Algebra

Subsets of the Real Numbers

Natural numbers = $\{1, 2, 3, \ldots\}$
Whole numbers = $\{0, 1, 2, 3, \ldots\}$

Integers = $\{\ldots -3, -2, -1, 0, 1, 2, 3, \ldots\}$

Rational = $\left\{\dfrac{a}{b} \,\middle|\, a \text{ and } b \text{ are integers with } b \neq 0\right\}$

Irrational = $\{x \mid x \text{ is not rational}\}$

Properties of the Real Numbers

For all real numbers a, b, and c

$a + b$ and ab are real numbers.	Closure
$a + b = b + a$; $a \cdot b = b \cdot a$	Commutative
$(a + b) + c = a + (b + c)$; $(ab)c = a(bc)$	Associative
$a(b + c) = ab + ac$; $a(b - c) = ab - ac$	Distributive
$a + 0 = a$; $1 \cdot a = a$	Identity
$a + (-a) = 0$; $a \cdot \dfrac{1}{a} = 1 \quad (a \neq 0)$	Inverse
$a \cdot 0 = 0$	Multiplication property of 0

Absolute Value

$$|a| = \begin{cases} a & \text{for } a \geq 0 \\ -a & \text{for } a < 0 \end{cases}$$

$\sqrt{x^2} = |x|$ for any real x

$|x| = k \Leftrightarrow x = k$ or $x = -k \quad (k > 0)$

$|x| < k \Leftrightarrow -k < x < k \quad (k > 0)$

$|x| > k \Leftrightarrow x < -k$ or $x > k \quad (k > 0)$

(The symbol $\Leftrightarrow$ means "if and only if.")

Interval Notation

$(a, b) = \{x \mid a < x < b\}$ $\quad [a, b] = \{x \mid a \leq x \leq b\}$

$(a, b] = \{x \mid a < x \leq b\}$ $\quad [a, b) = \{x \mid a \leq x < b\}$

$(-\infty, a) = \{x \mid x < a\}$ $\quad (a, \infty) = \{x \mid x > a\}$

$(-\infty, a] = \{x \mid x \leq a\}$ $\quad [a, \infty) = \{x \mid x \geq a\}$

Exponents

$a^n = a \cdot a \cdot \cdots \cdot a$ (n factors of a)

$a^0 = 1 \qquad a^{-n} = \dfrac{1}{a^n}$

$a^r a^s = a^{r+s} \qquad \dfrac{a^r}{a^s} = a^{r-s}$

$(a^r)^s = a^{rs} \qquad (ab)^r = a^r b^r$

$\left(\dfrac{a}{b}\right)^r = \dfrac{a^r}{b^r} \qquad \left(\dfrac{a}{b}\right)^{-r} = \left(\dfrac{b}{a}\right)^r$

Radicals

$a^{1/n} = \sqrt[n]{a} \qquad a^{m/n} = \left(\sqrt[n]{a}\right)^m = \sqrt[n]{a^m}$

$\sqrt[n]{ab} = \sqrt[n]{a} \cdot \sqrt[n]{b} \qquad \sqrt[n]{\dfrac{a}{b}} = \dfrac{\sqrt[n]{a}}{\sqrt[n]{b}}$

Factoring

$a^2 + 2ab + b^2 = (a + b)^2$

$a^2 - 2ab + b^2 = (a - b)^2$

$a^2 - b^2 = (a + b)(a - b)$

$a^3 - b^3 = (a - b)(a^2 + ab + b^2)$

$a^3 + b^3 = (a + b)(a^2 - ab + b^2)$

Rational Expressions

$\dfrac{ac}{bc} = \dfrac{a}{b} \qquad \dfrac{a}{b} + \dfrac{c}{d} = \dfrac{ad + bc}{bd}$

$\dfrac{a}{b} \cdot \dfrac{c}{d} = \dfrac{ac}{bd} \qquad \dfrac{a}{b} \div \dfrac{c}{d} = \dfrac{a}{b} \cdot \dfrac{d}{c}$

Quadratic Formula

The solutions to $ax^2 + bx + c = 0$ with $a \neq 0$ are

$$x = \frac{-b \pm \sqrt{b^2 - 4ac}}{2a}.$$

Distance Formula

The distance from (x_1, y_1) to (x_2, y_2), is

$$\sqrt{(x_2 - x_1)^2 + (y_2 - y_1)^2}.$$

COLLEGE ALGEBRA

SECOND EDITION

MARK DUGOPOLSKI
Southeastern Louisiana University

ADDISON-WESLEY
An imprint of Addison Wesley Longman, Inc.

Reading, Massachusetts • Menlo Park, California • New York • Harlow, England
Don Mills, Ontario • Sydney • Mexico City • Madrid • Amsterdam

Sponsoring Editor: Bill Poole

Senior Project Manager: Christine O'Brien

Managing Editor: Karen Guardino

Assistant Editor: Donna Bagdasarian

Senior Production Supervisor: Peggy McMahon

Design Supervisor: Barbara T. Atkinson

Art Editing Supervisor: Meredith Nightingale

Senior Marketing Manager: Brenda Bravener

Senior Manufacturing Manager: Ralph Mattivello

Cover Design: Barbara T. Atkinson

Text Design: Rebecca Lemna

Production Services: UG Production Services

Technical Art Illustration: Tech Graphics

Reflective Illustration: Jim Bryant, Gary Torisi, Howard Friedman

CREDITS: Cover Photograph: Sailboat: © Gary John Norman/Tony Stone Images, Water: © Tecmap/Westlight. Page xxiv: Painting of "Young America" courtesy of Jim DeWitt, DeWitt Studio, Point Richmond, CA. Pages 97, 105: Graph reprinted courtesy of *The Boston Globe*. Pages 262–263: Data from "Technical Preparation of the Airplane *Spirit of St. Louis*," by Donald A. Hall. Technical Note No. 257, National Advisory Committee for Aeronautics, Washington DC, July 1927. Pages 362–363: Based on "A New Look at Dinosaurs," by John H. Ostrom. *National Geographic*, August 1978, vol. 154, #2, pp. 152–185. Pages 803–804: Figure for Exercise 74 from *Sky and Telescope*, April 1997, used with permission. Guy Ottwell, Astronomical Calendar, 1997, p. 59.

LIBRARY OF CONGRESS CATALOGING-IN-PUBLICATION DATA

Dugopolski, Mark.
College algebra / Mark Dugopolski.—2nd ed.
p. cm.
ISBN 0-201-34711-3
1. Algebra. I. Title.
QA154.2.D82 1998
512.9—dc21 98-22932
CIP

ISBN: 0-201-34711-3

5 6 7 8 9 10 DOC 0100

CONTENTS

Preface *vii*
List of Supplements *xii*
Index of Applications *xvi*

P Prerequisites

P.1 Real Numbers and Their Properties *1*
P.2 Integral Exponents *16*
P.3 Rational Exponents and Radicals *29*
P.4 Polynomials *42*
P.5 Factoring Polynomials *54*
P.6 Rational Expressions *65*
Chapter P Highlights 76
Chapter P Review Exercises 78
Chapter P Test 80

1 Equations and Inequalities

1.1 Linear Equations *83*
1.2 Applications *94*
1.3 Complex Numbers *109*
1.4 Quadratic Equations *117*
1.5 Linear and Absolute Value Inequalities *132*
1.6 Quadratic and Rational Inequalities *145*
Chapter 1 Highlights 156
Chapter 1 Review Exercises 158
Chapter 1 Test 161
Tying It All Together 161

2 Functions and Graphs

2.1 The Cartesian Coordinate System *163*
2.2 Functions *176*
2.3 Graphs of Relations and Functions *191*
2.4 Transformations and Symmetry of Graphs *207*
2.5 Operations with Functions *222*
2.6 Inverse Functions *234*
2.7 Variation *247*
Chapter 2 Highlights 255
Chapter 2 Review Exercises 257
Chapter 2 Test 260
Tying It All Together 261

3 Polynomial and Rational Functions

3.1 Linear Functions *263*
3.2 Quadratic Functions *279*
3.3 Zeros of Polynomial Functions *291*
3.4 The Theory of Equations *303*
3.5 Miscellaneous Equations *313*
3.6 Graphs of Polynomial Functions *327*
3.7 Graphs of Rational Functions *339*
Chapter 3 Highlights 353
Chapter 3 Review Exercises 355
Chapter 3 Test 358
Tying It All Together 359

4 Exponential and Logarithmic Functions

4.1 Exponential Functions *361*
4.2 Logarithmic Functions *377*
4.3 Properties of Logarithms *389*
4.4 More Equations and Applications *402*
Chapter 4 Highlights 414
Chapter 4 Review Exercises 415
Chapter 4 Test 417
Tying It All Together 418

5 Systems of Equations and Inequalities

5.1 Systems of Linear Equations in Two Variables *421*
5.2 Systems of Linear Equations in Three Variables *434*
5.3 Nonlinear Systems of Equations *446*
5.4 Partial Fractions *455*
5.5 Inequalities and Systems of Inequalities in Two Variables *463*
5.6 Linear Programming *472*
Chapter 5 Highlights 480
Chapter 5 Review Exercises 481
Chapter 5 Test 482
Tying It All Together 483

6 Matrices and Determinants

6.1 Solving Linear Systems Using Matrices *485*
6.2 Operations with Matrices *500*
6.3 Multiplication of Matrices *507*
6.4 Inverses of Matrices *515*
6.5 Solution of Linear Systems in Two Variables Using Determinants *527*
6.6 Solution of Linear Systems in Three Variables Using Determinants *533*
Chapter 6 Highlights 543
Chapter 6 Review Exercises 545
Chapter 6 Test 546
Tying It All Together 547

7 The Conic Sections

7.1 The Parabola *549*
7.2 The Ellipse and the Circle *560*
7.3 The Hyperbola *574*
Chapter 7 Highlights 586
Chapter 7 Review Exercises 587
Chapter 7 Test 589
Tying It All Together 589

8 Sequences, Series, and Probability

8.1 Sequences *591*
8.2 Series *602*
8.3 Geometric Sequences and Series *610*
8.4 Counting and Permutations *623*
8.5 Combinations, Labeling, and the Binomial Theorem *630*
8.6 Probability *641*
8.7 Mathematical Induction *653*
Chapter 8 Highlights 659
Chapter 8 Review Exercises 661
Chapter 8 Test 663

Answers to Odd-numbered Exercises *A-1*
Index *I-1*

PREFACE

This text is designed for students who are pursuing further study in mathematics, as well as for those who are not. For those students who will take additional mathematics, the text will provide the skills, understanding, and insights necessary for success in future courses. For those students who will not pursue further mathematics, the extensive emphasis on applications and modeling will demonstrate the usefulness and applicability of mathematics in the world today. Additionally, the focus on problem solving that is a hallmark of this text provides numerous opportunities for students to reason and think their way through problem situations. Practicing such critical thinking skills will prove of future benefit to all students. I believe that the mathematics presented here is interesting, useful, and worth studying, and to get students to feel this way has been one of my goals in writing this text.

Use of Technology

I place increased emphasis on using graphical, numerical, and analytical points of view in discussions and in solving problems throughout the text. I also provide more opportunity for optional use of the graphing calculator as a tool to help students understand the concepts of algebra, and provide support for algebraic conclusions. While the use of technology is still optional, students who do not use a graphing calculator can still benefit from the technology discussions, as well as from the hundreds of calculator generated graphs that occur in the text. Any graphing utility or computer math package can be used in place of the graphing calculator.

I stress the difference between exact answers and the approximate answers given by a calculator, and discuss the limitations of a calculator graph. I also explain how a calculator drawn graph can assist students to draw or sketch a graph that clearly shows the important features of a function. In the exercises, the graphing calculator is often used as a means of making conjectures about new concepts and relationships. The text instructs students on how to use a graphing calculator, and more importantly, what to use it for and what not to use it for.

Graphing calculator screens generated from a TI-83 graphing calculator are included to give students the idea of the kinds of things a graphing calculator can do to aid in the learning process. But, because of the rapid changes occurring in technology, and the specific differences among brands of calculators, this text does not give specific instructions on how to use the various other machines available. However, the *Graphing Calculator Manual* that accompanies this text provides students with keystroke operations for many of the more popular graphing calculators, using specific examples drawn from this text to illustrate the various functions of those calculators.

Content Changes for the Second Edition

The entire first edition was carefully read for clarity of exposition, and rewritten where necessary. The major changes in this edition are as follows.

Chapter P This chapter contains prerequisite material on real numbers, rules of exponents, factoring, and simplifying expressions. Basic linear, quadratic, and absolute value equations and inequalities are covered in Chapter 1. Some sections from both of these chapters may be omitted depending on the preparation level of the students.

Revised Chapter 3 Quadratic type equations, equations with rational exponents or radicals, and more complicated absolute value equations now occur in Section 3.5, following the theory of polynomial equations, Section 3.4. Because some of these equations are polynomial equations, they will be better understood after the theory of polynomial equations has been studied.

New or Enhanced Features

Linking Concepts This **new feature** is located at the end of nearly every exercise set. It is a multi-part exercise or exploration that can be used for individual or group work. The idea of this feature is to use a concept from the current section along with concepts from preceding sections (including preceding chapters), and ask questions that help students see the links among various concepts. Some parts of these questions are open-ended, and require somewhat more thought than standard exercises. Answers to this feature are given only in the Instructor's Solutions Manual.

Applications Hundreds of **new exercises** have been added to the exercise sets, and most of them involve applications of real-world situations and often cite data sources. The emphasis of the new exercises is on understanding concepts and relationships.

Exercise Sets The exercise sets have been examined carefully to ensure that the exercises range from easy to challenging, and are arranged in order of increasing difficulty. Many new exercises require a graphing calculator.

Regression Problems Many **new regression problems** have been included in the text, so that students can start with real data, and use a calculator to obtain mathematical models of real problem situations.

Graphing Calculator Exercises Optional exercises that require a graphing calculator are now located in more natural positions in the exercises rather than at the end of the exercise sets as in the first edition. The exercises are optional and are marked with a graphing calculator icon.

Graphing Calculator Discussions Optional **new graphing calculator discussions** have been integrated into the text, and are set off by graphing calculator icons so that they can be easily skipped by those not using this technology.

Web Site A **new Web site** has been established—designed to increase student success in the course by offering section-by-section tutorial help, additional group projects, downloadable programs for TI graphing calculators, and author tips. This icon alerts students at times when this site would be helpful. The site will also be useful to instructors by providing dynamic resources for use in their course.

Continuing Features

Chapter Opener Each chapter begins with a Chapter Opener that discusses a real-world situation in which the mathematics of the chapter is used. Examples and exercises that relate back to the opener are included within the chapter.

Index of Applications The many applications contained within the text are listed in an Index of Applications that appears in the front of the text. The applications are page-referenced, and grouped by subject matter.

For Thought Each exercise set begins with a set of true or false questions that review the basic concepts in that section, help check student understanding before beginning the exercises, and offer opportunities for writing and/or discussion.

Writing/Discussion and Cooperative Learning Exercises These exercises deepen understanding by getting students to express mathematical ideas in writing, or to their classmates during small group or team discussions.

Highlights This end-of-chapter feature presents an overview of each section of the chapter, and is a useful summary of the basic information that students should have mastered in that chapter.

Chapter Review Exercises These exercises are designed to review the chapter without reference to the individual sections, and prepare students for the Chapter Test.

Chapter Test The problems in the Chapter Test are designed to help students measure their readiness for a classroom test. Instructors may also use them as a model for their own end-of-chapter tests.

Tying It All Together This reviews selected concepts from the present and prior chapters, and requires students to integrate multiple concepts and skills.

Acknowledgments

I would like to express my appreciation to the following reviewers whose comments and suggestions were invaluable in preparing this second edition:

Bill Ambrose, *West Texas A&M University*
Ronald Brent, *University of Massachusetts-Lowell*
LaJoycea Condry, *Tallahassee Community College*
William Grimes, *Central Missouri State University*
Dona Henderson, *University of South Dakota*
Julia Hicks, *University of Arkansas*
Dan Hostetler, *University of Cincinnati*
Heidi Howard, *Florida Community College*
Ben Neelley, *University of Texas Pan American*
Desley Plaisance, *Louisiana State University*
Allan Riveland, *Washburn University of Topeka*
Erlan Wheeler, *Carthage College*

I would also like to thank the following previous edition reviewers:

Victor A. Belfi
Daphne Bell
Donald W. Bellairs
Steven W. Blassberg
Robert R. Boltz
Jerry Burkhart
Roger Carlson
Floyd L. Downs
Rebecca Gehrke-Griswold
Carolyn A. Goldberg
Kenneth Grace
Mickey L. Hoffman
Carolyn T. Krause
Claire Krukenberg
Anita D. Lesmeister
Duncan J. Melville
Gael Mericle
Ricardo Moena
Robert E. Moyer
E. James Peake
James E. Peters II
Giles Wilson Maloof
Deborah J. Ritchie
Kenneth Seydel
Gerald M. Smith
Jan Vandever
Stewart Venit
Gerry Vidrine
Carol McRaney Walker
Christin Walker
Lyndon C. Weberg
S. K. Wyckoff
Mary F. Yorke

I also want to thank the many professors and students who have written to me with comments and suggestions. I am always glad to hear from users of my texts. You can write to me at the Department of Mathematics, Southeastern Louisiana University, Hammond, LA 70402, or send electronic mail to mdugopolski@selu.edu.

I also wish to express my thanks to Edgar Reyes, Southeastern Louisiana University, for working all exercises and writing the Solutions Manuals; Henry Smith, Southeastern Louisiana University, for working examples and exercises, and error checking the manuscript; Rebecca Muller, Southeastern Louisiana

University, for writing the Printed Test Bank; Margaret Donlan, University of Delaware, Dale Nauta, Orange Coast College, Patricia Schwarzkopf, University of Delaware, Burnette Thompson, Texas Southern University, and Lori Vrionis for presenting video lectures.

I give special thanks to my editor Bill Poole, project manager Christine O'Brien, and the entire Addison Wesley Longman team for their assistance, encouragement, and direction throughout the project: Barbara Atkinson, Donna Bagdasarian, Brenda Bravener, Karen Guardino, Peggy McMahon, Tricia Mescall, Karen Scott, and Greg Tobin.

I cannot put into words how much I appreciate the love and encouragement of my wife and daughters. Thanks to all.

M. D.
Hammond, Louisiana

SUPPLEMENTS

For the Student

Printed Supplements

Student's Solutions Manual

ISBN 0-201-38392-6
Edgar Reyes, Southeastern Louisiana University

- Complete solutions to all odd numbered graphical and numerical exercises.
- Ask your bookstore about ordering.

Graphing Calculator Manual

ISBN 0-201-38397-7
Stuart Moskowitz, Humboldt State University

- Graphing calculator usage instruction.
- Keystroke operations for the following calculator models: TI-82, TI-83, TI-85, TI-86, Casio Color Graphics 9850 Plus, HP38G.
- Worked-out examples taken directly from the text.

Media Supplements

InterAct Math Tutorial Software

Throughout the text, this icon 💻 indicates when this software would be helpful to students.

Windows ISBN 0-201-38381-0
Macintosh ISBN 0-201-38382-9

InterAct Math Tutorial Software has been developed and designed by professional software engineers working closely with a team of experienced math educators.

InterAct Math Tutorial Software includes exercises that are linked with every objective in the textbook and require the same computational and problem-solving skills as their companion exercises in the text. Each exercise has an example and an interactive guided solution that are designed to involve students in the solution process and to help them identify precisely where they are having trouble. The software recognizes common student errors and provides students with appropriate customized feedback.

With its sophisticated answer recognition capabilities, InterAct Math Tutorial Software recognizes appropriate forms of the same answer for any kind of input. It also tracks student activity and scores for each section, which can then be printed out.

The software is free to qualifying adopters or can be bundled with books for sale to students.

Videotape Series

Throughout the text, this icon indicates when these videotapes would be helpful to students.

ISBN 0-201-38403-5

- Keyed specifically to text.
- An engaging team of lecturers provide comprehensive coverage of each section and every topic.
- Uses worked-out examples, visual aids, and manipulatives to reinforce concepts.
- Emphasizes the relevance of material to the real world and relates mathematics to students' everyday lives.
- Can be ordered by mathematics instructors or departments.

SUPPLEMENTS

For the Instructor

Printed Supplements

Instructor's Solutions Manual

ISBN 0-201-38391-8
Edgar Reyes, Southeastern Louisiana University

- Complete solutions to all graphical and numerical exercises.
- Free to instructors with textbook adoption.

Instructor's Testing Manual

ISBN 0-201-38396-9
Rebecca Muller, Southeastern Louisiana University

- Contains six alternative forms of tests per chapter (two are multiple choice), and answer keys.
- Free to instructors with textbook adoption.

Media Supplements

TestGen-EQ with QuizMaster-EQ

Windows ISBN 0-201-38379-9
Macintosh ISBN 0-201-38380-2

TestGen-EQ is a computerized test generator with algorithmically defined problems organized specifically for this textbook. Its user-friendly graphical interface enables instructors to select, view, edit, and add test items, then print tests in a variety of fonts and forms. Seven question types are available, and search and sort features let the instructor quickly locate questions and arrange them in a preferred order. A built-in question editor gives the user the power to create graphs, import graphics, insert mathematical symbols and templates, and insert variable numbers or text. An ''Export to HTML'' feature lets instructors create practice tests that can be posted to a Web site. Tests created with TestGen-EQ can be used with QuizMaster-EQ, which enables students to take exams on a computer network. QuizMaster-EQ automatically grades the exams, stores results on disk, and allows the instructor to view or print a variety of reports for individual students, classes, or courses. This program is available in Windows and Macintosh formats, and is free to adopters of the text.

InterAct Math Plus Software

InterAct Math Plus combines course management and on-line testing with the features of the basic InterAct Math tutorial software to create an invaluable teaching resource.

Consult your Addison-Wesley representative for details.

INDEX OF APPLICATIONS

Archeology
Age of *Deinonychus*, 406
Ancient Egyptian trick, 15
Carbon 14 dating, 409
Dating a bone, 409
Dating a tree, 409
Dead sea scrolls, 410
Dinosaur speed, 259
Finding the half-life, 409, 416
Leakeys date Zinjanthropus, 400
Old clothes, 409
Radioactive decay, 374

Astronomy
Apogee, perigee, and eccentricity, 573
Comet Hale-Bopp, 571
Hale telescope, 558
Halley's Comet, 572
Hubble telescope, 556
Orbit of Mir, 572
Orbit of the moon, 572
Picturing earth, 572
Visual magnitude of Deneb, 411
Visual magnitude of a star, 411

Biology/Health/Life Sciences
Basic energy requirements, 176
Body-mass index, 26
Bungee jumping, 189
Cafeteria meals, 639
Carbon monoxide poisoning, 352
Cigarette consumption, 14
Contaminated chicken, 337
Dollars per death, 254
Drug testing, 302
Drug therapy, 609, 622
Efficiency for descending flight, 444
Excessive fat, 26
Extended drug therapy, 609
Free fall at Six Flags, 253
Growth rate for bacteria, 312
Having children, 640
Hiking time, 325
Hiking distance, 160
Human memory model, 401
Inspecting restaurants, 629
Learning curve, 416
Logistic growth, 412
Long-range therapy, 622
Male lung cancer death rate, 14
Nursing home care, 600
Nutrition, 498
Nutritional information, 514
Nutritional content, 506
Pediatric tuberculosis, 417
Percentage of body fat, 357
Physical fitness, 357
Poiseuille's Law, 245
Population of foxes, 411
Population of rabbits, 411
Prescribing drugs, 432
Protein and carbohydrates, 432
Rate of flu infection, 288
Recommended daily allowances, 506
Speed of a tortoise, 130
Spreading the flu, 639
Target heart rate, 14, 105

Business
Accounting problems, 446
Across-the-board raise, 220
Area of a lot, 107
Assigning vehicles, 640
Average cost, 129, 277
Balancing the costs, 352
Billboard advertising, 351
Bus routes, 629
Business expansion, 649
Capital cost and operating cost, 352
Capital investment, 453
Carpenters and helpers, 277
Choosing a team, 639
Company cars, 629
Comparing costs, 92
Computer design, 130
Computer passwords, 629
Computer recyclers, 338
Computers and printers, 276
Concert revenue, 189
Concert tickets, 288
Construction penalties, 105
Corporate taxes, 93
Cost accounting, 93
Cost of watermelons, 80
Cost of a car, 105
Costly campaign, 75
Daisies, carnations, and roses, 482
Demand and revenue, 156
Depreciation rate, 40, 245
Disaster relief, 622
Dogs and suds, 526
Doubles and singles, 432
Driving speed, 159
Economic forecast, 337
Economic impact, 400, 622
Economic order quantity, 39
Eggs and magazines, 526
Estimating costs, 14
Filing a tax return, 206
Filing invoices, 75
Free wax, 15
Furniture rental, 432
Gasoline sales, 542
Handling charge for a globe, 259
Hard drive capacity, 411
Harvesting wheat, 107
Imported and domestic cars, 482
Income tax reform, 278
In-house training, 276
Increasing revenue, 289
International communications, 92
Internet overcrowding, 401
Investment income, 105
Living comfortably, 108
Making a profit, 144
Marginal revenue, 401
Marketing plans, 357
Marking pickup trucks, 662
Maximizing revenue, 288, 478
Maximum profit, 337
Meeting between two cities, 159
Minimizing labor costs, 479

Minimizing operating costs, 478
Mixing breakfast cereal, 108
Mixing dried fruit, 108
Mowing a lawn, 151
National debt, 27
Numerous constraints, 479
On the Bayou, 527
Part-time and full-time workers, 479
Peeling apples, 159
Peppers, tomatoes, and eggplants, 482
Percentage of white meat, 130
Percentage of minority workers, 106
Phone extensions, 629
Placing advertisements, 662
Planting strawberries, 152
Predicting car sales, 546
Price per gigabyte, 411
Processing forms, 107
Production function, 220
Profit, 232
Profit from sales, 155
Profit for computer sales, 53
Pumping tomato soup, 452
Real Estate Commission, 105
Retail store profit, 312
Rising salary, 600
Sales goals, 622
Savings and loan bailout, 27
Scraping barnacles, 107
Selecting a team, 662
Shipping machinery, 206
Shortest route, 630
Small trucks and large trucks, 479
Start-up capital, 106
Stocking supplies, 527
Support for gambling, 159
Tacos and burritos, 478, 481
Tax reform, 432
Taxable income, 108
To buy or rent, 144
Total salary, 609
Traveling sales representative, 628
Triple feature, 662
Unemployment versus inflation, 220
Virtual pets, 526
Wages from two jobs, 498

Chemistry

Acidity of orange juice, 388
Acidity in your stomach, 388
Acidity of tomato juice, 388
Acidosis, 388
Celsius to Fahrenheit formula, 275
Controlling temperature, 143
Comparing pH, 416
Diluting antifreeze, 108
Diluting baneberry, 108
Drug buildup, 409
Formula for pH, 400
Lorazepam, 409
Mixing alloys, 472
Mixing alcohol solutions, 107
Mixing antifreeze in a radiator, 130
Nightshade-snakeroot drink, 15
Nitrogen gas shock absorber, 254
Weight of a hydrogen atom, 80

Construction

Area of a window, 232
Area of a foundation, 324
Bonus room, 358
Bordering a flower bed, 130
Bracing a gate, 130
Building costs, 514
Constructing an elliptical arch, 571
Countertop pricing, 600
Countertops, 600
Dimensions of a laundry room, 453
Fencing a rectangular area, 155
Heating and air, 312
Longest screwdriver, 40
Making an arch, 452
Minimizing construction cost, 326
Painting a house, 75
Painting a fence, 155
Plywood and insulation, 526
Size of a vent, 452
Spreading glue, 233
Stiff penalty, 600
Storing supplies, 326
Thickness of concrete, 160
Trimming a garage door, 106

Consumer

Admission to the zoo, 351
Average cost per drink, 75
Becoming a millionaire, 386
Book prices, 431
Budgeting, 506
Camaro depreciation, 277
Carpeting a room, 254
Choosing a pizza, 628
Coins in a vending machine, 108
College tuition, 290
Comparing investments, 416
Cost of window cleaning, 188
Cost of a newspaper, 108
Cost of business cards, 275
Cost of gravel, 188
Cost of copper tubing, 254
Cost of plastic sewer pipe, 254
Cost of a parking ticket, 375
Cost-of-living raise, 220
Delivering concrete, 205
Depreciation and inflation, 410
Depreciation of a Mustang, 188
Difference in prices, 143
Doubling your money, 386
Down payment, 622
Expensive models, 143
Fast food inflation, 445
Flat tax, 432
Further markdowns, 233
Gas mileage, 206
Getting fit, 431
Good planning, 600
Growing debits, 375
Have it your way, 628
High yields attract deposits, 374
Inflation of car prices, 155
Milk, coffee, and doughnuts, 445
More restrictions, 471
Optional equipment, 628
Parking charges, 205
Price of a burger, 142
Price range for a car, 142
Price range for a haircut, 160
Price of a car, 245
Product awareness, 156
Recursive pricing, 600
Renting a car, 351
Rising price of milk, 375
Saturating the market, 622
Scheduling departures, 662
Selling price of a home, 160
Simple interest, 104, 253
Size restrictions, 471
Social security benefits, 207
Ticket pricing, 275
Traffic jam, 387
Utility bills, 546
Volume discount, 275
Water bill, 205
Weight of a can, 254
Zoo admission prices, 429

Design

Accuracy of transducers, 325
Antique saw, 357
Area of a wing, 589
Capsize control, 175
Capsize screening value, 324
Cross section of a gutter, 287
Cross-sectional area of a well, 254
Designing a paper clip, 290

Designing fireworks, 312
Designing a crystal ball, 313
Designing a racing boat, 221
Displacement-length ratio, 27, 232
Finding the displacement, 129
Fitting a line to data points, 289
Folding sheet metal, 159
Giant teepee, 337
Golden rectangle, 160
Insulated carton, 324
Landing speed, 245
Lighter boat, 27
Limiting the beam, 175
Lindbergh's practical economic air speed, 276
Lindbergh's most economical air speed, 287
Making boxes, 338
Making a gas tank, 352
Making a glass tank, 352
Maximum volume, 337
Maximum volume of a cage, 287
Packing cheese, 337
Packing a triangular piece of glass, 232
Packing a square piece of glass, 232
Paint coverage, 338
Router bit, 436
Sail area-displacement ratio, 232, 325
Shipping carton, 324
Volume of a cubic box, 325

Environment

Arranging plants in a row, 259
Available habitat, 52
Average speed of an auto trip, 351
Catching cod, 288
Celsius temperature, 104
Cleaning up the river, 156
Costly cleanup, 75
Crowded or not, 28
Deforestation in Nigeria, 231, 386
Deforestation in El Salvador, 386
Deforestation, 189
Destruction of a rainforest, 231, 374
Elimination of tropical moist forest, 189
Energy consumption, 231
Energy consumption in U.S., 28
Fish population, 159
Global warming, 387
Gone fishing, 662
Hazardous waste, 74
Increasing gas mileage, 160
Inefficiency of Ivory Coast Mills, 260
Measuring ocean depth, 454
More congestion, 499
Municipal waste, 387
Noise pollution, 411
Nuclear power, 586
Paper consumption, 187
Radioactive waste, 409
Red fox, 52
Safe water, 387
Saving gasoline, 160
Saving energy, 233
Solid waste recovery, 109
Speed of an electric car, 130
Time for growth, 400
Traffic control, 499
Urban pollution, 189
Velocity of underground water, 254
Water pollution, 53, 74
World grain demand, 28
World population, 386

Geometry

Adjacent circles, 572
Area of a circle, 591
Area of a trapezoid, 63
Area of a rectangle, 53
Boston Molasses Disaster, 325
Cartridge box, 302
Circle inscribed in a square, 259
Circumference of a circle, 104
Computer case, 302
Depth of a reflecting pool, 108
Depth of a swimming pool, 107
Dimensions of a flag, 129
Dimensions of a picture frame, 160
Egyptian area formula, 53
Fencing dog pens, 106
Fencing a feed lot, 106
Increasing area of a field, 106
Legs of a right triangle, 452
Long shot, 129
Maximum girth, 142
Minimizing distance, 289
Missing circle, 205
Open-top box, 129, 302
Perimeter of a right triangle, 324
Pyramid builders, 64
Regular polygons, 472
Right triangle inscribed in a semicircle, 324
Right triangle, 324
Sides of a triangle, 452
Sides of a rectangle, 452
Surface area, 189
View from an airplane, 253
Volume of a rectangular box, 52, 63
Volume of a can of coke, 107
Volume of a cube, 245
Volume of a cylinder, 104

Investment

Annual growth rate, 245
Annual payments, 609
Annuities, 623
Ben's gift to Boston, 400
Ben's gift to Philadelphia, 400
Coin collecting, 432
Combining investments, 106
Comparing investments, 52
Compound interest, 622
Compounded quarterly, 609, 661
Doubling time with quarterly compounding, 416
Doubling time with continuous compounding, 416
Equality of investments, 410
Finding the rate, 386
Four-year investment, 52
Future value, 416
Income on investments, 431
Interest compounded continuously, 374
Interest compounded daily, 374
Interest compounded monthly, 374
Interest compounded quarterly, 374
Investment portfolio, 498
Large investment, 375
Market growth rate, 40
Miracle in Philadelphia, 386
Miracle in Boston, 386
Mixing investments, 526
Monthly payments, 661
Present value of a bond, 412
Present value of a CD, 412
Rule of 70, 386
Saving for retirement, 275, 622
Short term loans, 246
Stock market losses, 431
Stocks, bonds, and a mutual fund, 445
Thirty-year bonds, 80
Value of an annuity, 622

Miscellaneous

Age disclosure, 542
Arming the villagers, 506
Assigned reading, 609, 629
Assigning topics, 640
Atonement of Hercules, 629
Batman and Robin, 107
Bennie's coins, 542
Bridge hands, 639

Burgers, fries, and cokes, 445
Charleston earthquake, 129
Choosing a committee, 662
Choosing a prize, 629
Choosing a vacation, 662
Choosing a name, 629
Choosing songs, 629
Clark to the rescue, 629
Coffee and muffins, 633
Delicate balance, 472
Distribution of coins, 445
Distribution of coin types, 432
Doubling the bet, 417
Drawing cards from a deck, 628
Family tree, 622
Fantasy Five, 639
Fire code inspections, 639
Friends, 546
Grade on an algebra test, 254
Graduating seniors, 431
Granting tenure, 629
Job candidates, 639
Lindbergh's air speed over Newfoundland, 276
Making a dress, 130
Marching bands, 640
Mixed signals, 453
Modern American trick, 15
Mount of cans, 609
Parading in order, 629
Parking tickets, 640
Piano tuning, 39
Playing a lottery, 639
Poker hands, 628, 639
Political correctness, 472
Possible words, 629
Postal rates, 498
Prize-winning pigs, 639
Quality time, 431
Reading marathon, 600
Returning exam papers, 639
Rolling dice, 639
Rollover time, 400
Saving the best till last, 639
Scheduling radio shows, 629
Scratch and win, 649
Signal flags, 662
Sleepless night, 628
Television schedule, 639
Tourists and Alaskans, 108
Tummy masters, 661
Voyage of the whales, 453
Westward ho, 454
Working together, 109, 463

Navigation
Air navigation, 584
Marine navigation, 584
Uniform motion, 104

Number
Integers, 108
Lost numbers, 452
Misplaced house numbers, 445
More lost numbers, 452
Odd integers, 108
Recording experience, 108
Square roots, 324
Two more numbers, 155
Two numbers, 155

Physics
Air mobile, 452
Bird mobile, 432
Fish mobile, 445
Height of a ball, 129
Height of a flare, 155
Load on a spring, 259
Maximum height of a ball, 288
Maximum height of a football, 288
Newton's law of gravity, 260
Path of a missile, 444
Thermistor resistance, 412

Probability
Any card, 651
Checkmate, 651
Choosing a chairperson, 650
Colored marbles, 650
Drive defensively, 650
Dumping pennies, 652
Earth-crossing asteroids, 651
Electing Jones, 651
Family of five, 651
Five in five, 651
Foul play, 650
Four out of two, 651
Full house, 662
Future plans, 662
Hurricane alley, 650
Insurance categories, 651
Jelly beans, 662
Just guessing, 662
Killer asteroids, 650
Large family, 662
Lineup, 650
Lucky seven, 651
Morning line, 651
My three sons, 662
Numbered marbles, 650
On target, 650
Only one winner, 651
Pick a card, 651
Pick six, 651
Public opinion, 651
Read my lips, 650
Rolling a pair of dice, 649
Rolling fours, 650
Rolling a die, 649
Rolling a die twice, 649
Rolling once, 651
Rolling again, 662
Rolling dice, 662
Safe at last, 650
Selecting students, 651
Six or four, 651
Stock market rally, 650
Sum of six, 662
Tax time, 650
Ten or four, 651
Tossing a coin, 649
Tossing triplets, 650
Tossing one coin twice, 649
Tossing two coins once, 649
Two out of four, 651
Weather forecast, 651

Production
Acceptable bearings, 143
Acceptable targets, 143
Average cost, 75
Bird houses and mailboxes, 478
Break-even analysis, 324
Cookie time, 542
Cooling hot steel, 410
Cost of baking bread, 322
Demand equation, 128
Fluctuating costs, 482
Inventory control, 471
Manufacturing cost, 376
Marginal revenue, 190
Marginal cost, 190, 277
Pipeline or barge, 482
Processing oysters, 253
Production cost, 93

Science
Altitude and pressure, 289
Altitude of a rocket, 357
Bouncing ball, 622, 661
Challenger disaster, 375
Colombian earthquake of 1906, 400

Comparing exponential and linear growth, 376
Cooking a roast, 410
Distance to the sun, 27
Diversity index, 402
Doubling the sound level, 411
Eavesdropping, 558
Energy from the sun, 27
First-class diver, 253
Focus of an elliptical reflector, 571
Mass of the sun, 27
Masses of the planets, 255
Moon talk, 80
Orbit of the earth, 27
Parabolic mirror, 584
Radius of the earth, 27
Richter scale, 400
Room temperature, 410
Searchlight, 590
Second place, 253
Speed of light, 27
Telephoto lens, 584
Time of death, 410
Whispering gallery, 588

Sports

Adjusting the saddle, 105
Angle of completion, 590
Baseball statistics, 131
Basketball bucks, 53
Bicycle gear ratio, 142, 254
Changing speed of a dragster, 259
Choosing the right angle, 574
Diagonal of a football field, 129
Finding the freeboard, 104
Fly ball, 80
Green space, 108
Harmonic mean, 93
Hazards of altitude, 131
Hazards of depth, 188
Height of a sky diver, 129
Height of a baseball, 155
High altitude climbers, 156
High-income bracket, 14
Hockey money, 53
Initial velocity of a basketball player, 130
Injuries in the Boston Marathon, 96, 105
Laying out a track, 143
Less powerful boat, 40
Limiting velocity, 358
Maximum sail area, 54, 323
Minimum displacement for a yacht, 324
Olympic track, 108
Oxygen uptake, 94
Planning a race track, 107
Pole-vault principle, 254
Position of a football, 375
Racing speed, 106
Racing rules, 527
Ranking soccer teams, 515
Rowers and speed, 245
Rowing a boat, 106
Sail area-displacement ratio, 40
Selecting the cogs, 142
Shooting a target, 509
The 2.4-meter rule, 104
Throwing a javelin, 450
Time for a vertical leap, 156
Time swimming and running, 325
Track competition, 628
Velocity for a vertical leap, 155
Weight distribution, 445
Width of a football field, 107

Statistics/Demographics

Age groups, 445
Assigning volunteers, 640
Average age of vehicles, 445
Average speed, 75, 106
Basketball stats, 106
Black Death, 386
Bringing up your average, 142
Comparing growth, 190
Costly healthcare, 188
Crude exports, 143
Decline of the family, 375
Declining defense, 277
Dining in vs. dining out, 81
Electronic mail, 629
Enrollment data, 662
Final exam score, 142
Geometric mean, 325
Growing number of millionaires, 357
Health car expenditure, 174
Life expectancy, 431
Life expectancy in Lyon, 221
Life expectancy in Paris, 221
Linear extrapolation, 357
Linear interpolation, 357
Medicare and Medicaid, 188
Motor vehicle ownership, 206
Multiple-choice test, 629, 661
Other means, 610
Phone numbers, 629
Political party preference, 431
Population growth, 374
Possible families, 662
Poverty level, 410
Predicting population, 388
Raising a batting average, 142
Students, teachers, and pickup trucks, 445
Taking a test, 629
Two models, 453
Variance of the number of smokers, 288
Violent crime, 290
Weighted average with fractions, 142
Weighted average with whole numbers, 142
What a difference a weight makes, 542
Women on the board, 175
Women and marriage, 174

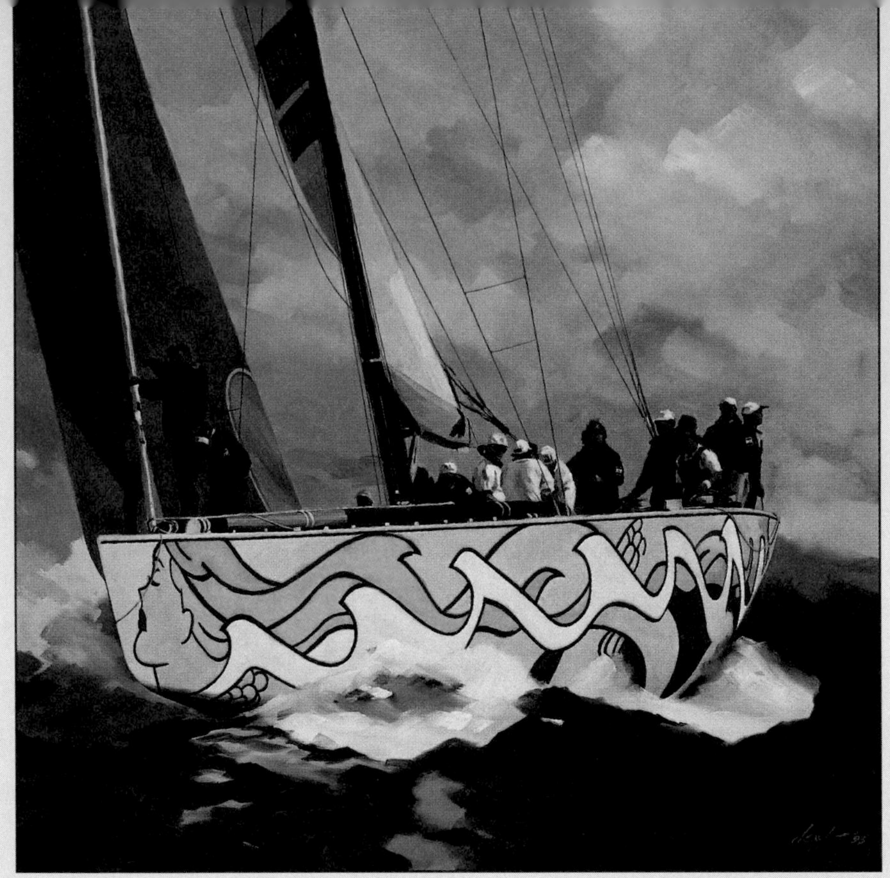

In the 1995 America's Cup the defending *Young America* kept up a good face, but from the start it was clear that the New Zealand entry, *Black Magic*, was sailing higher and faster in the 11-knot breeze and rolling swells. "Over the past two-and-a-half years, we've done everything we could to prepare for this moment," claimed New Zealand's team leader, Peter Blake, "You are seeing our best shot." And in the end a superbly prepared New Zealand crew overcame an American defense that was too little, too late.

The American/New Zealand rivalry intensified in 1988, when U.S. contender, Dennis Conner, stretched the Cup rules by audaciously sailing a nimble 60-foot catamaran against the Kiwi's 123-foot yacht. As a result, new rules were written to keep the race a sporting event. The new standards included a formula that established strict design boundaries for yacht sail area and hull shape, length, and displacement. It took New Zealand years to settle the score. But in the 1995 competition held off San Diego's shores, Blake and a hand-picked crew won the coveted trophy back from Conner.

Gone are the days when raw sailing ability and stamina won races like the America's Cup. In the modern sailing world, technology plays an ever increasing role in determining outcomes. For instance, all yachts in the 1995 Cup had a crewman on board to interpret computer data on sailing conditions. In addition, the New Zealander's *Black Magic* sported a mathematically arrived at design that optimized stability and speed specifically for San Diego's fickle wind conditions. By riding high in the water, *Black Magic* reduced its drag and optimized its ability to capture wind in its sail. Although the final design resulted in a boat that was trickier to handle, the crew's strenuous training carried the day.

Preparation is just as important in algebra as it was to the *Black Magic's* crew. In this chapter we will review the basic concepts that are necessary for success in algebra. Throughout the chapter you will see that even basic concepts have applications in business, science, engineering, and sailing. We begin by reviewing the real numbers and their properties.

P PREREQUISITES

P.1 Real Numbers and Their Properties
P.2 Integral Exponents
P.3 Rational Exponents and Radicals
P.4 Polynomials
P.5 Factoring Polynomials
P.6 Rational Expressions

P.1 Real Numbers and Their Properties

In arithmetic we learn facts about the real numbers and how to perform operations with them. Since algebra is an extension of arithmetic, we begin our study of algebra with a discussion of the real numbers and their properties.

The Real Numbers

The most basic set of numbers is the set of **counting** or **natural numbers**, N,

$$\{1, 2, 3, \ldots\}.$$

The natural numbers together with zero form the set of **whole numbers**, W,

$$\{0, 1, 2, 3, \ldots\}.$$

Negative numbers are used to represent losses or debts. The whole numbers together with the negative counting numbers are referred to as the set of **integers**, J,

$$\{\ldots, -3, -2, -1, 0, 1, 2, 3, \ldots\}.$$

The integers can be pictured as points on a line, the **number line**. To draw a number line, draw a line and label any convenient point with the number 0.

Now choose a convenient length, one **unit**, and use it to locate evenly spaced points corresponding to the positive integers to the right of zero and the negative integers to the left of zero as shown in Fig. P.1.

Figure P.1

The numbers corresponding to the points on the line are called the **coordinates** of the points.

The integers and their ratios form the set of **rational numbers**, Q. The rational numbers also correspond to points on the number line. For example, the rational number 1/2 is found halfway between 0 and 1 on the number line. In set notation the set of rational numbers is written as

$$\left\{\frac{a}{b}\,\middle|\, a \text{ and } b \text{ are integers with } b \neq 0\right\}.$$

This notation is read "The set of all numbers of the form a/b such that a and b are integers with b not equal to zero." In our set notation we used letters to represent integers. A letter that is used to represent a number is called a **variable**.

There are infinitely many rational numbers located between each pair of consecutive integers, yet there are infinitely many points on the number line that do not correspond to rational numbers. The numbers that correspond to those points are called **irrational** numbers. In decimal notation, the rational numbers are the numbers that are repeating or terminating decimals, and the irrational numbers are the nonrepeating nonterminating decimals. For example, the number 0.595959 . . . is a rational number because the pair 59 repeats indefinitely. By contrast, notice that in the number 5.010010001 . . . , each group of zeros contains one more zero than the previous group. Because no group of digits repeats, 5.010010001 . . . is an irrational number.

Numbers such as $\sqrt{2}$ or π are irrational also. We can visualize $\sqrt{2}$ as the length of the diagonal of a square whose sides are one unit in length. See Fig. P.2. In any circle, the ratio of the circumference c to the diameter d is π ($\pi = c/d$). See Fig. P.3. It is difficult to see that numbers like $\sqrt{2}$ and π are irrational because their decimal representations are not apparent. However, the irrationality of π was proven in 1767 by Johann Heinrich Lambert, and it can be shown that the square root of any positive integer that is not a perfect square is irrational.

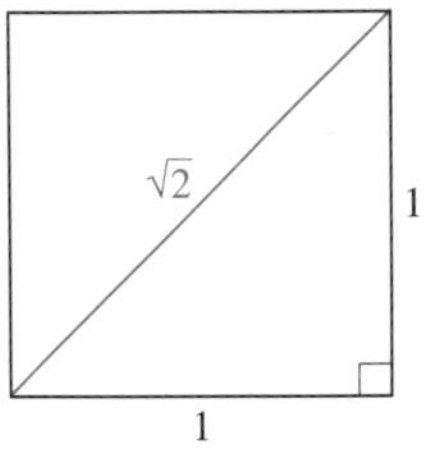

Figure P.2

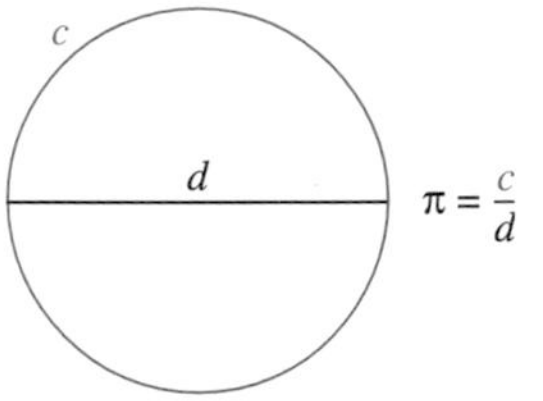

Figure P.3

Since a calculator operates with a fixed number of decimal places, it gives us a *rational approximation* for an irrational number such as $\sqrt{2}$ or π. See Fig. P.4. □

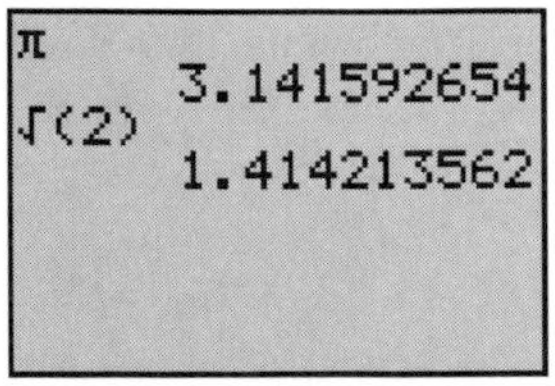

Figure P.4

The set of rational numbers, Q, together with the set of irrational numbers, I, is called the set of **real numbers**, R. Thc following are examples of real numbers:

$$-3, \quad -0.025, \quad \frac{1}{3}, \quad 0, \quad 0.595959\ldots, \quad \sqrt{2}, \quad \pi, \quad 5.010010001\ldots$$

These numbers are **graphed** on a number line in Fig. P.5.

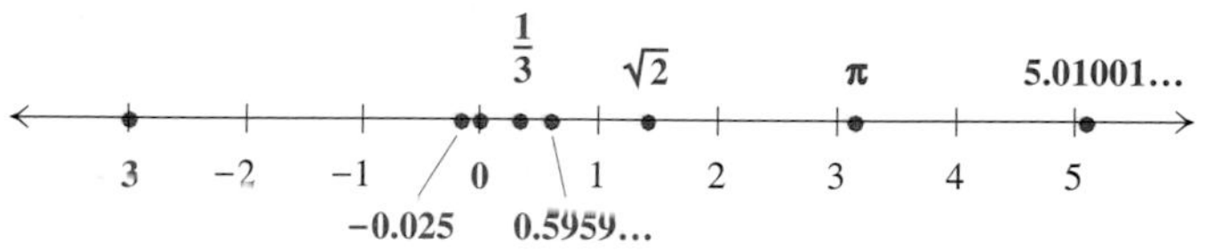

Figure P.5

Since there is a one-to-one correspondence between the points of the number line and the real numbers, we often refer to a real number as a point. Figure P.6 shows how the various subsets of the real numbers are related to one another.

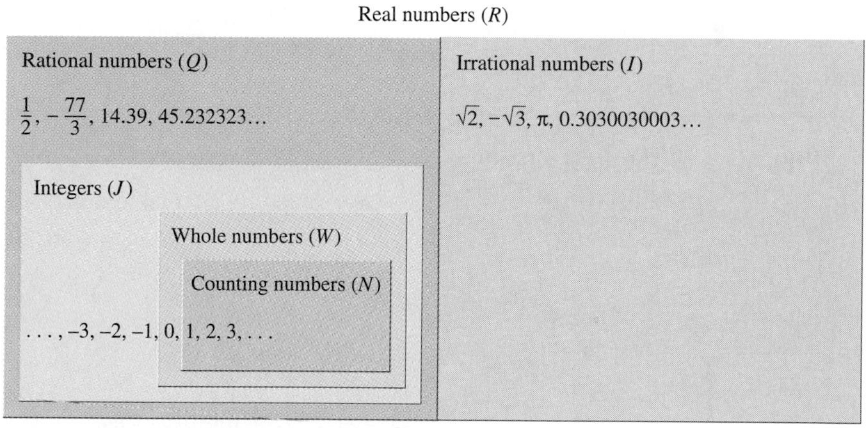

Figure P.6

To indicate that a number is a member of a set, we write $a \in A$, which is read "a is a member of set A." We write $a \notin A$ for "a is not a member of set A." Set A is a subset of set B ($A \subseteq B$) means that every member of set A is also a member of set B, and A is not a subset of B ($A \not\subseteq B$) means that there is at least one member of A that is not a member of B.

EXAMPLE 1 Classifying numbers and sets of numbers

Determine whether each statement is true or false and explain. See Fig. P.6.

a) $0 \in R$ **b)** $\pi \in Q$ **c)** $R \subseteq Q$ **d)** $I \not\subseteq Q$ **e)** $\sqrt{5} \in Q$

Solution

a) True, because 0 is a member of the set of whole numbers, a subset of the set of real numbers.

b) False, because π is irrational.

c) False, because every irrational number is a member of R but not Q.

d) True, because the irrational numbers and the rational numbers have no numbers in common.

e) False, because the square root of any integer that is not a perfect square is irrational.

Properties of the Real Numbers

In arithmetic we can observe that $3 + 4 = 4 + 3$, $6 + 9 = 9 + 6$, etc. We get the same sum when we add two real numbers in either order. This property of addition of real numbers is the **commutative property**. Using variables, the commutative property of addition is stated as $a + b = b + a$ for any real numbers a and b. There is also a commutative property of multiplication, which is written as $a \cdot b = b \cdot a$ or $ab = ba$. There are many properties concerning the operations of addition and multiplication on the real numbers that are useful in algebra.

Properties of the Real Numbers

For any real numbers a, b, and c:

$a + b$ and ab are real numbers	**Closure property**
$a + b = b + a$ and $ab = ba$	**Commutative properties**
$a + (b + c) = (a + b) + c$ and $a(bc) = (ab)c$	**Associative properties**
$a(b + c) = ab + ac$	**Distributive property**
$0 + a = a$ and $1 \cdot a = a$ (Zero is the **additive identity**, and 1 is the **multiplicative identity**.)	**Identity properties**
$0 \cdot a = 0$	**Multiplication property of zero**
For each real number a, there is a unique real number $-a$ such that $a + (-a) = 0$. ($-a$ is the **additive inverse** of a.)	**Additive inverse property**
For each nonzero real number a, there is a unique real number $1/a$ such that $a \cdot 1/a = 1$. ($1/a$ is the **multiplicative inverse** or **reciprocal** of a.)	**Multiplicative inverse property**

The closure property indicates that the sum and product of any pair of real numbers is a real number. The commutative properties indicate that we can add or multiply in either order and get the same result. Since we can add or multiply only a pair of numbers, the associative properties indicate two different ways to obtain the result when adding or multiplying three numbers. The operations within parentheses are performed first. Because of the commutative property, the distributive property can be used also in the form $(b + c)a = ab + ac$.

Note that the properties stated here involve only addition and multiplication, considered the basic operations of the real numbers. Subtraction and division are defined in terms of addition and multiplication. By definition $a - b = a + (-b)$ and $a \div b = a \cdot 1/b$ for $b \neq 0$. Note that $a - b$ is called the **difference** of a and b and $a \div b$ is called the **quotient** of a and b.

EXAMPLE 2 **Using the properties**

Complete each statement using the property named.

a) $a7 =$ ________, commutative

b) $2x + 4 =$ ________, distributive

c) $8($________$) = 1$, multiplicative inverse

d) $\frac{1}{3}(3x) =$ ________, associative

Solution

a) $a7 = 7a$

b) $2x + 4 = 2(x + 2)$

c) $8\left(\frac{1}{8}\right) = 1$

d) $\frac{1}{3}(3x) = \left(\frac{1}{3} \cdot 3\right)x$

Opposites and Negatives

The symbol $-$ has three meanings. In $a - b$ it means the operation of subtraction. In -7 it indicates the number negative seven. In the expression $-b$ it means the opposite or additive inverse of b. We read $-b$ as "the opposite of b" because $-b$ is not necessarily a negative number. If b is positive, then $-b$ is negative, but if b is negative, then $-b$ is positive.

Using two "opposite" signs has a cancellation effect. For example, $-(-5) = 5$ and $-(-(-3)) = -3$. Note that the additive inverse of a number can be obtained by multiplying the number by -1. For example, $-1 \cdot 3 = -3$. Calculators usually use the negative sign (-) to indicate opposite or negative and the subtraction sign $(-)$ for subtraction as shown in Fig. P.7.

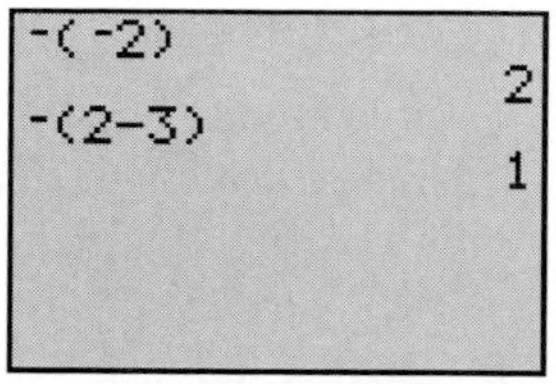

Figure P.7

We know that $a + b = b + a$ for any real numbers a and b, but is $a - b = b - a$ for any real numbers a and b? In general, $a - b$ is not equal to $b - a$. For example, $7 - 3 = 4$ and $3 - 7 = -4$. So subtraction is not commutative. Since

$a - b + b - a = 0$, we can conclude that $a - b$ and $b - a$ are opposites or additive inverses of each other. We summarize these properties of opposites as follows.

Properties of Opposites

For any real numbers a and b:

1. $-1 \cdot a = -a$ (The product of -1 and a is the opposite of a.)
2. $-(-a) = a$ (The opposite of the opposite of a is a.)
3. $-(a - b) = b - a$. (The opposite of $a - b$ is $b - a$.)

EXAMPLE 3 Using properties of opposites

Use the properties of opposites to complete each equation.

a) $-(-\pi) =$ ______ **b)** $-1(-2) =$ ______ **c)** $-1(x - h) =$ ______

Solution

a) $-(-\pi) = \pi$ **b)** $-1(-2) = -(-2) = 2$

c) $-1(x - h) = -(x - h) = h - x$

Relations

Symbols such as $<$, $>$, $=$, $\leq$, and $\geq$ are called **relations** because they indicate how numbers are related. We can visualize these relations by using a number line. For example, $\sqrt{2}$ is located to the right of 0 in Fig. P.5, so $\sqrt{2} > 0$. Since $\sqrt{2}$ is to the left of π in Fig. P.5, $\sqrt{2} < \pi$. In fact, if a and b are any two real numbers, we say that a is less than b (written $a < b$) provided that a is to the left of b on the number line. We say that a is greater than b (written $a > b$) if a is to the right of b on the number line. We say $a = b$ if a and b correspond to the same point on the number line. The fact that there are only three possibilities for ordering a pair of real numbers is called the **trichotomy property**.

Trichotomy Property

For any two real numbers a and b, exactly one of the following is true: $a < b$, $a = b$, or $a > b$.

The trichotomy property is very natural to use. For example, if we know that $r = t$ is false, then we can conclude (using the trichotomy property) that either $r > t$ or $r < t$ is true. If we know that $w + 6 > z$ is false, then we can conclude that $w + 6 \leq z$ is true. The following four properties of equality are also very natural to use, and we often use them without even thinking about them.

Properties of Equality

For any real numbers a, b, and c:

1. $a = a$	**Reflexive property**
2. If $a = b$, then $b = a$.	**Symmetric property**
3. If $a = b$ and $b = c$, then $a = c$.	**Transitive property**
4. If $a = b$, then a and b may be substituted for one another in any expression involving a or b.	**Substitution property**

Absolute Value

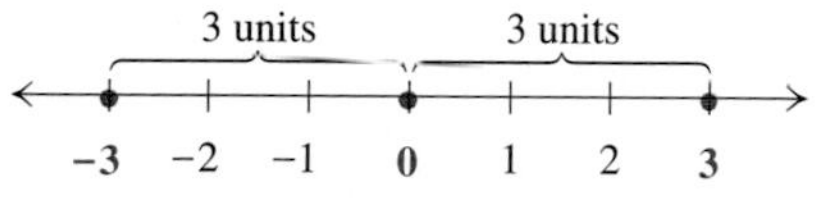

Figure P.8

The **absolute value** of a (in symbols, $|a|$) can be thought of as the distance from a to 0 on a number line. Since both 3 and -3 are three units from 0 on a number line as shown in Fig. P.8, $|3| = 3$ and $|-3| = 3$. A symbolic definition of absolute value is written as follows.

Definition: Absolute Value

For any real number a,

$$|a| = \begin{cases} a & \text{if } a \geq 0 \\ -a & \text{if } a < 0. \end{cases}$$

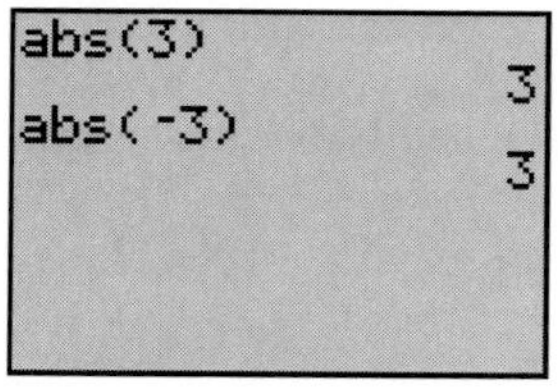

Figure P.9

A calculator typically uses **abs** for absolute value as shown in Fig. P.9. □

The symbolic definition of absolute value indicates that for $a \geq 0$ we use the equation $|a| = a$ (the absolute value of a is just a). For $a < 0$ we use the equation $|a| = -a$ (the absolute value of a is the opposite of a, a positive number).

EXAMPLE 4 Using the definition of absolute value

Use the symbolic definition of absolute value to simplify each expression.

a) $|5.6|$ **b)** $|0|$ **c)** $|-3|$

Solution

a) Since $5.6 \geq 0$, we use the equation $|a| = a$ to get $|5.6| = 5.6$.

b) Since $0 \geq 0$, we use the equation $|a| = a$ to get $|0| = 0$.

c) Since $-3 < 0$, we use the equation $|a| = -a$ to get $|-3| = -(-3) = 3$.

The definition of absolute value guarantees that the absolute value of any number is nonnegative. The definition also implies that additive inverses (or opposites) have the same absolute value. These properties of absolute value and two others are stated as follows.

Properties of Absolute Value

For any real numbers a and b:

1. $|a| \geq 0$ (The absolute value of any number is nonnegative.)
2. $|-a| = |a|$ (Additive inverses have the same absolute value.)
3. $|a \cdot b| = |a| \cdot |b|$ (The absolute value of a product is the product of the absolute values.)
4. $\left|\frac{a}{b}\right| = \frac{|a|}{|b|}$, $b \neq 0$ (The absolute value of a quotient is the quotient of the absolute values.)

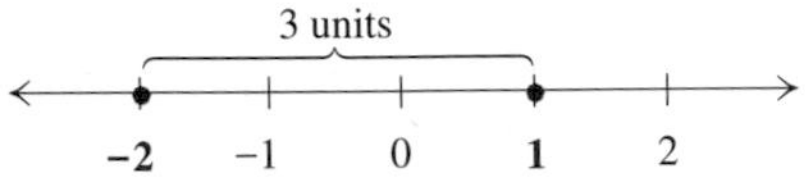

Figure P.10

Absolute value is used in finding the distance between points on a number line. Since 9 lies four units to the right of 5, the distance between 5 and 9 is 4. In symbols, $d(5, 9) = 4$. We can obtain 4 by $9 - 5 = 4$ or $|5 - 9| = 4$. In general, $|a - b|$ gives the distance between a and b for any values of a and b. For example, the distance between -2 and 1 in Fig. P.10 is three units and

$$d(-2, 1) = |-2 - 1| = |-3| = 3.$$

Distance Between Two Points on the Number Line

If a and b are any two points on the number line, then the distance between a and b is $|a - b|$. In symbols, $d(a, b) = |a - b|$.

Note that $d(a, 0) = |a - 0| = |a|$, which is consistent with the definition of absolute value of a as the distance between a and 0 on the number line.

EXAMPLE 5 Distance between two points on a number line

Find the distance between -3 and 5 on the number line.

Solution

The points corresponding to -3 and 5 are shown on the number line in Fig. P.11. The distance between these points is found as follows:

$$d(-3, 5) = |-3 - 5| = |-8| = 8$$

Notice that $d(-3, 5) = d(5, -3)$:

$$d(5, -3) = |5 - (-3)| = |8| = 8$$

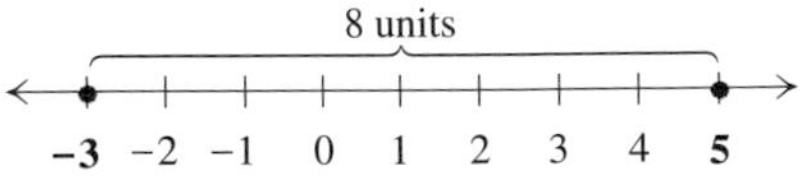

Figure P.11

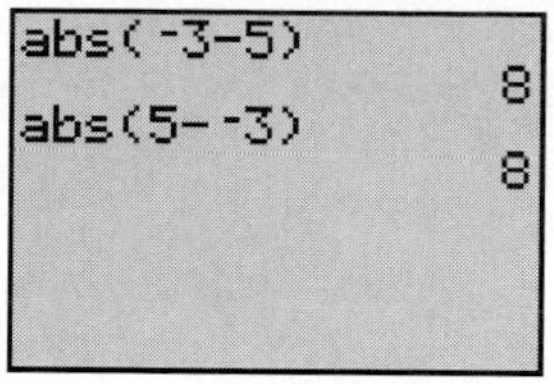

Figure P.12

When you use a calculator to find the absolute value of a difference or a sum, you must use parentheses as shown in Fig. P.12.

Arithmetic Expressions

The result of writing numbers in a meaningful combination with the ordinary operations of arithmetic is called an **arithmetic expression** or simply an expression. Some expressions are

$$\frac{1+3}{2-5},\quad (36+8)+2,\quad 3+5\cdot 6,\quad \text{and}\quad -1(7-9).$$

The **value** of an arithmetic expression is the real number obtained when all operations are performed. Symbols such as parentheses, brackets, braces, absolute value bars, and fraction bars are called **grouping symbols**. When no grouping symbols are used, we **evaluate** an expression (find its value) using the accepted order of operations.

The Order of Operations

1. Perform multiplication and division in order from left to right.
2. Perform addition and subtraction in order from left to right.

EXAMPLE 6 Using the order of operations

Evaluate each expression.

a) $3-4\cdot 2$ **b)** $5\cdot 8\div 4\cdot 2$ **c)** $3-4+9-2$

Solution

a) In an expression that has multiplication and subtraction, multiplication is done first. So $3-4\cdot 2=3-8=-5$.

b) In an expression that has only multiplication and division, the operations are performed from left to right. So $5\cdot 8\div 4\cdot 2=40\div 4\cdot 2=10\cdot 2=20$.

c) In an expression that has only addition and subtraction, the operations are performed from left to right. So $3-4+9-2=-1+9-2=8-2=6$.

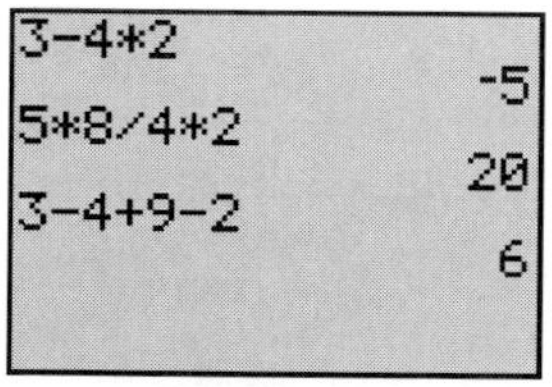

Figure P.13

A graphing calculator follows the order of operations as shown in Fig. P.13. However, some graphing calculators group products that are written without the multiplication symbol. For example, $8\div 4\cdot 2$ may be written on a graphing calculator as 8/4*2 or 8/4(2). Although we would give all three of these expressions the same value, some graphing calculators give the last two expressions different values. You should experiment with your calculator so that you know the order of operations your calculator is using. □

When grouping symbols are used in expressions, we first evaluate within grouping symbols, using the order of operations.

EXAMPLE 7 **Using grouping symbols**

Evaluate each expression. Check your answers with a calculator.

a) $\dfrac{3-9}{-2-(-5)}$ **b)** $2(3-|5-2\cdot 9|)$ **c)** $-17\cdot 3+(-17)\cdot 7$

Solution

a) Since the fraction bar acts as a grouping symbol, we evaluate the numerator and denominator before dividing:

$$\frac{3-9}{-2-(-5)}=\frac{-6}{3}=-2$$

b) First evaluate within the innermost grouping symbols, the absolute value symbols:

$$
\begin{aligned}
2(3-|5-2\cdot 9|) &= 2(3-|5-18|) && \text{Multiply before subtracting.}\\
&= 2(3-|-13|) && \text{Subtract within absolute value.}\\
&= 2(3-13) && \text{Evaluate absolute value.}\\
&= 2(-10) && \text{Subtract within parentheses.}\\
&= -20
\end{aligned}
$$

c) Using the order of operations, we would find each product before we add, but we can also use the distributive property to make the computation easier:

$$-17\cdot 3+(-17)\cdot 7=-17(3+7)=-17(10)=-170$$

All three expressions are shown on a graphing calculator in Fig. P.14. Note that the numerator and denominator must be enclosed in parentheses if the fraction is not in built-up form.

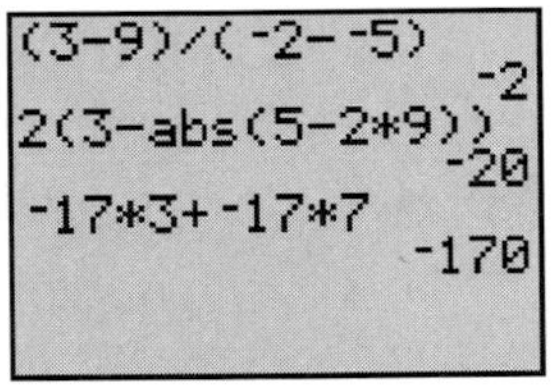

Figure P.14

Algebraic Expressions

When we write numbers and one or more variables in a meaningful combination with the ordinary operations of arithmetic, the result is called an **algebraic expression**, or simply an expression. Some examples are

$$7x,\quad \frac{1}{x},\quad \frac{a^2-b^2}{2},\quad \sqrt{x},\quad \text{and}\quad 3x+5x.$$

The **value of an algebraic expression** is the value of the arithmetic expression that is obtained when the variables are replaced by real numbers. Of course, we are not allowed to replace a variable by a number that will result in an expression that does not represent a real number. The **domain** of an algebraic expression is the set of real numbers that *are* allowed to be used for the variable. For example, the domain of $1/x$ is the set of nonzero real numbers. Zero is excluded because division by 0 is undefined. The domain of $\sqrt{x}$ is the set of nonnegative real numbers. The negative real numbers are excluded because the square root of a negative number is not a real number. Two algebraic expressions are **equivalent** if they have the same domain and if they have the same value for each x-value in the domain.

To **simplify** an expression means to find a simpler-looking equivalent expression. To simplify an expression such as $-1(7x)$ we use the associative property as follows:

$$-1(7x) = (-1 \cdot 7)x = -7x$$

Because the associative property is true for all real numbers, the equation $-1(7x) = -7x$ is true for any real number x. Therefore the expressions $-1(7x)$ and $-7x$ are equivalent.

A single number or the product of a number and one or more variables raised to powers is called a **term**. Expressions such as $3x$, $2kab^3$, and πr^2 are terms. In the term $3x$, 3 and x are called **factors**. The **coefficient** of any variable part of a term is the product of the remaining factors in the term. For example, the coefficient of x in $3x$ is 3. The coefficient of ab^3 in $2kab^3$ is $2k$ and the coefficient of b^3 is $2ka$. The coefficient of x in the equation $y = mx + b$ is m. If two terms contain the same variables with the same exponents, then they are called **like terms**. It is the distributive property that allows us to **combine like terms**:

$$\begin{aligned} -4x + 7x &= (-4 + 7)x && \text{Distributive property} \\ &= 3x \end{aligned}$$

Note that $-4x + 7x = 3x$ is true for any real number x. The expressions $-4x + 7x$ and $3x$ are equivalent.

The distributive property is used along with other properties to simplify expressions that have parentheses. We do not usually write down every step to simplify algebraic expressions, but we must be aware that we cannot take a step that is not justified by a property.

EXAMPLE 8 **Using properties to simplify an expression**

Simplify each expression.

a) $-4x - (6 - 7x)$ **b)** $\frac{1}{2}x - \frac{3}{4}x$ **c)** $-6(x - 3) - 3(5 - 7x)$

Solution

a) $-4x - (6 - 7x) = -4x + [-(6 + (-7x))]$ — Definition of subtraction

$= -4x + [-1(6 + (-7x))]$ — First property of opposites

$= -4x + [(-6) + 7x]$ — Distributive property

$= [-4x + 7x] + (-6)$ — Commutative and associative properties

$= 3x - 6$ — Combine like terms.

b) $\frac{1}{2}x - \frac{3}{4}x = \frac{2}{4}x - \frac{3}{4}x = -\frac{1}{4}x$ — Write $\frac{1}{2}$ as $\frac{2}{4}$ to obtain a common denominator.

c) $-6(x - 3) - 3(5 - 7x) = -6x + 18 - 15 + 21x$ — Distributive property

$= 15x + 3$ — Combine like terms.

It is often necessary to replace the variables in an expression with numbers and evaluate the expression.

EXAMPLE 9 Evaluating an expression

The expression $\frac{y_2 - y_1}{x_2 - x_1}$ is used to find the slope of a line. (We will explore this concept in Section 3.1.) Find the value when $x_1 = -3$, $x_2 = 5$, $y_1 = 6$, and $y_2 = -1$.

Solution

Use the substitution property to replace the variables by the appropriate numbers:

$$\frac{y_2 - y_1}{x_2 - x_1} = \frac{-1 - 6}{5 - (-3)} = \frac{-7}{8} = -\frac{7}{8}$$

FOR THOUGHT True or False? Explain.

1. Zero is the only number that is both rational and irrational.
2. Between any two distinct rational numbers there is another rational number.
3. Between any two distinct real numbers there is an irrational number.
4. Every real number has a multiplicative inverse.

5. If a is not less than and not equal to 3, then a is greater than 3.

6. If $a \leq w$ and $w \leq z$, then $a < z$.

7. For any real numbers a, b, and c, $a - (b - c) = (a - b) - c$.

8. If a and b are any two real numbers, then the distance between a and b on the number line is $a - b$.

9. Calculators give only rational answers.

10. For any real numbers a and b, the opposite of $a + b$ is $a - b$.

P.1 EXERCISES

Tape 1 Disk

Determine whether each statement is true or false. Explain. Sets are denoted by R for the real numbers, I for the irrational numbers, Q for the rational numbers, J for the integers, W for the whole numbers, and N for the natural numbers.

1. $\sqrt{2} \in R$
2. $\sqrt{3} \in Q$
3. $0 \notin I$
4. $-6 \notin J$
5. $J \subseteq R$
6. $I \subseteq Q$
7. $R \nsubseteq Q$
8. $N \nsubseteq W$

Determine which elements of the set $\{-3.5, -\sqrt{2}, -1, 0, 1, \sqrt{3}, 3.14, \pi, 4.3535\ldots, 5.090090009\ldots\}$ are members of the following sets.

9. Real numbers
10. Rational numbers
11. Irrational numbers
12. Integers
13. Whole numbers
14. Natural numbers

Complete each statement using the property named.

15. $7 + x =$ ________, commutative
16. $5(4y) =$ ________, associative
17. $5(x + 3) =$ ________, distributive
18. $-3(x - 4) =$ ________, distributive
19. $5x + 5 =$ ________, distributive
20. $-5x + 10 =$ ________, distributive
21. $-13 + (4 + x) =$ ________, associative
22. $yx =$ ________, commutative
23. $0.125($________$) = 1$, multiplicative inverse
24. $-(-3) + ($________$) = 0$, additive inverse

Use the properties of opposites to complete each equation.

25. $-(-\sqrt{3}) =$ ______
26. $-1(-6.4) =$ ______
27. $-1(x^2 - y^2) =$ ______
28. $-(1 - a^2) =$ ______

Use the symbolic definition of absolute value to simplify each expression.

29. $|7.2|$
30. $|0/3|$
31. $|-\sqrt{5}|$
32. $|-3/4|$

Find the distance on the number line between each pair of numbers.

33. 8, 13.5
34. −7.5, 0
35. −5, 17
36. −3.2, −14.9

Evaluate each expression without a calculator. Use a calculator to check.

37. $4 \div 5 \cdot 6 \cdot 3$
38. $4 \cdot 3 - 7(-6)$
39. $3 - 4 + 5 - 7 \cdot 6$
40. $64 \cdot 2 \div 8 \div 4 \cdot 6$
41. $(52 + 39) + (-39)$
42. $(7 - 15) - 13$
43. $49 \cdot 6 + 49 \cdot 4$
44. $-12 \cdot 39 + 2 \cdot 39$
45. $26 \cdot \frac{1}{5} \div \frac{1}{2} \cdot 5$
46. $\frac{4}{3} \cdot 50(0.75)2$
47. $|4| - |-3|$
48. $-8 - |-9|$
49. $|3 - 4.6| - 5$
50. $5 - |4 - 2.3|$
51. $(12 - 1)(12 + 1)$
52. $-2 - 3(5 - 2 \cdot 8)$
53. $2 - 3|3 - 4 \cdot 6|$
54. $1 - (3 - |1 - 2 \cdot 3|)$
55. $(2 - 8 \div 2)(8 \div 2 + 3)$
56. $2(2 - |-3| + 1)$
57. $\dfrac{-2 - (-6)}{-5 - (-9)}$
58. $\dfrac{4 - (-3)}{-3 - (-1)}$
59. $(598 + 432)(-3 \cdot 9 + 27)$
60. $(23 \cdot 5 + 99)(-3 \cdot 17 + 51)$

Use the properties of the real numbers to simplify each expression.

61. $-5x + 3x$
62. $-5x - (-8x)$
63. $x - 0.15x$
64. $x + 3 - 0.9x$
65. $-3(2xy)$
66. $\frac{1}{2}(8wz)$
67. $-8w(3x)$
68. $\frac{1}{2}x(-3y)$
69. $(3 - 4x) + (x - 9)$
70. $(9x - 3) + (4 - 6x)$
71. $x - 0.03(x + 200)$
72. $y - 0.9(y - 3000)$
73. $3 - 4zy - (5 - 6zy)$
74. $4x - 3z - (3x - 5z)$
75. $\frac{1}{3}x + \frac{1}{4}x$
76. $\frac{1}{3}y - \frac{1}{2}y$
77. $5x\left(\frac{3y}{20}\right)$
78. $-\frac{3}{4}z\left(\frac{7}{9}wt\right)$
79. $\frac{1}{2}(6 - 4x)$
80. $\frac{1}{4}(8x - 4)$
81. $\frac{6x - 2y}{2}$
82. $\frac{-9 - 6x}{-3}$
83. $-3\left(\frac{1}{2} - \frac{1}{4}y\right) + 6\left(\frac{3}{4} - \frac{1}{3}y\right)$
84. $\frac{1}{3}\left(\frac{3}{5}x - \frac{1}{3}\right) - \frac{1}{2}\left(\frac{1}{3}x - \frac{2}{5}\right)$

Evaluate $\frac{y_2 - y_1}{x_2 - x_1}$ in each of the following cases.

85. $x_1 = -2, x_2 = 3, y_1 = 5, y_2 = -7$
86. $x_1 = 3, x_2 = -4, y_1 = 6, y_2 = -1$
87. $x_1 = -8, x_2 = 5, y_1 = 3, y_2 = 3$
88. $x_1 = 3.9, x_2 = 4, y_1 = 5, y_2 = 8$

Solve each problem.

89. Arrange the following numbers in order from smallest to largest: $\frac{1}{2}, -\frac{1}{2}, \frac{1}{3}, -\frac{1}{3}, 0, \frac{5}{12}$, and $-\frac{5}{12}$. Do not use a calculator.

90. Use a calculator to help you arrange the following numbers in order from smallest to largest: $\frac{10}{3}, \sqrt{10}, \frac{22}{7}, \pi$, and $\frac{157}{50}$.

• CALCULUS •

91. *Cigarette Consumption* The accompanying graph shows cigarette consumption, male death rate, and female death rate in the U.S. since 1900 (Scientific American, June 1997). For what years did Americans consume more than 175 packs per person? In what year did cigarette consumption reach its maximum?

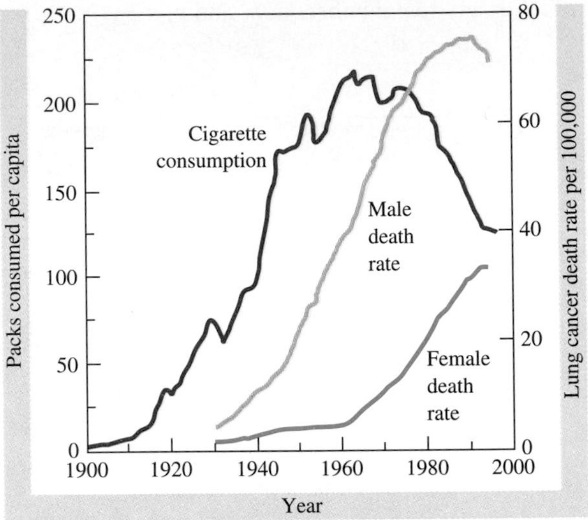

Figure for Exercises 91 and 92

92. *Male Lung Cancer Death Rate* For what years was the male death rate from lung cancer increasing? How many years are there between the year of maximum cigarette consumption and the year of maximum male death rate? Can you make any predictions about the female death rate?

93. *Target Heart Rate* If a is your age and r is your resting heart rate, then the expression

$$0.60(220 - a - r) + r$$

is used to find your target heart rate for a cardiovascular workout. (Cycling, Burkett and Darst, 1987). Simplify the expression. Find your target heart rate.

94. *High-Income Bracket* The annual salary for professional baseball player Cecil ''Big Daddy'' Fielder was \$9,237,500 in 1997 (ESPN). The expression

$$79{,}445 + 0.396(x - 263{,}750)$$

is used to find the federal income tax for a taxpayer with x dollars of taxable income, provided $x > 263{,}750$. Simplify the expression. Find the federal income tax on a taxable income of \$9,237,500.

95. *Estimating Costs* The accompanying graph shows the estimated cost E and the actual cost A for repairs done on five cars at Bill Poole Ford. Bill does not want the actual cost to differ from the estimated cost by more than 10% of the estimated cost. For which cars is $|A - E|$ greater than 10% of E?

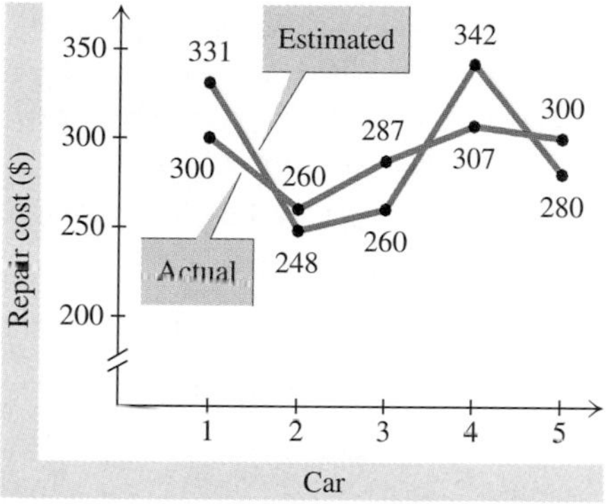

Figure for Exercises 95 and 96

96. *Free Wax* Bill Poole Ford does a free wax job on any car for which $|E - A| > 20$. Which of the cars listed in the accompanying graph get a free wax job?

97. *Modern American Trick* Think of a number between 1 and 10. Think of the product of your number and 9. Think of the sum of the digits in your answer. Think of the number that is 5 less than that sum. Think of the letter in the alphabet that corresponds to the number you are thinking about. (For example, if your number is seven, pick the seventh letter in the alphabet.) Think of a state that begins with the letter. Think of the second letter in the state. Think of a big animal that begins with that letter. Think of the color of that animal. Now check the answer in the back of this text to see what you are thinking about. Explain this trick.

Figure for Exercise 98

98. *Ancient Egyptian Trick* The following problem taken from the Rhind papyrus is believed to have originated in Egypt somewhere between 1849 and 1801 B.C.

Think of a number, and add 2/3 of this number to itself. From this sum, subtract 1/3 of its value and say what your answer is. From your answer subtract 1/10 of your answer. You will now have your original number. Try this procedure for yourself (with a multiple of 9 for simplicity). Verify that the following expression is what is described in the problem, then simplify the expression:

$$\left(x + \frac{2}{3}x\right) - \frac{1}{3}\left(x + \frac{2}{3}x\right) - \frac{1}{10}\left\{\left(x + \frac{2}{3}x\right) - \frac{1}{3}\left(x + \frac{2}{3}x\right)\right\}$$

For Writing/Discussion

99. *Nightshade-Snakeroot Drink* Slinky villainess, Poison Ivy, is demonstrating for Mr. Freeze her latest concoction. Ms. Ivy has one quart of deadly nightshade and one quart of extract of black snakeroot. She pours some nightshade into the snakeroot and mixes it up. She then pours the same amount of the mixture back into the nightshade flask so that she ends up with one quart of each mixture. Ms. Ivy challenges Freeze to determine whether there is now more nightshade in the snakeroot or more snakeroot in the nightshade. What do you think? Explain your answer.

100. *Number Line* To make a number line, we arbitrarily select a length for one unit and a starting point 0. We use the one-unit length to locate points for the integers. Explain how the figure shown here locates the point corresponding to $\sqrt{2}$.

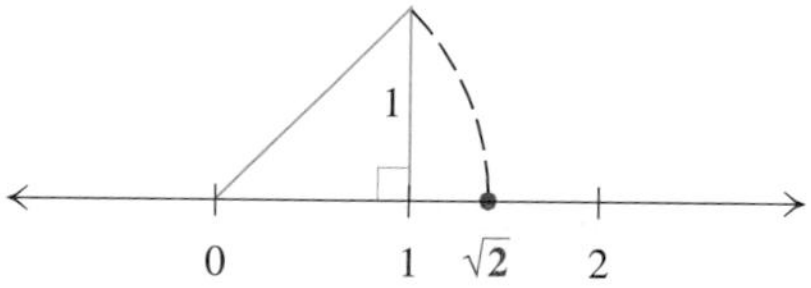

Figure for Exercise 100

101. *Locating Points* Explain how you can use geometry to locate a point on a number line corresponding to $\sqrt{3}$.

Linking Concepts

For Individual or Group Explorations

Rationals and Irrationals *In this section we studied many properties of the real numbers. A property that we did not mention is* density. *In simple terms, the rational numbers are a dense subset of the real numbers because they are found virtually everywhere. There is no escaping the rational numbers!*

a) Find a rational number between the rational numbers 0.156 and 0.157.

b) Find a rational number between 15/19 and 263/333.

c) Find a rational number between the irrational numbers $\sqrt{17}$ and $\sqrt[3]{70}$.

d) Write a detailed algorithm (or procedure) for finding a rational number between any two real numbers.

e) The irrational numbers are also a dense subset of the real numbers. Find an irrational number between each given pair of real numbers in parts (a), (b), and (c).

f) Write a detailed algorithm for finding an irrational number between any two real numbers.

P.2 Integral Exponents

One of the most convenient notations of algebra is the notation of exponents. Exponents are used to indicate repeated multiplication, and to represent very large and very small numbers. In this section we will review the definitions and rules concerning exponents and simplify expressions that have integral exponents.

Positive Integral Exponents

We use positive integral exponents to indicate the number of times a factor occurs in a product. For example, $2 \cdot 2 \cdot 2 \cdot 2$ is written as 2^4 using exponents. We read 2^4 as "the fourth power of 2" or "2 to the fourth power." We have the following definition.

Definition: Positive Integral Exponents

For any positive integer n,

$$a^n = \underbrace{a \cdot a \cdot a \cdot \cdots \cdot a}_{n \text{ factors of } a}.$$

We call a the **base**, n the **exponent** or **power**, and a^n an **exponential expression**.

We read a^n as "a to the nth power." For a^1 we usually omit the exponent and just write a. We refer to expressions having the exponents 2 and 3 as squares and cubes. For example, 3^2 is read "3 squared," 2^3 is read "2 cubed," x^4 is read "x to the fourth," b^5 is read "b to the fifth," and so on.

Exponential expressions are given the first priority in the order of operations. For example, $5 + 2 \cdot 3^2 = 5 + 2 \cdot 9 = 5 + 18 = 23$. Operations within grouping symbols are always performed before other operations. For example, $2(3 + 1)^2 = 2 \cdot 4^2 = 2 \cdot 16 = 32$. In an expression such as -3^2 we square 3 first, then take the opposite. So $-3^2 = -9$ and $(-3)^2 = (-3)(-3) = 9$.

EXAMPLE 1 **Evaluating arithmetic expressions that have exponents**

Evaluate each expression without using a calculator. Use a calculator to check.

a) $(-5)^2 - 4(-1)(-3)$ **b)** $(3 - (-2))^2 + (-1 - 5)^2$ **c)** $-2^4 - (-10)^3$

Solution

a) $(-5)^2 - 4(-1)(-3) = 25 - 12$ Powers and products before subtraction

$= 13$

b) $(3 - (-2))^2 + (-1 - 5)^2 = 5^2 + (-6)^2$ Evaluate within parentheses first.

$= 25 + 36$

$= 61$

c) $-2^4 - (-10)^3 = -16 - (-1000)$ Evaluate powers first.

$= -16 + 1000$

$= 984$

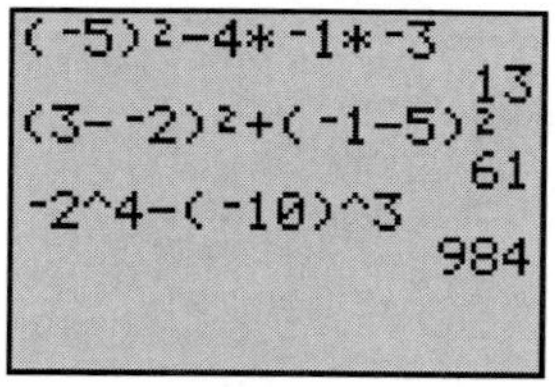

Figure P.15

Figure P.15 shows the three expressions evaluated with a graphing calculator. Notice the x^2 key is used to square a number and that the power key ^ is used for other powers.

EXAMPLE 2 **Evaluating algebraic expressions that have exponents**

Evaluate $3xy^4 - z^3$ if $x = 2$, $y = -1$, and $z = -3$.

Solution

Replace x, y, and z by 2, -1, and -3, respectively, and then evaluate:

$$3 \cdot 2 \cdot (-1)^4 - (-3)^3 = 3 \cdot 2 \cdot 1 - (-27) = 6 + 27 = 33$$

Negative Integral Exponents

We use a negative sign in an exponent to represent multiplicative inverses or reciprocals. For example, 3^{-2} represents the reciprocal of 3^2. Because $3^2 = 9$, $3^{-2} = \frac{1}{9}$. For negative exponents we do not allow the base to be zero because zero does not have a reciprocal. Negative exponents are used to represent reciprocals to make the rules for simplifying exponential expressions work smoothly.

Definition: Negative Integral Exponents

If a is a nonzero real number and n is a positive integer,

$$a^{-n} = \frac{1}{a^n}.$$

EXAMPLE 3 Evaluating expressions that have negative exponents

Simplify each expression without using a calculator, then check with a calculator.

a) $3^{-1} \cdot 5^{-2} \cdot 10^2$ **b)** $\left(\frac{2}{3}\right)^{-3}$ **c)** $\frac{6^{-2}}{2^{-3}}$

Solution

a) $3^{-1} \cdot 5^{-2} \cdot 10^2 = \frac{1}{3} \cdot \frac{1}{5^2} \cdot 100 = \frac{1}{3} \cdot \frac{1}{25} \cdot 100 = \frac{100}{75} = \frac{4}{3}$

b) $\left(\frac{2}{3}\right)^{-3} = \frac{1}{\left(\frac{2}{3}\right)^3} = \frac{1}{\frac{2}{3} \cdot \frac{2}{3} \cdot \frac{2}{3}} = \frac{1}{\frac{8}{27}} = \frac{27}{8}$ Note that $\left(\frac{2}{3}\right)^{-3} = \left(\frac{3}{2}\right)^3$.

c) $\frac{6^{-2}}{2^{-3}} = \frac{\frac{1}{6^2}}{\frac{1}{2^3}} = \frac{1}{6^2} \cdot \frac{2^3}{1} = \frac{8}{36} = \frac{2}{9}$ Note that $\frac{6^{-2}}{2^{-3}} = \frac{2^3}{6^2}$.

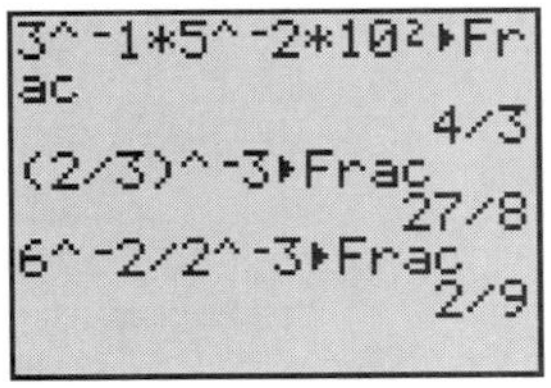

Figure P.16

These three expressions are shown on a graphing calculator in Fig. P.16. Note that the fractional base must be in parentheses. The fraction feature was used to get fractional answers.

Example 3(b) illustrates the fact that a fractional base can be inverted, if the sign of the exponent is changed. Example 3(c) illustrates the fact that a factor of the numerator or denominator can be moved from the numerator to the denominator or vice versa as long as we change the sign of the exponent. These rules follow from the definition of negative exponents.

Rules for Negative Exponents and Fractions

If a and b are nonzero real numbers and m and n are integers, then

$$\left(\frac{a}{b}\right)^{-m} = \left(\frac{b}{a}\right)^{m} \quad \text{and} \quad \frac{a^{-m}}{b^{-n}} = \frac{b^n}{a^m}.$$

Figure P.17

Using this rule, we could shorten Examples 3(b) and (c) as follows:

$$\left(\frac{2}{3}\right)^{-3} = \left(\frac{3}{2}\right)^{3} = \frac{27}{8} \quad \text{and} \quad \frac{6^{-2}}{2^{-3}} = \frac{2^3}{6^2} = \frac{8}{36} = \frac{2}{9}.$$

Note that we cannot apply these rules when addition or subtraction is involved.

$$\frac{1 + 3^{-2}}{2^{-3}} \neq \frac{1 + 2^3}{3^2}$$

Figure P.17 shows these expressions on a graphing calculator. □

Rules of Exponents

Consider the product $a^2 \cdot a^3$. Using the definition of exponents, we can simplify this product as follows:

$$a^2 \cdot a^3 = (a \cdot a)(a \cdot a \cdot a) = a^5$$

Similarly, if m and n are any positive integers we have

$$a^m \cdot a^n = \underbrace{\overbrace{a \cdot a \cdot \cdots \cdot a}^{m \text{ factors}} \cdot \overbrace{a \cdot a \cdot \cdots \cdot a}^{n \text{ factors}}}_{m + n \text{ factors}} = a^{m+n}.$$

This equation indicates that the product of exponential expressions *with the same base* is obtained by adding the exponents. This fact is called the **product rule**.

EXAMPLE 4 Using the product rule

Simplify each expression.

a) $(3x^8y^2)(-2xy^4)$ **b)** $2^3 \cdot 3^2$

Solution

a) Use the product rule to add the exponents when bases are identical:

$$(3x^8y^2)(-2xy^4) = -6x^9y^6$$

b) Since the bases are different we cannot use the product rule, but we can simplify the expression using the definition of exponents:

$$2^3 \cdot 3^2 = 8 \cdot 9 = 72$$

So far we have defined positive and negative integral exponents. The definition of zero as an exponent is given in the following box. Note that the zero power of zero is not defined.

Definition: Zero Exponent

If a is a nonzero real number, then $a^0 = 1$.

The definition of zero exponent allows us to extend the product rule to any integral exponents. For example, using the definition of negative exponents, we get

$$2^{-3} \cdot 2^3 = \frac{1}{2^3} \cdot 2^3 = 1.$$

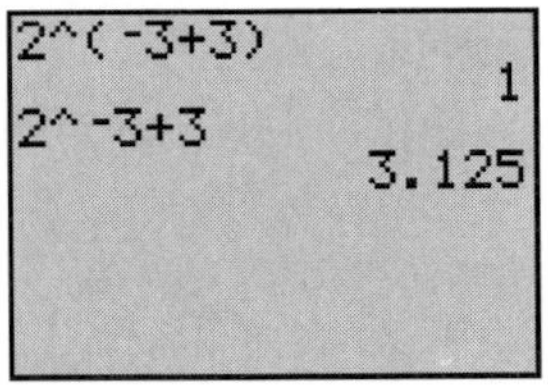

Figure P.18

Adding exponents, we get $2^{-3} \cdot 2^3 = 2^{-3+3} = 2^0$. The answer is the same because 2^0 is defined to be 1.

To evaluate 2^{-3+3} on a calculator, the expression $-3 + 3$ must be in parentheses as in Fig. P.18. □

Using the definitions of positive, negative, and zero exponents, we can show that the product rule and several other rules hold for any integral exponents. We list these rules in the following box.

Rules for Integral Exponents

If a and b are nonzero real numbers and m and n are integers, then

1. $a^m a^n = a^{m+n}$ **Product rule**
2. $\dfrac{a^m}{a^n} = a^{m-n}$ **Quotient rule**
3. $(a^m)^n = a^{mn}$ **Power of a power rule**
4. $(ab)^n = a^n b^n$ **Power of a product rule**
5. $\left(\dfrac{a}{b}\right)^n = \dfrac{a^n}{b^n}$ **Power of a quotient rule**

The rules for integral exponents are used to simplify expressions.

EXAMPLE 5 Simplifying expressions with integral exponents

Simplify each expression. Write your answer without negative exponents. Assume that all variables represent nonzero real numbers.

a) $(3x^2y^3)(-4x^{-2}y^{-5})$ **b)** $\dfrac{-6a^5b^{-1}}{2a^7b^{-3}}$

Solution

a) $(3x^2y^3)(-4x^{-2}y^{-5}) = -12x^{2+(-2)}y^{3+(-5)}$ Product rule

$= -12x^0y^{-2}$ Simplify the exponents.

$= -\frac{12}{y^2}$ Definition of negative and zero exponents

b) $\frac{-6a^5b^{-1}}{2a^7b^{-3}} = -3a^{5-7}b^{-1-(-3)}$ Quotient rule

$= -3a^{-2}b^2$ Simplify the exponents.

$= -\frac{3b^2}{a^2}$ Definition of negative exponents

In the next example, we use the rules of exponents to simplify expressions that have variables in the exponents.

EXAMPLE 6 **Simplifying expressions with variable exponents**

Simplify each expression. Assume that all bases are nonzero real numbers and all exponents are integers.

a) $(-3x^{a-5}y^{-3})^4$ b) $\left(\frac{3a^{2m-1}}{2a^{-3m}}\right)^{-3}$

Solution

a) $(-3x^{a-5}y^{-3})^4 = (-3)^4(x^{a-5})^4(y^{-3})^4$ Power of a product rule

$= 81x^{4a-20}y^{-12}$ Power of a power rule

$= \frac{81x^{4a-20}}{y^{12}}$ Definition of negative exponents

b) $\left(\frac{3a^{2m-1}}{2a^{-3m}}\right)^{-3} = \frac{3^{-3}(a^{2m-1})^{-3}}{2^{-3}(a^{-3m})^{-3}}$ Power of a quotient rule

$= \frac{2^3a^{-6m+3}}{3^3a^{9m}}$ Power of a power rule and definition of negative exponents

$= \frac{8a^{-15m+3}}{27}$ Quotient rule

Scientific Notation

Archimedes (287–212 B.C.) was a brilliant Greek inventor and mathematician who studied at the Egyptian city of Alexandria, then the center of the scientific world. Archimedes used his knowledge of mathematics to calculate for King

Gelon the number of grains of sand in the universe: 1 followed by 63 zeros. Of course, Archimedes' universe was different from our universe, and he performed his computations with Greek letter numerals, since the modern number system and scientific notation had not yet been invented. Although it is impossible to calculate the number of grains of sand in the universe, this story illustrates how long scientists have been interested in quantities ranging in size from the diameter of our galaxy to the diameter of an atom. Scientific notation offers a convenient way of expressing very large or very small numbers.

In scientific notation, a positive number is written as a product of a number between 1 and 10 and a power of 10. For example, 9.63×10^7 and 2.3×10^{-6} are numbers written in scientific notation.

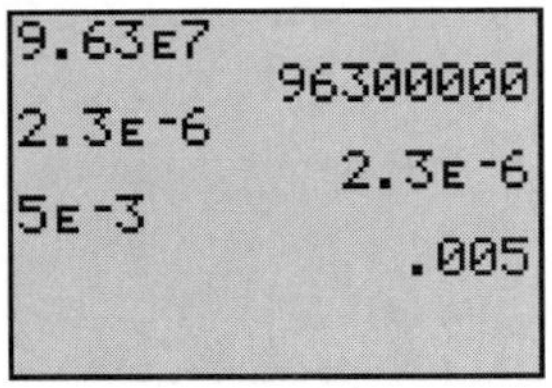

Figure P.19

These numbers are shown on a graphing calculator in scientific notation in Fig. P.19. Note that only the power of 10 shows in scientific notation on a calculator. The graphing calculator shown in Fig. P.19 converts to standard notation when the ENTER key is pressed (provided the number is neither too large nor too small. □

Conversion of numbers from scientific notation to standard notation is actually just multiplication.

EXAMPLE 7 Scientific notation to standard notation

Convert each number to standard notation.

a) 9.63×10^7 **b)** 2.3×10^{-6}

Solution

a) $9.63 \times 10^7 = 9.63 \times 10{,}000{,}000$ Evaluate 10^7.

$= 96{,}300{,}000$ Move decimal point seven places to the right.

b) $2.3 \times 10^{-6} = 2.3 \times \dfrac{1}{1{,}000{,}000}$ Evaluate 10^{-6}.

$= 2.3 \times 0.000001$

$= 0.0000023$ Move decimal point six places to the left.

Observe how the decimal point is relocated in Example 7. Converting from scientific notation to standard notation is simply a matter of moving the decimal point. To convert a number from scientific to standard notation, we move the decimal point the number of places indicated by the power of 10. Move the decimal point to the right for a positive power and to the left for a negative power. Note that in scientific notation a number greater than 0 is written with a positive power of 10 and a number less than 1 is written with a negative power of 10. Numbers between 1 and 10 are not written in scientific notation.

In the next example, we convert from standard notation to scientific notation by reversing the process used in Example 7.

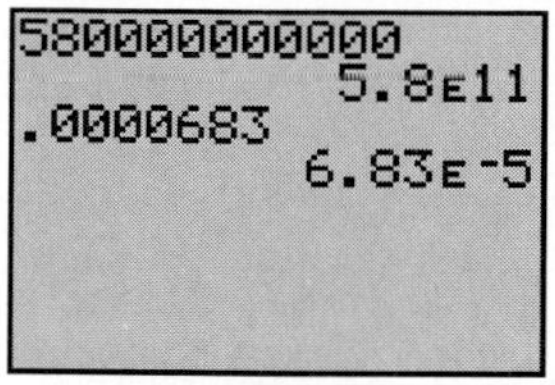

Figure P.20

A graphing calculator converts to scientific notation when you press ENTER as shown in Fig. P.20. □

EXAMPLE 8 Standard notation to scientific notation

Convert each number to scientific notation.

a) 580,000,000,000 **b)** 0.0000683

Solution

a) Determine the power of 10 by counting the number of places that the decimal must move so that there is a single digit to the left of the decimal point (11 places). Since 580,000,000,000 is larger than 10, we use a positive power of 10:

$$580{,}000{,}000{,}000 = 5.8 \times 10^{11}$$

b) Determine the power of 10 by counting the number of places the decimal must move so that there is a single digit to the left of the decimal point (five places). Since 0.0000683 is smaller than 1, we use a negative power of 10:

$$0.0000683 = 6.83 \times 10^{-5}$$

To convert a negative number to scientific notation, convert as you would a positive number and retain the negative sign:

$$-0.000000359 = -3.59 \times 10^{-7}$$

One advantage of scientific notation is that the rules of exponents can be used when performing certain computations involving scientific notation. Calculators can be used to perform computations with scientific notation, but it is good to practice some computation without a calculator.

EXAMPLE 9 Using scientific notation in computations

Perform the indicated operations without a calculator. Write your answers in scientific notation. Check your answers with a calculator.

a) $(4 \times 10^{13})(5 \times 10^{-9})$ **b)** $\dfrac{-1.2 \times 10^{-9}}{5 \times 10^{-7}}$ **c)** $\dfrac{(2{,}000{,}000{,}000)^3(0.00009)}{600{,}000{,}000}$

Solution

a)
$$\begin{aligned}(4 \times 10^{13})(5 \times 10^{-9}) &= 20 \times 10^{13+(-9)} && \text{Product rule for exponents}\\ &= 20 \times 10^{4} && \text{Simplify the exponent.}\\ &= 2 \times 10^{1} \times 10^{4} && \text{Write 20 in scientific notation.}\\ &= 2 \times 10^{5} && \text{Product rule for exponents}\end{aligned}$$

b) $\dfrac{-1.2 \times 10^{-9}}{4 \times 10^{-7}} = \dfrac{-1.2}{4} \times 10^{-9-(-7)}$ Quotient rule for exponents

$= -0.3 \times 10^{-2}$ Simplify the exponent.

$= -3 \times 10^{-3}$ Use $-0.3 = -3 \times 10^{-1}$.

c) First convert each number to scientific notation, then use the rules of exponents to simplify:

$$\frac{(2{,}000{,}000{,}000)^3(0.00009)}{600{,}000{,}000} = \frac{(2 \times 10^9)^3(9 \times 10^{-5})}{6 \times 10^8}$$

$$= \frac{(8 \times 10^{27})(9 \times 10^{-5})}{6 \times 10^8}$$

$$= \frac{72 \times 10^{22}}{6 \times 10^8} = 12 \times 10^{14} = 1.2 \times 10^{15}$$

These three computations are done on a calculator in Fig. P.21. Set the mode to scientific to get the answers in scientific notation.

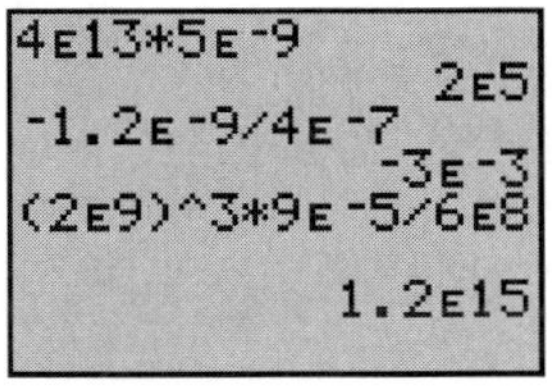

Figure P.21

In the next example we use scientific notation to perform the type of computation performed by Archimedes when he attempted to determine the number of grains of sand in the universe.

EXAMPLE 10 The number of grains of sand in Archimedes' earth

If the radius of the earth is approximately 6.38×10^3 kilometers and the radius of a grain of sand is approximately 1×10^{-3} meters, then what number of grains of sand have a volume equal to the volume of the earth?

Solution

Since the volume of a sphere is given by $V = \frac{4}{3}\pi r^3$, the volume of the earth is

$$\frac{4}{3}\pi(6.38 \times 10^3)^3 \text{ km}^3 = 1.09 \times 10^{12} \text{ km}^3.$$

Since 1 km $= 10^3$ m, the radius of a grain of sand is 1×10^{-6} km and its volume is

$$\frac{4}{3}\pi(1 \times 10^{-6})^3 \text{ km}^3 = 4.19 \times 10^{-18} \text{ km}^3.$$

To get the number of grains of sand, divide the volume of the earth by the volume of a grain of sand:

Figure P.22

$$\frac{1.09 \times 10^{12}\text{ km}^3}{4.19 \times 10^{-18}\text{ km}^3} = 2.60 \times 10^{29}$$

See Fig. P.22 for the computations.

For Thought True or False? Explain. Do Not Use a Calculator.

1. $2^{-1} + 2^{-1} = 1$ **2.** $2^{100} = 4^{50}$ **3.** $9^8 \cdot 9^8 = 81^8$

4. $(0.25)^{-1} = 4$ **5.** $\dfrac{5^{10}}{5^{-12}} = 5^{-2}$ **6.** $2 \cdot 2 \cdot 2 \cdot 2^{-1} = \dfrac{1}{16}$

7. $-3^{-3} = -\dfrac{1}{27}$ **8.** $\left(\dfrac{3}{4}\right)^{-2} = \left(\dfrac{4}{3}\right)^{2}$ **9.** $10^{-4} = 0.00001$

10. $98.6 \times 10^8 = 9.86 \times 10^7$

P.2 Exercises

Tape 1 Disk

Evaluate each expression without using a calculator. Check your answers with a calculator.

1. 4^3 **2.** -3^4 **3.** -4^2 **4.** $5 \cdot 10^4$

5. $\left(\dfrac{1}{2}\right)^3$ **6.** $\left(-\dfrac{3}{4}\right)^4$

7. $7^2 - 2(-3)(-6)$ **8.** $(-3)^2 - 4(-2)(-5)$

9. $(-2 - 3)^2 + (4 - (-1))^2$

10. $(5 - (-2))^2 + (6 - 7)^2$

Use $a = -2$, $b = 3$, and $c = -4$ to evaluate each algebraic expression.

11. $a^2 - b^2$ **12.** $a^2 + b^2$

13. $(a - b)(a + b)$ **14.** $(a + b)^2$

15. $(a - b)(a^2 + ab + b^2)$ **16.** $(a + b)(a^2 - ab + b^2)$

17. $a^b + c^b$ **18.** $(a + c)^b$

Evaluate each expression without using a calculator. Check your answers with a calculator.

19. 3^{-4} **20.** $\dfrac{1}{2^{-3}}$ **21.** $\dfrac{1}{5^{-2}}$

22. $-2 \cdot 10^{-3}$ **23.** $6^{-1} + 5^{-1}$ **24.** $2^0 + 2^{-1}$

25. $\dfrac{3^{-2}}{6^{-3}}$ **26.** $\dfrac{3^{-1}}{2^3}$

27. $\left(\dfrac{1}{2}\right)^{-3}$ **28.** $\left(-\dfrac{1}{10}\right)^{-4}$

29. $4 \cdot 4 \cdot 4 \cdot 4^{-1}$ **30.** $-1^{-1} \cdot (-2)^{-2}$

31. $5 \cdot 10^3 + 3 \cdot 10^2 + 6 \cdot 10^1$

32. $4 \cdot 10^{-2} + 3 \cdot 10^{-1} + 2 \cdot 10^0 + 7 \cdot 10^2 + 9 \cdot 10^3$

Simplify each expression.

33. $(-3x^2y^3)(2x^9y^8)$ **34.** $(-6a^7b^4)(3a^3b^5)$

35. $5^2 \cdot 3^2$ **36.** $x^2x^5 + x^3x^4$

Simplify each expression. Write answers without negative exponents. Assume that all variables represent nonzero real numbers.

37. $-1(2x^3)^2$ **38.** $(-3y^{-1})^{-1}$

39. $\left(\dfrac{-2x^2}{3}\right)^3$ **40.** $\left(\dfrac{-1}{2a}\right)^{-2}$

41. $\dfrac{6x^7}{2x^3}$ **42.** $\dfrac{-9x^2y}{3xy^2}$

43. $\left(\dfrac{y^2}{5}\right)^{-2}$ **44.** $\left(-\dfrac{y^2}{2a}\right)^4$

45. $\left(\frac{1}{2}x^{-4}y^{3}\right)\left(\frac{1}{3}x^{4}y^{-6}\right)$ **46.** $\left(\frac{1}{3}a^{-5}b\right)(a^{4}b^{-1})$

47. $\dfrac{-3m^{-1}n}{-6m^{-1}n^{-1}}$ **48.** $\dfrac{-p^{-1}q^{-1}}{-3pq^{-3}}$

49. $(2a^{2})^{3} + (-3a^{3})^{2}$ **50.** $(b^{-4})^{2} - (-b^{-2})^{4}$

51. $-3(3x^{-1}y^{3})^{-2}$ **52.** $6(-2a^{-1}b^{-3})^{-1}$

53. $\left(\dfrac{-2x^{4}y^{-4}}{3x^{-1}y^{-2}}\right)^{4}$ **54.** $\left(\dfrac{-3a^{3}b^{-5}}{6a^{-2}b^{-2}}\right)^{3}$

Simplify each expression. Assume that all bases are nonzero real numbers and all exponents are integers.

55. $(x^{b-1})^{3}(x^{b-4})^{-2}$ **56.** $(a^{2})^{m+2}(a^{3})^{4m}$

57. $(-5a^{2t}b^{-3t})^{3}$ **58.** $(-2x^{-5v}y^{3})^{2}$

59. $\dfrac{-9x^{3w}y^{9v}}{6x^{8w}y^{3v}}$ **60.** $\dfrac{6c^{9s}d^{4t}}{-9c^{3s}d^{8t}}$

61. $\left(\dfrac{a^{s+2}b^{t-3}}{a^{2s-3}b^{6-t}}\right)^{3}$ **62.** $\left(\dfrac{x^{2a-3}y^{-13}}{x^{-4a+1}y^{-9}}\right)^{-4}$

Convert each number given in standard notation to scientific notation and each number given in scientific notation to standard notation.

63. 4.3×10^{4} **64.** -5.98×10^{5}

65. -3.56×10^{-5} **66.** 9.333×10^{-9}

67. 5,000,000 **68.** −16,587,000

69. −0.0000672 **70.** 0.000000981

71. 7×10^{-9} **72.** -6×10^{-3}

73. −20,000,000,000 **74.** 0.00000000004

Perform the indicated operations without a calculator. Write your answers in scientific notation. Use a calculator to check.

75. $(5 \times 10^{8})(4 \times 10^{7})$ **76.** $(5 \times 10^{-10})(6 \times 10^{5})$

77. $\dfrac{8.2 \times 10^{-6}}{4.1 \times 10^{-3}}$ **78.** $\dfrac{9.3 \times 10^{12}}{3.1 \times 10^{-3}}$

79. $5(2 \times 10^{-10})^{3}$ **80.** $-2(2 \times 10^{8})^{-4}$

81. $\dfrac{(2{,}000{,}000)^{3}(0.000005)}{(0.00002)^{2}}$

82. $\dfrac{(-6{,}000{,}000)^{2}(-0.000003)^{-3}}{(2000)^{3}(1{,}000{,}000)}$

Use a calculator to perform the indicated operations. Give your answers in scientific notation.

83. $(4.32 \times 10^{-9})(-2.3 \times 10^{4})$

84. $(-2.33 \times 10^{23})(3.98 \times 10^{-9})$

85. $\dfrac{(5.63 \times 10^{-6})^{3}(3.5 \times 10^{7})^{-4}}{\pi(8.9 \times 10^{-4})^{2}}$

86. $\dfrac{\pi(2.39 \times 10^{-12})^{2}}{(6.75 \times 10^{-8})^{3}}$

87. $(2.7 \times 10^{9}) + (3.6 \times 10^{-5})$

88. $(-6.3 \times 10^{12}) - (7.25 \times 10^{-4})$

Solve each problem. Use your calculator.

89. *Body-Mass Index* The body-mass index is used to assess the relative amount of fat in a person's body (*New England Journal of Medicine*, September 14, 1995). The body-mass index, *BMI*, is given by the formula

$$BMI = 703wh^{-2},$$

where w is the person's weight in pounds and h is the person's height in inches. Find the *BMI*, rounded to the nearest whole number, for Charles Barkley, who played in the 1992 Olympic Games at 6 ft 6 in. and 250 pounds.

Figure for Exercises 89 and 90

90. *Excessive Fat* Some physicians consider your body fat excessive if your *BMI* (from the previous exercise) satisfies the inequality

$$|BMI - 23| > 3.$$

For the 1992 Olympic Games, Larry Bird was listed as 6 ft 9 in. and 220 pounds. Did Charles Barkley or Larry Bird have excessive fat in 1992?

91. *Displacement-Length Ratio* The displacement-length ratio D indicates whether a sail boat is relatively heavy or relatively light (*Sail*, September 1997). D is given by the formula

$$D = \frac{dL^{-3} \cdot 10^6}{2240},$$

where L is the length at the water line in feet, and d is the displacement in pounds. If $D > 300$, then a boat is considered relatively heavy. Find D for the USS Constitution, the nation's oldest floating ship, which has a displacement of 2200 tons and a length of 175 feet.

92. *A Lighter Boat* If $D < 150$ (from the previous exercise), then a boat is considered relatively light. Find D for the Open Class 50 boat shown in the figure. Its displacement is 11,500 pounds and its length is 50 feet.

Figure for Exercise 92

93. *National Debt* If the national debt is 1.3×10^{12} dollars and there are 2.53×10^8 people in the country, then how much does each person owe?

94. *Savings and Loan Bailout* It is estimated that the savings and loan bailout cost the United States \$500 billion. If there are 2.53×10^8 people in the United States, then what is the cost per person?

95. *Energy from the Sun* Merely an average star, our sun is a swirling mass of dense gases powered by thermonuclear reactions. The great solar furnace transforms 5 million tons of mass into energy every second. How many tons of mass will be transformed into energy during the sun's 10 billion-year lifetime?

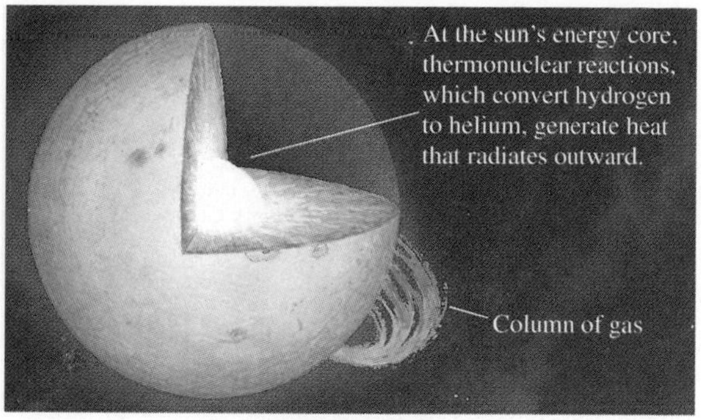

Figure for Exercise 95

96. *Orbit of the Earth* The earth orbits the sun in an approximately circular orbit with a radius of 1.495979×10^8 km. What is the area of the circle?

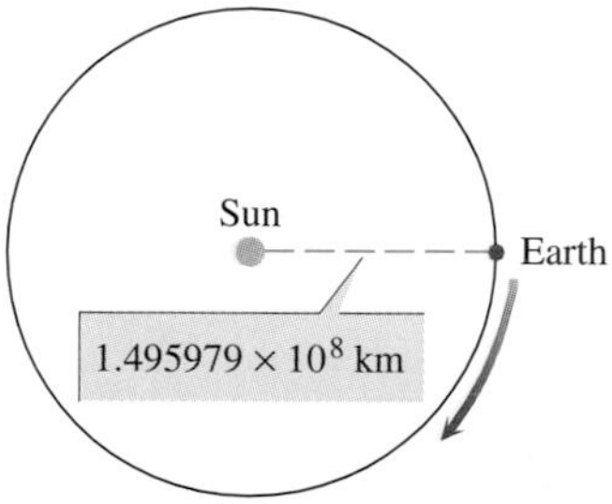

Figure for Exercises 96 and 98

97. *Radius of the Earth* The radius of the sun is 6.9599×10^5 km, which is 109.1 times the earth's radius. What is the radius of the earth?

98. *Distance to the Sun* The distance from the sun to the earth is 1.495979×10^8 km. Use the fact that 1 km = 0.621 mi to find the distance in miles from the earth to the sun.

99. *Mass of the Sun* The mass of the sun is 1.989×10^{30} kg and the mass of the earth is 5.976×10^{24} kg. How many times larger in mass is the sun than the earth?

100. *Speed of Light* If the speed of light is 3×10^8 m/sec, then how long does it take light from the sun to reach the earth? (See Exercise 98.)

101. *Energy Consumption in U.S.* According to the 1997 World Almanac, total world energy consumption for 1994 was 3.52×10^{17} Btu. Use the pie chart to determine the energy consumption for the United States.

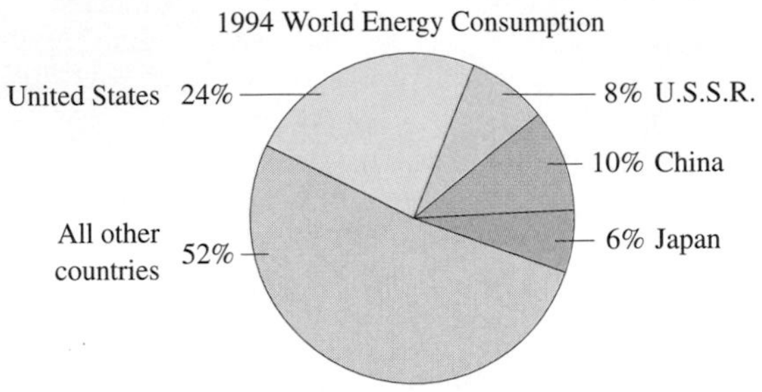

Figure for Exercise 101

102. *World Grain Demand* The accompanying figure shows Freeport McMoRan's projections for world grain demand through the year 2010 (*Forbes*, June 16, 1997). Use the figure to approximate (with scientific notation) the demand for grain in the year 2010. If the world population in 2010 is 8×10^9 and 1 metric ton = 1×10^3 kg, then what will be the world demand for grain per person?

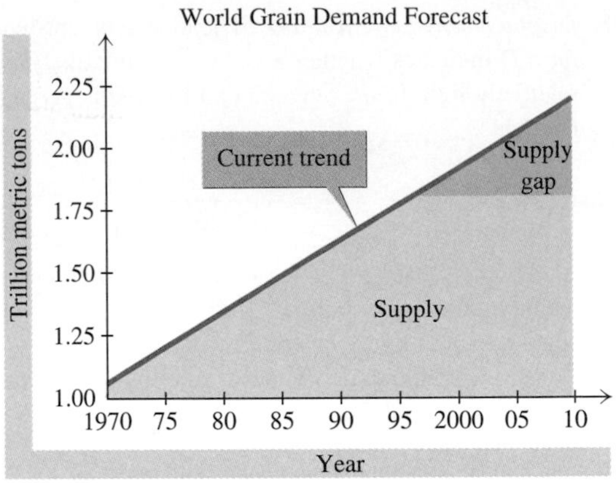

Figure for Exercise 102

Linking Concepts

For Individual or Group Explorations

Crowded or Not There is a lot of discussion on limiting the population of the Earth so that those who do inhabit the Earth can live in a comfortable manner. The experts disagree on the maximum number of people that the Earth can provide for. However, it does not take a lot of space to contain all of the people on Earth.

a) How many square feet does the average person occupy when standing?

b) How many people are there on the Earth at this time?

c) How many square miles would it take if all of the people on the Earth were to stand in one large crowd?

d) Would this crowd fit into the county in which you live?

e) If everyone on Earth lived in an average suburban house with a two-car garage with an average of three people per house, then would this huge suburb fit in the United States?

P.3 Rational Exponents and Radicals

Raising a number to a power is reversed by finding the root of a number. We indicate roots by using rational exponents or radicals. In this section we will review definitions and rules concerning rational exponents and radicals.

Roots

Since $2^4 = 16$ and $(-2)^4 = 16$, both 2 and -2 are fourth roots of 16. The nth root of a number is defined in terms of the nth power.

Definition: nth Roots

If n is a positive integer and $a^n = b$, then a is called an ***n*th root** of b. If $a^2 = b$, then a is a **square root** of b. If $a^3 = b$, then a is the **cube root** of b.

We also describe roots as even or odd, depending on whether the positive integer is even or odd. For example, if n is even (or odd) and a is an nth root of b, then a is called an **even** (or **odd**) **root** of b. Every positive real number has *two* real even roots, a positive root and a negative root. For example, both 5 and -5 are square roots of 25 because $5^2 = 25$ and $(-5)^2 = 25$. Moreover, every real number has exactly *one* real odd root. For example, because $2^3 = 8$ and 3 is odd, 2 is the only real cube root of 8. Because $(-2)^3 = -8$ and 3 is odd, -2 is the only real cube root of -8.

Finding an nth root is the reverse of finding an nth power, so we use the notation $a^{1/n}$ for the nth root of a. For example, since the positive square root of 25 is 5, we write $25^{1/2} = 5$.

Definition: Exponent $1/n$

If n is a positive even integer and a is positive, then $a^{1/n}$ denotes the **positive real *n*th root of *a*** and is called the **principal *n*th root of *a***.

If n is a positive odd integer and a is any real number, then $a^{1/n}$ denotes the real nth root of a.

If n is a positive integer, then $0^{1/n} = 0$.

EXAMPLE 1 Evaluating expressions involving exponent $1/n$

Evaluate each expression.

a) $4^{1/2}$ **b)** $8^{1/3}$ **c)** $(-8)^{1/3}$ **d)** $(-4)^{1/2}$

Solution

a) The expression $4^{1/2}$ represents the positive real square root of 4. So $4^{1/2} = 2$.

b) $8^{1/3} = 2$ **c)** $(-8)^{1/3} = -2$

d) Since the definition of nth root does not include an even root of a negative number, $(-4)^{1/2}$ has not yet been defined. Even roots of negative numbers do exist in the complex number system, which we define in Section 1.3. So an even root of a negative number is not a real number.

Rational Exponents

We have defined $a^{1/n}$ as the nth root of a. We extend this definition to $a^{m/n}$, which is defined as the mth power of the nth root of a. A rational exponent indicates both a root and a power.

Definition: Rational Exponents

If m and n are positive integers, then

$$a^{m/n} = (a^{1/n})^m$$

provided that $a^{1/n}$ is a real number.

According to the definition, expressions such as $(-25)^{-3/2}$, $(-43)^{1/4}$, and $(-1)^{2/2}$ are not real numbers because each of them involves an even root of a negative number. Of course a negative rational exponent indicates a reciprocal, just as a negative integral exponent does.

If $a^{1/n}$ is a real number, then $(a^{1/n})^m = (a^m)^{1/n}$. So the root and the power can be evaluated in either order. For example, to evaluate $8^{2/3}$, we find the square of the cube root of 8 or the cube of 8^2. In symbols,

$$8^{2/3} = (8^{1/3})^2 = 2^2 = 4 \quad \text{or} \quad 8^{2/3} = (8^2)^{1/3} = 64^{1/3} = 4.$$

To keep the numbers small we usually find the root and then the power.

We can evaluate $8^{-2/3}$ mentally as follows: The cube root of 8 is 2, 2 squared is 4, and the reciprocal of 4 is $\frac{1}{4}$. In written work it is usually best to eliminate the negative exponent first:

$$8^{-2/3} = \frac{1}{8^{2/3}} = \frac{1}{(8^{1/3})^2} = \frac{1}{2^2} = \frac{1}{4}$$

The three operations indicated by a negative rational exponent can be performed in any order. The simplest procedure for mental evaluation is summarized as follows.

PROCEDURE

Evaluating $a^{-m/n}$

To evaluate $a^{-m/n}$ mentally,

1. find the nth root of a, 2. raise it to the m power, 3. find the reciprocal.

Rational exponents can be reduced to lowest terms. For example, we can evaluate $2^{6/2}$ by first reducing the exponent:

$$2^{6/2} = 2^3 = 8$$

Exponents can be reduced only on expressions that are real numbers. For example, $(-1)^{2/2} \neq (-1)^1$ because $(-1)^{2/2}$ is not a real number, while $(-1)^1$ is a real number.

Your graphing calculator will probably evaluate $(-1)^{2/2}$ as -1, because it is not using our definition. Moreover, some calculators will not evaluate an expression with a negative base such as $(-8)^{2/3}$, but will evaluate the equivalent expression $((-8)^2)^{1/3}$. To use your calculator effectively, you must get to know it well.

EXAMPLE 2 **Evaluating expressions with rational exponents**

Evaluate each expression, then check with a calculator.

a) $(-8)^{2/3}$ **b)** $27^{-2/3}$ **c)** $100^{6/4}$

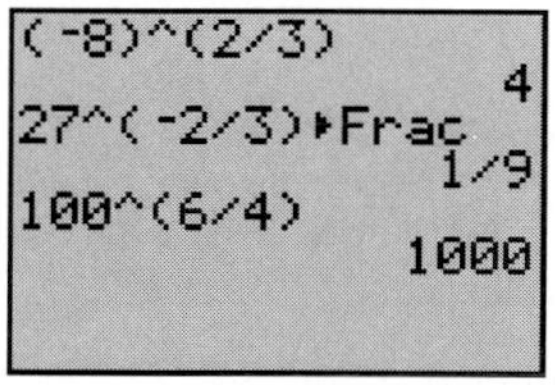

Figure P.23

Solution

a) $(-8)^{2/3} = ((-8)^{1/3})^2 = (-2)^2 = 4$

b) $27^{-2/3} = \dfrac{1}{27^{2/3}} = \dfrac{1}{(27^{1/3})^2} = \dfrac{1}{3^2} = \dfrac{1}{9}$

c) $100^{6/4} = 100^{3/2} = 10^3 = 1000$

The expressions are evaluated with a graphing calculator in Fig. P.23.

Rules for Rational Exponents

The rules for integral exponents from Section P.2 also hold for rational exponents.

Rules for Rational Exponents

The following rules are valid for all real numbers a and b and rational numbers r and s, provided that all indicated powers are real and no denominator is zero.

1. $a^r a^s = a^{r+s}$ **2.** $\dfrac{a^r}{a^s} = a^{r-s}$ **3.** $(a^r)^s = a^{rs}$

4. $(ab)^r = a^r b^r$ **5.** $\left(\dfrac{a}{b}\right)^r = \dfrac{a^r}{b^r}$ **6.** $\left(\dfrac{a}{b}\right)^{-r} = \left(\dfrac{b}{a}\right)^r$ **7.** $\dfrac{a^{-r}}{b^{-s}} = \dfrac{b^s}{a^r}$

When variable expressions involve even roots, we must be careful with signs. For example, $(x^2)^{1/2} = x$ is not correct for all values of x, because $((-5)^2)^{1/2} = 25^{1/2} = 5$. However, using absolute value we can write

$$(x^2)^{1/2} = |x| \quad \text{for every real number } x.$$

When finding an even root of an expression involving variables, remember that if n is even, $a^{1/n}$ is the *positive* nth root of a.

EXAMPLE 3 Using absolute value with rational exponents

Simplify each expression, using absolute value when necessary. Assume that the variables can represent any real numbers.

a) $(64a^6)^{1/6}$ **b)** $(x^9)^{1/3}$ **c)** $(a^8)^{1/4}$ **d)** $(y^{12})^{1/4}$

Solution

a) For any nonnegative real number a, we have $(64a^6)^{1/6} = 2a$. If a is negative, $(64a^6)^{1/6}$ is positive and $2a$ is negative. So we write

$$(64a^6)^{1/6} = |2a| = 2|a| \quad \text{for every real number } a.$$

b) For any nonnegative x, we have $(x^9)^{1/3} = x^{9/3} = x^3$. If x is negative, $(x^9)^{1/3}$ and x^3 are both negative. So we have

$$(x^9)^{1/3} = x^3 \quad \text{for every real number } x.$$

c) For nonnegative a, we have $(a^8)^{1/4} = a^2$. Since $(a^8)^{1/4}$ and a^2 are both positive if a is negative, no absolute value sign is needed. So

$$(a^8)^{1/4} = a^2 \quad \text{for every real number } a.$$

d) For nonnegative y, we have $(y^{12})^{1/4} = y^3$. If y is negative, $(y^{12})^{1/4}$ is positive but y^3 is negative. So

$$(y^{12})^{1/4} = |y^3| \quad \text{for every real number } y.$$

When simplifying expressions we will often assume that the variables represent positive real numbers so that we do not have to be concerned about undefined expressions or absolute value. In the following example we make that assumption as we use the rules of exponents to simplify expressions involving rational exponents.

EXAMPLE 4 Simplifying expressions with rational exponents

Use the rules of exponents to simplify each expression. Assume that the variables represent positive real numbers. Write answers without negative exponents.

a) $x^{2/3}x^{4/3}$ **b)** $(x^4y^{1/2})^{1/4}$ **c)** $\left(\dfrac{a^{3/2}b^{2/3}}{a^2}\right)^3$

Solution

a) $x^{2/3}x^{4/3} = x^{6/3}$ — Product rule

$= x^2$ — Simplify the exponent.

b) $(x^4y^{1/2})^{1/4} = (x^4)^{1/4}(y^{1/2})^{1/4}$ — Power of a product rule

$= xy^{1/8}$ — Power of a power rule

c) $\left(\dfrac{a^{3/2}b^{2/3}}{a^2}\right)^3 = \dfrac{(a^{3/2})^3(b^{2/3})^3}{(a^2)^3}$ — Power of a quotient rule

$= \dfrac{a^{9/2}b^2}{a^6}$ — Power of a power rule

$= a^{-3/2}b^2$ — Quotient rule $\left(\dfrac{9}{2} - 6 = -\dfrac{3}{2}\right)$

$= \dfrac{b^2}{a^{3/2}}$ — Definition of negative exponents

Radical Notation

The exponent $1/n$ and the **radical sign** $\sqrt[n]{}$ are both used to indicate nth root.

Definition: Radical

> If n is a positive integer and a is a number for which $a^{1/n}$ is defined, then the expression $\sqrt[n]{a}$ is called a **radical**, and
>
> $$\sqrt[n]{a} = a^{1/n}.$$
>
> If $n = 2$, we write $\sqrt{a}$ rather than $\sqrt[2]{a}$.

The number a is called the **radicand** and n is the **index** of the radical. Expressions such as $\sqrt{-3}$, $\sqrt[4]{-81}$, and $\sqrt[6]{-1}$ do not represent real numbers because each is an even root of a negative number.

EXAMPLE 5 Evaluating radicals

Evaluate each expression and check with a calculator.

a) $\sqrt{49}$ **b)** $\sqrt[3]{-1000}$ **c)** $\sqrt[4]{\dfrac{16}{81}}$

Solution

a) The symbol $\sqrt{49}$ indicates the positive square root of 49. So $\sqrt{49} = 49^{1/2} = 7$. Writing $\sqrt{49} = \pm 7$ is incorrect.

b) $\sqrt[3]{-1000} = (-1000)^{1/3} = -10$ — Check that $(-10)^3 = -1000$.

Figure P.24

c) $\sqrt[4]{\frac{16}{81}} = \left(\frac{16}{81}\right)^{1/4} = \frac{2}{3}$ Check that $\left(\frac{2}{3}\right)^4 = \frac{16}{81}$.

These expressions are evaluated with a calculator in Fig. P.24.

Since $a^{1/n} = \sqrt[n]{a}$, expressions involving rational exponents can be written with radicals.

Rule for Converting $a^{m/n}$ to Radical Notation

If a is a real number and m and n are integers for which $\sqrt[n]{a}$ is real, then

$$a^{m/n} = (\sqrt[n]{a})^m = \sqrt[n]{a^m}.$$

EXAMPLE 6 Writing rational exponents as radicals

Write each expression in radical notation. Assume that all variables represent positive real numbers. Simplify the radicand if possible.

a) $2^{2/3}$ **b)** $(3x)^{3/4}$ **c)** $2(x^2 + 3)^{-1/2}$

Solution

a) $2^{2/3} = \sqrt[3]{2^2} = \sqrt[3]{4}$ **b)** $(3x)^{3/4} = \sqrt[4]{(3x)^3} = \sqrt[4]{27x^3}$

c) $2(x^2 + 3)^{-1/2} = 2 \cdot \frac{1}{(x^2 + 3)^{1/2}} = \frac{2}{\sqrt{x^2 + 3}}$

The Product and Quotient Rules for Radicals

Using rational exponents we can write

$$(ab)^{1/n} = a^{1/n}b^{1/n} \quad \text{and} \quad \left(\frac{a}{b}\right)^{1/n} = \frac{a^{1/n}}{b^{1/n}}.$$

These equations say that the nth root of a product (or quotient) is the product (or quotient) of the nth roots. Using radical notation these rules are written as follows.

Rules for Radicals

For any positive integer n and real numbers a and b ($b \neq 0$),

1. $\sqrt[n]{ab} = \sqrt[n]{a} \cdot \sqrt[n]{b}$ **Product rule for radicals**
2. $\sqrt[n]{\frac{a}{b}} = \frac{\sqrt[n]{a}}{\sqrt[n]{b}}$ **Quotient rule for radicals**

provided that all of the roots are real.

An expression that is the square of a term that is free of radicals is called a **perfect square**. For example, $9x^6$ is a perfect square because $9x^6 = (3x^3)^2$. Likewise, $27y^{12}$ is a **perfect cube**. In general, an expression that is the nth power of an expression free of radicals is a **perfect *n*th power**. In the next example, the product and quotient rules for radicals are used to simplify radicals containing perfect squares, cubes, and so on.

EXAMPLE 7 Using the product and quotient rules for radicals

Simplify each radical expression. Assume that all variables represent positive real numbers.

a) $\sqrt[3]{125a^6}$ **b)** $\sqrt{\dfrac{3}{16}}$ **c)** $\sqrt[5]{\dfrac{-32y^5}{x^{20}}}$

Solution

a) Both 125 and a^6 are perfect cubes. So use the product rule to simplify:

$$\sqrt[3]{125a^6} = \sqrt[3]{125} \cdot \sqrt[3]{a^6} = 5a^2 \qquad \text{Since } \sqrt[3]{a^6} = a^{6/3} = a^2$$

b) Since 16 is a perfect square, use the quotient rule to simplify the radical:

$$\sqrt{\frac{3}{16}} = \frac{\sqrt{3}}{\sqrt{16}} = \frac{\sqrt{3}}{4}$$

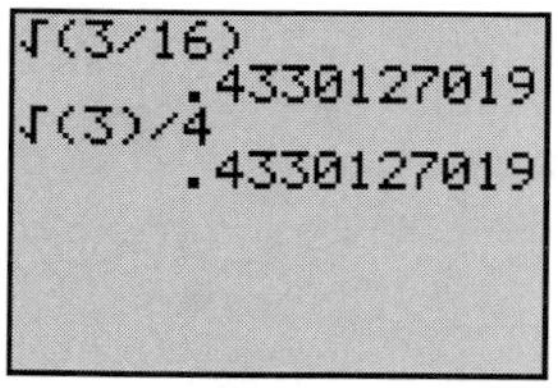

Figure P.25

We can check this answer by using a calculator as shown in Fig. P.25. Note that agreement in the first 10 decimal places supports our belief that the two expressions are equal, but does not prove it. The expressions are equal because of the quotient rule.

c) $$\sqrt[5]{\frac{-32y^5}{x^{20}}} = \frac{\sqrt[5]{-32y^5}}{\sqrt[5]{x^{20}}} = \frac{-2y}{x^4} \qquad \text{Since } \sqrt[5]{x^{20}} = x^{20/5} = x^4$$

Simplified Form and Rationalizing the Denominator

We have been simplifying radical expressions by just making them look simpler. However, a radical expression is in *simplified form* only if it satisfies the following three specific conditions. (You should check that the simplified expressions of Example 7 satisfy these conditions.)

Definition: Simplified Form for Radicals of Index *n*

A radical of index n in **simplified form** has

1. ***no*** perfect nth powers as factors of the radicand,
2. ***no*** fractions inside the radical, and
3. ***no*** radicals in a denominator.

The product rule is used to remove the perfect nth powers that are factors of the radicand, and the quotient rule is used when fractions occur inside the radical. The process of removing radicals from a denominator is called **rationalizing the denominator**. Radicals can be removed from the numerator by using the same type of procedure.

EXAMPLE 8 Simplified form of a radical expression

Write each radical expression in simplified form. Assume that all variables represent positive real numbers.

a) $\sqrt{20}$ **b)** $\sqrt{24x^8y^9}$ **c)** $\frac{9}{\sqrt{3}}$ **d)** $\sqrt[3]{\frac{3}{5a^4}}$

Solution

a) Since 4 is a factor of 20, $\sqrt{20}$ is not in its simplified form. Use the product rule for radicals to simplify it:

$$\sqrt{20} = \sqrt{4} \cdot \sqrt{5} = 2\sqrt{5}$$

b) Use the product rule to factor the radical, putting all perfect squares in the first factor:

$$\sqrt{24x^8y^9} = \sqrt{4x^8y^8} \cdot \sqrt{6y} \qquad \text{Product rule}$$
$$= 2x^4y^4\sqrt{6y} \qquad \text{Simplify the first radical.}$$

c) Since $\sqrt{3}$ appears in the denominator, we multiply the numerator and denominator by $\sqrt{3}$ to rationalize the denominator:

$$\frac{9}{\sqrt{3}} = \frac{9 \cdot \sqrt{3}}{\sqrt{3} \cdot \sqrt{3}} = \frac{9\sqrt{3}}{3} = 3\sqrt{3}$$

d) To rationalize this denominator, we must get a perfect cube in the denominator. The radicand $5a^4$ can be made into the perfect cube $125a^6$ by multiplying by $25a^2$:

$$\sqrt[3]{\frac{3}{5a^4}} = \frac{\sqrt[3]{3}}{\sqrt[3]{5a^4}} \qquad \text{Quotient rule for radicals}$$
$$= \frac{\sqrt[3]{3} \cdot \sqrt[3]{25a^2}}{\sqrt[3]{5a^4} \cdot \sqrt[3]{25a^2}} \qquad \text{Multiply numerator and denominator by } \sqrt[3]{25a^2}.$$
$$= \frac{\sqrt[3]{75a^2}}{\sqrt[3]{125a^6}} \qquad \text{Product rule for radicals}$$
$$= \frac{\sqrt[3]{75a^2}}{5a^2} \qquad \text{Since } (5a^2)^3 = 125a^6$$

Operations with Radical Expressions

Radical expressions with the same index can be added, subtracted, multiplied, or divided. For example, $2\sqrt{7} + 3\sqrt{7} = 5\sqrt{7}$ because $2x + 3x = 5x$ is true for any value of x. Because $2\sqrt{7}$ and $3\sqrt{7}$ are added in the same manner as like terms, they are called **like terms** or **like radicals**. Note that sums such as $\sqrt{3} + \sqrt{5}$ or $\sqrt[3]{2y} + \sqrt{2y}$ cannot be written as a single radical because the terms are not like terms. The next example further illustrates the basic operations with radicals.

EXAMPLE 9 Operations with radicals of the same index

Perform each operation and simplify each answer. Assume that each variable represents a positive real number.

a) $\sqrt{20} + \sqrt{5}$ **b)** $\sqrt[3]{24x} - \sqrt[3]{81x}$ **c)** $\sqrt[4]{4y^3} \cdot \sqrt[4]{12y^2}$ **d)** $\sqrt{40} \div \sqrt{5}$

Solution

a) $\sqrt{20} + \sqrt{5} = \sqrt{4} \cdot \sqrt{5} + \sqrt{5}$ Product rule for radicals

$= 2\sqrt{5} + \sqrt{5} = 3\sqrt{5}$ Simplify. Add like terms.

b) $\sqrt[3]{24x} - \sqrt[3]{81x} = \sqrt[3]{8} \cdot \sqrt[3]{3x} - \sqrt[3]{27} \cdot \sqrt[3]{3x}$ Product rule for radicals

$= 2\sqrt[3]{3x} - 3\sqrt[3]{3x} = -\sqrt[3]{3x}$ Simplify. Subtract like terms.

c) $\sqrt[4]{4y^3} \cdot \sqrt[4]{12y^2} = \sqrt[4]{48y^5}$ Product rule for radicals

$= \sqrt[4]{16y^4} \cdot \sqrt[4]{3y} = 2y\sqrt[4]{3y}$ Factor out the perfect fourth powers. Simplify.

d) $\sqrt{40} \div \sqrt{5} = \sqrt{\dfrac{40}{5}} = \sqrt{8}$ Quotient rule for radicals Divide.

$= \sqrt{4} \cdot \sqrt{2} = 2\sqrt{2}$ Product rule; simplify.

Radicals with different indices are not usually added or subtracted, but they can be combined in certain cases as shown in the next example.

EXAMPLE 10 Combining radicals with different indices

Write each expression using a single radical symbol. Assume that each variable represents a positive real number.

a) $\sqrt[3]{2} \cdot \sqrt{3}$ **b)** $\sqrt[3]{y} \cdot \sqrt[4]{2y}$ **c)** $\sqrt{\sqrt[3]{2}}$

Solution

a) $\sqrt[3]{2} \cdot \sqrt{3} = 2^{1/3} \cdot 3^{1/2}$ Rewrite radicals as rational exponents.

$= 2^{2/6} \cdot 3^{3/6}$ Write exponents with the least common denominator.

$= \sqrt[6]{2^2 \cdot 3^3}$ Rewrite in radical notation using the product rule.

$= \sqrt[6]{108}$ Simplify inside the radical.

b) $\sqrt[3]{y} \cdot \sqrt[4]{2y} = y^{1/3}(2y)^{1/4}$ Rewrite radicals as rational exponents.

$= y^{4/12}(2y)^{3/12}$ Write exponents with the LCD.

$= \sqrt[12]{y^4(2y)^3}$ Rewrite in radical notation using the product rule.

$= \sqrt[12]{8y^7}$ Simplify inside the radical.

c) $\sqrt{\sqrt[3]{2}} = (2^{1/3})^{1/2} = 2^{1/6} = \sqrt[6]{2}$

In Example 10(c) we found that the square root of a cube root is a sixth root. In general, an mth root of an nth root is an mnth root.

Theorem: mth Root of an nth Root

If m and n are positive integers for which all of the following roots are real, then

$$\sqrt[m]{\sqrt[n]{a}} = \sqrt[mn]{a}.$$

FOR THOUGHT True or False? Explain. Do Not Use a Calculator.

1. $8^{-1/3} = -2$
2. $16^{1/4} = 4^{1/2}$
3. $\sqrt{\frac{4}{6}} = \frac{2}{3}$
4. $(\sqrt{3})^3 = 3\sqrt{3}$
5. $(-1)^{2/2} = -1$
6. $\sqrt[3]{7^2} = 7^{3/2}$
7. $9^{1/2} = \sqrt{3}$
8. $\frac{1}{\sqrt{3}} = \frac{\sqrt{3}}{3}$
9. $\frac{2^{1/2}}{2^{1/3}} = \sqrt[6]{2}$
10. $\sqrt[3]{7^5} = 7\sqrt[3]{49}$

P.3 EXERCISES

Tape 1 Disk

Evaluate each expression. Use a calculator to check.

1. $-9^{1/2}$
2. $27^{1/3}$
3. $64^{1/2}$
4. $-144^{1/2}$
5. $(-64)^{1/3}$
6. $81^{1/4}$
7. $(-27)^{4/3}$
8. $125^{-2/3}$
9. $8^{-4/3}$
10. $4^{-3/2}$

Simplify each expression. Use absolute value when necessary.

11. $(x^6)^{1/6}$
12. $(x^{10})^{1/5}$
13. $(a^{15})^{1/5}$
14. $(y^2)^{1/2}$
15. $(a^8)^{1/4}$
16. $(z^{12})^{1/4}$
17. $(x^3y^6)^{1/3}$
18. $(16x^4y^8)^{1/4}$
19. $(a^2b^4)^{1/2}$
20. $(-x^3y^{12})^{1/3}$

Simplify each expression. Assume that all variables represent positive real numbers. Write your answers without negative exponents.

21. $(x^4y)^{1/2}$

22. $(a^{1/2}b^{1/3})^2$

23. $(2a^{1/2})(3a)$

24. $(-3y^{1/3})(-2y^{1/2})$

25. $\dfrac{6a^{1/2}}{2a^{1/3}}$

26. $\dfrac{-4y}{2y^{2/3}}$

27. $(a^2b^{1/2})(a^{1/3}b^{-1/2})$

28. $(4^{3/4}a^2b^3)(4^{3/4}a^{-2}b^{-5})$

29. $(a^{4m}b^{6n})^{1/2}b^n$

30. $(x^{2/n})^n(x^{n/6})^{1/n}$

31. $\left(\dfrac{x^6y^3}{z^9}\right)^{1/3}$

32. $\left(\dfrac{x^{1/2}y}{y^{1/2}}\right)^3$

Evaluate each radical expression. Use a calculator to check.

33. $\sqrt{900}$

34. $\sqrt{400}$

35. $\sqrt[3]{-8}$

36. $\sqrt[3]{64}$

37. $\sqrt[5]{-32}$

38. $\sqrt[6]{64}$

39. $\sqrt[3]{-\dfrac{8}{1000}}$

40. $\sqrt[4]{\dfrac{1}{625}}$

41. $\sqrt[4]{16^3}$

42. $\sqrt[3]{8^5}$

Write each expression involving rational exponents in radical notation and each expression involving radicals in exponential notation.

43. $10^{2/3}$

44. $-2^{3/4}$

45. $3y^{-3/5}$

46. $a(b^4+1)^{-1/2}$

47. $\dfrac{1}{\sqrt{x}}$

48. $-4\sqrt{x^3}$

49. $\sqrt[5]{x^3}$

50. $\sqrt[3]{x^3+y^3}$

Simplify each radical expression. Assume that all variables represent positive real numbers.

51. $\sqrt{16x^2}$

52. $\sqrt{121y^4}$

53. $\sqrt[3]{8y^9}$

54. $\sqrt[3]{125x^{18}}$

55. $\sqrt{\dfrac{xy}{100}}$

56. $\sqrt{\dfrac{t}{81}}$

57. $\sqrt[3]{\dfrac{-8a^3}{b^{15}}}$

58. $\sqrt[4]{\dfrac{16t^4}{y^8}}$

Write each radical expression in simplified form. Assume that all variables represent positive real numbers.

59. $\sqrt{28}$

60. $\sqrt{50}$

61. $\dfrac{1}{\sqrt{5}}$

62. $\dfrac{7}{\sqrt{7}}$

63. $\sqrt{\dfrac{x}{8}}$

64. $\sqrt{\dfrac{3y}{20}}$

65. $\sqrt[3]{40}$

66. $\sqrt[3]{54}$

67. $\sqrt[3]{-250x^4}$

68. $\sqrt[3]{24a^5}$

69. $\sqrt[3]{\dfrac{1}{2}}$

70. $\sqrt[3]{\dfrac{3x}{25}}$

Perform the indicated operations and simplify your answer. Assume that all variables represent positive real numbers. When possible use a calculator to verify your answer.

71. $\sqrt{8}+\sqrt{20}-\sqrt{12}$

72. $\sqrt{18}-\sqrt{50}+\sqrt{12}-\sqrt{75}$

73. $(-2\sqrt{3})(5\sqrt{6})$

74. $(-3\sqrt{2})(-2\sqrt{3})$

75. $(3\sqrt{5a})(4\sqrt{5a})$

76. $(-2\sqrt{6})(3\sqrt{6})$

77. $(-5\sqrt{3})^2$

78. $(3\sqrt{5})^2$

79. $\sqrt{18a}\div\sqrt{2a^4}$

80. $\sqrt{21x^7}\div\sqrt{3x^2}$

81. $5\div\sqrt{x}$

82. $a\div\sqrt{b}$

83. $\sqrt{20x^3}+\sqrt{45x^3}$

84. $\sqrt[3]{16a^4}+\sqrt[3]{54a^4}$

Write each expression using a single radical sign. Assume that all variables represent positive real numbers. Simplify the radicand where possible.

85. $\sqrt[3]{3}\cdot\sqrt{2}$

86. $\sqrt{5}\cdot\sqrt[3]{4}$

87. $\sqrt[3]{3}\cdot\sqrt[4]{x}$

88. $\sqrt[3]{3}\cdot\sqrt[4]{4}$

89. $\sqrt{xy}\cdot\sqrt[3]{2xy}$

90. $\sqrt[3]{2a}\cdot\sqrt{2a}$

91. $\sqrt[3]{\sqrt{7}}$

92. $\sqrt{\sqrt[3]{2a}}$

Solve each problem.

93. *Economic Order Quantity* Joseph Bursavich at Martin Marietta uses the formula

$$E=\sqrt{\frac{2AS}{I}}$$

to determine the most economic order quantity for parts used in the space shuttle fuel tanks. A is the quantity that the plant will use in one year, S is the cost of setup for making the part, and I is the cost of holding one unit in stock for one year. Find the most economic order quantity E if $S = \$6000$, $A = 25$, and $I = \$140$.

94. *Piano Tuning* The note middle C on a piano is tuned so that the string vibrates at 262 cycles per second, or 262 Hz (Hertz). The C note that is one octave higher is tuned to 524 Hz. Tuning for the 11 notes in-between using the using the method of *equal temperament* is $262\cdot 2^{n/12}$, where n takes the values 1 through 11. Find the tuning rounded to the nearest whole Hertz for those 11 notes.

95. *Sail Area-Displacement Ratio* The sail area-displacement ratio S measures the amount of power available to drive a sailboat in moderate to heavy winds (*Sail*, September 1997). Ratios typically range from about 15 to 25. With a high ratio indicating a powerful boat. S is given by the formula

$$S = \frac{16A}{d^{2/3}},$$

where A is the sail area in square feet and d is the displacement in pounds. Find S for the USS Constitution, which has a displacement of 2200 tons, a sail area of 42,700 ft^2, and 44 guns.

96. *A Less Powerful Boat* Find S (from the previous exercise) for the Ted Hood 51. It has a sail area of 1302 ft^2, a displacement of 49,400 pounds, a length of 51 feet, and no guns.

97. *Depreciation Rate* The formula $r = 1 - (S/C)^{1/n}$ for $n \geq 1$ gives the annual depreciation rate for a piece of equipment that has a useful life of n years, a salvage value S (in dollars), and an original cost C (in dollars). Find the annual depreciation rate for a computer that has a useful life of 5 years, a salvage value of \$200, and an original cost of \$5000.

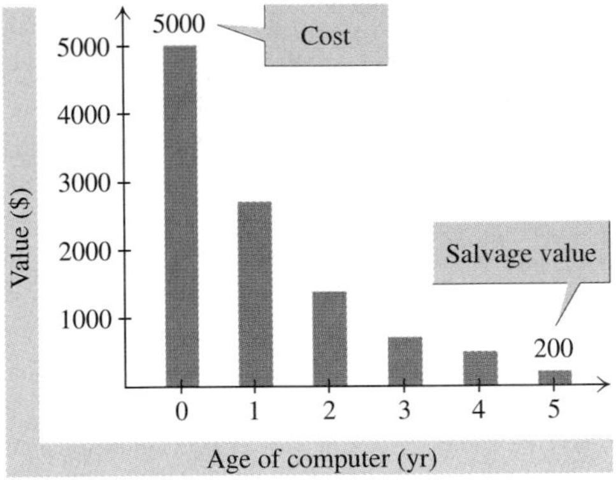

Figure for Exercise 97

98. *Market Growth Rate* The accompanying figure shows how \$100 invested in the stock market in 1801 grew to \$42.6 million in 1996 when adjusted for inflation (*Forbes*, May 5, 1997). The formula $r = (P/P_0)^{1/n} - 1$ gives the average annual growth rate r in terms of P, the value of the investment after n years, and the initial investment P_0. Find the average annual growth rate for the stock market for these years.

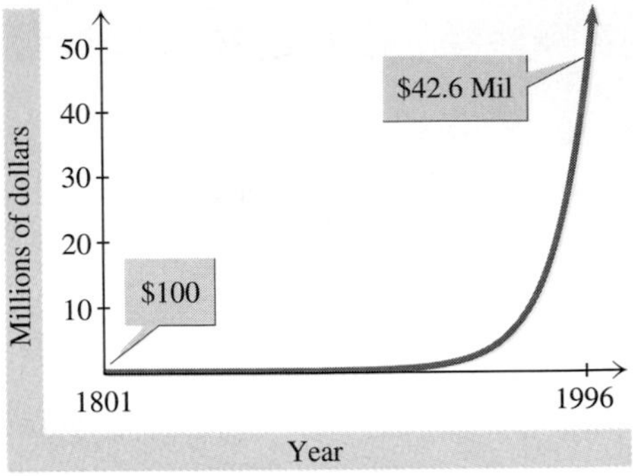

Figure for Exercise 98

99. *Longest Screwdriver* A toolbox has length L, width W, and height H. The length D of the longest screwdriver that will fit inside the box is given by

$$D = (L^2 + W^2 + H^2)^{1/2}.$$

Find the length of the longest screwdriver that will fit in a 4 in. by 6 in. by 12 in. box.

100. *Changing Radius* The radius of a sphere r is given in terms of its volume V by the formula

$$r = \left(\frac{0.75V}{\pi}\right)^{1/3}.$$

By how many inches has the radius of a spherical balloon increased when the amount of air in the balloon is increased from 4.2 ft^3 to 4.3 ft^3?

101. *Heron's Formula* If the lengths of the sides of a triangle are a, b, and c, and $s = (a + b + c)/2$, then the area A is given by the formula

$$A = \sqrt{s(s - a)(s - b)(s - c)}.$$

Find the area of a triangle whose sides are 6 ft, 7 ft, and 11 ft (see figure on facing page).

102. *Area of an Equilateral Triangle* Use Heron's formula from the previous exercise to find a formula for the area of an equilateral triangle with sides of length a and simplify it.

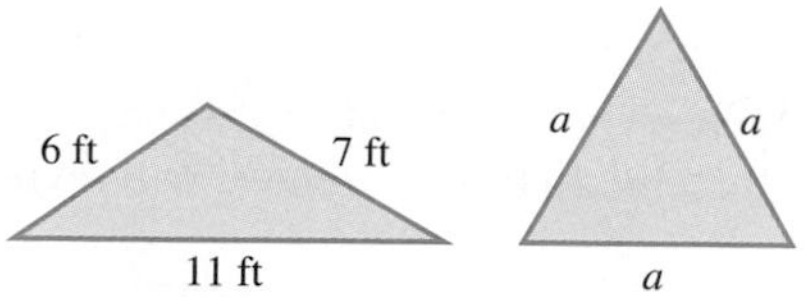

Figure for Exercises 101 and 102

For Writing/Discussion

103. *Roots or Powers* Which one of the following expressions is not equivalent to the others?
a) $(\sqrt[5]{t})^4$ **b)** $\sqrt[4]{t^5}$ **c)** $\sqrt[5]{t^4}$
d) $t^{4/5}$ **e)** $(t^{1/5})^4$

104. Which one of the following expressions is not equivalent to $\sqrt{a^2b^4}$?
a) $|a| \cdot b^2$ **b)** $|ab^2|$ **c)** ab^2
d) $(a^2b^4)^{1/2}$ **e)** $b^2\sqrt{a^2}$

105. *The Lost Rule?* Is it true that the square root of a sum is equal to the sum of the square roots? Explain.

106. *Technicalities* If m and n are real numbers and $m^2 = n$, then m is a square root of n, but if $m^3 = n$ then m is *the* cube root of n. How do we know when to use ''a'' or ''the''?

Linking Concepts

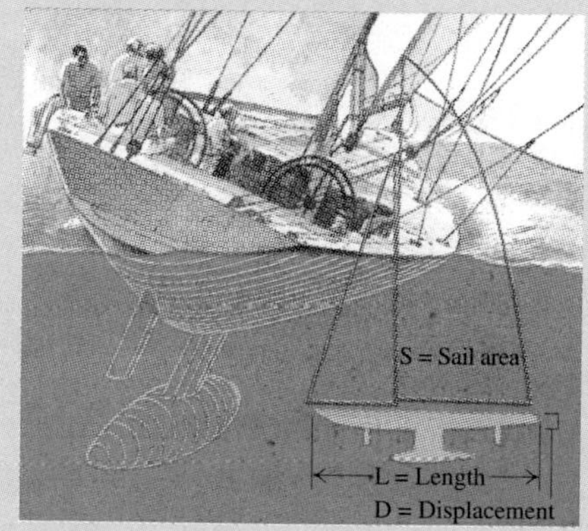

For Individual or Group Explorations

Yachting Rules The new International America's Cup Class rules, which took effect in 1989, establish the boundaries of a new class of longer, lighter, and faster yachts. In addition to satisfying 200 pages of other rules, the basic dimensions of any yacht competing for the silver trophy must satisfy the inequality

$$L + 1.25S^{1/2} - 9.8D^{1/3} \le 16.296$$

where L is the length in meters, S is the sail area in square meters, and D is the displacement in cubic meters. (Scientific American, May 1992)

a) Determine whether The New Zealand Challenge (considered a short, light boat) with a length of 20.95 m, a sail area of 277.3 m^2, and a displacement of 17.56 m^3 satisfies the inequality.

b) Verify that the published size of The Challenge Australia (considered a long, heavy boat) with a length of 21.87 m, a sail area of 311.78 m^2, and a displacement of 22.44 m^3 does not satisfy the inequality.

c) Is the inequality satisfied if the displacement of The Challenge Australia is increased by 0.01 m^3?

d) What does it mean to say that the displacement of a boat is 22.44 m^3?

e) What do you think could be done to The Challenge Australia to increase its displacement?

P.4 Polynomials

A polynomial is a type of algebraic expression that is as fundamental to algebra as the natural numbers are to arithmetic. In this section we will review some basic facts about polynomials.

Definitions

A polynomial is a single term or a finite sum of terms. Some examples of polynomials are

$$4, \quad -\frac{1}{2}x^5, \quad 2x^3 - \sqrt{3}x^2 + \pi x - 9, \quad \text{and} \quad a^3 + 3a^2b + 3ab^2 + b^3.$$

A term was defined in Section P.1 as a single number or the product of a number and one or more variables raised to powers. In a polynomial the powers of the variables must be whole numbers. We can give a formal definition of a polynomial in a single variable as follows.

Definition: Polynomial in x

If n is a nonnegative integer and $a_0, a_1, a_2, \ldots, a_n$ are real numbers, then

$$a_nx^n + a_{n-1}x^{n-1} + a_{n-2}x^{n-2} + \cdots + a_1x + a_0$$

is a **polynomial** in a single variable x.

In algebra a single number is often referred to as a **constant**. The last term, a_0, is called the **constant term**. When the coefficient of a term is negative, we write a difference rather than a sum.

EXAMPLE 1 Using the definition of polynomial

Determine whether each algebraic expression is a polynomial in x.

a) $x^3 - 4x + \sqrt{5}$ **b)** $4\sqrt{x} + 3$ **c)** $x^{-1} + 2$

Solution

The expression in (a) is a polynomial in x. The exponents on the variables in a polynomial must be nonnegative integers. So (b) and (c) are not polynomials.

A polynomial with only one term is called a **monomial**. A polynomial with two terms is called a **binomial**. A polynomial with three terms is called a **trinomial**. We usually write the terms of a polynomial in a single variable so that the exponents are in descending order from left to right. When a polynomial is written in this manner, the coefficient of the first term is called the **leading coefficient**. The **degree** of a polynomial in one variable is the highest power of the variable in the polynomial. The degree of $x^3 - 4x - \sqrt{5}$ is 3. A constant such as 5 is a polynomial with zero degree because $5 = 5x^0$. The number 0 is called the **zero polynomial**. It is a monomial without degree because $0 = 0x^n$ for any positive integer n. A first-degree polynomial is called a **linear polynomial**, a second-degree polynomial is called a **quadratic polynomial**, and a third-degree polynomial is called a **cubic polynomial**.

EXAMPLE 2 **Using the definitions of polynomial types**

Find the degree and leading coefficient of each polynomial and determine whether the polynomial is a monomial, binomial, or trinomial.

a) $\dfrac{x^3}{2} - \dfrac{1}{8}$ **b)** $5x^2 + x - 9$ **c)** $3x$

Solution

Polynomial (a) is a third-degree binomial, (b) is a second-degree trinomial, and (c) is a first-degree monomial. Because dividing by 2 is equivalent to multiplying by $\frac{1}{2}$, the leading coefficients are $\frac{1}{2}$, 5, and 3, respectively. We can also describe (a) as a cubic polynomial, (b) as a quadratic polynomial, and (c) as a linear polynomial.

We will mainly study polynomials in one variable, but you will also see polynomials in more than one variable. The degree of a term in more than one variable is the sum of the powers of the variables. For example, the degree of $5x^3y^4z$ is 8. The degree of a polynomial in more than one variable is equal to the highest degree of any of its terms. For example, the degree of $3x^2y^2 + x^2y - 5y$ is 4.

Addition and Subtraction of Polynomials

Since the coefficients of a polynomial are real numbers and the variables represent real numbers, the properties of the real numbers are valid for polynomials. We add or subtract polynomials by adding or subtracting the like terms. The results are the same whether you arrange your work horizontally, as in Example 3, or vertically, as in Example 4.

EXAMPLE 3 **Adding and subtracting polynomials horizontally**

Find each sum or difference.

a) $(3x^3 - x + 5) + (-8x^3 + 3x - 9)$ **b)** $(x^2 - 5x) - (3x^2 - 4x - 1)$

Solution

a) We use the commutative and associative properties of addition to rearrange the terms.

$$\begin{aligned}(3x^3 - x + 5) + (-8x^3 + 3x - 9) &= (3x^3 - 8x^3) + (-x + 3x) + (5 - 9)\\ &= -5x^3 + 2x - 4 \quad \text{Combine like terms.}\end{aligned}$$

b) The first step is to remove the parentheses and change the sign of every term in the second polynomial.

$$\begin{aligned}(x^2 - 5x) - (3x^2 - 4x - 1) &= x^2 - 5x - 3x^2 + 4x + 1 \quad \text{Distributive property}\\ &= -2x^2 - x + 1 \quad \text{Combine like terms.}\end{aligned}$$

EXAMPLE 4 **Adding and subtracting polynomials vertically**

Find each sum or difference.

a) $(3x^3 - x + 5) + (-8x^3 + 3x - 9)$ **b)** $(x^2 - 5x) - (3x^2 - 4x - 1)$

Solution

a) Add:
$$\begin{array}{r} 3x^3 - x + 5 \\ -8x^3 + 3x - 9 \\ \hline -5x^3 + 2x - 4 \end{array}$$

b) Subtract:
$$\begin{array}{r} x^2 - 5x \phantom{{}-1} \\ 3x^2 - 4x - 1 \\ \hline -2x^2 - x + 1 \end{array}$$

Multiplication of Polynomials

We multiplied monomials when we learned the product rule. For example, $(3x^2)(4x^5) = 12x^7$. To multiply a monomial and a binomial, we use the distributive property. For example,

$$3x(2x + 5) = (3x)(2x) + (3x)(5) = 6x^2 + 15x.$$

Note that the commutative and associative properties are also used to get $(3x)(2x) = 6x^2$.

We multiply two binomials by using the distributive property twice and combining like terms. For example,

$$\begin{aligned}(3x + 1)(2x + 5) &= 3x(2x + 5) + 1(2x + 5) \quad \text{Distributive property}\\ &= 6x^2 + 15x + 2x + 5 \quad \text{Distributive property}\\ &= 6x^2 + 17x + 5 \quad \text{Combine like terms.}\end{aligned}$$

When we multiply polynomials, we multiply each term of the first polynomial by every term of the second polynomial and then combine like terms. We can set up multiplication of polynomials vertically like multiplication of whole numbers:

Multiply:

$$\begin{array}{rl} 2x + 5 & \\ \underline{3x + 1} & \\ 2x + 5 & \text{Multiplying 1 and } 2x + 5 \text{ yields } 2x + 5. \\ \underline{6x^2 + 15x } & \text{Multiplying } 3x \text{ and } 2x + 5 \text{ yields } 6x^2 + 15x. \\ 6x^2 + 17x + 5 & \text{Add.} \end{array}$$

EXAMPLE 5 **Multiplying polynomials**

Use the distributive property to find each product.

a) $-3x(2x - 3)$ **b)** $(x + 5)(2x - 3)$ **c)** $(x^2 - 3x + 4)(2x - 3)$

Solution

a) $-3x(2x - 3) = -6x^2 + 9x$ Distributive property

b) $(x + 5)(2x - 3) = x(2x - 3) + 5(2x - 3)$ Distributive property

$= 2x^2 - 3x + 10x - 15$ Distributive property

$= 2x^2 + 7x - 15$ Combine like terms.

c) $(x^2 - 3x + 4)(2x - 3) = x^2(2x - 3) - 3x(2x - 3) + 4(2x - 3)$

$= 2x^3 - 3x^2 - 6x^2 + 9x + 8x - 12$

$= 2x^3 - 9x^2 + 17x - 12$

Using FOIL

The product of two binomials (as in Example 5b) results in four terms. Such products can be found quickly if we memorize where those four terms come from. The word FOIL is used as a memory aid. Consider the following product:

$$(a + b)(c + d) = a(c + d) + b(c + d) = \overbrace{ac}^{F} + \overbrace{ad}^{O} + \overbrace{bc}^{I} + \overbrace{bd}^{L}$$

(F: a and c; O: a and d; I: b and c; L: b and d)

The product of the two binomials consists of four terms:

the product of the *First* term of each (ac),

the product of the *Outer* terms (ad),

the product of the *Inner* terms (bc), and

the product of the *Last* term of each (bd).

Using FOIL to multiply two binomials is simply a way of speeding up the distributive property.

EXAMPLE 6 **Multiplying binomials using FOIL**

Find each product using FOIL.

a) $(x + 4)(2x - 3)$ **b)** $(x^2 - 3)(2x - 3)$

c) $(a^2b + 5)(a^2b - 5)$ **d)** $(a + b)^2$

Solution

a) $(x + 4)(2x - 3) = \overset{F}{\overbrace{2x^2}} - \overset{O}{\overbrace{3x}} + \overset{I}{\overbrace{8x}} - \overset{L}{\overbrace{12}} = 2x^2 + 5x - 12$

b) $(x^2 - 3)(2x - 3) = 2x^3 - 3x^2 - 6x + 9$

c) $(a^2b + 5)(a^2b - 5) = a^4b^2 - 5a^2b + 5a^2b - 25 = a^4b^2 - 25$

d) $(a + b)^2 = (a + b)(a + b) = a^2 + ab + ab + b^2 = a^2 + 2ab + b^2$

EXAMPLE 7 **Multiplying radicals using FOIL**

Use FOIL to find the product $(\sqrt{2} - 2\sqrt{3})(\sqrt{2} + 4\sqrt{3})$.

Solution

$$
\begin{aligned}
(\sqrt{2} - 2\sqrt{3})(\sqrt{2} + 4\sqrt{3}) &= \sqrt{2}\sqrt{2} + 4\sqrt{6} - 2\sqrt{6} - 2\sqrt{3} \cdot 4\sqrt{3} \\
&= 2 + 2\sqrt{6} - 24 \qquad 2\sqrt{3} \cdot 4\sqrt{3} = 8 \cdot 3 = 24 \\
&= -22 + 2\sqrt{6}
\end{aligned}
$$

Special Products

The products found in Examples 6(c) and 6(d) are called **special products** because of the way that the result is simplified. We used FOIL to find those special products, but it is better to memorize the rules given here so that the products can be found quickly.

The Special Products

$(a + b)^2 = a^2 + 2ab + b^2$	**The square of a sum**
$(a - b)^2 = a^2 - 2ab + b^2$	**The square of a difference**
$(a + b)(a - b) = a^2 - b^2$	**The product of a sum and a difference**

It is helpful to memorize the verbal statements of the special products. For example, the first special product would be: "The square of a sum is equal to the square of the first term, plus twice the product of the first and last terms, plus the square of the last term." Or, for the third special product: "The product of a sum and a difference (of the same terms) is equal to the difference of their squares."

EXAMPLE 8 Finding special products

Find each product by using the special product rules.

a) $(2x + 3)^2$ **b)** $(x^3 - 9)^2$ **c)** $(3x^m + 5)(3x^m - 5)$
d) $(3 - \sqrt{6})(3 + \sqrt{6})$

Solution

a) To find $(2x + 3)^2$ substitute $2x$ for a and 3 for b in $(a + b)^2 = a^2 + 2ab + b^2$:

$$(2x + 3)^2 = (2x)^2 + 2(2x)(3) + 3^2 = 4x^2 + 12x + 9$$

b) To find $(x^3 - 9)^2$ substitute x^3 for a and 9 for b in $(a - b)^2 = a^2 - 2ab + b^2$:

$$(x^3 - 9)^2 = (x^3)^2 - 2(x^3)(9) + 9^2 = x^6 - 18x^3 + 81$$

c) Substitute $3x^m$ for a and 5 for b in $(a + b)(a - b) = a^2 - b^2$:

$$(3x^m + 5)(3x^m - 5) = (3x^m)^2 - 5^2 = 9x^{2m} - 25$$

d) Substitute 3 for a and $\sqrt{6}$ for b in $(a - b)(a + b) = a^2 - b^2$:

$$(3 - \sqrt{6})(3 + \sqrt{6}) = 3^2 - (\sqrt{6})^2 = 9 - 6 = 3$$

In Example 8(d) the expressions $3 - \sqrt{6}$ and $3 + \sqrt{6}$ are called **conjugates**. Their product is a rational number. This fact is used to rationalize a denominator in the next example.

EXAMPLE 9 Using conjugates to rationalize a denominator

Simplify the expression and check with a calculator: $\dfrac{\sqrt{3}}{3 - \sqrt{6}}$

Solution

Multiply the numerator and denominator by $3 + \sqrt{6}$, the conjugate of $3 - \sqrt{6}$:

$$\frac{\sqrt{3}}{3 - \sqrt{6}} = \frac{\sqrt{3}(3 + \sqrt{6})}{(3 - \sqrt{6})(3 + \sqrt{6})} = \frac{3\sqrt{3} + \sqrt{18}}{3}$$

$$= \frac{3\sqrt{3} + 3\sqrt{2}}{3} = \sqrt{3} + \sqrt{2}$$

Checking the approximate values for the original expression and the answer shown in Fig. P.26 gives us confidence that our exact answer is correct.

Figure P.26

Division of Polynomials

We use the quotient rule for exponents to divide monomials. For example,

$$8x^5 \div (2x^2) = \frac{8x^5}{2x^2} = 4x^3.$$

We check division by multiplication, and since $4x^3 \cdot 2x^2 = 8x^5$, we have found the correct quotient. We use the distributive property to divide a polynomial by a monomial. For example,

$$(6x^3 - 9x^2 + 12x) \div (3x) = \frac{6x^3}{3x} - \frac{9x^2}{3x} + \frac{12x}{3x} = 2x^2 - 3x + 4.$$

Since $3x(2x^2 - 3x + 4) = 6x^3 - 9x^2 + 12x$, we have found the correct quotient. You should check all division by multiplication.

To divide the trinomial $x^2 - 3x - 10$ by the binomial $x - 5$, we use a process that looks like long division of whole numbers:

$$\begin{array}{r l}
 & \quad x + 2 \\
x - 5 \overline{)} & x^2 - 3x - 9 \\
 & \underline{x^2 - 5x} \\
 & \qquad 2x - 9 \\
 & \qquad \underline{2x - 10} \\
 & \qquad\qquad 1
\end{array}$$

$x^2 \div x = x$
$x(x - 5) = x^2 - 5x$
Subtract: $-3x - (-5x) = 2x$. Bring down -9.
$2x \div x = 2$, $2(x - 5) = 2x - 10$
Subtract: $-9 - (-10) = 1$.

Check: $(x + 2)(x - 5) + 1 = x^2 - 3x - 10 + 1 = x^2 - 3x - 9$. For this division, 1 is the remainder, $x - 5$ is the divisor, $x + 2$ is the quotient, and $x^2 - 3x + 9$ is the dividend. These quantities are related as follows.

The Division Algorithm

If the **dividend** $P(x)$ and the **divisor** $D(x)$ are polynomials such that $D(x) \neq 0$ and the degree of $P(x)$ is greater than or equal to the degree of $D(x)$, then there exist unique polynomials, the **quotient** $Q(x)$ and the **remainder** $R(x)$, such that

$$P(x) = Q(x)D(x) + R(x),$$

where $R(x) = 0$ or the degree of $R(x)$ is less than the degree of $D(x)$.

The division algorithm indicates that

$$\text{dividend} = (\text{quotient})(\text{divisor}) + \text{remainder}.$$

We can also express the relationship between these quantities as

$$\frac{\text{dividend}}{\text{divisor}} = \text{quotient} + \frac{\text{remainder}}{\text{divisor}}.$$

For the preceding division we can write

$$\frac{x^2 - 3x - 9}{x - 5} = x + 2 + \frac{1}{x - 5}.$$

EXAMPLE 10 Dividing polynomials

Find the quotient when the first polynomial is divided by the second.

a) $-6x^7 \div (-3x^2)$ **b)** $(8x^3 - 4x) \div (-2x)$ **c)** $(x^3 - 8) \div (x - 2)$

Solution

a) $-6x^7 \div (-3x^2) = \dfrac{-6x^7}{-3x^2} = 2x^5$

Check by multiplying $2x^5$ and $-3x^2$ to get $-6x^7$.

b) Notice that $-2x$ is a factor of each term in $8x^3 - 4x$:

$$(8x^3 - 4x) \div (-2x) = \frac{8x^3 - 4x}{-2x} = \frac{8x^3}{-2x} - \frac{4x}{-2x} = -4x^2 + 2$$

Check by multiplying $-2x$ and $-4x^2 + 2$ to get $8x^3 - 4x$.

c) To keep the division organized, insert $0x^2$ and $0x$ for the missing x^2 and x terms.

$$\begin{array}{r|l}
x^2 + 2x + 4 & \\
x - 2\overline{)x^3 + 0x^2 + 0x - 8} & x^3 \div x = x^2 \\
\underline{x^3 - 2x^2} & x^2(x - 2) = x^3 - 2x^2 \\
2x^2 + 0x & 0x^2 - (-2x^2) = 2x^2 \\
\underline{2x^2 - 4x} & 2x(x - 2) = 2x^2 - 4x \\
4x - 8 & 4(x - 2) = 4x - 8 \\
\underline{4x - 8} & \\
0 &
\end{array}$$

The quotient is $x^2 + 2x + 4$.

The Value of a Polynomial

A polynomial such as $x^2 + 3x + 1$ is an expression. It expresses operations to be performed on x, and it has no numerical value unless we choose a number for x. If we choose $x = 2$, then the value of the polynomial is $2^2 + 3(2) + 1$ or 11. Polynomials are often used to calculate quantities such as profit or revenue. A profit polynomial might be named $P(x)$ (read "P of x") and a revenue polynomial

(a)

Figure P.27(a)

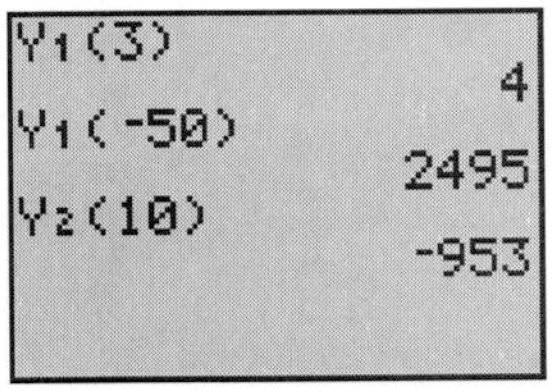

(b)

Figure P.27(b)

might be named $R(x)$. For example, if we refer to the polynomial $x^2 + 3x + 1$ as $P(x)$, then we write $P(x) = x^2 + 3x + 1$. $P(2)$ stands for the value of the polynomial $P(x)$ when $x = 2$. So we write $P(2) = 11$.

EXAMPLE 11 Evaluating a polynomial

Let $P(x) = x^2 - 5$ and $C(x) = -x^3 + 5x - 3$. Find the following.

a) $P(3)$ **b)** $P(-50)$ **c)** $C(10)$

Solution

a) $P(3) = 3^2 - 5 = 4$

b) $P(-50) = (-50)^2 - 5 = 2500 - 5 = 2495$

c) $C(10) = -10^3 + 5(10) - 3 = -1000 + 50 - 3 = -953$

To evaluate the polynomials in Example 11 on a calculator, define the polynomials with the $Y=$ key as in Fig. P.27(a), then use the y-variables to indicate which polynomials are to be evaluated as in Fig. P.27(b).

FOR THOUGHT True or False? Explain. Do Not Use a Calculator.

1. The expression $x^{-2} + 3x^{-1} + 9$ is a trinomial.
2. The degree of the polynomial $x^3 + 5x^2 - 6x^2y^2$ is 3.
3. $(3 + 5)^2 = 3^2 + 5^2$
4. $(50 + 1)(50 - 1) = 2499$
5. If the side of a square is $x + 3$ ft, then $x^2 + 9$ ft^2 is its area.
6. $(2x - 1) - (3x + 5) = -x + 4$ for any real number x.
7. $(a + b)^3 = a^3 + b^3$ for any real numbers a and b.
8. The dividend times the quotient plus the remainder equals the divisor.
9. $\dfrac{x^2 + 5x + 7}{x + 2} = x + 3 + \dfrac{1}{x + 2}$ for any real number x.
10. If $P(x) = 3x^2 + 7$, $Q(x) = 5x^2 - 9$, and $S(x) = 8x^2 - 2$, then $P(17) + Q(17) = S(17)$.

P.4 EXERCISES

Tape 2 Disk

Determine whether each algebraic expression is a polynomial in x. If it is a polynomial, find its degree and leading coefficient.

1. $x^3 - 4x^2 + \sqrt{5}$
2. $-x^7 - 6x^4$
3. $x^2 - 3x + 2 - \dfrac{1}{x}$
4. $x^{-1} + 5 + x^2$
5. 79
6. $\dfrac{x}{2} - \dfrac{x^2}{3} - \dfrac{x^3}{5} + \dfrac{1}{7}$

Find each sum or difference.

7. $(3x^2 - 4x) + (5x^2 + 7x - 1)$
8. $(-3x^2 - 4x + 2) + (5x^2 - 8x - 7)$
9. $(4x^2 - 3x) - (9x^2 - 4x + 3)$
10. $(x^2 + 2x + 4) - (x^2 + 4x + 4)$
11. $(4ax^3 - a^2x) - (5a^2x^3 - 3a^2x + 3)$
12. $(x^2y^2 - 3xy + 2x) - (6x^2y^2 + 4y - 6x)$

Find the sum or difference.

13. Add: $\begin{array}{r} 3x - 4 \\ \underline{-x + 3} \end{array}$

14. Add: $\begin{array}{r} -2x^2 - 5 \\ \underline{3x^2 - 6} \end{array}$

15. Subtract: $\begin{array}{r} x^2 \qquad\quad - 8 \\ \underline{-2x^2 + 3x - 2} \end{array}$

16. Subract: $\begin{array}{r} -2x^2 - 5x + 9 \\ \underline{4x^2 - 7x \qquad} \end{array}$

Use the distributive property to find each product.

17. $-3a^3(6a^2 - 5a + 2)$

18. $-2m(m^2 - 3m + 9)$

19. $(3b^2 - 5b + 2)(b - 3)$

20. $(-w^2 - 5w + 6)(w + 5)$

21. $(2x - 1)(4x^2 + 2x + 1)$

22. $(3x - 2)(9x^2 + 6x + 4)$

23. $(x - 4)(z + 3)$

24. $(a - 3)(b + c)$

25. $(a - b)(a^2 + ab + b^2)$

26. $(a + b)(a^2 - ab + b^2)$

Find each product using FOIL.

27. $(a + 9)(a - 2)$

28. $(z - 3)(z - 4)$

29. $(2y - 3)(y + 9)$

30. $(2y - 1)(3y + 4)$

31. $(2x - 9)(2x + 9)$

32. $(4x - 6y)(4x + 6y)$

33. $(2x + 5)^2$

34. $(5x - 3)^2$

35. $(2x^2 + 4)(3x^2 + 5)$

36. $(3x^3 - 2)(5x^3 + 6)$

Find each product using FOIL.

37. $(1 + \sqrt{2})(3 + \sqrt{2})$

38. $(5 + \sqrt{6})(2 + \sqrt{6})$

39. $(5 + \sqrt{2})(4 - 3\sqrt{2})$

40. $(2 - 3\sqrt{7})(5 - \sqrt{7})$

41. $(3\sqrt{2} + \sqrt{3})(2\sqrt{2} - \sqrt{3})$

42. $(2\sqrt{6} - \sqrt{3})(4\sqrt{6} - \sqrt{3})$

43. $(\sqrt{5} + \sqrt{3})^2$

44. $(\sqrt{8} - \sqrt{3})^2$

45. $(2 + \sqrt{x})^2$

46. $(1 + \sqrt{x - 1})^2$

Find each product using the special product rules.

47. $(3x + 5)^2$

48. $(x^3 - 2)^2$

49. $(x^n - 3)(x^n + 3)$

50. $(2z^b + 1)(2z^b - 1)$

51. $(\sqrt{2} - 5)(\sqrt{2} + 5)$

52. $(6 - \sqrt{3})(6 + \sqrt{3})$

53. $(3\sqrt{6} - 1)^2$

54. $(2\sqrt{5} + 2)^2$

55. $(3x^3 - 4)^2$

56. $(2x^2y^3 + 1)^2$

Simplify each expression by rationalizing the denominator.

57. $\dfrac{\sqrt{10}}{\sqrt{5} - 2}$

58. $\dfrac{\sqrt{3}}{\sqrt{2} - \sqrt{3}}$

59. $\dfrac{\sqrt{6}}{6 + \sqrt{3}}$

60. $\dfrac{\sqrt{2}}{\sqrt{8} + \sqrt{3}}$

Find each quotient.

61. $36x^6 \div (-4x^3)$

62. $-24y^{12} \div (3y^3)$

63. $(3x^2 - 6x) \div (-3x)$

64. $(6x^3 - 9x^2) \div (3x^2)$

65. $(x^2 + 6x + 9) \div (x + 3)$

66. $(x^2 - 3x - 54) \div (x - 9)$

67. $(a^3 - 1) \div (a - 1)$

68. $(b^6 + 8) \div (b^2 + 2)$

Find the quotient and remainder when the first polynomial is divided by the second.

69. $x^2 + 3x + 3,\ x - 2$

70. $3x^2 - x + 4,\ x + 2$

71. $2x^2 - 5,\ x + 3$

72. $-4x^2 + 1,\ x - 1$

73. $x^3 - 2x^2 - 2x - 3,\ x - 3$

74. $x^3 + 3x^2 - 3x + 4,\ x + 4$

75. $a^2b + 4ab + 3b,\ a + 3$

76. $3x^3y^2 - 4x^2y + x,\ xy - 1$

Use division to express each fraction in the form quotient + $\dfrac{\text{remainder}}{\text{divisor}}$.

77. $\dfrac{x^2 - 2}{x - 1}$

78. $\dfrac{x^2 + 4x + 5}{x + 1}$

79. $\dfrac{2x^2 - 3x + 1}{x}$

80. $\dfrac{2x^2 + 1}{x^2}$

81. $\dfrac{x^2 + x + 1}{x}$

82. $\dfrac{-x^2 + x - 1}{x}$

83. $\dfrac{x^2 - 2x + 1}{x - 2}$

84. $\dfrac{x^2 - x - 9}{x - 3}$

Let $P(x) = x^2 - 3x + 2$ and $M(x) = -x^3 + 5x^2 - x + 2$. Find the following.

85. $P(-2)$

86. $P(-1)$

87. $P(10)$

88. $P\left(\dfrac{1}{2}\right)$

89. $M(-3)$

90. $M(50)$

91. $M\left(\dfrac{1}{3}\right)$

92. $M\left(-\dfrac{1}{3}\right)$

Perform the indicated operations mentally. Write down only the answer.

93. $(x - 4)(x + 6)$

94. $(z^4 + 5)(z^4 - 4)$

95. $(2a^5 - 9)(a^5 + 3)$

96. $(3b^2 + 1)(2b^2 + 5)$

97. $(y - 3) - (2y + 6)$

98. $(a^2 + 3) - (a^2 - 6)$

99. $(7a - 9) + (3 - a)$

100. $(3m - 8) + (7 - m)$

101. $(w + 4)^2$

102. $(t - 2)^2$

103. $(4x - 9)(4x + 9)$

104. $(3x + 1)(3x - 1)$

105. $3y^2(y^3 - 3x)$

106. $a^4b(a^2b^2 + 1)$

107. $(6b^3 - 3b^2) \div (3b^2)$

108. $(3w - 8) \div (8 - 3w)$

109. $(x^3 + 8)(x^3 - 8)$

110. $(x^2 - 4)(x^2 + 4)$

111. $(3w^2 - 2n)^2$

112. $(7y^2 - 3x)^2$

113. $(\sqrt{5} - 2)(\sqrt{5} + 2)$

114. $(\sqrt{2} - \sqrt{3})(\sqrt{2} + \sqrt{3})$

Solve each problem.

115. *Area of a Rectangle* If the length of a field is $x + 3$ m and the width is $2x - 1$ m, then what trinomial represents the area of the field?

116. *Area of a Triangle* If the base of a triangular sign is $2x - 3$ cm and the height is $2x + 6$ cm, then what trinomial represents the area of the sign?

117. *Volume of a Rectangular Box* If the volume of a box is $x^3 + 9x^2 + 26x + 24$ cm^3 and the height is $x + 4$ cm, then what polynomial represents the area of the bottom?

Figure for Exercise 117

118. *Area of a Triangle* If the area of a triangle is $x^2 - x - 6$ m^2 and the base is $x - 3$ m, then what polynomial represents the height?

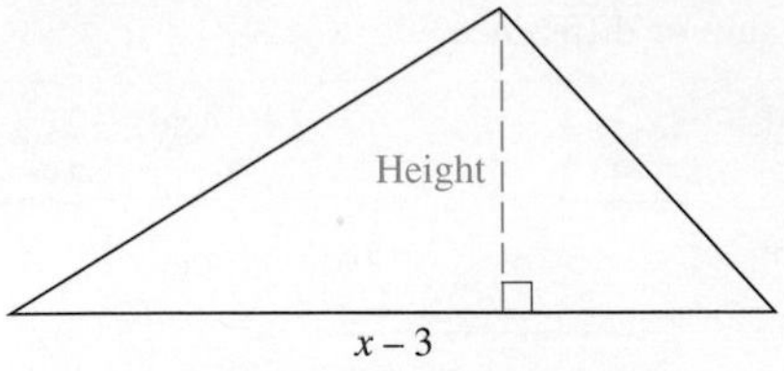

Figure for Exercise 118

119. *Available Habitat* A wild animal generally stays at least x mi from the edge of a forest. For a rectangular forest preserve that is 6 mi long and 4 mi wide, write a polynomial that represents the area of the available habitat for the wild animal.

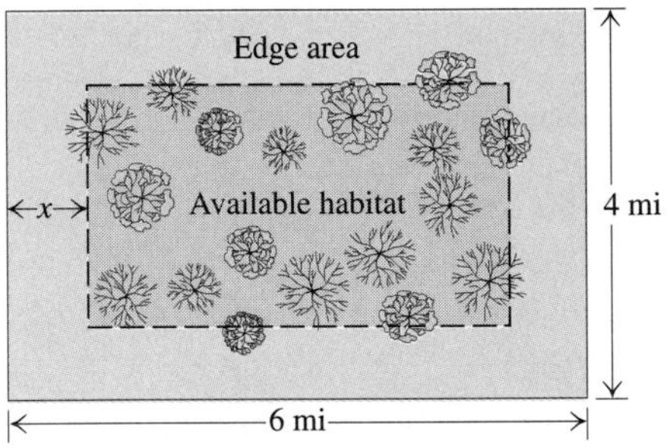

Figure for Exercise 119

120. *Red Fox* What percent of the forest in Exercise 119 is available habitat for the red fox if $x = 0.5$ mi for this wild animal?

121. *Comparing Investments* If you invest \$1 and it grows at an annual rate of $r\%$, then the amount of your investment in 3 years is given by the polynomial

$$A(r) = r^3 + 3r^2 + 3r + 1.$$

Fidelity's Short-Term Bond Fund has been averaging 5.02% annual growth while Fidelity's Equity-Income Fund has been averaging 17.24% annual growth. How much more would you have in 3 years if you invested your dollar in the higher yielding fund?

122. *Four-Year Investment* If you invest \$1 and it grows at an annual rate of $r\%$, then the amount of your investment in 4 years is given by the polynomial

$$A(r) = r^4 + 4r^3 + 6r^2 + 4r + 1.$$

Table for Exercises 125 and 126

	NHL	NBA	NFL	MLB
Average ticket price	\$38	\$32	\$35	\$9
Attendance	1.85×10^7	2.05×10^7	1.87×10^7	6.0×10^7
Merchandise	\$$1 \times 10^9$	\$$3 \times 10^9$	\$$3 \times 10^9$	\$$2.4 \times 10^9$
Broadcast	\$$8.5 \times 10^7$	\$$3.66 \times 10^8$	\$$1.1 \times 10^9$	\$$1.1 \times 10^6$

Fidelity's Magellan Fund, the world's largest fund, has averaged 14.22% for the past 5 years. How much would your dollar amount to in 4 years at 14.22%?

123. *Water Pollution* The polynomial $F(x) = 500 + 200x$ is used to determine the fine in dollars for violating water pollution standards, where x is the number of days that the violation persists after the citation. If a chemical plant persists in violating water pollution standards for 25 days after being cited, then how much is the fine?

124. *Profit for Computer Sales* The polynomial $P(n) = -2n^2 + 800n$ gives the profit in dollars on the sale of n computers in a single order. Find the profit for the sale of 200 computers.

125. *Hockey Money* The accompanying table shows the major sources of revenue for the 1995–96 season for major league hockey, basketball, football, and baseball (*Forbes*, March 24, 1997). Find the total revenue for the NHL by using the polynomial $R = 38a + m + b$, where a is the attendance, m is the merchandise revenue, and b is the broadcast revenue. What percent of total revenue comes from broadcasts for the NHL?

126. *Basketball Bucks* Write a polynomial as in the previous exercise that gives the total revenue for major league basketball. What percent of the total revenue comes from broadcasts for the NBA? Which sport has the highest percentage of its total revenue coming from broadcasts?

For Writing/Discussion

127. *Area of a Rectangle* Wilson thought that his lot was in the shape of a square. After he had it surveyed he discovered that it was rectangular in shape with the length 10 ft longer than he thought and the width 10 ft shorter than he thought. Does he have more or less area than he thought? How much? Explain.

128. *Egyptian Area Formula* Surrounded by thousands of square miles of arid desert, ancient Egypt was sustained by the green strips of land bordering the Nile. Villagers cooperated to control the annual flooding so that all might reap abundant harvests from a network of fields and orchards divided by irrigation canals. The Egyptians used the expression

$$\frac{(a + b)(c + d)}{4}$$

to approximate the area of a four-sided field, where a and b represent the lengths of two opposite sides and c and d represent the lengths of the other two opposite sides. Is the formula correct if the field is rectangular? Is it correct for every four-sided figure? Explain why the Egyptians used this formula.

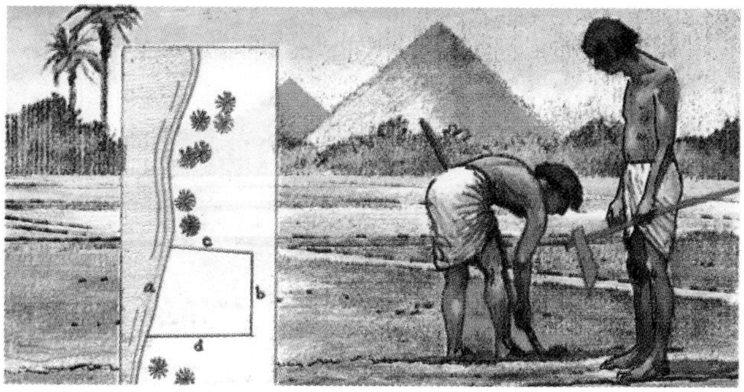

Figure for Exercise 128

129. *Cooperative Learning* Each student in your small group should write a polynomial $P(x)$ of a different degree. Use a calculator to evaluate each polynomial for $x = 5000$ and $x = -5000$. Repeat this procedure until you can make a conjecture about the relationship between the sign of x, the sign of the leading coefficient, the degree of $P(x)$, and the value of $P(x)$.

Linking Concepts

For Individual or Group Explorations

Maximum Sail Area *According to the International America's Cup Class rules, the polynomial*

$$S(L) = \left(\frac{16.296 + 9.8\sqrt[3]{18} - L}{1.25}\right)^2$$

gives the maximum sail area S (in square meters) in terms of the length L (in meters) for a boat with a fixed displacement of 18 m^3 and lengths less than 40 m.

a) What do you think is the purpose of rules such as this one?

b) Find the maximum sail area (to the nearest hundredth of a square meter) for lengths of 19.82 m, 20.56 m, and 21.24 m.

c) Is the maximum sail area increasing or decreasing as the length increases?

d) How does your answer to part (c) relate to your answer to part (a)?

e) Write the polynomial in the form $S(L) = aL^2 + bL + c$ where a, b, and c are real numbers rounded to the nearest hundredth.

f) Repeat part (b) using the polynomial that you found in part (e) and compare your answers to those found in part (b).

P.5 Factoring Polynomials

In Section P.4 we studied multiplication of polynomials. In this section we will factor polynomials. **Factoring** ''reverses'' multiplication. By factoring, we can express a complicated polynomial as a product of several simpler expressions, which are often easier to study.

Factoring Out the Greatest Common Factor

To factor $6x^2 - 3x$, notice that $3x$ is a monomial that can be divided evenly into each term. We can use the distributive property to write

$$6x^2 - 3x = 3x(2x - 1).$$

We call this process **factoring out** $3x$. Both $3x$ and $2x - 1$ are **factors** of $6x^2 - 3x$. Since 3 is a factor of $6x^2$ and $3x$, 3 is a **common factor** of the terms of the polynomial. The **greatest common factor (GCF)** is a monomial that includes every number and variable that is a factor of all terms of the polynomial. The monomial $3x$ is the greatest common factor of $6x^2 - 3x$. Usually the common factor has a positive coefficient, but at times it is useful to factor out a common factor with a negative coefficient. For example, we could factor out $-3x$ from $6x^2 - 3x$:

$$6x^2 - 3x = -3x(-2x + 1)$$

EXAMPLE 1 Factoring out the greatest common factor

Factor out the greatest common factor from each polynomial, first using the GCF with a positive coefficient and then using a negative coefficient.

a) $9x^4 - 6x^3 + 12x^2$ **b)** $x^2y + 10xy + 25y$ **c)** $-z - w$

Solution

a) $9x^4 - 6x^3 + 12x^2 = 3x^2(3x^2 - 2x + 4)$
$= -3x^2(-3x^2 + 2x - 4)$

b) $x^2y + 10xy + 25y = y(x^2 + 10x + 25)$
$= -y(-x^2 - 10x - 25)$

c) $-z - w = 1(-z - w)$
$= -1(z + w)$

Factoring by Grouping

Some four-term polynomials can be factored by **grouping** the terms in pairs and factoring out a common factor from each pair of terms. We show this technique in the next example. As we will see later, grouping can also be used to factor trinomials.

EXAMPLE 2 Factoring four-term polynomials

Factor each polynomial by grouping.

a) $x^3 + x^2 + 3x + 3$ **b)** $aw + bc - bw - ac$

Solution

a) Factor the common factor x^2 out of the first two terms and the common factor 3 out of the last two terms:

$x^3 + x^2 + 3x + 3 = x^2(x + 1) + 3(x + 1)$ **Factor out common factors.**
$= (x + 1)(x^2 + 3)$ **Factor out the common factor $(x + 1)$.**

b) We must first arrange the polynomial so that the first group of two terms has a common factor and the last group of two terms also has a common factor:

$$aw + bc - bw - ac = aw - bw - ac + bc \quad \text{Rearrange.}$$
$$= w(a - b) - c(a - b) \quad \text{Factor out common factors.}$$
$$= (w - c)(a - b) \quad \text{Factor out } (a - b).$$

Factoring $ax^2 + bx + c$

Factoring by grouping was used in Example 2 to factor polynomials with four terms. Factoring by grouping can be used also to factor a trinomial that is the product of two binomials, because there are four terms in the product of two binomials (FOIL). If we examine the multiplication of two binomials using the distributive property, we can develop a strategy for factoring trinomials.

$$(x + 3)(x + 5) = (x + 3)x + (x + 3)5 \quad \text{Distributive property}$$
$$= x^2 + 3x + 5x + 15 \quad \text{Distributive property}$$
$$= x^2 + 8x + 15 \quad \text{Combine like terms.}$$

To factor $x^2 + 8x + 15$ we simply reverse the steps in this multiplication. First write $8x$ as $3x + 5x$, and then factor by grouping. We could write $8x$ as $x + 7x$, $2x + 6x$, and so on, but we choose $3x + 5x$ because 3 and 5 have a product of 15. So to factor $x^2 + 8x + 15$ proceed as follows:

$$x^2 + 8x + 15 = x^2 + 3x + 5x + 15 \quad \text{Replace } 8x \text{ by } 3x + 5x.$$
$$= (x + 3)x + (x + 3)5 \quad \text{Factor out common factors.}$$
$$= (x + 3)(x + 5) \quad \text{Factor out } x + 3.$$

The key to factoring $ax^2 + bx + c$ with $a = 1$ is to find two numbers that have a product equal to c and a sum equal to b, then proceed as above. After you have practiced this procedure, you should skip the middle two steps shown here and just write the answer.

EXAMPLE 3 Factoring $ax^2 + bx + c$ with $a = 1$

Factor each trinomial.

a) $x^2 - 5x - 14$ **b)** $x^2 + 4x - 21$ **c)** $x^2 - 5x + 6$

Solution

a) Two numbers that have a product of -14 and a sum of -5 are -7 and 2.

$$x^2 - 5x - 14 = x^2 - 7x + 2x - 14 \quad \text{Replace } -5x \text{ by } -7x + 2x.$$
$$= (x - 7)x + (x - 7)2 \quad \text{Factor out common factors.}$$
$$= (x - 7)(x + 2) \quad \text{Factor out } (x - 7).$$

Check by using FOIL.

b) Two numbers that have a product of -21 and a sum of 4 are 7 and -3.

$$x^2 + 4x - 21 = (x + 7)(x - 3)$$

Check by using FOIL.

c) Two numbers that have a product of 6 and a sum of -5 are -2 and -3.

$$x^2 - 5x + 6 = (x - 2)(x - 3)$$

Check by using FOIL.

Before trying to factor $ax^2 + bx + c$ with $a \neq 1$, consider the product of two linear factors:

$$(Mx + N)(Px + Q) = MPx^2 + (MQ + NP)x + NQ$$

Observe that the product of MQ and NP (from the coefficient of x) is $MNPQ$. The product of MP (the leading coefficient) and NQ (the constant term) is also $MNPQ$. So if the trinomial $ax^2 + bx + c$ can be factored, there are two numbers that have a sum equal to b (the coefficient of x) and a product equal to ac (the product of the leading coefficient and the constant term). This fact is the key to the ***ac*-method** for factoring $ax^2 + bx + c$.

PROCEDURE

The *ac*-Method for Factoring

To factor $ax^2 + bx + c$ with $a \neq 1$:

1. Find two numbers whose sum is b and whose product is ac.
2. Replace b by the sum of these two numbers.
3. Factor the resulting four-term polynomial by grouping.

EXAMPLE 4 Factoring $ax^2 + bx + c$ with $a \neq 1$

Factor each trinomial.

a) $2x^2 + 5x + 2$ **b)** $6x^2 - x - 12$ **c)** $15x^2 - 14x + 3$

Solution

a) Since $ac = 2 \cdot 2 = 4$ and $b = 5$, we need two numbers that have a product of 4 and a sum of 5. The numbers are 4 and 1.

$$\begin{aligned} 2x^2 + 5x + 2 &= 2x^2 + 4x + x + 2 && \text{Replace } 5x \text{ by } 4x + x. \\ &= (x + 2)2x + (x + 2)1 && \text{Factor by grouping.} \\ &= (x + 2)(2x + 1) && \text{Check using FOIL.} \end{aligned}$$

b) Since $ac = 6(-12) = -72$ and $b = -1$, we need two numbers that have a product of -72 and a sum of -1. The numbers are 8 and -9.

$$\begin{aligned} 6x^2 - x - 12 &= 6x^2 - 9x + 8x - 12 && \text{Replace } -x \text{ by } -9x + 8x. \\ &= (2x - 3)3x + (2x - 3)4 && \text{Factor by grouping.} \\ &= (2x - 3)(3x + 4) && \text{Check using FOIL.} \end{aligned}$$

c) Since $ac = 15 \cdot 3 = 45$ and $b = -14$, we need two numbers that have a product of 45 and a sum of -14. The numbers are -5 and -9.

$$\begin{aligned} 15x^2 - 14x + 3 &= 15x^2 - 5x - 9x + 3 && \text{Replace } -14x \text{ by } -5x - 9x. \\ &= (3x - 1)5x + (3x - 1)(-3) && \text{Factor by grouping.} \\ &= (3x - 1)(5x - 3) && \text{Check using FOIL.} \end{aligned}$$

After factoring some polynomials by the grouping method of Example 3 and the ac-method of Example 4, you will find that you can often guess the factors without going through all of the steps in those methods. Guessing the factors is an acceptable method that is called the **trial-and-error** method. For this method you simply write down any factors that you think might work, use FOIL to check if they are correct, then try again if they are wrong.

Factoring the Special Products

In Section P.4 we learned how to find the special products: the square of a sum, the square of a difference, and the product of a sum and a difference. The trinomial that results from squaring a sum or a difference is called a **perfect square trinomial**. We can write a factoring rule for each special product.

Factoring the Special Products

$a^2 - b^2 = (a + b)(a - b)$	**Difference of two squares**
$a^2 + 2ab + b^2 = (a + b)^2$	**Perfect square trinomial**
$a^2 - 2ab + b^2 = (a - b)^2$	**Perfect square trinomial**

EXAMPLE 5 Factoring special products

Factor each polynomial.

a) $4x^2 - 1$ **b)** $x^2 - 6x + 9$ **c)** $9y^2 + 30y + 25$ **d)** $x^{2t} - 9$

Solution

a)
$$\begin{aligned} 4x^2 - 1 &= (2x)^2 - 1^2 && \text{Recognize the difference of two squares.} \\ &= (2x + 1)(2x - 1) \end{aligned}$$

b) $x^2 - 6x + 9 = x^2 - 2(3x) + 3^2$
$= (x - 3)^2$

Recognize the perfect square trinomial.

c) $9y^2 + 30y + 25 = (3y)^2 + 2(3y)(5) + 5^2$
$= (3y + 5)^2$

Recognize the perfect square trinomial.

d) $x^{2t} - 9 = (x^t - 3)(x^t + 3)$

Recognize the difference of two squares.

Factoring the Difference and Sum of Two Cubes

The following formulas are used to factor the difference of two cubes and the sum of two cubes. You should verify these formulas using multiplication.

Factoring the Difference and Sum of Two Cubes

$a^3 - b^3 = (a - b)(a^2 + ab + b^2)$ **Difference of two cubes**

$a^3 + b^3 = (a + b)(a^2 - ab + b^2)$ **Sum of two cubes**

EXAMPLE 6 Factoring differences and sums of cubes

Factor each polynomial.

a) $x^3 - 27$ **b)** $8w^6 + 125z^3$ **c)** $y^{3m} - 1$

Solution

a) Since $x^3 - 27 = x^3 - 3^3$, we use $a = x$ and $b = 3$ in the formula for factoring the difference of two cubes:

$$x^3 - 27 = (x - 3)(x^2 + 3x + 9)$$

b) Since $8w^6 + 125z^3 = (2w^2)^3 + (5z)^3$, we use $a = 2w^2$ and $b = 5z$ in the formula for factoring the sum of two cubes:

$$8w^6 + 125z^3 = (2w^2 + 5z)(4w^4 - 10w^2z + 25z^2)$$

c) Since $y^{3m} = (y^m)^3$, we can use the formula for the difference of two cubes with $a = y^m$ and $b = 1$:

$$y^{3m} - 1 = (y^m - 1)(y^{2m} + y^m + 1)$$

Factoring by Substitution

When a polynomial involves a complicated expression, we can use two substitutions to help us factor. First we replace the complicated expression by a single variable and factor the simpler-looking polynomial. Then we replace the single variable by the complicated expression. This method is called **substitution**.

EXAMPLE 7 Factoring higher-degree polynomials

Use substitution to factor each polynomial.

a) $w^4 - 6w^2 - 16$ **b)** $(a^2 - 3)^2 + 7(a^2 - 3) + 12$

Solution

a) Replace w^2 by x and w^4 by x^2.

$$w^4 - 6w^2 - 16 = x^2 - 6x - 16$$

$$= (x - 8)(x + 2) \quad \text{Factor the trinomial.}$$

$$= (w^2 - 8)(w^2 + 2) \quad \text{Replace } x \text{ by } w^2.$$

b) Replace $a^2 - 3$ by b in the polynomial.

$$(a^2 - 3)^2 + 7(a^2 - 3) + 12 = b^2 + 7b + 12$$

$$= (b + 3)(b + 4) \quad \text{Factor the trinomial.}$$

$$= (a^2 - 3 + 3)(a^2 - 3 + 4) \quad \text{Replace } b \text{ by } a^2 - 3.$$

$$= a^2(a^2 + 1) \quad \text{Simplify.}$$

Factoring Completely

A polynomial is factored when it is written as a product. Unless noted otherwise, our discussion of factoring will continue to be limited to polynomials whose factors have integral coefficients. Polynomials that cannot be factored using integral coefficients are called **prime** or **irreducible over the integers**. For example, $a^2 + 1$, $b^2 + b + 1$, and $x + 5$ are prime polynomials because they cannot be expressed as a product (in a nontrivial manner). Any monomial is also considered a prime polynomial. A polynomial is said to be **factored completely** when it is written as a product of prime polynomials. For example, $3x^2(2x - 3)$ is factored completely because it is a product of a monomial and a binomial that is prime.

EXAMPLE 8 Factoring completely

Factor each polynomial completely.

a) $2w^4 - 32$ **b)** $-6x^7 + 6x$

Solution

a) $2w^4 - 32 = 2(w^4 - 16)$ **Factor out the greatest common factor.**

$= 2(w^2 - 4)(w^2 + 4)$ **Difference of two squares**

$= 2(w - 2)(w + 2)(w^2 + 4)$ **Difference of two squares**

The polynomial is now factored completely because $w^2 + 4$ is prime.

b) $-6x^7 + 6x = -6x(x^6 - 1)$ **Greatest common factor**

$= -6x(x^3 - 1)(x^3 + 1)$ **Difference of two squares**

$= -6x(x - 1)(x^2 + x + 1)(x + 1)(x^2 - x + 1)$

Difference of two cubes; sum of two cubes

The polynomial is factored completely because all of the factors are prime.

If 7 is a factor of 147, then 7 is a divisor of 147. If we divide 147 by 7 to get 21 (with no remainder), then we have $147 = 7 \cdot 21$. By factoring 21, we factor 147 completely as $147 = 3 \cdot 7^2$. The same ideas hold for polynomials. If we know one factor of a polynomial, we can use division to find the other factor and then factor the polynomial completely.

EXAMPLE 9 Using division to factor completely

Factor $x^3 - 7x + 6$ completely, given that $x - 2$ is a factor.

Solution

Use long division to find the other factor:

$$\begin{array}{r} x^2 + 2x - 3 \\ x - 2\,\overline{)\,x^3 + 0x^2 - 7x + 6} \\ \underline{x^3 - 2x^2} \qquad\qquad \\ 2x^2 - 7x \qquad \\ \underline{2x^2 - 4x} \qquad \\ -3x + 6 \\ \underline{-3x + 6} \\ 0 \end{array}$$

Because the remainder is 0, we have $x^3 - 7x + 6 = (x - 2)(x^2 + 2x - 3)$. Factoring completely gives

$$x^3 - 7x + 6 = (x - 2)(x + 3)(x - 1).$$

For Thought Mark True If the Polynomial Is Factored Correctly and False Otherwise. Explain.

1. $x^2 + 6x - 16 = x(x + 6) - 16$
2. $2x^4 - 5x^2 = -x^2(5 - 2x^2)$
3. $2x^4 - 5x^2 - 3 = (x^2 - 3)(2x^2 + 1)$
4. $a^2 - 1 = (a - 1)^2$
5. $a^3 - 1 = (a - 1)(a^2 + 1)$
6. $x^3 - y^3 = (x - y)(x^2 + 2xy + y^2)$
7. $8a^3 + 27b^3 = (2a + 3b)(4a^2 + 6ab + 9b^2)$
8. $x^2 + 8x - 12 = (x + 2)(x - 6)$
9. $(a + 3)6 - (a + 3)x = (a + 3)(6 - x)$
10. $a - b = -1(b - a)$

P.5 Exercises

Tape 2 Disk

Factor out the greatest common factor from each polynomial, first using a positive coefficient on the GCF and then using a negative coefficient.

1. $6x^3 - 12x^2$
2. $12x^2 + 18x^3$
3. $4a - 8ab$
4. $3wm + 15wm^2$
5. $-ax^3 + 5ax^2 - 5ax$
6. $-sa^3 + sb^3 - sb$
7. $m - n$
8. $y - x$

Factor each polynomial by grouping.

9. $x^3 + 2x^2 + 5x + 10$
10. $2w^3 - 2w^2 + 3w - 3$
11. $y^3 - y^2 - 3y + 3$
12. $x^3 + x^2 - 7x - 7$
13. $ady - w + d - awy$
14. $xy + ab + by + ax$
15. $x^2y^2 + ab - ay^2 - bx^2$
16. $6yz - 3y - 10z + 5$

Factor each trinomial.

17. $x^2 + 10x + 16$
18. $x^2 + 7x + 12$
19. $x^2 - 4x - 12$
20. $y^2 + 3y - 18$
21. $m^2 - 12m + 20$
22. $n^2 - 8n + 7$
23. $t^2 + 5t - 84$
24. $s^2 - 6s - 27$
25. $2x^2 - 7x - 4$
26. $3x^2 + 5x - 2$
27. $8x^2 - 10x - 3$
28. $18x^2 - 15x + 2$
29. $6y^2 + 7y - 5$
30. $15x^2 - 14x - 8$

Factor each special product.

31. $t^2 - u^2$
32. $9t^2 - v^2$
33. $t^2 + 2t + 1$
34. $m^2 + 10m + 25$
35. $4w^2 - 4w + 1$
36. $9x^2 - 12xy + 4y^2$
37. $y^{4t} - 25$
38. $121w^{2t} - 1$
39. $9z^2x^2 + 24zx + 16$
40. $25t^2 - 20tw^3 + 4w^6$

Factor each sum or difference of two cubes.

41. $t^3 - u^3$
42. $m^3 + n^3$
43. $a^3 - 8$
44. $b^3 + 1$
45. $27y^3 + 8$
46. $1 - 8a^6$
47. $27x^3y^6 - 8z^9$
48. $8t^3h^3 + n^9$

Use substitution to factor each polynomial.

49. $y^6 + 10y^3 + 25$
50. $y^8 + 8y^4 + 12$
51. $4a^4b^8 - 8a^2b^4 - 5$
52. $3c^2m^{14} - 22cm^7 + 7$
53. $(2a + 1)^2 + 2(2a + 1) - 24$
54. $2(w - 3)^2 + 3(w - 3) - 2$
55. $(b^2 + 2)^2 - 5(b^2 + 2) + 4$
56. $(t^3 + 5)^2 + 5(t^3 + 5) + 4$

Factor each polynomial completely.

57. $-3x^3 + 27x$
58. $a^4b^2 - 16b^2$
59. $16t^4 + 54w^3t$
60. $8a^6 - a^3b^3$
61. $a^3 + a^2 - 4a - 4$
62. $2b^3 + 3b^2 - 18b - 27$
63. $x^4 - 2x^3 - 8x + 16$
64. $a^4 - a^3 + a - 1$
65. $-36x^3 + 18x^2 + 4x$
66. $-6a^4 - a^3 + 15a^2$
67. $a^7 - a^6 - 64a + 64$
68. $a^5 - 4a^4 - 4a + 16$
69. $-6x^2 - x + 15$
70. $-6x^2 - 9x + 42$
71. $(a^2 + 2)^2 - 4(a^2 + 2) + 3$
72. $(z^3 + 5)^2 - 10(z^3 + 5) + 24$

For each pair of polynomials, use long division to determine whether the first polynomial is a factor of the second.

73. 13, 1261

74. 7, 861

75. $x + 3$, $x^3 + 4x^2 + 4x + 3$

76. $x - 2$, $x^3 - 3x^2 + 5x - 6$

77. $x - 1$, $3x^3 + 5x^2 - 12x - 9$

78. $x + 1$, $2x^3 + 4x^2 - 9x + 2$

In each pair of polynomials, the second polynomial is a factor of the first. Factor the first polynomial completely.

79. $x^3 + 4x^2 + x - 6$, $x - 1$

80. $x^3 - 13x - 12$, $x + 1$

81. $x^3 - x^2 - 4x - 6$, $x - 3$

82. $2x^3 + 7x^2 + 5x + 1$, $2x + 1$

83. $x^4 + 5x^3 + 5x^2 - 5x - 6$, $x + 2$

84. $4x^4 + 4x^3 - 25x^2 - x + 6$, $x - 2$

Factor each polynomial. Assume that variables used as exponents represent positive integers.

85. $x^{2m} - 1$

86. $x^{10w} - y^{2z}$

87. $x^{2q} - 5x^q + 6$

88. $x^{2a+2} + 6x^{a+1} + 9$

89. $x^{3n} - 1$

90. $x^{12a} + 8$

91. $x^{2w+1} + x^{w+1} + 2x^w + 2$

92. $y^{3n} - 3y^{2n} - y^n + 3$

Solve each problem.

93. *Volume of a Rectangular Box* If the volume of a box is $x^3 - 1$ ft^3 and the height is $x - 1$ ft, then what polynomial represents the area of the bottom?

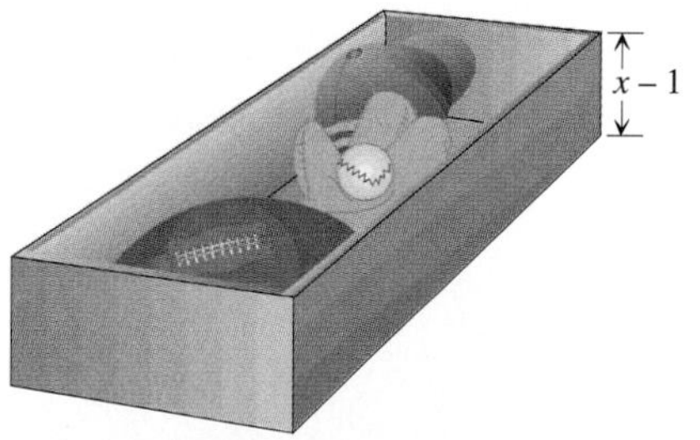

Figure for Exercise 93

94. *Area of a Trapezoid* If the area of a trapezoid is $2x^2 + 5x + 2$ square meters and the two parallel sides are x meters and $x + 1$ meters, then what polynomial represents the height of the trapezoid?

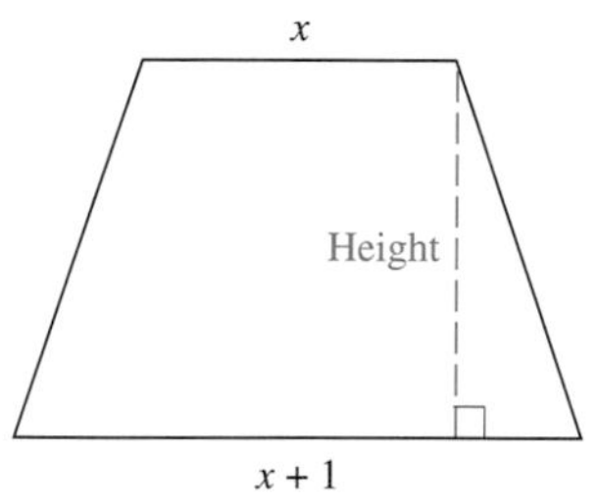

Figure for Exercise 94

95. *Making a Box* An open-top puzzle box is to be constructed from a 6 in. by 7 in. piece of cardboard by cutting out squares from each corner and folding up the sides as shown in the figure. Find the volume of the box for $x = 0.5$ in., 1 in., and 2 in. Which of these values for x produces the largest volume?

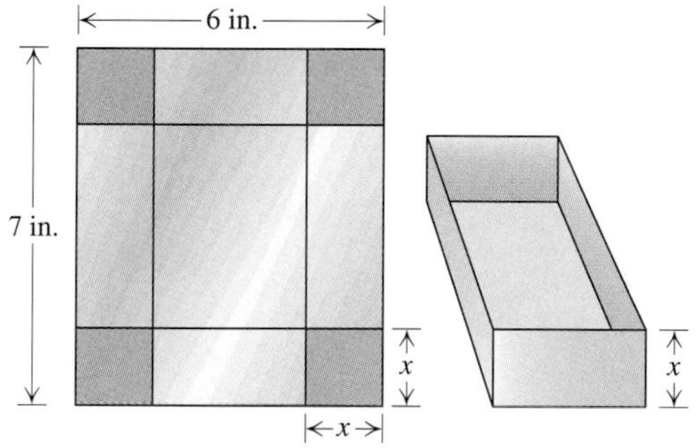

Figure for Exercises 95 and 96

• CALCULUS •

96. *Area and Volume* Find a polynomial that represents the area of the bottom of the box in the accompanying figure. Find a polynomial that represents the volume of the box.

For Writing/Discussion

97. Which of the following is not a perfect square trinomial? Explain.

a. $4x^8 - 20x^4y^3 + 25y^6$ **b.** $1000a^2 - 200ab + b^2$
c. $400w^4 - 40w^2 + 1$ **d.** $36a^{14} - 36a^7 + 9$

98. Which of the following is not a difference of two squares? Explain.

a. $196a^6b^4 - 289w^8$ **b.** $100x^{16} - 16h^{100}$
c. $25w^9 - 9y^{25}$ **d.** $1 - 9y^{36}$

99. *Another Lost Rule?* Is it true that the sum of two cubes factors as the cube of a sum? Explain.

100. *Cooperative Learning* Work in a small group to write a summary of the techniques used for factoring polynomials. When given a polynomial to factor completely, what should you look for first, second, third, and so on? Evaluate your summary by using it as a guide to factor some polynomials selected from the exercises.

Linking Concepts

For Individual or Group Explorations

Pyramid Builders *The Egyptians knew that the formula $V = \frac{1}{3}ha^2$ gives the volume of a pyramid that has a base with area a^2 and a height h. A pyramid missing its top (as it is during construction) is called a truncated pyramid. The ancient Egyptians used the formula $V = \frac{H}{3}(a^2 + ab + b^2)$ for the volume of a truncated pyramid with a square base of area a^2, a square top of area b^2, and height H as shown in the drawing. The Great Pyramid of Egypt was built with a square base of 755 feet on each side and a height of 480 feet.*

a) Find the volume of the Great Pyramid.

b) When the Great Pyramid was half as tall as its planned height (like a truncated pyramid), what volume of stone had been set in place?

c) Was the pyramid half finished when it was half as high as planned?

d) Assuming that the formula $V = \frac{1}{3}ha^2$ is correct, prove that the formula for the volume of a truncated pyramid is correct. (*Hint:* Subtract the volume of the missing top from the whole pyramid to get the volume of the truncated pyramid. You will need the fact that $a^2 + ab + b^2$ is a factor of $a^3 - b^3$ to get the given formula.)

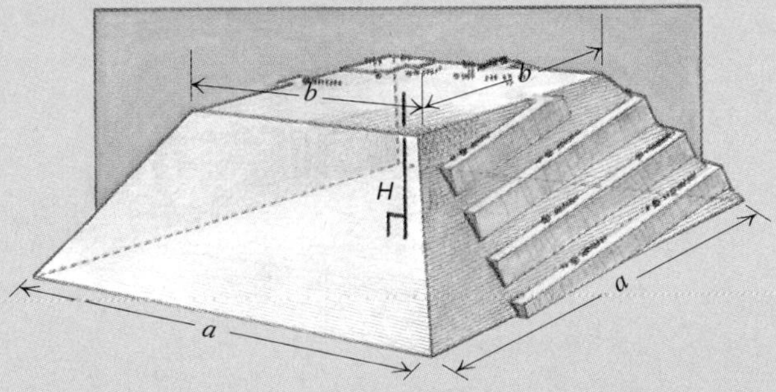

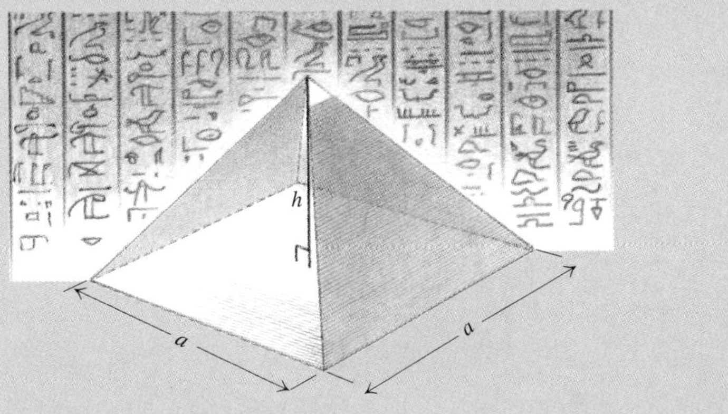

P.6 Rational Expressions

Rational expressions are to algebra what fractions are to arithmetic. In this section we will learn to reduce or build up rational expressions, and to perform basic operations with rational expressions.

Reducing

A **rational expression** is a ratio of two polynomials in which the denominator is not the zero polynomial. For example,

$$\frac{7}{2}, \quad \frac{2x - 1}{x + 3}, \quad 2a - 9, \quad \text{and} \quad \frac{2x + 4}{x^2 + 5x + 6}$$

are rational expressions. The **domain** of a rational expression is the set of all real numbers that can be used in place of the variable.

EXAMPLE 1 Domain of a rational expression

Find the domain of each rational expression.

a) $\dfrac{2x - 1}{x + 3}$ **b)** $\dfrac{2x + 4}{x^2 + 5x + 6}$ **c)** $\dfrac{1}{x^2 + 8}$

Solution

a) The domain is the set of all real numbers except those that cause $x + 3$ to have a value of 0. So -3 is excluded from the domain, because $x + 3$ has a value of 0 for $x = -3$. We write the domain in set notation as $\{x \mid x \neq -3\}$.

b) First factor $x^2 + 5x + 6$ as $(x + 2)(x + 3)$. The domain is the set of all real numbers except -2 and -3, because replacing x by either of these numbers would cause the denominator to be 0. The domain is written in set notation as $\{x \mid x \neq -2 \text{ and } x \neq -3\}$.

c) The value of $x^2 + 8$ is positive for any real number x. So the domain is the set of all real numbers, R.

In arithmetic we learned that each rational number has infinitely many equivalent forms. This fact is due to the **basic principle of rational numbers**.

Basic Principle of Rational Numbers

If a, b, and c are integers with $b \neq 0$ and $c \neq 0$, then

$$\frac{ac}{bc} = \frac{a}{b}.$$

The basic principle is actually true for all real numbers, but we use it most often with integers when we reduce fractions. For example, we reduce $\frac{3}{6}$ as follows:

$$\frac{3}{6} = \frac{1 \cdot 3}{2 \cdot 3} = \frac{1}{2}$$

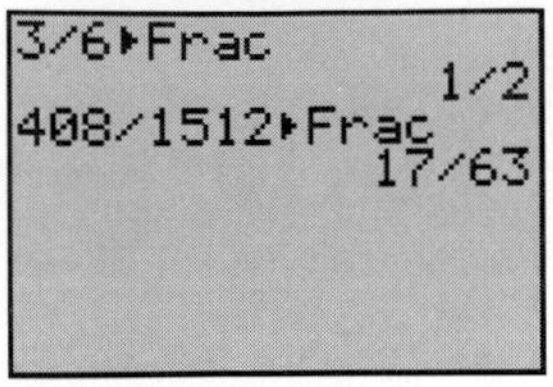

Figure P.28

Note that since 3 and 6 have a common factor of 3, we divide both the numerator and the denominator by 3 to reduce the fraction.

Your calculator reduces fractions as shown in Fig. P.28. □

To reduce a rational expression we factor the numerator and denominator completely, then *divide out* the common factors. A rational expression is in *lowest terms* when all common factors have been divided out.

EXAMPLE 2 Reducing to lowest terms

Reduce each rational expression to lowest terms.

a) $\dfrac{2x + 4}{x^2 + 5x + 6}$ **b)** $\dfrac{b - a}{a^3 - b^3}$ **c)** $\dfrac{x^2z^3}{x^5z}$

Solution

a) $\dfrac{2x + 4}{x^2 + 5x + 6} = \dfrac{2(x + 2)}{(x + 2)(x + 3)}$ **Factor the numerator and denominator.**

$= \dfrac{2}{x + 3}$ **Divide out the common factor $x + 2$.**

b) $\dfrac{b - a}{a^3 - b^3} = \dfrac{-1(a - b)}{(a - b)(a^2 + ab + b^2)}$ **Factor -1 out of $b - a$.**

$= \dfrac{-1}{a^2 + ab + b^2}$

c) $\dfrac{x^2z^3}{x^5z} = \dfrac{(x^2z)(z^2)}{(x^2z)(x^3)} = \dfrac{z^2}{x^3}$ **(The quotient rule yields the same result.)**

Be careful when reducing. The *only* way to reduce rational expressions is to factor and divide out the common *factors*. Identical terms that are not factors cannot be eliminated from a rational expression. For example,

$$\frac{x + 3}{3} \neq x$$

for all real numbers, because 3 is not a factor of the numerator.

Multiplication and Division

We multiply two rational numbers by multiplying their numerators and their denominators. For example,

$$\frac{2}{3} \cdot \frac{5}{7} = \frac{10}{21}.$$

Definition: Multiplication of Rational Numbers

If a/b and c/d are rational numbers, then

$$\frac{a}{b} \cdot \frac{c}{d} = \frac{ac}{bd}.$$

We multiply rational expressions in the same manner as rational numbers. Of course, any common factor can be divided out as we do when reducing rational expressions.

EXAMPLE 3 Multiplying rational expressions

Multiply the rational expressions.

a) $\dfrac{2a - 2b}{6} \cdot \dfrac{9a}{a^2 - b^2}$ **b)** $\dfrac{x - 1}{x^2 + 4x + 4} \cdot \dfrac{x + 2}{x^2 + 2x - 3}$

Solution

a)
$$\frac{2a - 2b}{6} \cdot \frac{9a}{a^2 - b^2} = \frac{2(a - b)}{2 \cdot 3} \cdot \frac{3 \cdot 3a}{(a - b)(a + b)} \quad \text{Factor completely.}$$
$$= \frac{3a}{a + b} \quad \text{Divide out common factors.}$$

b)
$$\frac{x - 1}{x^2 + 4x + 4} \cdot \frac{x + 2}{x^2 + 2x - 3} = \frac{x - 1}{(x + 2)^2} \cdot \frac{x + 2}{(x + 3)(x - 1)} \quad \text{Factor completely.}$$
$$= \frac{1}{x^2 + 5x + 6} \quad \text{Divide out common factors.}$$

We divide rational numbers by multiplying by the reciprocal of the divisor, or *invert and multiply*. For example, $6 \div \frac{1}{2} = 6 \cdot 2 = 12$.

Definition: Division of Rational Numbers

If a/b and c/d are rational numbers with $c \neq 0$, then

$$\frac{a}{b} \div \frac{c}{d} = \frac{a}{b} \cdot \frac{d}{c}.$$

Rational expressions are divided in the same manner as rational numbers.

EXAMPLE 4 **Dividing rational expressions**

Perform the indicated operations.

a) $\frac{9}{2x} \div \frac{3}{x}$ **b)** $\frac{4 - x^2}{6} \div \frac{x - 2}{2}$

Solution

a) $\frac{9}{2x} \div \frac{3}{x} = \frac{9}{2x} \cdot \frac{x}{3}$ Invert and multiply.

$= \frac{3 \cdot 3}{2x} \cdot \frac{x}{3}$ Factor completely.

$= \frac{3}{2}$ Divide out common factors.

b) $\frac{4 - x^2}{6} \div \frac{x - 2}{2} = \frac{-1(x - 2)(x + 2)}{2 \cdot 3} \cdot \frac{2}{x - 2}$ Invert and multiply.

$= \frac{-x - 2}{3}$ Divide out common factors.

Note that the division in Example 4(a) is valid only if $x \neq 0$. The division in Example 4(b) is valid only if $x \neq 2$ because $x - 2$ appears in the denominator after the rational expression is inverted.

Building Up the Denominator

The addition of fractions can be carried out only when their denominators are identical. To get a required denominator, we may **build up** the denominator of a fraction. To build up the denominator we use the basic principle of rational numbers in the reverse of the way we use it for reducing. We multiply the numerator and denominator of a fraction by the same nonzero number to get an equivalent fraction. For example, to get a fraction equivalent to $\frac{1}{3}$ with a denominator of 12, we multiply the numerator and denominator by 4:

$$\frac{1}{3} = \frac{1 \cdot 4}{3 \cdot 4} = \frac{4}{12}$$

To build up the denominator of a rational expression, we use the same procedure.

EXAMPLE 5 Writing equivalent rational expressions

Convert the first rational expression into an equivalent one that has the indicated denominator.

a) $\frac{3}{2a}, \frac{?}{6ab}$ **b)** $\frac{x-1}{x+2}, \frac{?}{x^2+6x+8}$ **c)** $\frac{a}{3b-a}, \frac{?}{a^2-9b^2}$

Solution

a) Compare the two denominators. Since $6ab = 2a(3b)$, we multiply the numerator and denominator of the first expression by $3b$:

$$\frac{3}{2a} = \frac{3 \cdot 3b}{2a \cdot 3b} = \frac{9b}{6ab}$$

b) Factor the second denominator and compare it to the first. Since $x^2 + 6x + 8 = (x + 2)(x + 4)$, we multiply the numerator and denominator by $x + 4$:

$$\frac{x-1}{x+2} = \frac{(x-1)(x+4)}{(x+2)(x+4)} = \frac{x^2+3x-4}{x^2+6x+8}$$

c) Factor the second denominator as

$$a^2 - 9b^2 = (a - 3b)(a + 3b) = -1(3b - a)(a + 3b).$$

Since $3b - a$ is a factor of $a^2 - 9b^2$, we multiply the numerator and denominator by $-1(a + 3b)$:

$$\frac{a}{3b-a} = \frac{a(-1)(a+3b)}{(3b-a)(-1)(a+3b)} = \frac{-a^2-3ab}{a^2-9b^2}$$

Addition and Subtraction

Fractions can be added or subtracted only if their denominators are identical. For example, $\frac{1}{3} + \frac{1}{3} = \frac{2}{3}$ and $\frac{7}{12} - \frac{2}{12} = \frac{5}{12}$.

Definition: Addition and Subtraction of Rational Numbers

If a/b and c/b are rational numbers, then

$$\frac{a}{b} + \frac{c}{b} = \frac{a+c}{b} \quad \text{and} \quad \frac{a}{b} - \frac{c}{b} = \frac{a-c}{b}.$$

For fractions with different denominators, we build up one or both denominators to get denominators that are equal to the least common multiple (LCM) of the denominators. The **least common denominator (LCD)** is the smallest number that is a multiple of all of the denominators. Use the following steps to find the LCD.

PROCEDURE

Finding the LCD

1. Factor each denominator completely.
2. Write a product using each factor that appears in a denominator.
3. For each factor, use the highest power of that factor that occurs in the denominators.

For example, to find the LCD for 10 and 12, we write $10 = 2 \cdot 5$ and $12 = 2^2 \cdot 3$. The LCD contains the factors 2, 3, and 5. Using the highest power of each, we get $2^2 \cdot 3 \cdot 5 = 60$ for the LCD. So, to add fractions with denominators of 10 and 12, we build up each fraction to a denominator of 60:

$$\frac{1}{10} + \frac{1}{12} = \frac{1 \cdot 6}{10 \cdot 6} + \frac{1 \cdot 5}{12 \cdot 5} = \frac{6}{60} + \frac{5}{60} = \frac{11}{60}$$

We use the same method to add or subtract rational expressions.

EXAMPLE 6 Adding and subtracting rational expressions

Perform the indicated operations.

a) $\dfrac{x}{x-1} + \dfrac{2x+3}{x^2-1}$ **b)** $\dfrac{x}{x^2+6x+9} - \dfrac{x-3}{x^2+5x+6}$

Solution

a)

$$\frac{x}{x-1} + \frac{2x+3}{x^2-1} = \frac{x}{x-1} + \frac{2x+3}{(x-1)(x+1)}$$ Factor denominators completely.

$$= \frac{x(x+1)}{(x-1)(x+1)} + \frac{2x+3}{(x-1)(x+1)}$$ Build up using the LCD, $(x-1)(x+1)$.

$$= \frac{x^2+x+2x+3}{(x-1)(x+1)}$$ Add the fractions.

$$= \frac{x^2+3x+3}{(x-1)(x+1)}$$ Simplify the numerator.

b)

$$\frac{x}{x^2+6x+9} - \frac{x-3}{x^2+5x+6} = \frac{x}{(x+3)^2} - \frac{x-3}{(x+2)(x+3)}$$

$$= \frac{x(x+2)}{(x+3)^2(x+2)} - \frac{(x-3)(x+3)}{(x+2)(x+3)(x+3)}$$

$$= \frac{x^2+2x}{(x+3)^2(x+2)} - \frac{x^2-9}{(x+3)^2(x+2)}$$

$$= \frac{2x+9}{(x+3)^2(x+2)}$$

Complex Fractions

A **complex fraction** is a fraction having rational expressions in the numerator, denominator, or both. Complex fractions can be simplified quickly by multiplying the numerator and denominator by the LCD of all of the denominators.

EXAMPLE 7 Simplifying a complex fraction

Simplify each complex fraction.

a) $\dfrac{4 - \dfrac{3}{4x}}{\dfrac{1}{x^2} - \dfrac{1}{6}}$ **b)** $\dfrac{\dfrac{x+6}{x^2-9}}{\dfrac{x}{x-3} - \dfrac{x+4}{x+3}}$

Solution

a) The LCD for the denominators 6, $4x$, and x^2 is $12x^2$. Multiply the numerator and denominator of the complex fraction by $12x^2$.

$$\frac{4 - \dfrac{3}{4x}}{\dfrac{1}{x^2} - \dfrac{1}{6}} = \frac{\left(4 - \dfrac{3}{4x}\right)12x^2}{\left(\dfrac{1}{x^2} - \dfrac{1}{6}\right)12x^2} = \frac{48x^2 - 9x}{12 - 2x^2}$$

b) The LCD for the denominators $x^2 - 9$, $x - 3$, and $x + 3$ is $x^2 - 9$, because $x^2 - 9 = (x - 3)(x + 3)$. Multiply the numerator and denominator by $(x - 3)(x + 3)$, or $x^2 - 9$:

$$\frac{\dfrac{x+6}{x^2-9}}{\dfrac{x}{x-3} - \dfrac{x+4}{x+3}} = \frac{\left(\dfrac{x+6}{x^2-9}\right)(x-3)(x+3)}{\left(\dfrac{x}{x-3} - \dfrac{x+4}{x+3}\right)(x-3)(x+3)}$$

$$= \frac{x+6}{x(x+3) - (x+4)(x-3)}$$

$$= \frac{x+6}{x^2 + 3x - (x^2 + x - 12)}$$

$$= \frac{x+6}{2x+12} = \frac{x+6}{2(x+6)} = \frac{1}{2}$$

The fractions in a complex fraction can be written with negative exponents. To simplify complex fractions with negative exponents we still multiply the numerator and denominator by the LCD.

EXAMPLE 8 A complex fraction with negative exponents

Simplify each complex fraction. Write answers with positive exponents only.

a) $\dfrac{a^{-1} + b^{-3}}{ab^{-2} + ba^{-4}}$ **b)** $pq + p^{-1}q^{-2}$

Solution

a) If we use the definition of negative exponent to rewrite each term, then the denominators would be a, b^3, b^2, and a^4. The LCD for these denominators is a^4b^3. Multiply the numerator and denominator by a^4b^3:

$$\frac{a^{-1} + b^{-3}}{ab^{-2} + ba^{-4}} = \frac{(a^{-1} + b^{-3})a^4b^3}{(ab^{-2} + ba^{-4})a^4b^3} = \frac{a^3b^3 + a^4}{a^5b + b^4}$$

Note that the exponents in a^4b^3 are just large enough to eliminate all negative exponents in the multiplication. Note also that we could have rewritten the complex fraction without negative exponents before multiplying by a^4b^3, but it is not necessary to do so.

b) Although this expression is not exactly a complex fraction, we can use the same technique to eliminate the negative exponents. Multiply the numerator and denominator by pq^2:

$$pq + p^{-1}q^{-2} = \frac{(pq + p^{-1}q^{-2})pq^2}{1 \cdot pq^2} = \frac{p^2q^3 + 1}{pq^2}$$

FOR THOUGHT True or False? Explain.

1. The rational expression $\dfrac{2x + 5}{2y}$ reduces to $\dfrac{x + 5}{y}$.

2. The rational expression $\dfrac{-3}{a - 5}$ is equivalent to $\dfrac{3}{5 - a}$.

3. The expressions $\dfrac{x(x + 2)}{x(x + 3)}$ and $\dfrac{x + 2}{x + 3}$ are equivalent.

4. The expression $\dfrac{a^2 - b^2}{a - b}$ reduced to lowest terms is $a - b$.

5. The LCD for the rational expressions $\dfrac{1}{x}$ and $\dfrac{1}{x + 1}$ is $x + 1$.

6. $\dfrac{2x - 1}{x - 3} + \dfrac{x + 5}{x - 3} = \dfrac{3x + 4}{x - 3}$ provided that $x \neq 3$.

7. $\dfrac{x}{2} = \dfrac{1}{2}x$ for all nonzero values of x.

8. $\dfrac{x}{3} - 1 = \dfrac{x - 3}{3}$ for any real number x.

9. If $x = 500$, then the approximate value of $\dfrac{2x + 1}{x - 3}$ is 2.

10. If $|x|$ is very large, then $\dfrac{5x + 1}{x}$ has an approximate value of 5.

P.6 EXERCISES

Tape 2 Disk

Find the domain of each rational expression.

1. $\dfrac{x-3}{x+2}$

2. $\dfrac{x^2-1}{x-5}$

3. $\dfrac{x^2-9}{(x-4)(x+2)}$

4. $\dfrac{2x-3}{(x+1)(x-3)}$

5. $\dfrac{x+1}{x^2-9}$

6. $\dfrac{x+2}{x^2+3x+2}$

7. $\dfrac{3x^2-2x+1}{x^2+3}$

8. $\dfrac{-2x^2-7}{3x^2+8}$

Reduce each rational expression to lowest terms.

9. $\dfrac{3x-9}{x^2-x-6}$

10. $\dfrac{-2x-4}{x^2-3x-10}$

11. $\dfrac{10a-8b}{12b-15a}$

12. $\dfrac{a^2-b^2}{b-a}$

13. $\dfrac{a^3b^6}{a^2b^3-a^4b^2}$

14. $\dfrac{18u^6v^5+24u^3v^3}{42u^2v^5}$

15. $\dfrac{x^4y^5z^2}{x^7y^3z}$

16. $\dfrac{t^3u^7}{-t^8u^5}$

17. $\dfrac{a^3-b^3}{a^2-b^2}$

18. $\dfrac{a^3+b^3}{a^2+b^2}$

Find the products or quotients.

19. $\dfrac{2a}{3b^2}\cdot\dfrac{9b}{14a^2}$

20. $\dfrac{14w}{51y}\cdot\dfrac{3w}{7y}$

21. $\dfrac{12a}{7}\div\dfrac{2a^3}{49}$

22. $\dfrac{20x}{y^3}\div\dfrac{30}{y^5}$

23. $\dfrac{a^2-9}{3a-6}\cdot\dfrac{a^2-4}{a^2-a-6}$

24. $\dfrac{6x^2+x-1}{6x+3}\cdot\dfrac{15}{9x^2-1}$

25. $\dfrac{x^2-y^2}{9}\div\dfrac{x^2+2xy+y^2}{18}$

26. $\dfrac{a^3-b^3}{a^2-2ab+b^2}\div\dfrac{2a^2+2ab+2b^2}{9a^2-9b^2}$

27. $\dfrac{x^2-y^2}{-3xy}\cdot\dfrac{6x^2y^3}{2y-2x}$

28. $\dfrac{a^2-a-2}{2}\cdot\dfrac{1}{4-a^2}$

In Exercises 29–34, convert the first rational expression into an equivalent one that has the indicated denominator.

29. $\dfrac{4}{3a},\ \dfrac{?}{12a^2}$

30. $\dfrac{a+2}{4a^2},\ \dfrac{?}{20a^3b}$

31. $\dfrac{x-5}{x+3},\ \dfrac{?}{x^2-9}$

32. $\dfrac{x+2}{x-8},\ \dfrac{?}{16-2x}$

33. $\dfrac{x}{x+5},\ \dfrac{?}{x^2+6x+5}$

34. $\dfrac{3a-b}{a+b},\ \dfrac{?}{9b^2-9a^2}$

Find the least common denominator (LCD) for each given pair of rational expressions.

35. $\dfrac{1}{4ab^2},\ \dfrac{7}{6a^2b^3}$

36. $\dfrac{3}{2x^2y},\ \dfrac{a}{5xy}$

37. $\dfrac{-7a}{3a+3b},\ \dfrac{5b}{2a+2b}$

38. $\dfrac{1}{3a-3b},\ \dfrac{2}{a^2-b^2}$

39. $\dfrac{2x}{x^2+5x+6},\ \dfrac{3x}{x^2-x-6}$

40. $\dfrac{x+7}{2x^2+7x-15},\ \dfrac{x-5}{2x^2-5x+3}$

Perform the indicated operations.

41. $\dfrac{3}{2x}+\dfrac{1}{6}$

42. $\dfrac{-7}{3a^2b}+\dfrac{4}{6ab^2}$

43. $\dfrac{x+3}{x-1}-\dfrac{x+4}{x+1}$

44. $\dfrac{x+2}{x-3}-\dfrac{x^2+3x-2}{x^2-9}$

45. $3+\dfrac{1}{a}$

46. $-1-\dfrac{3}{c}$

47. $t-1-\dfrac{1}{t+1}$

48. $w+\dfrac{1}{w-1}$

49. $\dfrac{x}{x^2+3x+2}+\dfrac{x-1}{x^2+5x+6}$

50. $\dfrac{x-1}{x^2+x-6}-\dfrac{x-2}{x^2+4x+3}$

51. $\dfrac{1}{x-3}-\dfrac{5}{6-2x}$

52. $\dfrac{5}{4 - x^2} - \dfrac{2x}{x - 2}$

53. $\dfrac{y^2}{x^3 - y^3} + \dfrac{x + y}{x^2 + xy + y^2}$

54. $\dfrac{ab}{a^3 + b^3} + \dfrac{a}{2a^2 - 2ab + 2b^2}$

55. $\dfrac{1}{x} + \dfrac{1}{x - 1} - \dfrac{1}{x + 1}$

56. $\dfrac{3}{x} - \dfrac{x - 1}{x^2 - 9} + \dfrac{1}{x - 3}$

Simplify each complex fraction.

57. $\dfrac{\dfrac{4}{a} - \dfrac{3}{b}}{\dfrac{1}{ab} + \dfrac{2}{b^2}}$

58. $\dfrac{\dfrac{2}{6xy} - \dfrac{1}{4x}}{\dfrac{1}{3y^2} + \dfrac{1}{2x}}$

59. $\dfrac{\dfrac{a}{ab^2} - \dfrac{b}{ab^3}}{\dfrac{3}{a^2} + \dfrac{a}{a^3b}}$

60. $\dfrac{\dfrac{1}{a^2b^3c}}{\dfrac{c}{ab^2} + \dfrac{a}{b^2c}}$

61. $\dfrac{a + \dfrac{4}{a + 4}}{a - \dfrac{4a + 4}{a + 4}}$

62. $\dfrac{y - \dfrac{y + 6}{y + 2}}{y - \dfrac{4y + 15}{y + 2}}$

63. $\dfrac{\dfrac{t + 2}{t - 1} - \dfrac{t - 3}{t}}{\dfrac{t + 4}{t} + \dfrac{t - 2}{t - 1}}$

64. $\dfrac{\dfrac{3}{2 + x} - \dfrac{4}{2 - x}}{\dfrac{1}{x + 2} - \dfrac{3}{x - 2}}$

Simplify. Write answers with positive exponents only.

65. $\dfrac{x^{-1} + 1}{x^{-1} - 1}$

66. $\dfrac{x^{-2} - 4}{x^{-1} - 2}$

67. $a^2 + a^{-1}b^{-3}$

68. $m + m^{-1} + m^{-2}$

69. $\dfrac{x^2 - y^2}{x^{-1} - y^{-1}}$

70. $\dfrac{x^{-2} - y^{-2}}{(x - y)^2}$

71. $(m^{-1} - n^{-1})^{-2}$

72. $(a^{-1} + y^{-1})^{-1}$

Let $R(x) = \dfrac{3}{x + 6}$, $S(x) = \dfrac{2x - 5}{2x - 9}$, and $T(x) = \dfrac{9x^2 - 1}{3x^2 - 2}$.

• CALCULUS • Find the following.

73. $R(1)$
74. $R(-1)$
75. $R(500)$
76. $R(-1000)$
77. $S(2)$
78. $S(-2)$
79. $S(600)$
80. $S(-600)$
81. $T(-4)$
82. $T(7)$
83. $T(-400)$
84. $T(500)$

Solve each problem.

85. *Hazardous Waste* High Tech Incineration charges a garbage hauler \$50 plus \$20 per ton to incinerate a truckload of hazardous waste. The average cost per ton is shown in the accompanying graph.
   a. Is the average cost per ton increasing or decreasing as the capacity of the truck increases?
   b. The average cost per ton (in dollars) for incinerating n tons of waste is given by the rational expression

$$R(n) = \frac{20n + 50}{n}.$$

   A garbage hauler has trucks with capacity of 7, 12, and 22 tons. What is the average cost per ton for each truck?

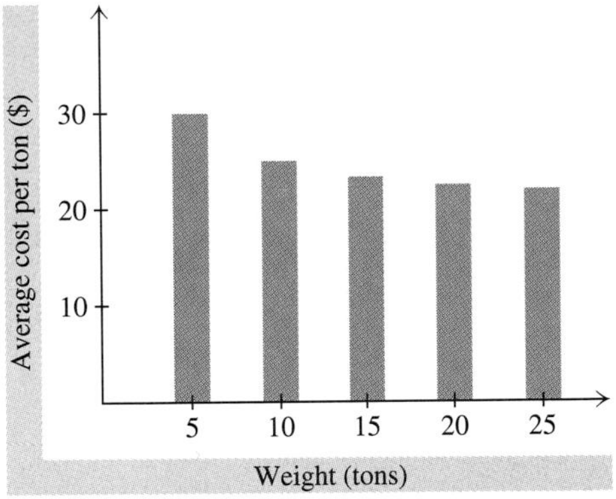

Figure for Exercise 85

86. *Water Pollution* High Tech Incineration (HTI) has been told by the EPA to stop polluting the Red River

immediately or it must pay \$5000 plus \$400 per day until it stops. Write a rational expression like that in the previous exercise for calculating the average fine per day F. What is the average fine per day if HTI stops polluting in 10 days, 20 days, or 30 days? Is the average fine per day increasing or decreasing as the number of days gets larger?

• CALCULUS •

87. *Costly Cleanup* The cost of cleaning up the ConChem hazardous waste site has been modeled by the rational expression

$$C(p) = \frac{6{,}000{,}000p}{100 - p}$$

where $C(p)$ is the cost in dollars for cleaning up $p\%$ of the pollution. Find the costs for cleaning up 50%, 75%, and 99% of the pollution. What is the domain of the rational expression?

88. *Costly Campaign* A political strategist used the rational expression

$$C(p) = \frac{2{,}000{,}000 + 1{,}000{,}000p}{50 - p}$$

to estimate the cost in dollars of a campaign that would get her candidate for governor $p\%$ of the votes. Find the costs for 30%, 40%, and 49%. Could this candidate raise enough money to get at least 50% of the vote?

89. *Filing Invoices* Gina can file all of the daily invoices in 4 hr and Bert can do the same job in 6 hr. If they work together, then what portion of the invoices can they file in 1 hr?

90. *Painting a House* Melanie can paint the entire house in x hours and Timothy can do the same job in y hours. Write a rational expression that represents the portion of the house that they can paint in 2 hr working together.

91. *Average Speed* Barry Allen, alias the Flash, finishes a meal at the Golden Buddha restaurant only to find he's left his wallet at home. Not wanting to reveal his secret identity to his date, he excuses himself and slips outside. Dashing home at 250 mph, he snatches his wallet from his nightstand and races back to the restaurant at 300 mph. Discounting the time it took to find his wallet, what was his average speed for the trip? (Average speed is total distance divided by total time for the trip.)

92. *Average Cost* Every day Denise spends the same amount on eggs for Denise's Diner. On Monday eggs were 50 cents per dozen, on Tuesday they were 60 cents per dozen, and on Wednesday they were 70 cents per dozen. What was her average cost for a dozen eggs for the three-day period? (The average cost is the total cost divided by the total number of dozens purchased.)

For Writing/Discussion

93. *Domain* Why is it important to know the domain of a rational expression?

94. *Cooperative Learning* Each student in your small group should write down a rational expression in which the degree of the denominator is greater than or equal to the degree of the numerator. Evaluate your rational expression for several ''very large'' values of x, using a calculator. Discuss your results. Can you predict the approximate value by looking at the rational expression?

95. Find the exact value of each expression.

a. $$\cfrac{1}{1 + \cfrac{1}{1 + \cfrac{1}{1 + \cfrac{1}{1 + \cfrac{1}{2}}}}}$$

b. $$\cfrac{1}{1 - \cfrac{1}{1 - \cfrac{1}{1 - \cfrac{1}{1 - \cfrac{1}{3}}}}}$$

96. Find the domain of each expression.

a. $$\cfrac{1}{1 + \cfrac{1}{1 + \cfrac{1}{x}}}$$

b. $$\cfrac{1}{1 + \cfrac{1}{1 + \cfrac{1}{1 + \cfrac{1}{x - 1}}}}$$

Linking Concepts

For Individual or Group Explorations

• CALCULUS •

Average Cost Per Drink *To save the environment, the Lion's Den Cafeteria sells an empty reusable 24-ounce plastic mug to students for \$2.99. Using the mug, a student can get a 24-ounce soft drink for a reduced price of 69 cents.*

Mug \$2.99
Refills 69¢

a) If a student buys the mug and uses it only once, then what is the cost for that one drink?

b) If a student buys the mug and uses it twice, then what is the average cost per drink?

c) Does the average cost per drink increase or decrease as the mug is used more and more?

d) If the regular price of a 24-ounce soft drink is 89 cents, then when does the student start saving money?

e) Write a rational expression for which the value of the expression is the average cost per drink and x is the number of times the mug is used.

f) If a student buys the mug and uses it every day for a semester, then what is the approximate average cost per drink?

g) If a professional student buys the mug and uses it every day for many years, then what value is the average cost per drink approaching?

h) Use a computer to create a graph showing the number of drinks purchased and the average cost per drink. Make sure that the graph illustrates your answer to part (g).

HIGHLIGHTS

• **Section P.1 Real Numbers and Their Properties**

1. Every real number is either rational (a ratio of integers) or irrational.
2. The properties of real numbers are the basic properties of algebra.
3. The absolute value of a real number a is a if $a \geq 0$ and $-a$ if $a < 0$.
4. The absolute value of a number indicates its distance from 0 on a number line.
5. The distance between a and b on a number line is $|a - b|$.
6. The order of operations is used for evaluating expressions.

• **Section P.2 Integral Exponents**

1. A positive integral exponent indicates the number of times the base is used as a factor.

2. An expression with a negative exponent is the reciprocal of the expression with a positive exponent.
3. If $a \neq 0$ then $a^0 = 1$.
4. In scientific notation, a positive number less than 1 or greater than 10 is written as a product of a number between 1 and 10 and a power of 10.

- **Section P.3 Rational Exponents and Radicals**

1. The fractional exponent $1/n$ indicates the nth root.
2. In a fractional exponent the numerator indicates the power and the denominator indicates the root.
3. An even root of a negative number is not a real number.
4. The same rules that apply to expressions with integral exponents apply to expressions with real numbers as exponents.
5. A radical expression of index n is in simplified form when it has no perfect nth powers as factors of the radicand, no fractions inside the radical, and no radicals in a denominator.

- **Section P.4 Polynomials**

1. If n is a nonnegative integer and $a_0, a_1, a_2, \ldots, a_n$ are real numbers, then a polynomial in x is an expression of the form $a_nx^n + a_{n-1}x^{n-1} + \cdots + a_1x + a_0$.
2. Polynomials can be added, subtracted, multiplied, and divided.
3. FOIL is a rule based on the distributive property that helps us find the product of two binomials quickly.
4. The value of a polynomial in x is the value of the expression obtained when x is replaced by a real number.

- **Section P.5 Factoring Polynomials**

1. To factor a polynomial means to write it as a product of two or more polynomials.
2. A perfect square trinomial is the square of a binomial.
3. Each special product rule is also a factoring rule.
4. All factoring can be checked by multiplication.

- **Section P.6 Rational Expressions**

1. A rational expression is a ratio of two polynomials, in which the denominator is not the zero polynomial.
2. The domain of a rational expression is the set of all real numbers for which the denominator does not have a value of zero.
3. Operations with rational expressions are done in the same manner as operations with fractions.
4. To simplify a complex fraction, multiply the numerator and denominator by the LCD of all denominators.

CHAPTER P REVIEW EXERCISES

Determine whether each statement is true or false and explain your answer.

1. Every real number is a rational number.
2. Zero is neither rational nor irrational.
3. There are no negative integers.
4. Every repeating decimal number is a rational number.
5. The terminating decimal numbers are irrational numbers.
6. The number $\sqrt{289}$ is a rational number.
7. Zero is a natural number.
8. The multiplicative inverse of 8 is 0.125.
9. The reciprocal of 0.333 is 3.
10. The real number π is irrational.
11. The additive inverse of 0.5 is 0.
12. The distributive property is used in adding like terms.

Simplify each expression.

13. $-3x - 4(3 - 5x)$
14. $x - 0.02(x - 9)$
15. $\dfrac{x}{5} + \dfrac{x}{10}$
16. $\dfrac{1}{3}x - \dfrac{1}{8}x$
17. $\dfrac{3x - 6}{9}$
18. $\dfrac{1}{2}(4x - 6)$
19. $\dfrac{-7 - (-1)}{3 - (-5)}$
20. $\dfrac{6 - (3 - x)}{2 - (-1)}$
21. $|-3| - |-5|$
22. $|5 - (-2)|$
23. $|3 - 7|$
24. $|-3 - (-4)|$
25. $8 - 9 \cdot 2 \div 3 + 5$
26. $3 - 4(2 - 3 \cdot 5^2)$
27. $12 \div 4 \cdot 3 \div 6 + 3^3$
28. $8 \cdot 3^2 - 3\sqrt{3^2 + 4^2}$

Simplify each expression. Assume that all variables represent positive real numbers.

29. 5^4
30. 2^{-4}
31. $(-2)^2 - 4(-2)(5)$
32. $6^2 - 4(-1)(-3)$
33. $2^{-1} + 2^0$
34. $\dfrac{3^{-1}}{-3^2}$
35. $\dfrac{-3^{-1}2^3}{2^{-1}}$
36. $\dfrac{-1}{-1^{-1}}$
37. $8^{-2/3}$
38. $-16^{-3/4}$
39. $(125x^6)^{1/3}$
40. $\dfrac{1}{(27t^{12})^{-1/3}}$
41. $\sqrt{121}$
42. $\sqrt[3]{-1000}$
43. $\sqrt{28s^3}$
44. $\sqrt{75a^2b^9}$
45. $\sqrt[3]{-2000}$
46. $\sqrt[3]{56w^4}$
47. $\sqrt{\dfrac{5}{2a}}$
48. $\sqrt{\dfrac{1}{18z^3}}$
49. $\sqrt[3]{\dfrac{2}{5}}$
50. $\sqrt[3]{\dfrac{3}{4y}}$
51. $\sqrt{18n^3} + \sqrt{50n^3}$
52. $\sqrt[3]{24} - \sqrt[3]{81}$
53. $\dfrac{2\sqrt{3}}{\sqrt{3} - 1}$
54. $\dfrac{2}{\sqrt{6} - 2}$
55. $\dfrac{\sqrt{6}}{\sqrt{8} + \sqrt{18}}$
56. $\dfrac{\sqrt{15}}{\sqrt{75} + \sqrt{20}}$

Convert each number given in scientific notation to standard notation and each number given in standard notation to scientific notation.

57. 3.2×10^8
58. -4.543×10^9
59. -1.85×10^{-4}
60. 9.44×10^{-5}
61. 0.000056
62. −0.000341
63. −2,340,000
64. 88,300,000,000

Perform the indicated operations. Write the answer in scientific notation.

65. $(5 \times 10^6)^3$
66. $\dfrac{(0.00000046)(3000)}{2{,}300{,}000}$
67. $\dfrac{(800)^2(0.00001)^{-3}}{(2{,}000{,}000)^3(0.00002)}$
68. $\dfrac{(5.1 \times 10^8)(-2 \times 10^{-3})}{1.7 \times 10^{-6}}$

Perform the indicated operations.

69. $(3x^2 - x - 2) + (-x^2 + 2x - 5)$
70. $(4y^3 - y^2 + 5y - 9) - (y^3 - 6y^2 + 3y - 2)$
71. $(-4x^4 - 3x^3 + x) - (x^4 - 6x^3 - 2x)$
72. $(3y^4 - 4y^2 - 6) + (-y^4 - 8y + 7)$

73. $(3a^2 - 2a + 5)(a - 2)$
74. $(w - 5)(w^2 + 5w + 25)$
75. $(b - 3y)^2$
76. $(x - 1)^3$
77. $(t - 3)(3t + 2)$
78. $(5y - 9)(5y + 9)$
79. $-35y^5 \div (7y^2)$
80. $(3x^3 - 6x^2) \div (3x)$
81. $(3 + \sqrt{2})(3 - \sqrt{2})$
82. $(2\sqrt{3} - 1)(3\sqrt{3} + 2)$
83. $(2\sqrt{5} + \sqrt{3})^2$
84. $(2\sqrt{x} + 3)^2$
85. $(1 + \sqrt{2x - 1})^2$
86. $(3 + \sqrt{y - 4})^2$

Find the quotient and remainder when the first polynomial is divided by the second.

87. $x^3 + 2x^2 - 9x + 3,\ x - 2$
88. $x^3 - 6x^2 + 3x - 9,\ x + 3$
89. $6x^2 + x + 2,\ 2x - 1$
90. $12x^2 - x - 21,\ 3x - 4$

Use division to express each fraction in the form

$$\text{quotient} + \frac{\text{remainder}}{\text{divisor}}.$$

91. $\dfrac{x^2 - 3}{x + 2}$
92. $\dfrac{a^2 + 5a + 2}{a}$
93. $\dfrac{2x + 3}{x - 5}$
94. $\dfrac{-3x + 2}{5x - 4}$

Factor each polynomial completely.

95. $6x^3 - 6x$
96. $4u^2 - 9v^2$
97. $9h^2 + 24ht + 16t^2$
98. $b^2 + 6b - 16$
99. $t^3 + y^3$
100. $8a^3 - 27$
101. $x^3 + 3x^2 - 9x - 27$
102. $3x - by + bx - 3y$
103. $t^6 - 1$
104. $y^4 - 625$
105. $18x^2 - 9x - 20$
106. $(x - 1)^2 - (x - 1) - 2$
107. $a^3b + 3a^2b - 18ab$
108. $x^3y^4 + 4x^2y^2 - 12x$
109. $2x^3 + x^2y - y - 2x$
110. $12x^3 - 2x^2y - 24xy^2$

Perform the indicated operations with the rational expressions.

111. $\dfrac{x - 2}{x + 3} + \dfrac{x + 8}{x + 3}$
112. $\dfrac{3x - 5}{x^2 - 4} - \dfrac{3 - x}{x^2 - 4}$
113. $\dfrac{x - 1}{x - 2} - \dfrac{x + 3}{x + 4}$
114. $\dfrac{1 - x}{x} + \dfrac{3 - x}{x - 2}$
115. $\dfrac{x^2 - 9}{x + 3} \cdot \dfrac{1}{6 - 2x}$
116. $\dfrac{x^3 - 8}{6} \cdot \dfrac{3x + 6}{x^2 - 4}$
117. $\dfrac{a^3bc^8}{a^9b^3c} \cdot \dfrac{(ab^3c^5)^2}{a^4b^3}$
118. $\dfrac{(x^3z^2)^5}{xz^4} \cdot \left(\dfrac{xz^2}{x^3}\right)^2$
119. $\dfrac{1}{x^2 - 4} + \dfrac{3}{x - 2}$
120. $\dfrac{3}{2x - 4} + \dfrac{5}{3x - 6}$
121. $\dfrac{1}{6x} - \dfrac{7}{10x^2}$
122. $\dfrac{1}{3y^2} - \dfrac{2}{9y}$
123. $\dfrac{a^2 - 25}{a^2 - 4a - 5} \div \dfrac{2a + 10}{a^2 - 1}$
124. $\dfrac{y^4 - 16}{y^2 + y - 2} \div \dfrac{y^3 + 4y}{y^3 - y}$
125. $\dfrac{x^2 - 16}{x^2 + 5x + 4} \div \dfrac{8 - 2x}{x^3 + 1}$
126. $\dfrac{x^2 + ax + bx + ab}{x^2 + 2bx + b^2} \div \dfrac{x^2 + 2ax + a^2}{x^3 + b^3}$
127. $\dfrac{a - 2}{a^2 + 6a + 5} + \dfrac{2a + 1}{a^2 - 1}$
128. $\dfrac{y - 1}{y^2 - 2y - 24} - \dfrac{y - 3}{y^2 + 2y - 8}$

Simplify.

129. $\dfrac{\dfrac{5}{2x} - \dfrac{3}{4x}}{\dfrac{1}{2} - \dfrac{2}{x}}$
130. $\dfrac{\dfrac{4}{4 - y^2} - \dfrac{5}{y - 2}}{\dfrac{1}{2 - y} - \dfrac{3}{y + 2}}$
131. $\dfrac{\dfrac{1}{y^2 - 2} - 3}{\dfrac{5}{y^2 - 2} + 4}$
132. $\dfrac{\dfrac{1}{6a^2b^3}}{\dfrac{5}{8a^3b} - \dfrac{3}{10a^3b^4}}$
133. $\dfrac{a^{-2} - b^{-3}}{a^{-1}b^{-1}}$
134. $\dfrac{x^{-1} - y^{-1}}{x^{-3} - y^{-3}}$
135. $p^{-1} + pq^{-3}$
136. $a^{-1} + x^{-1}$

Given that

$$P(x) = x^3 - 3x^2 + x - 9 \quad \text{and} \quad R(x) = \frac{3x - 1}{2x - 9},$$

find each of the following.

137. $P(2)$
138. $P(-1)$
139. $P(0)$
140. $P\left(\dfrac{1}{2}\right)$
141. $R(-1)$
142. $R(3)$
143. $R(50)$
144. $R(-40)$

Solve each problem.

145. *Fly Ball* The approximate distance in feet that a baseball will travel when hit at a 45° angle is given by the formula

$$d = \frac{(v_0)^2}{32},$$

where v_0 is the initial velocity of the ball in feet per second. If Troy O'Leary of the Red Sox hits a ball at a 45° angle at 126 ft/sec, then how far will the ball travel?

146. *Thirty-Year Bonds* In August of 1997, the thirty-year T-bond yield was 6.29% (*USA Today*, August 1, 1997). If you invested \$10,000 in T-bonds at that time, then in August of 2027 your investment would be worth

$$10{,}000(1.0629)^{30}.$$

Use a calculator to find this value.

147. *Weight of a Hydrogen Atom* One hydrogen atom weighs 1.7×10^{-24} g. Find the number of hydrogen atoms in 1 kg of hydrogen.

148. *Moon Talk* Radio waves travel at 3.0×10^8 m/sec (the same as the speed of light) and the distance from the earth to the moon is 3.84×10^8 m. How many seconds did it take for a radio wave to travel from mission control in Houston to the astronauts on the surface of the moon and then back to Houston? (A good demonstration of the speed of light actually occurred when Houston controllers heard an echo to their words that traveled to the moon and back.)

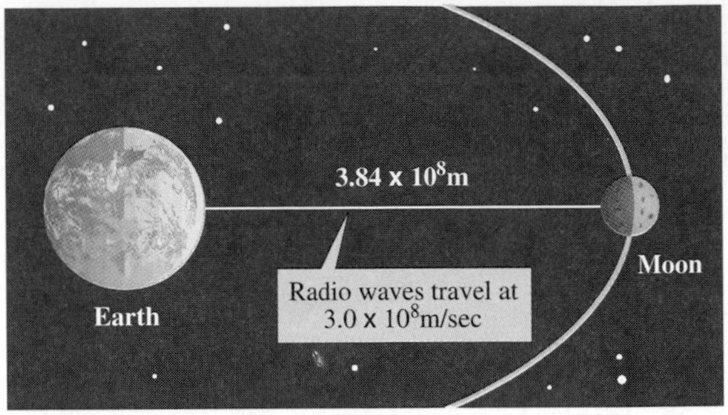

Figure for Exercise 148

149. *Distance* What is the distance between −2.35 and 8.77 on the number line?

150. *Absolute Value* Write an expression involving absolute value that gives the distance between the points *a* and −5 on the number line.

151. *Mowing a Lawn* Howard and Will get summer jobs doing yard work. If Howard can mow an entire lawn in 6 hr and Will can mow it in 4 hr, then what portion of it can they mow in 2 hr working together?

152. *Cost of Watermelons* Write a polynomial that expresses the total cost of $x + 3$ watermelons at \$2 apiece and $x + 9$ watermelons at \$3 apiece.

CHAPTER P TEST

Determine which elements of the set

$$\{-\pi, -\sqrt{3}, -1.22, -1, 0, 2, \sqrt{5}, 10/3, 6.020020002\ldots\}$$

are members of the following sets.

1. Real numbers

2. Rational numbers

3. Irrational numbers

4. Nonnegative integers

Evaluate each expression.

5. $|2 \cdot 3 - 5^2| - 6$

6. $\dfrac{1}{8^{-1/3}}$

7. $-27^{-2/3}$

8. $\dfrac{(-3)^2 - 4(-3) + 9}{(-3 - 2)(5 - (-1))}$

Simplify each expression. Assume that all variables represent positive numbers.

9. $(-2x^4y^2)(-3xy^5)$

10. $(4x^4)^{1/2} + (-8x^6)^{1/3}$

11. $\dfrac{(ab^2 + a^2b)^2}{a^2b^2}$

12. $\dfrac{(-2a^{-1}b^6)^3}{(8a^9b^{-12})^{-2/3}}$

Simplify each expression. Assume that all variables represent positive real numbers.

13. $\sqrt{27} - \sqrt{8} + \sqrt{32}$

14. $\dfrac{\sqrt{8}}{\sqrt{6} - \sqrt{2}}$

15. $\sqrt[3]{\dfrac{1}{4x^4}}$

16. $\sqrt{12x^3y^9z^0}$

Perform the indicated operations.

17. $(-x^3 - 5x) + (4x^3 + 3x^2 - 7x)$

18. $(-x^2 + 3x - 4) - (4x^2 - 6x + 9)$

19. $(x + 3)(x^2 - 2x - 1)$

20. $(8h^3 - 1) \div (2h - 1)$

21. $(x - 9y)(x + 3y)$

22. $(x^3 - 2x - 4) \div (x - 2)$

23. $(3x - 8)^2$

24. $(2t^4 - 1)(2t^4 + 1)$

Perform each operation.

25. $\dfrac{x^3 - 5x^2 + 6x}{2x^2 - 6x} \cdot \dfrac{4x^3 + 32}{2x^3 - 8x}$

26. $\dfrac{x + 5}{x^2 - 4x + 3} + \dfrac{x - 1}{x^2 + x - 12}$

27. $\dfrac{a - 1}{4a^2 - 9} - \dfrac{a - 2}{3 - 2a}$

28. $\dfrac{\dfrac{1}{2a^2b} - 2a}{\dfrac{1}{4ab^3} + \dfrac{1}{3b}}$

Factor completely.

29. $ax^2 - 11ax + 18a$

30. $m^5 - m$

31. $3x^2 + 14x - 5$

32. $bx^2 - 3bx + wx - 3w$

Solve each problem.

33. In 1989 (a record year), 2.9×10^6 people attended conventions in New York City and spent a total of \$1.1 billion. What amount was spent per person?

34. If one light year is equal to 6.3240×10^4 astronomical units and one astronomical unit is equal to 92.95582×10^6 mi, then what number of miles is equal to one light year?

35. The altitude in feet of an arrow t seconds after it is shot straight upward is calculated from the polynomial $A(t) = -16t^2 + 120t$. Find the altitude of the arrow 2 sec after it is shot upward.

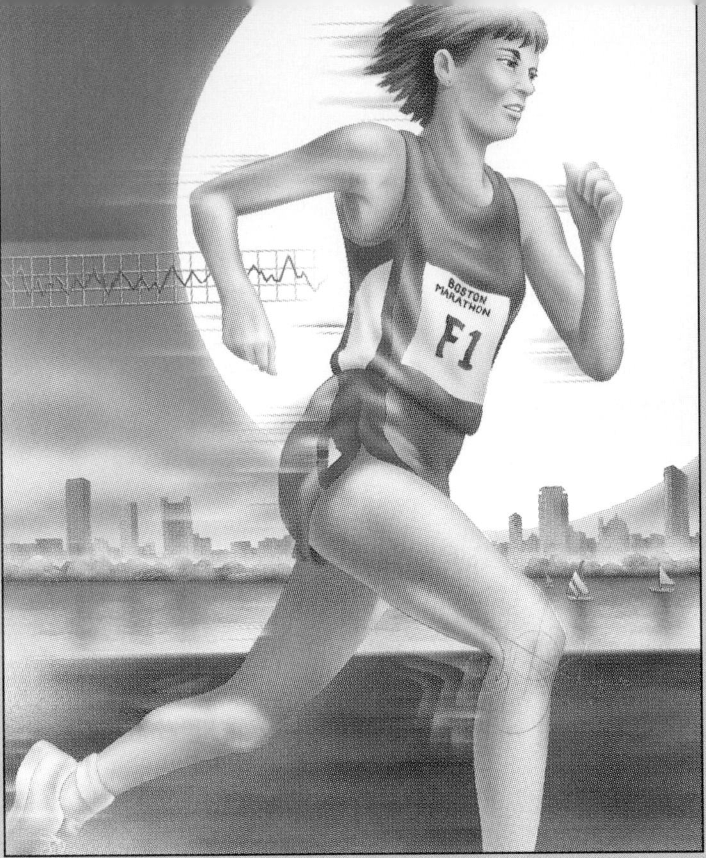

Even infamous Heartbreak Hill coundn't break the winning spirit of Russian runner Olga Markova, as she focused mind and muscle on outdistancing her competitors. It was the 96th running of the Boston Marathon, the world's oldest and most prestigious race of its kind, a 26.2-mile ordeal that one runner called "14 miles of fun, 8 miles of sweat, and 4 miles of hell!"

While Markova's rivals got off to a fast start, she shrewdly waited them out, breaking away at Mile 18 and never looking back until she burst through the finish line tape with a personal-best time of 2 hours, 23 minutes, 43 seconds—one of the best women's times ever in Boston.

Major sporting events like the Marathon have come a long way since the first Olympic Games, held in ancient Greece over 2500 years ago. In the 20th century, satellites beam the Games to a worldwide audience. And modern athletes and coaches often utilize both mathematics and computers to anlayze the critical variables that help competitors increase aerobic capacity, reduce air resistance, or strengthen muscles in order to jump higher, run faster, or throw farther than ever before. Scientists are particularly fascinated with long-distance running, and the Boston race has been studied more than most because of the area's many research centers. The Marathon also boasts a top medical-response team that's always searching for new ways to interpret the effects of the grueling pace on muscles, bones, heart, and other organs.

In this chapter you'll encounter various examples of sports applications while studying two of the most important skills in algebra: solving equations and creating mathematical models. In Chapter P we reviewed fundamentals of algebra. Now you're ready to learn how to use these tools to solve problems that can be expressed in the form of equations and inequalities. By the time you reach the finish line, you should be able to solve both linear and quadratic equations, as well as a few other special types of equations. You'll also be able to create and use several types of mathematical models.

1 EQUATIONS AND INEQUALITIES

1.1 Linear Equations
1.2 Applications
1.3 Complex Numbers
1.4 Quadratic Equations
1.5 Linear and Absolute Value Inequalities
1.6 Quadratic and Rational Inequalities

1.1 Linear Equations

One of our main goals in algebra is to develop techniques for solving a wide variety of equations. In this section we will solve linear equations and other similar equations.

Definitions

An **equation** is a statement (or sentence) indicating that two algebraic expressions are equal. The verb in an equation is the equality symbol. For example, $2x + 8 = 0$ is an equation. If we replace x by -4, we get $2 \cdot (-4) + 8 = 0$, a true statement. So we say that -4 is a **solution** or **root** to the equation or that -4 **satisfies** the equation. If we replace x by 3, we get $2 \cdot 3 + 8 = 0$, a false statement. Whether the equation $2x + 8 = 0$ is true or false depends on the value of x, and so it is called an **open sentence**. The equation is neither true nor false until we choose a value for x. The set of all solutions to an equation is called the **solution set** to the equation. To **solve** an equation means to find the solution set. The solution set for $2x + 8 = 0$ is $\{-4\}$. The equation $2x + 8 = 0$ is an example of a linear equation.

Definition: Linear Equation in One Variable

A **linear equation in one variable** is an equation of the form $ax + b = 0$, where a and b are real numbers, with $a \neq 0$.

Solving Equations

The equations $2x + 8 = 0$ and $2x = -8$ both have the solution set $\{-4\}$. Two equations with the same solution set are called **equivalent** equations. Adding the same real number to or subtracting the same real number from each side of an equation results in an equivalent equation. Multiplying or dividing each side of an equation by the same nonzero real number also results in an equivalent equation. These **properties of equality** are stated in symbols in the following box.

Properties of Equality

If A and B are algebraic expressions and C is a real number, then the following equations are equivalent to $A = B$:

$A + C = B + C$	**Addition property of equality**
$A - C = B - C$	**Subtraction property of equality**
$CA = CB \ (C \neq 0)$	**Multiplication property of equality**
$\frac{A}{C} = \frac{B}{C} \ (C \neq 0)$	**Division property of equality**

When C is a real number, the properties of equality allow us to perform the four basic operations of arithmetic on each side of an equation without changing the solution set. It is also correct to apply the properties of equality using an *algebraic expression* for C, because the value of an algebraic expression is a real number. However, if C involves a variable, we might obtain an equation that is not equivalent to the original equation. For example, if $1/x$ is subtracted from each side of the equation

$$x + \frac{1}{x} = 0 + \frac{1}{x}$$

we get $x = 0$. But replacing x by 0 in the original equation results in undefined expressions, so the equations are not equivalent. If an equation contains an expression whose domain excludes some real numbers, then we must check our solutions.

The strategy in solving equations is to use the properties of equality to simplify an equation until we get an equivalent equation whose solution set is obvious.

EXAMPLE 1 Using the properties of equality

Solve each equation.

a) $3x - 4 = 8$ **b)** $\frac{1}{2}x - 6 = \frac{3}{4}x - 9$ **c)** $3(4x - 1) = 4 - 6(x - 3)$

Solution

a)

$$3x - 4 = 8$$

$$3x - 4 + 4 = 8 + 4 \quad \text{Add 4 to each side.}$$

$$3x = 12 \quad \text{Simplify.}$$

$$\frac{3x}{3} = \frac{12}{3} \quad \text{Divide each side by 3.}$$

$$x = 4 \quad \text{Simplify.}$$

Since the last equation is equivalent to the original, the solution set to the original equation is $\{4\}$. We can check by replacing x by 4 in $3x - 4 = 8$. Since $3 \cdot 4 - 4 = 8$ is correct, we are confident that the solution set is $\{4\}$.

b)

$$\frac{1}{2}x - 6 = \frac{3}{4}x - 9$$

$$4\left(\frac{1}{2}x - 6\right) = 4\left(\frac{3}{4}x - 9\right) \quad \text{Multiply each side by 4, the LCD.}$$

$$2x - 24 = 3x - 36 \quad \text{Distributive property}$$

$$2x - 24 - 3x = 3x - 36 - 3x \quad \text{Subtract } 3x \text{ from each side.}$$

$$-x - 24 = -36 \quad \text{Simplify.}$$

$$-x = -12 \quad \text{Add 24 to each side.}$$

$$(-1)(-x) = (-1)(-12) \quad \text{Multiply each side by } -1.$$

$$x = 12 \quad \text{Simplify.}$$

Check 12 in the original equation. The solution set is $\{12\}$.

c) $3(4x - 1) = 4 - 6(x - 3)$

$$12x - 3 = 4 - 6x + 18 \quad \text{Distributive property}$$

$$12x - 3 = 22 - 6x \quad \text{Simplify.}$$

$$18x - 3 = 22 \quad \text{Add } 6x \text{ to each side.}$$

$$18x = 25 \quad \text{Add 3 to each side.}$$

$$x = \frac{25}{18} \quad \text{Divide each side by 18.}$$

Check 25/18 in the original equation. The solution set is $\{\frac{25}{18}\}$. A calculator check is shown in Fig. 1.1.

Figure 1.1

Note that checking the equation in Example 1(c) with a calculator did not prove that 25/18 is the correct solution. The properties of equality that were applied correctly in each step guarantee that we have the correct solution. The values of the two sides of the equation could agree for the 10 digits shown on the calculator and disagree for the digits not shown. Since that possibility is extremely unlikely, the calculator check does support our belief that we have the correct solution. □

Any linear equation, $ax + b = 0$, can be solved in two steps. Subtract b from each side and then divide each side by a ($a \neq 0$), to get $x = -b/a$. Although the equations of Example 1 are not exactly in the form $ax + b = 0$, they are often called linear equations because they are equivalent to linear equations. The next example of a linear equation involves radicals.

EXAMPLE 2 An equation involving radicals

Solve $\sqrt{2}x - 4 = \sqrt{6}$. Express the answer in simplified radical form.

Solution

We proceed as in any other linear equation.

$$\sqrt{2}x - 4 = \sqrt{6}$$

$$\sqrt{2}x = 4 + \sqrt{6} \qquad \text{Add 4 to each side.}$$

$$x = \frac{4 + \sqrt{6}}{\sqrt{2}} \qquad \text{Divide each side by } \sqrt{2}.$$

$$x = \frac{(4 + \sqrt{6})\sqrt{2}}{\sqrt{2} \cdot \sqrt{2}} \qquad \text{Rationalize the denominator.}$$

$$x = \frac{4\sqrt{2} + 2\sqrt{3}}{2}$$

$$x = 2\sqrt{2} + \sqrt{3}$$

Figure 1.2

The exact solution is $2\sqrt{2} + \sqrt{3}$. Using a calculator and rounding to three decimal places, we get $x \approx 4.560$ as shown in Fig. 1.2. The symbol $\approx$ means "is approximately equal to." The solution set is $\{2\sqrt{2} + \sqrt{3}\}$.

Use the ANS key to check as shown in Fig. 1.2. By subtracting the right side from the left side we can check in a single computation.

Identities, Conditional Equations, and Inconsistent Equations

An equation that is satisfied by every real number for which both sides are defined is an **identity**. Some examples of identities are

$$3x - 1 = 3x - 1, \quad 2x + 5x = 7x, \quad \text{and} \quad \frac{x}{x} = 1.$$

The solution set to the first two identities is the set of all real numbers, R. Since 0/0 is undefined, the solution set to $x/x = 1$ is the set of nonzero real numbers, $\{x | x \neq 0\}$.

A **conditional equation** is an equation that is satisfied by at least one real number but is not an identity. The equation $3x - 4 = 8$ is true only on condition that $x = 4$, and it is a conditional equation. The equations of Examples 1 and 2 are conditional equations.

An **inconsistent equation** is an equation that has no solution. Some inconsistent equations are

$$0 \cdot x + 1 = 2, \quad x + 3 = x + 5, \quad \text{and} \quad 9x - 9x = 8.$$

Note that each of these inconsistent equations is equivalent to a false statement: $1 = 2$, $3 = 5$, and $0 = 8$, respectively.

EXAMPLE 3 Classifying an equation

Determine whether the equation $3(x - 1) - 2x(4 - x) = (2x + 1)(x - 3)$ is an identity, an inconsistent equation, or a conditional equation.

Solution

$$\begin{aligned} 3(x - 1) - 2x(4 - x) &= (2x + 1)(x - 3) \\ 3x - 3 - 8x + 2x^2 &= 2x^2 - 5x - 3 \quad \text{Simplify each side.} \\ 2x^2 - 5x - 3 &= 2x^2 - 5x - 3 \end{aligned}$$

Since the last equation is equivalent to the original and the last equation is an identity, the original equation is an identity.

Equations Involving Rational Expressions

In Example 1(b) we solved an equation involving fractions. To simplify that equation, the first step was to multiply by the LCD of the fractions. The equations in the next example all involve rational expressions. Notice that multiplying each side of these equations by the LCD greatly simplifies the equations.

EXAMPLE 4 Equations involving rational expressions

Solve each equation and identify each as an identity, an inconsistent equation, or a conditional equation.

a) $\dfrac{y}{y - 3} + 3 = \dfrac{3}{y - 3}$ **b)** $\dfrac{1}{x - 1} - \dfrac{1}{x + 1} = \dfrac{2}{x^2 - 1}$ **c)** $\dfrac{1}{2} + \dfrac{1}{x - 1} = 1$

Solution

a) Since $y - 3$ is the denominator in each rational expression, $y - 3$ is the LCD.

$$(y-3)\left(\frac{y}{y-3}+3\right) = (y-3)\frac{3}{y-3} \quad \text{Multiply each side by the LCD.}$$

$$(y-3)\frac{y}{y-3} + (y-3)3 = 3 \quad \text{Distributive property}$$

$$y + 3y - 9 = 3$$

$$4y - 9 = 3$$

$$4y = 12 \quad \text{Add 9 to each side.}$$

$$y = 3 \quad \text{Divide each side by 4.}$$

If we replace y by 3 in the original equation, then we get two undefined expressions. So 3 is not a solution to the original equation. The original equation has no solution. The equation is inconsistent.

b) Since $x^2 - 1 = (x-1)(x+1)$, the LCD is $(x-1)(x+1)$.

$$\frac{1}{x-1} - \frac{1}{x+1} = \frac{2}{x^2-1}$$

$$(x-1)(x+1)\left(\frac{1}{x-1} - \frac{1}{x+1}\right) = (x-1)(x+1)\frac{2}{x^2-1} \quad \text{Multiply by the LCD.}$$

$$(x-1)(x+1)\frac{1}{x-1} - (x-1)(x+1)\frac{1}{x+1} = 2 \quad \text{Distributive property}$$

$$x + 1 - (x - 1) = 2$$

$$2 = 2$$

Since the last equation is an identity, the original equation is also an identity. The solution set is $\{x \mid x \neq 1 \text{ and } x \neq -1\}$, because 1 and -1 cannot be used for x in the original equation.

c)

$$\frac{1}{2} + \frac{1}{x-1} = 1$$

$$2(x-1)\left(\frac{1}{2} + \frac{1}{x-1}\right) = 2(x-1)1 \quad \text{Multiply by the LCD.}$$

$$x - 1 + 2 = 2x - 2$$

$$x + 1 = 2x - 2$$

$$3 = x$$

Check 3 in the original equation. The solution set is $\{3\}$, and the equation is a conditional equation.

In Example 4(a) the final equation had a root that did not satisfy the original equation, because the domain of the rational expression excluded the root. Such a root is called an **extraneous root**. If an equation has no solution, then its solution set is the **empty set** (the set with no members). The symbol ∅ is used to represent the empty set.

EXAMPLE 5 **Using a calculator in solving an equation**

Solve

$$\frac{7}{2.4x} + \frac{3}{5.9} = \frac{1}{8.2}$$

with the aid of a calculator. Round the answer to three decimal places.

Solution

We could multiply each side by the LCD, but since we are using a calculator, we can subtract 3/5.9 from each side to isolate x.

$$\frac{7}{2.4x} + \frac{3}{5.9} = \frac{1}{8.2}$$

$$\frac{7}{2.4x} = \frac{1}{8.2} - \frac{3}{5.9}$$

$$\frac{7}{2.4x} \approx -0.38652335 \qquad \text{Use a calculator to simplify.}$$

$$\frac{7}{2.4} \approx -0.38652335x \qquad \text{Multiply each side by } x.$$

$$\frac{7}{2.4(-0.38652335)} \approx x \qquad \text{Divide each side by } -0.38652335.$$

$$x \approx -7.546 \qquad \text{Round to three decimal places.}$$

Figure 1.3

The solution set is $\{-7.546\}$. Since -0.38652335 is an approximate value, the sign ≈ (for "approximately equal to") is used instead of the equal sign. To get three-decimal-place accuracy in the final answer, use as many digits as your calculator allows until you get to the final computation.

Figure 1.3 shows the computations and the check.

Equations Involving Absolute Value

To solve equations involving absolute value, remember that $|x| = x$ if $x \geq 0$, and $|x| = -x$ if $x < 0$. The absolute value of x is greater than or equal to 0 for any real number x. So an equation such as $|x| = -6$ has no solution. Since a number and its opposite have the same absolute value, $|x| = 4$ is equivalent to $x = 4$ or

$x = -4$. The only number that has 0 absolute value is 0. These ideas are summarized as follows.

SUMMARY **Basic Absolute Value Equations**

Absolute value equation	Equivalent statement	Solution set
$\lvert x \rvert = k \ (k > 0)$	$x = -k$ or $x = k$	$\{-k, k\}$
$\lvert x \rvert = 0$	$x = 0$	$\{0\}$
$\lvert x \rvert = k \ (k < 0)$		$\varnothing$

EXAMPLE 6 Equations involving absolute value

Solve each equation.

a) $|x - 5| = 4$ **b)** $2|x + 8| - 6 = 0$

Solution

a) First write an equivalent statement without using absolute value symbols.

$$|x - 5| = 4$$

$$x - 5 = -4 \quad \text{or} \quad x - 5 = 4$$

$$x = 1 \quad \text{or} \quad x = 9$$

The solution set is $\{1, 9\}$.

b) First isolate $|x + 8|$.

$$2|x + 8| - 6 = 0$$

$$2|x + 8| = 6$$

$$|x + 8| = 3$$

Now write an equivalent statement without absolute value symbols.

$$x + 8 = -3 \quad \text{or} \quad x + 8 = 3$$

$$x = -11 \quad \text{or} \quad x = -5$$

The solution set is $\{-11, -5\}$.

STRATEGY **Equation-Solving Overview**

In solving equations, there are usually many different sequences of correct steps that lead to the correct solution. If the approach that you try first does not work, *try another approach*, but learn from your failures as well as your successes. Solving equations successfully takes patience and practice.

FOR THOUGHT True or False? Explain.

1. The number 3 is in the solution set to $5(4 - x) = 2x - 1$.
2. The equation $3x - 1 = 8$ is equivalent to $3x - 2 = 7$.
3. The equation $x + \sqrt{x} = -2 + \sqrt{x}$ is equivalent to $x = -2$.
4. The solution set to $x - x = 7$ is the empty set.
5. The equation $12x = 0$ is an inconsistent equation.
6. The equation $x - 0.02x = 0.98x$ is an identity.
7. The equation $|x| = -8$ is equivalent to $x = 8$ or $x = -8$.
8. The equations $\frac{x}{x-5} = \frac{5}{x-5}$ and $x = 5$ are equivalent.
9. To solve $-\frac{2}{3}x = \frac{3}{4}$, we should multiply each side by $-\frac{2}{3}$.
10. If a and b are real numbers, then $ax + b = 0$ has a solution.

1.1 EXERCISES

Tape 3 Disk

Determine whether each given number is a solution to the equation following it.

1. $3,\ 2x - 4 = 9$
2. $-2,\ \frac{1}{x} - \frac{1}{2} = -1$
3. $-3,\ (x - 1)^2 = 16$
4. $4,\ \sqrt{3x + 4} = -4$

Solve each equation and check your answer.

5. $3x - 5 = 0$
6. $-2x + 3 = 0$
7. $-3x + 6 = 12$
8. $5x - 3 = -13$
9. $8x - 6 = 1 - 6x$
10. $4x - 3 = 6x - 1$
11. $7 + 3x = 4(x - 1)$
12. $-3(x - 5) = 4 - 2x$
13. $-\frac{3}{4}x = 18$
14. $\frac{2}{3}x = -9$
15. $\frac{x}{2} - 5 = -12 - \frac{2x}{3}$
16. $\frac{x}{4} - 3 = \frac{x}{2} + 3$

Find the exact solution to each equation. Use a calculator to check your answer.

17. $2x + 4 = \sqrt{8}$
18. $3x - 6 = \sqrt{18}$
19. $\sqrt{3}w - 5 = \sqrt{6}$
20. $\sqrt{2}z + 4 = \sqrt{10}$
21. $2\pi x - 3 = 4\pi x - 1$
22. $\pi r = 3 - 2\pi r$
23. $\sqrt[3]{2}x - 1 = 5$
24. $\sqrt[3]{4}x + 3 = 9$

Solve each equation. Identify each equation as an identity, an inconsistent equation, or a conditional equation.

25. $3(x - 6) = 3x - 18$
26. $2a + 3a = 6a$
27. $3(x - 6) = 3x + 18$
28. $2x + 3x = 5x$
29. $\frac{3x}{x} = 3$
30. $\frac{x(x + 2)}{x + 2} = x$

Solve each equation involving rational expressions. Identify each equation as an identity, an inconsistent equation, or a conditional equation.

31. $\frac{1}{w - 1} - \frac{1}{2w - 2} = \frac{1}{2w - 2}$
32. $\frac{1}{x - 3} - \frac{1}{x + 3} = \frac{6}{x^2 - 9}$
33. $\frac{1}{x} - \frac{1}{3x} = \frac{1}{2x} + \frac{1}{6x}$
34. $\frac{1}{5x} - \frac{1}{4x} + \frac{1}{3x} = -\frac{17}{60}$
35. $\frac{z + 2}{z - 3} = \frac{5}{-3}$
36. $\frac{2x - 3}{x - 4} = \frac{5}{x - 4}$
37. $\frac{1}{x} + \frac{1}{x - 3} = \frac{9}{x^2 - 3x}$
38. $\frac{4}{x - 1} - \frac{9}{x + 1} = \frac{3}{x^2 - 1}$
39. $4 + \frac{6}{y - 3} = \frac{2y}{y - 3}$
40. $\frac{x}{x + 6} - 3 = 1 - \frac{6}{x + 6}$

41. $\frac{t}{t+3} + 4 = \frac{2}{t+3}$

42. $\frac{3x}{x+1} - 5 = \frac{x-11}{x+1}$

Use a calculator to help you solve each equation. Round each approximate answer to three decimal places.

43. $0.27x - 3.9 = 0.48x + 0.29$

44. $0.06(x - 3.78) = 1.95$

45. $\sqrt{2}x - \sqrt{6} = \sqrt{3}$

46. $\sqrt[3]{2}x - 3 = 0$

47. $2a + 1 = -\sqrt{17}$

48. $3c + 4 = \sqrt{38}$

49. $\frac{0.001}{y - 0.333} = 3$

50. $1 + \frac{0.001}{t-1} = 0$

51. $\frac{x}{0.376} + \frac{x}{0.135} = 2$

52. $\frac{1}{x} + \frac{5}{6.72} = 10.379$

53. $(x + 3.25)^2 = (x - 4.1)^2$

54. $0.25(2x - 1.6)^2 = (x - 0.9)^2$

55. $(2.3 \times 10^6)x + 8.9 \times 10^5 = 1.63 \times 10^4$

56. $(-3.4 \times 10^{-9})x + 3.45 \times 10^{-8} = 1.63 \times 10^4$

Solve each absolute value equation.

57. $|x| = 9$

58. $|x| = 13.6$

59. $|2x - 3| = 7$

60. $|3x + 4| = 12$

61. $2|x + 5| - 10 = 0$

62. $6 - 4|x + 3| = -2$

63. $|3x - 2| = 0$

64. $5|6 - 3x| = 0$

65. $2|x| + 7 = 6$

66. $5 + 3|x - 4| = 0$

Solve each equation.

67. $x - 0.05x = 190$

68. $x + 0.1x = 121$

69. $0.1x - 0.05(x - 20) = 1.2$

70. $0.03x - 0.2 = 0.2(x + 0.03)$

71. $(x + 2)^2 = x^2 + 4$

72. $(x - 3)^2 = x^2 - 9$

73. $\frac{x}{2} + 1 = \frac{1}{4}(x - 6)$

74. $-\frac{1}{6}(x + 3) = \frac{1}{4}(3 - x)$

75. $\frac{y-3}{2} + \frac{y}{5} = 3 - \frac{y+1}{6}$

76. $\frac{y-3}{5} - \frac{y-4}{2} = 5$

77. $5 + 7|x + 6| = 19$

78. $9 - |2x - 3| = 6$

79. $\frac{3}{x-2} + \frac{4}{x+2} = \frac{7x-2}{x^2-4}$

80. $\frac{2}{x-1} - \frac{3}{x+2} = \frac{8-x}{x^2+x-2}$

81. $\frac{4}{x+3} - \frac{3}{2-x} = \frac{7x+1}{x^2+x-6}$

82. $\frac{3}{x} - \frac{4}{1-x} = \frac{7x-3}{x^2-x}$

83. $\frac{x-2}{x-3} = \frac{x-3}{x-4}$

84. $\frac{y-1}{y+4} = \frac{y+1}{y-2}$

Solve each problem.

85. *International Communications* In 1995, people spent 62 billion minutes phoning and faxing other countries (*Fortune*, September 8, 1997). The global investment in telecom infrastructure since 1990 can be modeled by the equation

$$I = 7.5t + 115,$$

where I is in billions of dollars and t is the number of years since 1990. Use the equation to predict the global investment in telecom infrastructure in the year 2003. Find the year in which the global investment reached \$150 billion by solving the equation $150 = 7.5t + 115$.

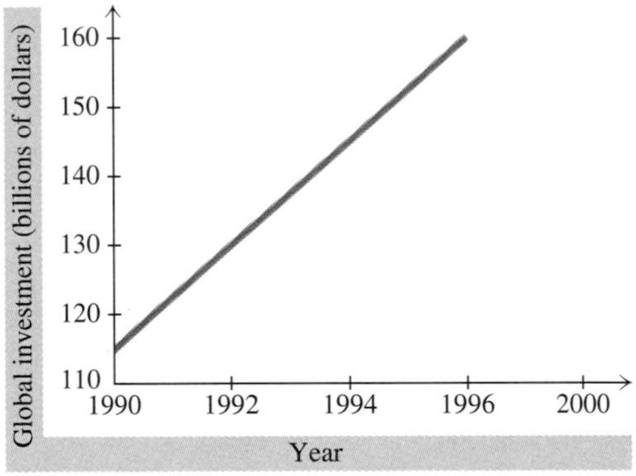

Figure for Exercise 85

86. *Comparing Costs* The total cost for renting a copy machine for 60 months at \$100 per month plus 5 cents per copy is $0.05x + 6000$ dollars, if x copies are made. If the same copy machine is purchased for \$2000, then it costs 7 cents per copy for supplies and maintenance, for a total cost of $2000 + 0.07x$ dollars for x copies. Solve the equation

$$0.05x + 6000 = 2000 + 0.07x$$

to find the number of copies for which the five-year cost is the same for renting or purchasing. If you plan to make 2000 copies per month for five years, is it cheaper to rent or purchase the copier? Assume that the copier has no value after five years. See the figure at the top of the next page.

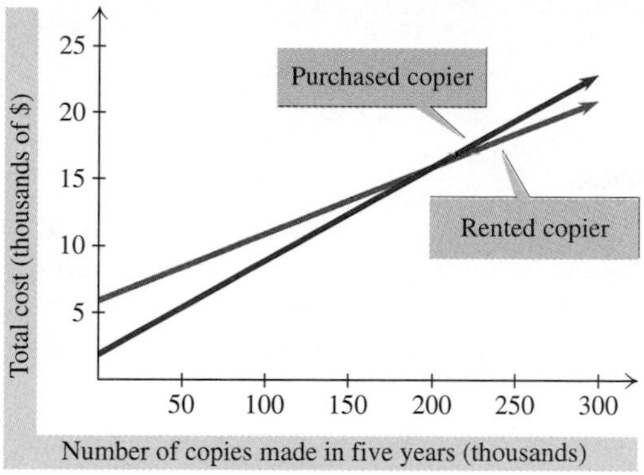

Figure for Exercise 86

87. *Cost Accounting* An accountant has been told to distribute a bonus to the employees that is 15% of the company's net income. Since the bonus is an expense to the accountant, it must be subtracted from the income to determine the net income. If the company has an income of \$140,000 before the bonus, then the accountant must solve

$$B = 0.15(140{,}000 - B).$$

to find the bonus B. Find B.

88. *Corporate Taxes* For a class C corporation in Louisiana, the amount of state income tax S is deductible on the federal return and the amount of federal income tax F is deductible on the state return. With \$200,000 taxable income and a 30% federal tax rate, the federal tax is $0.30(200{,}000 - S)$. If the state tax rate is 6% then the state tax satisfies

$$S = 0.06(200{,}000 - 0.30(200{,}000 - S)).$$

Find the state tax S and the federal tax F.

89. *Production Cost* An automobile manufacturer, who spent \$500 million to develop a new line of cars, wants the cost

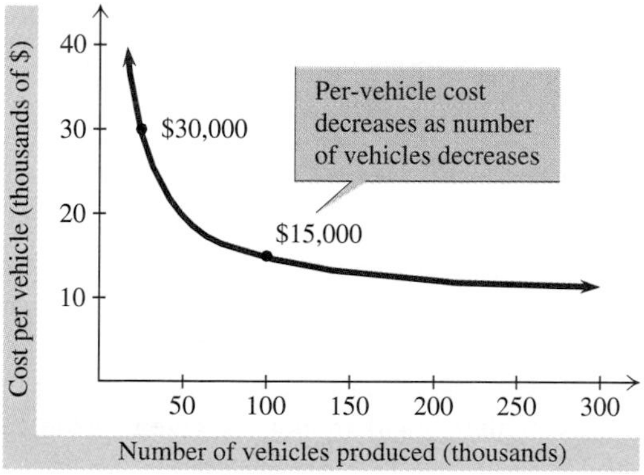

Figure for Exercise 89

of development and production to be \$12,000 per vehicle. If the production costs are \$10,000 per vehicle, then the cost per vehicle for development and production of x vehicles is $(10{,}000x + 500{,}000{,}000)/x$ dollars. Solve the equation

$$\frac{10{,}000x + 500{,}000{,}000}{x} = 12{,}000$$

to find the number of vehicles that must be sold so that the cost of development and production is \$12,000 per vehicle.

90. *Harmonic Mean* The *harmonic mean* of the numbers $x_1, x_2, \ldots, x_n$ is defined as

$$HM = \frac{n}{\dfrac{1}{x_1} + \dfrac{1}{x_2} + \cdots + \dfrac{1}{x_n}}.$$

The accompanying graph shows the winnings for Tiger Woods in the first four tournaments of 1997 (*The Boston Globe*, March 27, 1997).

a. What is the harmonic mean of the winnings for these four tournaments?

b. What would his winnings have to be for the fifth tournament for the harmonic mean for the first five tournaments to be \$55,000?

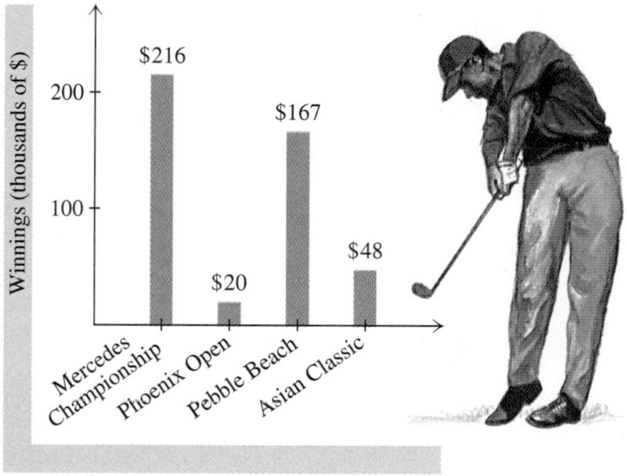

Figure for Exercise 90

For Writing/Discussion

91. *Definitions* Without looking back in the text, write the definitions of linear equation, identity, and inconsistent equation. Use complete sentences.

92. *Cooperative Learning* Each student in a small group should write a linear equation, an identity, an inconsistent equation, and an equation that has an extraneous root. The group should solve each equation and determine whether each equation is of the required type.

LINKING CONCEPTS

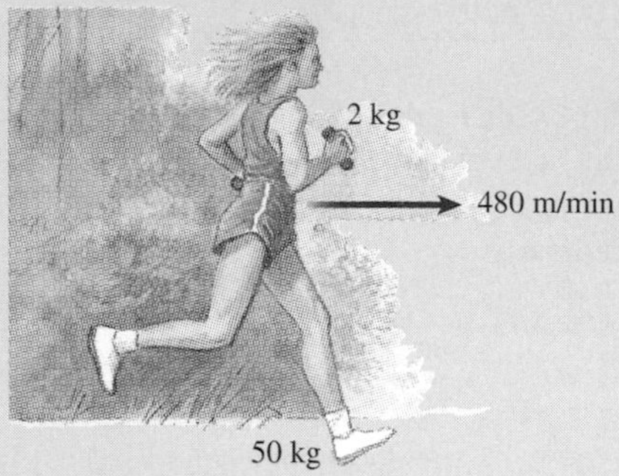

For Individual or Group Explorations

Oxygen Uptake *In a study of oxygen uptake rate for marathon runners* (Costill and Fox, *Medicine and Science in Sports,* Vol. 1) *it was found that the power expended in kilocalories per minute is given by the formula* $P = M(av - b)$ *where* M *is the mass of the runner in kilograms and* v *is the speed in meters per minute. The constants* a *and* b *have values*

$$a = 1.02 \times 10^{-3} \text{ kcal/kg m} \quad \text{and} \quad b = 2.62 \times 10^{-2} \text{ kcal/kg min}.$$

a) Runners with masses of 60 kg, 65 kg, and 70 kg are running together at 400 m/min. Find the power expenditure for each runner.

b) With a constant velocity, does the power expended increase or decrease as the mass of the runner increases?

c) Runners of 80 kg, 84 kg, and 90 kg, are all expending power at the rate of 38.7 kcal/min. Find the velocity at which each is running.

d) With a constant power expenditure, does the velocity increase or decrease as the mass of the runner increases?

e) A 50 kg runner in training has a velocity of 480 m/min while carrying weights of 2 kg. Assume that her power expenditure remains constant when the weights are removed and find her velocity without the weights.

f) Why do runners in training carry weights?

g) Use a computer to create graphs showing power expenditure versus mass with a velocity fixed at 400 m/min and velocity versus mass with a fixed power expenditure of 40 kcal/min.

1.2 Applications

In Section 1.1 we solved equations of various types. In this section we will put those new techniques to work in solving problems.

Formulas

A **formula** is an equation involving two or more variables. What distinguishes a formula from an ordinary equation is its application. For example, the equation $D = RT$ is a formula giving the relationship between distance, rate, and time in uniform motion. Formulas are often used to **model** real-life situations. A list of formulas commonly used as mathematical models is given inside the front cover of this text.

In general, a formula is a recipe for finding the value of one variable when the values of the other variables are known. For example, the formula $P = 2L + 2W$, which expresses the perimeter of a rectangle in terms of its length and width, is said to be **solved** for P. We can solve this formula for W as follows.

$$2L + 2W = P \qquad \text{Write the formula with } W \text{ on the left.}$$
$$2W = P - 2L \qquad \text{Subtract } 2L \text{ from each side.}$$
$$W = \frac{P - 2L}{2} \qquad \text{Divide each side by 2.}$$

When a formula is solved for a specified variable, that variable must not occur on both sides of the equal sign.

EXAMPLE 1 Solving a formula for a specified variable

Solve the formula $S = P + Prt$ for P.

Solution

$$P + Prt = S \qquad \text{Write the formula with } P \text{ on the left.}$$
$$P(1 + rt) = S \qquad \text{Factor out } P.$$
$$P = \frac{S}{1 + rt} \qquad \text{Divide each side by } 1 + rt.$$

In some situations we know the values of all variables except one. After we substitute values for those variables, the formula is an equation in one variable. We can then solve the equation to find the value of the remaining variable.

EXAMPLE 2 Finding the value of a variable in a formula

The total resistance R (in ohms) in the parallel circuit shown in Fig. 1.4 is modeled by the formula

$$\frac{1}{R} = \frac{1}{R_1} + \frac{1}{R_2}.$$

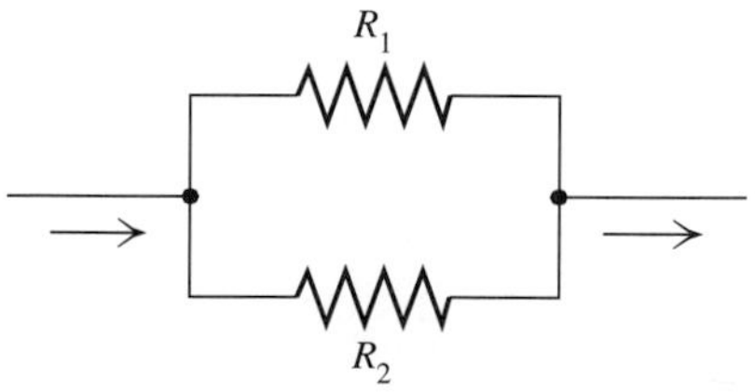

Figure 1.4

The subscripts 1 and 2 indicate that R_1 and R_2 represent the resistance for two different receivers. If $R = 7$ ohms and $R_1 = 10$ ohms, then what is the value of R_2?

Solution

Substitute the values for R and R_1 and solve for R_2.

$$\frac{1}{7} = \frac{1}{10} + \frac{1}{R_2} \qquad \text{The LCD of 7, 10, and } R_2 \text{ is } 70R_2.$$

$$70R_2 \cdot \frac{1}{7} = 70R_2\left(\frac{1}{10} + \frac{1}{R_2}\right) \qquad \text{Multiply each side by } 70R_2.$$

$$10R_2 = 7R_2 + 70$$

$$3R_2 = 70$$

$$R_2 = \frac{70}{3}$$

Check the solution in the original formula. The resistance R_2 is 70/3 ohms.

Linear Regression

When scientists discover relationships between variables in nature, they express the relationships with formulas. It is often the case that there appears to be a relationship between two variables, but there is no known formula. However, most graphing calculators are preprogrammed to perform a technique called **linear regression**, which will produce a formula that approximates the data. As you will see in the next example, getting a formula is relatively easy because the calculator performs all of the computations. The calculator even indicates how well the formula approximates or fits the data by giving a value r called the **correlation coefficient**. If r is 1 or -1 the fit is perfect, if r is near 1 or -1 the fit is good, but if r is near 0 the formula is not a good model for the data. You can get an idea of how linear regression works when you do Exercises 91 and 92 in Section 3.2.

• CALCULUS •

EXAMPLE 3 Injuries in the Boston Marathon

In the Boston Marathon there is a strong correlation between runner injuries and the temperature at the time of the race (*The Boston Globe*, April 20, 1992). See the bar graph in Fig. 1.5. When the Fahrenheit temperature at the time of the race is t, p% of the runners get injured.

a) Use linear regression to find a formula for p in terms of t.

b) Estimate the percentage of runners that would be injured if the marathon was run on a day when the temperature was 82°F.

c) If 9% of the runners are injured, then estimate the temperature at the start of the race.

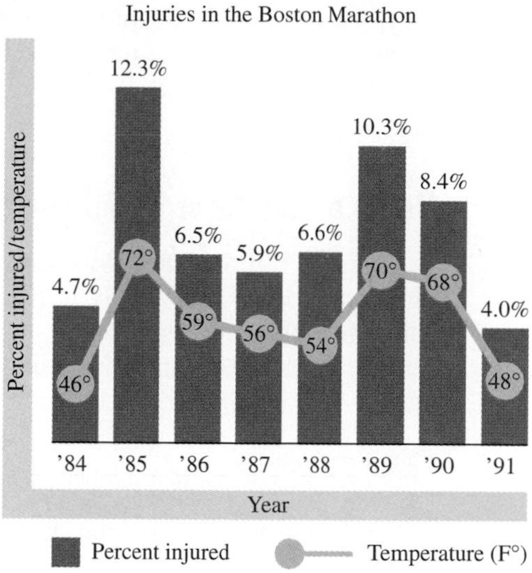

Figure 1.5

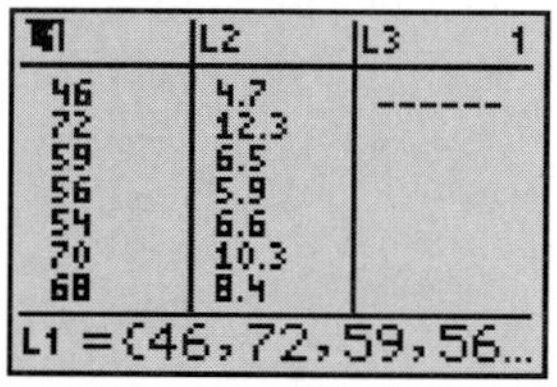

(a)

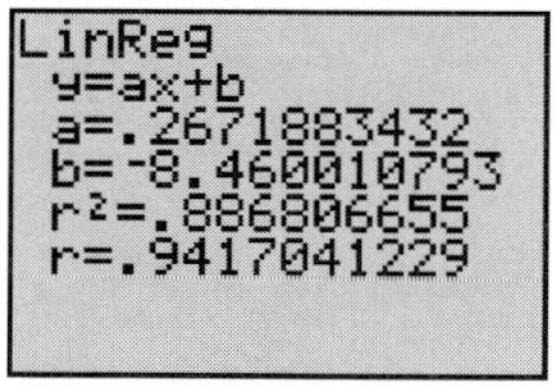

(b)

Figure 1.6

Solution

a) First enter the data into the calculator using the STAT EDIT feature as shown in Fig. 1.6(a). Enter the temperature in the first list or x-list and the percent injured in the second list or y-list. Consult your calculator manual if you have trouble or your calculator does not accept data in this manner. Use the STAT CALC feature and the choice LinReg, to find the formula as shown in Fig. 1.6(b). The calculator gives the formula in the form $y = ax + b$ where $a \approx 0.27$ and $b \approx -8.46$. Replace y by p and x by t to get the formula

$$p = 0.27t - 8.46.$$

Since $r \approx 0.94$, the formula is a good model for the data.

b) Use $t = 82$ in the formula $p = 0.27t - 8.46$:

$$p = 0.27(82) - 8.46 = 13.68$$

When the temperature is 82°F, we expect about 13.7% of the runners to be injured.

c) Now solve the formula for t:

$$0.27t - 8.46 = p$$

$$0.27t = p + 8.46$$

$$t = \frac{p + 8.46}{0.27}$$

Use $p = 9$ in this formula:

$$t = \frac{9 + 8.46}{0.27} \approx 65$$

If 9% of the runners are injured, then we would expect that the temperature was about 65°F. Note that we could have substituted $p = 9$ into the formula and then solved for t.

Using a Calculator in Solving Equations

In Example 5 of Section 1.1 we used a calculator in solving an equation, but in this section we use a slightly different approach. We solve the equation for x using the technique of solving for a specified variable. All computations are performed at the end, to minimize round-off errors.

EXAMPLE 4 Saving calculations until the end

Solve the equation for x, then use a calculator to find the value of x to three decimal places:

$$\frac{3.291}{x} + \frac{1}{2.35} = 4.76$$

Solution

$$2.35x\left(\frac{3.291}{x} + \frac{1}{2.35}\right) = 2.35x(4.76) \qquad \text{Multiply each side by the LCD.}$$

$$(2.35)(3.291) + x = (2.35)(4.76)x$$

$$x - (2.35)(4.76)x = -(2.35)(3.291) \qquad \text{Move all } x\text{-terms to the left side.}$$

$$x[1 - (2.35)(4.76)] = -(2.35)(3.291) \qquad \text{Factor out } x.$$

$$x = \frac{-2.35(3.291)}{1 - (2.35)(4.76)} \qquad \text{Divide each side by } 1 - (2.35)(4.76).$$

$$x \approx 0.759$$

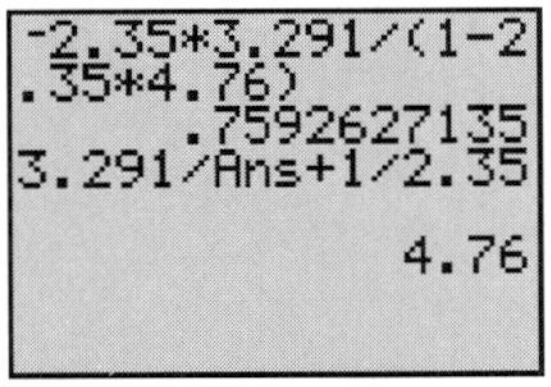

Figure 1.7

Check 0.759 in the original equation. The solution set is {0.759}. A graphing calculator check is shown in Fig. 1.7.

Problem-Solving Techniques

Applied problems in mathematics often involve solving an equation. The equations for some problems come from known formulas, as in Example 2, while in other problems we must write an equation that models a particular problem situation.

The best way to learn to solve problems is to study a few examples and then solve lots of problems. We will first look at an example of problem solving, then give a problem-solving strategy.

EXAMPLE 5 Solving a problem involving sales tax

Jeannie Fung bought a new, fire-engine red Mazda RX-7 for a total cost of \$18,966, including sales tax. If the sales tax rate is 9%, then what amount of tax did she pay?

Solution

There are two unknown quantities here, the price of the car and the amount of the tax. Represent the unknown quantities as follows:

$$x = \text{the price of the car}$$
$$0.09x = \text{amount of sales tax}$$

The price of the car plus the amount of sales tax is the total cost of the car. We can write this relationship as an equation and solve it for x.

$$x + 0.09x = 18{,}966$$
$$1.09x = 18{,}966$$
$$x = \frac{18{,}966}{1.09} = 17{,}400$$
$$0.09x = 1566$$

You can check this answer by adding \$17,400 and \$1566 to get \$18,966, the total cost. The amount of tax that Jeannie Fung paid was \$1566.

No two problems are exactly alike, but there are similarities. The following strategy will assist you in solving problems on your own.

STRATEGY

Problem Solving

1. Read the problem as many times as necessary to get an understanding of the problem.
2. If possible, draw a diagram to illustrate the problem.
3. Choose a variable, write down what it represents, and if possible, represent any other unknown quantities in terms of that variable.
4. Write an equation that models the situation. You may be able to use a known formula, or you may have to write an equation that models only that particular problem.
5. Solve the equation.
6. Check your answer by using it to solve the original problem (not just the equation).
7. Answer the question posed in the original problem.

In the next example we have a geometric situation. Notice how we are following the strategy for problem solving.

EXAMPLE 6 Solving a geometric problem

In 1974, Chinese workers found three pits that contained life-size sculptures of warriors, which were created to guard the tomb of an emperor (Nigel Hawkes, *Structures: The Way Things Are Built*, New York: Macmillan Co., 1990). The largest rectangular pit has a length that is 40 yards longer than three times the width. If its perimeter is 640 yards, what are the length and width?

Figure 1.8

Solution

First draw a diagram as shown in Fig. 1.8. Use the fact that the length is 40 yards longer than three times the width to represent the width and length as follows:

$$x = \text{the width in yards}$$
$$3x + 40 = \text{the length in yards}$$

The formula for the perimeter of a rectangle is $2L + 2W = P$. Replace W by x, L by $3x + 40$, and P by 640.

$$\begin{aligned} 2L + 2W &= P && \text{Perimeter formula} \\ 2(3x + 40) + 2x &= 640 && \text{Substitution} \\ 6x + 80 + 2x &= 640 \\ 8x + 80 &= 640 \\ 8x &= 560 \\ x &= 70 \end{aligned}$$

If $x = 70$, then $3x + 40 = 250$. Check that the dimensions of 70 yards and 250 yards really give a perimeter of 640 yards. We conclude that the length of the pit is 250 yards, and its width is 70 yards.

The next problem is called a **uniform-motion** problem because it involves motion at a constant rate. Of course, people do not usually move at a constant rate, but their average speed can be assumed to be constant over some time interval. The problem also illustrates how a table can be used as an effective technique for organizing information.

EXAMPLE 7 Solving a uniform-motion problem

A group of hikers from Tulsa hiked down into the Grand Canyon in 3 hours 30 minutes. Coming back up on a trail that was 4 miles shorter, they hiked 2 mph slower and it took them 1 hour longer. What was their rate going down?

Solution

Let x represent the rate going down into the canyon. Next make a table like Table 1.1 to show the distance, rate, and time for both the trip down and the trip back up. Once we fill in any two entries in a row of the table, we can use $D = RT$ to

Table 1.1

	Rate	Time	Distance
Down	x	3.5	$3.5x$
Up	$x - 2$	4.5	$4.5(x - 2)$

obtain an expression for the third entry. Using the fact that the distance up was 4 miles shorter, we can write the following equation.

$$4.5(x - 2) = 3.5x - 4$$
$$4.5x - 9 = 3.5x - 4$$
$$x - 9 = -4$$
$$x = 5$$

After checking that 5 mph satisfies the conditions given in the problem, we conclude that the hikers traveled at 5 mph going down into the canyon.

The next example involves mixing beverages with two different concentrations of orange juice. In other **mixture problems**, we may mix chemical solutions, candy, or even people.

EXAMPLE 8 Solving a mixture problem

A beverage producer makes two products, Orange Drink, containing 10% orange juice, and Orange Delight, containing 50% orange juice. How many gallons of Orange Delight must be mixed with 300 gallons of Orange Drink to create a new product containing 40% orange juice?

Solution

Let x represent the number of gallons of Orange Delight shown in Fig. 1.9. Table 1.2 (on the next page) shows three pertinent expressions for each product: the

Figure 1.9

Table 1.2

	Amount of product	Percent orange juice	Amount orange juice
Drink	300	10%	0.10(300)
Delight	x	50%	$0.50x$
Mixture	$x + 300$	40%	$0.40(x + 300)$

quantity of the product, the percentage of orange juice in the product, and the actual amount of orange juice in that quantity.

We write the following equation expressing the fact that the actual amount of orange juice in the mixture is the sum of the amounts of orange juice in the Orange Drink and in the Orange Delight.

$$\begin{aligned} 0.40(x + 300) &= 0.10(300) + 0.50x \\ 0.4x + 120 &= 30 + 0.5x \\ 90 &= 0.1x \\ 900 &= x \end{aligned}$$

Mix 900 gallons of Orange Delight with the 300 gallons of Orange Drink to obtain the proper mixture.

Work problems are problems in which people or machines are working together to accomplish a task. A typical situation might have two people painting a house at different rates. Suppose Joe and Frank are painting a house together for 2 hours and Joe paints at the rate of one-sixth of the house per hour while Frank paints at the rate of one-third of the house per hour. Note that

$$\left(\frac{1}{6} \text{ of house per hour}\right)(2 \text{ hr}) = \frac{1}{3} \text{ of house}$$

and

$$\left(\frac{1}{3} \text{ of house per hour}\right)(2 \text{ hr}) = \frac{2}{3} \text{ of house.}$$

The product of the rate and the time gives the fraction of the house completed by each person, and these fractions have a sum of 1 because the entire job is completed. Note how similar this situation is to a uniform-motion problem where $RT = D$.

EXAMPLE 9 Solving a work problem

Aboard the starship *Nostromo*, the human technician, Brett, can process the crew's medical history in 36 minutes. However, the android Science Officer, Ash, can process the same records in 24 minutes. After Brett worked on the records for 1 minute, Ash joined in and both crew members worked until the job was done. How long did Ash work on the records?

Solution

Let x represent the number of minutes that Ash worked and $x + 1$ represent the number of minutes that Brett worked. Ash works at the rate of 1/24 of the job per minute, while Brett works at the rate of 1/36 of the job per minute. Table 1.3 shows all of the pertinent quantities.

Table 1.3

	Rate	Time	Work completed
Ash	$\frac{1}{24}$	x	$\frac{1}{24}x$
Brett	$\frac{1}{36}$	$x + 1$	$\frac{1}{36}(x + 1)$

The following equation expresses the fact that the work completed together is the sum of the work completed by each worker alone.

$$\frac{1}{24}x + \frac{1}{36}(x + 1) = 1$$

$$72\left[\frac{1}{24}x + \frac{1}{36}(x + 1)\right] = 72 \cdot 1 \qquad \text{Multiply by the LCD 72.}$$

$$3x + 2x + 2 = 72$$

$$5x = 70$$

$$x = 14$$

Ash worked for 14 minutes.

Check 14 in the original equation as shown in Fig. 1.10.

Figure 1.10

For Thought True or False? Explain.

1. If we solve $P + Prt = S$ for P, we get $P = S - Prt$.
2. The perimeter of any rectangle is the product of its length and width.
3. If n is an odd integer, then $n + 1$ and $n + 3$ represent odd integers.
4. Solving $x - y = 1$ for y gives us $y = x - 1$.
5. Two numbers that have a sum of -3 can be represented by x and $-3 - x$.
6. If P is the number of professors and S is the number of students at the play, and there are twice as many professors as students, then $2P = S$.
7. If you need \$100,000 for your house, and the agent gets 9% of the selling price, then the agent gets \$9,000, and the house sells for \$109,000.
8. If John can mow the lawn in x hours, then he mows the lawn at the rate of $1/x$ of the lawn per hour.
9. If George hiked $3x$ miles and Anita hiked $4(x - 2)$ miles, and George hiked 5 more miles than Anita, then $3x + 5 = 4(x - 2)$.
10. Two numbers that differ by 9 can be represented as 9 and $x + 9$.

1.2 EXERCISES

Tape 3 Disk

Solve each formula for the specified variable.

1. $I = Prt$ for r (simple interest)
2. $D = RT$ for R (uniform motion)
3. $F = \frac{9}{5}C + 32$ for C (temperature)
4. $A = \frac{1}{2}bh$ for b (area of a triangle)
5. $C = 2\pi r$ for r (circumference of a circle)
6. $A = \frac{1}{2}h(b_1 + b_2)$ for h (area of a trapezoid)
7. $Ax + By = C$ for y (equation of a line)
8. $V = \pi r^2 h$ for h (volume of a cylinder)
9. $\frac{1}{R} = \frac{1}{R_1} + \frac{1}{R_2} + \frac{1}{R_3}$ for R_2 (resistance)
10. $S = \frac{a_1}{1 - r}$ for r (geometric series)
11. $a_n = a_1 + (n - 1)d$ for d (arithmetic sequence)
12. $S_n = \frac{n}{2}(a_1 + a_n)$ for a_1 (arithmetic series)
13. *The 2.4-Meter Rule* A 2.4-meter sailboat is a one-person boat that is about 13 feet in length, has a displacement of about 550 pounds, and a sail area of about 81 ft^2. To compete in the 2.4-meter class, a boat must satisfy the formula

$$2.4 = \frac{L + 2D - F\sqrt{S}}{2.37},$$

where L = length, F = freeboard, D = girth, and S = sail area. Solve the formula for D.

14. *Finding the Freeboard* Solve the formula in the previous exercise for F.

Use the appropriate formula to solve each problem.

15. *Simple Interest* If \$51.30 in interest is earned on a deposit of \$950 in one year, then what is the simple interest rate?
16. *Simple Interest* If you borrow \$100 and pay back \$105 at the end of one month, then what is the simple annual interest rate?
17. *Uniform Motion* How long does it take an SR-71 Blackbird, one of the fastest U.S. jets, to make a surveillance run of 5570 mi if it travels at an average speed of Mach 3 (2228 mph)?
18. *Circumference of a Circle* If the circumference of a circular sign is 72π in., then what is the radius?
19. *Volume of a Cylinder* If the volume of a cylinder is 126π in^3 and the radius is 6 in., then what is the height?
20. *Celsius Temperature* If the temperature at 1 P.M. on July 9 in Toronto was 30°C, then what was the temperature in degrees Fahrenheit?

21. *Injuries in the Boston Marathon* A study of the 4386 male runners in the 1985 Boston Marathon showed a high correlation between age and the percentage of injured runners for men aged 59 and below (*The Boston Globe*, April 20, 1992). The bar graph shows the percentage of injured men age 59 and below grouped into five categories. There appears to be a relationship between the age group A and the percentage p of those injured. Use the linear regression feature of a graphing calculator to find a formula that expresses p in terms of A. Find the percentage of runners that are predicted to be injured in age group 4 according to the formula and compare the answer to the actual percentage injured.

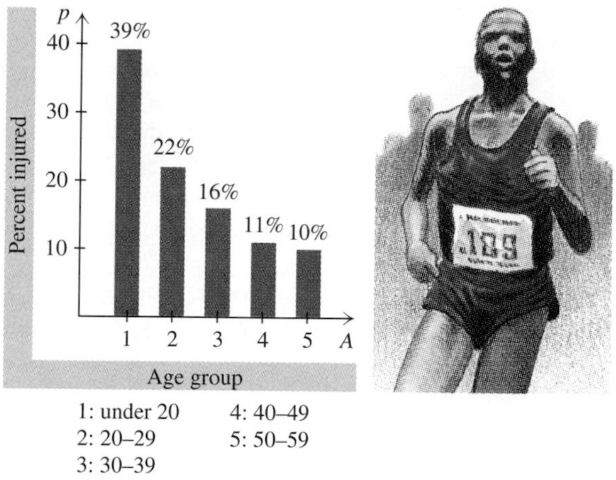

Figure for Exercise 21

22. *Injuries in the Boston Marathon* Use the formula of Exercise 21 to predict the percentage of runners injured in age group 6 (60-plus). In the 1985 race, 21% of the runners in the 60-plus age group were actually injured. Can you offer an explanation for the difference between the predicted and the actual figures?

Solve each equation for x, using your calculator only on the last step. Round answers to four decimal places.

23. $3.458x - 2.347 = 4.782$

24. $5.76 - 3.49x = 2.388x - 1.42$

25. $\dfrac{3.33}{x} + \dfrac{2.391}{3.4} = 9.876$

26. $\dfrac{4.57}{x} - \dfrac{1}{4.59} = \dfrac{1}{3.6}$

27. $\dfrac{x - 3.45}{x + 4.98} = 7.53$

28. $\dfrac{x + 2.41}{x - 3.98} = \dfrac{x - 2.31}{x + 4.55}$

Solve each problem.

29. *Cost of a Car* Jeff knows that his neighbor Sarah paid \$28,728, including sales tax, for a new Buick Park Avenue. If the sales tax rate is 8%, then what is the cost of the car before the tax?

30. *Real Estate Commission* To be able to afford the house of their dreams, Dave and Leslie must clear \$128,000 from the sale of their first house. If they must pay \$780 in closing costs and 6% of the selling price for the sales commission, then what is the minimum selling price for which they will get \$128,000?

31. *Adjusting the Saddle* The saddle height on a bicycle should be 109% of the inside leg measurement of the rider (Cycling, Burkett, and Darst, 1987). See the figure. If the saddle height is 37 in., then what is the inside leg measurement?

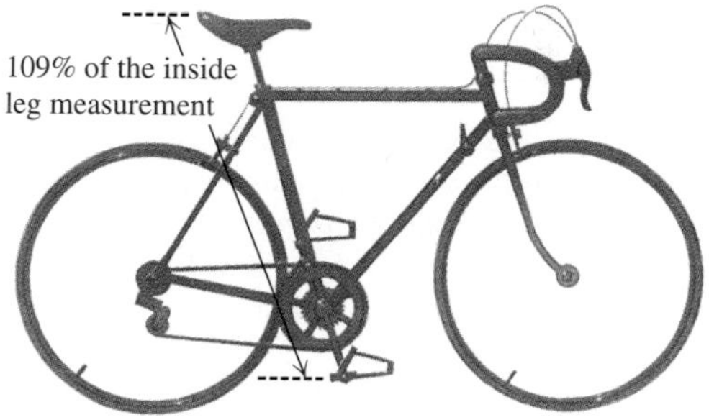

Figure for Exercise 31

32. *Target Heart Rate* For a cardiovascular work out, fitness experts recommend that you reach your target heart rate and stay at that rate for at least 20 minutes (Cycling, Burkett, and Darst, 1987). To find your target heart rate find the sum of your age and your resting heart rate, then subtract that sum from 220. Find 60% of that result and add it to your resting heart rate. If the target heart rate for a 30-year-old person is 144, then what is that person's resting heart rate?

33. *Investment Income* Tara paid one-half of her game-show winnings to the government for taxes. She invested one-third of her winnings in Jeff's copy shop at 14% interest and one-sixth of her winnings in Kaiser's German Bakery at 12% interest. If she earned a total of \$4000 on the investments in one year, then how much did she win on the game show?

34. *Construction Penalties* Gonzales Construction contracted Kentwood High and Memorial Stadium for a total cost of

$4.7 million. Because the construction was not completed on time, Gonzales paid 5% of the amount of the high school contract in penalties and 4% of the amount of the stadium contract in penalties. If the total penalty was $223,000, then what was the amount of each contract?

35. *Trimming a Garage Door* A carpenter used 30 ft of molding in three pieces to trim a garage door. If the long piece was 2 ft longer than twice the length of each shorter piece, then how long was each piece?

Figure for Exercise 35

36. *Increasing Area of a Field* Julia's soybean field is 3 m longer than it is wide. To increase her production, she plans to increase both the length and width by 2 m. If the new field is 46 m^2 larger than the old field, then what are the dimensions of the old field?

37. *Fencing a Feed Lot* Peter plans to fence off a square feed lot and then cross-fence to divide the feed lot into four smaller square feed lots. If he uses 480 ft of fencing, then how much area will be fenced in?

Figure for Exercise 37

38. *Fencing Dog Pens* Clint is constructing two adjacent rectangular dog pens. Each pen will be three times as long as it is wide, and the pens will share a common long side. If Clint has 65 ft of fencing, what are the dimensions of each pen?

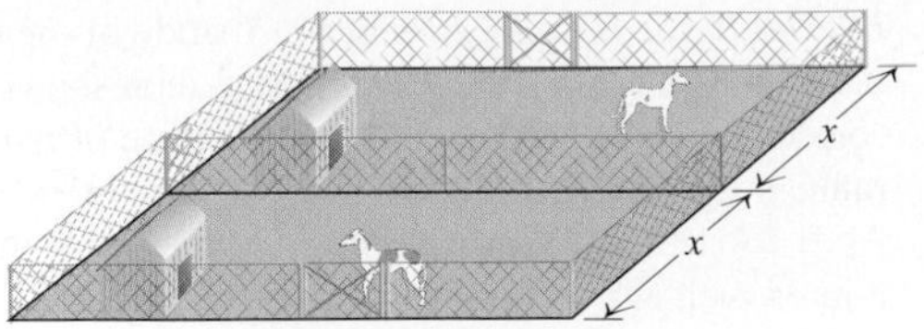

Figure for Exercise 38

39. *Racing Speed* Bobby and Rick are in a 10-lap race on a one-mile oval track. Bobby, averaging 90 mph, has completed two laps just as Rick is getting his car onto the track. What speed does Rick have to average to be even with Bobby at the end of the tenth lap?

40. *Rowing a Boat* Boudreaux rowed his pirogue from his camp on the bayou to his crab traps. Going down the bayou, he caught a falling tide that increased his normal speed by 2 mph, but coming back it decreased his normal speed by 2 mph. Going with the tide, the trip took only 10 min; going against the tide, the trip took 30 min. How far is it from Boudreaux's camp to his crab traps?

41. *Average Speed* Junior drove his rig on Interstate 10 from San Antonio to El Paso. At the halfway point he noticed that he had been averaging 80 mph, while his company requires his average speed to be 60 mph. What must be his speed for the last half of the trip so that he will average 60 mph for the trip?

42. *Basketball Stats* Two-thirds of the way through the 1997 basketball season, Houston Comets guard Cynthia Cooper has an average of 10 points per game. What must her point average be for the remaining games to average 20 points per game for the season?

43. *Start-Up Capital* Norma decided to delay starting her catering business for one year and invest her start-up capital in two banks. The annual percentage rate (APR) was 5% at one bank and 6% at the other. After one year she made $5880. If the amount that she invested at 6% was $10,000 larger than the amount invested at 5%, then what was the amount of her start-up capital?

44. *Combining Investments* Brent lent his brother Bob some money at 8% simple interest, and he lent his sister Betty half as much money at twice the interest rate. Both loans were for one year. If Brent made a total of 24 cents in interest, then how much did he lend to each one?

45. *Percentage of Minority Workers* At the Northside assembly plant, 5% of the workers were classified as minority, while at the Southside assembly plant, 80% of the workers were classified as minority. When Northside and Southside were closed, all workers transferred to the new Eastside plant to make up its entire work force. If 50% of the 1500 employees at Eastside are minority, then how many employees did Northside and Southside have originally?

46. *Mixing Alcohol Solutions* A pharmacist needs to obtain a 70% alcohol solution. How many ounces of a 30% alcohol solution must be mixed with 40 ounces of an 80% alcohol solution to obtain a 70% alcohol solution?

47. *Harvesting Wheat* With the old combine, Nikita's entire wheat crop can be harvested in 72 hr, but a new combine can do the same job in 48 hr. How many hours would it take to harvest the crop with both combines operating?

48. *Processing Forms* Rita can process a batch of insurance claims in 4 hr working alone. Eduardo can process a batch of insurance claims in 2 hr working alone. How long would it take them to process a batch of claims if they worked together?

49. *Batman and Robin* Batman can clean up all of the crime in Gotham City in 8 hr working alone. Robin can do the same job alone in 12 hr. If Robin starts crime-fighting at 8 A.M. and Batman joins him at 10 A.M., then at what time will they have all of the crime cleaned up?

50. *Scraping Barnacles* Della can scrape the barnacles from a 70-ft yacht in 10 hr using an electric barnacle scraper. Don can do the same job in 15 hr using a manual barnacle scraper. If Don starts scraping at noon and Della joins him at 3 P.M., then at what time will they finish the job?

Equations involving two variables x and y will be studied in later chapters. Solve each of the following equations for y.

51. $3x + 2y = -6$

52. $5x - 4y = 16$

53. $\frac{1}{2}x - \frac{1}{3}y = 3$

54. $x = \frac{5}{3}y + 10$

55. $y - 3 = \frac{3}{4}(x - 1)$

56. $y + 1 = -\frac{1}{2}(x - 3)$

57. $\frac{y - y_1}{x - x_1} = m$

58. $\frac{y - 3}{x + 5} = -\frac{2}{3}$

59. $y + xy = 3$

60. $xy + 3 = 2y - 9$

Solve each problem.

61. *Planning a Race Track* If Mario plans to develop a circular race track one mile in circumference on a square plot of land, then what is the minimum number of acres that he needs? (One acre is equal to 43,560 ft^2.)

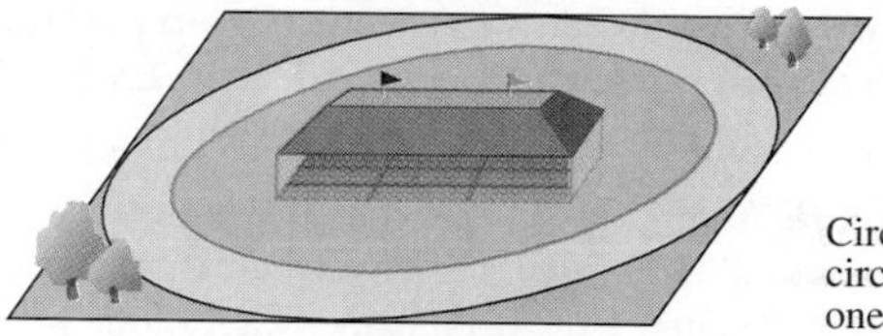

Circular track has circumference of one mile.

Figure for Exercise 61

62. *Volume of a Can of Coke* If a can of Coke contains 12 fluid ounces and the diameter of the can is 2.375 in., then what is the height of the can? (One fluid ounce equals approximately 1.8 in^3.)

63. *Area of a Lot* Julio owns a four-sided lot that lies between two parallel streets. If his 90,000-ft^2 lot has 500 ft frontage on one street and 300 ft frontage on the other, then how far apart are the streets?

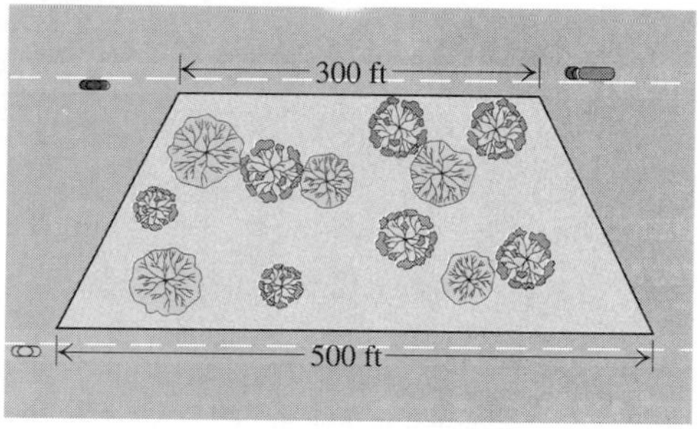

Figure for Exercise 63

64. *Width of a Football Field* If the perimeter of a football field in the NFL including the end zones is 1040 ft and the field is 120 yd long, then what is the width of the field in feet?

65. *Depth of a Swimming Pool* A circular swimming pool with a diameter of 30 ft and a horizontal bottom contains

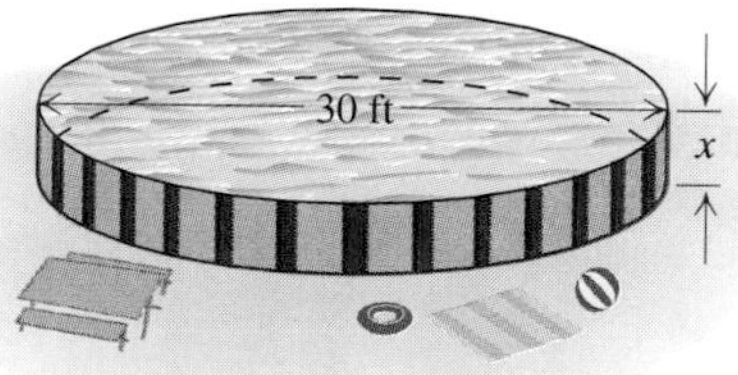

Figure for Exercise 65

22,000 gal of water. What is the depth of the water in the pool? (One cubic foot contains approximately 7.5 gal of water.)

66. *Depth of a Reflecting Pool* A rectangular reflecting pool with a horizontal bottom is 100 ft by 150 ft and contains 200,000 gal of water. How deep is the water in the pool?

67. *Olympic Track* To host the Summer Olympics, a city plans to build an eight-lane track. The track will consist of parallel 100-m straightaways with semicircular turns on either end as shown in the figure. The distance around the outside edge of the oval track is 514 m. If the track is built on a rectangular lot as shown in the drawing, then how many hectares (1 hectare = 10,000 m^2) of land are needed in the rectangular lot?

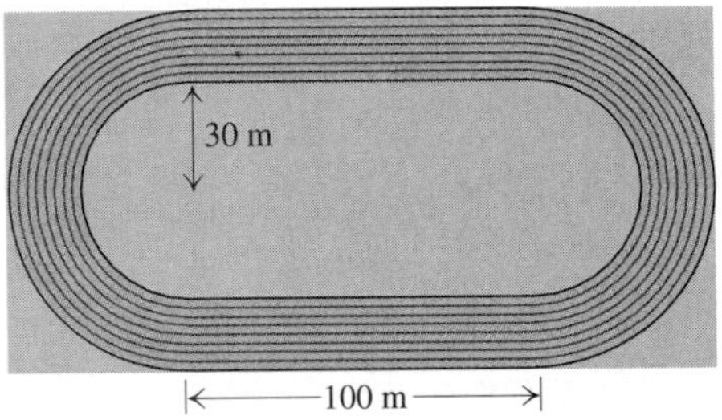

Figure for Exercises 67 and 68

68. *Green Space* If the inside radius of the turns is 30 m and grass is to be planted inside and outside the track of Exercise 67, then how many square meters of the rectangular lot will be planted in grass?

69. *Integers* If the sum of three consecutive integers is 105, then what are the integers?

70. *Odd Integers* If the sum of three consecutive odd integers is 87, then what are the integers?

71. *Taxable Income* According to the 1996 federal income tax rate schedule, a single taxpayer with taxable income over \$58,150 paid \$13,152 plus 31% of the amount over \$58,150 in federal income tax. If Lorinda paid \$18,731.10 in federal income tax in 1996, then how much was her taxable income that year?

72. *Living Comfortably* Glen, a single taxpayer in 1996, figured that he needed \$62,000 in after-federal-tax income to live comfortably. Use the information from Exercise 71 to find the amount of taxable income that would allow Glen to live comfortably in 1996.

73. *Diluting Baneberry* How much water must Poison Ivy add to a 4-liter solution that contains 5% extract of baneberry to get a solution that contains 3% extract of baneberry?

74. *Diluting Antifreeze* A mechanic is working on a car with a 20-quart radiator containing a 60% antifreeze solution. How much of the solution should he drain and replace with pure water to get a solution that is 50% antifreeze?

75. *Recording Experience* Kim has been in the recording business twice as long as Eric. In four years they will be able to advertise that their studio has 50 years of combined experience in recording. How much experience does Kim have now?

76. *Tourists and Alaskans* Twice as many tourists as native Alaskans were present at the Anchorage Gallery's opening of Inuit sculpture. Halfway through the evening, 14 tourists departed and 5 Alaskans arrived, making the number of Alaskans one greater than the number of tourists. How many tourists were in attendance at the start?

77. *Mixing Dried Fruit* The owner of a health-food store sells dried apples for \$1.20 per quarter-pound, and dried apricots for \$1.80 per quarter-pound. How many pounds of each must he mix together to get 20 lb of a mixture that sells for \$1.68 per quarter-pound?

78. *Mixing Breakfast Cereal* Raisins sell for \$4.50/lb, and bran flakes sell for \$2.80/lb. How many pounds of raisins should be mixed with 12 lb of bran flakes to get a mixture that sells for \$3.14/lb?

79. *Coins in a Vending Machine* Dana inserted eight coins, consisting of dimes and nickels, into a vending machine to purchase a Snickers bar for 55 cents. How many coins of each type did she use?

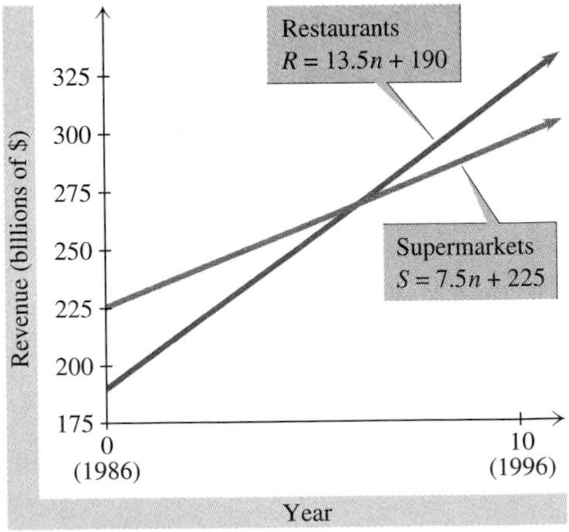

Figure for Exercise 81

80. *Cost of a Newspaper* Ravi took eight coins from his pocket, which contained only dimes, nickels, and quarters, and bought the Sunday Edition of *The Daily Star* for 75 cents. If the number of nickels he used was one more than the number of dimes, then how many of each type of coin did he use?

81. *Dining In vs. Dining Out* Revenue for both restaurants and supermarkets is growing as shown in the accompanying graph (*Forbes*, March 24, 1997). This revenue growth can be modeled by the equations $R = 13.5n + 190$ and $S = 7.5n + 225$, where n is the number of years since 1986. Use the equations to find the year in which restaurant revenue surpassed supermarket revenue. In what year will restaurant revenue be double the supermarket revenue?

82. *Working Together* If one lost hiker can pick a gallon of wild berries in 2 hr, then how long should it take two lost hikers working together to pick one gallon of wild berries? If one mechanic at Spee-Dee Oil Change can change the oil in a Saturn in 6 min, then how long should it take two mechanics working together to change the oil in a Saturn? How long would it take 60 mechanics? Are your answers reasonable? Explain.

Linking Concepts

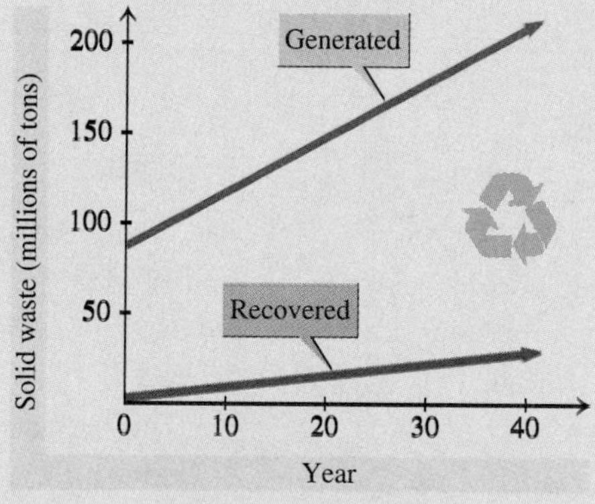

For Individual or Group Explorations

Solid Waste Recovery In 1960 the United States generated 87.1 million tons of municipal solid waste and recovered (or recycled) only 4.3% of it (U.S. EPA). The amount of municipal solid waste generated in the United States can be modeled by the formula $w = 3.14n + 87.1$*, while the amount recovered can be modeled by the formula* $w = 0.576n + 3.78$*, where w is in millions of tons and n is the number of years since 1960.*

a) Use the graph to estimate the first year in which the United States generated over 100 million tons of municipal solid waste.

b) Use the formulas to determine the year in which the United States generated 150 million more tons than it recovered.

c) Find the year in which 13% of the municipal solid waste generated will be recovered.

d) Find the years in which the recovery rates will be 14%, 15%, and 16%.

e) Will the recovery rate ever reach 25%?

f) According to this model, what is the maximum percentage of solid waste that will ever be recovered?

1.3 Complex Numbers

Our system of numbers developed as the need arose. Numbers were first used for counting. As society advanced, the rational numbers were formed to express fractional parts and ratios. Negative numbers were invented to express losses or debts. When it was discovered that the exact size of some very real objects could not be expressed with rational numbers, the irrational numbers were added to the system, forming the set of real numbers. Later still, there was a need for another expansion to the number system. In this section we study that expansion, the set of complex numbers.

Definitions

In Section 1.1 we solved equations of the form $ax + b = 0$ (linear equations). We learned that any linear equation has a real solution. In Section 1.4 we will study equations of the form $ax^2 + bx + c = 0$ (quadratic equations). For example, a simple quadratic equation such as $x^2 = 4$ has two solutions (2 and -2). However, $x^2 = -1$ has no real solution because the square of every real number is nonnegative. So that equations such as $x^2 = -1$ will have solutions, imaginary numbers are defined. They are combined with the set of real numbers to form the set of complex numbers.

The imaginary numbers are based on the solution of the equation $x^2 = -1$. Since no real number solves this equation, a solution is called an *imaginary number*. The imaginary number i is defined to be a solution to this equation.

Definition: Imaginary Number *i*

The **imaginary number *i*** is defined as

$$i = \sqrt{-1} \quad \text{and} \quad i^2 = -1.$$

A complex number is formed as a real number plus a real multiple of i.

Definition: Complex Numbers

The set of **complex numbers** is the set of all numbers of the form $a + bi$, where a and b are real numbers.

In the complex number $a + bi$, a is called the **real part** and b is called the **imaginary part**. If $b \neq 0$, then $a + bi$ is called an **imaginary number**. Two complex numbers $a + bi$ and $c + di$ are **equal** if and only if $a = c$ and $b = d$.

The form $a + bi$ is called the **standard form** of a complex number, but for convenience we use a few variations of that form. If $a = 0$, then a is omitted and only the imaginary part is written. If $b = 0$, then only a is written and the complex number is a real number. If b is a radical, then i is usually written before b. For example, we write $2 + i\sqrt{3}$ rather than $2 + \sqrt{3}i$, which could be confused with $2 + \sqrt{3i}$. If b is negative, we use the minus symbol to separate the real and imaginary parts. For example, instead of $3 + (-2)i$ we write $3 - 2i$. If both a and b are zero, the complex number $0 + 0i$ is the real number 0. For a complex number involving fractions, such as $\frac{1}{3} - \frac{2}{3}i$, we may write $(1 - 2i)/3$.

EXAMPLE 1 Standard form of a complex number

Determine whether each complex number is real or imaginary and write it in the standard form $a + bi$.

a) $3i$ **b)** 87 **c)** $4 - 5i$ **d)** 0 **e)** $\dfrac{1 + \pi i}{2}$

Solution

a) The complex number $3i$ is imaginary, and $3i = 0 + 3i$.

b) The complex number 87 is a real number, and $87 = 87 + 0i$.

c) The complex number $4 - 5i$ is imaginary, and $4 - 5i = 4 + (-5)i$.

d) The complex number 0 is real, and $0 = 0 + 0i$.

e) The complex number $\dfrac{1 + \pi i}{2}$ is imaginary, and $\dfrac{1 + \pi i}{2} = \dfrac{1}{2} + \dfrac{\pi}{2}i$.

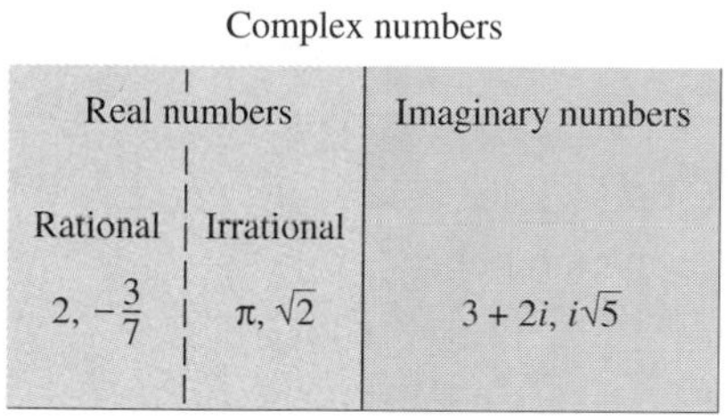

Figure 1.11

Just as there are two types of real numbers (rational and irrational), there are two types of complex numbers, the real numbers and the imaginary numbers. The relationship between these sets of numbers is shown in Fig. 1.11.

Addition, Subtraction, and Multiplication

Writing and thinking of complex numbers as binomials makes it easy to perform operations with complex numbers. The sum of $4 + 5i$ and $1 - 7i$ is found just as if we were adding binomials with i being the variable:

$$(4 + 5i) + (1 - 7i) = 5 - 2i$$

To find the sum, we add the real parts and add the imaginary parts. Subtraction is done similarly:

$$(9 - 2i) - (6 + 5i) = 3 - 7i$$

To multiply complex numbers, we use the FOIL method for multiplying binomials:

$$\begin{aligned}(2 + 3i)(5 - 4i) &= 10 - 8i + 15i - 12i^2 \\ &= 10 + 7i - 12(-1) \qquad \text{Replace } i^2 \text{ by } -1. \\ &= 22 + 7i\end{aligned}$$

In the box on the following page we give formal definitions of addition, subtraction, and multiplication of complex numbers.

Definition: Addition, Subtraction, and Multiplication

If $a + bi$ and $c + di$ are complex numbers, we define their sum, difference, and product as follows.

$$(a + bi) + (c + di) = (a + c) + (b + d)i$$
$$(a + bi) - (c + di) = (a - c) + (b - d)i$$
$$(a + bi)(c + di) = (ac - bd) + (bc + ad)i$$

Note that it is not necessary to memorize these definitions to perform these operations with complex numbers. We get the same results by working with complex numbers as if they were binomials in which i is the variable, replacing i^2 by -1 wherever it occurs.

EXAMPLE 2 Operations with complex numbers

Perform the indicated operations with the complex numbers.

a) $(-2 + 3i) + (-4 - 9i)$ **b)** $(-1 - 5i) - (3 - 2i)$ **c)** $2i(3 + i)$

d) $(3i)^2$ **e)** $(-3i)^2$ **f)** $(5 - 2i)(5 + 2i)$

Solution

a) $(-2 + 3i) + (-4 - 9i) = -6 - 6i$

b) $(-1 - 5i) - (3 - 2i) = -1 - 5i - 3 + 2i = -4 - 3i$

c) $2i(3 + i) = 6i + 2i^2 = 6i + 2(-1) = -2 + 6i$

d) $(3i)^2 = 3^2i^2 = 9(-1) = -9$

e) $(-3i)^2 = (-3)^2i^2 = 9(-1) = -9$

f) $(5 - 2i)(5 + 2i) = 25 - 4i^2 = 25 - 4(-1) = 29$

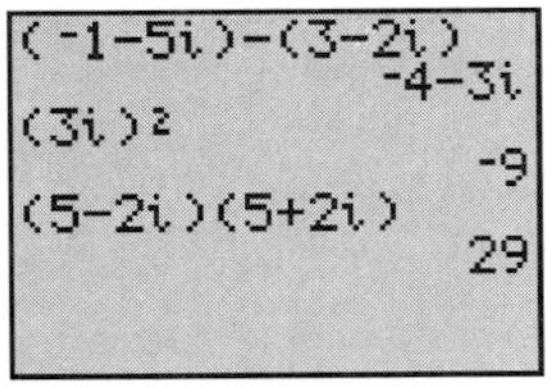

Figure 1.12

Check these results with a calculator that handles complex numbers as in Fig. 1.12.

Since $2^6 = 64$, 2^7 must be $2 \cdot 2^6$ or 128. Likewise, if we know any whole-number power of i, we can find the next higher power of i. Since $i^2 = -1$, we can find $i^3 = i \cdot i^2 = -i$. The first eight powers of i are listed here.

$$i^1 = i \quad i^2 = -1 \quad i^3 = -i \quad i^4 = 1$$
$$i^5 = i \quad i^6 = -1 \quad i^7 = -i \quad i^8 = 1$$

This list could be continued in this pattern, but any other whole-number power of i can be obtained from knowing the first four powers. We can simplify a power of i by using the fact that $i^4 = 1$ and $(i^4)^n = 1$ for any integer n.

EXAMPLE 3 Simplifying a power of i

Simplify i^{83}.

Solution

Divide 83 by 4 and write $83 = 4 \cdot 20 + 3$. So $i^{83} = (i^4)^{20} \cdot i^3 = 1 \cdot i^3 = -i$.

Division of Complex Numbers

The complex numbers $a + bi$ and $a - bi$ are called **complex conjugates** of each other.

EXAMPLE 4 Complex conjugates

Find the product of the given complex number and its conjugate.

a) $3 - i$ **b)** $4 + 2i$ **c)** $-i$

Solution

a) The conjugate of $3 - i$ is $3 + i$, and $(3 - i)(3 + i) = 9 - i^2 = 10$.

b) The conjugate of $4 + 2i$ is $4 - 2i$, and $(4 + 2i)(4 - 2i) = 16 - 4i^2 = 20$.

c) The conjugate of $-i$ is i, and $-i \cdot i = -i^2 = 1$.

In general we have the following theorem about complex conjugates.

Theorem: Complex Conjugates

If a and b are real numbers, then the product of $a + bi$ and its conjugate $a - bi$ is the real number $a^2 + b^2$. In symbols,

$$(a + bi)(a - bi) = a^2 + b^2.$$

We use the theorem about complex conjugates to divide imaginary numbers, in a process that is similar to rationalizing a denominator.

EXAMPLE 5 Dividing imaginary numbers

Write each quotient in the form $a + bi$.

a) $\dfrac{8 - i}{2 + i}$ **b)** $\dfrac{1}{5 - 4i}$ **c)** $\dfrac{3 - 2i}{i}$

Solution

a) Multiply the numerator and denominator by $2 - i$, the conjugate of $2 + i$:

$$\frac{8 - i}{2 + i} = \frac{(8 - i)(2 - i)}{(2 + i)(2 - i)} = \frac{16 - 10i + i^2}{4 - i^2} = \frac{15 - 10i}{5} = 3 - 2i$$

Check division using multiplication: $(3 - 2i)(2 + i) = 8 - i$.

b) $$\frac{1}{5 - 4i} = \frac{1(5 + 4i)}{(5 - 4i)(5 + 4i)} = \frac{5 + 4i}{25 + 16} = \frac{5 + 4i}{41} = \frac{5}{41} + \frac{4}{41}i$$

Check: $$\left(\frac{5}{41} + \frac{4}{41}i\right)(5 - 4i) = \frac{25}{41} + \frac{20}{41}i - \frac{20}{41}i - \frac{16}{41}i^2$$

$$= \frac{25}{41} + \frac{16}{41} = 1.$$

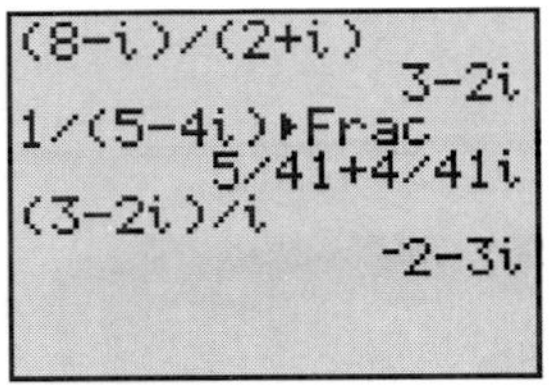

Figure 1.13

You can also check with a calculator that handles complex numbers as in Fig. 1.13. □

c) $$\frac{3 - 2i}{i} = \frac{(3 - 2i)(-i)}{i(-i)} = \frac{-3i + 2i^2}{-i^2} = \frac{-2 - 3i}{1} = -2 - 3i$$

Check: $(-2 - 3i)(i) = -3i^2 - 2i = 3 - 2i$.

Roots of Negative Numbers

In Examples 2(d) and 2(e), we saw that both $(3i)^2 = -9$ and $(-3i)^2 = -9$. This means that in the complex number system there are two square roots of -9, $3i$ and $-3i$. For any positive real number b, we have $(i\sqrt{b})^2 = -b$ and $(-i\sqrt{b})^2 = -b$. So there are two square roots of $-b$, $i\sqrt{b}$ and $-i\sqrt{b}$. We call $i\sqrt{b}$ the **principal square root** of $-b$ and make the following definition.

Definition: Square Root of a Negative Number

For any positive real number b, $\sqrt{-b} = i\sqrt{b}$.

In the real number system, $\sqrt{-2}$ and $\sqrt{-8}$ are undefined, but in the complex number system they are defined as $\sqrt{-2} = i\sqrt{2}$ and $\sqrt{-8} = i\sqrt{8}$. Even though we now have meaning for a symbol such as $\sqrt{-2}$, *all operations with complex numbers must be performed after converting to the $a + bi$ form.* If we perform operations with roots of negative numbers using properties of the real numbers, we can get contradictory results:

$$\sqrt{-2} \cdot \sqrt{-8} = \sqrt{(-2)(-8)} = \sqrt{16} = 4 \qquad \text{Incorrect.}$$

$$i\sqrt{2} \cdot i\sqrt{8} = i^2 \cdot \sqrt{16} = -4 \qquad \text{Correct.}$$

The product rule $\sqrt{a} \cdot \sqrt{b} = \sqrt{ab}$ is used *only* for nonnegative numbers a and b.

EXAMPLE 6 Square roots of negative numbers

Write each expression in the form $a + bi$, where a and b are real numbers.

a) $\sqrt{-8} + \sqrt{-18}$ **b)** $\dfrac{-4 + \sqrt{-50}}{4}$ **c)** $\sqrt{-27}(\sqrt{9} - \sqrt{-2})$

Solution

The first step in each case is to replace the square roots of negative numbers by expressions with i.

a) $\sqrt{-8} + \sqrt{-18} = i\sqrt{8} + i\sqrt{18} = 2i\sqrt{2} + 3i\sqrt{2}$
$= 5i\sqrt{2}$

b) $\dfrac{-4 + \sqrt{-50}}{4} = \dfrac{-4 + i\sqrt{50}}{4} = \dfrac{-4 + 5i\sqrt{2}}{4}$
$= -1 + \dfrac{5}{4}i\sqrt{2}$

c) $\sqrt{-27}(\sqrt{9} - \sqrt{-2}) = 3i\sqrt{3}(3 - i\sqrt{2}) = 9i\sqrt{3} - 3i^2\sqrt{6}$
$= 3\sqrt{6} + 9i\sqrt{3}$

Complex Solutions to Equations

In Section 1.4 we will be solving equations that have complex number solutions. We can use the operations for complex numbers defined in this section to determine whether a given complex number satisfies an equation.

EXAMPLE 7 Complex number solutions to equations

Determine whether the complex number $3 + i\sqrt{2}$ satisfies $x^2 - 6x + 11 = 0$.

Solution

Replace x by $3 + i\sqrt{2}$ in the polynomial $x^2 - 6x + 11$:

$$(3 + i\sqrt{2})^2 - 6(3 + i\sqrt{2}) + 11 = 9 + 6i\sqrt{2} + (i\sqrt{2})^2 - 18 - 6i\sqrt{2} + 11$$
$$= 9 + 6i\sqrt{2} - 2 - 18 - 6i\sqrt{2} + 11$$
$$= 0$$

Since the value of the polynomial is zero for $x = 3 + i\sqrt{2}$, the complex number $3 + i\sqrt{2}$ satisfies the equation $x^2 - 6x + 11 = 0$.

This polynomial is evaluated with a calculator as shown in Fig. 1.14.

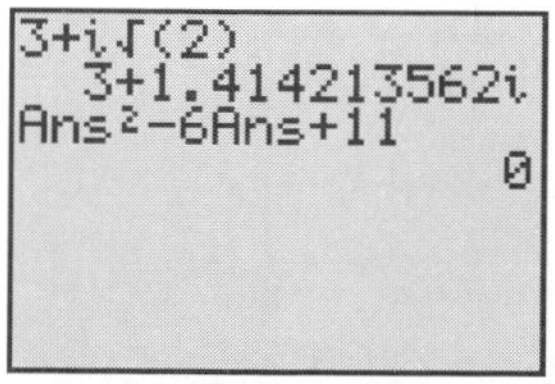

Figure 1.14

For Thought True or False? Explain.

1. The multiplicative inverse of i is $-i$.
2. The conjugate of i is $-i$.
3. The set of complex numbers is a subset of the set of real numbers.
4. $(\sqrt{3} - i\sqrt{2})(\sqrt{3} + i\sqrt{2}) = 5$
5. $(2 + 5i)(2 + 5i) = 4 + 25$
6. $5 - \sqrt{-9} = 5 - 9i$
7. If $P(x) = x^2 + 9$, then $P(3i) = 0$.
8. The imaginary number $-3i$ is a root to the equation $x^2 + 9 = 0$.
9. $i^4 = 1$
10. $i^{18} = 1$

1.3 Exercises

Tape 3 Disk

Determine whether each complex number is real or imaginary and write it in the standard form $a + bi$.

1. $6i$ **2.** $-3i + \sqrt{6}$ **3.** $\frac{1 + i}{3}$

4. -72 **5.** $\sqrt{7}$ **6.** $-i\sqrt{5}$

7. $\frac{\pi}{2}$ **8.** 0

Perform the indicated operations and write your answers in the form $a + bi$, where a and b are real numbers.

9. $(3 - 3i) + (4 + 5i)$ **10.** $(-3 + 2i) + (5 - 6i)$

11. $(1 - i) - (3 + 2i)$ **12.** $(6 - 7i) - (3 - 4i)$

13. $-6i(3 - 2i)$ **14.** $-3i(5 + 2i)$

15. $(2 - 3i)(4 + 6i)$ **16.** $(3 - i)(5 - 2i)$

17. $(5 - 2i)(5 + 2i)$ **18.** $(4 + 3i)(4 - 3i)$

19. $(\sqrt{3} - i)(\sqrt{3} + i)$

20. $(\sqrt{2} + i\sqrt{3})(\sqrt{2} - i\sqrt{3})$

21. $(3 + 4i)^2$ **22.** $(-6 - 2i)^2$

23. $(\sqrt{5} - 2i)^2$ **24.** $(\sqrt{6} + i\sqrt{3})^2$

25. i^{17} **26.** i^{24} **27.** i^{98} **28.** i^{19}

29. i^{-4} **30.** i^{-13} **31.** i^{-1} **32.** i^{-27}

Find the product of the given complex number and its conjugate.

33. $3 - 9i$ **34.** $4 + 3i$ **35.** $\frac{1}{2} + 2i$

36. $\frac{1}{3} - i$ **37.** i **38.** $-i\sqrt{5}$

39. $3 - i\sqrt{3}$ **40.** $\frac{5}{2} + i\frac{\sqrt{2}}{2}$

Write each quotient in the form $a + bi$.

41. $\frac{1}{2 - i}$ **42.** $\frac{1}{5 + 2i}$ **43.** $\frac{-3i}{1 - i}$

44. $\frac{3i}{-2 + i}$ **45.** $\frac{-2 + 6i}{2}$ **46.** $\frac{-6 - 9i}{-3}$

47. $\frac{-3 + 3i}{i}$ **48.** $\frac{-2 - 4i}{-i}$ **49.** $\frac{1 - i}{3 + 2i}$

50. $\frac{4 + 2i}{2 - 3i}$

Write each expression in the form $a + bi$, where a and b are real numbers.

51. $\sqrt{-4} - \sqrt{-9}$ **52.** $\sqrt{-16} + \sqrt{-25}$

53. $\sqrt{-4} - \sqrt{16}$ **54.** $\sqrt{-3} \cdot \sqrt{-3}$

55. $(\sqrt{-6})^2$ **56.** $(\sqrt{-5})^3$

57. $\sqrt{-2} \cdot \sqrt{-50}$ **58.** $\frac{-6 + \sqrt{-3}}{3}$

59. $\frac{-2 + \sqrt{-20}}{2}$ **60.** $\frac{9 - \sqrt{-18}}{-6}$

61. $-3 + \sqrt{3^2 - 4(1)(5)}$ **62.** $1 - \sqrt{(-1)^2 - 4(1)(1)}$

63. $\sqrt{-8}(\sqrt{-2} + \sqrt{8})$ **64.** $\sqrt{-6}(\sqrt{2} - \sqrt{-3})$

Evaluate the expression $\dfrac{-b + \sqrt{b^2 - 4ac}}{2a}$ for each choice of a, b, and c.

65. $a = 1, b = 2, c = 5$ **66.** $a = 5, b = -4, c = 1$

67. $a = 2, b = 4, c = 3$ **68.** $a = 2, b = -4, c = 5$

Evaluate the expression $\dfrac{-b - \sqrt{b^2 - 4ac}}{2a}$ for each choice of a, b, and c.

69. $a = 1, b = 6, c = 17$ **70.** $a = 1, b = -12, c = 84$

71. $a = -2, b = 6, c = 6$ **72.** $a = 3, b = 6, c = 8$

Let $P(x) = x^2 + 4x + 5$, $T(x) = 2x^2 + 1$, and $W(x) = x^2 - 6x + 14$. Find each of the following.

73. $P(-2 + i)$ **74.** $T\left(\dfrac{i\sqrt{2}}{2}\right)$ **75.** $P(-2 - i)$

76. $T\left(\dfrac{-i\sqrt{2}}{2}\right)$ **77.** $P(1 + i)$ **78.** $T(3 - i)$

79. $W(3 + i\sqrt{5})$ **80.** $W(3 - i\sqrt{5})$ **81.** $W(-1)$

82. $W(2 - i)$

Determine whether the given complex number satisfies the equation following it.

83. $2i$, $x^2 + 4 = 0$ **84.** $-4i$, $2x^2 + 32 = 0$

85. $1 - i$, $x^2 + 2x - 2 = 0$ **86.** $1 + i$, $x^2 - 2x + 2 = 0$

87. $3 - 2i$, $x^2 - 6x + 13 = 0$

88. $2 - i$, $x^2 - 4x - 6 = 0$

89. $\dfrac{i\sqrt{3}}{3}$, $3x^2 + 1 = 0$ **90.** $\dfrac{-i\sqrt{3}}{3}$, $3x^2 - 1 = 0$

Find the product of each pair of binomials. Note that in each case the product is a quadratic polynomial with real coefficients.

91. $(x - i)(x + i)$ **92.** $(x - 3i)(x + 3i)$

93. $(x - (1 + i))(x - (1 - i))$

94. $(x - (2 - i))(x - (2 + i))$

95. $(x - (2 + 3i))(x - (2 - 3i))$

96. $(x - (1 - 2i))(x - (1 + 2i))$

For Writing/Discussion

97. Explain in detail how to find i^n for any positive integer n.

98. Find a number $a + bi$ such that $a^2 + b^2$ is irrational.

99. Let $w = a + bi$ and $\overline{w} = a - bi$, where a and b are real numbers. Show that $w + \overline{w}$ is real and that $w - \overline{w}$ is imaginary. Write sentences (containing no mathematical symbols) stating these results.

100. Is it true that the product of a complex number and its conjugate is a real number? Explain.

101. Prove that the reciprocal of $a + bi$, where a and b are not both zero, is

$$\frac{a}{a^2 + b^2} - \frac{b}{a^2 + b^2}\,i.$$

102. *Cooperative Learning* Work in a small group to find the two square roots of 1 and two square roots of -1 in the complex number system. How many fourth roots of 1 are there in the complex number system and what are they? Explain how to find all of the fourth roots in the complex number system for any positive real number.

1.4 Quadratic Equations

One of our main goals is to solve polynomial equations. In Section 1.1 we learned to solve equations of the form $ax + b = 0$, the linear equations. Linear equations are first-degree polynomial equations. In this section we will solve second-degree polynomial equations, the quadratic equations.

Definition

Quadratic equations have a term that linear equations do not have, the x^2-term.

Definition: Quadratic Equation

A **quadratic equation** is an equation of the form

$$ax^2 + bx + c = 0,$$

where a, b, and c are real numbers with $a \neq 0$.

The condition that $a \neq 0$ in the definition ensures that the equation actually does have an x^2-term. There are several methods for solving quadratic equations. Which method is most appropriate depends on the type of quadratic equation that we are solving. We first consider a method for solving quadratic equations in which $b = 0$ and later consider methods for solving equations in which $b \neq 0$.

Solving $ax^2 + c = 0$

If $b = 0$ in the general form of the quadratic equation, then the equation is of the form $ax^2 + c = 0$. Equations of this form are solved by using the square root property.

The Square Root Property

For any real number k, the equation $x^2 = k$ is equivalent to $x = \pm\sqrt{k}$.

If $k > 0$, then $x^2 = k$ has two real solutions. If $k < 0$, then $x^2 = k$ has two imaginary solutions. If $k = 0$, then 0 is the only solution to $x^2 = k$.

EXAMPLE 1 Using the square root property

Solve each equation.

a) $x^2 - 9 = 0$ **b)** $2x^2 - 1 = 0$ **c)** $(x - 3)^2 = -8$

Solution

a) Before using the square root property, isolate x^2.

$$x^2 - 9 = 0$$

$$x^2 = 9$$

$$x = \pm\sqrt{9} = \pm 3 \qquad \text{Square root property}$$

Check: $3^2 - 9 = 0$ and $(-3)^2 - 9 = 0$. The solution set is $\{-3, 3\}$.

b) $2x^2 - 1 = 0$

$$2x^2 = 1$$

$$x^2 = \frac{1}{2} \qquad \textbf{Isolate } x^2.$$

$$x = \pm\sqrt{\frac{1}{2}} = \pm\frac{\sqrt{2}}{2} \qquad \textbf{Square root property}$$

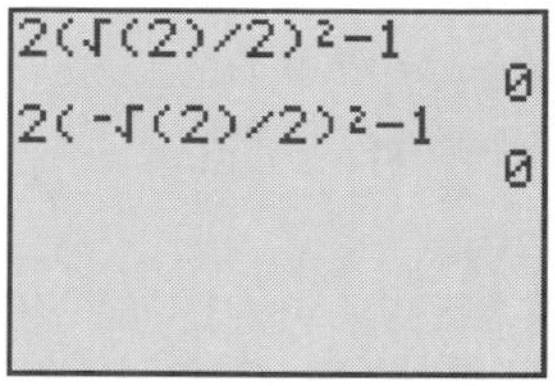

Figure 1.15

The solution set is $\left\{-\frac{\sqrt{2}}{2}, \frac{\sqrt{2}}{2}\right\}$.

Use a calculator to check as in Fig. 1.15. □

c) $(x - 3)^2 = -8$

$$x - 3 = \pm\sqrt{-8} \qquad \textbf{Square root property}$$

$$x = 3 \pm 2i\sqrt{2}$$

Check in the original equation. The solution set is $\{3 - 2i\sqrt{2}, 3 + 2i\sqrt{2}\}$.

The key idea in solving $ax^2 + c = 0$ is to solve for x^2 and then apply the square root property. We now turn our attention to equations of the form $ax^2 + bx + c = 0$ in which the trinomial can be factored.

Solving Quadratic Equations by Factoring

Many second-degree polynomials can be factored as a product of first-degree binomials. When one side of an equation is a product and the other side is 0, we can write an equivalent equation by setting each factor equal to zero. This idea is called the **zero factor property**.

The Zero Factor Property

If A and B are algebraic expressions, then the equation $AB = 0$ is equivalent to the compound statement $A = 0$ or $B = 0$.

EXAMPLE 2 Quadratic equations solved by factoring

Solve each equation by factoring.

a) $x^2 - x - 12 = 0$ **b)** $(x + 3)(x - 4) = 8$

Solution

a)

$$x^2 - x - 12 = 0$$

$$(x - 4)(x + 3) = 0 \qquad \text{Factor the left-hand side.}$$

$$x - 4 = 0 \quad \text{or} \quad x + 3 = 0 \qquad \text{Zero factor property}$$

$$x = 4 \quad \text{or} \quad x = -3$$

Check: $(-3)^2 - (-3) - 12 = 0$ and $4^2 - 4 - 12 = 0$. The solution set is $\{-3, 4\}$.

b) We must first rewrite the equation with 0 on one side, because the zero factor property applies only when the factors have a product of 0.

$$(x + 3)(x - 4) = 8$$

$$x^2 - x - 12 = 8 \qquad \text{Multiply on the left-hand side.}$$

$$x^2 - x - 20 = 0 \qquad \text{Get 0 on the right-hand side.}$$

$$(x - 5)(x + 4) = 0 \qquad \text{Factor.}$$

$$x - 5 = 0 \quad \text{or} \quad x + 4 = 0 \qquad \text{Zero factor property}$$

$$x = 5 \quad \text{or} \quad x = -4$$

Check in the original equation. The solution set is $\{-4, 5\}$.

Completing the Square

We cannot solve every quadratic equation by factoring because factoring methods are limited to integral coefficients. However, every quadratic equation can be written in the form of Example 1(c) and solved by using the square root property. This process is called **completing the square**.

Completing the square involves finding the third term of a perfect square trinomial when given the first two terms. For example, we can recognize $x^2 + 6x$ as the first two terms of the perfect square trinomial $x^2 + 6x + 9 = (x + 3)^2$. Note that one-half of 6 is 3 and 3^2 is 9. We can recognize $x^2 + 10x$ as the first two terms of the perfect square trinomial $x^2 + 10x + 25 = (x + 5)^2$. Note that one-half of 10 is 5, and 5^2 is 25. These examples suggest the following rule.

Rule for Finding the Last Term of $x^2 + bx + ?$

The last term of a perfect square trinomial (with $a = 1$) is the square of one-half of the coefficient of the middle term. In symbols, the perfect square trinomial whose first two terms are $x^2 + bx$ is

$$x^2 + bx + \left(\frac{b}{2}\right)^2.$$

A perfect square trinomial may have a leading coefficient that is not equal to 1. Since it is simpler to complete the square when $a = 1$, we rewrite any such polynomial so that $a = 1$ before completing the square.

EXAMPLE 3 **Solving a quadratic equation by completing the square**

Solve each equation by completing the square.

a) $x^2 + 6x + 7 = 0$ **b)** $2x^2 - 3x - 4 = 0$

Solution

a) Since $x^2 + 6x + 7$ is not a perfect square trinomial, we must find a perfect square trinomial that has $x^2 + 6x$ as its first two terms. Since one-half of 6 is 3 and $3^2 = 9$, our goal is to get $x^2 + 6x + 9$ on the left-hand side:

$$x^2 + 6x + 7 = 0$$

$$x^2 + 6x = -7 \qquad \text{Subtract 7 from each side.}$$

$$x^2 + 6x + 9 = -7 + 9 \qquad \text{Add 9 to each side.}$$

$$(x + 3)^2 = 2 \qquad \text{Factor the left-hand side.}$$

$$x + 3 = \pm\sqrt{2} \qquad \text{Square root property}$$

$$x = -3 \pm \sqrt{2}$$

Check in the original equation. The solution set is $\{-3 - \sqrt{2}, -3 + \sqrt{2}\}$.

b) $2x^2 - 3x - 4 = 0$

$$x^2 - \frac{3}{2}x - 2 = 0 \qquad \text{Divide each side by 2 to get } a = 1.$$

$$x^2 - \frac{3}{2}x = 2 \qquad \text{Add 2 to each side.}$$

$$x^2 - \frac{3}{2}x + \frac{9}{16} = 2 + \frac{9}{16} \qquad \frac{1}{2} \cdot \frac{3}{2} = \frac{3}{4} \text{ and } \left(\frac{3}{4}\right)^2 = \frac{9}{16}.$$

$$\left(x - \frac{3}{4}\right)^2 = \frac{41}{16} \qquad \text{Factor the left-hand side.}$$

$$x - \frac{3}{4} = \pm\frac{\sqrt{41}}{4} \qquad \text{Square root property}$$

$$x = \frac{3}{4} \pm \frac{\sqrt{41}}{4} = \frac{3 \pm \sqrt{41}}{4}$$

The solution set is $\left\{\frac{3 - \sqrt{41}}{4}, \frac{3 + \sqrt{41}}{4}\right\}$.

Check as shown in Figs. 1.16(a) and (b).

(a)

(b)

Figure 1.16

STRATEGY **Completing the Square**

In completing the square we want an *equivalent equation* that has a perfect square trinomial on one side. To get an equivalent equation, the number that completes the square must be added to *both sides* of the equation.

The Quadratic Formula

The method of completing the square can be applied to any quadratic equation

$$ax^2 + bx + c = 0, \quad \text{where } a \neq 0.$$

Assume for now that $a > 0$ and divide each side by a.

$$x^2 + \frac{b}{a}x + \frac{c}{a} = 0$$

$$x^2 + \frac{b}{a}x = -\frac{c}{a} \qquad \text{Subtract } \frac{c}{a} \text{ from each side.}$$

$$x^2 + \frac{b}{a}x + \frac{b^2}{4a^2} = -\frac{c}{a} + \frac{b^2}{4a^2} \qquad \frac{1}{2} \cdot \frac{b}{a} = \frac{b}{2a} \text{ and } \left(\frac{b}{2a}\right)^2 = \frac{b^2}{4a^2}.$$

Now factor the perfect square trinomial on the left-hand side. On the right-hand side get a common denominator and add.

$$\left(x + \frac{b}{2a}\right)^2 = \frac{b^2 - 4ac}{4a^2} \qquad \frac{c}{a} \cdot \frac{4a}{4a} = \frac{4ac}{4a^2}$$

$$x + \frac{b}{2a} = \pm\sqrt{\frac{b^2 - 4ac}{4a^2}} \qquad \text{Square root property}$$

$$x = -\frac{b}{2a} \pm \frac{\sqrt{b^2 - 4ac}}{2a} \qquad \text{Because } a > 0, \sqrt{4a^2} = 2a.$$

$$x = \frac{-b \pm \sqrt{b^2 - 4ac}}{2a}$$

We assumed that $a > 0$ so that $\sqrt{4a^2} = 2a$ would be correct. If a is negative, then $\sqrt{4a^2} = -2a$, and we get

$$x = -\frac{b}{2a} \pm \frac{\sqrt{b^2 - 4ac}}{-2a}.$$

However, the negative sign in $-2a$ can be deleted because of the $\pm$ symbol preceding it. For example, $5 \pm (-3)$ gives the same values as 5 ± 3. After deleting the negative sign on $-2a$, we get the same formula for the solution. It is called the **quadratic formula**. Its importance lies in its wide applicability. Any quadratic equation can be solved by using this formula.

The Quadratic Formula

The solution to $ax^2 + bx + c = 0$, with $a \neq 0$, is given by the formula

$$x = \frac{-b \pm \sqrt{b^2 - 4ac}}{2a}.$$

EXAMPLE 4 Using the quadratic formula

Solve each equation using the quadratic formula.

a) $x^2 + 8x + 6 = 0$ **b)** $5x^2 - 4x + 1 = 0$

Figure 1.17

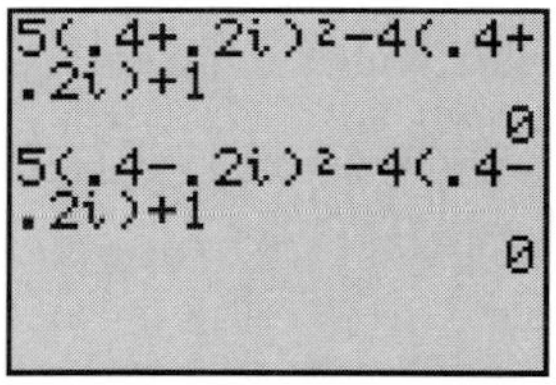

Figure 1.18

Solution

a) For $x^2 + 8x + 6 = 0$ we use $a = 1$, $b = 8$, and $c = 6$ in the formula:

$$x = \frac{-8 \pm \sqrt{8^2 - 4(1)(6)}}{2(1)} = \frac{-8 \pm \sqrt{40}}{2} = \frac{-8 \pm 2\sqrt{10}}{2} = -4 \pm \sqrt{10}$$

The solution set is $\{-4 - \sqrt{10}, -4 + \sqrt{10}\}$.

Check by using a calculator as in Fig. 1.17. □

b) For $5x^2 - 4x + 1 = 0$ we use $a = 5$, $b = -4$, and $c = 1$ in the formula:

$$x = \frac{-(-4) \pm \sqrt{(-4)^2 - 4(5)(1)}}{2(5)} = \frac{4 \pm \sqrt{-4}}{10} = \frac{4 \pm 2i}{10} = \frac{4}{10} \pm \frac{2}{10}i$$

$$= \frac{2}{5} \pm \frac{1}{5}i$$

You should check that these solutions are correct using operations with complex numbers. The solution set is $\{\frac{2}{5} - \frac{1}{5}i, \frac{2}{5} + \frac{1}{5}i\}$. Note that the solutions to this quadratic equation are complex conjugates.

You can check with a calculator as in Fig. 1.18.

To decide which of the four methods to use for solving a given quadratic equation, use the following strategy.

STRATEGY **Solving $ax^2 + bx + c = 0$**

1. If $b = 0$, solve $ax^2 + c = 0$ for x^2 and apply the square root property.
2. If $ax^2 + bx + c$ can be easily factored, then solve by factoring.
3. The quadratic formula or completing the square may be used on any quadratic equation, but the quadratic formula is usually easier to use.

The Discriminant

Some quadratic equations have two real solutions, some have two imaginary solutions, and some have only one real solution. If we examine the quadratic formula, we can see what determines the number and type of solutions. If $b^2 - 4ac > 0$ in the formula, then $\sqrt{b^2 - 4ac}$ is a real number and we get two real solutions. If $b^2 - 4ac < 0$, then $\sqrt{b^2 - 4ac}$ is an imaginary number and we get two imaginary solutions. If $b^2 - 4ac = 0$ in the quadratic formula, then there will be only one real solution, $x = -b/(2a)$. Because of the $\pm$ sign in the formula, the imaginary solutions always occur in conjugate pairs. Since $b^2 - 4ac$ determines the number and type of solutions, $b^2 - 4ac$ is called the **discriminant**. Table 1.4 summarizes this information.

Table 1.4 Number and type of solutions to a quadratic equation

Value of $b^2 - 4ac$	Type of solutions
Positive	Two real
Zero	One real
Negative	Two imaginary

EXAMPLE 5 Using the discriminant

For each equation, state the value of the discriminant, the number of solutions, and whether the solutions are real or imaginary.

a) $x^2 + 8x + 6 = 0$ **b)** $5x^2 - 4x + 1 = 0$ **c)** $4x^2 + 12x + 9 = 0$

Solution

a) Find the value of $b^2 - 4ac$ using $a = 1$, $b = 8$, and $c = 6$:

$$b^2 - 4ac = 8^2 - 4(1)(6) = 40$$

The value of the discriminant is 40 and the equation has two real solutions.

b) Find the value of the discriminant for the equation $5x^2 - 4x + 1 = 0$:

$$b^2 - 4ac = (-4)^2 - 4(5)(1) = -4$$

Because the discriminant is negative, the equation has two imaginary solutions.

c) For $4x^2 + 12x + 9 = 0$, we have $b^2 - 4ac = 12^2 - 4(4)(9) = 0$. So the equation has one real solution.

Applications

The problems that we solve in this section are very similar to those in Section 1.2. However, in this section the mathematical model of the situation results in a quadratic equation.

EXAMPLE 6 A problem solved with a quadratic equation

It took Susan 30 minutes longer to drive 275 miles on I-70 west of Green River, Utah, than it took her to drive 300 miles east of Green River. Because of a sand storm and a full load of cantaloupes, she averaged 10 mph less while traveling west of Green River. What was her average speed for each part of the trip?

Solution

Let x represent Susan's average speed east of Green River and $x - 10$ represent her average speed west of Green River. We can organize all of the given information as in Table 1.5. Since $D = RT$, the time is determined by $T = D/R$.

Table 1.5

	Distance	Rate	Time
East	300	x	$\frac{300}{x}$
West	275	$x - 10$	$\frac{275}{x - 10}$

The following equation expresses the fact that her time west of Green River was $\frac{1}{2}$ hour greater than her time east of Green River.

$$\frac{275}{x - 10} = \frac{300}{x} + \frac{1}{2}$$

$$2x(x - 10) \cdot \frac{275}{x - 10} = 2x(x - 10)\left(\frac{300}{x} + \frac{1}{2}\right)$$

$$550x = 600(x - 10) + x(x - 10)$$

$$-x^2 - 40x + 6000 = 0$$

$$x^2 + 40x - 6000 = 0$$

$$(x + 100)(x - 60) = 0 \quad \text{Factor.}$$

$$x = -100 \quad \text{or} \quad x = 60$$

The solution $x = -100$ is a solution to the equation, but not a solution to the problem. The other solution, $x = 60$, means that $x - 10 = 50$. Check that these two average speeds are a solution to the problem. Susan's average speed east of Green River was 60 mph, and her average speed west of Green River was 50 mph.

The next example involves the formula for the height of a rising or falling object, $S = -16t^2 + v_0 t + s_0$, where S is the height in feet, t is the time in seconds, v_0 is the initial velocity, and s_0 is the initial height. In this example, the solutions are irrational numbers. In this case it is usually best to give both exact and approximate answers to the problem.

EXAMPLE 7 Using the quadratic formula in a problem

Michael Chang, professional tennis player, skillfully uses the lob as both a defensive and an offensive shot. Chang knows that a well-placed lob can buy the time needed to prepare for the next shot. How long does it take for a tennis ball to hit the earth if it is hit straight upward with a velocity of 60 feet per second from a height of 5 feet?

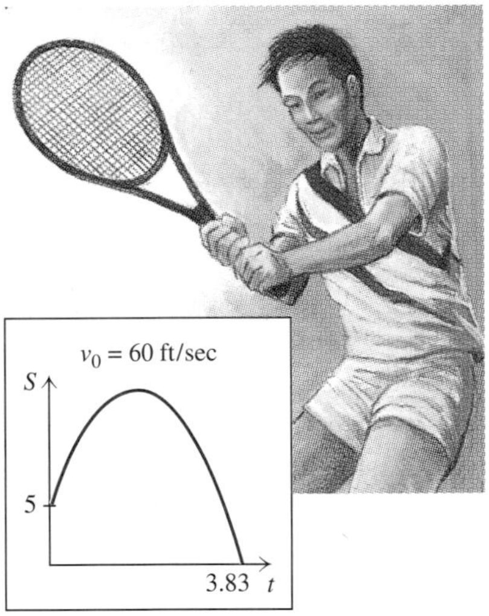

Solution

We are looking for the value of t for which the height S is 0. Use initial velocity $v_0 = 60$ feet per second, initial height $s_0 = 5$ feet, and $S = 0$ in the formula $S = -16t^2 + v_0 t + s_0$:

$$0 = -16t^2 + 60t + 5$$

Use the quadratic formula to solve the equation:

$$t = \frac{-60 \pm \sqrt{60^2 - 4(-16)(5)}}{2(-16)} = \frac{-60 \pm \sqrt{3920}}{-32} = \frac{-60 \pm 28\sqrt{5}}{-32}$$

$$= \frac{15 \pm 7\sqrt{5}}{8}$$

Now $(15 - 7\sqrt{5})/8 \approx -0.082$ and $(15 + 7\sqrt{5})/8 \approx 3.83$. Since the time at which the ball returns to the earth must be positive, it will take exactly $(15 + 7\sqrt{5})/8$ seconds or approximately 3.83 seconds for the ball to hit the earth.

Check as shown in Fig. 1.19. Note that due to round-off errors, the calculator did not get exactly zero.

Figure 1.19

Quadratic equations often arise from applications that involve the Pythagorean theorem from geometry: *A triangle is a right triangle if and only if the sum of the squares of the legs is equal to the square of the hypotenuse.*

EXAMPLE 8 Using the Pythagorean theorem

In the house shown in Fig. 1.20, the ridge of the roof at point B is 6 feet above point C. If the distance from A to C is 18 feet, then what is the length of a rafter from A to B?

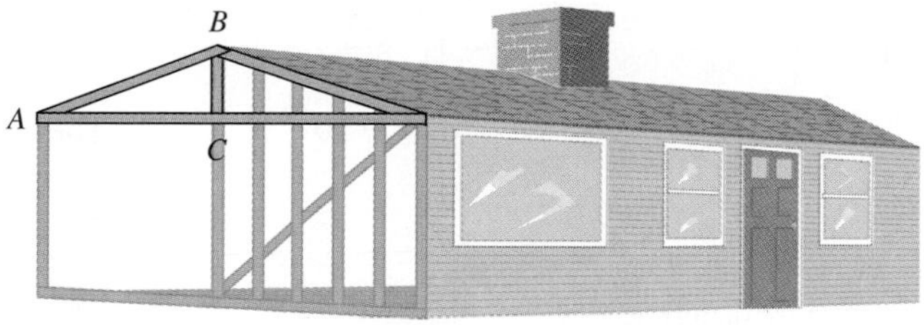

Figure 1.20

Solution

Let x be the distance from A to B. Use the Pythagorean theorem to write the following equation.

$$x^2 = 18^2 + 6^2$$
$$x^2 = 360$$
$$x = \pm\sqrt{360} = \pm 6\sqrt{10}$$

Since x must be positive in this problem, $x = 6\sqrt{10}$ feet or $x \approx 18.97$ feet.

FOR THOUGHT True or False? Explain.

1. The equation $(x - 3)^2 = 4$ is equivalent to $x - 3 = 2$.
2. Every quadratic equation can be solved by factoring.
3. The trinomial $x^2 + \frac{2}{3}x + \frac{4}{9}$ is a perfect square trinomial.
4. The equation $(x - 3)(2x + 5) = 0$ is equivalent to $x = 3$ or $x = \frac{5}{2}$.
5. All quadratic equations have two distinct solutions.
6. If m, n, and p are real numbers such that $mx^2 - nx + p = 0$ with $m \neq 0$, then $x = \frac{n \pm \sqrt{n^2 - 4mp}}{2m}$.
7. If $b = 0$, then $ax^2 + bx + c = 0$ cannot be solved by the quadratic formula.
8. All quadratic equations have at least one real solution.
9. The value of the discriminant is 0 for $4x^2 + 12x + 9 = 0$.
10. A quadratic equation with real coefficients can have one real and one imaginary solution.

1.4 EXERCISES

Tape 4 Disk

Solve each equation by using the square root property.

1. $x^2 - 5 = 0$ **2.** $x^2 - 8 = 0$ **3.** $3x^2 + 2 = 0$

4. $2x^2 + 16 = 0$ **5.** $(x - 3)^2 = 9$ **6.** $(x + 1)^2 = \frac{9}{4}$

7. $\left(x - \frac{1}{2}\right)^2 = \frac{25}{4}$ **8.** $(3x - 1)^2 = \frac{1}{4}$

9. $(x + 2)^2 = -4$ **10.** $(x - 3)^2 = -20$

11. $\left(x - \frac{2}{3}\right)^2 = -\frac{4}{9}$ **12.** $\left(x + \frac{3}{2}\right)^2 = -\frac{1}{2}$

Solve each equation by factoring.

13. $x^2 - x - 20 = 0$ **14.** $x^2 + 2x - 8 = 0$

15. $a^2 + 3a = -2$ **16.** $b^2 - 4b = 12$

17. $2x^2 - 5x - 3 = 0$ **18.** $2x^2 - 5x + 2 = 0$

19. $6x^2 - 7x + 2 = 0$ **20.** $12x^2 - 17x + 6 = 0$

21. $(y - 3)(y + 4) = 30$ **22.** $(w - 1)(w - 2) = 6$

23. $(2z - 1)(z + 3) = 15$ **24.** $(2t - 3)(2t + 1) = 5$

Find the perfect square trinomial whose first two terms are given.

25. $x^2 - 12x$ **26.** $y^2 + 20y$ **27.** $r^2 + 3r$

28. $t^2 - 7t$ **29.** $w^2 + \frac{1}{2}w$ **30.** $p^2 - \frac{2}{3}p$

Solve each equation by completing the square.

31. $x^2 + 6x + 1 = 0$ **32.** $x^2 - 10x + 5 = 0$

32. $n^2 - 2n - 1 = 0$ **34.** $m^2 - 12m + 33 = 0$

35. $h^2 + 3h - 1 = 0$ **36.** $t^2 - 5t + 2 = 0$

37. $x^2 + 2x + 5 = 0$ **38.** $x^2 - 4x + 5 = 0$

39. $2x^2 + 5x = 12$ **40.** $3x^2 + x = 2$

41. $3x^2 + 2x + 1 = 0$ **42.** $5x^2 + 4x + 3 = 0$

Solve each equation using the quadratic formula.

43. $x^2 + 3x - 4 = 0$ **44.** $x^2 + 8x + 12 = 0$

45. $2x^2 - 5x - 3 = 0$ **46.** $2x^2 + 3x - 2 = 0$

47. $9x^2 + 6x + 1 = 0$ **48.** $16x^2 - 24x + 9 = 0$

49. $2x^2 - 3 = 0$ **50.** $-2x^2 + 5 = 0$

51. $x^2 + 5 = 4x$ **52.** $x^2 = 6x - 13$

53. $9x^2 + 6x = 1$ **54.** $2x^2 + 3 = 6x$

Use a calculator and the quadratic formula to find all real and imaginary solutions to each equation. Round answers to two decimal places.

55. $3.2x^2 + 7.6x - 9 = 0$ **56.** $1.5x^2 - 6.3x = 10.1$

57. $3.25x^2 - 4.6x + 20 = 0$

58. $4.76x^2 + 6.12x + 55.3 = 0$

For each equation, state the value of the discriminant, the number of solutions, and whether the solutions are real or imaginary.

59. $9x^2 - 30x + 25 = 0$ **60.** $3x^2 - 7x + 3 = 0$

61. $5x^2 - 6x + 2 = 0$ **62.** $3x^2 + 5x + 5 = 0$

63. $7x^2 + 12x - 1 = 0$ **64.** $4x^2 + 28x + 49 = 0$

Solve each equation. Use the method of your choice.

65. $x^2 = \frac{4}{3}x - \frac{5}{9}$ **66.** $x^2 = \frac{2}{7}x - \frac{2}{49}$

67. $x^2 + \sqrt{2} = 0$ **68.** $\sqrt{2}x^2 - 1 = 0$

69. $12x^2 + x\sqrt{6} - 1 = 0$ **70.** $-10x^2 - x\sqrt{5} + 1 = 0$

71. $x(x + 6) = 72$ **72.** $x = \frac{96}{x + 4}$

73. $x = 1 + \frac{1}{x}$ **74.** $x = \frac{1}{x}$

75. $\frac{x - 12}{3 - x} = \frac{x + 4}{x + 7}$ **76.** $\frac{x - 9}{x - 2} = -\frac{x + 3}{x + 1}$

Use the methods for solving quadratic equations to solve each formula for the indicated variable.

77. $A = \pi r^2$ for r **78.** $S = 2\pi rh + 2\pi r^2$ for r

79. $x^2 + 2kx + 3 = 0$ for x **80.** $hy^2 - ky = p$ for y

81. $2y^2 + 4xy = x^2$ for y **82.** $\dfrac{\frac{1}{x + h} - \frac{1}{x}}{h} = 1$ for x

Find an exact solution to each problem. If the solution is irrational, then find an approximate solution also.

83. *Demand Equation* The demand equation for a certain product is $P = 40 - 0.001x$, where x is the number of units sold per week and P is the price in dollars at which each one is sold. The weekly revenue R is given by $R = xP$. What number of units sold produces a weekly revenue of \$175,000?

84. *Average Cost* The total cost in dollars of producing x items is given by $C = 0.02x^3 + 5x$. For what number of items is the average cost per item equal to \$5.50?

85. *Height of a Ball* A juggler can toss a ball into the air with a velocity of 40 ft/sec from a height of 4 ft. How long will it take for the ball to return to the height of 4 ft?

86. *Height of a Sky Diver* If a sky diver steps out of an airplane at 5000 ft, then how long does it take her to reach 4000 ft? What assumptions are you making to solve this problem?

87. *Diagonal of a Football Field* A football field is 100 yd long from goal line to goal line and 160 ft wide. If a player ran diagonally across the field from one goal line to the other, then how far did he run?

88. *Dimensions of a Flag* If the perimeter of a rectangular flag is 34 in. and the diagonal is 13 in., then what are the length and width?

89. *Long Shot* To avoid hitting the ball out, a tennis player in one corner of the 312 yd^2 court hits the ball to the farthest corner of the opponent's court as shown in the diagram. If the length L of the tennis court is 2 yd longer than twice the width W, then how far did the player hit the ball?

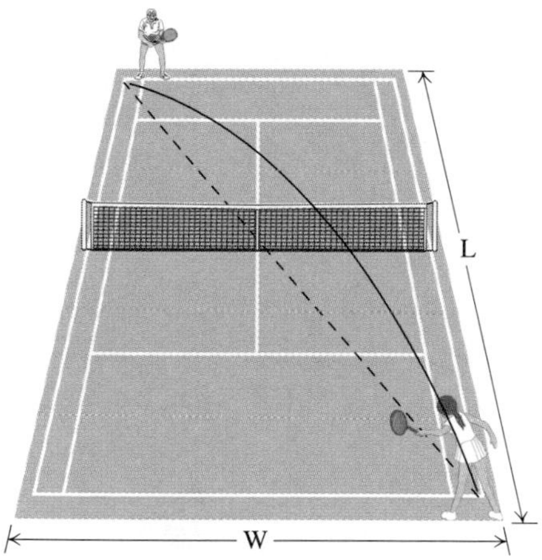

Figure for Exercise 89

90. *Open-Top Box* Imogene wants to make an open-top box for packing baked goods by cutting equal squares from each corner of an 11 in. by 14 in. piece of cardboard as shown in the diagram. She figures that for versatility the area of the bottom must be 80 in.^{2}. What size square should she cut from each corner?

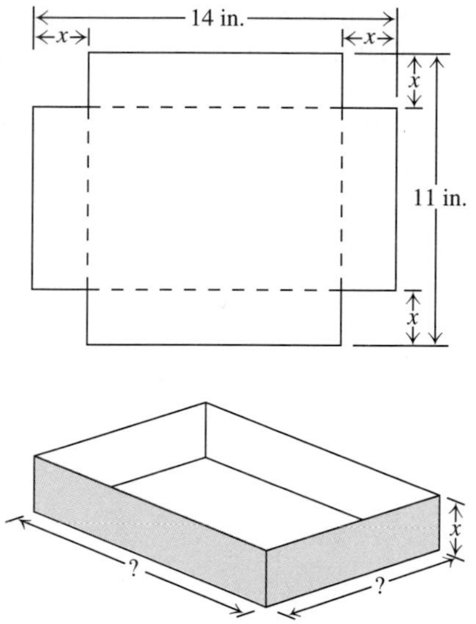

Figure for Exercise 90

91. *Finding the Displacement* The sail area-displacement ratio S measures the amount of power available to drive a sailboat in moderate to heavy winds (*Sail*, September 1997). For the Sabre 402 shown in the figure, the sail area A is 822 ft^2 and $S = 18.8$. The displacement d (in pounds) satisfies $2^{-12}d^2S^3 - A^3 = 0$. Find d.

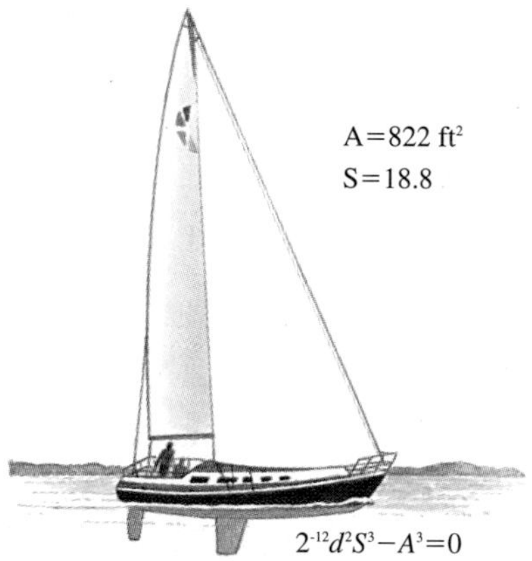

Figure for Exercise 91

92. *Charleston Earthquake* The Charleston, South Carolina earthquake of 1886 was a 7.6 on the Richter scale and it was felt over an area of 1.5 million square miles (Earthquakes and Volcanoes, U.S. Geological Survey). If the area in which it was felt was circular and centered at Charleston, then how far away was it felt?

93. *Bordering a Flower Bed* Juan, a landscaper in Albuquerque, is designing a spring display of tulips and daffodils. The city has requested a 6-ft by 8-ft area of tulips surrounded by a uniform border of daffodils. The total area of tulips and daffodils is to be 100 ft^2. How wide should the border be?

94. *Bracing a Gate* The width of a rectangular gate is 2 ft more than its height. If a diagonal board of length 8 ft is used for bracing, then what are the dimensions of the gate?

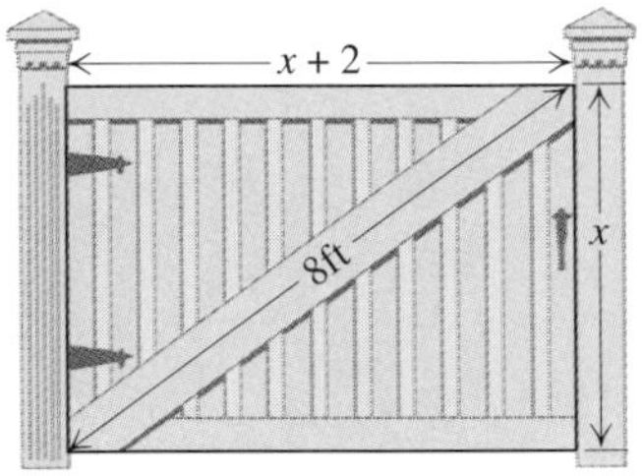

Figure for Exercise 94

95. *Speed of a Tortoise* When a tortoise crosses a highway, his speed is 2 ft/hr faster than normal. If he can cross a 24-ft lane in 24 min less time than he can travel that same distance off the highway, then what is his normal speed?

96. *Speed of an Electric Car* An experimental electric-solar car completed a 1000-mi race in 35 hr. For the 600 mi traveled during daylight the car averaged 20 mph more than it did for the 400 mi traveled at night. What was the average speed of the car during the daytime?

97. *Computer Design* Using a computer design package, Tina can write and design a direct-mail package in two days less time than it takes to create the same package using traditional design methods. If Tina uses the computer and her assistant Curt uses traditional methods, and together they complete the job in 3.5 days, then how long would it have taken Curt to do the job alone using traditional methods?

98. *Making a Dress* Rafael designed a sequined dress to be worn at the Academy Awards. His top seamstress, Maria, could sew on all the sequins in 10 hr less time than his next-best seamstress, Stephanie. To save time, he gave the job to both women and got all of the sequins attached in 17 hr. How long would it have taken Stephanie working alone?

99. *Percentage of White Meat* The Kansas Fried Chicken store sells a Party Size bucket that weighs 10 lb more than the Big Family Size bucket. The Party Size bucket contains 8 lb of white meat, while the Big Family Size bucket contains 3 lb of white meat. If the percentage of white meat in the Party Size is 10 percentage points greater than the percentage of white meat in the Big Family Size, then how much does the Party Size bucket weigh?

100. *Mixing Antifreeze in a Radiator* Steve's car had a large radiator that contained an unknown amount of pure water. He added two quarts of antifreeze to the radiator. After testing, he decided that the percentage of antifreeze in the radiator was not large enough. Not knowing how to solve mixture problems, Steve decided to add one quart of water and another quart of antifreeze to the radiator to see what he would get. After testing he found that the last addition increased the percentage of antifreeze by three percentage points. How much water did the radiator contain originally?

101. *Initial Velocity of a Basketball Player* Michael Jordan is famous for his high leaps and "hang time," especially while slam-dunking. If Jordan leaps for a dunk and reaches a peak height at the basket of 1.07 m, what is his upward velocity in meters per second at the moment his feet leave the floor? The formula $v_1^2 = v_0^2 + 2gS$ gives the relationship between final velocity v_1, initial velocity v_0, acceleration of gravity g, and his height S. Use $g = -9.8$ m/sec^2 and the fact that his final velocity is zero at his peak height. Use $S = \frac{1}{2}gt^2 + v_0t$ to find the amount of time he is in the air.

Figure for Exercise 101

Table for Exercise 102

Altitude (ft)	Atmospheric pressure (atm)	Comments
28,000	0.33	Mount Everest, Nepal
20,000	0.46	Mount Denali, Alaska
18,000		Climbers deteriorate
	0.52	Highest human settlements
10,000	0.69	Mountain sickness
0	1	Sea level

102. *Hazards of Altitude* The accompanying table shows height above sea level and atmospheric pressure (Encyclopedia of Sports Science, 1997). The atmospheric pressure a is related to altitude h by the equation

$$a = 3.89 \times 10^{-10}h^2 - 3.48 \times 10^{-5}h + 1.$$

Find the atmospheric pressure at the point where climbers deteriorate. Find the altitude of the highest human settlements.

For Writing/Discussion

103. Write down a quadratic equation that has nonintegral coefficients and two real solutions. Use the quadratic formula to solve it and a calculator to check your solutions.

104. *Cooperative Learning* Work in a small group to make up a problem involving uniform motion whose solution involves a quadratic equation. Give your problem to another group to solve.

Linking Concepts

For Individual or Group Explorations

Baseball Statistics *Baseball fans keep up with their favorite teams through charts where the teams are ranked according to the percentage of games won. The chart usually has a column indicating the games behind (GB) for each team. If the win-loss record of the number one team is (A, B), then the games behind of another team whose win-loss record is (a, b) is calculated by the formula*

$$GB = \frac{(A - a) + (b - B)}{2}.$$

a) Find GB for Atlanta and Montreal in the following table.

Team	Won	Lost	Pct.	GB
New York	38	24	.613	—
Atlanta	35	29	.547	?
Montreal	34	31	.523	?

b) In the following table Pittsburgh has a higher percentage of wins than Chicago, and so Pittsburgh is in first place. Find *GB* for Chicago.

Team	Won	Lost	Pct.	GB
Pittsburgh	18	13	.581	—
Chicago	22	16	.579	?

c) Is Chicago actually behind Pittsburgh in terms of the statistic *GB*?

d) Another measure of how far a team is from first place, called the deficit D, is the number of games that the two teams would have to play against each other to get equal percentages of wins, with the higher-ranked team losing all of the games. Find the deficits for Atlanta, Montreal, and Chicago.

e) Is it possible for *GB* or D to be negative? Does it make any sense if they are?

f) Compare the values of D and *GB* for each of the three teams. Which is a better measure of how far a team is from first place?

1.5 Linear and Absolute Inequalities

An equation states that two algebraic expressions are equal, while an **inequality** or **simple inequality** is a statement that two algebraic expressions are not equal in a particular way. Inequalities are stated using less than ($<$), less than or equal to ($\leq$), greater than ($>$), or greater than or equal to ($\geq$). In this section we study some basic inequalities. We will see more inequalities in Chapter 3.

Interval Notation

The solution set to an inequality is the set of all real numbers for which the inequality is true. The solution set to the inequality $x > 3$ is written $\{x | x > 3\}$ and consists of all real numbers to the right of 3 on the number line. This set is also called the **interval** of numbers greater than 3, and it is written in **interval notation** as $(3, \infty)$. The graph of the interval $(3, \infty)$ is shown in Fig. 1.21. A parenthesis is used next to the 3 to indicate that 3 is not in the interval or the solution set. The infinity symbol (∞) is not used as a number, but only to indicate that there is no bound on the numbers greater than 3.

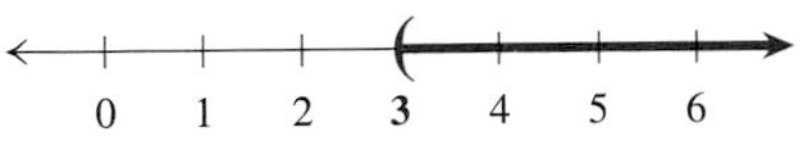

Figure 1.21

The solution set to $x \leq 4$ is written in set notation as $\{x | x \leq 4\}$ and in interval notation as $(-\infty, 4]$. The symbol $-\infty$ means that all numbers to the left of 4 on the number line are in the set and the bracket means that 4 is in the set. The graph of the interval $(-\infty, 4]$ is shown in Fig. 1.22. Intervals that use the infinity symbol are **unbounded** intervals. The following summary lists the different types of unbounded intervals used in interval notation and the graphs of those intervals on a number line. If only parentheses are used, the interval is

Figure 1.22

called **open**. If a bracket is used only on one end, the interval is called **half open** or **half closed**.

SUMMARY

Interval Notation for Unbounded Intervals

Set	Interval notation	Type	Graph
$\{x \mid x > a\}$	(a, ∞)	Open	
$\{x \mid x < a\}$	$(-\infty, a)$	Open	
$\{x \mid x \geq a\}$	$[a, \infty)$	Half open or half closed	
$\{x \mid x \leq a\}$	$(-\infty, a]$	Half open or half closed	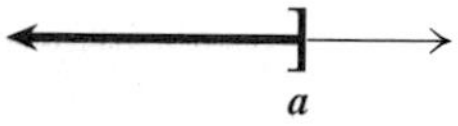
Real numbers	$(-\infty, \infty)$	Open	

We use a parenthesis when an endpoint of an interval is not included in the solution set and a bracket when an endpoint is included. A bracket is never used next to ∞ because infinity is not a number. On the graphs above, the number lines are shaded, showing that the solutions include all real numbers in the given interval.

EXAMPLE 1 Interval notation

Write an inequality whose solution set is the given interval.

a) $(-\infty, -9)$ **b)** $[0, \infty)$

Solution

a) The interval $(-\infty, -9)$ represents all real numbers less than -9. It is the solution set to $x < -9$.

b) The interval $[0, \infty)$ represents all real numbers greater than or equal to 0. It is the solution set to $x \geq 0$.

Linear Inequalities

Replacing the equal sign in the general linear equation $ax + b = 0$ by any of the symbols $<$, $\leq$, $>$, or $\geq$, gives a **linear inequality**. Two inequalities are **equivalent** if they have the same solution set. To solve linear inequalities, we use techniques similar to those used in solving linear equations. It can be shown that the following properties of inequality are valid. They allow us to write equivalent inequalities. These properties are stated for $<$, but they also hold for $>$, $\leq$, and $\geq$.

Properties of Inequality

If A and B are algebraic expressions and C is a real number, then the inequality $A < B$ is equivalent to

1. $A \pm C < B \pm C$,
2. $CA < CB$ (for C positive), $CA > CB$ (for C negative),
3. $\frac{A}{C} < \frac{B}{C}$ (for C positive), $\frac{A}{C} > \frac{B}{C}$ (for C negative).

Solving linear inequalities is very similar to solving linear equations, but pay particular attention to the second and third properties. *When an inequality is multiplied or divided by a negative number, the direction of the inequality symbol is reversed.* The inequality is reversed because of the rules for multiplying or dividing signed numbers. For example, multiplying the inequality $-4 < 7$ on each side by -2 yields $(-2)(-4) > (-2)(7)$, or $8 > -14$.

EXAMPLE 2 **Solving a linear inequality**

Solve $-3x - 5 < 4$. Write the solution set in interval notation and graph it.

Solution

Isolate the variable as is done in solving equations.

$$-3x - 5 < 4$$

$$-3x - 5 + 5 < 4 + 5 \quad \text{Add 5 to each side.}$$

$$-3x < 9$$

$$x > -3 \quad \text{Divide each side by } -3\text{, reversing the inequality.}$$

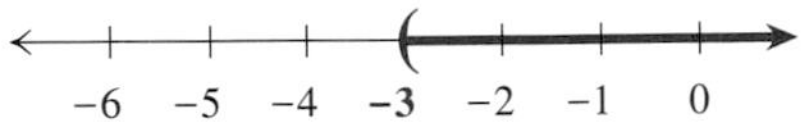

Figure 1.23

The solution set is the interval $(-3, \infty)$ and its graph is shown in Fig. 1.23. Checking the solution to an inequality is generally not as simple as checking an equation, because usually there are infinitely many solutions. We can do a "partial check" by checking one number in $(-3, \infty)$ and one number not in $(-3, \infty)$. For example, $0 > -3$ and $-3(0) - 5 < 4$ is correct, while $-6 < -3$ and $-3(-6) - 5 < 4$ is incorrect.

We can also perform operations on each side of an inequality using a variable expression. Addition or subtraction with variable expressions will give equivalent inequalities. However, we must always watch for undefined expressions. *Multiplication and division with a variable expression is usually avoided because we do not know whether the expression is positive or negative.*

EXAMPLE 3 Solving a linear inequality

Solve $\frac{1}{2}x - 3 \geq \frac{1}{4}x + 2$ and graph the solution set.

Solution

Multiply each side by the LCD to eliminate the fractions.

$$\frac{1}{2}x - 3 \geq \frac{1}{4}x + 2$$

$$4\left(\frac{1}{2}x - 3\right) \geq 4\left(\frac{1}{4}x + 2\right) \quad \text{Multiply each side by 4.}$$

$$2x - 12 \geq x + 8$$

$$x - 12 \geq 8$$

$$x \geq 20$$

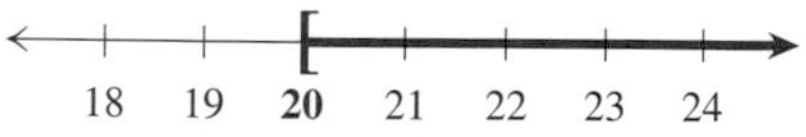

Figure 1.24

The solution set is the interval $[20, \infty)$. See Fig. 1.24 for the graph.

Compound Inequalities

A **compound inequality** is a sentence containing two simple inequalities connected with "and" or "or." The solution to a compound inequality can be an interval of real numbers that does not involve infinity, a **bounded** interval of real numbers. For example, the solution set to the compound inequality $x \geq 2$ and $x \leq 5$ is the set of real numbers between 2 and 5, inclusive. This inequality is also written as $2 \leq x \leq 5$. Its solution set is $\{x \mid 2 \leq x \leq 5\}$, which is written in interval notation as $[2, 5]$. Because $[2, 5]$ contains both of its endpoints, the interval is **closed**. The following summary lists the different types of bounded intervals used in interval notation and the graphs of those intervals on a number line.

SUMMARY Interval Notation for Bounded Intervals

Set	Interval notation	Type	Graph
$\{x \mid a < x < b\}$	(a, b)	Open	
$\{x \mid a \leq x \leq b\}$	$[a, b]$	Closed	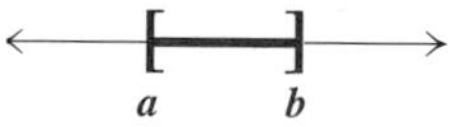
$\{x \mid a \leq x < b\}$	$[a, b)$	Half open or half closed	
$\{x \mid a < x \leq b\}$	$(a, b]$	Half open or half closed	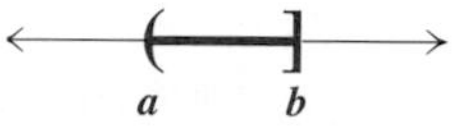

The notation $a < x < b$ is used only when x is actually between a and b, and a is less than b. We do not write $x > 5$ and $x < -3$ as $5 < x < -3$, nor do we write $a < x > b$ for $x > a$ and $x > b$.

The **intersection** of sets A and B is the set $A \cap B$, where $x \in A \cap B$ if and only if $x \in A$ and $x \in B$. The **union** of sets A and B is the set $A \cup B$, where $x \in A \cup B$ if and only if $x \in A$ or $x \in B$. In solving compound inequalities it is often necessary to find intersections and unions of intervals.

EXAMPLE 4 Intersections and unions of intervals

Let $A = (1, 5)$, $B = [3, 7)$, and $C = (6, \infty)$. Write $A \cup B$, $A \cap B$, $A \cap C$, and $A \cup C$ in interval notation.

Solution

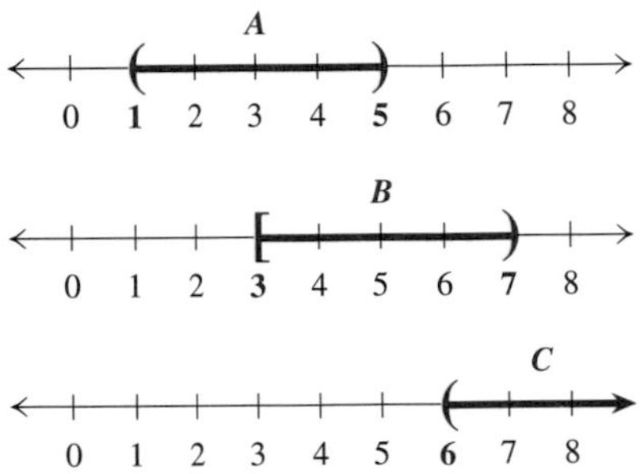

Figure 1.25

Figure 1.25 shows the intervals A, B, and C on the number line. Since $A \cup B$ consists of points that are either in A or in B, $A \cup B = (1, 7)$. Since $A \cap B$ consists of points that are in both A and B, $A \cap B = [3, 5)$. Since A and C have no points in common, $A \cap C = \emptyset$ and $A \cup C = (1, 5) \cup (6, \infty)$.

The solution set to a compound inequality using the connector "or" is the union of the two solution sets, and the solution set to a compound inequality using "and" is the intersection of the two solution sets.

EXAMPLE 5 Solving compound inequalities

Solve each compound inequality. Write the solution set using interval notation and graph it.

a) $2x - 3 > 5$ and $4 - x \leq 3$ **b)** $4 - 3x < -2$ or $3(x - 2) \leq -6$

c) $-4 \leq 3x - 1 < 5$

Solution

a)
$$\begin{aligned} 2x - 3 > 5 \quad &\text{and} \quad 4 - x \leq 3 \\ 2x > 8 \quad &\text{and} \quad -x \leq -1 \\ x > 4 \quad &\text{and} \quad x \geq 1 \end{aligned}$$

The intersection of the intervals $(4, \infty)$ and $[1, \infty)$ is $(4, \infty)$, because only numbers larger than 4 belong to both intervals. The solution set to the compound inequality is the open interval $(4, \infty)$. Its graph is shown in Fig. 1.26.

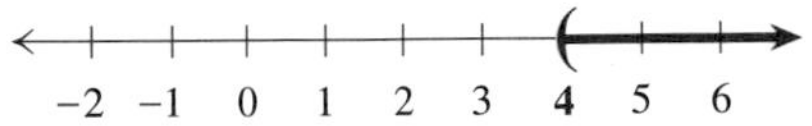

Figure 1.26

b)
$$\begin{aligned} 4 - 3x < -2 \quad &\text{or} \quad 3(x - 2) \leq -6 \\ -3x < -6 \quad &\text{or} \quad x - 2 \leq -2 \\ x > 2 \quad &\text{or} \quad x \leq 0 \end{aligned}$$

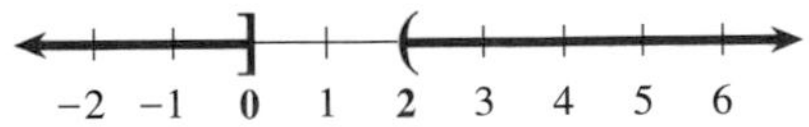

Figure 1.27

The solution set is $(-\infty, 0] \cup (2, \infty)$, and its graph is shown in Fig. 1.27.

c) We could write $-4 \leq 3x - 1 < 5$ as the compound inequality $-4 \leq 3x - 1$ and $3x - 1 < 5$, and then solve each simple inequality. Since each is solved using the same sequence of steps, we can solve the original inequality without separating it:

$$-4 \leq 3x - 1 < 5$$

$$-4 + 1 \leq 3x - 1 + 1 < 5 + 1 \quad \text{Add 1 to each part of the inequality.}$$

$$-3 \leq 3x < 6$$

$$\frac{-3}{3} \leq \frac{3x}{3} < \frac{6}{3} \quad \text{Divide each part by 3.}$$

$$-1 \leq x < 2$$

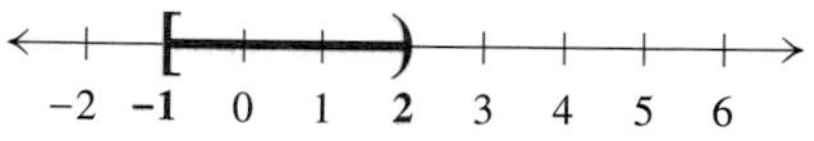

Figure 1.28

The solution set is the half-open interval $[-1, 2)$, graphed in Fig. 1.28.

It is possible that all real numbers satisfy a compound inequality or no real numbers satisfy a compound inequality.

EXAMPLE 6 Solving compound inequalities

Solve each compound inequality.

a) $3x - 9 \leq 9$ or $4 - x \leq 3$ **b)** $-\frac{2}{3}x < 4$ and $\frac{3}{4}x < -6$

Solution

a) Solve each simple inequality and find the union of their solution sets:

$$3x - 9 \leq 9 \quad \text{or} \quad 4 - x \leq 3$$

$$3x \leq 18 \quad \text{or} \quad -x \leq -1$$

$$x \leq 6 \quad \text{or} \quad x \geq 1$$

The union of $(-\infty, 6]$ and $[1, \infty)$ is the set of all real numbers, $(-\infty, \infty)$.

b) Solve each simple inequality and find the intersection of their solution sets:

$$-\frac{2}{3}x < 4 \quad \text{and} \quad \frac{3}{4}x < -6$$

$$\left(-\frac{3}{2}\right)\left(-\frac{2}{3}x\right) > \left(-\frac{3}{2}\right)4 \quad \text{and} \quad \left(\frac{4}{3}\right)\left(\frac{3}{4}x\right) < \left(\frac{4}{3}\right)(-6)$$

$$x > -6 \quad \text{and} \quad x < -8$$

Since $(-6, \infty) \cap (-\infty, -8) = \varnothing$, there is no solution to the compound inequality.

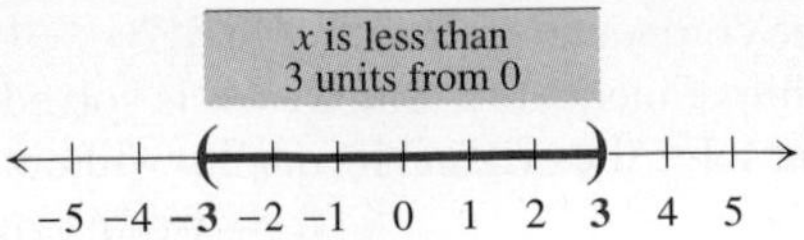

Figure 1.29

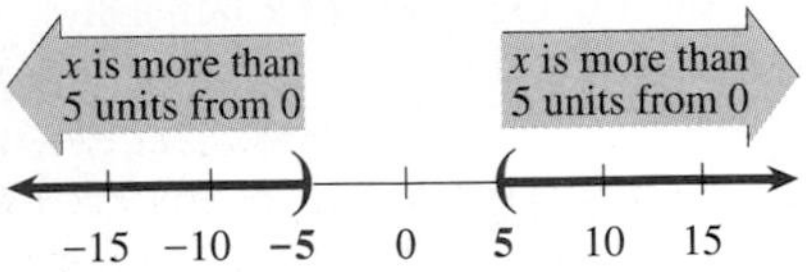

Figure 1.30

Absolute Value Inequalities

Recall that the absolute value of a number is the number's distance from 0 on the number line. The inequality $|x| < 3$ means that x is less than three units from 0. See Fig. 1.29. The real numbers that are less than three units from 0 are precisely the numbers that satisfy $-3 < x < 3$. So the solution set to $|x| < 3$ is the open interval $(-3, 3)$. The inequality $|x| > 5$ means that x is more than five units from 0 on the number line, which is equivalent to the compound inequality $x > 5$ or $x < -5$. See Fig. 1.30. So the solution to $|x| > 5$ is the union of two intervals, $(-\infty, -5) \cup (5, \infty)$. These ideas about absolute value inequalities are summarized as follows.

SUMMARY

Basic Absolute Value Inequalities (for $k > 0$)

Absolute value inequality	Equivalent statement	Solution set in interval notation	Graph of solution set
$\lvert x\rvert > k$	$x < -k$ or $x > k$	$(-\infty, -k) \cup (k, \infty)$	
$\lvert x\rvert \geq k$	$x \leq -k$ or $x \geq k$	$(-\infty, -k] \cup [k, \infty)$	
$\lvert x\rvert < k$	$-k < x < k$	$(-k, k)$	
$\lvert x\rvert \leq k$	$-k \leq x \leq k$	$[-k, k]$	

In the next example we use the rules for basic absolute value inequalities to solve more complicated absolute value inequalities.

EXAMPLE 7 Absolute value inequalities

Solve each absolute value inequality and graph the solution set.

a) $|3x + 2| < 7$ **b)** $-2|4 - x| \leq -4$ **c)** $|7x - 9| \geq -3$

Solution

a)
$$|3x + 2| < 7$$
$$-7 < 3x + 2 < 7 \quad \text{Write the equivalent compound inequality.}$$
$$-7 - 2 < 3x + 2 - 2 < 7 - 2 \quad \text{Subtract 2 from each part.}$$
$$-9 < 3x < 5$$
$$-3 < x < \frac{5}{3} \quad \text{Divide each part by 3.}$$

The solution set is the open interval $(-3, \frac{5}{3})$. The graph is shown in Fig. 1.31.

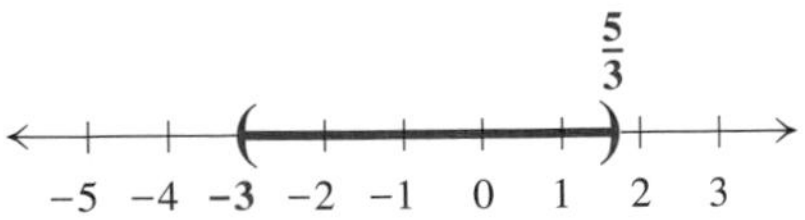

Figure 1.31

b)

$$-2|4 - x| \le -4$$

$$|4 - x| \ge 2 \qquad \text{Divide each side by } -2\text{, reversing the inequality.}$$

$$4 - x \le -2 \quad \text{or} \quad 4 - x \ge 2 \qquad \text{Write the equivalent compound inequality.}$$

$$-x \le -6 \quad \text{or} \quad -x \ge -2$$

$$x \ge 6 \quad \text{or} \quad x \le 2 \qquad \text{Multiply each side by } -1.$$

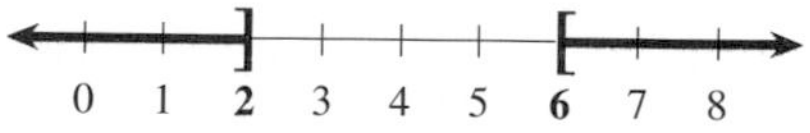

Figure 1.32

The solution set is $(-\infty, 2] \cup [6, \infty)$. Its graph is shown in Fig. 1.32. Check 8, 4, and 0 in the original inequality. If $x = 8$, we get $-2|4 - 8| \le -4$, which is true. If $x = 4$, we get $-2|4 - 4| \le -4$, which is false. If $x = 0$, we get $-2|4 - 0| \le -4$, which is true.

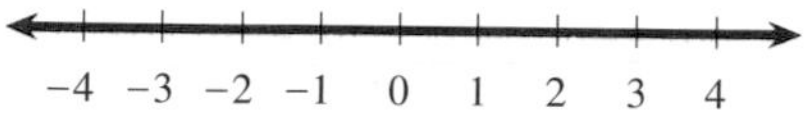

Figure 1.33

c) The expression $|7x - 9|$ has a nonnegative value for every real number x. So the inequality $|7x - 9| \ge -3$ is satisfied by every real number. The solution set is $(-\infty, \infty)$, and the graph is shown in Fig. 1.33.

Applications

Inequalities occur in applications just as equations do. In fact, in real life, equality in anything is usually the exception. The solution to a problem involving an inequality is generally an interval of real numbers. In this case we often ask for the range of values that solve the problem.

EXAMPLE 8 An application involving inequality

Remington scored 74 on his midterm exam in history. If he is to get a B, the average of his midterm and final exam must be between 80 and 89 inclusive. In what range must his final exam score lie for him to get a B in the course?

Solution

Let x represent Remington's final exam score. The average of his midterm and final must satisfy the following inequality.

$$80 \le \frac{74 + x}{2} \le 89$$

$$160 \le 74 + x \le 178$$

$$86 \le x \le 104$$

His final exam score must lie in the interval $[86, 104]$ if he is to get a B.

When discussing the error made in a measurement, we may refer to the *absolute error* or the *relative error*. For example, if L is the actual length of an object and x is the length determined by a measurement, then the absolute error is $|x - L|$ and the relative error is $|x - L|/L$.

EXAMPLE 9 **Application of absolute value inequality**

A technician is testing a scale with a 50 lb block of steel. The scale passes this test if the relative error when weighing this block is less than 0.1%. If x is the reading on the scale, then for what values of x does the scale pass this test?

Solution

If the relative error must be less than 0.1%, then x must satisfy the following inequality:

$$\frac{|x - 50|}{50} < 0.001$$

Solve the inequality for x:

$$|x - 50| < 0.05$$
$$-0.05 < x - 50 < 0.05$$
$$49.95 < x < 50.05$$

So the scale passes the test if it shows a weight in the interval (49.95, 50.05).

FOR THOUGHT True or False? Explain.

1. The inequality $-3 < x + 6$ is equivalent to $x + 6 > -3$.
2. The inequality $-2x < -6$ is equivalent to $\frac{-2x}{-2} < \frac{-6}{-2}$.
3. The smallest real number that satisfies $x > 12$ is 13.
4. The number -6 satisfies $|x - 6| > -1$.
5. $(-\infty, -3) \cap (-\infty, -2) = (-\infty, -2)$
6. $(5, \infty) \cap (-\infty, -3) = (-3, 5)$
7. All negative numbers satisfy $|x - 2| < 0$.
8. The compound inequality $x < -3$ or $x > 3$ is equivalent to $|x| < -3$.
9. The inequality $|x| + 2 < 5$ is equivalent to $-5 < x + 2 < 5$.
10. The fact that the difference between your age, y, and my age, m, is at most 5 years is written $|y - m| \leq 5$.

1.5 EXERCISES

Tape 4 Disk

For each given interval, write an inequality whose solution is the interval, and for each given inequality, write the solution set in interval notation.

1. $(-\infty, 12)$ **2.** $(-\infty, -3]$ **3.** $x \geq -8$

4. $x < 54$ **5.** $x < \pi/2$ **6.** $x \geq \sqrt{3}$

7. $[-7, \infty)$ **8.** $(1.2, \infty)$

Solve each inequality. Write the solution set using interval notation and graph it.

9. $3x - 6 > 9$ **10.** $2x + 1 < 6$

11. $7 - 5x \leq -3$ **12.** $-1 - 4x \geq 7$

13. $\frac{1}{2}x - 4 < \frac{1}{3}x + 5$ **14.** $\frac{1}{2} - x > \frac{x}{3} + \frac{1}{4}$

15. $\frac{7 - 3x}{2} \geq -3$ **16.** $\frac{5 - 3x}{-7} \leq 0$

17. $-2(3x - 2) \geq 4 - x$ **18.** $-5x \leq 3(x - 9)$

19. $\frac{2(76) + x}{3} \geq 80$ **20.** $x + 0.09x \leq 872$

Write as a single interval.

21. $(-3, \infty) \cup (5, \infty)$ **22.** $(-\infty, -2) \cap (-5, \infty)$

23. $(-\infty, -5) \cap (-2, \infty)$ **24.** $(-\infty, -3) \cup (-7, \infty)$

25. $(-\infty, 4) \cup [4, 5]$ **26.** $(4, 7) \cap (3, \infty)$

27. $(3, 5) \cup (-3, \infty)$ **28.** $(-\infty, 0) \cup (-\infty, 6)$

29. $[3, 5] \cup [5, 7]$ **30.** $(-3, \infty) \cap (2, \infty)$

31. $(3, 7) \cap [5, 7]$ **32.** $(-\infty, 0) \cap [-4, 3]$

Determine whether the given number satisfies the inequality following it.

33. -3, $4 - 2x \leq 9$ **34.** 4, $5 - 2x > -4x + 5$

35. -5, $7 - x > 6$ or $2x + 3 > 9$

36. -2, $3(x + 4) \leq 6$ and $x < 1$

37. 8, $2|x - 9| \leq 6$ **38.** 1, $-2 < \frac{5x - 9}{2} < 4$

Solve each compound inequality. Write the solution set using interval notation and graph it.

39. $5 > 8 - x$ and $1 + 0.5x < 4$

40. $\frac{3x + 4}{2} < -1$ and $5 - 2x < 5 - (4 + 3x)$

41. $\frac{2x - 5}{-2} < 2$ and $\frac{2x + 1}{3} > 0$

42. $1 - 2x \geq -2$ and $1 - x \leq \frac{1}{2} - 2x$

43. $x - 5 > -3$ or $3 - x > 6$

44. $5 - x > 2$ or $5 < x - 7$

45. $1 - x < 7 + x$ or $4x + 3 > x$

46. $1 - 2(x + 3) > -15$ or $4x < x - 3$

47. $\frac{1}{2}(x + 1) > 3$ or $0 < 7 - x$

48. $\frac{1}{2}(x + 6) > 3$ or $4(x - 1) < 3x - 4$

49. $1 - \frac{3}{2}x < 4$ and $\frac{1}{4}x - 2 \leq -3$

50. $\frac{3}{5}x - 1 > 2$ and $5 - \frac{2}{5}x \geq 3$

Solve each absolute value inequality. Write the solution set using interval notation and graph it.

51. $|3x - 1| < 2$ **52.** $|4x - 3| \leq 5$ **53.** $|5 - 4x| \leq 1$

54. $|6 - x| < 6$ **55.** $3 < |2x - 1|$ **56.** $5 \geq |4 - x|$

57. $|5 - 4x| < 0$ **58.** $|3x - 7| \geq -5$

59. $|2x - 8| \leq 0$ **60.** $|x - 6| > 0$

61. $|4 - 5x| < -1$ **62.** $|2 - 9x| \geq 0$

63. $|x - 2| + 6 > 9$ **64.** $3|x - 1| + 2 < 8$

65. $\left|\frac{x - 3}{2}\right| > 1$ **66.** $\left|\frac{9 - 4x}{2}\right| < 3$

Write an inequality of the form $|x - a| < k$ or of the form $|x - a| > k$ so that the inequality has the given solution set. (*Hint:* $|x - a| < k$ means that x is less than k units from a and $|x - a| > k$ means that x is more than k units from a on the number line.)

67. $(-\infty, 3) \cup (5, \infty)$ **68.** $(-\infty, -4) \cup (5, \infty)$

69. $(4, 8)$ **70.** $(-3, 8)$

71. $(-\infty, 4) \cup (4, \infty)$

72. $(-\infty, 0) \cup (0, \infty)$

73. $(-3, 7)$

74. $(3, 4)$

For each graph write an absolute value inequality that has the given solution set.

75.

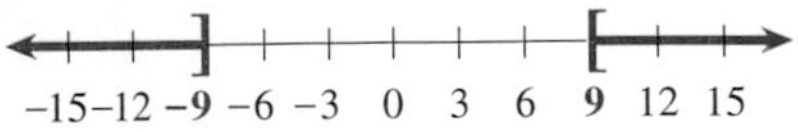

76.

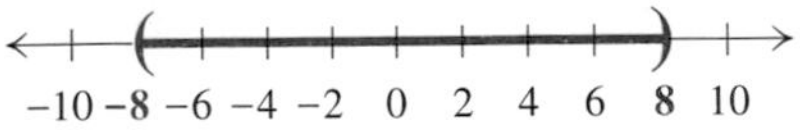

77.

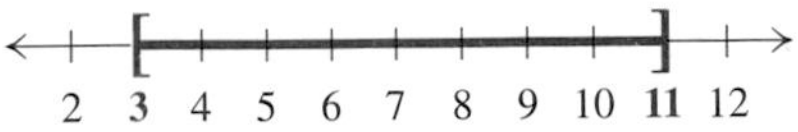

78.

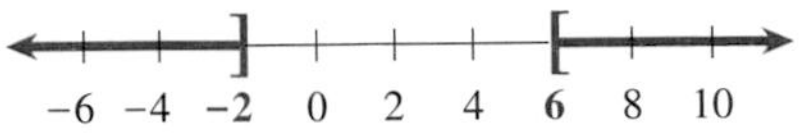

79.

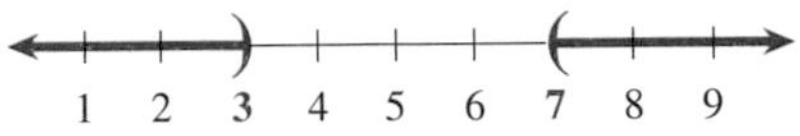

80.

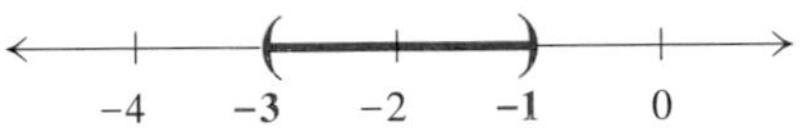

Find all values of x for which each expression is a real number.

81. $\sqrt{x - 2}$

82. $\sqrt{3x - 1}$

83. $\dfrac{1}{\sqrt{2 - x}}$

84. $\dfrac{5}{\sqrt{3 - 2x}}$

85. $\sqrt{|x| - 3}$

86. $\sqrt{5 - |x|}$

Solve each problem.

87. *Price Range for a Car* Yolanda is shopping for a car in a city where the sales tax is 10% and the title and license fee is \$300. If the maximum that she can spend is \$8000, then she should look at cars in what price range?

88. *Price of a Burger* The price of Elaine's favorite Big Salad at the corner restaurant is 10 cents more than the price of Jerry's hamburger. After treating a group of friends to lunch, Jerry is certain that for 10 hamburgers and 5 salads he spent more than \$9.14, but not more than \$13.19, including tax at 8% and a 50 cent tip. In what price range is a hamburger?

89. *Final Exam Score* Lucky scored 65 on his Psychology 101 midterm. If the average of his midterm and final must be between 79 and 90 for a B, then for what range of scores on the final exam would Lucky get a B?

90. *Bringing Up Your Average* Felix scored 52 and 64 on his first two tests in Sociology 212. What must he get on the third test to get an average for the three tests above 70?

91. *Weighted Average with Whole Numbers* Ingrid scored 65 on her calculus midterm. If her final exam counts twice as much as her midterm exam, then for what range of scores on her final would she get an average between 79 and 90?

92. *Weighted Average with Fractions* Elizabeth scored 62, 76, and 80 on three equally weighted tests in French. If the final exam score counts for two-thirds of the grade and the tests count for one-third, then what range of scores on the final exam would give her a final average over 70?

93. *Maximum Girth* United Parcel Service defines girth of a box as the sum of the length, twice the width, and twice the height. The maximum length that can be shipped with UPS is 108 in. and the maximum girth is 130 in. If a box has a length of 40 in. and a width of 30 in. then in what range must the height fall?

94. *Raising a Batting Average* At one point during the 1997 season, Javy Lopez of the Atlanta Braves had 93 hits in 317 times at bat for an average of 0.293. How many more times would he have to bat to get his average over 0.300 assuming he got a hit every time?

Figure for Exercise 94

95. *Bicycle Gear Ratio* The gear ratio r for a bicycle is defined by the following formula

$$r = \frac{Nw}{n},$$

where N is the number of teeth on the chainring (by the pedal), n is the number of teeth on the cog (by the wheel), and w is the wheel diameter in inches (Cycling, Burkett and Darst, 1987). The following chart gives uses for the various gear ratios.

Ratio	Use
$r > 90$	hard pedaling on level ground
$70 < r \le 90$	moderate effort on level ground
$50 < r \le 70$	mild hill climbing
$35 < r \le 50$	long hill with loaded bicycle

A bicycle with a 27-inch diameter wheel has 50 teeth on the chainring and 5 cogs with 14, 17, 20, 24, and 29 teeth. Find the gear ratio with each of the five cogs. Does this bicycle have a gear ratio for each of the four types of pedaling described in the table?

96. *Selecting the Cogs* Use the formula from the previous exercise to answer the following.

a. If a single-speed 27-inch bicycle has 17 teeth on the cog, then for what numbers of teeth on the chainring will the gear ratio be between 60 and 80?

b. If a 26-inch bicycle has 40 teeth on the chainring, then for what numbers of teeth on the cog will the gear ratio be between 60 and 75?

97. *Expensive Models* The prices of the 1998 BMW 740 iL and 750 iL differ by more than \$25,000. The list price of the 740 iL is \$66,070. Assuming that you do not know which model is more expensive, write an absolute value inequality that describes the price of the 750 iL. What are the possibilities for the price of the 750 iL?

98. *Differences in Prices* There is less than \$800 difference between the price of a \$14,200 Ford and a comparable Chevrolet. Express this as an absolute value inequality. What are the possibilities for the price of the Chevrolet?

• CALCULUS •

99. *Controlling Temperature* Michelle is trying to keep the water temperature in her chemistry experiment at 35° C. For the experiment to work, the relative error for the actual temperature must be less than 1%. Write an absolute value inequality for the actual temperature. Find the interval in which the actual temperature must lie.

100. *Laying Out a Track* Melvin is attempting to lay out a track for a 100 meter race. According to the rules of competition, the relative error of the actual length must be less than 0.5%. Write an absolute value inequality for the actual length. Find the interval in which the actual length must fall.

101. *Acceptable Bearings* A spherical bearing is to have a circumference of 7.2 cm with an error of no more than 0.1 cm. Use an absolute value inequality to find the acceptable range of values for the diameter of the bearing.

102. *Acceptable Targets* A manufacturer makes circular targets that have an area of 15 ft^2. According to competition rules, the area can be in error by no more than 0.5 ft^2. Use an absolute value inequality to find the acceptable range of values for the radius.

Figure for Exercise 102

103. *Crude Exports* The accompanying graph shows the total amount of crude and product exports per day for the top

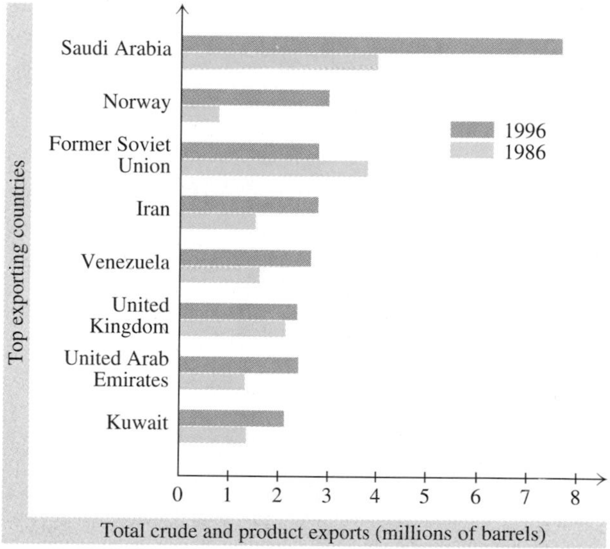

Figure for Exercise 103

oil exporting countries in 1986 and 1996 (*Forbes*, June 2, 1997). If a represents the amount for a country in 1986 and b represents the amount for that country in 1996, then for what countries is $|a - b| < 2$ million barrels per day? For what countries is $|a - b| > 0.5$ million barrels per day?

104. *Making a Profit* A strawberry farmer paid \$4200 for planting and fertilizing her strawberry crop. She must also pay \$2.40 per flat (12 pints) for picking and packing the berries and \$300 rent for space in a farmers' market where she sells the berries for \$11 per flat. For what number of flats will her revenue exceed her costs?

Linking Concepts

For Individual or Group Explorations

To Buy or Rent *A company can rent a copy machine for five years from American Business Machines for \$105 per month plus \$0.08 per copy. The same copy machine can be purchased for \$6500 with a per copy cost of \$0.04 plus \$25 per month for a maintenance contract. After five years the copier is worn out and worthless. To help the company make its choice, ABM sent the accompanying graph.*

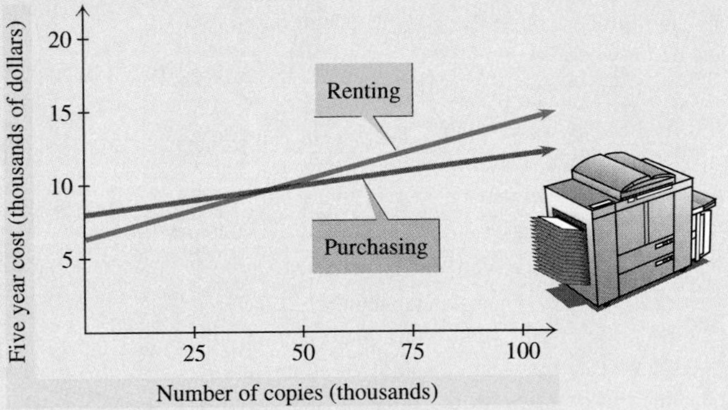

a) Write a formula for the total cost of renting and using the copier for five years in terms of the number of copies made during five years.

b) Write a formula for the total cost of buying and using the copier for five years in terms of the number of copies made during five years.

c) For what number of copies does the total cost of renting exceed \$10,000?

d) For what number of copies does the total cost of buying exceed \$10,000?

e) Use an absolute value inequality and solve it to find the number of copies for which the two plans differ by less than \$1000.

f) For what number of copies is the cost of renting equal to the cost of buying?

g) If the company estimates that it will make between 40,000 and 50,000 copies in five years, then which plan is better?

1.6 Quadratic and Rational Inequalities

In this section we continue the study of inequalities with quadratic inequalities and rational inequalities.

Quadratic Inequalities

If the equal sign in the quadratic equation $ax^2 + bx + c = 0$ is replaced by $<$, $\leq$, $>$, or $\geq$, we have a **quadratic inequality**. The method that we use for solving a quadratic inequality is very different from the methods used in Section 1.5.

Consider the quadratic inequality $x^2 - x - 6 > 0$. Since the left-hand side can be factored, we can write the inequality as $(x - 3)(x + 2) > 0$. A number x satisfies this inequality if the product of $x - 3$ and $x + 2$ is positive. To analyze this situation, we make a **sign graph** showing where each factor is positive, negative, or zero. The value of the factor $x - 3$ is positive for $x > 3$, negative for $x < 3$, and zero for $x = 3$. So on the sign graph of Fig. 1.34 we put positive signs above the interval $(3, \infty)$, negative signs above the interval $(-\infty, 3)$, and 0 above 3.

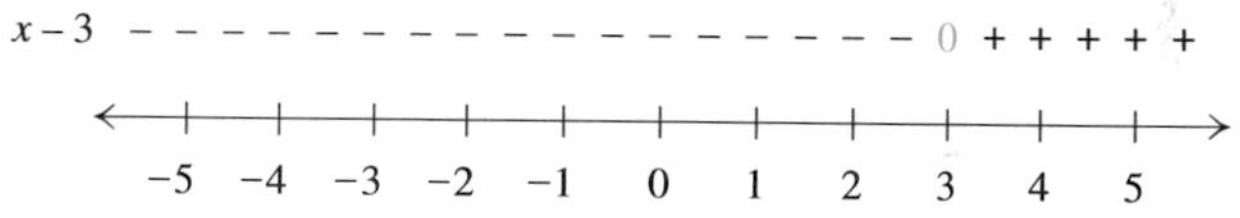

Figure 1.34

Now analyze the factor $x + 2$. The value of $x + 2$ is positive for $x > -2$, negative for $x < -2$, and zero for $x = -2$. So we add a line for the factor $x + 2$ to the sign graph of Fig. 1.34. In Fig. 1.35 we put positive signs above the interval $(-2, \infty)$, negative signs above $(-\infty, -2)$, and 0 above -2.

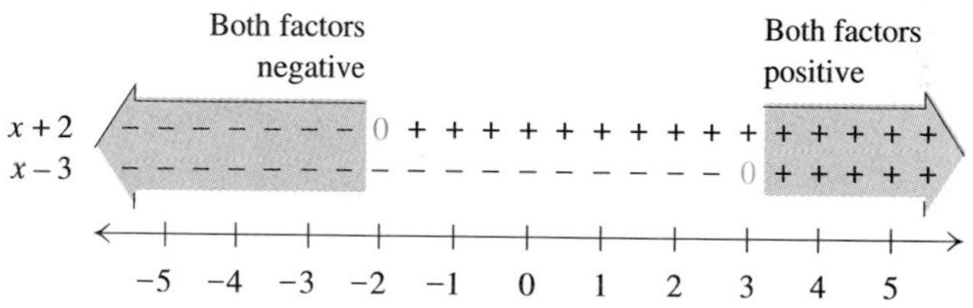

Figure 1.35

The product $(x - 3)(x + 2)$ is positive where both factors are positive or both factors are negative. From Fig. 1.35 we see that the solution set to the inequality $(x - 3)(x + 2) > 0$ is $(-\infty, -2) \cup (3, \infty)$.

EXAMPLE 1 Solving quadratic inequalities

Solve each inequality. Write the solution set in interval notation and graph it.

a) $x^2 - x \leq 6$ **b)** $9 - x^2 < 0$

Solution

a) Write the inequality with 0 on the right-hand side before factoring.

$$x^2 - x \leq 6$$
$$x^2 - x - 6 \leq 0$$
$$(x - 3)(x + 2) \leq 0$$

The solution set to $(x - 3)(x + 2) \leq 0$ includes any value of x for which the product of $x - 3$ and $x + 2$ is negative or zero. Note that prior to this example we solved $(x - 3)(x + 2) > 0$. The signs of the factors $x - 3$ and $x + 2$ are already shown in Fig. 1.35. The product is negative where one factor is positive and the other is negative, between -2 and 3. The product is zero if $x = -2$ or $x = 3$. The solution set is $[-2, 3]$ and its graph is shown in Fig. 1.36. Note that the solution to $(x - 3)(x + 2) \leq 0$ includes -2 and 3, while the solution to $(x - 3)(x + 2) > 0$ did not include them.

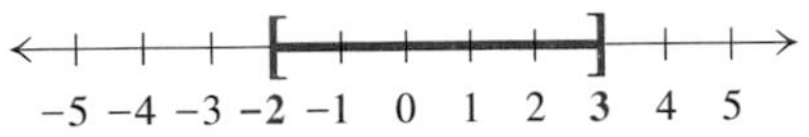

Figure 1.36

b) If we factor the left-hand side of $9 - x^2 < 0$, we get $(3 - x)(3 + x) < 0$. The factor $3 - x$ is positive if $x < 3$, negative if $x > 3$, and 0 if $x = 3$. The factor $3 + x$ is positive if $x > -3$, negative if $x < -3$, and 0 if $x = -3$. This information is shown on the sign graph in Fig. 1.37.

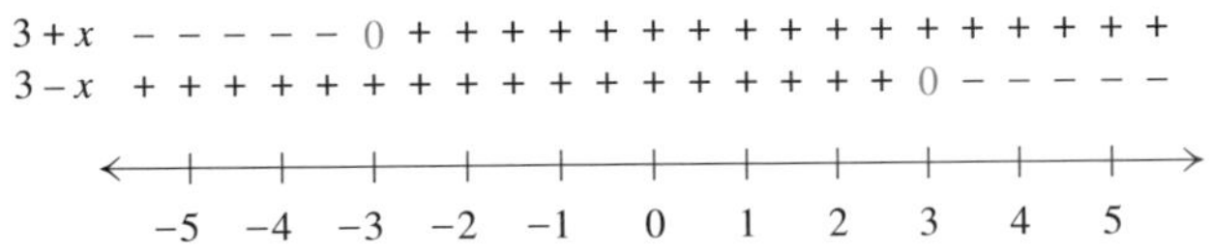

Figure 1.37

The solution set includes the values of x for which the product of the two factors is negative. Since the product is negative when the factors have opposite signs, the solution set to the inequality is $(-\infty, -3) \cup (3, \infty)$. The graph of the solution set is shown in Fig. 1.38.

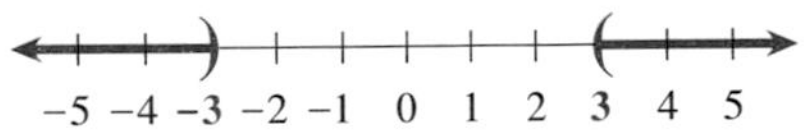

Figure 1.38

The inequality in Example 1(b) is of the type $x^2 > k$, where k is a positive real number. We can use the same steps as in Example 1(b) to show that the solution set to $x^2 > k$ is $(-\infty, -\sqrt{k}) \cup (\sqrt{k}, \infty)$. Likewise, the solution set to $x^2 < k$ is $(-\sqrt{k}, \sqrt{k})$. See Fig. 1.39.

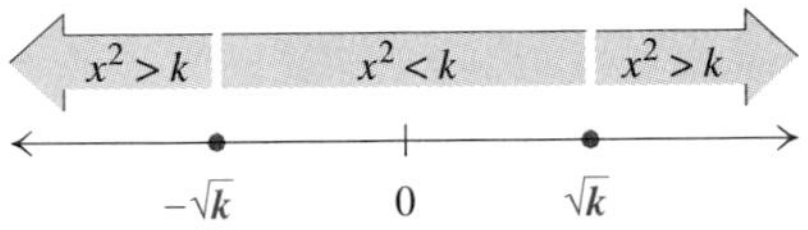

Figure 1.39

Rational Inequalities

Rational inequalities are inequalities that involve rational expressions. Some examples are

$$\frac{x+3}{x-2} \leq 2, \quad \frac{x^2-4}{x-3} \geq 0, \quad \text{and} \quad \frac{2}{x-1} \geq \frac{1}{x+1}.$$

When we multiply each side of an inequality by a real number, we must know whether that number is positive or negative. *So we generally do not multiply each side of a rational inequality by an expression that involves a variable.* As we did for quadratic inequalities, we can use sign graphs to solve rational inequalities.

EXAMPLE 2 **Solving rational inequalities**

Solve the inequality. Write the solution set using interval notation and graph it.

$$\frac{x+3}{x-2} \le 2$$

Solution

To use the sign graph method, we must have a 0 on one side of the inequality.

$$\frac{x+3}{x-2} \le 2$$

$$\frac{x+3}{x-2} - 2 \le 0 \qquad \text{Subtract 2 from each side to get 0 on the right-hand side.}$$

$$\frac{x+3}{x-2} - \frac{2(x-2)}{x-2} \le 0 \qquad \text{Get a common denominator.}$$

$$\frac{x+3-2(x-2)}{x-2} \le 0 \qquad \text{Subtract.}$$

$$\frac{7-x}{x-2} \le 0 \qquad \text{Get a single rational expression on the left.}$$

The denominator $x - 2$ is positive if $x > 2$, negative if $x < 2$, and 0 if $x = 2$. The numerator $7 - x$ is positive if $x < 7$, negative if $x > 7$, and 0 if $x = 7$. This information is shown on the sign graph in Fig. 1.40.

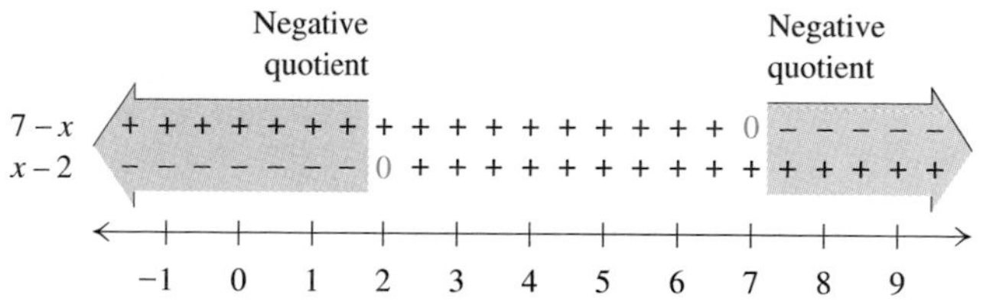

Figure 1.40

To discuss this sign graph, let

$$R(x) = \frac{7-x}{x-2}.$$

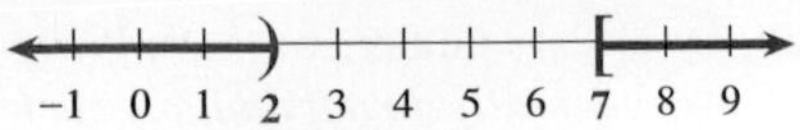

Figure 1.41

We want the values of x for which $R(x)$ is negative or zero. Since $R(x)$ is a quotient, $R(x)$ will be zero only if the numerator is zero (the denominator may not be zero), and $R(x)$ will be negative where $7 - x$ and $x - 2$ have opposite signs on the sign graph. The solution set to the inequality is $(-\infty, 2) \cup [7, \infty)$. Note that 2 is not in the solution set, because $R(2)$ is undefined; 7 is in the solution set because $R(7) = 0$. The graph of the solution set is shown in Fig. 1.41.

You can use the table feature of a graphing calculator to check the solution to the inequality of Example 2. Let $y_1 = (7 - x)/(x - 2)$ as shown in Fig. 1.42(a). Now use the table feature to scroll through a table of values for y_1. Figure 1.42(b) verifies that there is a change in sign of y_1 at $x = 2$. Figure 1.42(c) verifies that there is another change in sign of y_1 at $x = 7$. These results provide numerical support for our algebraic solution in Example 2. Note that these tables do not give us an accurate method for solving inequalities.

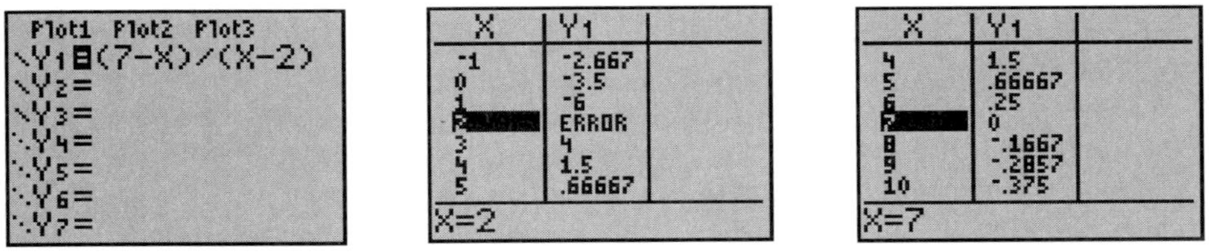

(a) (b) (c)

Figure 1.42

EXAMPLE 3 Rational inequality with three factors

Solve the inequality. Write the solution set in interval notation and graph it.

$$\frac{2}{x-1} \geq \frac{1}{x+1}$$

Solution

Instead of multiplying by the LCD, we use subtraction to get both rational expressions on the same side of the inequality.

$$\frac{2}{x-1} - \frac{1}{x+1} \geq 0$$

$$\frac{2(x+1)}{(x-1)(x+1)} - \frac{1(x-1)}{(x+1)(x-1)} \geq 0 \qquad \text{Get a common denominator.}$$

$$\frac{2x+2-1(x-1)}{(x-1)(x+1)} \geq 0 \qquad \text{Subtract.}$$

$$\frac{x+3}{(x-1)(x+1)} \geq 0$$

Let

$$R(x) = \frac{x + 3}{(x - 1)(x + 1)}.$$

The solution set includes any value of x for which $R(x)$ is positive or zero. The signs for each binomial in $R(x)$ are shown in Fig. 1.43.

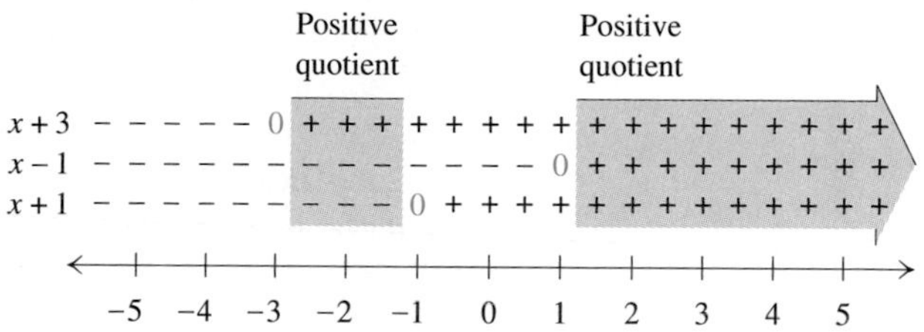

Figure 1.43

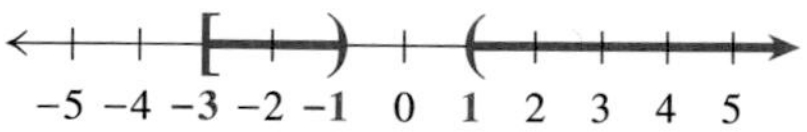

Figure 1.44

Since $R(x)$ involves only multiplication and division of the three binomials on the sign graph, $R(x)$ is positive if all three binomials are positive or if exactly one is positive. $R(x) = 0$ only if $x = -3$, because the denominator must not be 0. From the sign graph the solution set is $[-3, -1) \cup (1, \infty)$. The graph of the solution is shown in Fig. 1.44.

We checked Example 2 using the table feature of a graphing calculator, but we will check Example 3 by using the test feature. On a graphing calculator an inequality is treated like an expression. It has a value of 0 if it is not satisfied and a value of 1 if it is satisfied. Setting y_1 equal to the inequality of Example 3 as shown in Fig. 1.45(a) means that y_1 will have a value of 0 or 1. (The inequality symbols are found in the test menu.) Now examine the graph of y_1 in Fig. 1.45(b). The horizontal line appears only above intervals on which the inequality is satisfied. This result supports the algebraic solution in Example 3. □

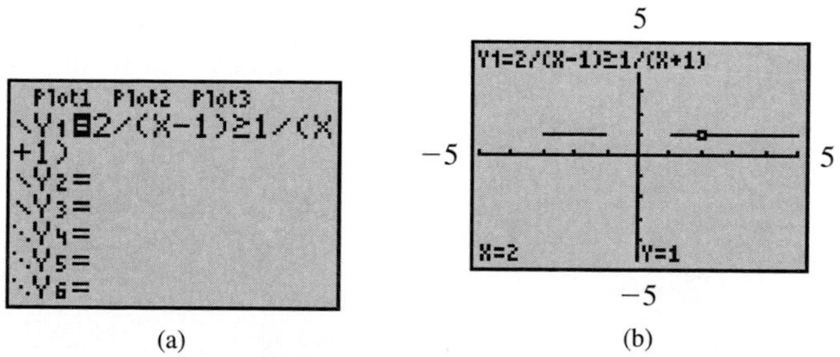

Figure 1.45

The Test Point Method

In Example 3 we worked with the rational expression

$$R(x) = \frac{x + 3}{(x - 1)(x + 1)}.$$

From Fig. 1.43 we see that $R(x)$ is negative for x in $(-\infty, -3)$, positive for x in $(-3, -1)$, negative for x in $(-1, 1)$, and positive for x in $(1, \infty)$. The value of $R(x)$ changes sign only at a value of x for which the expression is zero or undefined. This idea is used in solving inequalities by the **test point method**. In this method, we determine all of the points at which the expression is zero or undefined. These points divide the line into intervals. We test a number from each interval in the expression to determine the sign of the expression for that interval. Then we write the solution set from the interval or intervals that satisfy the inequality.

EXAMPLE 4 Solving an inequality with test points

Solve the inequality. State the solution set in interval notation and graph it.

$$\frac{x^2 - 6x + 8}{x^2 + 4x + 4} \geq 0$$

Solution

$$\frac{(x - 2)(x - 4)}{(x + 2)^2} \geq 0 \qquad \text{Factor the numerator and denominator.}$$

Let

$$R(x) = \frac{(x - 2)(x - 4)}{(x + 2)^2}.$$

We want to solve $R(x) \geq 0$. Now $R(x) = 0$ if $x = 2$ or $x = 4$. The denominator is 0 if $x = -2$, but -2 is not in the domain of the rational expression. So $R(x)$ is undefined for -2. Mark these points on a number line with a "U" for undefined or a zero as shown in Fig. 1.46. Then select one test point in each of the intervals determined by these points. We have selected -3, 0, 3, and 6.

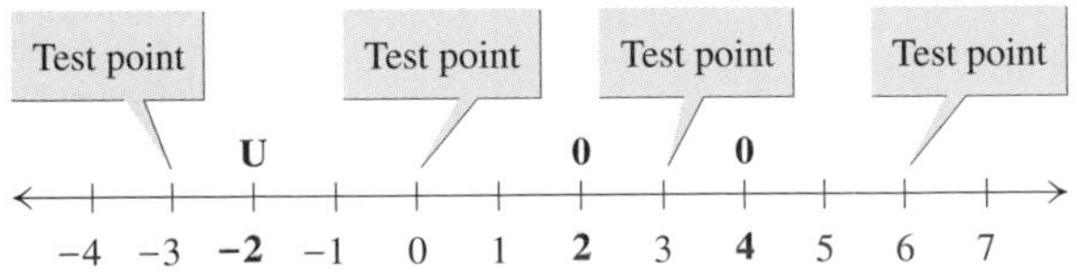

Figure 1.46

Table 1.6

x	$R(x)$	Conclusion
-3	35	$R(x) > 0$ on $(-\infty, -2)$
0	2	$R(x) > 0$ on $(-2, 2)$
3	$-1/25$	$R(x) < 0$ on $(2, 4)$
6	$1/8$	$R(x) > 0$ on $(4, \infty)$

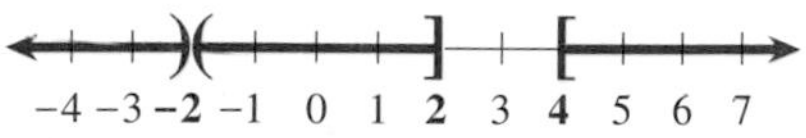

Figure 1.47

The value of $R(x)$ for each test point is listed in Table 1.6. If $R(x)$ is positive for a test point, we conclude that $R(x)$ is positive on the corresponding interval. From the table, we see three intervals on which $R(x) > 0$. Since $R(x) = 0$ for $x = 2$ and 4, those points are included in the solution set. So the solution set to $R(x) \geq 0$ is $(-\infty, -2) \cup (-2, 2] \cup [4, \infty)$. Its graph is shown in Fig. 1.47.

In the next example we use the test point method to solve a quadratic inequality that involves a prime polynomial.

EXAMPLE 5 Quadratic inequality with test points

Solve $x^2 < x + 1$. State the solution set in interval notation and graph it.

Solution

Write the inequality as $x^2 - x - 1 < 0$. Let $P(x) = x^2 - x - 1$. Since $P(x)$ is a prime polynomial, use the quadratic formula to solve $x^2 - x - 1 = 0$:

$$x = \frac{1 \pm \sqrt{(-1)^2 - 4(1)(-1)}}{2(1)} = \frac{1 \pm \sqrt{5}}{2}$$

Since $\frac{1 + \sqrt{5}}{2} \approx 1.6$ and $\frac{1 - \sqrt{5}}{2} \approx -0.6$, we mark these two points on a number line and choose a test point in each of the resulting three regions (Fig. 1.48). Table 1.7 lists the value of $P(x) = x^2 - x - 1$ at test points -2, 0, and 4.

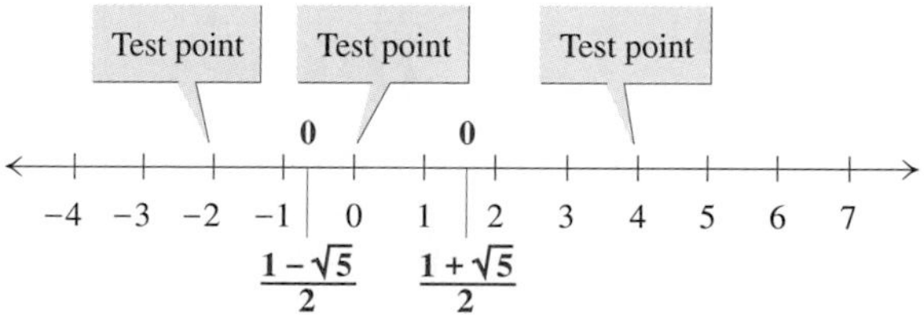

Figure 1.48

Table 1.7

x	$P(x)$	Conclusion
-2	5	$P(x) > 0$ on $\left(-\infty, \frac{1-\sqrt{5}}{2}\right)$
0	-1	$P(x) < 0$ on $\left(\frac{1-\sqrt{5}}{2}, \frac{1+\sqrt{5}}{2}\right)$
4	11	$P(x) > 0$ on $\left(\frac{1+\sqrt{5}}{2}, \infty\right)$

We conclude that the sign of $P(x)$ on each corresponding interval is the same as the sign of $P(x)$ at the test point. The original inequality is equivalent to $P(x) < 0$ and there is only one interval for which $P(x) < 0$. So the solution set is

$$\left(\frac{1-\sqrt{5}}{2}, \frac{1+\sqrt{5}}{2}\right)$$

and its graph is shown in Fig. 1.49.

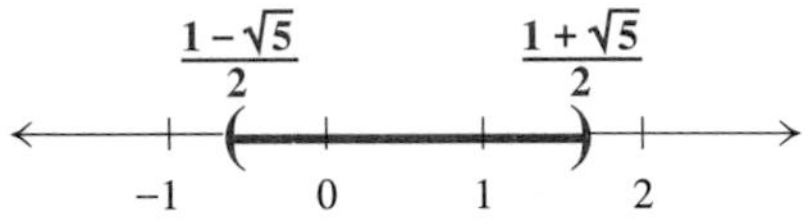

Figure 1.49

The polynomial equation $P(x) = 0$ in Example 5 had two real solutions. If the equation $P(x) = 0$ has no real solutions, then there are no points on the real line where $P(x)$ changes sign. So $P(x) < 0$ will be satisfied by either all real numbers or no real numbers.

Applications

In Section 1.4 we introduced the equation that is used to model the motion of a projectile: $S = -16t^2 + v_0t + s_0$, where S is the height in feet at time t in seconds, v_0 is the initial velocity in feet per second, and s_0 is the initial height in feet. In the next example, we write a quadratic inequality based on this model for the motion of a projectile.

EXAMPLE 6 Application of a quadratic inequality

During the summer of 1992, Mark Lenzi of Virginia won a gold medal in the springboard diving competition at Barcelona. The complex maneuvers that he performs require split-second timing. If Mark springs into the air at 20 feet per second from a diving board 10 feet above the water, then for how many seconds is he at least 15 feet above the water?

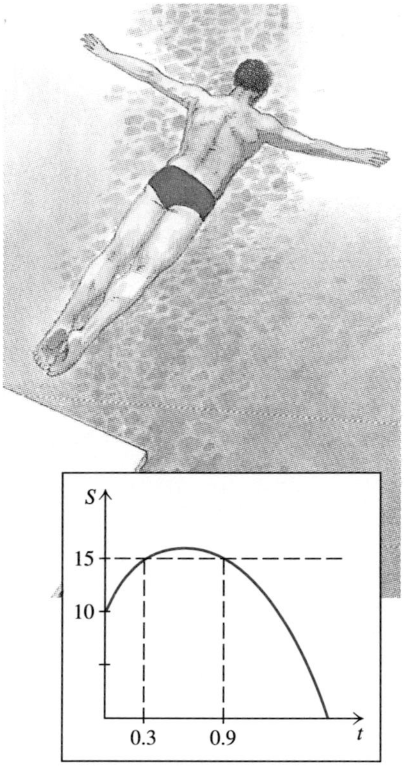

Figure 1.50

Solution

Use an initial velocity of 20 feet per second and an initial height of 10 feet. The height of the diver at any time t is given by the formula $S = -16t^2 + 20t + 10$. See Fig. 1.50. We want the values of t that satisfy $-16t^2 + 20t + 10 \geq 15$, or $-16t^2 + 20t - 5 \geq 0$. Use the quadratic formula to solve $-16t^2 + 20t - 5 = 0$:

$$t = \frac{-20 \pm \sqrt{20^2 - 4(-16)(-5)}}{2(-16)} = \frac{-5 \pm \sqrt{5}}{-8} = \frac{5 \pm \sqrt{5}}{8} \approx 0.3, 0.9$$

Pick test points 0, 0.5, and 1. If $P(t) = -16t^2 + 20t - 5$, then

$$P(0) = -5, \quad P(0.5) = 1, \quad \text{and} \quad P(1) = -1.$$

So $-16t^2 + 20t - 5 \geq 0$ is satisfied only for t in the interval

$$\left[\frac{5 - \sqrt{5}}{8}, \frac{5 + \sqrt{5}}{8}\right].$$

The length of this interval is the amount of time for which the diver is at least 15 feet above the water:

$$\frac{5 + \sqrt{5}}{8} - \frac{5 - \sqrt{5}}{8} = \frac{\sqrt{5}}{4} \approx 0.6$$

He is at least 15 feet above the water for exactly $\sqrt{5}/4$ seconds, or approximately 0.6 seconds.

FOR THOUGHT **True or False? Explain.**

1. The solution set to $x^2 > 9$ is the interval $(3, \infty)$.
2. The inequality $\frac{3}{x} > 2$ is equivalent to $3 > 2x$.
3. The inequality $x(2x - 1)(x - 3) < 0$ is equivalent to $x < 0$ or $2x - 1 < 0$ or $x - 3 < 0$.
4. We can solve any quadratic inequality.
5. If $\frac{-3}{a} > 0$, then $a < 0$.
6. The solution set to $(x - 2)(x + 5) > 0$ is $(-\infty, -5) \cup (2, \infty)$.
7. Any real number a satisfies $(a - 3)^2 > 0$.
8. The solution set to $\frac{x - 1}{(x + 3)^2} > 0$ is $(1, \infty)$.
9. Every real number satisfies $(x - (2 + i))(x - (2 - i)) > 0$.
10. The solution to $(x - 3)(x + 1)(x - 1) < 0$ consists of all values of x for which exactly one factor is negative or all three factors are negative.

1.6 EXERCISES

Tape 4 Disk

Solve each inequality. State the solution set in interval notation and graph it.

1. $2x - 3 > 0$ **2.** $3x + 2 < 0$ **3.** $w + 1 < 0$

4. $w - 2 > 0$ **5.** $5 - x < 0$ **6.** $1 - x > 0$

7. $z + 3 < 0$ **8.** $z - 4 < 0$

Solve using a sign graph. State the solution set in interval notation and graph it.

9. $2x^2 - x < 3$ **10.** $3x^2 - 4x \le 4$

11. $2x + 15 < x^2$ **12.** $5x - x^2 < 4$

13. $w^2 - 4w - 12 \ge 0$ **14.** $y^2 + 8y + 15 \le 0$

15. $t^2 \le 16$ **16.** $36 \le h^2$

17. $a^2 + 6a + 9 \le 0$ **18.** $c^2 + 4 \le 4c$

19. $4z^2 - 12z + 9 > 0$ **20.** $9s^2 + 6s + 1 \ge 0$

Find the indicated value of each polynomial or rational expression in Exercises 21–28.

21. $P(3)$ if $P(x) = x^2 - 3x + 1$

22. $P(-2)$ if $P(x) = 2x^2 - 5x + 6$

23. $R(-1)$ if $R(x) = \dfrac{x - 1}{x + 3}$ **24.** $R(-3)$ if $R(x) = \dfrac{4 - x}{2x - 1}$

25. $P(2)$ if $P(x) = (x - 1)(x - 5)(x - 6)$

26. $P(4)$ if $P(x) = (3 - x)(x - 2)(x + 2)$

27. $R(-4)$ if $R(x) = \dfrac{(x - 3)(5 - x)}{(x + 2)(x - 1)}$

28. $R(3)$ if $R(x) = \dfrac{x^2 + 2x + 7}{(x - 5)(x - 1)}$

Solve using a sign graph. State the solution set in interval notation and graph it.

29. $\dfrac{x - 4}{x + 2} \le 0$ **30.** $\dfrac{x + 3}{x + 5} \ge 0$ **31.** $\dfrac{x - 7}{x + 7} < 0$

32. $\dfrac{x + 3}{x - 1} > 0$ **33.** $\dfrac{1}{y} \le 0$ **34.** $\dfrac{1}{y^2} \le 0$

35. $\dfrac{w^2 - w - 6}{w - 6} \ge 0$ **36.** $\dfrac{z - 5}{z^2 + 2z - 8} \le 0$

37. $\dfrac{3 - t}{2t^2 - 9t - 5} > 0$ **38.** $\dfrac{m^2 - 5m + 6}{1 - 2m} < 0$

39. $\dfrac{q - 2}{q + 3} < 2$ **40.** $\dfrac{p + 1}{2p - 1} \ge 1$

41. $\dfrac{1}{x + 2} > \dfrac{1}{x - 3}$ **42.** $\dfrac{1}{x + 1} > \dfrac{2}{x - 1}$

43. $x < \dfrac{3x - 8}{5 - x}$ **44.** $\dfrac{2}{x + 3} \ge \dfrac{1}{x - 1}$

Solve using the test point method. State the solution set in interval notation and graph it.

45. $\dfrac{(x - 3)(x + 1)}{x - 5} \ge 0$ **46.** $\dfrac{(x + 2)(x - 1)}{(x + 4)^2} \le 0$

47. $\dfrac{x^2 - 25}{9 - x^2} \le 0$ **48.** $\dfrac{x^2 + 2x + 1}{x^2 - 2x - 15} \ge 0$

49. $x^2 - 2x - 24 > 0$ **50.** $4x + 60 \le x^2$

51. $4x^2 - 9 < 0$ **52.** $2x^2 - 7 > 0$

53. $y^2 + 23 > 10y$ **54.** $y^2 + 3 \ge 6y$

55. $6t < t^2 + 10$ **56.** $8t < t^2 + 25$

57. $\dfrac{1}{w} > \dfrac{1}{w^2}$ **58.** $\dfrac{1}{w} > w^2$

59. $w > \dfrac{w - 5}{w - 3}$ **60.** $w < \dfrac{w - 2}{w + 1}$

Solve by the method of your choice. State the solution set in interval notation and graph it.

61. $(x - 5)(3x + 2)(x - 1) < 0$

62. $(3 - x)(x + 4)(3x - 2) \ge 0$

63. $(y - 1)(y - 3)(y - 5)(y - 7) > 0$

64. $y(y + 2)(y - 3)(y - 6) < 0$

65. $y^3 - 9y \ge 0$

66. $2y^3 - 5y \le 0$

67. $t + 5t^2 > 4t + 20$ **68.** $t^3 - t^2 < 16t - 16$

69. $\dfrac{5z - 12}{z - 3} \le z$ **70.** $\dfrac{1}{z} \ge z$

71. $\dfrac{d - 1}{d + 2} > \dfrac{d + 3}{d - 4}$ **72.** $\dfrac{v + 1}{v - 2} \le \dfrac{v - 1}{v + 2}$

73. $\dfrac{u^2 - 3u + 2}{u^2 - 7u + 12} \le 0$ **74.** $\dfrac{4b^2 - 1}{b^2 - 5} \le 0$

75. $\dfrac{2t}{t-1} \geq \dfrac{t+2}{t-2}$

76. $\dfrac{2q+3}{q-1} \leq \dfrac{q+2}{q-4}$

Find the values of w for which each expression represents a real number. Use interval notation.

77. $\sqrt{2w-1}$

78. $\sqrt{9-w}$

79. $\sqrt{9-w^2}$

80. $\sqrt{w^2-4}$

81. $\dfrac{1}{\sqrt{w^2-w-6}}$

82. $\dfrac{1}{\sqrt{w^2-9}}$

83. $\sqrt{\dfrac{w-1}{w+2}}$

84. $\sqrt{\dfrac{1-w}{w-2}}$

Solve each problem.

85. *Fencing a Rectangular Area* Hector has 100 ft of fencing to fence a rectangular area. What are the possible lengths for the north side that would give an area of less than or equal to 400 ft^2?

86. *Profit from Sales* Shelley's weekly profit in dollars for selling x bottles of skin toner is given by the formula $P(x) = -2x^2 + 50x + 20$. For what values of x is her profit at least \$250?

87. *Height of a Flare* A flare is fired straight into the air from a height of 6 ft above the water. If the flare leaves the barrel of the gun at 80 ft/sec, then for how long is the flare more than 100 ft above the water?

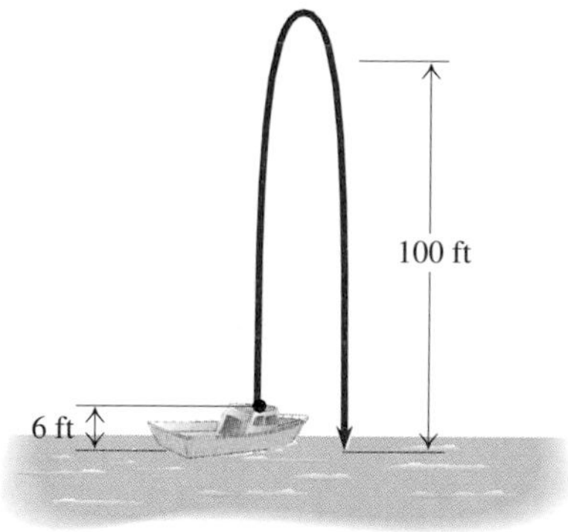

Figure for Exercise 87

88. *Height of a Baseball* St. Louis Cardinals' first baseman, Mark McGwire, is a power hitter whose swing produces a lot of fly balls. If McGwire's bat connects with the ball at a height of 3 ft and sends it straight up with a velocity of 50 ft/sec, how long is the baseball more than 20 ft above the ground?

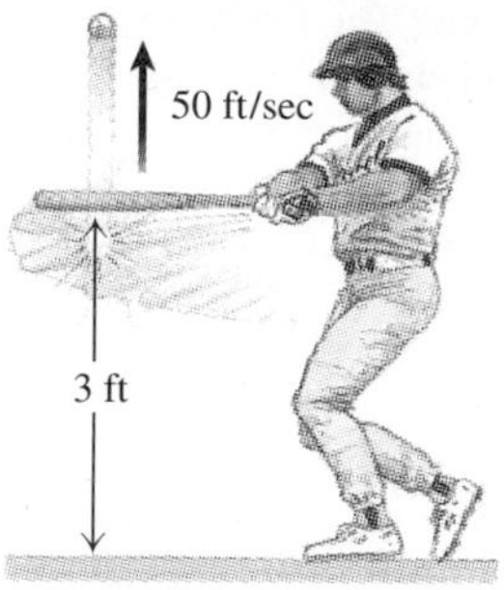

Figure for Exercise 88

89. *Two Numbers* If two numbers must have a sum of 12, then what are the possible values for the first number so that their product is at least 10?

90. *Two More Numbers* If two numbers must have a difference of 6, then what are the possible values for the smaller number so that their product is at least 9?

91. *Inflation of Car Prices* The cost of a BMW in two years is given by the formula $C = P(1 + r)^2$, where P is the present cost and r is the annual inflation rate. According to a market analyst, today's \$38,000 BMW will cost at least \$43,000 in two years. For what possible values for r would this statement be true?

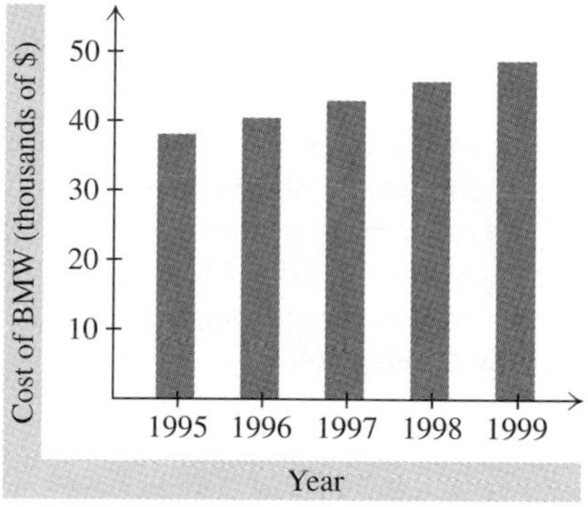

Figure for Exercise 91

92. *Painting a Fence* Joe can paint a fence working alone in 40 hr. Joe plans to work with a helper who could do the job alone in x hours. For what values of x could Joe and the helper get the fence done in less than 30 hr?

93. *Velocity for a Vertical Leap* Toronto Raptors' player Dee Brown, at 6 ft and 160 lb, may have the highest vertical leap in the NBA. If Brown makes a vertical leap of 1.17 m, what is his initial velocity at the time his feet leave the floor? The formula $v_1^2 = v_0^2 + 2gS$ gives the relationship between terminal velocity v_1, initial velocity v_0, the acceleration of gravity g, and his height S. Use $g = -9.8$ m/sec^2 and the fact that his terminal velocity at his peak height is zero.

Figure for Exercise 93

94. *Time for a Vertical Leap* Use the initial velocity from Exercise 93 to find the amount of time that Dee Brown is more than 0.7 m above the court when he makes his vertical leap. Use the formula $S = -4.9t^2 + v_0 t$, where S is height in meters, t is time in seconds, and v_0 is initial velocity in meters per second.

95. *High Altitude Climbers* Mountain climbers know that when the atmospheric pressure is less than 0.50 atm they must exercise extreme caution because of the lack of oxygen in the air (Encyclopedia of Sports Science, 1997). If the atmospheric pressure a is related to the altitude h (in feet) by the equation

$$a = 3.89 \times 10^{-10}h^2 - 3.48 \times 10^{-5}h + 1,$$

then for what altitudes is the atmospheric pressure less than 0.50?

96. *Demand and Revenue* The weekly demand d for the Acme Virtual Pet is given by the equation

$$d = 3000 - 60p$$

where p is the price of each in dollars. The weekly revenue is the product of the demand and the price each,

$$(3000 - 60p)p.$$

For what prices (to the nearest cent) would the weekly revenue be greater than $30,000?

97. *Cleaning Up the River* Pollution in the Tangipahoa River has been blamed primarily on the area dairy farmers. The formula

$$C = \frac{400{,}000p}{100 - p}$$

is used to model the cost in dollars to area dairy farmers for removing p% of the coliform bacteria from the river. If less than $1.2 million is spent, then what percentage of the coliform bacteria can be removed?

98. *Product Awareness* An advertising agency estimates that the cost in dollars of an advertising company that reaches p% of the consumers in Chicago is given by

$$C = \frac{100{,}000 + 500{,}000p}{100 - p}.$$

If less than $752,500 is spent on the campaign, then what percentage of the consumers will be reached?

For Writing/Discussion

99. *General Summary* Write a general summary concerning the solution set to the quadratic inequality $ax^2 + bx + c > 0$, based on the value of the discriminant.

100. *Extended Summary* Extend the summary of Exercise 99 to include quadratic inequalities using $\geq$, $<$, and $\leq$.

101. *Cooperative Learning* Each student in your small group should write a linear inequality, an absolute value inequality, a quadratic inequality, and a rational inequality. The group should solve each inequality and determine whether it is the required type.

HIGHLIGHTS

- **Section 1.1 Linear Equations**

1. An equation is a statement indicating that two algebraic expressions are equal.
2. A linear equation is an equation of the form $ax + b = 0$, where a and b are real numbers, with $a \neq 0$.

3. To solve an equation means to find the set of all numbers for which the equation is true.
4. The properties of equality are used to simplify an equation until we get an equivalent equation whose solution set is obvious.
5. An equation may be an identity, an inconsistent equation, or a conditional equation.
6. An equation involving absolute value may have more than one solution.

- **Section 1.2 Applications**

1. A formula is an equation in two or more variables that is usually of an applied nature.
2. The value of one variable can usually be found if values are known for all of the other variables in the formula.
3. The strategy for problem solving presented gives a step-by-step procedure for solving verbal problems.
4. Three main categories of applied problems studied here are uniform-motion problems, work problems, and mixture problems.

- **Section 1.3 Complex Numbers**

1. Complex numbers are of the form $a + bi$, where a and b are real numbers and where $i = \sqrt{-1}$ and $i^2 = -1$.
2. Operations with complex numbers can be performed as if they are binomials with i being a variable, and using $i^2 = -1$ to simplify.
3. Division of complex numbers is performed by multiplying numerator and denominator by the complex conjugate of the denominator.
4. Expressions involving square roots of negative numbers must be converted to standard form using $\sqrt{-b} = i\sqrt{b}$ $(b > 0)$ before we can perform any operations.

- **Section 1.4 Quadratic Equations**

1. A quadratic equation is an equation of the form $ax^2 + bx + c = 0$, where a, b, and c are real numbers with $a \neq 0$.
2. A quadratic equation can have two distinct real solutions, one real solution, or two imaginary solutions that are conjugates.
3. The four methods presented for solving quadratic equations are factoring, the square root property, completing the square, and the quadratic formula.

- **Section 1.5 Linear and Absolute Value Inequalities**

1. The inequality symbol is reversed if the inequality is multiplied or divided by a negative number.
2. We usually do not multiply or divide an inequality by a variable expression.
3. Interval notation is used to state the solution sets of most inequalities.
4. To solve an absolute value inequality, remember that the absolute value of x is the distance from x to 0 on the number line.

5. Most absolute value inequalities can be written as equivalent compound inequalities.

- **Section 1.6 Quadratic and Rational Inequalities**

1. Polynomial and rational inequalities are solved by sign graphs and the rules for multiplying and dividing signed numbers. The test point method is also used for solving polynomial or rational inequalities.
2. The test point method is based on the fact that the value of a polynomial or rational expression can change sign only at a point on the number line where it has value 0 or is undefined.

CHAPTER 1 REVIEW EXERCISES

Find all real and imaginary solutions to each equation.

1. $3x - 2 = 0$

2. $3x - 5 = 5(x + 7)$

3. $3x^2 - 2 = 0$

4. $x^2 - 16 = 0$

5. $\frac{1}{2}y - \frac{1}{3} = \frac{1}{4}y + \frac{1}{5}$

6. $\frac{1}{2} - \frac{w}{5} = \frac{w}{4} - \frac{1}{8}$

7. $\frac{2}{x} = \frac{3}{x - 1}$

8. $\frac{5}{x + 1} = \frac{2}{x - 3}$

9. $\frac{x + 1}{x - 1} = \frac{x + 2}{x - 3}$

10. $\frac{x + 3}{x - 8} = \frac{x + 7}{x - 4}$

11. $\frac{x + 2}{x + 4} = \frac{x - 2}{x - 4}$

12. $\frac{2x - 5}{x + 2} = \frac{x - 3}{x - 2}$

13. $\frac{1}{x} + \frac{1}{x - 1} = \frac{3}{2}$

14. $\frac{2}{x - 2} - \frac{3}{x + 2} = \frac{1}{2}$

15. $b^2 + 10 = 6b$

16. $4t^2 + 17 = 16t$

17. $s^2 - 4s + 1 = 0$

18. $3z^2 - 2z - 1 = 0$

19. $|3q - 4| = 2$

20. $|2v - 1| = 3$

21. $|2h - 3| = 0$

22. $|3y - 1| = -2$

Solve each inequality. State the solution set using interval notation and graph the solution set.

23. $4x - 1 > 3x + 2$

24. $6(x - 3) < 5(x + 4)$

25. $5 - 2x > -3$

26. $7 - x > -6$

27. $\frac{1}{2}x - \frac{1}{3} > x + 2$

28. $0.06x + 1000 > x + 60$

29. $-2 < \frac{x - 3}{2} \leq 5$

30. $-1 \leq \frac{3 - 2x}{4} < 3$

31. $3 - 4x < 1$ and $5 + 3x < 8$

32. $-3x < 6$ and $2x + 1 > -1$

33. $-2x < 8$ or $3x > -3$

34. $1 - x < 6$ or $-5 + x < 1$

35. $|x - 3| > 2$

36. $|4 - x| \leq 3$

37. $|2x - 7| \leq 0$

38. $|6 - 5x| < 0$

39. $|7 - 3x| > -4$

40. $|4 - 3x| \geq 1$

Solve each inequality using a sign graph. State the solution set using interval notation.

41. $8x^2 + 1 < 6x$

42. $x^2 + 2x \leq 63$

43. $(3 - x)(x + 5) \geq 0$

44. $-x^2 - 2x + 15 < 0$

45. $\frac{x - 3}{5 - x} \geq 0$

46. $\frac{4 - x}{x + 7} \leq 0$

Solve each inequality using the test point method. State the solution set using interval notation.

47. $\frac{x + 10}{x + 2} < 5$

48. $\frac{x - 6}{2x + 1} \geq 1$

49. $\frac{12 - 7x}{x^2} > -1$

50. $x - \frac{2}{x} \leq -1$

51. $x^2 \leq 2x + 4$

52. $x^2 - x \leq \frac{1}{2}$

Solve each inequality using the method of your choice. State the solution set using interval notation and graph it.

53. $\frac{x^2 - 3x + 2}{x^2 - 7x + 12} \geq 0$

54. $\frac{x^2 + 4x + 3}{x^2 - 2x - 15} \leq 0$

55. $\frac{x + 3}{x + 3} \leq 0$

56. $\frac{1}{(x - 2)^2} > 1$

57. $\frac{51}{(x - 1)^2} \geq 0$

58. $\frac{-1}{x} > 2$

Solve each equation for y.

59. $2x - 3y = 6$

60. $x(y - 2) = 1$

61. $xy = 1 + 3y$

62. $x^2y = 1 + 9y$

63. $ax + by = c$

64. $y^2 + my + n = 0$

65. $\frac{1}{y} = \frac{1}{x} + \frac{1}{2}$

66. $y^2 + by = ay + ab$

Write each expression in the form $a + bi$, where a and b are real numbers.

67. $(3 - 7i) + (-4 + 6i)$

68. $(-6 - 3i) - (3 - 2i)$

69. $(4 - 5i)^2$

70. $7 - i(2 - 3i)^2$

71. $(1 - 3i)(2 + 6i)$

72. $(0.3 + 2i)(0.3 - 2i)$

73. $(2 - 3i) \div i$

74. $(-2 + 4i) \div (-i)$

75. $(1 - i) \div (2 + i)$

76. $(3 + 6i) \div (4 - i)$

77. $\frac{1 + i}{2 - 3i}$

78. $\frac{3 - i}{4 - 3i}$

79. $\frac{6 + \sqrt{-8}}{2}$

80. $\frac{-2 - \sqrt{-18}}{2}$

81. $\frac{-6 + \sqrt{(-2)^2 - 4(-1)(-6)}}{-8}$

82. $\frac{-9 - \sqrt{(-9)^2 - 4(-3)(-9)}}{-6}$

83. $i^{34} + i^{19}$

84. $\sqrt{6} + \sqrt{-3}\sqrt{-2}$

Find the values for x for which each expression represents a real number. Use interval notation.

85. $\sqrt{3x - 4}$

86. $\frac{1}{\sqrt{x - 1}}$

87. $\sqrt{x^2 - 25}$

88. $\sqrt{\frac{x^2 - x - 2}{4 - x}}$

Evaluate the discriminant for each equation, and use it to determine the type and number of solutions to the equation.

89. $x^2 + 2 = 4x$

90. $y^2 + 2 = 3y$

91. $4w^2 = 20w - 25$

92. $2x^2 - 3x + 10 = 0$

Solve each problem. Use an equation or an inequality as appropriate. For problems in which the answer involves an irrational number, find the exact answer and an approximate answer.

• CALCULUS •

93. *Folding Sheet Metal* A square is to be cut from each corner of an 8-in. by 11-in. rectangular piece of copper and the sides are to be folded up to form a box as shown in the figure. If the area of the bottom is to be 50 in.^{2}, what is the length of the side of the square to be cut from the corner?

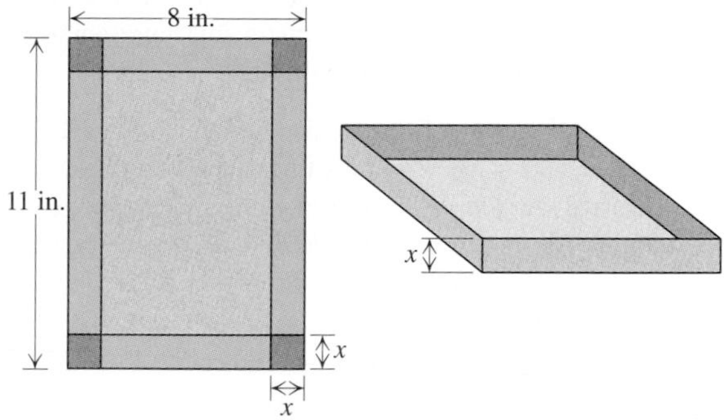

Figure for Exercise 93

94. *Peeling Apples* Bart can peel a batch of 3000 lb of apples in 12 hr with the old machine, while Mona can do the same job in 8 hr using the new machine. If Bart starts peeling at 8 A.M. and Mona joins him at 9 A.M., then at what time will they finish the batch working together?

95. *Meeting Between Two Cities* Lisa, an architect from Huntsville, has arranged a business luncheon with her client, Taro Numata from Norwood. They plan to meet at a restaurant off the 300-mi highway connecting their two cities. If they leave their offices simultaneously and arrive at the restaurant simultaneously, and Lisa averages 50 mph while Taro averages 60 mph, then how far from Huntsville is the restaurant?

96. *Driving Speed* Lisa and Taro agree to meet at the construction site located 100 mi from Norwood on the 300-mi highway connecting Huntsville and Norwood. Lisa leaves Huntsville at noon, while Taro departs from Norwood one hour later. If they arrive at the construction site simultaneously and Lisa's driving speed averaged 10 mph faster than Taro's, then how fast did she drive?

97. *Fish Population* A channel was dug to connect Homer and Mirror Lakes. Before the channel was dug, a biologist estimated that 20% of the fish in Homer Lake and 30% of the fish in Mirror Lake were bass. After the lakes were joined, the biologist made another estimate of the bass population in order to set fishing quotas. If she decided that 28% of the total fish population of 8000 were bass, then how many fish were in Homer Lake originally?

98. *Support for Gambling* Eighteen pro-gambling representatives in the state house of representatives bring up a gambling bill every year. After redistricting, four new representatives are added to the house, causing the percentage of pro-gambling representatives to increase by 5 percentage points. If all four of the new representatives are pro-gambling and they still do not constitute a majority of the house, then how many representatives are there in the house after redistricting?

99. *Hiking Distance* The distance between Marjorie and the dude ranch was $4\sqrt{34}$ mi straight across a rattlesnake-infested canyon. Instead of crossing the canyon, she hiked due north for a long time and then hiked due east for the shorter leg of the journey to the ranch. If she averaged 4 mph and it took her 8 hr to get to the ranch, then how far did she hike in a northerly direction?

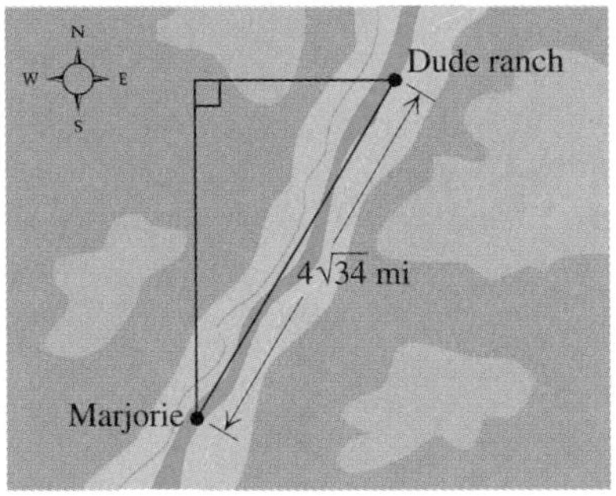

Figure for Exercise 99

100. *Golden Rectangle* The *golden rectangle* of the ancient Greeks was thought to be the rectangle with the shape that was most pleasing to the eye. The golden rectangle was defined as a rectangle that would retain its original shape after removal of a W by W square from one end, as shown in the figure. So the length and width of the original rectangle must satisfy

$$\frac{L}{W} = \frac{W}{L - W}.$$

If the length of a golden rectangle is 20 m, then what is its width? If the width of a golden rectangle is 8 m, then what is its length?

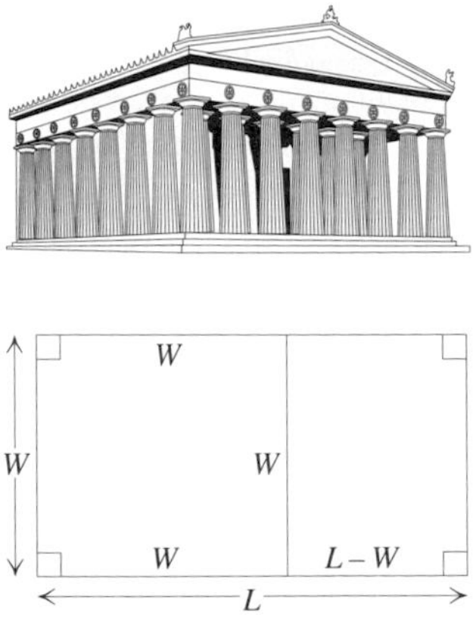

Figure for Exercise 100

101. *Price Range for a Haircut* A haircut at Joe's Barber Shop costs \$50 less than a haircut at Renee's French Salon. In fact, you can get five haircuts at Joe's for less than the cost of one haircut at Renee's. What is the price range for a haircut at Joe's?

102. *Selling Price of a Home* Elena must get at least \$120,000 for her house in order to break even with the original purchase price. If the real estate agent gets 6% of the selling price, then what should the selling price be?

103. *Dimensions of a Picture Frame* The length of a picture frame must be 2 in. longer than the width. If Reginald can use between 32 and 50 in. of frame molding for the frame, then what are the possibilities for the width?

104. *Saving Gasoline* If Americans drive 10^{12} mi per year and the average gas mileage is raised from 27.5 mpg to 29.5 mpg, then how many gallons of gasoline are saved?

105. *Increasing Gas Mileage* If Americans continue to drive 10^{12} mi per year and the average gas mileage is raised from 29.5 to 31.5 mpg, then how many gallons of gasoline are saved? If the average mpg is presently 29.5, then what should it be increased to in order to achieve the same savings in gallons as the increase from 27.5 to 29.5 mpg?

106. *Thickness of Concrete* Alfred used 40 yd^3 of concrete to pour a section of interstate highway that was 12 ft wide and 54 ft long. How thick was the concrete?

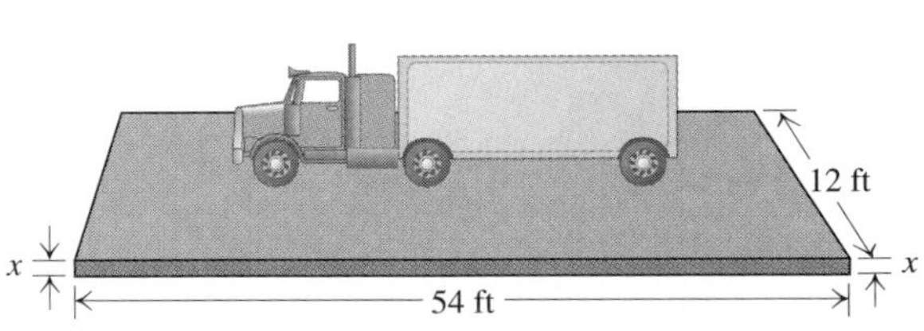

Figure for Exercise 106

CHAPTER 1 TEST

Find all real and imaginary solutions to each equation.

1. $\frac{1}{2}x - \frac{1}{6} = \frac{1}{3}x$

2. $3x^2 - 2 = 0$

3. $x^2 + 1 = 6x$

4. $x^2 + 14 = 9x$

5. $\frac{x - 1}{x + 3} = \frac{x + 2}{x - 6}$

6. $x^2 = 2x - 5$

Solve each inequality. State the solution set using interval notation and graph it.

7. $3 - 2x > 7$

8. $\frac{x}{2} > 3$ and $5 < x$

9. $|2x - 1| \le 3$

10. $2|x - 3| + 1 > 5$

Perform the indicated operations and write the answer in the form $a + bi$, where a and b are real.

11. $(4 - 3i)^2$

12. $\frac{2 - i}{3 + i}$

13. $i^6 - i^{35}$

14. $\sqrt{-8}(\sqrt{-2} + \sqrt{6})$

Solve each inequality using the method of your choice.

15. $x^2 - 2x < 8$

16. $\frac{x + 2}{x - 3} > -1$

17. $\frac{x + 3}{(4 - x)(x + 1)} \ge 0$

Solve each problem.

18. Find the value of the discriminant for $x^2 - 5x + 9 = 0$.

19. Find the product of $7 - 2i$ and its conjugate.

20. Solve $5 - 2y = 4 + 3xy$ for y.

21. For what values of x is $\sqrt{x - 5}$ a real number?

22. Terry had a square patio. After expanding the length by 20 ft and the width by 10 ft the area was 999 ft^2. What was the original area?

23. How many gallons of 20% alcohol solution must be mixed with 10 gal of a 50% alcohol solution to obtain a 30% alcohol solution?

TYING IT ALL TOGETHER

Chapters P–1

Perform the indicated operations.

1. $3x + 4x$

2. $5x \cdot 6x$

3. $\frac{1}{x} + \frac{1}{2x}$

4. $(x + 3)^2$

5. $(2x - 1)(3x + 2)$

6. $\frac{(x + h)^2 - x^2}{h}$

7. $\frac{1}{x - 1} + \frac{1}{x + 1}$

8. $\left(x + \frac{3}{2}\right)^2$

Solve each equation.

9. $3x + 4x = 7x$

10. $5x \cdot 6x = 11x$

11. $\frac{1}{x} + \frac{1}{2x} = \frac{3}{2x}$

12. $(x + 3)^2 = x^2 + 9$

13. $6x^2 + x - 2 = 0$

14. $\frac{1}{x - 1} + \frac{1}{x + 1} = \frac{5}{8}$

15. $3x + 4x = 7x^2$

16. $3x + 4x = 7$

Given that $P(x) = x^3 - 3x^2 + 2$ and $T(x) = -x^2 + 3x - 4$, find the value of each polynomial for the indicated values of x.

17. $P(1)$

18. $P(-1)$

19. $P(-1/2)$

20. $P(-1/3)$

21. $T(-1)$

22. $T(2)$

23. $T(0)$

24. $T(0.5)$

Fast forward to the year 2200 . . . A spacecraft streaks toward Earth on its return from the Outer Galaxy. As the astronauts approach the blue planet, they spot a band of vegetation encircling the equator. They recognize this emerald belt as tropical rainforest, but its ragged outline—like a scar on Earth's surface—appears as an ominous sign. "A little over two centuries ago," remarks the Science Officer, "that green line was practically unbroken. Now look at it. If only people had cared enough to do more while they still had the chance!"

There may still be time to preserve the rainforests, a lush habitat for life that thrives in a warm, wet climate. Although covering less than an eighth of a percent of Earth's surface, rainforests are home to over half our animal and plant species. Destroying only a percentage of these millions of species could irrevocably change our planet's future. Rainforests are also a source of many common foods, such as chocolate, vanilla, pineapples, and cinnamon, as well as raw products like latex, resins, dyes, and lumber. In addition, they provide us with twenty to twenty-five percent of all medicines. (One or more rainforest plants may hold the cure for cancer.)

Why are we destroying this richest of resources? Why do we cut down or burn rainforests in Africa, Asia, and Latin America at a rate of some 25,000 square miles per year? This destruction is rooted in a complex combination of social, economic, and biological causes: Loggers looking for easy profits open up the rainforests; poor migrant farmers follow, seeking fertile soil for their crops. And, as habitats vanish, animals disappear at the rate of a thousand species a year.

Depletion of the rainforests is only one of many hot ecological issues that include damage to the ozone layer, effects of radiation, and disposal of hazardous wastes. In assessing the impact humans have on the earth, scientists are looking for relationships between variables. In mathematics such relationships are called functions. This chapter discusses functions from both an algebraic and a geometric (graphing) point of view. We begin by looking at the two-dimensional coordinate system.

2 FUNCTIONS AND GRAPHS

2.1 The Cartesian Coordinate System
2.2 Functions
2.3 Graphs of Relations and Functions
2.4 Transformations and Symmetry of Graphs
2.5 Operations with Functions
2.6 Inverse Functions
2.7 Variation

2.1 The Cartesian Coordinate System

In Chapter 1 we studied equations and inequalities in one variable, whose solution sets are found on the number line. In this section we begin studying pairs of variables. For example, p might represent the price of gasoline and n the number of gallons that you consume in a month; x might be the number of toppings on a medium pizza and y the cost of that pizza; or h might be the height of a two-year-old child and w the weight. Since we cannot study relationships between pairs of variables on a one-dimensional number line, in this section we develop a two-dimensional coordinate system.

Ordered Pairs

Table 2.1

Toppings x	Cost y
0	\$ 5
1	7
2	9
3	11
4	13

Suppose that the cost of a medium pizza is \$5 plus \$2 per topping. Table 2.1 shows a chart in which x is the number of toppings and y is the cost. To indicate that $x = 3$ and $y = 11$ go together, we can use the **ordered pair** (3, 11). The ordered pair (2, 9) indicates that a two-topping pizza costs \$9. The **first coordinate** of the ordered pair represents the number of toppings and the **second coordinate** represents the cost. The assignment of what the coordinates represent is arbitrary, but once made it is kept fixed. For example, in the present context the ordered pair (11, 3) indicates that an 11-topping pizza costs $3, and it is not the same as (3, 11). This is the reason they are called "ordered pairs." Recall that in Chapter 1 we used parentheses to indicate intervals of real numbers, and now we are using the same notation for ordered pairs. However, the meaning of this notation should always be clear from the discussion.

Just as sets of real numbers are pictured on a number line, sets of ordered pairs are pictured on a plane in the **rectangular coordinate system** or **Cartesian coordinate system**, named after the French mathematician René Descartes (1596–1650). The Cartesian coordinate system consists of two number lines drawn perpendicular to one another, intersecting at zero on each number line as in Fig. 2.1. The point at which they intersect is called the **origin**. On the horizontal number line, the ***x*-axis**, the positive numbers are to the right of the origin, and on the vertical number line, the ***y*-axis**, the positive numbers are above the origin. The two number lines divide the plane into four regions called **quadrants**, numbered as in Fig. 2.1. The quadrants do not include any points on the axes.

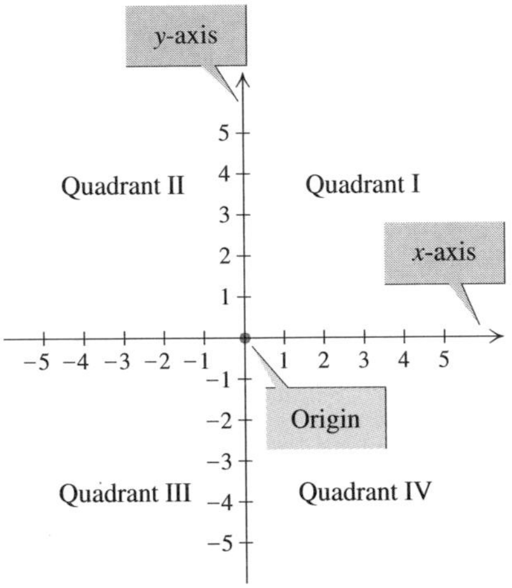

Figure 2.1

We call a plane with a rectangular coordinate system the **coordinate plane** or the ***xy*-plane**. Just as every real number corresponds to a point on the number line, every ordered pair of real numbers (a, b) corresponds to a point P in the xy-plane. For this reason, ordered pairs of numbers are often called **points**. The numbers a and b are called the **coordinates** of point P. Locating the point P that corresponds to (a, b) in the xy-plane is referred to as **plotting** or **graphing** the point, and P is called the **graph** of (a, b).

EXAMPLE 1 Plotting points

Plot the points (3, 5), (4, −5), (−3, 4), (−2, −5), and (0, 2) in the xy-plane.

Solution

The point (3, 5) is located three units to the right of the origin and five units above the x-axis as shown in Fig. 2.2. The point (4, −5) is located four units to the right of the origin and five units below the x-axis. The point (−3, 4) is located three units to the left of the origin and four units above the x-axis. The point (−2, −5) is located two units to the left of the origin and five units below the x-axis. The point (0, 2) is on the y-axis because its first coordinate is zero.

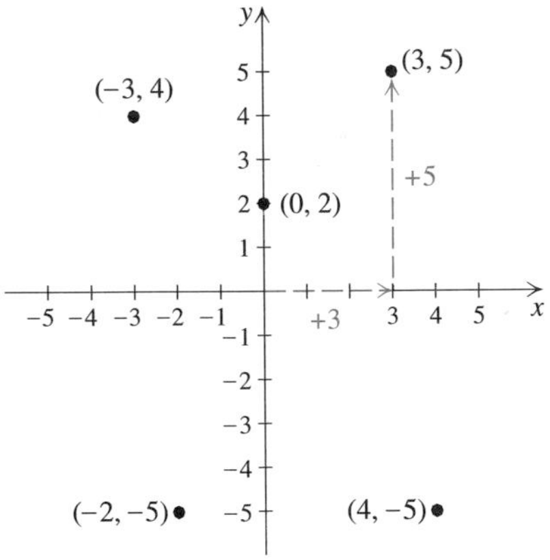

Figure 2.2

Note that for points in quadrant I, both coordinates are positive. In quadrant II the first coordinate is negative and the second is positive, while in quadrant III, both coordinates are negative. In quadrant IV the first coordinate is positive and the second is negative.

Graphing a Line

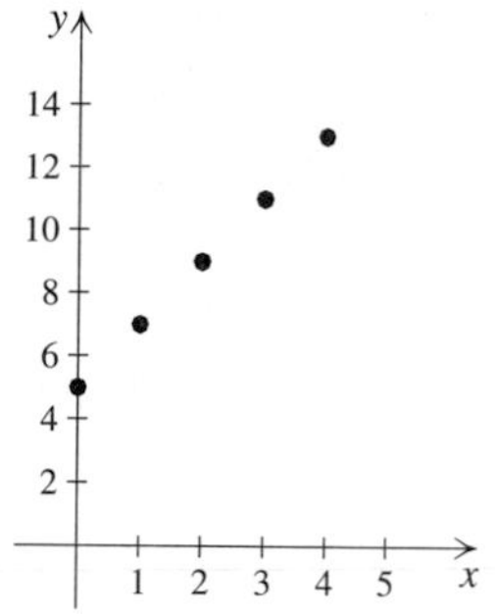

Figure 2.3

If a medium pizza costs \$5 plus \$2 per topping, then the equation $y = 2x + 5$ expresses the cost in terms of the number of toppings. If $x = 3$ and $y = 11$, then the equation becomes $11 = 2 \cdot 3 + 5$, a true statement. The ordered pair (3, 11) **satisfies** the equation. If $x = 4$, then $y = 2 \cdot 4 + 5 = 13$. So (4, 13) satisfies the equation. The following table shows five ordered pairs that satisfy $y = 2x + 5$.

x	0	1	2	3	4
$y = 2x + 5$	5	7	9	11	13

These ordered pairs are graphed in Fig. 2.3.

If we think of $y = 2x + 5$ as an equation without its application to the cost of a pizza, then there are infinitely many ordered pairs that satisfy $y = 2x + 5$, all lying along a line through the points of Fig. 2.3. For example, (0.5, 6), $(-1, 3)$, and $(-2, 1)$ satisfy $y = 2x + 5$ and lie on the same line as the other points. The set of all ordered pairs that satisfy $y = 2x + 5$ is the **solution set** of the equation. The solution set is expressed in set notation as $\{(x, y) \mid y = 2x + 5\}$, and its graph is shown in Fig. 2.4. The graph of the solution set to an equation is called the **graph of the equation.**

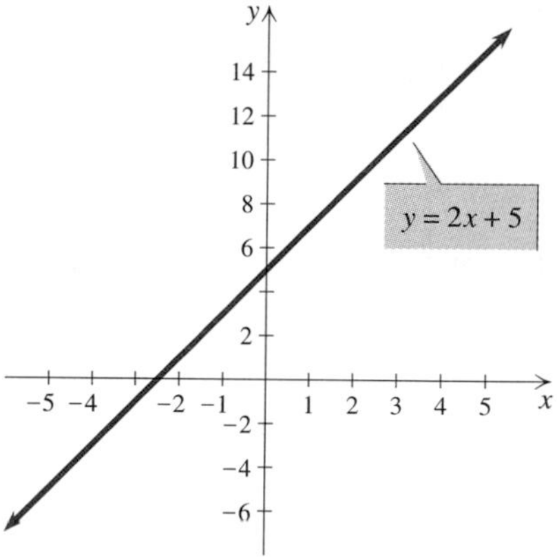

Figure 2.4

To graph an equation such as $y = 2x + 5$ on a graphing calculator, define the equation with the $Y=$ key, set the viewing window, then press GRAPH as shown in Figs. 2.5(a), (b), and (c). Because of the resolution, a calculator graph does not usually look as good as a hand-drawn graph and might not show some important features. However, the speed and accuracy of a calculator more than compensate for its lack of artistic ability. A typical calculator can instantly plot about 100 points that satisfy an equation. But the points shown on a calculator can be misleading and often need some interpretation. In the sections ahead we will see how to best use the graphing capabilities of a graphing calculator. □

In this chapter we will learn that graphs of equations come in many different shapes. It will save a great deal of time and effort in graphing if we can identify the type of graph an equation has before plotting any points. For example, the graph of any linear equation in two variables is a straight line, and that is why it is called a *linear equation.*

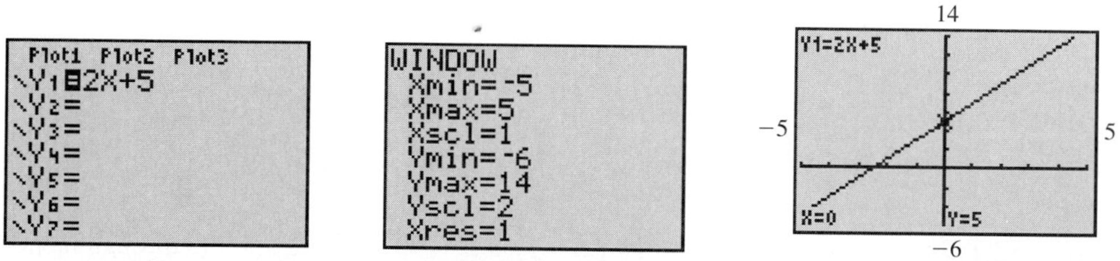

(a) (b) (c)

Figure 2.5

Definition: Linear Equation in Two Variables

A **linear equation** in two variables x and y is an equation of the form

$$Ax + By = C,$$

where A, B, and C are real numbers and A and B are not both zero.

The form $Ax + By = C$ is the **standard form** for a linear equation. So the graphs of the equations

$$x + y = 5, \quad y = \frac{1}{2}x - 9, \quad x = 4, \quad \text{and} \quad y = 5$$

are straight lines, because they are all variations of the standard form.

There is only one line containing any two distinct points in the xy-plane. So to graph a linear equation, we need locate only two points that satisfy the equation and draw a line through them. However, it is good practice to find three points as a check, because it is unlikely that all three points will be on the same line if an error has been made.

EXAMPLE 2 Graphing a linear equation

Sketch the graph of each linear equation.

a) $x - 2y = 4$ **b)** $y = 40 - x$

Solution

a) It is easier to find ordered pairs that satisfy the equation if the equation is solved for y.

$$x - 2y = 4$$
$$-2y = -x + 4$$
$$y = \frac{1}{2}x - 2$$

Arbitrarily select $x = 0$, 2, and 4 and find the corresponding y-values shown in the table.

x	0	2	4
$y = \frac{1}{2}x - 2$	-2	-1	0

Plot $(0, -2)$, $(2, -1)$, and $(4, 0)$ and draw a line through them as in Fig. 2.6. The calculator graph shown in Fig. 2.7 is consistent with the graph in Fig. 2.6. □

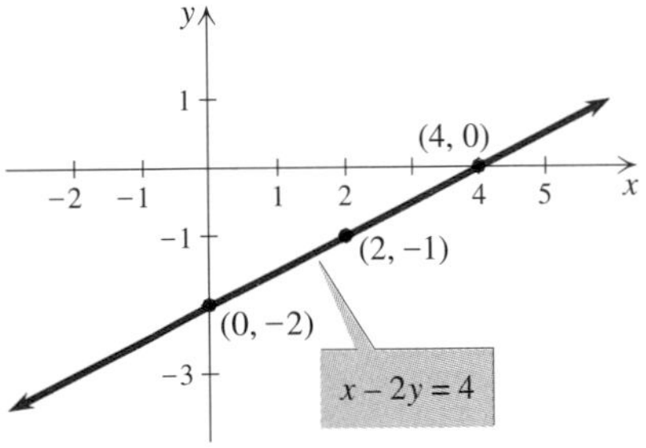

Figure 2.6

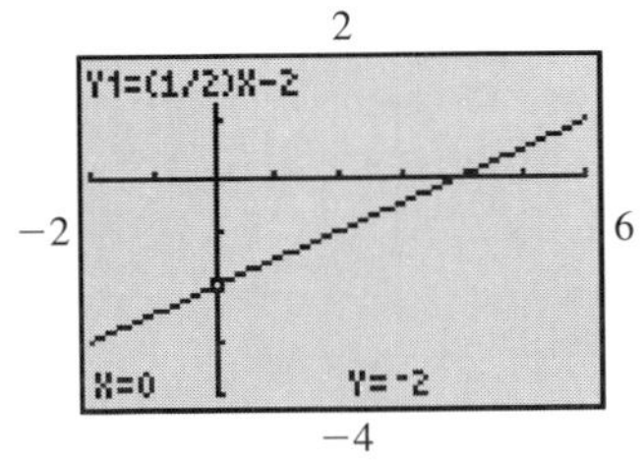

Figure 2.7

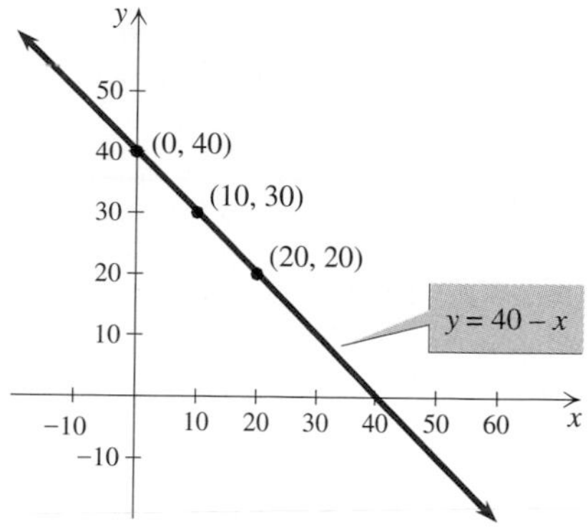

Figure 2.8

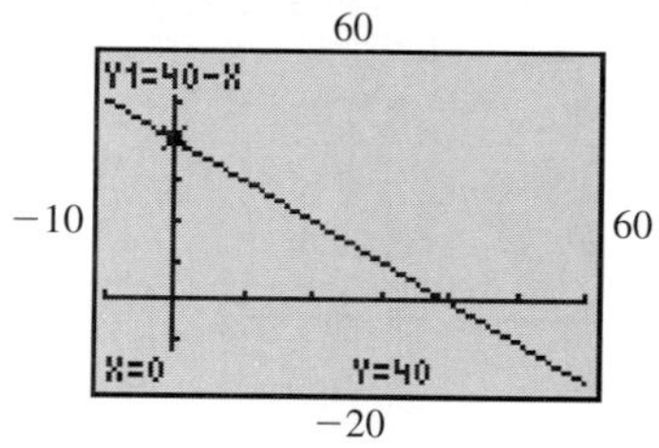

Figure 2.9

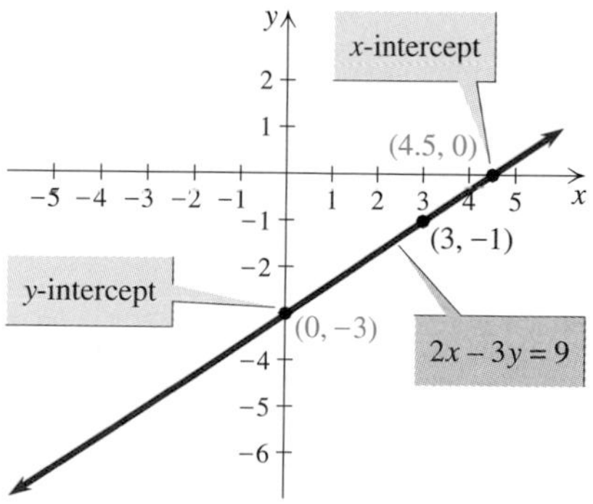

Figure 2.10

b) Arbitrarily select $x = 0$, 10, and 20 and make a table of ordered pairs.

x	0	10	20
$y = 40 - x$	40	30	20

Label the axes to accommodate these ordered pairs. Plot (0, 40), (10, 30), and (20, 20) and draw a line through them as shown in Fig. 2.8.

The calculator graph shown in Fig. 2.9 is consistent with the graph in Fig. 2.8. Note that the viewing window must be adjusted to accommodate the graph.

Graphing Lines Using the Intercepts

A point at which a graph intersects the x-axis is an ***x*-intercept**, and a point at which a graph intersects the y-axis is a ***y*-intercept**. Since these points have 0 for one of the coordinates, they are usually easier to find than other points on the line. We often graph a linear equation by finding the intercepts.

EXAMPLE 3 Using the intercepts to graph a line

Graph the equation $2x - 3y = 9$ by finding the intercepts.

Solution

Since the y-coordinate of the x-intercept is 0, we replace y by 0 in the equation:

$$2x - 3(0) = 9$$
$$x = 4.5$$

Since the x-coordinate of the y-intercept is 0, we replace x by 0 in the equation:

$$2(0) - 3y = 9$$
$$y = -3$$

The x-intercept is (4.5, 0), and the y-intercept is (0, −3). Locate the intercepts and draw the line as shown in Fig. 2.10. To check, locate a point such as (3, −1), which also satisfies the equation, and see if the line goes through it.

The x-coordinate of the x-intercept for $y = ax + b$ satisfies the equation $ax + b = 0$. For that reason it is called a **zero** or **root**.

To solve $ax + b = 0$ with a graphing calculator, first graph $y = ax + b$, then use the ZERO or ROOT feature of the calculator to find the x-intercept. □

EXAMPLE 4 Using a graph to solve an equation

Use a graphing calculator to solve $0.55(x - 3.45) + 13.98 = 0$.

Solution

First graph $y = 0.55(x - 3.45) + 13.98$ using a viewing window that shows the x-intercept as in Fig. 2.11(a). Next press ZERO on the CALC menu. The calculator can find a zero between a *left bound* and a *right bound*, which you must enter. The calculator also asks you to make a guess. See Fig. 2.11(b). The more accurate the guess, the faster the calculator will find the zero. The solution to the equation rounded to two decimal places is -21.97. See Fig. 2.11(c).

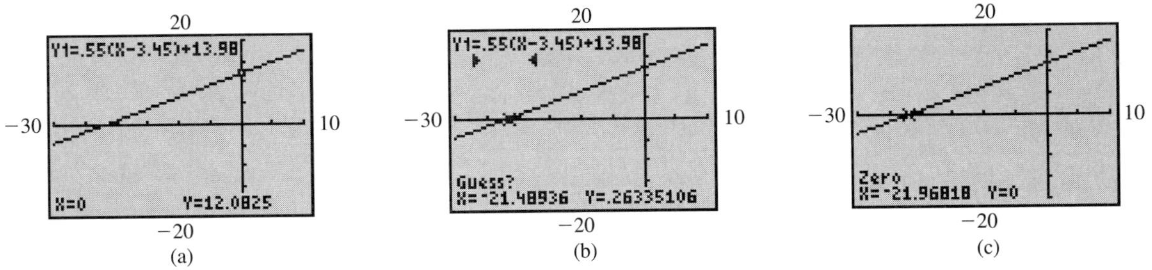

Figure 2.11

EXAMPLE 5 Graphing horizontal and vertical lines

Sketch the graph of each equation in the rectangular coordinate system.

a) $y = 3$ **b)** $x = 4$

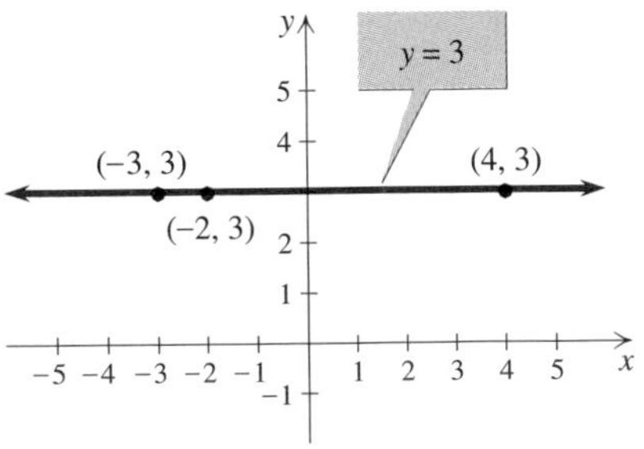

Figure 2.12

Solution

a) The equation $y = 3$ is equivalent to $0 \cdot x + y = 3$. Because x is multiplied by 0, we can choose any value for x as long as we choose 3 for y. So ordered pairs such as $(-3, 3)$, $(-2, 3)$, and $(4, 3)$ satisfy the equation $y = 3$. The graph of $y = 3$ is the horizontal line shown in Fig. 2.12.

b) The equation $x = 4$ is equivalent to $x + 0 \cdot y = 4$. Because y is multiplied by 0, we can choose any value for y as long as we choose 4 for x. So ordered

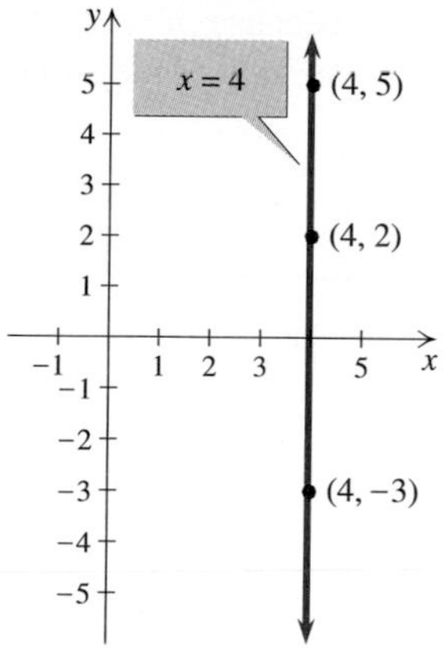

Figure 2.13

pairs such as (4, −3), (4, 2), and (4, 5) satisfy the equation $x = 4$. The graph of $x = 4$ is the vertical line shown in Fig. 2.13.

Note that most graphing calculators graph only equations that are solved for y. So you cannot graph $x = 4$ on many graphing calculators.

Distance Between Two Points

We have been studying sets of points that satisfy linear equations. Their graphs all have the same geometric shape, the straight line. We now turn our attention to some other geometric figures.

Consider the points $A(-1, 5)$ and $B(3, 2)$ as shown in Fig. 2.14. We use the notation $\overline{AB}$ for the line segment with endpoints A and B and AB for the length of that line segment. $\overline{AB}$ is the hypotenuse of the right triangle with a right angle at $C(3, 5)$ shown in Fig. 2.14. If A and C are thought of as lying on a horizontal number line, then we can use the formula for the distance between two points on a number line to get $AC = |-1 - 3| = 4$. Likewise $CB = |5 - 2| = 3$. The Pythagorean theorem says that a triangle is a right triangle if and only if the sum of the squares of the legs is equal to the hypotenuse squared. So by the Pythagorean theorem

$$AB = \sqrt{3^2 + 4^2} = \sqrt{25} = 5.$$

The Pythagorean theorem can also be used to find a general formula for the distance between two arbitrary points $A(x_1, y_1)$ and $B(x_2, y_2)$ shown in Fig. 2.15. BC depends only on the y-coordinates of points B and C, and $BC = |y_2 - y_1|$. Similarly, $AC = |x_2 - x_1|$. Using the Pythagorean theorem with d being the distance between A and B, we have

$$d^2 = |x_2 - x_1|^2 + |y_2 - y_1|^2.$$

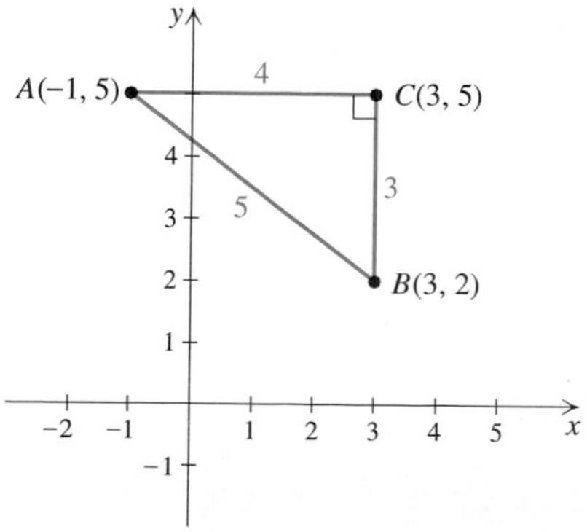

Figure 2.14

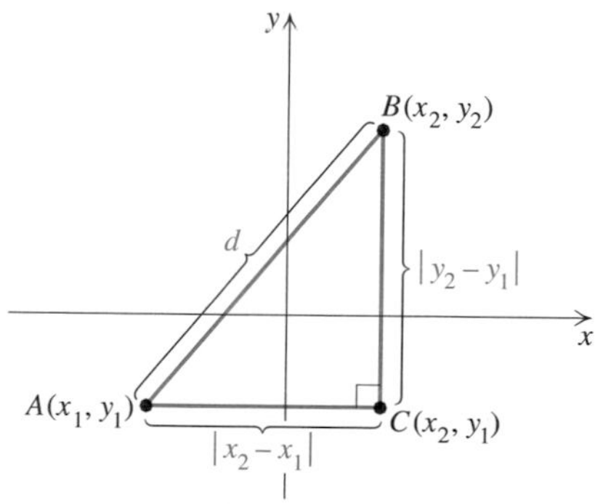

Figure 2.15

The absolute value symbols can be replaced by parentheses, because $|a - b|^2 = (a - b)^2$ for any values of a and b. Since the distance d must be a nonnegative number, we have the following result.

The Distance Formula

The distance d between the points (x_1, y_1) and (x_2, y_2) is given by the formula

$$d = \sqrt{(x_2 - x_1)^2 + (y_2 - y_1)^2}.$$

EXAMPLE 6 Finding the distance between two points

Find the distance between $(5, -3)$ and $(-1, -6)$.

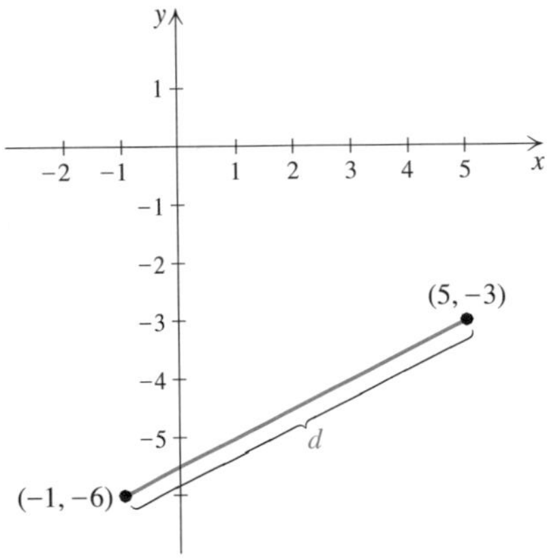

Figure 2.16

Solution

Let $(x_1, y_1) = (5, -3)$ and $(x_2, y_2) = (-1, -6)$. These points are shown on the graph in Fig. 2.16. Substitute these values into the distance formula:

$$\begin{aligned} d &= \sqrt{(-1 - 5)^2 + (-6 - (-3))^2} \\ &= \sqrt{(-6)^2 + (-3)^2} \\ &= \sqrt{36 + 9} = \sqrt{45} = 3\sqrt{5} \end{aligned}$$

The exact distance between the points is $3\sqrt{5}$.

Note that the distance between two points is the same regardless of which point is chosen as (x_1, y_1) or (x_2, y_2).

The distance formula can be used to establish properties of geometric figures drawn in the coordinate plane. The study of geometric figures in the coordinate plane is called **coordinate geometry** or **analytical geometry**. The main purpose of the Cartesian coordinate system is to help us visualize abstract concepts. In every problem involving coordinate geometry, you should first plot the points, draw the picture, and get a visual idea of what is going on.

EXAMPLE 7 Deciding whether a triangle is a right triangle

Determine whether the triangle with vertices $(-3, -5)$, $(3, -3)$, and $(0, 6)$ is a right triangle.

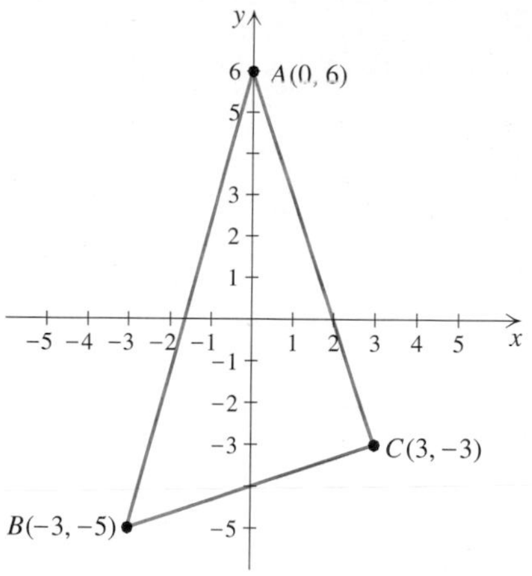

Figure 2.17

Solution

The Pythagorean theorem can be used to determine whether a triangle is a right triangle. Plot the three points and find the length of each side of the triangle of Fig. 2.17:

$$AC = \sqrt{(3 - 0)^2 + (-3 - 6)^2} = \sqrt{90}$$
$$BC = \sqrt{(3 - (-3))^2 + (-3 - (-5))^2} = \sqrt{40}$$
$$AB = \sqrt{(-3 - 0)^2 + (-5 - 6)^2} = \sqrt{130}$$

Since $(\sqrt{90})^2 + (\sqrt{40})^2 = (\sqrt{130})^2$, we have $(AC)^2 + (BC)^2 = (AB)^2$. So by the Pythagorean theorem, the triangle with the given vertices is a right triangle and $\overline{AB}$ is the hypotenuse.

The Midpoint Formula

With geometric figures we are often interested in knowing whether one line segment bisects another. This question may be easy to decide if we know the location of the midpoint of a line segment. For the line segment with endpoints -3 and 9 on the number line shown in Fig. 2.18, the midpoint is 3. We find the midpoint just as we would average two test scores, $(-3 + 9)/2 = 3$.

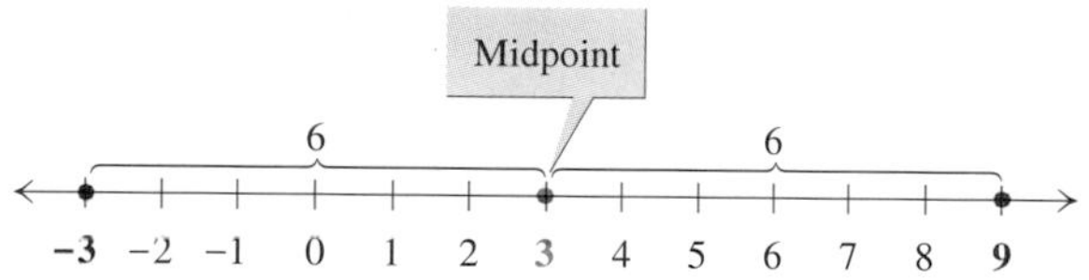

Figure 2.18

In general, $(a + b)/2$ is the midpoint of the line segment with endpoints a and b shown in Fig. 2.19. (See Exercises 89 and 90.)

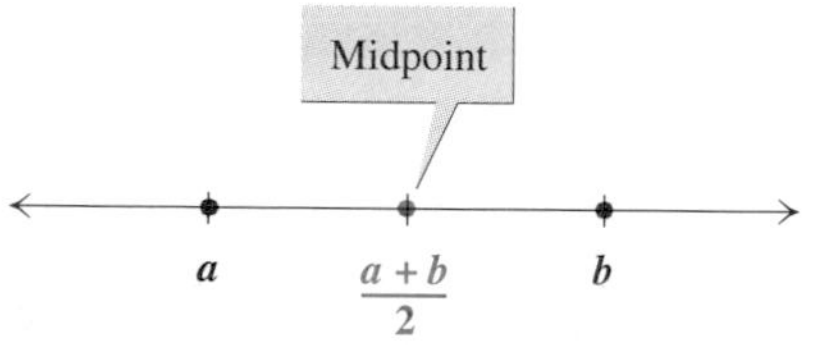

Figure 2.19

The formula for the midpoint on the number line can be extended to a similar formula for the midpoint of a line segment in the xy-plane.

Theorem: The Midpoint Formula

The midpoint of the line segment with endpoints (x_1, y_1) and (x_2, y_2) is

$$\left(\frac{x_1 + x_2}{2}, \frac{y_1 + y_2}{2}\right).$$

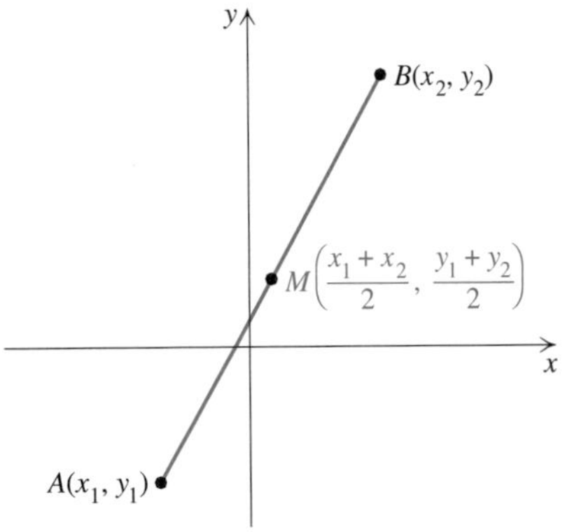

Figure 2.20

Proof To prove that M is the midpoint of $\overline{AB}$ as shown in Fig. 2.20, we will show that $AM = MB$ and that A, M, and B are colinear (on the same line). To prove that they are colinear, it is sufficient to show that $AM + MB = AB$. (If $AM + MB > AB$, then they form a triangle.) First we use the distance formula to find AM and MB:

$$AM = \sqrt{\left(\frac{x_1 + x_2}{2} - x_1\right)^2 + \left(\frac{y_1 + y_2}{2} - y_1\right)^2} = \sqrt{\left(\frac{x_2 - x_1}{2}\right)^2 + \left(\frac{y_2 - y_1}{2}\right)^2}$$

$$MB = \sqrt{\left(x_2 - \frac{x_1 + x_2}{2}\right)^2 + \left(y_2 - \frac{y_1 + y_2}{2}\right)^2} = \sqrt{\left(\frac{x_2 - x_1}{2}\right)^2 + \left(\frac{y_2 - y_1}{2}\right)^2}$$

Since $AM = MB$, we can conclude that M is equidistant from A and B. The expressions for AM and MB can be further simplified by factoring 2^2 out of each denominator:

$$AM = MB = \frac{1}{2}\sqrt{(x_2 - x_1)^2 + (y_2 - y_1)^2}$$

Since $AB = \sqrt{(x_2 - x_1)^2 + (y_2 - y_1)^2}$, we have $AM + MB = AB$. So A, M, and B are colinear. ■

EXAMPLE 8 Using the midpoint formula

Prove that the diagonals of the parallelogram with vertices (0, 0), (1, 3), (5, 0), and (6, 3) bisect each other.

Solution

The parallelogram is shown in Fig. 2.21. The midpoint of the diagonal from (0, 0) to (6, 3) is

$$\left(\frac{0 + 6}{2}, \frac{0 + 3}{2}\right),$$

or (3, 1.5). The midpoint of the diagonal from (1, 3) to (5, 0) is

$$\left(\frac{1 + 5}{2}, \frac{3 + 0}{2}\right),$$

or (3, 1.5). Since the diagonals have the same midpoint, they bisect each other. ■

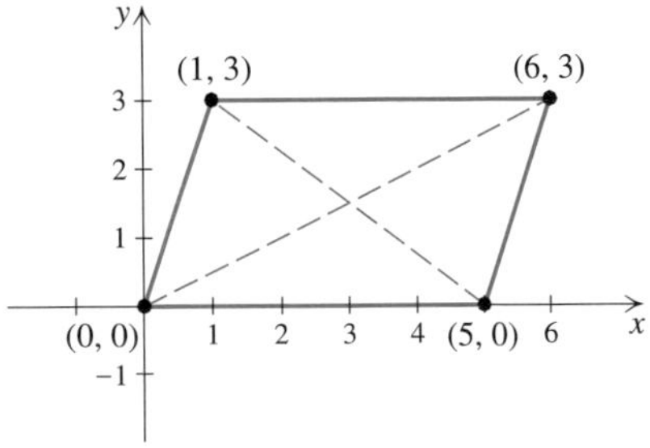

Figure 2.21

For Thought True or False? Explain.

1. The point $(2, -3)$ is in quadrant II.
2. The point $(4, 0)$ is in quadrant I.
3. The distance between (a, b) and (c, d) is $\sqrt{(a-b)^2 + (c-d)^2}$.
4. The equation $3x^2 + y = 5$ is a linear equation.
5. The solution to $7x - 9 = 0$ is the x-coordinate of the x-intercept of $y = 7x - 9$.
6. $\sqrt{7^2 + 9^2} = 7 + 9$
7. The origin lies midway between $(1, 3)$ and $(-1, -3)$.
8. The distance between $(3, -7)$ and $(3, 3)$ is 10.
9. The x-intercept for the graph of $3x - 2y = 7$ is $(7/3, 0)$.
10. For a and b on a number line, $\frac{a+b}{3}$ is a point between a and b.

2.1 Exercises

Tape 5 Disk

In Exercises 1–10, for each point shown in the xy-plane, write the corresponding ordered pair and name the quadrant in which it lies or the axis on which it lies.

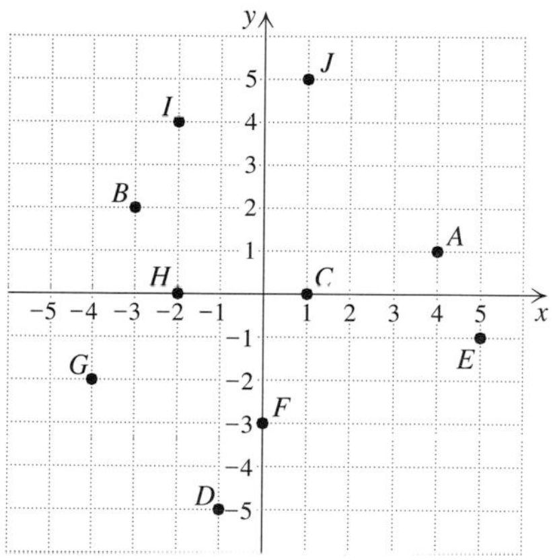

1. A 2. B 3. C 4. D
5. E 6. F 7. G 8. H
9. I 10. J

Graph each set of ordered pairs.

11. $\{(-1, 0), (0, 1), (1, 2)\}$
12. $\{(-3, 3), (0, 0), (3, 3)\}$
13. $\{(x, y) \mid y = x\}$
14. $\{(x, y) \mid y = -x\}$
15. $\{(2, 2), (-2, 2), (-2, -2), (2, -2)\}$
16. $\{(-2, 4), (-1, 1), (0, 0), (1, 1), (2, 4)\}$
17. $\{(x, y) \mid y = -2x + 3\}$
18. $\{(x, y) \mid y = -3x + 5\}$
19. $\{(1980, 100), (1981, 200), (1982, 400)\}$
20. $\{(1, 300), (5, 400), (3, 100)\}$
21. $\{(x, y) \mid x + 0 \cdot y = 3\}$
22. $\{(x, y) \mid 0 \cdot x + y = -2\}$

Sketch the graph of each linear equation.

23. $y = 3x - 4$
24. $y = 5x - 5$
25. $3x - y = 6$
26. $5x - 2y = 10$
27. $x + y = 80$
28. $2x + y = -100$
29. $x = 3y - 90$
30. $x = 80 - 2y$
31. $\frac{1}{2}x - \frac{1}{3}y = 600$
32. $\frac{2}{3}y - \frac{1}{2}x = 400$

Graph each equation by finding the x- and y-intercepts.

33. $5x + 4y = 20$
34. $3x - 2y = -6$
35. $y = 3x - 2$
36. $y = -2x + 5$
37. $2x + 4y = 0.01$
38. $3x - 5y = 1.5$
39. $4y - 3x = 60$
40. $2y - 4x = 1$
41. $\frac{1}{2}x + \frac{1}{3}y = 20$
42. $\frac{1}{4}x - \frac{2}{3}y = 60$
43. $0.03x + 0.06y = 150$
44. $0.09x - 0.06y = 54$

Use a graphing calculator to estimate the solution to each equation to two decimal places. Then find the solution algebraically and compare it with your estimate.

45. $1.2x + 3.4 = 0$
46. $3.2x - 4.5 = 0$
47. $1.23x - 687 = 0$
48. $-2.46x + 1500 = 0$
49. $0.03x - 3497 = 0$
50. $0.09x + 2000 = 0$

51. $4.3 - 3.1(2.3x - 9.9) = 0$

52. $3.7(2.3x + 9.4) - 4.37(3.5x - 9.76) = 0$

Graph each equation in the rectangular coordinate system.

53. $x = 5$ **54.** $y = -2$ **55.** $y = 4$

56. $x = -3$ **57.** $x = -4$ **58.** $y = 5$

59. $y - 1 = 0$ **60.** $5 - x = 4$

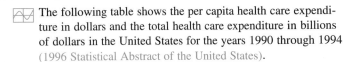

The following table shows the per capita health care expenditure in dollars and the total health care expenditure in billions of dollars in the United States for the years 1990 through 1994 (1996 Statistical Abstract of the United States).

Table for Exercises 61 and 62

Year	Per capita expenditure ($)	Total expenditure (billions of $)
1990	2688	697.5
1991	2902	761.3
1992	3144	833.6
1993	3331	892.3
1994	3510	949.4

61. a. Use linear regression with a graphing calculator to find a linear equation that expresses the per capita expenditure E in terms of the year y and graph the equation. See Example 3 in Section 1.2.

b. Use your equation to predict the per capita health care expenditure for 1995.

c. Find the actual per capita health care expenditure for 1995 in the Statistical Abstract of the United States and compare it to your answer to part (b). (Try the Internet.)

d. Do you think that in the year 2001 a self-employed person could buy a good health insurance policy for $500 per month? Explain your answer.

62. a. Use linear regression with a graphing calculator to find a linear equation that expresses the total health expenditure T in terms of the year y and graph the equation.

b. Use your equation to predict the total health care expenditure in the U.S. for the year 2000.

For each pair of points find the distance between them and the midpoint of the line segment joining them.

63. $(1, 3), (4, 7)$ **64.** $(-3, -2), (9, 3)$

65. $(-1, -2), (1, 0)$ **66.** $(-1, 0), (1, 2)$

67. $(-1, 1), (-1 + 3\sqrt{3}, 4)$

68. $(1 + \sqrt{2}, -2), (1 - \sqrt{2}, 2)$

69. $(1.2, 4.8), (-3.8, -2.2)$ **70.** $(-2.3, 1.5)(4.7, -7.5)$

71. $(a, 0), (b, 0)$ **72.** $(a, 0), \left(\frac{a + b}{2}, 0\right)$

73. $(\pi, 0), (\pi/2, 1)$ **74.** $(0, 0), (\pi/2, 1)$

Solve each problem.

75. Determine whether the points $(-6, 3)$, $(-2, -2)$, and $(3, 1)$ are the vertices of a right triangle.

76. Determine whether the points $(-5, 3)$, $(-1, -2)$, and $(4, 2)$ are the vertices of a right triangle.

77. Determine whether the points $(-2, 2)$, $(-2, -3)$, and $(1, 1)$ are the vertices of an isosceles triangle.

78. Determine whether the points $(-1, -1)$, $(-1, 5)$, and $(4, 2)$ are the vertices of an equilateral triangle.

79. Find the value of k so that the distance between $(3, -2)$ and $(5, k)$ is 6.

80. Find all points on the x-axis that are eight units from $(2, 6)$.

81. Find (a, b) such that $(2, 5)$ is the midpoint of the line segment joining $(-1, 3)$ and (a, b).

82. Find a and b such that $(2, 5)$ is the midpoint of the line segment joining $(a, 3)$ and $(7, b)$.

83. Show that the points $(0, 0)$, $(2, -2)$, $(-2, -3)$, and $(-4, -1)$ are the vertices of a parallelogram.

84. Show that the points $(-4, -3)$, $(-2, 2)$, $(3, 4)$, and $(1, -1)$ are the vertices of a rhombus.

85. *Women and Marriage* The percentage of women in the 20–24 age group who have married went from 64.2% in 1970 to 33.3% in 1996 as shown in the figure (Census Bureau). Find the midpoint of the line segment in the figure and interpret your result.

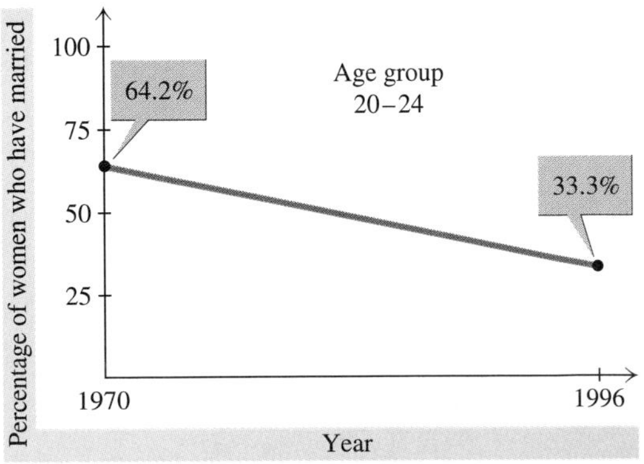

Figure for Exercise 85

86. *Women on the Board* The percentage of companies with at least one woman on the board is growing steadily as shown in the figure (*Forbes*, February 10, 1997). The percentage can be approximated with the linear equation $p = 2n + 36$, where n is the number of years since 1980. Find and interpret the p-intercept for the line. Find and interpret the n-intercept for the line.

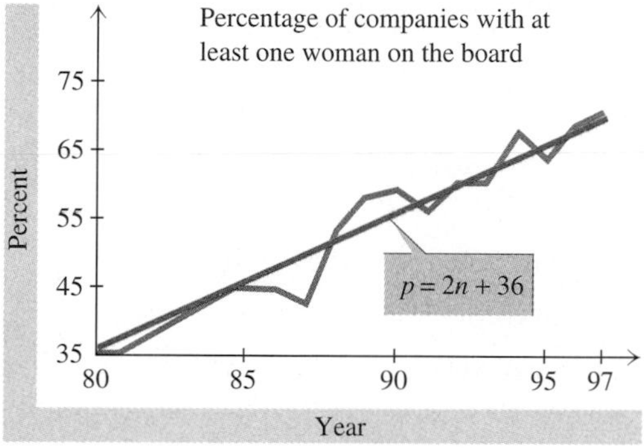

Figure for Exercise 86

87. *Capsize Control* The capsize screening value C is an indicator of a sailboat's suitability for extended offshore sailing (*Sail*, September 1997). C is determined by the formula

$$C = 4D^{-1/3}B,$$

where D is the displacement of the boat in pounds and B is its beam (or width) in feet. Sketch the graph of this equation for B ranging from 0 to 20 ft assuming that D is fixed at 22,800 lbs. Find C for the Island Packet 40, which has a displacement of 22,800 pounds and a beam of 12 feet 11 inches.

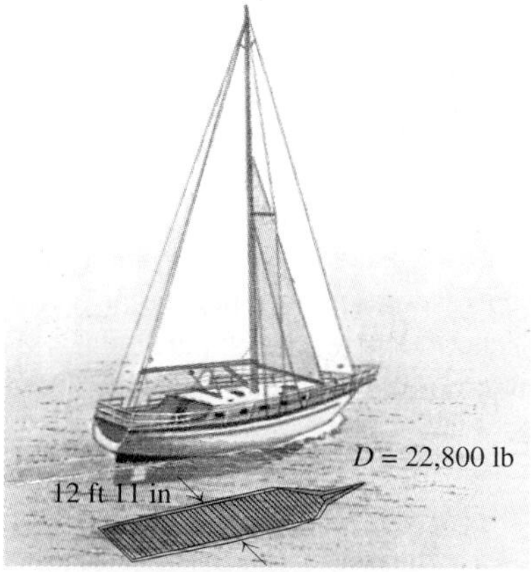

Figure for Exercises 87 and 88

88. *Limiting the Beam* The International Offshore Rules require that the capsize screening value C (from the previous exercise) be less than or equal to 2 for safety. What is the maximum allowable beam (to the nearest inch) for a boat with a displacement of 22,800 lbs? For a fixed displacement, is a boat more or less likely to capsize as its beam gets larger?

89. Prove that if a and b are real numbers such that $a < b$, then $a < (a + b)/2 < b$.

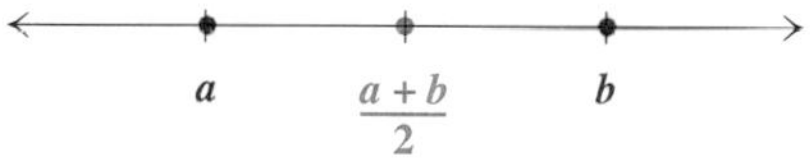

Figure for Exercises 89 and 90

90. Prove that the point $(a + b)/2$ is equidistant from a and b on the number line for any real numbers a and b. The proof of this statement together with the previous exercise proves that $(a + b)/2$ is the midpoint of the segment with endpoints a and b on a number line.

91. Show that the points $A(-4, -5)$, $B(1, 1)$, and $C(6, 7)$ are colinear. (*Hint:* Points A, B, and C lie on a straight line if $AB + BC = AC$.)

92. Show that the midpoint of the hypotenuse of any right triangle is equidistant from all three vertices.

For Writing/Discussion

93. *Finding Points* Can you find two points such that their coordinates are integers and the distance between them is 10? $\sqrt{10}$? $\sqrt{19}$? Explain.

94. *Plotting Points* Plot at least five points in the xy-plane that satisfy the inequality $y > 2x$. Give a verbal description of the solution set to $y > 2x$.

95. *Cooperative Learning* Work in a small group to plot the points $(-1, 3)$ and $(4, 1)$ on graph paper. Assuming that these two points are adjacent vertices of a square, find the other two vertices. Now select your own pair of adjacent vertices and ''complete the square.'' Next, start with the points (x_1, y_1) and (x_2, y_2) as adjacent vertices of a square and write expressions for the coordinates of the other two vertices.

96. *Cooperative Learning* Repeat the previous exercise assuming that the first two points are opposite vertices of a square.

Linking Concepts

For Individual or Group Explorations

Basic Energy Requirements *Clinical dietitians must design diets for patients that meet their basic energy requirements and are suitable for the condition of their health. The basic energy requirement B (in calories) for a male is given by the formula*

$$B = 655.096 + 9.563W + 1.85H - 4.676A,$$

where W is the patient's weight in kilograms, H is the height in centimeters, and A is the age in years. For a female the formula is

$$B = 66.473 + 13.752W + 5.003H - 6.755A.$$

a) Find your basic energy requirement.

b) Replace H and A with your actual height and age. Now draw a graph showing how B depends on weight for suitable values of W.

c) For a fixed height and age, does the basic energy requirement increase or decrease as weight increases?

d) Replace W and H with your actual weight and height. Now draw a graph showing how B depends on age for suitable values of A.

e) For a fixed weight and height, does the basic energy requirement increase or decrease as a person gets older?

f) Replace W and A with your actual weight and height. Now draw a graph showing how B depends on height for suitable values of H.

g) For a fixed weight and age, does the basic energy requirement increase or decrease for taller persons?

h) For the three graphs that you have drawn, explain how to determine whether the graph is increasing or decreasing from the formula for the graph.

2.2 Functions

In Section 2.1 we studied ordered pairs of numbers and graphed sets of ordered pairs. For some sets of ordered pairs we used an equation such as $y = 2x + 5$ to determine the second coordinate corresponding to a particular first coordinate. In this section we continue to study sets of ordered pairs, particularly those in which the second coordinate is determined uniquely by the first coordinate.

The Idea of a Function

Any set of ordered pairs is called a **relation**. Consider the relations defined by Tables 2.2 and 2.3. This data is from the 1996 Statistical Abstract of the United

Table 2.2 U.S. iron ore production

Number of mines	Price per ton
37	$36.56
32	40.39
27	41.72
20	43.61
20	44.17
20	42.03
19	35.63
19	31.88

Table 2.3 U.S. primary energy production

Year	Quadrillion Btu
1989	66
1990	71
1991	70
1992	70
1993	68
1994	70

States. Table 2.2 shows the number of iron ore mines in operation in the United States for each year beginning in 1980 and the average price per ton at the mine. Table 2.3 shows the energy production of the United States for various years. Each table defines a relation, but there is an important difference between the tables. In Table 2.2 the second coordinate, price per ton, is not uniquely determined by the first coordinate, number of mines, because when 20 mines were in operation the price was either $43.61, $44.17, or $42.03. Energy production is determined by the year in Table 2.3, because each year corresponds to a single production level. For example, in 1989 the production was 66 quadrillion Btu of energy. Note that in Table 2.3 no first coordinate is paired with two different second coordinates. In this case we say that the second coordinate *is a function of* the first coordinate.

The Definition of Function

To keep the definition of function simple and precise, we state it in terms of ordered pairs.

Definition: Function

A **function** is a set of ordered pairs in which no two ordered pairs have the same first coordinate and different second coordinates.

The values of the first coordinate of an ordered pair are thought of as values of a first or **independent variable**, while the values of the second coordinate are thought of as values of a second or **dependent variable**. A function is thought of as a rule that indicates how values of the independent variable are paired with values of the dependent variable. We say that the dependent variable **is a function of** the independent variable if and only if the set of ordered pairs determined by these variables is a function. Ordered pairs in the following example are given in set notation and as a table, but as we will see later, ordered pairs may also be given by a verbal description, a formula, or a graph.

EXAMPLE 1 Using the definition of function

Determine whether each relation is a function.

a) $\{(1, 3), (2, 3), (4, 9)\}$ **b)** $\{(9, -3), (9, 3), (4, 2), (0, 0)\}$

c)

Quantity	Price each
1–5	\$9.40
5–10	\$8.75

Solution

a) This set of ordered pairs is a function because no two ordered pairs have the same first coordinate and different second coordinates.

b) This set of ordered pairs is not a function because both $(9, 3)$ and $(9, -3)$ are in the set and they have the same first coordinate and different second coordinates.

c) The quantity 5 corresponds to a price of \$9.40 and also to a price of \$8.75. Assuming that quantity is the first coordinate, the ordered pairs (5, \$9.40) and (5, \$8.75) both belong to this relation. If you are purchasing items whose price was determined from this table, you would certainly say that something is wrong, the table has a mistake in it, or the price you should pay is not clear. The price is not a function of the quantity purchased.

Certain keys on a nongraphing calculator are referred to as the **function keys**. These are the keys marked $\sqrt{x}$, x^2, $x!$, $1/x$, 10^x, e^x, $\sin(x)$, $\cos(x)$, $\tan(x)$, etc. If an appropriate number for x is on the display and one of these keys is pressed, then the corresponding y-coordinate is calculated and displayed. The ordered pairs certainly satisfy the definition of function because the calculator will not produce two different second coordinates corresponding to one first coordinate.

We have seen and used many functions as formulas. For example, the formula $c = \pi d$ defines a set of ordered pairs in which the first coordinate is the diameter of a circle and the second coordinate is the circumference. Since each diameter corresponds to a unique circumference, the circumference is a function of the diameter. If the set of ordered pairs satisfying an equation is a function, then we say that the equation is a function or the equation defines a function. Other well-known formulas such as $C = \frac{5}{9}(F - 32)$, $A = \pi r^2$, and $V = \frac{4}{3}\pi r^3$ are also functions, but, as we will see in the next example, not every equation defines a function. The variables in the next example and all others in this text represent real numbers unless indicated otherwise.

EXAMPLE 2 Using the definition of function

Determine whether each equation defines y as a function of x.

a) $|y| = x$ **b)** $y = x^2 - 3x + 2$ **c)** $x^2 + y^2 = 1$ **d)** $3x - 4y = 8$

Solution

a) Since $|2| = 2$ and $|-2| = 2$, the ordered pairs $(2, 2)$ and $(2, -2)$ both satisfy the equation $|y| = x$. So this equation does *not* define y as a function of x.

b) Since y is determined by $y = x^2 - 3x + 2$, for every x-coordinate there is only one y-coordinate. So this equation *does* define y as a function of x.

c) Since the ordered pairs $(0, 1)$ and $(0, -1)$ both satisfy $x^2 + y^2 = 1$, this equation does *not* define y as a function of x. Note that $x^2 + y^2 = 1$ is equivalent to $y = \pm\sqrt{1 - x^2}$.

d) The equation $3x - 4y = 8$ is equivalent to $y = \frac{3}{4}x - 2$. Since each x determines only one y, this equation *does* define y as a function of x.

In the next example we find a formula for a function relating two measurements in a geometric figure.

EXAMPLE 3 Finding a formula for a function

Given that a square has diagonal of length d and side of length s, write the area A as a function of the length of the diagonal.

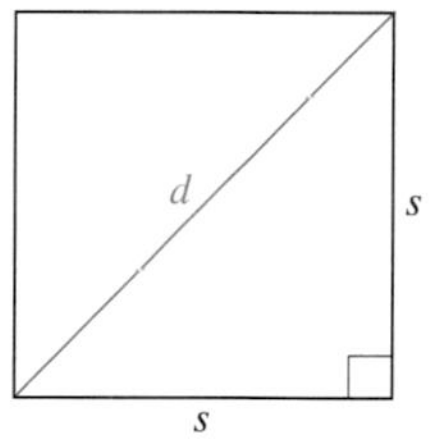

Figure 2.22

Solution

The area of any square is given by $A = s^2$. The diagonal is the hypotenuse of a right triangle as shown in Fig. 2.22. By the Pythagorean theorem, $d^2 = s^2 + s^2$, $d^2 = 2s^2$, or $s^2 = d^2/2$. Since $A = s^2$ and $s^2 = d^2/2$, we get the formula

$$A = \frac{d^2}{2}$$

expressing the area of the square as a function of the length of the diagonal.

Domain and Range

A relation is a set of ordered pairs. The **domain** of a relation is the set of all first coordinates of the ordered pairs. The **range** of a relation is the set of all second coordinates of the ordered pairs. The relation

$$\{(19, 2.4), (27, 3.0), (19, 3.6), (22, 2.4), (36, 3.8)\}$$

shows the ages and grade point averages of five randomly selected students. The domain of this relation is the set of ages, {19, 22, 27, 36}. The range is the set of grade point averages, {2.4, 3.0, 3.6, 3.8}. This relation matches elements of the domain (ages) with elements of the range (grade point averages), as shown in Fig. 2.23. Is this relation a function?

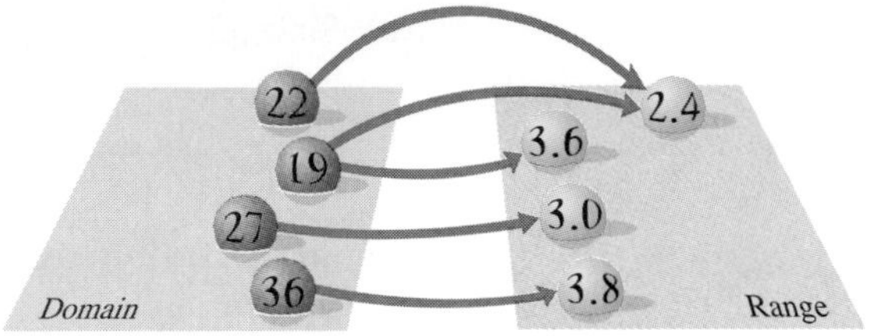

Figure 2.23

For some relations, all of the ordered pairs are listed, but for others, only an equation is given for determining the ordered pairs. When the domain of the relation is not stated, it is understood that the domain consists of only values of the independent variable that can be used in the expression defining the relation. When we use x and y for the variables, we always assume that x is the independent variable and y is the dependent variable.

EXAMPLE 4 Determining domain and range

State the domain and range of each relation and whether the relation is a function.

a) $\{(-1, 1), (3, 9), (3, -9)\}$ **b)** $y = \sqrt{2x - 1}$ **c)** $x = |y|$

Solution

a) The domain is the set of first coordinates $\{-1, 3\}$, and the range is the set of second coordinates $\{1, 9, -9\}$. Note that an element of a set is not listed more than once. Since (3, 9) and (3, −9) are in the relation, the relation is not a function.

b) Since y is determined uniquely from x by the formula $y = \sqrt{2x - 1}$, y is a function of x. We are discussing only real numbers. So $\sqrt{2x - 1}$ is a real number provided that $2x - 1 \geq 0$, or $x \geq 1/2$. So the domain of the function is the interval $[1/2, \infty)$. If $2x - 1 \geq 0$ and $y = \sqrt{2x - 1}$, we have $y \geq 0$. So the range of the function is the interval $[0, \infty)$.

The graph of $y = \sqrt{2x - 1}$ shown in Fig. 2.24 supports these answers because the points that are plotted appear to have $x \geq 1/2$ and $y \geq 0$. □

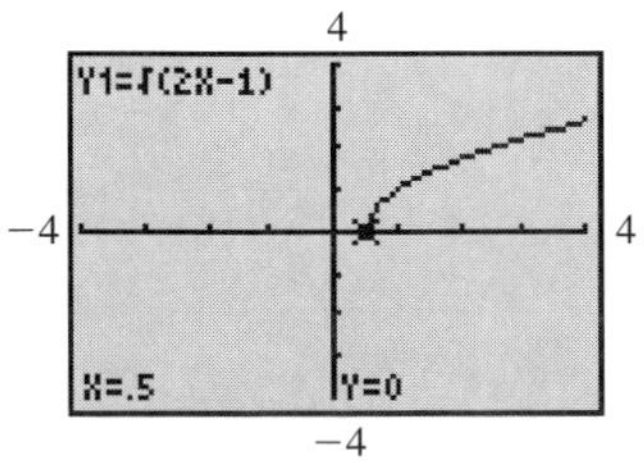

Figure 2.24

c) The expression $|y|$ is defined for any real number y. So the range is the interval of all real numbers, $(-\infty, \infty)$. Since $|y|$ is nonnegative, the values of x must be nonnegative. So the domain is $[0, \infty)$. Since ordered pairs such as (2, 2) and (2, −2) satisfy $x = |y|$, this equation does not give y as a function of x.

In Example 4(b) and (c) we found the domain and range by examining an equation defining a relation. If the relation is a function as in Example 4(b), you can easily draw the graph with a graphing calculator and use it to support your answer. However, to choose an appropriate viewing window you must know the domain and range to begin with. So it is best to use a calculator graph to support your conclusions about domain and range rather than to make conclusions about domain and range. □

The *f*-Notation

If y is a function of x, we may use the expression $f(x)$, rather than y, for the dependent variable. The notation $f(x)$ is read "the value of f at x" or simply "f of x." So y and $f(x)$ are both symbols for the value of the second coordinate when the first coordinate is x, and we may write $y = f(x)$. For example, we could write $f(x) = \sqrt{2x - 1}$ for the function $y = \sqrt{2x - 1}$ from Example 4(b). The notation $f(x)$ is called **function notation** or ***f*-notation**.

EXAMPLE 5 Using *f*-notation

Find each of the following for the function $f(x) = \sqrt{2x - 1}$.

a) $f(5)$ **b)** x, if $f(x) = 0$ **c)** $f(-3)$ **d)** $f(a + 2)$

Solution

a) The expression $f(5)$ is the second coordinate when the first coordinate is 5. To find the value of $f(5)$ replace x by 5 in the formula $f(x) = \sqrt{2x - 1}$:

$$f(5) = \sqrt{2 \cdot 5 - 1} = \sqrt{9} = 3$$

b) To find a value of x so that the second coordinate is 0, we must solve $\sqrt{2x - 1} = 0$. Since 0 is the only number whose square root is 0, we have $2x - 1 = 0$ or $x = 1/2$.

c) The expression $f(-3)$ is undefined, because in Example 4 we found that the domain of $f(x) = \sqrt{2x - 1}$ is the interval $[1/2, \infty)$.

d) The expression $f(a + 2)$ is the second coordinate when the first coordinate is $a + 2$. So replace x by $a + 2$ in the formula $f(x) = \sqrt{2x - 1}$:

$$f(a + 2) = \sqrt{2(a + 2) - 1} = \sqrt{2a + 3}$$

In Example 5(d) the binomial $a + 2$ was used as the x-coordinate. In fact, any expression (whose values are in the domain of the function) can be used in

Table 2.4

First coordinate	Find second coordinate using $f(x) = 3x + 1$	Ordered pair
4	$f(4) = 3(4) + 1 = 13$	$(4, 13)$
a	$f(a) = 3a + 1$	$(a, 3a + 1)$
$a + 2$	$f(a + 2) = 3(a + 2) + 1 = 3a + 7$	$(a + 2, 3a + 7)$
$x - 5$	$f(x - 5) = 3(x - 5) + 1 = 3x - 14$	$(x - 5, 3x - 14)$

place of x. The function notation such as $f(x) = 3x + 1$ provides a rule for finding the second coordinate: Multiply the first coordinate (whatever it is) by 3 and then add 1. The x in this notation is called a **dummy variable** because the letter used is unimportant. We could write $f(t) = 3t + 1$,

$$f(\text{first coordinate}) = 3(\text{first coordinate}) + 1,$$

or even $f(\quad) = 3(\quad) + 1$ to convey the same idea. It is often necessary to even use expressions involving x as the first coordinate. For example, if $f(x) = 3x + 1$, then to find $f(x - 4)$, replace x by $x - 4$:

$$f(x - 4) = 3(x - 4) + 1 = 3x - 11$$

Study the additional examples given in Table 2.4.

The f-notation is especially convenient for naming functions when more than one function is under discussion. For clarity, we name different functions with different letters. For example, let

$$f = \{(1, 4), (2, 5), (3, 6)\} \quad \text{and} \quad g = \{(1, 9), (5, 7)\}.$$

The expression $f(1)$ is the value of the second coordinate when the first coordinate is 1 in the function named f. So $f(1) = 4$, while $g(1) = 9$.

EXAMPLE 6 Using f-notation

Given that $f(x) = x^2 - 2$ and $g(x) = 2x - 3$, find and simplify each of the following expressions.

a) $f(-3)$ **b)** $g(5)$ **c)** $f(a + 1)$ **d)** $g(x + h)$ **e)** $f(g(x))$

Solution

a) $f(-3) = (-3)^2 - 2 = 7$

b) $g(5) = 2 \cdot 5 - 3 = 7$

c) $f(a + 1) = (a + 1)^2 - 2$ Replace x by $a + 1$ in $f(x) = x^2 - 2$.

$= a^2 + 2a + 1 - 2$

$= a^2 + 2a - 1$

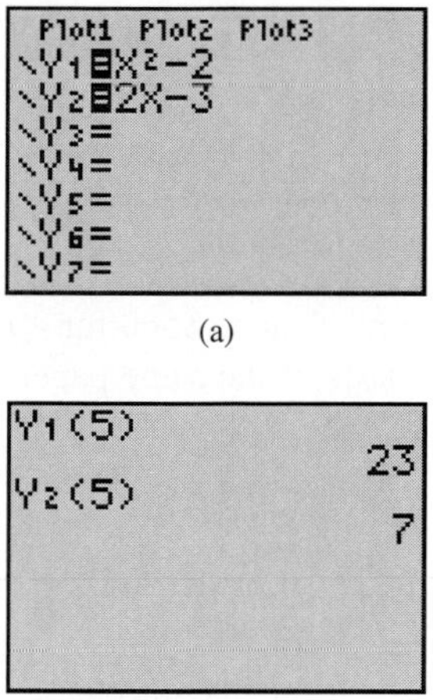

(a)

(b)

Figure 2.25

d) $g(x + h) = 2(x + h) - 3$ Replace x by $x + h$ in $g(x) = 2x - 3$.

$= 2x + 2h - 3$

e) $f(g(x)) = f(2x - 3)$ Replace $g(x)$ by $2x - 3$, because $g(x) = 2x - 3$.

$= (2x - 3)^2 - 2$ Use $2x - 3$ in placce of x in $f(x) = x^2 - 2$.

$= 4x^2 - 12x + 7$ Simplify.

A graphing calculator uses subscripts to indicate different functions. For example, if $y_1 = x^2 - 2$ and $y_2 = 2x - 3$, then $y_1(5) = 23$ and $y_2(5) = 7$ as shown in Figs. 2.25(a) and (b).

If a function describes some real application, then a letter that fits the situation is usually used. For example, if watermelons are three dollars apiece, then the cost of x watermelons is given by the function $C(x) = 3x$. The cost of five watermelons is $C(5) = 3 \cdot 5 = \$15$. In trigonometry the abbreviations sin, cos, and tan are used rather than a single letter to name the trigonometric functions. The dependent variables are written as $\sin(x)$, $\cos(x)$, and $\tan(x)$.

• CALCULUS •

The Average Rate of Change of a Function

At the 1996 Atlanta Olympics, Michael Johnson ran the fastest 200-meter race in the history of sports (*Newsweek*, August 12, 1996). At the first 100 meters, Johnson's time was 10.12 seconds; at 200 meters, as he crossed the finish line, his time was 19.32 seconds. There is a function here pairing times to locations. Two ordered pairs of this function are (10.12, 100) and (19.32, 200). Johnson's average speed over the time interval from 10.12 seconds to 19.32 seconds is his change in location divided by his change in time:

$$\frac{\text{change in location}}{\text{change in time}} = \frac{200 - 100}{19.32 - 10.12} = \frac{100}{9.20} \approx 10.87 \text{ m/s}$$

Johnson did not run at a constant 10.87 meters per second. This rate is the average rate at which his location changed for the time between 10.12 seconds and 19.32 seconds.

Definition: Average Rate of Change from x_1 to x_2

If (x_1, y_1) and (x_2, y_2) are two ordered pairs of a function, we define the **average rate of change** of the function as x varies from x_1 to x_2 as

$$\frac{y_2 - y_1}{x_2 - x_1}.$$

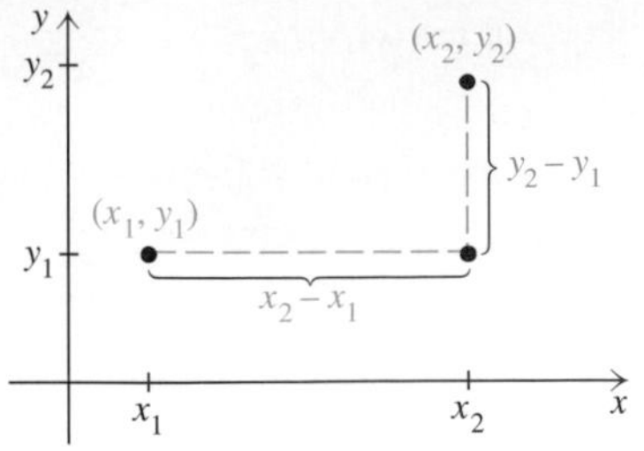

Figure 2.26

The average rate of change is the ratio of the vertical change to the horizontal change between the points (x_1, y_1) and (x_2, y_2). See Fig. 2.26.

EXAMPLE 7 Average rate of change of a function

A wallpaper company specializing in antique reproductions charges \$560 for 40 yards of hand silk-screened wallpaper and \$780 for 60 yards of the same paper. Find the average rate of change of the cost as the length varies from 40 to 60 yards.

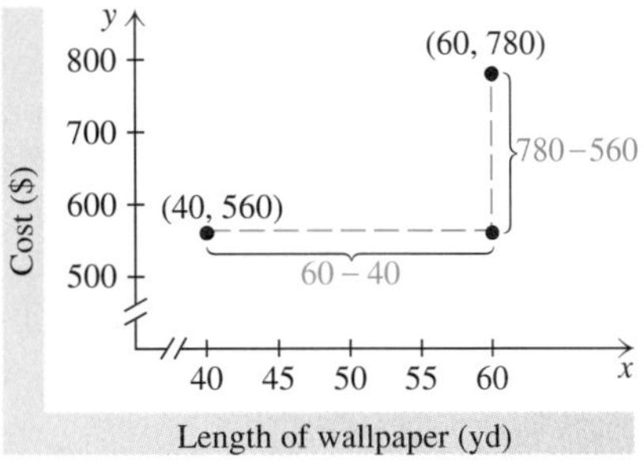

Figure 2.27

Solution

Figure 2.27 shows the ordered pairs (40, 560) and (60, 780). Note that the change in cost is the vertical distance between the points and the change in length is the horizontal distance. The average rate of change of the cost is the change in cost divided by the change in length.

$$\frac{\text{change in cost}}{\text{change in length}} = \frac{y_2 - y_1}{x_2 - x_1} = \frac{780 - 560}{60 - 40} = \frac{220}{20} = 11$$

The average rate of change of the cost is \$11 per yard as the length varies from 40 yards to 60 yards. Note that the company does not charge \$11 per yard for 40 yards of wallpaper, but the charge changes by \$11 per yard for every yard between 40 and 60.

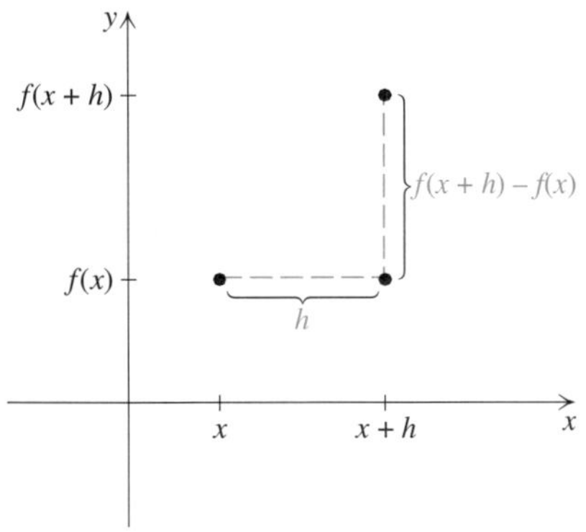

Figure 2.28

In calculus we are often interested in the average rate of change of a function f from $x_1 = x$ to $x_2 = x + h$, where $h \neq 0$. The function contains the points $(x, f(x))$ and $(x + h, f(x + h))$. See Fig. 2.28. In this case, $x_2 - x_1 = x + h - x = h$ and $y_2 - y_1 = f(x + h) - f(x)$. With this notation the average rate of change is written as

$$\frac{y_2 - y_1}{x_2 - x_1} = \frac{f(x + h) - f(x)}{h}.$$

Definition: Difference Quotient

The expression

$$\frac{f(x + h) - f(x)}{h}$$

for $h \neq 0$ is called the **difference quotient**.

In the difference quotient, h is usually thought of as a number very close to 0, so that the average rate of change can be found for a very short interval.

• CALCULUS •

EXAMPLE 8 Finding a difference quotient

Find and simplify the difference quotient for each of the following functions.

a) $j(x) = 3x + 2$ **b)** $f(x) = x^2 - 2x$ **c)** $g(x) = \sqrt{x}$

Solution

a)
$$\begin{aligned}\frac{j(x+h) - j(x)}{h} &= \frac{3(x+h) + 2 - (3x+2)}{h}\\ &= \frac{3x + 3h + 2 - 3x - 2}{h}\\ &= \frac{3h}{h} = 3\end{aligned}$$

b)
$$\begin{aligned}\frac{f(x+h) - f(x)}{h} &= \frac{[(x+h)^2 - 2(x+h)] - (x^2 - 2x)}{h}\\ &= \frac{x^2 + 2xh + h^2 - 2x - 2h - x^2 + 2x}{h}\\ &= \frac{2xh + h^2 - 2h}{h}\\ &= 2x + h - 2\end{aligned}$$

c)
$$\begin{aligned}\frac{g(x+h) - g(x)}{h} &= \frac{\sqrt{x+h} - \sqrt{x}}{h}\\ &= \frac{(\sqrt{x+h} - \sqrt{x})(\sqrt{x+h} + \sqrt{x})}{h(\sqrt{x+h} + \sqrt{x})} \quad \text{Rationalize the numerator.}\\ &= \frac{x + h - x}{h(\sqrt{x+h} + \sqrt{x})}\\ &= \frac{1}{\sqrt{x+h} + \sqrt{x}}\end{aligned}$$

Note that in Examples 8(b) and 8(c) the average rate of change of the function depends on the values of x and h, while in Example 8(a) the average rate of change of the function is constant. In Example 8(c) the expression does not look much different after rationalizing the numerator than it did before. However, we did remove h as a factor of the denominator, and in calculus it is often necessary to perform this step.

FOR THOUGHT True or False? Explain.

1. Any set of ordered pairs is a function.
2. If $f = \{(1, 1), (2, 4), (3, 9)\}$, then $f(5) = 25$.
3. The domain of $f(x) = 1/x$ is $(-\infty, 0) \cup (0, \infty)$.
4. Each student's exam grade is a function of the student's IQ.
5. If $f(x) = x^2$, then $f(x + h) = x^2 + h$.
6. The domain of $g(x) = |x - 3|$ is $[3, \infty)$.
7. The range of $y = 8 - x^2$ is $(-\infty, 8]$.
8. The equation $x = y^2$ does not define y as a function of x.
9. If $f(t) = \dfrac{t-2}{t+2}$, then $f(0) = -1$.
10. The set $\{(\frac{3}{8}, 8), (\frac{4}{7}, 7), (0.16, 6), (0.375, 5)\}$ is a function.

2.2 EXERCISES

Tape 5 Disk

Determine whether each relation is a function.

1. $\{(-1, -1), (2, 2), (3, 3)\}$
2. $\{(0.5, 7), (0, 7), (1, 7), (9, 7)\}$
3. $\{(25, 5), (25, -5), (0, 0)\}$
4. $\{(1, \pi), (30, \pi/2), (60, \pi/4)\}$

5.

x	y
3	6
4	9
3	12

6.

x	y
1	6.98
5	5.98
9	6.98

7. $\{(x, y) | y = (x + 2)^2\}$
8. $\{(x, y) | x = y^2 + 1\}$
9. $\{(x, y) | x = |2y|\}$
10. $\{(x, y) | 2x + y = 5\}$
11. $\{(x, y) | x = 6y\}$
12. $\{(x, y) | y^2 = x^4\}$
13. $\{(x, y) | x = y^2 \text{ and } y \geq 0\}$
14. $\{(x, y) | x + 0 \cdot y = 5\}$

Determine whether each equation defines y as a function of x.

15. $y = ax + b$
16. $y = ax^2 + bx + c$
17. $x = 3y - 9$
18. $x = y^2 - 6y + 9$
19. $x^2 = y^2$
20. $y^2 - x^2 = 9$
21. $y^3 = x$
22. $x = \sqrt{y}$
23. $y + 2 = |x|$
24. $\dfrac{x^2}{4} + \dfrac{y^2}{9} = 1$

For each pair of variables determine whether a is a function of b, b is a function of a, or neither.

25. a is the radius of any U.S. coin and b is its circumference.
26. a is the length of any rectangle with a width of 5 in. and b is its perimeter.
27. a is the length of any piece of U.S. paper currency and b is its denomination.
28. a is the diameter of any U.S. coin and b is its value.
29. a is the universal product code for an item at Wal Mart and b is its price.
30. a is the final exam score for a student in your class and b is his/her semester grade.
31. a is the time spent studying for the final exam for a student in your class and b is the student's final exam score.
32. a is the age of an adult male and b is his shoe size.
33. a is the height of a car in inches and b is its height in centimeters.
34. a is the cost for mailing a first class letter and b is its weight.

Determine the domain and range of each relation.

35. $\{(-3, 1), (\pi, \sqrt{2}), (-3, 6), (5, 6)\}$

36. $\{(1, 1/2), (2, 1/4), (3, 1/8), (4, 1/16)\}$

37. $y = |x| + 5$ **38.** $y = x^2 + 8$ **39.** $x = -y^2$

40. $x + 3 = |y|$ **41.** $y = 5 + \sqrt{x}$ **42.** $y = x^{1/3}$

43. $y = \sqrt{2x - 1}$ **44.** $y = \sqrt{5 - 2x}$

45. $\{(x, y) | 2x + 3y = 6\}$ **46.** $\{(x, y) | x = 5\}$

47. $\{(x, y) | 0 < x < 3 \text{ and } y = x + 2\}$

48. $\{(x, y) | y = x^2 \text{ and } |x| < 3\}$

Let $f(x) = 3x^2 - x$, $g(x) = 4x - 2$, and $k(x) = |x + 3|$. Find the following.

49. $f(2)$ **50.** $f(-4)$ **51.** $g(-1)$ **52.** $g(\ \ 2)$

53. $k(5)$ **54.** $k(-4)$ **55.** $f(0.56)$ **56.** $g(-0.5)$

57. $f(-1) + g(-1)$ **58.** $f(3) \cdot k(3)$

59. $g(-5) - k(-5)$ **60.** $\dfrac{f(1)}{g(1)}$

61. $f(a)$ **62.** $g(b)$ **63.** $f(a + 3)$

64. $g(b + 2)$ **65.** $f(x + h)$ **66.** $g(x + h)$

67. $f(x + h) - f(x)$ **68.** $g(x + h) - g(x)$

69. $f(x) + g(x)$ **70.** $f(x) - g(x)$ **71.** $f(x) \cdot g(x)$

72. $g\left(\dfrac{x + 2}{4}\right)$ **73.** x, if $f(x) = 0$ **74.** x, if $g(x) = 3$

75. a, if $k(a) = 4$ **76.** t, if $f(t) = 10$ **77.** $f(g(-1))$

78. $g(f(2))$ **79.** $f(k(-4))$ **80.** $k(g(-2))$

• CALCULUS •

Find the difference quotient for each function and simplify it.

81. $f(x) = 3x + 5$ **82.** $f(x) = -2x + 3$

83. $g(x) = 3x^2 + 1$ **84.** $g(x) = -2x^2 - 4$

85. $y = -x^2 + x - 2$ **86.** $y = x^2 - x + 3$

87. $f(x) = \sqrt{x + 2}$ **88.** $f(x) = \sqrt{\dfrac{x}{2}}$

89. $g(x) = \dfrac{1}{x}$ **90.** $g(x) = 3 + \dfrac{2}{x - 1}$

Consider a square with side of length s, diagonal of length d, perimeter P, and area A.

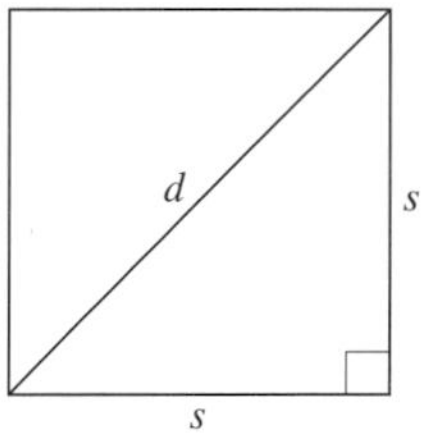

Figure for Exercises 91–98

91. Write A as a function of s.

92. Write s as a function of A.

93. Write s as a function of d.

94. Write d as a function of s.

95. Write P as a function of s.

96. Write s as a function of P.

97. Write A as a function of P.

98. Write d as a function of A.

99. *Aerobics* A woman in an aerobic dance class burns 353 calories per hour. Express the number of calories burned, C, as a function of the number of hours danced, n.

100. *Paper Consumption* The average American uses 580 lb of paper annually. Express the total annual paper consumption of the United States, P, as a function of the size of the population, n.

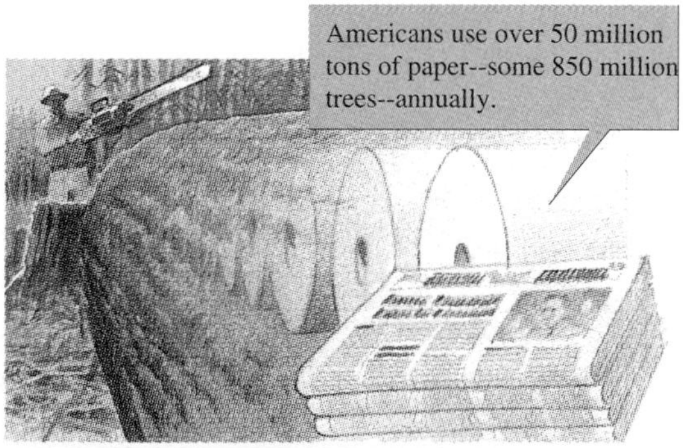
Americans use over 50 million tons of paper--some 850 million trees--annually.

Figure for Exercise 100

101. *Cost of Window Cleaning* A window cleaner charges \$50 per visit plus \$35 per hour. Express the total charge as a function of the number of hours worked, n.

102. *Hazards of Depth* The accompanying table shows depth below sea level and atmospheric pressure (Encyclopedia of Sports Science, 1997). The equation

$$A(d) = 0.03d + 1$$

expresses the atmospheric pressure as a function of depth d. Find the atmospheric pressure at the depth where nitrogen narcosis begins. Find the maximum depth for intermediate divers.

Table for Exercise 102

Depth (ft)	Atmospheric pressure (atm)	Comments
21	1.63	Bends are a danger
60	2.8	Maximum for beginners
100		Nitrogen narcosis begins
	4.9	Maximum for intermediate
200	7.0	Severe nitrogen narcosis
250	8.5	Extremely dangerous depth

103. *Medicare and Medicaid* The payment in billions by Medicare (health care for elderly) and Medicaid (health care for poor) can be modeled by the functions

$$E(n) = 15.6n + 99.8$$

and

$$P(n) = 12.6n + 81.3$$

respectively, where n is the number of years since 1990 (Healthcare Financing Administration).

a. Find $E(9) + P(9)$ and explain what it represents.

b. In what year will Medicaid spending reach \$250 billion?

104. *Costly Healthcare* The accompanying graph shows the Medicare and Medicaid spending from 1990 through 1998.

a. Do we spend less on Medicare or Medicaid?

b. Which program is growing faster?

c. In what year will total spending reach \$500 billion?

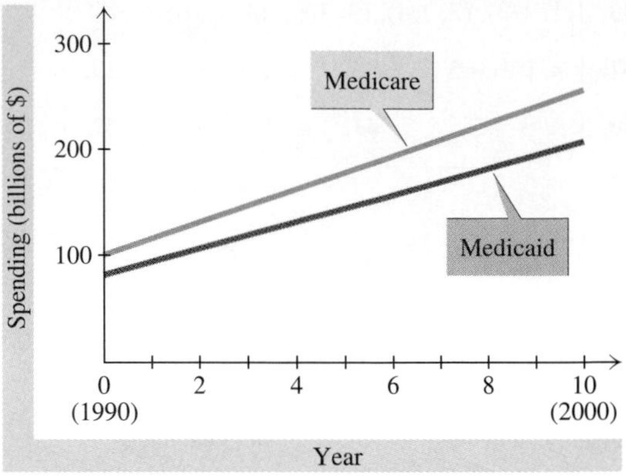

Figure for Exercises 103 and 104

105. *Depreciation of a Mustang* If a new Mustang LX is valued at \$16,000 and five years later it is valued at \$4000, then what is the average rate of change of its value during those five years?

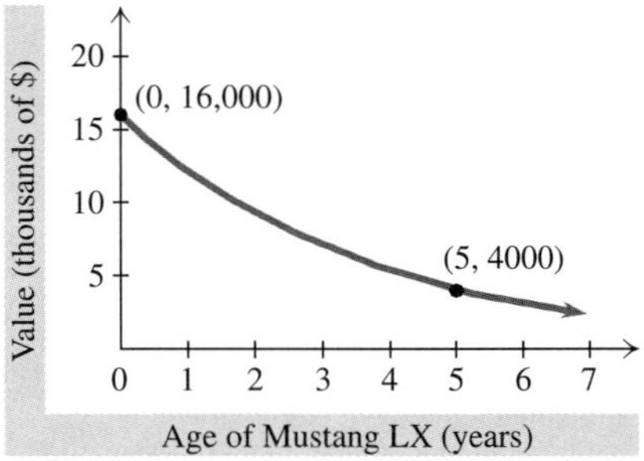

Figure for Exercise 105

106. *Cost of Gravel* Wilson's Sand and Gravel will deliver 12 yd^3 of gravel for \$240, 30 yd^3 for \$528, and 60 yd^3 for \$948. What is the average rate of change of the cost as the number of cubic yards varies from 12 to 30? What is the average rate of change as cubic yards varies from 30 to 60?

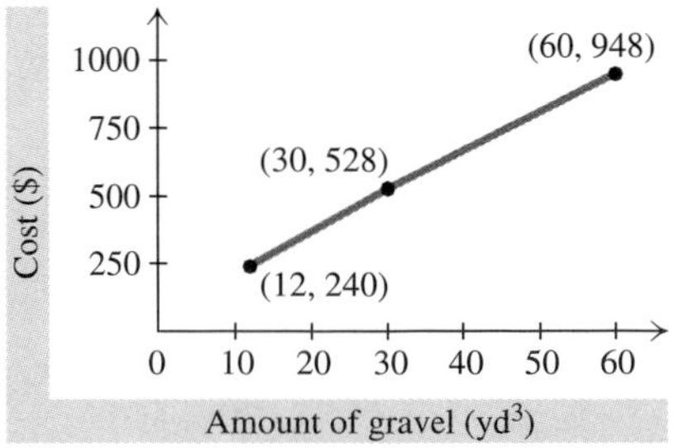

Figure for Exercise 106

107. *Urban Pollution* The accompanying graph shows how urban concentration of sulfur dioxide in the air is related to income throughout the world (World Resources 1996–97). Find the average rates of change of sulfur dioxide concentration as income varies from \$100 to \$2000 per capita, \$2000 to \$31,000 per capita, and \$100 to \$31,000 per capita. Which of these average rates of change makes the most sense considering the graph? Explain.

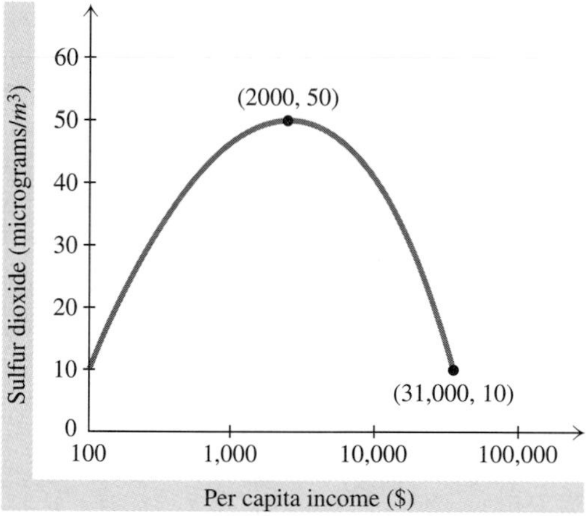

Figure for Exercise 107

108. *Bungee Jumping* Billy Joe McCallister jumped off the Tallahatchie Bridge, 70 ft above the water, with a bungee cord tied to his legs. If he was 6 ft above the water 2 sec after jumping, then what was the average rate of change of his altitude as the time varied from 0 to 2 sec?

109. *Deforestation* In 1984 the world's tropical moist forest covered approximately 1056 million hectares (1 hectare = 10,000 m^2). In 1996 the world's tropical moist forest covered approximately 982 million hectares. What was the average rate of change of the area of tropical moist forest over those 12 years?

110. *Elimination of Tropical Moist Forest* If the deforestation described in Exercise 109 continues at the same amount per year, then in which year will the tropical moist forest be totally eliminated?

• CALCULUS •

111. *Concert Revenue* The revenue in dollars from the sale of concert tickets at x dollars each is given by the function

$$R(x) = 20{,}000x - 500x^2.$$

Find the difference quotient when $x = 18$ and $h = 0.1$ and when $x = 22$ and $h = 0.1$. Interpret your answers.

• CALCULUS •

112. *Surface Area* The amount of tin A (in square inches) needed to make a tin can with radius r inches and volume 36 in.^{3} can be found by the function

$$A(r) = \frac{72}{r} + 2\pi r^2.$$

Find the difference quotient when $r = 1.4$ and $h = 0.1$ and when $r = 2$ and $h = 0.1$. Interpret your answers.

The notations $C(x)$, $R(x)$, and $P(x)$ are used in business for functions that give the cost, revenue, and profit for the production of x units. The **marginal cost** function $MC(x)$ is the function that gives the change in cost if a production level of x is raised to $x + 1$. $MC(x)$ is equal to the difference quotient for $C(x)$ with

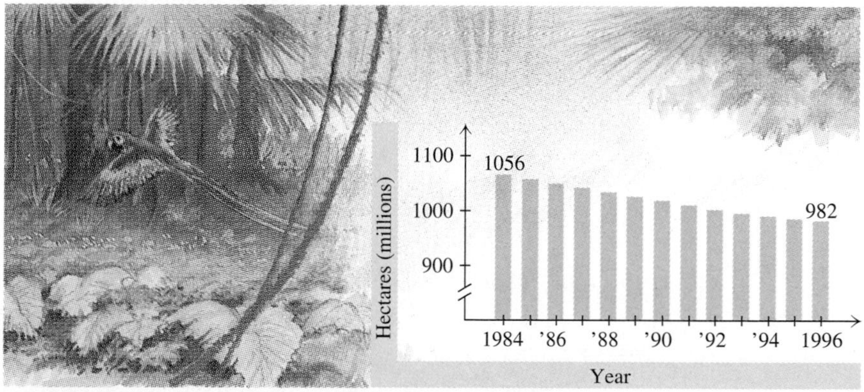

Figure for Exercises 109 and 110

$h = 1$. The definitions of **marginal revenue** $MR(x)$ and **marginal profit** $MP(x)$ are similar.

113. *Marginal Cost* A dairy's weekly cost in dollars for producing x hundred gallons of milk can be modeled by the function $C(x) = 0.03x^2 + 40x + 6000$. Find $MC(x)$ and simplify it. What is the marginal cost when production is at 10,000 gallons per week?

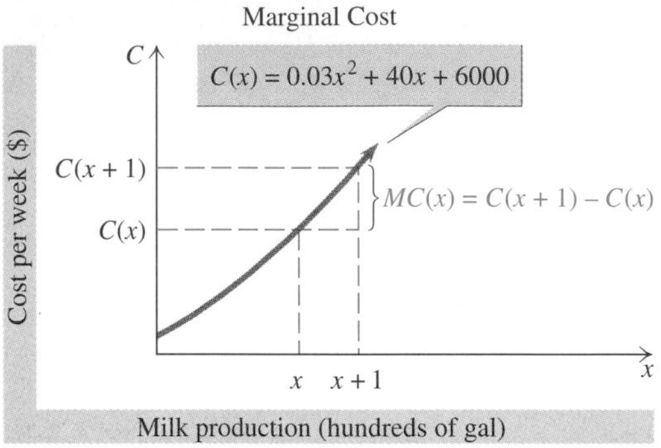

Figure for Exercise 113

114. *Marginal Revenue* A manufacturer's weekly revenue in dollars for producing x items can be modeled by the function $R(x) = 200x - x^2$. Find $MR(x)$ and simplify it. What is the marginal revenue when production is at 30 units per week?

For Writing/Discussion

115. *Find a Function* Give an example of a function and an example of a relation that is not a function, from situations that you have encountered outside of this textbook.

116. *Cooperative Learning* Work in a small group to consider the equation $y^n = x^m$ for any integers n and m. For which integers n and m does the equation define y as a function of x?

LINKING CONCEPTS

For Individual or Group Explorations

• CALCULUS •

Comparing Growth The following tables give the U.S. federal debt in billions of dollars and the population in millions of people for the years 1940 through 1990 (from the 1995 Statistical Abstract of the United States).

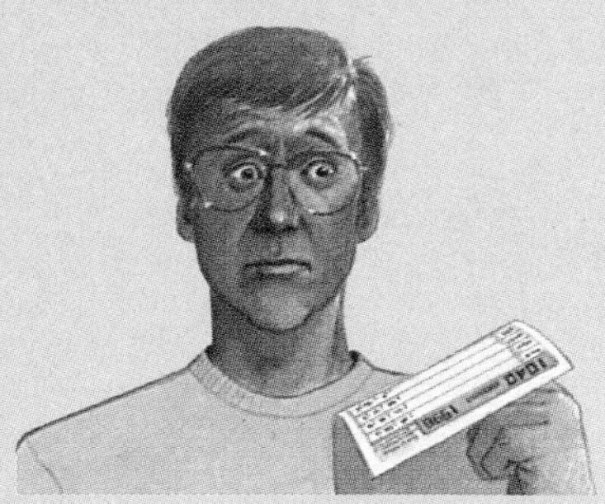

Year	Debt
1940	51
1950	257
1960	291
1970	381
1980	909
1990	3207

Year	Pop
1940	131.7
1950	150.7
1960	179.3
1970	203.3
1980	226.5
1990	248.7

Do parts (a) through (e) for each function given by the tables.

a) Draw an accurate graph of the function.

b) Find the average rate of change of the function over each ten-year period.

c) Take the average rate of change for each ten-year period (starting with 1950–1960) and subtract from it the average rate of change for the previous ten-year period.

d) Are the average rates of change for the function positive or negative?

e) Are the answers to part (c) mostly positive or mostly negative?

f) Judging from the graphs and the average rates of change, which is growing out of control, the federal debt or the population?

g) Explain the relationship between your answer to part (f) and your answer to part (e).

2.3 Graphs of Relations and Functions

When we graph the set of ordered pairs that satisfy an equation, we are combining algebra with geometry. We saw in Section 2.1 that any equation of the form $Ax + By = C$ has a graph that is a straight line. In this section we will see that graphs of equations can take on many different geometric shapes and see how to determine whether a relation is a function by its graph.

The Circle

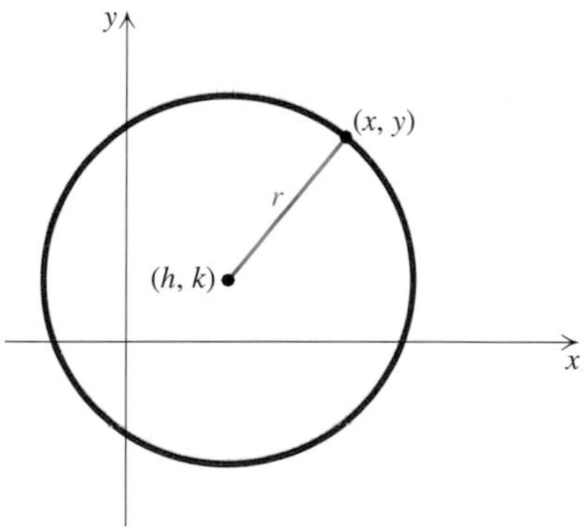

Figure 2.29

A **circle** is the set of all points in a plane that lie a fixed distance from a given point in the plane. The fixed distance is called the **radius**, and the given point is the **center**. The distance formula of Section 2.1 can be used to write an equation for the circle shown in Fig. 2.29 with center (h, k) and radius r for $r > 0$. A point (x, y) is on the circle if and only if it satisfies the equation

$$\sqrt{(x - h)^2 + (y - k)^2} = r.$$

Using the definition of square root, we can write the following equation.

Theorem: Equation for a Circle

The equation for a circle with center (h, k) and radius r for $r > 0$ is

$$(x - h)^2 + (y - k)^2 = r^2.$$

Note how we applied the definition of square root to rewrite the equation of the circle. This step is also thought of as **squaring both sides** of the equation. If we square both sides of $\sqrt{x} = 3$ we get $x = 9$, an equivalent equation. However, if negative numbers are involved, then squaring both sides might not yield an equivalent equation. Squaring both sides of $\sqrt{x} = -3$ yields $x = 9$. But $\sqrt{9} \neq -3$, because the square root symbol always represents the nonnegative square root. We will study the idea of raising both sides of an equation to a power in greater detail in Section 3.5.

The form $(x - h)^2 + (y - k)^2 = r^2$ is called the **standard form** for the equation of a circle. If we have an equation in this form (or one that can be rewritten in this form), we can conclude that its graph is a circle centered at (h, k) with radius r. On the other hand, if we know the center and radius of a circle, we can write an equation for it in the standard form. Note that if a circle is centered at the origin, then its standard equation is $x^2 + y^2 = r^2$.

EXAMPLE 1 Graphing a circle

Sketch the graph of the equation $(x - 1)^2 + (y + 2)^2 = 9$ and state the domain and range of the relation.

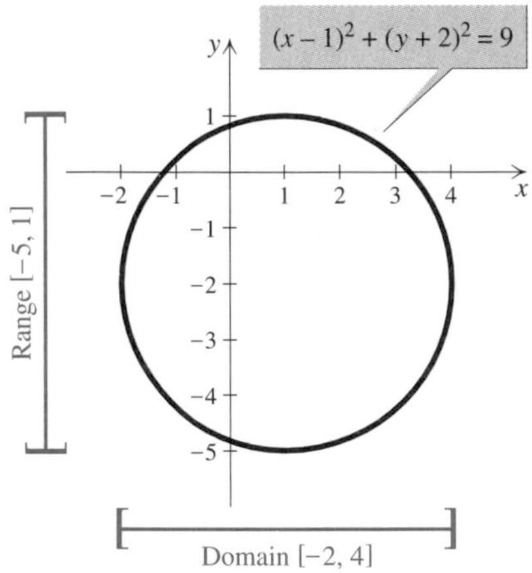

Figure 2.30

Solution

The equation is in the standard form of the equation of a circle. Its center is $(1, -2)$ and its radius is 3. The graph is shown in Fig. 2.30. To find the domain, observe from the graph that all x-coordinates on the graph are within three units of the x-coordinate of the center. So the domain is the closed interval $[-2, 4]$. Likewise, all y-coordinates are within three units of the y-coordinate of the center. So the range is the closed interval $[-5, 1]$.

To support these results with a graphing calculator you must first solve the equation for y:

$$\begin{aligned}(x - 1)^2 + (y + 2)^2 &= 9\\ (y + 2)^2 &= 9 - (x - 1)^2\\ y + 2 &= \pm\sqrt{9 - (x - 1)^2}\\ y &= -2 \pm \sqrt{9 - (x - 1)^2}\end{aligned}$$

Now enter the functions y_1 and y_2 as in Fig. 2.31(a). Set the viewing window using the domain $[-2, 4]$ and range $[-5, 1]$. The graph in Fig. 2.31(b) supports our previous answers. A circle will look round only if the same unit distance is used on both axes.

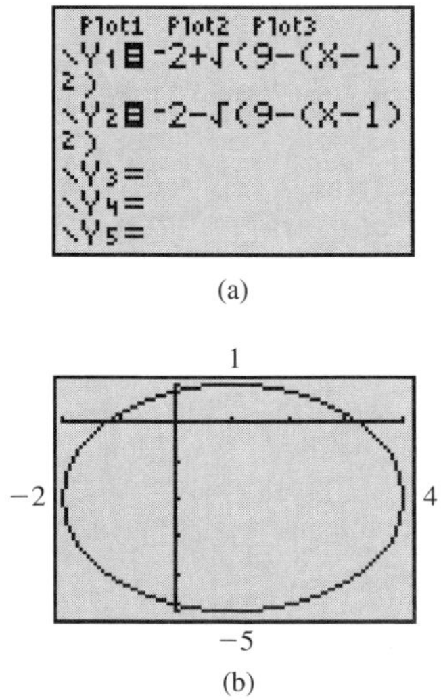

Figure 2.31

Note that an equation such as $(x - 1)^2 + (y + 2)^2 = -9$ is not satisfied by any pair of real numbers, because the left-hand side is a nonnegative real number, while the right-hand side is negative. The equation $(x - 1)^2 + (y + 2)^2 = 0$ is satisfied only by $(1, -2)$. Since only one point satisfies $(x - 1)^2 + (y + 2)^2 = 0$, its graph is sometimes called a degenerate circle with radius zero. We will study circles again later in this text when we study the conic sections.

In Example 1 we were given an equation and we sketched its graph. In the next example, we are given a description of the circle and we are asked to find its equation.

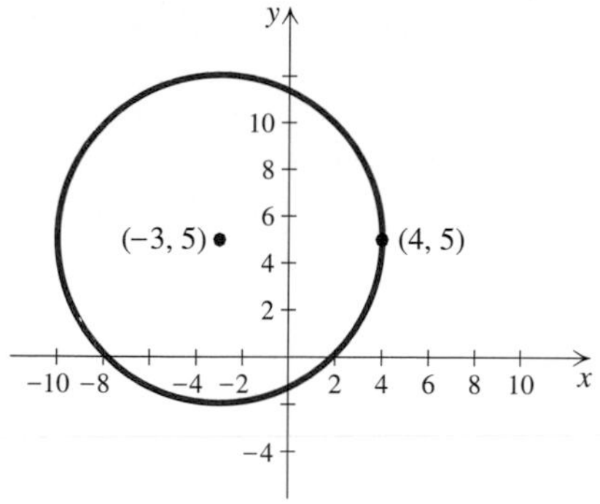

Figure 2.32

EXAMPLE 2 **Writing an equation of a circle**

Write the standard equation for the circle with the center $(-3, 5)$ and passing through $(4, 5)$ as shown in Fig. 2.32.

Solution

Since the distance between the center $(-3, 5)$ and $(4, 5)$ is seven units, the radius of the circle is 7. Use the center $h = -3$, $k = 5$, and radius $r = 7$ in the standard equation of a circle:

$$(x - (-3))^2 + (y - 5)^2 = 7^2$$

So the equation of the circle is $(x + 3)^2 + (y - 5)^2 = 49$.

If the equation of a circle is not given in the standard form, we can still recognize it by completing the square.

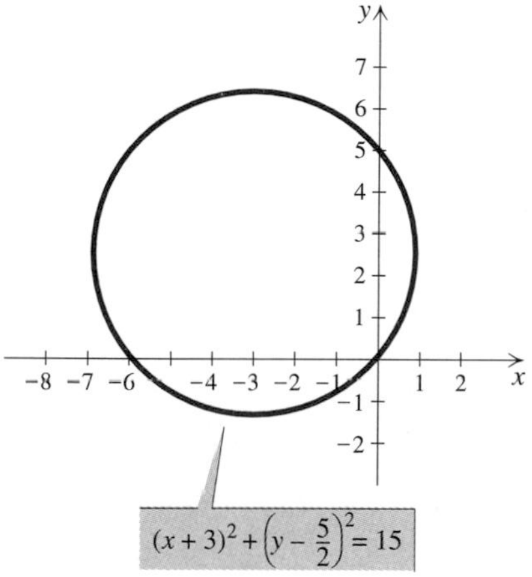

Figure 2.33

EXAMPLE 3 **Changing an equation of a circle to standard form**

Graph the equation $x^2 + 6x + y^2 - 5y = -\frac{1}{4}$.

Solution

Complete the square for both x and y to get the standard form.

$$x^2 + 6x + 9 + y^2 - 5y + \frac{25}{4} = -\frac{1}{4} + 9 + \frac{25}{4} \qquad \left(\frac{1}{2}\cdot 6\right)^2 = 9, \quad \left(\frac{1}{2}\cdot 5\right)^2 = \frac{25}{4}$$

$$(x + 3)^2 + \left(y - \frac{5}{2}\right)^2 = 15$$

Factor the trinomials on the left side.

The graph is a circle with center $(-3, \frac{5}{2})$ and radius $\sqrt{15}$. See Fig. 2.33.

Graphing Equations

The circle and the line provide nice examples of how algebra and geometry are interrelated. When you see an equation that you recognize as the equation of a circle or a line, sketching a graph is easy to do. Other equations have graphs that are not such familiar shapes. Until we learn to recognize the kinds of graphs that other equations have, we graph other equations by calculating enough ordered pairs to determine the shape of the graph. When you graph equations, try to anticipate what the graph will look like, and after the graph is drawn, pause to reflect on the shape of the graph and the type of equation that produced it.

Of course a graphing calculator can speed up this process. Remember that a graphing calculator shows only finitely many points and a graph consists of infinitely many points. After looking at the display of a graphing calculator, you must still decide what the entire graph looks like. □

EXAMPLE 4 **Graphing by plotting ordered pairs**

Graph each equation and state the domain and range of the relation.

a) $y = x^2$ **b)** $x = y^2$ **c)** $y = |x|$ **d)** $y = \sqrt{x}$

Solution

a) Make a table of ordered pairs that satisfy $y = x^2$:

x	0	1	-1	2	-2
$y = x^2$	0	1	1	4	4

These ordered pairs indicate a graph in the shape shown in Fig. 2.34. This curve is called a **parabola**. We still study parabolas in greater detail in Chapter 3. The domain is $(-\infty, \infty)$ because any real number can be used for x in $y = x^2$. Since all y-coordinates are nonnegative, the range is $[0, \infty)$.

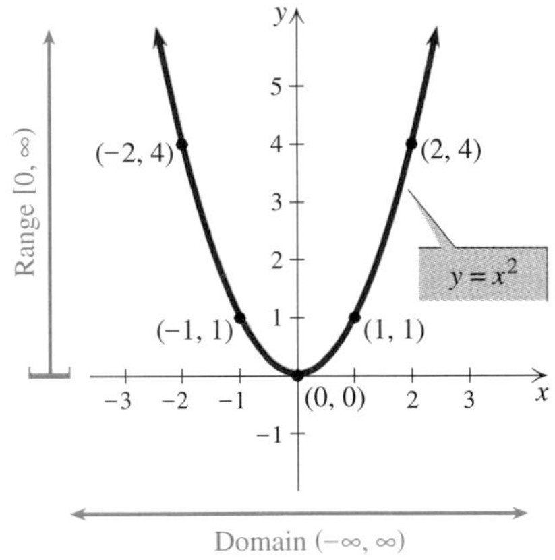

Figure 2.34

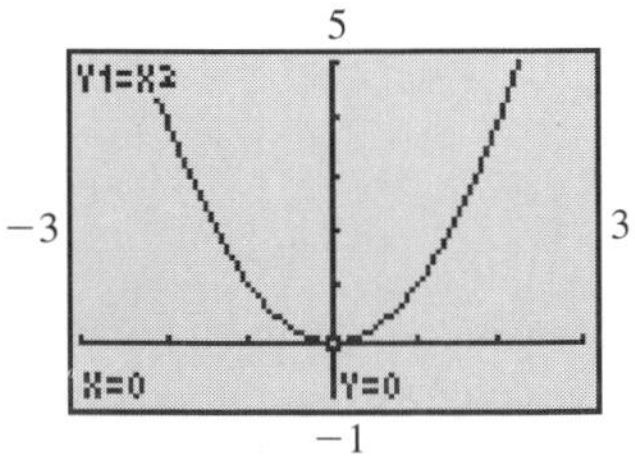

Figure 2.35

The calculator graph shown in Fig. 2.35 supports these conclusions. □

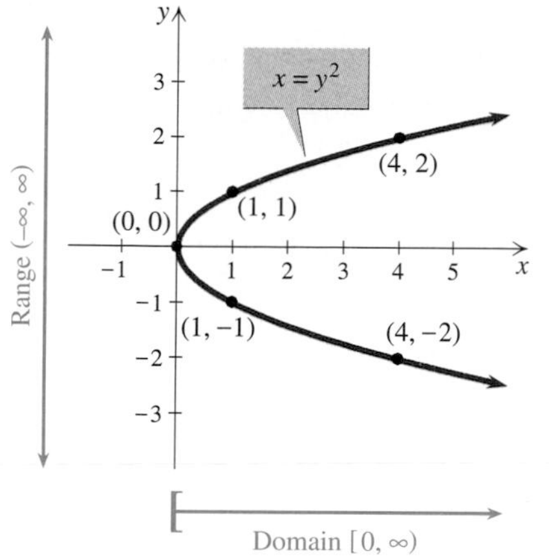

Figure 2.36

b) Make a table of ordered pairs that satisfy $x = y^2$. In this case choose y and calculate x.

$x = y^2$	0	1	1	4	4
y	0	1	−1	2	−2

The ordered pairs (0, 0), (1, 1), (1, −1), (4, 2), and (4, −2) satisfy $x = y^2$. Note that these ordered pairs could be obtained from the ordered pairs that satisfy $y = x^2$ by interchanging the x- and y-coordinates. For this reason, the graph of $x = y^2$ in Fig. 2.36 has the same shape as the parabola in Fig. 2.34, and it is also a parabola. The domain of $x = y^2$ is $[0, \infty)$, and the range is $(-\infty, \infty)$.

To support these conclusions with a graphing calculator, graph $y_1 = \sqrt{x}$ and $y_2 = -\sqrt{x}$ as shown in Fig. 2.37. □

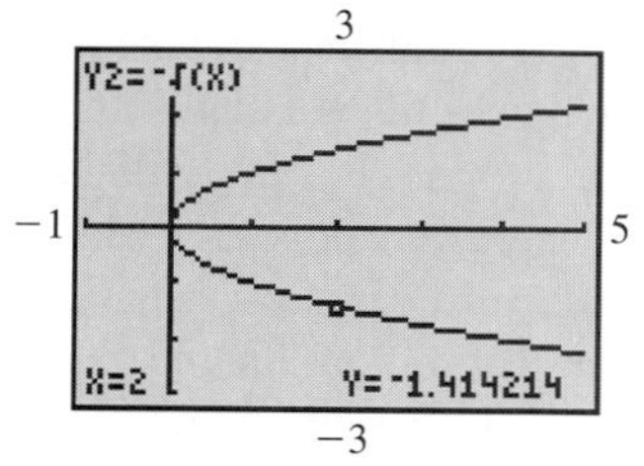

Figure 2.37

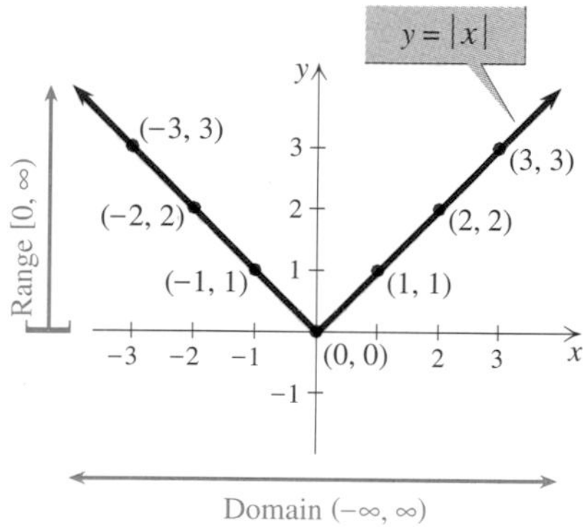

Figure 2.38

c) Make a table of ordered pairs that satisfy $y = |x|$:

x	0	1	−1	2	−2	3	−3
$y = \|x\|$	0	1	1	2	2	3	3

These ordered pairs suggest the V-shaped graph of Fig. 2.38. The function $y = |x|$ is known as the **absolute value function**. The domain is $(-\infty, \infty)$, and the range is $[0, \infty)$.

To support these conclusions with a graphing calculator, graph $y_1 =$ abs(x) as shown in Fig. 2.39. □

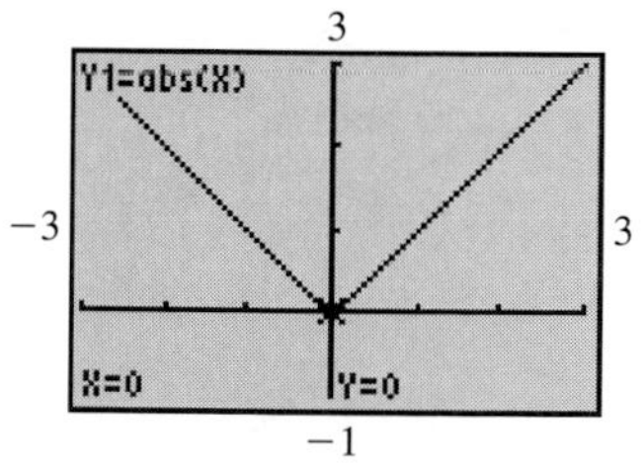

Figure 2.39

d) Make a table of ordered pairs that satisfy $y = \sqrt{x}$:

x	0	1	4	9
$y = \sqrt{x}$	0	1	2	3

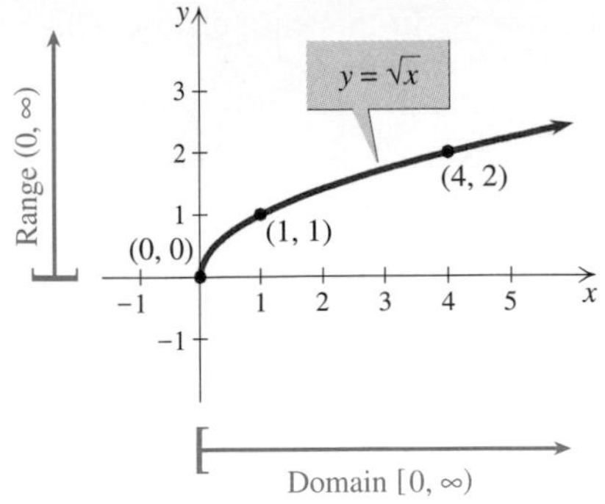

Figure 2.40

Plotting these ordered pairs suggests the graph shown in Fig. 2.40. The domain is $[0, \infty)$, and the range is $[0, \infty)$. Note that $x = y^2$ is equivalent to $y = \pm\sqrt{x}$. The graph of $y = \sqrt{x}$ is the same as the top half of the parabola of Fig. 2.36, and $y = -\sqrt{x}$ is the same as the bottom half.

The Vertical Line Test

The graphs shown in Figs. 2.32, 2.34, 2.36, 2.38, and 2.40 are all graphs of relations. But which of these relations are functions? The graph of $x = y^2$ is not the graph of a function because the points (4, 2) and (4, -2) are both on the graph. The graph of the circle in Fig. 2.32 is not the graph of a function, because there are also ordered pairs on this graph with the same x-coordinate and different y-coordinates. Note that on these graphs at least one of the points is directly above another. If a vertical line can be drawn that crosses the graph at more than one point, then there are at least two ordered pairs with the same x-coordinate and different y-coordinates, and the graph is not the graph of a function. This criterion is referred to as the **vertical line test**. In general, the vertical line test can be used on any graph that has the independent variable on the horizontal axis, which is always the case in this text. Because the accuracy of all graphs is limited, the vertical line test cannot serve as a proof of whether a graph represents a function. However, the vertical line test helps us understand the visual difference between a graph that represents a function and one that does not.

EXAMPLE 5 Applying the vertical line test

Determine which of the graphs shown in Fig. 2.41 are graphs of functions.

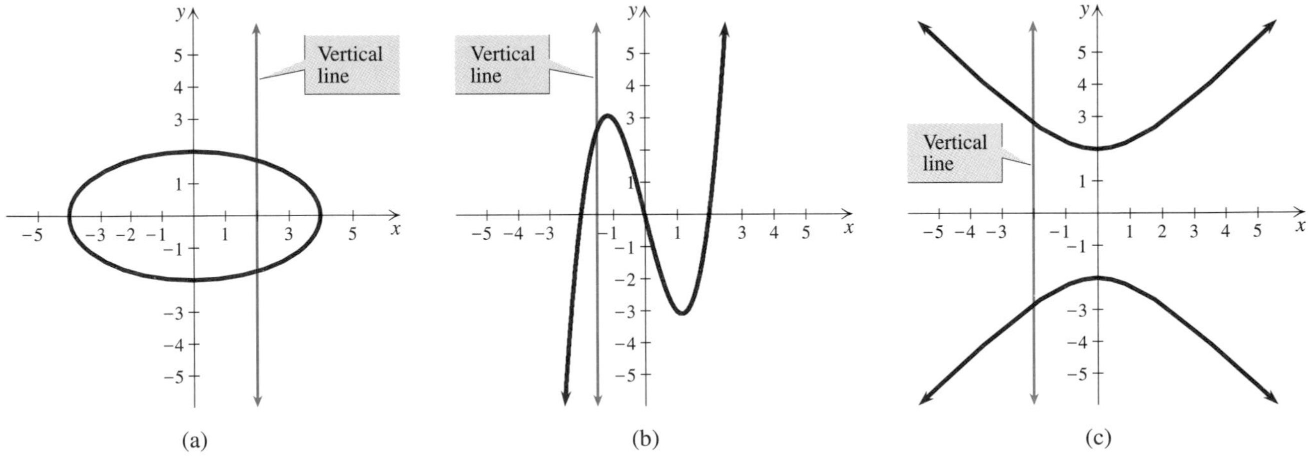

Figure 2.41

Solution

Neither (a) nor (c) is the graph of a function because we can draw a vertical line that crosses the graph more than once. The graph in Fig. 2.41(b) is the graph of a function because every vertical line appears to cross the graph at most once.

Semicircles

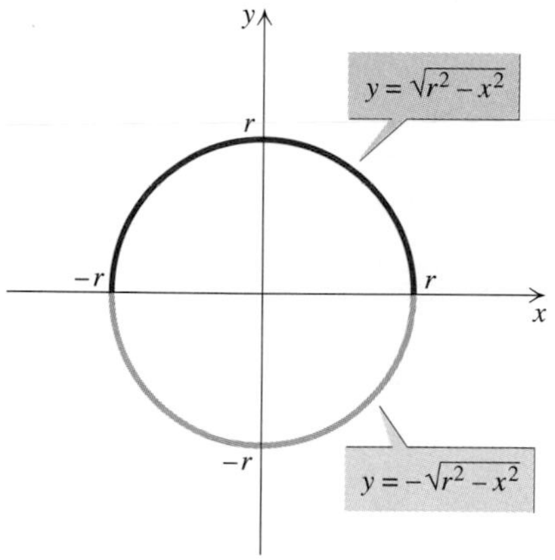

Figure 2.42

The graph of $x^2 + y^2 = r^2$ $(r > 0)$ is a circle centered at the origin of radius r. The circle does not pass the vertical line test, and a circle is not the graph of a function. We can find an equivalent equation by solving for y:

$$x^2 + y^2 = r^2$$
$$y^2 = r^2 - x^2$$
$$y = \pm\sqrt{r^2 - x^2}$$

The equation $y = \sqrt{r^2 - x^2}$ does define y as a function of x. Because y is nonnegative in this equation, the graph is the top semicircle in Fig. 2.42. The top semicircle passes the vertical line test. Likewise, the equation $y = -\sqrt{r^2 - x^2}$ defines y as a function of x, and its graph is the bottom semicircle in Fig. 2.42.

EXAMPLE 6 Graphing a semicircle

Sketch the graph of each function and state the domain and range of the function.

a) $y = -\sqrt{4 - x^2}$ **b)** $y = \sqrt{9 - x^2}$

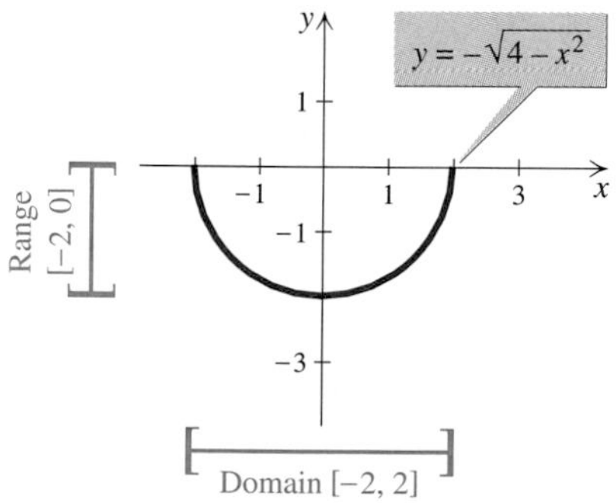

Figure 2.43

Solution

a) Rewrite the equation in the standard form for a circle:

$$y = -\sqrt{4 - x^2}$$
$$y^2 = 4 - x^2 \quad \text{Square each side.}$$
$$x^2 + y^2 = 4 \quad \text{Standard form for the equation of a circle.}$$

The graph of $x^2 + y^2 = 4$ is a circle of radius 2 centered at (0, 0). Since y must be negative in $y = -\sqrt{4 - x^2}$, the graph of $y = -\sqrt{4 - x^2}$ is the semicircle shown in Fig. 2.43. We can see from the graph that the domain is $[-2, 2]$ and the range is $[-2, 0]$.

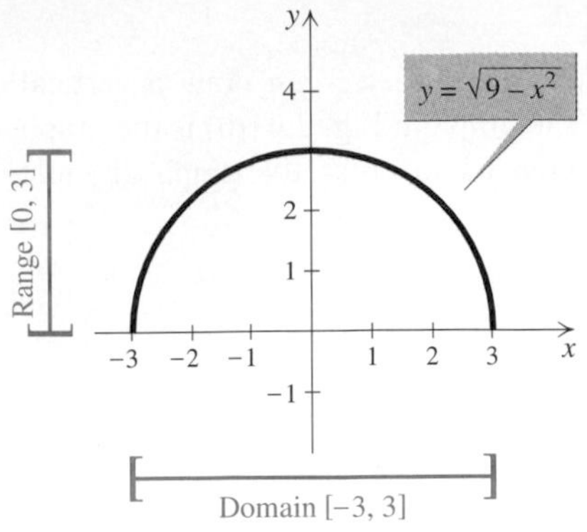

Figure 2.44

b) Rewrite the equation in the standard form for a circle:

$$y = \sqrt{9 - x^2}$$
$$y^2 = 9 - x^2 \qquad \text{Square each side.}$$
$$x^2 + y^2 = 9$$

The graph of $x^2 + y^2 = 9$ is a circle with center (0, 0) and radius 3. But this equation is not equivalent to the original. The value of y in $y = \sqrt{9 - x^2}$ is nonnegative. So the graph of the original equation is the semicircle shown in Fig. 2.44. We can read the domain $[-3, 3]$ and the range $[0, 3]$ from the graph.

Piecewise Functions

For some functions, different formulas are used in different regions of the domain. Such functions are called **piecewise functions**. The simplest example of such a function is the absolute value function $f(x) = |x|$, which can be written as

$$f(x) = \begin{cases} x & \text{for } x \ge 0 \\ -x & \text{for } x < 0. \end{cases}$$

For $x \ge 0$ the equation $f(x) = x$ is used to obtain the second coordinate, and for $x < 0$ the equation $f(x) = -x$ is used. The graph of the absolute value function was drawn in Fig. 2.38.

EXAMPLE 7 Graphing a piecewise function

Sketch the graph of each function and state the domain and range.

a) $f(x) = \begin{cases} 1 & \text{for } x < 0 \\ -1 & \text{for } x \ge 0 \end{cases}$ **b)** $f(x) = \begin{cases} x^2 - 4 & \text{for } -2 \le x \le 2 \\ x - 2 & \text{for } x > 2 \end{cases}$

Solution

a) For $x < 0$ the graph is the horizontal line $y = 1$. For $x \ge 0$ the graph is the horizontal line $y = -1$. Note that $(0, -1)$ is on the graph shown in Fig. 2.45, but $(0, 1)$ is not, because when $x = 0$ we have $y = -1$. The domain is the interval $(-\infty, \infty)$ and the range consists of only two numbers, -1 and 1. The range is not an interval. It is written in set notation as $\{-1, 1\}$.

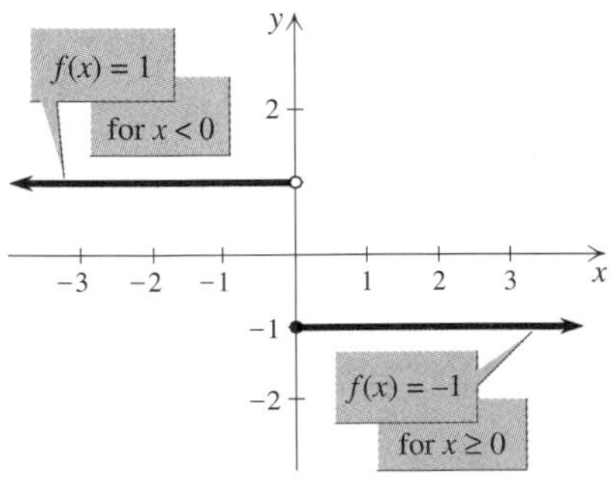

Figure 2.45

b) Make a table of ordered pairs using $y = x^2 - 4$ for x between -2 and 2 and $y = x - 2$ for $x > 2$.

x	−2	−1	0	1	2
$y = x^2 - 4$	0	−3	−4	−3	0

x	2.1	3	4	5
$y = x - 2$	0.1	1	2	3

For x in the interval $[-2, 2]$ the graph is a portion of a parabola as shown in Fig. 2.46. For $x > 2$, the graph is a straight line through $(3, 1)$, $(4, 2)$, and $(5, 3)$. The domain is $[-2, \infty)$, and the range is $[-4, \infty)$.

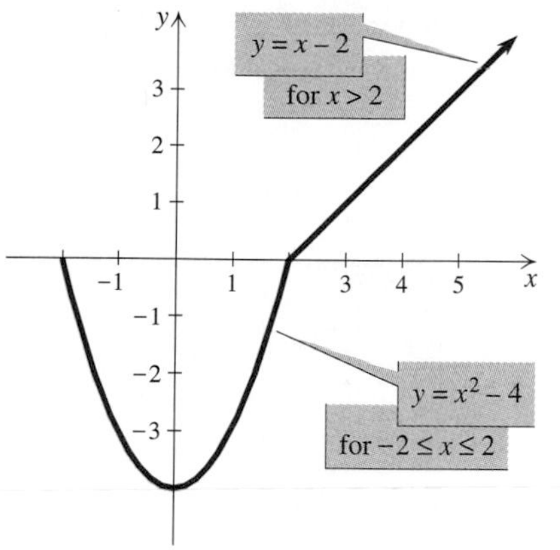

Figure 2.46

Figure 2.47(a) shows how to use the inequality symbols from the TEST menu to enter a piecewise function on a calculator. The inequality $x \geq -2$ is not treated as a normal inequality, but instead the calculator gives it a value of 1 when it is satisfied and 0 when it is not satisfied. So, dividing $x^2 - 4$ by the inequalities $x \geq -2$ and $x \leq 2$ will cause $x^2 - 4$ to be graphed only when they are both satisfied and not graphed when they are not both satisfied. The calculator graph in Fig. 2.47(b) is consistent with the conclusions that we have made.

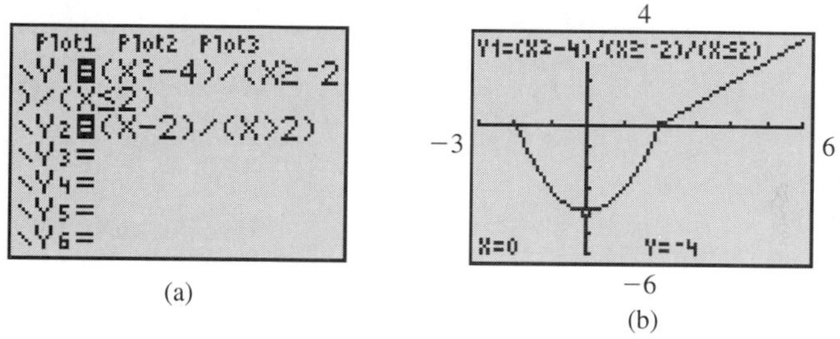

Figure 2.47

Piecewise functions are often found in shipping charges. For example, if the weight in pounds of an order is in the interval $(0, 1]$, the shipping and handling charge is \$3. If the weight is in the interval $(1, 2]$, the shipping and handling charge is \$4, and so on. The next example is a function that is similar to a shipping and handling charge. This function is referred to as the **greatest integer function** and is written $f(x) = [\![x]\!]$ or $f(x) = \text{int}(x)$. The symbol $[\![x]\!]$ is defined to be the largest integer that is less than or equal to x. For example, $[\![5.01]\!] = 5$, because the greatest integer less than or equal to 5.01 is 5. Likewise, $[\![3.2]\!] = 3$, $[\![-2.2]\!] = -3$, and $[\![7]\!] = 7$.

EXAMPLE 8 Graphing the greatest integer function

Sketch the graph of $f(x) = [\![x]\!]$ and state the domain and range.

Solution

For any x in the interval $[0, 1)$ the greatest integer less than or equal to x is 0. For any x in $[1, 2)$ the greatest integer less than or equal to x is 1. For any x in $[-1, 0)$ the greatest integer less than or equal to x is -1. The definition of $[\![x]\!]$

causes the function to be constant between the integers and to "jump" at each integer. The graph of $f(x) = [\![x]\!]$ is shown in Fig. 2.48. The domain is $(-\infty, \infty)$, and the range is the set of integers.

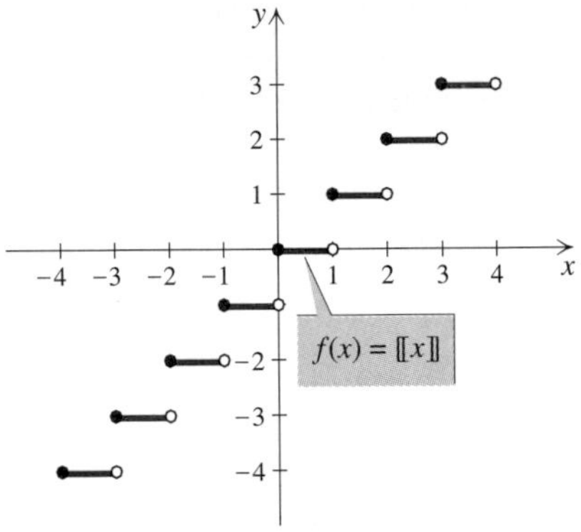

Figure 2.48

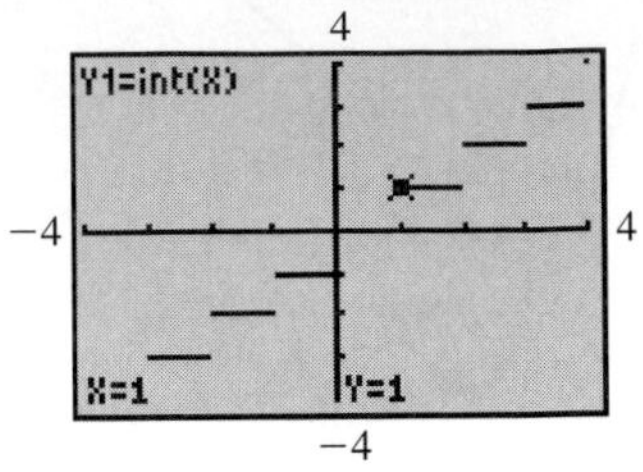

Figure 2.49

The calculator graph of $y_1 = \text{int}(x)$ looks best in dot mode as in Fig. 2.49, because in connected mode the calculator connects the disjoint pieces of the graph. The calculator graph in Fig. 2.49 supports our conclusion that the graph of this function looks like the one drawn in Fig. 2.48. Note that the calculator is incapable of showing whether the endpoints of the line segments are included.

In the next example we vary the form of the greatest integer function, but the graph is still similar to the graph in Fig. 2.48.

EXAMPLE 9 A variation of the greatest integer function

Sketch the graph of $f(x) = [\![x - 2]\!]$ for $0 \leq x \leq 5$.

Solution

If $x = 0$, $f(0) = [\![-2]\!] = -2$. If $x = 0.5$, $f(0.5) = [\![-1.5]\!] = -2$. In fact, $f(x) = -2$ for any x in the interval $[0, 1)$. Similarly, $f(x) = -1$ for any x in the interval $[1, 2)$. This pattern continues with $f(x) = 2$ for any x in the interval $[4, 5)$, and $f(x) = 3$ for $x = 5$. The graph of $f(x) = [\![x - 2]\!]$ is shown in Fig. 2.50.

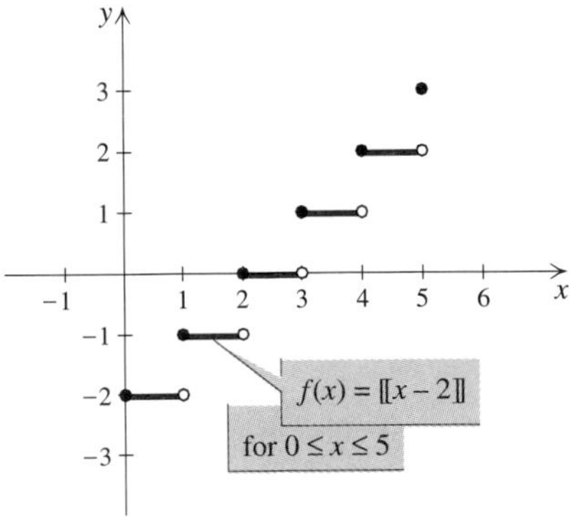

Figure 2.50

Increasing, Decreasing, and Constant

A graph is often used to get a general idea of the relationship between two variables. From a graph we can determine at a glance approximately where the temperature is increasing, the population is decreasing, or the cost is constant.

EXAMPLE 10 Reading a graph

The graph in Fig. 2.51 shows the population of rabbits in a forest preserve from 1980 through 1990, where 0 corresponds to 1980, 1 to 1981, and so on. When was the population increasing, decreasing, or constant?

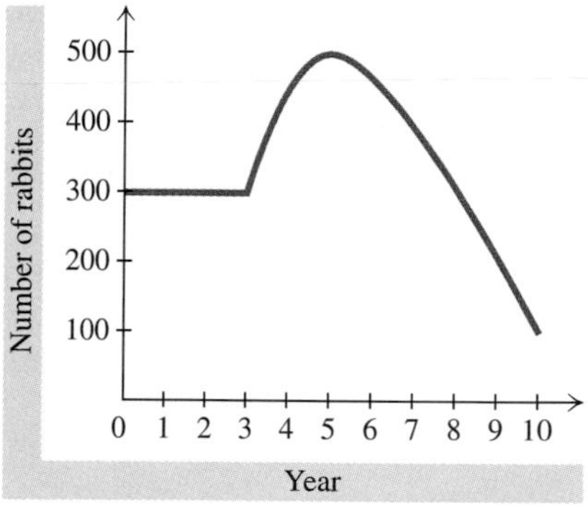

Figure 2.51

Solution

From the graph we see that the population was constant from 1980 through 1983, increasing from 1983 through 1985, and decreasing from 1985 through 1990. Using interval notation we say that the population was constant on the interval (1980, 1983), increasing on the interval (1983, 1985), and decreasing on the interval (1985, 1990).

To better understand the ideas of increasing, decreasing, and constant, imagine a point moving from left to right along the graph of the function. In an interval where the function is increasing, the point will be rising; in an interval where the function is decreasing, the point will be falling; in an interval where the function is constant, the point will be moving horizontally. A formal definition of these ideas is given as follows.

Definitions: Increasing, Decreasing, and Constant

1. If $f(x_1) < f(x_2)$ for every x_1 and x_2 in (a, b) with $x_1 < x_2$, then f is **increasing** on (a, b).
2. If $f(x_1) > f(x_2)$ for every x_1 and x_2 in (a, b) with $x_1 < x_2$, then f is **decreasing** on (a, b).
3. If $f(x_1) = f(x_2)$ for every x_1 and x_2 in (a, b), then f is **constant** on (a, b).

In this text we will not work with the inequalities shown in the definition of increasing, decreasing, and constant. We will simply identify the intervals on which a function is increasing, decreasing, or constant, by inspecting a graph as we did in the case of the rabbit population. Note that we have defined these terms only for open intervals (a, b). Remember that the interval (a, b) gives the x values for which the y values are increasing, decreasing, or constant.

EXAMPLE 11 **Increasing, decreasing, or constant**

Sketch the graph of each function and identify any open intervals on which the function is increasing, decreasing, or constant.

a) $f(x) = 4 - x^2$ **b)** $g(x) = \begin{cases} 0 & \text{for } x \leq 0 \\ \sqrt{x} & \text{for } 0 < x < 4 \\ 2 & \text{for } x \geq 4 \end{cases}$

Solution

a) The graph of $f(x) = 4 - x^2$ includes the points $(-2, 0)$, $(-1, 3)$, $(0, 4)$, $(1, 3)$, and $(2, 0)$. The graph is shown in Fig. 2.52. The function is increasing on the interval $(-\infty, 0)$ and decreasing on $(0, \infty)$.

b) The graph of g is shown in Fig. 2.53. The function g is constant on the intervals $(-\infty, 0)$ and $(4, \infty)$ and increasing on $(0, 4)$.

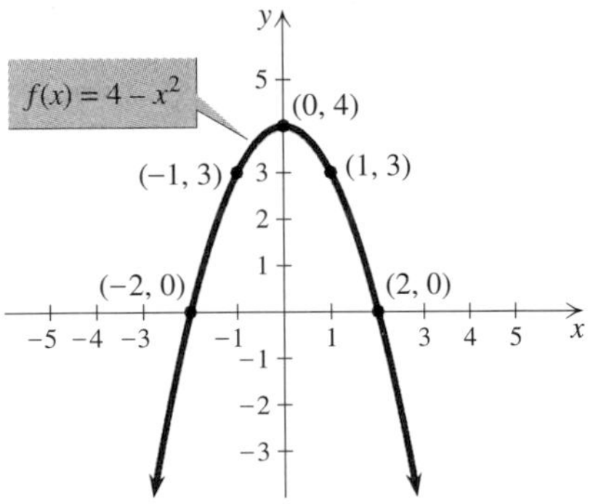

Figure 2.52

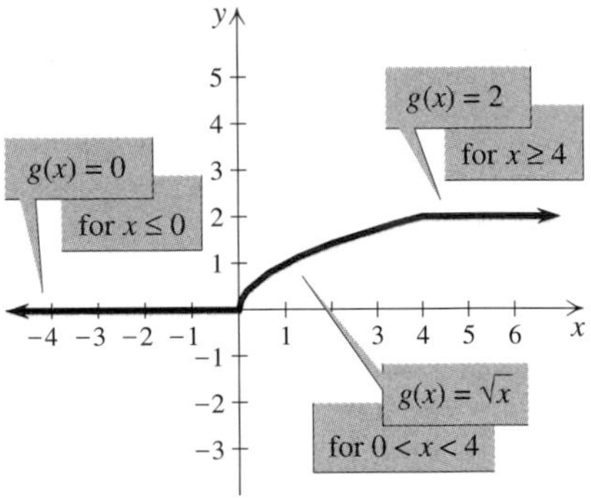

Figure 2.53

FOR THOUGHT **True or False? Explain.**

1. The graph of $\sqrt{(x - 2)^2 + (y + 3)^2} = 9$ is a circle with center $(2, -3)$ and radius 9.
2. The graph of $(y - 2)^2 + (x - 3)^2 = 25$ is a circle with center $(2, 3)$ and radius 5.
3. The graph of $(x + 1)^2 + (y + 1)^2 = -16$ is a circle centered at $(-1, -1)$ with radius 4.
4. If $f(x) = [\![x + 3]\!]$, then $f(-4.5) = -2$.
5. The range of the function $f(x) = \dfrac{|x|}{x}$ is the interval $(-1, 1)$.
6. The range of $f(x) = [\![x - 1]\!]$ is the set of integers.
7. The only ordered pair that satisfies $(x - 5)^2 + (y + 6)^2 = 0$ is $(5, -6)$.
8. The domain of the function $y = \sqrt{4 - x^2}$ is the interval $[-2, 2]$.
9. The range of the function $y = \sqrt{16 - x^2}$ is the interval $[0, \infty)$.
10. The function $y = \sqrt{4 - x^2}$ is increasing on $(-2, 0)$ and decreasing on $(0, 2)$.

2.3 EXERCISES

Tape 5 Disk

Determine the center and radius of each circle and sketch the graph.

1. $x^2 + y^2 = 16$
2. $x^2 + y^2 = 1$
3. $(x + 6)^2 + y^2 = 36$
4. $x^2 = 9 - (y - 3)^2$
5. $(x - 2)^2 = 8 - (y + 2)^2$
6. $(y + 2)^2 = 20 - (x - 4)^2$

Write the standard equation for each circle.

7. Center at (0, 0) with radius $\sqrt{7}$
8. Center at (0, 0) with radius $2\sqrt{3}$
9. Center at $(-2, 5)$ with radius 1/2
10. Center at $(-1, -6)$ with radius 1/3
11. Center at (3, 5) and passing through the origin
12. Center at $(-3, 9)$ and passing through the origin
13. Center at $(5, -1)$ and passing through (1, 3)
14. Center at $(-2, -3)$ and passing through (2, 5)

Determine the center and radius of each circle and sketch the graph.

15. $x^2 + y^2 + 6y = 0$
16. $x^2 + y^2 = 4x$
17. $x^2 - 6x + y^2 - 8y = 0$
18. $x^2 + 10x + y^2 - 8y = -40$
19. $x^2 + y^2 = 4x + 3y$
20. $x^2 + y^2 = 5x - 6y$
21. $x^2 + y^2 = \frac{x}{2} - \frac{y}{3} - \frac{1}{16}$
22. $x^2 + y^2 = x - y + \frac{1}{2}$

Graph each equation by plotting ordered pairs of numbers. Determine the domain and range, and whether the relation is a function.

23. $x = \sqrt{y}$
24. $x = |y|$
25. $y = x^2 - 1$
26. $y = 1 + \sqrt{x}$
27. $y = 5$
28. $y = 2x$
29. $x = 2y$
30. $x - y = 0$
31. $x - y = 2$
32. $x = 3$
33. $y = 2|x|$
34. $y = 2x^2$
35. $x = y^2 + 1$
36. $x = 1 - y^2$
37. $y = |x - 1|$
38. $x = |y| + 1$
39. $x = |y + 2|$
40. $y = |x| - 3$
41. $y = \sqrt{1 - x^2}$
42. $y = -\sqrt{25 - x^2}$
43. $y^2 = 1 - x^2$
44. $y = x^3$
45. $x + y^2 = 0$
46. $x^2 + y^2 = 0$
47. $y = 1 - x^2$
48. $y^2 = -1 - x^2$
49. $y = -|x|$
50. $y = \frac{|x|}{x}$

Use the vertical line test to determine whether each graph in Exercises 51–56 is the graph of a function.

51.
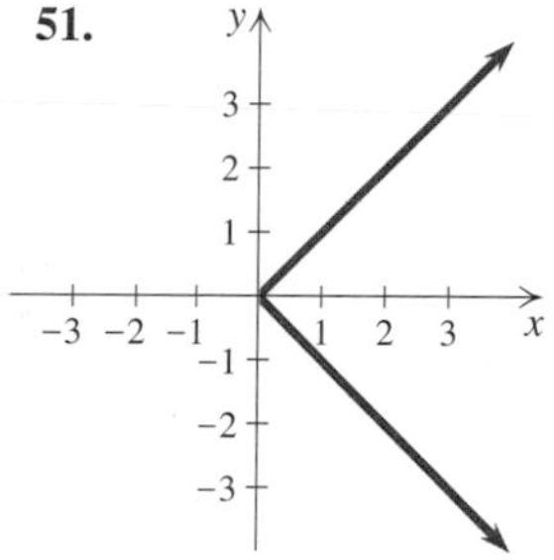

52.
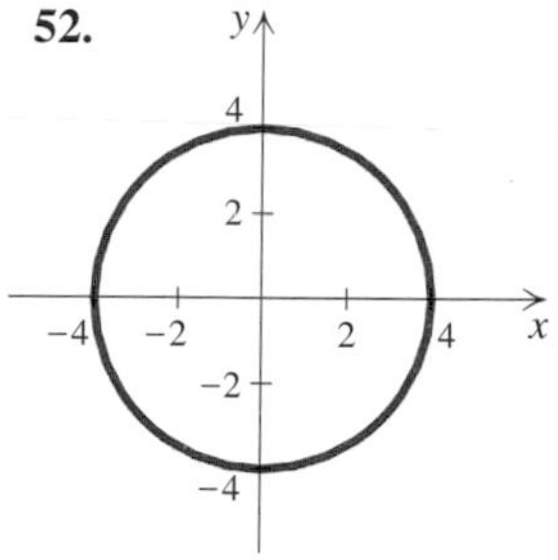

53.
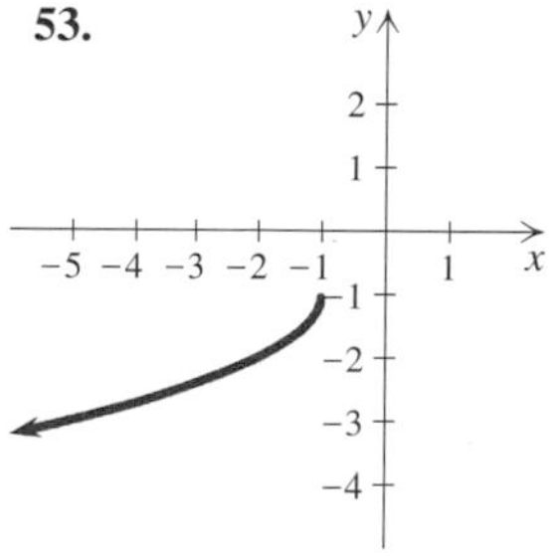

54.
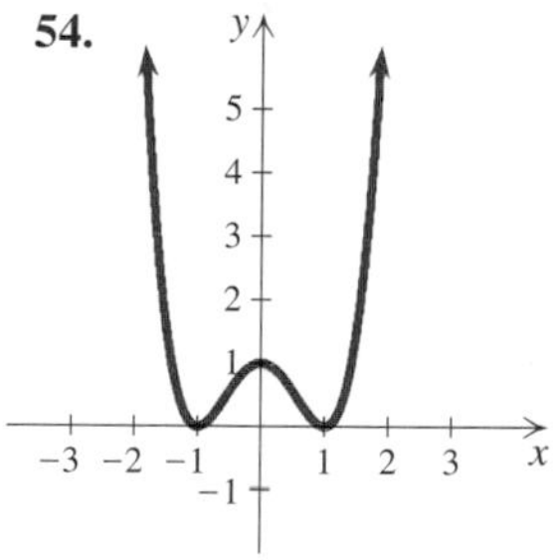

55.
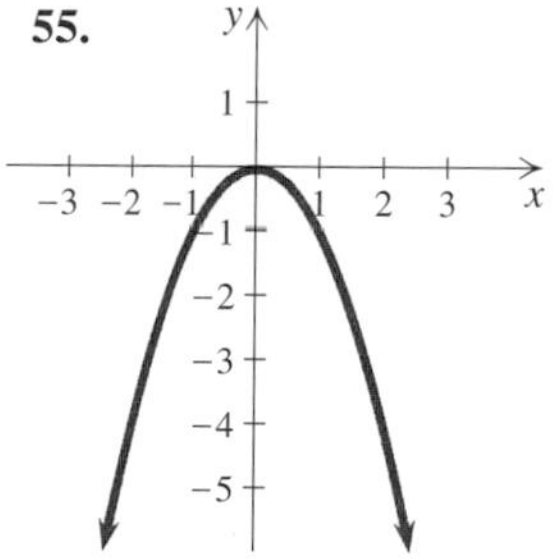

56.
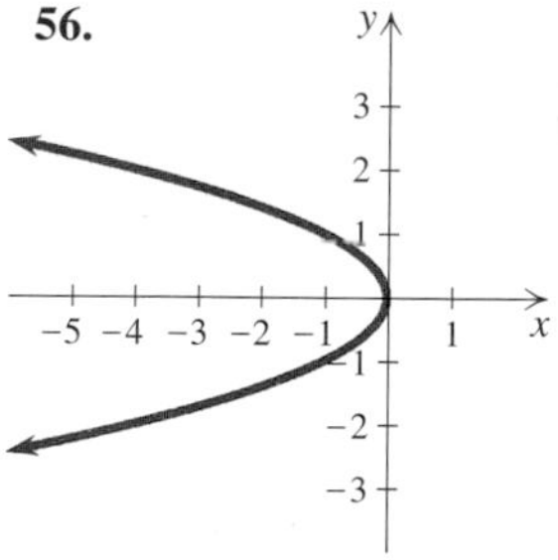

Sketch the graph of each function and state the domain and range.

57. $f(x) = \begin{cases} 2 & \text{for } x < -1 \\ -2 & \text{for } x \geq -1 \end{cases}$

58. $f(x) = \begin{cases} 3 & \text{for } x < 2 \\ x & \text{for } x \geq 2 \end{cases}$

59. $f(x) = \begin{cases} \sqrt{x+2} & \text{for } -2 \le x \le 2 \\ 4 - x & \text{for } x > 2 \end{cases}$

60. $f(x) = \begin{cases} \sqrt{x} & \text{for } x \ge 1 \\ -x & \text{for } x < 1 \end{cases}$

61. $f(x) = \begin{cases} \sqrt{-x} & \text{for } x < 0 \\ \sqrt{x} & \text{for } x \ge 0 \end{cases}$

62. $f(x) = \begin{cases} x & \text{for } x < -1 \\ -x & \text{for } x \ge -1 \end{cases}$

63. $f(x) = \begin{cases} 4 - x^2 & \text{for } -2 \le x \le 2 \\ x - 2 & \text{for } x > 2 \end{cases}$

64. $f(x) = \begin{cases} 3 & \text{for } x < 0 \\ 3 + \sqrt{x} & \text{for } x \ge 0 \end{cases}$

65. $f(x) = [\![x + 1]\!]$

66. $f(x) = 2[\![x]\!]$

67. $f(x) = [\![x]\!] + 2 \quad \text{for } 0 \le x < 4$

68. $f(x) = [\![x - 3]\!] \quad \text{for } 0 < x \le 5$

From the graph of each function in Exercises 69–74, state the domain, the range, and the intervals on which the function is increasing, decreasing, or constant.

69.

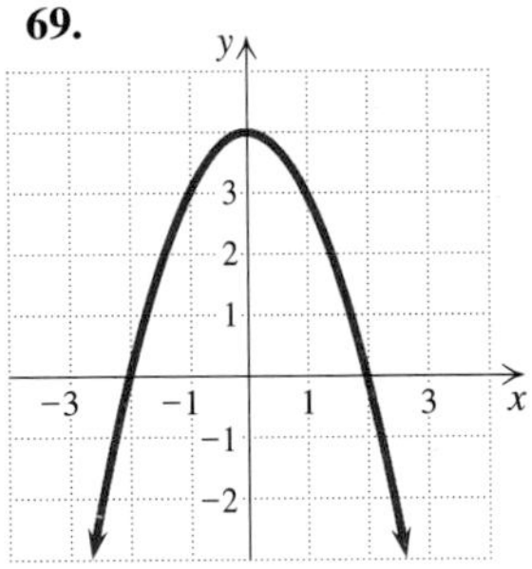

70.

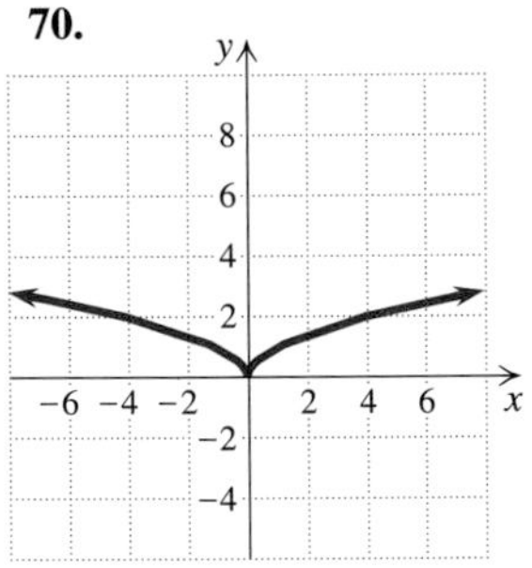

71.

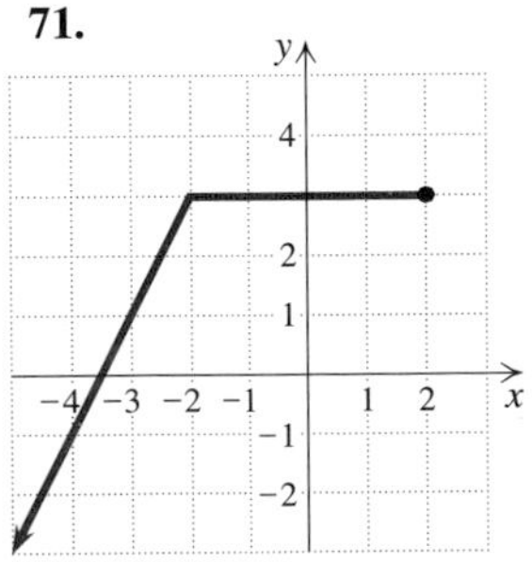

72.

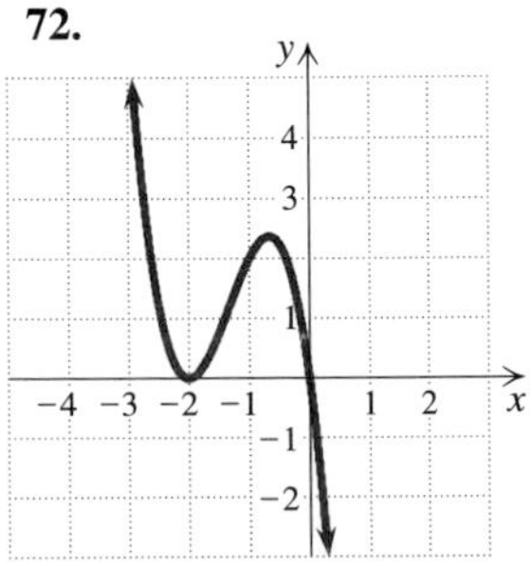

73.

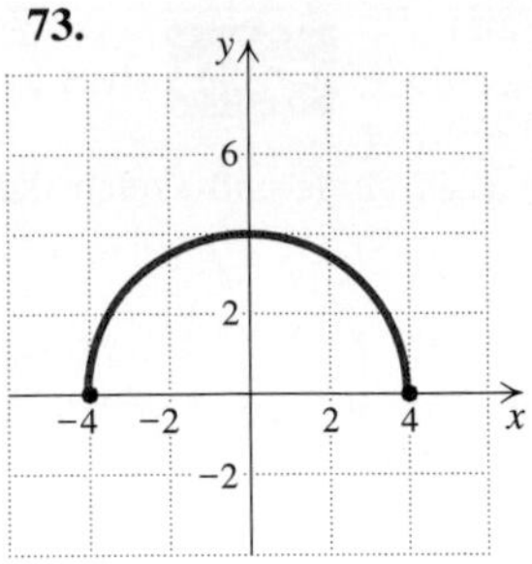

74.

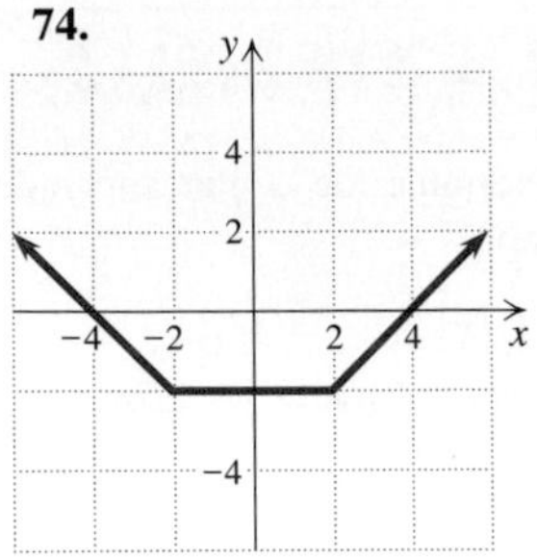

Select the graph (a, b, c, or d) that best depicts each of the following scenarios.

75. Profits for Cajun Drilling Supplies soared during the seventies, but went flat when the bottom fell out of the oil industry in the eighties. Cajun saw a period of moderate growth in the nineties.

76. To attain a minimum therapeutic level of Flexeril in her blood, Millie takes three Flexeril tablets, one every four hours. Because the half-life of Flexeril is four hours, one-half of the Flexeril in her blood is eliminated after four hours.

77. The bears dominated the market during the first quarter with massive sell-offs. The second quarter was an erratic period in which investors could not make up their minds. The bulls returned during the third quarter sending the market to record highs.

78. Medicare spending soared during the eighties. Congress managed to slow the rate of growth of Medicare during the nineties and actually managed to decrease Medicare spending in the first decade of the twenty-first century.

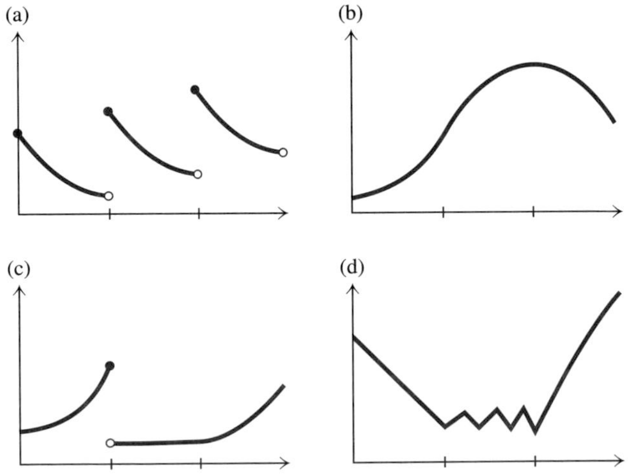

Figure for Exercises 75 to 78

Draw a graph that pictures each situation. Explain any choices that you make. Identify the independent and dependent variables. Determine the intervals on which the dependent variable is increasing, decreasing, or constant. Answers may vary.

• CALCULUS •

79. Captain Janeway left the holodeck at 7:45 to meet Tuvok, her Chief of Security, on the Bridge. After walking for 3 minutes, she realized she had forgotten her tricorder and returned to get it. She picked up the tricorder and resumed her walk, arriving on the Bridge at 8:00. After 15 minutes the discussion was over, and Janeway returned to the holodeck. Graph Janeway's distance from the holodeck as a function of time.

80. Starting from the pit, Helen made three laps around a circular race track at 40 seconds per lap. She then made a 30 second pit stop and two and a half laps before running off the track and getting stuck in the mud for the remainder of the five-minute race. Graph Helen's distance from the pit as a function of time.

81. Winona deposited \$30 per week into her cookie jar. After 2 years she spent half of her savings on a stereo. After spending 6 months in a coma she proceeded to spend \$15 per week on CDs until all of the money in the jar was gone. Graph the amount in her cookie jar as a function of time.

82. Michael started buying Navajo crafts with \$8,000 in his checking account. He spent \$200 per day for 10 days on pottery, then \$400 per day for the next 10 days on turquoise and silver jewelry. For the next 20 days, he spent \$50 per day on supplies while he set up his retail shop. He rested for 5 days then took in \$800 per day for the next 20 days from the resale of his collection. Graph the amount in his checking account as a function of time.

Sketch the graph of each function. Determine the domain and range. Identify any intervals on which f is increasing, decreasing, or constant.

83. $f(x) = 2x + 1$

84. $f(x) = x^2 - 3$

85. $f(x) = |x - 1|$

86. $f(x) = |x| + 1$

87. $f(x) = \dfrac{2x}{|x|}$

88. $f(x) = -3x$

89. $f(x) = -\sqrt{1 - x^2}$

90. $f(x) = \sqrt{9 - x^2}$

91. $f(x) = \begin{cases} x + 1 & \text{for } x \geq 3 \\ x + 2 & \text{for } x < 3 \end{cases}$

92. $f(x) = \begin{cases} \sqrt{-x} & \text{for } x < 0 \\ -\sqrt{x} & \text{for } x \geq 0 \end{cases}$

93. $f(x) = \begin{cases} x + 3 & \text{for } x \leq -2 \\ \sqrt{4 - x^2} & \text{for } -2 < x < 2 \\ -x + 3 & \text{for } x \geq 2 \end{cases}$

94. $f(x) = \begin{cases} 8 + 2x & \text{for } x \leq -2 \\ x^2 & \text{for } -2 < x < 2 \\ 8 - 2x & \text{for } x \geq 2 \end{cases}$

Use the minimum and maximum features of a graphing calculator to find approximately the intervals on which each function is increasing or decreasing. Round answers to two decimal places.

95. $y = 3x^2 - 5x - 4$

96. $y = -6x^2 + 2x - 9$

97. $y = x^3 - 3x$

98. $y = x^4 - 11x^2 + 18$

99. $y = 2x^4 - 12x^2 + 25$

100. $y = x^5 - 13x^3 + 36x$

Solve each problem.

101. *Missing Circle* Find the standard equation of the circle that has (4, 1) and (−6, 5) as endpoints of a diameter.

102. *Missing Circle* Find the standard equation of the circle that has (−3, −2) and (4, 5) as endpoints of a diameter.

103. *Parking Charges* A garage charges \$4/hr up to 3 hr, with any fraction of an hour charged as a whole hour. Any time over 3 hr is charged at the all-day rate of \$15. Use function notation to write the charge as a function of the number of hours x, where $0 < x \leq 8$, and graph this function.

104. *Delivering Concrete* A concrete company charges \$150 for delivering less than 3 yd^3 of concrete. For 3 yd^3 and more, the charge is \$50/yd^3 with a fraction of a yard charged as a fraction of \$50. Use function notation to write the charge as a function of the number x of cubic yards delivered, where $0 < x \leq 10$, and graph this function.

105. *Water Bill* The monthly water bill in Hammond is a function of the number of gallons used. The cost is \$10.30 for 10,000 gal or less. Over 10,000 gal, the cost is \$10.30 plus \$1 for each 1000 gal over 10,000 with any fraction of 1000 gal charged at a fraction of \$1. On what interval is the cost constant? On what interval is the cost increasing?

106. *Gas Mileage* The number of miles per gallon obtained with a new Firebird is a function of the speed at which it is driven. Is this function increasing or decreasing on its domain? Explain.

107. *Filing a Tax Return* An accountant determines the charge for filing a tax return by using the function $C = 50 + 40[\![t]\!]$ for $t > 0$, where C is in dollars and t is in hours. Sketch the graph of this function. For what values of t is the charge over \$235?

108. *Shipping Machinery* The cost in dollars of shipping a machine is given by the function $C = 200 + 37[\![w/100]\!]$ for $w > 0$, where w is the weight of the machine in pounds. For which values of w is the cost less than \$862?

109. *Motor Vehicle Ownership* World motor vehicle ownership in developed countries can be modeled by the function

$$M(t) = \begin{cases} 17.5t + 250 & 0 \le t \le 20 \\ 10t + 400 & 20 < t \le 40 \end{cases}$$

where t is the number of years since 1970 and $M(t)$ is in millions of vehicles (World Resources, 1996–1997). See the accompanying figure. How many vehicles were there in developed countries in 1988? How many will there be in 2002? What was the average rate of change of motor vehicle ownership from 1984 to 1994?

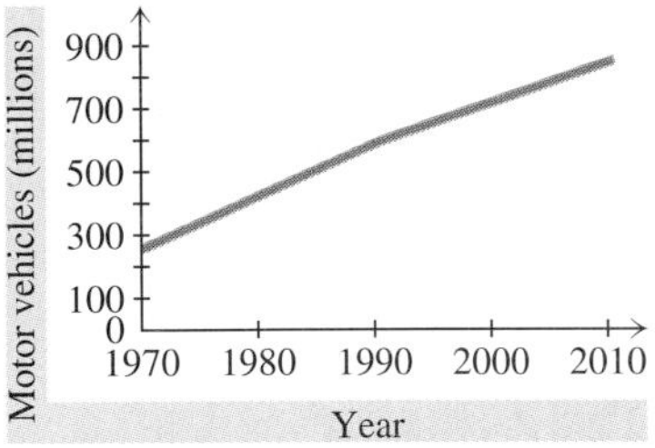

Figure for Exercise 109

110. *Motor Vehicle Ownership* World motor vehicle ownership in developing countries and Eastern Europe can be modeled by the function

$$M(t) = 6.25t + 50$$

where t is the number of years since 1970 and $M(t)$ is in millions of vehicles (World Resources, 1996–1997). Graph this function. What is the expected average rate of change of motor vehicle ownership from 1990 through 2010? Is motor vehicle ownership expected to grow faster in developed or developing countries in the period 1990 through 2010? (See the previous exercise.)

For Writing/Discussion

111. *Steps* Find an example of a function in real life whose graph has ''steps'' like that of the greatest integer function. Graph your function and find a formula for it if possible.

112. *Cooperative Learning* Select two numbers a and b. Then define a piecewise function using different formulas on the intervals $(-\infty, a]$, (a, b), and $[b, \infty)$ so that the graph does not ''jump'' at a or b. Give your function to a classmate to graph and check.

113. Define x and y and make up a situation so that the given graph represents your situation.

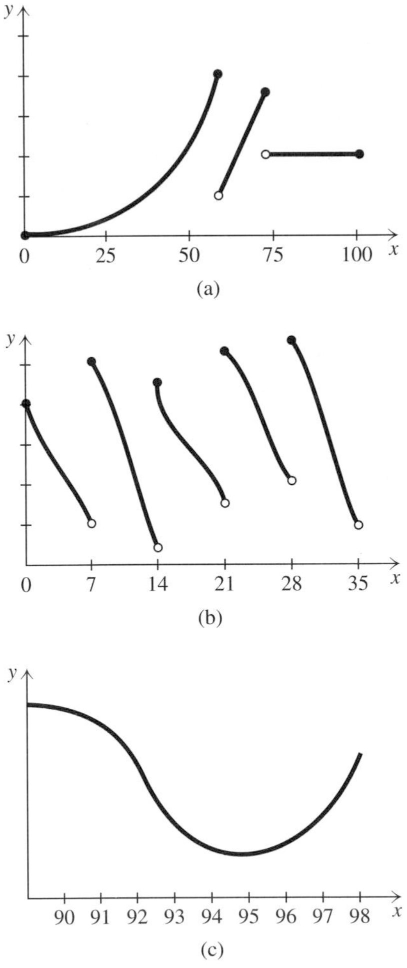

Figure for Exercise 113

Linking Concepts

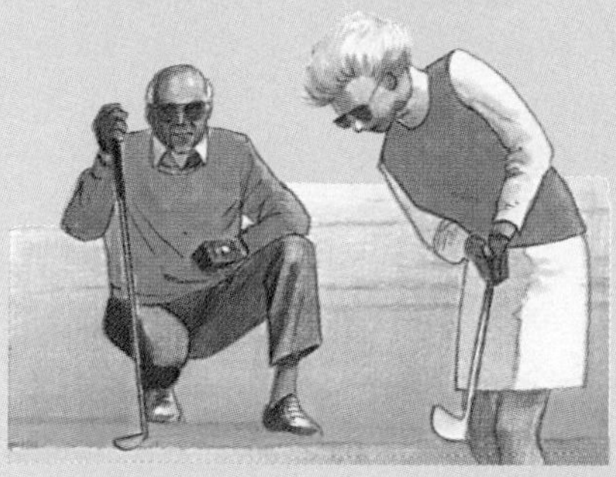

For Individual or Group Explorations

Social Security Benefits *The annual Social Security benefit of a retiree who earned a lifetime average annual salary of \$25,000 depends on the age at the time of retirement. The following function gives the annual benefit in dollars for persons retiring at ages 62 through 70 in the year 2005 or later based on current legislation.*

$$B = \begin{cases} 500a - 24{,}000 & 62 \le a < 64 \\ \dfrac{2000a - 104{,}000}{3} & 64 \le a < 67 \\ 800a - 43{,}600 & 67 \le a \le 70 \end{cases}$$

a) Graph the benefit function.

b) What is the annual benefit for a person who retires at age 64?

c) At what age does a person receive an annual benefit of \$11,600?

d) What is the average rate of change of the benefit for the ages 62 through 64? Ages 64 through 67? Ages 67 through 70?

e) Do the answers to part (d) appear in the annual benefit formula?
The life expectancy L for a U.S. white male with present age a can be modeled by the formula

$$L = 67.0166(1.00308)^a.$$

f) If Bob is a white male retiring in 2005 at age 62, then what total amount can he be expected to draw from Social Security before he dies?

g) If Bill is a white male retiring in 2005 at age 70, then what total amount can he be expected to draw from Social Security before he dies?

h) Why does Bill draw more than Bob?

2.4 Transformations and Symmetry of Graphs

In Section 2.3 we studied the graphs of relations and functions. In this section we continue that study by discussing how the graphs of some functions are transformed into the graphs of other functions. We also study the idea of symmetry of a graph.

Reflection

The graph of $f(x) = x^2$ goes through (0, 0), (1, 1), and (2, 4). The graph of $g(x) = -x^2$ goes through (0, 0), (1, −1), and (2, −4). Consider the graphs of $f(x) = x^2$ and $g(x) = -x^2$ shown in Fig. 2.54. Notice that the graph of g is a

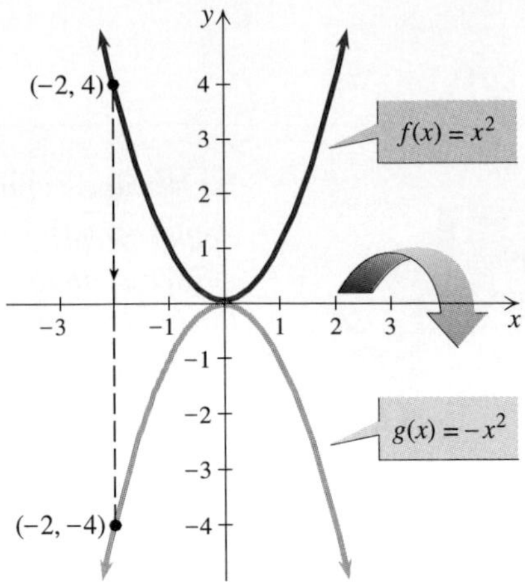

Figure 2.54

mirror image (or reflection) of the graph of f. For every ordered pair (x, y) on the graph of f, the ordered pair $(x, -y)$ is on the graph of g.

Definition: Reflection

The graph of $y = -f(x)$ is a **reflection** in the x-axis of the graph of $y = f(x)$.

Knowing that a graph is a reflection of a familiar graph makes graphing easier. We reflect the familiar graph in the x-axis to obtain a new graph.

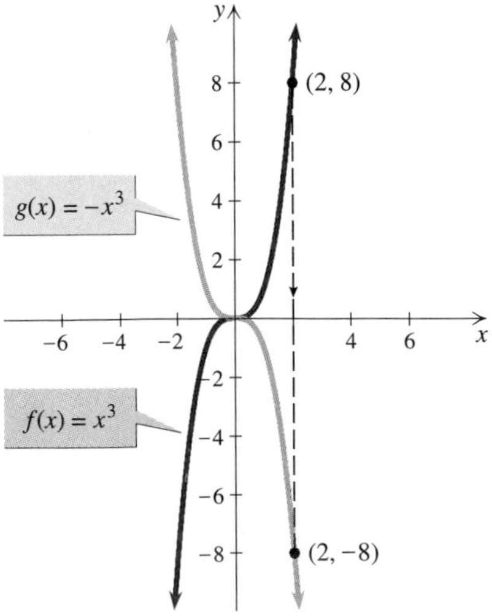

Figure 2.55

EXAMPLE 1 Graphing using reflection

Sketch the graphs of each pair of functions on the same coordinate plane.

a) $f(x) = x^3$, $g(x) = -x^3$ **b)** $f(x) = |x|$, $g(x) = -|x|$

Solution

a) Make a table of ordered pairs for f as follows:

x	-2	-1	0	1	2
$f(x) = x^3$	-8	-1	0	1	8

Sketch the graph of f through these ordered pairs as shown in Fig. 2.55. Since $g(x) = -f(x)$, the graph of g can be obtained by reflecting the graph of f in the x-axis. Each point on the graph of f corresponds to a point on the graph of g with the opposite y-coordinate. For example, $(2, 8)$ on the graph of f corresponds to $(2, -8)$ on the graph of g. Both graphs are shown in Fig. 2.55.

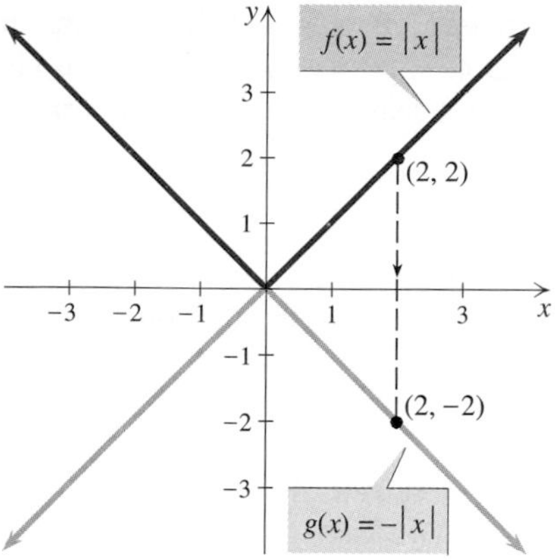

Figure 2.56

b) The graph of f is the familiar V-shaped graph of the absolute value function as shown in Fig. 2.56. Since $g(x) = -f(x)$, the graph of g can be obtained by reflecting the graph of f in the x-axis. Each point on the graph of f corresponds to a point on the graph of g with the opposite y-coordinate. For example, (2, 2) on f corresponds to (2, −2) on g. Both graphs are shown in Fig. 2.55. ■

Translation

Consider the functions $f(x) = \sqrt{x}$, $g(x) = \sqrt{x} + 3$, and $h(x) = \sqrt{x} - 5$. Because these functions are similar, their graphs ought to be similar also. In the expression $\sqrt{x} + 3$, adding 3 is the last operation performed to obtain the y-coordinate. So every point on the graph of g is exactly three units above a corresponding point on the graph of f. This translating of points gives the graph of g the same shape as the graph of f. Likewise, every point on the graph of h is exactly five units below a corresponding point on the graph of f. See Fig. 2.57.

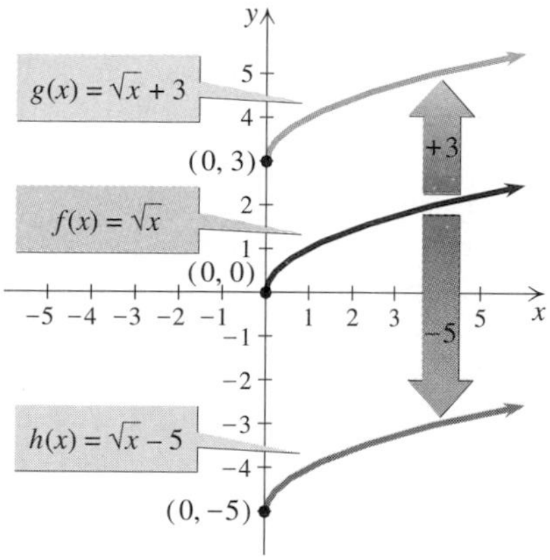

Figure 2.57

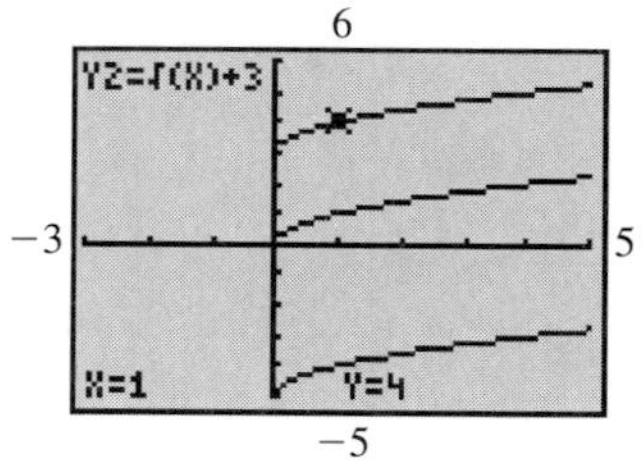

Figure 2.58

The relationship between the graphs of f, g, and h can be seen with a graphing calculator in Fig. 2.58. With a graphing calculator you can easily experiment with other functions to see how a slight change in the formula changes the graph. □

Definition: Translation Upward or Downward

If $k > 0$, then the graph of $y = f(x) + k$ is an **upward translation** of the graph of $y = f(x)$ and the graph of $y = f(x) - k$ is a **downward translation** of the graph of $y = f(x)$.

Knowing that the graph of one function is a translation of the graph of a familiar function makes graphing easier. If you use a graphing calculator to assist you in graphing, the facts that we discover in this section will help you to avoid errors.

EXAMPLE 2 Graphing using translation

Graph the functions $f(x) = x^2$, $g(x) = x^2 + 2$, and $h(x) = x^2 - 3$ on the same coordinate plane.

Solution

First sketch the familiar graph of $f(x) = x^2$ through $(-2, 4)$, $(-1, 1)$, $(0, 0)$, $(1, 1)$, and $(2, 4)$, as shown in Fig. 2.59. Since $g(x) = f(x) + 2$, the graph of g can be obtained by translating the graph of f upward two units. Since $h(x) = f(x) - 3$, the graph of h can be obtained by translating the graph of f downward three units. For example, the point $(-1, 1)$ on the graph of f moves up to $(-1, 3)$ on the graph of g and down to $(-1, -2)$ on the graph of h. All three graphs are shown in Fig. 2.59.

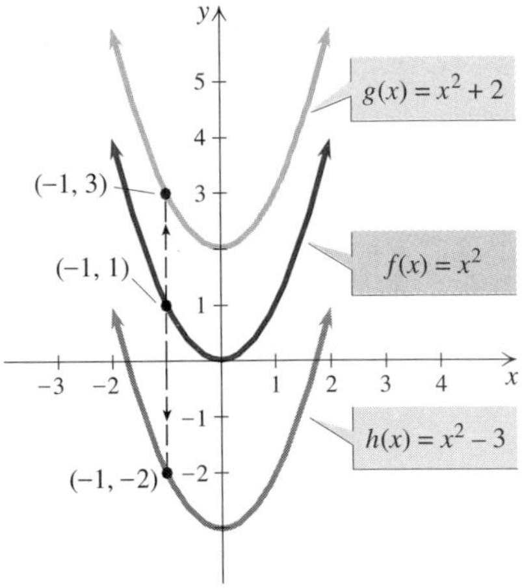

Figure 2.59

Consider the graphs of $f(x) = \sqrt{x}$, $g(x) = \sqrt{x - 3}$, and $h(x) = \sqrt{x + 5}$ in Fig. 2.60. In the expression $\sqrt{x - 3}$, subtracting 3 is the first operation to perform. So every point on the graph of g is exactly three units to the right of a corresponding point on the graph of f. (We must start with a larger value of x to get the same y-coordinate, because we first subtract 3.) Every point on the graph of h is exactly five units to the left of a corresponding point on the graph of f. See Fig. 2.60. Compare the graphs in Figs. 2.57 and 2.60. Notice that a slight difference in the formula makes a big difference in the location of the graph.

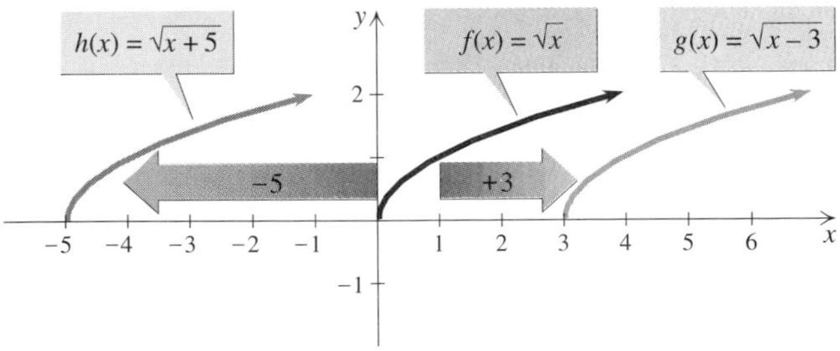

Figure 2.60

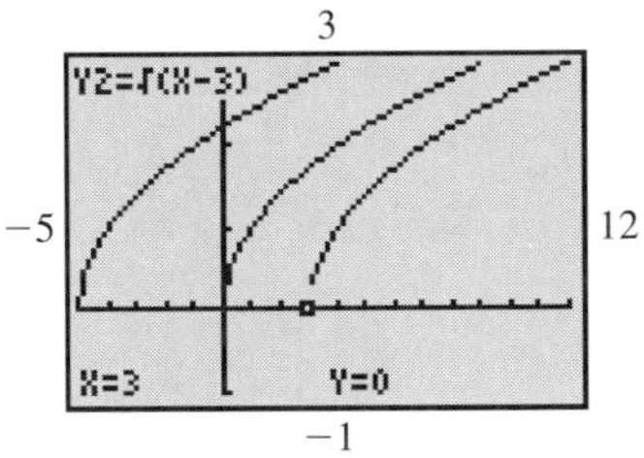

Figure 2.61

The calculator graphs of f, g, and h are shown in Fig. 2.61. Be sure to note the difference between $y = \sqrt{}(x) - 3$ and $y = \sqrt{}(x - 3)$ on a calculator.

Definition: Translation to the Right or Left

If $h > 0$, then the graph of $y = f(x - h)$ is a **translation to the right** of the graph of $y = f(x)$ and the graph of $y = f(x + h)$ is a **translation to the left** of the graph of $y = f(x)$.

EXAMPLE 3 Graphing using translation

Sketch the graph of each function.

a) $f(x) = |x - 1|$ **b)** $f(x) = (x + 3)^2$

Solution

a) The graph of $f(x) = |x - 1|$ is a translation one unit to the right of the familiar graph of $f(x) = |x|$. Calculate a few ordered pairs to get an accurate graph. The points (0, 1), (1, 0), and (2, 1) are on the graph of $f(x) = |x - 1|$ shown in Fig. 2.62.

b) The graph of $f(x) = (x + 3)^2$ is a translation three units to the left of the familiar graph of $f(x) = x^2$. Calculate a few ordered pairs to get an accurate graph. The points (−3, 0), (−2, 1), and (−4, 1) are on the graph shown in Fig. 2.63.

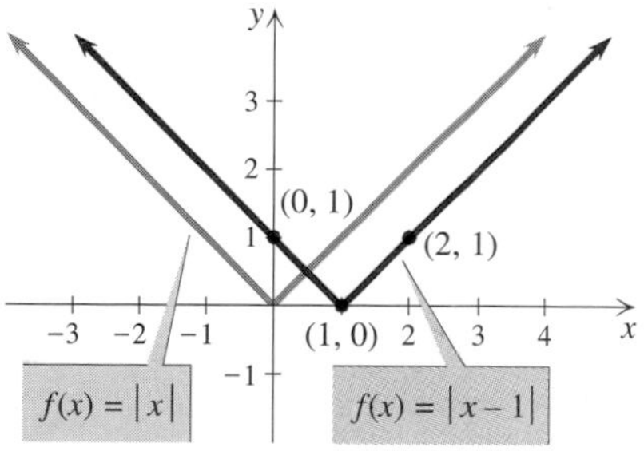

Figure 2.62

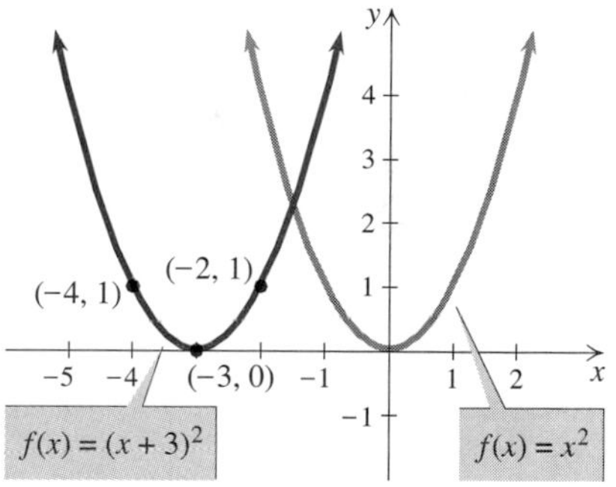

Figure 2.63

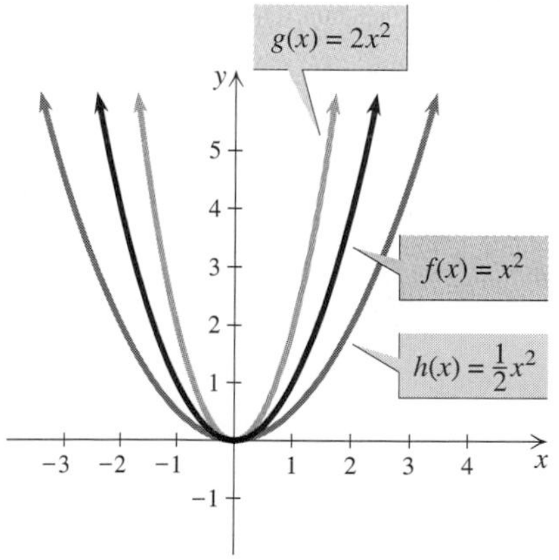

Figure 2.64

Nonrigid Transformation

Consider the graphs of the functions $f(x) = x^2$, $g(x) = 2x^2$, and $h(x) = \frac{1}{2}x^2$ shown in Fig. 2.64. For any given x-coordinate, the y-coordinate on g is twice as large as the corresponding y-coordinate on f, and the y-coordinate on h is one-half as large as the corresponding y-coordinate on f. Such transformations are referred to as stretching and shrinking transformations.

Definitions: Stretching and Shrinking

The graph of $y = af(x)$ is obtained from the graph of $y = f(x)$ by

1. **stretching** the graph of $y = f(x)$ when $a > 1$, or
2. **shrinking** the graph of $y = f(x)$ when $0 < a < 1$.

Note that the last operation to be performed in stretching or shrinking is multiplication by a. The function $f(x) = 2\sqrt{x}$ is obtained by stretching $f(x) = \sqrt{x}$ by a factor of 2. However, $f(x) = \sqrt{2x}$ is obtained from $f(x) = \sqrt{x}$ by stretching not by a factor of 2, but by a factor of $\sqrt{2}$.

Reflecting, translating, stretching, and shrinking are all referred to as **transformations** because they transform one graph into another. Since reflection and translation do not change the shape of the graph, they are **rigid transformations**. Stretching and shrinking change the shape of a graph, so they are **nonrigid transformations**.

EXAMPLE 4 Graphing using stretching and shrinking

Graph the functions $f(x) = \sqrt{x}$, $g(x) = 2\sqrt{x}$, and $h(x) = \frac{1}{2}\sqrt{x}$ on the same coordinate plane.

Solution

The graph of g is obtained by stretching the graph of f, and the graph of h is obtained by shrinking the graph of f. The graph of f includes the points (0, 0), (1, 1), and (4, 2). The graph of g includes the points (0, 0), (1, 2), and (4, 4). The graph of h includes the points (0, 0), (1, 0.5), and (4, 1). The graphs are shown in Fig. 2.65.

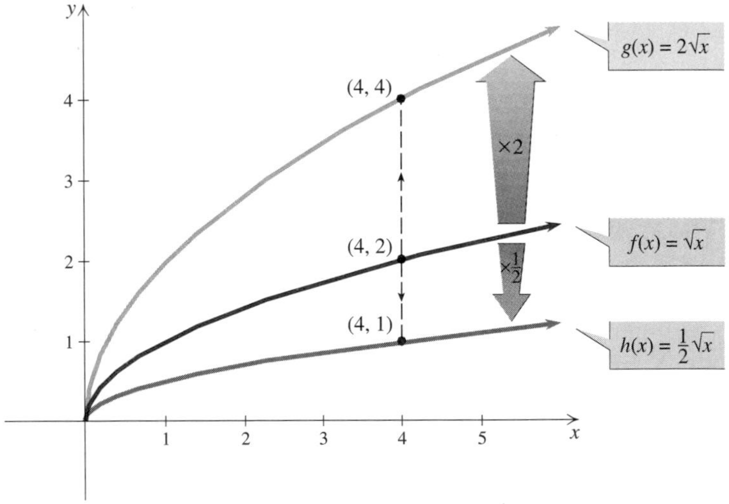

Figure 2.65

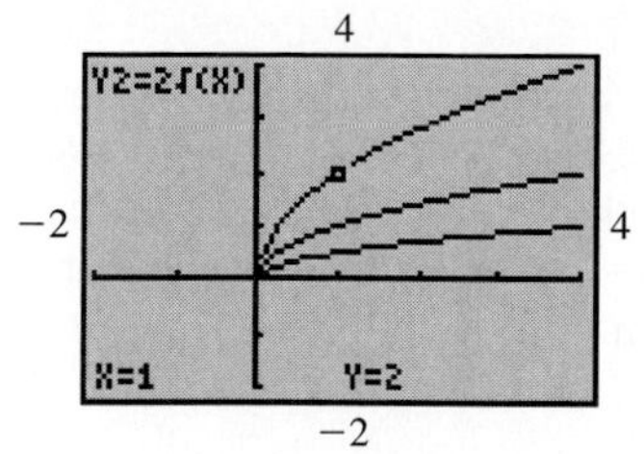

Figure 2.66

The functions f, g, and h are shown on a graphing calculator in Fig. 2.66. Note how the viewing window affects the shape of the graph. They do not appear as separated on the calculator as they do in Fig. 2.65.

A function involving more than one transformation may be graphed using the following procedure.

PROCEDURE **Multiple Transformations**

Graph a function involving more than one transformation in the following order:

1. **Horizontal translation**
2. **Stretching or shrinking**
3. **Reflecting**
4. **Vertical translation**

The function in the next example involves all four of the above transformations.

EXAMPLE 5 **Graphing using several transformations**

Graph the function $y = 4 - 2\sqrt{x + 1}$.

Solution

The graph of $y = \sqrt{x + 1}$ is a *horizontal* translation one unit to the left of the graph of $y = \sqrt{x}$. The graph of $y = 2\sqrt{x + 1}$ is obtained from $y = \sqrt{x + 1}$ by *stretching* it by a factor of 2. *Reflect* $y = 2\sqrt{x + 1}$ in the x-axis to obtain the graph of $y = -2\sqrt{x + 1}$. Finally, the graph of $y = 4 - 2\sqrt{x + 1}$ is a *vertical* translation of $y = -2\sqrt{x + 1}$, four units upward. All of these graphs are shown in Fig. 2.67.

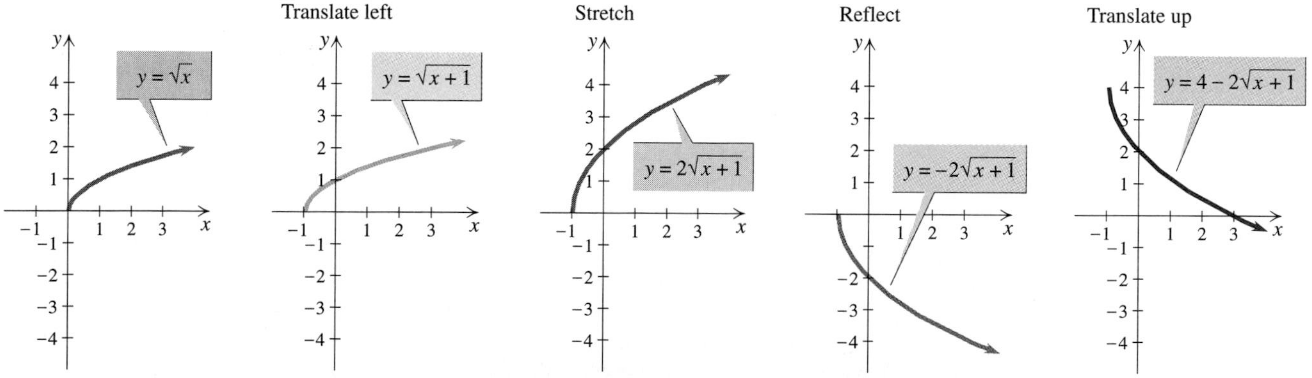

Figure 2.67

Linear Functions

A **linear function** is a function of the form $y = mx + b$, where m and b are real numbers. The simplest linear function is $y = x$, and the graphs of all others are obtained by transforming $y = x$. The function $y = x$ or $f(x) = x$ is called the **identity function**. We will study linear functions in more detail in Section 3.1.

EXAMPLE 6 Graphing linear functions using transformations

Sketch the graphs of $y = x$, $y = 2x$, $y = -2x$, and $y = -2x - 3$.

Solution

The graph of $y = x$ is a line through (0, 0), (1, 1), and (2, 2) as shown in Fig. 2.68. Stretch the graph of $y = x$ by a factor of 2 to get the graph of $y = 2x$ shown in Fig. 2.68. The graph of $y = -2x - 3$ is obtained by reflecting $y = 2x$ in the x-axis, and then translating downward three units as shown in Fig. 2.68.

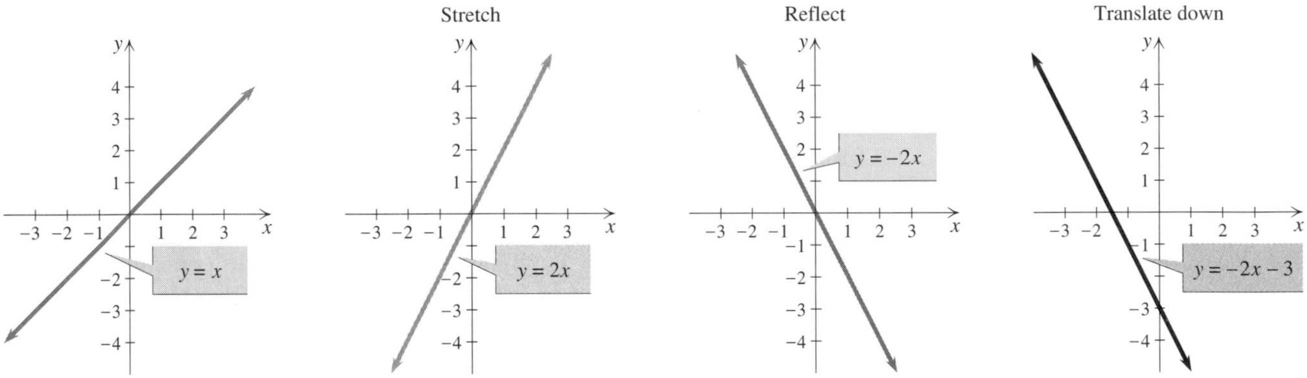

Figure 2.68

Symmetry

The graph of $g(x) = -x^2$ is a reflection in the x-axis of the graph of $f(x) = x^2$. If the paper were folded along the x-axis, the graphs would coincide. See Fig. 2.69. The symmetry that we call reflection occurs between two functions, but the graph of $f(x) = x^2$ has a symmetry within itself. Points such as (2, 4) and (−2, 4) are on the graph and are equidistant from the y-axis. Folding the paper along the y-axis brings all such pairs of points together. See Fig. 2.70. The reason for this symmetry about the y-axis is the fact that $f(x) = f(-x)$ for every value of x. We get the same y-coordinate whether we evaluate the function at a number or at its opposite.

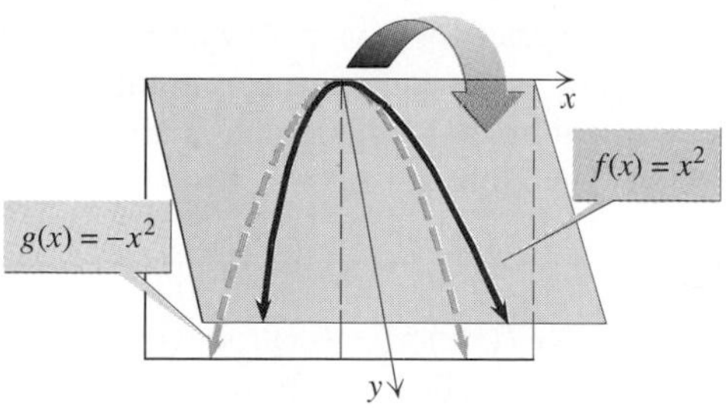

Figure 2.69

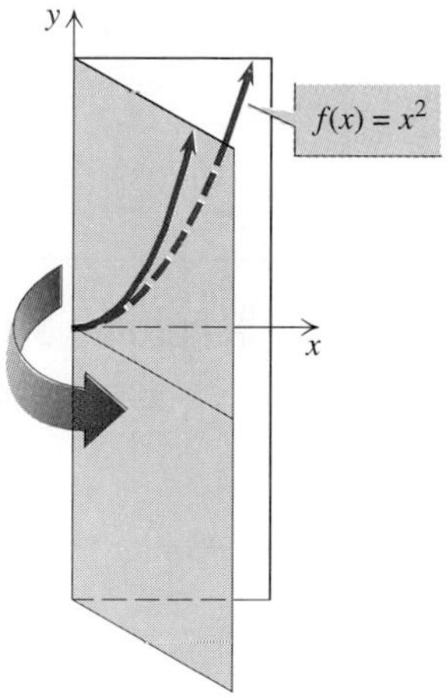

Figure 2.70

Definition: Symmetric about the *y*-Axis

If $f(x) = f(-x)$ for every value of x in the domain of the function f, then f is called an **even function** and its graph is **symmetric about the *y*-axis**.

The function $f(x) = x^2$ is an example of an even function and its graph is symmetric about the y-axis.

Consider the graph of $f(x) = x^3$ shown in Fig. 2.71. It is not symmetric about the y-axis like the graph of $f(x) = x^2$, but it has another kind of symmetry. On

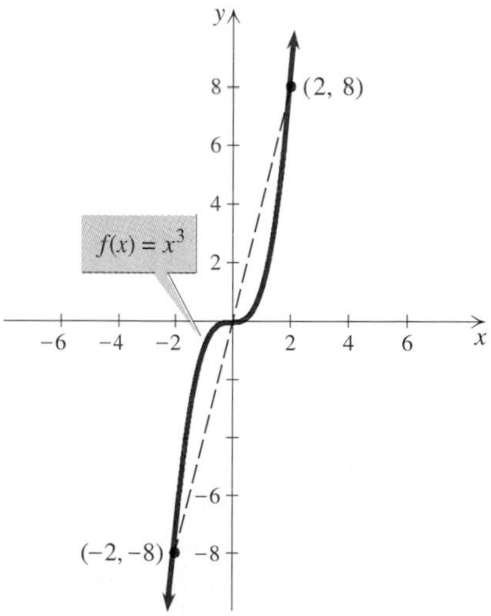

Figure 2.71

the graph of $f(x) = x^3$ we find pairs of points such as (2, 8) and (−2, −8). These points are equidistant from the origin and on opposite sides of the origin. So the symmetry of this graph is about the origin. In this case, $f(x)$ and $f(-x)$ are not equal, but $f(-x) = -f(x)$.

Definition: Symmetric about the Origin

If $f(-x) = -f(x)$ for every value of x in the domain of the function f, then f is called an **odd function** and its graph is **symmetric about the origin**.

We can look at a graph and see if it is symmetric about the y-axis or the origin, but it makes graphing easier and more accurate if we can identify symmetry *before* graphing. Using the definitions, we can determine whether a function is even, odd, or neither even nor odd from the formula defining the function. In this way we know the type of graph to expect before we start plotting points.

EXAMPLE 7 Determining symmetry in a graph

Discuss the symmetry of the graph of each function.

a) $f(x) = 5x^3 - x$ **b)** $f(x) = |x| + 3$ **c)** $f(x) = x^2 - 3x + 6$

Solution

a) Replace x by $-x$ in the formula for $f(x)$ and simplify:

$$f(-x) = 5(-x)^3 - (-x) = -5x^3 + x$$

Is $f(-x)$ equal to $f(x)$ or the opposite of $f(x)$? Since $-f(x) = -5x^3 + x$, we have $f(-x) = -f(x)$. So f is an odd function and the graph is symmetric about the origin.

b) Since $|-x| = |x|$ for any x, we have $f(-x) = |-x| + 3 = |x| + 3$. Because $f(-x) = f(x)$, the function is even and the graph is symmetric about the y-axis.

c) In this case, $f(-x) = (-x)^2 - 3(-x) + 6 = x^2 + 3x + 6$. So $f(-x) \neq f(x)$, and $f(-x) \neq -f(x)$. This function is neither odd nor even and its graph has neither type of symmetry.

Using Graphs to Solve Inequalities

In Chapter 1 we solved a variety of inequalities using various techniques. We can solve many inequalities just by examining a graph of a function. Consider the inequality $x^2 - 4 < 0$. The graph of $y = x^2 - 4$ is shown in Fig. 2.72. For any value of x for which $x^2 - 4 < 0$, the y-coordinate on the graph of $y = x^2 - 4$ is negative. The y-coordinate is negative precisely for x in the open interval $(-2, 2)$ as shown in Fig. 2.72. The solution set to $x^2 - 4 < 0$ is the interval $(-2, 2)$. The solution set to $x^2 - 4 \geq 0$ is also read from the graph of $y = x^2 - 4$ as $(-\infty, -2] \cup [2, \infty)$.

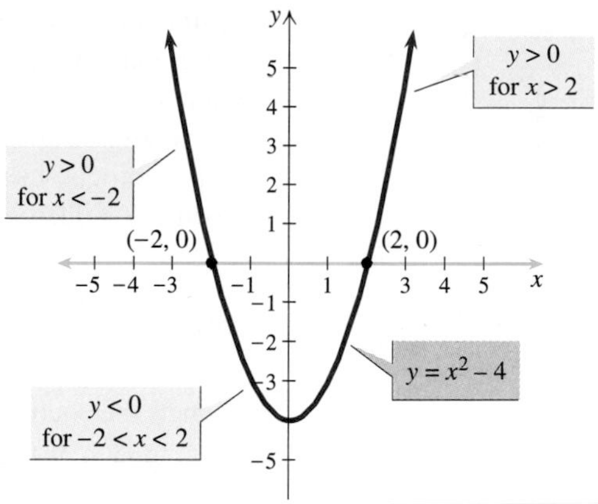

Figure 2.72

Since the x-intercepts are the points where the y-coordinate may change sign, they are critical for solving an inequality by inspecting a graph of a function.

EXAMPLE 8 Using a graph to solve an inequality

Solve the inequality $(x - 1)^2 - 2 < 0$ by graphing.

Solution

The graph of $y = (x - 1)^2 - 2$ is obtained by translating the graph of $y = x^2$ one unit to the right and two units downward. See Fig. 2.73. To find the x-intercepts we solve $(x - 1)^2 - 2 = 0$:

$$(x - 1)^2 = 2$$
$$x - 1 = \pm\sqrt{2}$$
$$x = 1 \pm \sqrt{2}$$

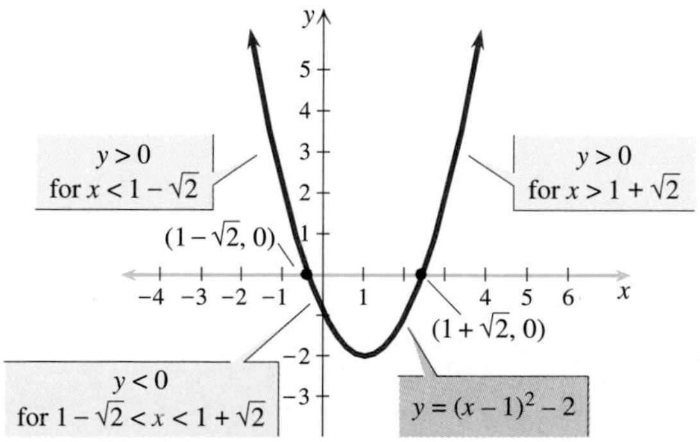

Figure 2.73

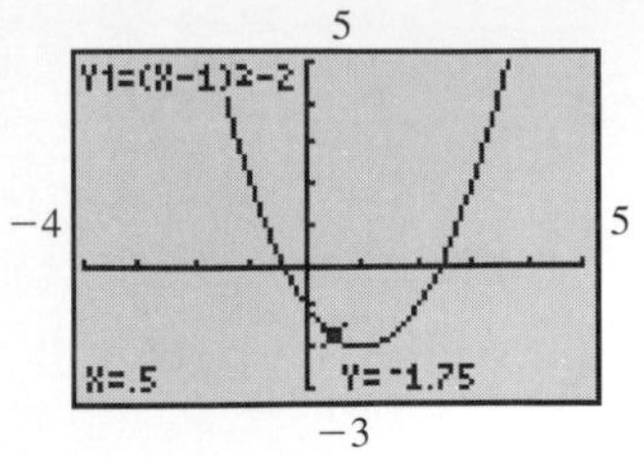

Figure 2.74

The x-intercepts are $(1 - \sqrt{2}, 0)$ and $(1 + \sqrt{2}, 0)$. If the y-coordinate of a point on the graph is negative, then the x-coordinate satisfies $(x - 1)^2 - 2 < 0$. So the solution set to $(x - 1)^2 - 2 < 0$ is the open interval $(1 - \sqrt{2}, 1 + \sqrt{2})$.

Although a graphing calculator will not find the exact solution to this inequality, you can use TRACE to support the answer and see that y is negative between the x-intercepts. See Fig. 2.74.

Note that the solution set to $(x - 1)^2 - 2 \geq 0$ can also be obtained from the graph in Fig. 2.73. If the y-coordinate of a point on the graph is positive or zero, then the x-coordinate satisfies $(x - 1)^2 - 2 \geq 0$. So the solution set to $(x - 1)^2 - 2 \geq 0$ is $(-\infty, 1 - \sqrt{2}] \cup [1 + \sqrt{2}, \infty)$.

FOR THOUGHT True or False? Explain.

1. The graph of $f(x) = (-x)^4$ is a reflection in the x-axis of the graph of $g(x) = x^4$.
2. The graph of $f(x) = x - 4$ lies four units to the right of the graph of $f(x) = x$.
3. The graph of $y = |x + 2| + 3$ is a translation two units to the right and three units upward of the graph of $y = |x|$.
4. The graph of $f(x) = -3$ is a reflection in the x-axis of the graph of $g(x) = 3$.
5. The functions $y = x^2 + 4x + 1$ and $y = (x + 2)^2 - 3$ have the same graph.
6. The graph of $y = -(x - 3)^2 - 4$ can be obtained by moving $y = x^2$ three units to the right and down four units, and then reflecting in the x-axis.
7. If $f(x) = -x^3 + 2x^2 - 3x + 5$, then $f(-x) = x^3 + 2x^2 + 3x + 5$.
8. The graphs of $f(x) = -\sqrt{x}$ and $g(x) = \sqrt{-x}$ are identical.
9. If $f(x) = x^3 - x$, then $f(-x) = -f(x)$.
10. The solution set to $|x| - 1 \leq 0$ is $[-1, 1]$.

2.4 EXERCISES

Tape 5 Disk

Sketch the graphs of each pair of functions on the same coordinate plane.

1. $f(x) = \sqrt{x}$, $g(x) = -\sqrt{x}$
2. $f(x) = x^2 + 1$, $g(x) = -x^2 - 1$
3. $y = x$, $y = -x$
4. $y = \sqrt{4 - x^2}$, $y = -\sqrt{4 - x^2}$
5. $f(x) = |x|$, $g(x) = |x| - 4$
6. $f(x) = \sqrt{x}$, $g(x) = \sqrt{x} + 3$
7. $f(x) = x$, $g(x) = x + 3$
8. $f(x) = x^2$, $g(x) = x^2 - 5$
9. $y = x^2$, $y = (x - 3)^2$
10. $y = |x|$, $y = |x + 2|$
11. $y = \sqrt{x}$, $y = 3\sqrt{x}$
12. $y = |x|$, $y = \frac{1}{3}|x|$
13. $y = x^2$, $y = \frac{1}{4}x^2$
14. $y = \sqrt{1 - x^2}$, $y = 4\sqrt{1 - x^2}$

Match each function in Exercises 15–22 with its graph (a)–(h).

15. $y = x^2$
16. $y = (x - 4)^2 + 2$
17. $y = (x + 4)^2 - 2$
18. $y = -2(x - 2)^2$
19. $y = -2(x + 2)^2$
20. $y = -\frac{1}{2}x^2 - 4$
21. $y = \frac{1}{2}(x + 4)^2 + 2$
22. $y = -2(x - 4)^2 - 2$

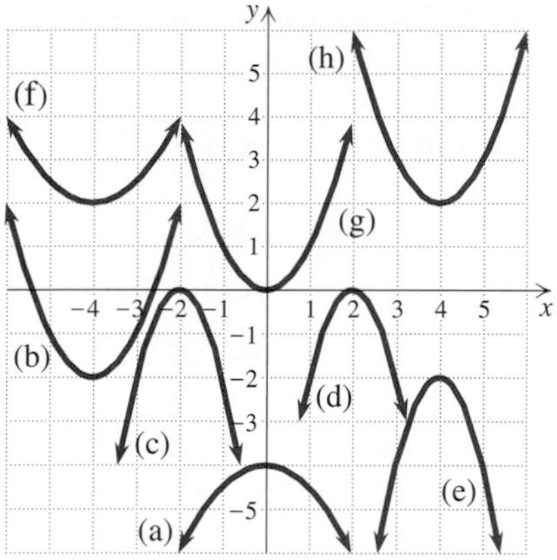

Figure for Exercises 15 to 22

Write the equation of each graph in its final resting place.

23. The graph of $y = x^2$ is translated 10 units to the right and 4 units upward.

24. The graph of $y = \sqrt{x}$ is translated 5 units to the left and 12 units downward.

25. The graph of $y = |x|$ is reflected in the x-axis, stretched by a factor of 3, then translated 7 units to the right and 9 units upward.

26. The graph of $y = x$ is stretched by a factor of 2, reflected in the x-axis, then translated 8 units downward and 6 units to the left.

27. The graph of $y = \sqrt{x}$ is stretched by a factor of 3, translated 5 units upward, then reflected in the x-axis.

28. The graph of $y = x^2$ is translated 13 units to the right and 6 units downward, then reflected in the x-axis.

Use transformations to graph each function.

29. $y = (x - 1)^2 + 2$ **30.** $y = \sqrt{x + 5} - 4$

31. $y = |x - 1| + 3$ **32.** $y = |x + 3| - 4$

33. $y = 3x - 40$ **34.** $y = -4x + 200$

35. $y = \frac{1}{2}x - 20$ **36.** $y = -\frac{1}{2}x + 40$

37. $y = -\frac{1}{2}|x| + 40$ **38.** $y = 3|x| - 200$

39. $y = -\frac{1}{2}|x + 4|$ **40.** $y = 3|x - 2|$

41. $y = -\sqrt{x - 3} + 1$ **42.** $y = -(x + 2)^2 - 4$

43. $y = -2\sqrt{x + 3} + 2$ **44.** $y = -\frac{1}{2}\sqrt{x + 2} + 4$

Discuss the symmetry of the graph of each function and determine whether the function is even, odd, or neither.

45. $f(x) = x^4$ **46.** $f(x) = x^4 - 2x^2$

47. $f(x) = x^4 - x^3$ **48.** $f(x) = x^3 - x$

49. $f(x) = (x + 3)^2$ **50.** $f(x) = (x - 1)^2$

51. $f(x) = \sqrt{x}$ **52.** $f(x) = |x| - 9$

53. $f(x) = x$ **54.** $f(x) = -x$

55. $f(x) = 3x + 2$ **56.** $f(x) = x - 3$

57. $f(x) = x^3 - 5x + 1$ **58.** $f(x) = x^6 - x^4 + x^2$

59. $f(x) = |x - 2|$ **60.** $f(x) = (x^2 - 2)^3$

61. $f(x) = 1 + \frac{1}{x^2}$ **62.** $f(x) = \sqrt{9 - x^2}$

63. $f(x) = |x^2 - 3|$ **64.** $f(x) = \sqrt{x^2 + 3}$

Match each function with its graph (a)–(h) on pages 219–220.

65. $y = 2 + \sqrt{x}$ **66.** $y = \sqrt{2 + x}$

67. $y = \sqrt{x^2}$ **68.** $y = \sqrt{\frac{x}{2}}$

69. $y = \frac{1}{2}\sqrt{x}$ **70.** $y = 2 - \sqrt{x - 2}$

71. $y = -2\sqrt{x}$ **72.** $y = -\sqrt{-x}$

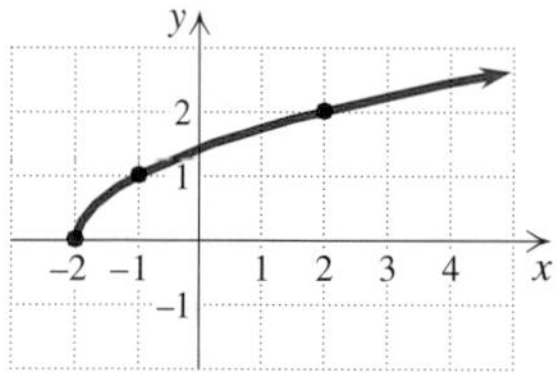

(a)

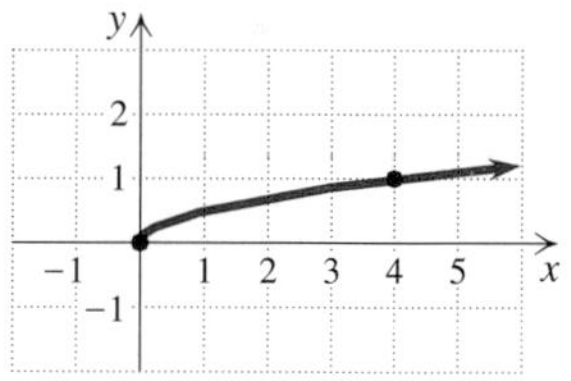

(b)

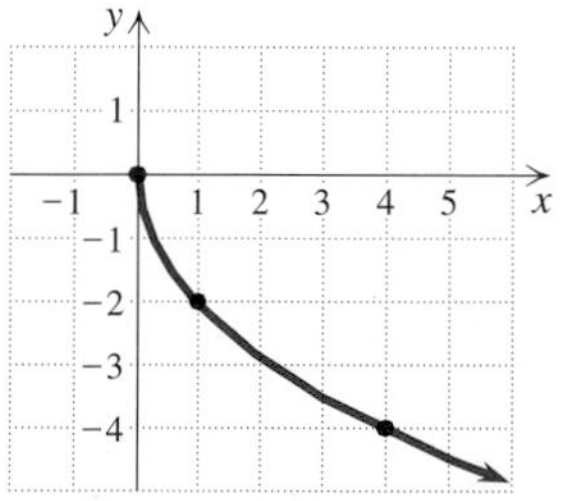

(c)

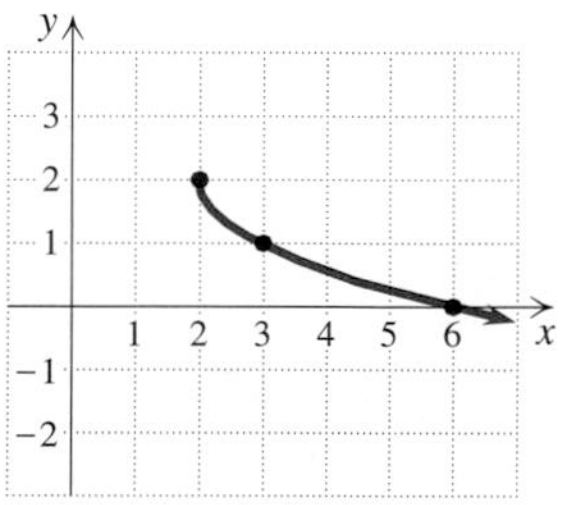

(d)

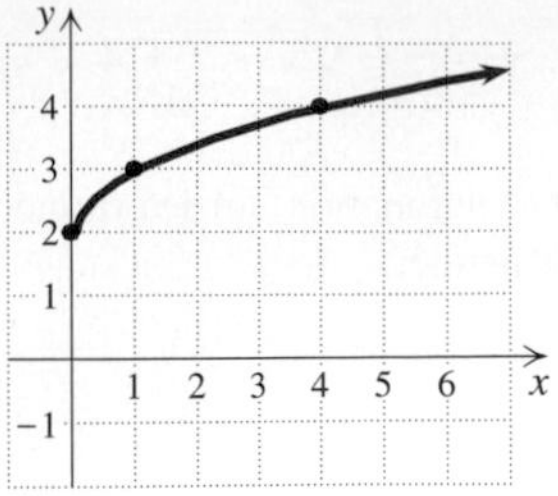

(e)

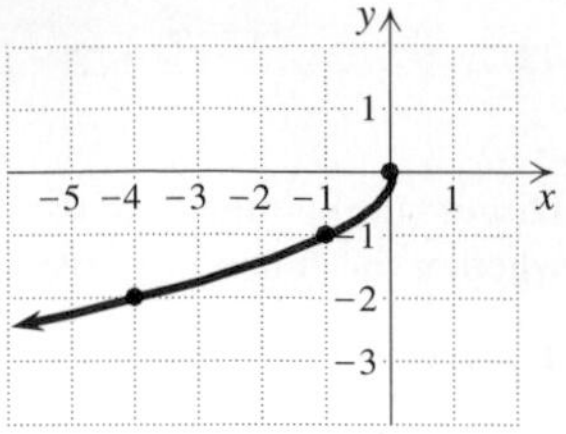

(f)

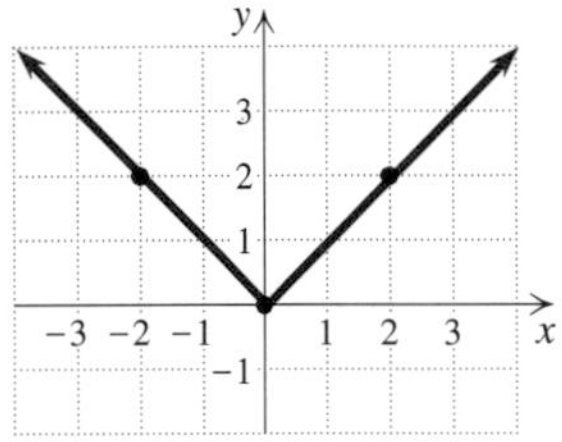

(g)

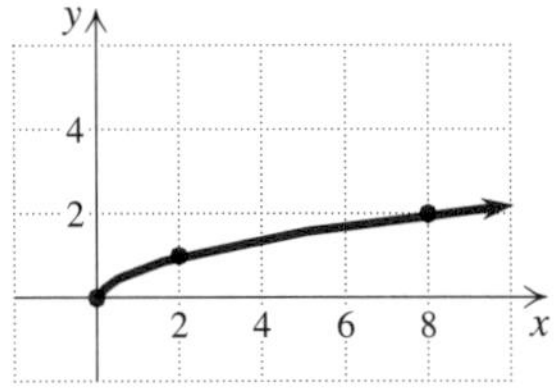

(h)

Figure for Exercises 65 to 72

Solve each inequality by reading the corresponding graph.

73. $x^2 - 1 \geq 0$

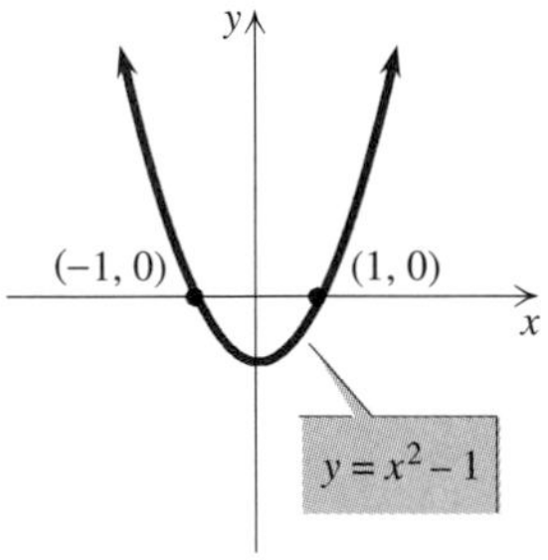

74. $2x^2 - 3 < 0$

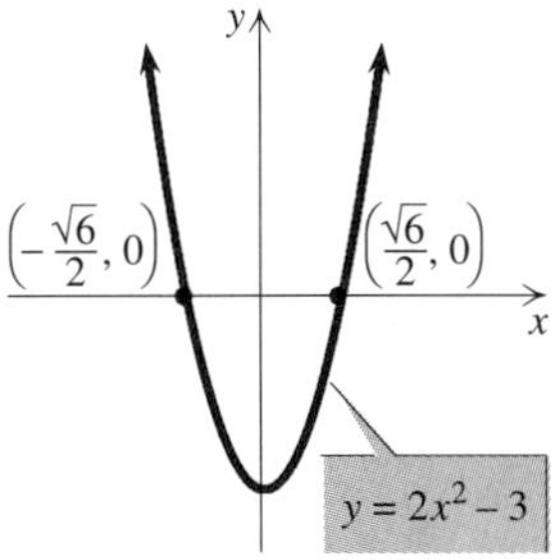

75. $|x - 2| - 3 > 0$

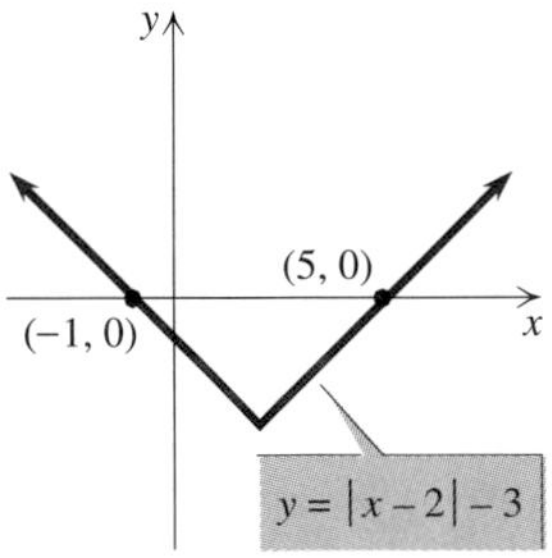

76. $2 - |x + 1| \geq 0$

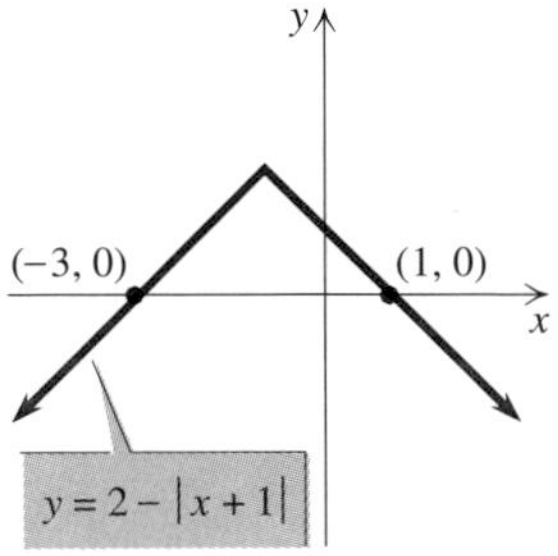

Solve each inequality by graphing an appropriate function. State the solution set using interval notation.

77. $(x - 1)^2 - 9 < 0$

78. $\left(x - \dfrac{1}{2}\right)^2 - \dfrac{9}{4} \geq 0$

79. $5 - \sqrt{x} \geq 0$

80. $\sqrt{x + 3} - 2 \geq 0$

81. $(x - 2)^2 > 3$

82. $(x - 1)^2 < 4$

83. $\sqrt{25 - x^2} > 0$

84. $\sqrt{4 - x^2} \geq 0$

Use a graphing calculator to find an approximate solution to each inequality by reading the graph of an appropriate function. Round to two decimal places.

85. $\sqrt{3}x^2 + \pi x - 9 < 0$

86. $x^3 - 5x^2 + 6x - 1 > 0$

Solve each problem.

87. *Across-the-Board Raise* Each teacher at C. F. Gauss Elementary School is given an across-the-board raise of \$2000. Write a function that *transforms* each old salary x into a new salary $N(x)$.

88. *Cost-of-Living Raise* Each registered nurse at Blue Hills Memorial Hospital is first given a 5% cost-of-living raise and then a \$3000 merit raise. Write a function that *transforms* each old salary x into a new salary $N(x)$. Does it make any difference in which order these raises are given? Explain.

89. *Unemployment Versus Inflation* The Phillips curve shows the relationship between the unemployment rate x and the inflation rate y. If the equation of the curve is $y = 1 - \sqrt{x}$ for a certain Third World country, then for what values of x is the inflation rate less than 50%?

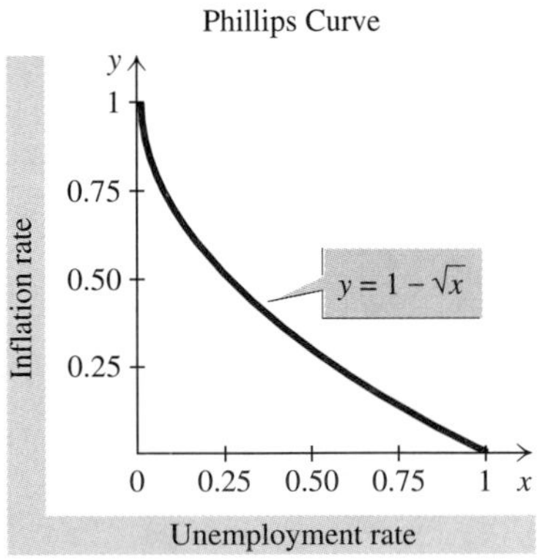

Figure for Exercise 89

90. *Production Function* The production function shows the relationship between inputs and outputs. A manufacturer of custom windows produces y windows per week using x hours of labor per week, where $y = 1.75\sqrt{x}$. How many hours of labor are required to keep production at or above 28 windows per week?

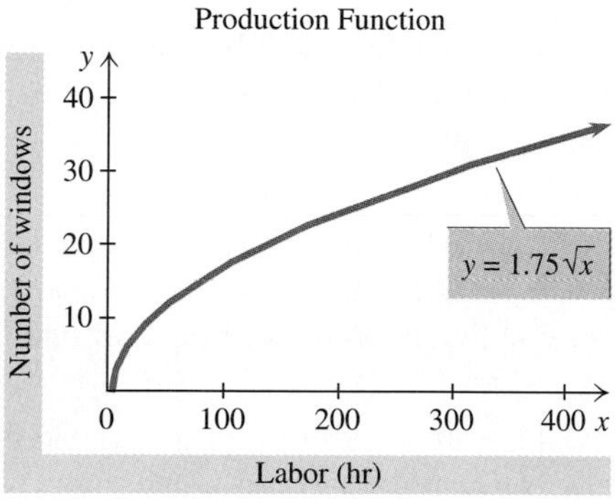

Figure for Exercise 89

91. *Life Expectancy in Lyon* Lyon, France began improving its water supply and sanitation around 1840 while Paris began around 1850. The accompanying graph shows how water and sanitation improvements corresponded to increased life expectancy for those cities (World Resources, 1996–1997). Life expectancy for Lyon is modeled by the equation

$$y = 0.005x^2 + 32,$$

where x is the number of years since 1840 and y is the life expectancy in years. Determine the year in which life expectancy in Lyon surpassed 46 years.

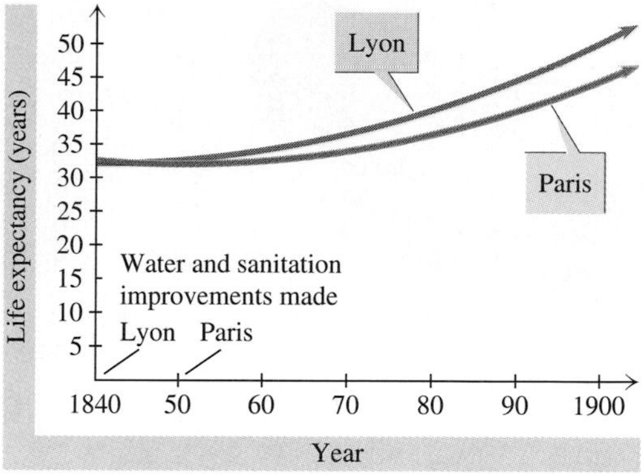

Figure for Exercises 91 and 92

92. *Life Expectancy in Paris* Life expectancy for Paris is modeled by the equation

$$y = 0.005(x - 10)^2 + 32,$$

where x is the number of years since 1840 and y is the life expectancy in years. Determine the year in which life expectancy in Paris surpassed 46 years.

For Writing/Discussion

93. Graph each pair of functions (without simplifying the second function) on the same screen of a graphing calculator and explain what each exercise illustrates.

a. $y = x^4 - x^2$, $y = (-x)^4 - (-x)^2$
b. $y = x^3 - x$, $y = (-x)^3 - (-x)$
c. $y = x^4 - x^2$, $y = (x + 1)^4 - (x + 1)^2$
d. $y = x^3 - x$, $y = (x - 2)^3 - (x - 2) + 3$

94. Graph $y = x^3 + 6x^2 + 12x + 8$ on a graphing calculator. Explain how this graph could be obtained as a transformation of a simpler function.

Linking Concepts

For Individual or Group Explorations

Designing a Racing Boat *According to International America's Cup Class rules the basic dimensions of any yacht competing for the silver trophy must satisfy the inequality*

$$16.96 + 9.8D^{1/3} - L - 1.25S^{1/2} \geq 0,$$

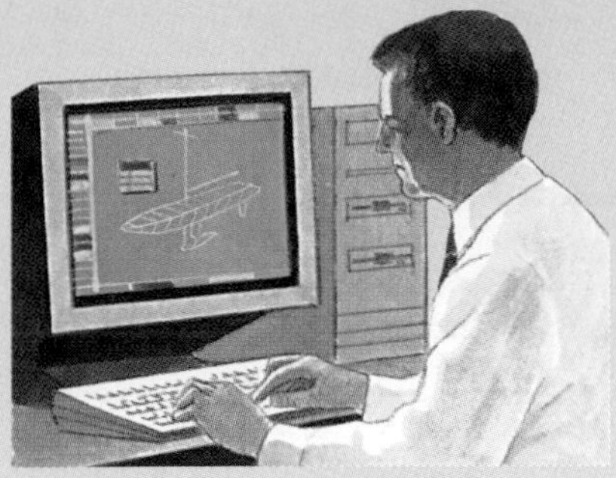

where L is the length in meters, S is the sail area in square meters, and D is the displacement in cubic meters (Scientific American, May 1992). Use a graphing utility to get approximate answers to the following questions.

a) A team of British designers wants their new boat to have a length of 20.85 m and a displacement of 17.67 m^3. Write the inequality that must be satisfied by the sail area S. Use the graphing technique described in this section to find the interval in which S must lie.

b) A team of Australian designers wants their new boat to have a sail area of 312.54 m^2 and length of 21.45 m. Write the inequality that must be satisfied by the displacement D. Use the graphing technique to find the interval in which D must lie.

c) A team of American designers wants their new boat to have a sail area of 310.28 m^2 and a displacement of 17.26 m^3. Write the inequality that must be satisfied by the length L. Use the graphing technique to find the interval in which L must lie.

d) When two of the three variables are fixed, the third variable either has a maximum or minimum value. Explain how you can determine whether it is a maximum or minimum by looking at the original inequality.

2.5 Operations with Functions

In Sections 2.3 and 2.4 we studied the graphs of functions to see the relationships between types of functions and their graphs. In this section we will study various ways in which two or more functions can be combined to make new functions. The emphasis here will be on formulas that define functions.

Basic Operations with Functions

A college student is hired to deliver new telephone books and collect the old ones for recycling. She is paid \$4 per hour plus \$0.30 for each old phone book she collects. Her salary for a 40-hour week is a function of the number of phone books collected. If x represents the number of phone books collected in one week, then the function $S(x) = 0.30x + 160$ gives her salary in dollars. However, she must use her own car for this job. She figures that her car expenses average \$0.20 per phone book collected plus a fixed cost of \$20 per week for insurance. We can write her expenses as a function of the number of phone books collected, $E(x) = 0.20x + 20$. Her profit for one week is her salary minus her expenses:

$$\begin{aligned} P(x) &= S(x) - E(x) \\ &= 0.30x + 160 - (0.20x + 20) \\ &= 0.10x + 140 \end{aligned}$$

By subtracting, we get $P(x) = 0.10x + 140$. Her weekly profit is written as a function of the number of phone books collected. In this example we obtained a new function by subtracting two functions. In general, there are four basic arithmetic operations defined for functions.

Definition: Sum, Difference, Product, and Quotient Functions

For two functions f and g, the **sum**, **difference**, **product**, and **quotient functions**, functions $f + g$, $f - g$, $f \cdot g$, and f/g, respectively, are defined as follows:

$$(f + g)(x) = f(x) + g(x)$$
$$(f - g)(x) = f(x) - g(x)$$
$$(f \cdot g)(x) = f(x) \cdot g(x)$$
$$(f/g)(x) = f(x)/g(x) \quad \text{provided that } g(x) \neq 0.$$

EXAMPLE 1 Evaluating functions

Let $f(x) = 3\sqrt{x} - 2$ and $g(x) = x^2 + 5$. Find and simplify each expression.

a) $(f + g)(4)$ **b)** $(f - g)(x)$ **c)** $(f \cdot g)(0)$ **d)** $\left(\frac{f}{g}\right)(9)$

(a)

(b)

Figure 2.75

Solution

a) $(f + g)(4) = f(4) + g(4) = 3\sqrt{4} - 2 + 4^2 + 5 = 25$

b) $(f - g)(x) = f(x) - g(x) = 3\sqrt{x} - 2 - (x^2 + 5) = 3\sqrt{x} - x^2 - 7$

c) $(f \cdot g)(0) = f(0) \cdot g(0) = (3\sqrt{0} - 2)(0^2 + 5) = (-2)(5) = -10$

d) $\left(\frac{f}{g}\right)(9) = \frac{f(9)}{g(9)} = \frac{3\sqrt{9} - 2}{9^2 + 5} = \frac{7}{86}$

Parts (a), (c), and (d) can be checked with a graphing calculator as shown in Figs. 2.75(a) and (b).

Think of $f + g$, $f - g$, $f \cdot g$, and f/g as generic names for the sum, difference, product, and quotient of the functions f and g. If any of these functions has a particular meaning, as in the phone book example, we can use a new letter to identify it. The domain of $f + g$, $f - g$, $f \cdot g$, or f/g is the intersection of the domain of f with the domain of g. Of course, we exclude from the domain of f/g any number for which $g(x) = 0$.

EXAMPLE 2 The sum, product, and quotient functions

Let $f = \{(1, 3), (2, 8), (3, 6), (5, 9)\}$ and $g = \{(1, 6), (2, 11), (3, 0), (4, 1)\}$. Find $f + g$, $f \cdot g$, and f/g. State the domain of each function.

Solution

The domain of $f + g$ and $f \cdot g$ is $\{1, 2, 3\}$ because that is the intersection of the domains of f and g. The ordered pair $(1, 9)$ belongs to $f + g$ because

$$(f + g)(1) = f(1) + g(1) = 3 + 6 = 9.$$

The ordered pair $(2, 19)$ belongs to $f + g$ because $(f + g)(2) = 19$. The pair $(3, 6)$ also belongs to $f + g$. So

$$f + g = \{(1, 9), (2, 19), (3, 6)\}.$$

Since $(f \cdot g)(1) = f(1) \cdot g(1) = 3 \cdot 6 = 18$, the pair $(1, 18)$ belongs to $f \cdot g$. Likewise, $(2, 88)$ and $(3, 0)$ also belong to $f \cdot g$. So

$$f \cdot g = \{(1, 18), (2, 88), (3, 0)\}.$$

The domain of f/g is $\{1, 2\}$ because $g(3) = 0$. So

$$\frac{f}{g} = \left\{\left(1, \frac{1}{2}\right), \left(2, \frac{8}{11}\right)\right\}.$$

In Example 2 the functions are given as sets of ordered pairs, and the results of performing operations with these functions are sets of ordered pairs. In the next example the sets of ordered pairs are defined by means of equations, so the result of performing operations with these functions will be new equations that determine the ordered pairs of the function.

EXAMPLE 3 The sum, quotient, product, and difference functions

Let $f(x) = \sqrt{x}$, $g(x) = 3x + 1$, and $h(x) = x - 1$. Find each function and state its domain.

a) $f + g$ **b)** $\frac{g}{f}$ **c)** $g \cdot h$ **d)** $g - h$

Solution

a) Since the domain of f is $[0, \infty)$ and the domain of g is $(-\infty, \infty)$, the domain of $f + g$ is $[0, \infty)$. Since $(f + g)(x) = f(x) + g(x) = \sqrt{x} + 3x + 1$, the equation defining the function $f + g$ is

$$(f + g)(x) = \sqrt{x} + 3x + 1.$$

b) The number 0 is not in the domain of g/f because $f(0) = 0$. So the domain of g/f is $(0, \infty)$. The equation defining g/f is

$$\left(\frac{g}{f}\right)(x) = \frac{3x + 1}{\sqrt{x}}.$$

c) The domain of both g and h is $(-\infty, \infty)$. So the domain of $g \cdot h$ is $(-\infty, \infty)$. Since $(3x + 1)(x - 1) = 3x^2 - 2x - 1$, the equation defining the function $g \cdot h$ is

$$(g \cdot h)(x) = 3x^2 - 2x - 1.$$

d) The domain of both g and h is $(-\infty, \infty)$. So the domain of $g - h$ is $(-\infty, \infty)$. Since $(3x + 1) - (x - 1) = 2x + 2$, the equation defining $g - h$ is

$$(g - h)(x) = 2x + 2.$$

Composition of Functions

The area of a circle is a function of the radius, $A = \pi r^2$. The radius is a function of the diameter, $r = d/2$. So the area is also a function of the diameter. The formula for A as a function of the diameter is obtained by substituting $d/2$ for r:

$$A = \pi \frac{d^2}{4}$$

This example illustrates the **composition of functions**.

EXAMPLE 4 Composition of functions: rainforest application

Tropical rainforests are one of the earth's most ancient and complex ecosystems, yet each year deforestation claims an area the size of West Virginia. In "Deforestation in the Tropics" (*Scientific American*, April 1990) it is stated that loggers typically harvest only 10% of the area in a rainforest. However, the loggers destroy 50% of the area not harvested just to get at the valuable trees. In a forest of x square kilometers, the function $u = 0.9x$ gives the unharvested area when the valuable timber is removed. The function $w = 0.5u$ gives the amount of unharvested area that is destroyed and wasted. Write the wasted area w as a function of the original area x.

Solution

Since w is a function of u and u is a function of x, w is a function of x. Writing the composition of the two functions is simply a matter of replacing u by $0.9x$ in the formula for w:

$$w = 0.5u = 0.5(0.9x) = 0.45x$$

The composition function $w = 0.45x$ indicates that 45% of the area of a rainforest is wasted when the valuable timber is harvested.

Example 4 illustrates how composition is handled using functions defined by formulas. The salary of the phone book collector mentioned at the start of this section was given in the f-notation as $S(x) = 0.30x + 160$, where x is the number of phone books collected. Now suppose that 20% of her salary is withheld for income taxes. The withholding function is $W(x) = 0.20x$, where x is her salary. The salary is a function of the number of phone books, and the withholding is a function of the salary, so the withholding is a function of the number of phone books, and we can write

$$W(S(x)) = W(0.30x + 160) = 0.20(0.30x + 160) = 0.06x + 32.$$

We have a new function pairing the number of phone books collected with the amount of withholding. We have found the composition of the functions W and S. The notation $W \circ S$ is used for the composition of W and S. $W \circ S$ is read "W compose S" or "W of S." In this case, $(W \circ S)(x) = 0.06x + 32$. For example, if the student collects 100 phone books, the amount of withholding is $(W \circ S)(100) = 0.06(100) + 32 = \$38$.

Definition: Composition of Functions

If f and g are two functions, the **composition** of f and g, written $f \circ g$, is defined by the equation

$$(f \circ g)(x) = f(g(x)),$$

provided that $g(x)$ is in the domain of f. The composition of g and f, written $g \circ f$, is defined by

$$(g \circ f)(x) = g(f(x)),$$

provided that $f(x)$ is in the domain of g.

Note that $f \circ g$ is not the same function as $f \cdot g$, the product of f and g.

EXAMPLE 5 Evaluating compositions defined by equations

Let $f(x) = \sqrt{x}$, $g(x) = 2x - 1$, and $h(x) = x^2$. Find the value of each expression.

a) $(f \circ g)(5)$ **b)** $(g \circ f)(5)$ **c)** $(h \circ g \circ f)(9)$

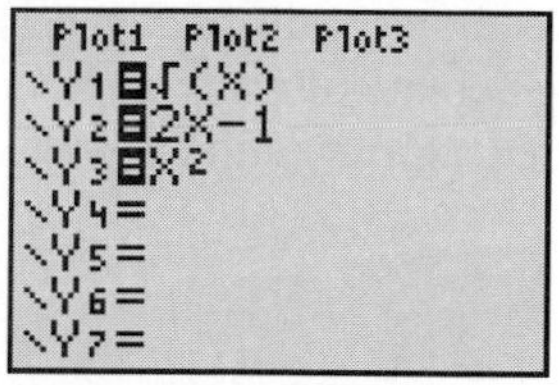

(a)

(b)

Figure 2.76

Solution

a) $(f \circ g)(5) = f(g(5))$ — Definition of composition

$= f(9)$ — $g(5) = 2 \cdot 5 - 1 = 9$

$= \sqrt{9}$

$= 3$

b) $(g \circ f)(5) = g(f(5)) = g(\sqrt{5}) = 2\sqrt{5} - 1$

c) $(h \circ g \circ f)(9) = h(g(f(9))) = h(g(3)) = h(5) = 5^2 = 25$

You can check these answers with a graphing calculator as shown in Figs. 2.76(a) and (b).

Note that for the composition $f \circ g$ to be defined at x, $g(x)$ must be in the domain of f. So the domain of $f \circ g$ is the set of all values of x in the domain of g for which $g(x)$ is in the domain of f. The diagram shown in Fig. 2.77 will help you to understand the composition of functions. In Example 5 we found specific values of compositions. In the next example we will find the entire composition function.

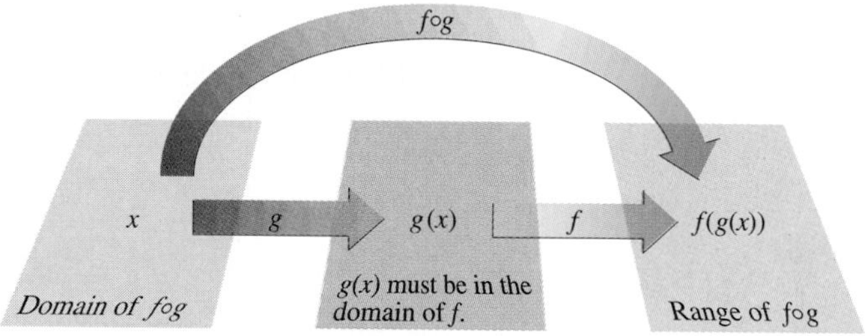

Figure 2.77

EXAMPLE 6 Composition of functions defined by sets

Let $g = \{(1, 4), (2, 5), (3, 6)\}$ and $f = \{(3, 8), (4, 9), (5, 10)\}$. Find $f \circ g$.

Solution

Since $g(1) = 4$ and $f(4) = 9$, $(f \circ g)(1) = 9$. So the ordered pair $(1, 9)$ is in $f \circ g$. Since $g(2) = 5$ and $f(5) = 10$, $(f \circ g)(2) = 10$. So $(2, 10)$ is in $f \circ g$. Now $g(3) = 6$, but 6 is not in the domain of f. So there are only two ordered pairs in $f \circ g$:

$$f \circ g = \{(1, 9), (2, 10)\}$$

In Example 6, the domain of g is $\{1, 2, 3\}$ while the domain of $f \circ g$ is $\{1, 2\}$. To find the domain of $f \circ g$, we remove from the domain of g any number x such that $g(x)$ is not in the domain of f. In the next example we find compositions of functions defined by equations.

EXAMPLE 7 **Composition of functions defined by equations**

Let $f(x) = \sqrt{x}$, $g(x) = 2x - 1$, and $h(x) = x^2$. Find each composition function and state its domain.

a) $f \circ g$ **b)** $g \circ f$ **c)** $h \circ g$

Solution

a) In the composition $f \circ g$, $g(x)$ must be in the domain of f. The domain of f is $[0, \infty)$. If $g(x)$ is in $[0, \infty)$, then $2x - 1 \geq 0$, or $x \geq \frac{1}{2}$. So the domain of $f \circ g$ is $[\frac{1}{2}, \infty)$. Since

$$(f \circ g)(x) = f(g(x)) = f(2x - 1) = \sqrt{2x - 1},$$

the function $f \circ g$ is defined by the equation $(f \circ g)(x) = \sqrt{2x - 1}$, for $x \geq \frac{1}{2}$.

b) Since the domain of g is $(-\infty, \infty)$, $f(x)$ is certainly in the domain of g. So the domain of $g \circ f$ is the same as the domain of f, $[0, \infty)$. Since

$$(g \circ f)(x) = g(f(x)) = g(\sqrt{x}) = 2\sqrt{x} - 1,$$

the function $g \circ f$ is defined by the equation $(g \circ f)(x) = 2\sqrt{x} - 1$. Note that $g \circ f$ is generally not equal to $f \circ g$, but in Section 2.6 we will study special types of functions for which they are equal.

c) Since the domain of h is $(-\infty, \infty)$, $g(x)$ is certainly in the domain of h. So the domain of $h \circ g$ is the same as the domain of g, $(-\infty, \infty)$. Since

$$(h \circ g)(x) = h(g(x)) = h(2x - 1) = (2x - 1)^2 = 4x^2 - 4x + 1,$$

the function $h \circ g$ is defined by $(h \circ g)(x) = 4x^2 - 4x + 1$.

When studying rates of change of functions in calculus, we often need to look at a function that has several operations as a composition of simpler functions. We can also better understand the transformations of graphs in Section 2.4 when we think of a complicated function as a composition of simpler functions. For example, to find the second coordinate of an ordered pair for the function $H(x) = (x + 3)^2$, we start with x, add 3 to x, then square the result. These two operations can be accomplished by composition, using $f(x) = x + 3$ followed by $g(x) = x^2$:

$$(g \circ f)(x) = g(f(x)) = g(x + 3) = (x + 3)^2$$

So the function H is the same as the composition of g and f, $H = g \circ f$. Notice that $(f \circ g)(x) = x^2 + 3$ and it is not the same as $H(x)$.

• CALCULUS •

EXAMPLE 8 **Writing a function as a composition**

Let $f(x) = \sqrt{x}$, $g(x) = x - 3$, and $h(x) = 2x$. Write each given function as a composition of appropriate functions chosen from f, g, and h.

a) $F(x) = \sqrt{x - 3}$ **b)** $G(x) = x - 6$ **c)** $H(x) = 2\sqrt{x} - 3$

Solution

a) The function F consists of subtracting 3 from x and finding the square root of that result. These two operations can be accomplished by composition, using g followed by f. So $F = f \circ g$. Check this answer as follows:

$$(f \circ g)(x) = f(g(x)) = f(x - 3) = \sqrt{x - 3} = F(x)$$

b) Subtracting 6 from x can be accomplished by subtracting 3 from x and then subtracting 3 from that result, so $G = g \circ g$. Check as follows:

$$(g \circ g)(x) = g(g(x)) = g(x - 3) = (x - 3) - 3 = x - 6 = G(x)$$

c) For the function H, find the square root of x, then multiply by 2, and finally subtract 3. These three operations can be accomplished by composition, using f, then h, and then g. So $H = g \circ h \circ f$. Check as follows:

$$(g \circ h \circ f)(x) = g(h(f(x))) = g(h(\sqrt{x})) = g(2\sqrt{x}) = 2\sqrt{x} - 3 = H(x)$$

FOR THOUGHT True or False? Explain.

1. If $f = \{(2, 4)\}$ and $g = \{(1, 5)\}$, then $f + g = \{(3, 9)\}$.
2. If $f = \{(1, 6), (9, 5)\}$ and $g = \{(1, 3), (9, 0)\}$, then $f/g = \{(1, 2)\}$.
3. If $f = \{(1, 6), (9, 5)\}$ and $g = \{(1, 3), (9, 0)\}$, then $f \cdot g = \{(1, 18), (9, 0)\}$.
4. If $f(x) = x + 2$ and $g(x) = x - 3$, then $(f \cdot g)(5) = 14$.
5. If $s = P/4$ and $A = s^2$, then A is a function of P.
6. If $f(3) = 19$ and $g(19) = 99$, then $(g \circ f)(3) = 99$.
7. If $f(x) = \sqrt{x}$ and $g(x) = x - 2$, then $(f \circ g)(x) = \sqrt{x} - 2$.
8. If $f(x) = 5x$ and $g(x) = x/5$, then $(f \circ g)(x) = (g \circ f)(x) = x$.
9. If $F(x) = (x - 9)^2$, $g(x) = x^2$, and $h(x) = x - 9$, then $F = h \circ g$.
10. If $f(x) = \sqrt{x}$ and $g(x) = x - 2$, then the domain of $f \circ g$ is $[2, \infty)$.

2.5 EXERCISES

Tape 6 Disk

Let $f(x) = x - 3$ and $g(x) = x^2 - x$. Find and simplify each expression.

1. $(f + g)(2)$ **2.** $(g + f)(3)$ **3.** $(f - g)(-2)$

4. $(g - f)(-6)$ **5.** $(f \cdot g)(-1)$ **6.** $(g \cdot f)(0)$

7. $(f/g)(4)$ **8.** $(g/f)(4)$ **9.** $(f + g)(a)$

10. $(f - g)(b)$ **11.** $(f \cdot g)(a)$ **12.** $(f/g)(b)$

Use the two given functions to write y as a function of x.

13. $y = 2a - 3,\ a = 3x + 1$

14. $y = -4d - 1,\ d = -3x - 2$

15. $y = w^2 - 2,\ w = x + 3$

16. $y = 3t^2 - 3,\ t = x - 1$

17. $y = 3m - 1,\ m = \dfrac{x + 1}{3}$

18. $y = 2z + 5,\ z = \dfrac{1}{2}x - \dfrac{5}{2}$

19. $y = 2k^3 - 1,\ k = \sqrt[3]{\dfrac{x + 1}{2}}$

20. $y = \dfrac{s^3 + 1}{5},\ s = \sqrt[3]{5x - 1}$

Let $f = \{(-3, 1), (0, 4), (2, 0)\}$, $g = \{(-3, 2), (1, 2), (2, 6), (4, 0)\}$, and $h = \{(2, 4), (1, 0)\}$. Find each function.

21. $f + g$ **22.** $f - g$ **23.** g/f **24.** f/g

25. $f \cdot g$ **26.** $g - f$ **27.** $f \circ g$ **28.** $g \circ f$

29. $f \circ h$ **30.** $h \circ f$ **31.** $h \circ g$ **32.** $g \circ h$

Let $f(x) = \sqrt{x}$, $g(x) = x - 4$, and $h(x) = \dfrac{1}{x - 2}$.

Find an equation defining each function and state the domain of the function.

33. $f + g$ **34.** $f + h$ **35.** $g \cdot h$ **36.** g/f

37. f/g **38.** $f - h$ **39.** g/h **40.** $f \cdot h$

Let $f(x) = 3x - 1$, $g(x) = x^2 + 1$, and $h(x) = \dfrac{x + 1}{3}$.

Find and simplify each expression.

41. $f(g(-1))$ **42.** $g(f(-1))$ **43.** $(f \circ h)(5)$

44. $(h \circ f)(-7)$ **45.** $(f \circ g)(4.39)$ **46.** $(g \circ h)(-9.87)$

47. $f(g(x))$ **48.** $g(f(x))$ **49.** $(g \circ f)(x)$

50. $(f \circ g)(x)$ **51.** $(h \circ g)(x)$ **52.** $(g \circ h)(x)$

53. $(f \circ f)(x)$ **54.** $(g \circ g)(x)$ **55.** $(h \circ g \circ f)(x)$

56. $(f \circ g \circ h)(x)$

Let $f(x) = \sqrt{x}$, $g(x) = 2x - 1$, and $h(x) = \dfrac{1}{x - 3}$.

Find an equation defining each function and state the domain of the function.

57. $f \circ g$ **58.** $h \circ g$ **59.** $h \circ f$ **60.** $f \circ h$

61. $h \circ h$ **62.** $g \circ h$ **63.** $f \circ f$ **64.** $f \circ h \circ g$

Show that $(f \circ g)(x) = x$ and $(g \circ f)(x) = x$ for each given pair of functions.

65. $f(x) = 2x - 1,\ g(x) = \dfrac{x + 1}{2}$

66. $f(x) = x^3 + 5,\ g(x) = \sqrt[3]{x - 5}$

67. $f(x) = \dfrac{x - 1}{x + 1},\ g(x) = \dfrac{x + 1}{1 - x}$

68. $f(x) = \dfrac{x + 1}{x - 3},\ g(x) = \dfrac{3x + 1}{x - 1}$

69. $f(x) = \sqrt[3]{2x^5 - 1},\ g(x) = \sqrt[5]{\dfrac{x^3 + 1}{2}}$

70. $f(x) = x^{3/5} - 3,\ g(x) = (x + 3)^{5/3}$

Graph each function on a graphing calculator without simplifying the given expression. Examine the graph and write a function for the graph. Then simplify the original function and see whether it is the same as your function. Is the domain of the original function equal to the domain of the function after it is simplified?

71. $y = ((x - 1)/(x + 1) + 1)/((x - 1)/(x + 1) - 1)$

72. $y = ((3x + 1)/(x - 1) + 1)/((3x + 1)/(x - 1) - 3)$

Define $y_1 = \sqrt{x + 1}$ and $y_2 = 3x - 4$ on your graphing calculator. For each function y_3, defined in terms of y_1 and y_2, determine the domain and range of y_3 from its graph on your calculator and explain what each graph illustrates.

73. $y_3 = y_1 + y_2$

74. $y_3 = 3y_1 - 4$

75. $y_3 = \sqrt{y_2 + 1}$

76. $y_3 = \sqrt{y_1 + 1}$

Define $y_1 = \sqrt[3]{x}$, $y_2 = \sqrt{x}$, and $y_3 = x + 4$. For each function y_4, determine the domain and range of y_4 from its graph on your calculator and explain what each graph illustrates.

77. $y_4 = y_1 + y_2 + y_3$

78. $y_4 = \sqrt{y_1 + 4}$

Find each function from the given verbal description of the function.

79. If m is n minus 4 and y is the square of m, then write y as a function of n.

80. If u is the sum of t and 9 and v is u divided by 3, then write v as a function of t.

81. If w is equal to the sum of x and 16, z is the square root of w, and y is z divided by 8, then write y as a function of x.

82. If a is the cube of b, c is the sum of a and 25, and d is the square root of c, then write d as a function of b.

• CALCULUS •

Let $f(x) = |x|$, $g(x) = x - 7$, and $h(x) = x^2$. Write each of the following functions as a composition of functions chosen from f, g, and h.

83. $F(x) = x^2 - 7$

84. $G(x) = |x| - 7$

85. $H(x) = |x^2 - 7|$

86. $M(x) = x^2 - 14x + 49$

87. $N(x) = (|x| - 7)^2$

88. $R(x) = |x - 7|$

89. $P(x) = |x - 7| - 7$

90. $C(x) = (x^2 - 7)^2$

Solve each problem.

91. Write the area A of a square with a side of length s as a function of its diagonal d.

92. Write the perimeter of a square P as a function of the area A.

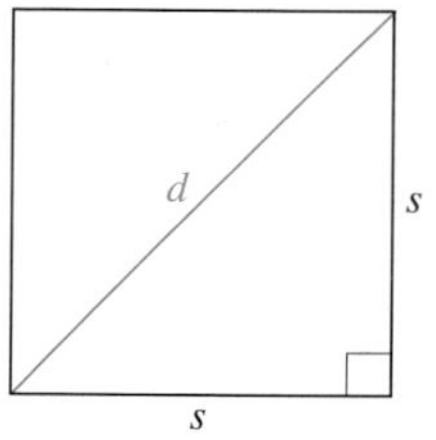

Figure for Exercises 91 and 92

93. *Destruction of a Rainforest* The function $f(x) = 0.1x$ gives the area used when a rainforest of area x is harvested. The function $g(x) = 0.45x$ found in Example 4 gives the area wasted as a function of the initial area x. Find and interpret the function $f + g$.

Figure for Exercise 93

94. *Deforestation in Nigeria* In Nigeria deforestation occurs at the rate of about 5.2% per year. If x is the total forest area of Nigeria at the start of 1996, then $f(x) = 0.948x$ is the amount of forest land in Nigeria at the start of 1997. Find and interpret $(f \circ f)(x)$. Find and interpret $(f \circ f \circ f)(x)$.

95. *Energy Consumption* Energy consumption for OECD (Organization for Economic Cooperation and Development) countries, developing countries, and transition countries (former Soviet Union and Central

Europe) is shown in the accompanying graph (World Resources, 1996–1997). Let $F(x) = (f + g + h)(x)$. Find a formula for $F(x)$ and explain what $F(x)$ represents. Find $F(20)$.

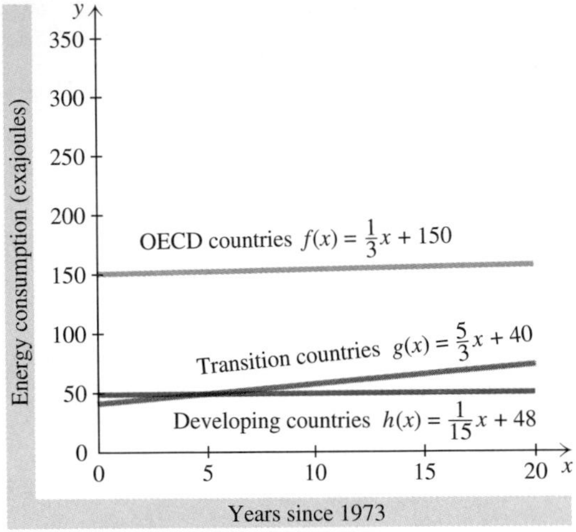

Figure for Exercise 95

• CALCULUS •

96. *Profit* The revenue in dollars that a company receives for installing x alarm systems per month is given by $R(x) = 3000x - 20x^2$, while the cost in dollars is given by $C(x) = 600x + 4000$. The function $P(x) = R(x) - C(x)$ gives the profit for installing x alarm systems per month. Find $P(x)$ and simplify it. Find the marginal profit function $MP(x)$, marginal revenue function $MR(x)$, and marginal cost function $MC(x)$. (See Exercises 113 and 114 of Section 2.2.) Is $MP(x) = MR(x) - MC(x)$?

97. *Displacement-Length Ratio* The displacement-length ratio D indicates whether a sailboat is relatively heavy or relatively light (*Sail*, September 1997). According to *Sail*,

$$D = (d \div 2240) \div x$$

where d is the displacement in pounds and

$$x = (L \div 100)^3$$

where L is the length at the waterline in feet. Assuming that the displacement is 26,000 pounds, write D as a function of L and simplify it.

98. *Sail Area-Displacement Ratio* The sail area-displacement ratio S measures the sail power available to drive a sailboat (*Sail*, September 1997). According to *Sail*,

$$S = A \div y$$

where A is the sail area in square feet and

$$y = (d \div 64)^{2/3}$$

where d is the displacement in pounds. Assuming that the sail area is 6500 square feet, write S as a function of d and simplify it.

99. *Area of a Window* A window is in the shape of a square with a side of length s, with a semicircle of diameter w adjoining the top of the square. Write the total area of the window W as a function of s.

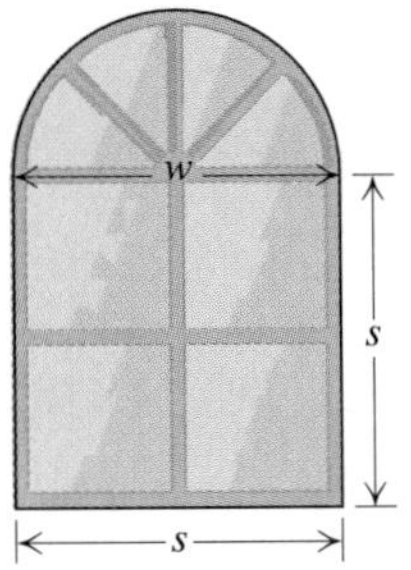

Figure for Exercises 99 and 100

100. *Area of a Window* Using the window of Exercise 99, write the area of the square A as a function of the area of the semicircle S.

101. *Packing a Square Piece of Glass* A glass prism with a square cross section is to be shipped in a cylindrical cardboard tube that has an inside diameter of d inches. Given that s is the length of a side of the square and the glass fits snugly into the tube, write s as a function of d.

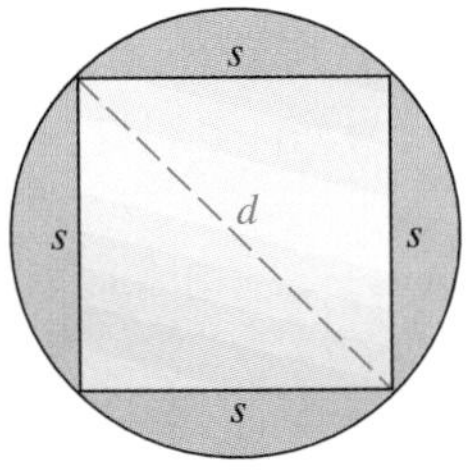

Figure for Exercise 101

102. *Packing a Triangular Piece of Glass* A glass prism is to be shipped in a cylindrical cardboard tube that has an inside diameter of d inches. Given that the cross section of the prism is an equilateral triangle with side of length p and the prism fits snugly into the tube, write p as a function of d.

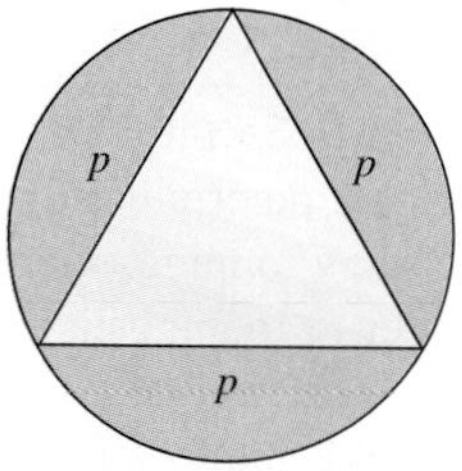

Figure for Exercise 102

103. *Further Markdowns* A clothing store marked down all of its summer merchandise 30%. Let x be the original price and s be the sale price of an item.

a. Write s as a function of x.

b. For the final clearance of the season the prices are reduced 25% off of the sale prices. Let c be the clearance price and write c as a function of s.

c. Write c as a function of x.

d. If you buy an item at the clearance sale are you getting 55% off?

104. *Saving Energy* Caulking for sealing cracks around windows and doors is sold in a tube containing 20 in^3 of caulking. The contents of one tube will make a cylindrical bead with diameter d that is n inches long. Write n as a function of d.

LINKING CONCEPTS

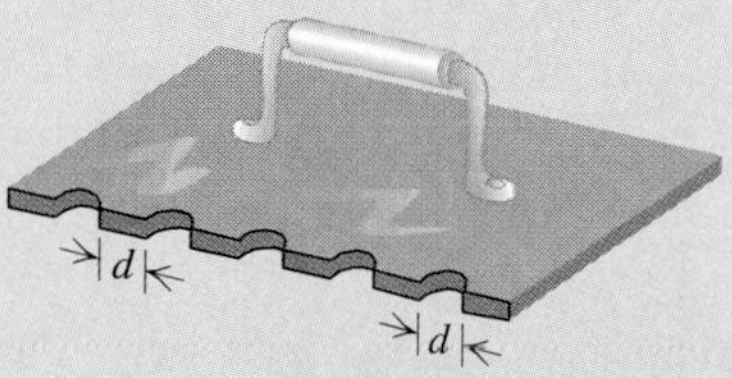

For Individual or Group Explorations

Spreading Glue Frank is designing a notched trowel for spreading glue. The notches on the trowel are in the shape of semicircles. The diameter of each notch and the space between consecutive notches is d inches as shown in the figure. Suppose that the trowel is used to make parallel beads of glue on one square foot of floor.

a) Write the number of parallel beads made as a function of d.

b) The cross section of each parallel bead is a semicircle with diameter d. Write the area of a cross section as a function of d.

c) Write the volume of one of the parallel beads as a function of d.

d) Write the volume of glue on one square foot of floor as a function of d.

e) Write the number of square feet that one gallon of glue will cover as a function of d. (Use 1 ft^3 = 7.5 gal)

f) Suppose that the trowel has square notches where the sides of the squares are d inches in length and the distance between consecutive notches is d inches. Write the number of square feet that one gallon of glue will cover as a function of d.

2.6 Inverse Functions

It is possible for one function to undo what another function does. For example, squaring undoes the operation of taking a square root. The composition of two such functions is the identity function. In this section we explore this idea in detail.

Definitions

Consider again the medium pizza that costs \$5 plus \$2 per topping. Table 2.5 shows the ordered pairs of the function that determines the cost. If we know the cost of a pizza, can we determine how many toppings it has? We can, and Table 2.6 shows ordered pairs where the number of toppings is determined by the cost. Of course we could just read Table 2.5 backwards, but we make a new table because we want a function for finding the number of toppings depending on the cost. The function in Table 2.6 is the **inverse function** of the function in Table 2.5.

Table 2.5

Toppings x	Cost y
0	\$ 5
1	7
2	9
3	11
4	13

Table 2.6

Cost x	Toppings y
\$ 5	0
7	1
9	2
11	3
13	4

A function is a set of ordered pairs in which no two ordered pairs have the same first coordinates and different second coordinates. If we interchange the x- and y-coordinates in each ordered pair of a function, as in Tables 2.5 and 2.6, the resulting set of ordered pairs might or might not be a function. For example, if $f = \{(1, 3), (2, 4)\}$, then the set $\{(3, 1), (4, 2)\}$ is a function. However, if $f = \{(4, 6), (5, 6)\}$, then the set $\{(6, 4), (6, 5)\}$ is *not* a function. If the set obtained by interchanging the coordinates of each ordered pair of a function is a function, then the function is called an **invertible function**.

Definition: Inverse Function

The **inverse** of an invertible function f is the function f^{-1} (read "f inverse"), where the ordered pairs of f^{-1} are obtained by interchanging the coordinates in each ordered pair of f.

In this notation, the number -1 in f^{-1} does not represent a negative exponent. It is merely a symbol for denoting the inverse function.

EXAMPLE 1 **Finding an inverse function**

Let $f = \{(1, 3), (2, 4), (5, 7)\}$. Find f^{-1}, $f^{-1}(3)$, and $(f^{-1} \circ f)(1)$.

Solution

Interchange the x- and y-coordinates of each ordered pair of f to find f^{-1}:

$$f^{-1} = \{(3, 1), (4, 2), (7, 5)\}$$

To find the value of $f^{-1}(3)$, notice that $f^{-1}(3)$ is the second coordinate when the first coordinate is 3 in the function f^{-1}. So $f^{-1}(3) = 1$. To find the composition, use the definition of composition of functions:

$$(f^{-1} \circ f)(1) = f^{-1}(f(1)) = f^{-1}(3) = 1$$

Since the coordinates in the ordered pairs are interchanged, the domain of f^{-1} is the range of f, and the range of f^{-1} is the domain of f. If f^{-1} is the inverse function of f, then certainly f is the inverse of f^{-1}. The functions f and f^{-1} are inverses of each other.

Whether a function is invertible depends on the ordered pairs of the function f. If f has ordered pairs with different first coordinates and the same second coordinate, then f is not invertible because the definition of function is violated when the ordered pairs are reversed.

Definition: One-to-One Function

If a function has no two ordered pairs with different first coordinates and the same second coordinate, then the function is called **one-to-one**.

In a one-to-one function, each member of the domain corresponds to a unique member of the range. The following theorem describes the invertible functions.

Theorem: Invertible Functions

A function is invertible if and only if it is a one-to-one function.

EXAMPLE 2 Finding an inverse function

For each function, determine whether it is invertible. If it is invertible, then find the inverse.

a) $f = \{(-2, 3), (4, 5), (2, 3)\}$

b) $g = \{(3, 1), (5, 2), (7, 4), (9, 8)\}$

Solution

a) This function f is *not* one-to-one because of the ordered pairs $(-2, 3)$ and $(2, 3)$. So f is not invertible.

b) The function g is one-to-one, and so g is invertible. The inverse of g is the function $g^{-1} = \{(1, 3), (2, 5), (4, 7), (8, 9)\}$.

If the ordered pairs of an invertible function are defined by means of an equation, we can obtain the inverse function by simply interchanging x and y in the equation. Interchanging x and y in the equation causes the first and second coordinates of each ordered pair to be interchanged.

EXAMPLE 3 Finding an inverse function

Given that $f = \{(x, y) \mid y = 2x\}$, find f^{-1}.

Solution

To interchange the coordinates in all of the ordered pairs, interchange x and y in the equation:

$$f^{-1} = \{(x, y) \mid x = 2y\} = \left\{(x, y) \mid y = \frac{1}{2}x\right\}$$

For an ordered pair such as $(3, 6)$ in f, $(6, 3)$ is in f^{-1}.

Determining Whether a Function Is One-to-One

If a function is given as a short list of ordered pairs, then it is easy to determine whether the function is one-to-one. A graph of a function can also be used to determine whether a function is one-to-one using the **horizontal line test**.

Horizontal Line Test

If each horizontal line crosses the graph of a function at no more than one point, then the function is one-to-one.

The graph of a one-to-one function never has the same y-coordinate for two different x-coordinates on the graph. So if it is possible to draw a horizontal line that crosses the graph of a function two or more times, then the function is not one-to-one.

EXAMPLE 4 The horizontal line test

Use the horizontal line test to determine whether the functions shown in Fig. 2.78 are one-to-one.

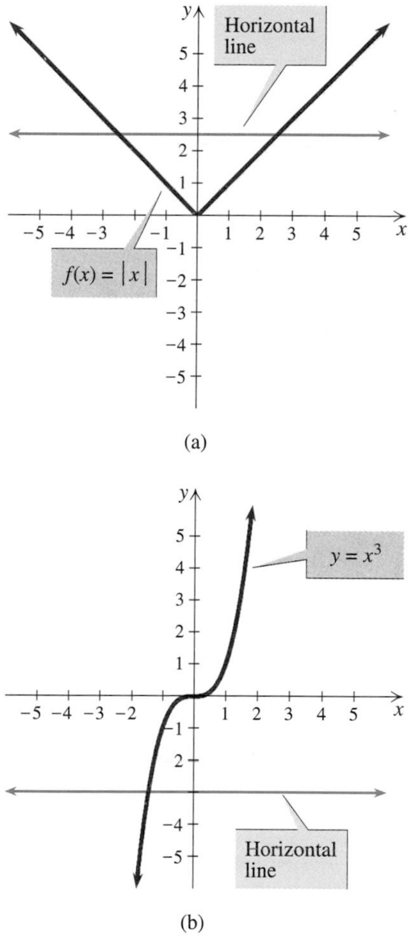

Figure 2.78

Solution

The function $f(x) = |x|$ is not one-to-one because it is possible to draw a horizontal line that crosses the graph twice as shown in Fig. 2.78(a). The function $y = x^3$ in Fig. 2.78(b) is one-to-one because it appears to be impossible to draw a horizontal line that crosses the graph more than once.

The horizontal line test explains the visual difference between the graph of a one-to-one function and the graph of a function that is not one-to-one. Because no graph of a function is perfectly accurate, conclusions made from a graph alone may not be correct. For example, from the graph of $y = x^3 - 0.01x$ shown in Fig. 2.79(a) we would conclude that the function is one-to-one and invertible. However, another view of the same function in Fig. 2.79(b) shows that the function is not one-to-one.

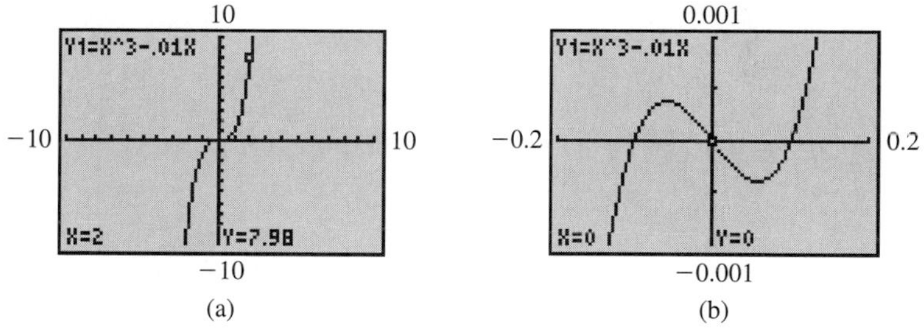

Figure 2.79

Using the equation for a function we can often determine conclusively whether the function is one-to-one. A function f is *not* one-to-one if it is possible to find two different numbers x_1 and x_2 such that $f(x_1) = f(x_2)$. For example, for the function $f(x) = x^2$, it is possible to find two different numbers, 2 and -2, such that $f(2) = f(-2)$. So $f(x) = x^2$ is not one-to-one. To prove that a function f is one-to-one, we must show that $f(x_1) = f(x_2)$ implies that $x_1 = x_2$.

EXAMPLE 5 **Using the definition of one-to-one**

Determine whether each function is one-to-one.

a) $f(x) = \dfrac{2x + 1}{x - 3}$ **b)** $g(x) = |x|$

Solution

a) If $f(x_1) = f(x_2)$, then we have the following equation:

$$\frac{2x_1 + 1}{x_1 - 3} = \frac{2x_2 + 1}{x_2 - 3}$$

$$(2x_1 + 1)(x_2 - 3) = (2x_2 + 1)(x_1 - 3) \quad \text{Multiply by the LCD.}$$

$$2x_1x_2 + x_2 - 6x_1 - 3 = 2x_2x_1 + x_1 - 6x_2 - 3$$

$$7x_2 = 7x_1$$

$$x_2 = x_1$$

Since $f(x_1) = f(x_2)$ implies that $x_1 = x_2$, $f(x)$ is a one-to-one function.

b) Since $g(3) = |3| = 3$ and $g(-3) = 3$, the function g is *not* one-to-one.

Inverse Functions Using *f*-Notation

The function $f(x) = 2x + 5$ gives the cost of a pizza where \$5 is the basic cost and x is the number of toppings at \$2 each. If we assume that x could be any real number, this function defines the set

$$f = \{(x, y) | y = 2x + 5\}.$$

Since the coordinates of each ordered pair are interchanged for f^{-1}, the ordered pairs of f^{-1} must satisfy $x = 2y + 5$:

$$f^{-1} = \{(x, y) | x = 2y + 5\}.$$

Since $x = 2y + 5$ is equivalent to $y = \dfrac{x - 5}{2}$,

$$f^{-1} = \left\{(x, y) \middle| y = \frac{x - 5}{2}\right\}.$$

The function f^{-1} is described in f-notation as $f^{-1}(x) = \dfrac{x - 5}{2}$. The function f^{-1} gives the number of toppings as a function of cost.

A one-to-one function f pairs members of the domain of f with members of the range of f, and its inverse function f^{-1} exactly reverses those pairings as shown in Fig. 2.80. The above example suggests the following steps for finding an inverse function using f-notation.

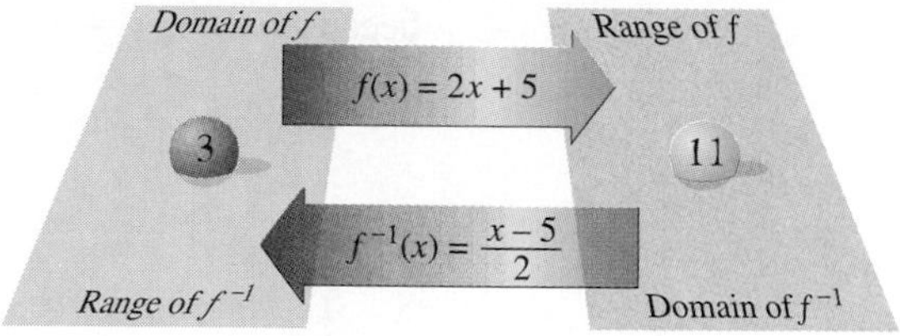

Figure 2.80

PROCEDURE

Finding $f^{-1}(x)$

To find the inverse of a one-to-one function given in f-notation:

1. Replace $f(x)$ by y.
2. Interchange x and y.
3. Solve the equation for y.
4. Replace y by $f^{-1}(x)$.
5. Check that the domain of f is the range of f^{-1} and the range of f is the domain of f^{-1}.

EXAMPLE 6 Finding an inverse function

Find the inverse of the function $f(x) = \dfrac{2x + 1}{x - 3}$.

Solution

In Example 5(a) we showed that f is a one-to-one function. So we can find f^{-1} by interchanging x and y and solving for y:

$$y = \frac{2x + 1}{x - 3} \qquad \text{Replace } f(x) \text{ by } y.$$

$$x = \frac{2y + 1}{y - 3} \qquad \text{Interchange } x \text{ and } y.$$

$$x(y - 3) = 2y + 1 \qquad \text{Solve for } y.$$

$$xy - 3x = 2y + 1$$

$$xy - 2y = 3x + 1$$

$$y(x - 2) = 3x + 1$$

$$y = \frac{3x + 1}{x - 2}$$

Replace y by $f^{-1}(x)$ to get $f^{-1}(x) = \dfrac{3x + 1}{x - 2}$. The domain of f is all real numbers except 3. The range of f^{-1} is the set of all real numbers except 3. We exclude 3 because $\dfrac{3x + 1}{x - 2} = 3$ has no solution. Check that the range of f is equal to the domain of f^{-1}.

To find an inverse function according to the definition, we simply interchange the coordinates in all of the ordered pairs of the given function. When a function is defined by a formula, it might not be easy to see that a "possible" formula for the inverse really does this. However, if the conditions in the following theorem are satisfied, then we can conclude that g is the inverse of f (without checking all of the ordered pairs).

Theorem: Determining Whether g Is the Inverse of f

A function g is the inverse of a one-to-one function f if and only if

1. the domain of g is equal to the range of f and
2. $(g \circ f)(x) = x$ for any x in the domain of f.

The equation $(g \circ f)(x) = x$ means that the composition $g \circ f$ is the identity function. This condition is often used as the starting point for finding a formula for an inverse function. That is, we look for a function g that undoes the operations of f.

EXAMPLE 7 Using composition to determine an inverse

For $f(x) = \sqrt{x}$ and $g(x) = x^2$, find $(g \circ f)(x)$ and determine whether g is the inverse of f.

Solution

The function $f(x) = \sqrt{x}$ is one-to-one because $\sqrt{x_1} = \sqrt{x_2}$ implies that $x_1 = x_2$. The domain of f is $[0, \infty)$ and $(g \circ f)(x) = g(\sqrt{x}) = (\sqrt{x})^2 = x$ for any $x \geq 0$. So g satisfies the second condition of the theorem. Since the range of f is $[0, \infty)$ and the domain of g is $(-\infty, \infty)$, g fails the first condition of the theorem and g is *not* the inverse of f.

Note that the calculator graph in Fig. 2.81 supports the conclusion that f is one-to-one.

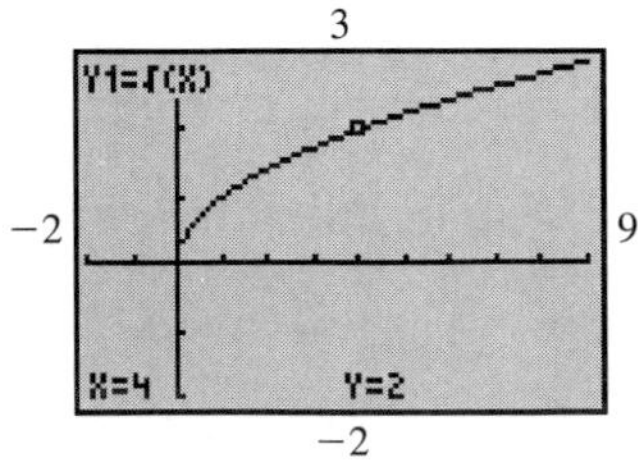

Figure 2.81

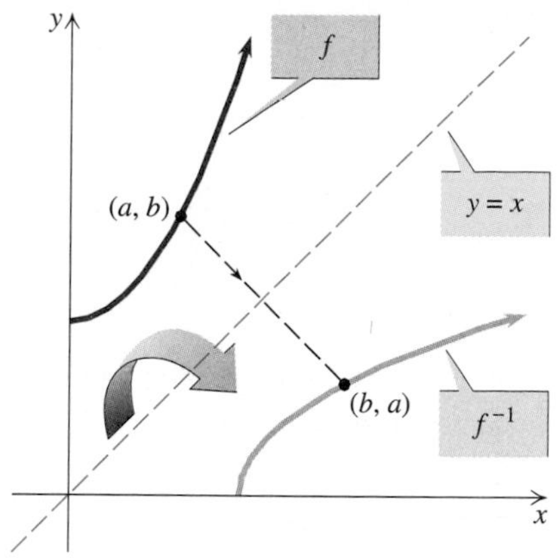

Figure 2.82

Even though $g(x) = x^2$ undoes what $f(x) = \sqrt{x}$ does, it is not the inverse of f because its domain is not the range of f. This problem is easily corrected by restricting the domain of the squaring function. Since the inverse of $f(x) = \sqrt{x}$ must have domain $[0, \infty)$, we have $f^{-1}(x) = x^2$ for $x \geq 0$.

Graphs of f and f^{-1}

If a point (a, b) is on the graph of an invertible function f, then (b, a) is on the graph of f^{-1}. See Fig. 2.82. Since the points (a, b) and (b, a) are symmetric with respect to the line $y = x$, the graph of f^{-1} is a reflection of the graph of f with respect to the line $y = x$.

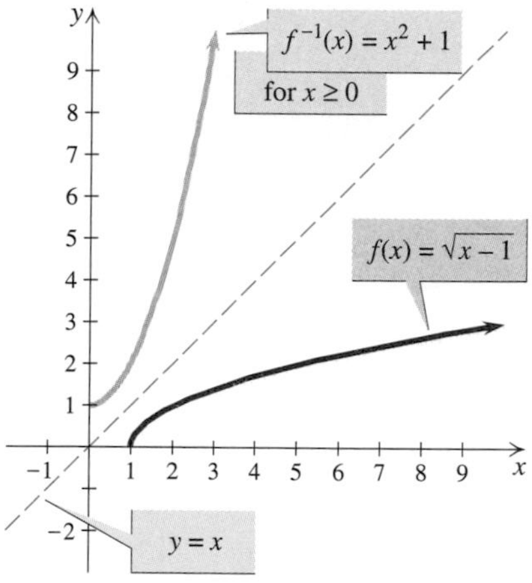

Figure 2.83

EXAMPLE 8 Graphing a function and its inverse

Find the inverse of the function $f(x) = \sqrt{x - 1}$ and graph both f and f^{-1} on the same coordinate axes.

Solution

If $\sqrt{x_1 - 1} = \sqrt{x_2 - 1}$, then $x_1 = x_2$. So f is one-to-one. Since the range of $f(x) = \sqrt{x - 1}$ is $[0, \infty)$, f^{-1} must be a function with domain $[0, \infty)$. To find a formula for the inverse, interchange x and y in $y = \sqrt{x - 1}$ to get $x = \sqrt{y - 1}$. Solving this equation for y gives $y = x^2 + 1$. So $f^{-1}(x) = x^2 + 1$ for $x \geq 0$. The graph of f is a translation one unit to the right of the graph of $y = \sqrt{x}$, and the graph of f^{-1} is a translation one unit upward of the graph of $y = x^2$ for $x \geq 0$. Both graphs are shown in Fig. 2.83 along with the line $y = x$. Notice that the graph of f^{-1} is a reflection of the graph of f.

Finding Inverse Functions Mentally

It is no surprise that the inverse of $f(x) = x^2$ for $x \geq 0$ is the function $f^{-1}(x) = \sqrt{x}$. For nonnegative numbers, taking the square root undoes what squaring does. It is also no surprise that the inverse of $f(x) = 3x$ is $f^{-1}(x) = x/3$ or that the inverse of $f(x) = x + 9$ is $f^{-1}(x) = x - 9$. If an invertible function involves a single operation, it is usually easy to write the inverse function because for most operations there is an inverse operation. See Table 2.7. If an invertible function involves more than one operation, we find the inverse function by applying the inverse operations in the opposite order from the order in which they appear in the original function.

Table 2.7

Function	Inverse
$f(x) = 2x$	$f^{-1}(x) = x/2$
$f(x) = x - 5$	$f^{-1}(x) = x + 5$
$f(x) = \sqrt{x}$	$f^{-1}(x) = x^2 \quad (x \geq 0)$
$f(x) = x^3$	$f^{-1}(x) = \sqrt[3]{x}$
$f(x) = 1/x$	$f^{-1}(x) = 1/x$
$f(x) = -x$	$f^{-1}(x) = -x$

EXAMPLE 9 Finding inverses mentally

Find the inverse of each function mentally.

a) $f(x) = 2x + 1$ **b)** $g(x) = \dfrac{x^3 + 5}{2}$

Solution

a) The function $f(x) = 2x + 1$ is a composition of multiplying x by 2 and then adding 1. So the inverse function is a composition of subtracting 1 and then dividing by 2: $f^{-1}(x) = (x - 1)/2$. See Fig. 2.84.

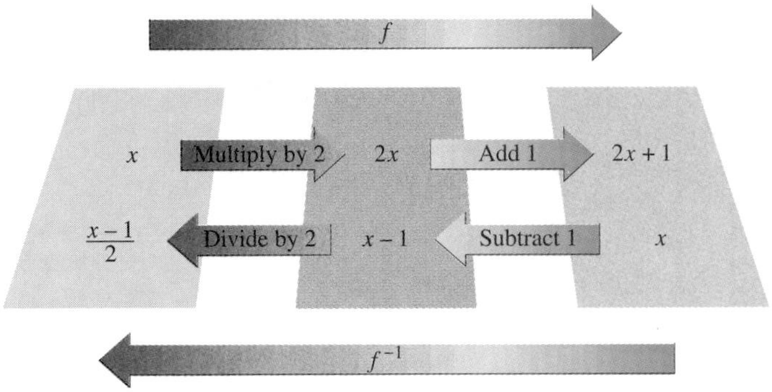

Figure 2.84

b) The function $g(x) = (x^3 + 5)/2$ is a composition of cubing x, adding 5, and dividing by 2. The inverse is a composition of multiplying by 2, subtracting 5, and then taking the cube root: $g^{-1}(x) = \sqrt[3]{2x - 5}$. See Fig. 2.85.

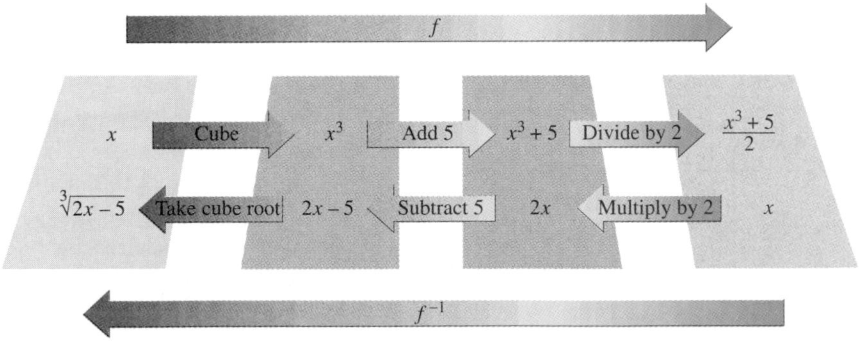

Figure 2.85

FOR THOUGHT True or False? Explain.

1. The inverse of the function $\{(2, 3), (5, 5)\}$ is $\{(5, 2), (5, 3)\}$.
2. The function $f(x) = -2$ is an invertible function.
3. If $g(x) = x^2$, then $g^{-1}(x) = \sqrt{x}$.
4. The only functions that are invertible are the one-to-one functions.
5. Every function has an inverse function.
6. The function $f(x) = x^4$ is invertible.
7. If $f(x) = 3\sqrt{x - 2}$, then $f^{-1}(x) = \dfrac{(x + 2)^2}{3}$ for $x \geq 0$.
8. If $f(x) = |x - 3|$, then $f^{-1}(x) = |x| + 3$.
9. According to the horizontal line test, $y = |x|$ is one-to-one.
10. The function $g(x) = -x$ is the inverse of the function $f(x) = -x$.

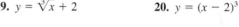

2.6 EXERCISES

Tape 6 Disk

For each function f, find f^{-1}, $f^{-1}(5)$, and $(f^{-1} \circ f)(2)$.

1. $f = \{(2, 1), (3, 5)\}$
2. $f = \{(-1, 5), (0, 0), (2, 6)\}$
3. $f = \{(-3, -3), (0, 5), (2, -7)\}$
4. $f = \{(3.2, 5), (2, 1.99)\}$

Determine whether each function is invertible. If it is invertible, find the inverse.

5. $\{(-1, 0), (0, 0), (1, 0)\}$
6. $\{(0, 0)\}$
7. $\{(3, 0), (2, 5), (4, 6), (7, 9)\}$
8. $\{(-1, -\sqrt{3}), (1, -\sqrt{3})\}$
9. $\{(1, 1), (2, 2), (4.5, 4.5)\}$
10. $\{(-3, 2), (3, 0), (6, 0)\}$
11. $\{(x, y) \mid y = x + 2\}$
12. $\{(x, y) \mid y = 3x - 1\}$
13. $\{(x, y) \mid y = 2x + 7\}$
14. $\{(x, y) \mid y = 1/x^2\}$
15. $\{(x, y) \mid y = |x|\}$
16. $\{(x, y) \mid y = \sqrt{x}\}$

Use the horizontal line test to determine whether each function is one-to-one.

17. $f(x) = x^2 - 3x$

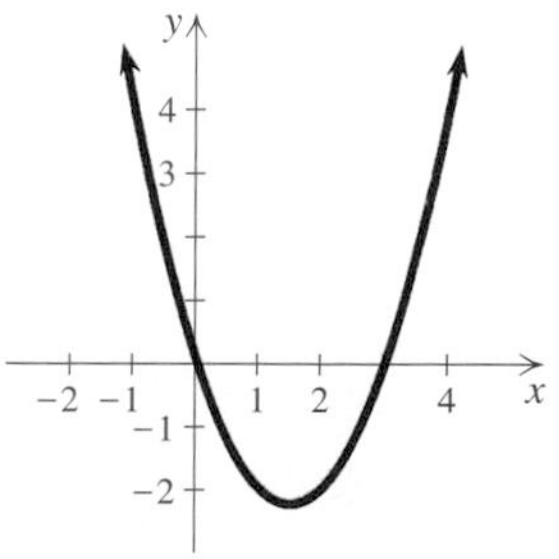

18. $g(x) = |x - 2| + 1$

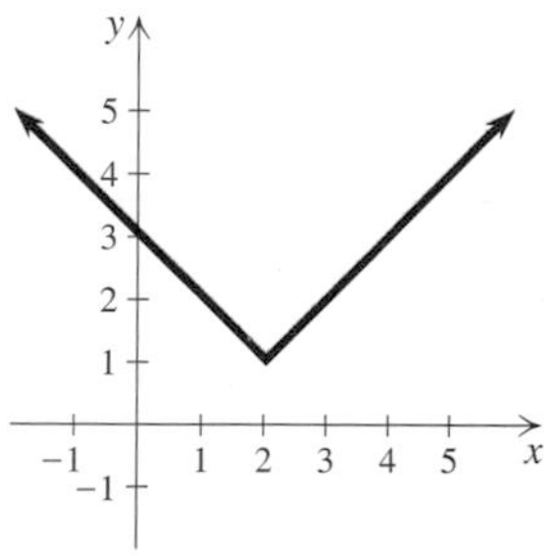

19. $y = \sqrt[3]{x} + 2$

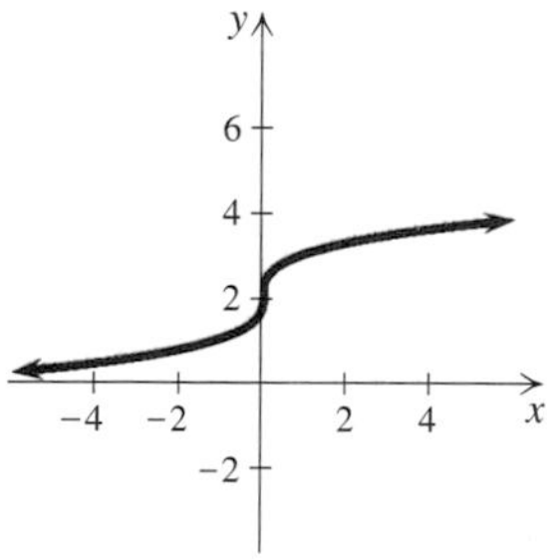

20. $y = (x - 2)^3$

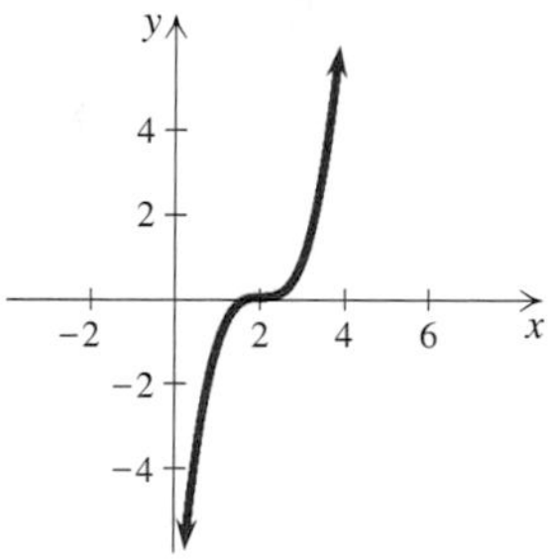

21. $y = x^3 - x$

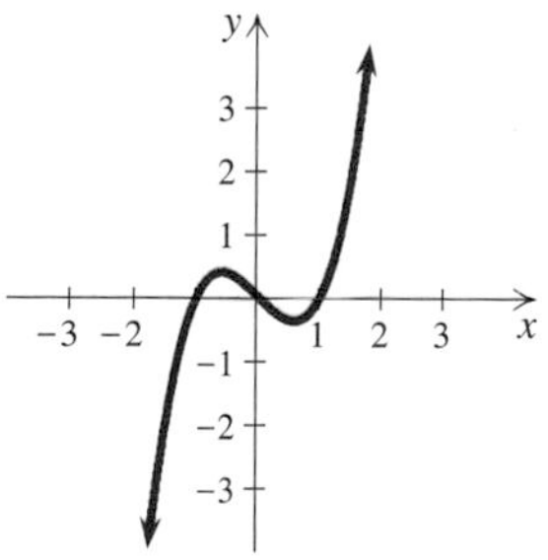

22. $y = \dfrac{1}{x}$

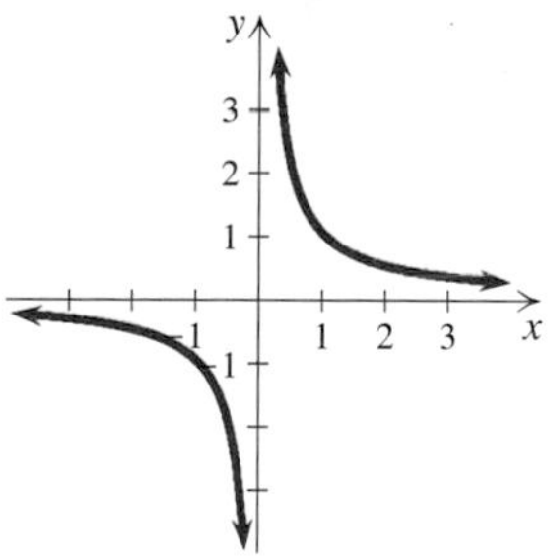

Determine whether each function is one-to-one.

23. $\{(2, 3), (3, 3), (4, 6)\}$
24. $\{(5, \pi), (6, \pi/2), (7, \pi/3)\}$
25. $\{(x, y) \mid y = |x + 1|\}$
26. $\{(x, y) \mid y = x^6\}$
27. $y = 3 + \dfrac{1 - x}{x - 5}$
28. $y = \dfrac{-3x}{x + 7}$

29. $y = x^2 + 3$

30. $y = 2x^2 - 4$

31. $y = x^2 - x$

32. $f(x) = 5 + \sqrt[3]{x + 9}$

Determine whether each function is invertible and explain your answer.

33. The function that pairs the universal product code of an item at Sears with a price.

34. The function that pairs the number of days since your birth with your age in years.

35. The function that pairs the length of a VCR tape in feet with the playing time in minutes.

36. The function that pairs the speed of your car in miles per hour with the speed in kilometers per hour.

37. The function that pairs the number of days of a hotel stay with the total cost for the stay.

38. The function that pairs the number days that a deposit of \$100 earns interest at 6% compounded daily with the amount of interest.

Determine whether each function is invertible by inspecting its graph on a graphing calculator.

39. $f(x) = (x + 0.01)(x + 0.02)(x + 0.03)$

40. $f(x) = x^3 - 0.6x^2 + 0.11x - 0.006$

41. $f(x) = |x - 2| - |5 - x|$

42. $f(x) = \sqrt[3]{0.1x + 3} + \sqrt[3]{-0.1x}$

Find the inverse of each function.

43. $f(x) = 3x - 7$

44. $f(x) = -2x + 5$

45. $f(x) = 2 + \sqrt{x - 3}$

46. $f(x) = \sqrt{3x - 1}$

47. $f(x) = -x - 9$

48. $f(x) = -x + 3$

49. $f(x) = \dfrac{x + 3}{x - 5}$

50. $f(x) = \dfrac{2x - 1}{x - 6}$

51. $f(x) = -\dfrac{1}{x}$

52. $f(x) = x$

53. $f(x) = \sqrt[3]{x - 9} + 5$

54. $f(x) = \sqrt[3]{\dfrac{x}{2}} + 5$

55. $f(x) = (x - 2)^2$ for $x \geq 2$

56. $f(x) = x^2$ for $x \leq 0$

In each case, find $(g \circ f)(x)$ and determine whether g is the inverse of f.

57. $f(x) = 4x + 4$, $g(x) = 0.25x - 1$

58. $f(x) = 20 - 5x$, $g(x) = -0.2x + 4$

59. $f(x) = x^2 + 1$, $g(x) = \sqrt{x - 1}$

60. $f(x) = \sqrt[4]{x}$, $g(x) = x^4$

61. $f(x) = \dfrac{1}{x} + 3$, $g(x) = \dfrac{1}{x - 3}$

62. $f(x) = 4 - \dfrac{1}{x}$, $g(x) = \dfrac{1}{4 - x}$

63. $f(x) = \sqrt[3]{\dfrac{x - 2}{5}}$, $g(x) = 5x^3 + 2$

64. $f(x) = x^3 - 27$, $g(x) = \sqrt[3]{x} + 3$

For each exercise, graph the three functions on the same screen of a graphing calculator. Enter these functions as shown without simplifying any expressions. Explain what these exercises illustrate.

65. $y_1 = \sqrt[3]{x} - 1$, $y_2 = (x + 1)^3$, $y_3 = (\sqrt[3]{x} - 1 + 1)^3$

66. $y_1 = (2x - 1)^{1/3}$, $y_2 = (x^3 + 1)/2$, $y_3 = (2((x^3 + 1)/2) - 1)^{1/3}$

Each of the following functions is invertible. Find the inverse mentally.

67. $f(x) = 5x$

68. $f(x) = 0.5x$

69. $f(x) = x - 88$

70. $f(x) = x + 99$

71. $f(x) = 3x - 7$

72. $f(x) = 5x + 1$

73. $f(x) = 4 - 3x$

74. $f(x) = 5 - 2x$

75. $f(x) = \dfrac{1}{2}x - 9$

76. $f(x) = \dfrac{x}{3} + 6$

77. $f(x) = -x$

78. $f(x) = \dfrac{1}{x}$

79. $f(x) = \sqrt[3]{x} - 9$

80. $f(x) = \sqrt[3]{x - 9}$

81. $f(x) = 2x^3 - 7$

82. $f(x) = -x^3 + 4$

Determine whether each pair of functions f and g are inverses of each other.

83.

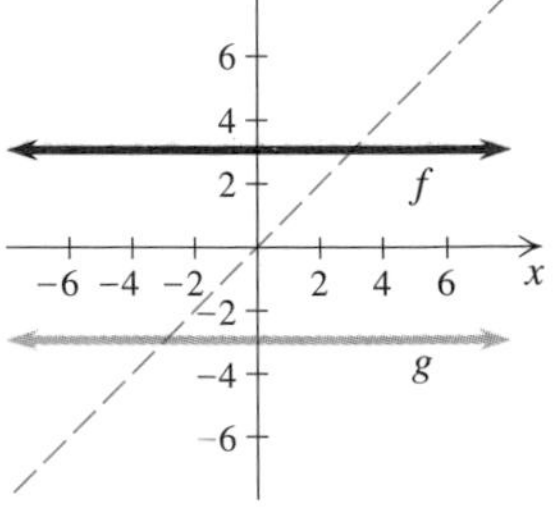

84.

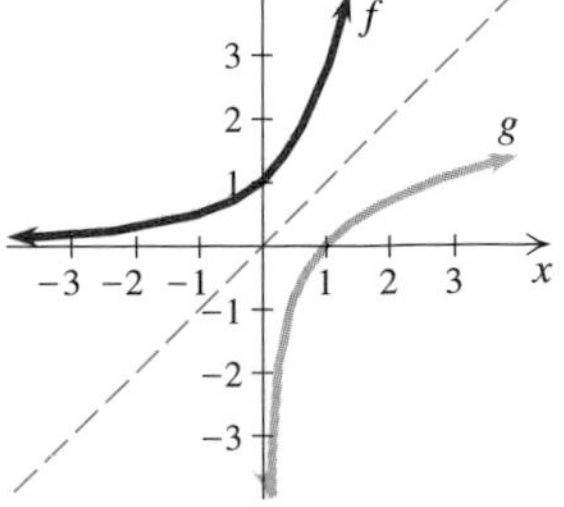

85.

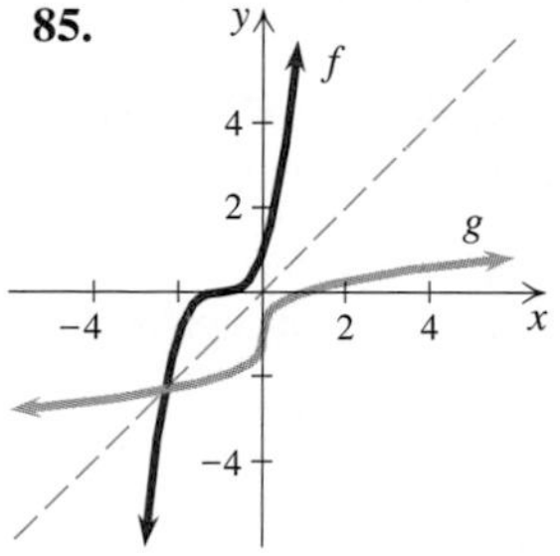

86.

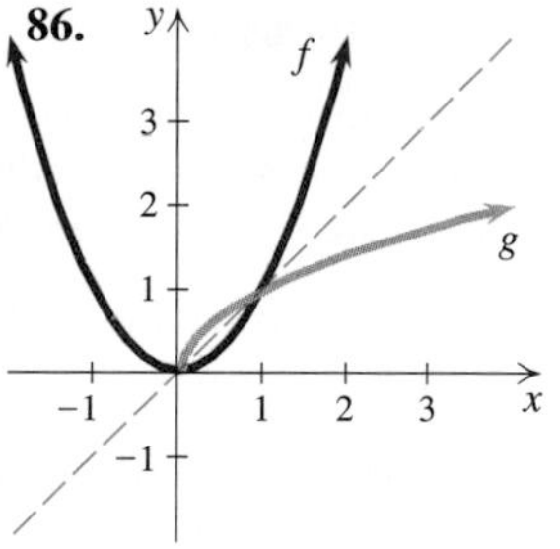

Find the inverse of each function and graph both f and f^{-1} on the same coordinate plane.

87. $f(x) = 3x + 2$

88. $f(x) = -x - 8$

89. $f(x) = x^2 - 4$ for $x \geq 0$

90. $f(x) = 1 - x^2$ for $x \geq 0$

91. $f(x) = x^3$

92. $f(x) = -x^3$

93. $f(x) = \sqrt{x} - 3$

94. $f(x) = \sqrt{x - 3}$

Solve each problem.

95. *Price of a Car* The tax on a new car is 8% of the purchase price P. Express the total cost C as a function of the purchase price. Express the purchase price P as a function of the total cost C.

96. *Volume of a Cube* Express the volume of a cube $V(x)$ as a function of the length of a side x. Express the length of a side of a cube $S(x)$ as a function of the volume x.

97. *Rowers and Speed* The world record times in the 2000-meter race are a function of the number of rowers as shown in the accompanying figure (Encyclopedia of Sports Science, 1997). If r is the number of rowers and t is the time in minutes, then the formula $t = -0.39r + 7.89$ models this relationship. Is this function invertible?

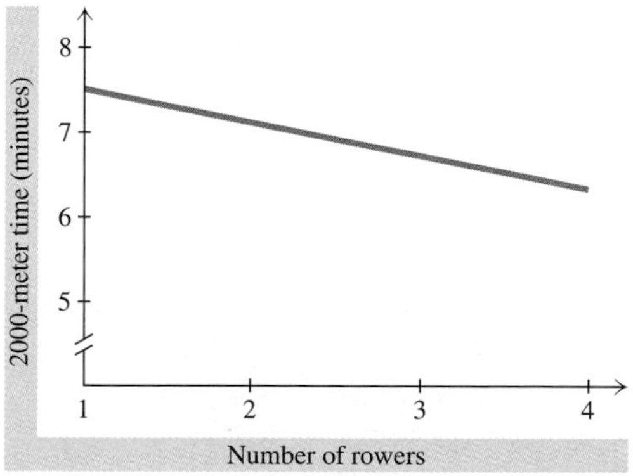

Figure for Exercise 97

Find the inverse. If the time for a 2000-meter race was 5.55 minutes, then how many rowers were probably in the boat?

98. *Temperature* The function $C = \frac{5}{9}(F - 32)$ expresses the Celsius temperature as a function of the Fahrenheit temperature. Find the inverse function. What is it used for?

99. *Landing Speed* The function $V = \sqrt{1.496w}$ expresses the landing speed V (in feet per second) as a function of the gross weight (in pounds) of the Piper Cheyenne aircraft. Find the inverse function. Use the inverse function to find the gross weight for a Piper Cheyenne for which the proper landing speed is 115 ft/sec.

100. *Poiseuille's Law* Under certain conditions, the velocity V of blood in a vessel at distance r from the center of the vessel is given by $V = 500(5.625 \times 10^{-5} - r^2)$ where $0 \leq r \leq 7.5 \times 10^{-3}$. Write r as a function of V.

101. *Depreciation Rate* A new Lexus 400 SC sells for about \$50,000. The function $r = 1 - \left(\frac{V}{50{,}000}\right)^{1/5}$ expresses the depreciation rate r for a five-year-old car as a function of the value V of a five-year-old car. What is the depreciation rate for a car that is worth \$28,000 after five years? Write V as a function of r.

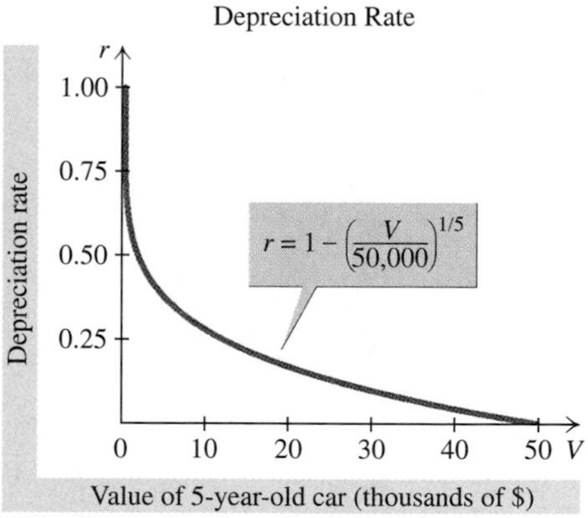

Figure for Exercise 101

102. *Annual Growth Rate* One measurement of the quality of a mutual fund is its average annual growth rate over the last 10 years. The function $r = \left(\frac{P}{10{,}000}\right)^{1/10} - 1$ expresses the average annual growth rate r as a function of the present value P of an investment of \$10,000 made 10 years ago. An investment of \$10,000 in Fidelity's Contrafund in 1987 was worth \$67,583 in 1997 (Fidelity Contrafund Prospectus). What was the annual growth rate for that period? Write P as a function of r.

Annual Growth Rate

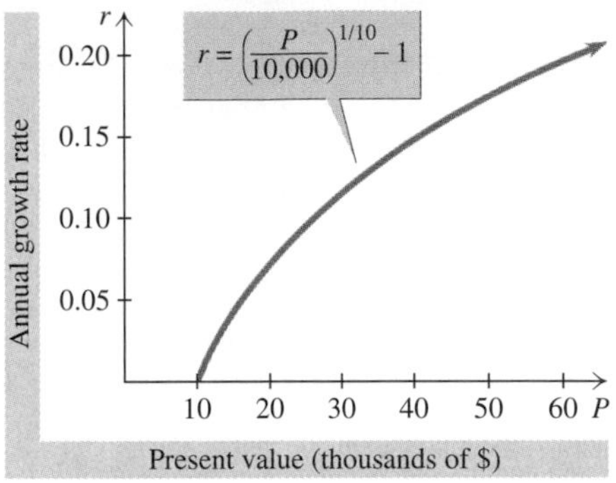

Figure for Exercise 102

For Writing/Discussion

103. Use a geometric argument to prove that if $a \neq b$, then (a, b) and (b, a) lie on a line perpendicular to the line $y = x$ and are equidistant from $y = x$.

104. Why is it difficult to find the inverse of a function such as $f(x) = (x - 3)/(x + 2)$ mentally?

105. Find an expression equivalent to $(x - 3)/(x + 2)$ in which x appears only once. Explain.

106. Use the result of Exercise 105 to find the inverse of $f(x) = (x - 3)/(x + 2)$ mentally. Explain.

107. If $f(x) = 2x + 1$ and $g(x) = 3x - 5$, verify the equation $(f \circ g)^{-1} = g^{-1} \circ f^{-1}$.

108. Explain why $(f \circ g)^{-1} = g^{-1} \circ f^{-1}$ for any invertible functions f and g. Discuss any restrictions on the domains and ranges of f and g for this equation to be correct.

LINKING CONCEPTS

Quick Cash PAY DAY Loans
BORROW YOU PAY BACK
$200.00 $229.00
$300.00 $339.00
$400.00 $449.00
WE SPECIALIZE IN ALL TYPES OF LOANS
* If loan is paid out within 30 days

For Individual or Group Explorations

Short-Term Loans *Star Financial Corporation offers short-term loans as shown in the accompanying advertisement. If you borrow P dollars for n days at the annual percentage rate r, then you pay back A dollars where*

$$A = P\left(1 + \frac{r}{365}\right)^n.$$

a) Solve the formula for r in terms of A, P, and n.

b) Find the annual percentage rate (APR) for each of the loans described in the advertisement.

c) If you borrowed \$200, then how much would you have to pay back in one year using the APR that you determined in part (b)?

d) Find the current annual percentage rates for a car loan, house mortgage, and a credit card.

e) Why do you think Star Financial Corporation charges such high rates?

2.7 Variation

The area of a circle is a function of the radius, $A = \pi r^2$. As the values of r change or vary, the values of A vary also. Certain relationships can be indicated by describing how one variable changes according to the values of another variable. In this section we will learn to write formulas for certain types of functions that are traditionally described in terms of variation.

Direct Variation

If we save 3 cubic yards of landfill space for every ton of paper that we recycle (Earth Works Group, 1989), then the volume V of space saved is a linear function of the number of tons recycled n, $V = 3n$. If n is 20 tons, then V is 60 cubic yards. If n is doubled to 40 tons, then V is also doubled to 120 cubic yards. If n is tripled, V is also tripled. This relationship is called **direct variation**.

Definition: Direct Variation

The statement **y varies directly as x** or **y is directly proportional to x** means that

$$y = kx$$

for some fixed nonzero real number k. The constant k is called the **variation constant** or **proportionality constant**.

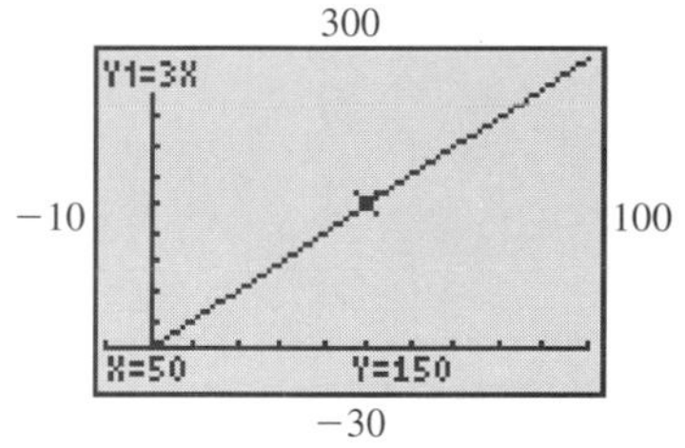

Figure 2.86

Compare the landfill savings to a pizza that costs \$5 plus \$2 per topping. The cost is a linear function of the number of toppings n, $C = 2n + 5$. If $n = 2$, then $C = \$9$ and if $n = 4$, then $C = \$13$. The cost increases as n increases, but doubling n does not double C. The savings in landfill space is directly proportional to the number of tons of paper that is recycled, but the cost of the pizza is not directly proportional to the number of toppings. The graph of $V = 3n$ in Fig. 2.86 is a straight line through the origin. In general, if y varies directly as x, then y is a linear function of x whose graph is a straight line through the origin. We are merely introducing some new terms to describe an old idea.

EXAMPLE 1 Direct variation

The cost, C, of a house in Wedgewood Estates is directly proportional to the size of the house, s. If a 2850-square-foot house costs \$182,400, then what is the cost of a 3640-square-foot house?

Solution

Because C is directly proportional to s, there is a constant k such that

$$C = ks.$$

We can find the constant by using $C = \$182{,}400$ when $s = 2850$:

$$\begin{aligned} 182{,}400 &= k(2850) \\ 64 &= k \end{aligned}$$

The cost for a house is \$64 per square foot. To find the cost of a 3640-square-foot house, use the formula $C = 64s$:

$$\begin{aligned} C &= 64(3640) \\ &= 232{,}960 \end{aligned}$$

The 3640-square-foot house costs \$232,960.

Cruising speed: 580 mph

Cruising speed: 1350 mph

Inverse Variation

The time that it takes to make the 3470-mile trip by air from New York to London is a function of the rate of the airplane, $T = 3470/R$. A Boeing 747 averaging 675 mph can make the trip in about 5 hours. The Concorde, traveling twice as fast at 1350 mph, can make it in half the time, about 2.5 hours. If you travel three times as fast as the 747 you could make the trip in one-third the time. This relationship is called **inverse variation**.

Definition: Inverse Variation

The statement **y varies inversely as x** or **y is inversely proportional to x** means that

$$y = \frac{k}{x}$$

for a fixed nonzero real number k.

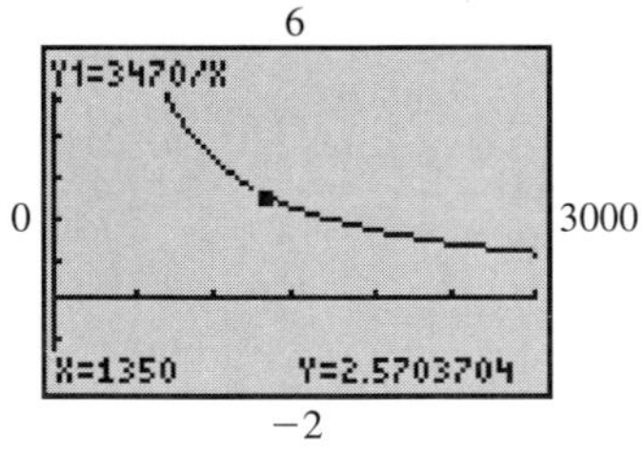

Figure 2.87

The graph of $T = 3470/R$ is shown in Fig. 2.87. As the speed increases the time decreases, but does not reach zero for any speed.

EXAMPLE 2 Inverse variation

The time it takes to remove all of the campaign signs in Bakersfield after an election varies inversely with the number of volunteers working on the job. If 12 volunteers could complete the job in 7 days, then how many days would it take for 15 volunteers?

Solution

Since the time T varies inversely as the number of volunteers n, we have

$$T = \frac{k}{n}.$$

Since $T = 7$ when $n = 12$, we can find k:

$$7 = \frac{k}{12}$$

$$84 = k$$

So the formula is $T = 84/n$. Now if $n = 15$, we get

$$T = \frac{84}{15} = 5.6.$$

It would take 15 volunteers 5.6 days to remove all of the political signs.

Joint Variation

We have been studying functions of one variable, but situations often arise in which one variable depends on the values of two or more variables. For example, the area of a rectangular room depends on the length *and* the width, $A = LW$. If carpeting costs \$25 per square yard, then the total cost is a function of the length and width, $C = 25LW$. This example illustrates **joint variation.**

Definition: Joint Variation

The statement **y varies jointly as x and z** or **y is jointly proportional to x and z** means that

$$y = kxz$$

for a fixed nonzero real number k.

EXAMPLE 3 Joint variation

The cost of constructing a 9-foot by 12-foot patio is \$734.40. If the cost varies jointly as the length and width, then what does an 8-foot by 14-foot patio cost?

Solution

Since the cost varies jointly as L and W, we have $C = kLW$ for some constant k. Use $W = 9$, $L = 12$, and $C = \$734.40$ to find k:

$$734.40 = k(12)(9)$$

$$6.80 = k$$

So the formula is $C = 6.80LW$. Use $W = 8$ and $L = 14$ to find C:

$$C = 6.80(14)(8)$$
$$= 761.60$$

The cost of constructing an 8-foot by 14-foot patio is \$761.60.

Combined Variation

Some relationships are combinations of direct and inverse variation. For example, the formula $V = kT/P$ expresses the volume of a gas in terms of its temperature T and its pressure P. The volume varies directly as the temperature and inversely as the pressure. The next example gives more illustrations of combining direct, inverse, and joint variation.

EXAMPLE 4 Writing formulas for combined variation

Write each sentence as a formula involving a constant of variation k.

a) y varies directly as x and inversely as z.

b) y varies jointly as x and the square root of z.

c) y varies jointly as x and the square of z and inversely as the cube of w.

Solution

a) $y = \frac{kx}{z}$ **b)** $y = kx\sqrt{z}$ **c)** $y = \frac{kxz^2}{w^3}$

The language of variation evolved as a means of describing functions that involve only multiplication and/or division. The terms "varies directly," "directly proportional," and "jointly proportional" indicate multiplication. The term "inverse" indicates division. The term "variation" is *never* used in this text to refer to a formula involving addition or subtraction. Note that the formula for the cost of a pizza $C = 2n + 5$ is *not* the kind of formula that is described in terms of variation.

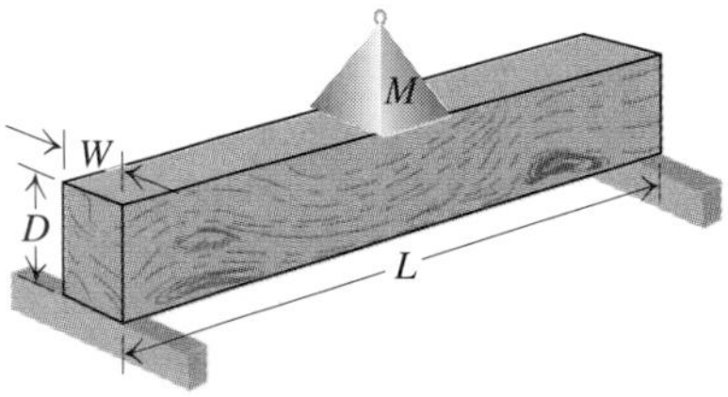

Figure 2.88

EXAMPLE 5 Solving a combined variation problem

The recommended maximum load for the ceiling joist with rectangular cross section shown in Fig. 2.88 varies directly as the product of the width and the square of the depth of the cross section, and inversely as the length of the joist.

If the recommended maximum load is 800 pounds for a 2-inch by 6-inch joist that is 12 feet long, then what is the recommended maximum load for a 2-inch by 12-inch joist that is 16 feet long?

Solution

The maximum load M varies directly as the product of the width W and the square of the depth D, and inversely as the length L:

$$M = \frac{kWD^2}{L}$$

Because the depth is squared, the maximum load is always obtained by positioning the joist so that the larger dimension is the depth. Use $W = 2$, $D = 6$, $L = 12$, and $M = 800$ to find k:

$$800 = \frac{k \cdot 2 \cdot 6^2}{12}$$

$$800 = 6k$$

$$\frac{400}{3} = k$$

Now use $W = 2$, $D = 12$, $L = 16$, and $k = 400/3$ in $M = \dfrac{kWD^2}{L}$ to find M:

$$M = \frac{400 \cdot 2 \cdot 12^2}{3 \cdot 16} = 2400$$

The maximum load on the joist is 2400 pounds. Notice that feet and inches were used in the formula without converting one to the other. If we solve the problem using only feet or only inches, then the value of k is different but we get the same value for M.

FOR THOUGHT True or False? Explain.

1. If the cost of an 800 number is \$29.95 per month plus 28 cents per minute, then the monthly cost varies directly as the number of minutes.
2. If bananas are \$0.39 per pound, then your cost for a bunch of bananas varies directly with the number of bananas purchased.
3. The area of a circle varies directly as the square of the radius.
4. The area of a triangle varies jointly as the length of the base and the height.
5. The number π is a constant of proportionality.

6. If y is inversely proportional to x, then for $x = 0$, we get $y = 0$.
7. If the cost of potatoes varies directly with the number of pounds purchased, then the proportionality constant is the price per pound.
8. The height of a student in centimeters is directly proportional to the height of the same student in inches.
9. The surface area of a cube varies directly as the square of the length of an edge.
10. The surface area of a rectangular box varies jointly as its length, width, and height.

2.7 EXERCISES

Tape 6 Disk

Write a formula that expresses the relationship described by each statement. Use k as the constant of variation.

1. Your grade on the next test, G, varies directly with the number of hours, n, that you study for it.
2. The volume of a gas in a cylinder, V, is inversely proportional to the pressure on the gas, P.
3. For two lines that are perpendicular and have slopes, the slope of one is inversely proportional to the slope of the other.
4. The amount of sales tax on a new car is directly proportional to the purchase price of the car.
5. The cost of constructing a silo varies jointly as the height and the radius.
6. The volume of grain that a silo can hold varies jointly as the height and the square of the radius.
7. Y varies directly as x and inversely as the square root of z.
8. W varies directly with r and t and inversely with v.

For each formula that illustrates a type of variation, write the formula in words using the vocabulary of variation. The letters k and π represent constants, and all other letters represent variables.

9. $A = \pi r^2$
10. $C = \pi D$
11. $a = kzw$
12. $m_1 = \dfrac{-1}{m_2}$
13. $T = 3W + 5$
14. $y = k\sqrt{x - 3}$
15. $y = \dfrac{1}{x}$
16. $V = LWH$
17. $H = \dfrac{k\sqrt{t}}{s}$
18. $B = \dfrac{y^2}{2\sqrt{x}}$
19. $D = \dfrac{2LJ}{W}$
20. $E = mc^2$

Find the constant of variation and write the formula that is expressed by the indicated variation.

21. y varies directly as x, and $y = 5$ when $x = 9$.
22. h is directly proportional to z, and $h = 210$ when $z = 200$.
23. T is inversely proportional to y, and $T = -30$ when $y = 5$.
24. H is inversely proportional to n, and $H = 9$ when $n = -6$.
25. m varies directly as the square of t, and $m = 54$ when $t = 3\sqrt{2}$.
26. p varies directly as the cube root of w, and $p = \sqrt[3]{2}/2$ when $w = 4$.
27. y varies directly as x and inversely as the square root of z, and $y = 2.192$ when $x = 2.4$ and $z = 2.25$.
28. n is jointly proportional to x and the square root of b, and $n = -18.954$ when $x = -1.35$ and $b = 15.21$.

Solve each variation problem.

29. If y varies directly as x, and $y = 9$ when $x = 2$, what is y when $x = -3$?
30. If y is directly proportional to z, and $y = 6$ when $z = \sqrt{12}$, what is y when $z = \sqrt{75}$?
31. If P is inversely proportional to w, and $P = 2/3$ when $w = 1/4$, what is P when $w = 1/6$?
32. If H varies inversely as q, and $H = 0.03$ when $q = 0.01$, what is H when $q = 0.05$?
33. If A varies jointly as L and W, and $A = 30$ when $L = 3$ and $W = 5\sqrt{2}$, what is A when $L = 2\sqrt{3}$ and $W = 1/2$?
34. If J is jointly proportional to G and V, and $J = \sqrt{3}$ when $G = \sqrt{2}$ and $V = \sqrt{8}$, what is J when $G = \sqrt{6}$ and $V = 8$?

35. If y is directly proportional to u and inversely proportional to the square of v, and $y = 7$ when $u = 9$ and $v = 6$, what is y when $u = 4$ and $v = 8$?

36. If q is directly proportional to the square root of h and inversely proportional to the cube of j, and $q = 18$ when $h = 9$ and $j = 2$, what is q when $h = 16$ and $j = 1/2$?

In each case, determine whether the first variable varies directly or inversely with the other variable. Write a formula that describes the variation using the appropriate variation constant.

37. The length of a car in inches, the length of the same car in feet

38. The time in seconds to pop a bag of microwave popcorn, the time in minutes to pop the same bag of popcorn

39. The price for each person sharing a \$20 pizza, the number of hungry people

40. The number of identical steel rods with a total weight of 40,000 lb, the weight of one of those rods

41. The speed of a car in miles per hour, the speed of the same car at the same time in kilometers per hour

42. The weight of a sumo wrestler in pounds, the weight of the same sumo wrestler in kilograms. (Chad Rowan of the United States, at 455 lb or 206 kg, recently became the first-ever foreign grand champion in Tokyo's annual tournament.)

Figure for Exercise 42

43. The temperature in degrees Celsius, the temperature at the same time and place in degrees Fahrenheit

44. The perimeter of a rectangle with a length of 4 ft, the width of the same rectangle

45. The area of a rectangle with a length of 30 in., the width of the same rectangle

46. The area of a triangle with a height of 10 cm, the base of the same triangle

47. The number of gallons of gasoline that you can buy for \$5, the price per gallon of that same gasoline

48. The length of Yolanda's 40 ft^2 closet, the width of that closet

Solve each problem.

49. *First-Class Diver* The pressure exerted by water at a point below the surface varies directly with the depth. The pressure is 4.34 pounds per square inch at a depth of 10 feet. What pressure does the sperm whale (the deepest diver among the air-breathing mammals) experience when it dives 6000 feet below the surface?

50. *Second Place* The elephant seal (second only to the sperm whale) experiences a pressure of 2170 pounds per square inch when it makes its deepest dives. At what depth does the elephant seal experience this pressure? See the previous exercise.

51. *Processing Oysters* The time required to process a shipment of oysters varies directly with the number of pounds in the shipment and inversely with the number of workers assigned. If 3000 lb can be processed by 6 workers in 8 hr, then how long would it take 5 workers to process 4000 lb?

52. *View from an Airplane* The view V from the air is directly proportional to the square root of the altitude A. If the view from horizon to horizon at an altitude of 16,000 ft is approximately 154 mi, then what is the view from 36,000 ft?

53. *Simple Interest* The amount of simple interest on a deposit varies jointly with the principal and the time in days. If \$20.80 is earned on a deposit of \$4000 for 16 days, how much interest would be earned on a deposit of \$6500 for 24 days?

54. *Free Fall at Six Flags* Visitors at Six Flags in Atlanta get a brief thrill in the park's "free fall" ride. Passengers seated in a small cab drop from a 10-story tower down a track that eventually curves level with the ground. The distance the cab falls varies directly with the square of the time it is falling. If the cab falls 16 ft in the first second, then how far has it fallen after 2 sec?

55. *Cost of Plastic Sewer Pipe* The cost of a plastic sewer pipe varies jointly as its diameter and its length. If a 20-ft pipe with a diameter of 6 in. costs \$18.60, then what is the cost of a 16-ft pipe with a diameter of 8 in.?

56. *Cost of Copper Tubing* A plumber purchasing materials for a renovation observed that the cost of copper tubing varies jointly as the length and the diameter of the tubing. If 20 ft of $\frac{1}{2}$-in.-diameter tubing costs \$36.60, then what is the cost of 100 ft of $\frac{3}{4}$-in.-diameter tubing?

57. *Weight of a Can* The weight of a can of baked beans varies jointly with the height and the square of the diameter. If a 4-in.-high can with a 3-in. radius weighs 14.5 oz, then what is the weight of a 5-in.-high can with a diameter of 6 in.?

58. *Nitrogen Gas Shock Absorber* The volume of gas in a nitrogen gas shock absorber varies directly with the temperature of the gas and inversely with the amount of weight on the piston. If the volume is 10 in.^{3} at 80° F with a weight of 600 lb, then what is the volume at 90° F with a weight of 800 lb?

59. *Velocity of Underground Water* Darcy's law states that the velocity V of underground water through sandstone varies directly as the head h and inversely as the length l of the flow. The head is the vertical distance between the point of intake into the rock and the point of discharge such as a spring, and the length is the length of the flow from intake to discharge. In a certain sandstone a velocity of 10 ft per year has been recorded with a head of 50 ft and length of 200 ft. What would we expect the velocity to be if the head is 60 ft and the length is 300 ft?

60. *Cross-Sectional Area of a Well* The rate of discharge of a well, V, varies jointly as the hydraulic gradient, i, and the cross-sectional area of the well wall, A. Suppose that a well with a cross-sectional area of 10 ft^2 discharges 3 gal of water per minute in an area where the hydraulic gradient is 0.3. If we dig another well nearby where the hydraulic gradient is 0.4, and we want a discharge of 5 gal/min, then what should be the cross-sectional area for the well?

61. *Dollars Per Death* The accompanying table shows the annual amount of money that the National Institute of Health spends on disease research and the annual number of deaths for three diseases (*Forbes*, January 27, 1997). Is the annual spending directly proportional to the annual number of deaths? Should it be?

Table for Exercise 61

Disease	Research spending	Annual deaths
AIDS	\$1.34 billion	42,506
Diabetes	\$295 million	59,085
Heart Disease	\$958 million	738,781

62. *Bicycle Gear Ratio* The gear ratio for a bicycle varies jointly with the number of teeth on the chainring by the pedals and the diameter of the wheel w, and inversely with the number of teeth on the cog by the wheel (Cycling, Burket and Darst, 1987). Find the missing entries in the following table.

Table for Exercise 62

Teeth on chainring	Teeth on cog	Wheel diameter	Gear ratio
50	25	27 in.	54
40	13	26 in.	
45		27 in.	67.5

63. *Grade on an Algebra Test* Calvin believes that his grade on a college algebra test varies directly with the number of hours spent studying during the week prior to the test and inversely with the number of hours spent at the Beach Club playing volleyball during the week prior to the test. If he scored 76 on a test when he studied 12 hr and played 10 hr during the week prior to the test, then what score should he expect if he studies 9 hr and plays 15 hr?

64. *Carpeting a Room* The cost of carpeting a room varies jointly as the length and width. Julie advertises a price of \$263.40 to carpet a 9-ft by 12-ft room with Dupont Stainmaster. Julie gave a customer a price of \$482.90 for carpeting a room with a width of 4 yd with that same carpeting, but she has forgotten the length of the room. What is the length of the room?

65. *Pole-Vault Principle* Due to increasingly flexible poles that return virtually all of the energy athletes put into them, the deciding factor in today's pole-vaulting competitions is speed down the runway (*Newsweek*, July 1992). The height a pole-vaulter attains is directly proportional to the square of his velocity on the runway. Given that a speed of 32 ft/sec will loft a vaulter to 16 ft, find the speed necessary for World Record Holder Sergei Bubka to reach a record height of 20 ft 2.5 in.

For Writing/Discussion

66. *Cooperative Learning* Working in groups, drop a ball from various heights and measure the distance that the ball rebounds after hitting the floor. Enter your pairs of data into a graphing calculator and calculate the regression line. What is the value of r? Do you think that the rebound distance is directly proportional to the drop distance? Compare results with other groups.

67. *Moving Melons* To move more watermelons, a retailer decided to make the price of each watermelon purchased by a single customer inversely proportional to the number of watermelons purchased by the customer. Is this a good idea? Explain.

68. *Common Units* In general we do not perform computations involving inches and feet until they are converted to a common unit of measure. Explain how it is possible to use inches and feet in Example 5 without converting to a common unit of measure.

LINKING CONCEPTS

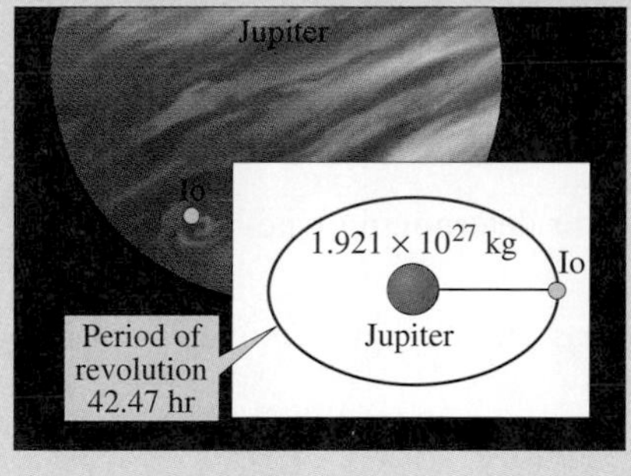

For Individual or Group Explorations

Masses of the Planets A result of Kepler's harmonic law is that the mass of a planet with a satellite is directly proportional to the cube of the mean distance from the satellite to the planet, and inversely proportional to the square of the period of revolution. Early astronomers estimated the mass of the Earth to be 5.976×10^{24} kg and observed that the Moon orbited the Earth with a period of 27.322 days at a mean distance of 384.4×10^3 km.

a) Write a formula for the mass of a planet according to Kepler.

b) Find the proportionality constant using the observations and estimates for the Earth and the Moon.

c) Extend your knowledge across the solar system to find the mass of Mars based on observations that Phobos orbits Mars in 7.65 hr at a mean distance of 9330 km.

d) What is the approximate ratio of the mass of the Earth to the mass of Mars?

e) If the mass of Jupiter is 1.921×10^{27} kg and the period of revolution of Io around Jupiter is 42.47 hr, then what is the mean distance between Io and Jupiter?

HIGHLIGHTS

- **Section 2.1 The Cartesian Coordinate System**

1. The Cartesian coordinate system is a means of picturing sets of ordered pairs.
2. Any equation of the form $Ax + By = C$, where A and B are not both zero, has a graph that is a straight line.
3. The distance between (x_1, y_1) and (x_2, y_2) is $\sqrt{(x_2 - x_1)^2 + (y_2 - y_1)^2}$.

4. The midpoint for a line segment with endpoints (x_1, y_1) and (x_2, y_2) is $\left(\frac{x_1 + x_2}{2}, \frac{y_1 + y_2}{2}\right)$.

- **Section 2.2**
  Functions

1. Any set of ordered pairs is called a relation.
2. A function is a set of ordered pairs in which no two ordered pairs have the same first coordinate and different second coordinates.
3. The ordered pairs of a function may be defined by a table, a list, an equation, or a rule.
4. The average rate of change of a function is a measure of the rate of change of the dependent variable over some interval of the independent variable.
5. The difference quotient is the average rate of change of a function over the interval $(x, x + h)$.

- **Section 2.3**
  Graphs of Relations and Functions

1. The set of ordered pairs satisfying $(x - h)^2 + (y - k)^2 = r^2$ for $r > 0$ forms a circle centered at (h, k) with radius r.
2. The equation of a circle can be solved for y, giving the equations of two semicircles.
3. The vertical line test is a visual method for determining whether a graph is a graph of a function.
4. For piecewise functions, different formulas are used in different regions of the domain.
5. A function is increasing, decreasing, or constant on an interval, depending on whether its graph is rising, falling, or staying level as you move from left to right along the graph.

- **Section 2.4**
  Transformations and Symmetry of Graphs

1. If we know what some basic graphs look like, we can graph a wide variety of functions and equations by using the ideas of transformations and symmetry.
2. Any function of the form $f(x) = mx + b$ has a graph that is a straight line.
3. The graph of an even function $(f(-x) = f(x))$ is symmetric about the y-axis and the graph of an odd function $(f(-x) = -f(x))$ is symmetric about the origin.
4. The solution sets to certain inequalities can be determined by visually inspecting the graphs of functions.

- **Section 2.5**
  Operations with Functions

1. Two functions can be combined to form the sum, difference, product, and quotient functions.
2. In the composition $f \circ g$, the function g is followed by f (the input for f is the output of g).

- **Section 2.6 Inverse Functions**

1. If a function has no two ordered pairs with different first coordinates and the same second coordinate, then the function is a one-to-one function.
2. Only one-to-one functions are invertible.
3. The horizontal line test is a visual method of determining whether a function is one-to-one.
4. The inverse function of a one-to-one function is the function formed by interchanging the coordinates in each ordered pair.
5. The graph of f^{-1} can be obtained by reflecting the graph of f about the line $y = x$.

- **Section 2.7 Variation**

1. Certain functions of one and several variables are customarily described using the ideas of direct, inverse, and joint variation.
2. The terms ''varies directly,'' ''directly proportional,'' and ''jointly proportional'' indicate multiplication.
3. The terms ''varies inversely'' and ''inversely proportional'' indicate division.

CHAPTER 2 REVIEW EXERCISES

Graph each set of ordered pairs. State the domain and range of each relation. Determine whether each set is a function.

1. $\{(0, 0), (1, 1), (-2, -2)\}$

2. $\{(0, -3), (1, -1), (2, 1), (2, 3)\}$

3. $\{(x, y) \mid y = 3 - x\}$

4. $\{(x, y) \mid 2x + y = 5\}$

5. $\{(x, y) \mid x = 2\}$

6. $\{(x, y) \mid y = 3\}$

7. $\{(x, y) \mid x^2 + y^2 = 0.01\}$

8. $\{(x, y) \mid x^2 + y^2 = 2x - 4y\}$

9. $\{(x, y) \mid x = y^2 + 1\}$

10. $\{(x, y) \mid y = |x - 2|\}$

11. $\{(x, y) \mid y = \sqrt{x} - 3\}$

12. $\{(x, y) \mid x = \sqrt{y}\}$

Let $f(x) = x^2 + 3$ and $g(x) = 2x - 7$. Find and simplify each expression.

13. $f(-3)$

14. $g(3)$

15. $g(12)$

16. $f(-1)$

17. x, if $f(x) = 19$

18. x, if $g(x) = 9$

19. $(g \circ f)(-3)$

20. $(f \circ g)(3)$

21. $(f + g)(2)$

22. $(f - g)(-2)$

23. $(f \cdot g)(-1)$

24. $(f/g)(4)$

25. $f(g(2))$

26. $g(f(-2))$

27. $(f \circ g)(x)$

28. $(g \circ f)(x)$

29. $(f \circ f)(x)$

30. $(g \circ g)(x)$

31. $f(a + 1)$

32. $g(a + 2)$

33. $\dfrac{f(3 + h) - f(3)}{h}$

34. $\dfrac{g(5 + h) - g(5)}{h}$

35. $\dfrac{f(x + h) - f(x)}{h}$

36. $\dfrac{g(x + h) - g(x)}{h}$

37. $g\left(\dfrac{x + 7}{2}\right)$

38. $f(\sqrt{x - 3})$

39. $g^{-1}(x)$

40. $g^{-1}(-3)$

Use transformations to graph each pair of functions on the same coordinate plane.

41. $f(x) = \sqrt{x}$, $g(x) = 2\sqrt{x + 3}$

42. $f(x) = \sqrt{x}$, $g(x) = -2\sqrt{x} + 3$

43. $f(x) = |x|$, $g(x) = -2|x + 2| + 4$

44. $f(x) = |x|$, $g(x) = \dfrac{1}{2}|x - 1| - 3$

45. $f(x) = x^2,\ g(x) = \frac{1}{2}(x - 2)^2 + 1$

46. $f(x) = x^2,\ g(x) = -2x^2 + 4$

Let $f(x) = \sqrt[3]{x}$, $g(x) = x - 4$, $h(x) = x/3$, and $j(x) = x^2$. Write each function as a composition of the appropriate functions chosen from f, g, h, and j.

47. $F(x) = \sqrt[3]{x - 4}$

48. $G(x) = -4 + \sqrt[3]{x}$

49. $H(x) = \sqrt[3]{\dfrac{x^2 - 4}{3}}$

50. $M(x) = \left(\dfrac{x - 4}{3}\right)^2$

51. $N(x) = \frac{1}{3}x^{2/3}$

52. $P(x) = x^2 - 16$

53. $R(x) = \dfrac{x^2}{3} - 4$

54. $Q(x) = x^2 - 8x + 16$

For each pair of points, find the distance between them and the midpoint of the line segment joining them.

55. $(-3, 5), (2, -6)$

56. $(-1, 1), (-2, -3)$

57. $(1/2, 1/3), (1/4, 1)$

58. $(0.5, 0.2), (-1.2, 2.1)$

Find the difference quotient for each function and simplify it.

59. $f(x) = -5x + 9$

60. $f(x) = \sqrt{x - 7}$

61. $f(x) = \dfrac{1}{2x}$

62. $f(x) = -5x^2 + x$

Sketch the graph of each function and state its domain and range. Determine the intervals on which the function is increasing, decreasing, or constant.

63. $f(x) = \sqrt{100 - x^2}$

64. $f(x) = -\sqrt{7 - x^2}$

65. $f(x) = \begin{cases} -x^2 & \text{for } x \le 0 \\ x^2 & \text{for } x > 0 \end{cases}$

66. $f(x) = \begin{cases} x^2 & \text{for } x \le 0 \\ x & \text{for } 0 < x \le 4 \end{cases}$

67. $f(x) = \begin{cases} -x - 4 & \text{for } x \le -2 \\ -|x| & \text{for } -2 < x < 2 \\ x - 4 & \text{for } x \ge 2 \end{cases}$

68. $f(x) = \begin{cases} (x + 1)^2 - 1 & \text{for } x \le -1 \\ -1 & \text{for } -1 < x < 1 \\ (x - 1)^2 - 1 & \text{for } x \ge 1 \end{cases}$

Each of the following graphs is the graph of a function involving absolute value. Write an equation for each graph and state the domain and range of the function.

69.

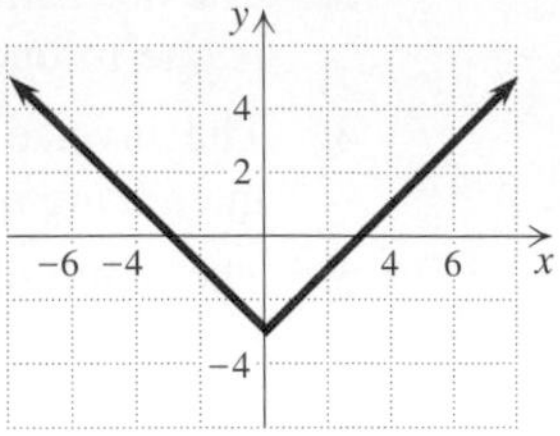

70.

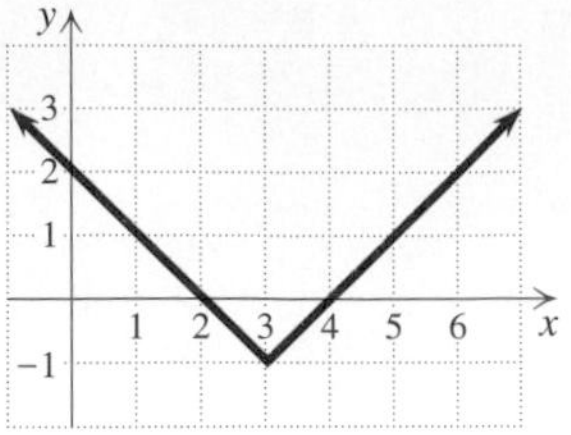

71.

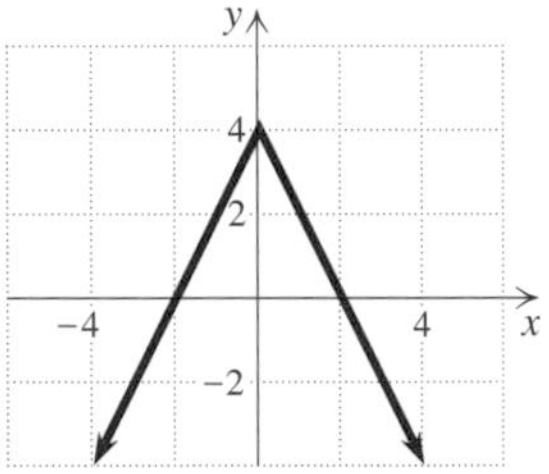

72.

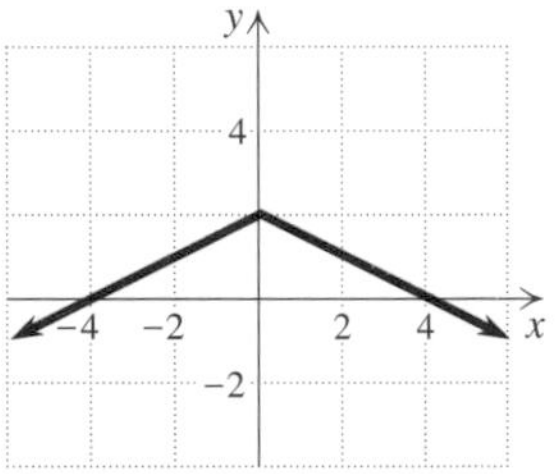

73.

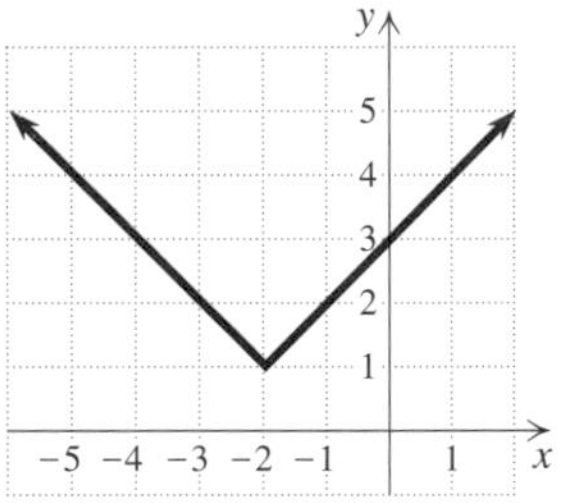

74.

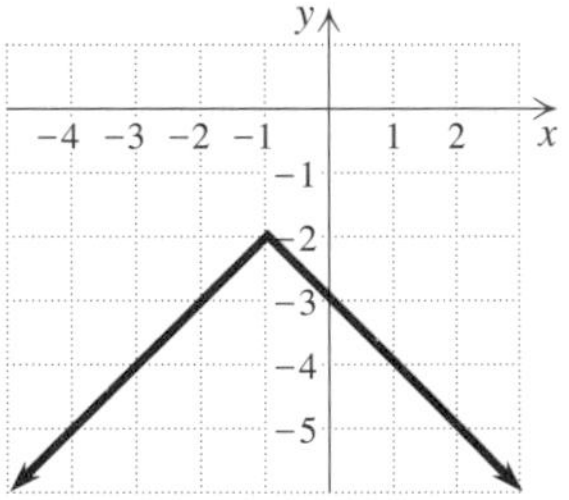

Discuss the symmetry of the graph of each function.

75. $f(x) = x^4 - 9x^2$

76. $f(x) = |x| - 99$

77. $f(x) = -x^3 - 5x$

78. $f(x) = \dfrac{15}{x}$

79. $f(x) = -x + 1$

80. $f(x) = |x - 1|$

81. $f(x) = \sqrt{x^2}$

82. $f(x) = \sqrt{16 - x^2}$

Use the graph of an appropriate function to find the solution set to each inequality.

83. $|x - 3| \ge 1$

84. $|x + 2| + 1 < 0$

85. $-2x^2 + 4 > 0$

86. $-\frac{1}{2}x^2 + 2 \le 0$

87. $-\sqrt{x + 1} - 2 > 0$

88. $\sqrt{x} - 3 < 0$

Graph each pair of functions on the same coordinate plane. What is the relationship between the functions in each pair?

89. $f(x) = \sqrt{x + 3}$, $g(x) = x^2 - 3$ for $x \geq 0$

90. $f(x) = (x - 2)^3$, $g(x) = \sqrt[3]{x} + 2$

91. $f(x) = 2x - 4$, $g(x) = \frac{1}{2}x + 2$

92. $f(x) = -\frac{1}{2}x + 4$, $g(x) = -2x + 8$

Determine whether each function is invertible. If the function is invertible, then find the inverse function, and state the domain and range of the inverse function.

93. $\{(\pi, 0), (\pi/2, 1), (2\pi/3, 0)\}$

94. $\{(-2, 1/3), (-3, 1/4), (-4, 1/5)\}$

95. $f(x) = 3x - 21$

96. $f(x) = 3|x|$

97. $y = \sqrt{9 - x^2}$

98. $y = 7 - x$

99. $f(x) = \sqrt{x - 9}$

100. $f(x) = \sqrt{x} - 9$

101. $f(x) = \dfrac{x - 7}{x + 5}$

102. $f(x) = \dfrac{2x - 3}{5 - x}$

103. $f(x) = x^2 + 1$ for $x \leq 0$

104. $f(x) = (x + 3)^4$ for $x \geq 0$

Solve each problem.

105. Find the x- and y-intercepts for the graph of $3x - 4y = 9$.

106. Given that D is directly proportional to W and $D = 9$ when $W = 25$, find D when $W = 100$.

107. Given that t varies directly as u and inversely as v, and $t = 6$ when $u = 8$ and $v = 2$, find t when $u = 19$ and $v = 3$.

108. Write in standard form the equation of the circle that has center $(-3, 5)$ and radius $\sqrt{3}$.

109. Use completing the square to write the equation $x^2 + y^2 = x - 2y + 1$ in the form $(x - h)^2 + (y - k)^2 = r^2$, and identify the center and radius of this circle.

110. Write the diameter of a circle, d, as a function of its area, A.

111. *Circle Inscribed in a Square* A cylindrical pipe with an outer radius r must fit snugly through a square hole in a wall. Write the area of the square as a function of the radius of the pipe.

112. *Load on a Spring* When a load of 5 lb is placed on a spring, its length is 6 in., and when a load of 9 lb is placed on the spring, its length is 8 in. What is the average rate of change of the length of the spring as the load varies from 5 lb to 9 lb?

113. *Changing Speed of a Dragster* Suppose that 2 sec after starting, a dragster is traveling 40 mph, and 5 sec after starting, the dragster is traveling 130 mph. What is the average rate of change of the speed of the dragster over the time interval from 2 sec to 5 sec? What are the units for this measurement?

114. *Dinosaur Speed* R. McNeill Alexander, a British paleontologist, uses observations of living animals to estimate the speed of dinosaurs (*Scientific American*, April 1991). Alexander believes that two animals of different sizes but geometrically similar shapes will move in a similar fashion. For animals of similar shapes, their velocity is directly proportional to the square root of their hip height. Alexander has compared a white rhinoceros, with a hip height of 1.5 m, and a member of the genus Triceratops, with a hip height of 2.8 m. If a white rhinoceros can move at 45 km/hr, then what is the estimated velocity of the dinosaur?

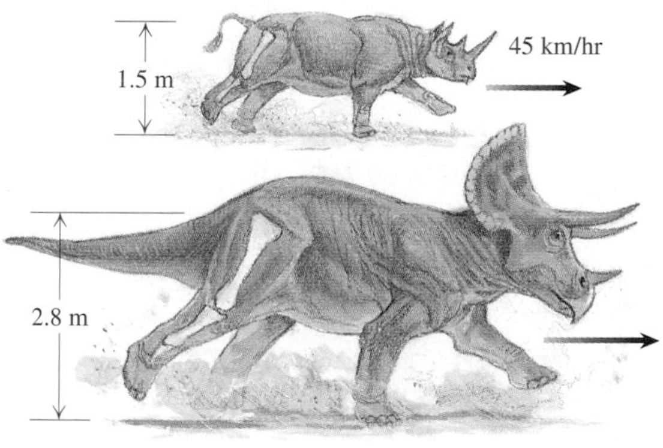

Figure for Exercise 114

115. *Arranging Plants in a Row* Timothy has a shipment of strawberry plants to plant in his field, which is prepared with rows of equal length. He knows that the number of rows that he will need varies inversely with the number of plants per row. If he plants 240 plants per row, then he needs 21 rows. How many rows does he need if he plants 224 plants per row?

116. *Handling Charge for a Globe* The shipping and handing charge for a globe from Mercator's Map Company varies directly as the square of the diameter. If the charge is \$4.32 for a 6-in.-diameter globe, then what is the charge for a 16-in.-diameter globe?

117. *Newton's Law of Gravity* Newton's law of gravity states that the force of gravity acting between two objects varies directly as the product of the masses of the two objects and inversely as the square of the distance between their centers. Write an equation for Newton's law of gravity.

118. *Inefficiency of Ivory Coast Mills* Not only are the tropical forests being consumed at an alarming rate, but the inefficiency of the lumber mills is contributing to the problem (*Scientific American*, April 1990). For example, Ivory Coast mills require 30% more logs to produce the same output as more efficient mills. Given that the output of an Ivory Coast mill varies directly with the input, and a mill produces 75 tons of lumber out of 130 tons of logs, write the output y as a function of the input x.

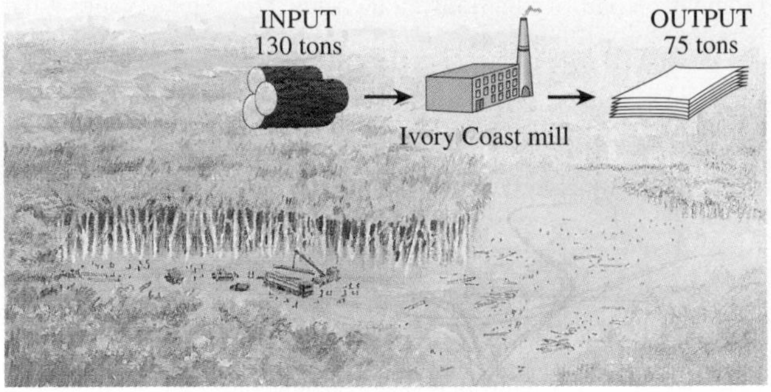

Figure for Exercise 118

For Writing/Discussion

119. Explain what the vertical and horizontal line tests are and how to apply them.

120. Is a vertical line the graph of a linear function? Explain.

CHAPTER 2 TEST

Determine whether each equation defines y as a function of x.

1. $x^2 + y^2 = 25$

2. $3x - 5y = 20$

3. $x = |y + 2|$

4. $y = x^3 - 3x^2 + 2x - 1$

State the domain and range of each relation.

5. $\{(2, -3), (2, -4), (5, 7)\}$

6. $y = \sqrt{x - 9}$

7. $x = |y - 1|$

Sketch the graph of each function.

8. $3x - 4y = 12$

9. $y = 2x - 3$

10. $y = \sqrt{25 - x^2}$

11. $y = -(x - 2)^2 + 5$

12. $y = 2|x| - 4$

13. $y = \sqrt{x + 3} - 5$

14. $f(x) = \begin{cases} x & \text{for } x < 0 \\ 2 & \text{for } x \geq 0 \end{cases}$

Let $f(x) = \sqrt{x + 2}$ and $g(x) = 3x - 1$. Find and simplify each of the following expressions.

15. $f(7)$

16. $(f \circ g)(2)$

17. $(f \circ g)(x)$

18. $g^{-1}(x)$

19. $(f + g)(14)$

20. $\dfrac{g(x + h) - g(x)}{h}$

Solve each problem.

21. Find the midpoint of the line segment whose endpoints are $(3, 5)$ and $(-3, 6)$.

22. Find the distance between the points $(-2, 4)$ and $(3, -1)$.

23. Write the equation in standard form for a circle with center $(-4, 1)$ and radius $\sqrt{7}$.

24. State the intervals on which $f(x) = (x - 3)^2$ is increasing and those on which $f(x)$ is decreasing.

25. Discuss the symmetry of the graph of the function $f(x) = x^4 - 3x^2 + 9$.

26. State the solution set to the inequality $|x - 1| - 3 < 0$ using interval notation.

27. Find the inverse of the function $g(x) = 3 + \sqrt[3]{x - 2}$.

28. Jang's Postal Service charges \$35 for addressing 200 envelopes, and \$60 for addressing 400 envelopes. What is the average rate of change of the cost as the number of envelopes goes from 200 to 400?

29. The intensity of illumination I from a light source varies inversely as the square of the distance d from the source. If a flash on a camera has an intensity of 300 candlepower at a distance of 2 m, then what is the intensity of the flash at a distance of 10 m?

30. Write a formula expressing the volume V of a cube as a function of the length of the diagonal d of a side of the cube.

TYING IT ALL TOGETHER

Chapters P–2

Perform the indicated operations.

1. $2x + 3x$

2. $(3 - x)^2$

3. $\sqrt{x^2}$

4. $(x - 2)(x + 2)$

5. $x^2(x - 2)(x + 3)$

6. $(3 - 2i)(3 + 2i)$

7. $\dfrac{-4 \pm \sqrt{4^2 - 4(1)(2)}}{2}$

8. $\dfrac{-6 \pm \sqrt{(-6)^2 - 4(3)(5)}}{2(3)}$

Solve each equation.

9. $2x + 3x = 5x$

10. $\sqrt{9 - x^2} = 2$

11. $|x| - 1 = 0$

12. $x^2 - 4 = 0$

13. $x^2 - 4x + 5 = 0$

14. $4x^2 - 4x - 1 = 0$

15. $x^2 - 9x = 0$

16. $\dfrac{1}{x} + \dfrac{1}{x - 1} = \dfrac{3}{x^2 - x}$

Graph each function.

17. $y = 5x$

18. $y = \sqrt{9 - x^2}$

19. $y = |x| - 1$

20. $y = x^2 - 4$

21. $y = 3 - x$

22. $y = 3 - \sqrt{x}$

23. $y = 2(x + 2)^2 - 5$

24. $y = (x - 1)(x^2 + x + 1)$

Write each expression as a single rational expression.

25. $x + \dfrac{1}{x}$

26. $x + 1 + \dfrac{1}{x - 1}$

27. $3 + \dfrac{2}{x - 4}$

28. $-2 + \dfrac{-1}{x + 3}$

"Which way is Ireland?" A shout cut the air and hovered briefly over a fishing boat trawling the dark ocean waters below. The voice belonged to American aviator Charles Lindbergh, then in the 26th hour of his nonstop, solo flight across the Atlantic. A man stuck his head out of a porthole but did not reply. In exasperation the weary pilot gave up and headed his plane east.

Others had died attempting this aviation feat, and in May 1927, the $25,000 prize offered by a New York businessman was still unclaimed. Lindbergh, disoriented by hours of flying without sleep and by thick fog that obscured his view, desperately wanted to sight land. He feared he had drifted hundreds of miles off course. When no answer came from the boat, he knew he must make another vital determination of air speed: If he'd strayed too far south, he should throttle down to save fuel in order to fly the additional thousand miles it would take to reach Europe. But, if he was on course to Ireland and slowed his air speed, it was unlikely he'd sight land before dark.

For much of his flight, Lindbergh's calculations of the interrelationships among wind velocity, air speed, fuel consumption, and time elapsed held life-or-death implications. The longer he remained in the air, the wearier he became and the greater the risk of something going wrong. But the faster he flew, the more rapidly the Spirit of St. Louis burned fuel and the greater the risk he'd fail to reach Paris, his ultimate destination.

In this chapter we explore polynomial and rational functions and their graphs. We'll learn how to use these basic functions of algebra to model many real-world situations, and we'll see how Lindbergh used them in planning several aspects of his 3610-mile flight between New York and Paris. In fact, as you'll discover, polynomial and rational functions have many common real-world applications. After reading this chapter, perhaps a better understanding of these functions will both extend your knowledge and—as with Lindbergh—expand your own horizons.

3 Polynomial and Rational Functions

3.1 Linear Functions
3.2 Quadratic Functions
3.3 Zeros of Polynomial Functions
3.4 The Theory of Equations
3.5 Miscellaneous Equations
3.6 Graphs of Polynomial Functions
3.7 Graphs of Rational Functions

3.1 Linear Functions

In this chapter we will study the polynomial functions. The linear functions, first introduced and graphed in Section 2.1, are the simplest polynomial functions. In this section, we'll study different forms of their equations, see how graphs of lines are related to equations, and use linear functions in a variety of applications.

Definitions and Review

If a polynomial is used to define a function, then the function is called a **polynomial function**. Some examples of polynomial functions are

$$h(x) = 2, \quad g(x) = 3x - 6, \quad P(x) = 5x^2 - 2x + 6, \quad \text{and} \quad T(x) = x^3.$$

Definition: Polynomial Function

Let $a_n, a_{n-1}, \ldots, a_0$ be real numbers with $a_n \neq 0$. The function defined by

$$f(x) = a_n x^n + a_{n-1}x^{n-1} + \cdots + a_1 x + a_0$$

is called a **polynomial function of degree *n***. The **leading coefficient** is a_n.

The degree of a polynomial function is the degree of the polynomial used to define it. The polynomial functions h, g, P, and T given as examples have degrees 0, 1, 2, and 3, respectively.

A linear function is a polynomial function of degree 1.

Definition: Linear Function

If m and b are real numbers with $m \neq 0$, then

$$f(x) = mx + b$$

is called a **linear function**.

If $m = 0$, then $f(x) = b$ has degree 0 (provided $b \neq 0$), and it is a **constant function**. The graph of $f(x) = b$ is a horizontal line. We learned in Chapter 2 that the graph of any linear function is a straight line.

Slope of a Line

In Section 2.2 we defined the average rate of change of a function. For a linear function the average rate of change is called the *slope* of the line.

Definition: Slope

The **slope** of a line through (x_1, y_1) and (x_2, y_2) with $x_1 \neq x_2$ is

$$\frac{y_2 - y_1}{x_2 - x_1}.$$

If (x_1, y_1) and (x_2, y_2) are two distinct points with $x_1 = x_2$, then the points lie on a vertical line. So slope is undefined for vertical lines. Equations of vertical lines were discussed in Chapter 2. Since a vertical line is not the graph of a function, it is not the graph of a linear function.

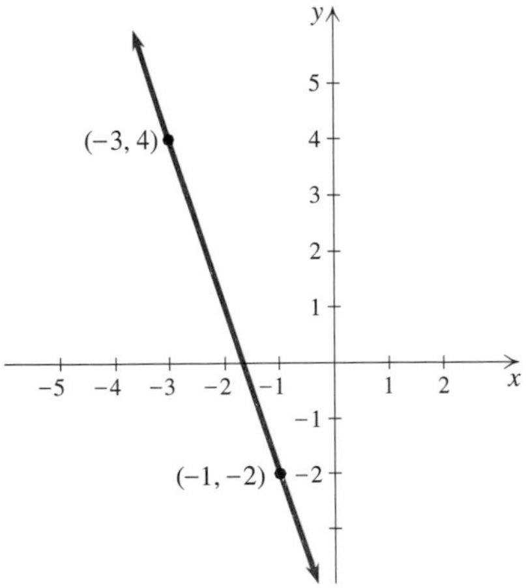

Figure 3.1

EXAMPLE 1 Finding the slope of a line through two points

Find the slope of the line that contains the points $(-3, 4)$ and $(-1, -2)$.

Solution

The line through $(-3, 4)$ and $(-1, -2)$ is shown in Fig. 3.1. Use $(x_1, y_1) = (-3, 4)$ and $(x_2, y_2) = (-1, -2)$ in the expression for slope:

$$\text{slope} = \frac{y_2 - y_1}{x_2 - x_1} = \frac{-2 - 4}{-1 - (-3)} = \frac{-6}{2} = -3$$

The slope of the line is -3.

If (x_1, y_1) and (x_2, y_2) are *any* two points that satisfy $y = mx + b$, then $y_1 = mx_1 + b$ and $y_2 = mx_2 + b$. Using the definition of slope, we have

$$\text{slope} = \frac{y_2 - y_1}{x_2 - x_1} = \frac{(mx_2 + b) - (mx_1 + b)}{x_2 - x_1} = \frac{m(x_2 - x_1)}{x_2 - x_1} = m$$

provided that $x_1 \neq x_2$. This computation proves that the slope is the same no matter which two points of a line are used to calculate it and the slope actually appears in the formula $y = mx + b$ as m. Since $y = mx + b$ has y-intercept $(0, b)$ and slope m, $y = mx + b$ is called the **slope-inercept form** of the equation of a line.

Theorem: Slope-Intercept Form

The graph of $y = mx + b$ is a line with slope m and intercept $(0, b)$. Furthermore,

$$m = \frac{y_2 - y_1}{x_2 - x_1}$$

for any points (x_1, y_1) and (x_2, y_2) on the line $y = mx + b$ with $x_1 \neq x_2$.

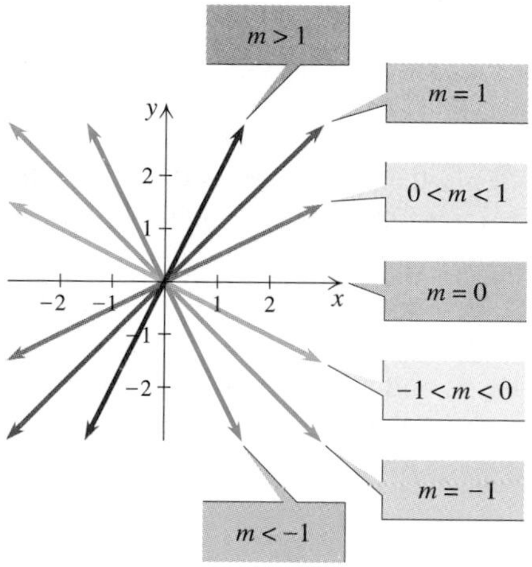

Graphs of $y = mx$

Figure 3.2

In Section 2.4 we considered $y = x$ as the basic linear function, with the graph of $y = mx$ obtained by transforming the graph of $y = x$. For example, if $m > 1$, each y-coordinate of $y = x$ is increased or stretched to a y-coordinate of $y = mx$. If $0 < m < 1$, each y-coordinate of $y = x$ is decreased or shrunk to a y-coordinate of $y = mx$. Graphs of $y = mx$ are shown in Fig. 3.2 for various values of m.

From Fig. 3.2 we can see that the slope is a measure of the steepness of the line. For a positive slope, the linear function is increasing on its entire domain, and it is called an *increasing function*. For a negative slope, $y = mx$ is a *decreasing function*. Since the average rate of change of a linear function is the constant m, the slope is the rate at which the linear function is increasing or decreasing.

EXAMPLE 2 Slope-intercept form

Write the equation $2x - 3y = 6$ in slope-intercept form and identify the slope and y-intercept. Is the linear function increasing or decreasing?

Solution

Write $2x - 3y = 6$ in slope-intercept form by solving for y:

$$-3y = -2x + 6$$

$$y = \frac{2}{3}x - 2$$

The slope is 2/3 and the y-intercept is $(0, -2)$. The function $y = \frac{2}{3}x - 2$ is an increasing function because the slope is positive.

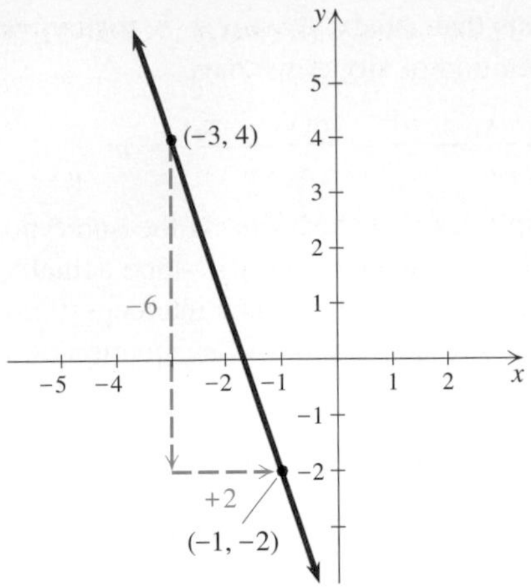

Figure 3.3

Using Slope in Graphing Lines

In Example 1 we found the slope of the line through $(-3, 4)$ and $(-1, -2)$. In this case, $y_2 - y_1 = -6$ and $x_2 - x_1 = 2$. We can move from $(-3, 4)$ to $(-1, -2)$ by going down six units and then moving two units to the right as shown in Fig. 3.3. The numerator in the slope formula, $y_2 - y_1$, is called the **rise**, and it gives the vertical change in going from (x_1, y_1) to (x_2, y_2) as shown in Fig. 3.4. The denominator $x_2 - x_1$ is called the **run**, and it gives the horizontal change. The slope of a line is the ratio of the rise to the run:

$$m = \frac{y_2 - y_1}{x_2 - x_1} = \frac{\text{rise}}{\text{run}}$$

Note that we can think of a positive rise as motion upward and a negative rise as motion downward. Likewise, a positive run indicates motion to the right and a negative run indicates motion to the left.

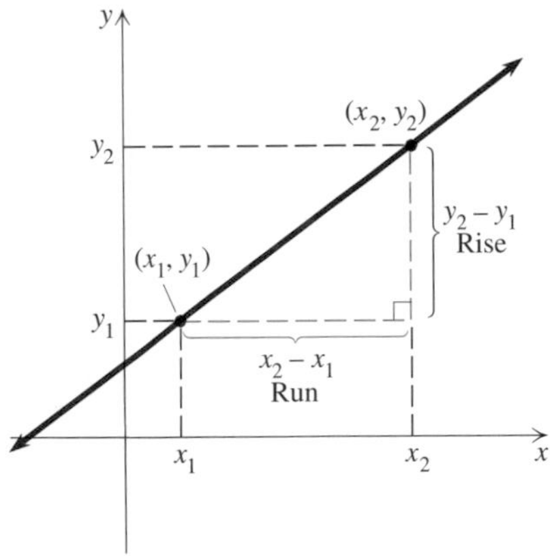

Figure 3.4

EXAMPLE 3 Graphing a linear function using slope and *y*-intercept

Graph each linear function.

a) $f(x) = \frac{2}{3}x - 1$ **b)** $y = -3x + 4$

Solution

a) In the slope-intercept form $f(x) = mx + b$, the y-intercept is $(0, b)$ and the slope is m. So for $f(x) = \frac{2}{3}x - 1$, the y-intercept is $(0, -1)$ and the slope is $2/3$. The line goes through $(0, -1)$. To find a second point on the line, start at $(0, -1)$ and rise 2 and run 3 to get to the point $(3, 1)$ as shown in Fig. 3.5. The graph of $f(x)$ is a line through $(0, -1)$ and $(3, 1)$.

b) For $y = -3x + 4$, the y-intercept is $(0, 4)$ and the slope is -3. The graph goes through $(0, 4)$. The slope of -3 can be expressed as the rise/run ratio

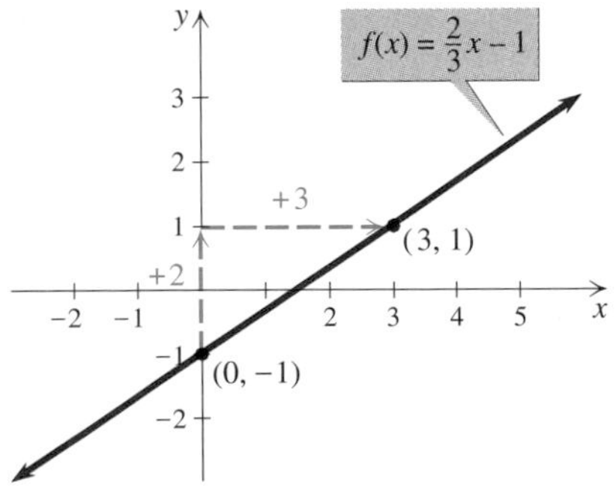

Figure 3.5

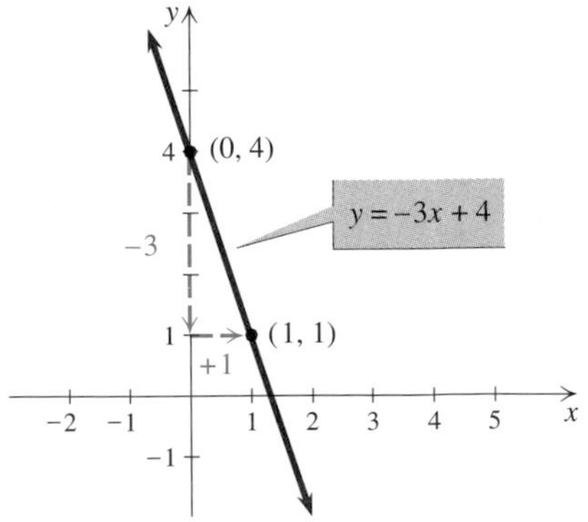

Figure 3.6

$-3/1$. To find a second point on the line, start at (0, 4), move downward three units, and then one unit to the right. Draw a line through (0, 4) and (1, 1), as in Fig. 3.6.

Point-Slope Form of a Line

In Example 3 we started with the equation of a line and graphed the line. We can also start with a description of a line or even a graph of the line and write the equation for the line. Suppose a line has slope m and goes through the point (x_1, y_1). Since the slope is a constant, any other point (x, y) on the line must satisfy

$$\frac{y - y_1}{x - x_1} = m.$$

Multiply each side of this equation by $x - x_1$ to get $y - y_1 = m(x - x_1)$, which is called the **point-slope form** of the equation of a line. We have proved the following theorem.

Theorem: Point-Slope Form

The equation of a line through (x_1, y_1) with slope m is

$$y - y_1 = m(x - x_1).$$

EXAMPLE 4 Using the point-slope form

Find the equation of the line shown in Fig. 3.7 and write it in slope-intercept form. Assume that the line goes through the two points shown in Fig. 3.7.

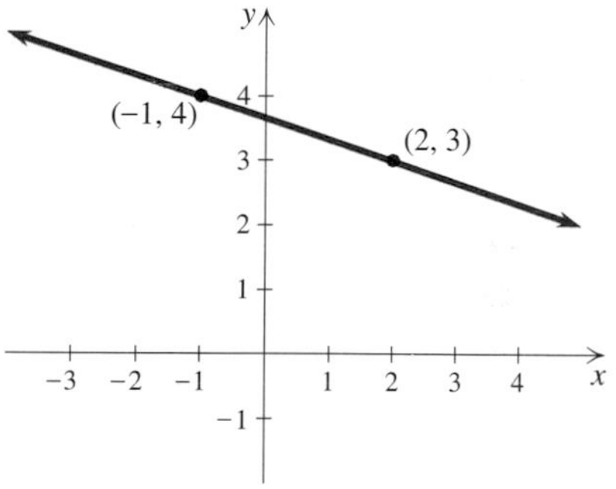

Figure 3.7

Solution

We can use any two points on the line to find its slope, and then use the point-slope form for the equation of the line. The points in Fig. 3.7 are $(-1, 4)$ and $(2, 3)$. To get from $(-1, 4)$ to $(2, 3)$ we have a rise of -1 and a run of 3. So the slope is $-1/3$. Let $m = -1/3$ and $(x_1, y_1) = (-1, 4)$ in the equation $y - y_1 = m(x - x_1)$:

$$y - 4 = -\frac{1}{3}(x - (-1)) \quad \text{The equation in point-slope form}$$

$$y - 4 = -\frac{1}{3}(x + 1)$$

$$y - 4 = -\frac{1}{3}x - \frac{1}{3}$$

$$y = -\frac{1}{3}x + \frac{11}{3} \quad \text{The equation in slope-intercept form}$$

Check that $(-1, 4)$ and $(2, 3)$ both satisfy the last equation. All four of the equations are equations for the line in Fig. 3.7, but only the last one is in slope-intercept form as required.

Standard Form of the Equation of a Line

Since vertical lines are the only lines that have no slope and are not the graphs of functions, they are the only lines that cannot be described by the linear function $y = mx + b$. However, all lines including vertical lines can be described by an equation in standard form.

Definition: Standard Form

If A, B, and C are real numbers with A and B not both zero, then

$$Ax + By = C$$

is called the **standard form** of the equation of a line.

If $B = 0$ and $A = 1$ in the standard form, the graph of $x = C$ would be a vertical line with x-intercept $(C, 0)$.

EXAMPLE 5 Equation of a line through two points

Write the standard form of the equation of the line through the given pair of points.

a) $(-1, 3)$ and $(2, 4)$ **b)** $(3, 5)$ and $(3, -6)$ **c)** $(2, 7)$ and $(-3, 7)$

Solution

a) First find the slope:

$$m = \frac{4 - 3}{2 - (-1)} = \frac{1}{3}$$

Use the point $(2, 4)$ and the slope $1/3$ in the point-slope form:

$$y - 4 = \frac{1}{3}(x - 2)$$

$$3y - 12 = x - 2 \qquad \text{Multiply each side by 3 to get integers.}$$

$$-x + 3y = 10$$

$$x - 3y = -10$$

The equation in standard form is $x - 3y = -10$. Both ordered pairs satisfy $x - 3y = -10$, because $-1 - 3(3) = -10$ and $2 - 3(4) = -10$.

b) The points (3, 5) and (3, −6) are two points on a vertical line through (3, 0). Its equation is $x = 3$. Note that slope cannot be used to find this equation because a vertical line does not have a slope.

c) Because (2, 7) and (−3, 7) have the same y-coordinate, they lie on a horizontal line that is 7 units above the x-axis. Its equation is $y = 7$.

Parallel Lines

Two lines in a plane are said to be **parallel** if they have no points in common. Any two vertical lines are parallel, and slope can be used to determine whether nonvertical lines are parallel. For example, the lines $y = 3x - 4$ and $y = 3x + 1$ are parallel because their slopes are equal.

Theorem: Parallel Lines

Two nonvertical lines in the coordinate plane are parallel if and only if their slopes are equal.

A proof to this theorem is outlined in Exercises 97 and 98.

EXAMPLE 6 Parallel lines

Find the equation in slope-intercept form of the line through (1, −4) that is parallel to $y = 3x + 2$.

Solution

Since $y = 3x + 2$ has slope 3, any line parallel to it also has slope 3. Write the equation of the line through (1, −4) with slope 3 in point-slope form:

$$y - (-4) = 3(x - 1) \qquad \text{Point-slope form}$$
$$y + 4 = 3x - 3$$
$$y = 3x - 7 \qquad \text{Slope-intercept form}$$

The line $y = 3x - 7$ goes through (1, −4) and is parallel to $y = 3x + 2$. The graphs of $y_1 = 3x - 7$ and $y_2 = 3x + 2$ in Fig. 3.8 support the answer.

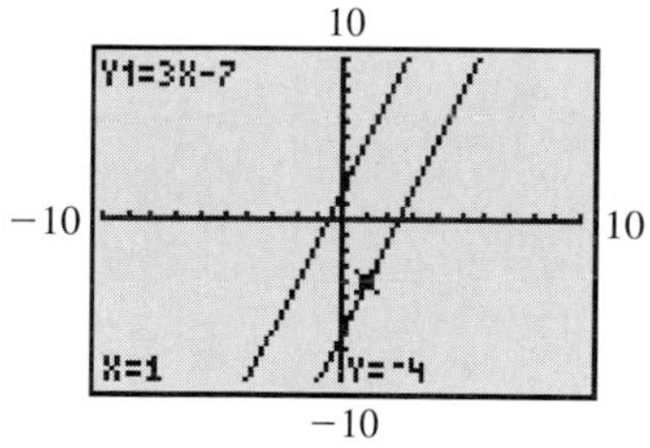

Figure 3.8

Perpendicular Lines

Two lines are **perpendicular** if they intersect at a right angle. Slope can be used to determine whether lines are perpendicular. For example, lines with slopes such as 2/3 and −3/2 are perpendicular. The slope −3/2 is the opposite of the reciprocal of 2/3. In the following theorem we use the equivalent condition that the product of the slopes of two perpendicular lines is -1, provided they both have slopes.

Theorem: Perpendicular Lines

Two lines with slopes m_1 and m_2 are perpendicular if and only if $m_1m_2 = -1$.

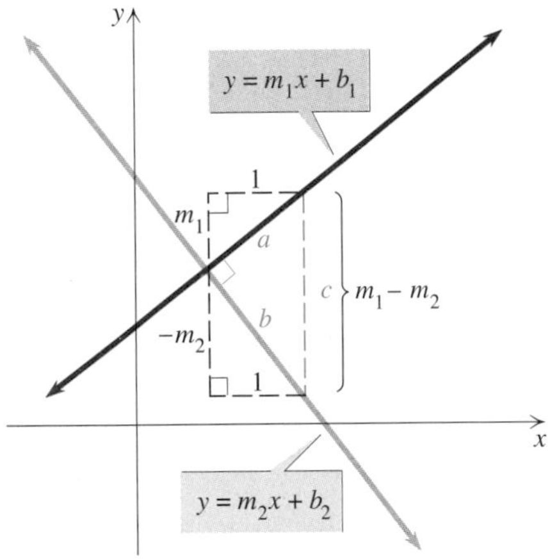

Figure 3.9

Proof The phrase "if and only if" means that there are two statements to prove. First we prove that if two lines are perpendicular, then the product of their slopes is -1. Suppose $y = m_1x + b_1$ and $y = m_2x + b_2$ are perpendicular as shown in Fig. 3.9. Assuming that $m_1 > 0$ and $m_2 < 0$, start at the intersection of the lines and form a triangle by a rise of m_1 and a run of 1 to indicate a slope of m_1. Use the slope m_2 as a negative rise of m_2 and a run of 1. By the Pythagorean theorem, $a = \sqrt{1 + m_1^2}$ and $b = \sqrt{1 + m_2^2}$. Applying the Pythagorean theorem to the third right triangle with $c = m_1 - m_2$, we get

$$(m_1 - m_2)^2 = (\sqrt{1 + m_1^2})^2 + (\sqrt{1 + m_2^2})^2 \quad \text{Since } c^2 = a^2 + b^2$$

$$(m_1 - m_2)^2 = 1 + m_1^2 + 1 + m_2^2 \quad \text{Simplify.}$$

$$m_1^2 - 2m_1m_2 + m_2^2 = 2 + m_1^2 + m_2^2$$

$$-2m_1m_2 = 2$$

$$m_1m_2 = -1$$

For the second half of the proof, we show that if $m_1m_2 = -1$, then the lines are perpendicular. If $m_1m_2 = -1$, we can reverse the order of the preceding equations to conclude that the triangle with sides a, b, and c is a right triangle and that the lines are perpendicular. Thus we have proved the theorem. ■

EXAMPLE 7 Writing equations of perpendicular lines

Find the equation of the line perpendicular to the line $3x - 4y = 8$ and containing the point $(-2, 1)$. Write the answer in slope-intercept form.

Solution

Rewrite $3x - 4y = 8$ in slope-intercept form:

$$-4y = -3x + 8$$

$$y = \frac{3}{4}x - 2 \quad \text{Slope of this line is 3/4.}$$

Since the product of the slopes of perpendicular lines is -1, the slope of the line that we seek is $-4/3$. Use the slope $-4/3$ and the point $(-2, 1)$ in the point-slope form:

$$y - 1 = -\frac{4}{3}(x - (-2))$$

$$y - 1 = -\frac{4}{3}x - \frac{8}{3}$$

$$y = -\frac{4}{3}x - \frac{5}{3}$$

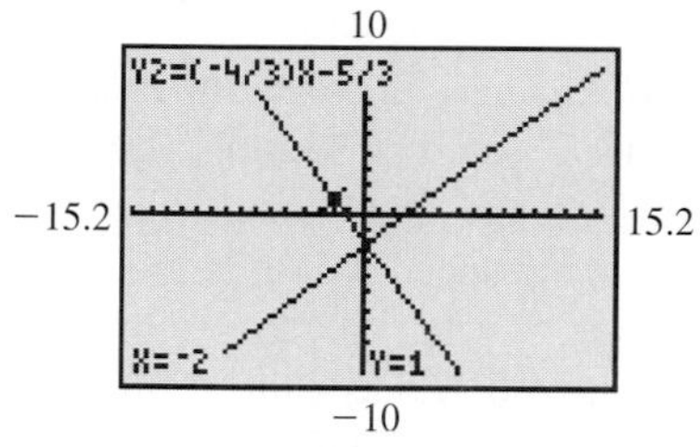

Figure 3.10

The last equation is the required equation in slope-intercept form. The graphs of these two equations should look perpendicular.

If you graph $y_1 = \frac{3}{4}x - 2$ and $y_2 = -\frac{4}{3}x - \frac{5}{3}$ with a graphing calculator, the graphs will not appear perpendicular in the standard viewing window because each axis has a different unit length. The graphs appear perpendicular in Fig. 3.10 because the unit lengths were made equal with the ZSquare feature of the TI-83.

Geometric Application of Slope

In Section 2.1 the distance formula was used to establish some facts about geometric figures in the coordinate plane. We can also use slope to prove that lines are parallel or perpendicular.

EXAMPLE 8 The diagonals of a rhombus are perpendicular

Given a rhombus with vertices (0, 0), (5, 0), (3, 4), and (8, 4), prove that the diagonals of this rhombus are perpendicular.

Solution

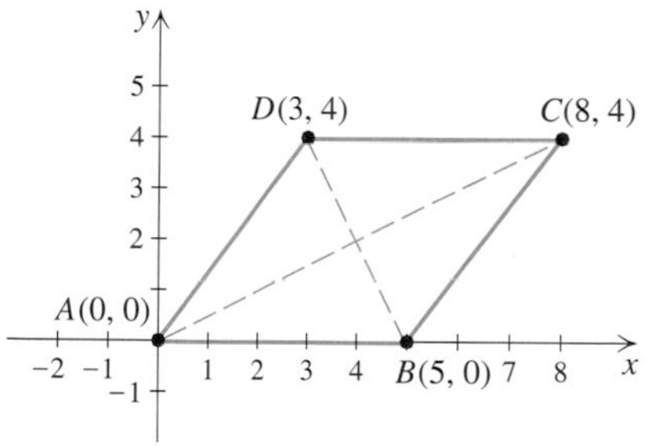

Figure 3.11

Plot the four points A, B, C, and D as shown in Fig. 3.11. You should show that each side of this figure has length 5, verifying that the figure is a rhombus. Find the slope of each diagonal and their product:

$$m_{AC} = \frac{4 - 0}{8 - 0} = \frac{1}{2} \qquad m_{BD} = \frac{0 - 4}{5 - 3} = -2$$

$$m_{AC} \cdot m_{BD} = \frac{1}{2}(-2) = -1$$

Since the product of the slopes of the diagonals is -1, the diagonals are perpendicular.

Applications of Linear Functions

If one variable is a linear function of another and we know two ordered pairs of that function, then we can write a formula for the linear function using the point-slope form of the equation of a line. In the next example we will see one of the linear functions that Charles Lindbergh used in planning his record-breaking flight across the Atlantic (from *The Spirit of St. Louis* by Charles Lindbergh).

EXAMPLE 9 An application of linear functions

Lindbergh and Don Hall, chief engineer at Ryan Airlines, calculated that at the beginning of the flight the most economical air speed would be 97 mph, at which Lindbergh would get 1.2 miles per pound of fuel. At the flight's end, when the plane was almost out of fuel, the most economical air speed would be 67 mph, at which he would get 2.3 miles per pound of fuel. See Fig. 3.12. Through repeated testing, Lindbergh knew that the miles per pound M was a linear function of the most economical air speed A. Write a formula for that function.

Solution

The slope of the linear function through the points (97, 1.2) and (67, 2.3) is

$$m = \frac{2.3 - 1.2}{67 - 97} = \frac{1.1}{-30} = -\frac{11}{300}.$$

Use the point-slope form to get the equation of the line:

$$M - 1.2 = -\frac{11}{300}(A - 97)$$

$$M = -0.0366A + 4.7566$$

The number of miles per pound of fuel is determined from the most economical air speed A by the function $M = -0.0366A + 4.7566$, where $67 \leq A \leq 97$.

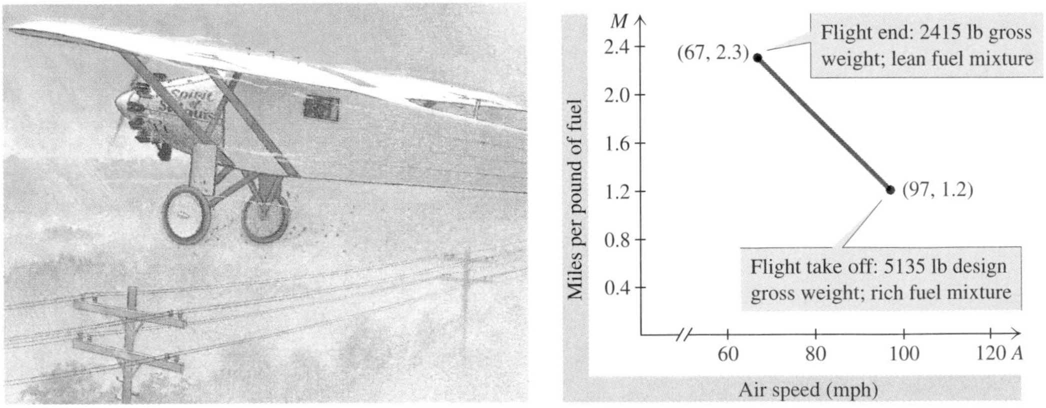

Figure 3.12

Lindbergh planned to adjust his air speed to stay on the line shown in Fig. 3.12 from the start to the end of his flight. From knowledge of how much fuel he had used, he could estimate what portion of the flight he had completed and adjust his air speed accordingly.

In the next example we do not need the slope to write the equation of the line because we are given values for A, B, and C that we can use in standard form $Ax + By = C$.

EXAMPLE 10 Interpreting slope

A manager for a country market will spend a total of \$80 on apples at \$0.25 each and pears at \$0.50 each. Write the number of apples she can buy as a function of the number of pears. Find the slope and interpret your answer. Is the function increasing or decreasing?

Solution

Let a represent the number of apples and p represent the number of pears. Write an equation about the total amount spent:

$$\begin{aligned} 0.25a + 0.50p &= 80 \\ 0.25a &= 80 - 0.50p \\ a &= 320 - 2p \end{aligned}$$

The equation $a = 320 - 2p$ expresses the number of apples as a function of the number of pears. Because the slope is -2 the function is decreasing. Since the dependent variable is a and the independent variable is p, the slope is actually -2 apples per pear. So if the number of pears is increased by one, the number of apples is decreased by two. Which makes sense because the pears cost twice as much as the apples.

FOR THOUGHT True or False? Explain.

1. The slope of the line through (2, 2) and (3, 3) is 3/2.
2. A linear function with positive slope is increasing on $(-\infty, \infty)$.
3. Any two distinct parallel lines have equal slopes.
4. The graph of $x = 3$ in the coordinate plane is the single point (3, 0).
5. Two lines with slopes m_1 and m_2 are perpendicular if $m_1 = -1/m_2$.
6. Every line in the coordinate plane has an equation in slope-intercept form.
7. The leading coefficient of the polynomial function $f(x) = 2x - x^3$ is 2.
8. Every line in the coordinate plane has an equation in standard form.
9. The line $y = 3x$ is parallel to the line $y = -3x$.
10. The line $x - 3y = 4$ contains the point $(1, -1)$ and has slope 1/3.

3.1 EXERCISES

Tape 7 Disk

Find the slope of the line containing each pair of points.

1. $(-2, 3), (4, 5)$
2. $(-1, 2), (3, 6)$
3. $(1, 3), (3, -5)$
4. $(2, -1), (5, -3)$
5. $(5, 2), (-3, 2)$
6. $(0, 0), (5, 0)$
7. $(\sqrt{2}, 1), (3\sqrt{2}, -3)$
8. $(3, \sqrt{3}), (7, -\sqrt{3})$
9. $(3, \pi/2), (1, \pi/4)$
10. $(-3, \pi/2), (-5, \pi/3)$

Write each equation in slope-intercept form and identify the slope and y-intercept of the line.

11. $3x - 5y = 10$
12. $2x - 2y = 1$
13. $y - 3 = 2(x - 4)$
14. $y + 5 = -3(x - (-1))$
15. $y + 1 = \frac{1}{2}(x - (-3))$
16. $y - 2 = -\frac{3}{2}(x + 5)$
17. $y - 4 = 0$
18. $x - y + 5 = 0$
19. $y - 0.4 = 0.03(x - 100)$
20. $0.02x - y = 0.6$

Use the y-intercept and slope to sketch the graph of each equation.

21. $y = \frac{1}{2}x - 2$
22. $y = \frac{2}{3}x + 1$
23. $f(x) = -3x + 1$
24. $f(x) = -x + 3$
25. $y = -\frac{3}{4}x - 1$
26. $y = -\frac{3}{2}x$
27. $x - y = 3$
28. $2x - 3y = 6$
29. $y - 5 = 0$
30. $6 - y = 0$
31. $y = 3x - 4$
32. $y = 5x - 5$
33. $y = \sqrt{2}x - \sqrt{2}$
34. $3x - 4y = 5$

Write an equation in slope-intercept form for each of the lines shown.

35.

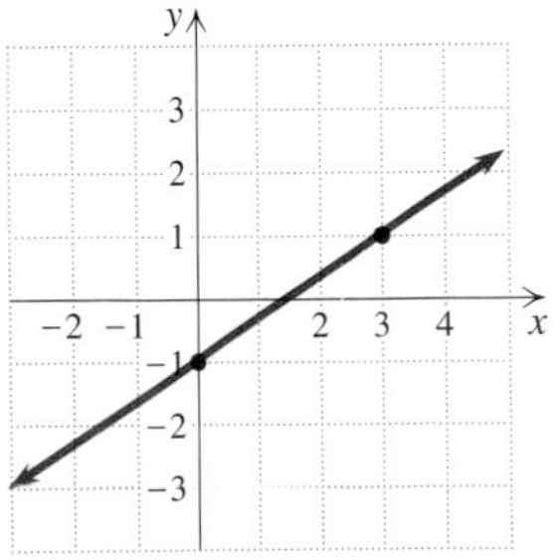

36.

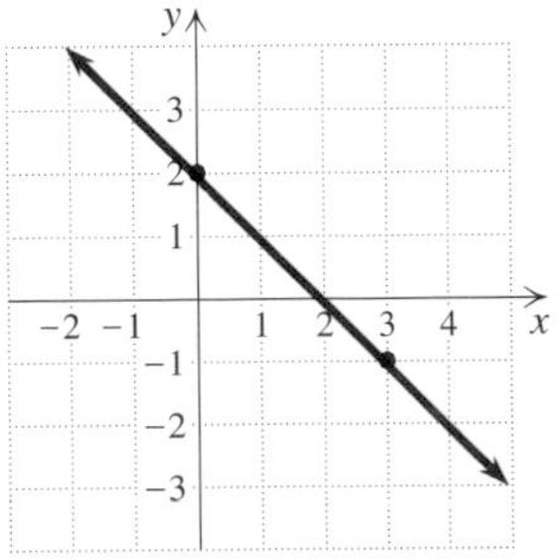

37.

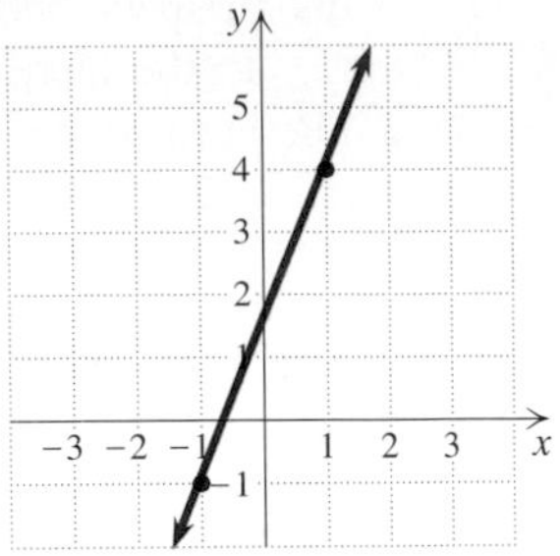

38.

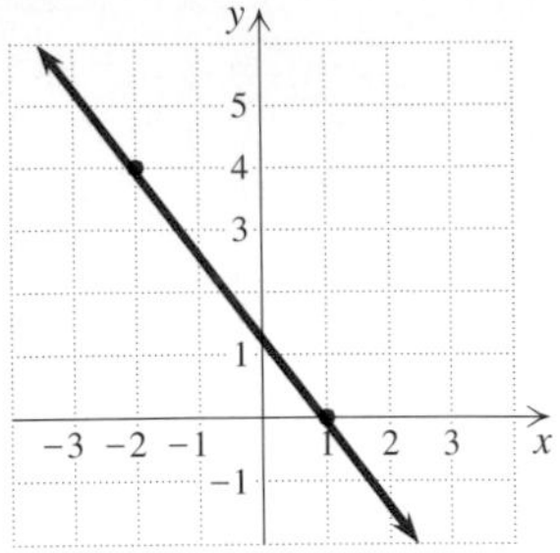

39.

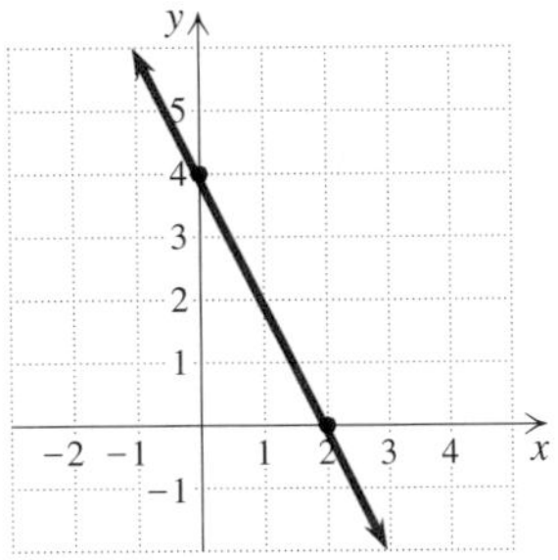

40.

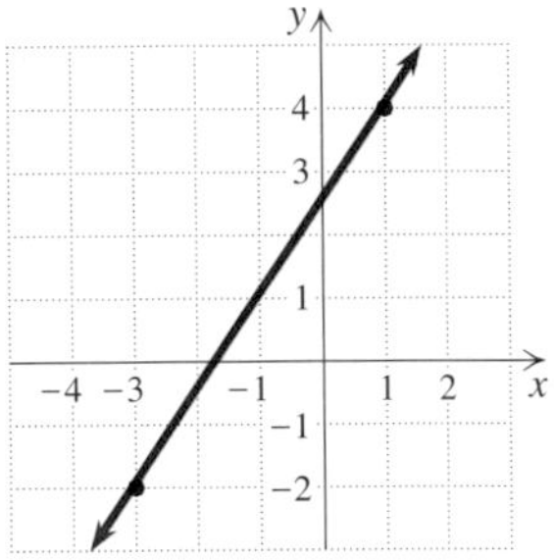

Write the standard form of the equation of the line through the given pair of points.

41. $(3, 0)$ and $(0, -4)$
42. $(-2, 0)$ and $(0, 3)$
43. $(2, 3)$ and $(-3, -1)$
44. $(4, -1)$ and $(-2, -6)$
45. $(-4, 2)$ and $(-4, 5)$
46. $(-3, 6)$ and $(9, 6)$
47. $(1, -6)$ and $(3, -6)$
48. $(5, -3)$ and $(5, 2)$

Find the slope of each line described.

49. A line parallel to $y = 0.5x - 9$
50. A line parallel to $3x - 9y = 4$
51. A line perpendicular to $3y - 3x = 7$
52. A line perpendicular to $2x - 3y = 8$
53. A line perpendicular to the line $x = 4$
54. A line parallel to $y = 5$

Write an equation in standard form with integral coefficients for each of the lines described. In each case make a sketch.

55. The line with slope 2, going through $(1, -2)$

56. The line with slope -3, going through $(-3, 4)$

57. The line through $(1, 4)$, parallel to $y = -3x$

58. The line through $(-2, 3)$ parallel to $y = \frac{1}{2}x + 6$

59. The line perpendicular to $x - 2y = 3$ and containing $(-3, 1)$

60. The line perpendicular to $3x - y = 9$ and containing $(0, 0)$

61. The line perpendicular to $x = 4$ and containing $(2, 5)$

62. The line perpendicular to $y = 9$ and containing $(-1, 3)$

63. The line through $(0, 3)$ running parallel to the line through $(-1, -2)$ and $(4, 1)$

64. The line through $(0, -2)$ running perpendicular to the line through $(-2, 1)$ and $(3, 5)$

Find the value of a in each case.

65. The line through $(3, 4)$ and $(7, a)$ has slope $2/3$.

66. The line through $(-2, a)$ and $(a, 3)$ has slope $-1/2$.

67. The line through $(-2, 3)$ and $(8, 5)$ is perpendicular to $y = ax + 2$.

68. The line through $(-1, a)$ and $(3, -4)$ is parallel to $y = ax$.

Either prove or disprove each statement. Use a graph only as a guide. Your proof should rely on algebraic calculations.

69. The points $(-1, 2)$, $(2, -1)$, $(3, 3)$, and $(-2, -2)$ are the vertices of a parallelogram.

70. The points $(-1, 1)$, $(-2, -5)$, $(2, -4)$, and $(3, 2)$ are the vertices of a parallelogram.

71. The points $(-5, -1)$, $(-3, -4)$, $(3, 0)$, and $(1, 3)$ are the vertices of a rectangle.

72. The points $(-5, -1)$, $(1, -4)$, $(4, 2)$, $(-1, 5)$ are the vertices of a square.

73. The points $(-5, 1)$, $(-2, -3)$, and $(4, 2)$ are the vertices of a right triangle.

74. The points $(-4, -3)$, $(1, -2)$, $(2, 3)$, and $(-3, 2)$ are the vertices of a rhombus.

75. The diagonals of the quadrilateral with vertices $(-3, 2)$, $(-1, -2)$, $(6, -1)$, and $(1, 4)$ are perpendicular.

76. The points $(-5, 1)$, $(-5, -4)$, $(4, 2)$, and $(-2, 3)$ are the vertices of a trapezoid.

Use a graphing calculator to solve each problem.

77. Graph $y_1 = (x - 5)/3$ and $y_2 = x - 0.67(x + 4.2)$. Do the lines appear to be parallel? Are the lines parallel?

78. Graph $y_1 = 99x$ and $y_2 = -x/99$. Do the lines appear to be perpendicular? Should they appear perpendicular?

79. Graph $y = (x^3 - 8)/(x^2 + 2x + 4)$. Use TRACE to examine points on the graph. Write a linear function for the graph. Explain why this function is linear.

80. Graph $y = (x^3 + 2x^2 - 5x - 6)/(x^2 + x - 6)$. Use TRACE to examine points on the graph. Write a linear function for the graph. Factor $x^3 + 2x^2 - 5x - 6$ completely.

Solve each problem.

81. *Celsius to Fahrenheit Formula* Fahrenheit temperature F is a linear function of Celsius temperature C. The ordered pair $(0, 32)$ is an ordered pair of this function because 0°C is equivalent to 32°F, the freezing point of water. The ordered pair $(100, 212)$ is also an ordered pair of this function because 100°C is equivalent to 212°F, the boiling point of water. Use the two given points and the point-slope formula to write F as a function of C. Find the Fahrenheit temperature of an oven at 150°C.

82. *Cost of Business Cards* Speedy Printing charges \$23 for 200 deluxe business cards and \$35 for 500 deluxe business cards. Given that the cost is a linear function of the number of cards printed, find a formula for that function and find the cost of 700 business cards.

83. *Volume Discount* Mona Kalini gives a walking tour of Honolulu to one person for \$49. To increase her business, she advertised at the National Orthodontist Convention that she would lower the price by \$1 per person for each additional person. Write the cost per person c as a function of the number of people n on the tour. How much does she make for a tour with 40 people?

84. *Ticket Pricing* At \$10 per ticket, Willie Williams and the Wranglers will fill all 8000 seats in the Assembly Center. The manager knows that for every \$1 increase in the price, 500 tickets will go unsold. Write the number of tickets sold, n, as a function of the ticket price, p. How much money will be taken in if the tickets are \$20 each?

85. *Saving for Retirement* Financial advisors at Fidelity Investments, Boston, use the accompanying table as a measure of whether a client is on the road to a comfortable retirement.

Age	Years of salary saved
35	0.5
40	1.0
45	1.5
50	2.0
55	3.0
60	4.0
65	5.0

Write the years of salary saved y as a piecewise function of age a, assuming that each piece is linear. How many years of salary should a person have saved at age 47? At age 63?

86. *In-House Training* The accompanying graph shows that the percentage of U.S. workers receiving training by their employers went from 5% in 1981 to 16% in 1995 to a projected 25% in 2000. Assuming that the percentage is growing linearly as shown in the figure, write the percentage as a piecewise function of the year.

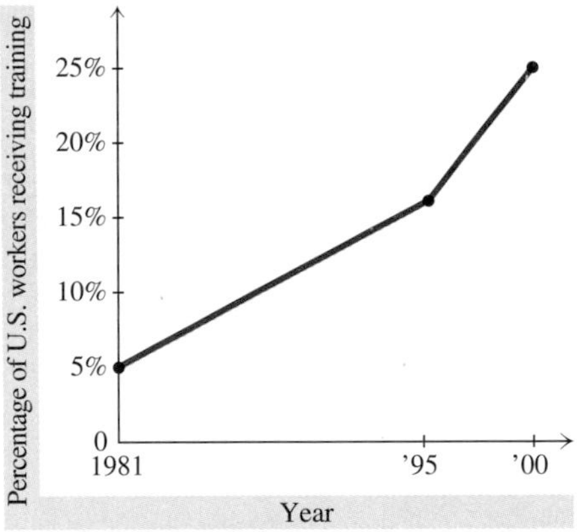

Figure for Exercise 86

87. *Lindbergh's Practical Economical Air Speed* Lindbergh and Hall spent hours graphing variables such as flight speed, fuel consumption, and weight of the plane. In a graph like the one shown here, they experimented with the relationship between *practical economical air speed* and distance, as measured from the starting point of the flight. They estimated that at the starting point the practical economical air speed was 95 mph and at 4000 statute miles from the starting point it was 75 mph. Assume that the practical economical air speed S is a linear function of the distance D from the starting point and find a formula for that function. (The moral of the story: Start out fast when you have maximum load to lift, and decrease air speed as fuel burns off and load lightens.)

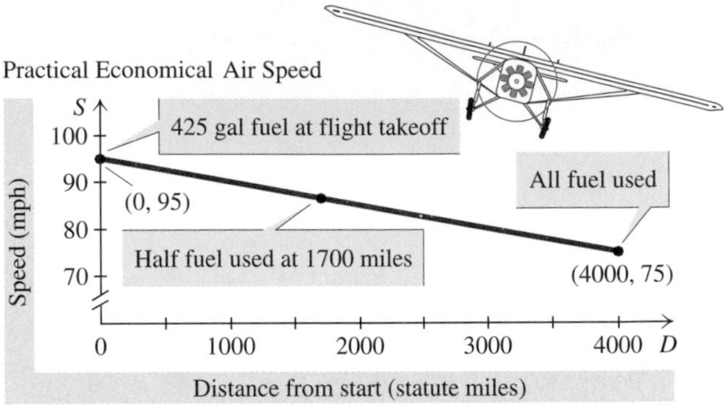

Figure for Exercise 87

88. *Lindbergh's Air Speed Over Newfoundland* Lindbergh did not stick precisely to his calculations during his flight, but used them as a guide. By the time he reached Newfoundland, he had flown 11 hr and covered 1100 mi. Based on the formula in Exercise 87, what should his air speed have been at that point? (In his flying log, Lindbergh recorded his actual air speed over Newfoundland as 98 mph and estimated that with the wind at his tail he was making a mile every 30 sec.)

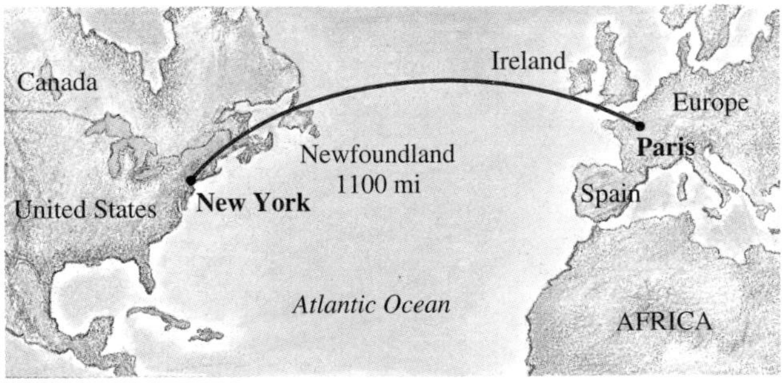

Figure for Exercise 88

89. *Computers and Printers* An office manager will spend a total of \$60,000 on computers at \$2000 each and printers at \$1500 each. Write the number of computers purchased as a function of the number of printers purchased. Find and interpret the slope.

90. *Carpenters and Helpers* Because a job was finished early, a contractor will distribute a total of \$2400 in bonuses to 9 carpenters and 3 helpers. The carpenters all get the same amount and the helpers all get the same amount. Write the amount of a helper's bonus as a function of the amount of a carpenter's bonus. Find and interpret the slope.

91. *Camaro Depreciation* The average base retail price in the spring of 1997 for a used Camaro Z28 that is either one, two, three, four, or five-years-old is given in the following table. (Edmund's 1997 Used Car Prices)

Age (years)	Retail price (\$)
1	15,400
2	11,940
3	10,825
4	10,065
5	8,115

a. Use linear regression on a graphing calculator to find the equation of the line that best fits the given data and graph the line. See Example 3 in Section 1.2.
b. Find and interpret the slope of the line.
c. Use the regression line to predict the average retail price for a seven-year-old Z28.
d. At what age is a Z28 worthless?

92. *Declining Defense* The total manpower in all branches of the U.S. military for the years 1990 through 1995 is given in the following table. (Department of Defense)

Year	Manpower (in thousands)
1990	2044
1991	1986
1992	1807
1993	1705
1994	1610
1995	1518

a. Use linear regression on a graphing calculator to find the equation of the line that best fits the given data and graph the line.
b. Find and interpret the slope of the regression line.
c. If the reduction in military manpower continues, then in what year will the U.S. military be reduced to one-half of its 1995 level?
d. How does the average rate of change of the manpower of the U.S. military from 1990 to 1995 compare to the slope of the regression line?

• CALCULUS •

93. *Marginal Cost* A company estimates that its weekly cost $C(x)$ for manufacturing x items is \$8000 plus \$50 per item. Write a formula for $C(x)$. The marginal cost function $MC(x)$ is the difference quotient for the function $C(x)$ with $h = 1$. Find $MC(x)$. For what values of x is $MC(x)$ increasing, decreasing, or constant?

94. *Average Cost* The cost to a company for producing x cellular telephones per day is given by the function $C(x) = 125x + 1000$.
a. The function $AC(x) = C(x)/x$ gives the average cost of production per telephone. Find $AC(10)$ and $AC(50)$. What happens to the average cost as x gets large?
b. The marginal average cost function, $MAC(x)$, is the difference quotient for the function $AC(x)$ with $h = 1$. Find $MAC(x)$, $MAC(5)$, and $MAC(20)$. What happens to the marginal average cost as x gets large?

For Writing/Discussion

95. *Summing Angles* Consider the angles shown in the accompanying figure. Show that the degree measure of angle A plus the degree measure of angle B is equal to the degree measure of angle C. (This could be tricky.)

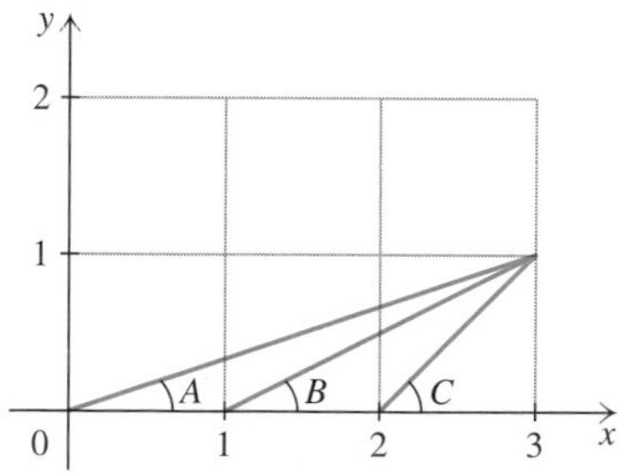

Figure for Exercise 95

96. *Radius of a Circle* Find the radius of the circle shown in the figure on the next page. (This is tricky.)

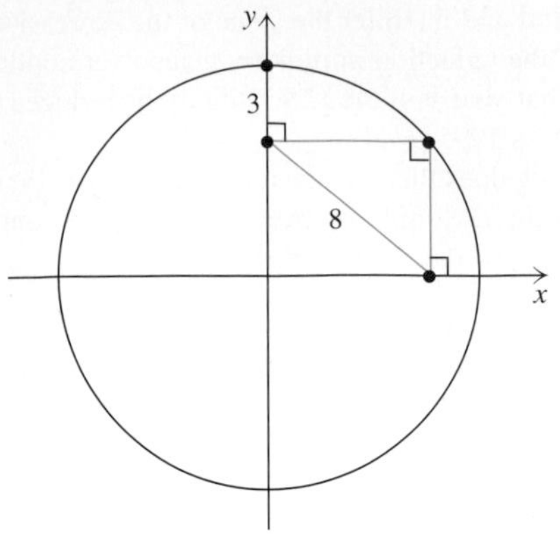

Figure for Exercise 96

97. *Equal Slopes* Show that if $y = mx + b_1$ and $y = mx + b_2$ are equations of lines with equal slopes, but $b_1 \neq b_2$, then they have no point in common. (*Hint:* Assume that they have a point in common and see that this assumption leads to a contradiction.)

98. *Unequal Slopes* Show that the lines $y = m_1x + b_1$ and $y = m_2x + b_2$ intersect at a point with x-coordinate $(b_2 - b_1)/(m_1 - m_2)$ provided $m_1 \neq m_2$. Explain how this exercise and the previous exercise prove the theorem that two nonvertical lines are parallel if and only if they have equal slopes.

99. *Cooperative Learning* Work in a small group to find and simplify the difference quotient for the function $f(x) = x^2$. Evaluate the difference quotient for $h = 0$ and $x = -1, 0, 1, 2$, and 3, and draw the graph of $f(x) = x^2$. Do these values give any indication of how fast the curve is increasing or decreasing at the points with x-coordinates $-1, 0, 1, 2$, and 3? Use the difference quotient to define the concept of "slope" of a curve $y = f(x)$ at any point on the curve.

Linking Concepts

For Individual or Group Explorations

Income Tax Reform *One idea for income tax reform is that of a negative income tax. Under this plan families with an income below a certain level receive a payment from the government in addition to their income while families above a certain level pay taxes. In one proposal a family with an earned income of \$6000 would receive a \$4000 payment, giving the family a disposable income of \$10,000. A family with an earned income of \$24,000 would pay \$2000 in taxes, giving the family a disposable income of \$22,000. A family's disposable income D is a linear function of the family's earned income E.*

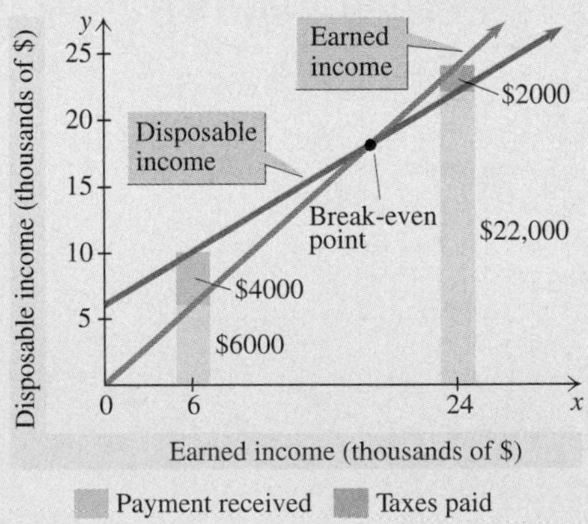

a) Use the given data to write a formula for D as a linear function of E.

b) What is the disposable income for a family with an earned income of \$60,000?

c) Find the break-even point, the income at which a family pays no taxes and receives no payment.

d) What percentage of their earned income is paid in taxes for a family with an earned income of \$25,000? \$100,000? \$2,000,000?

e) What is the maximum percentage of earned income that anyone will pay in taxes?

f) Compare your answer to part (e) with current tax laws. Does this negative income tax favor the rich, poor, or middle class?

3.2 Quadratic Functions

In Section 3.1 we studied linear functions, which are polynomial functions of degree 1. Quadratic functions are polynomial functions of degree 2.

Definition: Quadratic Function

A **quadratic function** is any function of the form

$$f(x) = ax^2 + bx + c,$$

where a, b, and c are real numbers and $a \neq 0$.

In Chapter 2 we graphed some quadratic functions, but we did not call them quadratic functions. In this section we will study quadratic functions in greater detail.

Using Transformations to Graph a Quadratic Function

We saw in Section 2.4 that the graph of any function of the form $f(x) = a(x - h)^2 + k$ is a transformation of the graph of $f(x) = x^2$. By completing the square, we can write any quadratic function in the form $f(x) = a(x - h)^2 + k$, and then quickly sketch the graph using transformations.

EXAMPLE 1 Graphing a quadratic function

Rewrite $f(x) = -2x^2 - 4x + 3$ in the form $f(x) = a(x - h)^2 + k$ and sketch its graph.

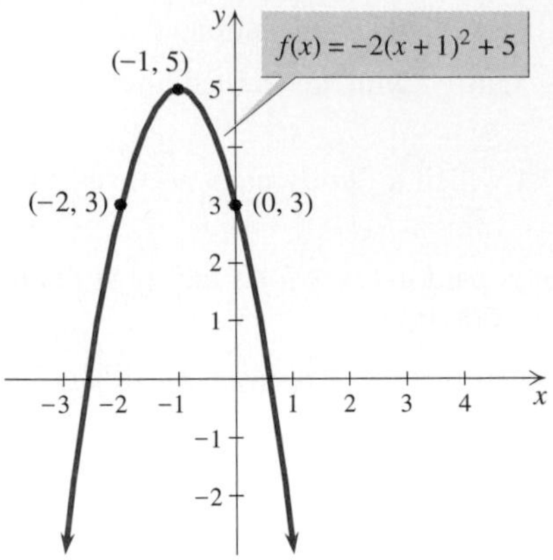

Figure 3.13

Solution

Start by completing the square:

$$\begin{aligned} f(x) &= -2(x^2 + 2x) + 3 && \text{Factor out } -2 \text{ from first two terms.} \\ &= -2(x^2 + 2x + 1 - 1) + 3 && \text{Complete the square for } x^2 + 2x. \\ &= -2(x^2 + 2x + 1) + 2 + 3 && \text{Remove } -1 \text{ from the parentheses.} \\ &= -2(x + 1)^2 + 5 \end{aligned}$$

The function is now in the form $f(x) = a(x - h)^2 + k$. The number 1 indicates that the graph of $f(x) = x^2$ is translated one unit to the left. Since $a = -2$, the graph is stretched by a factor of 2 and reflected below the x-axis. Finally, the graph is translated five units upward. The graph shown in Fig. 3.13 includes the points $(0, 3)$, $(-1, 5)$, and $(-2, 3)$.

We can use completing the square as in Example 1 to write any quadratic function $f(x) = ax^2 + bx + c$ in the form $f(x) = a(x - h)^2 + k$:

$$\begin{aligned} f(x) &= ax^2 + bx + c \\ &= a\left(x^2 + \frac{b}{a}x\right) + c && \text{Factor } a \text{ out of the first two terms.} \end{aligned}$$

To complete the square for $x^2 + \frac{b}{a}x$, add and subtract $\frac{b^2}{4a^2}$ inside the parentheses:

$$\begin{aligned} f(x) &= a\left(x^2 + \frac{b}{a}x + \frac{b^2}{4a^2} - \frac{b^2}{4a^2}\right) + c \\ &= a\left(x^2 + \frac{b}{a}x + \frac{b^2}{4a^2}\right) - \frac{b^2}{4a} + c && \text{Remove } -\frac{b^2}{4a^2} \text{ from the parentheses.} \\ &= a\left(x + \frac{b}{2a}\right)^2 + \frac{4ac - b^2}{4a} && \text{Factor and get a common denominator.} \\ &= a(x - h)^2 + k && \text{Let } h = -\frac{b}{2a} \text{ and } k = \frac{4ac - b^2}{4a}. \end{aligned}$$

Since $y = ax^2 + bx + c$ is equivalent to $y = a(x - h)^2 + k$, *the graph of any quadratic function is a transformation of the graph of* $y = x^2$. The graph of any quadratic function is called a **parabola**.

Opening, Vertex, and Axis of Symmetry

If $a > 0$, the graph of $f(x) = a(x - h)^2 + k$ **opens upward**; if $a < 0$, the graph **opens downward** as shown in Fig. 3.14. Notice that h determines the amount of horizontal translation and k determines the amount of vertical translation of the graph of $f(x) = x^2$. Because of the translations, the point $(0, 0)$ on the graph of

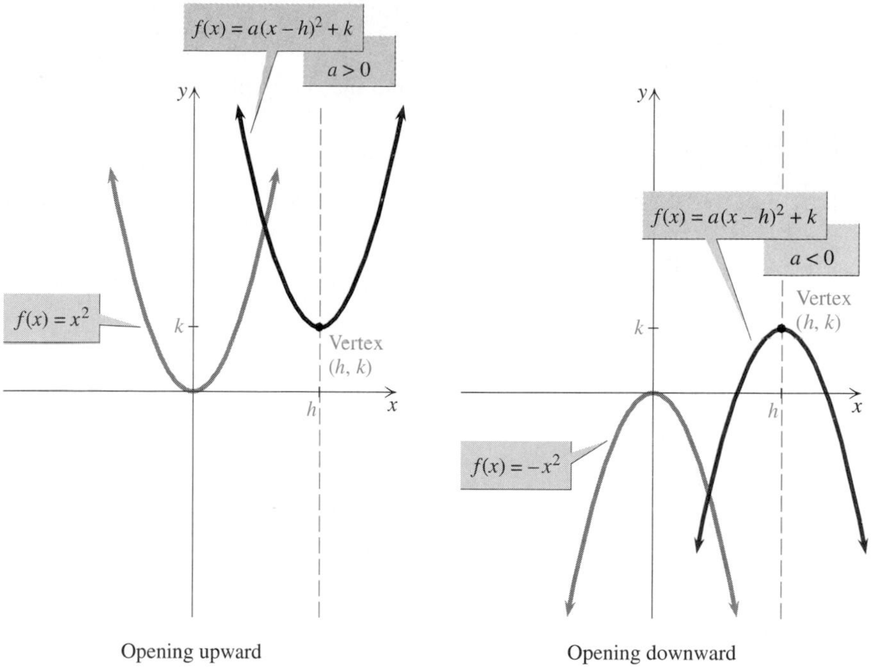

Figure 3.14

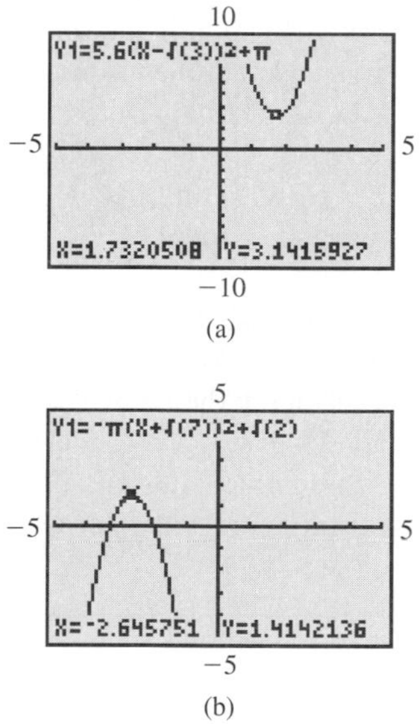

Figure 3.15

$f(x) = x^2$ moves to the point (h, k) on the graph of $f(x) = a(x - h)^2 + k$. Since $(0, 0)$ is the lowest point on the graph of $f(x) = x^2$, the lowest point on any parabola that opens upward is (h, k). Since $(0, 0)$ is the highest point on the graph of $f(x) = -x^2$, the highest point on any parabola that opens downward is (h, k). The point (h, k) is called the **vertex** of the parabola.

You can use a graphing calculator to experiment with different values for a, h, and k. Two possibilities are shown in Fig. 3.15(a) and (b). The graphing calculator allows you to see results quickly and to use values that you would not use otherwise. □

We can also find the vertex of the parabola when the function is written in the form $f(x) = ax^2 + bx + c$. The x-coordinate of the vertex is $-b/(2a)$ because we used $h = -b/(2a)$ when we obtained $y = a(x - h)^2 + k$ from $y = ax^2 + bx + c$. The y-coordinate is found by substitution.

SUMMARY Vertex of a Parabola

1. For a quadratic function in the form $f(x) = a(x - h)^2 + k$, the vertex of the parabola is (h, k).
2. For the form $f(x) = ax^2 + bx + c$, the x-coordinate of the vertex is $-b/(2a)$. The y-coordinate is $f(-b/(2a))$.

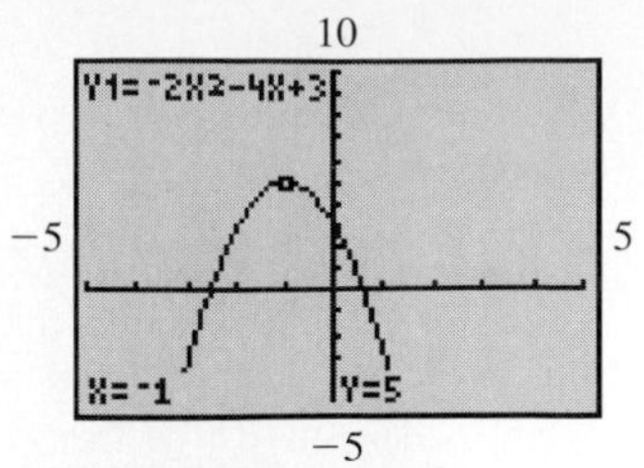

Figure 3.16

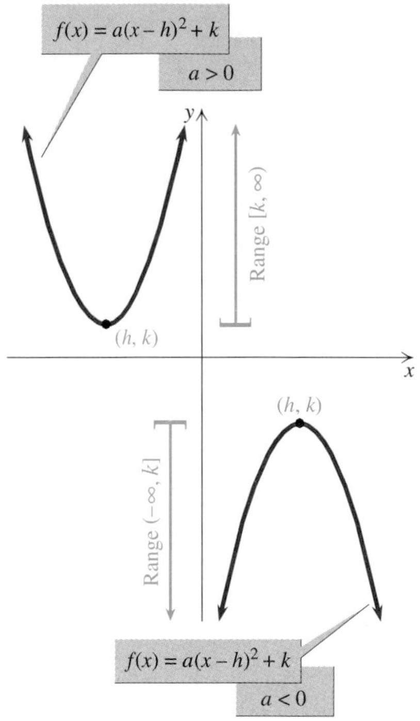

Figure 3.17

EXAMPLE 2 Finding the vertex

Find the vertex of the graph of $f(x) = -2x^2 - 4x + 3$.

Solution

Use $x = -b/(2a)$ to find the x-coordinate of the vertex:

$$x = -\frac{b}{2a} = -\frac{-4}{2(-2)} = -1$$

Now find $f(-1)$:

$$f(-1) = -2(-1)^2 - 4(-1) + 3 = 5$$

The vertex is $(-1, 5)$. In Example 1, $f(x) = -2x^2 - 4x + 3$ was rewritten as $f(x) = -2(x + 1)^2 + 5$. In this form we see immediately that the vertex is $(-1, 5)$.

The graph in Fig. 3.16 supports these conclusions.

The domain of every quadratic function $f(x) = a(x - h)^2 + k$ is the set of real numbers, $(-\infty, \infty)$. The range of a quadratic function is determined from the second coordinate of the vertex. If $a > 0$, the range is $[k, \infty)$ and k is called the **minimum value of the function**. The function is decreasing on $(-\infty, h)$ and increasing on (h, ∞). See Fig. 3.17. If $a < 0$, the range is $(-\infty, k]$ and k is called the **maximum value of the function**. The function is increasing on $(-\infty, h)$ and decreasing on (h, ∞).

The graph of $f(x) = x^2$ is symmetric about the y-axis, which runs vertically through the vertex of the parabola. Since this symmetry is preserved in transformations, the graph of any quadratic function is symmetric about the vertical line through its vertex. The vertical line $x = -b/(2a)$ is called the **axis of symmetry** for the graph of $f(x) = ax^2 + bx + c$. Identifying these characteristics of a parabola before drawing a graph makes graphing easier and more accurate.

• CALCULUS •

EXAMPLE 3 Identifying the characteristics of a parabola

For each parabola, determine whether the parabola opens upward or downward, and find the vertex, axis of symmetry, and range of the function. Find the maximum or minimum value of the function and the intervals on which the function is increasing or decreasing.

a) $y = -2(x + 4)^2 - 8$ **b)** $y = 2x^2 - 4x - 9$

Solution

a) Since $a = -2$, the parabola opens downward. In $y = a(x - h)^2 + k$, the vertex is (h, k). So the vertex is $(-4, -8)$. The axis of symmetry is the vertical line through the vertex, $x = -4$. Since the parabola opens downward from

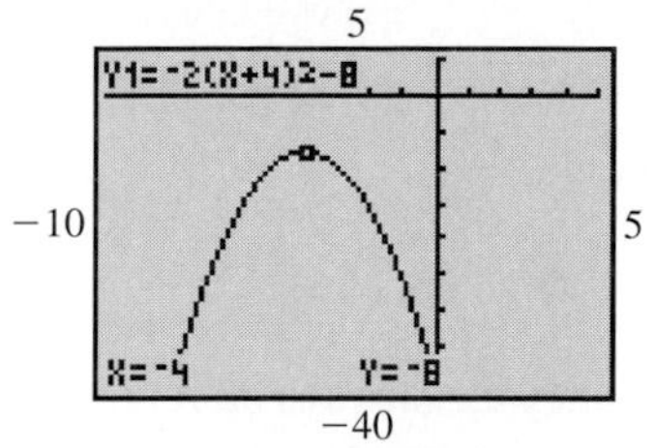

Figure 3.18

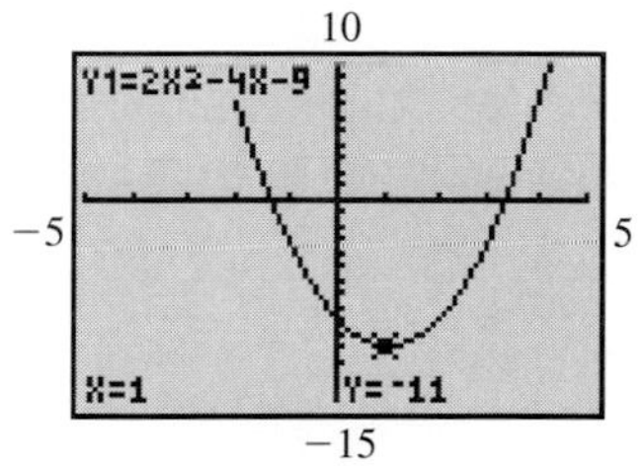

Figure 3.19

$(-4, -8)$, the range of the function is $(-\infty, -8]$. The maximum value of the function is -8, and the function is increasing on $(-\infty, -4)$ and decreasing on $(-4, \infty)$.

The graph in Fig. 3.18 supports these results. □

b) Since $a = 2$, the parabola opens upward. In the form $y = ax^2 + bx + c$, the x-coordinate of the vertex is $x = -b/(2a)$. In this case,

$$x = \frac{-b}{2a} = \frac{-(-4)}{2(2)} = 1.$$

Use $x = 1$ to find $y = 2(1)^2 - 4(1) - 9 = -11$. So the vertex is $(1, -11)$, and the axis of symmetry is the vertical line $x = 1$. Since the parabola opens upward, the vertex is the lowest point and the range is $[-11, \infty)$. The function is decreasing on $(-\infty, 1)$ and increasing on $(1, \infty)$. The minimum value of the function is -11.

The graph in Fig. 3.19 supports these results.

Intercepts

The x-intercepts and the y-intercept are important points on the graph of a parabola. The x-intercepts are used in solving quadratic inequalities and the y-intercept is the starting point on the graph of a function whose domain is the nonnegative real numbers. The y-intercept is easily found by letting $x = 0$. The x-intercepts are found by letting $y = 0$ and solving the resulting quadratic equation.

EXAMPLE 4 Finding the intercepts

Find the y-intercept and the x-intercepts for each parabola and sketch the graph of each parabola.

a) $y = -2(x + 4)^2 - 8$ **b)** $y = 2x^2 - 4x - 9$

Solution

a) If $x = 0$, $y = -2(0 + 4)^2 - 8 = -40$. The y-intercept is $(0, -40)$. Because the graph is symmetric about the line $x = -4$, the point $(-8, -40)$ is also on the graph. Since the parabola opens downward from $(-4, -8)$, which is below the x-axis, there are no x-intercepts. If we try to solve $-2(x + 4)^2 - 8 = 0$ to find the x-intercepts, we get $(x + 4)^2 = -4$, which has no real solution. The graph is shown in Fig. 3.20.

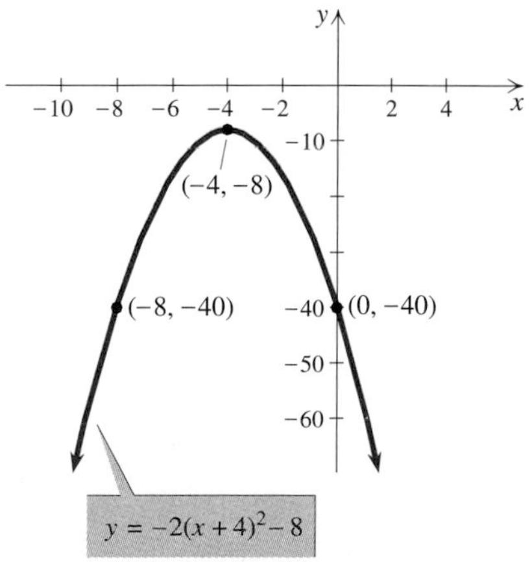

Figure 3.20

b) If $x = 0$, $y = 2(0)^2 - 4(0) - 9 = -9$. The y-intercept is $(0, -9)$. From Example 3(b), the vertex is $(1, -11)$. Because the graph is symmetric about the line $x = 1$, the point $(2, -9)$ is also on the graph. The x-intercepts are found by solving $2x^2 - 4x - 9 = 0$:

$$x = \frac{4 \pm \sqrt{(-4)^2 - 4(2)(-9)}}{2(2)} = \frac{4 \pm \sqrt{88}}{4} = \frac{2 \pm \sqrt{22}}{2}$$

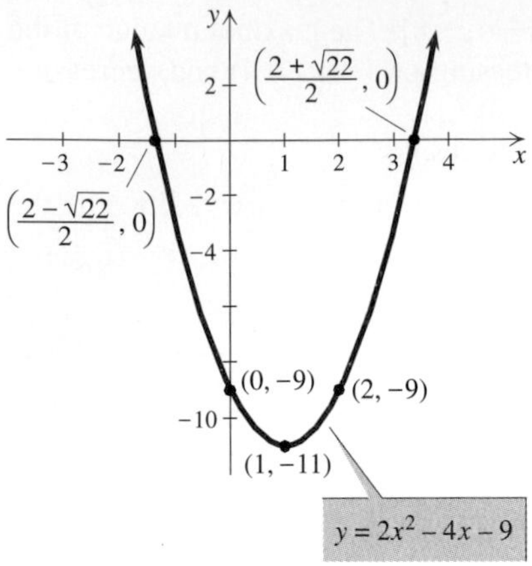

Figure 3.21

The x-intercepts are $\left(\frac{2 + \sqrt{22}}{2}, 0\right)$ and $\left(\frac{2 - \sqrt{22}}{2}, 0\right)$. See Fig. 3.21.

Note that if $y = ax^2 + bx + c$ has x-intercepts, they can always be found by using the quadratic formula. The x-coordinates of the x-intercepts are

$$x = \frac{-b \pm \sqrt{b^2 - 4ac}}{2a} = \frac{-b}{2a} \pm \frac{\sqrt{b^2 - 4ac}}{2a}.$$

Note how the axis of symmetry, $x = -b/(2a)$, appears in this formula. The x-intercepts are on opposite sides of the graph and are equidistant from the axis of symmetry.

Solving Quadratic Inequalities

In Section 2.4 we saw that inequalities could be solved by graphing. Since we can graph any quadratic function, we can solve any quadratic inequality by graphing. As shown in the next example, the key points for solving a quadratic inequality are the x-intercepts.

EXAMPLE 5 Solving a quadratic inequality by graphing

Solve each inequality by graphing an appropriate function.

a) $-2(x + 4)^2 > 8$ **b)** $2x^2 - 4x - 9 < 0$

Solution

a) The inequality is equivalent to $-2(x + 4)^2 - 8 > 0$. The graph of $y = -2(x + 4)^2 - 8$ is shown in Fig. 3.20. Since no y-coordinate is greater than 0 on this graph, there is no solution to this inequality.

b) The graph of $y = 2x^2 - 4x - 9$ is shown in Fig. 3.21. The y-coordinates on the graph are less than zero provided that the x-coordinates are between the x-intercepts. So the solution to the inequality is the interval

$$\left(\frac{2 - \sqrt{22}}{2}, \frac{2 + \sqrt{22}}{2}\right).$$

Applications of Maximum and Minimum

In applications we often seek to minimize cost, maximize profit, or maximize area. If one variable is a quadratic function of another, then the maximum or minimum value of the dependent variable occurs at the vertex of the parabola.

• CALCULUS •

EXAMPLE 6 Maximizing area of a rectangle

If 100 meters of fencing will be used to fence a rectangular region, then what dimensions for the rectangle will maximize the area of the region?

Solution

Since $P = 2(L + W)$ for a rectangle, the length plus the width is 50 meters. If we let x represent the length of the rectangular region, then $50 - x$ represents the width. See Fig. 3.22. Since $A = LW$ for a rectangle, $A = x(50 - x) = -x^2 + 50x$. So the area is a quadratic function of the length. The graph of this function is the parabola in Fig. 3.23.

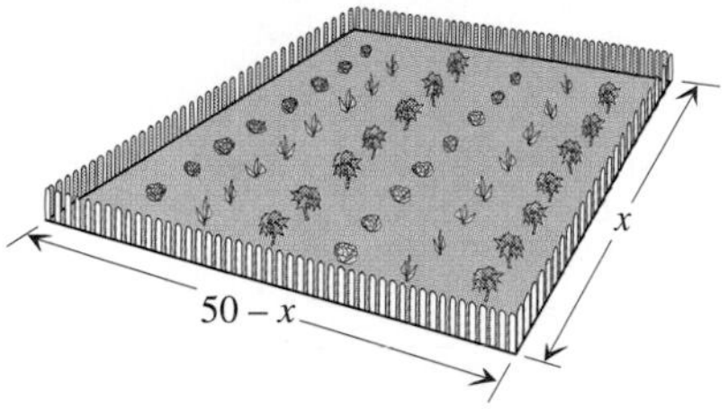

Figure 3.22

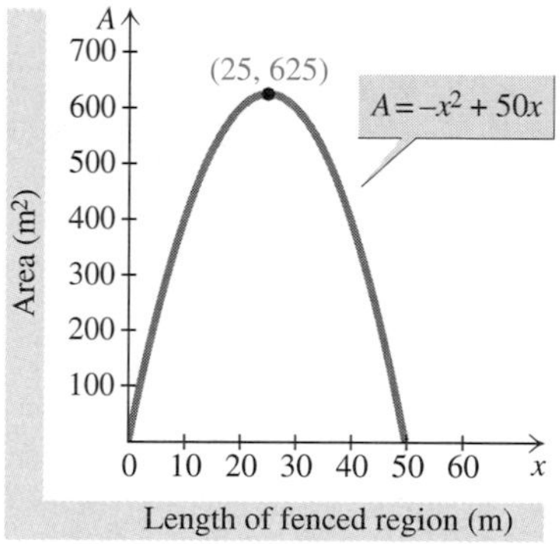

Figure 3.23

Since the parabola opens downward, the maximum value of A occurs when

$$x = \frac{-b}{2a} = \frac{-50}{2(-1)} = 25.$$

So the length should be 25 meters and the width should be 25 meters. The rectangle that gives the maximum area is actually a square with an area of 625 m^2.

FOR THOUGHT True or False? Explain.

1. The domain and range of a quadratic function are $(-\infty, \infty)$.
2. The vertex of the graph of $y = 2(x - 3)^2 - 1$ is $(3, 1)$.
3. The graph of $y = -3(x + 2)^2 - 9$ has no x-intercepts.
4. The maximum value of y in the function $y = -4(x - 1)^2 + 9$ is 9.
5. For $y = 3x^2 - 6x + 7$, the value of y is at its minimum when $x = 1$.
6. The graph of $f(x) = 9x^2 + 12x + 4$ has one x-intercept and one y-intercept.

7. The graph of every quadratic function has exactly one y-intercept.

8. The inequality $\pi(x - \sqrt{3})^2 + \pi/2 \leq 0$ has no solution.

9. The maximum area of a rectangle with fixed perimeter p is $p^2/16$.

10. The function $f(x) = (x - 3)^2$ is increasing on the interval $[-3, \infty)$.

3.2 EXERCISES

Tape 7 Disk

Write each quadratic function in the form $y = a(x - h)^2 + k$ and sketch its graph.

1. $y = x^2 - 3x$

2. $y = -x^2 + x$

3. $y = 2x^2 - 12x + 22$

4. $y = -2x^2 - 4x - 5$

5. $y = -\frac{1}{2}x^2 + x + \frac{5}{2}$

6. $y = \frac{1}{2}x^2 - x + 1$

7. $y = x^2 + 3x + \frac{5}{2}$

8. $y = x^2 - x + 1$

9. $y = -2x^2 + 3x - 1$

10. $y = -2x^2 + x$

11. $y = -3x^2 + 2x$

12. $y = 3x^2 + 4x + 2$

Find the vertex of the graph of each quadratic function.

13. $f(x) = 3x^2 - 12x + 1$

14. $f(x) = -2x^2 - 8x + 9$

15. $f(x) = -3(x - 4)^2 + 1$

16. $f(x) = \frac{1}{2}(x + 6)^2 - \frac{1}{4}$

17. $y = -\frac{1}{2}x^2 - \frac{1}{3}x$

18. $y = \frac{1}{4}x^2 + \frac{1}{2}x - 1$

From the graph of each parabola, determine whether the parabola opens upward or downward, and find the vertex, axis of symmetry, and range of the function. Find the maximum or minimum value of the function and the intervals on which the function is increasing or decreasing.

19.

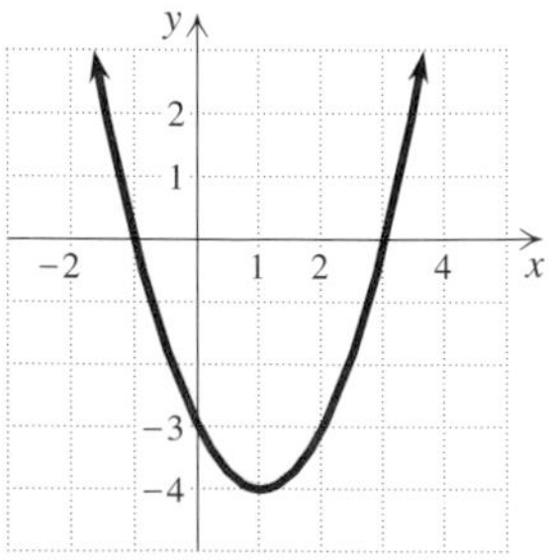

20.

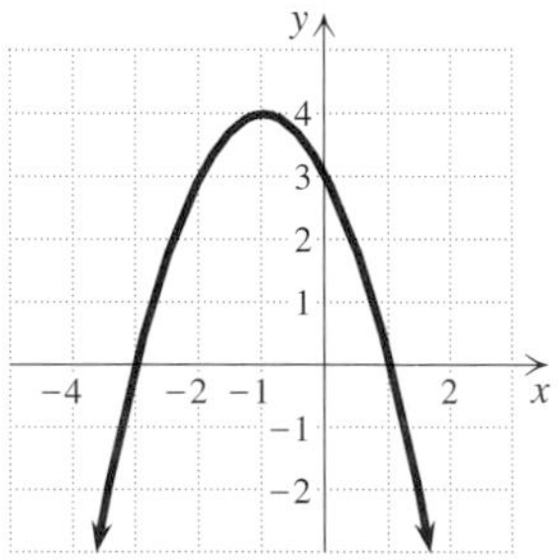

Find the range of each quadratic function and the maximum or minimum value of the function. Identify the intervals on which each function is increasing or decreasing.

21. $y = (x - 1)^2 - 1$

22. $y = (x + 3)^2 + 4$

23. $y = \sqrt{3} - x^2$

24. $y = \pi - x^2$

25. $f(x) = \frac{1}{2}(x - 3)^2 + 4$

26. $f(x) = -\frac{1}{3}(x + 6)^2 + 37$

27. $y = -\frac{3}{4}\left(x - \frac{1}{2}\right)^2 + 9$

28. $y = \frac{3}{2}\left(x - \frac{1}{3}\right)^2 - 6$

29. $f(x) = -2x^2 + 6x + 9$

30. $f(x) = -3x^2 - 9x + 4$

31. $y = \sqrt{2}x^2 + \sqrt{8}x - 2$

32. $f(x) = x^2 - \sqrt{2}x - 3$

Identify the vertex, axis of symmetry, y-intercept, x-intercepts, and opening of each parabola, then sketch the graph.

33. $y = x^2 - 3$

34. $y = 8 - x^2$

35. $f(x) = x^2 - \pi x$

36. $f(x) = 2\pi x - x^2$

37. $f(x) = x^2 + 6x + 9$

38. $f(x) = x^2 - 6x$

39. $f(x) = (x - 3)^2 - 4$

40. $f(x) = (x + 1)^2 - 9$

41. $y = -3(x - 2)^2 + 12$

42. $y = -2(x + 3)^2 + 8$

43. $y = -2x^2 + 4x + 1$

44. $y = -x^2 + 2x - 6$

Identify the solution set to each quadratic inequality by inspecting the graphs of $y = x^2 - 2x - 3$ and $y = -x^2 - 2x + 3$ shown here.

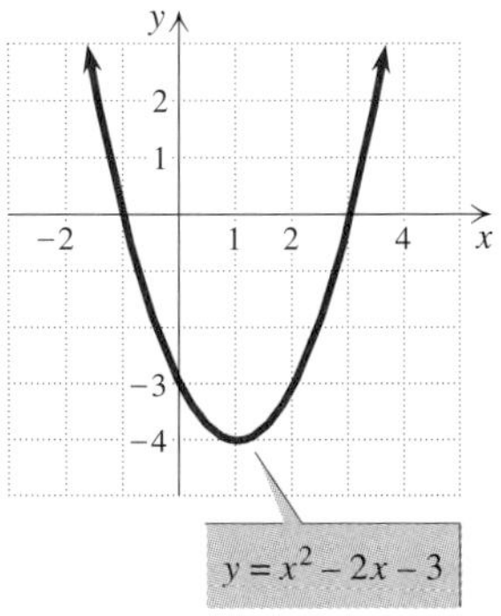

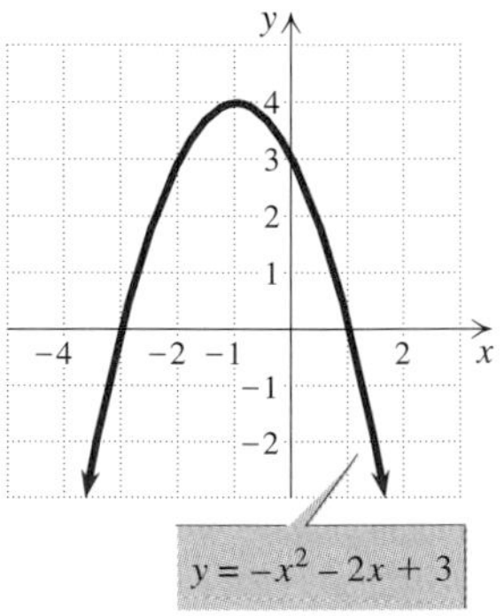

45. $x^2 - 2x - 3 \geq 0$

46. $x^2 - 2x - 3 < 0$

47. $-x^2 - 2x + 3 > 0$

48. $-x^2 - 2x + 3 \leq 0$

49. $x^2 + 2x \leq 3$

50. $x^2 \leq 2x + 3$

Solve each quadratic inequality by graphing an appropriate quadratic function.

51. $x^2 - 5x + 6 \leq 0$

52. $x^2 + 3x < 4$

53. $x^2 + 7 > 6x$

54. $2x - x^2 \leq -2$

55. $x^2 - 6x \geq -10$

56. $-2x^2 + 4x > 3$

Solve each inequality by using a graphing calculator to find approximate x-intercepts. Round to two decimal places.

57. $x^2 + 0.1x - 0.021 < 0$

58. $x^2 + 60x + 899.5 < 0$

For each given quadratic polynomial, use a graphing calculator to graph the corresponding quadratic function and locate the x-intercepts. Then factor the quadratic polynomial. (*Hint:* $y = (x - m)(x - n)$ has x-intercepts $(m, 0)$ and $(n, 0)$.)

59. $x^2 - 24x - 3456$

60. $2x^2 - 503x + 750$

61. $2x^2 - 39x - 2970$

62. $24x^2 - 85x - 550$

Find the axis of symmetry for each parabola whose equation is given. Use the axis of symmetry to name a point on the parabola that has the same y-coordinate as the given point.

63. $y = x^2$, $(2, 4)$

64. $y = 5 - x^2$, $(3, -4)$

65. $y = 3(x - 1)^2 + 5$, $(2, 8)$

66. $y = -2(x + 3)^2 + 1$, $(0, -17)$

67. $y = 3x^2 - 6x + 7$, $(-1, 16)$

68. $y = 5x^2 - 5x + 12$, $(2, 22)$

69. $y = -6x^2 - 6x - 1$, $(-2, -13)$

70. $y = 2x^2 - x - 3$, $(-1/2, -2)$

Solve each problem.

71. Find the value of b so that the vertex of the parabola $y = -3x^2 + bx - 16$ is $(-2, -4)$.

72. Find the value of a so that the vertex of the parabola $y = ax^2 + 10x - 3$ is $(1, 2)$.

73. Write the equation of the parabola that has vertex $(1, -3)$ and y-intercept $(0, 2)$.

74. Write the equation of the parabola that has x-intercepts $(1, 0)$ and $(5, 0)$ and y-intercept $(0, 6)$.

75. Write the equation of the parabola that has vertex $(-3, 2)$ and x-intercept $(-1, 0)$.

76. Write the equation of the parabola that has vertex $(3, -5)$ and goes through the point $(5, -2)$.

77. Find two numbers that have the maximum possible product and a sum of 7.

78. What is the maximum possible product that can be attained by two numbers with a sum of -8?

• CALCULUS •

79. *Cross Section of a Gutter* Seth has a piece of aluminum that is 10 in. wide and 12 ft long. He plans to form a gutter with a rectangular cross section and an open top by folding up the sides as shown in the figure. What dimensions of the gutter would maximize the amount of water that it can hold?

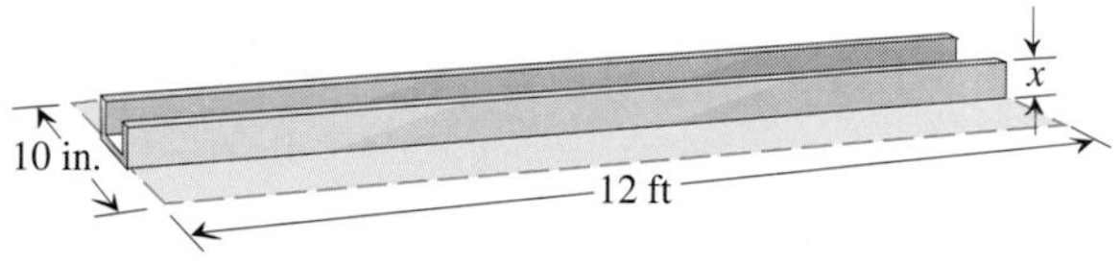

Figure for Exercise 79

80. *Maximum Volume of a Cage* Sharon has a 12-ft board that is 12 in. wide. She wants to cut it into five pieces to make a cage for two pigeons, as shown in the figure. The front and back will be covered with chicken wire. What should be the dimensions of the cage to maximize the volume and use all of the 12-ft board?

Figure for Exercise 80

81. *Lindbergh's Most Economical Air Speed* Lindbergh and Hall used a graph similar to the one shown here to determine the air speed at which the number of miles per pound of fuel, M, would be at a maximum for a fully loaded plane. The curve shown was determined by testing the loaded plane at Camp Kearney, San Diego, in 1927. Using empirical data, Lindbergh figured that the most economical air speed would occur at the highest point on the curve, at roughly 97 mph. In the figure, the curve

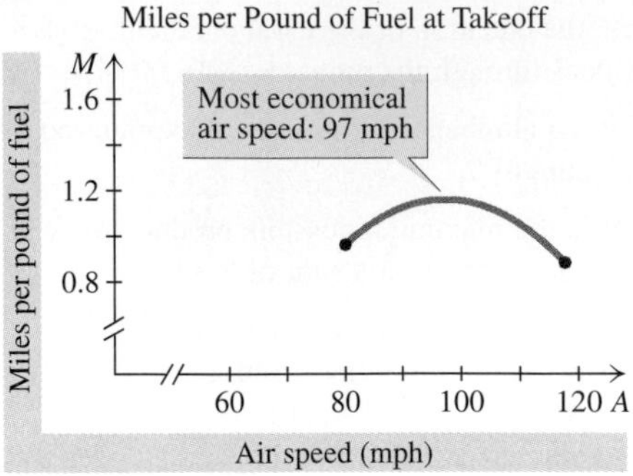

Figure for Exercise 81

appears to be a parabola. If we assume that M is a quadratic function of air speed A and is determined by the equation $M = -0.000653A^2 + 0.127A - 5.01$, then what value of A would maximize M? How many gallons of fuel did Lindbergh burn per hour if he flew at 97 mph and got 1.2 mi/lb of fuel, which weighed 6.12 lb/gal?

82. *Catching Cod* The nominal catch of Atlantic cod has been declining for the last three decades as shown in the accompanying figure (World Resources, 1996–1997). We can model the catch with the quadratic function

$$y = -0.003x^2 + 0.120x + 1.909$$

where x is the number of years since 1950 and y is the catch in millions of metric tons.

a. Find the year in which the catch reached its maximum.
b. Find the year in which the catch will be 0.
c. For what years was the catch increasing? Decreasing?
d. When was the catch greater than 2.5 million metric tons?

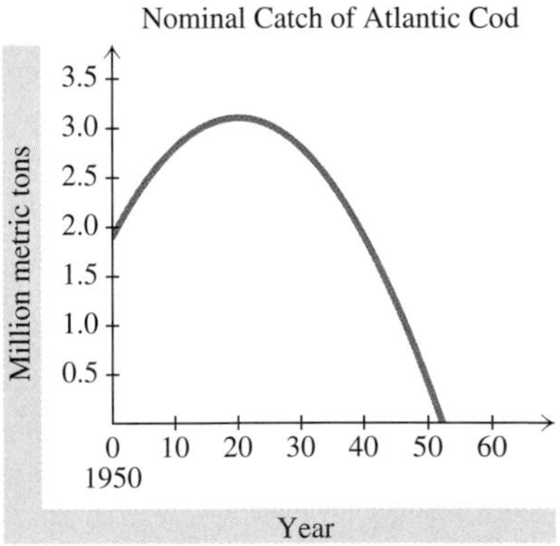

Figure for Exercise 82

83. *Maximum Height of a Football* If a football is kicked straight up with an initial velocity of 128 ft/sec from a height of 5 ft, then its height above the earth is a function of time given by $h(t) = -16t^2 + 128t + 5$. What is the maximum height reached by this ball?

84. *Maximum Height of a Ball* If a juggler can toss a ball into the air at a velocity of 64 ft/sec from a height of 6 ft, then what is the maximum height reached by the ball?

• CALCULUS •

85. *Maximizing Revenue* Mona Kalini gives a walking tour of Honolulu to one person for \$49. To increase her business, she advertised at the National Orthodontist Convention that she would lower the price by \$1 per person for each additional person, up to 49 people. Write the price per person p as a function of the number of people n. See Exercise 83 of Section 3.1. Her revenue is the product of n and p. Write her revenue as a function of the number of people on the tour. What number of people on the tour would maximize her revenue? What is the maximum revenue for her tour?

86. *Concert Tickets* At \$10 per ticket, Willie Williams and the Wranglers will fill all 8000 seats in the Assembly Center. The manager knows that for every \$1 increase in the price, 500 tickets will go unsold. Write the number of tickets sold n as a function of ticket price p. See Exercise 84 of Section 3.1. Write the total revenue as a function of the ticket price. What ticket price will maximize the revenue?

87. *Variance of the Number of Smokers* If p is the probability that a randomly selected person in Chicago is a smoker, then $1 - p$ is the probability that the person is not a smoker. The variance of the number of smokers in a random sample of 50 Chicagoans is $50p(1 - p)$. What value of p maximizes the variance?

88. *Rate of Flu Infection* In a town of 5000 people the daily rate of infection with a flu virus varies directly with the product of the number of people who have been infected and the number of people not infected. When 1000 people have been infected, the flu is spreading at a rate of 40 new cases per day. For what number of people infected is the daily rate of infection at its maximum?

89. *Altitude and Pressure* At an altitude of h feet the atmospheric pressure a (atm) is given by

$$a = 3.89 \times 10^{-10}h^2 - 3.48 \times 10^{-5}h + 1$$

(Encyclopedia of Sports Science, 1997).

a. Graph this quadratic function.

b. In 1953, Edmund Hillary of New Zealand was the first person to climb Mount Everest (see accompanying figure). Was the atmospheric pressure increasing or decreasing as he went to the 29,029 foot summit?

c. Where is the function decreasing? Increasing?

d. Does your answer to part (c) make sense?

e. For what altitudes do you think the function is valid?

Figure for Exercise 89

90. *Increasing Revenue* A company's weekly revenue in dollars is given by $R(x) = 2000x - 2x^2$, where x is the number of items produced during a week.

a. For what x is $R(x) > 0$?

b. On what interval is $R(x)$ increasing? Decreasing?

c. Find the marginal revenue function $MR(x)$, where $MR(x)$ is defined by

$$MR(x) = R(x + 1) - R(x).$$

d. On what interval is $MR(x)$ positive? Negative?

A line that runs as close as possible to a set of data points is called the **line of best fit** or the **regression line**. If we require that the line goes through the origin, then we need to determine only the slope, which can be done with a quadratic function as in Exercises 91 and 92. To find the slope and y-intercept for a regression line requires techniques beyond the scope of this course. In this case we will use the built-in linear regression feature of a graphing calculator as in Exercises 93 and 94.

91. *Minimizing Distance* A caravan starting at (0, 0) must travel in a line ($y = mx$) to come as close as possible to outposts located at (2, 3) and (6, 5) as shown in the figure. One measure of the total distance from the caravan's path to the outposts is the sum of the squares of the vertical distances from the path to each outpost. In this case we get $d = (3 - 2m)^2 + (6m - 5)^2$. What value for m would minimize d?

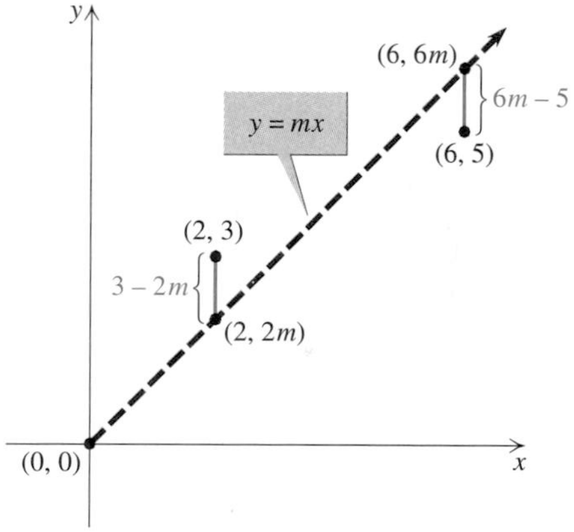

Figure for Exercise 91

92. *Fitting a Line to Data Points* In a test of a new type of brake pad, stopping distances from 30, 40, and 50 mph were measured at 40, 50, and 70 ft, respectively. Assuming that the stopping distance y is directly proportional to the speed of the vehicle x, find the equation of the form $y = mx$ that "best fits" the data points shown in the figure.

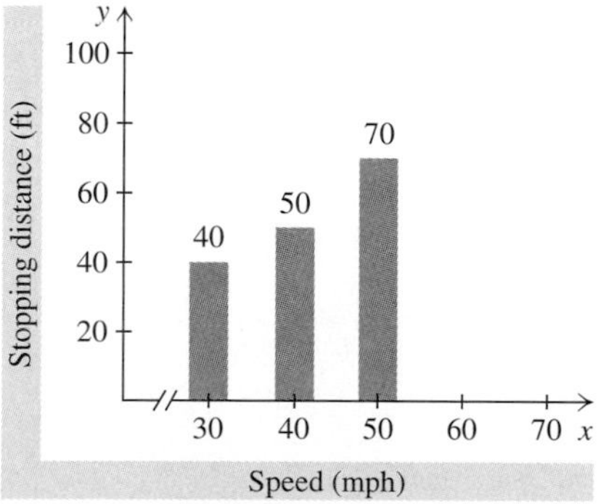

Figure for Exercise 92

Measure the fit of the line as in Exercise 91. Use your equation to predict the stopping distance from 65 mph.

93. *College Tuition* The following table gives the consumer price index for college tuition for the years 1985 through 1994. According to the table, it took \$129.60 in 1986 to buy the same tuition that you could get for \$119.90 in 1985. (U.S. Bureau of Labor Statistics)

Year	Index	Year	Index
1985	119.9	1990	175.0
1986	129.6	1991	192.8
1987	139.4	1992	213.5
1988	150.0	1993	233.5
1989	161.9	1994	249.8

a. Use a graphing calculator and linear regression to express the consumer price index for tuition as a linear function of the year. The procedure that a calculator uses for linear regression is similar to that used in Exercise 91. See also Example 3 in Section 1.2.
b. Use the function from part (a) to predict the consumer price index for tuition in the year 1998.
c. If you paid \$1489 for tuition in 1994, then what amount would you expect to pay in 1998?

94. *Violent Crime* The following table gives the number of violent crimes per 100,000 population in the U.S. in the years 1985 through 1994. (Crime in the United States, U.S. FBI)

Year	Crimes	Year	Crimes
1985	557	1990	732
1986	618	1991	758
1987	610	1992	758
1988	637	1993	747
1989	663	1994	716

a. Use a graphing calculator and quadratic regression to express the number of violent crimes per 100,000 as a quadratic function of the year.
b. Use the equation from part (a) to find the year in which the number of violent crimes per 100,000 reached its maximum value.
c. Use the table to determine when violent crime reached its maximum value.
d. Use the function from part (a) to predict the number of violent crimes in 1995.
e. Graph the function on a graphing calculator to find the year in which there will be no more violent crime.

LINKING CONCEPTS

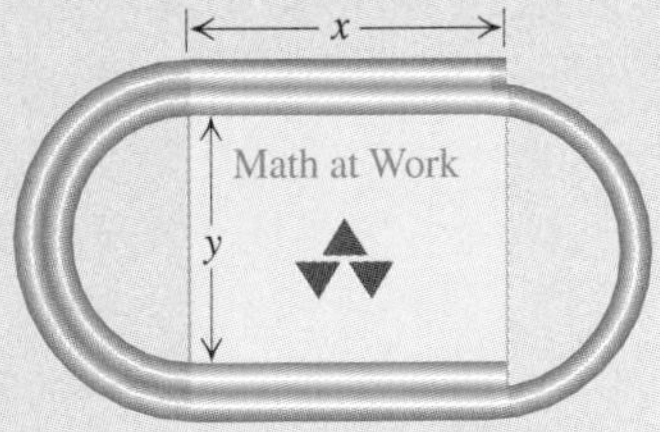

For Individual or Group Explorations

Designing a Paper Clip Sylvia is designing a paper clip that will carry a company logo in the rectangular center section as shown in the accompanying figure. The length of the wire used to make the clip will be 8 in.

a) Assume that the radii of the three semicircles on the ends are all $y/2$. Find the dimensions for x and y that will maximize the area of the rectangular center section.

b) Assume that the wire has a 1/16-in. diameter and do not assume that the radii of the semicircles are equal. Find the dimensions for x and y that will maximize the area of the rectangular center section. Explain your choices for the three radii.

c) Instead of maximizing the area of the rectangular region, find the dimensions for x and y that will maximize the total area enclosed by the wire. State any assumptions that you make.

3.3 Zeros of Polynomial Functions

In Sections 3.1 and 3.2 we studied polynomial functions of degree 2 or less. In this section we concentrate on polynomial functions in general, particularly those of degree 3 or higher.

The Remainder Theorem

If $y = P(x)$ is a polynomial function, then a value of x that satisfies $P(x) = 0$ is called a **zero** of the polynomial function or a zero of the polynomial. For example, 3 and -3 are zeros of the function $P(x) = x^2 - 9$, because $P(3) = 0$ and $P(-3) = 0$. Note that the zeros of $P(x) = x^2 - 9$ are the same as the solutions to the equation $x^2 - 9 = 0$. The zeros of a polynomial function appear on the graph of the function as the x coordinates of the x-intercepts. The x-intercepts of the graph of $P(x) = x^2 - 9$ shown in Fig. 3.24 are $(-3, 0)$ and $(3, 0)$.

For polynomial functions of degree 2 or less, the zeros can be found by solving quadratic or linear equations. Our goal in this section is to find all of the zeros of a polynomial function when possible. For polynomials of degree higher than 2, the difficulty of this task ranges from easy to impossible, but we have some theorems to assist us. The remainder theorem relates evaluating a polynomial to division of polynomials.

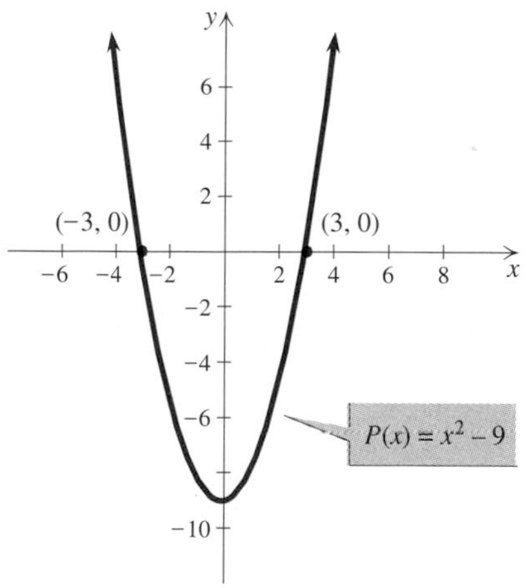

Figure 3.24

The Remainder Theorem

If R is the remainder when a polynomial $P(x)$ is divided by $x - c$, then $R = P(c)$.

Proof Let $Q(x)$ be the quotient and R be the remainder when $P(x)$ is divided by $x - c$. Since the dividend is equal to the divisor times the quotient plus the remainder, we have

$$P(x) = (x - c)Q(x) + R$$

This statement is true for any value of x, and so it is also true for $x = c$:

$$\begin{aligned} P(c) &= (c - c)Q(c) + R \\ &= 0 \cdot Q(c) + R \\ &= R \end{aligned}$$

So $P(c)$ is equal to the remainder when $P(x)$ is divided by $x - c$. ■

EXAMPLE 1 **Using the remainder theorem to evaluate a polynomial**

Use the remainder theorem to find $P(3)$ if $P(x) = 2x^3 - 5x^2 + 4x - 6$.

Solution

By the remainder theorem $P(3)$ is the remainder when $P(x)$ is divided by $x - 3$:

$$\begin{array}{r} 2x^2 + x + 7 \\ x - 3\overline{)2x^3 - 5x^2 + 4x - 6} \\ \underline{2x^3 - 6x^2} \\ x^2 + 4x \\ \underline{x^2 - 3x} \\ 7x - 6 \\ \underline{7x - 21} \\ 15 \end{array}$$

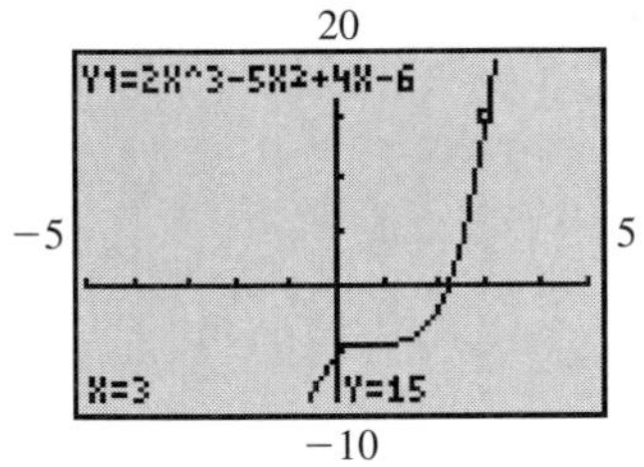

Figure 3.25

The remainder is 15 and therefore $P(3) = 15$. We can check by finding $P(3)$ in the usual manner:

$$P(3) = 2 \cdot 3^3 - 5 \cdot 3^2 + 4 \cdot 3 - 6 = 54 - 45 + 12 - 6 = 15$$

You can check this with a graphing calculator as shown in Fig. 3.25. ■

Synthetic Division

In Example 1 we found $P(3) = 15$ in two different ways. Certainly, evaluating $P(x)$ for $x = 3$ in the usual manner is faster than dividing $P(x)$ by $x - 3$ using the ordinary method of dividing polynomials. However, there is a faster method, called **synthetic division**, for dividing by $x - 3$. Compare the two methods side by side, both showing $2x^3 - 5x^2 + 4x - 6$ divided by $x - 3$:

Ordinary Division of Polynomials

$$\begin{array}{r}
2x^2 + x + 7 \leftarrow \text{Quotient} \\
x - 3\overline{)2x^3 - 5x^2 + 4x - 6} \\
\underline{2x^3 - 6x^2} \quad\quad\quad\quad \\
x^2 + 4x \quad\quad \\
\underline{x^2 - 3x} \quad\quad \\
7x - 6 \\
\underline{7x - 21} \\
15 \leftarrow \text{Remainder}
\end{array}$$

Synthetic Division

$$\begin{array}{r|rrrr}
3 & 2 & -5 & 4 & -6 \\
 & & 6 & 3 & 21 \\
\hline
 & 2 & 1 & 7 & 15 \leftarrow \text{Remainder}
\end{array}$$

$\underbrace{2 \quad 1 \quad 7}_{\text{Quotient}}$

Synthetic division certainly looks easier than ordinary division, and in general it is faster than evaluating the polynomial by substitution. Synthetic division is used as a quick means of dividing a polynomial by a binomial of the form $x - c$.

In synthetic division we write just the necessary parts of the ordinary division. Instead of writing $2x^3 - 5x^2 + 4x - 6$, write the coefficients 2, -5, 4, and -6. For $x - 3$, write only the 3. The bottom row in synthetic division gives the coefficients of the quotient and the remainder.

To actually perform the synthetic division, start with the following arrangement of coefficients:

$$\begin{array}{r|rrrr}
3 & 2 & -5 & 4 & -6 \\
\hline
\end{array}$$

Bring down the first coefficient, 2. Multiply 2 by 3 and write the answer beneath -5. Then add:

$$\begin{array}{rr|rrrr}
 & 3 & 2 & -5 & 4 & -6 \\
 & & \downarrow & 6 & & \\
\hline
\text{Multiply} \rightarrow & & 2 & 1 & & \\
 & & & \uparrow & & \\
 & & & \text{Add} & &
\end{array}$$

Using 3 rather than -3 when dividing by $x - 3$ allows us to multiply and add rather than multiply and subtract as in ordinary division. Now multiply 1 by 3 and write the answer beneath 4. Then add. Repeat the multiply-and-add step for the remaining column:

$$\begin{array}{r|rrrr}
3 & 2 & -5 & 4 & -6 \\
 & & 6 & 3 & 21 \\
\hline
 & 2 & 1 & 7 & 15
\end{array}$$

To perform this arithmetic on a graphing calculator, start with the leading coefficient 2 as the answer. Then repeatedly multiply the answer by 3 and add the next coefficient as shown in Fig. 3.26. □

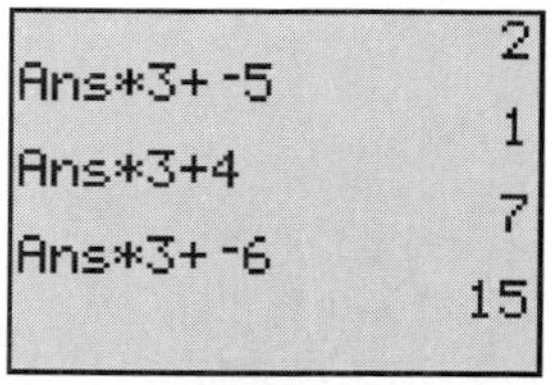

Figure 3.26

The quotient is $2x^2 + x + 7$, and the remainder is 15. Since the divisor in synthetic division is of the form $x - c$, the degree of the quotient is always one less than the degree of the dividend.

EXAMPLE 2 Synthetic division

Use synthetic division to find the quotient and remainder when $x^4 - 14x^2 + 5x - 9$ is divided by $x + 4$.

Solution

Since $x + 4 = x - (-4)$, we use -4 in the synthetic division. Use 1, 0, -14, 5, and -9 as the coefficients of the polynomial. We use 0 for the coefficient of the missing x^3-term, as we would in ordinary division of polynomials.

$$\begin{array}{r|rrrrr} -4 & 1 & 0 & -14 & 5 & -9 \\ & & -4 & 16 & -8 & 12 \\ \hline \text{Multiply} \rightarrow & 1 & -4 & 2 & -3 & 3 \\ & & \uparrow & & & \\ & & \text{Add} & & & \end{array}$$

The quotient is $x^3 - 4x^2 + 2x - 3$ and the remainder is 3.

In the next example, we evaluate a polynomial using the remainder theorem, with synthetic division replacing the ordinary division used in Example 1.

EXAMPLE 3 Using synthetic division to evaluate a polynomial

Let $f(x) = x^3$ and $g(x) = x^3 - 3x^2 + 5x - 12$. Use synthetic division to find the following function values.

a) $f(-2)$ **b)** $g(4)$

Solution

a) To find $f(-2)$, divide the polynomial x^3 by $x - (-2)$ or $x + 2$ using synthetic division. Write x^3 as $x^3 + 0x^2 + 0x + 0$, and use 1, 0, 0, and 0 as the coefficients. We use a zero for each power of x below x^3.

$$\begin{array}{r|rrrr} -2 & 1 & 0 & 0 & 0 \\ & & -2 & 4 & -8 \\ \hline & 1 & -2 & 4 & -8 \end{array}$$

The remainder is -8, so $f(-2) = -8$. To check, find $f(-2) = (-2)^3 = -8$.

b) To find $g(4)$, use synthetic division to divide $x^3 - 3x^2 + 5x - 12$ by $x - 4$:

$$\begin{array}{r|rrrr} 4 & 1 & -3 & 5 & -12 \\ & & 4 & 4 & 36 \\ \hline & 1 & 1 & 9 & 24 \end{array}$$

The remainder is 24, so $g(4) = 24$. Check this answer by finding $g(4) = 4^3 - 3(4^2) + 5(4) - 12 = 24$.

There are two ways to find the value of a polynomial $P(x)$ for $x = c$: Find the remainder when the polynomial is divided by $x - c$, or directly evaluate the polynomial by substituting $x = c$. In Example 3(a), it is easier to directly evaluate $(-2)^3$ than to do synthetic division. In Example 3(b), synthetic division is easier because there are fewer arithmetic operations in the synthetic division than in substituting $x = 4$ and computing the result. For polynomials of higher degree there may be an even bigger difference between the number of operations in the two methods. To gain a better understanding of why synthetic division gives the value of the polynomial, see the Linking Concepts at the end of this section.

The Factor Theorem

Consider the polynomial function $P(x) = x^2 - x - 6$. We can find the zeros of the function by solving $x^2 - x - 6 = 0$ by factoring:

$$(x - 3)(x + 2) = 0$$

$$x - 3 = 0 \quad \text{or} \quad x + 2 = 0$$

$$x = 3 \quad \text{or} \quad x = -2$$

Both 3 and -2 are zeros of the function $P(x) = x^2 - x - 6$. Note how each factor of the polynomial corresponds to a zero of the function. This example suggests the following theorem.

The Factor Theorem

The number c is a zero of the polynomial function $y = P(x)$ if and only if $x - c$ is a factor of the polynomial $P(x)$.

Proof If c is a zero of the polynomial function $y = P(x)$, then $P(c) = 0$. If $P(x)$ is divided by $x - c$, we get a quotient $Q(x)$ and a remainder R such that

$$P(x) = (x - c)Q(x) + R.$$

By the remainder theorem, $R = P(c)$. Since $P(c) = 0$, we have $R = 0$ and $P(x) = (x - c)Q(x)$, which proves that $x - c$ is a factor of $P(x)$.

In Exercise 83 you will be asked to prove that if $x - c$ is a factor of $P(x)$, then c is a zero of the polynomial function. These two arguments together establish the truth of the factor theorem. ∎

Synthetic division can be used in conjunction with the factor theorem. If the remainder of dividing $P(x)$ by $x - c$ is 0, then $P(c) = 0$ and c is a zero of the polynomial function. By the factor theorem, $x - c$ is a factor of $P(x)$.

EXAMPLE 4 Using the factor theorem to factor a polynomial

Determine whether $x + 4$ is a factor of the polynomial $P(x) = x^3 - 13x + 12$. If it is a factor, then factor $P(x)$ completely.

Solution

By the factor theorem, $x + 4$ is a factor of $P(x)$ if and only if $P(-4) = 0$. We can find $P(-4)$ using synthetic division:

$$\begin{array}{r|rrrr} -4 & 1 & 0 & -13 & 12 \\ & & -4 & 16 & -12 \\ \hline & 1 & -4 & 3 & 0 \end{array}$$

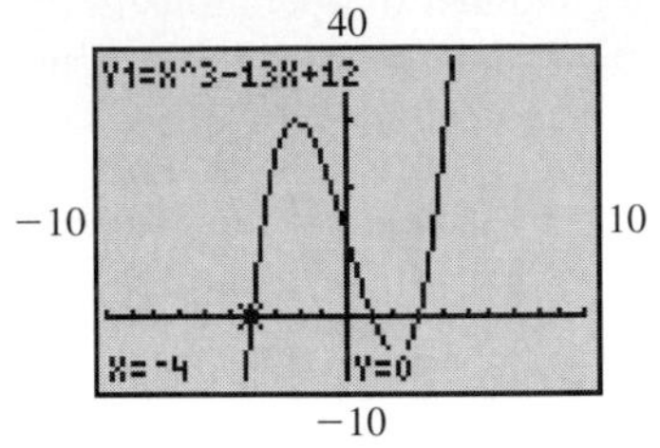

Figure 3.27

Since $P(-4)$ is equal to the remainder, $P(-4) = 0$ and -4 is a zero of $P(x)$. By the factor theorem, $x + 4$ is a factor of $P(x)$. Since the other factor is the quotient from the synthetic division, $P(x) = (x + 4)(x^2 - 4x + 3)$. Factor the quadratic polynomial to get $P(x) = (x + 4)(x - 1)(x - 3)$.

You can check this result by examining the calculator graph of $y = x^3 - 13x + 12$ shown in Fig. 3.27. The graph appears to cross the x-axis at -4, 1, and 3, supporting the conclusion that $P(x) = (x + 4)(x - 1)(x - 3)$.

The Fundamental Theorem of Algebra

Whether a number is a zero of a polynomial function can be determined by synthetic division. But does every polynomial function have a zero? This question was answered in the affirmative by Carl F. Gauss when he proved the fundamental theorem of algebra in his doctoral thesis in 1799 at the age of 22.

The Fundamental Theorem of Algebra

If $y = P(x)$ is a polynomial function of positive degree, then $y = P(x)$ has at least one zero in the set of complex numbers.

Gauss also proved the n-root theorem of Section 3.4 which says that the number of zeros of a polynomial (or polynomial function) of degree n is at most n. For example, a fifth degree polynomial function has at least one zero and at most five.

Note that the zeros guaranteed by Gauss are in the set of complex numbers. So the zero might be real or imaginary. The theorem applies only to polynomial functions of degree 1 or more, because a polynomial function of zero degree such as $P(x) = 7$ has no zeros. A polynomial function of degree 1, $f(x) = ax + b$, has exactly one zero which is found by solving $ax + b = 0$. A polynomial function of degree 2, $f(x) = ax^2 + bx + c$, has one or two zeros which can be found by the quadratic formula. For higher degree polynomials the situation is not as simple. The fundamental theorem tells us that a polynomial function has at least one zero but not how to find it. For this purpose we have some other theorems.

The Rational Zero Theorem

Zeros that are rational numbers, the **rational zeros**, are generally the easiest to find. The polynomial function $f(x) = 6x^2 - x - 35$ has two rational zeros which can be found as follows:

$$6x^2 - x - 35 = 0$$

$$(2x - 5)(3x + 7) = 0 \qquad \text{Factor.}$$

$$x = \frac{5}{2} \quad \text{or} \quad x = -\frac{7}{3}$$

Note that in 5/2, 5 is a factor of -35 (the constant term) and 2 is a factor of 6 (the leading coefficient). For the zero $-7/3$, -7 is a factor of -35 and 3 is a factor of 6. Of course, these observations are not surprising, because we used these facts to factor the quadratic polynomial in the first place. Note that there are a lot of other factors of -35 and 6 for which the ratio is *not* a zero of this function. This example illustrates the rational zero theorem.

The Rational Zero Theorem

If $f(x) = a_nx^n + a_{n-1}x^{n-1} + a_{n-2}x^{n-2} + \cdots + a_1x + a_0$ is a polynomial function with integral coefficients ($a_n \neq 0$ and $a_0 \neq 0$) and p/q (in lowest terms) is a rational zero of $f(x)$, then p is a factor of the constant term a_0 and q is a factor of the leading coefficient a_n.

Proof If p/q is a zero of $f(x)$, then $f(p/q) = 0$:

$$a_n\left(\frac{p}{q}\right)^n + a_{n-1}\left(\frac{p}{q}\right)^{n-1} + a_{n-2}\left(\frac{p}{q}\right)^{n-2} + \cdots + a_1\frac{p}{q} + a_0 = 0$$

Subtract a_0 from each side of this equation.

$$a_n\left(\frac{p}{q}\right)^n + a_{n-1}\left(\frac{p}{q}\right)^{n-1} + a_{n-2}\left(\frac{p}{q}\right)^{n-2} + \cdots + a_1\frac{p}{q} = -a_0$$

Since p/q is in lowest terms, p is not a factor of q. So p appears at least once in each term on the left-hand side of the equation, and p is a factor of the left-hand side. Since the left-hand side is equal to the integer $-a_0$, p is also a factor of a_0.

To prove that q is a factor of a_n, multiply each side of the original equation by $(q/p)^n$, then subtract a_n from each side:

$$a_{n-1}\left(\frac{q}{p}\right) + a_{n-2}\left(\frac{q}{p}\right)^2 + a_{n-3}\left(\frac{q}{p}\right)^3 + \cdots + a_0\left(\frac{q}{p}\right)^n = -a_n$$

Now q is a factor of a_n by the same argument that we used above. ■

The rational zero theorem does not identify exactly which rational numbers are zeros of a function; it only gives *possibilities* for the rational zeros.

EXAMPLE 5 **Using the rational zero theorem**

Find all possible rational zeros for each polynomial function.

a) $f(x) = 2x^3 - 3x^2 - 11x + 6$ **b)** $g(x) = 3x^3 - 8x^2 - 8x + 8$

Solution

a) If the rational number p/q is a zero of $f(x)$, then p is a factor of 6 and q is a factor of 2. The positive factors of 6 are 1, 2, 3, and 6. The positive factors of 2 are 1 and 2. Take each factor of 6 and divide by 1, to get 1/1, 2/1, 3/1, and 6/1. Take each factor of 6 and divide by 2, to get 1/2, 2/2, 3/2, and 6/2. Simplify the ratios, eliminate duplications, and put in the negative factors to get

$$\pm 1, \quad \pm 2, \quad \pm 3, \quad \pm 6, \quad \pm\frac{1}{2}, \quad \text{and} \quad \pm\frac{3}{2}$$

as the possible rational zeros to the function $f(x)$.

b) If the rational number p/q is a zero of $g(x)$, then p is a factor of 8 and q is a factor of 3. The factors of 8 are 1, 2, 4, and 8. The factors of 3 are 1 and 3. If we take all possible ratios of a factor of 8 over a factor of 3, we get

$$\pm 1, \quad \pm 2, \quad \pm 4, \quad \pm 8, \quad \pm\frac{1}{3}, \quad \pm\frac{2}{3}, \quad \pm\frac{4}{3}, \quad \text{and} \quad \pm\frac{8}{3}$$

as the possible rational zeros of the function $g(x)$.

Our goal is to find all of the zeros to a polynomial function. The zeros to a polynomial function might be rational, irrational, or imaginary. We can determine the rational zeros by simply evaluating the polynomial function for every number in the list of possible rational zeros. If the list is long, looking at a graph of the function can speed up the process. We will use synthetic division to evaluate the polynomial, because synthetic division gives the quotient polynomial as well as the value of the polynomial.

EXAMPLE 6 **Finding all zeros of a polynomial function**

Find all of the real and imaginary zeros for each polynomial function of Example 5.

a) $f(x) = 2x^3 - 3x^2 - 11x + 6$ **b)** $g(x) = 3x^3 - 8x^2 - 8x + 8$

Solution

a) The possible rational zeros of $f(x)$ are listed in Example 5(a). Use synthetic division to check each possible zero to see whether it is actually a zero. Try 1 first.

$$\begin{array}{r|rrrr} 1 & 2 & -3 & -11 & 6 \\ & & 2 & -1 & -12 \\ \hline & 2 & -1 & -12 & -6 \end{array}$$

Since the remainder is -6, 1 is not a zero of the function. Keep on trying numbers from the list of possible rational zeros. To save space, we will not show any more failures. So try 1/2 next.

$$\begin{array}{r|rrrr} \frac{1}{2} & 2 & -3 & -11 & 6 \\ & & 1 & -1 & -6 \\ \hline & 2 & -2 & -12 & 0 \end{array}$$

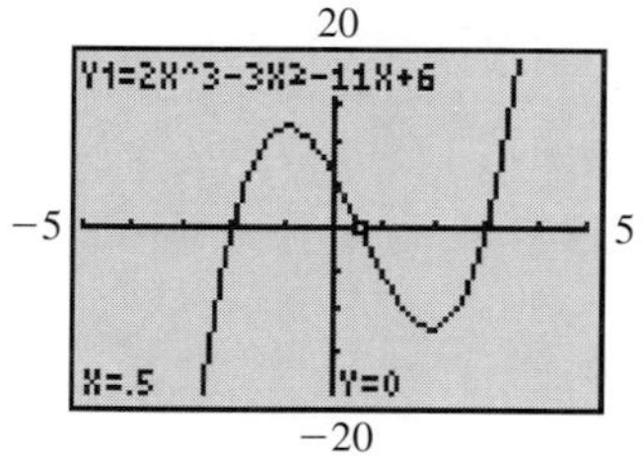

Figure 3.28

You could speed up this process of finding a zero with the calculator graph shown in Fig. 3.28. It is not too hard to discover that 1/2 is a zero by looking at the graph and the list of possible rational zeros that we listed in Example 5. If you use a graph to find that 1/2 is a zero, you still need to do synthetic division to factor the polynomial. □

Since the remainder in the synthetic division is 0, 1/2 is a zero of $f(x)$. By the factor theorem, $x - 1/2$ is a factor of the polynomial. The quotient is the other factor.

$$\begin{aligned} 2x^3 - 3x^2 - 11x + 6 &= 0 & & \\ \left(x - \frac{1}{2}\right)(2x^2 - 2x - 12) &= 0 & & \text{Factor.} \\ (2x - 1)(x^2 - x - 6) &= 0 & & \text{Factor 2 out of the second ``factor'' and distribute it into the first factor.} \\ (2x - 1)(x - 3)(x + 2) &= 0 & & \text{Factor completely.} \end{aligned}$$

$$2x - 1 = 0 \quad \text{or} \quad x - 3 = 0 \quad \text{or} \quad x + 2 = 0$$

$$x = \frac{1}{2} \quad \text{or} \quad x = 3 \quad \text{or} \quad x = -2$$

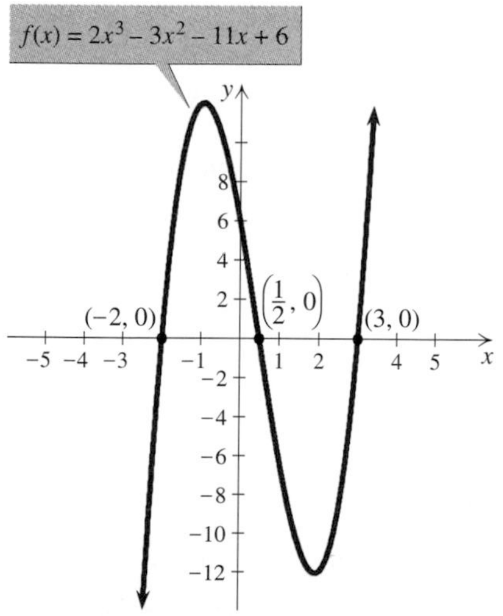

Figure 3.29

The zeros of the function f are 1/2, 3, and -2. Note that each zero of f corresponds to an x-intercept on the graph of f shown in Fig. 3.29. Because this polynomial had three rational zeros, we could have found them all by using synthetic division or by examining the calculator graph. However, it is good to factor the polynomial to see the correspondence between the three zeros, the three factors, and the three x-intercepts.

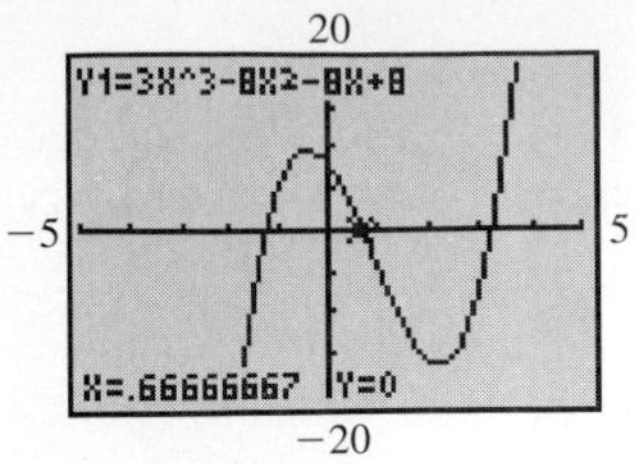

Figure 3.30

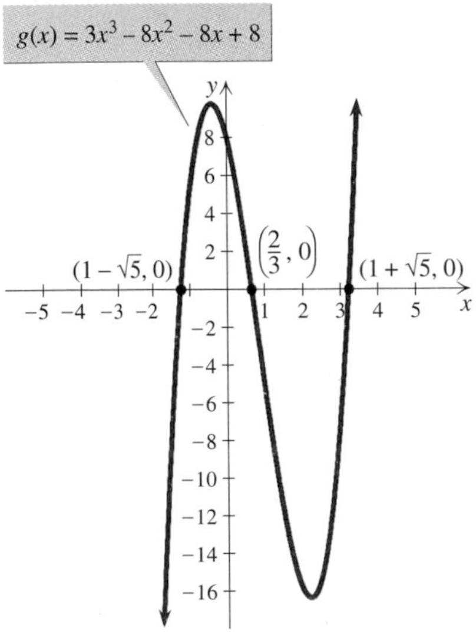

Figure 3.31

b) The possible rational zeros of $g(x)$ are listed in Example 5(b). First check 2/3 to see whether it produces a remainder of 0.

$\frac{2}{3}$	3	−8	−8	8
		2	−4	−8
	3	−6	−12	0

You could graph the function with a calculator as shown in Fig. 3.30. Keeping in mind the list of possible rational zeros, it is not too hard to discover that 2/3 is a zero. □

Since the remainder in the synthetic division is 0, 2/3 is a zero of $g(x)$. By the factor theorem, $x - 2/3$ is a factor of the polynomial. The quotient is the other factor.

$$3x^3 - 8x^2 - 8x + 8 = 0$$

$$\left(x - \frac{2}{3}\right)(3x^2 - 6x - 12) = 0$$

$$(3x - 2)(x^2 - 2x - 4) = 0$$

$$3x - 2 = 0 \quad \text{or} \quad x^2 - 2x - 4 = 0$$

$$x = \frac{2}{3} \quad \text{or} \quad x = \frac{2 \pm \sqrt{20}}{2}$$

$$x = \frac{2}{3} \quad \text{or} \quad x = 1 \pm \sqrt{5}$$

There are one rational and two irrational roots to the equation. So the zeros of the function g are 2/3, $1 + \sqrt{5}$, and $1 - \sqrt{5}$. Each zero corresponds to an x-intercept on the graph of g shown in Fig. 3.31. ■

Note that in Example 6(a) all of the zeros were rational. All three could have been found by continuing to check the possible rational zeros using synthetic division. In Example 6(b) we would be wasting time if we continued to check the possible rational zeros, because there is only one. When we get to a quadratic polynomial, it is best to either factor the quadratic polynomial or use the quadratic formula to find the remaining zeros.

FOR THOUGHT True or False? Explain.

1. The function $f(x) = 1/x$ has at least one zero.
2. If $P(x) = x^4 - 6x^2 - 8$ is divided by $x^2 - 2$, then the remainder is $P(\sqrt{2})$.
3. If $1 - 2i$ and $1 + 2i$ are zeros of $P(x) = x^3 - 5x^2 + 11x - 15$, then $x - 1 - 2i$ and $x - 1 + 2i$ are factors of $P(x)$.
4. If we divide $x^5 - 1$ by $x - 2$, then the remainder is 31.
5. Every polynomial function has at least one zero.
6. If $P(x) = x^3 - x^2 + 4x - 5$ and b is the remainder from division of $P(x)$ by $x - c$, then $b^3 - b^2 + 4b - 5 = c$.

7. If $P(x) = x^3 - 5x^2 + 4x - 15$, then $P(4) = 0$.

8. The equation $\pi^2 x^4 - \frac{1}{\sqrt{2}} x^3 + \frac{1}{\sqrt{7} + \pi} = 0$ has at least one complex solution.

9. The binomial $x - 1$ is a factor of $x^5 + x^4 - x^3 - x^2 - x + 1$.

10. The binomial $x + 3$ is a factor of $3x^4 - 5x^3 + 7x^2 - 9x - 2$.

3.3 EXERCISES

Tape 7 Disk

Use ordinary division of polynomials to find the quotient and remainder when the first polynomial is divided by the second.

1. $x^2 - 5x + 7,\ x - 2$
2. $x^2 - 3x + 9,\ x - 4$
3. $-2x^3 + 4x - 9,\ x + 3$
4. $-4w^3 + 5w^2 - 7,\ w - 3$
5. $s^4 - 3s^2 + 6,\ s^2 - 5$
6. $h^4 + 3h^3 + h - 5,\ h^2 - 3$

Use synthetic division to find the quotient and remainder when the first polynomial is divided by the second.

7. $x^2 + 4x + 1,\ x - 2$
8. $2x^2 - 3x + 6,\ x - 5$
9. $-x^2 + 4x + 9,\ x + 3$
10. $-3x^2 + 4x - 1,\ x + 1$
11. $4x^3 - 5x + 2,\ x - \frac{1}{2}$
12. $-6x^3 - 25x^2 - 9,\ x - \frac{3}{2}$
13. $2a^3 - 3a^2 + 4a + 3,\ a + \frac{1}{2}$
14. $-3b^3 - b^2 - 3b - 1,\ b + \frac{1}{3}$
15. $x^4 - 3,\ x - 1$
16. $x^4 - 16,\ x - 2$
17. $x^2 - 3x + 1,\ x - \frac{1}{2}$
18. $x^3 - x^2 + x - 1,\ x - \frac{1}{2}$

Let $f(x) = x^5 - 1$, $g(x) = x^3 - 4x^2 + 8$, and $h(x) = 2x^4 + x^3 - x^2 + 3x + 3$. Find the following function values by using synthetic division. Check by using substitution.

19. $f(1)$
20. $f(-1)$
21. $f(-2)$
22. $f(3)$
23. $g(1)$
24. $g(-1)$
25. $g\left(-\frac{1}{2}\right)$
26. $g\left(\frac{1}{2}\right)$
27. $h(-1)$
28. $h(2)$
29. $h(1)$
30. $h(-3)$

Determine whether the given binomial is a factor of the polynomial following it. If it is a factor, then factor the polynomial completely.

31. $x + 3,\ x^3 + 4x^2 + x - 6$
32. $x + 5,\ x^3 + 8x^2 + 11x - 20$
33. $x - 4,\ x^3 + 4x^2 - 17x - 60$
34. $x - 2,\ x^3 - 12x^2 + 44x - 48$

Determine whether each given number is a zero of the polynomial function following the number.

35. $3,\ f(x) = 2x^3 - 5x^2 - 4x + 3$
36. $-2,\ g(x) = 3x^3 - 6x^2 - 3x - 19$
37. $-2,\ g(d) = d^3 + 2d^2 + 3d + 1$
38. $-1,\ w(x) = 3x^3 + 2x^2 - 2x - 1$
39. $-1,\ P(x) = x^4 + 2x^3 + 4x^2 + 6x + 3$
40. $3,\ G(r) = r^4 + 4r^3 + 5r^2 + 3r + 17$
41. $\frac{1}{2},\ H(x) = x^3 + 3x^2 - 5x + 7$
42. $-\frac{1}{2},\ T(x) = 2x^3 + 3x^2 - 3x - 2$

Use the rational zero theorem to find all possible rational zeros for each polynomial function.

43. $f(x) = x^3 - 9x^2 + 26x - 24$
44. $g(x) = x^3 - 2x^2 - 5x + 6$
45. $h(x) = x^3 - x^2 - 7x + 15$
46. $m(x) = x^3 + 4x^2 + 4x + 3$
47. $P(x) = 8x^3 - 36x^2 + 46x - 15$
48. $T(x) = 18x^3 - 9x^2 - 5x + 2$
49. $M(x) = 18x^3 - 21x^2 + 10x - 2$
50. $N(x) = 4x^3 - 10x^2 + 4x + 5$

Find all of the real and imaginary zeros for each polynomial function.

51. $f(x) = x^3 - 9x^2 + 26x - 24$
52. $g(x) = x^3 - 2x^2 - 5x + 6$

53. $h(x) = x^3 - x^2 - 7x + 15$

54. $m(x) = x^3 + 4x^2 + 4x + 3$

55. $P(a) = 8a^3 - 36a^2 + 46a - 15$

56. $T(b) = 18b^3 - 9b^2 - 5b + 2$

57. $M(t) = 18t^3 - 21t^2 + 10t - 2$

58. $N(t) = 4t^3 - 10t^2 + 4t + 5$

59. $S(w) = w^4 + w^3 - w^2 + w - 2$

60. $W(v) = 2v^4 + 5v^3 + 3v^2 + 15v - 9$

61. $V(x) = x^4 + 2x^3 - x^2 - 4x - 2$

62. $U(x) = x^4 - 4x^3 + x^2 + 12x - 12$

63. $f(x) = 24x^3 - 26x^2 + 9x - 1$

64. $f(x) = 30x^3 - 47x^2 - x + 6$

65. $y = 16x^3 - 33x^2 + 82x - 5$

66. $y = 15x^3 - 37x^2 + 44x - 14$

67. $f(x) = 21x^4 - 31x^3 - 21x^2 - 31x - 42$

68. $f(x) = 119x^4 - 5x^3 + 214x^2 - 10x - 48$

Use division to write each rational expression in the form quotient + remainder/divisor. Use synthetic division when possible.

69. $\dfrac{2x + 1}{x - 2}$ **70.** $\dfrac{x - 1}{x + 3}$ **71.** $\dfrac{a^2 - 3a + 5}{a - 3}$

72. $\dfrac{2b^2 - 3b + 1}{b + 2}$ **73.** $\dfrac{c^2 - 3c - 4}{c^2 - 4}$ **74.** $\dfrac{2h^2 + h - 2}{h^2 - 1}$

75. $\dfrac{4t - 5}{2t + 1}$ **76.** $\dfrac{6y - 1}{3y - 1}$

Solve each problem.

77. *Drug Testing* The concentration of a drug (in parts per million) in a patient's bloodstream t hours after administration of the drug is given by the function

$$P(t) = -t^4 + 12t^3 - 58t^2 + 132t.$$

How many hours after administration will the drug be totally eliminated from the bloodstream?

78. *Open-Top Box* Joan intends to make an 18-in.^{3} open-top box out of a 6 in. by 7 in. piece of copper by cutting equal squares (x in. by x in.) from the corners and folding up the sides. Write the difference between the intended volume and the actual volume as a function of x. For what value of x is there no difference between the intended volume and the actual volume?

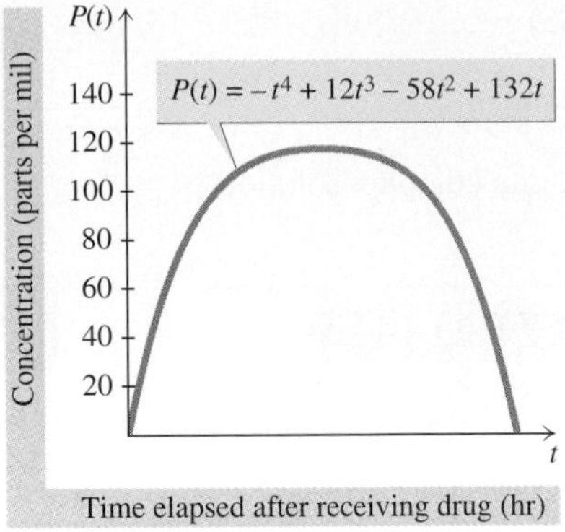

Figure for Exercise 77

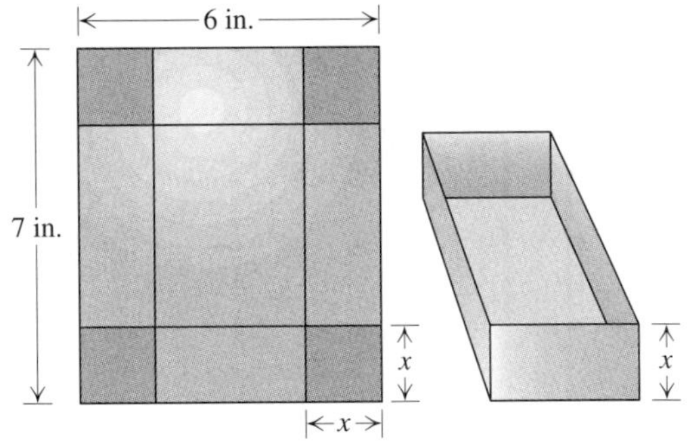

Figure for Exercise 78

79. *Cartridge Box* The height of a box containing an HP Laser Jet III printer cartridge is 4 in. more than the width and the length is 9 in. more than the width. If the volume of the box is 630 in.^{3}, then what are the dimensions of the box?

80. *Computer Case* The width of the case for a 133 megahertz Pentium computer is 4 in. more than twice the height and the depth is 1 in. more than the width. If the volume of the case is 1632 in.^{3}, then what are the dimensions of the case?

81. For what value of c is the equation $\dfrac{x^2 - 2x + 7}{x - 3} = x + 1 + \dfrac{c}{x - 3}$ an identity?

82. *Synthetic Division* Explain how synthetic division can be used to find the quotient and remainder when $3x^3 + 4x^2 + 2x - 4$ is divided by $3x - 2$.

83. Prove that if $x - c$ is a factor of $P(x)$, then c is a zero of the polynomial function.

LINKING CONCEPTS

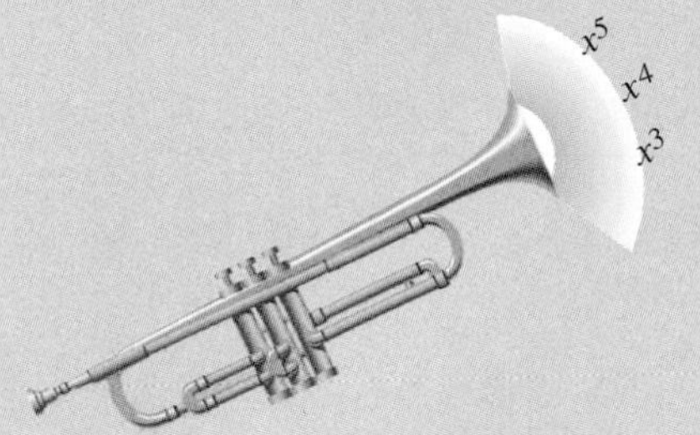

For Individual or Group Explorations

Horner's Method *A fourth degree polynomial in x such as* $3x^4 + 5x^3 + 4x^2 + 3x + 1$ *contains all of the powers of x from the first through the fourth. However, any polynomial can be written without powers of x. Evaluating a polynomial without powers of x (Horner's method) is somewhat easier than evaluating a polynomial with powers.*

a) Show that $\{[(3x + 5)x + 4]x + 3\}x + 1 = 3x^4 + 5x^3 + 4x^2 + 3x + 1$ is an identity.

b) Rewrite the polynomial $P(x) = 6x^5 - 3x^4 + 9x^3 + 6x^2 - 8x + 12$ without powers of x as in part (a).

c) Find $P(2)$ without a calculator using both forms of the polynomial.

d) For which form did you perform fewer arithmetic operations?

e) Explain in detail how to rewrite any polynomial without powers of x.

f) Explain how this new form relates to synthetic division and the remainder theorem.

3.4 The Theory of Equations

One of the main goals in algebra is to keep expanding our knowledge of solving equations. The solutions (roots) of a polynomial equation $P(x) = 0$ are precisely the zeros of a polynomial function $y = P(x)$. Therefore the theorems of Section 3.3 concerning zeros of polynomial functions apply also to the roots of polynomial equations. In this section we study several additional theorems that are useful in solving polynomial equations.

The Number of Roots of a Polynomial Equation

When a polynomial equation is solved by factoring, a factor may occur more than once. For example, $x^2 - 10x + 25 = 0$ is equivalent to $(x - 5)^2 = 0$. Since the factor $x - 5$ occurs twice, we say that 5 is a root of the equation with *multiplicity* 2.

Definition: Multiplicity

If the factor $x - c$ occurs k times in the complete factorization of the polynomial $P(x)$, then c is called a root of $P(x) = 0$ with **multiplicity** k.

If a quadratic equation has a single root, as in $x^2 - 10x + 25 = 0$, then that root has multiplicity 2. If a root with multiplicity 2 is counted as two roots, then every quadratic equation has two roots in the set of complex numbers. This situation is generalized in the following theorem, where the phrase ''when multiplicity is considered'' means that a root with multiplicity k is counted as k individual roots.

n-Root Theorem

If $P(x) = 0$ is a polynomial equation with real or complex coefficients and positive degree n, then, when multiplicity is considered, $P(x) = 0$ has n roots.

Proof By the fundamental theorem of algebra, the polynomial equation $P(x) = 0$ with degree n has at least one complex root c_1. By the factor theorem, $P(x) = 0$ is equivalent to

$$(x - c_1)Q_1(x) = 0,$$

where $Q_1(x)$ is a polynomial with degree $n - 1$ (the quotient when $P(x)$ is divided by $x - c_1$). Again, by the fundamental theorem of algebra, there is at least one complex root c_2 of $Q_1(x) = 0$. By the factor theorem, $P(x) = 0$ can be written as

$$(x - c_1)(x - c_2)Q_2(x) = 0,$$

where $Q_2(x)$ is a polynomial with degree $n - 2$. Reasoning in this manner n times, we get a quotient polynomial that has 0 degree, n factors for $P(x)$, and n complex roots, not necessarily all different. ■

EXAMPLE 1 Finding all roots of a polynomial equation

State the degree of each polynomial equation. Find all real and imaginary roots of each equation, stating multiplicity when it is greater than one.

a) $6x^5 + 24x^3 = 0$ **b)** $(x - 3)^2(x + 14)^5 = 0$

Solution

a) This fifth-degree equation can be solved by factoring:

$$6x^3(x^2 + 4) = 0$$

$$6x^3 = 0 \quad \text{or} \quad x^2 + 4 = 0$$

$$x^3 = 0 \quad \text{or} \quad x^2 = -4$$

$$x = 0 \quad \text{or} \quad x = \pm 2i$$

The roots are $\pm 2i$ and 0. Since there are two imaginary roots and 0 is a root with multiplicity 3, there are five roots when multiplicity is considered.

Because 0 is the only real root, the graph of $y = 6x^5 + 24x^3$ has only one x-intercept at (0, 0) as shown in Fig. 3.32. □

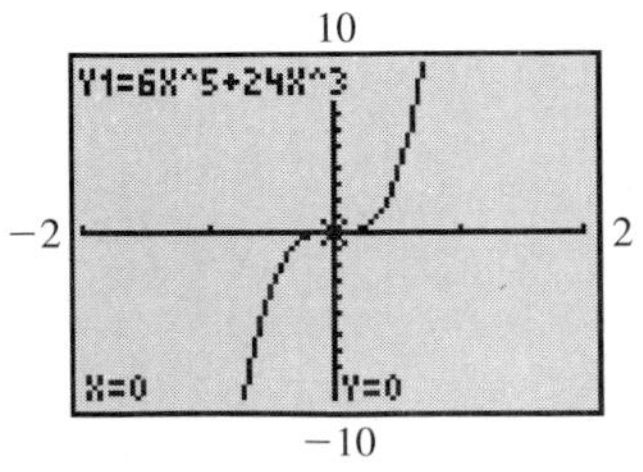

Figure 3.32

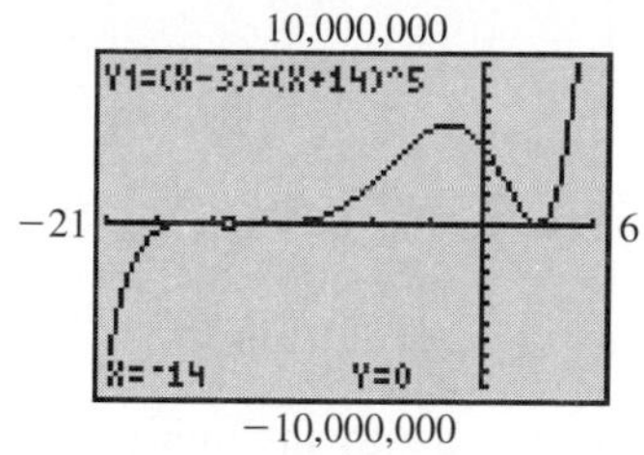

Figure 3.33

b) The highest power of x in $(x - 3)^2$ is 2, and in $(x + 14)^5$ is 5. By the product rule for exponents, the highest power of x in this equation is 7. The only roots of this seventh-degree equation are 3 and -14. The root 3 has multiplicity 2, and -14 has multiplicity 5. So there are seven roots when multiplicity is considered.

Because the equation has two real solutions, the graph of $y = (x - 3)^2(x + 14)^5$ has two x-intercepts at $(3, 0)$ and $(-14, 0)$ as shown in Fig. 3.33.

Note that graphing polynomial functions and solving polynomial equations go hand in hand. The solutions to the equation can help us find an appropriate viewing window for the graph as they did in Example 1, and the graph can help us find solutions to the equation.

The Conjugate Pairs Theorem

For second-degree polynomial equations, we know that the roots occur in pairs. For example, the roots of $x^2 - 2x + 5 = 0$ are

$$x = \frac{2 \pm \sqrt{(-2)^2 - 4(1)(5)}}{2} = 1 \pm 2i.$$

The roots $1 - 2i$ and $1 + 2i$ are complex conjugates. The $\pm$ symbol in the quadratic formula causes the complex solutions of a quadratic equation with real coefficients to occur in conjugate pairs. The conjugate pairs theorem indicates that this situation occurs also for polynomial equations of higher degree.

Conjugate Pairs Theorem

If $P(x) = 0$ is a polynomial equation with real coefficients and the complex number $a + bi$ $(b \neq 0)$ is a root, then $a - bi$ is also a root.

The proof for this theorem is left for the exercises.

EXAMPLE 2 Using the conjugate pairs theorem

Find a polynomial equation with real coefficients that has 2 and $1 - i$ as roots.

Solution

If the polynomial has real coefficients, then its imaginary roots occur in conjugate pairs. So a polynomial with these two roots must actually have at least three roots: 2, $1 - i$, and $1 + i$. Since each root of the equation corresponds to a factor of the polynomial, we can write the following equation.

$$(x - 2)[x - (1 - i)][x - (1 + i)] = 0$$
$$(x - 2)[x^2 - x(1 + i) - x(1 - i) + (1 - i)(1 + i)] = 0$$
$$(x - 2)(x^2 - 2x + 2) = 0 \quad (1 - i)(1 + i) = 1 - i^2 = 2$$
$$x^3 - 4x^2 + 6x - 4 = 0$$

This equation has the required roots and the smallest degree. Any multiple of this equation would also have the required roots but would not be as simple.

Descartes's Rule of Signs

None of the theorems in this chapter tells us how to find all of the n roots to a polynomial equation of degree n. However, the theorems and rules presented here add to our knowledge of polynomial equations and help us to predict the type and number of solutions to expect for a particular equation. Descartes's rule of signs is a method for determining the number of positive, negative, and imaginary solutions.

When a polynomial is written in descending order, a **variation of sign** occurs when the signs of consecutive terms change. For example, if

$$P(x) = 3x^5 - 7x^4 - 8x^3 - x^2 + 3x - 9,$$

there are sign changes in going from the first to the second term, from the fourth to the fifth term, and from the fifth to the sixth term. So there are three variations of sign for $P(x)$. This information determines the number of positive real solutions to $P(x) = 0$. Descartes's rule requires that we look at $P(-x)$ and also count the variations of sign after it is simplified:

$$\begin{aligned} P(-x) &= 3(-x)^5 - 7(-x)^4 - 8(-x)^3 - (-x)^2 + 3(-x) - 9 \\ &= -3x^5 - 7x^4 + 8x^3 - x^2 - 3x - 9 \end{aligned}$$

In $P(-x)$ the signs of the terms change from the second to the third term and again from the third to the fourth term. So there are two variations of sign for $P(-x)$. This information determines the number of negative real solutions to $P(x) = 0$.

Descartes's Rule of Signs

Suppose $P(x) = 0$ is a polynomial equation with real coefficients and with terms written in descending order.

- The number of positive real roots of the equation is either equal to the number of variations of sign of $P(x)$ or less than that by an even number.
- The number of negative real roots of the equation is either equal to the number of variations of sign of $P(-x)$ or less than that by an even number.

The proof of Descartes's rule of signs is beyond the scope of this text, but we can apply the rule to polynomial equations. Descartes's rule of signs is especially helpful when the number of variations of sign is 0 or 1.

EXAMPLE 3 Using Descartes's rule of signs

Discuss the possibilities for the roots to $2x^3 - 5x^2 - 6x + 4 = 0$.

Solution

The number of variations of sign in

$$P(x) = 2x^3 - 5x^2 - 6x + 4$$

is 2. By Descartes's rule, the number of positive real roots is either 2 or 0. Since

$$\begin{aligned} P(-x) &= 2(-x)^3 - 5(-x)^2 - 6(-x) + 4 \\ &= -2x^3 - 5x^2 + 6x + 4 \end{aligned}$$

there is one variation of sign in $P(-x)$. So there is exactly one negative real root.

The equation must have three roots, because it is a third-degree polynomial equation. Since there must be three roots and one is negative, the other two roots must be either both imaginary numbers or both positive real numbers. Table 3.1 summarizes these two possibilities.

Table 3.1 Number of roots

Positive	Negative	Imaginary
2	1	0
0	1	2

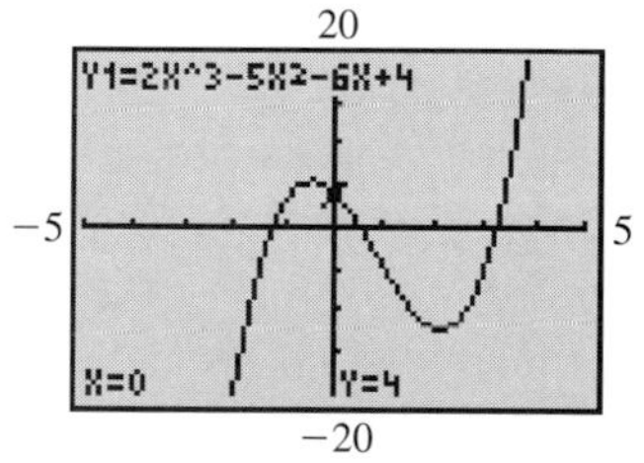

Figure 3.34

The graph of $y = 2x^3 - 5x^2 - 6x + 4$ shown in Fig. 3.34 crosses the positive x-axis twice and the negative x-axis once. So the first case in Table 3.1 is actually correct.

EXAMPLE 4 Using Descartes's rule of signs

Discuss the possibilities for the roots to $3x^4 - 5x^3 - x^2 - 8x + 4 = 0$.

Solution

There are two variations of sign in the polynomial

$$P(x) = 3x^4 - 5x^3 - x^2 - 8x + 4.$$

According to Descartes's rule, there are either two or zero positive real roots to the equation. Since

$$\begin{aligned} P(-x) &= 3(-x)^4 - 5(-x)^3 - (-x)^2 - 8(-x) + 4 \\ &= 3x^4 + 5x^3 - x^2 + 8x + 4 \end{aligned}$$

there are two variations of sign in $P(-x)$. So the number of negative real roots is either two or zero. Since the degree of the polynomial is 4, there must be four roots. Each line of Table 3.2 gives a possible distribution of the type of those four roots. Note that the number of imaginary roots is even in each case, as we would expect from the conjugate pairs theorem.

Table 3.2 **Number of roots**

Positive	Negative	Imaginary
2	2	0
2	0	2
0	2	2
0	0	4

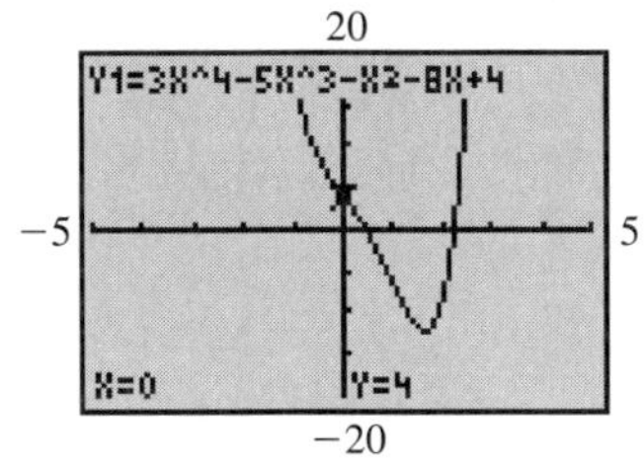

Figure 3.35

The calculator graph of $y = 3x^4 - 5x^3 - x^2 - 8x + 4$ shown in Fig. 3.35 shows two positive intercepts and no negative intercepts. However, we might not have the appropriate viewing window. The negative intercepts might be less than -5. In this case the graph did not allow us to conclude which line in Table 3.2 is correct.

Bounds on the Roots

If a polynomial equation has no roots greater than c, then c is called an **upper bound** for the roots. If there are no roots less than c, then c is called a **lower bound** for the roots. The next theorem is used to determine upper and lower bounds for the roots of a polynomial equation. We will not prove this theorem.

Theorem on Bounds

Suppose that $P(x)$ is a polynomial with real coefficients and a positive leading coefficient, and synthetic division with $x - c$ is performed.

- If $c > 0$ and all terms in the bottom row are nonnegative, then c is an upper bound for the roots of $P(x) = 0$.
- If $c < 0$ and the terms in the bottom row alternate in sign, then c is a lower bound for the roots of $P(x) = 0$.

If 0 appears in the bottom row of the synthetic division, then it may be assigned either a positive or negative sign in determining whether the signs alternate. For example, the numbers 3, 0, 5, and -6 would be alternating in sign if we assign a negative sign to 0. The numbers -7, 5, -8, 0, and -2 would be alternating in sign if we assign a positive sign to 0.

EXAMPLE 5 Finding bounds for the roots

Use the theorem on bounds to establish the best integral bounds for the roots of $2x^3 - 5x^2 - 6x + 4 = 0$.

Solution

Try synthetic division with the integers 1, 2, 3, and so on. The first integer for which all terms on the bottom row are nonnegative is the best upper bound for the roots according to the theorem on bounds. Note that we are not trying fractions because we are looking for integral bounds for the roots and not the actual roots.

$$\begin{array}{r|rrrr} 1 & 2 & -5 & -6 & 4 \\ & & 2 & -3 & -9 \\ \hline & 2 & -3 & -9 & -5 \end{array} \qquad \begin{array}{r|rrrr} 2 & 2 & -5 & -6 & 4 \\ & & 4 & -2 & -16 \\ \hline & 2 & -1 & -8 & -12 \end{array}$$

$$\begin{array}{r|rrrr} 3 & 2 & -5 & -6 & 4 \\ & & 6 & 3 & -9 \\ \hline & 2 & 1 & -3 & -5 \end{array} \qquad \begin{array}{r|rrrr} 4 & 2 & -5 & -6 & 4 \\ & & 8 & 12 & 24 \\ \hline & 2 & 3 & 6 & 28 \end{array}$$

By the theorem on bounds, no number greater than 4 can be a root to the equation. Now try synthetic division with the integers -1, -2, -3, and so on. The first negative integer for which the terms on the bottom row alternate in sign is the best lower bound for the roots.

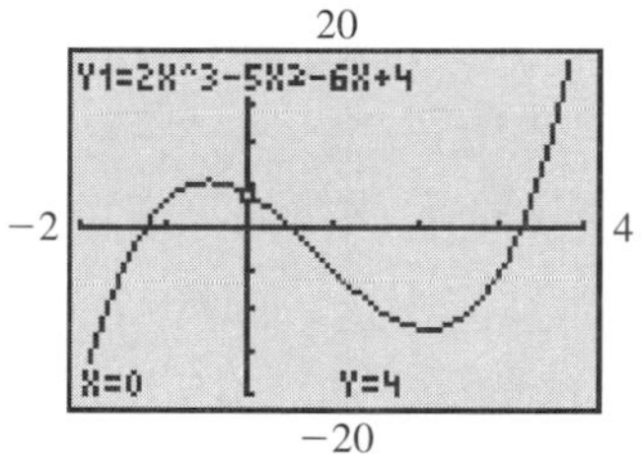

Figure 3.36

$$\begin{array}{r|rrrr} -1 & 2 & -5 & -6 & 4 \\ & & -2 & 7 & -1 \\ \hline & 2 & -7 & 1 & 3 \end{array} \qquad \begin{array}{r|rrrr} -2 & 2 & -5 & -6 & 4 \\ & & -4 & 18 & -24 \\ \hline & 2 & -9 & 12 & -20 \end{array}$$

By the theorem on bounds, no number less than -2 can be a root to the equation. So all of the real roots to this equation are between -2 and 4.

The graph of $y = 2x^3 - 5x^2 - 6x + 4$ in Fig. 3.36 shows three x-intercepts between -2 and 4, which supports our conclusion about the bounds for the roots.

In the next example we use all of the available information about roots.

EXAMPLE 6 Using all of the theorems about roots

Find all of the solutions to $2x^3 - 5x^2 - 6x + 4 = 0$.

Solution

In Example 3 we used Descartes's rule of signs on this equation to determine that it has either two positive roots and one negative root or one negative root and two imaginary roots. In Example 5 we used the theorem on bounds to determine that all of the real roots to this equation are between -2 and 4. From the rational zero theorem, the possible rational roots are ± 1, ± 2, ± 4, and $\pm 1/2$. Since there must be one negative root and it must be greater than -2, the only possible numbers from the list are -1 and $-1/2$. So start by checking $-1/2$ and -1 with synthetic division.

$$\begin{array}{r|rrrr} -\frac{1}{2} & 2 & -5 & -6 & 4 \\ & & -1 & 3 & \frac{3}{2} \\ \hline & 2 & -6 & -3 & \frac{11}{2} \end{array} \qquad \begin{array}{r|rrrr} -1 & 2 & -5 & -6 & 4 \\ & & -2 & 7 & -1 \\ \hline & 2 & -7 & 1 & 3 \end{array}$$

Since neither -1 nor $-1/2$ is a root, the negative root must be irrational. The only rational possibilities for the two positive roots smaller than 4 are 1/2, 1, and 2.

$$\begin{array}{r|rrrr} \frac{1}{2} & 2 & -5 & -6 & 4 \\ & & 1 & -2 & -4 \\ \hline & 2 & -4 & -8 & 0 \end{array}$$

Since 1/2 is a root of the equation, $x - 1/2$ is a factor of the polynomial. The last line in the synthetic division indicates that the other factor is $2x^2 - 4x - 8$.

$$\left(x - \frac{1}{2}\right)(2x^2 - 4x - 8) = 0$$

$$(2x - 1)(x^2 - 2x - 4) = 0$$

$$2x - 1 = 0 \quad \text{or} \quad x^2 - 2x - 4 = 0$$

$$x = \frac{1}{2} \quad \text{or} \quad x = \frac{2 \pm \sqrt{4 - 4(1)(-4)}}{2} = 1 \pm \sqrt{5}$$

There are two positive roots, 1/2 and $1 + \sqrt{5}$. The negative root is $1 - \sqrt{5}$. Note that the roots guaranteed by Descartes's rule of signs are real numbers but not necessarily rational numbers.

The graph of $y = 2x^3 - 5x^2 - 6x + 4$ in Fig. 3.37 supports these conclusions, because its x-intercepts appear to be $(1 - \sqrt{5}, 0)$, $(1/2, 0)$, and $(1 + \sqrt{5}, 0)$.

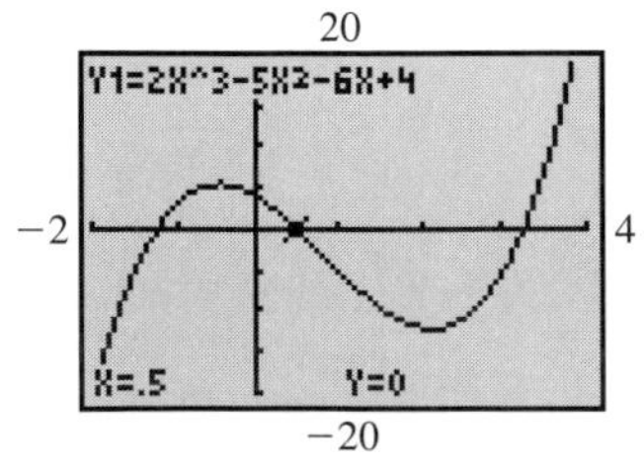

Figure 3.37

FOR THOUGHT True or False? Explain.

1. The number 1 is a root of $x^3 - 1 = 0$ with multiplicity 3.
2. The equation $x^3 = 125$ has three complex number solutions.
3. For $(x + 1)^3(x^2 - 2x + 1) = 0$, -1 is a root with multiplicity 3.
4. For $(x - 5)^3(x^2 - 3x - 10) = 0$, 5 is a root with multiplicity 3.
5. If $4 - 5i$ is a solution to a polynomial equation with real coefficients, then $5i - 4$ is also a solution to the equation.
6. If $P(x) = 0$ is a polynomial equation with real coefficients and i, $2 - 3i$, and $5 + 7i$ are roots, then the degree of $P(x)$ is at least 6.
7. Both $-3 - i\sqrt{5}$ and $3 - i\sqrt{5}$ are solutions to $5x^3 - 9x^2 + 17x - 23 = 0$.
8. Both 3/2 and 2 are solutions to $2x^5 - 4x^3 - 6x^2 - 3x - 6 = 0$.
9. The equation $x^3 - 5x^2 + 6x - 1 = 0$ has no negative roots.
10. The equation $5x^3 - 171 = 0$ has two imaginary solutions.

3.4 EXERCISES

Tape 8 Disk

State the degree of each polynomial equation. Find all of the real and imaginary roots of each equation, stating multiplicity when it is greater than one.

1. $x^2 - 10x + 25 = 0$ **2.** $x^2 - 18x + 81 = 0$

3. $x^5 - 9x^3 = 0$ **4.** $x^6 + x^4 = 0$

5. $x^4 - 2x^3 + x^2 = 0$ **6.** $x^5 - 6x^4 + 9x^3 = 0$

7. $(2x - 3)^2(3x + 4)^2 = 0$ **8.** $(2x^2 + x)^2(3x - 1)^4 = 0$

9. $x^3 - 4x^2 - 6x = 0$ **10.** $-x^3 + 8x^2 - 14x = 0$

Find each product.

11. $(x - 3i)(x + 3i)$ **12.** $(x + 6i)(x - 6i)$

13. $[x - (1 + \sqrt{2})][x - (1 - \sqrt{2})]$

14. $[x - (3 - \sqrt{5})][x - (3 + \sqrt{5})]$

15. $[x - (3 + 2i)][x - (3 - 2i)]$

16. $[x - (3 - i)][x - (3 + i)]$

17. $(x - 2)[x - (3 + 4i)][x - (3 - 4i)]$

18. $(x + 1)(x - (1 - i))(x - (1 + i))$

Find a polynomial equation with real coefficients that has the given roots.

19. $-3, 5$ **20.** $6, -1$ **21.** $-4i, 4i$ **22.** $-9i, 9i$

23. $3 - i$ **24.** $4 + i$ **25.** $-2, i$ **26.** $4, -i$

27. $0, i\sqrt{3}$ **28.** $-2, i\sqrt{2}$ **29.** $3, 1 - i$

30. $5, 4 - 3i$ **31.** $1, 2, 3$ **32.** $-1, 2, -3$

33. $1, 2 - 3i$ **34.** $-1, 4 - 2i$ **35.** $\frac{1}{2}, \frac{1}{3}, \frac{1}{4}$

36. $-\frac{1}{2}, -\frac{1}{3}, 1$ **37.** $i, 1 + i$ **38.** $3i, 3 - i$

Use Descartes's rule of signs to discuss the possibilities for the roots of each equation. Do not solve the equation.

39. $x^3 + 5x^2 + 7x + 1 = 0$ **40.** $2x^3 - 3x^2 + 5x - 6 = 0$

41. $-x^3 - x^2 + 7x + 6 = 0$ **42.** $-x^4 - 5x^2 - x + 7 = 0$

43. $y^4 + 5y^2 + 7 = 0$ **44.** $-3y^4 - 6y^2 + 7 = 0$

45. $t^4 - 3t^3 + 2t^2 - 5t + 7 = 0$

46. $-5r^4 + 4r^3 + 7r - 16 = 0$

47. $x^5 + x^3 + 5x = 0$ **48.** $x^4 - x^2 + 1 = 0$

Use the theorem on bounds to establish the best integral bounds for the roots of each equation.

49. $2x^3 - 5x^2 + 6 = 0$ **50.** $2x^3 - x^2 - 5x + 3 = 0$

51. $4x^3 + 8x^2 - 11x - 15 = 0$

52. $6x^3 + 5x^2 - 36x - 35 = 0$

53. $w^4 - 5w^3 + 3w^2 + 2w - 1 = 0$

54. $3z^4 - 7z^2 + 5 = 0$

55. $-2x^3 + 5x^2 - 3x + 9 = 0$

56. $-x^3 + 8x - 12 = 0$

Use the rational zero theorem, Descartes's rule of signs, and the theorem on bounds as aids in finding all real and imaginary roots to each equation.

57. $x^3 - 4x^2 - 7x + 10 = 0$

58. $x^3 + 9x^2 + 26x + 24 = 0$

59. $x^3 - 10x - 3 = 0$

60. $2x^3 - 7x^2 - 16 = 0$

61. $x^4 + 2x^3 - 7x^2 + 2x - 8 = 0$

62. $x^4 - 4x^3 + 7x^2 - 16x + 12 = 0$

63. $6x^3 + 25x^2 - 24x + 5 = 0$

64. $6x^3 - 11x^2 - 46x - 24 = 0$

65. $x^4 + 2x^3 - 3x^2 - 4x + 4 = 0$

66. $x^5 + 3x^3 + 2x = 0$

67. $x^4 - 6x^3 + 12x^2 - 8x = 0$

68. $x^4 + 9x^3 + 27x^2 + 27x = 0$

69. $x^6 - x^5 - x^4 + x^3 - 12x^2 + 12x = 0$

70. $2x^7 - 2x^6 + 7x^5 - 7x^4 - 4x^3 + 4x^2 = 0$

For each of the following functions use synthetic division and the theorem on bounds to find integers a and b, such that the interval $[a, b]$ contains all real zeros of the function. This method does not necessarily give the shortest interval containing all real zeros. By inspecting the graph of each function, find the shortest interval $[c, d]$ that contains all real zeros of the function with c and d integers. The second interval should be a subinterval of the first.

71. $y = 2x^3 - 3x^2 - 50x + 18$

72. $f(x) = x^3 - 33x - 58$

73. $f(x) = x^4 - 26x^2 + 153$

74. $y = x^4 + x^3 - 16x^2 - 10x + 60$

75. $y = 4x^3 - 90x^2 - 2x + 45$

76. $f(x) = x^4 - 12x^3 + 27x^2 - 6x - 8$

Solve each problem.

77. *Growth Rate for Bacteria* The instantaneous growth rate r of a colony of bacteria t hours after the start of an experiment is given by the function $r = 0.01t^3 - 0.08t^2 + 0.11t + 0.20$ for $0 \le t \le 7$. Find the times for which the instantaneous growth rate is zero.

78. *Retail Store Profit* The manager of a retail store has figured that her monthly profit P (in thousands of dollars) is determined by her monthly advertising expense x (in tens of thousands of dollars) according to the formula

$$P = x^3 - 20x^2 + 100x \quad \text{for} \quad 0 \le x \le 4.$$

For what value of x does she get \$147,000 in profit?

79. *Designing Fireworks* Marshall is designing a rocket for the Red Rocket Fireworks Company. The rocket will consist of a cardboard circular cylinder with a height that is four times as large as the radius. On top of the cylinder will be a cone with a height of 2 in. and a radius equal to the radius of the base as shown in the figure. If he wants to fill the cone and the cylinder with a total of 114π in.^{3} of powder, then what should be the radius of the cylinder?

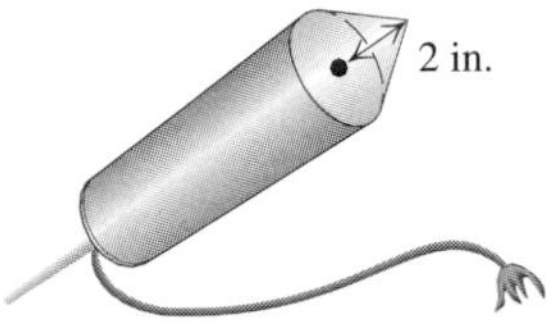

Figure for Exercise 79

80. *Heating and Air* An observatory is built in the shape of a right circular cylinder with a hemispherical roof as shown in the figure. The heating and air contractor has figured the volume of the structure as 3168π ft^3. If the height of the cylindrical walls are 2 ft more than the radius of the building, then what is the radius of the building?

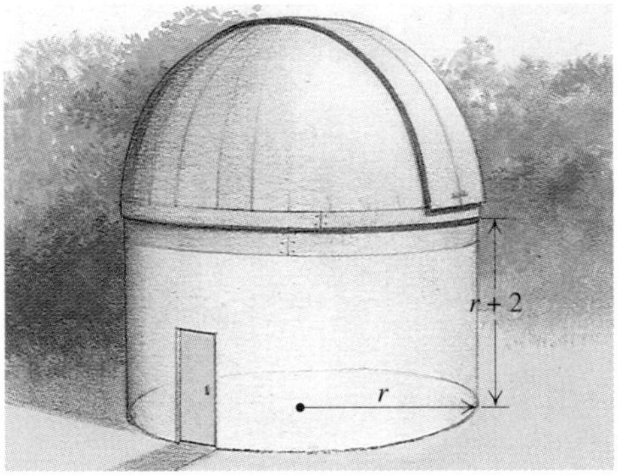

Figure for Exercise 80

For Writing/Discussion

81. *Conjugate of a Sum* Show that the conjugate of the sum of two complex numbers is equal to the sum of their conjugates.

82. *Conjugate of a Product* Show that the conjugate of the product of two complex numbers is equal to the product of their conjugates.

83. *Conjugate of a Real Number* Show that $\bar{a} = a$ for any real number a.

84. *Conjugate Pairs* Assume that $a + bi$ is a root of $a_n x^n + a_{n-1}x^{n-1} + \cdots + a_1 x + a_0 = 0$ and substitute $a + bi$ for x. Take the conjugate of each side of the resulting equation and use the results of Exercises 81–83 to simplify it. Your final equation should show that $a - bi$ is a root of $a_n x^n + a_{n-1}x^{n-1} + \cdots + a_1 x + a_0 = 0$. This proves the conjugate pairs theorem.

85. *Missing Polynomials* Find a third-degree polynomial function such that $f(0) = 3$ and whose zeros are 1, 2, and 3. Explain how you found it.

86. *Missing Polynomials* Is there a third-degree polynomial function such that $f(0) = 6$ and $f(-1) = 12$ and whose zeros are 1, 2, and 3? Explain.

LINKING CONCEPTS

For Individual or Group Explorations

Designing a Crystal Ball *The wizard Gandalf is creating a massive crystal ball. To achieve maximum power, the orb must be a perfect solid sphere of clear crystal mounted on a square solid silver base, as shown in the figure.*

a) Write a formula for the total volume of material used in terms of the radius of the sphere, the thickness of the base, and the length of the side of the base.

b) The wizard has determined that the diameter of the sphere must equal the length of the side of the square base, and the thickness of the base must be π in. Find the exact radius of the sphere if the total amount of material used in the sphere and base must be 1296π in.^{3}.

c) If it turns out that the sphere in part (b) is not powerful enough, Gandalf plans to make a sphere and base of solid dilitheum crystal. For this project the diameter of the sphere must be 2 in. less than the length of the side of the base and the thickness of the base must be 2 in. Find the approximate radius of the sphere if the total volume of material used in the sphere and base must be 5000 in.^{3}.

3.5 Miscellaneous Equations

In Section 3.4 we learned that an nth-degree polynomial equation has n roots. However, it is not always obvious how to find them. In this section we will solve polynomial equations using some new techniques and we will solve several other types of equations. Unfortunately, we cannot generally predict the number of roots to nonpolynomial equations.

Factoring Higher-Degree Equations

We usually use factoring to solve quadratic equations. However, since we can also factor many higher-degree polynomials, we can solve many higher-degree equations by factoring. Factoring is often the fastest method for solving an equation.

EXAMPLE 1 Solving an equation by factoring

Solve $x^3 + 3x^2 + x + 3 = 0$.

Solution

Factor the polynomial on the left-hand side by grouping.

$$x^2(x + 3) + 1(x + 3) = 0 \qquad \text{Factor by grouping.}$$
$$(x^2 + 1)(x + 3) = 0 \qquad \text{Factor out } x + 3.$$
$$x^2 + 1 = 0 \quad \text{or} \quad x + 3 = 0 \qquad \text{Zero factor property}$$
$$x^2 = -1 \quad \text{or} \quad x = -3$$
$$x = \pm i \quad \text{or} \quad x = -3$$

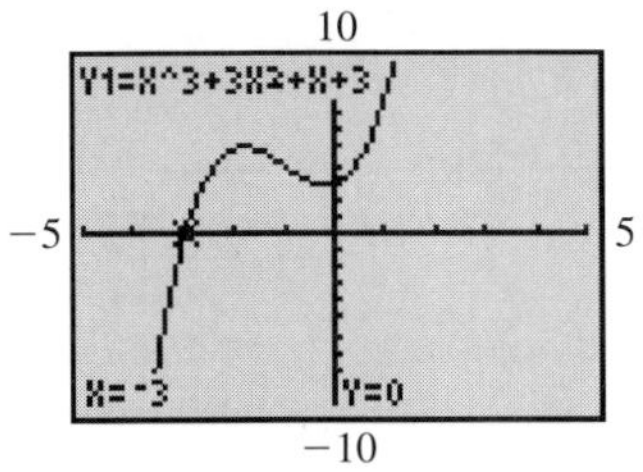

Figure 3.38

The solution set is $\{-3, -i, i\}$.
The graph in Fig. 3.38 supports these solutions. Because there is only one real solution the graph crosses the x-axis only once.

EXAMPLE 2 Solving an equation by factoring

Solve $2x^5 = 16x^2$.

Solution

Write the equation with 0 on the right-hand side, then factor completely.

$$2x^5 - 16x^2 = 0$$
$$2x^2(x^3 - 8) = 0 \qquad \text{Factor out greatest common factor.}$$
$$2x^2(x - 2)(x^2 + 2x + 4) = 0 \qquad \text{Factor difference of two cubes.}$$
$$2x^2 = 0 \text{ or } x - 2 = 0 \text{ or } x^2 + 2x + 4 = 0$$
$$x = 0 \text{ or } \quad x = 2 \text{ or } \quad x = \frac{-2 \pm \sqrt{-12}}{2} = -1 \pm i\sqrt{3}$$

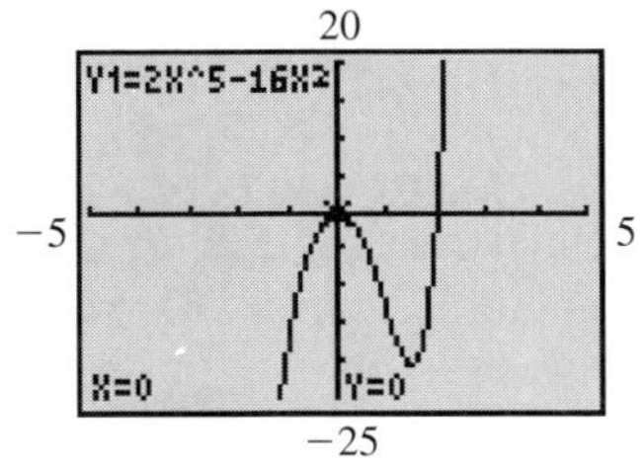

Figure 3.39

The solution set is $\{0, 2, -1 \pm i\sqrt{3}\}$. Since 0 is a root with multiplicity 2, there are five roots, counting multiplicity, to this fifth-degree equation.
The graph in Fig. 3.39 supports this solution.

Note that in Example 2, if we had divided each side by x^2 as our first step, we would have lost the solution $x = 0$. *We do not usually divide each side of an equation by a variable expression.* Instead, bring all expressions to the same side and factor out the common factors.

Equations Involving Square Roots

Recall that $\sqrt{x}$ represents the nonnegative square root of x. To solve $\sqrt{x} = 3$ we can use the definition of square root. Since the nonnegative square root of 9 is 3, the solution to $\sqrt{x} = 3$ is 9. To solve $\sqrt{x} = -3$ we again use the definition of square root. Because $\sqrt{x}$ is nonnegative while -3 is negative, this equation has no solution.

More complicated equations involving square roots are usually solved by squaring both sides. However, squaring both sides does not always lead to an equivalent equation. If we square both sides of $\sqrt{x} = 3$ we get $x = 9$, which is equivalent to $\sqrt{x} = 3$. But if we square both sides of $\sqrt{x} = -3$ we also get $x = 9$, which is not equivalent to $\sqrt{x} = -3$. Because 9 appeared in the attempt to solve $\sqrt{x} = -3$, but does not satisfy the equation, it is called an *extraneous root*. This same situation can occur with an equation involving a fourth root or any other even root. So if you raise each side of an equation to an even power, you must check for extraneous roots.

EXAMPLE 3 **Squaring each side to solve an equation**

Solve $\sqrt{x} + 2 = x$.

Solution

Isolate the radical before squaring each side.

$$\sqrt{x} = x - 2$$

$$(\sqrt{x})^2 = (x - 2)^2 \qquad \text{Square each side.}$$

$$x = x^2 - 4x + 4 \qquad \text{Use the special product } (a - b)^2 = a^2 - 2ab + b^2.$$

$$0 = x^2 - 5x + 4 \qquad \text{Write in the form } ax^2 + bx + c = 0.$$

$$0 = (x - 4)(x - 1) \qquad \text{Factor the quadratic polynomial.}$$

$$x - 4 = 0 \quad \text{or} \quad x - 1 = 0 \qquad \text{Zero factor property}$$

$$x = 4 \quad \text{or} \quad x = 1$$

Checking $x = 4$, we get $\sqrt{4} + 2 = 4$, which is correct. Checking $x = 1$, we get $\sqrt{1} + 2 = 1$, which is incorrect. So 1 is an extraneous root and the solution set is $\{4\}$.

The graph in Fig. 3.40 supports this solution.

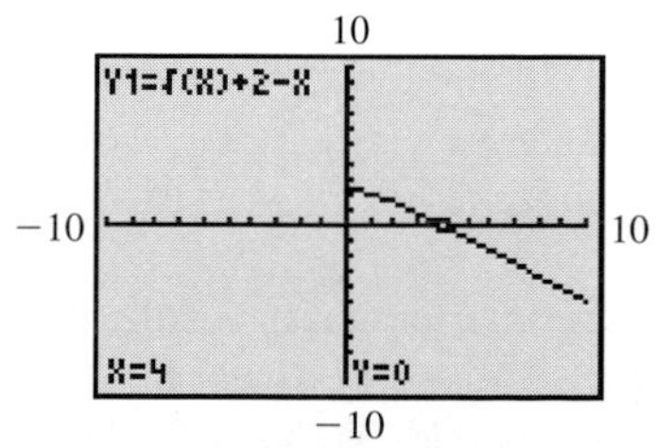

Figure 3.40

The next example involves two radicals. In this example we will isolate the more complicated radical before squaring each side. But not all radicals are eliminated upon squaring each side. So we isolate the remaining radical and square each side again.

EXAMPLE 4 **Squaring each side twice**

Solve $\sqrt{2x+1} - \sqrt{x} = 1$.

Solution

First we write the equation so that the more complicated radical is isolated. Then we square each side. On the left side, when we square $\sqrt{2x+1}$ we get $2x + 1$. On the right side, when we square $1 + \sqrt{x}$, we use the special product rule $(a+b)^2 = a^2 + 2ab + b^2$.

$$\sqrt{2x+1} = 1 + \sqrt{x}$$

$$(\sqrt{2x+1})^2 = (1+\sqrt{x})^2 \quad \text{Square each side.}$$

$$2x + 1 = 1 + 2\sqrt{x} + x$$

$$x = 2\sqrt{x} \quad \text{All radicals are not eliminated by the first squaring.}$$

$$x^2 = (2\sqrt{x})^2 \quad \text{Square each side a second time.}$$

$$x^2 = 4x$$

$$x^2 - 4x = 0$$

$$x(x-4) = 0$$

$$x = 0 \quad \text{or} \quad x - 4 = 0$$

$$x = 0 \quad \text{or} \quad x = 4$$

1

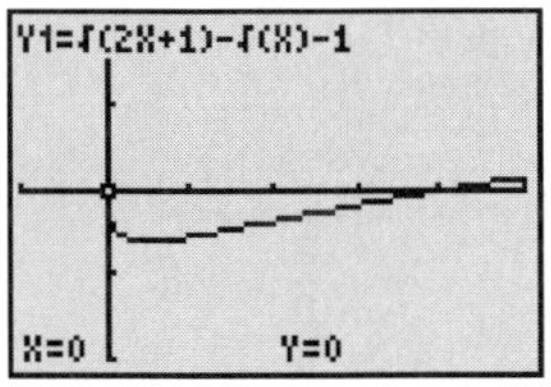

−1 5 −1

Figure 3.41

Both 0 and 4 satisfy the original equation. So the solution set is $\{0, 4\}$. The graph in Fig. 3.41 supports this solution.

Equations with Rational Exponents

To solve equations of the form $x^{m/n} = k$ in which m and n are positive integers and m/n is in lowest terms, we adapt the methods of Examples 3 and 4 of raising each side to a power. Cubing each side of $x^{2/3} = 4$, yields $(x^{2/3})^3 = 4^3$ or $x^2 = 64$. By the square root property, $x = \pm 8$. We can shorten this solution by raising each side of the equation to the power 3/2 (the reciprocal of 2/3) and inserting the $\pm$ symbol to obtain the two square roots.

$$x^{2/3} = 4$$

$$(x^{2/3})^{3/2} = \pm 4^{3/2} \quad \text{Raise each side to the power 3/2 and insert } \pm.$$

$$x = \pm 8$$

The equation $x^{2/3} = 4$ has two solutions because the numerator of the exponent 2/3 is an even number. An equation such as $x^{-3/2} = 1/8$ has only one solution because the numerator of the exponent $-3/2$ is odd. To solve $x^{-3/2} = 1/8$, raise each side to the power $-2/3$ (the reciprocal of $-3/2$).

$$x^{-3/2} = \frac{1}{8}$$

$$(x^{-3/2})^{-2/3} = \left(\frac{1}{8}\right)^{-2/3} \quad \text{Raise each side to the power } -2/3.$$

$$x = 4$$

In the next example we solve two more equations of this type by raising each side to a fractional power. Note that we use the ± symbol only when the numerator of the original exponent is even.

EXAMPLE 5 **Equations with rational exponents**

Solve each equation.

a) $x^{4/3} = 625$ **b)** $(y - 2)^{-5/2} = 32$

Solution

a) Raise each side of the equation to the power 3/4. Use the ± symbol because the numerator of 4/3 is even.

$$x^{4/3} = 625$$

$$(x^{4/3})^{3/4} = \pm 625^{3/4}$$

$$x = \pm 125$$

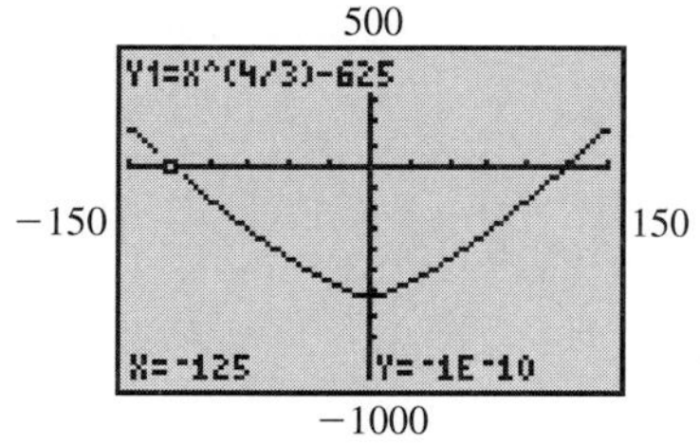

Figure 3.42

Check in the original equation. The solution set is $\{-125, 125\}$.

The graph in Fig. 3.42 supports this solution. □

b) Raise each side to the power −2/5. Because the numerator in −5/2 is an odd number, there is only one solution.

$$(y - 2)^{-5/2} = 32$$

$$((y - 2)^{-5/2})^{-2/5} = 32^{-2/5} \quad \text{Raise each side to the power } -2/5.$$

$$y - 2 = \frac{1}{4}$$

$$y = 2 + \frac{1}{4} = \frac{9}{4}$$

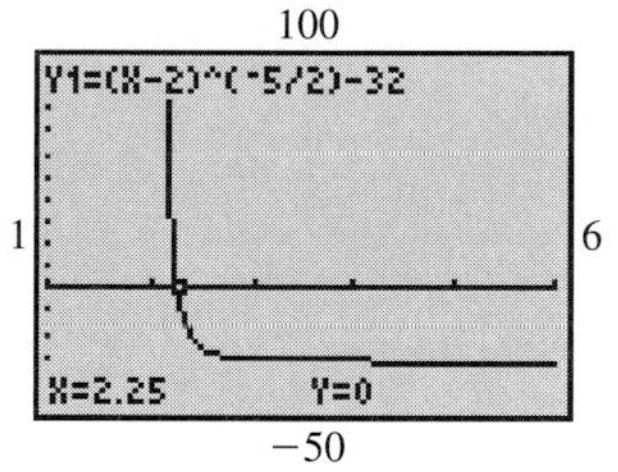

Figure 3.43

Check 9/4 in the original equation. The solution set is $\{\frac{9}{4}\}$.

The graph in Fig. 3.43 supports this solution.

Equations of Quadratic Type

In some cases, an equation can be converted to a quadratic equation by substituting a single variable for a more complicated expression. Such equations are called **equations of quadratic type**. An equation of quadratic type has the form $au^2 + bu + c = 0$, where $a \neq 0$ and u is an algebraic expression.

In the next example, the expression x^2 in a fourth-degree equation is replaced by u, yielding a quadratic equation. After the quadratic equation is solved, u is replaced by x^2 so that we find values for x that satisfy the original fourth-degree equation.

EXAMPLE 6 Solving a fourth-degree polynomial equation

Solve $x^4 - 14x^2 + 45 = 0$.

Solution

We let $u = x^2$ so that $u^2 = (x^2)^2 = x^4$.

$$(x^2)^2 - 14x^2 + 45 = 0$$

$$u^2 - 14u + 45 = 0 \qquad \text{Replace } x^2 \text{ by } u.$$

$$(u - 9)(u - 5) = 0$$

$$u - 9 = 0 \quad \text{or} \quad u - 5 = 0$$

$$u = 9 \quad \text{or} \quad u = 5$$

$$x^2 = 9 \quad \text{or} \quad x^2 = 5 \qquad \text{Replace } u \text{ by } x^2.$$

$$x = \pm 3 \quad \text{or} \quad x = \pm\sqrt{5}$$

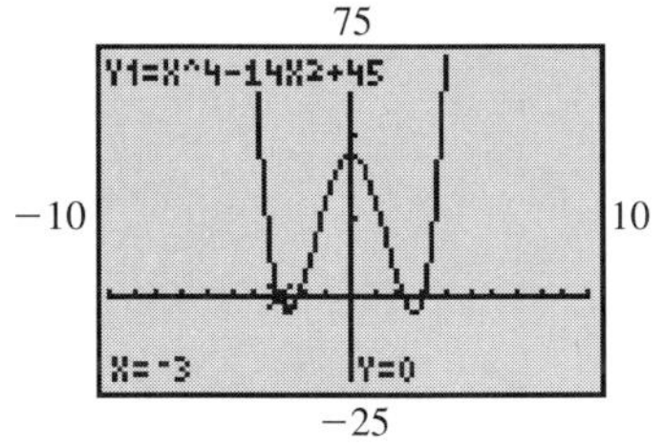

Figure 3.44

Check in the original equation. The solution set is $\{-3, -\sqrt{5}, \sqrt{5}, 3\}$. The graph in Fig. 3.44 supports this solution.

Note that the equation of Example 6 could be solved by factoring without doing substitution, because $x^4 - 14x^2 + 45 = (x^2 - 9)(x^2 - 5)$. Since the next example involves a more complicated algebraic expression, we use substitution to simplify it, although it too could be solved by factoring, without substitution.

EXAMPLE 7 Another equation of quadratic type

Solve $(x^2 - x)^2 - 18(x^2 - x) + 72 = 0$.

Solution

If we let $u = x^2 - x$, then the equation becomes a quadratic equation.

$$(x^2 - x)^2 - 18(x^2 - x) + 72 = 0$$

$$u^2 - 18u + 72 = 0 \qquad \text{Replace } x^2 - x \text{ by } u.$$

$$(u - 6)(u - 12) = 0$$

$$\begin{aligned} u - 6 &= 0 & &\text{or} & u - 12 &= 0 \\ u &= 6 & &\text{or} & u &= 12 \\ x^2 - x &= 6 & &\text{or} & x^2 - x &= 12 \end{aligned}$$

Replace u by $x^2 - x$.

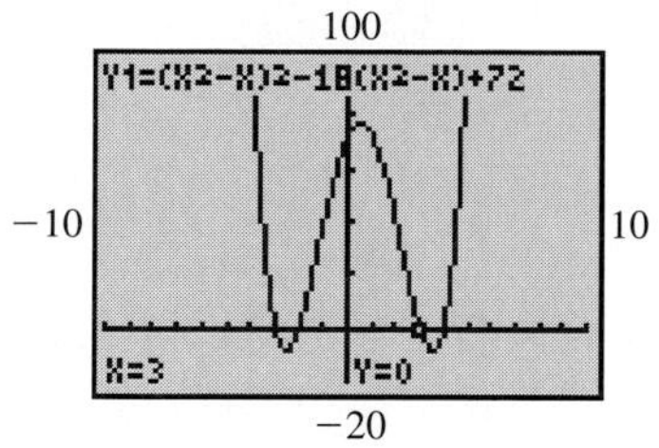

Figure 3.45

$$\begin{aligned} x^2 - x - 6 &= 0 & &\text{or} & x^2 - x - 12 &= 0 \\ (x - 3)(x + 2) &= 0 & &\text{or} & (x - 4)(x + 3) &= 0 \end{aligned}$$

$$x = 3 \quad \text{or} \quad x = -2 \quad \text{or} \quad x = 4 \quad \text{or} \quad x = -3$$

Check in the original equation. The solution set is $\{-3, -2, 3, 4\}$.
The graph in Fig. 3.45 supports this solution.

The next equations of quadratic type have rational exponents.

EXAMPLE 8 Quadratic type and rational exponents

Find all real solutions to each equation.

a) $x^{2/3} - 9x^{1/3} + 8 = 0$ **b)** $(11x^2 - 18)^{1/4} = x$

Solution

a) If we let $u = x^{1/3}$, then $u^2 = (x^{1/3})^2 = x^{2/3}$.

$$\begin{aligned} u^2 - 9u + 8 &= 0 \\ (u - 8)(u - 1) &= 0 \end{aligned}$$

Replace $x^{2/3}$ by u^2 and $x^{1/3}$ by u.

$$\begin{aligned} u &= 8 & &\text{or} & u &= 1 \\ x^{1/3} &= 8 & &\text{or} & x^{1/3} &= 1 \\ (x^{1/3})^3 &= 8^3 & &\text{or} & (x^{1/3})^3 &= 1^3 \\ x &= 512 & &\text{or} & x &= 1 \end{aligned}$$

Replace u by $x^{1/3}$.

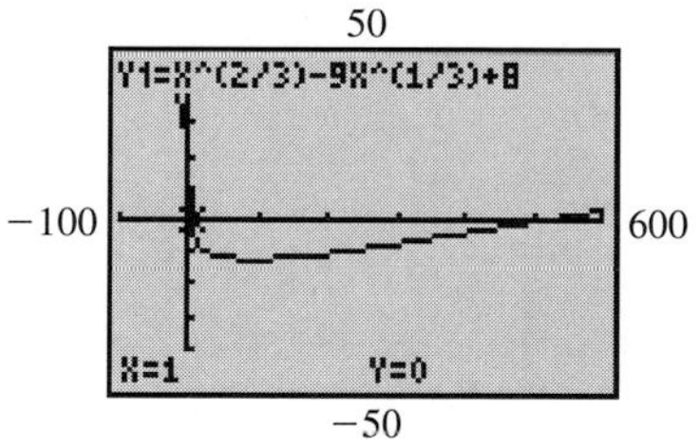

Figure 3.46

Check in the original equation. The solution set is $\{1, 512\}$.
The graph in Fig. 3.46 supports this solution. □

b)

$$\begin{aligned} (11x^2 - 18)^{1/4} &= x \\ ((11x^2 - 18)^{1/4})^4 &= x^4 \\ 11x^2 - 18 &= x^4 \\ x^4 - 11x^2 + 18 &= 0 \\ (x^2 - 9)(x^2 - 2) &= 0 \end{aligned}$$

Raise each side to the power 4.

$$\begin{aligned} x^2 &= 9 & &\text{or} & x^2 &= 2 \\ x &= \pm 3 & &\text{or} & x &= \pm\sqrt{2} \end{aligned}$$

Since the exponent 1/4 means principal fourth root, the right-hand side of the equation cannot be negative. So -3 and $-\sqrt{2}$ are extraneous roots. Since 3 and $\sqrt{2}$ satisfy the original equation, the solution set is $\{\sqrt{2}, 3\}$.
The graph in Fig. 3.47 supports this solution.

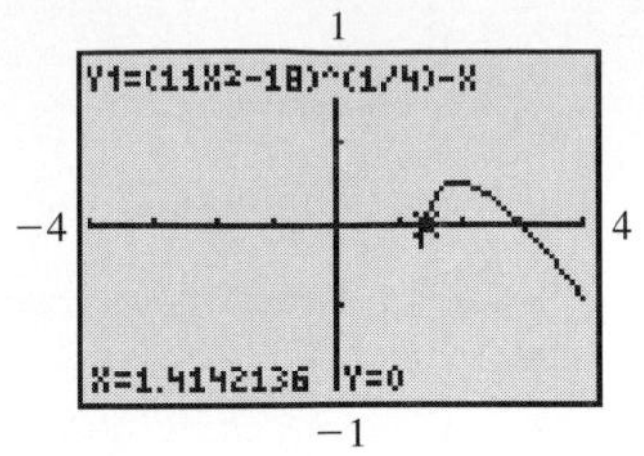

Figure 3.47

Equations Involving Absolute Value

We solved basic absolute value equations in Section 1.1. In the next two examples we solve some more complicated absolute value equations.

EXAMPLE 9 An equation involving absolute value

Solve $|x^2 - 2x - 16| = 8$.

Solution

First write an equivalent statement without using absolute value symbols.

$$\begin{array}{lcl} x^2 - 2x - 16 = 8 & \text{or} & x^2 - 2x - 16 = -8 \\ x^2 - 2x - 24 = 0 & \text{or} & x^2 - 2x - 8 = 0 \\ (x - 6)(x + 4) = 0 & \text{or} & (x - 4)(x + 2) = 0 \\ x = 6 \quad \text{or} \quad x = -4 & \text{or} & x = 4 \quad \text{or} \quad x = -2 \end{array}$$

The solution set is $\{-4, -2, 4, 6\}$.
The graph in Fig. 3.48 supports this solution.

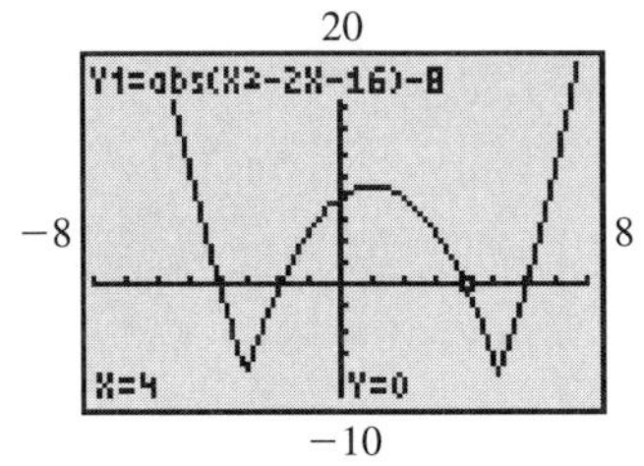

Figure 3.48

In the next example we have an equation in which an absolute value expression is equal to an expression that could be positive or negative and an equation with two absolute value expressions.

EXAMPLE 10 More equations involving absolute value

Solve each equation.

a) $|x^2 - 6| = 5x$ **b)** $|a - 1| = |2a - 3|$

Solution

a) Since $|x^2 - 6|$ is nonnegative for any value of x, $5x$ must be nonnegative. Write the equivalent statement assuming that $5x$ is nonnegative:

$$\begin{array}{lcl} x^2 - 6 = 5x & \text{or} & x^2 - 6 = -5x \\ x^2 - 5x - 6 = 0 & \text{or} & x^2 + 5x - 6 = 0 \\ (x - 6)(x + 1) = 0 & \text{or} & (x + 6)(x - 1) = 0 \\ x = 6 \quad \text{or} \quad x = -1 & \text{or} & x = -6 \quad \text{or} \quad x = 1 \end{array}$$

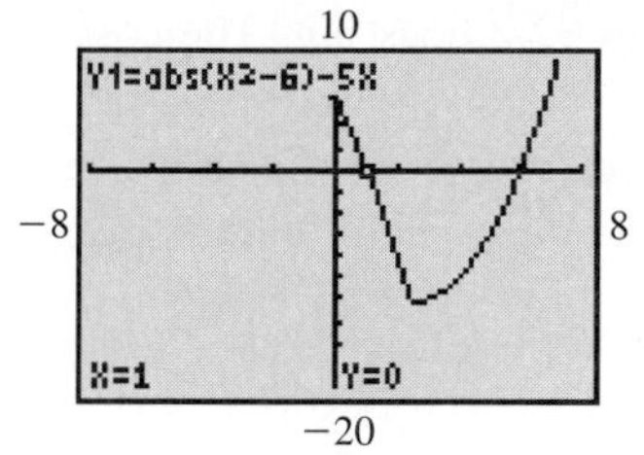

Figure 3.49

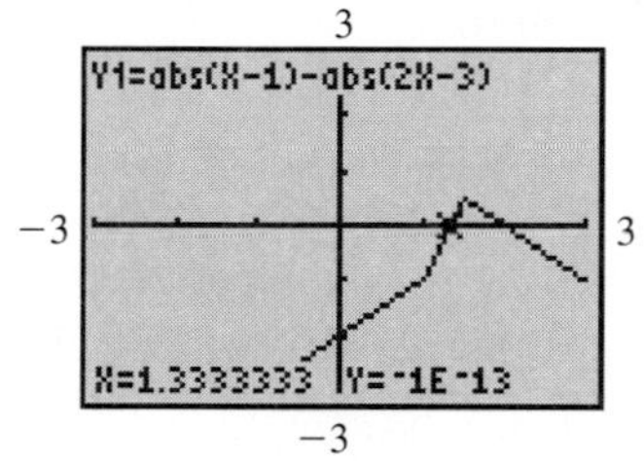

Figure 3.50

The expression $|x^2 - 6|$ is nonnegative for any real number x. But $5x$ is negative if $x = -1$ or if $x = -6$. So -1 and -6 are extraneous roots. They do not satisfy the original equation. The solution set is $\{1, 6\}$.
The graph in Fig. 3.49 supports this solution.

b) The equation $|a - 1| = |2a - 3|$ indicates that $a - 1$ and $2a - 3$ have the same absolute value. If two quantities have the same absolute value, they are either equal or opposites. Use this fact to write an equivalent statement without absolute value signs.

$$a - 1 = 2a - 3 \quad \text{or} \quad a - 1 = -(2a - 3)$$
$$a + 2 = 2a \quad \text{or} \quad a - 1 = -2a + 3$$
$$2 = a \quad \text{or} \quad a = \frac{4}{3}$$

Check that both 2 and $\frac{4}{3}$ satisfy the original absolute value equation. The solution set is $\{\frac{4}{3}, 2\}$.
The graph in Fig. 3.50 supports this solution.

Applications

The **break-even point** for a business is the point at which the cost of doing business is equal to the revenue generated by the business. The business is profitable when the revenue is greater than the cost.

EXAMPLE 11 Break-even point for a bus tour

A tour operator uses the equation $C = 3x + \sqrt{50x + 9000}$ to find his cost in dollars for taking x people on a tour of San Francisco.

a) For what value of x is the cost \$160?

b) If he charges \$10 per person for the tour, then what is his break-even point?

Solution

a) Replace C by \$160 and solve the equation.

$$160 = 3x + \sqrt{50x + 9000}$$
$$160 - 3x = \sqrt{50x + 9000} \qquad \text{Isolate the radical.}$$
$$25{,}600 - 960x + 9x^2 = 50x + 9000 \qquad \text{Square each side.}$$
$$9x^2 - 1010x + 16{,}600 = 0$$
$$x = \frac{-(-1010) \pm \sqrt{1010^2 - 4(9)(16{,}600)}}{2(9)}$$
$$= 20 \quad \text{or} \quad 92.2$$

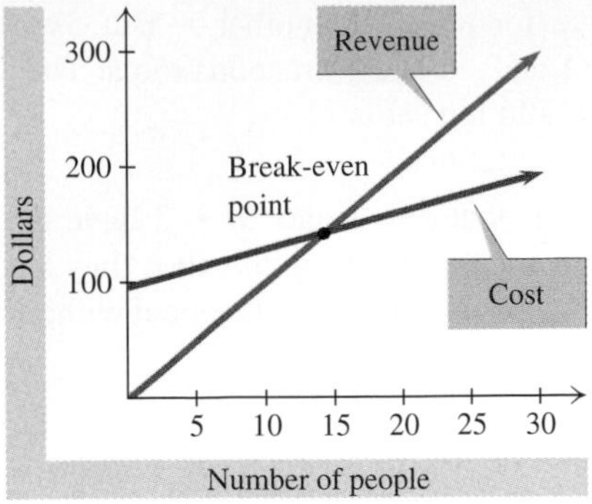

Figure 3.51

Check that 20 satisfies the original equation but 92.2 does not. The cost is \$160 when 20 people take the tour.

b) At \$10 per person, the revenue in dollars is given by $R = 10x$. When the revenue is equal to the cost as shown in Fig. 3.51, the operator breaks even.

$$10x = 3x + \sqrt{50x + 9000}$$
$$7x = \sqrt{50x + 9000}$$
$$49x^2 = 50x + 9000 \quad \text{Square each side.}$$
$$49x^2 - 50x - 9000 = 0$$
$$x = \frac{50 \pm \sqrt{50^2 - 4(49)(-9000)}}{2(49)}$$
$$= -13.052 \quad \text{or} \quad 14.072$$

If 14.072 people took the tour, the operator would break even. Since the break-even point is not a whole number, the operator actually needs 15 people to make a profit.

For Thought True or False? Explain.

1. Squaring each side of $\sqrt{x - 1} + \sqrt{x} = 6$ yields $x - 1 + x = 36$.
2. The equations $(2x - 1)^2 = 9$ and $2x - 1 = 3$ are equivalent.
3. The equations $x^{2/3} = 9$ and $x = 27$ have the same solution set.
4. To solve $2x^{1/4} - x^{1/2} + 3 = 0$, we let $u = x^{1/2}$ and $u^2 = x^{1/4}$.
5. If $(x - 1)^{-2/3} = 4$, then $x = 1 \pm 4^{-3/2}$.
6. No negative number satisfies $x^{-2/5} = 4$.
7. The solution set to $|2x + 10| = 3x$ is $\{-2, 10\}$.
8. No negative number satisfies $|x^2 - 3x + 2| = 7x$.
9. The equation $|2x + 1| = |x|$ is equivalent to $2x + 1 = x$ or $2x + 1 = -x$.
10. The equation $x^9 - 5x^3 + 6 = 0$ is an equation of quadratic type.

3.5 Exercises

Tape 8 Disk

Find all real and imaginary solutions to each equation. Check your answers.

1. $x^3 + 3x^2 - 4x - 12 = 0$
2. $x^3 - x^2 - 5x + 5 = 0$
3. $2x^3 + 1000x^2 - x - 500 = 0$
4. $3x^3 - 1200x^2 - 2x + 800 = 0$
5. $a^3 + 5a = 15a^2$
6. $b^3 + 20b = 9b^2$
7. $3y^4 - 12y^2 = 0$
8. $5m^4 - 10m^3 + 5m^2 = 0$
9. $a^4 - 16 = 0$
10. $w^4 + 8w = 0$

Find all real solutions to each equation. Check your answers.

11. $\sqrt{x + 1} = x - 5$
12. $\sqrt{x - 1} = x - 7$
13. $\sqrt{x} - 2 = x - 22$
14. $3 + \sqrt{x} = 1 + x$
15. $w = \dfrac{\sqrt{1 - 3w}}{2}$
16. $t = \dfrac{\sqrt{2 - 3t}}{3}$
17. $\dfrac{1}{z} = \dfrac{3}{\sqrt{4z + 1}}$
18. $\dfrac{1}{p} - \dfrac{2}{\sqrt{9p + 1}} = 0$
19. $\sqrt{x^2 - 2x - 15} = 3$
20. $\sqrt{3x^2 + 5x - 3} = x$
21. $\sqrt{x + 40} - \sqrt{x} = 4$
22. $\sqrt{x} + \sqrt{x - 36} = 2$

23. $\sqrt{n+4} + \sqrt{n-1} = 5$

24. $\sqrt{y+10} - \sqrt{y-2} = 2$

25. $\sqrt{2x+5} + \sqrt{x+6} = 9$

26. $\sqrt{3x-2} - \sqrt{x-2} = 2$

Find all real solutions to each equation. Check your answers.

27. $x^{2/3} = 2$ **28.** $x^{2/3} = \frac{1}{2}$ **29.** $w^{-4/3} = 16$

30. $w^{-3/2} = 27$ **31.** $t^{-1/2} = 7$ **32.** $t^{-1/2} = \frac{1}{2}$

33. $(s-1)^{-1/2} = 2$ **34.** $(s-2)^{-1/2} = \frac{1}{3}$

Find all real and imaginary solutions to each equation. Check your answers.

35. $x^4 - 12x^2 + 27 = 0$ **36.** $x^4 + 10 = 7x^2$

37. $\left(\frac{2c-3}{5}\right)^2 + 2\left(\frac{2c-3}{5}\right) = 8$

38. $\left(\frac{b-5}{6}\right)^2 - \left(\frac{b-5}{6}\right) - 6 = 0$

39. $\frac{1}{(5x-1)^2} + \frac{1}{5x-1} - 12 = 0$

40. $\frac{1}{(x-3)^2} + \frac{2}{x-3} - 24 = 0$

41. $(v^2 - 4v)^2 - 17(v^2 - 4v) + 60 = 0$

42. $(u^2 + 2u)^2 - 2(u^2 + 2u) - 3 = 0$

43. $x - 4\sqrt{x} + 3 = 0$ **44.** $2x + 3\sqrt{x} - 20 = 0$

45. $q - 7q^{1/2} + 12 = 0$ **46.** $h + 1 = 2h^{1/2}$

47. $x^{2/3} + 10 = 7x^{1/3}$ **48.** $x^{1/2} - 3x^{1/4} + 2 = 0$

Solve each absolute value equation.

49. $|w^2 - 4| = 3$ **50.** $|a^2 - 1| = 1$

51. $|v^2 - 3v| = 5v$ **52.** $|z^2 - 12| = z$

53. $|x^2 - x - 6| = 6$ **54.** $|2x^2 - x - 2| = 1$

55. $|x + 5| = |2x + 1|$ **56.** $|3x - 4| = |x|$

Solve each equation. Find imaginary solutions when possible.

57. $\sqrt{16x+1} - \sqrt{6x+13} = -1$

58. $\sqrt{16x+1} - \sqrt{6x+13} = 1$

59. $v^6 - 64 = 0$ **60.** $t^4 - 1 = 0$

61. $(7x^2 - 12)^{1/4} = x$ **62.** $(10x^2 - 1)^{1/4} = 2x$

63. $\sqrt[3]{2 + x - 2x^2} = x$

64. $\sqrt{48 + \sqrt{x}} - 4 = \sqrt[4]{x}$

65. $\left(\frac{x-2}{3}\right)^2 - 2\left(\frac{x-2}{3}\right) + 10 = 0$

66. $\frac{1}{(x+1)^2} - \frac{2}{x+1} + 2 = 0$

67. $(3u - 1)^{2/5} = 2$

68. $(2u + 1)^{2/3} = 3$

69. $x^2 - 11\sqrt{x^2+1} + 31 = 0$

70. $2x^2 - 3\sqrt{2x^2-3} - 1 = 0$

71. $|x^2 - 2x| = |3x - 6|$ **72.** $|x^2 + 5x| = |3 - x^2|$

73. $(3m + 1)^{-3/5} = -\frac{1}{8}$ **74.** $(1 - 2m)^{-5/3} = -\frac{1}{32}$

75. $|x^2 - 4| = x - 2$ **76.** $|x^2 + 7x| = x^2 - 4$

The International America's Cup Rules, which took effect in 1989, establish the boundaries of a new class of longer, lighter, and faster yachts. In addition to 200 pages of other rules, the basic dimensions are governed by an equation that balances length L in meters, sail area S in square meters, and displacement D in cubic meters. (*Scientific American*, May 1992)

77. *Maximum Sail Area for a Yacht* The maximum sail area S for a boat with length 21.24 m and a displacement of 18.34 m^3 is determined by the equation $L + 1.25S^{1/2} - 9.8D^{1/3} = 16.296$. Find the maximum sail area for this boat.

Figure for Exercises 77 and 78

78. *Minimum Displacement for a Yacht* The minimum displacement D for a boat with length 21.52 m and a sail area of 310.64 m^2 is determined by the equation $L + 1.25S^{1/2} - 9.8D^{1/3} = 16.296$. Find this boat's minimum displacement.

Solve each problem.

79. *Cost of Baking Bread* The daily cost for baking x loaves of bread at Juanita's Bakery is given in dollars by $C = 0.5x + \sqrt{8x + 5000}$. Find the number of loaves for which the cost is \$83.50.

80. *Break-Even Analysis* If the bread in Exercise 79 sells for \$1.10 per loaf, then what is the minimum number of loaves that Juanita must bake and sell to make a profit?

81. *Square Roots* Find two numbers that differ by 6 and whose square roots differ by 1.

82. *Right Triangle* One leg of a right triangle is 1 cm longer than the other leg. What is the length of the short leg if the total length of the hypotenuse and the short leg is 10 cm?

83. *Perimeter of a Right Triangle* A sign in the shape of a right triangle has one leg that is 7 in. longer than the other leg. What is the length of the shorter leg if the perimeter is 30 in.?

84. *Right Triangle Inscribed in a Semicircle* One leg of a right triangle is 1 ft longer than the other leg. If the triangle is inscribed in a circle and the hypotenuse of the triangle is the diameter of the circle, then what is the length of the radius for which the length of the radius is equal to the length of the shortest side?

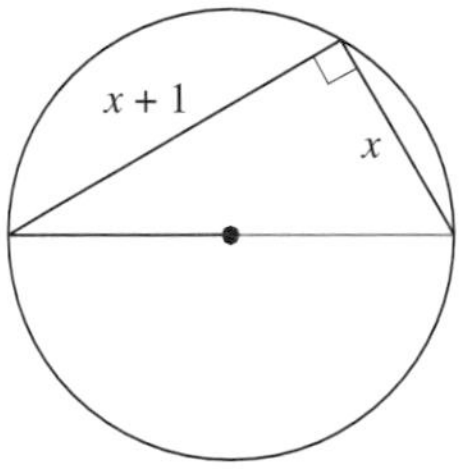

Figure for Exercise 84

85. *Area of a Foundation* The original plans for Jennifer's house called for a square foundation. After increasing one side by 30 ft and decreasing the other by 10 ft, the area of the rectangular foundation was 2100 ft^2. What was the area of the original square foundation?

86. *Shipping Carton* Heloise designed a cubic box for shipping paper. The height of the box was acceptable, but the paper would not fit into the box. After increasing the length by 0.5 in. and the width by 6 in. the area of the bottom was 119 in.^{2} and the paper would fit. What was the original volume of the cubic box?

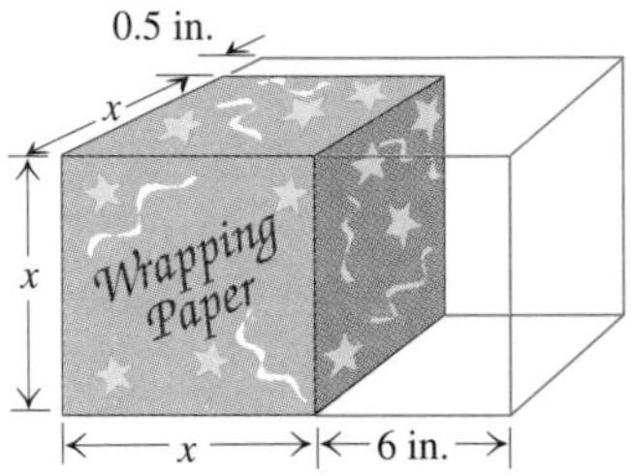

Figure for Exercise 86

87. *Sail Area-Displacement Ratio* The sail area-displacement ratio S is defined as

$$S = A\left(\frac{d}{64}\right)^{-2/3},$$

where A is the sail area in square feet and d is the displacement in pounds. The Oceanis 381 is a 39 ft sailboat with a sail area-displacement ratio of 14.26 and a sail area of 598.9 ft^2. Find the displacement for the Oceanis 381.

88. *Capsize Screening Value* The capsize screening value C is defined as

$$C = b\left(\frac{d}{64}\right)^{-1/3},$$

where b is the beam (or width) in feet and d is the displacement in pounds. The Bahia 46 is a 46 ft catamaran with a capsize screening value of 3.91 and a beam of 26.1 ft. Find the displacement for the Bahia 46.

89. *Insulated Carton* Nina is designing a box for shipping frozen shrimp. The box is to have a square base and a height that is 2 in. greater than the width of the base. The box will be surrounded with a 1-in. thick layer of styrofoam. If the volume of the inside of the box must be equal to the volume of the styrofoam used, then what volume of shrimp can be shipped in the box?

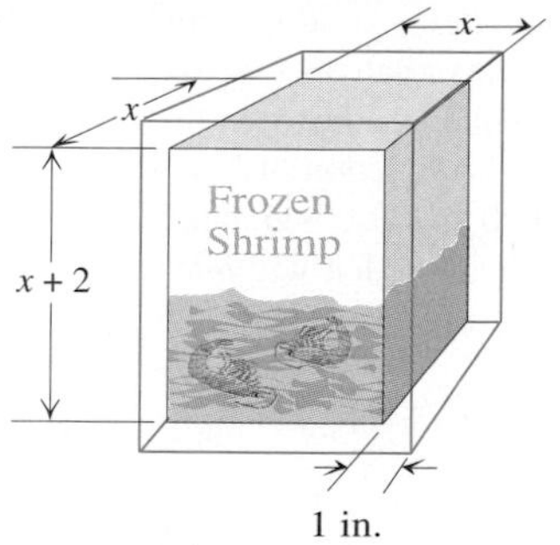

Figure for Exercise 89

90. *Volume of a Cubic Box* If the width of the base of a cubic container is increased by 3 m and the length of the base decreased by 1 m, then the volume of the new container is 6 m^3. What is the height of the cubic container?

91. *Hiking Time* William, Nancy, and Edgar met at the lodge at 8 A.M., and William began hiking west at 4 mph. At 10 A.M., Nancy began hiking north at 5 mph and Edgar went east on a three-wheeler at 12 mph. At what time was the distance between Nancy and Edgar 14 mi greater than the distance between Nancy and William?

92. *Accuracy of Transducers* Setra Systems Inc., of Acton, MA, calculates the accuracy A of its pressure transducers using the formula

$$A = \sqrt{(NL)^2 + (HY)^2 + (NR)^2}$$

where NL represents nonlinearity, HY represents hysteresis, and NR represents nonrepeatability. If the nonlinearity is 0.1%, the hysteresis is 0.05%, and the nonrepeatability is 0.02%, then what is the accuracy? Solve the formula for hysteresis.

93. *Geometric Mean* The *geometric mean* of the numbers $x_1, x_2, \ldots, x_n$ is defined as

$$GM = \sqrt[n]{x_1 \cdot x_2 \cdot \cdots \cdot x_n}.$$

The accompanying graph shows the profits for Compaq Computer for the years 1995 through 1997 (*Fortune*, June 23, 1997).

a. Find the geometric mean of the profits for these three years.

b. What would the profits for Compaq have to be in 1998 so that the geometric mean for 1995 through 1998 would be $1.6 billion?

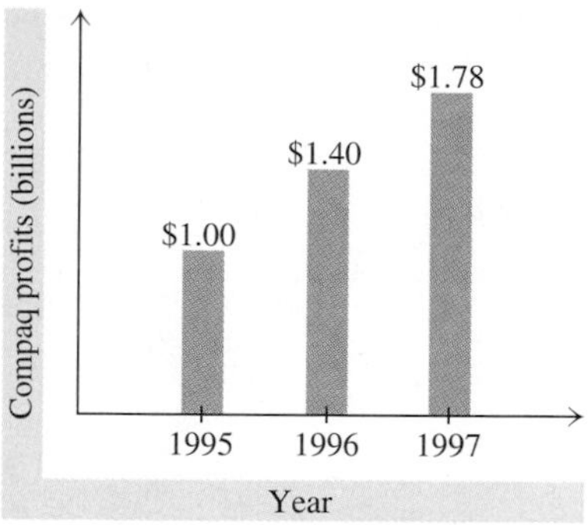

Figure for Exercise 93

• CALCULUS •

94. *Time Swimming and Running* Lauren is competing in her town's cross-country competition. Early in the event, she must race from point A on the Greenbriar River to point B, which is 5 mi downstream and on the opposite bank. The Greenbriar is 1 mi wide. In planning her strategy, Lauren knows she can use any combination of running and swimming. She can run 10 mph and swim 8 mph. How long would it take if she ran 5 mi downstream and then swam across? Find the time it would take if she swam diagonally from A to B. Find x so that she could run x miles along the bank, swim diagonally to B, and complete the race in 36 min. (Ignore the current in the river.)

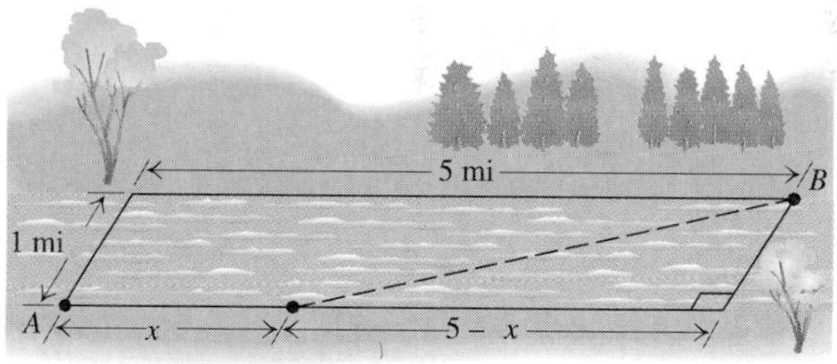

Figure for Exercise 94

95. *Boston Molasses Disaster* In the city of Boston, during the afternoon of January 15, 1919, a cylindrical metal tank containing 25,850,000 kg of molasses ruptured. The sticky liquid poured into the streets in a 9-m-deep stream that knocked down buildings and killed pedestrians and horses (*The Boston Globe*, Sept. 19, 1991). If the diameter of the

tank was equal to its height and the weight of molasses is 1600 kg/m^3, then what was the height of the tank in meters?

Figure for Exercise 95

96. *Storing Supplies* An army sergeant wants to use a 20 ft by 40 ft piece of canvas to make a two-sided tent for holding supplies as shown in the figure. Write the volume of the tent as a function of b. For what value of b is the volume 1600 ft^3? Use a graphing calculator to find the values for b and h that will maximize the volume of the tent.

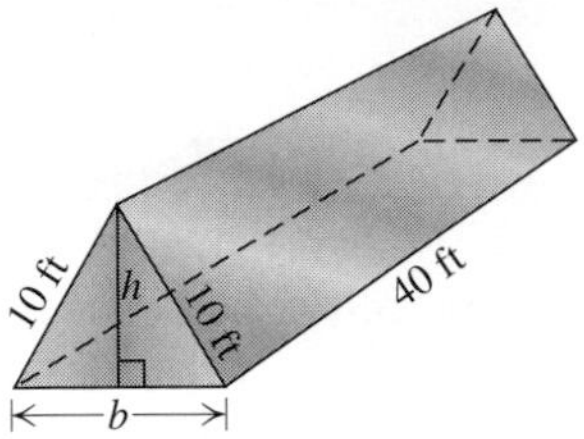

Figure for Exercise 96

LINKING CONCEPTS

For Individual or Group Explorations

• CALCULUS •

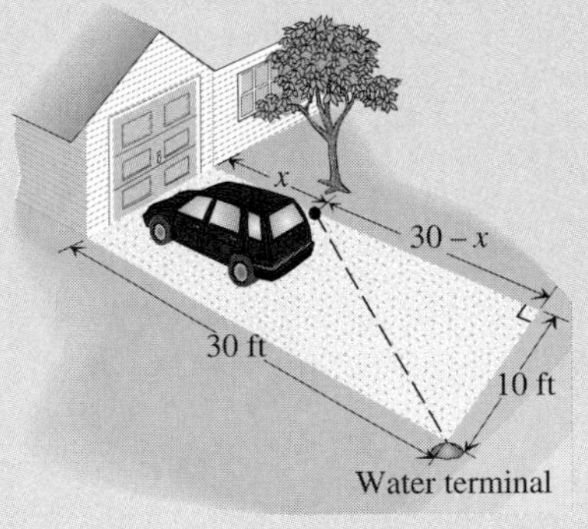

Minimizing Construction Cost *A homeowner needs to run a new water pipe from his house to a water terminal as shown in the accompanying diagram. The terminal is 30 ft down the 10-ft-wide driveway and on the other side. A contractor charges \$3/ft alongside the driveway and \$4/ft for underneath the driveway.*

a) What will it cost if the contractor runs the pipe entirely under the driveway along the diagonal of the 30-ft by 10-ft rectangle?

b) What will it cost if the contractor runs the pipe 30 ft alongside the driveway and then 10 ft straight across?

c) The contractor claims that he can do the job for \$120 by going alongside the driveway for some distance and then going under the drive diagonally to the terminal. Find x, the distance alongside the driveway.

d) Write the cost as a function of x and sketch the graph of the function.

e) Use the minimum feature of a graphing calculator to find the approximate value for x that will minimize the cost.

f) What is the minimum cost (to the nearest cent) for which the job can be done?

3.6 Graphs of Polynomial Functions

In Chapter 2 we learned that the graph of a polynomial function of degree 0 or 1 is a straight line. In Section 3.2 we learned that the graph of a second-degree polynomial function is a parabola. In this section we will concentrate on graphs of polynomial functions of degree greater than 2.

Drawing Good Graphs

A graph of an equation is a picture of all of the ordered pairs that satisfy the equation. However, it is impossible to draw a perfect picture of any set of ordered pairs. We usually find a few important features of the graph and make sure that our picture brings out those features. For example, you can make a good graph of a linear function by drawing a line through the intercepts using a ruler and a sharp pencil. A good parabola should look smooth and symmetric and pass through the vertex and intercepts.

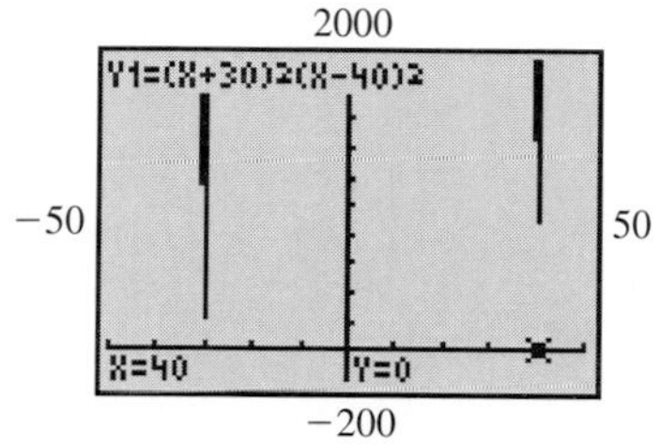

Figure 3.52

A graphing calculator is a tremendous aid in graphing because it can quickly plot a lot of points. However, the calculator does not know if it has drawn a graph that shows the important features. For example, the graph of $y = (x + 30)^2(x - 40)^2$ has x-intercepts at $(-30, 0)$ and $(40, 0)$, but they do not appear on the graph in Fig. 3.52. □

To draw good graphs of polynomial functions we must understand symmetry, the behavior of the graph at its x-intercepts, and how the leading coefficient affects the shape of the graph.

Symmetry

Symmetry is a very special property of graphs of some functions but not others. Recognizing that the graph of a function has some symmetry usually cuts in half the work required to obtain the graph and also helps cut down on errors in graphing. So far we have discussed the following types of symmetry.

SUMMARY **Types of Symmetry**

1. The graph of a function $f(x)$ is *symmetric about the y-axis* if $f(-x) = f(x)$ for any value of x in the domain of the function. (Section 2.4)
2. The graph of a function $f(x)$ is *symmetric about the origin* if $f(-x) = -f(x)$ for any value of x in the domain of the function. (Section 2.4)
3. The graph of a quadratic function $f(x) = ax^2 + bx + c$ is symmetric about its *axis of symmetry*, $x = -b/(2a)$. (Section 3.2)

The graphs of $f(x) = x^2$ and $f(x) = x^3$ shown in Figs. 3.53 and 3.54 are nice examples of symmetry about the y-axis and symmetry about the origin, respec-

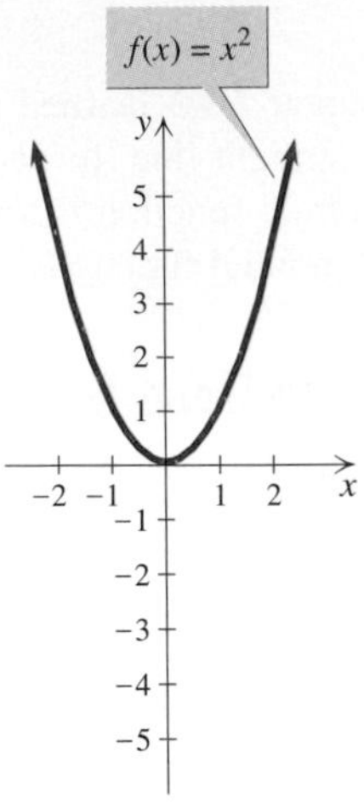

Symmetry about y-axis

Figure 3.53

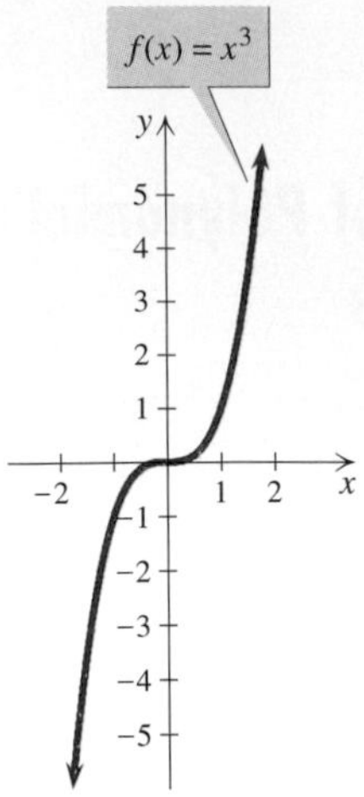

Symmetry about origin

Figure 3.54

tively. The axis of symmetry of $f(x) = x^2$ is the y-axis. The symmetry of the other quadratic functions comes from the fact that the graph of every quadratic function is a transformation of the graph of $f(x) = x^2$.

EXAMPLE 1 Determining the symmetry of a graph

Discuss the symmetry of the graph of each polynomial function.

a) $f(x) = 5x^3 - x$ **b)** $g(x) = 2x^4 - 3x^2$ **c)** $h(x) = x^2 - 3x + 6$
d) $j(x) = x^4 - x^3$

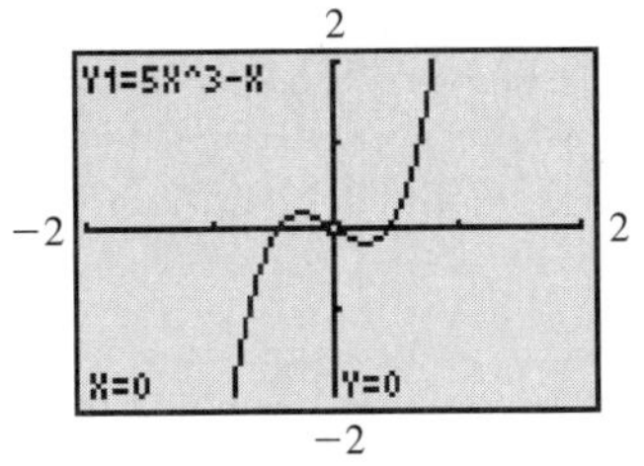

Figure 3.55

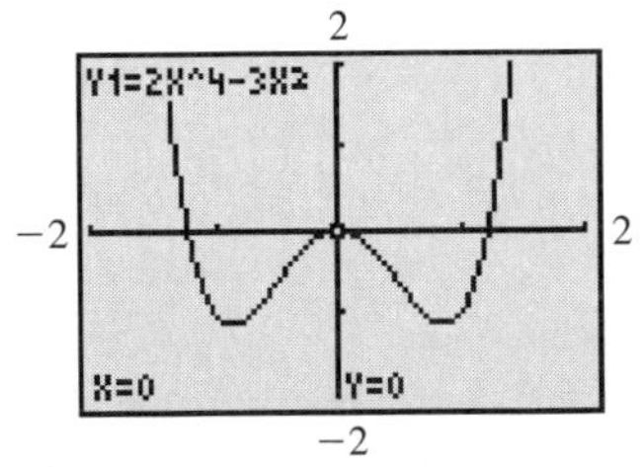

Figure 3.56

Solution

a) Replace x by $-x$ in $f(x) = 5x^3 - x$ and simplify:

$$f(-x) = 5(-x)^3 - (-x)$$
$$= -5x^3 + x$$

Since $f(-x)$ is the opposite of $f(x)$, the graph is symmetric about the origin. Fig. 3.55 supports this conclusion. □

b) Replace x by $-x$ in $g(x) = 2x^4 - 3x^2$ and simplify:

$$g(-x) = 2(-x)^4 - 3(-x)^2$$
$$= 2x^4 - 3x^2$$

Since $g(-x) = g(x)$, the graph is symmetric about the y-axis. Fig. 3.56 supports this conclusion. □

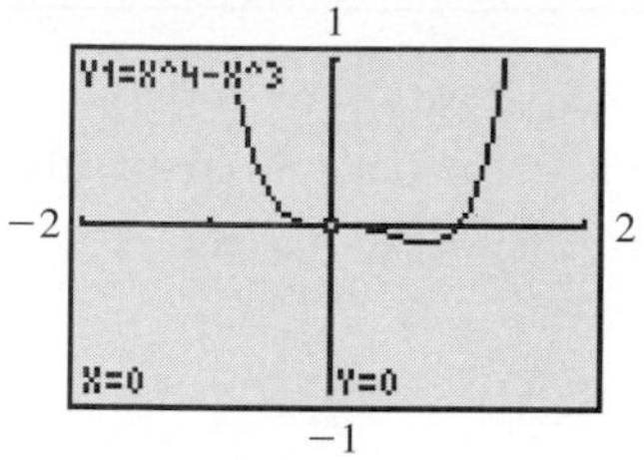

Figure 3.57

c) Because h is a quadratic function, its graph is symmetric about the line $x = -b/(2a)$, which in this case is the line $x = 3/2$.

d) In this case

$$j(-x) = (-x)^4 - (-x)^3$$
$$= x^4 + x^3$$

So $j(-x) \neq j(x)$ and $j(-x) \neq -j(x)$. The graph of j has neither type of symmetry.

Fig. 3.57 supports this conclusion.

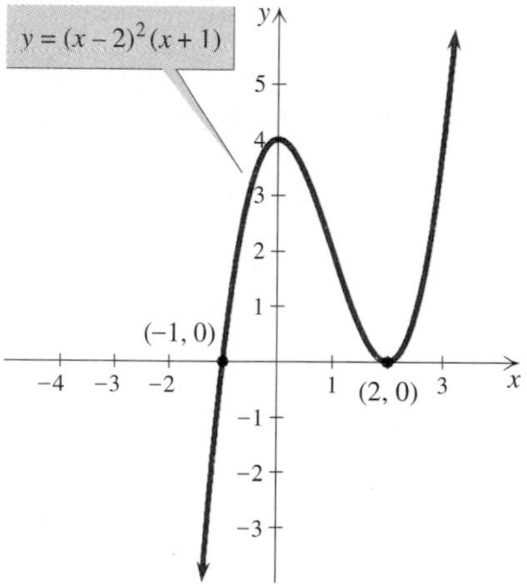

Figure 3.58

Behavior at the *x*-Intercepts

The x-intercepts are key points for the graph of a polynomial function, as they are for any function. Consider the graph of $y = (x - 2)^2(x + 1)$ in Fig. 3.58. Near 2 the values of y are positive as shown in Fig. 3.59 and the graph does not cross the x-axis. Near -1 the values of y are negative before -1 and positive after -1 as shown in Fig. 3.60 and the graph crosses the x-axis. The reason for this behavior is the exponents in $(x - 2)^2$ and $(x + 1)^1$. Because $(x - 2)^2$ has an even exponent, $(x - 2)^2$ cannot be negative. Because $(x + 1)^1$ has an odd exponent, $(x + 1)^1$ changes sign at -1 but does not change sign at 2. So the product $(x - 2)^2(x + 1)$ does not change sign at 2, but does change sign at -1.

X	Y1
1.7	.243
1.8	.112
1.9	.029
2	0
2.1	.031
2.2	.128
2.3	.297

X=2

Figure 3.59

X	Y1
-1.3	-3.267
-1.2	-2.048
-1.1	-.961
-1	0
-.9	.841
-.8	1.568
-.7	2.187

X= -1

Figure 3.60

Every x-intercept corresponds to a factor of the polynomial. Whether that factor occurs an odd or even number of times determines the behavior of the graph at that intercept. *The graph crosses the x-axis at an x-intercept if the factor corresponding to that intercept is raised to an odd power, and the graph touches but does not cross if the factor is raised to an even power.* As another example, consider the graphs of $f(x) = x^2$ and $f(x) = x^3$ shown in Figs. 3.53 and 3.54. Each has only one x-intercept. Note that $f(x) = x^2$ does not cross the x-axis at (0, 0), but $f(x) = x^3$ does cross the x-axis at (0, 0).

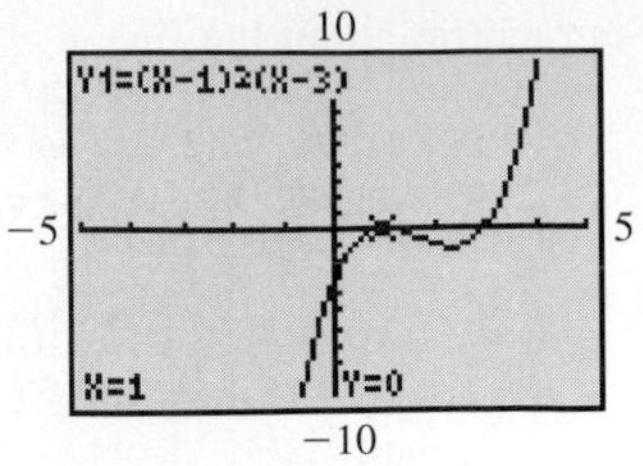

Figure 3.61

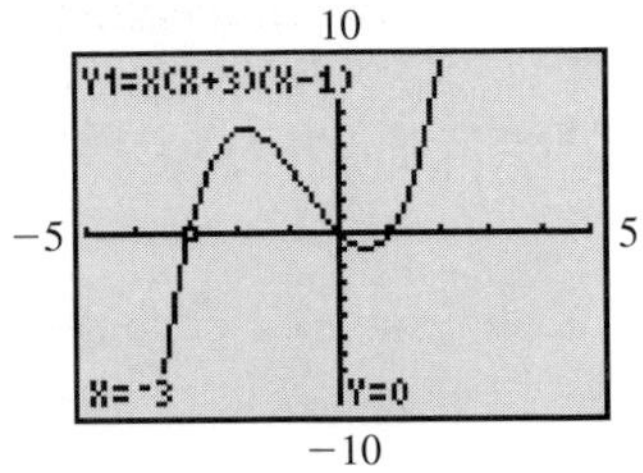

Figure 3.62

EXAMPLE 2 Crossing at the x-intercepts

Find the x-intercepts and determine whether the graph of the function crosses the x-axis at each x-intercept.

a) $f(x) = (x - 1)^2(x - 3)$ **b)** $f(x) = x^3 + 2x^2 - 3x$

Solution

a) The x-intercepts are found by solving $(x - 1)^2(x - 3) = 0$. The x-intercepts are $(1, 0)$ and $(3, 0)$. The graph does not cross the x-axis at $(1, 0)$ because the factor $x - 1$ occurs to an even power. The graph crosses the x-axis at $(3, 0)$ because $x - 3$ occurs to an odd power.
The graph in Fig. 3.61 supports these conclusions.

b) The x-intercepts are found by solving $x^3 + 2x^2 - 3x = 0$. By factoring, we get $x(x + 3)(x - 1) = 0$. The x-intercepts are $(0, 0)$, $(-3, 0)$, and $(1, 0)$. Since each factor occurs an odd number of times (once), the graph crosses the x-axis at each of the x-intercepts.
The graph in Fig. 3.62 supports these conclusions.

• CALCULUS •

The Leading Coefficient Test

Polynomial functions of degree 2 or less have graphs that are either parabolas or lines. We have seen many graphs of polynomial functions of degree 3 and higher. There is a lot of variety to such graphs, but they do have similarities. All of these graphs are continuous curves with no breaks like those in the greatest integer function, and they are smooth curves without sharp corners like that in the graph of the absolute value function. We know why a curve crosses or does not cross the x-axis at an x-intercept. Now we see what determines the behavior of the curve as x takes on larger and larger values (or smaller and smaller values).

The graph of any polynomial function either rises or falls as x gets larger without bound, $x \to \infty$ (read "x goes to ∞"), or as x gets smaller without bound, $x \to -\infty$ (read "x goes to $-\infty$"). This behavior is determined by the degree of the polynomial and the sign of the leading coefficient. Because the first term of the polynomial has the largest degree, it "dominates" all other terms of the polynomial as $x \to \infty$ or as $x \to -\infty$. The four types of behavior are illustrated in the next example.

EXAMPLE 3 Behavior as $x \to \infty$ or $x \to -\infty$

Determine the behavior of the graph of each function as $x \to \infty$ or $x \to -\infty$.

a) $y = x^3 - x$ **b)** $y = -x^3 + 1$ **c)** $y = x^4 - 4x^2$

d) $y = -x^4 + 4x^2 + x$

Solution

a) The graph of $y = x^3 - x$ is shown in Fig. 3.63. As x gets larger and larger ($x \to \infty$), y increases without bound ($y \to \infty$). As x gets smaller and smaller ($x \to -\infty$), y decreases without bound ($y \to -\infty$). Notice that the degree of the polynomial is odd and the sign of the leading coefficient is positive.

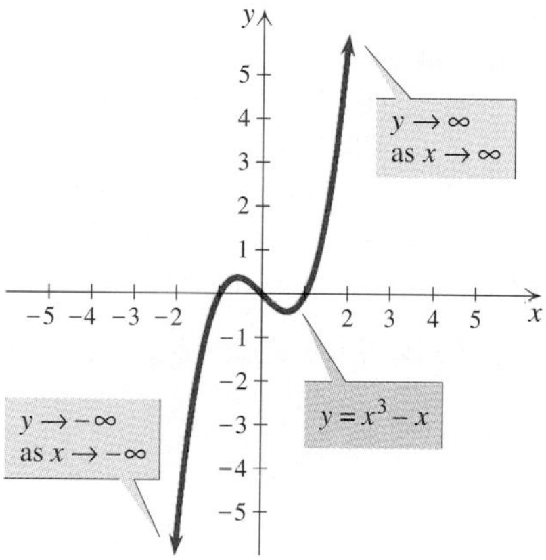

Figure 3.63

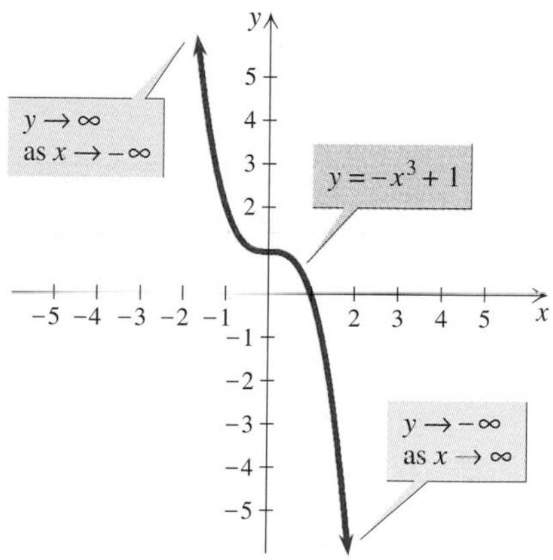

Figure 3.64

b) The graph of $y = -x^3 + 1$ is shown in Fig. 3.64. As x gets larger and larger ($x \to \infty$), y decreases without bound ($y \to -\infty$). As x gets smaller and smaller ($x \to -\infty$), y increases without bound ($y \to \infty$). Notice that the degree of this polynomial is odd and the sign of the leading coefficient is negative.

c) The graph of $y = x^4 - 4x^2$ is shown in Fig. 3.65. As $x \to \infty$ or $x \to -\infty$, y increases without bound ($y \to \infty$). Notice that the degree of this polynomial is even and the sign of the leading coefficient is positive.

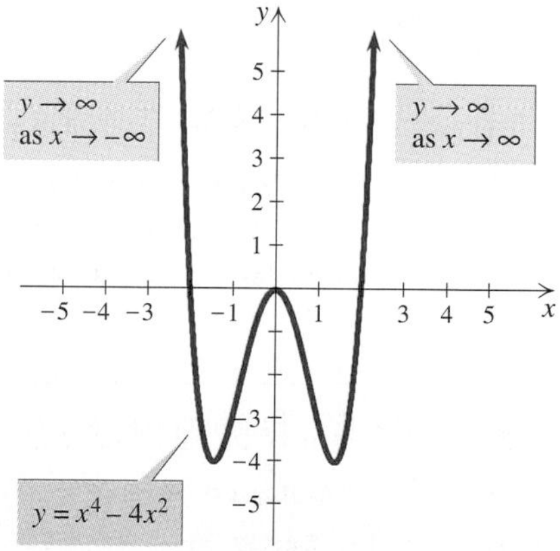

Figure 3.65

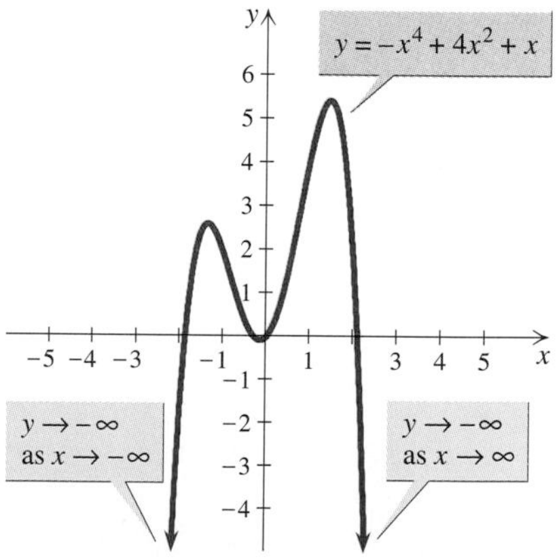

Figure 3.66

d) The graph of $y = -x^4 + 4x^2 + x$ is shown in Fig. 3.66 on page 331. As $x \to \infty$ or $x \to -\infty$, y decreases without bound ($y \to -\infty$). Notice that the degree of this polynomial is even and the sign of the leading coefficient is negative.

The observations that we made in Example 3 are generalized in the **leading coefficient test**.

Leading Coefficient Test

If $f(x) = a_n x^n + a_{n-1}x^{n-1} + \cdots + a_1 x + a_0$, the behavior of the graph of f to the left and right is determined as follows:

For n odd and $a_n > 0$, $y \to \infty$ as $x \to \infty$, and $y \to -\infty$ as $x \to -\infty$.

For n odd and $a_n < 0$, $y \to -\infty$ as $x \to \infty$, and $y \to \infty$ as $x \to -\infty$.

For n even and $a_n > 0$, $y \to \infty$ as $x \to \infty$ or $x \to -\infty$.

For n even and $a_n < 0$, $y \to -\infty$ as $x \to \infty$ or $x \to -\infty$.

• CALCULUS •

Sketching Graphs of Polynomial Functions

A good graph of a polynomial function should include the features that we have been discussing. The following strategy will help you graph polynomial functions.

STRATEGY

Graphing a Polynomial Function

1. Check for symmetry of any type.
2. Find all real zeros of the polynomial function.
3. Determine the behavior at the corresponding x-intercepts.
4. Determine the behavior as $x \to \infty$ and as $x \to -\infty$.
5. Calculate several ordered pairs including the y-intercept to verify your suspicions about the shape of the graph.
6. Draw a smooth curve through the points to make the graph.

EXAMPLE 4 Graphing polynomial functions

Sketch the graph of each polynomial function.

a) $f(x) = x^3 - 5x^2 + 7x - 3$ **b)** $f(x) = x^4 - 200x^2 + 10{,}000$

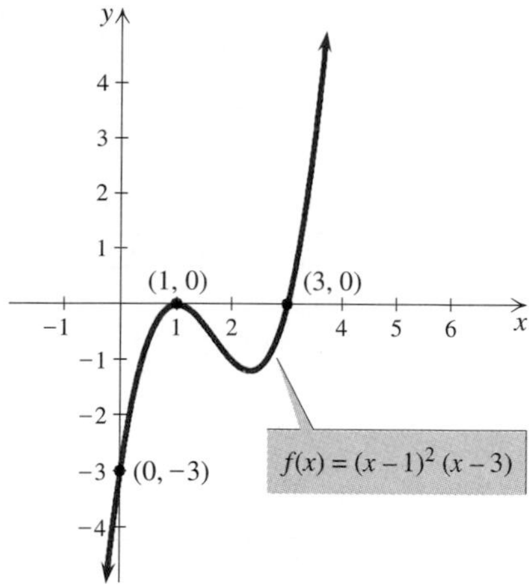

Figure 3.67

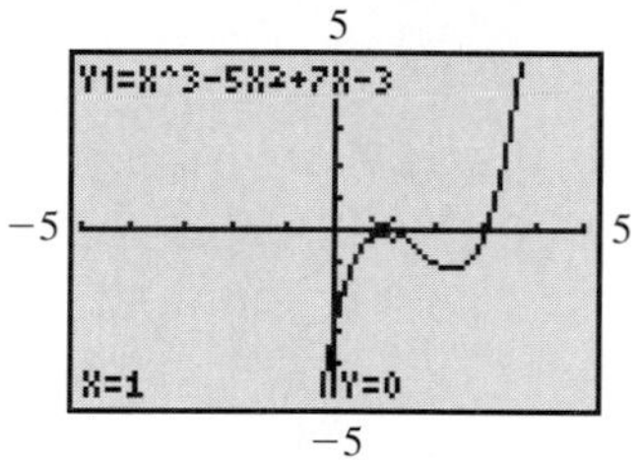

Figure 3.68

Solution

a) First find $f(-x)$ to determine symmetry and the number of negative roots.

$$\begin{aligned} f(-x) &= (-x)^3 - 5(-x)^2 + 7(-x) - 3 \\ &= -x^3 - 5x^2 - 7x - 3 \end{aligned}$$

From $f(-x)$, we see that the graph has neither type of symmetry. Because $f(-x)$ has no sign changes, $x^3 - 5x^2 + 7x - 3 = 0$ has no negative roots by Descartes's rule of signs. The only possible rational roots are 1 and 3.

$$\begin{array}{r|rrrr} 1 & 1 & -5 & 7 & -3 \\ & & 1 & -4 & 3 \\ \hline & 1 & -4 & 3 & 0 \end{array}$$

From the synthetic division we know that 1 is a root and we can factor $f(x)$:

$$\begin{aligned} f(x) &= (x - 1)(x^2 - 4x + 3) \\ &= (x - 1)^2(x - 3) \qquad \text{Factor completely.} \end{aligned}$$

The x-intercepts are $(1, 0)$ and $(3, 0)$. The graph of f does not cross the x-axis at $(1, 0)$ because $x - 1$ occurs to an even power, while the graph crosses at $(3, 0)$ because $x - 3$ occurs to an odd power. The y-intercept is $(0, -3)$. Since the leading coefficient is positive and the degree is odd, $y \to \infty$ as $x \to \infty$ and $y \to -\infty$ as $x \to -\infty$. Calculate two more ordered pairs for accuracy, say $(2, -1)$ and $(4, 9)$. Draw a smooth curve as in Fig. 3.67.

The calculator graph shown in Fig. 3.68 supports these conclusions. □

b) First find $f(-x)$:

$$\begin{aligned} f(-x) &= (-x)^4 - 200(-x)^2 + 10{,}000 \\ &= x^4 - 200x^2 + 10{,}000 \end{aligned}$$

Since $f(x) = f(-x)$, the graph is symmetric about the y-axis. We can factor the polynomial as follows.

$$\begin{aligned} f(x) &= x^4 - 200x^2 + 10{,}000 \\ &= (x^2 - 100)(x^2 - 100) \\ &= (x - 10)(x + 10)(x - 10)(x + 10) \\ &= (x - 10)^2(x + 10)^2 \end{aligned}$$

The x-intercepts are $(10, 0)$ and $(-10, 0)$. Since each factor for these intercepts has an even power, the graph does not cross the x-axis at the intercepts. The y-intercept is $(0, 10{,}000)$. Since the leading coefficient is positive and the degree is even, $y \to \infty$ as $x \to \infty$ or as $x \to -\infty$. The graph also goes through $(-20, 90{,}000)$ and $(20, 90{,}000)$. Draw a smooth curve through these points and the intercepts as shown in Fig. 3.69.

The calculator graph shown in Fig. 3.70 supports these conclusions.

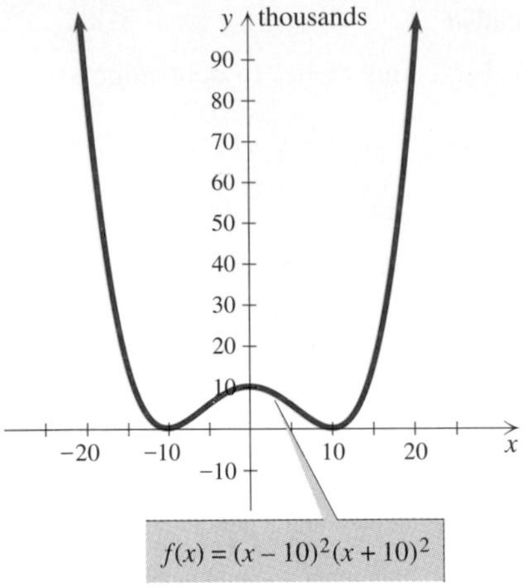

Figure 3.69

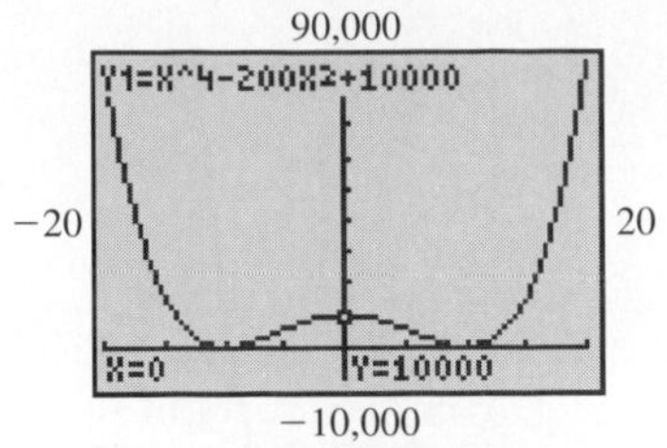

Figure 3.70

Polynomial Inequalities

We solved quadratic inequalities in Section 3.2 by examining the graphs of the corresponding quadratic functions. We can now use graphs to solve higher-degree polynomial inequalities in the same manner.

EXAMPLE 5 **Solve the inequality $x^4 + x^3 - 15x^2 - 3x + 36 < 0$.**

Solution

Use synthetic division to see that 3 and −4 are zeros of the function $y = x^4 + x^3 - 15x^2 - 3x + 36$:

3	1	1	−15	−3	36
		3	12	−9	−36
−4	1	4	−3	−12	0
		−4	0	12	
	1	0	−3	0	

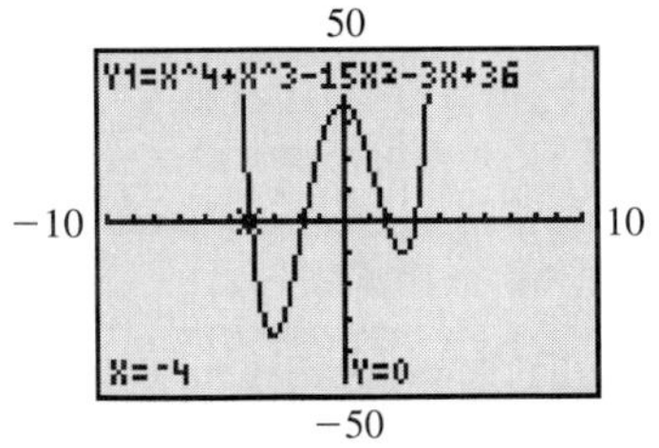

Figure 3.71

Since $x^4 + x^3 - 15x^2 - 3x + 36 = (x - 3)(x + 4)(x^2 - 3)$, the other two zeros are $\pm\sqrt{3}$. Since the multiplicity of each zero is one, the graph crosses the x-axis at each of the four intercepts as shown in Fig. 3.71. The inequality is satisfied for any value of x for which the graph is below the x-axis. So the solution set is $(-4, -\sqrt{3}) \cup (\sqrt{3}, 3)$.

The method used in Example 5 is a variation of the test point method. By examining the graph we are testing points between the intercepts to see whether the graph is above or below the x-axis. The graph actually shows more points than is necessary. We could determine whether the graph is above or below the x-axis by simply testing one point between two adjacent x-intercepts.

FOR THOUGHT True or False? Explain.

1. If P is a function for which $P(2) = 8$ and $P(-2) = -8$, then the graph of P is symmetric about the origin.
2. If $y = -3x^3 + 4x^2 - 6x + 9$, then $y \to -\infty$ as $x \to \infty$.
3. If the graph of $y = P(x)$ is symmetric about the origin and $P(8) = 4$, then $-P(-8) = 4$.
4. If $f(x) = x^3 - 3x$, then $f(x) = f(-x)$ for any value of x.
5. If $f(x) = x^4 - x^3 + x^2 - 6x + 7$, then $f(-x) = x^4 + x^3 + x^2 + 6x + 7$.
6. The graph of $f(x) = x^2 - 6x + 9$ has only one x-intercept.
7. The x-intercepts for $P(x) = (x - 1)^2(x + 1)$ are $(0, 1)$ and $(0, -1)$.
8. The y-intercept for $P(x) = 4(x - 3)^2 + 2$ is $(0, 2)$.
9. The graph of $f(x) = x^2(x + 8)^2$ has no points in quadrants III and IV.
10. The graph of $f(x) = x^3 - 1$ has three x-intercepts.

3.6 EXERCISES

Tape 8 Disk

For each graph discuss its symmetry, indicate whether the graph crosses the x-axis at each x-intercept, and determine whether $y \to \infty$ or $y \to -\infty$ as $x \to \infty$ and $x \to -\infty$.

1.

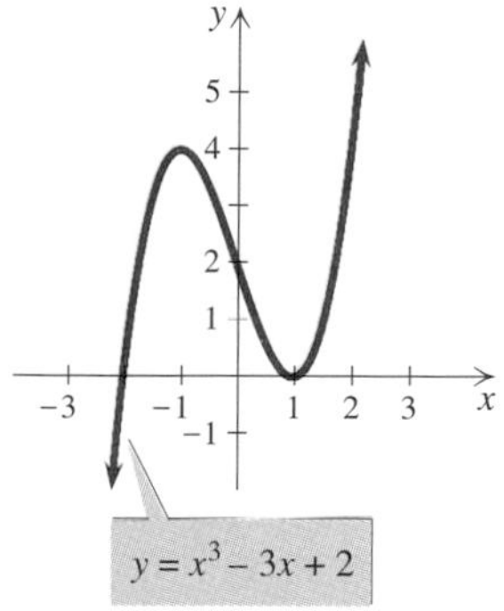

2.

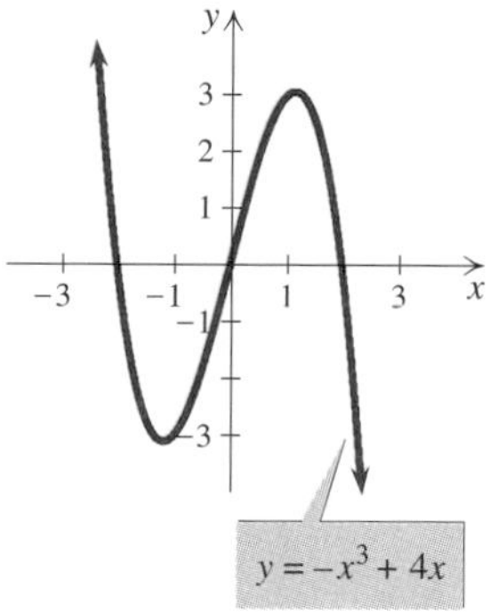

3.

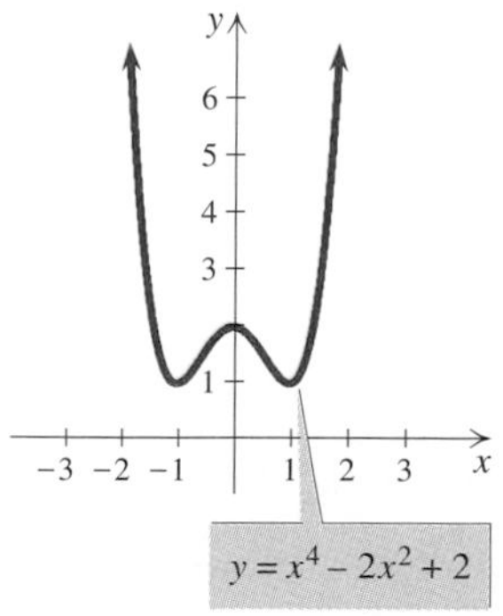

4.

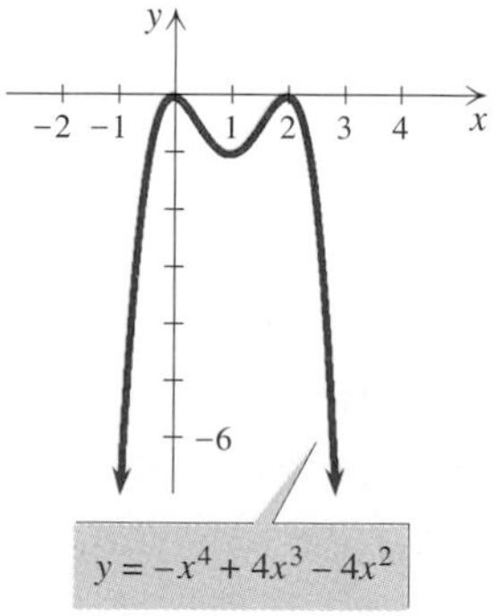

Discuss the symmetry of the graph of each polynomial function.

5. $f(x) = x^6$
6. $f(x) = x^5 - x$
7. $f(x) = x^2 - 3x + 5$
8. $f(x) = 5x^2 + 10x + 1$
9. $f(x) = 3x^6 - 5x^2 + 3x$
10. $f(x) = x^6 - x^4 + x^2 - 8$
11. $f(x) = 4x^3 - x$
12. $f(x) = 7x^3 + x^2$
13. $f(x) = (x - 5)^2$
14. $f(x) = (x^2 - 1)^2$
15. $f(x) = -x$
16. $f(x) = 3x$

Find the x-intercepts and discuss the behavior of the graph of each polynomial function at its x-intercepts.

17. $f(x) = (x - 4)^2$
18. $f(x) = (x - 1)^2(x + 3)^2$
19. $f(x) = (2x - 1)^3$
20. $f(x) = x^6$

21. $f(x) = 4x - 1$

22. $f(x) = x^2 - 5x - 6$

23. $f(x) = x^2 - 3x + 10$

24. $f(x) = x^4 - 16$

25. $f(x) = x^3 - 3x^2$

26. $f(x) = x^3 - x^2 - x + 1$

27. $f(x) = 2x^3 - 5x^2 + 4x - 1$

28. $f(x) = x^3 - 3x^2 + 4$

29. $f(x) = -2x^3 - 8x^2 + 6x + 36$

30. $f(x) = -x^3 + 7x - 6$

For each function use the leading coefficient test to determine whether $y \to \infty$ or $y \to -\infty$ as $x \to \infty$.

31. $y = 2x^3 - x^2 + 9$

32. $y = -3x + 7$

33. $y = -3x^4 + 5$

34. $y = 6x^4 - 5x^2 - 1$

35. $y = x - 3x^3$

36. $y = 5x - 7x^4$

For each function use the leading coefficient test to determine whether $y \to \infty$ or $y \to -\infty$ as $x \to -\infty$.

37. $y = -2x^5 - 3x^2$

38. $y = x^3 + 8x + \sqrt{2}$

39. $y = 3x^6 - 999x^3$

40. $y = -12x^4 - 5x$

Match each polynomial function with its graph (a)–(h).

41. $f(x) = -2x + 1$

42. $f(x) = -2x^2 + 1$

43. $f(x) = -2x^3 + 1$

44. $f(x) = -2x^2 + 4x - 1$

45. $f(x) = -2x^4 + 6$

46. $f(x) = -2x^4 + 6x^2$

47. $f(x) = x^3 + 4x^2 - x - 4$

48. $f(x) = (x - 2)^4$

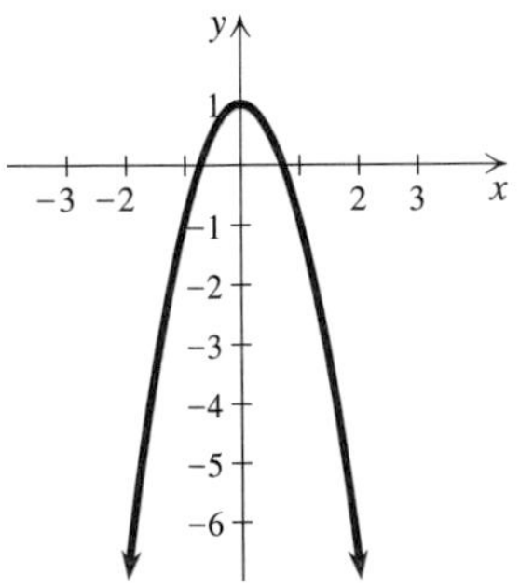

(a)

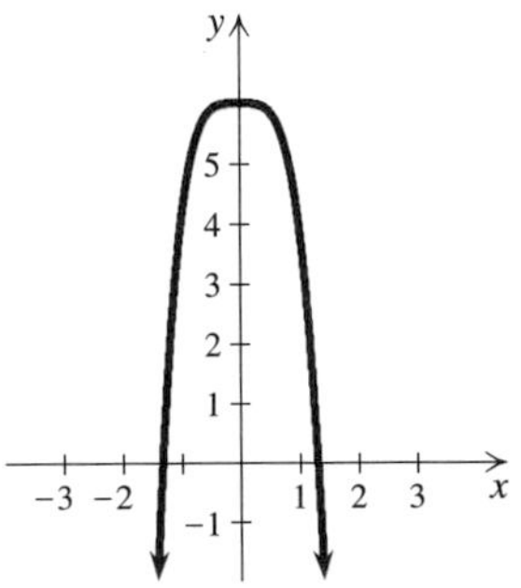

(b)

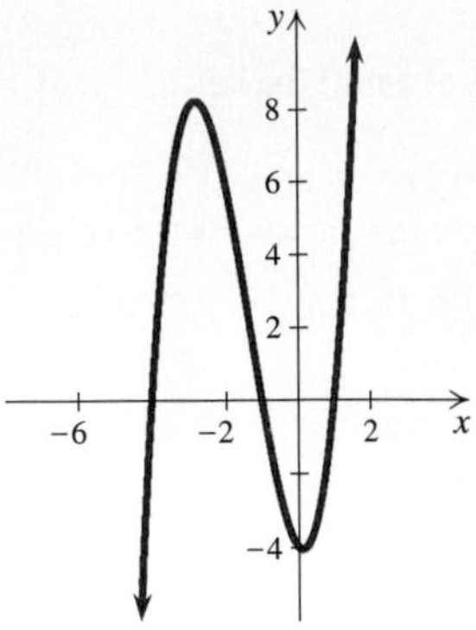

(c)

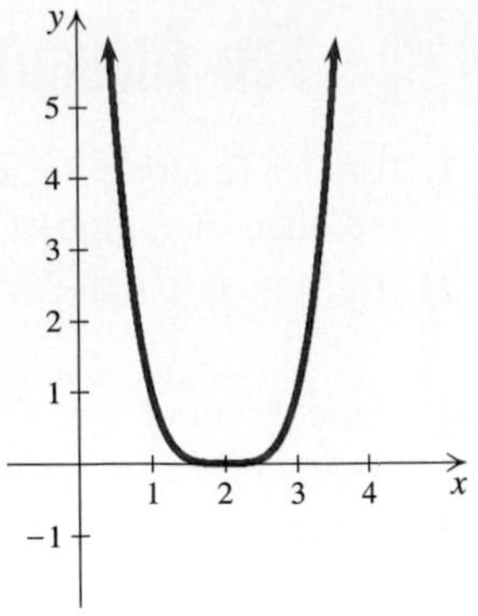

(d)

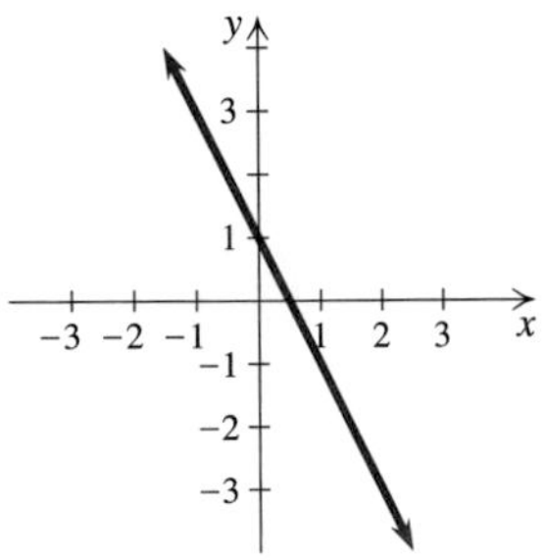

(e)

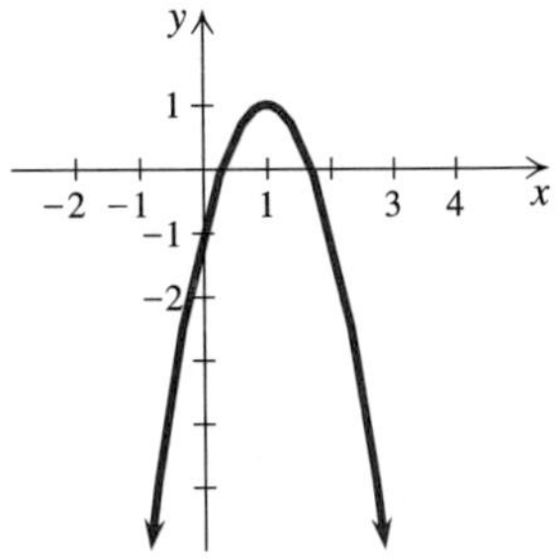

(f)

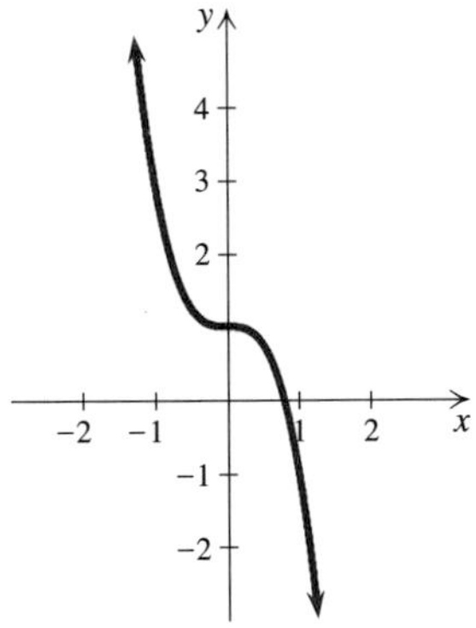

(g)

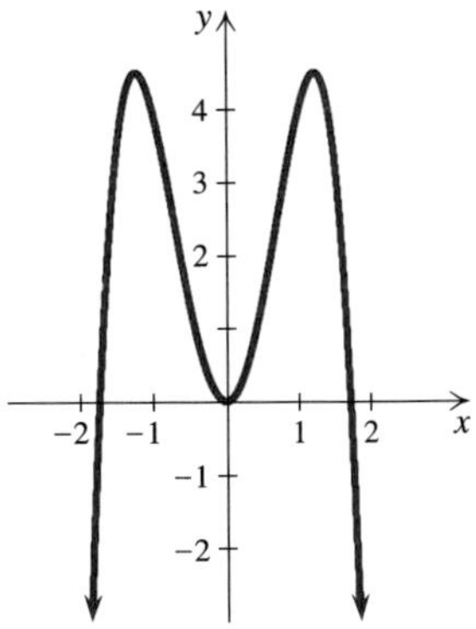

(h)

Sketch the graph of each polynomial function.

49. $f(x) = x - 30$

50. $f(x) = 40 - x$

51. $f(x) = (x - 30)^2$

52. $f(x) = (40 - x)^2$

53. $f(x) = x^3 - 40x^2$

54. $f(x) = x^3 - 900x$

55. $f(x) = (x - 20)^2(x + 20)^2$

56. $f(x) = (x - 20)^2(x + 12)$

57. $f(x) = -x^3 - x^2 + 5x - 3$

58. $f(x) = -x^4 + 6x^3 - 9x^2$

59. $f(x) = x^3 - 10x^2 - 600x$

60. $f(x) = -x^4 + 24x^3 - 144x^2$

61. $f(x) = x^3 + 18x^2 - 37x + 60$

62. $f(x) = x^3 - 7x^2 - 25x - 50$

63. $f(x) = -x^4 + 196x^2$

64. $f(x) = -x^4 + x^2 + 12$

65. $f(x) = x^3 + 3x^2 + 3x + 1$

66. $f(x) = -x^3 + 3x + 2$

Solve each polynomial inequality.

67. $x^3 - 3x > 0$

68. $-x^3 + 3x + 2 < 0$

69. $2x^2 - x^4 \leq 0$

70. $-x^4 + x^2 + 12 \geq 0$

71. $x^3 + 4x^2 - x - 4 > 0$

72. $x^3 + 2x^2 - 2x - 4 < 0$

Each of the following third-degree polynomial functions is increasing on $(-\infty, a)$, decreasing on (a, b), and increasing on (b, ∞), where $a < b$. There is no maximum or minimum value for this type of function, but we say that the function has a **local maximum** value of $f(a)$ at $x = a$ and a **local minimum** value of $f(b)$ at $x = b$. Graph each function with a graphing calculator and estimate the local maximum and local minimum values of the function to the nearest hundredth. (Using calculus, exact answers could be found for these questions.)

73. $f(x) = x^3 + x^2 - 2x + 1$

74. $y = x^3 - x^2 - 20x - 5$

75. $y = x^3 - 9x^2 - 12x + 20$

76. $f(x) = x^3 - x^2 - 4x + 6$

77. $f(x) = 2x^3 - 3x^2 - 4x + 20$

78. $y = x^3 - 7x^2 + 12x - 14$

• **CALCULUS** • Solve each problem.

79. *Maximum Profit* A company's weekly profit (in thousands of dollars) is given by the function $P(x) = x^3 - 3x^2 + 2x + 3$, where x is the amount (in thousands of dollars) spent per week on advertising. Use a graphing calculator to estimate the local maximum and the local minimum value for the profit. What amount must be spent on advertising to get the profit higher than the local maximum profit?

80. *Maximum Volume* An open-top box is to be made from a 6 in. by 7 in. piece of copper by cutting equal squares (x in. by x in.) from each corner and folding up the sides. Write the volume of the box as a function of x. Use a graphing calculator to find the maximum possible volume to the nearest hundredth of a cubic inch.

81. *Economic Forecast* The total annual profit (in thousands of dollars) for a department store chain is determined by $P = 90x - 6x^2 + 0.1x^3$, where x is the number of stores in the chain. The company now has 10 stores in operation and plans to pursue an aggressive expansion program. What happens to the total annual profit as the company opens more and more stores? For what values of x is profit increasing?

82. *Contaminated Chicken* The number of bacteria of a certain type found on a chicken t minutes after processing in a contaminated plant is given by the function

$$N = 30t + 25t^2 + 44t^3 - 0.01t^4.$$

Find N for $t = 30$, 300, and 3000. What happens to N as t keeps increasing?

83. *Packing Cheese* Workers at the Green Bay Cheese Factory are trying to cover a block of cheese with an 8 in. by 12 in. piece of foil paper as shown in the figure. The ratio of the length and width of the block must be 4 to 3 to accommodate the label. Find a polynomial function that gives the volume of the block of cheese covered in this manner as a function of the thickness x. Use a graphing calculator to find the dimensions of the block that will maximize the volume.

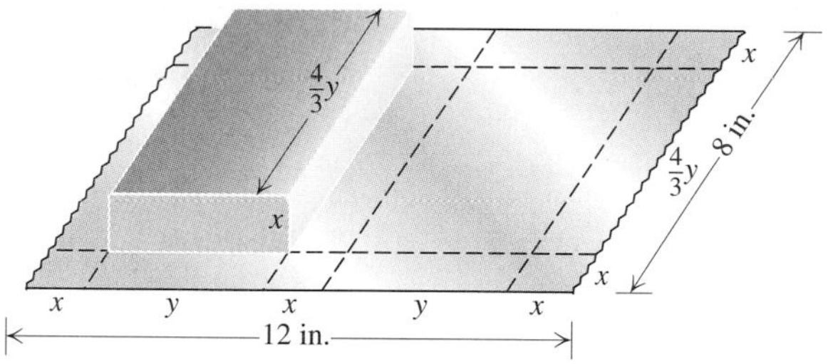

Figure for Exercise 83

84. *Giant Teepee* A casino designer is planning a giant teepee that is 80 ft in diameter and 120 ft high as shown in the figure. Inside the teepee is to be a cylindrical room for slot machines. Write the volume of the cylindrical room as a function of its radius. Use a graphing calculator to find the radius that maximizes the volume of the cylindrical room.

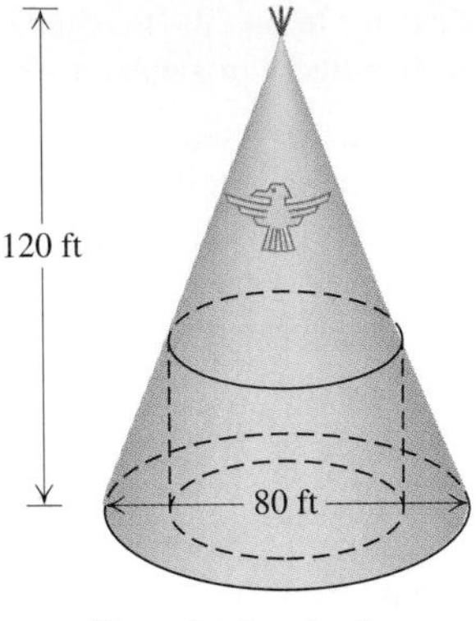

Figure for Exercise 84

85. *Paint Coverage* A one pint can of spray paint will cover 50 ft^2. Of course a small amount of paint coats the inside of the cylindrical can and it is not used. What would happen if the inside of the can had a surface area of 50 ft^2? Find the dimensions for the two one pint cans that have a surface area of 50 ft^2. You will need to know that there are about 7.5 gallons in a cubic foot, and for a cylinder $V = \pi r^2 h$ and $S = 2\pi r^2 + 2\pi rh$.

Figure for Exercise 85

86. *Booming Business* When Computer Recyclers opened its doors business started booming. After a few months, there was a temporary slowdown in sales, after which sales took off again. We can model sales for this business with the function

$$N = 8t^3 - 133t^2 + 653t,$$

where N is the number of computers sold in month t ($t = 0$ corresponds to the opening of the business).

a. Use a graphing calculator to estimate the month in which the temporary slowdown was the worst.

b. What percentage drop in sales occurred at the bottom of the slowdown compared to the previous high point in sales?

Linking Concepts

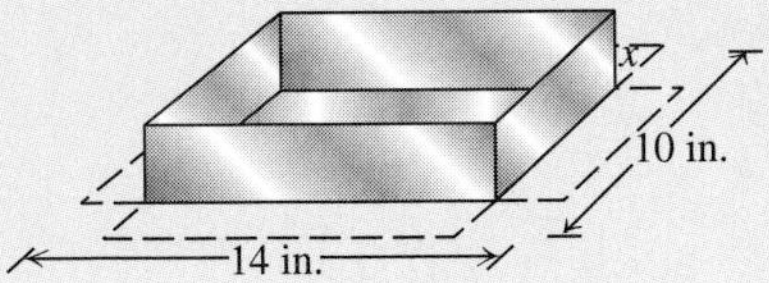

For Individual or Group Explorations

Making Boxes *A sheet metal worker is planning to make an open-top box by cutting equal squares (x-in. by x-in.) from the corners of a 10-in. by 14-in. piece of copper. A second box is to be made in the same manner from an 8-in. by 10-in. piece of aluminum, but its height is to be one-half that of the first box.*

a) Find polynomial functions for the volume of each box.

b) Find the values of x for which the copper box is 72 $in.^3$ larger than the aluminum box.

c) Write the difference between the two volumes d as a function of x and graph it.

d) Find d for $x = 1.5$ in. and $x = 8$ in.

e) For what value of x is the difference between the two volumes the largest?

f) For what value of x is the total volume the largest?

3.7 Graphs of Rational Functions

A rational expression was defined in Chapter P as a ratio of two polynomials. In this section we will study functions that are defined by rational expressions. Since polynomials are used to form rational functions, we will be able to use some of the facts that we have learned about polynomial functions.

Domain

If a ratio of two polynomials is used to define a function, then the function is called a rational function, provided the denominator is not the zero polynomial. Functions such as

$$y = \frac{1}{x}, \quad f(x) = \frac{x - 3}{x - 1}, \quad \text{and} \quad g(x) = \frac{2x - 3}{x^2 - 4}$$

are rational functions.

Definition: Rational Function

If $P(x)$ and $Q(x)$ are polynomials, then a function of the form

$$f(x) = \frac{P(x)}{Q(x)}$$

is called a **rational function**, provided that $Q(x)$ is not the zero polynomial.

To simplify discussions of rational functions we will assume that $f(x)$ is in lowest terms ($P(x)$ and $Q(x)$ have no common factors) unless it is stated otherwise.

The domain of a polynomial function is the set of all real numbers, while the domain of a rational function is restricted to real numbers that do not cause the denominator to have a value of 0. The domain of $y = 1/x$ is the set of all real numbers except 0.

EXAMPLE 1 The domain of a rational function

Find the domain of each rational function.

a) $f(x) = \dfrac{x - 3}{x - 1}$ **b)** $g(x) = \dfrac{2x - 3}{x^2 - 4}$

Solution

a) Since $x - 1 = 0$ only for $x = 1$, the domain of f is the set of all real numbers except 1. The domain is written in interval notation as $(-\infty, 1) \cup (1, \infty)$.

b) Since $x^2 - 4 = 0$ for $x = \pm 2$, any real number except 2 and -2 can be used for x. So the domain of g is $(-\infty, -2) \cup (-2, 2) \cup (2, \infty)$.

Horizontal and Vertical Asymptotes

• CALCULUS •

The graph of a rational function such as $f(x) = 1/x$ does not look like the graph of a polynomial function. The domain of $f(x) = 1/x$ is the set of all real numbers except 0. However, 0 is an important number for the graph of this function because of the behavior of the graph when x is close to 0. Figure 3.72 shows ordered pairs in which x is close to 0. Notice that the closer x is to 0, the farther y is from 0. In symbols, $y \to \infty$ as $x \to 0$ from the right and $y \to -\infty$ as $x \to 0$ from the left. Plotting these ordered pairs suggests a curve that gets closer and closer to the vertical line $x = 0$ (the y-axis) but never touches it as shown in Fig. 3.73. The y-axis is called a *vertical asymptote* for this curve.

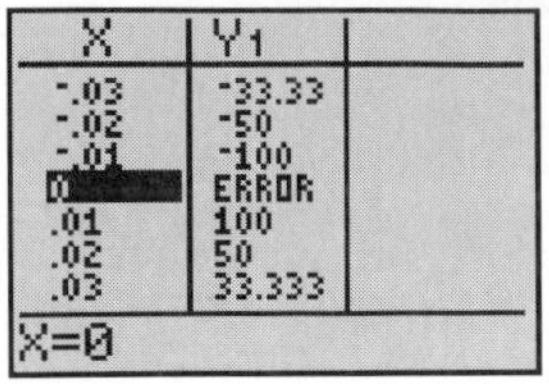

X	Y1
-.03	-33.33
-.02	-50
-.01	-100
0	ERROR
.01	100
.02	50
.03	33.333

X=0

Figure 3.72

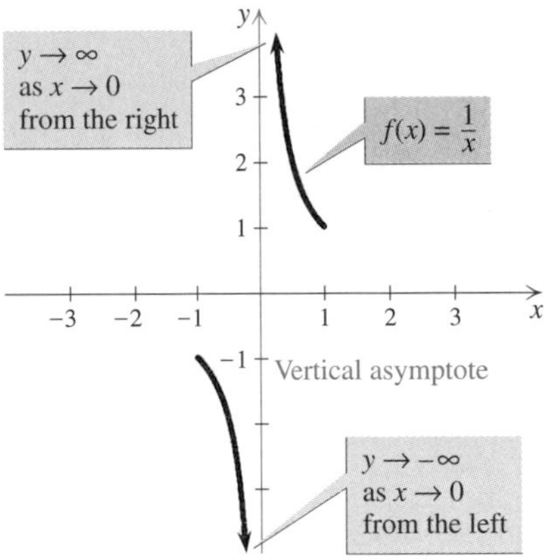

Figure 3.73

Figure 3.74 shows ordered pairs in which x is far from 0. Notice that the farther x is from 0, the closer y is to 0. These ordered pairs suggest a curve that lies just above the positive x-axis and just below the negative x-axis. The x-axis is called a *horizontal asymptote* for the graph of f. The complete graph of $f(x) = 1/x$ is shown in Fig. 3.75. Since $f(-x) = 1/(-x) = -f(x)$, the graph is symmetric about the origin.

For any rational function expressed in lowest terms, a horizontal asymptote is determined by the value approached by the rational expression as $|x| \to \infty$ ($x \to \infty$ or $x \to -\infty$). A vertical asymptote occurs for every number that causes the denominator of the function to have a value of 0, provided the rational function is in lowest terms. As we will learn shortly, not every rational function has a vertical and a horizontal asymptote. We can give a formal definition of asymptotes as described below.

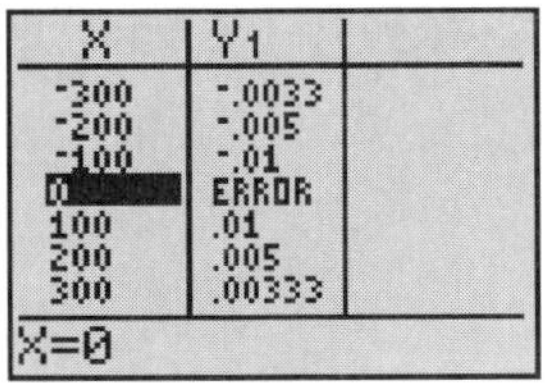

X	Y1
-300	-.0033
-200	-.005
-100	-.01
0	ERROR
100	.01
200	.005
300	.00333

X=0

Figure 3.74

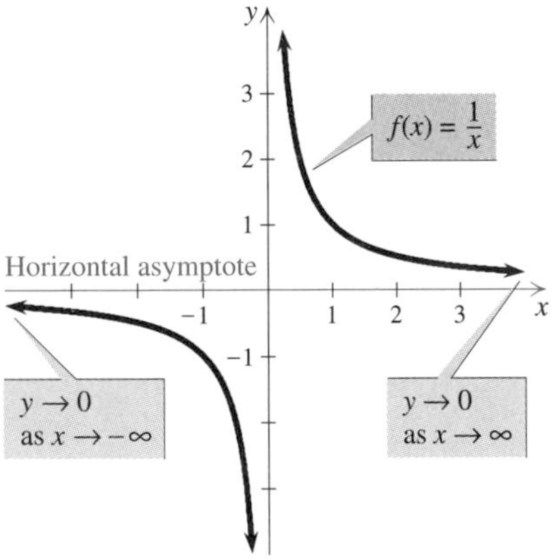

Figure 3.75

Definition: Vertical and Horizontal Asymptotes

Let $f(x) = P(x)/Q(x)$ be a rational function written in lowest terms.

If $|f(x)| \to \infty$ as $x \to a$, then the vertical line $x = a$ is a **vertical asymptote**.

If $f(x) \to a$ as $|x| \to \infty$, then the horizontal line $y = a$ is a **horizontal asymptote**.

For a rational function, the asymptotes (if there are any) *must* be determined before the function is graphed. In the next example, we concentrate on finding the asymptotes of rational functions. Later, in Example 4 of this section, we will sketch the graphs of the functions of Example 2. If you wish, you may look ahead to the graphs in Figs. 3.78, 3.80, and 3.82 to see how important the asymptotes are to the graphs of these functions.

EXAMPLE 2 **Identifying horizontal and vertical asymptotes**

Find the horizontal and vertical asymptotes for each rational function.

a) $f(x) = \dfrac{3}{x^2 - 1}$ **b)** $g(x) = \dfrac{x}{x^2 - 4}$ **c)** $h(x) = \dfrac{2x + 1}{x + 3}$

Solution

a) The denominator $x^2 - 1$ has a value of 0 if $x = \pm 1$. So the lines $x = 1$ and $x = -1$ are vertical asymptotes. As $x \to \infty$ or $x \to -\infty$, $x^2 - 1$ gets larger, making $3/(x^2 - 1)$ the ratio of 3 and a large number. Thus $3/(x^2 - 1) \to 0$ and the x-axis is a horizontal asymptote.

b) The denominator $x^2 - 4$ has a value of 0 if $x = \pm 2$. So the lines $x = 2$ and $x = -2$ are vertical asymptotes. As $x \to \infty$, $x/(x^2 - 4)$ is a ratio of two large numbers. The approximate value of this ratio is not clear. However, if we divide the numerator and denominator by x^2, the highest power of x, we can get a clearer picture of the value of this ratio:

$$g(x) = \frac{x}{x^2 - 4} = \frac{\frac{x}{x^2}}{\frac{x^2}{x^2} - \frac{4}{x^2}} = \frac{\frac{1}{x}}{1 - \frac{4}{x^2}}$$

As $|x| \to \infty$, the values of $1/x$ and $4/x^2$ go to 0. So

$$g(x) \to \frac{0}{1 - 0} = 0.$$

So the x-axis is a horizontal asymptote.

c) The denominator $x + 3$ has a value of 0 if $x = -3$. So the line $x = -3$ is a vertical asymptote. To find any horizontal asymptotes, rewrite the rational expression by dividing the numerator and denominator by the highest power of x:

$$h(x) = \frac{2x + 1}{x + 3} = \frac{\frac{2x}{x} + \frac{1}{x}}{\frac{x}{x} + \frac{3}{x}} = \frac{2 + \frac{1}{x}}{1 + \frac{3}{x}}$$

As $x \to \infty$ or $x \to -\infty$ the values of $1/x$ and $3/x$ approach 0. So

$$h(x) \to \frac{2 + 0}{1 + 0} = 2.$$

So the line $y = 2$ is a horizontal asymptote.

If the degree of the numerator of a rational function is less than the degree of the denominator, as in Examples 2(a) and 2(b), then the x-axis is a horizontal asymptote for the graph of the function. If the degree of the numerator is equal to the degree of the denominator, as in Example 2(c), then the x-axis is not a horizontal asymptote. We can see this clearly if we use division to rewrite the expression as quotient + remainder/divisor. For the function of Example 2(c), we get

$$h(x) = \frac{2x + 1}{x + 3} = 2 + \frac{-5}{x + 3}.$$

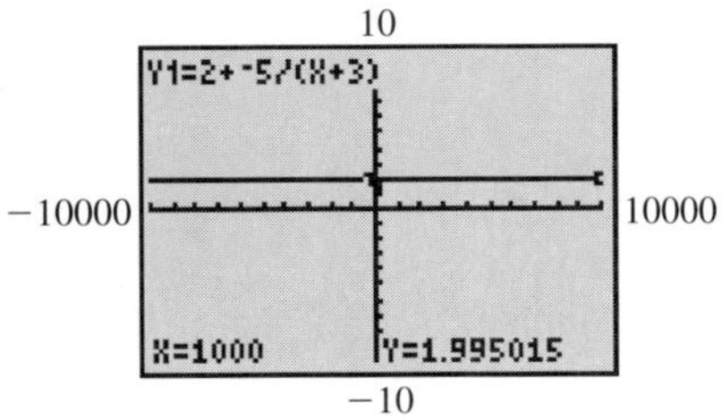

Figure 3.76

The graph of h is a translation two units upward of the graph of $y = -5/(x + 3)$, which has the x-axis as a horizontal asymptote. So $y = 2$ is a horizontal asymptote for h. Note that 2 is simply the ratio of the leading coefficients.

Because the graph gets very close to $y = 2$, the wide view of h in Fig. 3.76 looks like $y = 2$. □

Oblique Asymptotes

Each rational function of Example 2 had one horizontal asymptote and had a vertical asymptote for each zero of the polynomial in the denominator. The horizontal asymptote $y = 0$ occurs because the y-coordinate gets closer and closer to 0 as $x \to \infty$ or $x \to -\infty$. Some rational functions have a nonhorizontal line for an asymptote. An asymptote that is neither horizontal nor vertical is called an **oblique asymptote** or **slant asymptote**.

EXAMPLE 3 A rational function with an oblique asymptote

Determine all of the asymptotes for $g(x) = \dfrac{2x^2 + 3x - 5}{x + 2}$.

Solution

If $x + 2 = 0$, then $x = -2$. So the line $x = -2$ is a vertical asymptote. Because the degree of the numerator is larger than the degree of the denominator, we use long division to rewrite the function as quotient + remainder/divisor (dividing the numerator and denominator by x^2 will not work in this case):

$$g(x) = \frac{2x^2 + 3x - 5}{x + 2} = 2x - 1 + \frac{-3}{x + 2}$$

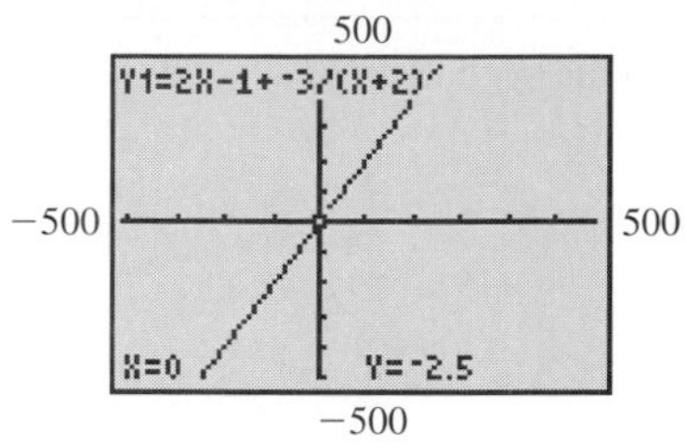

Figure 3.77

If $|x| \to \infty$, then $-3/(x+2) \to 0$. So the value of $g(x)$ approaches $2x - 1$ as $|x| \to \infty$. The line $y = 2x - 1$ is an oblique asymptote for the graph of g. You may look ahead to Fig. 3.84 to see the graph of this function with its oblique asymptote.

Note that the wide view of g in Fig. 3.77 looks like the line $y = 2x - 1$.

If the degree of $P(x)$ is 1 greater than the degree of $Q(x)$ and the degree of $Q(x)$ is at least 1, then the rational function has an oblique asymptote. In this case use division to rewrite the function as quotient + remainder/divisor. The graph of the equation formed by setting y equal to the quotient is an oblique asymptote. We conclude this discussion of asymptotes with a summary.

SUMMARY

Finding Asymptotes for a Rational Function

Let $f(x) = P(x)/Q(x)$ be a rational function in lowest terms with the degree of $Q(x)$ at least 1.

1. The graph of f has a vertical asymptote corresponding to each root of $Q(x) = 0$.
2. If the degree of $P(x)$ is less than the degree of $Q(x)$, then the x-axis is a horizontal asymptote.
3. If the degree of $P(x)$ equals the degree of $Q(x)$, then the horizontal asymptote is determined by the ratio of the leading coefficients.
4. If the degree of $P(x)$ is greater than the degree of $Q(x)$, then use division to rewrite the function as quotient + remainder/divisor. The graph of the equation formed by setting y equal to the quotient is an asymptote.

Sketching Graphs of Rational Functions

We now use asymptotes and symmetry to help us sketch the graphs of the rational functions discussed in Examples 2 and 3. Use the following steps to graph a rational function.

PROCEDURE

Graphing a Rational Function

To graph a rational function in lowest terms:

1. Determine the asymptotes and draw them as dashed lines.
2. Check for symmetry.
3. Find any intercepts.
4. Plot several selected points to determine how the graph approaches the asymptotes.
5. Draw curves through the selected points, approaching the asymptotes.

EXAMPLE 4 Functions with horizontal and vertical asymptotes

Sketch the graph of each rational function.

a) $f(x) = \dfrac{3}{x^2 - 1}$ **b)** $g(x) = \dfrac{x}{x^2 - 4}$ **c)** $h(x) = \dfrac{2x + 1}{x + 3}$

Solution

a) From Example 2(a), $x = 1$ and $x = -1$ are vertical asymptotes, and the x-axis is a horizontal asymptote. Draw the vertical asymptotes using dashed lines. Since all of the powers of x are even, $f(-x) = f(x)$ and the graph is symmetric about the y-axis. The y-intercept is $(0, -3)$. There are no x-intercepts because $f(x) = 0$ has no solution. Evaluate the function at $x = 0.9$ and $x = 1.1$ to see how the curve approaches the vertical asymptote at $x = 1$. Evaluate at $x = 2$ and $x = 3$ to see whether the curve approaches the horizontal asymptote from above or below. We get $(0.9, -15.789)$, $(1.1, 14.286)$, $(2, 1)$, and $(3, 3/8)$. From these points you can see that the curve is going downward toward $x = 1$ from the left and upward toward $x = 1$ from the right. It is approaching its horizontal asymptote from above. Now use the symmetry with respect to the y-axis to draw the curve approaching its asymptotes as shown in Fig. 3.78.

The calculator graph in dot mode in Fig. 3.79 supports these conclusions. □

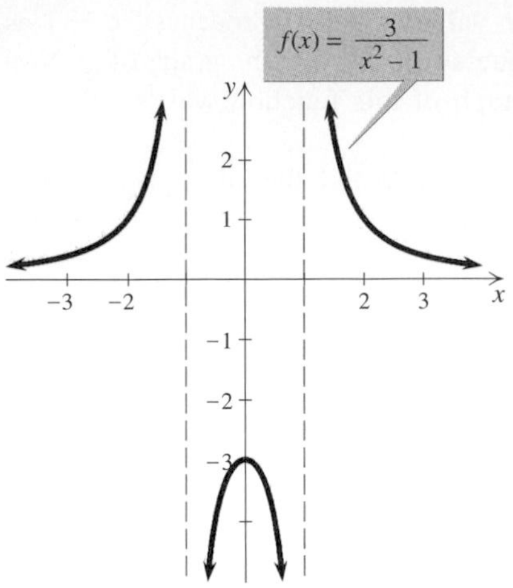

Figure 3.78

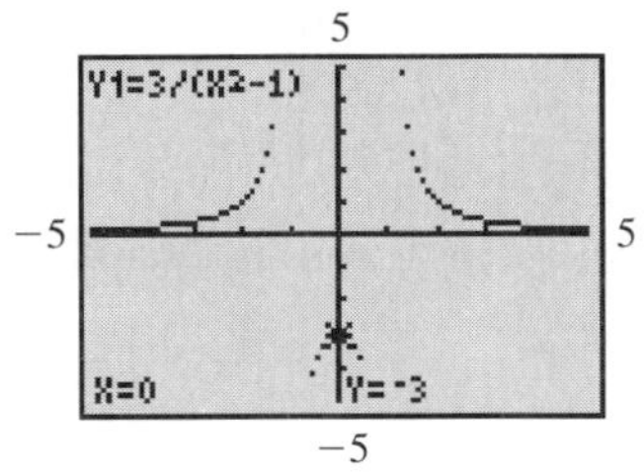

Figure 3.79

b) Draw the vertical asymptotes $x = 2$ and $x = -2$ from Example 2(b) as dashed lines. The x-axis is a horizontal asymptote. Because $f(-x) = -f(x)$, the graph is symmetric about the origin. The x-intercept is $(0, 0)$. Evaluate the function near the vertical asymptote $x = 2$, and for larger values of x to see how the curve approaches the horizontal asymptote. We get $(1.9, -4.872)$, $(2.1, 5.122)$, $(3, 3/5)$, and $(4, 1/3)$. From these points you can see that the curve is going downward toward $x = 2$ from the left and upward toward $x = 2$ from the right. It is approaching its horizontal asymptote from above. Now use the symmetry with respect to the origin to draw the curve approaching its asymptotes as shown in Fig. 3.80.

The calculator graph in connected mode in Fig. 3.81 confirms these conclusions. Note how the calculator appears to draw the vertical asymptotes

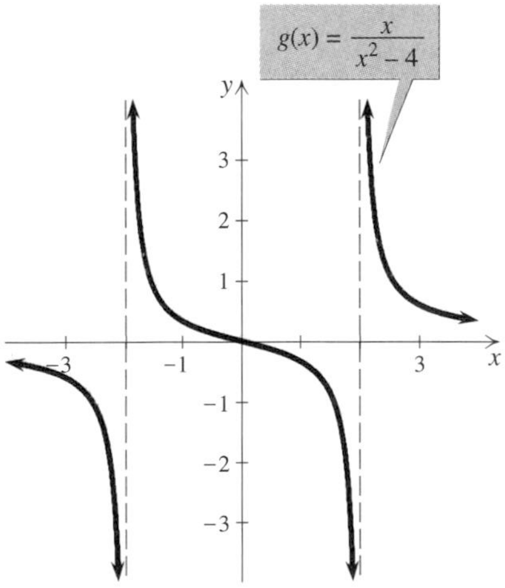

Figure 3.80

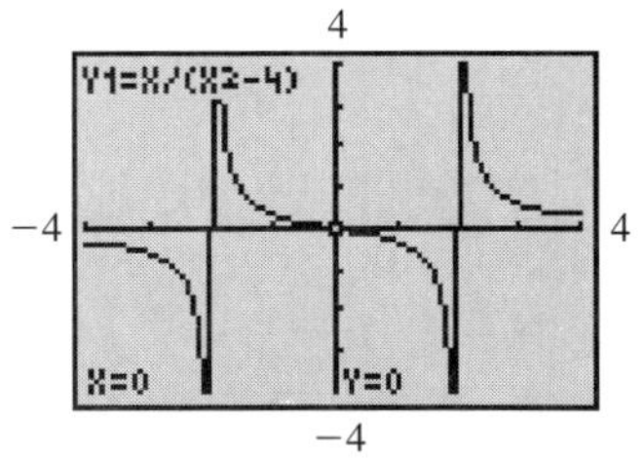

Figure 3.81

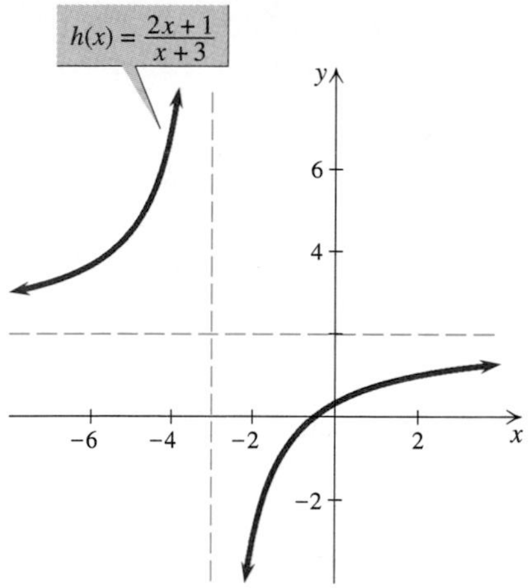

Figure 3.82

as it connects the points that are close to but on opposite sides of the asymptotes. □

c) Draw the vertical asymptote $x = -3$ and the horizontal asymptote $y = 2$ from Example 2(c) as dashed lines. The x-intercept is $(-1/2, 0)$ and the y-intercept is $(0, 1/3)$. The points $(-2, -3)$, $(7, 1.5)$, $(-4, 7)$, and $(-13, 2.5)$ are also on the graph. From these points we can conclude that the curve goes upward toward $x = -3$ from the left and downward toward $x = -3$ from the right. It approaches $y = 2$ from above as x goes to $-\infty$ and from below as x goes to ∞. Draw the graph approaching its asymptotes as shown in Fig. 3.82.

The calculator graph in Fig. 3.83 supports these conclusions.

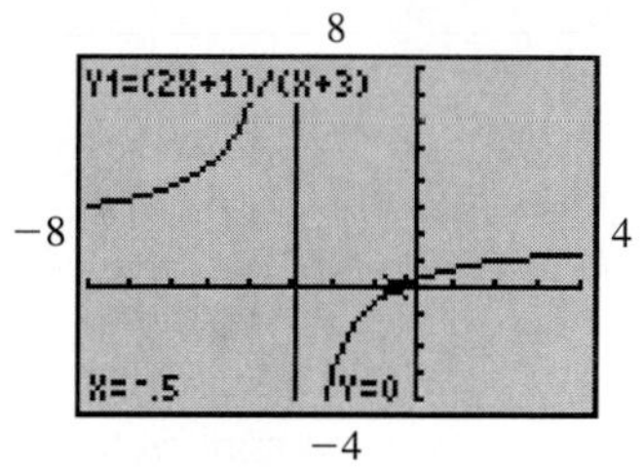

Figure 3.83

Although the graphing calculator is very helpful for graphing rational functions, the calculator has two serious limitations. The graph on a calculator crosses its vertical asymptotes in connected mode and it usually appears to touch its horizontal asymptotes.

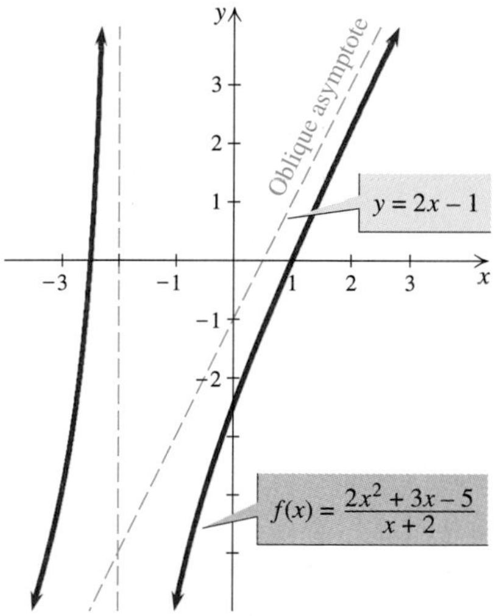

Figure 3.84

EXAMPLE 5 Graphing a function with an oblique asymptote

Sketch the graph of $k(x) = \dfrac{2x^2 + 3x - 5}{x + 2}$.

Solution

Draw the vertical asymptote $x = -2$ and the oblique asymptote $y = 2x - 1$ determined in Example 3 as dashed lines. The x-intercepts, $(1, 0)$ and $(-2.5, 0)$, are found by solving $2x^2 + 3x - 5 = 0$. The y-intercept is $(0, -2.5)$. The points $(-1, -6)$, $(4, 6.5)$, and $(-3, -4)$ are also on the graph. From these points we can conclude that the curve goes upward toward $x = -2$ from the left and downward toward $x = -2$ from the right. It approaches $y = 2x - 1$ from above as x goes to $-\infty$ and from below as x goes to ∞. Draw the graph approaching its asymptotes as shown in Fig. 3.84.

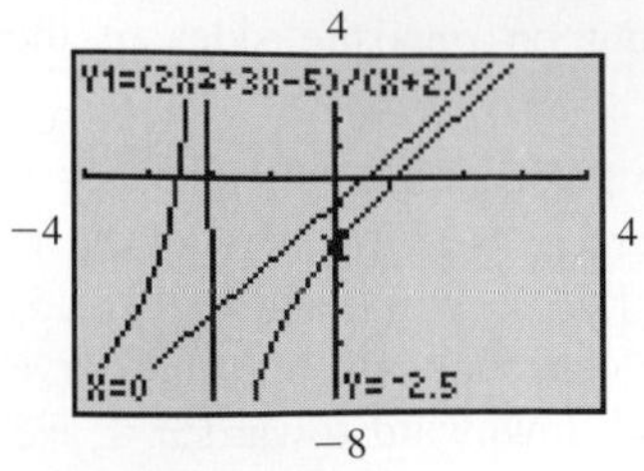

Figure 3.85

The calculator graph in Fig. 3.85 supports these conclusions.

The graph of a rational function cannot cross a vertical asymptote, but it can cross a nonvertical asymptote. The graph of a rational function gets closer and closer to a nonvertical asymptote as $|x| \to \infty$, but it can also cross a nonvertical asymptote at a "small" value of x, as illustrated in the next example.

EXAMPLE 6 A rational function that crosses an asymptote

Sketch the graph of $f(x) = \dfrac{4x}{(x + 1)^2}$.

Solution

The graph of f has a vertical asymptote at $x = -1$, and since the degree of the numerator is less than the degree of the denominator, the x-axis is a horizontal asymptote. If $x > 0$ then $f(x) > 0$, and if $x < 0$ then $f(x) < 0$. So the graph approaches the horizontal axis from above for $x > 0$, and from below for $x < 0$. However, the x-intercept is (0, 0). So the graph crosses its horizontal asymptote at (0, 0). The graph shown in Fig. 3.86 goes through (1, 1), (2, 8/9), (−2, −8), and (−3, −3).

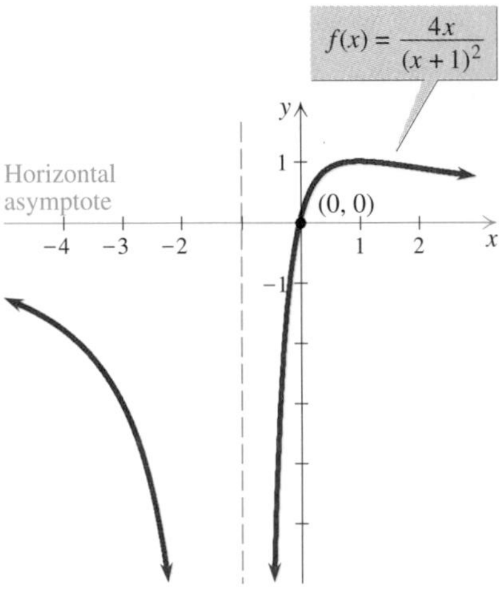

Figure 3.86

All rational functions so far have been given in lowest terms. In the next example the numerator and denominator have a common factor. In this case the graph does not have as many vertical asymptotes as you might expect. The graph is almost identical to the graph of the rational function obtained by reducing the expression to lowest terms.

• CALCULUS •

EXAMPLE 7 A graph with a hole in it

Sketch the graph of $f(x) = \dfrac{x - 2}{x^2 - 4}$.

Solution

Since $x^2 - 4 = (x - 2)(x + 2)$, the domain of f is the set of all real numbers except 2 and −2. Since the rational expression can be reduced, the function f could also be defined as

$$f(x) = \frac{1}{x + 2} \quad \text{for } x \neq 2 \text{ and } x \neq -2.$$

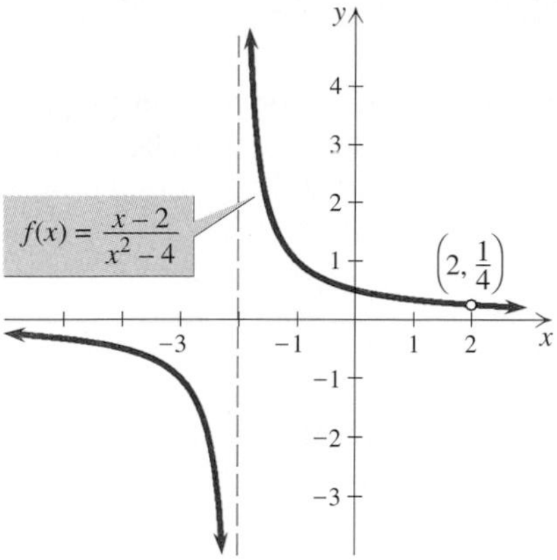

Figure 3.87

The graph of $y = 1/(x + 2)$ has only one vertical asymptote $x = -2$. In fact, the graph of $y = 1/(x + 2)$ is a translation two units to the left of $y = 1/x$. Since $f(2)$ is undefined, the point (2, 1/4) that would normally be on the graph of $y - 1/(x + 2)$ is omitted. The missing point is indicated on the graph in Fig. 3.87 as a small open circle. Note that the graph made by a graphing calculator will usually not show the missing point.

Rational Inequalities

The only way that the graph of a polynomial function can get from one side of the x-axis to the other is by crossing it. So to solve polynomial inequalities in Section 3.6, we had to find all x-intercepts on the graph of the corresponding polynomial functions. The graph of a rational function can go from one side of the x-axis to the other at an x-intercept or at a vertical asymptote. So to solve a rational inequality, we must find all x-intercepts and vertical asymptotes for the graph of the corresponding rational function.

EXAMPLE 8 Solving a rational inequality

Solve the inequality $\dfrac{2x^2 + 7x - 16}{x^2 - 5} \leq 2$

Solution

First rewrite the inequality with 0 on the right and a single rational expression on the left:

$$\frac{2x^2 + 7x - 16}{x^2 - 5} - 2 \leq 0$$

$$\frac{2x^2 + 7x - 16}{x^2 - 5} - \frac{2(x^2 - 5)}{x^2 - 5} \leq 0$$

$$\frac{7x - 6}{x^2 - 5} \leq 0$$

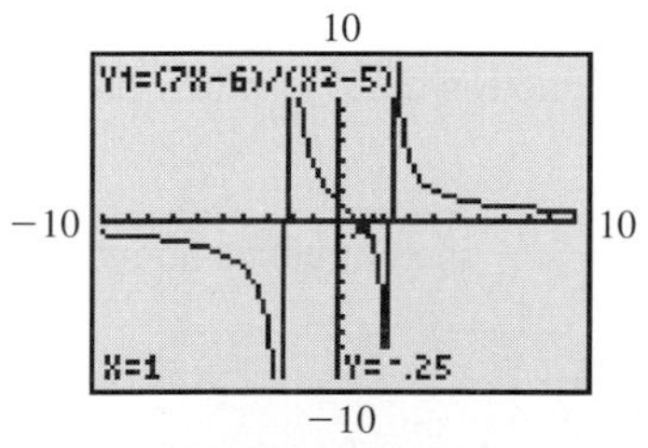

Figure 3.88

Now consider the graph of the rational function $y = \dfrac{7x - 6}{x^2 - 5}$. This graph can be obtained with a graphing calculator as in Fig. 3.88 or by hand. The graph crosses the x-axis at 6/7 and has vertical asymptotes at $\pm\sqrt{5}$. In Fig. 3.88 we see where the graph is below or on the x-axis. The solution set is $(-\infty, -\sqrt{5}) \cup [6/7, \sqrt{5})$. Note that 6/7 is included in the solution set because $y = 0$ when $x = 6/7$.

Applications

Rational functions can occur in many applied situations. A horizontal asymptote might indicate that the average cost of producing a product approaches a fixed value in the long run. A vertical asymptote might show that the cost of a project goes up astronomically as a certain barrier is approached.

EXAMPLE 9 Average cost of a handbook

Eco Publishing spent \$5000 in producing an environmental handbook and \$8 each for printing. Write a function that gives the average cost to the company per printed handbook. Graph the function for $0 < x \le 500$. What happens to the average cost if the book becomes very, very popular?

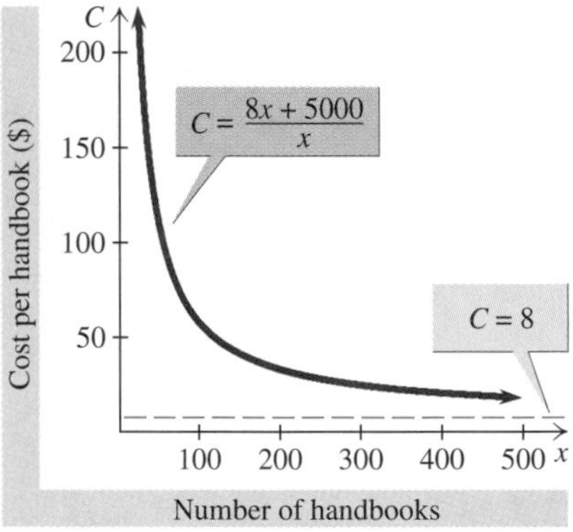

Figure 3.89

Solution

The total cost of producing and printing x handbooks is $8x + 5000$. To find the average cost per book, divide the total cost by the number of books:

$$C = \frac{8x + 5000}{x}$$

This rational function has a vertical asymptote at $x = 0$ and a horizontal asymptote $C = 8$. The graph is shown in Fig. 3.89. As x gets larger and larger, the \$5000 production cost is spread out over more and more books, and the average cost per book approaches \$8, as shown by the horizontal asymptote.

FOR THOUGHT True or False? Explain.

1. The function $f(x) = \dfrac{1}{\sqrt{x} - 3}$ is a rational function.
2. The domain of $f(x) = \dfrac{x + 2}{x - 2}$ is $(-\infty, -2) \cup (-2, 2) \cup (2, \infty)$.
3. The number of vertical asymptotes for a rational function equals the degree of the denominator.
4. The graph of $f(x) = \dfrac{1}{x^4 - 4x^2}$ has four vertical asymptotes.
5. The x-axis is a horizontal asymptote for $f(x) = \dfrac{x^2 - 2x + 7}{x^3 - 4x}$.
6. The x-axis is a horizontal asymptote for the graph of $f(x) = \dfrac{5x - 1}{x + 2}$.
7. The line $y = x - 3$ is an asymptote for the graph of $f(x) = x - 3 + \dfrac{1}{x - 1}$.
8. The graph of a rational function cannot intersect its asymptote.
9. The graph of $f(x) = \dfrac{4}{x^2 - 16}$ is symmetric about the y-axis.
10. The graph of $f(x) = \dfrac{2x + 6}{x^2 - 9}$ has only one vertical asymptote.

3.7 EXERCISES

Tape 8 Disk

Find the domain of each rational function.

1. $f(x) = \dfrac{4}{x+2}$

2. $f(x) = \dfrac{-1}{x-2}$

3. $f(x) = \dfrac{-x}{x^2-4}$

4. $f(x) = \dfrac{2}{x^2-x-2}$

5. $f(x) = \dfrac{2x+3}{x-3}$

6. $f(x) = \dfrac{4-x}{x+2}$

7. $f(x) = \dfrac{x^2-2x+4}{x}$

8. $f(x) = \dfrac{x^3+2}{x^2}$

9. $f(x) = \dfrac{3x^2-1}{x^3-x}$

10. $f(x) = \dfrac{x^2+1}{8x^3-2x}$

11. $f(x) = \dfrac{-x^2+x}{x^2+5x+6}$

12. $f(x) = \dfrac{-x^2+2x-3}{x^2+x-12}$

Determine the domain and the equations of the asymptotes for the graph of each rational function.

13.

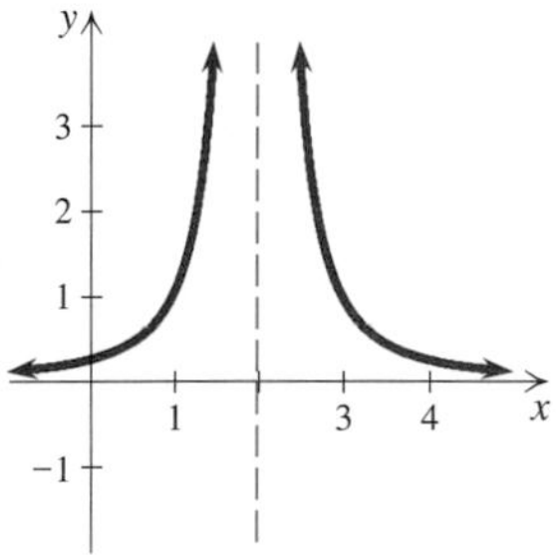

14.

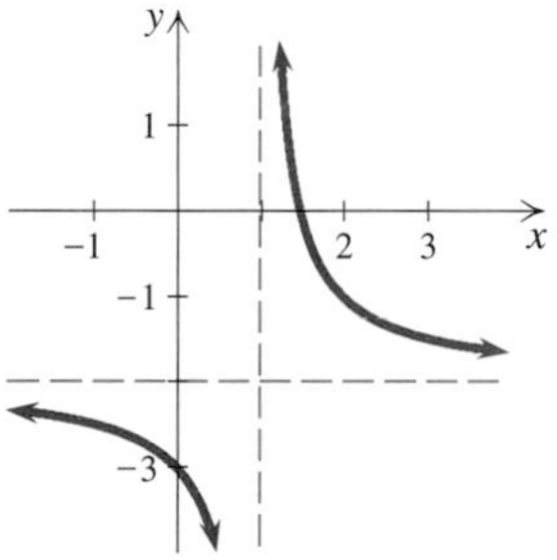

15.

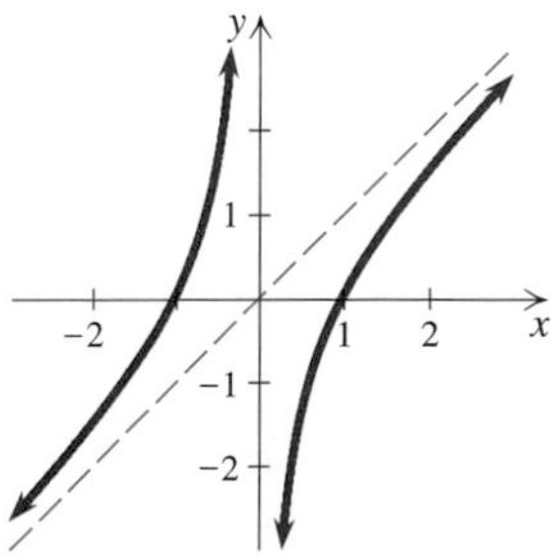

16.

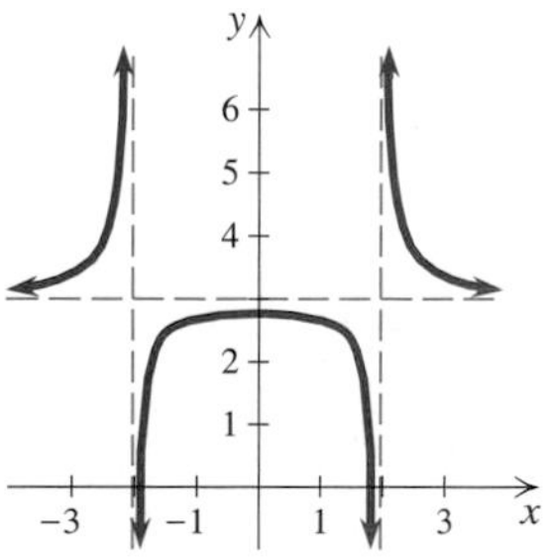

Determine the equations of all asymptotes for the graph of each function.

17. $f(x) = \dfrac{5}{x-2}$

18. $f(x) = \dfrac{-1}{x+12}$

19. $f(x) = \dfrac{-x}{x^2-9}$

20. $f(x) = \dfrac{-2}{x^2-5x+6}$

21. $f(x) = \dfrac{2x+4}{x-1}$

22. $f(x) = \dfrac{5-x}{x+5}$

23. $f(x) = \dfrac{x^2-2x+1}{x}$

24. $f(x) = \dfrac{x^3-8}{x^2}$

25. $f(x) = \dfrac{3x^2+4}{x+1}$

26. $f(x) = \dfrac{x^2}{x-9}$

27. $f(x) = \dfrac{-x^2+4x}{x+2}$

28. $f(x) = \dfrac{-x^2+3x-7}{x-3}$

Find all asymptotes, x-intercepts, and y-intercepts for the graph of each rational function and sketch the graph of the function.

29. $f(x) = \dfrac{-1}{x}$

30. $f(x) = \dfrac{1}{x^2}$

31. $f(x) = \dfrac{1}{x-2}$

32. $f(x) = \dfrac{-1}{x+1}$

33. $f(x) = \dfrac{1}{x^2-4}$

34. $f(x) = \dfrac{1}{x^2-2x+1}$

35. $f(x) = \dfrac{-1}{(x+1)^2}$

36. $f(x) = \dfrac{-2}{x^2-9}$

37. $f(x) = \dfrac{2x+1}{x-1}$

38. $f(x) = \dfrac{3x-1}{x+1}$

39. $f(x) = \dfrac{x-3}{x+2}$

40. $f(x) = \dfrac{2-x}{x+2}$

41. $f(x) = \dfrac{x}{x^2-1}$

42. $f(x) = \dfrac{-x}{x^2-9}$

43. $f(x) = \dfrac{4x}{x^2-2x+1}$

44. $f(x) = \dfrac{-2x}{x^2+6x+9}$

45. $f(x) = \dfrac{8-x^2}{x^2-9}$

46. $f(x) = \dfrac{2x^2+x-8}{x^2-4}$

47. $f(x) = \dfrac{2x^2+8x+2}{x^2+2x+1}$

48. $f(x) = \dfrac{-x^2+7x-9}{x^2-6x+9}$

Find the oblique asymptote and sketch the graph of each rational function.

49. $f(x) = \dfrac{x^2 + 1}{x}$

50. $f(x) = \dfrac{x^2 - 1}{x}$

51. $f(x) = \dfrac{x^3 - 1}{x^2}$

52. $f(x) = \dfrac{x^3 + 1}{x}$

53. $f(x) = \dfrac{x^2}{x + 1}$

54. $f(x) = \dfrac{x^2}{x - 1}$

55. $f(x) = \dfrac{2x^2 - x}{x - 1}$

56. $f(x) = \dfrac{-x^2 + x + 1}{x + 1}$

Match each rational function with its graph (a)–(h), without using a graphing calculator.

57. $f(x) = -\dfrac{3}{x}$

58. $f(x) = \dfrac{1}{3 - x}$

59. $f(x) = \dfrac{x}{x - 3}$

60. $f(x) = \dfrac{x - 3}{x}$

61. $f(x) = \dfrac{1}{x^2 - 3x}$

62. $f(x) = \dfrac{x^2}{x^2 - 9}$

63. $f(x) = \dfrac{x^2 - 3}{x}$

64. $f(x) = \dfrac{-x^3 + 1}{x^2}$

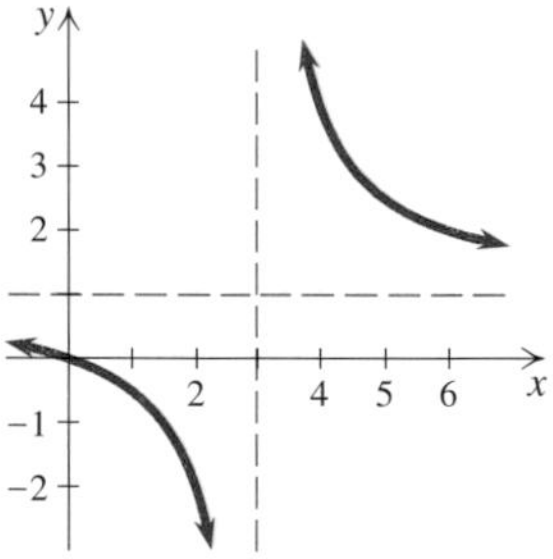
(a)

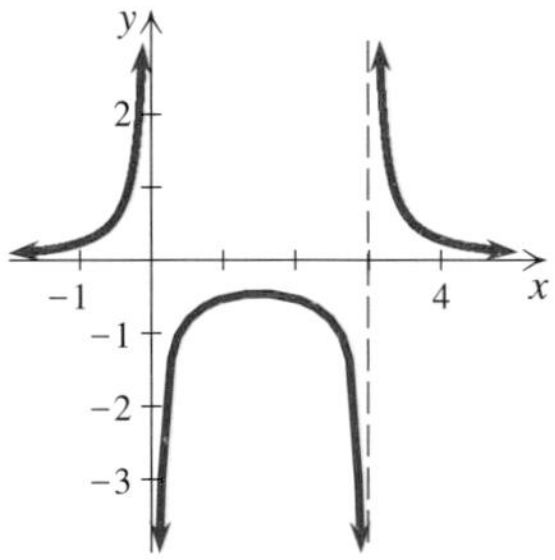
(b)

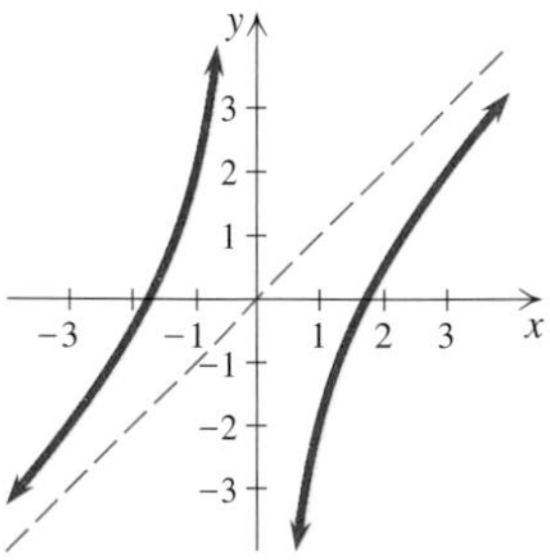
(c)

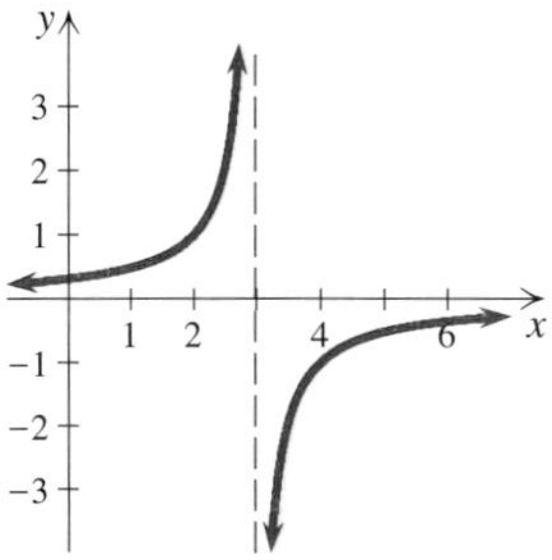
(d)

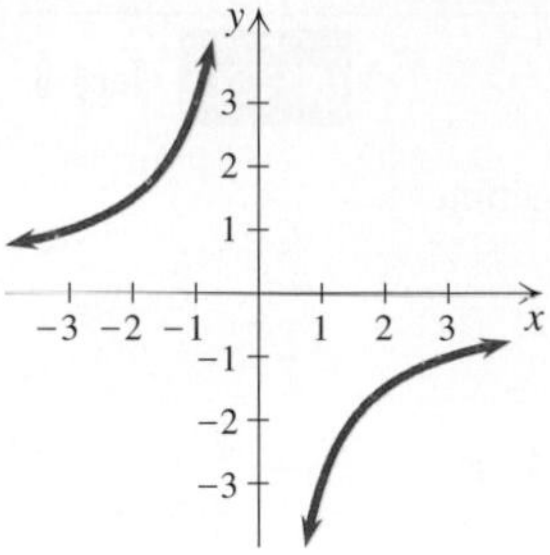
(e)

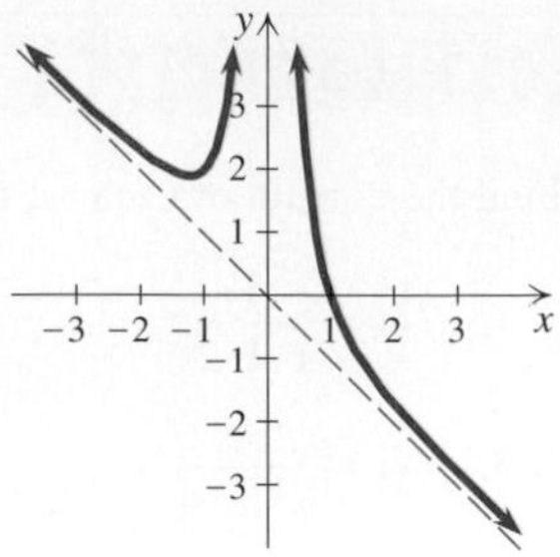
(f)

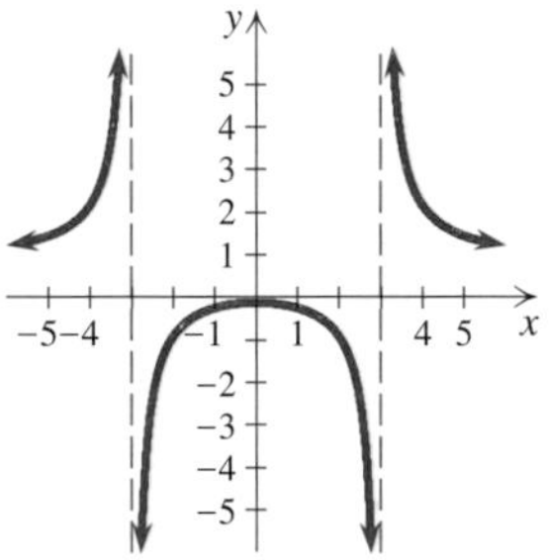
(g)

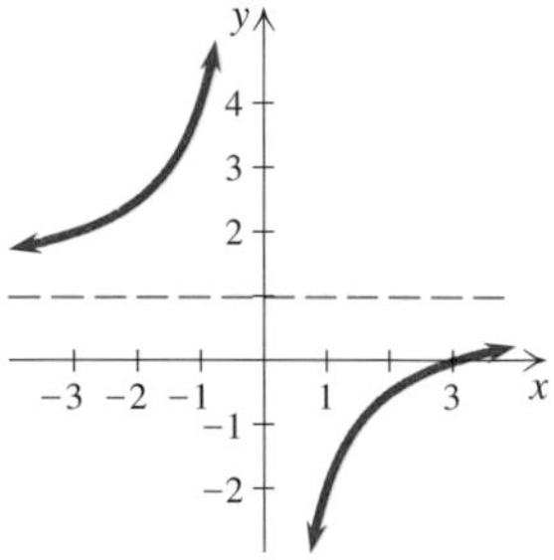
(h)

Sketch the graph of each rational function. Note that the functions are not in lowest terms.

65. $f(x) = \dfrac{x + 1}{x^2 - 1}$

66. $f(x) = \dfrac{x}{x^2 + 2x}$

67. $f(x) = \dfrac{x^2 - 1}{x - 1}$

68. $f(x) = \dfrac{x^2 - 5x + 6}{x - 2}$

Sketch the graph of each rational function.

69. $f(x) = \dfrac{2}{x^2 + 1}$

70. $f(x) = \dfrac{x}{x^2 + 1}$

71. $f(x) = \dfrac{x - 1}{x^3 - 9x}$

72. $f(x) = \dfrac{x^2 + 1}{x^3 - 4x}$

73. $f(x) = \dfrac{x + 1}{x^2}$

74. $f(x) = \dfrac{x - 1}{x^2}$

For each function find $f(5000)$ to the nearest tenth. Then find $f(5000) - k$ where $y = k$ is the horizontal asymptote of the graph.

75. $f(x) = \dfrac{3x - 1}{2x - 9}$

76. $f(x) = \dfrac{x - 9}{2 - x}$

77. $f(x) = \dfrac{4}{3x + 40}$

78. $f(x) = \dfrac{2x + 5}{x^2 - 1}$

For each function find $f(-5000)$ to the nearest tenth. Then find $f(-5000) - k$ where $y = k$ is the horizontal asymptote of the graph.

79. $f(x) = \dfrac{9x^2 - 1}{3x^2 - 2}$

80. $f(x) = \dfrac{5 - 4x^2}{10x^2 + 7}$

81. $f(x) = \dfrac{-3x + 4}{2x^3 - 9x^2}$

82. $f(x) = \dfrac{-7x + 9}{-10x^2 - 4x + 2}$

Use a graphing calculator to help you graph each function. Why do these graphs fail to have vertical asymptotes? What is the domain of each function? Estimate the range of each function from the graph. What is the equation of the horizontal asymptote?

83. $y = \dfrac{100}{x^2 + 1}$

84. $y = \dfrac{9x}{x^2 + 1}$

85. $y = \dfrac{5}{x^2 + 2x + 6}$

86. $y = \dfrac{-6x^2 + 12x - 29}{x^2 - 2x + 5}$

87. $f(x) = \dfrac{2x - 3}{x^2 + x + 1}$

88. $f(x) = \dfrac{x^3 + x + 5}{x^2 + 1}$

Solve each rational inequality. State the solution set in interval notation.

89. $\dfrac{3}{9x + 2} < 0$

90. $\dfrac{1}{6 - 7x} > 0$

91. $\dfrac{6x - 5}{x^2 - 5} > 0$

92. $\dfrac{22x - 3}{x^2 - 6} < 0$

93. $\dfrac{2x}{x + 2} \le \dfrac{6}{x - 3}$

94. $\dfrac{2}{x + 3} \ge \dfrac{1}{x - 1}$

Solve each problem.

95. *Admission to the Zoo* Winona paid \$100 for a lifetime membership to Friends of the Zoo, so that she could gain admittance to the zoo for only \$1 per visit. Write Winona's average cost per visit C as a function of the number of visits when she has visited x times. What is her average cost per visit when she has visited the zoo 100 times? Graph the function for $x > 0$. What happens to her average cost per visit if she starts when she is young and visits the zoo every day?

96. *Renting a Car* The cost of renting a car for one day is \$19 plus 30 cents per mile. Write the average cost per mile C as a function of the number of miles driven in one day x. Graph the function for $x > 0$. What happens to C as the number of miles gets very large?

97. *Average Speed of an Auto Trip* A 200-mi trip by an electric car must be completed in 4 hr. Let x be the number of hours it takes to travel the first half of the distance and write the average speed for the second half of the trip as a function of x. Graph this function for $0 < x < 4$. What is the significance of the vertical asymptote?

98. *Billboard Advertising* An Atlanta marketing agency figures that the monthly cost of a billboard advertising campaign depends on the fraction of the market p that the client wishes to reach. For $0 \le p < 1$ the cost in dollars is determined by the formula $C = (4p - 1200)/(p - 1)$. What is the monthly cost for a campaign intended to reach 95% of the market? Graph this function for $0 \le p < 1$. What happens to the cost for a client who wants to reach 100% of the market?

99. *Carbon Monoxide Poisoning* The accompanying graph shows the symptoms associated with various levels of carbon monoxide (CO) for various amounts of time. (American Sensors, Carbon Monoxide Detector Owner's Manual)

a. An exposure to 800 parts per million (PPM) for what amount of time will cause drowsiness?

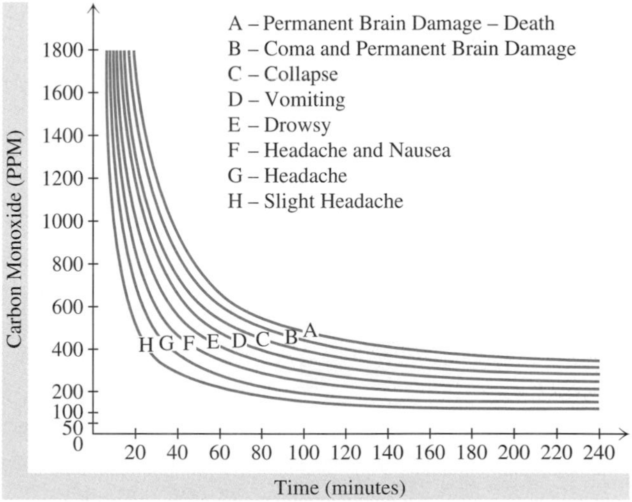

Figure for Exercise 99

b. If the worst symptom a person exhibits after exposure to at least 400 PPM of carbon monoxide is collapse, then for what amount of time was the person exposed?
c. If a person is exposed to less than 800 PPM of CO for less than 40 minutes, then what is the worst symptom possible?
d. What are the horizontal and vertical asymptotes for curve A? Explain what these asymptotes mean in simple terms.

100. *Balancing the Costs* A furniture maker buys foam rubber x times per year. The delivery charge is \$400 per purchase regardless of the amount purchased. The annual cost of storage is figured as \$10,000/x, because the more frequent the purchase, the less it costs for storage. So the annual cost of delivery and storage is given by

$$C = 400x + \frac{10,000}{x}.$$

a. Graph the function with a graphing calculator.
b. Find the number of purchases per year that minimizes the annual cost of delivery and storage.

101. *Making a Gas Tank* An engineer is designing a cylindrical metal tank that is to hold 500 ft^3 of gasoline.
a. The volume of the tank is given by $V = \pi r^2 h$, where r is the radius and h is the height. Write h as a function of r.
b. The surface area of a tank is given by $S = 2\pi r^2 + 2\pi rh$. Use the result of part (a) to write S as a function of r and graph it.
c. Use the minimum feature of a graphing calculator to find the radius to the nearest tenth of a foot that minimizes the surface area. Ignore the thickness of the metal.
d. If the tank costs \$8 per square foot to construct, then what is the minimum cost for making the tank?

102. *Making a Glass Tank* An architect for the Aquarium of the Americas is designing a cylindrical fish tank to hold 1000 ft^3 of water. The bottom and side of the tank will be 3 in.-thick glass and the tank will have no top.
a. Write the inside surface area of the tank as a function of the inside radius and graph the function.
b. Use the minimum feature of a graphing calculator to find the inside radius that minimizes that inside surface area.
c. If Owens Corning will build the tank for \$250 per cubic foot of glass used, then what will be the cost for the minimal tank?

For Writing/Discussion

103. *Approximate Value* Given an arbitrary rational function $R(x)$, explain how you can find the approximate value of $R(x)$ when the absolute value of x is very large.

104. *Cooperative Learning* Each student in your small group should write the equations of the horizontal, vertical, and/or oblique asymptotes of an unknown rational function. Then work as a group to find a rational function that has the specified asymptotes. Is there a rational function with any specified asymptotes?

Linking Concepts

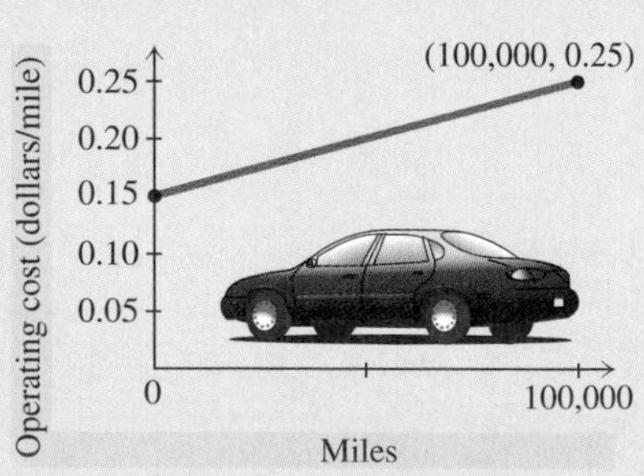

For Individual or Group Explorations

Capital Cost and Operating Cost To decide when to replace company cars, an accountant looks at two cost components: capital cost and operating cost. The capital cost is the difference between the original cost and the salvage value and the operating cost is the cost of operating the vehicle.

a) For a new Ford Taurus, the capital cost is \$3000 plus \$0.12 for each mile that the car is driven. Write the capital cost C as a function of the number of miles the car is driven x.

b) If a Ford Taurus is purchased for \$19,846 and driven 80,000 miles, then what is its salvage value?

c) The operating cost P for the Taurus starts at \$0.15 per mile when the car is new and increases linearly to \$0.25 cents per mile when the car reaches 100,000 miles. Write P as a function of x.

d) What is the operating cost at 70,000 miles?

e) The total cost per mile is given by $T = \frac{C}{x} + P$. Identify all asymptotes and graph T as a function of x.

f) Explain each asymptote in terms of this application.

g) Find T for $x = 20{,}000$, 30,000, and 90,000 miles.

h) The senior vice president has told the accountant that the goal is to keep the total operating cost less than \$0.38 per mile. Is this goal practical?

HIGHLIGHTS

- **Section 3.1 Linear Functions**

1. The graph of every linear function $y = mx + b$ is a line with slope m and y-intercept $(0, b)$.
2. The slope of a line through (x_1, y_1) and (x_2, y_2) with $x_1 \neq x_2$ is $(y_2 - y_1)/(x_2 - x_1)$.
3. The slope-intercept form of the equation of a line through (x_1, y_1) with slope m is $y - y_1 = m(x - x_1)$.
4. Two nonvertical lines are parallel if and only if their slopes are equal.
5. Two lines with slopes m_1 and m_2 are perpendicular if and only if $m_1 m_2 = -1$.

- **Section 3.2 Quadratic Functions**

1. The graph of every quadratic function $y = ax^2 + bx + c$ $(a \neq 0)$ is a parabola, which opens upward if $a > 0$ and downward if $a < 0$.
2. The graph of $y = ax^2 + bx + c$ $(a \neq 0)$ is a transformation of the graph of $y = x^2$.
3. The x-coordinate of the vertex for $y = ax^2 + bx + c$ is $-b/(2a)$.
4. The vertex for the parabola $y = a(x - h)^2 + k$ is (h, k).
5. The minimum or maximum value of a quadratic function occurs at the vertex of the parabola.

- **Section 3.3 Zeros of Polynomial Functions**

1. The remainder when a polynomial $P(x)$ is divided by $x - c$ is $P(c)$.
2. Synthetic division provides a quick means of dividing $P(x)$ by $x - c$.
3. The number c is a zero of $y = P(x)$ if and only if $x - c$ is a factor of the polynomial $P(x)$.
4. A polynomial function of positive degree has at least one complex zero.

5. If p/q is a rational zero in lowest terms for a polynomial function with integral coefficients, then p is a factor of the constant term and q is a factor of the leading coefficient.

- **Section 3.4 The Theory of Equations**

1. A polynomial equation with positive degree n has n roots, when multiplicity is counted.
2. The roots of an equation $P(x) = 0$ are the same as the zeros of the function $y = P(x)$.
3. In a polynomial equation with real coefficients, the imaginary roots occur in conjugate pairs.
4. Descartes's rule of signs and the theorem on bounds give us clues about the size and type of roots to a polynomial equation.

- **Section 3.5 Miscellaneous Equations**

1. Equations of higher degree can be solved by factoring.
2. Squaring each side of an equation can produce extraneous roots.
3. Equations of quadratic type are not quadratic equations, but they can be solved by the techniques used for quadratic equations.
4. To solve an equation involving absolute value, we first write an equivalent equation without absolute value.

- **Section 3.6 Graphs of Polynomial Functions**

1. The graph of a polynomial function can be symmetric about the y-axis ($f(-x) = f(x)$), symmetric about the origin ($f(-x) = -f(x)$), or have neither type of symmetry.
2. The graph of a quadratic function is symmetric about a vertical line through its vertex.
3. The graph of a polynomial function crosses the x-axis at an x-intercept if the factor corresponding to that intercept is raised to an odd power, and the graph touches but does not cross if the factor is raised to an even power.
4. The leading coefficient and the degree of the function determine the behavior of a polynomial function as $x \to \infty$ or $x \to -\infty$.

- **Section 3.7 Graphs of Rational Functions**

1. The graph of a rational function in lowest terms has a vertical asymptote for each value of x for which the denominator has a value of zero.
2. If the degree of the numerator of a rational function is less than the degree of the denominator, then the x-axis is a horizontal asymptote.
3. If the degree of the numerator equals the degree of the denominator, then the horizontal asymptote is determined by the ratio of the leading coefficients.
4. If the degree of the numerator of a rational function is greater than the degree of the denominator, then use division to discover the oblique asymptote.
5. Rational inequalities are solved by locating all zeros and vertical asymptotes for the graph of a rational function.

CHAPTER 3 REVIEW EXERCISES

Solve each problem.

1. Find the slope of the line $3x - 5y = 8$.
2. Find the slope of the line through $(-3, 6)$ and $(5, -4)$.
3. Find the slope-intercept form of the equation of the line that goes through $(1, -2)$ and is perpendicular to the line $2x - 3y = 6$.
4. Find the equation of the line in slope-intercept form that goes through $(3, -4)$ and is parallel to the line through $(0, 2)$ and $(3, -1)$.
5. Write the function $f(x) = 3x^2 - 2x + 1$ in the form $f(x) = a(x - h)^2 + k$.
6. Write the function $f(x) = -4\left(x - \frac{1}{3}\right)^2 - \frac{1}{2}$ in the form $f(x) = ax^2 + bx + c$.
7. Find the vertex, axis of symmetry, x-intercepts, and y-intercept for the parabola $y = 2x^2 - 4x - 1$.
8. Find the maximum value of the function $y = -x^2 + 3x - 5$.
9. Write the equation of a parabola that has x-intercepts $(-1, 0)$ and $(3, 0)$ and y-intercept $(0, 6)$.
10. Write the equation of the parabola that has vertex $(1, 2)$ and y-intercept $(0, 5)$.
11. Solve the inequality $2x^2 + 5x < 3$.
12. Solve the inequality $-x^2 + 4x - 7 < 0$.

Find all real and imaginary zeros for each polynomial function.

13. $f(x) = 3x - 1$
14. $g(x) = 7$
15. $h(x) = x^2 - 8$
16. $m(x) = x^3 - 8$
17. $n(x) = 8x^3 - 1$
18. $C(x) = 3x^2 - 2$
19. $P(t) = t^4 - 100$
20. $S(t) = 25t^4 - 1$
21. $R(s) = 8s^3 - 4s^2 - 2s + 1$
22. $W(s) = s^3 + s^2 + s + 1$
23. $f(x) = x^3 + 2x^2 - 6x$
24. $f(x) = 2x^3 - 4x^2 + 3x$

For each polynomial, find the indicated value in two different ways.

25. $P(x) = 4x^3 - 3x^2 + x - 1$, $P(3)$
26. $P(x) = 2x^3 + 5x^2 - 3x - 2$, $P(-2)$
27. $P(x) = -8x^5 + 2x^3 - 6x + 2$, $P\left(-\frac{1}{2}\right)$
28. $P(x) = -4x^4 + 3x^2 + 1$, $P\left(\frac{1}{2}\right)$

Use the rational zero theorem to list all possible rational zeros for each polynomial function.

29. $f(x) = -3x^3 + 6x^2 + 5x - 2$
30. $f(x) = 2x^4 + 9x^2 - 8x - 3$
31. $f(x) = 6x^4 - x^2 - 9x + 3$
32. $f(x) = 4x^3 - 5x^2 - 13x - 8$

Find a polynomial equation with integral coefficients (and lowest degree) that has the given roots.

33. $-\frac{1}{2}, 3$
34. $\frac{1}{2}, -5$
35. $3 - 2i$
36. $4 + 2i$
37. $2, 1 - 2i$
38. $-3, 3 - 4i$
39. $2 - \sqrt{3}$
40. $1 + \sqrt{2}$

Use Descartes's rule of signs to discuss the possibilities for the roots to each equation. Do not solve the equation.

41. $x^8 + x^6 + 2x^2 = 0$
42. $-x^3 - x - 3 = 0$
43. $4x^3 - 3x^2 + 2x - 9 = 0$
44. $5x^5 + x^3 + 5x = 0$
45. $x^3 + 2x^2 + 2x + 1 = 0$
46. $-x^4 - x^3 + 3x^2 + 5x - 8 = 0$

Establish the best integral bounds for the roots of each equation according to the theorem on bounds.

47. $6x^2 + 5x - 50 = 0$
48. $4x^2 - 12x - 27 = 0$
49. $2x^3 - 15x^2 + 31x - 12 = 0$
50. $x^3 + 6x^2 + 11x + 7 = 0$
51. $12x^3 - 4x^2 - 3x + 1 = 0$
52. $4x^3 - 4x^2 - 9x + 10 = 0$

Find all real and imaginary solutions to each equation, stating multiplicity when it is greater than one.

53. $x^3 - 6x^2 + 11x - 6 = 0$
54. $x^3 + 7x^2 + 16x + 12 = 0$
55. $6x^4 - 5x^3 + 7x^2 - 5x + 1 = 0$
56. $6x^4 + 5x^3 + 25x^2 + 20x + 4 = 0$
57. $x^3 - 9x^2 + 28x - 30 = 0$
58. $x^3 - 4x^2 + 6x - 4 = 0$
59. $x^3 - 4x^2 + 7x - 6 = 0$

60. $2x^3 - 5x^2 + 10x - 4 = 0$

61. $2x^4 - 5x^3 - 2x^2 + 2x = 0$

62. $2x^5 - 15x^4 + 26x^3 - 12x^2 = 0$

Find all real solutions to each equation.

63. $|2v - 1| = 3v$

64. $|2h - 3| = |h|$

65. $x^4 + 7x^2 = 18$

66. $2x^{-2} + 5x^{-1} = 12$

67. $\sqrt{x + 6} - \sqrt{x - 5} = 1$

68. $\sqrt{2x - 1} = \sqrt{x - 1} + 1$

69. $\sqrt{y} + \sqrt[4]{y} = 6$

70. $\sqrt[3]{x^2} + \sqrt[3]{x} = 2$

71. $x^4 - 3x^2 - 4 = 0$

72. $(y - 1)^2 - (y - 1) = 2$

73. $(x - 1)^{2/3} = 4$

74. $(2x - 3)^{-1/2} = \frac{1}{2}$

75. $(x + 3)^{-3/4} = -8$

76. $\left(\frac{1}{x - 3}\right)^{-1/4} = \frac{1}{2}$

77. $\sqrt[3]{3x - 7} = \sqrt[3]{4 - x}$

78. $\sqrt[3]{x + 1} = \sqrt[6]{4x + 9}$

Discuss the symmetry of the graph of each function.

79. $f(x) = 2x^2 - 3x + 9$

80. $f(x) = -3x^2 + 12x - 1$

81. $f(x) = -3x^4 - 2$

82. $f(x) = \frac{-x^3}{x^2 - 1}$

83. $f(x) = \frac{x}{x^2 + 1}$

84. $f(x) = 2x^4 + 3x^2 + 1$

Find the domain of each rational function.

85. $f(x) = \frac{x^2 - 4}{2x + 5}$

86. $f(x) = \frac{4x + 1}{x^2 - x - 6}$

87. $f(x) = \frac{1}{x^2 + 1}$

88. $f(x) = \frac{x - 9}{x^2 - 1}$

Find the x-intercepts, y-intercept, and asymptotes for the graph of each function and sketch the graph.

89. $f(x) = -\frac{1}{2}x + 4$

90. $f(x) = 3x - 5$

91. $f(x) = x^2 - x - 2$

92. $f(x) = -2(x - 1)^2 + 6$

93. $f(x) = x^3 - 3x - 2$

94. $f(x) = x^3 - 3x^2 + 4$

95. $f(x) = \frac{1}{2}x^3 - \frac{1}{2}x^2 - 2x + 2$

96. $f(x) = \frac{1}{2}x^3 - 3x^2 + 4x$

97. $f(x) = \frac{1}{4}x^4 - 2x^2 + 4$

98. $f(x) = \frac{1}{2}x^4 + 2x^3 + 2x^2$

99. $f(x) = \frac{2}{x + 3}$

100. $f(x) = \frac{1}{2 - x}$

101. $f(x) = \frac{2x}{x^2 - 4}$

102. $f(x) = \frac{2x^2}{x^2 - 4}$

103. $f(x) = \frac{x^2 - 2x + 1}{x - 2}$

104. $f(x) = \frac{-x^2 + x + 2}{x - 1}$

105. $f(x) = \frac{2x - 1}{2 - x}$

106. $f(x) = \frac{1 - x}{x + 1}$

107. $f(x) = \frac{x^2 - 4}{x - 2}$

108. $f(x) = \frac{x^3 + x}{x}$

Solve each inequality. State the solution set using interval notation.

109. $4x^3 - 400x^2 - x + 100 \geq 0$

110. $x^3 - 49x^2 - 52x + 100 < 0$

111. $\frac{x + 10}{x + 2} < 5$

112. $\frac{x - 6}{2x + 1} \geq 1$

113. $\frac{12 - 7x}{x^2} > -1$

114. $x - \frac{2}{x} \leq -1$

115. $\frac{x^2 - 3x + 2}{x^2 - 7x + 12} \geq 0$

116. $\frac{x^2 + 4x + 3}{x^2 - 2x - 15} \leq 0$

117. $\frac{x + 3}{x + 3} \leq 0$

118. $\frac{1}{(x - 2)^2} > 1$

119. $\frac{51}{(x - 1)^2} \geq 0$

120. $\frac{-1}{x} > 2$

Solve each problem.

121. Find the quotient and remainder when $x^3 - 6x^2 + 9x - 15$ is divided by $x - 3$.

122. Find the quotient and remainder when $3x^3 + 4x^2 + 2x - 4$ is divided by $3x - 2$.

123. *Linear Interpolation* The accompanying graph shows that enrollment in U.S. higher education went from 12.2

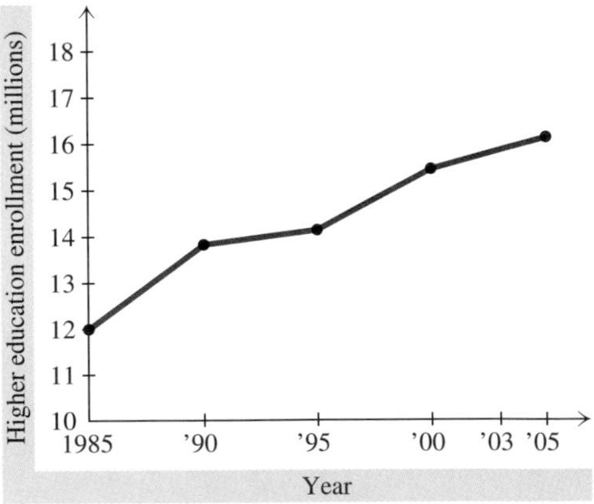

Figure for Exercise 123

million in 1985 to 13.9 million in 1990 (*Fortune*, June 23,1997). The process of using a linear equation to estimate a point between two known data points is called *linear interpolation.* Find the equation of the line through these two points and use the equation to estimate the enrollment in 1988.

124. *Linear Extrapolation* The accompanying graph shows how IBM's share of total computer-industry market capitalization has decreased from 45.9% in 1988 to 14.2% in 1997 (*Fortune*, June 23, 1997). The process of using a linear equation to estimate a point that is not between two known data points is called *linear extrapolation.* Find the equation of the line through the two given points and use the equation to estimate IBM's share in the year 2000.

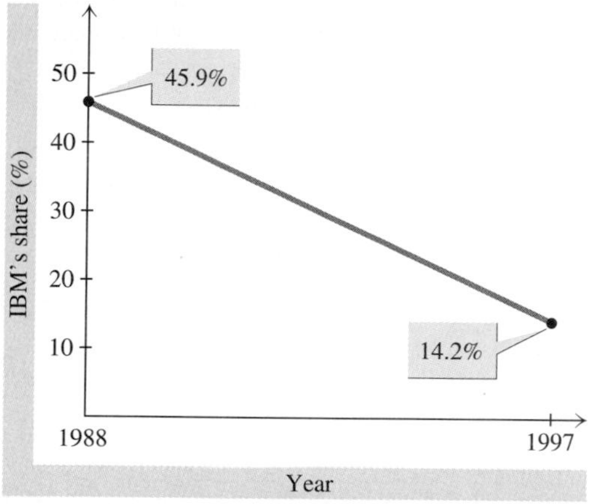

Figure for Exercise 124

125. *Percent of Body Fat* The average 20-year-old woman has 23% body fat and the average 50-year-old woman has 47% body fat (*Boston Globe*, August 4, 1997). Assuming that the percentage of body fat is a linear function of age, find the function. Use the function to determine the percentage of body fat in the average 65-year-old woman.

126. *Growing Number of Millionaires* According to IRS figures (*Newsweek*, August 4, 1997), 13,505 taxpayers declared income of over \$1 million in 1979 and 69,935 did in 1994. Assuming that the number of taxpayers declaring income of over \$1 million is growing linearly, express this number as a linear function of the year. Use the function you found to predict the number of taxpayers who will declare income of over \$1 million in 2002.

127. *Altitude of a Rocket* If the altitude in feet of a model rocket is given by the equation $S = -16t^2 + 156t$, where t is the time in seconds after ignition, then what is the maximum height attained by the rocket?

128. *Antique Saw* Willard is making a reproduction of an antique saw. The handle consists of two pieces of wood joined at a right angle, with the blade being the hypotenuse of the right triangle as shown in the figure. If the total length of the handle is to be 36 in., then what length for each piece would minimize the square of the length of the blade?

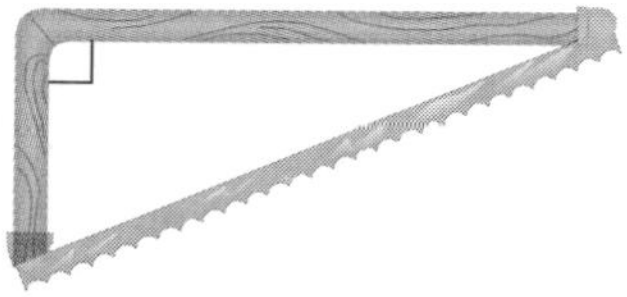

Figure for Exercise 128

129. *Physical Fitness* For a 25-year-old woman to be considered in the high fitness category, the maximum time allowed for a one-mile walk at a pulse rate of 110 is 19.1 min ("Monitoring Your Health," *Reader's Digest*, 1991). For walking at a pulse rate of 170, the maximum time allowed is 16.2 min. If the maximum time allowed is a linear function of the pulse rate and 25-year-old Tina walks one mile in 17.5 min with a pulse rate of 145, is she in the high fitness category?

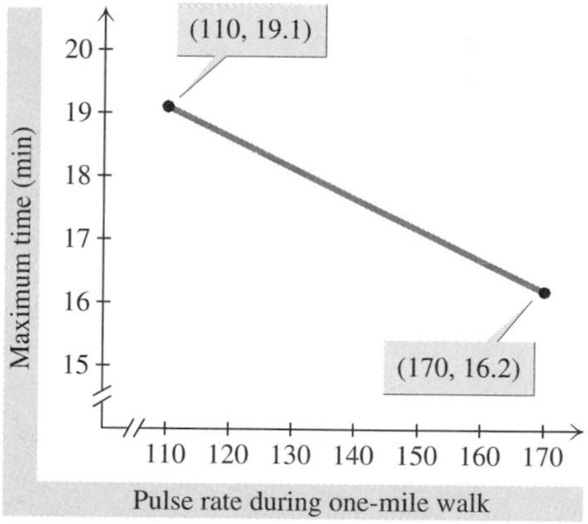

Figure for Exercise 129

130. *Marketing Plans* A company has determined through market research that the demand equation for a new product is $P = 30 - \sqrt{0.002x + 3}$, where x is the number of units sold per month and P is the price in dollars of one unit. If the company sets the price of the product at \$27 per unit, then how many units can it expect to sell per month?

131. *Bonus Room* A homeowner wants to put a room in the attic of her house. The house is 48 ft wide and the roof has a 5–12 pitch. (The roof rises 5 ft in a horizontal distance of 12 ft.) Find the dimensions of the room that will maximize the area of the cross section shown in the figure.

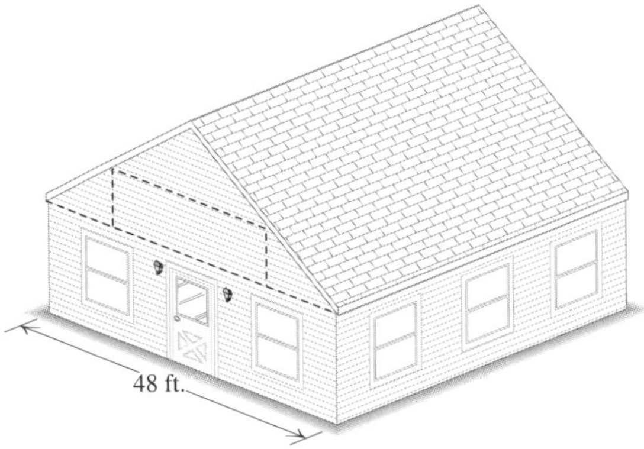

Figure for Exercise 131

132. *Maximizing Area* An isosceles triangle has one vertex at the origin and a horizontal base below the x-axis with its endpoints on the curve $y = x^2 - 16$. See the figure. Let (a, b) be the vertex in the fourth quadrant and write the area of the triangle as a function of a. Use a graphing calculator to find the point (a, b) for which the triangle has the maximum possible area.

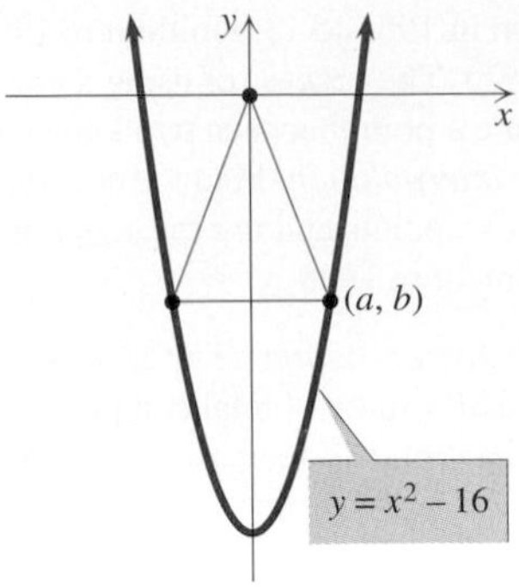

Figure for Exercise 132

133. *Limiting Velocity* As a skydiver falls, his velocity keeps increasing. However, because of air resistance, the rate at which the velocity increases keeps decreasing and there is a limit velocity that the skydiver cannot exceed. To see this behavior, graph

$$V = \frac{1000t}{5t + 8},$$

where V is the velocity in feet per second and t is the time in seconds.

a. What is the velocity at time $t = 10$ sec?
b. What is the horizontal asymptote for this graph?
c. What is the limiting velocity that cannot be exceeded?

CHAPTER 3 TEST

Solve each problem.

1. Find the slope of the line that goes through $(3, -6)$ and $(-1, 2)$.
2. Find the slope of the line $3x - 4y = 9$.
3. Find the equation of the line through $(2, -4)$ that is perpendicular to the line $y = -3x - 5$. Write the answer in slope-intercept form.
4. Write $y = 3x^2 - 12x + 1$ in the form $y = a(x - h)^2 + k$.
5. Identify the vertex, axis of symmetry, y-intercept, x-intercepts, and range for $y = 3x^2 - 12x + 1$.
6. What is the minimum value of y in the function $y = 3x^2 - 12x + 1$?
7. Use synthetic division to find the quotient and remainder when $2x^3 - 4x + 5$ is divided by $x + 3$.
8. What is the remainder when
$$x^{98} - 19x^{73} + 17x^{44} - 12x^{23} + 2x^9 - 3$$
is divided by $x - 1$?
9. List the possible rational zeros for the function $f(x) = 3x^3 - 4x^2 + 5x - 6$ according to the rational zero theorem.
10. Find a polynomial equation with real coefficients that has the roots -3 and $4i$.
11. Use Descartes's rule of signs to discuss the possibilities for the roots of $x^3 - 3x^2 + 5x + 7 = 0$.
12. The altitude in feet of a toy rocket t seconds after launch is given by the function $S(t) = -16t^2 + 128t$. Find the maximum altitude reached by the rocket.

13. The median price of a new home in Springfield in 1994 was \$88,000 and in 1997 it was \$92,800. Assuming that the price is a linear function of the year, write a formula for that function. Use the formula to predict the median price in the year 2003.

Find all real and imaginary zeros for each polynomial function.

14. $f(x) = x^2 - 9$ **15.** $f(x) = x^4 - 16$
16. $f(x) = x^3 - 4x^2 - x + 10$

Find all real and imaginary roots of each equation. State the multiplicity of a root when it is greater than one.

17. $x^4 + 2x^2 + 1 = 0$ **18.** $(x^3 - 2x^2)(2x + 3)^3 = 0$
19. $2x^3 - 9x^2 + 14x - 5 = 0$

Sketch the graph of each function.

20. $y = -\frac{1}{2}x + 3$ **21.** $y = 2(x - 3)^2 + 1$

22. $y = (x - 2)^2(x + 1)$ **23.** $y = x^3 - 4x$

Find all asymptotes and sketch the graph of each function.

24. $y = \frac{1}{x - 2}$ **25.** $y = \frac{2x - 3}{x - 2}$

26. $f(x) = \frac{x^2 + 1}{x}$ **27.** $y = \frac{4}{x^2 - 4}$

28. $y = \frac{x^2 - 2x + 1}{x - 1}$

Solve each inequality.

29. $x^2 - 2x < 8$ **30.** $\frac{x + 2}{x - 3} > -1$

31. $\frac{x + 3}{(4 - x)(x + 1)} \geq 0$

Find all real solutions to each equation.

32. $(x - 3)^{-2/3} = \frac{1}{3}$ **33.** $\sqrt{x} - \sqrt{x - 7} = 1$

TYING IT ALL TOGETHER

Chapters P–3

Find all real and imaginary solutions to each equation.

1. $2x + 1 = 3x - 1$ **2.** $(2x + 1)(3x - 1) = 0$ **3.** $\frac{2x + 1}{3x - 1} = 0$

4. $\frac{1}{2x + 1} = \frac{1}{3x - 1}$ **5.** $\frac{1}{3x - 1} - \frac{1}{2x + 1} = \frac{1}{6}$ **6.** $6x^4 + x^3 + 5x^2 + x - 1 = 0$

7. $\sqrt{2x + 1} = \sqrt{3x - 1}$ **8.** $3x^2 + 4 = 2x^2 + 1$ **9.** $\sqrt{3x + 4} - \sqrt{2x + 1} = 1$

10. $|2x + 1| = 3$ **11.** $(2x + 1)^{2/3} = 9$ **12.** $|2x + 1| = |3x - 1|$

Sketch the graph of each function and state the domain and range.

13. $y = 3 - x$ **14.** $y = |3 - x|$ **15.** $y = 3 - x^2$ **16.** $y = \frac{x - 3}{|x - 3|}$

17. $y = 3 - \sqrt{x}$ **18.** $y = \frac{1}{3 - x}$ **19.** $y = x^2 - 6x + 9$ **20.** $y = \frac{1}{(x - 3)^2}$

21. $y = \frac{x^2 - 3}{x}$ **22.** $y = \sqrt{3 - x^2}$ **23.** $y = \sqrt{3 - x}$ **24.** $y = x^3 - 3x^2$

Simplify each expression.

25. 10^3 **26.** 3^0 **27.** $4^{-1/2}$ **28.** 10^{-3}

29. 3^{-2} **30.** $9^{1/2}$ **31.** $27^{-2/3}$ **32.** $4^{3/2}$

Vast, lonesome, remote—the Montana badlands are no man's land. In the summer of 1964, paleontologist John Ostrom was exploring this region when he spotted some weathered fragments of fossil bone protruding from a hillside. "Mine were the first human eyes to see those fragments," he wrote. "They had lain exposed there for centuries. But I knew at once that they were unique and that this was the most important discovery our expedition had made all summer."

Back at Yale University, as Ostrom carefully reassembled the skeletons, an extraordinary carnivore emerged. If fleshed out, this new dinosaur would have weighed about 170 pounds and measured 9 feet in length. Its structure suggested that it ran swiftly on its hind legs, much like a large bird. But the striking feature that inspired its name, Deinonychus or "terrible claw," was a 6-inch sickle-like claw on a toe of each hind foot. Ostrom believed these talons served as deadly weapons with which Deinonychus slashed its prey. Later, when he unearthed several Deinonychus specimens entombed with the remains of a 1000-pound plant-eater, Ostrom deduced that the agile killer hunted in packs in order to overcome large quarry.

Ongoing studies of animal behavior and new information on dinosaur anatomy continue to challenge old assumptions about what dinosaurs were like. Since imagination must often take over where scientific evidence ends, many of the new theories have sparked controversy. Our search for the truth remains an ongoing detective hunt in which we interpret data extracted from mute bones and petrified footprints.

In this chapter we learn how scientists use exponential functions, which measure the rate of growth and decay over time, to date archaeological finds that are millions of years old. And through applications ranging from population growth to the growth of financial investments, we'll see how eponential functions can help us both predict the future and discover the past.

4 Exponential and Logarithmic Functions

4.1 Exponential Functions
4.2 Logarithmic Functions
4.3 Properties of Logarithms
4.4 More Equations and Applications

4.1 Exponential Functions

The functions that involve some combination of basic arithmetic operations, powers, or roots are called **algebraic functions**. Most of the functions studied so far are algebraic functions. In this chapter we turn to the exponential and logarithmic functions. These functions are used to describe phenomena ranging from growth of investments to the decay of radioactive materials, which cannot be described with algebraic functions. Since the exponential and logarithmic functions transcend what can be described with algebraic functions, they are called **transcendental functions**.

The Definition

In algebraic functions such as

$$j(x) = x^2, \quad p(x) = x^5, \quad \text{and} \quad m(x) = x^{1/3}$$

the base is a variable and the exponent is constant. For the exponential functions the base is constant and the exponent is a variable. The functions

$$f(x) = 2^x, \quad g(x) = 5^x, \quad \text{and} \quad h(x) = \left(\frac{1}{3}\right)^x$$

are exponential functions.

Definition: Exponential Function

An **exponential function** with **base *a*** is a function of the form

$$f(x) = a^x$$

where a and x are real numbers such that $a > 0$ and $a \neq 1$.

We rule out the base $a = 1$ in the definition because $f(x) = 1^x$ is the constant function $f(x) = 1$. Negative numbers are not used as bases because powers such as $(-4)^{1/2}$ are not real numbers. Functions that have variable exponents but are not exactly in the form of the definition may also be called exponential functions.

To evaluate an exponential function, we use our knowledge of exponents.

EXAMPLE 1 Evaluating exponential functions

Let $f(x) = 4^x$, $g(x) = 5^{2-x}$, and $h(x) = \left(\frac{1}{3}\right)^x$. Find the following values. Check your answers with a calculator.

a) $f(3/2)$ **b)** $g(3)$ **c)** $h(-2)$

Solution

a) $f(3/2) = 4^{3/2} = (\sqrt{4})^3 = 2^3 = 8$ **Take the square root, then cube.**

b) $g(3) = 5^{2-3} = 5^{-1} = \dfrac{1}{5}$

c) $h(-2) = \left(\dfrac{1}{3}\right)^{-2} = 3^2 = 9$ **Find the reciprocal, then square.**

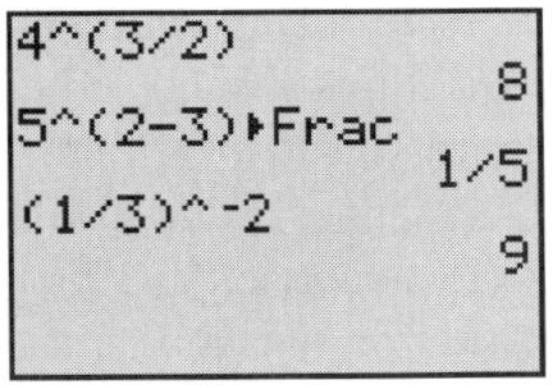

Figure 4.1

Evaluating these functions with a calculator as in Fig. 4.1 yields the same results.

Domain of Exponential Functions

The domain of the exponential function $f(x) = 2^x$ is the set of all real numbers. To understand how this function can be evaluated for any *real* number, first note that the function is evaluated for any *rational* number using powers and roots. For example,

$$f(1.7) = f(17/10) = 2^{17/10} = \sqrt[10]{2^{17}} \approx 3.2490095985.$$

Until now, we have not considered *irrational* exponents.

Consider the expression $2^{\sqrt{3}}$ and recall that the irrational number $\sqrt{3}$ is an infinite nonterminating nonrepeating decimal number:

$$\sqrt{3} = 1.7320508075\ldots$$

If we use rational approximations to $\sqrt{3}$ as exponents, we see a pattern:

$$2^{1.7} = 3.249009585\ldots$$
$$2^{1.73} = 3.317278183\ldots$$
$$2^{1.732} = 3.321880096\ldots$$
$$2^{1.73205} = 3.321995226\ldots$$
$$2^{1.7320508} = 3.321997068\ldots$$

As the exponents get closer and closer to $\sqrt{3}$ we get results that are approaching some number. We define $2^{\sqrt{3}}$ to be that number. Of course, it is impossible to write the exact value of $\sqrt{3}$ or $2^{\sqrt{3}}$ as a decimal, but you can use a calculator to get $2^{\sqrt{3}} \approx 3.321997085$. Since any irrational number can be approximated by rational numbers in this same manner, 2^x is defined similarly for any irrational number. This idea extends to any exponential function.

Domain of an Exponential Function

The domain of $f(x) = a^x$ for $a > 0$ and $a \neq 1$ is the set of all real numbers.

Graphing Exponential Functions

Even though the domain of an exponential function is the set of real numbers, for ease of computation, we generally choose only rational numbers for x to find ordered pairs on the graph of the function.

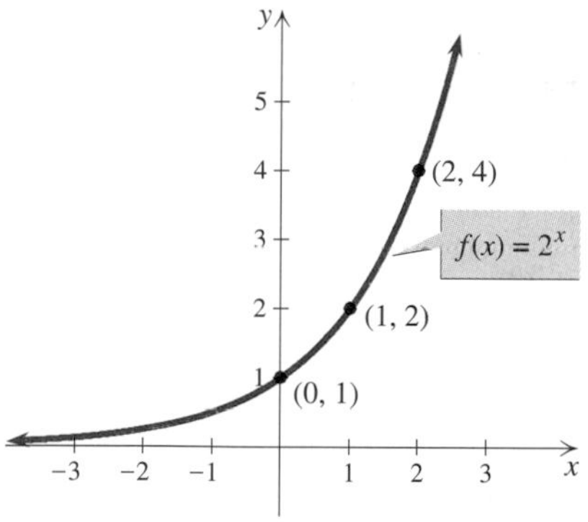

Figure 4.2

EXAMPLE 2 Graphing an exponential function ($a > 1$)

Sketch the graph of each exponential function and state the domain and range.

a) $f(x) = 2^x$ **b)** $g(x) = 10^x$

Solution

a) Find some ordered pairs satisfying $f(x) = 2^x$ as follows:

x	-2	-1	0	1	2	3
$y = 2^x$	1/4	1/2	1	2	4	8

Plot these points. As x gets larger, so does 2^x. As x approaches $-\infty$, 2^x approaches but never reaches 0. So the x-axis is a horizontal asymptote for the curve. It can be shown that the graph of $f(x) = 2^x$ increases in a continuous manner, with no "jumps" or "breaks" in the graph. This fact allows us to draw a smooth curve through the points as shown in Fig. 4.2. The domain of $f(x) = 2^x$ is $(-\infty, \infty)$. The range is $(0, \infty)$.

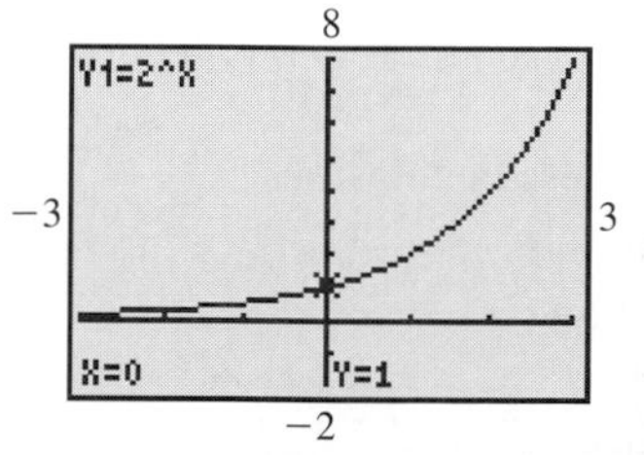

Figure 4.3

The calculator graph shown in Fig. 4.3 supports these conclusions. Note that the calculator graph appears to touch its horizontal asymptote. When

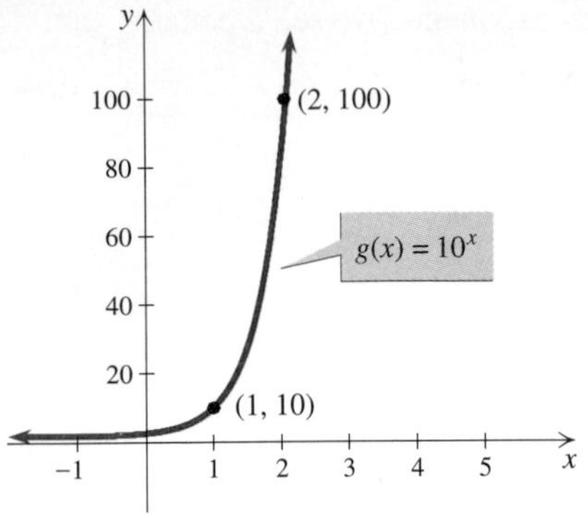

Figure 4.4

drawing any curve with an asymptote by hand, we draw it so that it does not touch its asymptote. □

b) Find some ordered pairs satisfying $g(x) = 10^x$ as follows:

x	-2	-1	0	1	2	3
$y = 10^x$	0.01	0.1	1	10	100	1000

These points indicate that the graph of $g(x) = 10^x$ increases in the same manner as the graph of $f(x) = 2^x$. The domain of $g(x) = 10^x$ is $(-\infty, \infty)$. The range is $(0, \infty)$. Plot some points and sketch the curve shown in Fig. 4.4. Make sure that your hand-drawn curve does not touch its asymptote, the x-axis.

The calculator graph in Fig. 4.5 supports these conclusions.

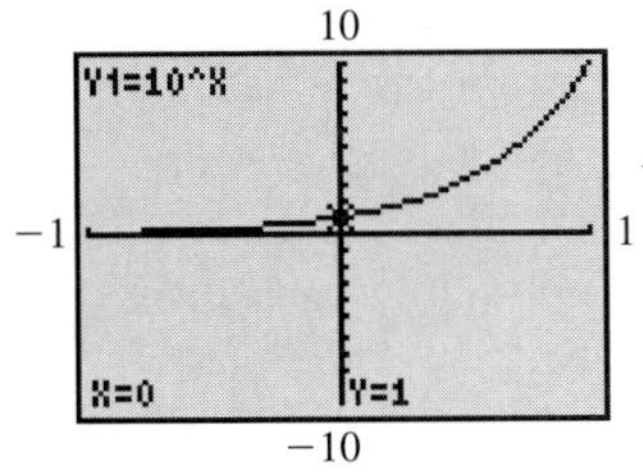

Figure 4.5

Note the similarities between the graphs shown in Figs. 4.2 and 4.4. Both pass through the point (0, 1), both functions are increasing, and both have the x-axis as a horizontal asymptote. Since a horizontal line can cross these graphs only once, the functions $f(x) = 2^x$ and $g(x) = 10^x$ are one-to-one by the horizontal line test. All functions of the form $f(x) = a^x$ for $a > 1$ have graphs similar to those in Figs. 4.2 and 4.4. In the next example we graph functions of the form $f(x) = a^x$ for $0 < a < 1$.

EXAMPLE 3 Graphing an exponential function ($0 < a < 1$)

Sketch the graph of each function and state the domain and range of the function.

a) $f(x) = (1/2)^x$ **b)** $g(x) = 3^{-x}$

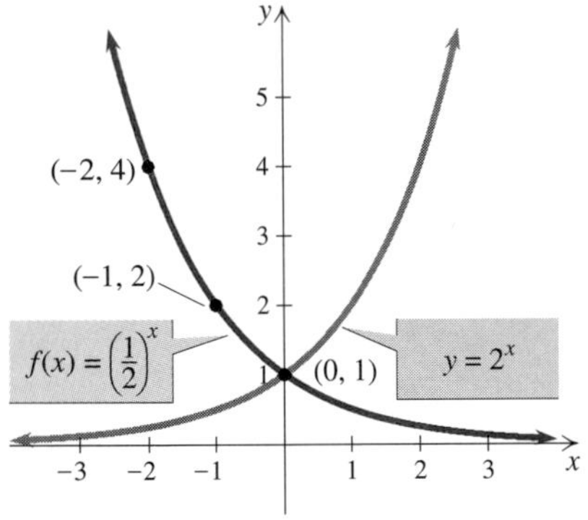

Figure 4.6

Solution

a) Find some ordered pairs satisfying $f(x) = (1/2)^x$ as follows:

x	-2	-1	0	1	2	3
$y = \left(\frac{1}{2}\right)^x$	4	2	1	1/2	1/4	1/8

Plot these points and draw a smooth curve through them as shown in Fig. 4.6. From the graph in Fig. 4.6 we see that the domain is $(-\infty, \infty)$ and the range is $(0, \infty)$. Because $(1/2)^x = 2^{-x}$ the graph of $f(x) = (1/2)^x$ is a reflection in the y-axis of the graph of $y = 2^x$.

The calculator graph in Fig. 4.7 supports these conclusions. □

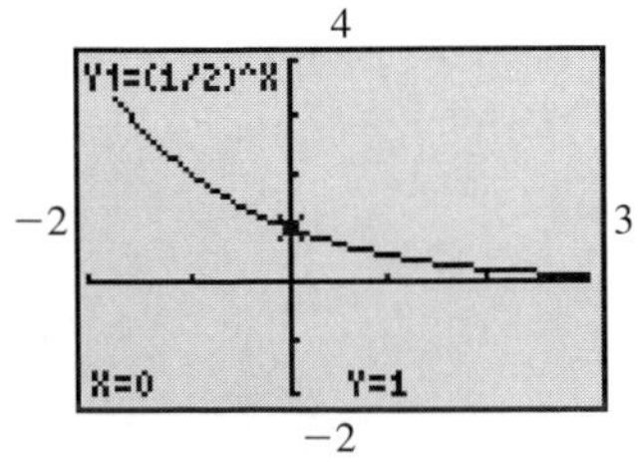

Figure 4.7

b) Since $3^{-x} = (1/3)^x$, this function is of the form $y = a^x$ for $0 < a < 1$. Find some ordered pairs satisfying $g(x) = 3^{-x}$ as follows:

x	-2	-1	0	1	2	3
$y = 3^{-x}$	9	3	1	1/3	1/9	1/27

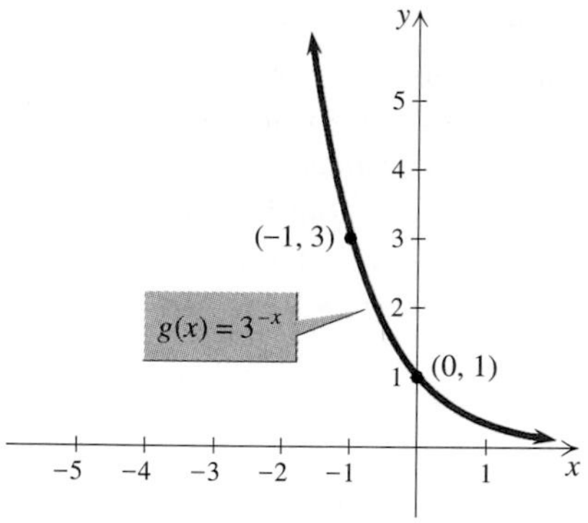

Figure 4.8

Plot these points and draw a smooth curve through them as shown in Fig. 4.8. The domain is $(-\infty, \infty)$ and the range is $(0, \infty)$. The x-axis is again a horizontal asymptote.

The calculator graph in Fig. 4.9 supports these conclusions.

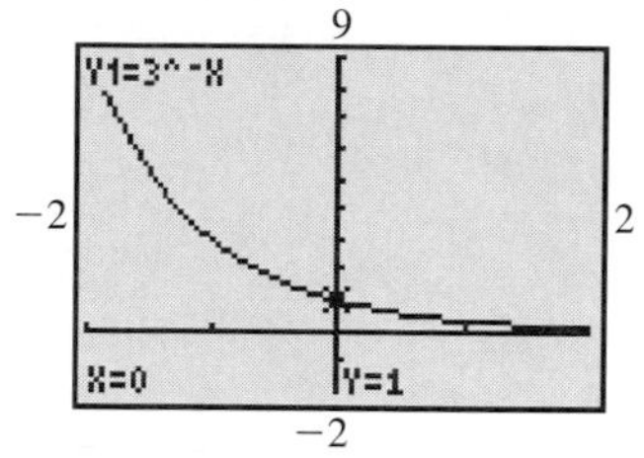

Figure 4.9

Again note the similarities between the graphs in Figs. 4.6 and 4.8. Both pass through (0, 1), both functions are decreasing, and both have the x-axis as a horizontal asymptote. By the horizontal line test, these functions are also one-to-one. The base determines whether the exponential function is increasing or decreasing. In general, we have the following properties of exponential functions.

Properties of Exponential Functions

The exponential function $f(x) = a^x$ has the following properties:

1. The function f is increasing for $a > 1$ and decreasing for $0 < a < 1$.
2. The y-intercept of the graph of f is (0, 1).
3. The graph has the x-axis as a horizontal asymptote.
4. The range of f is $(0, \infty)$.
5. The function f is one-to-one.

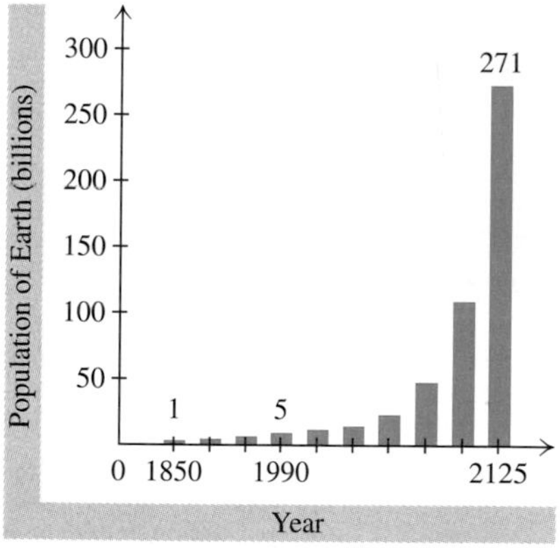

Figure 4.10

The phrase ''growing exponentially'' is often used to describe a population or other quantity that is increasing in the manner of an exponential function. For example, the earth's population is said to be growing exponentially because of the shape of the graph shown in Fig. 4.10. For an exponential function such as $f(x) = 2^x$, a small increase in x can cause a relatively large growth in y. If x increases from 9 to 10, $f(10) - f(9) = 512$. Compare this increase with the growth obtained in the algebraic function $g(x) = x^2$ when x increases from 9 to 10: $g(10) - g(9) = 19$.

Transformation of Exponential Functions

The graphs of exponential functions can be translated, stretched, shrunk, or reflected using the ideas of Section 2.4. Just as $y = (x + 2)^2$ is a translation two units to the left of $y = x^2$, so $y = 3^{x+2}$ is a translation two units to the left of

$y = 3^x$. If the graph is translated upward or downward, the horizontal asymptote will move also. If you have a graphing calculator, use it to graph the functions presented in the examples and exercises, and also experiment with it. Experimenting with a graphing calculator can lead you to an understanding of exponential functions much faster than doing all of the calculations by hand. Remember that the intention here is not simply to draw a picture, but to understand how the size of the base affects the shape of the graph and how transformations determine the location of the graph.

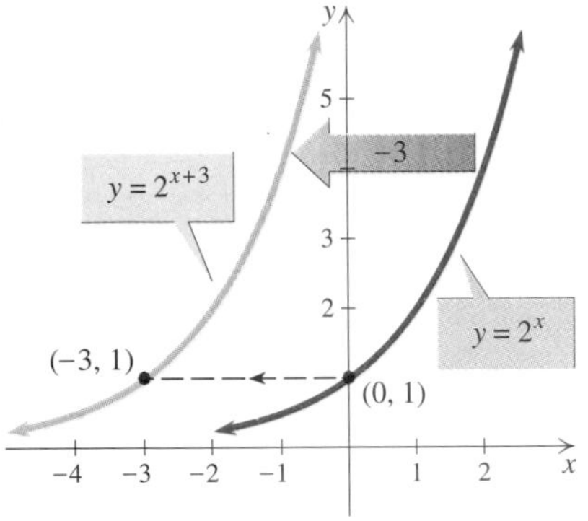

Figure 4.11

EXAMPLE 4 Graphing an exponential function with translations

Sketch the graph of each function and state its domain and range.

a) $y = 2^{x+3}$ **b)** $f(x) = -4 + 3^{x+2}$

Solution

a) The graph of $y = 2^{x+3}$ is a translation three units to the left of the graph of $y = 2^x$. For accuracy, find a few ordered pairs that satisfy $y = 2^{x+3}$:

x	-3	-2	-1
$y = 2^{x+3}$	1	2	4

Draw the graph of $y = 2^{x+3}$ through these points as shown in Fig. 4.11. The domain of $y = 2^{x+3}$ is $(-\infty, \infty)$, and the range is $(0, \infty)$.

The calculator graph in Fig. 4.12 supports these conclusions. □

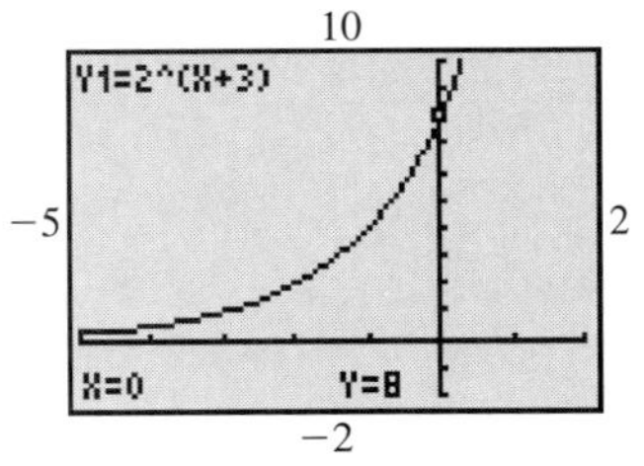

Figure 4.12

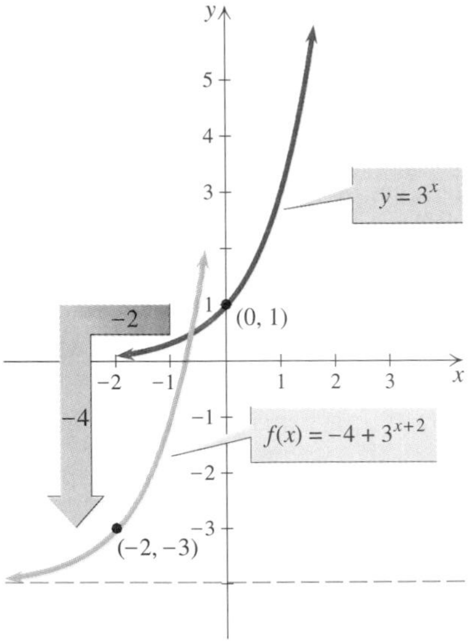

Figure 4.13

b) The graph of $f(x) = -4 + 3^{x+2}$ is obtained by translating $y = 3^x$ to the left two units and downward four units. For accuracy, find a few points on the graph of $f(x) = -4 + 3^{x+2}$:

x	-3	-2	-1	0
$y = -4 + 3^{x+2}$	$-11/3$	-3	-1	5

Sketch a smooth curve through these points as shown in Fig. 4.13. The horizontal asymptote is the line $y = -4$. The domain of f is $(-\infty, \infty)$, and the range is $(-4, \infty)$.

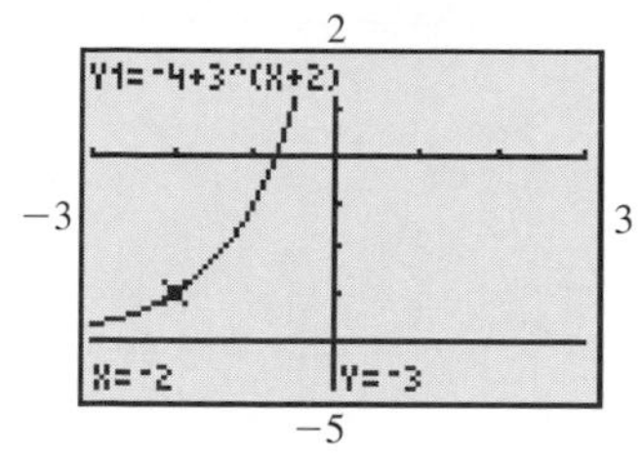

Figure 4.14

The calculator graph of $y = -4 + 3^{x+2}$ and the asymptote $y = -4$ in Fig. 4.14 supports these conclusions.

If you recognize that a function is a transformation of a simpler function, then you know what its graph looks like. This information along with a few ordered pairs will help you to make accurate graphs.

EXAMPLE 5 Graphing an exponential function with a reflection

Sketch the graph of $f(x) = -\frac{1}{2} \cdot 3^{-x}$ and state the domain and range.

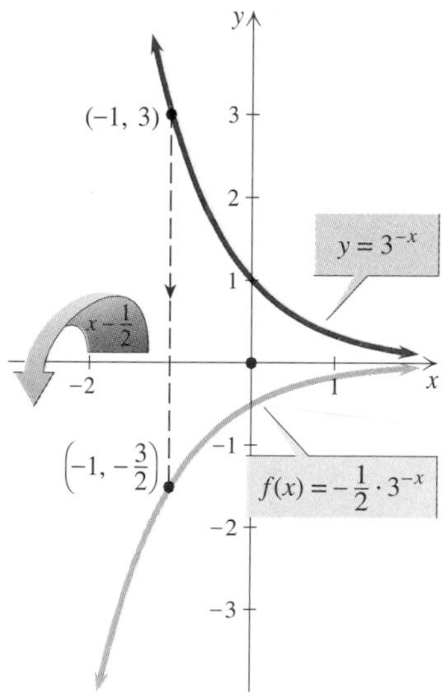

Figure 4.15

Solution

The graph of $y = 3^{-x}$ is shown in Fig. 4.8. Multiplying by $-1/2$ shrinks the graph and reflects it below the x-axis. The x-axis is the horizontal asymptote for the graph of f. Find a few ordered pairs as follows:

x	-2	-1	0	1	2
$y = -\frac{1}{2} \cdot 3^{-x}$	$-9/2$	$-3/2$	$-1/2$	$-1/6$	$-1/18$

Sketch the curve through these points as in Fig. 4.15. The domain is $(-\infty, \infty)$ and the range is $(-\infty, 0)$.

Exponential Equations

From the graphs of exponential functions, we observed that they are one-to-one. For an exponential function, one-to-one means that if two exponential expressions with the same base are equal, then the exponents are equal.

One-to-One Property of Exponential Functions

For $a > 0$ and $a \neq 1$,

$$\text{if} \quad a^{x_1} = a^{x_2}, \quad \text{then} \quad x_1 = x_2.$$

The one-to-one property is used in solving simple exponential equations. For example, to solve $2^x = 8$, we recall that $8 = 2^3$. So the equation becomes $2^x = 2^3$. By the one-to-one property, $x = 3$ is the only solution. The one-to-one property applies only to equations in which each side is a power of the same base.

EXAMPLE 6 **Solving exponential equations**

Solve each exponential equation.

a) $4^x = \frac{1}{4}$ **b)** $\left(\frac{1}{10}\right)^x = 100$

Solution

a) Since $1/4 = 4^{-1}$, we can write the right-hand side as a power of 4 and use the one-to-one property:

$$4^x = \frac{1}{4} = 4^{-1}$$

$$x = -1 \qquad \text{One-to-one property}$$

b) Since $(1/10)^x = 10^{-x}$ and $100 = 10^2$, we can write each side as a power of 10:

$$\left(\frac{1}{10}\right)^x = 100$$

$$10^{-x} = 10^2$$

$$-x = 2 \qquad \text{One-to-one property}$$

$$x = -2$$

The type of equation that we solved in Example 6 arises naturally when we try to find the first coordinate of an ordered pair of an exponential function when given the second coordinate.

EXAMPLE 7 **Finding the first coordinate given the second**

Let $f(x) = 5^{2-x}$. Find x such that $f(x) = 125$.

Solution

To find x such that $f(x) = 125$, we must solve $5^{2-x} = 125$:

$$5^{2-x} = 5^3 \qquad \text{Since } 125 = 5^3$$

$$2 - x = 3 \qquad \text{One-to-one property}$$

$$x = -1$$

Compound Interest

Exponential functions are used to model phenomena such as population growth, radioactive decay, and compound interest. Here we will show how these functions are used to determine the amount of an investment earning compound interest.

If P dollars are deposited in an account with a simple annual interest rate r for t years, then the amount A in the account at the end of t years is found by using the formula $A = P + Prt$ or $A = P(1 + rt)$. If \$1000 is deposited at 6% annual rate for one quarter of a year, then at the end of three months the amount is

$$A = 1000\left(1 + 0.06 \cdot \frac{1}{4}\right) = 1000(1.015) = \$1015.$$

If the account begins the next quarter with \$1015, then at the end of the second quarter we again multiply by 1.015 to get the amount

$$A = 1000(1.015)^2 = \$1030.23.$$

This process is referred to as compound interest because interest is put back into the account and the interest also earns interest. At the end of 20 years, or 80 quarters, the amount is

$$A = 1000(1.015)^{80} = \$3290.66.$$

Compound interest can be thought of as simple interest computed over and over. The general **compound interest formula** follows.

Compound Interest Formula

If a principal P is invested for t years at an annual rate r compounded n times per year, then the amount A, or ending balance, is given by

$$A = P\left(1 + \frac{r}{n}\right)^{nt}.$$

The principal P is also called **present value** and the amount A is called **future value**.

EXAMPLE 8 Using the compound interest formula

Find the amount of \$20,000 at 6% compounded daily for three years.

Figure 4.16

Solution

Use $P = \$20{,}000$, $r = 0.06$, $n = 365$, and $t = 3$ in the compound interest formula:

$$A = \$20{,}000\left(1 + \frac{0.06}{365}\right)^{365 \cdot 3} = \$23{,}943.99$$

Figure 4.16 shows this expression on a graphing calculator.

When interest is compounded daily, financial institutions usually use the exact number of days. To keep our discussion of interest simple, we assume that all years have 365 days and ignore leap years. When discussing months we assume that all months have 30 days. There may be other rules involved in computing interest that are specific to the institution doing the computing. One popular method is to compound quarterly and only give interest on money that is on deposit for full quarters. With this rule you could have money in an account for nearly six months and receive no interest.

• CALCULUS •

Continuous Compounding and the Number *e*

The more often that interest is figured during the year, the more interest an investment will earn. The first five lines of Table 4.1 show the future value of \$10,000 invested at 12% for one year for more and more frequent compounding. The last line shows the limiting amount \$11,274.97, which we cannot exceed no matter how often we compound the interest on this investment for one year.

To better understand the last line of Table 4.1, we need to use a fact from calculus. The expression $[1 + (r/n)]^{nt}$ used in calculating the first five values in the table can be shown to approach e^{rt} as $n \to \infty$. The number e (like π) is an irrational number that occurs in many areas of mathematics. The factor e^{rt} is used in the last line of the table to find the limit to the values of the investment obtained by more and more frequent compounding. Using e^{rt} to find the future value is called **continuous compounding**.

Table 4.1

Compounding	Future value in one year
Annually	$\$10{,}000\left(1 + \frac{0.12}{1}\right)^{1} = \$11{,}200$
Quarterly	$\$10{,}000\left(1 + \frac{0.12}{4}\right)^{4} = \$11{,}255.09$
Monthly	$\$10{,}000\left(1 + \frac{0.12}{12}\right)^{12} = \$11{,}268.25$
Daily	$\$10{,}000\left(1 + \frac{0.12}{365}\right)^{365} = \$11{,}274.75$
Hourly	$\$10{,}000\left(1 + \frac{0.12}{8760}\right)^{8760} = \$11{,}274.96$
Continuously	$\$10{,}000e^{0.12(1)} = \$11{,}274.97$

Figure 4.17

You can find approximate values for powers of e using a calculator with an e^x key. To find an approximate value for e itself, find e^1 on a calculator:

$$e \approx 2.718281828459$$

Figure 4.17 shows the graphing calculator computation of the future value of \$10,000 at 12% compounded continuously for one year and the value of e^1. □

Continuous Compounding Formula

If a principal P is invested for t years at an annual rate r compounded continuously, then the amount A, or ending balance, is given by

$$A = P \cdot e^{rt}.$$

EXAMPLE 9 Interest compounded continuously

Find the amount when a principal of \$5600 is invested at $6\frac{1}{4}\%$ annual rate compounded continuously for 5 years and 9 months.

Solution

Convert 5 years and 9 months to 5.75 years. Use $r = 0.0625$, $t = 5.75$, and $P = \$5600$ in the continuous compounding formula:

$$A = 5600 \cdot e^{(0.0625)(5.75)} = \$8021.63$$

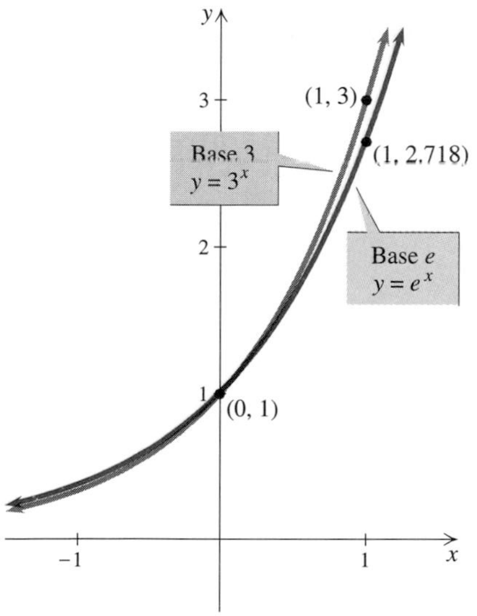

Exponential Functions

Figure 4.18

The function $f(x) = e^x$ is called the **base-e exponential function**. Variations of this function are used to model many types of growth and decay. The graph of $y = e^x$ looks like the graph of $y = 3^x$, because the value of e is close to 3. Use your calculator to check that the graph of $y = e^x$ shown in Fig. 4.18 goes approximately through the points $(-1, 0.368)$, $(0, 1)$, $(1, 2.718)$, and $(2, 7.389)$.

Radioactive Decay

The mathematical model of radioactive decay is based on the formula

$$A = A_0 e^{rt},$$

which gives the amount A of a radioactive substance remaining after t years, where A_0 is the initial amount present and r is the annual rate of decay. The only difference between this formula and the continuous compounding formula is that when decay is involved the rate r is negative.

EXAMPLE 10 Radioactive decay of carbon-14

Finely chipped spear points shaped by nomadic hunters in the Ice Age were found at the scene of a large kill in Lubbock, Texas. From the amount of decay of carbon-14 in the bone fragments of the giant bison, scientists determined that the kill took place about 9833 years ago (*National Geographic*, December 1955).

Carbon-14 is a radioactive substance that is absorbed by an organism while it is alive but begins to decay upon the death of the organism. The number of grams of carbon-14 remaining in a fragment of charred bone from the giant bison after t years is given by the formula $A = 3.6e^{rt}$, where $r = -1.21 \times 10^{-4}$. Find the amount present initially and after 9833 years.

Solution

To find the initial amount, let $t = 0$ and $r = -1.21 \times 10^{-4}$ in the formula:

$$A = 3.6e^{(-1.21\times10^{-4})(0)} = 3.6 \text{ grams}$$

To find the amount present after 9833 years, let $t = 9833$ and $r = -1.21 \times 10^{-4}$:

$$\begin{aligned} A &= 3.6e^{(-1.21\times10^{-4})(9833)} \\ &\approx 1.1 \text{ grams} \end{aligned}$$

The initial amount of carbon-14 was 3.6 grams, and after 9833 years, 1.1 grams of carbon-14 remained. See Fig. 4.19.

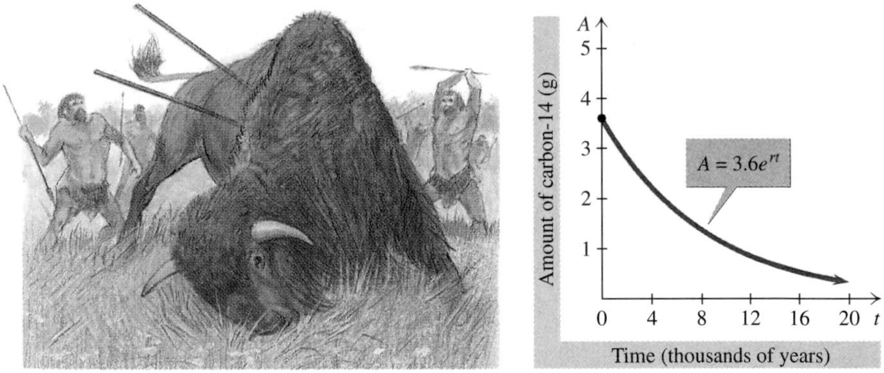

Figure 4.19

The calculator graph in Fig. 4.20 can be used to find the amount of radioactive substance present after 9833 years.

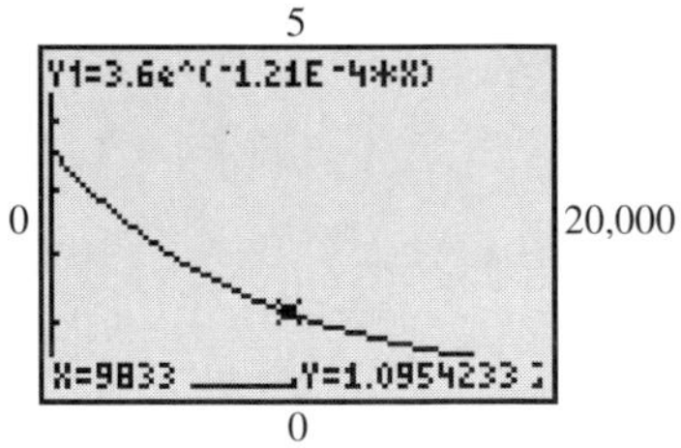

Figure 4.20

The formula for radioactive decay is used to determine the age of ancient objects such as bones, campfire charcoal, and meteorites. To date objects, we must solve the radioactive decay formula for t, which will be done in Section 4.4 after we have studied logarithms.

FOR THOUGHT True or False? Explain.

1. The function $f(x) = (-2)^x$ is an exponential function.
2. The function $f(x) = 2^x$ is invertible.
3. If $2^x = -\frac{1}{8}$, then $x = -3$.
4. If $f(x) = 3^x$, then $f(0.5) = \sqrt{3}$.
5. If $f(x) = e^x$, then $f(0) = 1$.
6. If $f(x) = e^x$ and $f(t) = e^2$, then $t = 2$.
7. The x-axis is a horizontal asymptote for the graph of $y = e^x$.
8. The function $f(x) = (0.5)^x$ is increasing.
9. The functions $f(x) = 4^{x-1}$ and $g(x) = (0.25)^{1-x}$ have the same graph.
10. $2^{1.73} = \sqrt[100]{2^{173}}$

4.1 EXERCISES

Tape 9 Disk

Let $f(x) = 3^x$, $g(x) = 2^{1-x}$, and $h(x) = (1/4)^x$. Find the following values.

1. $f(2)$ **2.** $f(4)$ **3.** $f(-2)$ **4.** $f(-3)$
5. $g(2)$ **6.** $g(1)$ **7.** $g(-2)$ **8.** $g(-3)$
9. $h(-1)$ **10.** $h(-2)$ **11.** $h(-1/2)$ **12.** $h(3/2)$

Sketch the graph of each function by finding at least three ordered pairs on the graph. State the domain, the range, and whether the function is increasing or decreasing.

13. $f(x) = 5^x$ **14.** $f(x) = 4^x$ **15.** $y = 10^{-x}$
16. $y = e^{-x}$ **17.** $f(x) = (1/4)^x$ **18.** $f(x) = (0.2)^x$

Use transformations to help you graph each function. Identify the horizontal asymptote.

19. $f(x) = 2^x - 3$ **20.** $f(x) = 3^{-x} + 1$
21. $y = -2^{-x}$ **22.** $y = -10^{-x}$
23. $y = 1 - 2^x$ **24.** $y = -1 - 2^{-x}$
25. $f(x) = 0.5 \cdot 3^{x-2}$ **26.** $f(x) = -0.1 \cdot 5^{x+4}$
27. $y = 500(0.5)^x$ **28.** $y = 100 \cdot 2^x$
29. $f(x) = 2^{x+3} - 5$ **30.** $f(x) = 3^{1-x} - 4$

Write the equation of each graph in its final resting place.

31. The graph of $y = 2^x$ is translated 5 units to the right and then 2 units downward.

32. The graph of $y = e^x$ is translated 3 units to the left and then 1 unit upward.

33. The graph of $y = (1/4)^x$ is translated 1 unit to the right, reflected in the x-axis, and then translated 2 units downward.

34. The graph of $y = 10^x$ is translated 3 units upward, 2 units to the left, and then reflected in the x-axis.

Solve each equation.

35. $2^x = 64$ **36.** $5^x = 1$ **37.** $10^x = 0.1$
38. $10^{2x} = 1000$ **39.** $-3^x = -27$ **40.** $-2^x = -\frac{1}{2}$
41. $3^{-x} = 9$ **42.** $2^x = \frac{1}{8}$ **43.** $8^x = 2$
44. $9^x = 3$ **45.** $e^x = \frac{1}{e^2}$ **46.** $e^{-x} = \frac{1}{e}$
47. $\left(\frac{1}{2}\right)^x = 8$ **48.** $\left(\frac{2}{3}\right)^x = \frac{9}{4}$
49. $10^{x-1} = 0.01$ **50.** $10^{|x|} = 1000$

Let $f(x) = 2^x$, $g(x) = (1/3)^x$, $h(x) = 10^x$, $m(x) = e^x$. Find the value of x in each equation.

51. $f(x) = 4$ **52.** $f(x) = 32$ **53.** $f(x) = \frac{1}{2}$
54. $f(x) = 1$ **55.** $g(x) = 1$ **56.** $g(x) = 9$
57. $g(x) = 27$ **58.** $g(x) = \frac{1}{9}$ **59.** $h(x) = 1000$
60. $h(x) = 10^5$ **61.** $h(x) = 0.1$ **62.** $h(x) = 0.0001$

63. $m(x) = e$

64. $m(x) = e^3$

65. $m(x) = \frac{1}{e}$

66. $m(x) = 1$

Fill in the missing coordinate in each ordered pair so that the pair is a solution to the given equation.

67. $y = 3^x$ (2,), (, 3), (−1,), (, 1/9)

68. $y = 10^x$ (3,), (, 1), (−2,), (, 0.01)

69. $f(x) = 5^{-x}$ (0,), (, 25), (−1,), (, 1/5)

70. $f(x) = e^{-x}$ (1,), (, e), (0,), (, e^2)

71. $f(x) = -2^x$ (4,), (, −1/4), (−1,), (, −32)

72. $f(x) = -(1/4)^{x-1}$ (3,), (, −4), (−1,), (, −1/16)

Find an approximate solution to each equation by graphing an appropriate function on a graphing calculator and reading from the graph. Round answers to two decimal places.

73. $3^x - 5 = 0$

74. $2^{-x} - 3 = 0$

75. $2^x - 3^{x+1} = 0$

76. $e^x - 2^{-x+1} = 0$

77. $e^{x+1} = 9$

78. $2^x = 3^{-x}$

79. $200e^{0.06x} = 400$

80. $300(1.05)^x = 600$

• CALCULUS •

Use a graphing calculator to help you sketch the graph of each function. Identify the domain and range.

81. $f(x) = 2^{|x|}$

82. $f(x) = 3^{-|x|}$

83. $y = 2^{-x} + 2^x$

84. $y = x2^x$

85. $y = 2^{-x^2}$

86. $y = x^2e^{-x}$

Solve each problem.

87. *Interest Compounded Quarterly* If \$4500 is deposited in a bank account paying 8% compounded quarterly, then what amount will be in the account at the end of 6 years? How much interest will be earned during the 6 years?

88. *Interest Compounded Monthly* If Melinda invests her \$80,000 winnings from Publishers Clearing House at 9% compounded monthly, then what will the investment be worth at the end of 20 years? How much interest will be earned during the 20 years?

89. *Interest Compounded Daily* If a credit union pays 6.5% annual interest compounded daily, then what will a deposit of \$2300 be worth after 5 years and 4 months? Use 30 days per month and 365 days per year.

90. *High Yields Attract Deposits* To attract funds, the financially troubled Commercial Federal Savings and Loan offered $9\frac{3}{4}$% annual interest compounded daily on certificates of deposit. At this rate how much interest would a deposit of \$30,000 earn in 1 year and 6 months? Use 30 days per month and 365 days per year.

91. *Interest Compounded Continuously* The Lakewood Savings Bank pays 5% annual interest compounded continuously on deposits. How much will a deposit of \$100,000 amount to in 5 years and 45 days at 5% compounded continuously? Compounded daily?

92. *Interest Compounded Continuously* If one million dollars is deposited in an account paying 6% compounded continuously, then will it earn more or less than minimum wage in its first hour on deposit? How much interest is earned during its 500th hour on deposit?

93. *Population Growth* The population of a small country in Central America is growing according to the function $P = 2{,}400{,}000e^{0.03t}$, where t is the number of years since 1990. What was the population in 1990? What will the population be in 2001 according to the formula?

94. *Radioactive Decay* The number of grams of a certain radioactive substance present at time t is given by the formula $A = 200e^{-0.001t}$, where t is the number of years. How many grams are present at time $t = 0$? How many grams are present at time $t = 500$?

95. *Destruction of the Rain Forests* With the value of timber growing at 5% annually and interest rates as high as 24%, owners of the rain forests are finding it hard to delay turning their timber into cash (*Scientific American*, April 1997). The accompanying graph compares the growth in

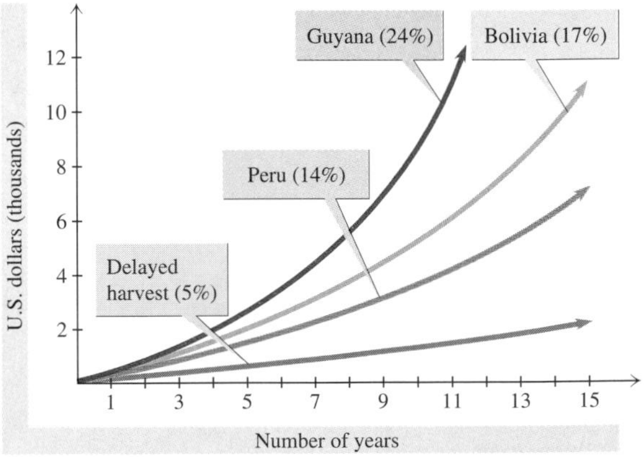

Figure for Exercise 95

value of \$1000 worth of timber in the forest to the growth of \$1000 in cash in Guyana, Bolivia, and Peru.

a. How much more does a landowner in Guyana have by investing \$1000 for 10 years compared to delaying the harvest of \$1000 worth of timber for 10 years? (Use $A = Pe^{rt}$.)

b. How much more does a landowner in Bolivia have by investing \$20,000 for 13 years compared to delaying the harvest of \$20,000 worth of timber for 13 years?

c. Enter the four equations for these curves into a graphing calculator. Set the viewing window to match that of the accompanying figure. Your graph should look like the one in the figure.

96. *Growing Debits* The number of debit card transactions in the U.S. has been growing exponentially as shown in the accompanying graph (Forbes, June 16, 1997). The number of debit card purchases can be approximated by the equation $y = 0.08e^{0.42x}$, where $x = 0$ corresponds to 1990 and y is in billions.

a. How many debit card purchases were there in 1995?

b. How many will there be in 2001?

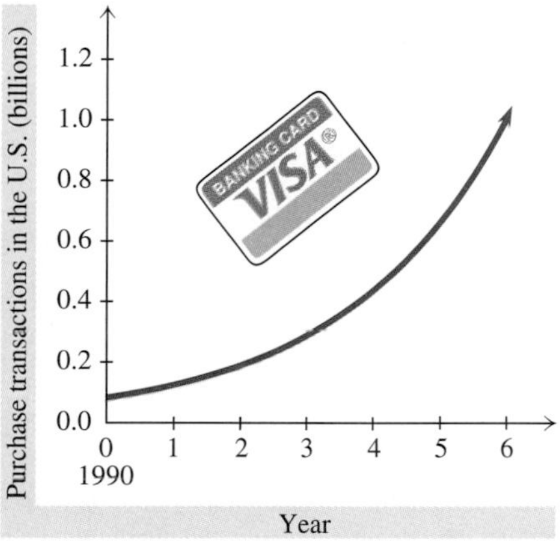

Figure for Exercise 96

97. *Decline of the Family* The Census Bureau defines a household as a *family* if its members are related by marriage, birth, or adoption. The percentage of families among U.S. households can be modeled by the function

$$p = 89.7e^{-0.0058t}$$

where t is the number of years since 1950 (Census Bureau).

a. Is the percentage of households that are families increasing or decreasing?

b. Find the percentage in 1950, 1990, and 1998.

c. Predict the percentage in the year 2020.

98. *Rising Price of Milk* In the second quarter of 1997, the annual rate of inflation was 1.4%, the lowest rate in 34 years (Boston Globe, August 1, 1997). If the price of a gallon of skim milk was \$2.59 in 1997, then what will be the price of a gallon of skim milk in 2015 if its price increases at a 2% annual rate until then? What will the price be if the price increases at a 4% annual rate until 2015? What assumption(s) are you making to solve this problem?

99. *Position of a Football* The football is on the 10-yard line. Several penalties in a row are given, and each moves the ball half the distance to the closer goal line. Write a formula that gives the ball's position P after the nth such penalty.

100. *Cost of a Parking Ticket* The cost of a parking ticket on campus is \$15 for the first offense. Given that the cost doubles for each additional offense, write a formula for the cost C as a function of the number of tickets n.

• CALCULUS •

101. *The Value of e* Make a table showing the values of $(1 + 1/n)^n$ for $n = 20$, 200, 2000, and 20,000. What is the difference between the last entry in your table and the value of e found on your calculator?

102. *A Large Investment* Find the amount of \$80 million earning interest for one year at 9% compounded annually, monthly, daily, hourly, and every minute. What amount are these values approaching?

103. *Challenger Disaster* Using data on O-ring damage from 24 previous space shuttle launches, Professor Edward R. Tufte of Yale University concluded that the number of O-rings damaged per launch is an exponential function of the temperature at the time of the launch. If NASA had used a model such as $n = 644e^{-0.15t}$, where t is the

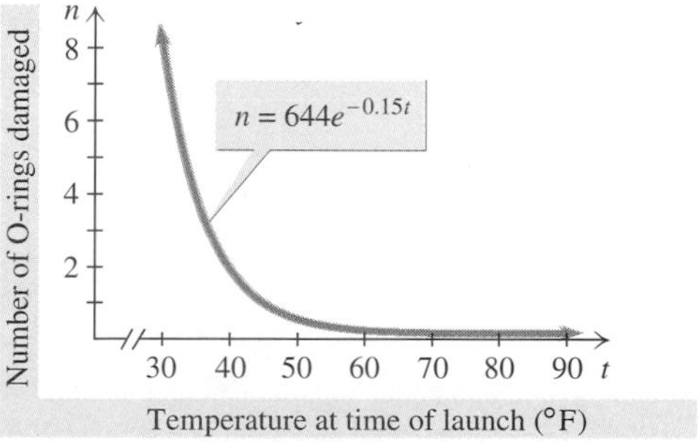

Figure for Exercise 103

Fahrenheit temperature at the time of launch and n is the number of O-rings damaged, then the tragic end to the flight of the space shuttle Challenger might have been avoided. Using this model, find the number of O-rings that would be expected to fail at 31°F, the temperature at the time of the Challenger launch on January 28, 1986.

104. *Manufacturing Cost* The cost in dollars for manufacturing x units of a certain drug is given by the function $C(x) = xe^{0.001x}$. Find the cost for manufacturing 500 units. Find the function $AC(x)$ that gives the average cost per unit for manufacturing x units. What happens to the average cost per unit as x gets larger and larger?

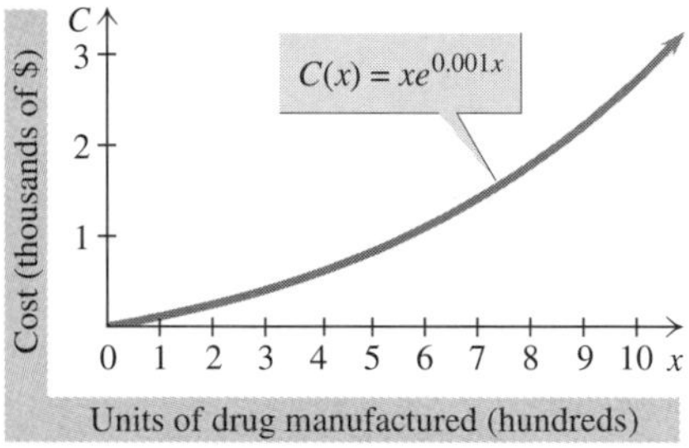

Figure for Exercise 104

For Writing/Discussion

105. *Inverse Function* The function $f(x) = 10^x$ is a one-to-one function, and so it has an inverse function. What is the value of $f^{-1}(100)$? Explain your answer.

106. *Another Inverse Function* The function $f(x) = e^x$ is a one-to-one function, and so it has an inverse function. What is the value of $f^{-1}(1)$? Explain your answer.

107. *Cooperative Learning* Work in a small group to write a summary (including drawings) of the types of graphs that can be obtained for exponential functions of the form $y = a^x$ for $a > 0$ and $a \neq 1$.

108. *Group Toss* Have all students in your class stand and toss a coin. Those that get heads sit down. Those that are left standing toss again and all who obtain heads must sit down. Repeat until no one is left standing. For each toss record the number of coins that are tossed. For example, (1, 30), (2, 14), and so on. Do not include a pair with zero coins tossed. Enter the data into a graphing calculator and use exponential regression to find an equation of the form $y = a \cdot b^x$ that fits the data. What percent of those standing did you expect would sit down after each toss? Is the value of b close to this number?

LINKING CONCEPTS

• CALCULUS •

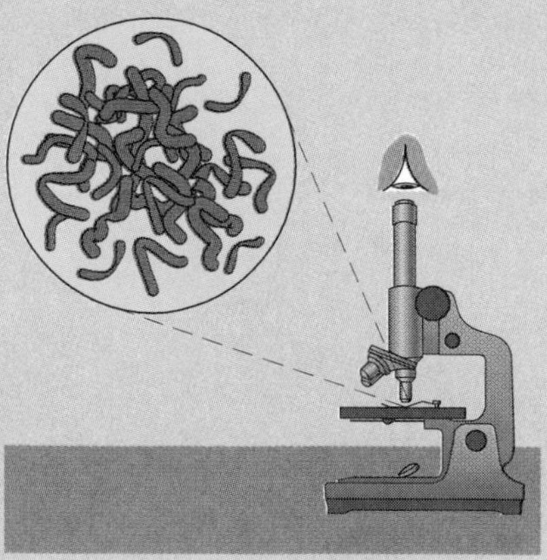

For Individual or Group Explorations

Comparing Exponential and Linear Growth *The function $f(t) = 300e^{0.5t}$ gives the number of bacteria present in a culture t hours after the start of an experiment in which the bacteria are growing continuously at a rate of 50% per hour. Let A[a, b] represent the average rate of change of f on the time interval [a, b].*

a) Fill in the following table.

Interval $[a, b]$	$f(a)$	$A[a, b]$	$A[a, b]/f(a)$
[3.00, 3.05] [7.50, 7.51] [8.623, 8.624]			

b) What are the units for the quantity $A[a, b]$?

c) What can you conjecture about the ratio $A[a, b]/f(a)$?

d) Test your conjecture on a few more intervals of various lengths. Explain how the length of the interval affects the ratio.

e) State your conjecture in terms of variation (Section 2.7). What is the constant of proportionality?

f) Suppose the bacteria were growing in a linear manner, say $f(t) = 800t + 300$. Make a table like the given table and make a conjecture about the ratio $A[a, b]/f(a)$.

g) Explain how your conclusions about average rate of change can be used to justify an exponential model as better than a linear model for modeling growth of a bacteria (or human) population.

4.2 Logarithmic Functions

Since exponential functions are one-to-one functions (Section 4.1), they are invertible. In this section we will study the inverses of the exponential functions.

The Definition

The inverses of the exponential functions are called **logarithmic functions**. Since f^{-1} is a general name for an inverse function, we adopt a more descriptive notation for these inverses. If $f(x) = a^x$, then instead of $f^{-1}(x)$, we write $\log_a(x)$ for the inverse of the base-a exponential function. We read $\log_a(x)$ as "log of x with base a," and we call the expression $\log_a(x)$ a **logarithm**.

The meaning of $\log_a(x)$ will be clearer if we consider the exponential function

$$f(x) = 2^x$$

as an example. Since $f(3) = 2^3 = 8$, the base-2 exponential function pairs the exponent 3 with the value of the exponential expression 8. Since the function $\log_2(x)$ reverses that pairing, we have $\log_2(8) = 3$. So $\log_2(8)$ is the exponent that is used on the base 2 to obtain 8. *In general, $\log_a(x)$ is the exponent that is used on the base a to obtain the value x.*

Definition: Logarithmic Function

For $a > 0$ and $a \neq 1$, the **logarithmic function with base a** is denoted $f(x) = \log_a(x)$, where

$$y = \log_a(x) \quad \text{if and only if} \quad a^y = x.$$

EXAMPLE 1 Evaluating logarithmic functions

Find the indicated values of the logarithmic functions.

a) $\log_3(9)$ **b)** $\log_2\left(\frac{1}{4}\right)$ **c)** $\log_{1/2}(8)$

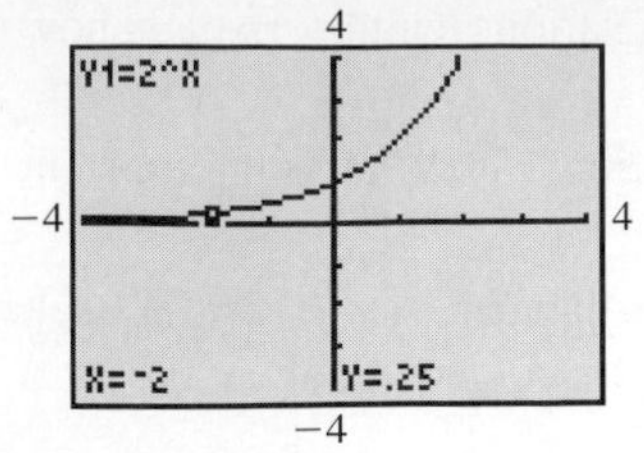

Figure 4.21

Solution

a) By the definition of logarithm, $\log_3(9)$ is the exponent that is used on the base 3 to obtain 9. Since $3^2 = 9$, we have $\log_3(9) = 2$.

b) Since $\log_2(1/4)$ is the exponent that is used on the base 2 to obtain 1/4, we try various powers of 2 until we find $2^{-2} = 1/4$. So $\log_2(1/4) = -2$.
You can use the graph of $y = 2^x$ as in Fig. 4.21 to check that $2^{-2} = 1/4$. □

c) Since $\log_{1/2}(8)$ is the exponent that is used on a base of 1/2 to obtain 8, we try various powers of 1/2 until we find $(1/2)^{-3} = 8$. So $\log_{1/2}(8) = -3$.

Since the exponential function $f(x) = a^x$ has domain $(-\infty, \infty)$ and range $(0, \infty)$, the logarithmic function $f(x) = \log_a(x)$ has domain $(0, \infty)$ and range $(-\infty, \infty)$. So there are no logarithms of negative numbers or zero. Expressions such as $\log_2(-4)$ and $\log_3(0)$ are undefined. Note that $\log_a(1) = 0$ for any base a, because $a^0 = 1$ for any base a.

There are two bases that are used more frequently than the others; they are 10 and e. The notation $\log_{10}(x)$ is abbreviated $\log(x)$, and $\log_e(x)$ is abbreviated $\ln(x)$. Most scientific calculators have function keys for the exponential functions 10^x and e^x and their inverses $\log(x)$ and $\ln(x)$, which are called **common logarithms** and **natural logarithms**, respectively. Natural logarithms are also called **Napierian logarithms** after John Napier (1550–1617).

Figure 4.22

Note that $\log(76)$ is approximately 1.8808 and $10^{1.8808}$ is approximately 76 as shown in Fig. 4.22. If you use more digits for $\log(76)$ as the power of 10, then the calculator gets closer to 76. □

You can use a calculator to find common or natural logarithms, but you should know how to find the values of logarithms such as those in Examples 1 and 2 without using a calculator.

EXAMPLE 2 Evaluating logarithmic functions

Find the indicated values of the logarithmic functions without a calculator. Use a calculator to check.

a) $\log(1000)$ **b)** $\ln(1)$ **c)** $\ln(-6)$

Solution

a) To find $\log(1000)$, we must find the exponent that is used on the base 10 to obtain 1000. Since $10^3 = 1000$, we have $\log(1000) = 3$.

b) Since $e^0 = 1$, $\ln(1) = 0$.

c) The expression $\ln(-6)$ is undefined because -6 is not in the domain of the natural logarithm function. There is no power of e that results in -6.

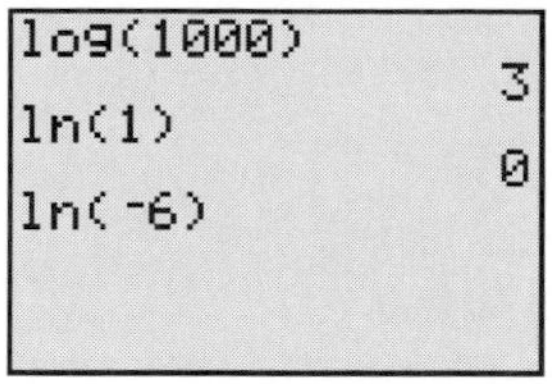

Figure 4.23

The calculator results for parts (a) and (b) are shown in Fig. 4.23. If you ask a calculator for $\ln(-6)$, it will give you an error message.

Graphs of Logarithmic Functions

The functions $y = a^x$ and $y = \log_a(x)$ for $a > 0$ and $a \neq 1$ are inverse functions. So the graph of $y = \log_a(x)$ is a reflection about the line $y = x$ of the graph of $y = a^x$. The graph of $y = a^x$ has the x-axis as its horizontal asymptote, while the graph of $y = \log_a(x)$ has the y-axis as its vertical asymptote.

EXAMPLE 3 Graph of a base-a logarithmic function with $a > 1$

Sketch the graphs of $y = 2^x$ and $y = \log_2(x)$ on the same coordinate system. State the domain and range of each function.

Solution

Since these two functions are inverses of each other, the graph $y = \log_2(x)$ is a reflection of the graph of $y = 2^x$ about the line $y = x$. Make a table of ordered pairs for each function:

x	-1	0	1	2
$y = 2^x$	1/2	1	2	4

x	1/2	1	2	4
$y = \log_2(x)$	-1	0	1	2

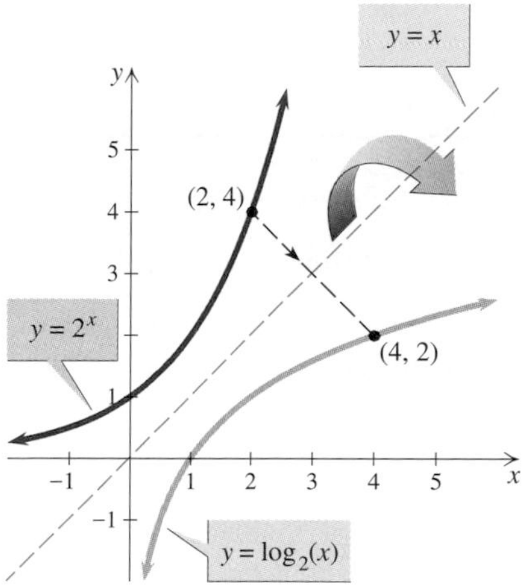

Figure 4.24

Sketch $y = 2^x$ as in Section 4.1. Then sketch $y = \log_2(x)$ through the points given in the table, keeping in mind that it is a reflection of $y = 2^x$. Both curves are shown in Fig. 4.24 along with the line $y = x$. The domain of $y = 2^x$ is $(-\infty, \infty)$, and its range is $(0, \infty)$. The domain of $y = \log_2(x)$ is $(0, \infty)$, and its range is $(-\infty, \infty)$.

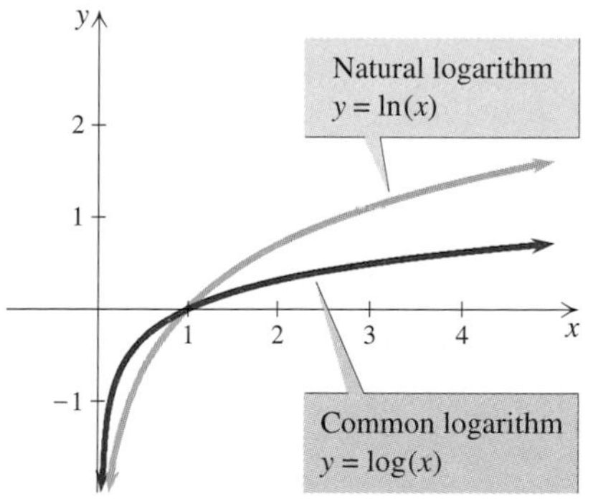

Figure 4.25

Note that the function $y = \log_2(x)$ graphed in Fig. 4.24 is one-to-one by the horizontal line test and it is an increasing function. The graphs of the common logarithm function $y = \log(x)$ and the natural logarithm function $y = \ln(x)$ are shown in Fig. 4.25 and they are also increasing functions. The function $y = \log_a(x)$ is increasing if $a > 1$. By contrast, if $0 < a < 1$, the function $y = \log_a(x)$ is decreasing, as illustrated in the next example.

EXAMPLE 4 Graph of a base-a logarithmic function with $0 < a < 1$

Sketch the graph of $f(x) = \log_{1/3}(x)$ and state its domain and range.

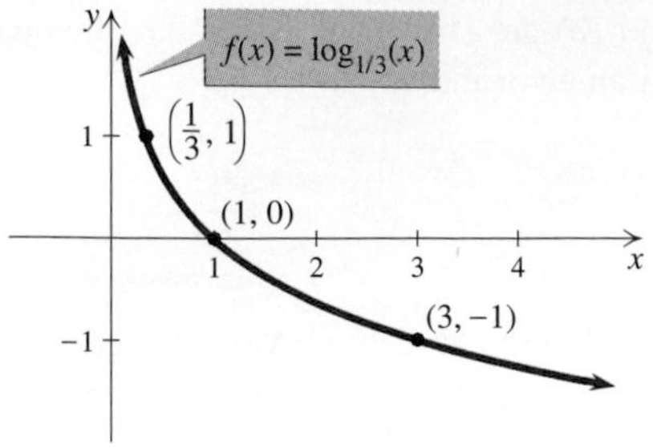

Figure 4.26

Solution

Make a table of ordered pairs for the function:

x	1/9	1/3	1	3	9
$y = \log_{1/3}(x)$	2	1	0	−1	−2

Sketch the curve through these points as shown in Fig. 4.26. The y-axis is a vertical asymptote for the curve. The domain of $f(x) = \log_{1/3}(x)$ is $(0, \infty)$, and its range is $(-\infty, \infty)$.

The logarithmic functions have properties corresponding to the properties of the exponential functions stated in Section 4.1.

Properties of Logarithmic Functions

The logarithmic function $f(x) = \log_a(x)$ has the following properties:

1. The function f is increasing for $a > 1$ and decreasing for $0 < a < 1$.
2. The x-intercept of the graph of f is (1, 0).
3. The graph has the y-axis as a vertical asymptote.
4. The domain of f is $(0, \infty)$ and the range of f is $(-\infty, \infty)$.
5. The function f is one-to-one.

Transformation of Logarithmic Functions

The graphs of logarithmic functions can be transformed in the same manner as the graphs of other functions. For example, the graph of $y = \log_a(x + 2)$ lies two units to the left of the graph of $y = \log_a(x)$, while $y = -3 + \log_a(x)$ is three units below $y = \log_a(x)$, and $y = -\log_a(x)$ is a reflection of $y = \log_a(x)$.

EXAMPLE 5 Transformation of the graph of a logarithmic function

Sketch the graph of each function and state its domain and range.

a) $y = \log_2(x - 1)$ **b)** $f(x) = -\frac{1}{2}\log_2(x + 3)$

Solution

a) The graph of $y = \log_2(x - 1)$ is obtained by translating the graph of $y = \log_2(x)$ to the right one unit. Since the domain of $y = \log_2(x)$ is $(0, \infty)$, the domain of $y = \log_2(x - 1)$ is $(1, \infty)$. The line $x = 1$ is the vertical asymptote. Calculate a few ordered pairs to get an accurate graph.

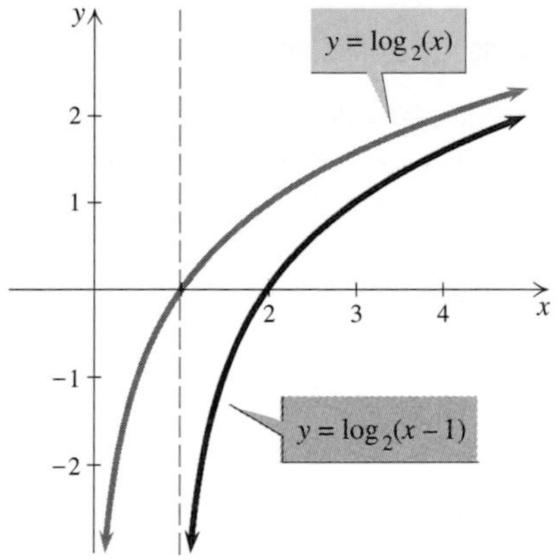

Figure 4.27

x	1.5	2	3	5
$y = \log_2(x - 1)$	-1	0	1	2

Sketch the curve as shown in Fig. 4.27. The range is $(-\infty, \infty)$.

b) The graph of f is obtained by translating the graph of $y = \log_2(x)$ to the left three units. The vertical asymptote for f is the line $x = -3$. Multiplication by $-1/2$ shrinks the graph and reflects it in the x-axis. Calculate a few ordered pairs for accuracy.

x	-2	-1	1	5
$y = -\frac{1}{2}\log_2(x + 3)$	0	$-1/2$	-1	$-3/2$

Sketch the curve through these points as shown in Fig. 4.28. The domain of f is $(-3, \infty)$, and the range is $(-\infty, \infty)$.

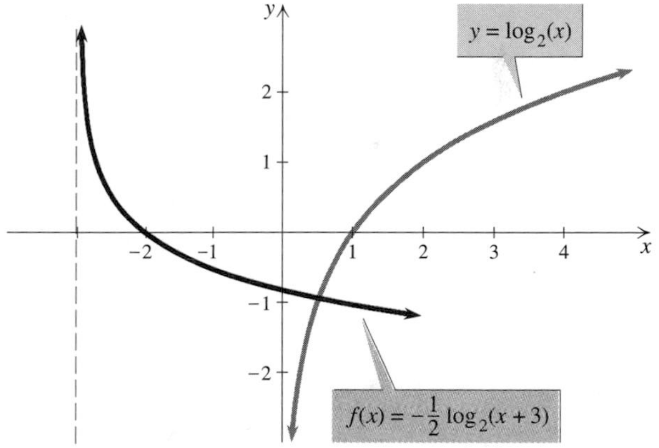

Figure 4.28

Note that logarithmic functions involving common or natural logarithms can be graphed with a graphing calculator because the calculator has keys for log and ln. To graph logarithmic functions involving other bases, we need the base-change formula that is discussed in Section 4.3.

Logarithmic and Exponential Equations

Some equations involving logarithms can be solved by writing an equivalent exponential equation. On the other hand, some equations involving exponents can be solved by writing an equivalent logarithmic equation. Rewriting logarithmic and exponential equations is possible because of the definition of logarithms:

$$y = \log_a(x) \quad \text{if and only if} \quad a^y = x$$

EXAMPLE 6 Rewriting logarithmic and exponential equations

Write each equation involving logarithms as an equivalent exponential equation, and write each equation involving exponents as an equivalent logarithmic equation.

a) $\log_5(625) = 4$ **b)** $\log_3(n - 1) = 5$ **c)** $3^{2x+3} = 50$ **d)** $e^{x-1} = 9$

Solution

a) $\log_5(625) = 4$ is equivalent to $5^4 = 625$.

b) $\log_3(n - 1) = 5$ is equivalent to $3^5 = n - 1$.

c) $3^{2x+3} = 50$ is equivalent to $2x + 3 = \log_3(50)$.

d) $e^{x-1} = 9$ is equivalent to $x - 1 = \ln(9)$.

The one-to-one property of exponential functions was used to solve exponential equations in Section 4.1. Likewise, the one-to-one property of logarithmic functions is used in solving logarithmic equations. The one-to-one property says that *if two quantities have the same base-a logarithm, then the quantities are equal.*

One-to-One Property of Logarithms

For $a > 0$ and $a \neq 1$,

$$\text{if} \quad \log_a(x_1) = \log_a(x_2), \quad \text{then} \quad x_1 = x_2.$$

The one-to-one properties and the definition of logarithm are at present the only new tools that we have for solving equations. Later in this chapter we will develop more properties of logarithms and solve more complicated equations.

EXAMPLE 7 Solving equations involving logarithms

Solve each equation.

a) $\log_3(x) = -2$ **b)** $\log_x(5) = 2$ **c)** $5^x = 9$ **d)** $\ln(x^2) = \ln(3x)$

Solution

a) Use the definition of logarithm to write the equivalent exponential equation:

$$\begin{aligned} \log_3(x) &= -2 \\ x &= 3^{-2} && \text{Definition of logarithm} \\ &= \frac{1}{9} \end{aligned}$$

Since $\log_3(1/9) = -2$ is correct, the solution to the equation is $1/9$.

b) $\log_x(5) = 2$

$x^2 = 5$ Definition of logarithm

$x = \pm\sqrt{5}$

Since the base of a logarithm is always nonnegative, the only solution is $\sqrt{5}$.

c) $5^x = 9$

$x = \log_5(9)$ Definition of logarithm

The exact solution to $5^x = 9$ is the irrational number $\log_5(9)$. In the next section we will learn how to find a rational approximation for $\log_5(9)$.

d) $\ln(x^2) = \ln(3x)$

$x^2 = 3x$ One-to-one property of logarithms

$x^2 - 3x = 0$

$x(x - 3) = 0$

$x = 0$ or $x = 3$

Checking $x = 0$ in the original equation, we get the undefined expression $\ln(0)$. So the only solution to the equation is 3.

EXAMPLE 8 Equations involving common and natural logarithms

Use a calculator to find the value of x rounded to four decimal places.

a) $e^x = 125$ **b)** $x \cdot \log(2) + 3x \cdot \log(17) = 6$

Solution

a) $e^x = 125$

$x = \ln(125)$ Definition of logarithm

≈ 4.8283

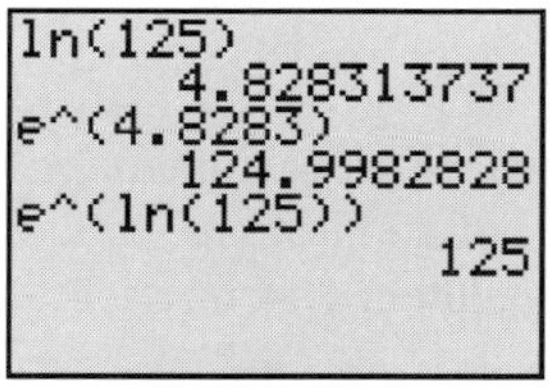

Figure 4.29

Figure 4.29 shows how to find the approximate value of $\ln(125)$ and how to check the approximate answer and exact answer.

b) $x \cdot \log(2) + 3x \cdot \log(17) = 6$

$x[\log(2) + 3 \cdot \log(17)] = 6$ Factor out the common factor.

$$x = \frac{6}{\log(2) + 3 \cdot \log(17)}$$

≈ 1.5029

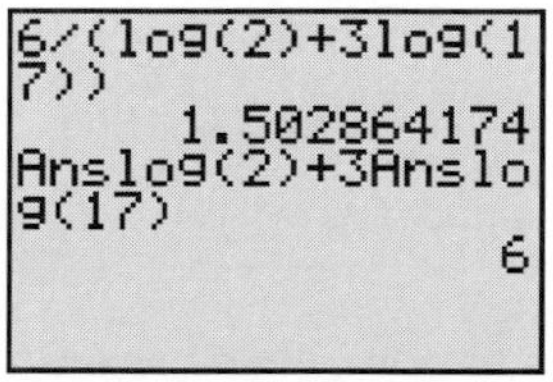

Figure 4.30

Figure 4.30 shows how to find the approximate value of x and to check it in the original equation.

Applications

We saw in Section 4.1 that if a principal of P dollars earns interest at an annual rate r compounded continuously, then the amount after t years is given by

$$A = Pe^{rt}.$$

If either the rate or the time is the only unknown in this formula, then the definition of logarithms can be used to solve for the rate or the time.

EXAMPLE 9 **Finding the time in a continuous compounding problem**

If \$8000 is invested at 9% compounded continuously, then how long will it take for the investment to grow to \$20,000?

Solution

Use the formula $A = Pe^{rt}$ with $A = \$20{,}000$, $P = \$8000$, and $r = 0.09$:

$$20{,}000 = 8000e^{0.09t}$$

$$2.5 = e^{0.09t} \qquad \text{Divide each side by 8000.}$$

$$0.09t = \ln(2.5) \qquad \text{Definition of logarithm: } y = a^x \text{ if and only if } x = \log_a(y)$$

$$t = \frac{\ln(2.5)}{0.09}$$

$$\approx 10.181 \text{ years}$$

Figure 4.31

We can multiply 365 by 0.181 to get approximately 66 days. So the investment grows to \$20,000 in approximately 10 years and 66 days.

You can use a graphing calculator to check as shown in Fig. 4.31.

The formula $A = Pe^{rt}$ was introduced to model the continuous growth of money. However, this type of formula is used in a wide variety of applications. In the following exercises you will find problems involving population growth, declining forests, and global warming.

FOR THOUGHT True or False? Explain.

1. The first coordinate of an ordered pair in an exponential function is a logarithm.
2. $\log_{100}(10) = 2$
3. If $f(x) = \log_3(x)$, then $f^{-1}(x) = 3^x$.
4. $10^{\log(1000)} = 1000$
5. The domain of $f(x) = \ln(x)$ is $(-\infty, \infty)$.
6. $\ln(e^{2.451}) = 2.451$
7. For any positive real number x, $e^{\ln(x)} = x$.
8. For any base a, where $a > 0$ and $a \neq 1$, $\log_a(0) = 1$.
9. $\log(10^3) + \log(10^5) = \log(10^8)$
10. $\log_2(32) - \log_2(8) = \log_2(4)$

4.2 EXERCISES

Tape 9 Disk

Let $f(x) = 5^x$. Evaluate the following expressions.

1. $f(1)$ and $f^{-1}(5)$

2. $f(0)$ and $f^{-1}(1)$

3. $f(-1)$ and $f^{-1}(1/5)$

4. $f(-2)$ and $f^{-1}(0.04)$

5. $f^{-1}(125)$

6. $f^{-1}(1/625)$

Find the indicated values of the logarithmic functions.

7. $\log_2(64)$ **8.** $\log_2(16)$ **9.** $\log_3(1/81)$

10. $\log_3(1)$ **11.** $\log_{16}(2)$ **12.** $\log_{16}(16)$

13. $\log_{1/5}(125)$ **14.** $\log_{1/5}(1/125)$ **15.** $\log(0.1)$

16. $\log(10^6)$ **17.** $\log(1)$ **18.** $\log(10)$

19. $\ln(e)$ **20.** $\ln(0)$

21. $\ln(e^{-5})$ **22.** $\ln(e^9)$

Sketch the graph of each function, and state the domain and range of each function.

23. $y = \log_3(x)$

24. $y = \log_4(x)$

25. $f(x) = \log_5(x)$

26. $g(x) = \log_8(x)$

27. $y = \log_{1/2}(x)$

28. $y = \log_{1/4}(x)$

29. $h(x) = \log_{1/5}(x)$

30. $s(x) = \log_{1/10}(x)$

31. $f(x) = \ln(x - 1)$

32. $f(x) = \log_3(x + 2)$

33. $f(x) = -3 + \log(x + 2)$

34. $f(x) = 4 - \log(x + 6)$

35. $f(x) = -\frac{1}{2}\log(x - 1)$

36. $f(x) = \log_2(-x)$

Write the equation of each graph in its final resting place.

37. The graph of $y = \ln(x)$ is translated 3 units to the right and then 4 units downward.

38. The graph of $y = \log(x)$ is translated 5 units to the left and then 7 units upward.

39. The graph of $y = \log_2(x)$ is translated 5 units to the right, reflected in the x-axis, and then translated 1 unit downward.

40. The graph of $y = \log_3(x)$ is translated 4 units upward, 6 units to the left, and then reflected in the x-axis.

Write each equation involving logarithms as an equivalent exponential equation, and write each equation involving exponents as an equivalent logarithmic equation.

41. $\log(30) = y$ **42.** $\log(t) = 9$ **43.** $5^y = 7$

44. $10^s = 13$ **45.** $\log_2(1) = 0$ **46.** $\log_3(h) = k$

47. $e^{0.09t} = 2$

48. $e^{x-1} = y$

49. $\log_b(N/M) = 3$

50. $\log_a(NM) = t$

51. $b^{3x} = y$

52. $(1/2)^{5-y} = z$

Solve each equation. Find the exact solutions.

53. $\log_2(x) = 8$

54. $\log_5(x) = 3$

55. $\log_3(x) = \frac{1}{2}$

56. $\log_4(x) = \frac{1}{3}$

57. $\log_x(16) = 2$

58. $\log_x(16) = 4$

59. $3^x = 77$

60. $\frac{1}{2^x} = 5$

61. $\ln(x - 3) = \ln(2x - 9)$

62. $\log_2(4x) = \log_2(x + 6)$

63. $\log_x(18) = 2$

64. $\log_x(9) = \frac{1}{2}$

65. $3^{x+1} = 7$

66. $5^{3-x} = 12$

67. $\log(x) = \log(6 - x^2)$

68. $\log_3(2x) = \log_3(24 - x^2)$

69. $100^{-3/2} = x$

70. $\left(\frac{1}{8}\right)^{-4/3} = x$

71. $\log_x\left(\frac{1}{9}\right) = -\frac{2}{3}$

72. $\log_x\left(\frac{1}{16}\right) = \frac{4}{3}$

73. $4^{2x-1} = \frac{1}{2}$

74. $e^{3x-4} = 1$

Use a calculator to find the approximate solution rounded to four decimal places.

75. $10^x = 25$

76. $e^x = 2$

77. $x \cdot \ln(3) = \ln(7)$

78. $x \cdot \log(3) = \log(7)$

79. $x \cdot \ln(8) - 5 = \ln(20)$

80. $1.27 - x \cdot \log(23) = \log(54)$

81. $x \cdot \ln(2) - x \cdot \ln(3) = 5$

82. $x \cdot \log(17) + 3x \cdot \log(4) = 7$

83. $(x - 1)\log(5) = (x - 2)\log(9)$

84. $(x + 2)\ln(3) = (2x - 1)\ln(12)$

Find an approximate solution to each equation by graphing an appropriate function on a graphing calculator and reading from the graph. Round answers to two decimal places.

85. $\ln(x - 2) = 3.2$

86. $\log(x) = -\log(x + 2)$

87. $\log(x + 1) = -\ln(x + 2)$

88. $\log|x| = x^2 - 4$

For each function, find f^{-1}.

89. $f(x) = 2^x$

90. $f(x) = 5^x$

91. $f(x) = \log_7(x)$

92. $f(x) = \log(x)$

93. $f(x) = \ln(x - 1)$

94. $f(x) = \log(x + 4)$

95. $f(x) = 3^{x+2}$

96. $f(x) = 6^{x-1}$

Graph each function and state the domain and range.

97. $y = \log_3(|x|)$

98. $y = \log_2(2^{|x|+1})$

99. $y = \log_2(1/x)$

100. $y = |\log_3(x)|$

Solve each problem.

101. *Becoming a Millionaire* Find the amount of time, to the nearest day, that it would take for a deposit of \$1000 to grow to \$1 million at 14% compounded continuously.

102. *Doubling Your Money* How long does it take for a deposit of \$1000 to double at 8% compounded continuously?

103. *Finding the Rate* Solve the equation $A = Pe^{rt}$ for r, then find the rate at which a deposit of \$1000 would double in 3 years compounded continuously.

104. *Finding the Rate* At what interest rate would a deposit of \$30,000 grow to \$2,540,689 in 40 years with continuous compounding?

105. *Rule of 70* a) Find the time that it takes for an investment to double at 10% compounded continuously. b) The time that it takes for an investment to double is approximately 70 divided by the interest rate (the rule of 70). So at 10%, an investment will double in about 7 years. Explain why this rule works.

106. *Using the Rule of 70* Find approximate answers to these questions without using a calculator.

a. Connie deposits \$1000 in a bank at an annual interest rate of 2% (typical bank interest rate in 1997). How long does it take her money to double?

b. Celeste invests \$1000 in the stock market and her money grows at an annual rate of 10% (the historical average rate of return for the stock market). How long does it take her money to double?

c. What is the ratio of the value of Celeste's investment after 35 years to the value of Connie's investment after 35 years?

107. *Miracle in Boston* To illustrate the "miracle" of compound interest, Ben Franklin bequeathed \$4000 to the city of Boston in 1790. The fund grew to \$4.5 million in 200 years. Find the annual rate compounded continuously that would cause this "miracle" to happen.

108. *Miracle in Philadelphia* Ben Franklin's gift of \$4000 to the city of Philadelphia in 1790 was not managed as well as his gift to Boston. The Philadelphia fund grew to only \$2 million in 200 years. Find the annual rate compounded continuously that would yield this total value.

109. *Deforestation in Nigeria* In Nigeria, deforestation occurs at the rate of about 5.25% per year. Assuming that the amount of forest remaining is determined by the function

$$F = F_0e^{-0.052t},$$

where F_0 is the present acreage of forest land and t is the time in years from the present. In how many years will there be only 60% of the present acreage remaining?

110. *Deforestation in El Salvador* It is estimated that at the present rate of deforestation in El Salvador, in 20 years only 53% of the present forest will be remaining. Use the exponential model $F = F_0e^{rt}$ to determine the annual rate of deforestation in El Salvador.

111. *World Population* The population of the world doubled from 1950 to 1987, going from 2.5 billion to 5 billion people. Using the exponential model,

$$P = P_0e^{rt},$$

find the annual growth rate r for that period. Although the annual growth rate has declined slightly to 1.63% annually, the population of the world is still growing at a tremendous rate. Using the initial population of 5 billion in 1987 and an annual rate of 1.63%, estimate the world population in the year 2000.

112. *Black Death* Because of the Black Death, or plague, the only substantial period in recorded history when the

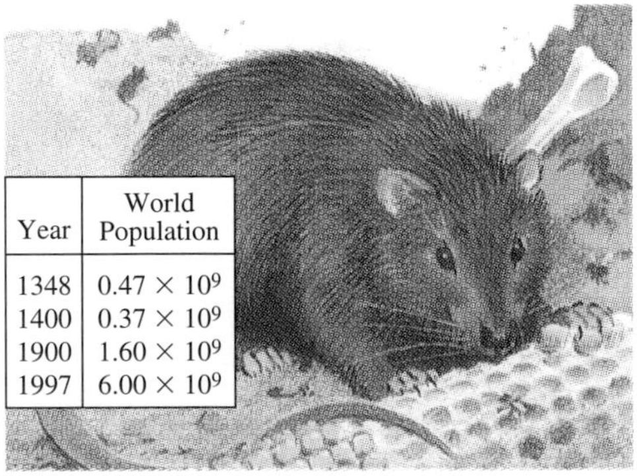

Year	World Population
1348	0.47×10^9
1400	0.37×10^9
1900	1.60×10^9
1997	6.00×10^9

Figure for Exercise 112

earth's population was not increasing was from 1348 to 1400. During that period the world population decreased by about 100 million people. Use the exponential model $P = P_0e^{rt}$ and the data from the accompanying table to find the annual growth rate for the period 1400 to 1997. If the 100 million people had not been lost, then how many people would they have grown to in 600 years using the growth rate that you just found?

113. *Traffic Jam* The amount of U.S. long-distance traffic for data transmission is expected to grow exponentially in the coming years. See the accompanying figure. The expected growth can be modeled by the function

$$d = 0.1e^{0.46t},$$

where t is the number of years since 1994 and d is measured in billions of gigabits per year.

a. What will be the amount of long-distance data transmission in 2001?

b. In what year will long-distance data transmission reach 14 billion gigabits per year?

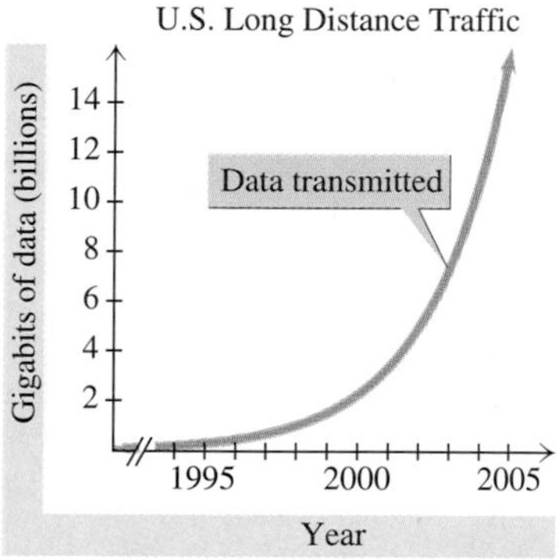

Figure for Exercise 113

114. *Global Warming* The increasing global temperature can be modeled by the function

$$I = 0.1e^{0.02t},$$

where I is the increase in global temperature in degrees Celsius since 1900, and t is the number of years since 1900 (Scientific American, May 1997).

a. How much warmer will it be in 2000 than it was in 1950?

b. In what year will the global temperature be 4° greater than the global temperature in 2000?

115. *Safe Water* The accompanying graph shows the percentage of population without safe water p as a function of per capita income I (World Resources 1996–1997). Because the horizontal axis has a *log scale* (each mark is 10 times the previous mark) the relationship looks linear but it is not.

a. Find a formula for this function by first finding the equation of the line through (2, 100) and (4, 10) in the form $p = mx + b$, then replace x by $\log(I)$.

b. Try to graph the function that you found in part (a) without using a log scale. Explain why a log scale is better for graphing this function.

c. What percent of the population would be expected to be without safe drinking water in a city with a per capita income of $100,000?

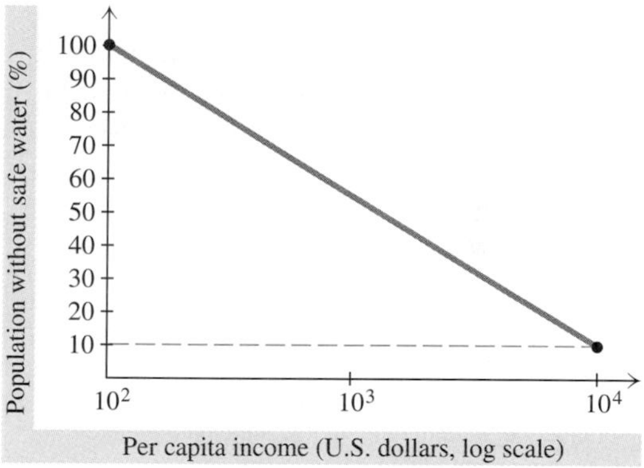

Figure for Exercise 115

116. *Municipal Waste* The accompanying graph shows the amount of municipal waste per capita w as a function of per capita income I (World Resources 1996–1997).

a. Use the technique of the previous exercise to find a formula for w as a function of I.

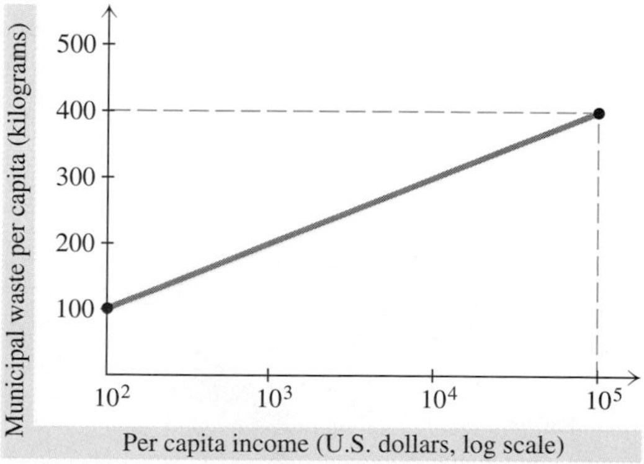

Figure for Exercise 116

b. How much municipal waste per capita would you expect in a city where the per capita income is \$200,000?

Please note the following for Exercises 117–120. In chemistry, the pH of a solution is defined to be

$$\text{pH} = -\log[H^+],$$

where H^+ is the hydrogen ion concentration of the solution in moles per liter. Distilled water has a pH of approximately 7. A substance with a pH under 7 is called an acid and one with a pH over 7 is called a base.

117. *Acidity of Tomato Juice* Tomato juice has a hydrogen ion concentration of $10^{-4.1}$ moles per liter. Find the pH of tomato juice.

118. *Acidity in Your Stomach* The gastric juices in your stomach have a hydrogen ion concentration of 10^{-1} moles per liter. Find the pH of your gastric juices.

119. *Acidity of Orange Juice* The hydrogen ion concentration of orange juice is $10^{-3.7}$ moles per liter. Find the pH of orange juice.

120. *Acidosis* A healthy body maintains the hydrogen ion concentration of human blood at $10^{-7.4}$ moles per liter. What is the pH of normal healthy blood? The condition of low blood pH is called acidosis. Its symptoms are sickly sweet breath, headache, and nausea.

For Writing/Discussion

121. *Different Models* Calculators that perform exponential regression often use $y = a \cdot b^x$ as the exponential growth model instead of $y = a \cdot e^{cx}$. For what value of c is $a \cdot b^x = a \cdot e^{cx}$? If a calculator gives $y = 500(1.036)^x$ for a growth model, then what is the continuous growth rate to the nearest hundredth of a percent?

122. *Making Conjectures* Consider the function $y = \log(10^n \cdot x)$ where n is an integer. Use a graphing calculator to graph this function for several choices of n. Make a conjecture about the relationship between the graph of $y = \log(10^n \cdot x)$ and the graph of $y = \log(x)$. Save your conjecture and attempt to prove it after you have studied the properties of logarithms, which are coming in Section 4.3. Repeat this exercise with $y = \log(x^n)$ where n is an integer.

123. *Increasing or Decreasing* Which exponential and logarithmic functions are increasing? Decreasing? Is the inverse of an increasing function increasing or decreasing? Is the inverse of a decreasing function increasing or decreasing? Explain.

124. *Cooperative Learning* Work in a small group to write a summary (including drawings) of the types of graphs that can be obtained for logarithmic functions of the form $y = \log_a(x)$ for $a > 0$ and $a \neq 1$.

LINKING CONCEPTS

For Individual or Group Explorations

Predicting Population *To effectively plan for the future one must attempt to predict the future. Government agencies use data about the past to construct a model and predict the future. The following table gives the population of the U.S. in millions every ten years since 1900 (from the 1995 Statistical Abstract of the United States).*

Year	Pop.	Year	Pop.
1900	76	1950	152
1910	92	1960	180
1920	106	1970	204
1930	123	1980	227
1940	132	1990	249

a) Draw a bar graph of the data in the accompanying table. Use a computer graphics program if one is available.

b) Enter the data into your calculator and use exponential regression to find an exponential model of the form $y = a \cdot b^x$.

c) Make a table (like the given table) that shows the predicted population according to the exponential model rather than the actual population.

d) Plot the points from part (c) on your bar graph and sketch an exponential curve through the points.

e) Use the given population data with linear regression to find a linear model of the form $y = ax + b$ and graph the line on your bar graph.

f) Predict the population in the year 2000 using the exponential model and the linear model.

g) Judging from your exponential curve and your line on your bar graph, in which prediction do you have the most confidence?

h) Use only the fact that the population grew from 76 million at time $t = 0$ (the year 1900) to 249 million at time $t = 90$ (the year 1990) to find a formula of the form $P(t) = P_0 \cdot e^{rt}$ for the population at any time t. Do not use regression.

i) Use the formula from part (h) to predict the population in the year 2000. Compare this prediction to the prediction that you found using exponential regression.

4.3 Properties of Logarithms

The properties of logarithms are closely related to the properties of exponents, because logarithms are exponents. The properties of exponents in Chapter P were given only for rational exponents. It can also be shown that the same properties hold if the exponents are real numbers. In this section we use the properties of exponents to develop some properties of logarithms. With these properties of logarithms we will be able to solve more equations involving exponents and logarithms.

The Logarithm of a Product

By the product rule for exponents, we add exponents when multiplying exponential expressions having the same base. To find a corresponding rule for logarithms, let's examine the equation $2^3 \cdot 2^2 = 2^5$. Notice that the exponents 3, 2, and 5 are the base-2 logarithms of 8, 4, and 32, respectively.

$$\underbrace{2^3}_{\log_2(8)} \cdot \underbrace{2^2}_{\log_2(4)} = \underbrace{2^5}_{\log_2(32)}$$

When we add the exponents 3 and 2 to get 5, we are adding logarithms and getting another logarithm as the result. So the base-2 logarithm of 32 (the product of 8 and 4) is the sum of the base-2 logarithms of 8 and 4:

$$\log_2(8 \cdot 4) = \log_2(8) + \log_2(4)$$

This example suggests the **product rule for logarithms**.

Product Rule for Logarithms

For $M > 0$ and $N > 0$,

$$\log_a(MN) = \log_a(M) + \log_a(N).$$

Proof If $M = a^x$ and $N = a^y$, then

$$MN = a^x a^y = a^{x+y}.$$

By the definition of logarithm, $MN = a^{x+y}$ is equivalent to

$$\log_a(MN) = x + y.$$

Since $M = a^x$, we have $x = \log_a(M)$, and since $N = a^y$, we have $y = \log_a(N)$. So

$$\log_a(MN) = \log_a(M) + \log_a(N).$$

Figure 4.32

The product rule for logarithms says that *the logarithm of a product of two numbers is equal to the sum of their logarithms*, provided that all of the logarithms are defined and all have the same base. There is no rule about the logarithm of a sum, and the logarithm of a sum is generally *not* equal to the sum of the logarithms. For example, $\log_2(8 + 8) \neq \log_2(8) + \log_2(8)$ because $\log_2(8 + 8) = 4$ while $\log_2(8) + \log_2(8) = 6$.

You can use a calculator to illustrate the product rule as in Fig. 4.32.

EXAMPLE 1 Using the product rule for logarithms

Write each expression as a single logarithm. All variables represent positive real numbers.

a) $\log_3(x) + \log_3(6)$ **b)** $\ln(3) + \ln(x^2) + \ln(y)$

Solution

a) By the product rule for logarithms, $\log_3(x) + \log_3(6) = \log_3(6x)$.

b) By the product rule for logarithms, $\ln(3) + \ln(x^2) + \ln(y) = \ln(3x^2y)$.

4.3 EXERCISES

Tape 7 Disk

Rewrite each expression as a single logarithm.

1. $\log(5) + \log(3)$

2. $\ln(6) + \ln(2)$

3. $\log_2(x - 1) + \log_2(x)$

4. $\log_3(x + 2) + \log_3(x - 1)$

5. $\log_4(12) - \log_4(2)$

6. $\log_2(25) - \log_2(5)$

7. $\ln(x^8) - \ln(x^3)$

8. $\log(x^2 - 4) - \log(x - 2)$

Rewrite each expression in terms of log(5).

9. $\log(25)$

10. $\log(1/5)$

11. $\log(1/125)$

12. $\log(5^x)$

13. $\log(\sqrt[3]{5})$

14. $\log(\sqrt[4]{125})$

Simplify each expression.

15. $e^{\ln(\sqrt{y})}$

16. $10^{\log(3x+1)}$

17. $\log(10^{y+1})$

18. $\ln(e^{2k})$

19. $7^{\log_7(999)}$

20. $\log_4(2^{300})$

Rewrite each expression in terms of log(2) and log(5).

21. $\log(10)$

22. $\log(0.4)$

23. $\log(2.5)$

24. $\log(250)$

25. $\log(\sqrt{20})$

26. $\log(0.0005)$

27. $\log(4/25)$

28. $\log(0.1)$

Rewrite each expression as a sum or difference of multiples of logarithms.

29. $\log_3(5x)$

30. $\log_2(xyz)$

31. $\log_2\left(\frac{5}{2y}\right)$

32. $\log\left(\frac{4a}{3b}\right)$

33. $\log(3\sqrt{x})$

34. $\ln(\sqrt{x/4})$

35. $\log(3 \cdot 2^{x-1})$

36. $\ln\left(\frac{5^{-x}}{2}\right)$

37. $\ln\left(\frac{\sqrt[3]{xy}}{t^{4/3}}\right)$

38. $\log\left(\frac{3x^2}{(ab)^{2/3}}\right)$

39. $\ln\left(\frac{6\sqrt{x-1}}{5x^3}\right)$

40. $\log_4\left(\frac{3x\sqrt{y}}{\sqrt[3]{x-1}}\right)$

Rewrite each expression as a single logarithm.

41. $\log_2(5) + 3 \cdot \log_2(x)$

42. $\log(x) + 5 \cdot \log(x)$

43. $\log_7(x^5) - 4 \cdot \log_7(x^2)$

44. $\frac{1}{3}\ln(6) - \frac{1}{3}\ln(2)$

45. $\log(2) + \log(x) + \log(y) - \log(z)$

46. $\ln(2) + \ln(3) + \ln(5) - \ln(7)$

47. $\frac{1}{2}\log(x) - \log(y) + \log(z) - \frac{1}{3}\log(w)$

48. $\frac{5}{6}\log_2(x) + \frac{2}{3}\log_2(y) - \frac{1}{2}\log_2(x) - \log_2(y)$

49. $3 \cdot \log_4(x^2) - 4 \cdot \log_4(x^{-3}) + 2 \cdot \log_4(x)$

50. $\frac{1}{2}[\log(x) + \log(y)] - \log(z)$

Find an approximate rational solution to each equation. Round answers to four decimal places.

51. $2^x = 9$

52. $3^x = 12$

53. $(0.56)^x = 8$

54. $(0.23)^x = 18.4$

55. $(1.06)^x = 2$

56. $(1.09)^x = 3$

57. $(0.73)^x = 0.5$

58. $(0.62)^x = 0.25$

Use a calculator and the base-change formula to find each logarithm to four decimal places.

59. $\log_4(9)$

60. $\log_3(4.78)$

61. $\log_{9.1}(2.3)$

62. $\log_{1.2}(13.7)$

Solve each equation. Round answers to four decimal places.

63. $(1.02)^{4t} = 3$

64. $(1.025)^{12t} = 3$

65. $(1.0001)^{365t} = 3.5$

66. $(1.00012)^{365t} = 2.4$

67. $(1 + r)^3 = 2.3$

68. $\left(1 + \frac{r}{4}\right)^{20} = 3$

69. $2\left(1 + \frac{r}{12}\right)^{360} = 8.4$

70. $5\left(1 + \frac{r}{360}\right)^{720} = 12$

Solve each equation. Round answers to four decimal places.

71. $\log_x(33.4) = 5$

72. $\log_x(12.33) = 2.3$

73. $\log_x(0.546) = -1.3$

74. $\log_x(0.915) = -3.2$

Use the properties of logarithms to determine whether each equation is true or false. Explain your answer.

75. $\log_3(81) = \log_3(9) \cdot \log_3(9)$

76. $\log(81) = \log(9) \cdot \log(9)$

77. $\ln(3^2) = (\ln(3))^2$

78. $\ln\left(\frac{5}{8}\right) = \frac{\ln(5)}{\ln(8)}$

79. $\log_2(8^4) = 12$

80. $\log(8.2 \times 10^{-9}) = -9 + \log(8.2)$

81. $\log(1006) = 3 + \log(6)$

82. $\dfrac{\log_2(16)}{\log_2(4)} = \log_2(16) - \log_2(4)$

83. $\dfrac{\log_2(8)}{\log_2(16)} = \log_2(8) - \log_2(16)$

84. $\ln(e^{(e^x)}) = e^x$

85. $\log_2(25) = 2 \cdot \log(5)$

86. $\log_3\left(\dfrac{5}{7}\right) = \log(5) - \log(7)$

87. $\log_2(7) = \dfrac{\log(7)}{\log(2)}$

88. $\dfrac{\ln(17)}{\ln(3)} = \dfrac{\log(17)}{\log(3)}$

For Exercises 89–94 find the time or rate required for each investment given in the table to grow to the specified amount. The letter W represents an unknown principal.

	Principal	Ending balance	Rate	Compounded	Time
89.	\$800	\$2000	8%	Daily	?
90.	\$10,000	\$1,000,000	7.75%	Annually	?
91.	\$$W$	\$3W	10%	Quarterly	?
92.	\$$W$	\$2W	12%	Monthly	?
93.	\$500	\$2000	?	Annually	25 yr
94.	\$1000	\$2500	?	Monthly	8 yr

95. *Ben's Gift to Boston* Ben Franklin's gift of \$4000 to Boston grew to \$4.5 million in 200 years. At what interest rate compounded annually would this growth occur?

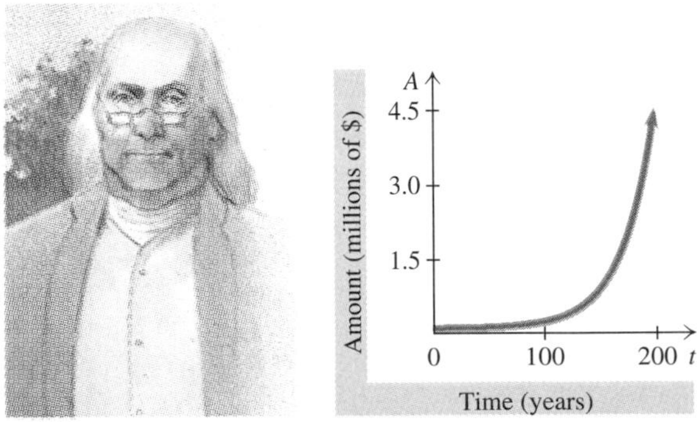

Figure for Exercise 95

96. *Ben's Gift to Philadelphia* Ben Franklin's gift of \$4000 to Philadelphia grew to \$2 million in 200 years. At what interest rate compounded monthly would this growth occur?

97. *Richter Scale* The common logarithm is used to measure the intensity of an earthquake on the Richter scale. The Richter scale rating of an earthquake of intensity I is given by $\log(I) - \log(I_0)$, where I_0 is the intensity of a small "benchmark" earthquake. Write the Richter scale rating as a single logarithm. What is the Richter scale rating of an earthquake for which $I = 1000 \cdot I_0$?

98. *Colombian Earthquake of 1906* At 8.6 on the Richter scale, the Colombian earthquake of January 31, 1906, was one of the strongest earthquakes on record. Use the formula from Exercise 97 to write I as a multiple of I_0 for this earthquake.

99. *Time for Growth* The time in years for a population of size P_0 to grow to size P at the annual growth rate r is given by $t = \ln((P/P_0)^{1/r})$. Use the properties of logarithms to express t in terms of $\ln(P)$ and $\ln(P_0)$.

100. *Formula for pH* The pH of a solution is given by $\text{pH} = \log(1/H^+)$, where H^+ is the hydrogen ion concentration of the solution. Use the properties of logarithms to express the pH in terms of $\log(H^+)$.

101. *Rollover Time* The probability that a \$1 ticket does not win the Louisiana Lottery is $\dfrac{7{,}059{,}051}{7{,}059{,}052}$. The probability p that n independently sold tickets are all losers and the lottery rolls over is given by

$$p = \left(\frac{7{,}059{,}051}{7{,}059{,}052}\right)^n.$$

a. As n increases is p increasing or decreasing?

b. For what number of tickets sold is the probability of a rollover greater than 50%?

102. *Economic Impact* An economist estimates that 75% of the money spent in Chattanooga is respent in four days on the average in Chattanooga. So if P dollars are spent, then the economic impact I in dollars after n respendings is given by

$$I = P(0.75)^n.$$

When $I < 0.02P$, then P no longer has an impact on the economy. If the Telephone Psychics Convention brings

\$1.3 million to the city, then in how many days will that money no longer have an impact on the city?

103. *Marginal Revenue* The revenue in dollars from the sale of x items is given by the function $R(x) = 500 \cdot \log(x + 1)$. The marginal revenue function $MR(x)$ is the difference quotient for $R(x)$ when $h = 1$. Find $MR(x)$ and write it as a single logarithm. What happens to the marginal revenue as x gets larger and larger?

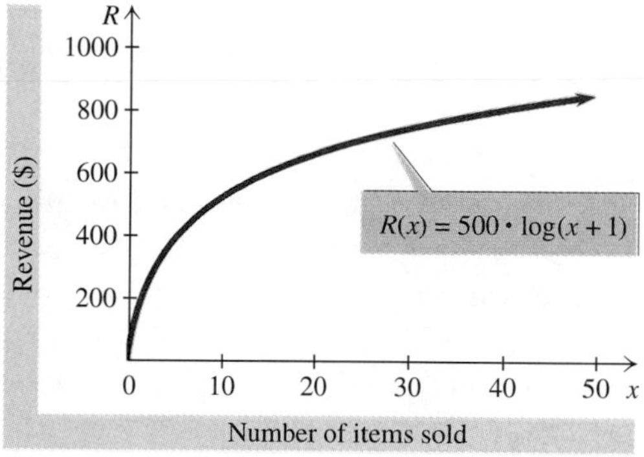

Figure for Exercise 103

104. *Human Memory Model* A class of college algebra students was given a test on college algebra concepts every month for one year after completing a college algebra course. The mean score for the class t months after completing the course can be modeled by the function $m = \ln[e^{80}/(t + 1)^7]$ for $0 \le t \le 12$. Find the mean score of the class for $t = 0$, 5, and 12. Use the properties of logarithms to rewrite the function.

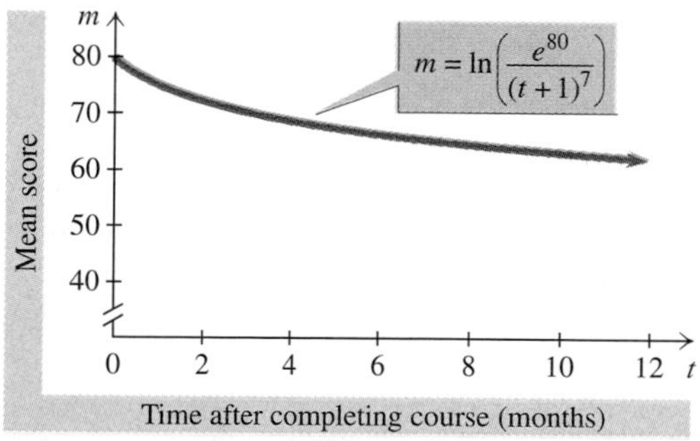

Figure for Exercise 104

105. *Internet Overcrowding* The growth of Internet hosts from 1991 through 1997 is shown in the accompanying graph (Nature, May 1997).

a. Use exponential regression on a graphing calculator to find the best fitting curve of the form $y = a \cdot b^x$.

b. Write your equation in the form $y = a \cdot e^{cx}$.

c. Assuming that the net is growing continuously, what is the annual percentage rate?

d. Use the equation to predict the number of Internet hosts in the year 2002.

e. In what year will the number of Internet hosts reach one billion? When will the number of Internet hosts surpass the population of the Earth?

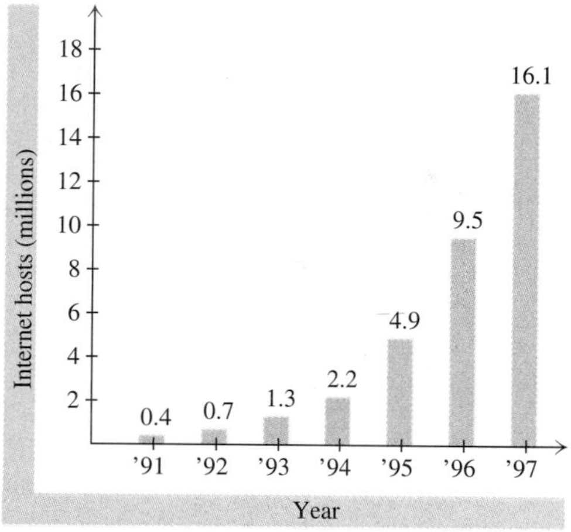

Figure for Exercise 105

For Writing/Discussion

106. *Quotient Rule* Write a proof for the quotient rule for logarithms that is similar to the proof given in the text for the product rule for logarithms.

107. *Cooperative Learning* Work in a small group to write a proof for the power rule for logarithms.

108. *Finding Relationships* Graph each of the following pairs of functions on the same screen of a graphing calculator. (Use the base-change formula to graph with bases other than 10 or e.) Explain how the functions in each pair are related.

a. $y_1 = \log_3(\sqrt{3}x)$, $y_2 = 0.5 + \log_3(x)$

b. $y_1 = \log_2(1/x)$, $y_2 = -\log_2(x)$

c. $y_1 = 3^{x-1}$, $y_2 = \log_3(x) + 1$

d. $y_1 = 3 + 2^{x-4}$, $y_2 = \log_2(x - 3) + 4$

Linking Concepts

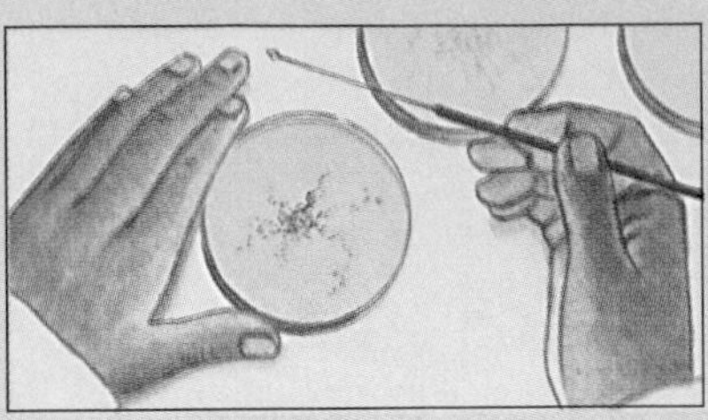

For Individual or Group Explorations

Diversity Index *The United States Geological Survey measures the quality of a water sample by using the diversity index d, given by*

$$d = -[p_1 \cdot \log_2(p_1) + p_2 \cdot \log_2(p_2) + \cdots + p_n \cdot \log_2(p_n)]$$

where n is the number of different taxons (biological classifications) represented in the sample and p_1 through p_n are the percentages of organisms in each of the n taxons. For example, if 10% of the organisms in a water sample are E-coli and 90% are fecal coliform, then

$$d = -[0.1 \cdot \log_2(0.1) + 0.9 \cdot \log_2(0.9)] \approx 0.5.$$

a) Find the value of d when the only organism found in a sample is *E coli* bacteria.

b) Let $n = 3$ in the formula and write the diversity index as a single logarithm.

c) If a water sample is found to contain 20% of its organisms of one type, 30% of another type, and 50% of a third type, then what is the diversity index for the water sample?

d) If the organisms in a water sample are equally distributed among 100 different taxons, then what is the diversity index?

e) If 99% of the organisms in a water sample are from one taxon and the other 1% are equally distributed among 99 other taxons, then what is the diversity index?

f) The diversity index can be found for populations other than organisms in a water sample. Find the diversity index for the dogs in the movie *101 Dalmations* (cartoon version).

g) Identify a population of your choice and different classifications within the population. (For example, the trees on campus can be classified as pine, maple, spruce, or elm.) Gather real data and calculate the diversity index for your population.

4.4 More Equations and Applications

The properties of Section 4.3 combined with the techniques that we have already used in Sections 4.1 and 4.2 allow us to solve several new types of equations involving exponents and logarithms.

Logarithmic Equations

An equation involving a single logarithm can usually be solved by using the definition of logarithm as we did in Section 4.2.

EXAMPLE 1 An equation involving a single logarithm

Solve the equation $\log(x - 3) = 4$.

Solution

Write the equivalent equation using the definition of logarithm:

$$x - 3 = 10^4$$
$$x = 10{,}003$$

Check this number in the original equation. The solution is 10,003.

When more than one logarithm is present, we can use the one-to-one property as in Section 4.2 or use the other properties of logarithms to combine logarithms.

EXAMPLE 2 Equations involving more than one logarithm

Solve each equation.

a) $\log_2(x) + \log_2(x + 2) = \log_2(6x + 1)$ **b)** $\log(x) - \log(x - 1) = 2$

Solution

a) Since the sum of two logarithms is equal to the logarithm of a product, we can rewrite the left-hand side of the equation.

$$\log_2(x) + \log_2(x + 2) = \log_2(6x + 1)$$
$$\log_2(x(x + 2)) = \log_2(6x + 1) \quad \text{Product rule for logarithms}$$
$$x^2 + 2x = 6x + 1 \quad \text{One-to-one property of logarithms}$$
$$x^2 - 4x - 1 = 0 \quad \text{Solve quadratic equation.}$$
$$x = \frac{4 \pm \sqrt{16 - 4(-1)}}{2} = 2 \pm \sqrt{5}$$

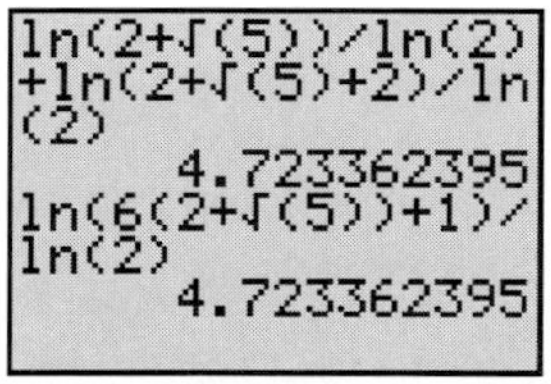

Figure 4.39

Since $2 - \sqrt{5}$ is a negative number, $\log_2(2 - \sqrt{5})$ is undefined and $2 - \sqrt{5}$ is not a solution. The only solution to the equation is $2 + \sqrt{5}$. Check this solution by using a calculator and the base-change formula. The check is shown with a graphing calculator in Fig. 4.39. □

b) Since the difference of two logarithms is equal to the logarithm of a quotient, we can rewrite the left-hand side of the equation:

$$\log(x) - \log(x - 1) = 2$$

$$\log\left(\frac{x}{x-1}\right) = 2 \qquad \text{Quotient rule for logarithms}$$

$$\frac{x}{x-1} = 10^2 \qquad \text{Definition of logarithm}$$

$$x = 100x - 100 \qquad \text{Solve for } x.$$

$$-99x = -100$$

$$x = \frac{100}{99}$$

Check 100/99 in the original equation as shown in Fig. 4.40.

Figure 4.40

All solutions to logarithmic and exponential equations should be checked in the original equation, because extraneous roots can occur, as in Example 2(a). Use your calculator to check every solution and you will increase your proficiency with your calculator.

Exponential Equations

An exponential equation with a single exponential expression can usually be solved by using the definition of logarithm, as in Section 4.2.

EXAMPLE 3 Equations involving a single exponential expression

Solve the equation $(1.02)^{4t-1} = 5$.

Solution

Write an equivalent equation, using the definition of logarithm.

$$4t - 1 = \log_{1.02}(5)$$

$$t = \frac{1 + \log_{1.02}(5)}{4} \qquad \text{The exact solution}$$

$$= \frac{1 + \dfrac{\ln(5)}{\ln(1.02)}}{4} \qquad \text{Base-change formula}$$

$$\approx 20.5685$$

The approximate solution is 20.5685.

Check with a graphing calculator as shown in Fig. 4.41.

Figure 4.41

We could have solved Example 3 by taking the common or natural logarithm of each side and using the power rule for logarithms. If an equation has an exponential expression on each side, as in Example 4, then we must take the common or natural logarithm of each side to solve it.

EXAMPLE 4 **Equations involving two exponential expressions**

Find the exact and approximate solutions to $3^{2x-1} = 5^x$.

Solution

$$\begin{aligned} \ln(3^{2x-1}) &= \ln(5^x) && \text{Take the natural logarithm of each side.} \\ (2x - 1)\ln(3) &= x \cdot \ln(5) && \text{Power rule for logarithms} \\ 2x \cdot \ln(3) - \ln(3) &= x \cdot \ln(5) && \text{Distributive property} \\ 2x \cdot \ln(3) - x \cdot \ln(5) &= \ln(3) \\ x[2 \cdot \ln(3) - \ln(5)] &= \ln(3) \\ x &= \frac{\ln(3)}{2 \cdot \ln(3) - \ln(5)} && \text{Exact solution} \\ &\approx 1.8691 \end{aligned}$$

Figure 4.42

Had we used common logarithms, similar steps would give

$$\begin{aligned} x &= \frac{\log(3)}{2 \cdot \log(3) - \log(5)} \\ &\approx 1.8691 \end{aligned}$$

Check the solution in the original equation as shown in Fig. 4.42.

The technique of Example 4 can be used on any equation of the form $a^M = b^N$. Take the natural logarithm of each side and apply the power rule, to get an equation of the form $M \cdot \ln(a) = N \cdot \ln(b)$. This last equation usually has no exponents and is relatively easy to solve.

Strategy for Solving Equations

We solved equations involving exponential and logarithmic functions in Sections 4.1 through 4.4. There is no formula that will solve every exponential or logarithmic equation, but the following box summarizes the procedures we have been using.

STRATEGY **Solving Exponential and Logarithmic Equations**

1. If the equation involves a single logarithm or a single exponential expression, then use the definition of logarithm: $y = \log_a(x)$ if and only if $a^y = x$.
2. Use the one-to-one properties when applicable:
   a) if $a^M = a^N$, then $M = N$.
   b) if $\log_a(M) = \log_a(N)$, then $M = N$.
3. If an equation has several logarithms with the same base, then combine them using the product and quotient rules:
   a) $\log_a(M) + \log_a(N) = \log_a(MN)$
   b) $\log_a(M) - \log_a(N) = \log_a(M/N)$
4. If an equation has only exponential expressions with different bases on each side, then take the natural or common logarithm of each side and use the power rule: $a^M = b^N$ is equivalent to $\ln(a^M) = \ln(b^N)$ or $M \cdot \ln(a) = N \cdot \ln(b)$.

Radioactive Dating

In Section 4.1 we stated that the amount A of a radioactive substance remaining after t years is given by

$$A = A_0 e^{rt},$$

where A_0 is the initial amount present and r is the annual rate of decay for that particular substance. A standard measurement of the speed at which a radioactive substance decays is its **half-life**. The half-life of a radioactive substance is the amount of time that it takes for one-half of the substance to decay. Of course, when one-half has decayed, one-half remains.

Now that we have studied logarithms we can use the formula for radioactive decay to determine the age of an ancient object that contains a radioactive substance. One such substance is potassium-40, which is found in rocks. Once the rock is formed, the potassium-40 begins to decay. The amount of time that has passed since the formation of the rock can be determined by measuring the amount of potassium-40 that has decayed into argon-40. Dating rocks using potassium-40 is known as **potassium-argon dating**.

EXAMPLE 5 Finding the age of *Deinonychus* ("terrible claw")

Our chapter opening case described the 1964 find of *Deinonychus* (*National Geographic*, August 1978). Since dinosaur bones are too old to contain enough organic material for radiocarbon dating, paleontologists often estimate the age of bones by dating volcanic debris in the surrounding rock. The age of *Deinonychus* was determined from the age of surrounding rocks by using potassium-argon dating. The half-life of potassium-40 is 1.31 billion years. If 94.5% of the original amount of potassium-40 is still present in the rock, then how old are the bones of *Deinonychus*?

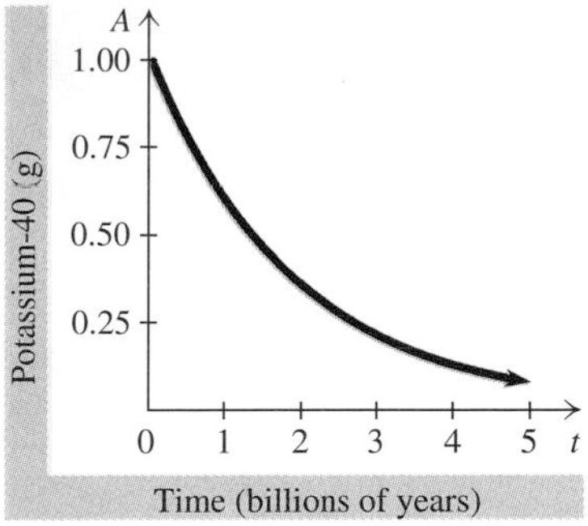

Solution

The half-life is the amount of time that it takes for 1 gram to decay to 0.5 gram. Use $A_0 = 1$, $A = 0.5$, and $t = 1.31 \times 10^9$ in the formula $A = A_0e^{rt}$ to find r.

$$0.5 = 1 \cdot e^{(1.31\times10^9)(r)}$$

$$(1.31 \times 10^9)(r) = \ln(0.5) \quad \textbf{Definition of logarithm}$$

$$r = \frac{\ln(0.5)}{1.31 \times 10^9}$$

$$\approx -5.29 \times 10^{-10}$$

Now we can find the amount of time that it takes for 1 gram to decay to 0.945 gram. Use $r \approx -5.29 \times 10^{-10}$, $A_0 = 1$, and $A = 0.945$ in the formula.

$$0.945 = 1 \cdot e^{(-5.29\times10^{-10})(t)}$$

$$(-5.29 \times 10^{-10})(t) = \ln(0.945) \quad \textbf{Definition of logarithm}$$

$$t = \frac{\ln(0.945)}{-5.29 \times 10^{-10}}$$

$$\approx 107 \text{ million years}$$

The dinosaur *Deinonychus* lived about 107 million years ago.

Newton's Law of Cooling

Newton's law of cooling states that when a warm object is placed in colder surroundings or a cold object is placed in warmer surroundings, then the difference between the two temperatures decreases in an exponential manner. If D_0 is the initial difference in temperature, then the difference D at time t is given by the formula

$$D = D_0e^{kt},$$

where k is a constant that depends on the object and the surroundings. In the next example we use Newton's law of cooling to answer a question that you might have asked yourself as the appetizers were running low.

EXAMPLE 6 Using Newton's law of cooling

A turkey with a temperature of 40°F is moved to a 350° oven. After 4 hours the internal temperature of the turkey is 170°F. If the turkey is done when its temperature reaches 185°, then how much longer must it cook?

Solution

The initial difference of 310° has dropped to a difference of 180° after 4 hours. See Fig. 4.43. With this information we can find k:

$$\begin{aligned} 180 &= 310e^{4k} \\ e^{4k} &= \frac{180}{310} && \text{Isolate } k \text{ by dividing by 310.} \\ 4k &= \ln(18/31) && \text{Definition of logarithm} \\ k &= \frac{\ln(18/31)}{4} \approx -0.1359 \end{aligned}$$

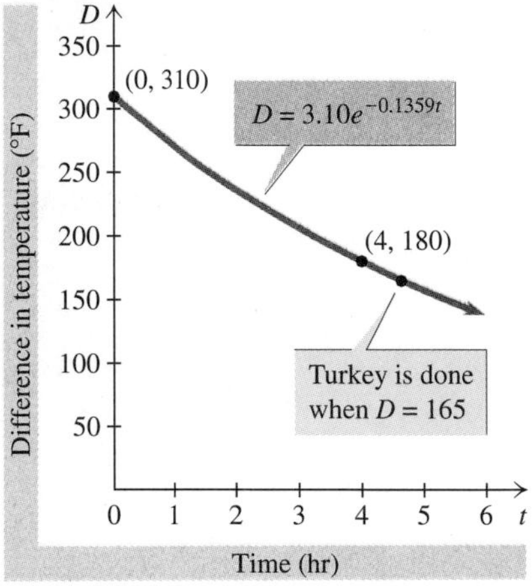

Figure 4.43

The turkey is done when the difference in temperature between the turkey and the oven is 165° (the oven temperature 350° minus the turkey temperature 185°). Now find the time for which the difference will be 165°:

$$\begin{aligned} 165 &= 310e^{-0.1359t} \\ e^{-0.1359t} &= \frac{165}{310} \\ -0.1359t &= \ln(165/310) && \text{Definition of logarithm} \\ t &= \frac{\ln(165/310)}{-0.1359} \approx 4.6404 \end{aligned}$$

The difference in temperature between the turkey and the oven will be 165° when the turkey has cooked 4.6404 hours. So the turkey must cook approximately 0.6404 hour (38.4 minutes) longer.

FOR THOUGHT True or False? Explain.

1. The equation $3(1.02)^x = 21$ is equivalent to $x = \log_{1.02}(7)$.
2. If $x - x \cdot \ln(3) = 8$, then $x = \dfrac{8}{1 - \ln(3)}$.
3. The solution to $\ln(x) - \ln(x - 1) = 6$ is $1 - \sqrt{6}$.
4. If $2^{x-3} = 3^{2x+1}$, then $x - 3 = \log_2(3^{2x+1})$.
5. The exact solution to $3^x = 17$ is 2.5789.
6. The equation $\log(x) + \log(x - 3) = 1$ is equivalent to $\log(x^2 - 3x) = 1$.
7. The equation $4^x = 2^{x-1}$ is equivalent to $2x = x - 1$.
8. The equation $(1.09)^x = 2.3$ is equivalent to $x \cdot \ln(1.09) = \ln(2.3)$.
9. $\ln(2) \cdot \log(7) = \log(2) \cdot \ln(7)$
10. $\log(e) \cdot \ln(10) = 1$

4.4 EXERCISES

Tape 7 Disk

Solve each equation.

1. $\log(x + 20) = 2$
2. $\log(x^2 - 15) = 1$
3. $2 = \log_x(9)$
4. $-2 = \log_x(4)$
5. $\log_2(x + 2) + \log_2(x - 2) = 5$
6. $\log(x + 1) - \log(x) = 3$
7. $\log(5) = 2 - \log(x)$
8. $\log(4) = 1 + \log(x - 1)$
9. $\ln(x) + \ln(x + 2) = \ln(8)$
10. $\log_3(x) = \log_3(2) - \log_3(x - 2)$
11. $\log(4) + \log(x) = \log(5) - \log(x)$
12. $\ln(x) - \ln(x + 1) = \ln(x + 3) - \ln(x + 5)$
13. $\log_2(x) - \log_2(3x - 1) = 0$
14. $\log_3(x) + \log_3(1/x) = 0$
15. $x \cdot \ln(3) = 2 - x$
16. $x \cdot \log(5) + x \cdot \log(7) = \log(9)$

Solve each equation. If the exact solution is an irrational number, then find an approximate solution rounded to four decimal places.

17. $2^{x-1} = 7$
18. $5^{3x} = 29$
19. $(1.09)^{4x} = 3.4$
20. $(1.04)^{2x} = 2.5$
21. $3^{-x} = 30$
22. $10^{-x+3} = 102$
23. $9 = e^{-3x^2}$
24. $25 = 10^{-2x}$
25. $6^x = 3^{x+1}$
26. $2^x = 3^{x-1}$
27. $e^{x+1} = 10^x$
28. $e^x = 2^{x+1}$
29. $2^{x-1} = 4^{3x}$
30. $3^{3x-4} = 9^x$
31. $6^{x+1} = 12^x$
32. $2^x \cdot 2^{x+1} = 4^{x^2+x}$
33. $e^{-\ln(w)} = 3$
34. $10^{2 \cdot \log(y)} = 4$
35. $(\log(z))^2 = \log(z^2)$
36. $\ln(e^x) - \ln(e^6) = \ln(e^2)$
37. $4(1.02)^x = 3(1.03)^x$
38. $500(1.06)^x = 400(1.02)^{4x}$
39. $e^{3 \cdot \ln(x^2) - 2 \cdot \ln(x)} = \ln(e^{16})$
40. $\sqrt{\log(x) - 3} = \log(x) - 3$
41. $\left(\frac{1}{2}\right)^{2x-1} = \left(\frac{1}{4}\right)^{3x+2}$
42. $\left(\frac{2}{3}\right)^{x+1} = \left(\frac{9}{4}\right)^{x+2}$

Find the approximate solution to each equation by graphing an appropriate function on a graphing calculator and locating the x-intercept. Note that these equations cannot be solved by the techniques that we have learned in this chapter.

43. $2^x = 3^{x-1} + 5^{-x}$
44. $2^x = \log(x + 4)$
45. $\ln(x + 51) = \log(-48 - x)$
46. $2^x = 5 - 3^{x+1}$
47. $x^2 = 2^x$
48. $x^3 = e^x$

Solve each problem.

49. *Dating a Bone* A piece of bone from an organism is found to contain 10% of the carbon-14 that it contained when the organism was living. If the half-life of carbon-14 is 5730 years, then how long ago was the organism alive?

50. *Old Clothes* If only 15% of the carbon-14 in a remnant of cloth has decayed, then how old is the cloth?

51. *Dating a Tree* How long does it take for 12 g of carbon-14 in a tree trunk to be reduced to 10 g of carbon-14 by radioactive decay?

52. *Carbon-14 Dating* How long does it take for 2.4 g of carbon-14 to be reduced to 1.3 g of carbon-14 by radioactive decay?

53. *Radioactive Waste* If 25 g of radioactive waste reduces to 20 g of radioactive waste after 8000 years, then what is the half-life for this radioactive element?

54. *Finding the Half-Life* If 80% of a radioactive element remains radioactive after 250 million years, then what percent remains radioactive after 600 million years? What is the half-life of this element?

55. *Lorazepam* The drug lorazepam, used to relieve anxiety and nervousness, has a half-life of 14 hours. Its chemical structure is shown in the accompanying figure. If a doctor prescribes one 2.5 milligram tablet every 24 hours, then what percentage of the last dosage remains in the patient's body when the next dosage is taken?

Lorazepam

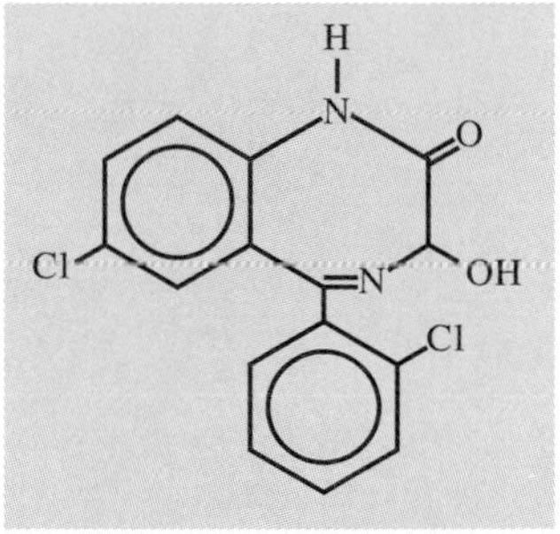

Figure for Exercise 55

56. *Drug Build-Up* The level of a prescription drug in the human body over time can be found using the formula

$$L = \frac{D}{1 - (0.5)^{n/h}},$$

where D is the amount taken every n hours and h is the drug's half-life in hours.

a. If 2.5 milligrams of lorazepam with a half-life of 14 hours is taken every 24 hours, then to what level does the drug build up over time?

b. If a doctor wants the level of lorazepam to build up to a level of 5.58 milligrams in a patient taking 2.5 milligram doses, then how often should the doses be taken?

c. What is the difference between taking 2.5 milligrams of lorazepam every 12 hours and taking 5 milligrams every 24 hours?

57. *Dead Sea Scrolls* Willard Libby, a nuclear chemist from the University of Chicago, developed radiocarbon dating in the 1940s. This dating method, effective on specimens up to about 40,000 years old, works best on objects like shells, charred bones, and plants that contain organic matter (carbon). Libby's first great success came in 1951 when he dated the Dead Sea Scrolls. Carbon-14 has a half-life of 5730 years. If Libby found 79.3% of the original carbon-14 still present, then in about what year were the scrolls made?

58. *Leakeys Date Zinjanthropus* In 1959, archaeologists Louis and Mary Leakey were exploring Olduvai Gorge, Tanzania, when they uncovered the remains of *Zinjanthropus*, an early hominid with traits of both ape and man. Dating of the volcanic rock revealed that 91.2% of the original potassium had not decayed into argon. What age was assigned to the rock and the bones of *Zinjanthropus*? The radioactive decay of potassium-40 to argon-40 occurs with a half-life of 1.31 billion years.

Figure for Exercise 58

59. *Cooking a Roast* James knows that to get well-done beef, it should be brought to a temperature of 170°F. He placed a sirloin tip roast with a temperature of 35°F in an oven with a temperature of 325°, and after 3 hr the temperature of the roast was 140°. How much longer must the roast be in the oven to get it well done? If the oven temperature is set at 170°, how long will it take to get the roast well done?

60. *Room Temperature* Marlene brought a can of polyurethane varnish that was stored at 40°F into her shop, where the temperature was 74°. After 2 hr the temperature of the varnish was 58°. If the varnish must be 68° for best results, then how much longer must Marlene wait until she uses the varnish?

61. *Time of Death* A detective discovered a body in a vacant lot at 7 A.M. and found that the body temperature was 80°F. The county coroner examined the body at 8 A.M. and found that the body temperature was 72°. Assuming that the body temperature was 98° when the person died and that the air temperature was a constant 40° all night, what was the approximate time of death?

62. *Cooling Hot Steel* A blacksmith immersed a piece of steel at 600°F into a large bucket of water with a temperature of 65°. After 1 min the temperature of the steel was 200°. How much longer must the steel remain immersed to bring its temperature down to 100°?

63. *Poverty Level* In a certain country the number of people below the poverty level, B, is currently 20 million and growing according to the formula $B = 20e^{0.07t}$, where B is in millions of people and t is the time in years from the present. The number of people above the poverty level, A, is 30 million and decreasing according to the formula $A = 30e^{-0.05t}$, where A is in millions of people and t is the time in years from the present. In how many years will the numbers of people above and below the poverty level be equal?

64. *Equality of Investments* Tasha invested \$400 at 5% compounded annually. At the same time, Eunice invested \$500 at 4% compounded annually. In how many years will their investment be equal in value?

65. *Equality of Investments* Fiona invested \$1000 at 6% compounded continuously. At the same time, Maria invested \$1100 at 6% compounded daily. How long will it take (to the nearest day) for their investments to be equal in value?

66. *Depreciation and Inflation* Boris won a \$35,000 luxury car on Wheel of Fortune. He plans to keep it until he can trade it evenly for a new compact car that currently costs

\$10,000. If the value of the luxury car decreases by 8% each year and the cost of the compact car increases by 5% each year, then in how many years will he be able to make the trade?

67. *Population of Rabbits* The population of rabbits in a certain national forest appears to be growing according to the formula $P = 12{,}300 + 1000 \cdot \ln(t + 1)$, where t is the time in years from the present. How many rabbits are now in the forest? In how many years will there be 15,000 rabbits?

68. *Population of Foxes* The population of foxes in the forest of Exercise 67 appears to be growing according to the formula $P = 400 + 50 \cdot \ln(90t + 1)$. When the population of foxes is equal to 5% of the population of rabbits, the system is considered to be out of ecological balance. In how many years will the system be out of balance?

69. *Visual Magnitude of a Star* If all stars were at the same distance, it would be a simple matter to compare their brightness. However, the brightness that we see, the apparent visual magnitude m, depends on a star's intrinsic brightness, or absolute visual magnitude M_V, and the distance d from the observer in parsecs (1 parsec = 3.262 light years), according to the formula $m = M_V - 5 + 5 \cdot \log(d)$. The values of M_V range from -8 for the intrinsically brightest stars to $+15$ for the intrinsically faintest stars. The nearest star to the sun, Alpha Centauri, has an apparent visual magnitude of 0 and an absolute visual magnitude of 4.39. Find the distance d in parsecs to Alpha Centauri.

70. *Visual Magnitude of Deneb* The star Deneb is 490 parsecs away and has an apparent visual magnitude of 1.26, which means that it is harder to see than Alpha Centauri. Use the formula from Exercise 69 to find the absolute visual magnitude of Deneb. Is Deneb intrinsically very bright or very faint? If Deneb and Alpha Centauri were both 490 parsecs away, then which would appear brighter?

71. *Price Per Gigabyte* According to Moore's Law, the performance of electronic parts grows exponentially while the cost of those parts declines exponentially (Forbes, March 24, 1997). The accompanying figure shows the decline in price P for a gigabyte of hard drive storage for the years 1982 through 2000. The relationship looks linear because the price axis has a log scale.

a. Find a formula for P in terms of x, where x is the number of years since 1982. See Exercise 115 of Section 4.2.

b. Find P in 1991.

c. Find the year in which the price will be \$10.

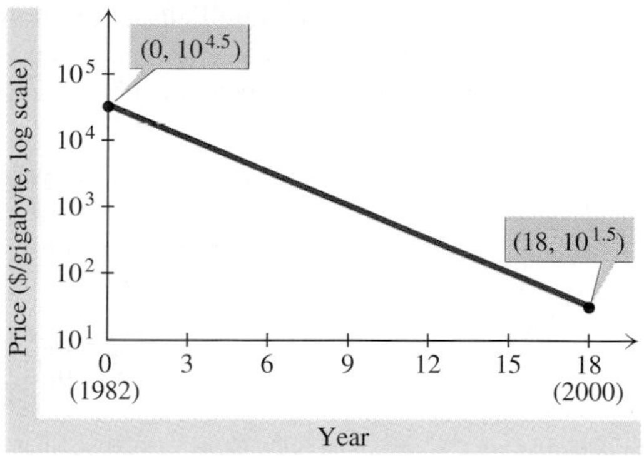

Figure for Exercise 71

72. *Hard Drive Capacity* The accompanying figure shows how hard drive capacity C has increased for the years 1982 through 2000. The relationship looks linear because the capacity axis has a log scale.

a. Find a formula for C in terms of x, where x is the number of years since 1982. See the previous exercise.

b. Find C in 1994.

c. Find the year in which the capacity will be 100 gigabytes.

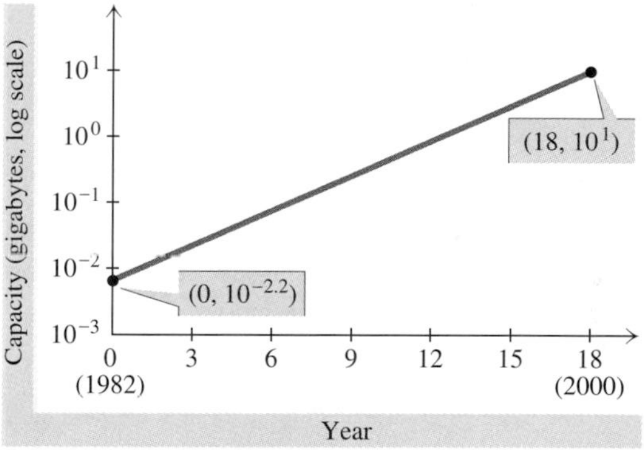

Figure for Exercise 72

73. *Noise Pollution* The level of a sound in decibels (db) is determined by the formula

$$\text{sound level} = 10 \cdot \log(I \times 10^{12}) \text{ db},$$

where I is the intensity of the sound in watts per square meter. To combat noise pollution, a city has an ordinance prohibiting sounds above 90 db on a city street. What value of I gives a sound of 90 db?

74. *Doubling the Sound Level* A small stereo amplifier produces a sound of 50 db at a distance of 20 ft from the speakers. Use the formula from Exercise 73 to find the intensity of the sound at this point in the room. If the intensity just found is doubled, what happens to the sound level? What must the intensity be to double the sound level to 100 db?

75. *Present Value of a CD* What amount (present value) must be deposited today in a certificate of deposit so that the investment will grow to \$20,000 in 18 years at 6% compounded continuously?

76. *Present Value of a Bond* A \$50 U.S. Savings Bond paying 6.22% compounded monthly matures in 11 years 2 months. What is the present value of the bond?

77. *Thermistor Resistance* A Thermistor is a resistor whose resistance varies with the temperature as shown in the figure. The relationship between temperature T in °C and resistance R in Ohms for a BetaTHERM Thermistor model 100K6A1 is given by the Steinhart-Hart equation

$$T = \frac{1}{A + B \cdot \ln(R) + C \cdot [\ln(R)]^3} - 273.15$$

where

$$A = 8.27153 \times 10^{-4}$$
$$B = 2.08796 \times 10^{-4}$$
$$C = 8.060985 \times 10^{-8}.$$

At what temperature is the resistance 1×10^5 Ohms? Use a graphing calculator to find the resistance when $T = 35°C$.

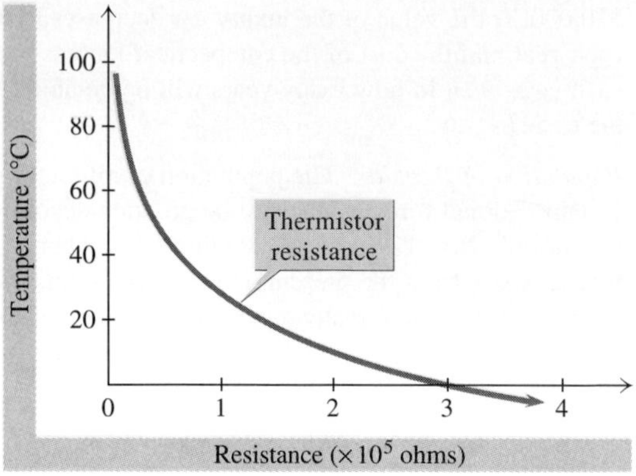

Figure for Exercise 77

• CALCULUS •

78. *Infinite Series for e^x* The following formula from calculus is used to compute values of e^x:

$$e^x = 1 + x + \frac{x^2}{2!} + \frac{x^3}{3!} + \frac{x^4}{4!} + \cdots + \frac{x^n}{n!} + \cdots,$$

where $n! = 1 \cdot 2 \cdot 3 \cdot \cdots \cdot n$ for any positive integer n. The notation $n!$ is read "n factorial." For example, $3! = 1 \cdot 2 \cdot 3 = 6$. In calculating e^x, the more terms that we use from the formula, the closer we get to the true value of e^x. Use the first five terms of the formula to estimate the value of $e^{0.1}$ and compare your result to the value of $e^{0.1}$ obtained using the e^x-key on your calculator.

79. *Infinite Series for Logarithms* The following formula from calculus can be used to compute values of natural logarithms:

$$\ln(1 + x) = x - \frac{x^2}{2} + \frac{x^3}{3} - \frac{x^4}{4} + \cdots$$

where $-1 < x < 1$. The more terms that we use from the formula, the closer we get to the true value of $\ln(1 + x)$. Find $\ln(1.4)$ by using the first five terms of the series and compare your result to the calculator value for $\ln(1.4)$.

LINKING CONCEPTS

For Individual or Group Explorations

• CALCULUS •

Logistic Growth *When a virus injects a finite population of size P in which no one is immune, the virus spreads slowly at first, then more rapidly as more people are infected, and finally slows down when nearly everyone has been infected. This situation is modeled by a* ***logistic curve*** *of the form*

$$n = \frac{P}{1 + (P - 1) \cdot e^{-ct}}$$

where n is the number of people who have caught the virus on or before day t and c is a positive constant.

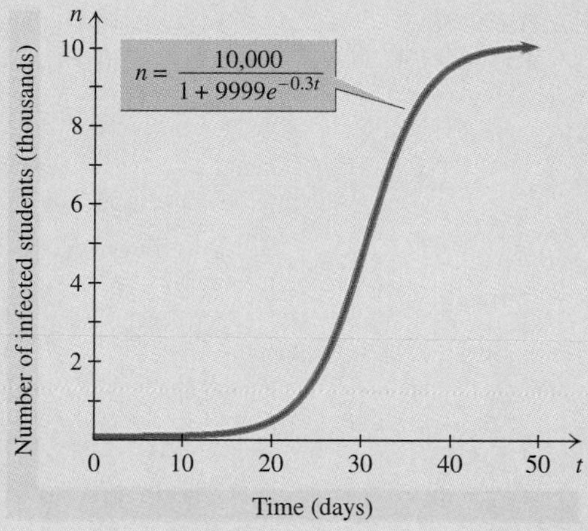

a) According to the model, how many people have caught the virus at time $t = 0$?

b) Now consider what happens when one student carrying a flu virus returns from spring break to a university of 10,000 students. For $c = 0.1$, 0.2, and so on through $c = 0.9$, graph the logistic curve

$$y_1 = 10000/(1 + 9999e\text{^}(-cx))$$

and the daily number of new cases

$$y_2 = y_1(x) - y_1(x - 1).$$

For each value of c use the graph of y_2 to find the day on which the flu is spreading most rapidly.

c) The Health Center estimated that the greatest number of new cases of the flu occurred on the 19th day after the students returned from spring break. What value of c should be used to model this situation? How many new cases occurred on the 19th day?

d) Algebraically find the day on which the number of infected students reached 9000.

e) The Health Center has a team of doctors from Atlanta arriving on the 30th day to help with this three-day flu epidemic. What do you think of this plan?

HIGHLIGHTS

• Section 4.1
Exponential Functions

1. If $a > 0$ and $a \neq 1$, then $f(x) = a^x$ is an exponential function. Its domain is all real numbers.
2. Exponential functions are one-to-one: if $a^{x_1} = a^{x_2}$ then $x_1 = x_2$.
3. If $a > 1$, then $f(x) = a^x$ is increasing, and if $0 < a < 1$, then $f(x) = a^x$ is decreasing.
4. The x-axis is a horizontal asymptote for the graph of $f(x) = a^x$.
5. P dollars invested at annual rate r compounded n times per year amounts to $A = P\left(1 + \frac{r}{n}\right)^{nt}$ in t years.
6. P dollars invested at an annual rate r compounded continuously amounts to $A = Pe^{rt}$ in t years, where $e \approx 2.718$.

• Section 4.2
Logarithmic Functions

1. The inverse of $f(x) = a^x$ is $f^{-1}(x) = \log_a(x)$, where $y = \log_a(x)$ if and only if $a^y = x$.
2. The common logarithm function (base 10) is $y = \log(x)$, and the natural logarithm function (base e) is $y = \ln(x)$.
3. The domain of $f(x) = \log_a(x)$ is $(0, \infty)$, and the y-axis is a vertical asymptote for its graph.
4. If $a > 1$, $f(x) = \log_a(x)$ is an increasing function, and if $0 < a < 1$, it is a decreasing function.
5. Logarithmic functions are one-to-one: if $\log_a(x_1) = \log_a(x_2)$, then $x_1 = x_2$.

• Section 4.3
Properties of Logarithms

1. The logarithm of a product is the sum of two logarithms: $\log_a(MN) = \log_a(M) + \log_a(N)$.
2. The logarithm of a quotient is the difference of two logarithms: $\log_a(M/N) = \log_a(M) - \log_a(N)$.
3. The logarithm of a power of a number is the power times the logarithm of the number: $\log_a(M^N) = N \cdot \log_a(M)$.
4. Because exponential and logarithmic functions are inverses, $\log_a(a^x) = x$ and $a^{\log_a(x)} = x$.
5. The base-change formula $\log_a(M) = \dfrac{\log_b(M)}{\log_b(a)}$ is used to evaluate logarithms with a calculator.

• Section 4.4
More Equations and Applications

1. The definition of logarithm, one-to-one properties, product rule, quotient rule, and power rule are all used in solving equations involving exponential functions and logarithmic functions.

2. Exponential and logarithmic functions have a wide variety of applications, including population growth, compound interest, radioactive dating, and Newton's law of cooling.

CHAPTER 4 REVIEW EXERCISES

Simplify each expression.

1. 2^6 **2.** $\ln(e^2)$ **3.** $\log_2(64)$

4. $3 + 2 \cdot \log(10)$ **5.** $\log_9(1)$ **6.** $5^{\log_5(99)}$

7. $\log_2(2^{17})$ **8.** $\log_2(\log_2(16))$

Let $f(x) = 2^x$, $g(x) = 10^x$, and $h(x) = \log_2(x)$. Simplify each expression.

9. $f(5)$ **10.** $g(-1)$ **11.** $\log(g(3))$ **12.** $g(\log(5))$

13. $(h \circ f)(9)$ **14.** $(f \circ h)(7)$ **15.** $g^{-1}(1000)$

16. $g^{-1}(1)$ **17.** $h(1/8)$ **18.** $f(1/2)$

19. $f^{-1}(8)$ **20.** $(f^{-1} \circ f)(13)$

Rewrite each expression as a single logarithm.

21. $\log(x - 3) + \log(x)$ **22.** $(1/2)\ln(x) - 2 \cdot \ln(y)$

23. $2 \cdot \ln(x) + \ln(y) + \ln(3)$

24. $3 \cdot \log_2(x) - 2 \cdot \log_2(y) + \log_2(z)$

Rewrite each expression as a sum or difference of multiples of logarithms.

25. $\log(3x^4)$ **26.** $\ln\left(\frac{x^5}{y^3}\right)$

27. $\log_3\left(\frac{5\sqrt{x}}{y^4}\right)$ **28.** $\log_2(\sqrt{xy^3})$

Rewrite each expression in terms of ln(2) and ln(5).

29. $\ln(10)$ **30.** $\ln(0.4)$ **31.** $\ln(50)$ **32.** $\ln(\sqrt{20})$

Find the exact solution to each equation.

33. $\log(x) = 10$ **34.** $\log_3(x + 1) = -1$

35. $\log_x(81) = 4$ **36.** $\log_x(1) = 0$

37. $\log_{1/3}(27) = x + 2$ **38.** $\log_{1/2}(4) = x - 1$

39. $3^{x+2} = \frac{1}{9}$ **40.** $2^{x-1} = \frac{1}{4}$

41. $e^{x-2} = 9$ **42.** $\frac{1}{2^{1-x}} = 3$

43. $4^{x+3} = \frac{1}{2^x}$ **44.** $3^{2x-1} \cdot 9^x = 1$

45. $\log(x) + \log(2x) = 5$

46. $\log(x + 90) - \log(x) = 1$

47. $\log_2(x) + \log_2(x - 4) = \log_2(x + 24)$

48. $\log_5(x + 18) + \log_5(x - 6) = 2 \cdot \log_5(x)$

49. $2 \cdot \ln(x + 2) = 3 \cdot \ln(4)$

50. $x \cdot \log_2(12) = x \cdot \log_2(3) + 1$

51. $x \cdot \log(4) = 6 - x \cdot \log(25)$

52. $\log(\log(x)) = 1$

Find the missing coordinate so that each ordered pair satisfies the given equation.

53. $y = \left(\frac{1}{3}\right)^x$: $(-1, \quad)$, $(\quad, 27)$, $(-1/2, \quad)$, $(\quad, 1)$

54. $y = \log_9(x - 1)$: $(2, \quad)$, $(4, \quad)$, $(\quad, 3/2)$, $(\quad, -1)$

Match each equation to one of the graphs (a)–(h).

55. $y = 2^x$ **56.** $y = 2^{-x}$

57. $y = \log_2(x)$ **58.** $y = \log_{1/2}(x)$

59. $y = 2^{x+2}$ **60.** $y = \log_2(x + 2)$

61. $y = 2^x + 2$ **62.** $y = 2 + \log_2(x)$

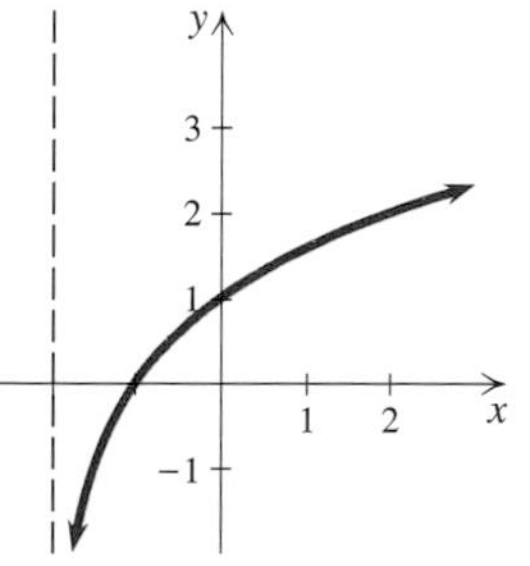

(a)

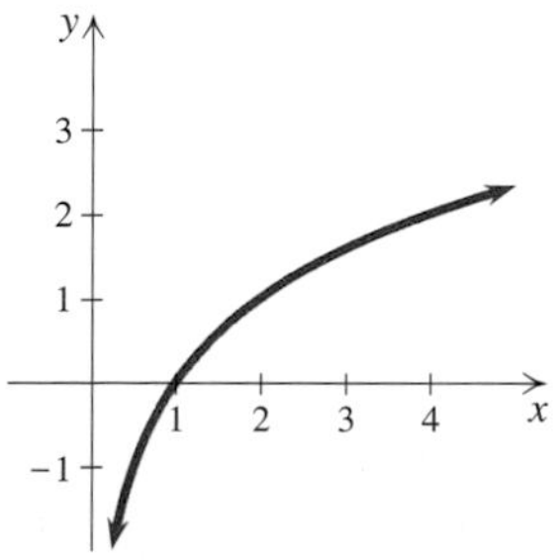

(b)

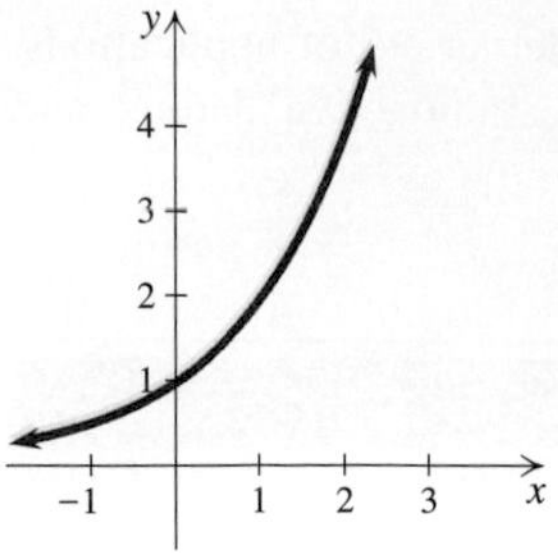

(c)

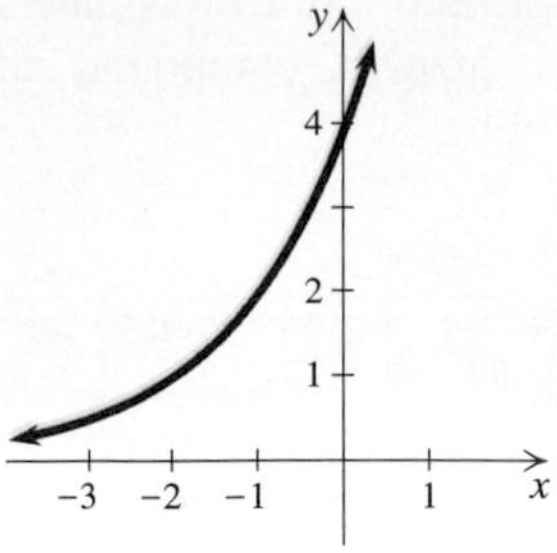

(d)

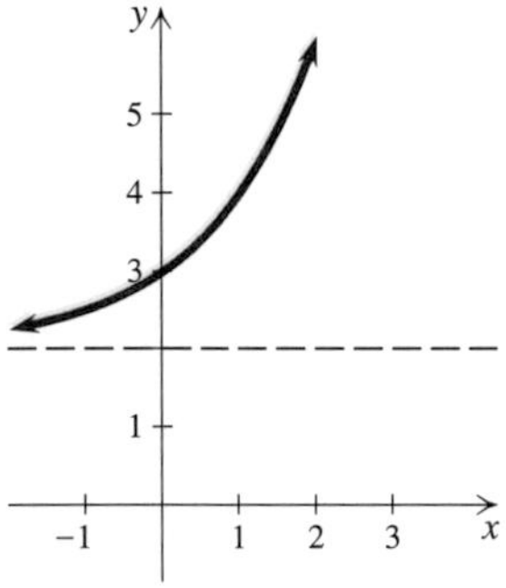

(e)

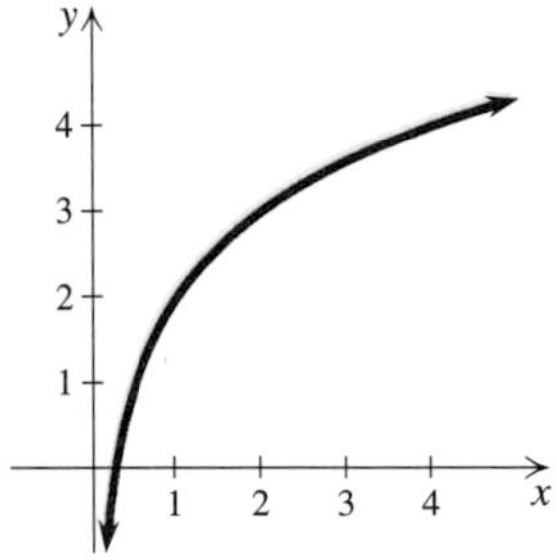

(f)

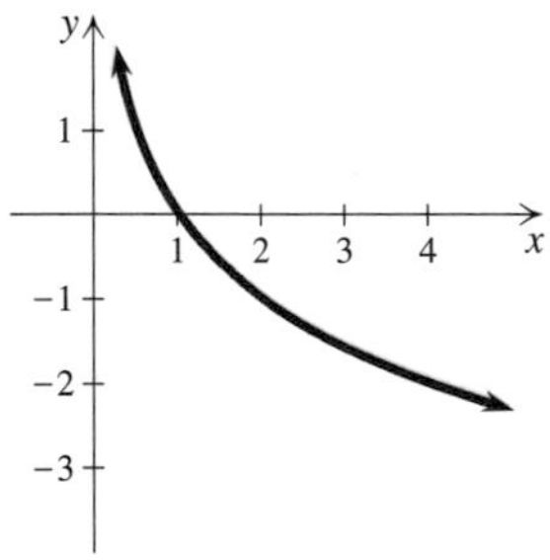

(g)

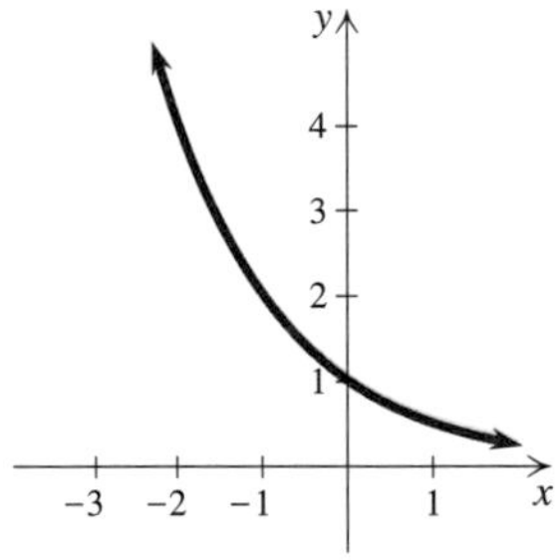

(h)

Sketch the graph of each function. State the domain, the range, and whether the function is increasing or decreasing. Identify any asymptotes.

63. $f(x) = 5^x$ **64.** $f(x) = e^x$

65. $f(x) = 10^{-x}$ **66.** $f(x) = (1/2)^x$

67. $y = \log_3(x)$ **68.** $y = \log_5(x)$

69. $y = 1 + \ln(x + 3)$ **70.** $y = 3 - \log_2(x)$

71. $f(x) = 1 + 2^{x-1}$ **72.** $f(x) = 3 - 2^{x+1}$

73. $y = \log_3(-x + 2)$ **74.** $y = 1 + \log_3(x + 2)$

For each function f, find f^{-1}.

75. $f(x) = 7^x$ **76.** $f(x) = \log_8(x)$

77. $f(x) = \log_5(x)$ **78.** $f(x) = 3^x$

Use a calculator to find an approximate solution to each equation. Round answers to four decimal places.

79. $3^x = 10$ **80.** $4^{2x} = 12$

81. $\log_3(x) = 1.876$ **82.** $\log_5(x + 2) = 2.7$

83. $5^x = 8^{x+1}$ **84.** $3^x = e^{x+1}$

Solve each problem.

85. *Comparing pH* If the hydrogen ion concentration of liquid A is 10 times the hydrogen ion concentration of liquid B, then how does the pH of A compare with the pH of B?

86. *Solving a Formula* Solve the formula $A = P + Ce^{-kt}$ for t.

87. *Future Value* If \$50,000 is deposited in a bank account paying 5% compounded quarterly, then what will be the value of the account at the end of 18 years?

88. *Future Value* If \$30,000 is deposited in First American Savings and Loan in an account paying 6.18% compounded continuously, then what will be the value of the account after 12 years and 3 months?

89. *Doubling Time with Quarterly Compounding* How long (to the nearest quarter) will it take for the investment of Exercise 87 to double?

90. *Doubling Time with Continuous Compounding* How long (to the nearest day) will it take for the investment of Exercise 88 to double?

91. *Finding the Half-Life* The number of grams A of a certain radioactive substance present at time t is given by the formula $A = 25e^{-0.00032t}$, where t is the number of years from the present. How many grams are present initially? How many grams are present after 1000 years? What is the half-life of this substance?

92. *Comparing Investments* Shinichi invested \$800 at 6% compounded continuously. At the same time Toshio invested \$1000 at 5% compounded monthly. How long (to the nearest month) will it take for their investments to be equal in value?

93. *Learning Curve* According to an educational psychologist, the number of words learned by a student of a foreign language after t hours in the language laboratory is given by $f(t) = 40{,}000(1 - e^{-0.0001t})$. How many hours would it take to learn 10,000 words?

94. *Pediatric Tuberculosis* Health researchers divided the state of Maryland into regions according to per capita income and found the relationship between per capita income i (in thousands) and the percentage of pediatric TB cases P shown in the accompanying figure for 1986–1993 (Public Health Reports, April 1997). The equation $P = 31.5(0.935)^i$ can be used to model the data.

a. What percentage of TB cases would you expect to find in the region with a per capita income of \$8000?

b. What per capita income would you expect in the region where only 2% of the TB cases occurred?

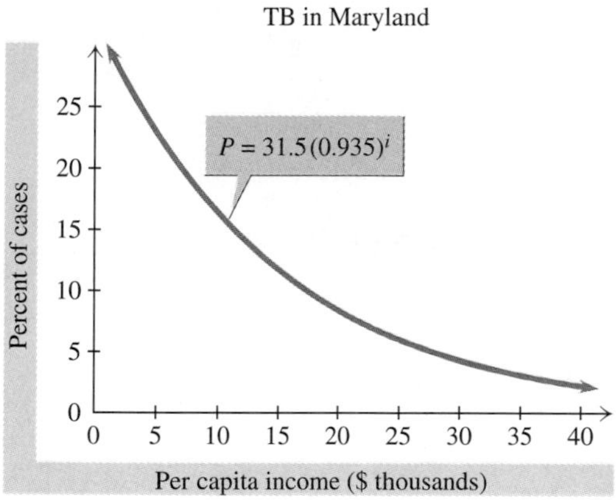

Figure for Exercise 94

95. *Doubling the Bet* A strategy used by some gamblers when they lose is to play again and double the bet. So if a gambler loses \$2, then he bets \$4, then \$8, and so on.

a. Why would the gambler do this?

b. If he keeps losing, then he bets 2^n dollars on his nth bet. Use logarithms to find the first value of n for which his bet is more than \$1 million.

Three numbers that are used for judging a boat's sea worthiness are the sail area-displacement ratio

$$A(d/64)^x,$$

the displacement-length ratio

$$(d/2240)(L/100)^y,$$

and thc capsize screening value

$$b(d/64)^z,$$

where A is the sail area in square feet, d is the displacement in pounds, L is the length at the water line in feet, and b is the beam in feet. The Freedom 40 is a \$291,560 sailboat with a sail area of 1026 ft^2, a displacement of 25,005 pounds, a beam of 13 ft 6 in., and a length at the water line of 35 ft 1 in.

96. If the sail area-displacement ratio is 19.20 for the Freedom 40, then what is the value of x?

97. If the displacement-length ratio for the Freedom 40 is 258.51, then what is the value of y?

98. If the capsize screening value for the Freedom 40 is 1.85, then what is the value of z?

CHAPTER 4 TEST

Simplify each expression.

1. $\log(1000)$

2. $\log_7(1/49)$

3. $3^{\log_3(6.47)}$

4. $\ln(e^{\sqrt{2}})$

Find the inverse of each function.

5. $f(x) = \ln(x)$

6. $f(x) = 8^x$

Write each expression as a single logarithm.

7. $\log(x) + 3 \cdot \log(y)$

8. $\frac{1}{2} \cdot \ln(x - 1) - \ln(33)$

Write each expression in terms of $\log_a(2)$ and $\log_a(7)$.

9. $\log_a(28)$

10. $\log_a(3.5)$

Find the exact solution to each equation.

11. $\log_2(x) + \log_2(x - 2) = 3$

12. $\log(10x) - \log(x + 2) = 2 \cdot \log(3)$

Find the exact solution and an approximate solution (to four decimal places) for each equation.

13. $3^x = 5^{x-1}$

14. $\log_3(x - 1) = 5.46$

Graph each equation in the xy-plane. State the domain and range of the function and whether the function is increasing or decreasing. Identify any asymptotes.

15. $f(x) = 2^x + 1$

16. $y = \log_{1/2}(x - 1)$

Solve each problem.

17. What is the only ordered pair that satisfies both $y = \log_2(x)$ and $y = \ln(x)$?

18. A student invested \$2000 at 8% annual percentage rate for 20 years. What is the amount of the investment if the interest is compounded quarterly? Compounded continuously?

19. A satellite has a radioisotope power supply. The power supply in watts is given by the formula $P = 50e^{-t/250}$, where t is the time in days. How much power is available at the end of 200 days? What is the half-life of the power supply? If the equipment aboard the satellite requires 9 watts of power to operate properly, then what is the operational life of the satellite?

20. If \$4000 is invested at 6% compounded quarterly, then how long will it take for the investment to grow to \$10,000?

21. An educational psychologist uses the model $t = -50 \cdot \ln(1 - p)$ for $0 \le p < 1$ to predict the number of hours t that it will take for a child to reach level p in the new video game Mario Goes to Mars. Level $p = 0$ means that the child knows nothing about MGM. What is the predicted level of a child for 100 hr of playing MGM? If $p = 1$ corresponds to mastery of MGM, then is it possible to master MGM?

TYING IT ALL TOGETHER

Chapters P–4

Solve each equation.

1. $(x - 3)^2 = 4$ **2.** $2 \cdot \log(x - 3) = \log(4)$ **3.** $\log_2(x - 3) = 4$

4. $2^{x-3} = 4$ **5.** $\sqrt{x - 3} = 4$ **6.** $|x - 3| = 4$

7. $x^2 - 4x = -2$ **8.** $2^{x-3} = 4^x$ **9.** $\sqrt{x} + \sqrt{x - 5} = 5$

10. $2^x = 3$ **11.** $\log(x - 3) + \log(4) = \log(x)$ **12.** $x^3 - 4x^2 + x + 6 = 0$

Sketch the graph of each function.

13. $y = x^2$ **14.** $y = (x - 2)^2$ **15.** $y = 2^x$ **16.** $y = x^{-2}$

17. $y = \log_2(x - 2)$ **18.** $y = x - 2$ **19.** $y = 2x$ **20.** $y = \log(2^x)$

21. $y = e^2$ **22.** $y = 2 - x^2$ **23.** $y = \dfrac{2}{x}$ **24.** $y = \dfrac{1}{x - 2}$

Find the inverse of each function.

25. $f(x) = \dfrac{1}{3}x$ **26.** $f(x) = \dfrac{1}{3^x}$ **27.** $f(x) = \sqrt{x - 2}$ **28.** $f(x) = 2 + (x - 5)^3$

29. $f(x) = \log(\sqrt{x} - 3)$ **30.** $f = \{(3, 1), (5, 4)\}$ **31.** $f(x) = 3 + \dfrac{1}{x - 5}$ **32.** $f(x) = 3 - e^{\sqrt{x}}$

Find a formula for each composition function given that $p(x) = e^x$, $m(x) = x + 5$, $q(x) = \sqrt{x}$, and $r(x) = \ln(x)$. State the domain and range of the composition function.

33. $p \circ m$ **34.** $p \circ q$ **35.** $q \circ p \circ m$ **36.** $m \circ r \circ q$

37. $p \circ r \circ m$ **38.** $r \circ q \circ p$

Express each of the following functions as a composition of the functions f, g, and h, where $f(x) = \log_2(x)$, $g(x) = x - 4$, and $h(x) = x^3$.

39. $F(x) = \log_2(x^3 - 4)$ **40.** $H(x) = (\log_2(x))^3 - 4$ **41.** $G(x) = (\log_2(x) - 4)^3$ **42.** $M(x) = (\log_2(x - 4))^3$

SOME BASIC FUNCTIONS OF ALGEBRA

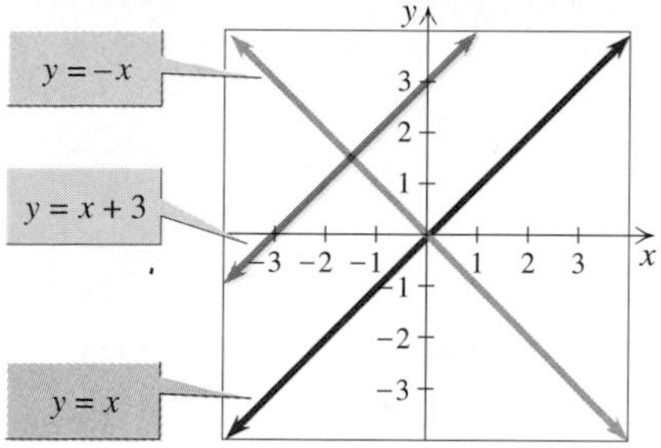

Linear

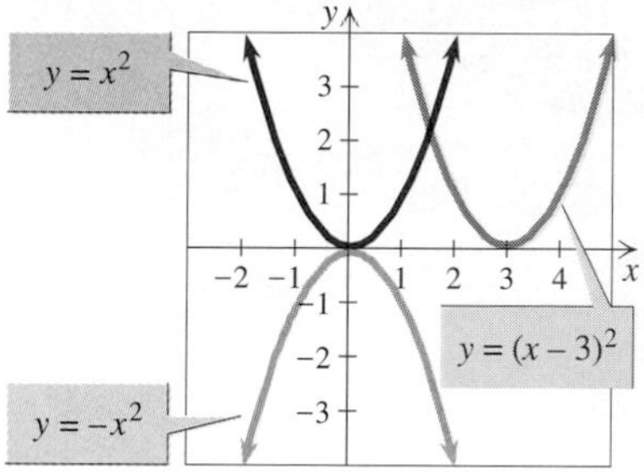

Quadratic

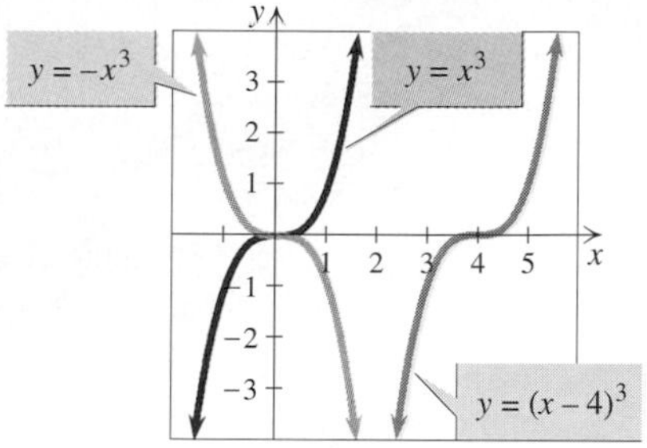

Cubic

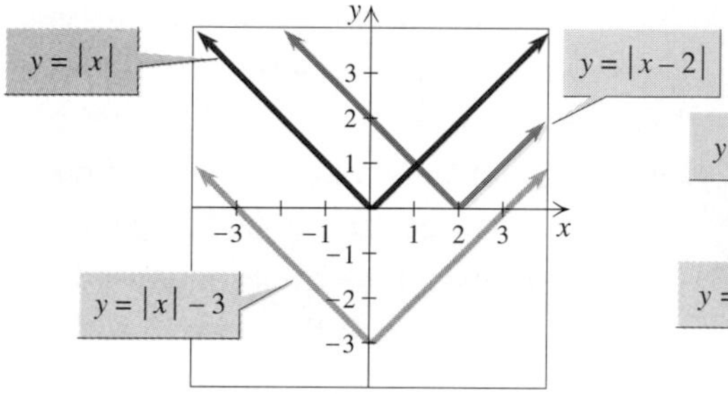

Absolute value

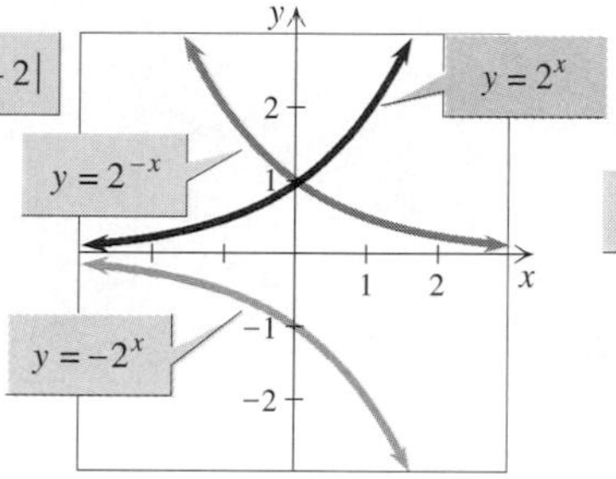

Exponential

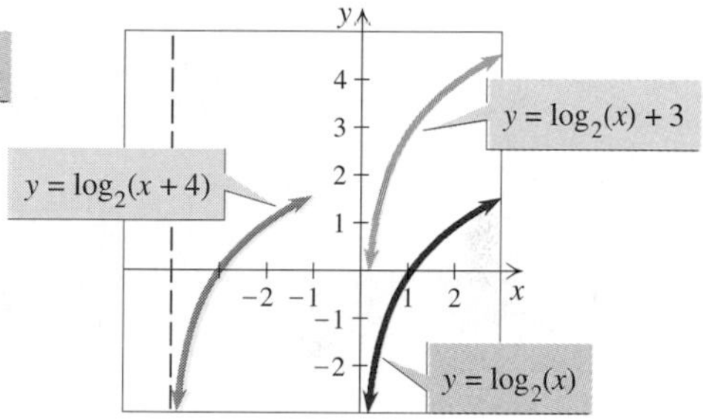

Logarithmic

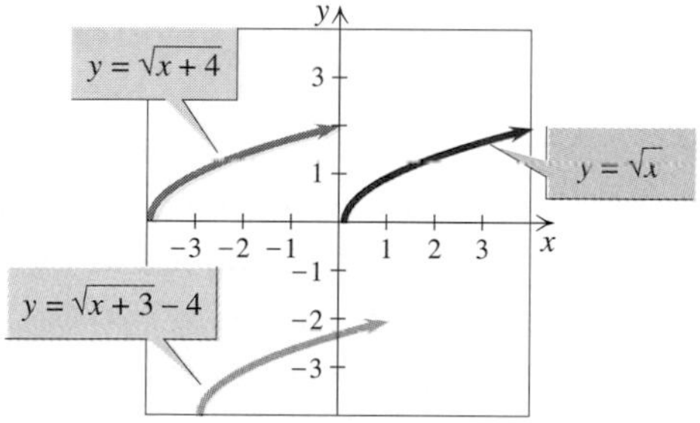

Square root

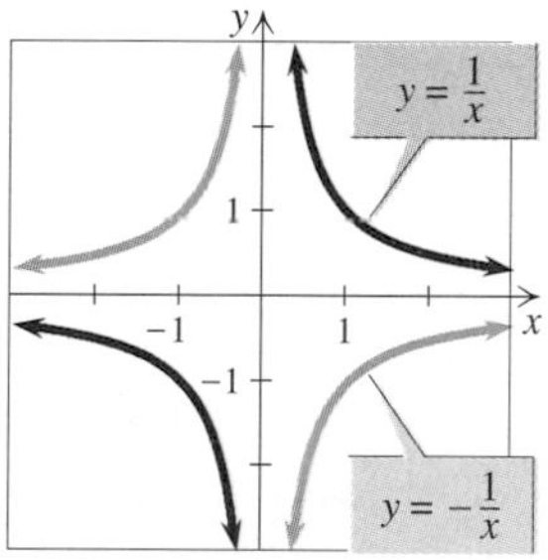

Reciprocal

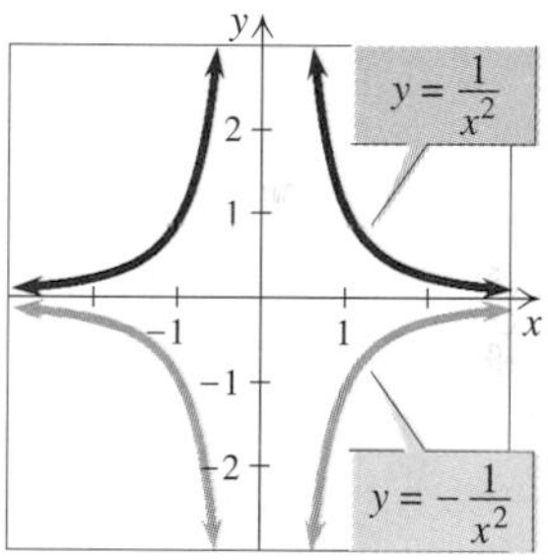

Rational

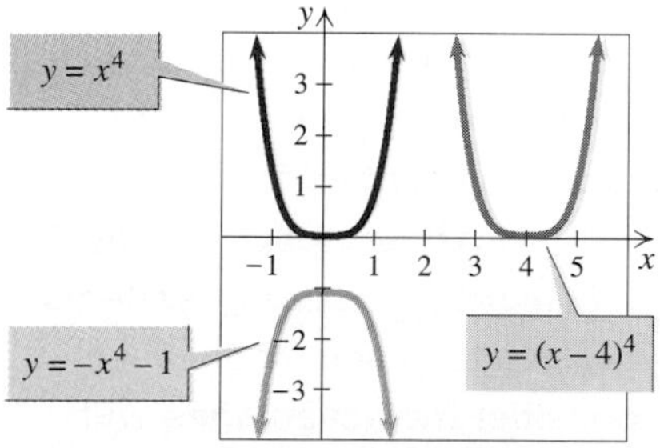

Fourth degree

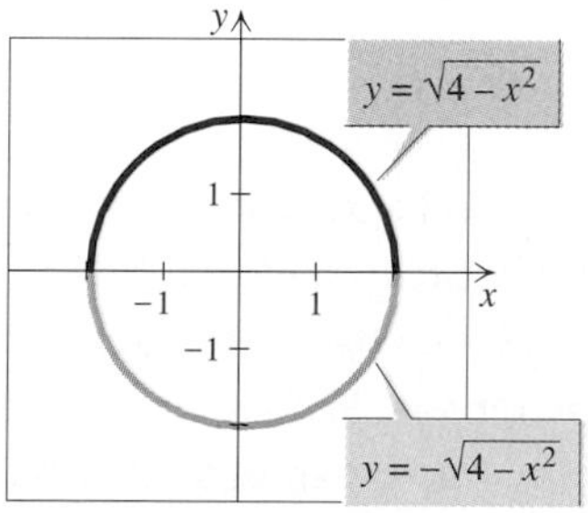

Semicircle

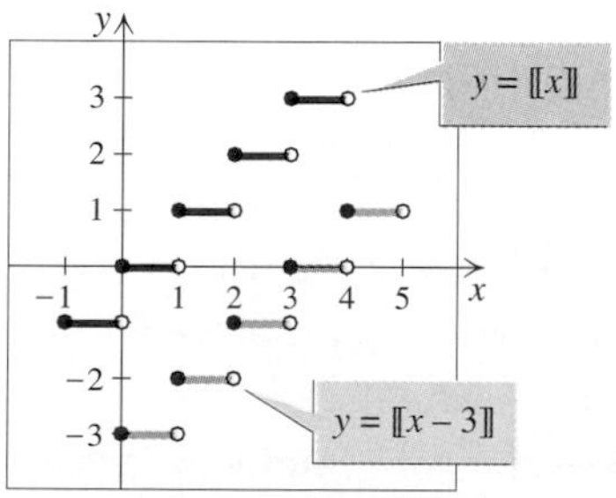

Greatest integer

The autumn day is sultry, with humid winds drifting inland from the sea. Over a field of jewelweed and crimson cardinal flowers, a cloud of ruby-throated hummingbirds dart and hover, feeding on nectar in the brilliant blossoms. These tiny travelers—many weighing only 1/8 ounce—have flown over 1000 miles since leaving their summer range in New England. Now they must double their body weight before undertaking the final leg of their journey. Before them lies the perilous 500-mile, nonstop flight across the Gulf of Mexico.

Blessed with relatively large wing muscles, hummingbirds can flap 3000 beats per minute in level flight. In the process, the little ruby-throat migrates over 2000 miles—from southern Canada to Central America and the Caribbean. Intelligent and resourceful, hummers wait for favorable weather conditions and a strong tailwind before attempting to cross the Gulf—a marathon journey completed in an amazing 18 hours.

Scientists have been investigating the mysteries of bird migration for decades, yet they still do not fully understand how birds can navigate such long distances. We know some species rely on smells, sounds, and visual landmarks; others set their course by the sun, stars, and Earth's magnetic field. Experiments indicate that birds' stamina comes from an ability to efficiently burn body fat for fuel.

Researcher Vance Tucker set up a wind tunnel experiment at Duke University. After training parakeets and gulls to fly into an airstream, he then measured the rate at which they converted proteins, carbohydrates, and fats into usable energy. His data show that birds are expert at choosing a flight speed that will minimize the number of calories they burn. In this chapter we'll learn how to pinpoint the most economic flight speed using systems of equations to fit a parabolic curve to Tucker's data.

In earlier chapters we solved equations that contained one variable. However, many business, engineering, and science applications require solving problems containing two or more variables. As you study the techniques for solving systems of equations and inequalities, consider the advantages and disadvantages of each method.

5 Systems of Equations and Inequalities

5.1 Systems of Linear Equations in Two Variables
5.2 Systems of Linear Equations in Three Variables
5.3 Nonlinear Systems of Equations
5.4 Partial Fractions
5.5 Inequalities and Systems of Inequalities in Two Variables
5.6 Linear Programming

5.1 Systems of Linear Equations in Two Variables

In Section 2.1 we defined a linear equation in two variables as an equation of the form

$$Ax + By = C$$

where A and B are not both zero, and we discussed numerous applications of linear equations. There are infinitely many ordered pairs that satisfy a single linear equation. In applications, however, we are often interested in finding a single ordered pair that satisfies a *pair* of linear equations. In this section we discuss several methods for solving that problem.

Solving a System by Graphing

Any collection of two or more equations is called a **system of equations**. For example, the system of equations consisting of $x + 2y = 6$ and $2x - y = -8$ is written as follows:

$$\begin{aligned} x + 2y &= 6 \\ 2x - y &= -8 \end{aligned}$$

The **solution set** of a system of two linear equations in two variables is the set of all ordered pairs that satisfy *both* equations of the system. The graph of an equation shows all ordered pairs that satisfy it, so we can solve some systems by

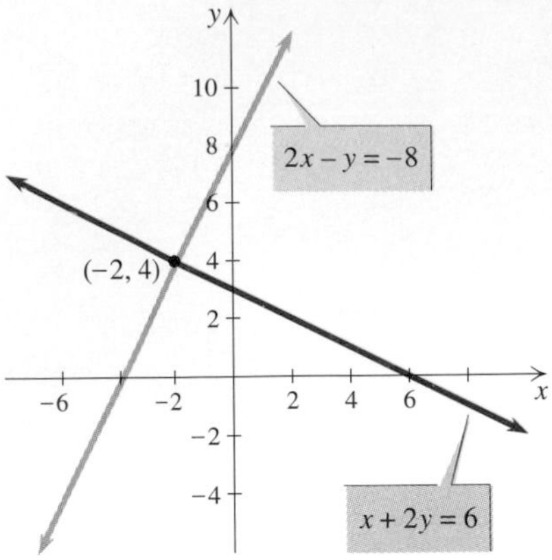

Figure 5.1

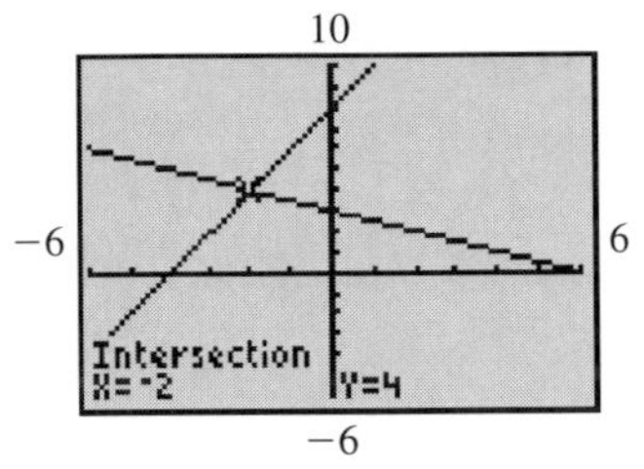

Figure 5.2

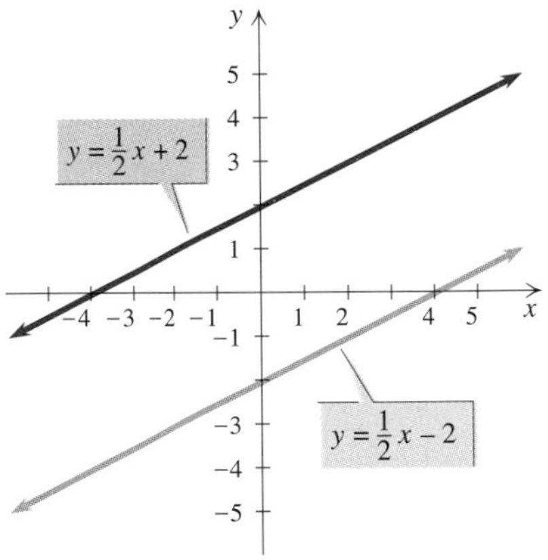

Figure 5.3

graphing the equations and observing which points (if any) satisfy all of the equations.

EXAMPLE 1 **Solving a system by graphing**

Solve each system by graphing.

a) $x + 2y = 6$
$2x - y = -8$

b) $y = \frac{1}{2}x + 2$
$x - 2y = 4$

Solution

a) Graph the straight line

$$x + 2y = 6$$

by using its intercepts, (0, 3) and (6, 0). Graph the straight line

$$2x - y = -8$$

by using its intercepts, (0, 8) and (−4, 0). The graphs are shown in Fig. 5.1. The lines appear to intersect at (−2, 4). Check (−2, 4) in both equations. Since

$$-2 + 2(4) = 6 \quad \text{and} \quad 2(-2) - 4 = -8$$

are both correct, we can be certain that (−2, 4) satisfies both equations. The solution set of the system is $\{(-2, 4)\}$.

You can check by graphing the equations on a calculator and finding the intersection as shown in Fig. 5.2. □

b) Use the y-intercept (0, 2) and the slope 1/2 to graph

$$y = \frac{1}{2}x + 2$$

as shown in Fig. 5.3. Since $x - 2y = 4$ is equivalent to

$$y = \frac{1}{2}x - 2,$$

its graph has y-intercept (0, −2) and is parallel to the first line. Since the lines are parallel, there is no point that satisfies both equations of the system. In fact, if you substitute any value of x in the two equations, the corresponding y-values will differ by 4.

Independent, Inconsistent, and Dependent Equations

Most of the systems of equations that occur in applications correspond to pairs of lines that intersect in a single point, as in Fig. 5.1. In this case the equations are called **independent** or the system is called an independent system. If the two lines corresponding to the equations are parallel, as in Fig. 5.3, then there is no solution to the system and the equations are called **inconsistent** or the system is called inconsistent. The third possibility is that two equations are equivalent and have the same graph. In this case the equations are called **dependent**. For example, the equations of the system

$$\begin{aligned} x + y &= 5 \\ 2x + 2y &= 10 \end{aligned}$$

have the same graph (a line) and the system is dependent. Any ordered pair that satisfies one of these equations, satisfies both of these equations. So the solution set to the system consists of all ordered pairs that satisfy one of the equations, described in set notation as

$$\{(x, y) \mid x + y = 5\}.$$

Since the equations are equivalent, we could use either equation in this set notation, but we usually use the simpler one. Figure 5.4 illustrates typical graphs for independent, inconsistent, and dependent systems.

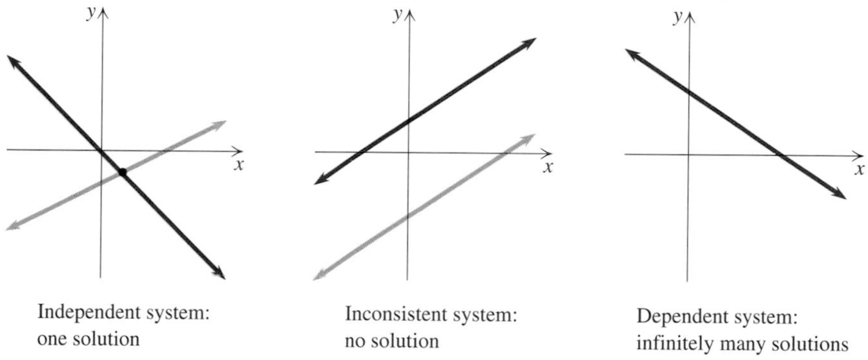

Figure 5.4

The Substitution Method

Graphing the equations of a system helps us to visualize the system and determine how many solutions it has. However, solving systems of linear equations by graphing is not very accurate unless the solution is fairly simple. The accuracy of graphing can be improved with a graphing calculator, but even with a graphing calculator, we generally get only approximate solutions. For example, the solution to $y_1 = (28 - 7x)/13$ and $y_2 = (29 + 7x)/26$ is $(9/7, 19/13)$, but the graphing calculator solution in Fig. 5.5 does not give this exact answer. However, by using an algebraic technique such as the substitution method, we can get exact solutions quickly. In this method, shown in Example 2, we eliminate a variable from one equation by substituting an expression for that variable from the other equation.

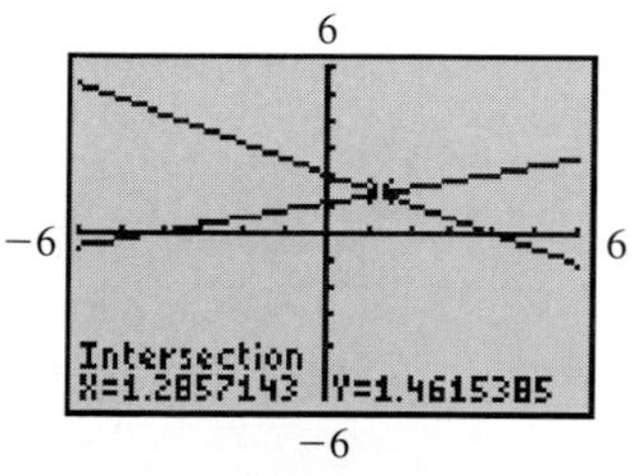

Figure 5.5

EXAMPLE 2 Solving a system by substitution

Solve each system by substitution.

a) $3x - y = 6$
$6x + 5y = -23$

b) $y = 2x + 1000$
$0.05x + 0.06y = 400$

Solution

a) Since y occurs with coefficient -1 in $3x - y = 6$, it is simpler to isolate y in this equation than to isolate any other variable in the system.

$$-y = -3x + 6$$
$$y = 3x - 6$$

Use $3x - 6$ in place of y in the equation $6x + 5y = -23$:

$$6x + 5(3x - 6) = -23 \quad \text{Substitution}$$
$$6x + 15x - 30 = -23$$
$$21x = 7$$
$$x = \frac{1}{3}$$

The x-coordinate of the solution is $\frac{1}{3}$. To find y, use $x = \frac{1}{3}$ in $y = 3x - 6$:

$$y = 3\left(\frac{1}{3}\right) - 6$$
$$y = -5$$

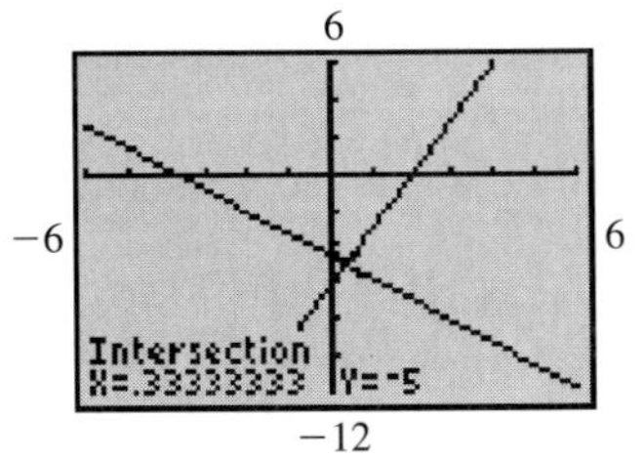

Figure 5.6

Check that $(\frac{1}{3}, -5)$ satisfies both of the original equations. The solution set is $\{(\frac{1}{3}, -5)\}$.

The graphs of $y_1 = 3x - 6$ and $y_2 = (-23 - 6x)/5$ in Fig. 5.6 support this solution. □

b) The first equation, $y = 2x + 1000$, already has one variable isolated. So we can replace y by $2x + 1000$ in $0.05x + 0.06y = 400$:

$$0.05x + 0.06(2x + 1000) = 400 \quad \text{Substitution}$$
$$0.05x + 0.12x + 60 = 400$$
$$0.17x = 340$$
$$x = 2000$$

Use $x = 2000$ in $y = 2x + 1000$ to find y:

$$y = 2(2000) + 1000$$
$$y = 5000$$

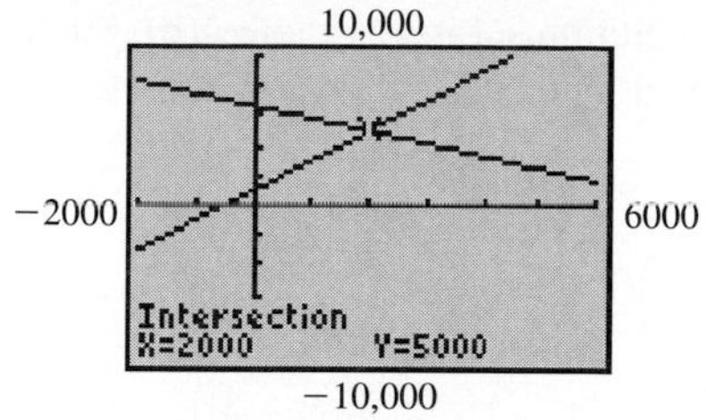

Figure 5.7

Check (2000, 5000) in the original equations. The solution set is $\{(2000, 5000)\}$.

The graphs of $y_1 = 2x + 1000$ and $y_2 = (400 - 0.05x)/0.06$ in Fig. 5.7 support this solution.

If substitution results in a false statement, then the system is inconsistent. If substitution results in an identity, then the system is dependent. In the next example we solve an inconsistent system and a dependent system by substitution.

EXAMPLE 3 Inconsistent and dependent systems

Solve each system by substitution.

a) $3x - y = 9$
$2y - 6x = 7$

b) $\frac{1}{2}x - \frac{2}{3}y = -2$
$4y = 3x + 12$

Solution

a) Solve $3x - y = 9$ for y to get $y = 3x - 9$. Replace y by $3x - 9$ in the equation $2y - 6x = 7$:

$$2(3x - 9) - 6x = 7$$
$$6x - 18 - 6x = 7$$
$$-18 = 7$$

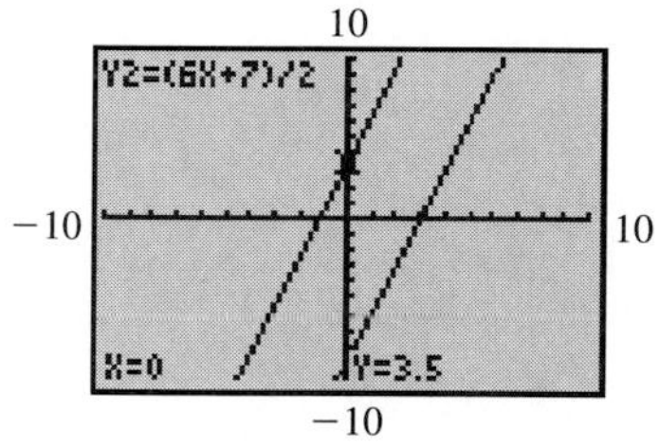

Figure 5.8

Since the last statement is false, the system is inconsistent and has *no solution*.

The graphs of $y_1 = 3x - 9$ and $y_2 = (6x + 7)/2$ in Fig. 5.8 appear to be parallel lines and support the conclusion that the system is inconsistent. □

b) Solve $4y = 3x + 12$ for y to get $y = \frac{3}{4}x + 3$. Replace y by $\frac{3}{4}x + 3$ in the first equation:

$$\frac{1}{2}x - \frac{2}{3}\left(\frac{3}{4}x + 3\right) = -2$$
$$\frac{1}{2}x - \frac{1}{2}x - 2 = -2$$
$$-2 = -2$$

Since the last statement is an identity, the system is dependent. The solution set is $\{(x, y) \mid 4y = 3x + 12\}$.

Note that there are many ways of writing the solution set in Example 3(b). Since $4y = 3x + 12$ is equivalent to $y = \frac{3}{4}x + 3$ or to $x = \frac{4}{3}y - 4$, we could write the solution set as

$$\left\{\left(x, \frac{3}{4}x + 3\right) \middle| x \text{ is any real number}\right\} \quad \text{or}$$

$$\left\{\left(\frac{4}{3}y - 4, y\right) \middle| y \text{ is any real number}\right\}.$$

The Addition Method

In the substitution method we eliminate a variable in one equation by substituting from the other equation. In the addition method we eliminate a variable by adding the two equations. It might be necessary to multiply each equation by an appropriate number so that a variable will be eliminated by the addition.

EXAMPLE 4 Solving systems by addition

Solve each system by addition.

a) $3x - y = 9$ **b)** $2x - 3y = -2$
$\quad 2x + y = 1$ $\quad 3x - 2y = 12$

Solution

a) Add the equations to eliminate the y-variable:

$$\begin{aligned} 3x - y &= 9 \\ \underline{2x + y} &\underline{= 1} \\ 5x \quad &= 10 \\ x &= 2 \end{aligned}$$

Use $x = 2$ in $2x + y = 1$ to find y:

$$\begin{aligned} 2(2) + y &= 1 \\ y &= -3 \end{aligned}$$

Substituting the values $x = 2$ and $y = -3$ in the original equations yields $3(2) - (-3) = 9$ and $2(2) + (-3) = 1$, which are both correct. So $(2, -3)$ satisfies both equations and the solution set to the system is $\{(2, -3)\}$.

b) To eliminate x upon addition, we multiply the first equation by 3 and the second equation by -2:

$$\begin{aligned} 3(2x - 3y) &= 3(-2) \\ -2(3x - 2y) &= -2(12) \end{aligned}$$

This multiplication produces $6x$ in one equation and $-6x$ in the other. So the x-variable is eliminated upon addition of the equations.

$$\begin{aligned} 6x - 9y &= -6 \\ -6x + 4y &= -24 \\ \hline -5y &= -30 \\ y &= 6 \end{aligned}$$

Use $y = 6$ in $2x - 3y = -2$ to find x:

$$\begin{aligned} 2x - 3(6) &= -2 \\ 2x - 18 &= -2 \\ 2x &= 16 \\ x &= 8 \end{aligned}$$

If $y = 6$ is used in the other equation, $3x - 2y = 12$, we would also get $x = 8$. Substituting $x = 8$ and $y = 6$ in both of the original equations yields $2(8) - 3(6) = -2$ and $3(8) - 2(6) = 12$, which are both correct. So the solution set to the system is $\{(8, 6)\}$.

In Example 4(b), we started with the given system and multiplied the first equation by 3 and the second equation by -2 to get

$$\begin{aligned} 6x - 9y &= -6 \\ -6x + 4y &= -24. \end{aligned}$$

Since each equation of the new system is equivalent to an equation of the old system, the solution sets to these systems are identical. Two systems with the same solution set are **equivalent systems**. If we had multiplied the first equation by 2 and the second by -3, we would have obtained the equivalent system

$$\begin{aligned} 4x - 6y &= -4 \\ -9x + 6y &= -36 \end{aligned}$$

and we would have eliminated y by adding the equations.

When we have a choice of which method to use for solving a system, we generally avoid graphing because it is often inaccurate. Substitution or addition both yield exact solutions, but sometimes one method is easier to apply than the other. Substitution is usually used when one equation gives one variable in terms of the other as in Example 2. Addition is usually used when both equations are in the form $Ax + By = C$ as in Example 4. By doing the exercises, you will soon discover which method works best on a given system.

When a system is solved by the addition method, an inconsistent system results in a false statement and a dependent system results in an identity, just as they did for the substitution method.

EXAMPLE 5 Inconsistent and dependent systems

Solve each system by addition.

a) $0.2x - 0.4y = 0.5$
$x - 2y = 1.3$

b) $\frac{1}{2}x - \frac{2}{3}y = -2$
$-3x + 4y = 12$

Solution

a) It is usually a good idea to eliminate the decimals in the coefficients, so we multiply the first equation by 10:

$$\begin{aligned} 2x - 4y &= 5 \qquad \text{First equation multiplied by 10} \\ x - 2y &= 1.3 \end{aligned}$$

Now multiply the second equation by -2 and add to eliminate x:

$$\begin{aligned} 2x - 4y &= 5 \\ -2x + 4y &= -2.6 \\ \hline 0 &= 2.4 \end{aligned}$$

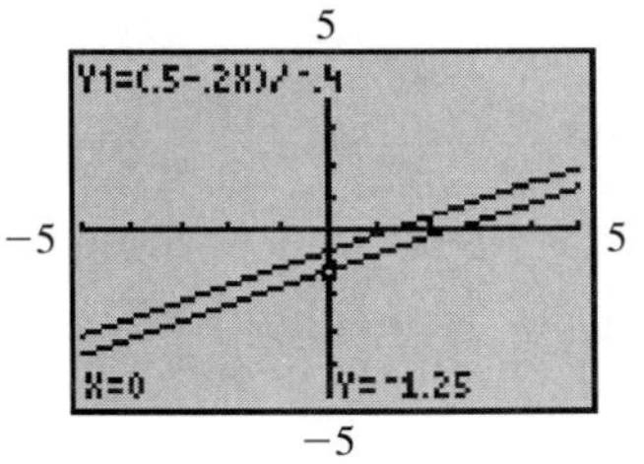

Figure 5.9

Since $0 = 2.4$ is false, there is no solution to the system.

The graphs of $y_1 = (0.5 - 0.2x)/(-0.4)$ and $y_2 = (1.3 - x)/(-2)$ in Fig. 5.9 appear to be parallel lines and support the conclusion that there is no solution to the system. □

b) To eliminate fractions in the coefficients, multiply the first equation by the LCD 6:

$$\begin{aligned} 3x - 4y &= -12 \\ -3x + 4y &= 12 \\ \hline 0 &= 0 \end{aligned}$$

Since $0 = 0$ is an identity, the solution set is $\{(x, y) \mid -3x + 4y = 12\}$.

Note that there are many ways to solve a system by addition. In Example 5(a), we could have multiplied the first equation by -5 or the second equation by -0.2. In either case, x would be eliminated upon addition. Try this for yourself.

Applications

We solved many problems involving linear equations in the past, but we always wrote all unknown quantities in terms of a single variable. Now that we can solve systems of equations, we can solve problems involving two unknown quantities by using two variables and a system of at least two equations.

EXAMPLE 6 Problem solving using a system of equations

While on a stakeout protecting Earth from the scum of the universe, Agent K observed five men in black enter a coffee shop and get three doughnuts and five coffees for \$3.30. Next, three aliens in disguise entered the shop and got four doughnuts and three coffees for \$2.75. Finally, Agent Zed from the home office entered and got a doughnut and a cup of coffee. How much was Zed's bill?

Solution

Let x be the cost of one doughnut and y be the cost of one cup of coffee. We can write a system of equations about x and y:

$$3x + 5y = 3.30$$
$$4x + 3y = 2.75$$

To eliminate x, multiply the first equation by -4 and the second by 3:

$$-4(3x + 5y) = -4(3.30)$$
$$3(4x + 3y) = 3(2.75)$$

Add the two resulting equations:

$$\begin{aligned} -12x - 20y &= -13.20 \\ 12x + 9y &= 8.25 \\ \hline -11y &= -4.95 \\ y &= 0.45 \end{aligned}$$

Use $y = 0.45$ in $3x + 5y = 3.30$ to find x:

$$\begin{aligned} 3x + 5(0.45) &= 3.30 \\ 3x + 2.25 &= 3.30 \\ 3x &= 1.05 \\ x &= 0.35 \end{aligned}$$

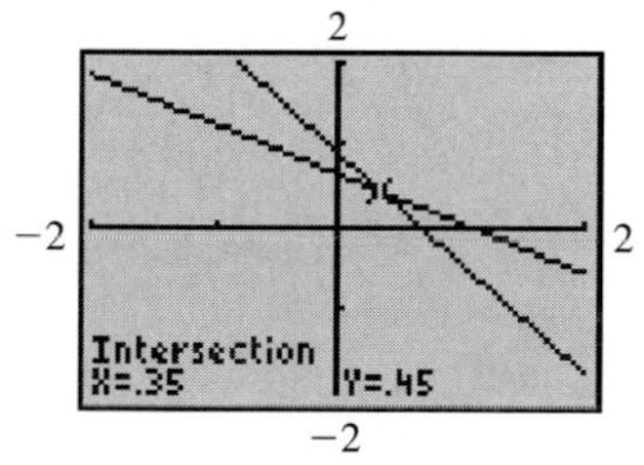

Figure 5.10

Check that 35 cents for a doughnut and 45 cents for a cup of coffee satisfy the statements in the original problem. Given these prices, Agent Zed's bill is 80 cents.

The graphs of $y_1 = (3.30 - 3x)/5$ and $y_2 = (2.75 - 4x)/3$ in Fig. 5.10 support this solution.

FOR THOUGHT True or False? Explain.

The following systems are referenced in these statements.

a) $x + y = 5$, $x - y = 1$ **b)** $x - 2y = 4$, $3x - 6y = 8$ **c)** $x = 5 + 3y$, $9y - 3x = -15$

1. The ordered pair (2, 3) is in the solution set to $x + y = 5$.

2. The ordered pair (2, 3) is in the solution set to system (a).

3. System (a) is inconsistent.
4. There is no solution to system (b).
5. Adding the equations in system (a) would eliminate y.
6. To solve system (c) we could substitute $5 + 3y$ for x in $9y - 3x = -15$.
7. System (c) is inconsistent.
8. The solution set to system (c) is the set of all real numbers.
9. The graphs of the equations of system (c) intersect at a single point.
10. The graphs of the equations of system (b) are parallel.

5.1 EXERCISES

Tape 16 Disk

Solve each system by inspecting the graphs of the equations.

1. $2x - 3y = -4$
$y = -2x + 4$

2. $x + 2y = -1$
$2x + 3y = -3$

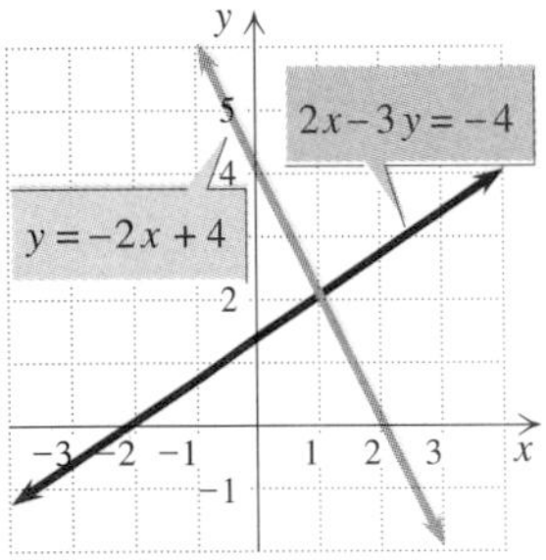

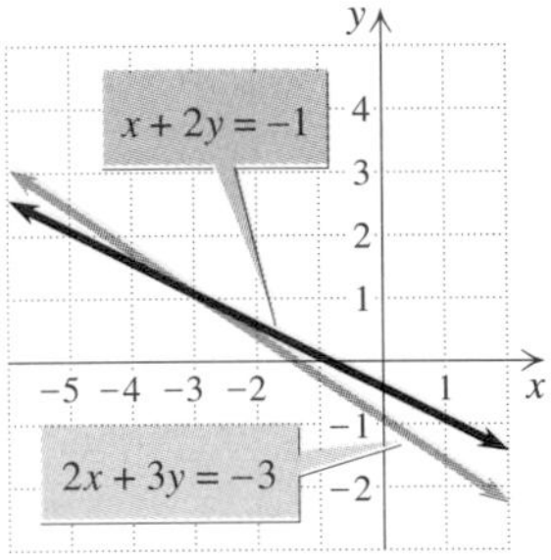

3. $3x - 4y = 0$
$y = \frac{3}{4}x + 2$

4. $x - 2y = -3$
$y = \frac{1}{2}x + \frac{3}{2}$

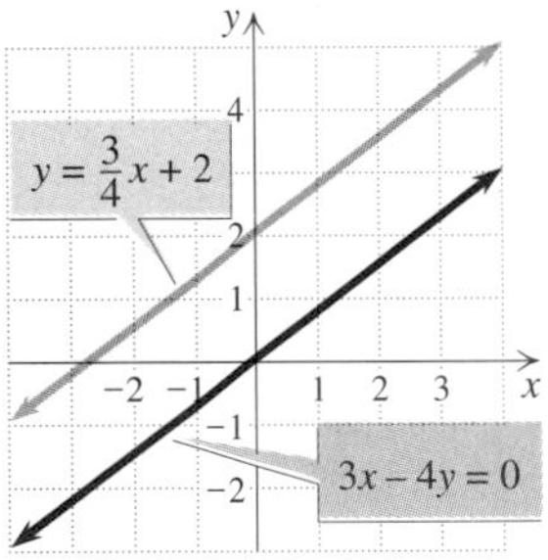

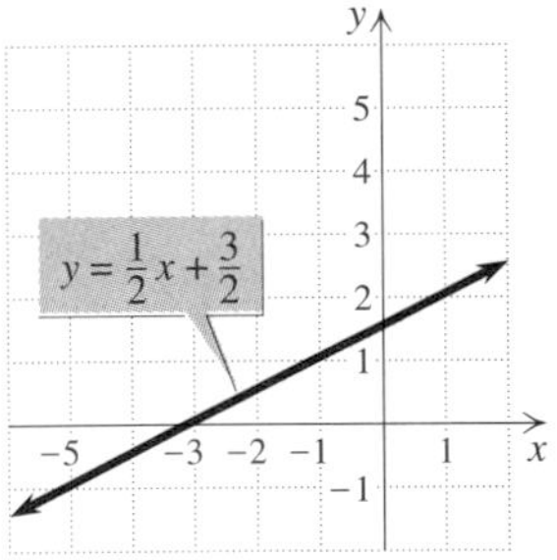

Solve each system by graphing.

5. $x + y = 5$
$x - y = 1$

6. $2x + y = -1$
$y - x = 5$

7. $y = x - 2$
$y = -x + 4$

8. $y = -3x$
$x - 2y = 7$

9. $3y + 2x = 6$
$y = -\frac{2}{3}x - 1$

10. $2x + 4y = 12$
$2y = 6 - x$

11. $y = \frac{1}{2}x - 3$
$2x - 4y = 12$

12. $y = -2x + 6$
$4x + 2y = 8$

Solve each system by substitution. Determine whether each system is independent, inconsistent, or dependent.

13. $y = 2x + 1$
$3x - 4y = 1$

14. $5x - 6y = 23$
$x = 6 - 3y$

15. $x + y = 1$
$2x - 3y = 8$

16. $x + 2y = 3$
$2x + y = 5$

17. $y - 3x = 5$
$3(x + 1) = y - 2$

18. $2y = 1 - 4x$
$2x + y = 0$

19. $2y = 6 - 3x$
$\frac{1}{2}x + \frac{1}{3}y = 3$

20. $2x = 10 - 5y$
$\frac{1}{5}x + \frac{1}{2}y = 1$

21. $x + y = 200$
$0.05x + 0.06y = 10.50$

22. $2x + y = 300$
$\frac{1}{2}x + \frac{1}{3}y = 80$

23. $y = 3x + 1$
$y = 3x - 7$

24. $2x + y = 9$
$4x + 2y = 10$

25. $\frac{1}{2}x - \frac{1}{3}y = 12$
$\frac{1}{4}x - \frac{1}{2}y = 1$

26. $0.05x + 0.1y = 10$
$0.06x + 0.2y = 16$

Solve each system by addition. Determine whether each system is independent, inconsistent, or dependent.

27. $x + y = 20$
$x - y = 6$

28. $3x - 2y = 7$
$-3x + y = 5$

29. $x - y = 5$
$3x + 2y = 10$

30. $x - 4y = -3$
$-3x + 5y = 2$

31. $x - y = 7$
$y - x = 5$

32. $2x - y = 6$
$-4x + 2y = 9$

33. $\begin{aligned} 2x + 3y &= 1 \\ 3x - 5y &= -8 \end{aligned}$

34. $\begin{aligned} -2x + 5y &= 14 \\ 7x + 6y &= -2 \end{aligned}$

35. $\begin{aligned} 0.05x + 0.1y &= 0.6 \\ x + 2y &= 12 \end{aligned}$

36. $\begin{aligned} 0.02x - 0.04y &= 0.08 \\ x - 2y &= 4 \end{aligned}$

37. $\begin{aligned} \frac{x}{2} + \frac{y}{2} &= 5 \\ \frac{3x}{2} - \frac{2y}{3} &= 2 \end{aligned}$

38. $\begin{aligned} \frac{x}{4} + \frac{y}{3} &= 0 \\ \frac{x}{8} - \frac{y}{6} &= 2 \end{aligned}$

39. $\begin{aligned} 3x - 2.5y &= -4.2 \\ 0.12x + 0.09y &= 0.4932 \end{aligned}$

40. $\begin{aligned} 1.5x - 2y &= 8.5 \\ 3x + 1.5y &= 6 \end{aligned}$

Classify each system as independent, inconsistent, or dependent without doing any written work.

41. $\begin{aligned} y &= 5x - 6 \\ y &= -5x - 6 \end{aligned}$

42. $\begin{aligned} y &= 5x - 6 \\ y &= 5x + 4 \end{aligned}$

43. $\begin{aligned} 5x - y &= 6 \\ y &= 5x - 6 \end{aligned}$

44. $\begin{aligned} 5x - y &= 6 \\ y &= -5x + 6 \end{aligned}$

Solve each system by graphing the equations on a graphing calculator and estimating the point of intersection.

45. $\begin{aligned} y &= 0.5x + 3 \\ y &= 0.499x + 2 \end{aligned}$

46. $\begin{aligned} y &= 2x - 3 \\ y &= 1.9999x - 2 \end{aligned}$

47. $\begin{aligned} 0.23x + 0.32y &= 1.25 \\ 0.47x - 1.26y &= 3.58 \end{aligned}$

48. $\begin{aligned} 342x - 78y &= 474 \\ 123x + 145y &= 397 \end{aligned}$

Solve each problem using two variables and a system of two equations. Solve the system by the method of your choice. Note that some of these problems lead to dependent or inconsistent systems.

49. *Income on Investments* Carmen made \$25,000 profit on the sale of her condominium. She lent part of the profit to Jim's Orange Grove at 10% interest and the remainder to Ricky's Used Cars at 8% interest. If she received \$2200 in interest after one year, then how much did she lend to each business?

50. *Stock Market Losses* In 1997 Gerhart lost twice as much in the futures market as he did in the stock market. If his losses totaled \$18,630, then how much did he lose in each market?

51. *Zoo Admission Prices* The Lincoln Park Zoo has different admission prices for adults and children. When Mr. and Mrs. Weaver went with their five children, the bill was \$33. If Mrs. Wong and her three children got in for \$18.50, then what is the price of an adult's ticket and what is the price of a child's ticket?

52. *Book Prices* At the Book Exchange, all paperbacks sell for one price and all hardbacks sell for another price. Tanya got six paperbacks and three hardbacks for \$8.25, while Gretta got four paperbacks and five hardbacks for \$9.25. What was Todd's bill for seven paperbacks and nine hardbacks?

53. *Getting Fit* The Valley Health Club sold a dozen memberships in one week for a total of \$6000. If male memberships cost \$500 and female memberships cost \$500, then how many male memberships and how many female memberships were sold?

54. *Quality Time* Mr. Thomas and his three children paid a total of \$65.75 for admission to Water World. Mr. and Mrs. Li and their six children paid a total of \$131.50. What is the price of an adult's ticket and what is the price of a child's ticket?

55. *Coffee and Muffins* On Monday the office staff paid a total of \$7.77 including tax for 3 coffees and 7 muffins. On Tuesday the bill was \$14.80 including tax for 6 coffees and 14 muffins. If the sales tax rate is 7%, then what is the price of a coffee and what is the price of a muffin?

56. *Graduating Seniors* In Sociology 410 there are 55 more males than there are females. Two-thirds of the males and two-thirds of the females are graduating seniors. If there are 30 more graduating senior males than graduating senior females, then how many males and how many females are in the class?

57. *Political Party Preference* The results of a survey of students at Central High School concerning political party preference are given in the accompanying table. If 230 students preferred the Democratic party and 260 students preferred the Republican party, then how many students are there at CHS?

Table for Exercise 57

	Male	Female
Democratic	50%	30%
Republican	20%	60%
Other	30%	10%

58. *Protein and Carbohydrates* Nutritional information for Rice Krispies and Grape-nuts is given in the accompanying table. How many servings of each would it take to get exactly 23 g of protein and 215 g of carbohydrates?

Table for Exercise 58

	Rice Krispies	Grape-nuts
Protein (g/serving)	2	3
Carbohydrates (g/serving)	25	23

59. *Distribution of Coin Types* Isabelle paid for her $1.75 lunch with 87 coins. If all of the coins were nickels and pennies, then how many were there of each type?

60. *Coin Collecting* Theodore has a collection of 166 old coins consisting of quarters and dimes. If he figures that each coin is worth two and a half times its face value, then his collection is worth $61.75. How many of each type of coin does he have?

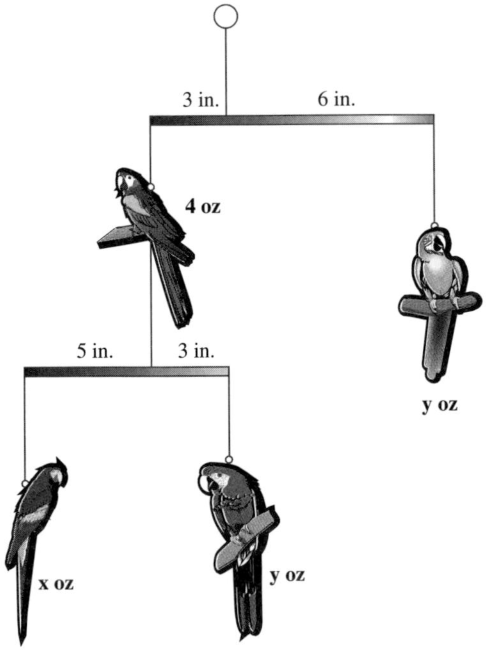

Figure for Exercise 61

61. *Bird Mobile* A wood carver is making a bird mobile as shown in the figure below. The weights of the horizontal bars and strings are negligible. The mobile will balance if the product of the weight and distance on one side of the balance point is equal to the product of the weight and distance on the other side. For what values of x and y will the mobile be balanced?

62. *Doubles and Singles* The Executive Inn rents a double room for $10 more per night than a single. One night the motel took in $2159 by renting 15 doubles and 26 singles. What is the rental price for each type of room?

63. *Furniture Rental* A civil engineer has a choice of two plans for renting furniture for her new office. Under Plan A, she pays $800 plus $150 per month, while under Plan B she pays $200 plus $200 per month. For each plan, write the cost as a function of the number of months. Which plan is cheaper in the long run? For what number of months do the two plans cost the same?

64. *Flat Tax* The 1996 tax rate schedule for a single taxpayer is given in the accompanying table. Suppose the federal tax were simplified to be $100 plus 25% of taxable income. Use a system of equations to find the taxable income at which a single taxpayer would pay the same amount of tax under the simplified plan as under the 1996 schedule.

Table for Exercise 64
1996 Tax Rate Schedule—Single taxpayers

If taxable income is over	but not over	your tax is	of the amount over
$0	$24,000	15%	$0
24,000	58,150	3,600 + 28%	24,000
58,150	121,300	13,162 + 31%	58,150
121,300	263,750	32,738.50 + 36%	121,300
263,750	—	84,020.50 + 39.6%	263,750

65. *Prescribing Drugs* Doctors often prescribe the same drugs for children as they do for adults. If a is the age of a child and D is the adult dosage, then to find the child's dosage d, doctors can use the formula $d = 0.08aD$ (Friend's rule) or $d = D(a + 1)/24$ (Cowling's rule). For what age do the two formulas give the same child's dosage?

66. *Tax Reform* One plan for federal income tax reform is to tax an individual's income in excess of $15,000 at a 17% rate. Another plan is to institute a national retail sales tax of 15%. If an individual spends 75% of his or her income in retail stores where it is taxed at 15%, then for what income would the amount of tax be the same under either plan?

A system of equations can be used to find the equation of a line that goes through two points. For example, if $y = ax + b$ goes through (3, 5), then a and b must satisfy $3a + b = 5$. For each given pair of points, find the equation of the line $y = ax + b$ that goes through the points by solving a system of equations.

67. $(-3, 9), (2, -1)$

68. $(1, -1), (3, 7)$

69. $(-2, 3), (4, -7)$

70. $(-3, -1), (4, 9)$

For Writing/Discussion

71. *Number of Solutions* Explain how you can tell (without graphing) whether a system of linear equations has one solution, no solutions, or infinitely many solutions. Be sure to account for linear equations that are not functions.

72. *Cooperative Learning* Write a step-by-step procedure (or algorithm) based on the addition method that will solve any system of two equations of the form $Ax + By = C$. Ask a classmate to solve a system using your procedure.

73. *Cooperative Learning* Write an independent system of two linear equations for which $(2, -3)$ is the solution. Ask a classmate to solve your system.

74. *Cooperative Learning* Write a dependent system of two linear equations for which $\{(t, t + 5) \mid t \text{ is any real number}\}$ is the solution set. Ask a classmate to solve your system.

Linking Concepts

For Individual or Group Explorations

Life Expectancy The accompanying table gives the life expectancy at birth for U.S. men and women (Reader's Digest, Monitoring Your Health, 1991)

Year of birth	Life expectancy (Male)	Life expectancy (Female)
1930	58.1	61.6
1940	60.8	65.2
1950	65.6	71.1
1960	66.6	73.1
1970	67.1	74.7
1980	70.0	77.4
1990	71.7	78.6

a) Use linear regression on your graphing calculator to find the life expectancy for men as a function of the year of birth.

b) Use linear regression on your graphing calculator to find the life expectancy for women as a function of the year of birth.

c) Graph the functions that you found in parts (a) and (b) on the same coordinate system.

d) According to this model, will men ever catch up to women in life expectancy?

e) For what year of birth did men and women have the same life expectancy?

f) Interpret the slope of these two lines.

g) Why do you think that life expectancy for women is increasing at a greater rate than life expectancy for men?

5.2 Systems of Linear Equations in Three Variables

Systems of many linear equations in many variables are used to model a variety of situations ranging from airline scheduling to allocating resources in manufacturing. The same techniques that we are studying with small systems can be extended to much larger systems. In this section we use the techniques of substitution and addition from Section 5.1 to solve systems of linear equations in three variables.

Definitions

A **linear equation in three variables** x, y, and z is an equation of the form

$$Ax + By + Cz = D$$

where A, B, C, and D are real numbers with A, B, and C not all equal to zero. For example,

$$x + y + 2z = 9$$

is a linear equation in three variables. The equation is called *linear* because its form is similar to that of a linear equation in two variables. A solution to a linear equation in three variables is an **ordered triple** of real numbers in the form (x, y, z) that satisfies the equation. For instance, the ordered triple $(1, 2, 3)$ is a solution to

$$x + y + 2z = 9$$

because $1 + 2 + 2(3) = 9$. Other ordered triples such as $(4, 5, 0)$ or $(3, 4, 1)$ are also in the solution set to $x + y + 2z = 9$. In fact, there are infinitely many ordered triples in the solution set to a linear equation in three variables.

The graph of the solution set of a linear equation in three variables requires a three-dimensional coordinate system. A three-dimensional coordinate system has a z-axis through the origin of the xy-plane, as shown in Fig. 5.11. The third coordinate of a point indicates its distance above or below the xy-plane. The point $(1, 2, 3)$ is shown in Fig. 5.11.

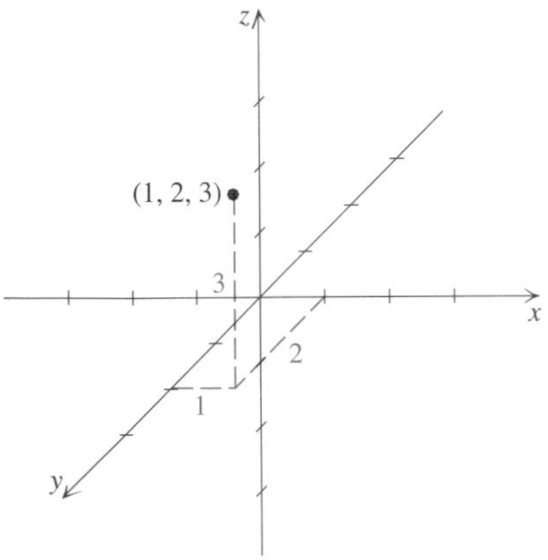

Figure 5.11

The graph of a linear equation in three variables is a plane and not a line as the name might suggest. We think of a plane as an infinite sheet of paper (with no edges), but that is difficult to draw. One way to draw a representation of a plane in a three-dimensional coordinate system is to draw a triangle whose vertices are the points of intersection of the plane and the axes.

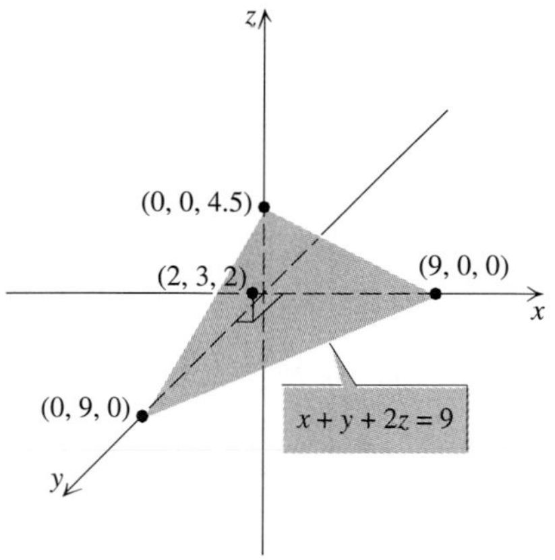

Figure 5.12

EXAMPLE 1

Sketch the graph of $x + y + 2z = 9$ in a three-dimensional coordinate system by locating the three intercepts. Find one additional point that satisfies the equation and plot it.

Solution

Let x and y both be zero in the equation:

$$0 + 0 + 2z = 9$$
$$z = 4.5$$

So (0, 0, 4.5) is the z-intercept of the plane. If x and z are both zero, then we get $y = 9$. So (0, 9, 0) is the y-intercept. If y and z are both zero, then we get $x = 9$. So (9, 0, 0) is the x-intercept. Plot the three intercepts and draw a triangle as shown in Fig. 5.12. If $x = 2$ and $y = 3$, then $z = 2$. So (2, 3, 2) satisfies the equation and appears to be on the plane when it is plotted as in Fig. 5.12. Of course there is more to a plane than the triangle in Fig. 5.12, but the triangle gives us an idea of the location of the plane.

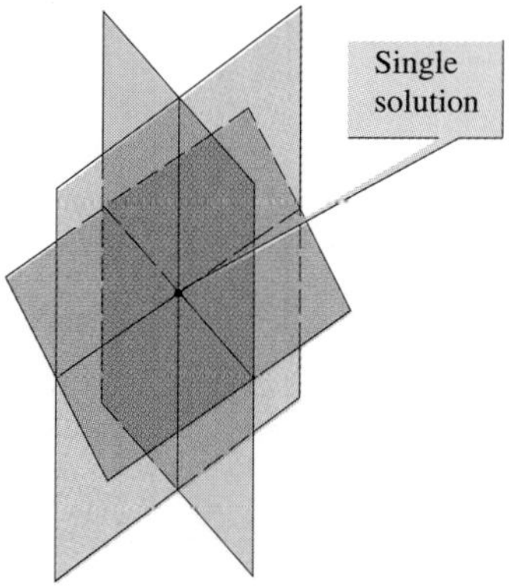

Independent System of Three Equations

Figure 5.13

It is fairly easy to draw a triangle through the intercepts to represent a plane as we did in Example 1. However, if we were to draw two or three planes in this manner, it could be very difficult to see how they intersect. So we will often refer to the graphs of the equations to aid in understanding a system, but we will not attempt to solve a system by graphing. We will solve systems by using the algebraic methods of substitution, addition, or a combination of both.

Independent Systems

As with linear systems in two variables, a linear system in three variables can have one, infinitely many, or zero solutions. If the three equations correspond to three planes that intersect at a single point, as in Fig. 5.13, then the solution to the system is a single ordered triple. In this case, the system is called **independent**. Note that for simplicity, Fig. 5.13 is drawn without showing the coordinate axes.

EXAMPLE 2 An independent system of equations

Solve the system.

(1) $x + y - z = 0$

(2) $3x - y + 3z = -2$

(3) $x + 2y - 3z = -1$

Solution

Look for a variable that is easy to eliminate by addition. Since y occurs in Eq. (1) and $-y$ occurs in Eq. (2), we can eliminate y by adding Eqs. (1) and (2):

$$\begin{array}{lrcl} & x + y - z &=& 0 \\ & 3x - y + 3z &=& -2 \\ \hline (4) & 4x \quad + 2z &=& -2 \end{array}$$

Now repeat the process to eliminate y from Eqs. (1) and (3). Multiply Eq. (1) by -2 and add the result to Eq. (3):

$$\begin{array}{lrcll} & -2x - 2y + 2z &=& 0 & \text{Eq. (1) multiplied by } -2 \\ & x + 2y - 3z &=& -1 & \text{Eq. (3)} \\ \hline (5) & -x \quad - z &=& -1 & \end{array}$$

Equations (4) and (5) are a system of two linear equations in two variables. We could solve this system by substitution or addition. To solve by addition, multiply Eq. (5) by 2 and add to Eq. (4) to eliminate z:

$$\begin{array}{rcll} 4x + 2z &=& -2 & \text{Eq. (4)} \\ -2x - 2z &=& -2 & \text{Eq. (5) multiplied by 2} \\ \hline 2x &=& -4 & \\ x &=& -2 & \end{array}$$

Use $x = -2$ in $4x + 2z = -2$ to find z:

$$\begin{aligned} 4(-2) + 2z &= -2 \\ 2z &= 6 \\ z &= 3 \end{aligned}$$

To find y, use $x = -2$ and $z = 3$ in $x + y - z = 0$ (Eq. 1):

$$\begin{aligned} -2 + y - 3 &= 0 \\ y &= 5 \end{aligned}$$

Check that the ordered triple $(-2, 5, 3)$ satisfies all three of the original equations:

$$\begin{array}{lrcll} (1) & -2 + 5 - 3 &=& 0 & \text{Correct.} \\ (2) & 3(-2) - 5 + 3(3) &=& -2 & \text{Correct.} \\ (3) & -2 + 2(5) - 3(3) &=& -1 & \text{Correct.} \end{array}$$

The solution set is $\{(-2, 5, 3)\}$.

When solving a system of three equations in three variables, we try to get a system of two equations in two variables by eliminating one of the variables. Any one of the three variables can be eliminated first, but we usually look for the

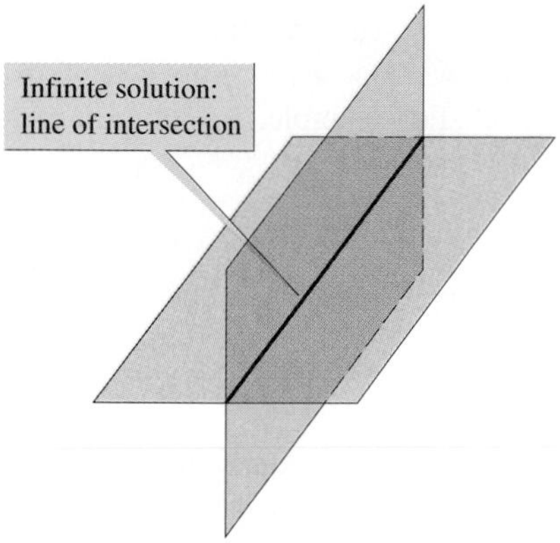

Dependent System of Two Equations

Figure 5.14

easiest one. We may eliminate that variable from the first and second, the first and third, or the second and third equations. For example, we could have started Example 3 by adding Eqs. (2) and (3) to eliminate z. We then repeat the process, eliminating the same variable from another pair of equations in the system. After getting two equations in two unknowns, we solve that system using methods for systems involving two variables.

Systems with Infinite Solution Sets

Two planes in three-dimensional space either are parallel or intersect along a line as shown in Fig. 5.14. If the planes are parallel, there is no common point and no solution to the system. If the two planes intersect along a line, then there are infinitely many points on that line that satisfy both equations of the system. In our next example we solve a system of two equations whose graphs are planes that intersect along a line.

EXAMPLE 3 Two linear equations in three variables

Solve the system

$$(1) \qquad -2x + 3y - z = -1$$

$$(2) \qquad x - 2y + z = 3$$

Solution

Add the equations to eliminate z:

$$\begin{array}{rl} & -2x + 3y - z = -1 \\ & \underline{\quad x - 2y + z = 3} \\ & -x + y \qquad = 2 \\ (3) & \qquad y = x + 2 \end{array}$$

Equation (3) indicates that (x, y, z) satisfies both equations if and only if $y = x + 2$. Write Eq. (2) as $z = 3 - x + 2y$ and substitute $y = x + 2$ into this equation:

$$z = 3 - x + 2(x + 2)$$

$$z = x + 7$$

Now (x, y, z) satisfies (1) and (2) if and only if $y = x + 2$ and $z = x + 7$. So the solution set to the system could be written as

$$\{(x, y, z) \mid y = x + 2 \text{ and } z = x + 7\}$$

or more simply

$$\{(x, x + 2, x + 7) \mid x \text{ is any real number}\}.$$

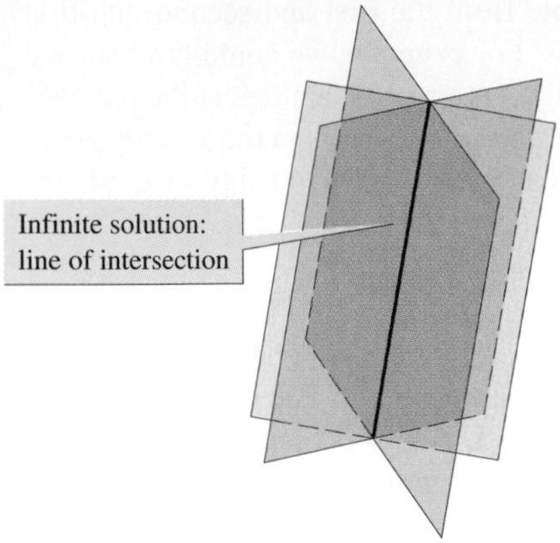

Dependent System of Three Equations

Figure 5.15

This system has infinitely many solutions. Note that we can write the solution set in terms of x, y, or z, depending on which variable we choose to eliminate in the first step and which variable we solve for after that. For example, the solution set could also be written as

$$\{(z - 7, z - 5, z) \mid z \text{ is any real number}\},$$

because $z = x + 7$.

A line in three-dimensional space does not have a simple equation like a line in a two-dimensional coordinate system. In Example 3, the triple $(x, x + 2, x + 7)$ is a point on the line of intersection of the two planes, for any real number x. For example, the points $(0, 2, 7)$, $(1, 3, 8)$, and $(-2, 0, 5)$ all satisfy both equations of the system and lie on the line of intersection of the planes.

In the next example we solve a system consisting of three planes that intersect along a single line, as shown in Fig. 5.15. This example is similar to Example 3.

EXAMPLE 4 A dependent system of three equations

Solve the system.

$$\begin{aligned}
&(1) \qquad & 2x + y - z &= 1\\
&(2) \qquad & -3x - 3y + 2z &= 1\\
&(3) \qquad & -10x - 14y + 8z &= 10
\end{aligned}$$

Solution

Examine the system and decide which variable to eliminate. If Eq. (1) is multiplied by 2, then $-2z$ will appear in Eq. (1) and $2z$ in Eq. (2). So, multiply Eq. (1) by 2 and add the result to Eq. (2) to eliminate z:

$$\begin{aligned}
& & 4x + 2y - 2z &= 2 & &\text{Eq. (1) multiplied by 2}\\
& & -3x - 3y + 2z &= 1 & &\text{Eq. (2)}\\
&(4) & x - y &= 3
\end{aligned}$$

Now eliminate z from Eqs. (2) and (3) by multiplying Eq. (2) by -4 and adding the result to Eq. (3):

$$\begin{aligned}
& & 12x + 12y - 8z &= -4 & &\text{Eq. (2) multiplied by } -4\\
& & -10x - 14y + 8z &= 10 & &\text{Eq. (3)}\\
&(5) & 2x - 2y &= 6
\end{aligned}$$

Note that Eq. (5) is a multiple of Eq. (4). Since Eqs. (4) and (5) are dependent, the original system has infinitely many solutions. We can describe all solutions in terms of the single variable x. To do this, get $y = x - 3$ from Eq. (4). Then substitute $x - 3$ for y in Eq. (1) to find z in terms of x:

$$z = 2x + y - 1 \qquad \textbf{Eq. (1) solved for } z$$
$$z = 2x + (x - 3) - 1 \qquad \textbf{Replace } y \textbf{ by } x - 3.$$
$$z = 3x - 4$$

An ordered triple (x, y, z) satisfies the system provided $y = x - 3$ and $z = 3x - 4$. So the solution set to the system is $\{(x, x - 3, 3x - 4) | x \text{ is any real number}\}$.

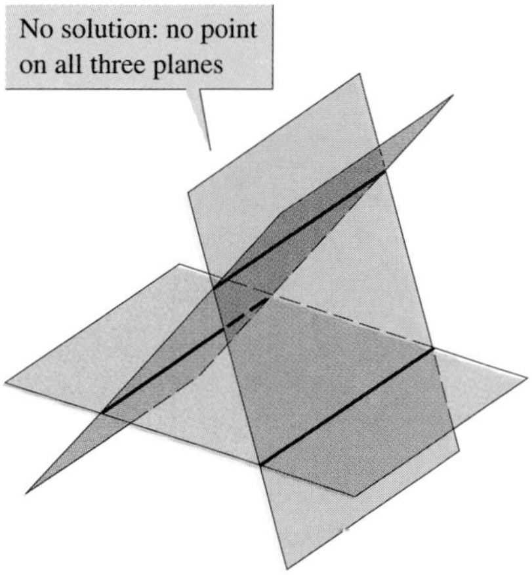

Inconsistent System of Three Equations

Figure 5.16

If all of the original equations are equivalent, then the solution set to the system is the set of all points that satisfy one of the equations. For example, the system

$$x + y + z = 1$$
$$2x + 2y + 2z = 2$$
$$3x + 3y + 3z = 3$$

has solution set $\{(x, y, z) | x + y + z = 1\}$.

Inconsistent Systems

There are several ways that three planes can be positioned so that there is no single point that is on all three planes. For example, Fig. 5.16 corresponds to a system where there are ordered triples that satisfy two equations, but no ordered triple that satisfies all three equations. Whatever the configuration of the planes, if there are no points in common to all three, the corresponding system is called **inconsistent** and has no solution. It is easy to identify a system that has no solution because a false statement will occur when we try to solve the system.

EXAMPLE 5 A system with no solution

Solve the system.

$$(1) \qquad x + y - z = 5$$
$$(2) \qquad x + 2y - 3z = 9$$
$$(3) \qquad x - y + 3z = 3$$

Solution

Multiply Eq. (1) by -2 and add the result to Eq. (2):

$$-2x - 2y + 2z = -10 \qquad \textbf{Eq. (1) multiplied by } -2$$
$$x + 2y - 3z = 9 \qquad \textbf{Eq. (2)}$$
$$(4) \qquad -x \qquad - z = -1$$

Add Eq. (1) and Eq. (3) to eliminate y:

$$\begin{array}{rll} x + y - z = 5 & & \text{Eq. (1)} \\ x - y + 3z = 3 & & \text{Eq. (3)} \\ \hline 2x + 2z = 8 & & \\ (5) \quad x + z = 4 & & \end{array}$$

Now add Eq. (4) and Eq. (5):

$$\begin{array}{rl} -x - z = -1 & \text{Eq. (4)} \\ x + z = 4 & \text{Eq. (5)} \\ \hline 0 = 3 & \end{array}$$

Since $0 = 3$ is false, the system is inconsistent. There is no solution.

Applications

It is a fairly simple matter to write the equation of a line when given two points on the line, but it is a bit more complicated to write the equation of a parabola when given three points on the parabola. Surprisingly, the equation of a parabola through three points can be found by solving a system of three linear equations. The process of finding a curve that goes through some given points is called **curve fitting**.

In the next example, we fit a parabola to some given data points to determine the most economic flight speed for a parakeet in level flight.

EXAMPLE 6 Power expenditure of birds in flight

In studying how migratory birds can fly thousands of miles without rest, Vance Tucker gathered data on parakeets and sea gulls in a wind tunnel (*Scientific American*, May 1969). Tucker found that a parakeet in level flight at 12 mph has a *power expenditure* of 152. (That is, it requires 152 calories to move each gram of body weight at that speed for one hour.) At 22 mph the parakeet's power

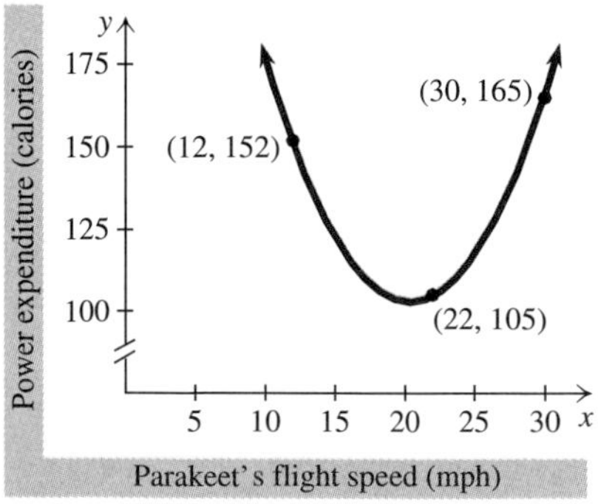

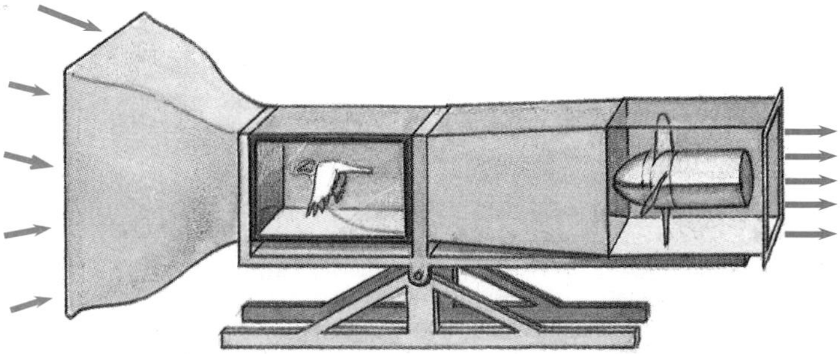

Figure 5.17

expenditure is 105, and at 30 mph its power expenditure is 165. These points are graphed in Fig. 5.17. Tucker's data suggest that power expenditure is a quadratic function of the flying speed. Find the quadratic function whose graph goes through the three given points and find the speed at which the power expenditure is a minimum.

Solution

The ordered pairs (12, 152), (22, 105), and (30, 165) satisfy $y = ax^2 + bx + c$, where x is the speed and y is the power expenditure. We get three equations by using values of x and y from the ordered pairs in the equation $y = ax^2 + bx + c$.

(1) $144a + 12b + c = 152$ Use $x = 12$ and $y = 152$.

(2) $484a + 22b + c = 105$ Use $x = 22$ and $y = 105$.

(3) $900a + 30b + c = 165$ Use $x = 30$ and $y = 165$.

The easiest variable to eliminate in this system is c. Multiplying Eq. (1) by -1 and adding it to Eq. (2) yields

(4) $340a + 10b = -47.$

Multiplying Eq. (2) by -1 and adding it to Eq. (3) yields

$$416a + 8b = 60.$$

Dividing each side of this equation by 4 (to simplify it) yields

(5) $104a + 2b = 15.$

We now solve the system consisting of Eqs. (4) and (5). Multiplying Eq. (5) by -5 and adding it to Eq. (4), we get

$$-180a = -122$$
$$a \approx 0.6778.$$

Substituting $a = 0.6778$ into Eq. (5) to find b, we get

$$b \approx -27.7446.$$

Using the values of a and b in Eq. (1) gives

$$c \approx 387.3349.$$

The equation

$$y = 0.6778x^2 - 27.7446x + 387.3349$$

expresses power expenditure y as a function of speed x. The minimum value of y occurs at the vertex of the parabola, at $x = -b/(2a)$:

$$x = \frac{27.7446}{2(0.6778)} \approx 20.5$$

So the parakeet expends the least amount of energy flying at about 20.5 mph. In fact, other research has confirmed that many migrating birds fly (without the influence of the wind) at speeds of 20 to 25 mph.

The equation for the parabola in Example 6 could be found with a graphing calculator. See Exercise 52.

The next example involves three unknown prices, but we are given enough information to write three linear equations concerning those prices.

EXAMPLE 7 A problem involving three unknowns

Lionel delivers milk, bread, and eggs to Marcie's Camp Store. On Monday the bill for eight half-gallons of milk, four loaves of bread, and six dozen eggs was \$14.20. On Tuesday the bill for five half-gallons of milk, ten loaves of bread, and three dozen eggs was \$14.50. On Wednesday the bill for two half-gallons of milk, five loaves of bread, and seven dozen eggs was \$9.50. How much is Thursday's bill for one half-gallon of milk, two loaves of bread, and one dozen eggs?

Solution

Let x represent the price of a half-gallon of milk, y represents the price of a loaf of bread, and z represent the price of a dozen eggs. We can write an equation for the bill on each of the three days:

$$\begin{array}{ll} (1) & 8x + 4y + 6z = 14.20 \\ (2) & 5x + 10y + 3z = 14.50 \\ (3) & 2x + 5y + 7z = 9.50 \end{array}$$

Multiply Eq. (2) by -2 and add the result to Eq. (1):

$$\begin{array}{lll} & 8x + 4y + 6z = 14.20 & \text{Eq. (1)} \\ & -10x - 20y - 6z = -29 & \text{Eq. (2) multiplied by } -2 \\ \hline & -2x - 16y = -14.80 & \\ (4) & x + 8y = 7.40 & \end{array}$$

Multiply Eq. (2) by -7, multiply Eq. (3) by 3, and add the results:

$$\begin{array}{lll} & -35x - 70y - 21z = -101.50 & \text{Eq. (2) multiplied by } -7 \\ & 6x + 15y + 21z = 28.50 & \text{Eq. (3) multiplied by } 3 \\ \hline (5) & -29x - 55y = -73 & \end{array}$$

Multiply Eq. (4) by 29 and add the result to Eq. (5) to eliminate x:

$$\begin{array}{ll} 29x + 232y = 214.60 & \text{Eq. (4) multiplied by 29} \\ -29x - 55y = -73 & \text{Eq. (5)} \\ \hline 177y = 141.60 & \\ y = 0.80 & \end{array}$$

Use $y = 0.80$ in $x + 8y = 7.40$ (Eq. 4):

$$x + 8(0.80) = 7.40$$
$$x + 6.40 = 7.40$$
$$x = 1$$

Use $x = 1$ and $y = 0.80$ in $8x + 4y + 6z = 14.20$ (Eq. 1) to find z:

$$8(1) + 4(0.80) + 6z = 14.20$$
$$6z = 3$$
$$z = 0.50$$

So milk is \$1.00 per half-gallon, bread is 80 cents per loaf, and eggs are 50 cents per dozen. Thursday's bill should be \$3.10.

FOR THOUGHT True or False? Explain.

The following systems are referenced in statements 1–7:

(a) $\begin{aligned} x + y - z &= 2 \\ -x - y + z &= 4 \\ x - 2y + 3z &= 9 \end{aligned}$

(b) $\begin{aligned} x + y - z &= 6 \\ x + y + z &= 4 \\ x - y - z &= 8 \end{aligned}$

(c) $\begin{aligned} x - y + z &= 1 \\ -x + y - z &= -1 \\ 2x - 2y + 2z &= 2 \end{aligned}$

1. The point $(1, 1, 0)$ is in the solution set to $x + y - z = 2$.
2. The point $(1, 1, 0)$ is in the solution set to system (a).
3. System (a) is inconsistent.
4. The point $(2, 3, -1)$ satisfies all equations of system (b).
5. The point $(6, -1, -1)$ satisfies all equations of system (b).
6. The solution set to system (c) is $\{(x, y, z) | x - y + z = 1\}$.
7. System (c) is dependent.
8. The solution set to $y = 2x + 3$ is $\{(x, 2x + 3) | x \text{ is any real number}\}$.
9. $(3, 1, 0) \in \{(x + 2, x, x - 1) | x \text{ is any real number}\}$.
10. x nickels, y dimes, and z quarters are worth $5x + 10y + 25z$ dollars.

5.2 EXERCISES

Tape 16 Disk

Sketch the graph of each equation in a three-dimensional coordinate system.

1. $x + y + z = 5$
2. $x + 2y + z = 6$
3. $x + y - z = 3$
4. $2x + y - z = 6$

Solve each system of equations.

5. $\begin{aligned} x + y + z &= 6 \\ 2x - 2y - z &= -5 \\ 3x + y - z &= 2 \end{aligned}$

6. $\begin{aligned} 3x - y + 2z &= 14 \\ x + y - z &= 0 \\ 2x - y + 3z &= 18 \end{aligned}$

7. $\begin{aligned} 3x + 2y + z &= 1 \\ x + y - 2z &= -4 \\ 2x - 3y + 3z &= 1 \end{aligned}$

8. $\begin{aligned} 4x - 2y + z &= 13 \\ 3x - y + 2z &= 13 \\ x + 3y - 3z &= -10 \end{aligned}$

9. $\begin{aligned} 2x + y - 2z &= -15 \\ 4x - 2y + z &= 15 \\ x + 3y + 2z &= -5 \end{aligned}$

10. $\begin{aligned} x - 2y - 3z &= 4 \\ 2x - 4y + 5z &= -3 \\ 5x - 6y + 4z &= -7 \end{aligned}$

11. $\begin{aligned} x - 2y + 3z &= 5 \\ 2x - 4y + 6z &= 3 \\ 2x - 3y + z &= 9 \end{aligned}$

12. $\begin{aligned} -2x + y - 3z &= 6 \\ 4x - y + z &= 2 \\ 2x - y + 3z &= 1 \end{aligned}$

13. $\begin{aligned} x + 2y - 3z &= -17 \\ 3x - 2y - z &= -3 \end{aligned}$

14. $\begin{aligned} x + 2y + z &= 4 \\ 2x - y - z &= 3 \end{aligned}$

15. $\begin{aligned} x + 2y - 3z &= 5 \\ -x - 2y + 3z &= -5 \\ 2x + 4y - 6z &= 10 \end{aligned}$

16. $\begin{aligned} 2x - 6y + 4z &= 8 \\ 3x - 9y + 6z &= 12 \\ 5x - 15y + 10z &= 20 \end{aligned}$

17. $\begin{aligned} x + y - z &= 2 \\ 2x - y + z &= 4 \end{aligned}$

18. $\begin{aligned} -2x + 2y - z &= 4 \\ 2x - y + z &= 1 \end{aligned}$

19. $\begin{aligned} x + y &= 5 \\ y - z &= 2 \\ x + z &= 3 \end{aligned}$

20. $\begin{aligned} 2x - y &= -1 \\ -2x + z &= 1 \\ y - z &= 0 \end{aligned}$

21. $\begin{aligned} x - y + z &= 7 \\ 2y - 3z &= -13 \\ 3x - 2z &= -3 \end{aligned}$

22. $\begin{aligned} 2x + y - z &= 5 \\ 2y + 3z &= -14 \\ -3y - 2z &= 11 \end{aligned}$

23. $\begin{aligned} x + y + 2z &= 7.5 \\ 3x + 4y + z &= 12 \\ 5x + 2y + 5z &= 21 \end{aligned}$

24. $\begin{aligned} 100x + 200y + 500z &= 47 \\ 350x + 5y + 250z &= 33.9 \\ 200x + 80y + 100z &= 23.4 \end{aligned}$

25. $\begin{aligned} x + y + z &= 9000 \\ 0.05x + 0.06y + 0.09z &= 710 \\ z &= 3y \end{aligned}$

26. $\begin{aligned} x + y + z &= 200{,}000 \\ 0.09x + 0.08y + 0.12z &= 20{,}200 \\ z &= x + y \end{aligned}$

27. $\begin{aligned} x &= 2y - 1 \\ y &= 3z + 2 \\ z &= 2x - 3 \end{aligned}$

28. $\begin{aligned} x + 2y - 3z &= 0 \\ 2x - y + z &= 0 \\ 3x + y - 4z &= 0 \end{aligned}$

Use a system of equations to find the parabola of the form $y = ax^2 + bx + c$ that goes through the three given points.

29. $(-1, -2), (2, 1), (-2, 1)$

30. $(1, 2), (2, 3), (3, 6)$

31. $(0, 0), (1, 3), (2, 2)$

32. $(0, -6), (1, -3), (2, 6)$

33. $(0, 4), (-2, 0), (-3, 1)$

34. $(0, 6), (3, 0), (-1, 12)$

Write a linear equation in three variables that is satisfied by all three of the given ordered triples.

35. $(0, 0, 1), (0, 1, 0), (1, 0, 0)$

36. $(0, 0, 2), (0, 1, 0), (1, 0, 0)$

37. $(1, 1, 1), (0, 2, 0), (1, 0, 0)$

38. $(1, 0, 1), (2, 1, 0), (0, 2, 1)$

Solve each problem by using a system of three linear equations in three variables.

39. *Efficiency for Descending Flight* In studying flight of birds (*Scientific American*, May 1969), Vance Tucker measured the efficiency (the relationship between power input and power output) for parakeets flying at various speeds in a descending flight pattern. He recorded an efficiency of 0.18 at 12 mph, 0.23 at 22 mph, and 0.14 at 30 mph. Tucker's measurements suggest that efficiency E is a quadratic function of the speed s. Find the quadratic function whose graph goes through the three given ordered pairs, and find the speed that gives the maximum efficiency for descending flight.

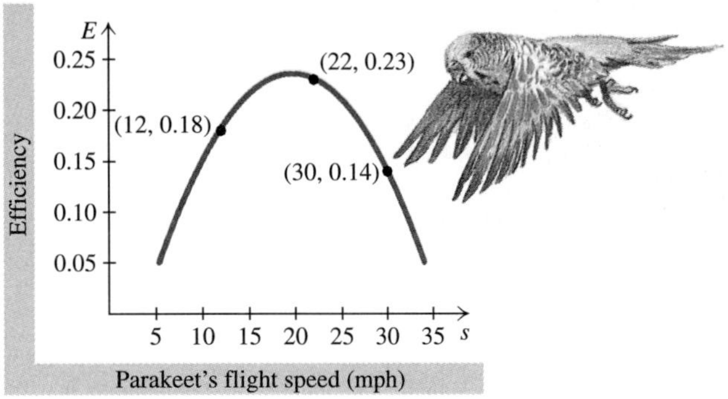

Figure for Exercise 39

40. *Path of a Missile* A missile is fired from the origin in the direction of the positive x-axis as shown in the drawing. Radar shows the missile at 4000 ft when it is 1 mi from the origin and 7000 ft when it is 2 mi from the origin. Assuming that the path of the missile is parabolic, find the

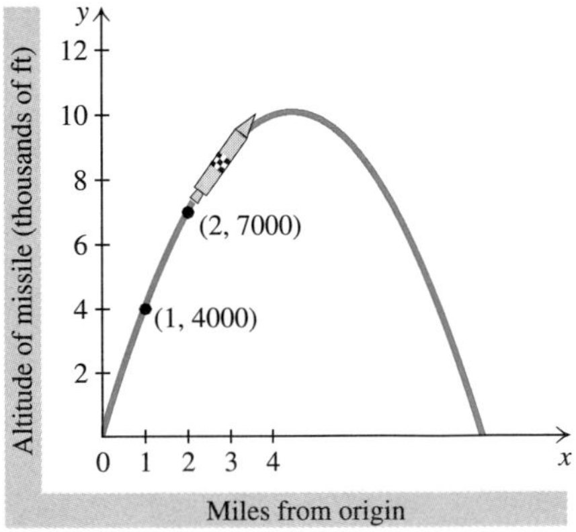

Figure for Exercise 40

equation of the parabola. What is the highest altitude reached by the missile? How many miles from the origin will the missile strike?

41. *Stocks, Bonds, and a Mutual Fund* Marita invested a total of \$25,000 in stocks, bonds, and a mutual fund. In one year she earned 8% on her stock investment, 10% on her bond investment, and 6% on her mutual fund, with a total return of \$1860. Unfortunately, the amount she invested in the mutual fund was twice as large as the amount she invested in the bonds. How much did she invest in each?

42. *Age Groups* In 1980 the population of Springfield was 1911. In 1990 the number of people under 20 years old increased 10%, while the number of people in the 20 to 60 category decreased by 8%, and the number of people over 60 increased by one-third. If the 1990 population was 2136 and in 1990 the number of people over 60 was equal to the number of people 60 and under, then how many were in each age group in 1980?

43. *Fast Food Inflation* Last year you could get a hamburger, fries, and a Coke at Francisco's Drive-In for \$1.90. Since the price of a hamburger has increased 10%, the price of fries has increased 20%, and the price of a Coke has increased 50%, the same meal now costs \$2.37. If the price of a Coke is now 9 cents more than that of a hamburger, then what was the price of each item last year?

44. *Misplaced House Numbers* Angelo on Elm Street removed his three-digit house number for painting and noticed that the sum of the digits was 9 and that the units digit was three times as large as the hundreds digit. When the painters put the house number back up, they reversed the digits. The new house number was now 396 larger than the correct house number. What is Angelo's correct address?

45. *Weight Distribution* A driver of a 1200-pound race car wants to have 51% of the car's weight on the left front and left rear tires and 48% of the car's weight on the left rear and right rear tires. If there must be at least 280 pounds on every tire, then find three possible weight distributions. Answers may vary.

46. *Burgers, Fries, and Cokes* Jennifer bought 5 burgers, 7 orders of fries, and 6 Cokes for \$11.25. Marylin bought 6 burgers, 8 orders of fries, and 7 Cokes for \$13.20. John wants to buy an order of fries from Marylin for \$0.80, but Marylin says that she paid more than that for the fries. What do you think? Find three possibilities for the prices of the burgers, fries, and Cokes. Answers may vary.

47. *Distribution of Coins* Emma paid the \$10.36 bill for her lunch with 232 coins consisting of pennies, nickels, and dimes. If the number of nickels plus the number of dimes was equal to the number of pennies, then how many coins of each type did she use?

48. *Students, Teachers, and Pickup Trucks* Among the 564 students and teachers at Jefferson High School, 128 drive to school each day. One-fourth of the male students, one-sixth of the female students, and three-fourths of the teachers drive. Among those who drive to school, there are 41 who drive pickup trucks. If one-half of the driving male students, one-tenth of the driving female students, and one-third of the driving teachers drive pickups, then how many male students, female students, and teachers are there?

49. *Milk, Coffee, and Doughnuts* The employees from maintenance go for coffee together every day at 9 A.M. On Monday, Hector paid \$5.45 for three cartons of milk, four cups of coffee, and seven doughnuts. On Tuesday, Guillermo paid \$5.30 for four milks, two coffees, and eight doughnuts. On Wednesday, Anna paid \$5.15 for two milks, five coffees, and six doughnuts. On Thursday, Alphonse had to pay for five milks, two coffees, and nine doughnuts. How much change did he get back from his \$10 bill?

50. *Average Age of Vehicles* The average age of the Johnsons' cars is eight years. Three years ago the Toyota was twice as old as the Ford. Two years ago the sum of the Buick's and the Ford's ages was equal to the age of the Toyota. How old is each car now?

51. *Fish Mobile* A sculptor is designing a fish mobile as shown in the figure. The weights of the horizontal bars and strings are negligible compared to the cast iron fish. The

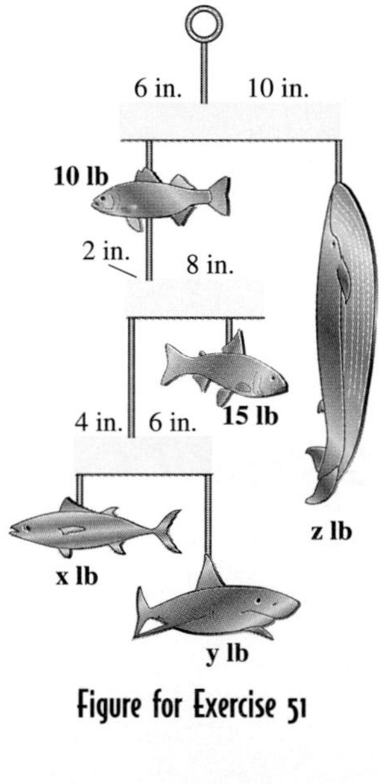

Figure for Exercise 51

mobile will balance if the product of the weight and distance on one side of the balance point is equal to the product of the weight and distance on the other side. How much must the bottom three fish weigh for the mobile to be balanced? (Adapted from Discover, May 1997)

52. *Quadratic Regression* Use the data from Example 6 and quadratic regression on a graphing calculator to find the equation of the parabola that passes through the given points. Then use your calculator to find the x-coordinate corresponding to the minimum y-coordinate on the graph of a parabola. Compare your results to those in Example 6.

For Writing/Discussion

53. *Cooperative Learning* Write a system of three linear equations in three unknowns for which (1/2, 1/3, 1/4) is the only solution. Ask a classmate to solve the system.

54. *Cooperative Learning* Write a word problem for which a system of three linear equations in three unknowns can be used to find the solution. Ask a classmate to solve the problem.

LINKING CONCEPTS

For Individual or Group Explorations

Accounting Problems *For Class C corporations in Louisiana the amount of state income tax is deductible on the federal income tax return and the amount of federal income tax is deductible on the state return. Assume that the state tax rate is 5%, the federal tax rate is 30%, and the corporation has an income of $200,000 before taxes.*

a) Write the amount of state tax as a function of the amount of federal tax.

b) Write the amount of federal tax as a function of the amount of state tax.

c) Solve your system to find the amount of state tax and the amount of federal tax.

d) If the corporation wants to give a 20% bonus to employees, the accountant deducts the amount of the bonus, the state tax, and the federal tax from the $200,000 income to get the amount on which a 20% bonus is computed. The bonus and sate tax are deductible before the federal tax is computed, and the federal tax and bonus are deductible before the state tax is computed. Find the amount of the bonus, the federal tax, and the state tax for this corporation.

5.3 Nonlinear Systems of Equations

In Sections 5.1 and 5.2 we solved linear systems of equations, but we have seen many equations in two variables that are not linear. Equations such as

$$y = 3x^2, \quad y = \sqrt{x}, \quad y = |x|, \quad y = 10^x, \quad \text{and} \quad y = \log(x)$$

are called *nonlinear equations*, because their graphs are not straight lines. If a system has at least one nonlinear equation, it is called a **nonlinear system**. Systems of nonlinear equations arise in applications just as systems of linear equations do. In this section we will use the techniques that we learned for linear systems to solve nonlinear systems.

Solving by Elimination of Variables

To solve nonlinear systems, we combine equations to eliminate variables just as we did for linear systems. However, since the graphs of nonlinear equations are not straight lines, the graphs might intersect at more than one point, and the solution set might contain more than one point. The next examples show systems whose solution sets contain two points, four points, and one point.

EXAMPLE 1 A parabola and a line

Solve the system of equations and sketch the graph of each equation on the same coordinate plane.

$$y = x^2 + 1$$
$$y - x = 2$$

Solution

The graph of $y = x^2 + 1$ is a parabola opening upward. The graph of $y = x + 2$ is a line with y-intercept $(0, 2)$ and slope 1. The graphs are shown in Fig. 5.18. To find the exact coordinates of the points of intersection of the graphs, we solve the system by substitution. Substitute $y = x^2 + 1$ into $y - x = 2$:

$$x^2 + 1 - x = 2$$
$$x^2 - x - 1 = 0$$

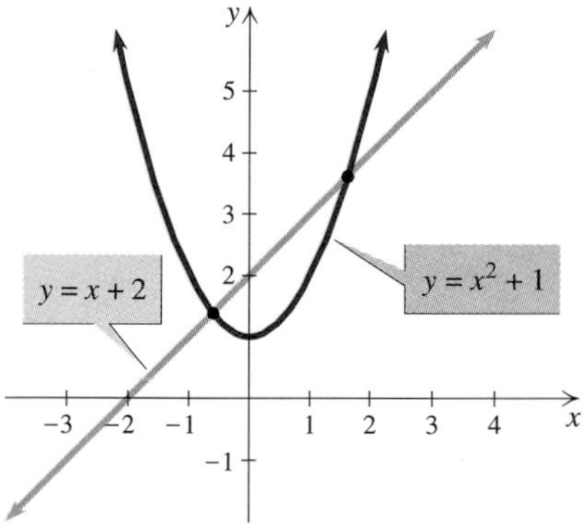

Figure 5.18

Solve this quadratic equation using the quadratic formula:

$$x = \frac{1 \pm \sqrt{1 - 4(1)(-1)}}{2(1)} = \frac{1 \pm \sqrt{5}}{2}$$

Use $x = (1 \pm \sqrt{5})/2$ in $y = x + 2$ to find y:

$$y = \frac{1 \pm \sqrt{5}}{2} + 2 = \frac{5 \pm \sqrt{5}}{2}$$

The solution set to the system is

$$\left\{\left(\frac{1 + \sqrt{5}}{2}, \frac{5 + \sqrt{5}}{2}\right), \left(\frac{1 - \sqrt{5}}{2}, \frac{5 - \sqrt{5}}{2}\right)\right\}.$$

To check the solution, use a calculator to find the approximations (1.62, 3.62) and (−0.62, 1.38). These points of intersection are consistent with Fig. 5.18. You should also check the decimal approximations in the original equations. Note that we could have solved this system by solving the second equation for x (or for y) and then substituting into the first.

In the next example we find the points of intersection for the graph of an absolute value function and a parabola.

EXAMPLE 2 An absolute value function and a parabola

Solve the system of equations and sketch the graph of each equation on the same coordinate plane.

$$y = |3x|$$
$$y = x^2 + 2$$

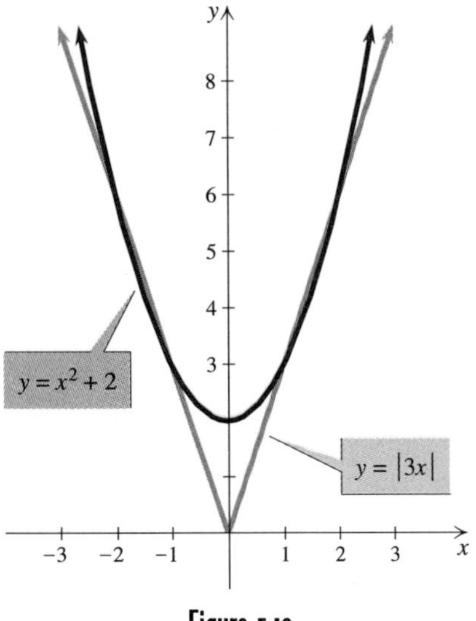

Figure 5.19

Solution

The graph of $y = x^2 + 2$ is a parabola opening upward with vertex at (0, 2). The graph of $y = |3x|$ is V-shaped and passes through (0, 0) and $(\pm 1, 3)$. Both graphs are shown in Fig. 5.19. To eliminate y, we can substitute $|3x|$ for y in the second equation:

$$|3x| = x^2 + 2$$

Write an equivalent compound equation without absolute value and solve:

$$\begin{aligned} 3x &= x^2 + 2 & \text{or} && 3x &= -(x^2 + 2) \\ x^2 - 3x + 2 &= 0 & \text{or} && x^2 + 3x + 2 &= 0 \\ (x - 2)(x - 1) &= 0 & \text{or} && (x + 2)(x + 1) &= 0 \\ x = 2 \quad \text{or} \quad x &= 1 & \text{or} && x = -2 \quad \text{or} \quad x &= -1 \end{aligned}$$

Using each of these values for x in the equation $y = |3x|$ yields the solution set $\{(-2, 6), (-1, 3), (1, 3), (2, 6)\}$. This solution set is consistent with what we see on the graphs in Fig. 5.19.

It is not necessary to draw the graphs of the equations of a nonlinear system to solve it. However, the graphs give us an idea of how many solutions to expect for the system. So graphs can be used to support our solutions.

EXAMPLE 3 Solving a nonlinear system

Solve the system of equations.

$$(1) \qquad x^2 + y^2 = 25$$
$$(2) \qquad \frac{x^2}{18} + \frac{y^2}{32} = 1$$

Solution

Write Eq. (1) as $y^2 = 25 - x^2$ and substitute into Eq. (2):

$$\frac{x^2}{18} + \frac{25 - x^2}{32} = 1$$

$$288\left(\frac{x^2}{18} + \frac{25 - x^2}{32}\right) = 288 \cdot 1 \qquad \text{The LCD for 18 and 32 is 288.}$$

$$16x^2 + 9(25 - x^2) = 288$$
$$7x^2 + 225 = 288$$
$$7x^2 = 63$$
$$x^2 = 9$$
$$x = \pm 3$$

Use $x = 3$ in $y^2 = 25 - x^2$: $\quad y^2 = 25 - 3^2$

$$y^2 = 16$$
$$y = \pm 4$$

Use $x = -3$ in $y^2 = 25 - x^2$: $\quad y^2 = 25 - (-3)^2$

$$y^2 = 16$$
$$y = \pm 4$$

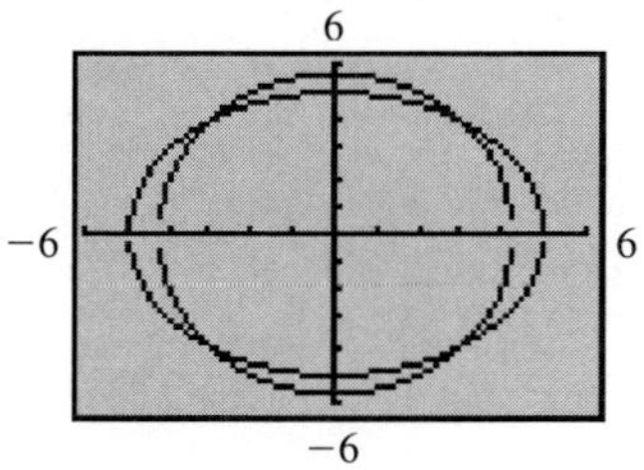

Figure 5.20

The solution set is $\{(-3, 4), (-3, -4), (3, 4), (3, -4)\}$. Check that all four ordered pairs satisfy both equations of the original system.

The graphs of $y = \pm\sqrt{25 - x^2}$ and $y = \pm\sqrt{32 - 32x^2/18}$ in Fig. 5.20 intersect at four points and support the conclusion that there are four solutions to the system.

EXAMPLE 4 Solving a nonlinear system

Solve the system of equations.

(1) $\quad \dfrac{2}{x} + \dfrac{3}{y} = \dfrac{1}{2}$

(2) $\quad \dfrac{4}{x} + \dfrac{1}{y} = \dfrac{2}{3}$

Solution

Since these equations have the same form, we can use addition to eliminate a variable. Multiply Eq. (2) by -3 and add the result to Eq. (1):

$$\frac{2}{x} + \frac{3}{y} = \frac{1}{2}$$
$$\frac{-12}{x} + \frac{-3}{y} = -2$$
$$\frac{-10}{x} = -\frac{3}{2}$$
$$3x = 20$$
$$x = \frac{20}{3}$$

Use $x = 20/3$ in Eq. (1) to find y:

$$\frac{2}{20/3} + \frac{3}{y} = \frac{1}{2}$$

$$\frac{3}{10} + \frac{3}{y} = \frac{1}{2}$$

$$3y + 30 = 5y \qquad \text{Multiply each side by } 10y.$$

$$30 = 2y$$

$$15 = y$$

The solution set is $\{(20/3, 15)\}$. Check this solution in the original system.

Applications

In the next example we solve a problem using a nonlinear system.

EXAMPLE 5 Application of a nonlinear system

A 10-inch diagonal measure television is advertised as having a viewing area of 48 square inches. What are the width and height of the screen?

Solution

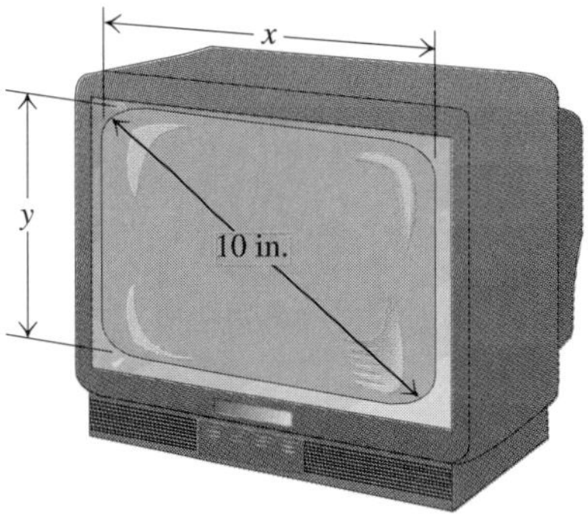

Figure 5.21

Let x be the width and y be the height as shown in Fig. 5.21. Use the Pythagorean theorem to write the first equation and the formula $A = LW$ to write the second:

$$x^2 + y^2 = 10^2$$

$$xy = 48$$

Write $xy = 48$ as $y = 48/x$ and replace y by $48/x$ in the first equation:

$$x^2 + \left(\frac{48}{x}\right)^2 = 100$$

$$x^2 + \frac{2304}{x^2} = 100$$

$$x^4 + 2304 = 100x^2$$

$$x^4 - 100x^2 + 2304 = 0 \qquad \text{An equation of quadratic type}$$

$$(x^2 - 36)(x^2 - 64) = 0 \qquad \text{Solve by factoring.}$$

$$x^2 = 36 \quad \text{or} \quad x^2 = 64$$

$$x = \pm 6 \quad \text{or} \quad x = \pm 8$$

Since x cannot be negative in this situation, x is either 6 or 8. If $x = 6$ then $y = 8$; and if $x = 8$, then $y = 6$. The width of a television screen is usually larger than the height, so we conclude that the screen is 8 inches wide and 6 inches high.

FOR THOUGHT True or False? Explain.

1. The line $y = x$ intersects the circle $x^2 + y^2 = 1$ at two points.
2. A line and a circle intersect at two points or not at all.
3. A parabola and a circle can intersect at three points.
4. The parabolas $y = x^2 - 1$ and $y = 1 - x^2$ do not intersect.
5. The line $y = x$ intersects $x^2 + y^2 = 2$ at $(1, 1)$, $(1, -1)$, $(-1, 1)$, and $(-1, -1)$.
6. Two distinct circles can intersect at more than two points.
7. The area of a right triangle is half the product of the lengths of its legs.
8. The surface area of a rectangular solid with length L, width W, and height H is $2LW + 2LH + 2WH$.
9. Two numbers with a sum of 6 and a product of 7 are $3 - \sqrt{2}$ and $3 + \sqrt{2}$.
10. It is impossible to find two numbers with a sum of 7 and a product of 1.

5.3 EXERCISES

Tape 16 Disk

Solve each system. Then graph both equations on the same coordinate system to support your solution.

1. $5x - y = 6$, $y = x^2$
2. $2x^2 - y = 8$, $7x + y = -4$
3. $y = |x| - 1$, $2y - x = 1$
4. $y = x + 3$, $y = |x|$
5. $y = \sqrt{x}$, $y = 2x$
6. $y = \sqrt{x + 3}$, $x - 4y = -7$
7. $y = x^3$, $y = 4x$
8. $y = x^2$, $x = y^2$
9. $x^2 + (y - 2)^2 = 4$, $2y = x^2$
10. $y = 2^x$, $y = 3^{-x}$
11. $y = \log_2(x)$, $y = \log_{1/3}(x)$
12. $y = x^3 - x$, $y = x$
13. $y = x^4 - x^2$, $y = x^2$
14. $x^2 + y^2 = 25$, $2x - 3y = -6$

Solve each system.

15. $x + y = -4$, $xy = 1$
16. $x + y = 10$, $xy = 21$
17. $2x^2 - y^2 = 1$, $x^2 - 2y^2 = -1$
18. $xy - 2x = 2$, $2x - y = 1$
19. $\frac{3}{x} - \frac{1}{y} = \frac{13}{10}$, $\frac{1}{x} + \frac{2}{y} = \frac{9}{10}$
20. $\frac{2}{x} + \frac{3}{2y} = \frac{11}{4}$, $\frac{5}{2x} - \frac{2}{y} = \frac{3}{2}$
21. $x^2 + xy - y^2 = -5$, $x + y = 1$
22. $x^2 + xy + y^2 = 12$, $x + y = 2$
23. $x^2 + 2xy - 2y^2 = -11$, $-x^2 - xy + 2y^2 = 9$
24. $-3x^2 + 2xy - y^2 = -9$, $3x^2 - xy + y^2 = 15$

Solve each system of exponential or logarithmic equations.

25. $y = 2^{x+1}$, $y = 4^{-x}$
26. $y = 3^{2x+1}$, $y = 9^{x}$
27. $y = \log_2(x)$, $y = \log_4(x + 2)$
28. $y = \log_2(-x)$, $y = \log_2(x + 4)$
29. $y = \log_2(x + 2)$, $y = 3 - \log_2(x)$
30. $y = \log(2x + 4)$, $y = 1 + \log(x - 2)$
31. $y = 3^x$, $y = 2^x$
32. $y = 6^{x-1}$, $y = 2^{x+1}$

Using a graphing calculator, we can solve systems that are too difficult to solve algebraically. Solve each system by graphing its equations on a graphing calculator and finding points of intersection to the nearest tenth.

33. $y = \log_2(x)$, $y = 2^x - 3$
34. $x^2 - y^2 = 1$, $y = 2^{x-3}$
35. $x^2 + y^2 = 4$, $y = \log_3(x)$
36. $y = \frac{x^3}{6} + \frac{x^2}{2} + x + 1$, $y = e^x$
37. $y = x^2$, $y = 2^x$
38. $y = e^x$, $y = x^3 + 1$

Solve each problem using a system of two equations in two unknowns.

39. *Legs of a Right Triangle* Find the lengths of the legs of a right triangle whose hypotenuse is 15 m and whose area is 54 m^2.

40. *Sides of a Rectangle* What are the length and width of a rectangle that has a perimeter of 98 cm and a diagonal of 35 cm?

41. *Sides of a Triangle* Find the lengths of the sides of a triangle whose perimeter is 12 ft and whose angles are 30°, 60°, and 90°. (*Hint:* Use a, $a/2$, and b to represent the lengths of the sides.)

42. *Size of a Vent* Kwan is constructing a triangular vent in the gable end of a house, as shown in the diagram. If the pitch of the roof is 6–12 (run 12 ft and rise 6 ft) and the vent must have an area of 4.5 ft^2, then what size should he make the base and height of the triangle?

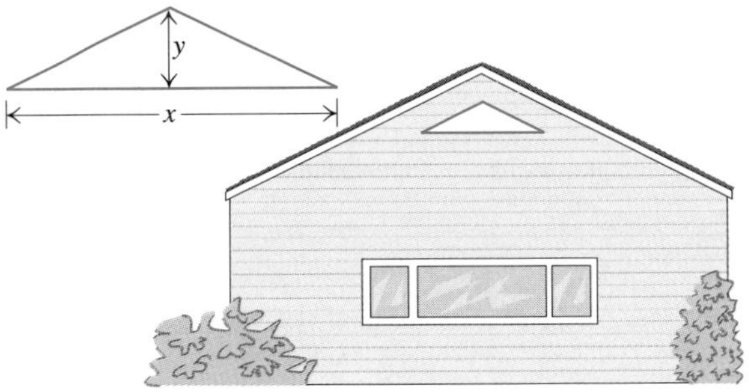

Figure for Exercise 42

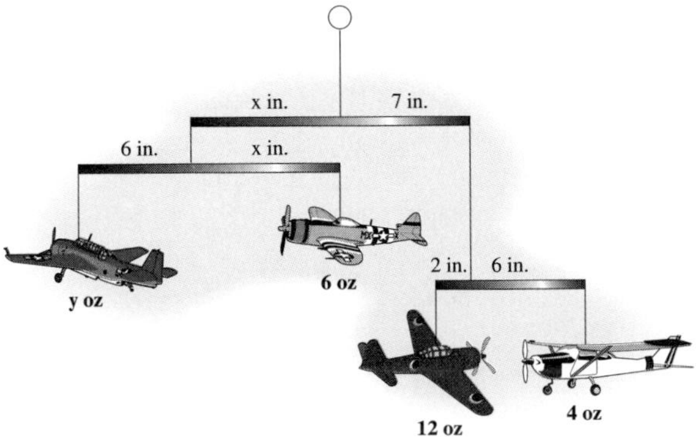

Figure for Exercise 43

43. *Air Mobile* A hobbyist is building a mobile out of model airplanes as shown in the figure. Find x and y so that the mobile will balance. Ignore the weights of the horizontal bars and the strings. The mobile will be in balance if the product of the weight and distance on one side of the balance point is equal to the product of the weight and distance on the other side.

44. *Making an Arch* A bricklayer is constructing a circular arch with a radius of 9 feet as shown in the figure. If the height h must be 1.5 times as large as the width w, then what are h and w?

Figure for Exercise 44

45. *Pumping Tomato Soup* At the Acme Soup Company a large vat of tomato soup can be filled in 8 min by pump A and pump B working together. Pump B can be reversed so that it empties the vat at the same rate at which it fills the vat. One day a worker filled the vat in 12 min using pumps A and B, but accidentally ran pump B in reverse. How long does it take each pump to fill the vat working alone?

46. *Planting Strawberries* Blanche and Morris can plant an acre of strawberries in 8 hr working together. Morris takes 2 hr longer to plant an acre of strawberries working alone than it takes Blanche working alone. How long does it take Morris to plant an acre by himself?

47. *Lost Numbers* Find two complex numbers whose sum is 6 and whose product is 10.

48. *More Lost Numbers* Find two complex numbers whose sum is 1 and whose product is 5.

49. *Voyage of the Whales* In one of Captain James Kirk's most challenging missions, he returned to late twentieth century San Francisco to bring back a pair of humpback whales. Chief Engineer Scotty built a tank of transparent aluminum to hold the time-traveling cetaceans. If Scotty's 20-ft-high tank had a volume of 36,000 ft^3, and it took 7,200 ft^2 of transparent aluminum to cover all six sides, then what were the length and width of the tank?

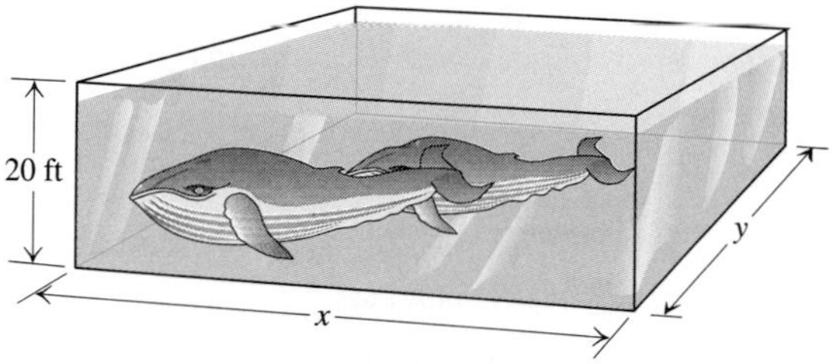

Figure for Exercise 49

50. *Dimensions of a Laundry Room* The plans call for a rectangular laundry room in the Wilsons' new house. Connie wants to increase the width by 1 ft and the length by 2 ft, which will increase the area by 30 ft^2. Christopher wants to increase the width by 2 ft and decrease the length by 3 ft, which will decrease the area by 6 ft^2. What are the original dimensions for the laundry room?

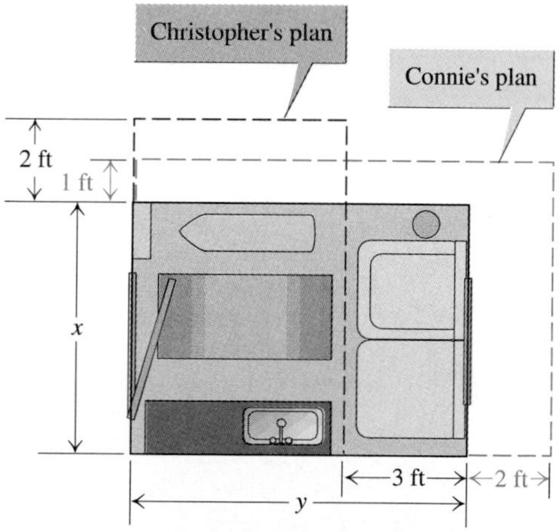

Figure for Exercise 50

51. *Two Models* One demographer believes that the population growth of a certain country is best modeled by the function $P(t) = 20e^{0.07t}$, while a second demographer believes that the population growth of that same country is best modeled by the function $P(t) = 20 + 2t$. In each case t is the number of years from the present and $P(t)$ is given in millions of people. For what values of t do these two models give the same population? In how many years is the population predicted by the exponential model twice as large as the population predicted by the linear model?

52. *Capital Investment* The equation $p = 0.13(1.1)^x$ can be used to model the percentage of gross domestic product spent on capital investment in information technology and $p = 0.03x + 7$ to model the percentage of GDP spent on other types of capital investment as shown in the accompanying figure, where x is the number of years since 1960 (Scientific American, July 1997). Use a graphing calculator to find the year in which investment in information technology will amount to the same percentage of GDP as other types of capital investment.

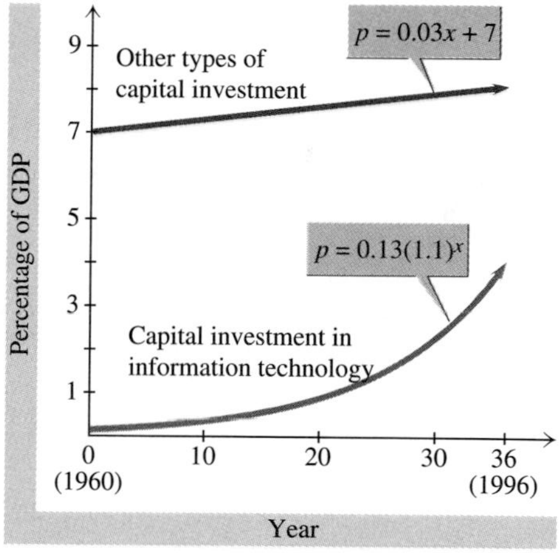

Figure for Exercise 52

53. *Mixed Signals* Sally and Bob live quite far from each other on Interstate 75 and decide to get together on Saturday. However, they get their signals crossed and they each start out at sunrise for the other's house. At noon they pass each other on the freeway. At 4 P.M. Sally arrives at Bob's house and at 9 P.M. Bob arrives at Sally's house. At what time did the sun rise? See the figure on the next page.

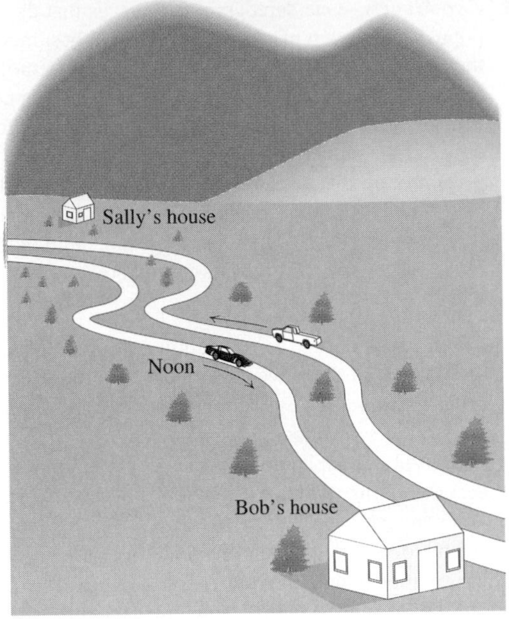

Figure for Exercise 53

54. *Westward Ho* A wagon train that is one mile long advances one mile at a constant rate. At the same time, the wagon master rides his horse at a constant rate from the front of the wagon train to the rear, and then back to the front. How far did the wagon master ride?

For Writing/Discussion

55. *Cooperative Learning* Write a nonlinear system of two equations that has three points in its solution set and a nonlinear system of two equations that has no solution. Ask a classmate to solve your systems.

56. *How Many?* Explain why the system

$$y = 3 + \log_2(x - 1)$$
$$8x - 8 = 2^y$$

has infinitely many solutions.

57. *Line and Circle* For what values of b does the solution set to $y = x + b$ and $x^2 + y^2 = 1$ consist of one point, two points, and no points? Explain.

LINKING CONCEPTS

For Individual or Group Explorations

Measuring Ocean Depth *Geophysicists who map the ocean floor use sound reflection to measure the depth of the ocean. The problem is complicated by the fact that the speed of sound in water is not constant, but depends on the temperature and other conditions of the water. Let v be the velocity of sound through the water and d_1 be the depth of the ocean below the ship as shown in the figure.*

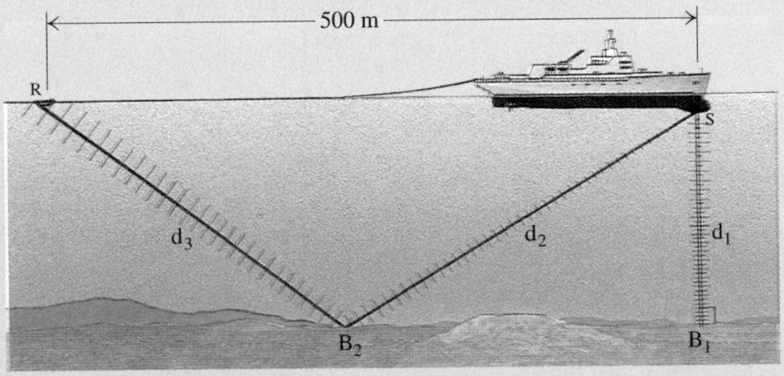

a) The time that it takes for sound to travel to the ocean floor at point B_1 and back to the ship at point S is measured as 0.280 sec. Write d_1 as a function of v.

b) It takes 0.446 sec for sound to travel from point S to point B_2 and then to a receiver at R, which is towed 500 meters behind the ship. Assuming that $d_2 = d_3$, write d_2 as a function of v.

c) Write d_1 as a function of d_2.

d) Solve your system to find the ocean depth d_1.

5.4 Partial Fractions

• CALCULUS •

In algebra we usually learn a process and then learn to reverse it. For example, we multiply two binomials, and then we factor trinomials into a product of two binomials. We solve polynomial equations, and we write polynomial equations with given solutions. After we studied functions, we learned about their inverses. In Section P.6 we learned how to add rational expressions. Now we will reverse the process of addition. We start with a rational expression and write it as a sum of two or more simpler rational expressions. This technique is useful in calculus, and it also shows a nice application of systems of equations.

The Basic Idea

Before trying to reverse addition of rational expressions, recall how to add them.

EXAMPLE 1 Adding rational expressions

Perform the indicated operation.

$$\frac{3}{x-3} + \frac{-2}{x+1}$$

Solution

The least common denominator (LCD) for $x - 3$ and $x + 1$ is $(x - 3)(x + 1)$. We convert each rational expression or fraction into an equivalent fraction with this denominator:

$$\begin{aligned}\frac{3}{x-3} + \frac{-2}{x+1} &= \frac{3(x+1)}{(x-3)(x+1)} + \frac{-2(x-3)}{(x+1)(x-3)} \\ &= \frac{3x+3}{(x-3)(x+1)} + \frac{-2x+6}{(x-3)(x+1)} \\ &= \frac{x+9}{(x-3)(x+1)}\end{aligned}$$

The following rational expression is similar to the result in Example 1.

$$\frac{8x - 7}{(x + 1)(x - 2)}$$

Thus, it is possible that this fraction is the sum of two fractions with denominators $x + 1$ and $x - 2$. In the next example, we will find those two fractions.

EXAMPLE 2 **Reversing the addition of rational expressions**

Write the following rational expression as a sum of two rational expressions.

$$\frac{8x - 7}{(x + 1)(x - 2)}$$

Solution

To write the given expression as a sum, we need numbers A and B such that

$$\frac{8x - 7}{(x + 1)(x - 2)} = \frac{A}{x + 1} + \frac{B}{x - 2}.$$

Simplify this equation by multiplying each side by the LCD, $(x + 1)(x - 2)$:

$$(x + 1)(x - 2)\frac{8x - 7}{(x + 1)(x - 2)} = (x + 1)(x - 2)\left(\frac{A}{x + 1} + \frac{B}{x - 2}\right)$$

$$8x - 7 = A(x - 2) + B(x + 1)$$

$$8x - 7 = Ax - 2A + Bx + B$$

$$8x - 7 = (A + B)x - 2A + B \qquad \text{Combine like terms.}$$

Since the last equation is an identity, the coefficient of x on one side equals the coefficient of x on the other, and the constant on one side equals the constant on the other. So A and B satisfy the following two equations.

$$A + B = 8$$

$$-2A + B = -7$$

We can solve this system of two linear equations in two unknowns by addition:

$$\begin{aligned} A + B &= 8 \\ 2A - B &= 7 \qquad \text{Second equation multiplied by } -1 \\ \hline 3A \quad &= 15 \\ A &= 5 \end{aligned}$$

If $A = 5$, then $B = 3$, and we have

$$\frac{8x - 7}{(x + 1)(x - 2)} = \frac{5}{x + 1} + \frac{3}{x - 2}.$$

Check by adding the fractions on the right-hand side of the equation.

Each of the two fractions on the right-hand side of the equation

$$\frac{8x - 7}{(x + 1)(x - 2)} = \frac{5}{x + 1} + \frac{3}{x - 2}$$

is called a **partial fraction**. This equation shows the **partial fraction decomposition** of the rational expression on the left-hand side.

General Decomposition

In general, let $N(x)$ be the polynomial in the numerator and $D(x)$ be the polynomial in the denominator of the fraction that is to be decomposed. *We will decompose only fractions for which the degree of the numerator is smaller than the degree of the denominator.* If the degree of $N(x)$ is not smaller than the degree of $D(x)$, we can use long division to write the rational expression as quotient + remainder/divisor. For example,

$$\frac{x^3 + x^2 - x + 5}{x^2 + x - 6} = x + \frac{5x + 5}{x^2 + x - 6} = x + \frac{A}{x + 3} + \frac{B}{x - 2}.$$

You should find the values of A and B in the above equation as we did in Example 2, and check.

If a factor of $D(x)$ is repeated n times, then all powers of the factor from 1 through n might occur as denominators in the partial fractions. To understand the reason for this statement, look at

$$\frac{7}{8} = \frac{1}{2} + \frac{1}{4} + \frac{1}{8} = \frac{1}{2} + \frac{1}{2^2} + \frac{1}{2^3}.$$

Here the factor 2 is repeated three times in $D(x) = 8$. Notice that each of the powers of 2 (2^1, 2^2, 2^3) occurs in the denominators of the partial fractions.

Now consider a fraction $N(x)/D(x)$, where $D(x) = (x - 1)(x + 3)^2$. Since the factor $x + 3$ occurs to the second power, both $x + 3$ and $(x + 3)^2$ occur in the partial fraction decomposition of $N(x)/D(x)$. For example, we write the partial fraction decomposition for $(3x^2 + 17x + 12)/D(x)$ as follows:

$$\frac{3x^2 + 17x + 12}{(x - 1)(x + 3)^2} = \frac{A}{x - 1} + \frac{B}{x + 3} + \frac{C}{(x + 3)^2}$$

It is possible that we do not need the fraction $B/(x + 3)$, but we do not know this until we find the value of B. If $B = 0$, then the decomposition does not include a fraction with denominator $x + 3$. This decomposition is completed in Example 3.

To find the partial fraction decomposition of a rational expression, the denominator must be factored into a product of prime polynomials. *If a quadratic prime polynomial occurs in the denominator, then the numerator of the partial fraction for that polynomial is of the form $Ax + B$.* For example, if $D(x) = x^3 +$

$x^2 + 4x + 4$, then $D(x) = (x^2 + 4)(x + 1)$. The partial fraction decomposition for $(5x^2 + 3x + 13)/D(x)$ is written as follows:

$$\frac{5x^2 + 3x + 13}{x^3 + x^2 + 4x + 4} = \frac{Ax + B}{x^2 + 4} + \frac{C}{x + 1}$$

This decomposition is completed in Example 4.

The main points to remember for partial fraction decomposition are summarized as follows.

STRATEGY

Decomposition into Partial Fractions

To decompose a rational expression $N(x)/D(x)$ into partial fractions, use the following strategies:

1. If the degree of the numerator $N(x)$ is greater than or equal to the degree of the denominator $D(x)$, use division to express $N(x)/D(x)$ as quotient + remainder/divisor and decompose the resulting fraction.
2. If the degree of $N(x)$ is less than the degree of $D(x)$, factor the denominator completely into prime factors that are either linear $(ax + b)$ or quadratic $(ax^2 + bx + c)$.
3. For each linear factor of the form $(ax + b)^n$, the partial fraction decomposition must include the following fractions:

$$\frac{A_1}{ax + b} + \frac{A_2}{(ax + b)^2} + \cdots + \frac{A_n}{(ax + b)^n}$$

4. For each quadratic factor of the form $(ax^2 + bx + c)^m$, the partial fraction decomposition must include the following fractions:

$$\frac{B_1x + C_1}{ax^2 + bx + c} + \frac{B_2x + C_2}{(ax^2 + bx + c)^2} + \cdots + \frac{B_mx + C_m}{(ax^2 + bx + c)^m}$$

5. Set up and solve a system of equations involving the As, Bs, and/or Cs.

The strategy for decomposition applies to very complicated rational expressions. To actually carry out the decomposition, we must be able to solve the system of equations that arises. Theoretically, we can solve systems of many equations in many unknowns. Practically, we are limited to fairly simple systems of equations. Large systems of equations are generally solved by computers or calculators using techniques that we will develop in the next chapter.

EXAMPLE 3 Repeated linear factor

Find the partial fraction decomposition for the rational expression

$$\frac{3x^2 + 17x + 12}{(x - 1)(x + 3)^2}.$$

Solution

Since the factor $x + 3$ occurs twice in the original denominator, it might occur in the partial fractions with powers 1 and 2:

$$\frac{3x^2 + 17x + 12}{(x - 1)(x + 3)^2} = \frac{A}{x - 1} + \frac{B}{x + 3} + \frac{C}{(x + 3)^2}$$

Multiply each side of the equation by the LCD, $(x - 1)(x + 3)^2$.

$$\begin{aligned} 3x^2 + 17x + 12 &= A(x + 3)^2 + B(x - 1)(x + 3) + C(x - 1) \\ &= Ax^2 + 6Ax + 9A + Bx^2 + 2Bx - 3B + Cx - C \\ &= (A + B)x^2 + (6A + 2B + C)x + 9A - 3B - C \end{aligned}$$

Next, write a system of equations by equating the coefficients of like terms from opposite sides of the equation. The corresponding coefficients are highlighted above.

$$\begin{aligned} A + B \quad\quad &= 3 \\ 6A + 2B + C &= 17 \\ 9A - 3B - C &= 12 \end{aligned}$$

We can solve this system of three equations in the variables A, B, and C by first eliminating C. Add the last two equations to get $15A - B = 29$. Add this equation to $A + B = 3$:

$$\begin{aligned} 15A - B &= 29 \\ A + B &= 3 \\ \hline 16A \quad\quad &= 32 \\ A &= 2 \end{aligned}$$

If $A = 2$ and $A + B = 3$, then $B = 1$. Use $A = 2$ and $B = 1$ in the equation $6A + 2B + C = 17$:

$$\begin{aligned} 6(2) + 2(1) + C &= 17 \\ C &= 3 \end{aligned}$$

The partial fraction decomposition is written as follows:

$$\frac{3x^2 + 17x + 12}{(x - 1)(x + 3)^2} = \frac{2}{x - 1} + \frac{1}{x + 3} + \frac{3}{(x + 3)^2}$$

EXAMPLE 4 Single prime quadratic factor

Find the partial fraction decomposition for the rational expression

$$\frac{5x^2 + 3x + 13}{x^3 + x^2 + 4x + 4}.$$

Solution

Factor the denominator by grouping:

$$x^3 + x^2 + 4x + 4 = x^2(x + 1) + 4(x + 1) = (x^2 + 4)(x + 1)$$

Write the partial fraction decomposition:

$$\frac{5x^2 + 3x + 13}{x^3 + x^2 + 4x + 4} = \frac{Ax + B}{x^2 + 4} + \frac{C}{x + 1}$$

Note that $Ax + B$ is used over the prime quadratic polynomial $x^2 + 4$. Multiply each side of this equation by the LCD, $(x^2 + 4)(x + 1)$:

$$\begin{aligned} 5x^2 + 3x + 13 &= (Ax + B)(x + 1) + C(x^2 + 4) \\ &= Ax^2 + Bx + Ax + B + Cx^2 + 4C \\ &= (A + C)x^2 + (A + B)x + B + 4C \end{aligned}$$

Write a system of equations by equating the coefficients of like terms from opposite sides of the last equation:

$$\begin{aligned} A + C &= 5 \\ A + B &= 3 \\ B + 4C &= 13 \end{aligned}$$

One way to solve the system is to substitute $C = 5 - A$ and $B = 3 - A$ into $B + 4C = 13$:

$$\begin{aligned} 3 - A + 4(5 - A) &= 13 \\ 3 - A + 20 - 4A &= 13 \\ 23 - 5A &= 13 \\ -5A &= -10 \\ A &= 2 \end{aligned}$$

Since $A = 2$, we get $C = 5 - 2 = 3$ and $B = 3 - 2 = 1$. So the partial fraction decomposition is written as follows:

$$\frac{5x^2 + 3x + 13}{x^3 + x^2 + 4x + 4} = \frac{2x + 1}{x^2 + 4} + \frac{3}{x + 1}$$

EXAMPLE 5 **Repeated prime quadratic factor**

Find the partial fraction decomposition for the rational expression

$$\frac{4x^3 - 2x^2 + 7x - 6}{4x^4 + 12x^2 + 9}.$$

Solution

The denominator factors as $(2x^2 + 3)^2$, and $2x^2 + 3$ is prime. Write the partial fractions using denominators $2x^2 + 3$ and $(2x^2 + 3)^2$.

$$\frac{4x^3 - 2x^2 + 7x - 6}{(2x^2 + 3)^2} = \frac{Ax + B}{2x^2 + 3} + \frac{Cx + D}{(2x^2 + 3)^2}$$

Multiply each side of the equation by the LCD, $(2x^2 + 3)^2$, to get the following equation:

$$\begin{aligned} 4x^3 - 2x^2 + 7x - 6 &= (Ax + B)(2x^2 + 3) + Cx + D \\ &= 2Ax^3 + 2Bx^2 + 3Ax + 3B + Cx + D \\ &= 2Ax^3 + 2Bx^2 + (3A + C)x + 3B + D \end{aligned}$$

Equating the coefficients produces the following system of equations:

$$\begin{aligned} 2A &= 4 \\ 2B &= -2 \\ 3A + C &= 7 \\ 3B + D &= -6 \end{aligned}$$

From the first two equations $2A = 4$ and $2B = -2$, we get $A = 2$ and $B = -1$. Using $A = 2$ in $3A + C = 7$ gives $C = 1$. Using $B = -1$ in $3B + D = -6$ gives $D = -3$. So the partial fraction decomposition is written as follows:

$$\frac{4x^3 - 2x^2 + 7x - 6}{(2x^2 + 3)^2} = \frac{2x - 1}{2x^2 + 3} + \frac{x - 3}{(2x^2 + 3)^2}$$

FOR THOUGHT True or False? Explain.

1. $\dfrac{1}{x} + \dfrac{3}{x + 1} = \dfrac{4x + 1}{x^2 + x}$ for any real number x except 0 and -1.

2. $x + \dfrac{3x}{x^2 - 1} = \dfrac{x^3 + 2x}{x^2 - 1}$ for any real number x except -1 and 1.

3. The partial fraction decomposition of $\dfrac{x^2}{x^2 - 9}$ is $\dfrac{x^2}{x^2 - 9} = \dfrac{A}{x - 3} + \dfrac{B}{x + 3}$.

4. In the decomposition $\dfrac{5}{8} = \dfrac{A}{2} + \dfrac{B}{2^2} + \dfrac{C}{2^3}$, $A = 1$, $B = 0$, and $C = 1$.

5. The partial fraction decomposition of $\dfrac{3x - 1}{x^3 + x}$ is $\dfrac{3x - 1}{x^3 + x} = \dfrac{A}{x} + \dfrac{B}{x^2 + 1}$.

6. $\dfrac{1}{x^2 - 1} = \dfrac{1}{x - 1} + \dfrac{1}{x + 1}$ for any real number except 1 and -1.

7. $\dfrac{x^3 + 1}{x^2 + x - 2} = x - 1 + \dfrac{3x - 1}{x^2 + x - 2}$ for any real number except 1 and -2.

8. $x^3 - 8 = (x - 2)(x^2 + 4x + 4)$ for any real number x.

9. $\dfrac{x^2 + 2x}{x^3 - 1} = \dfrac{1}{x - 1} + \dfrac{1}{x^2 + x + 1}$ for any real number except 1.

10. There is no partial fraction decomposition for $\dfrac{2x}{x^2 + 9}$.

5.4 EXERCISES

Tape 17 Disk

Perform the indicated operations.

1. $\dfrac{3}{x - 2} + \dfrac{4}{x + 1}$

2. $\dfrac{-1}{x + 5} + \dfrac{-3}{x - 4}$

3. $\dfrac{1}{x - 1} + \dfrac{-3}{x^2 + 2}$

4. $\dfrac{x + 3}{x^2 + x + 1} + \dfrac{1}{x - 1}$

5. $\dfrac{2x + 1}{x^2 + 3} + \dfrac{x^3 + 2x + 2}{(x^2 + 3)^2}$

6. $\dfrac{3x - 1}{x^2 + x - 3} + \dfrac{x^3 + x - 1}{(x^2 + x - 3)^2}$

7. $\dfrac{1}{x - 1} + \dfrac{2x + 3}{(x - 1)^2} + \dfrac{x^2 + 1}{(x - 1)^3}$

8. $\dfrac{3}{x + 2} + \dfrac{x - 1}{(x + 2)^2} + \dfrac{1}{x^2 + 2}$

Find A and B for each partial fraction decomposition.

9. $\dfrac{12}{x^2 - 9} = \dfrac{A}{x - 3} + \dfrac{B}{x + 3}$

10. $\dfrac{5x + 2}{x^2 - 4} = \dfrac{A}{x - 2} + \dfrac{B}{x + 2}$

Find the partial fraction decomposition for each rational expression.

11. $\dfrac{5x - 1}{(x + 1)(x - 2)}$

12. $\dfrac{-3x - 5}{(x + 3)(x - 1)}$

13. $\dfrac{2x + 5}{x^2 + 6x + 8}$

14. $\dfrac{x + 2}{x^2 + 12x + 32}$

15. $\dfrac{2}{x^2 - 9}$

16. $\dfrac{1}{9x^2 - 1}$

17. $\dfrac{1}{x^2 - x}$

18. $\dfrac{2}{x^2 - 2x}$

Find A, B, and C for each partial fraction decomposition.

19. $\dfrac{x^2 + x - 31}{(x + 3)^2(x - 2)} = \dfrac{A}{x + 3} + \dfrac{B}{(x + 3)^2} + \dfrac{C}{x - 2}$

20. $\dfrac{x^2 - x - 7}{(x + 1)(x^2 + 4)} = \dfrac{A}{x + 1} + \dfrac{Bx + C}{x^2 + 4}$

Find each partial fraction decomposition.

21. $\dfrac{-2x - 7}{x^2 + 4x + 4}$

22. $\dfrac{-3x + 2}{x^2 - 2x + 1}$

23. $\dfrac{6x^2 - x + 1}{x^3 + x^2 + x + 1}$

24. $\dfrac{3x^2 - 2x + 8}{x^3 + 2x^2 + 4x + 8}$

25. $\dfrac{3x^3 - x^2 + 19x - 9}{x^4 + 18x^2 + 81}$

26. $\dfrac{-x^3 - 10x - 3}{x^4 + 10x^2 + 25}$

27. $\dfrac{3x^2 + 17x + 14}{x^3 - 8}$

28. $\dfrac{2x^2 + 17x - 21}{x^3 + 27}$

29. $\dfrac{2x^3 + x^2 + 3x - 2}{x^2 - 1}$

30. $\dfrac{2x^3 - 19x - 9}{x^2 - 9}$

31. $\dfrac{3x^3 - 2x^2 + x - 2}{(x^2 + x + 1)^2}$

32. $\dfrac{x^3 - 8x^2 - 5x - 33}{(x^2 + x + 4)^2}$

33. $\dfrac{3x^3 + 4x^2 - 12x + 16}{x^4 - 16}$

34. $\dfrac{9x^2 + 3}{81x^4 - 1}$

35. $\dfrac{5x^3 + x^2 + x - 3}{x^4 - x^3}$

36. $\dfrac{2x^3 + 3x^2 - 8x + 4}{x^4 - 4x^2}$

37. $\dfrac{6x^2 - 28x + 33}{(x - 2)^2(x - 3)}$

38. $\dfrac{7x^2 + 45x + 58}{(x + 3)^2(x + 1)}$

39. $\dfrac{9x^2 + 21x - 24}{x^3 + 4x^2 - 11x - 30}$

40. $\dfrac{3x^2 + 24x + 6}{2x^3 - x^2 - 13x - 6}$

41. $\dfrac{x^2 - 2}{x^3 - 3x^2 + 3x - 1}$

42. $\dfrac{2x^2}{x^3 + 3x^2 + 3x + 1}$

Find the partial fraction decomposition for each rational expression. Assume that a, b, and c are nonzero constants.

43. $\dfrac{x}{(ax + b)^2}$

44. $\dfrac{x^2}{(ax + b)^3}$

45. $\dfrac{x + c}{ax^2 + bx}$

46. $\dfrac{1}{x^3(ax + b)}$

47. $\dfrac{1}{x^2(ax + b)}$

48. $\dfrac{1}{x^2(ax + b)^2}$

LINKING CONCEPTS

For Individual or Group Explorations

Working Together *Suppose that Andrew can paint a garage by himself in a hours and Betty can paint the same garage by herself in b hours. Working together they can paint the garage in 2 hr 24 min.*

a) What is the sum of $1/a$ and $1/b$?

b) Write a as a function of b and graph it.

c) What are the horizontal and vertical asymptotes? Interpret the asymptotes in the context of this situation.

d) Use a graphing calculator to make a table for a and b. If a and b are positive integers, find all possible values for a and b from the table.

e) Now suppose that the time it takes for Andrew, Betty, or Carl to paint a garage working alone is a, b, or c hours respectively, where a, b, and c are positive integers. If they paint the garage together in 1 hr 15 min, then what are the possible values of a, b, and c? Explain your solution.

5.5 Inequalities and Systems of Inequalities in Two Variables

Earlier in this chapter we solved systems of equations in two variables. In this section we turn to linear and nonlinear inequalities in two variables and systems of inequalities in two variables. In the next section systems of linear inequalities will be used to solve linear programming problems.

Linear Inequalities

Three hamburgers and two Cokes cost at least \$5.50. If a hamburger costs x dollars and a Coke costs y dollars, then this sentence can be written as the linear inequality

$$3x + 2y \geq 5.50.$$

A linear inequality in two variables is simply a linear equation in two variables with the equal sign replaced by an inequality symbol.

Definition: Linear Inequality

If A, B, and C are real numbers with A and B not both zero, then

$$Ax + By < C$$

is called a **linear inequality in two variables**. In place of $<$ we can also use the symbols $\leq$, $>$, or $\geq$.

Some examples of linear inequalities are

$$y > x - 3, \quad 2x - y \geq 1, \quad x + y < 0, \quad \text{and} \quad y \leq 3.$$

The inequality $y \leq 3$ is considered an inequality in two variables because it can be written in the form $0 \cdot x + y \leq 3$.

An ordered pair (a, b) is a **solution to an inequality** if the inequality is true when x is replaced by a and y is replaced by b. Ordered pairs such as (9, 1) and (7, 3) satisfy $y \leq 3$. Ordered pairs such as (4, 2) and $(-1, -2)$ satisfy $y > x - 3$.

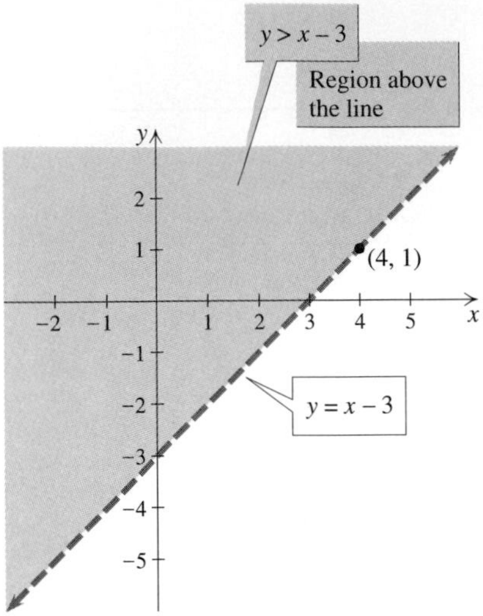

Figure 5.22

The ordered pairs satisfying $y = x - 3$ form a line in the rectangular coordinate system, but what does the solution set to $y > x - 3$ look like? An ordered pair satisfies $y > x - 3$ whenever the y-coordinate is *greater than* the x-coordinate minus 3. For example, (4, 1) satisfies the equation $y = x - 3$ and is on the line, while (4, 1.001) satisfies $y > x - 3$ and is above the line. In fact, any point in the coordinate plane directly above (4, 1) satisfies $y > x - 3$ (and any point directly below (4, 1) satisfies $y < x - 3$). Since this fact holds true for every point on the line $y = x - 3$, the solution set to the inequality is the set of all points *above* the line. The graph of $y > x - 3$ is indicated by shading the region above the line as shown in Fig. 5.22. We draw a dashed line for $y = x - 3$ because it is not part of the solution set to $y > x - 3$. However, the graph of $y \geq x - 3$ in Fig. 5.23 has a solid boundary line because the line is included in its solution set. Notice also that the region below the line is the solution set to $y < x - 3$.

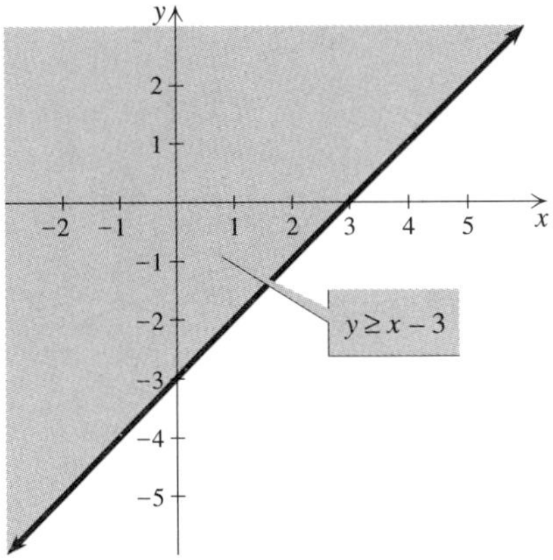

Figure 5.23

EXAMPLE 1 Graphing linear inequalities

Graph the solution set to each inequality.

a) $y < -\frac{1}{2}x + 2$ **b)** $2x - y \geq 1$ **c)** $y > 2$ **d)** $x \leq 3$

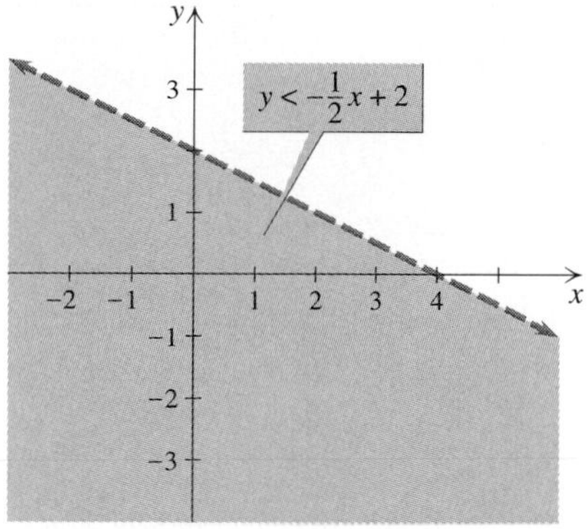

Figure 5.24

Solution

a) Graph the corresponding equation

$$y = -\frac{1}{2}x + 2$$

as a dashed line by using its intercept (0, 2) and slope $-\frac{1}{2}$. Since the inequality symbol is $<$, the solution set to the inequality is the region below this line, as shown in Fig. 5.24.

b) Solve the inequality for y:

$$2x - y \geq 1$$
$$-y \geq -2x + 1$$
$$y \leq 2x - 1 \quad \text{Multiply by } -1 \text{ and reverse the inequality.}$$

Graph the line $y = 2x - 1$ by using its intercept $(0, -1)$ and its slope 2. Because of the symbol $\leq$, we use a solid line and shade the region below it, as shown in Fig. 5.25.

c) Every point in the region above the horizontal line $y = 2$ satisfies $y > 2$. See Fig. 5.26.

d) Every point on or to the left of the vertical line $x = 3$ satisfies $x \leq 3$. See Fig. 5.27.

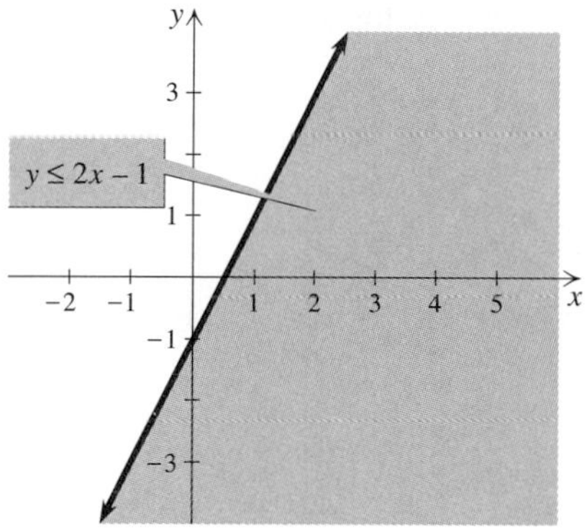

Figure 5.25

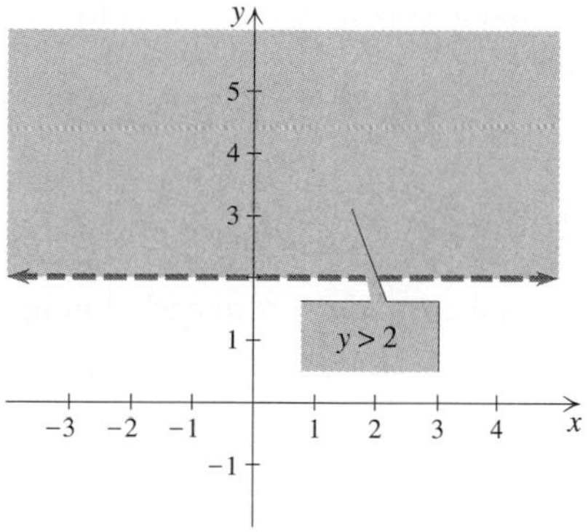

Figure 5.26

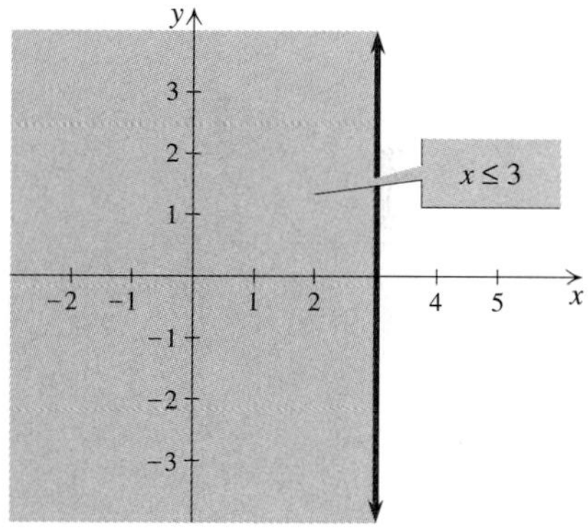

Figure 5.27

The graph in Example 1(b) is the region below the line because $2x - y \geq 1$ is equivalent to $y \leq 2x - 1$. The graph of $y > mx + b$ is above the line and $y < mx + b$ is below the line. The symbol $>$ corresponds to "above the line" and $<$ corresponds to "below the line" *only if the inequality is solved for y.*

With the **test point method** it is not necessary to solve the inequality for y. The graph of $Ax + By = C$ divides the plane into two regions. On one side, $Ax + By > C$; and on the other, $Ax + By < C$. We can simply test a single point in one of the regions to see which is which.

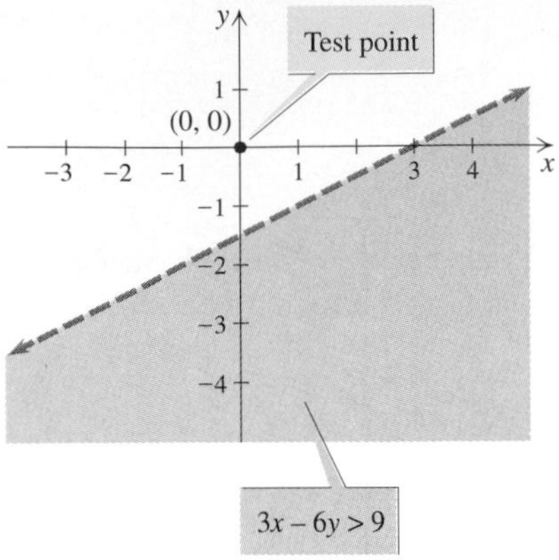

Figure 5.28

EXAMPLE 2 **Graphing an inequality using test points**

Use the test point method to graph $3x - 6y > 9$.

Solution

First graph $3x - 6y = 9$, using a dashed line going through its intercepts $(0, -\frac{3}{2})$ and $(3, 0)$. Select a test point that is not on the line, say, $(0, 0)$. Test $(0, 0)$ in $3x - 6y > 9$:

$$3 \cdot 0 - 6 \cdot 0 > 9$$
$$0 > 9 \quad \text{Incorrect.}$$

Since $(0, 0)$ does not satisfy $3x - 6y > 9$, any point on the *other* side of the line satisfies $3x - 6y > 9$. So we shade the region that does not contain $(0, 0)$, as shown in Fig. 5.28.

Nonlinear Inequalities

Any nonlinear equation in two variables becomes a nonlinear inequality in two variables when the equal sign is replaced by an inequality symbol. The test point method is generally the easiest to use for graphing nonlinear inequalities in two variables.

EXAMPLE 3 **Graphing nonlinear inequalities in two variables**

Graph the solution set to each nonlinear inequality.

a) $y > x^2$ **b)** $x^2 + y^2 \leq 4$ **c)** $y < \log_2(x)$

Solution

a) The graph of $y = x^2$ is a parabola opening upward with vertex at $(0, 0)$. Draw the parabola dashed, as shown in Fig. 5.29, and select a point that is not on the parabola, say, $(0, 5)$. Since $5 > 0^2$ is correct, the region of the plane containing $(0, 5)$ is shaded.

b) The graph of $x^2 + y^2 = 4$ is a circle of radius 2 centered at $(0, 0)$. Select $(0, 0)$ as a test point. Since $0^2 + 0^2 \leq 4$ is correct, we shade the region inside the circle, as shown in Fig. 5.30.

c) The graph of $y = \log_2(x)$ is a curve through the points $(1, 0)$, $(2, 1)$, and $(4, 2)$ as shown in Fig. 5.31. Select $(4, 0)$ as a test point. Since $0 < \log_2(4)$ is correct, shade the region shown in Fig. 5.31.

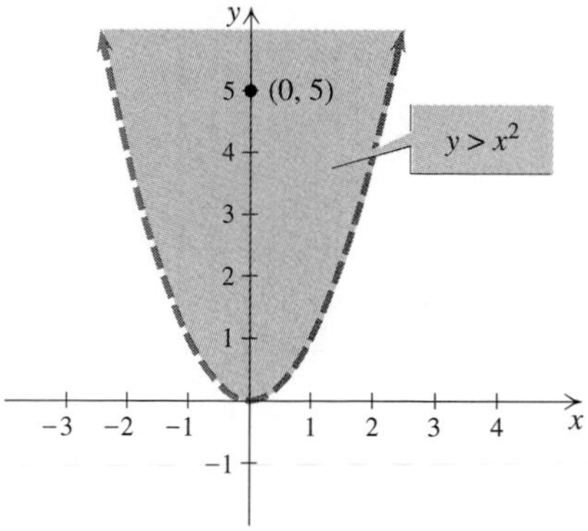

Figure 5.29

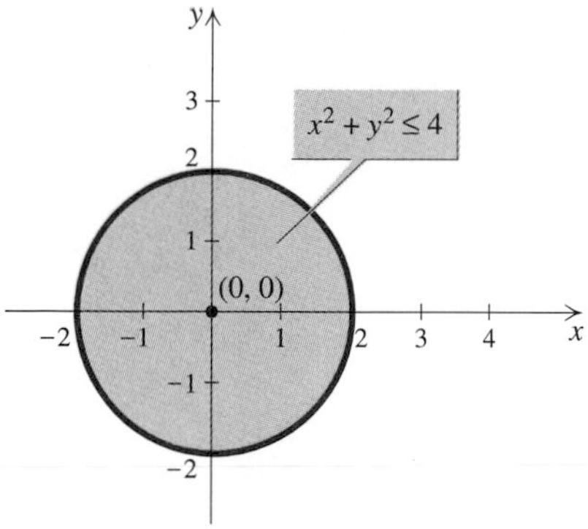

Figure 5.30

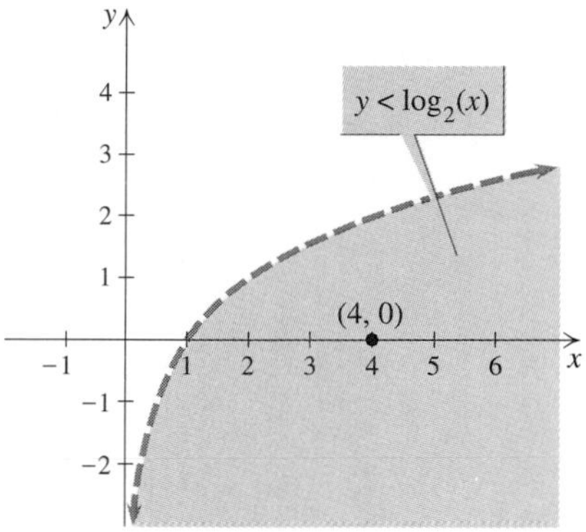

Figure 5.31

Systems of Inequalities

The solution set to a system of inequalities in two variables consists of all ordered pairs that satisfy *all* of the inequalities in the system. For example, the system

$$x + y > 5$$
$$x - y < 9$$

has (4, 2) as a solution because $4 + 2 > 5$ and $4 - 2 < 9$. There are infinitely many solutions to this system.

The solution set to a system is generally a region of the coordinate plane. It is the intersection of the solution sets to thc individual inequalities. To find the solution set to a system, we graph the equation corresponding to each inequality in the system and then test a point in each region to see whether it satisfies all inequalities of the system.

EXAMPLE 4 Solving a system of linear inequalities

Graph the solution set to the system.

$$x + 2y \leq 4$$
$$y \geq x - 3$$

Solution

The graph of $x + 2y = 4$ is a line through (0, 2) and (4, 0). The graph of $y = x - 3$ is a line through $(0, -3)$ with slope 1. These two lines divide the plane into four regions, as shown in Fig. 5.32 on the next page. Select a test point in each of the four regions. We use (0, 0), (0, 5), $(0, -5)$, and (5, 0) as test points. Only (0, 0) satisfies both of the inequalities of the system. So we shade the region containing (0, 0) including its boundaries, as shown in Fig. 5.33.

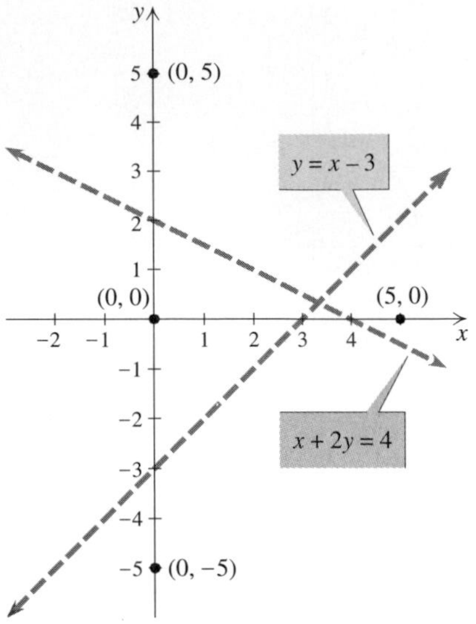

Figure 5.32

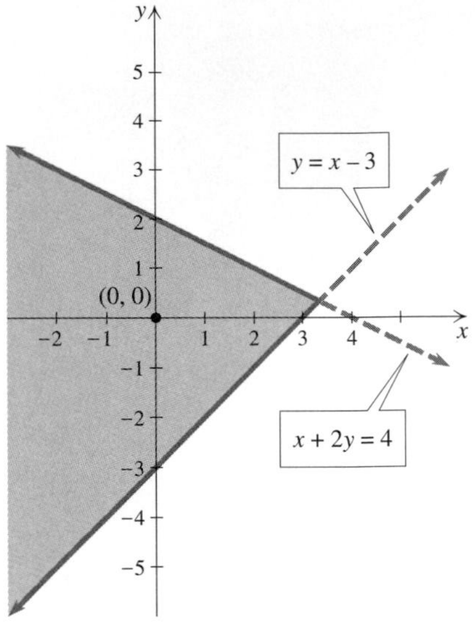

Figure 5.33

Note that the set of points indicated in Fig. 5.33 is the intersection of the set of points on or below the line $x + 2y = 4$ with the set of points on or above the line $y = x - 3$. If you can visualize how these regions intersect, then you can find the solution set without test points.

EXAMPLE 5 Solving a system of nonlinear inequalities

Graph the solution set to the system.

$$x^2 + y^2 \leq 16$$
$$y > 2^x$$

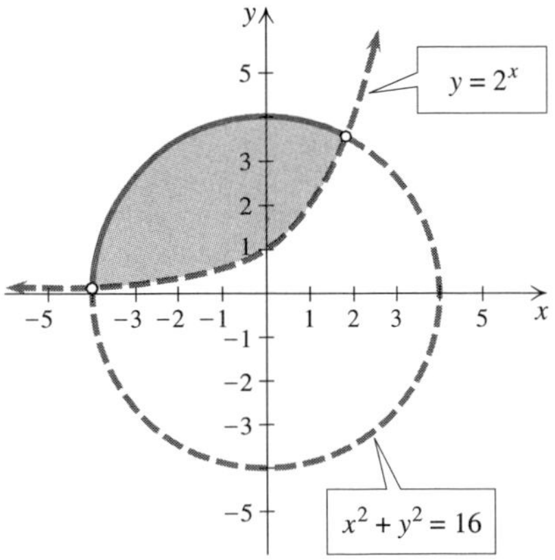

Figure 5.34

Solution

Points that satisfy $x^2 + y^2 \leq 16$ are on or inside the circle of radius 4 centered at (0, 0). Points that satisfy $y > 2^x$ are above the curve $y = 2^x$. Points that satisfy both inequalities are on or inside the circle and above the curve $y = 2^x$ as shown in Fig. 5.34. Note that the circular boundary of the solution set is drawn as a solid curve because of the $\leq$ symbol. We could also find the solution set by using test points.

EXAMPLE 6 Solving a system of three inequalities

Graph the solution set to the system

$$y > x^2$$
$$y < x + 6$$
$$y < -x + 6$$

Solution

Points that satisfy $y > x^2$ are above the parabola $y = x^2$. Points that satisfy $y < x + 6$ are below the line $y = x + 6$. Points that satisfy $y < -x + 6$ are below the line $y = -x + 6$. Points that satisfy all three inequalities lie in the region shown in Fig. 5.35.

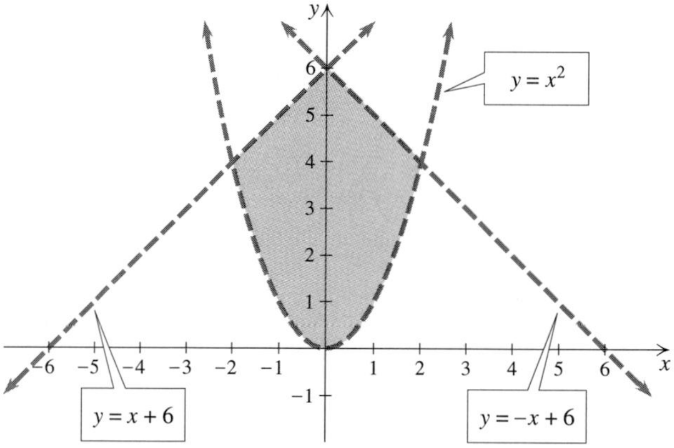

Figure 5.35

FOR THOUGHT True or False? Explain.

1. The point (1, 3) satisfies the inequality $y > x + 2$.
2. The graph of $x - y < 2$ is the region below the line $x - y = 2$.
3. The graph of $x + y > 2$ is the region above the line $x + y = 2$.
4. The graph of $x^2 + y^2 > 5$ is the region outside the circle of radius 5.

The following systems are referenced in statements 5–10:

a) $x - 2y < 3$
$y - 3x > 5$

b) $x^2 + y^2 > 9$
$y < x + 2$

c) $y \geq x^2 - 4$
$y < x + 2$

5. The point $(-2, 1)$ is in the solution set to system (a).
6. The solution to system (b) consists of points outside a circle and below a line.
7. The point $(-2, 0)$ is in the solution set to system (c).
8. The point $(-1, 2)$ is a test point for system (a).
9. The origin is in the solution set to system (c).
10. The point $(2, 3)$ is in the solution set to system (b).

5.5 EXERCISES

Tape 17 Disk

Match each inequality with one of the graphs (a)–(d).

1. $y > x - 2$
2. $x < 2 - y$
3. $x - y > 2$
4. $x + y > 2$

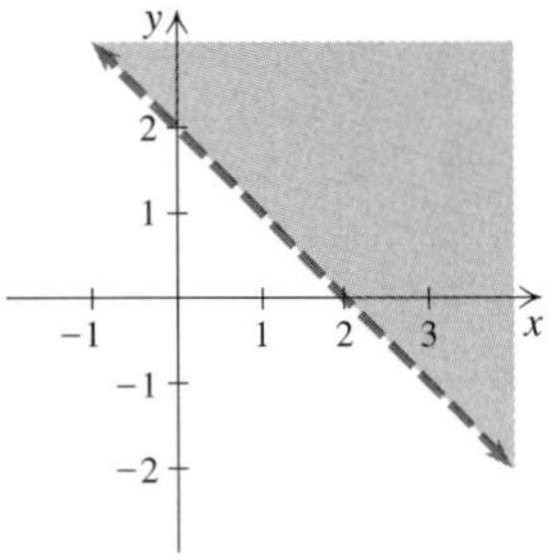

(a)

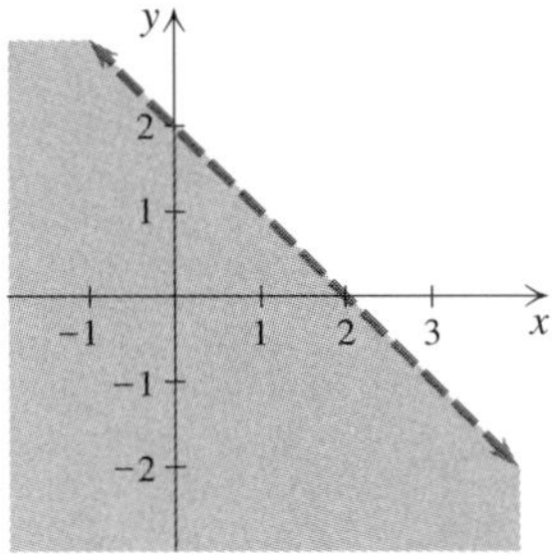

(b)

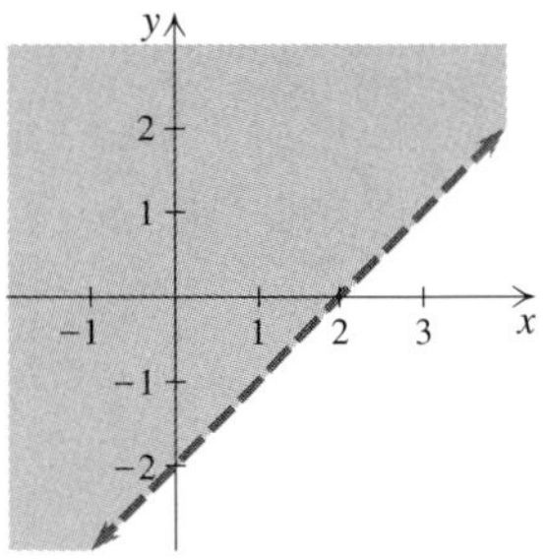

(c)

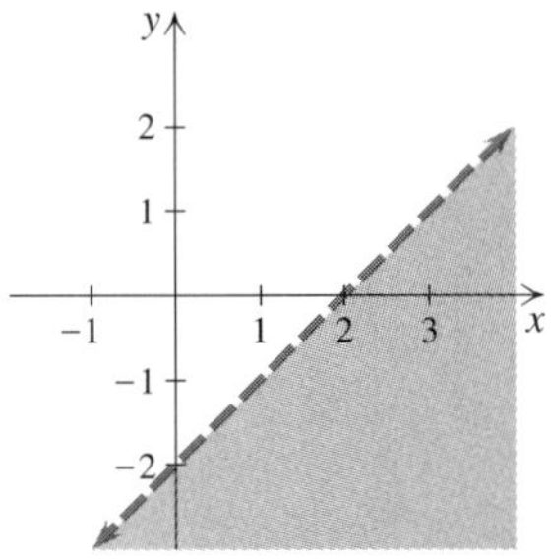

(d)

Sketch the graph of the solution set to each linear inequality in the rectangular coordinate system.

5. $y < 2x$
6. $x > y$
7. $x + y > 3$
8. $2x + y < 1$
9. $2x - y \le 4$
10. $x - 2y \ge 6$
11. $y < -3x - 4$
12. $y > \frac{2}{3}x - 3$
13. $x - 3 \ge 0$
14. $y + 1 \le 0$
15. $20x - 30y \le 6000$
16. $30y + 40x > 1200$
17. $y < 3$
18. $x > 0$

Sketch the graph of each nonlinear inequality.

19. $y > -x^2$
20. $y < 4 - x^2$
21. $x^2 + y^2 \ge 1$
22. $x^2 + y^2 < 36$
23. $x > |y|$
24. $y < x^{1/3}$
25. $x \ge y^2$
26. $y^2 > x - 1$
27. $y \ge x^3$
28. $y < |x - 1|$
29. $y > 2^x$
30. $y < \log_2(x)$

Sketch the graph of the solution set to each system of inequalities.

31. $y > x - 4$
    $y < -x - 2$
32. $y < \frac{1}{2}x + 1$
    $y < -\frac{1}{3}x + 1$
33. $3x - 4y \le 12$
    $x + y \ge -3$
34. $2x + y \ge -1$
    $y - 2x \le -3$
35. $3x - y < 4$
    $y < 3x + 5$
36. $x - y > 0$
    $y + 4 > x$
37. $y + x < 0$
    $y > 3 - x$
38. $3x - 2y \le 6$
    $2y - 3x \le -8$
39. $x + y < 5$
    $y \ge 2$
40. $x \le 2$
    $y > -2$
41. $y < x - 3$
    $x \le 4$
42. $y > 0$
    $y \le x$

Sketch the graph of the solution set to each nonlinear system of inequalities.

43. $y > x^2 - 3$
    $y < x + 1$
44. $y < 5 - x^2$
    $y > (x - 1)^2$
45. $x^2 + y^2 \ge 4$
    $x^2 + y^2 \le 16$
46. $x^2 + y^2 \le 9$
    $y \ge x - 1$
47. $(x - 3)^2 + y^2 < 25$
    $(x + 3)^2 + y^2 < 25$
48. $x^2 + y^2 \ge 64$
    $x^2 + y^2 \le 16$
49. $x^2 + y^2 > 4$
    $|x| \le 4$
50. $x^2 + y^2 < 36$
    $|y| < 3$

51. $y > |2x| - 4$
$y \leq \sqrt{4 - x^2}$

52. $y < 4 - |x|$
$y \geq |x| - 4$

53. $|x - 1| < 2$
$|y - 1| < 4$

54. $x \geq y^2 - 1$
$(x + 1)^2 + y^2 \geq 4$

Solve each system of inequalities.

55. $x \geq 0$
$y \geq 0$
$x + y \leq 4$

56. $x \geq 0$
$y \geq 0$
$y \geq -\frac{1}{2}x + 2$

57. $x \geq 0,\ y \geq 0$
$x + y \geq 4$
$y \geq -2x + 6$

58. $x \geq 0,\ y \geq 0$
$4x + 3y \leq 12$
$3x + 4y \leq 12$

59. $x^2 + y^2 \geq 9$
$x^2 + y^2 \leq 25$
$y \geq |x|$

60. $x - 2 < y < x + 2$
$x^2 + y^2 < 16$
$x > 0,\ y > 0$

61. $y > (x - 1)^3$
$y > 1$
$x + y > -2$

62. $x \geq |y|$
$y \geq -3$
$2y - x \leq 4$

63. $y > 2^x$
$y < 6 - x^2$
$x + y > 0$

64. $x^2 + y \leq 5$
$y \geq x^3 - x$
$y \leq 4$

Write a system of inequalities whose solution set is the region shown.

65.

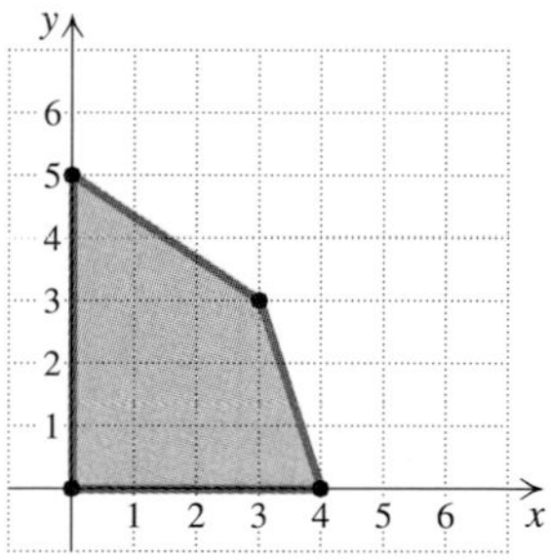

66.

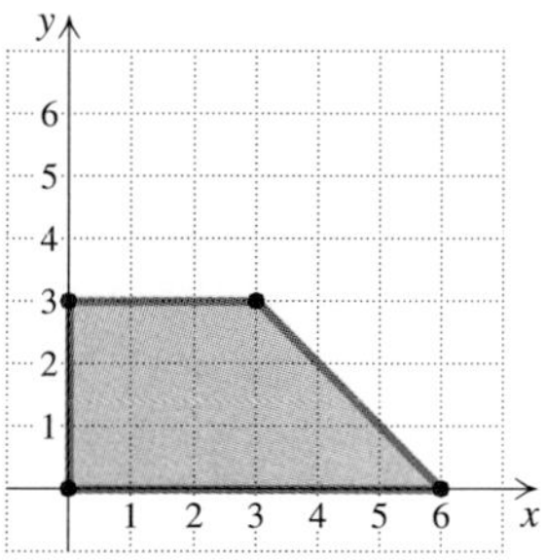

67.

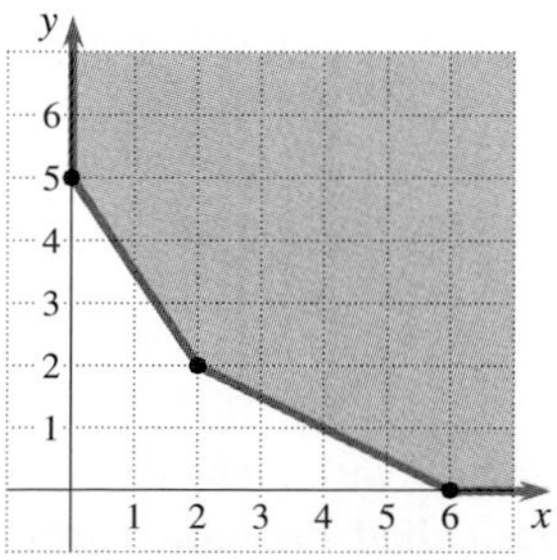

68.

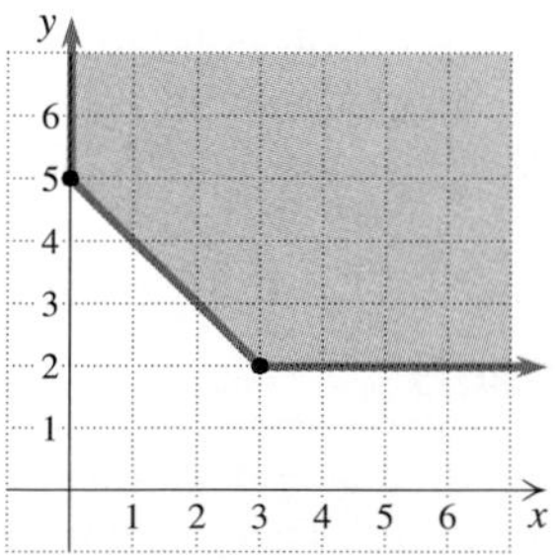

Find a system of inequalities to describe the given region.

69. Points inside the square that has vertices (2, 2), (−2, 2), (−2, −2), and (2, −2).

70. Points inside the triangle that has vertices (0, 0), (0, 6), and (3, 0).

71. Points in the first quadrant less than nine units from the origin.

72. Points that are closer to the x-axis than they are to the y-axis.

Use a graphing calculator to graph the equation corresponding to each inequality. From the display of the graphing calculator, locate one ordered pair in the solution set to the system and check that it satisfies all inequalities of the system.

73. $y > 2^{x+2}$
$y < x^3 - 3x$

74. $y > e^{x-0.8}$
$y < \log(x + 2.5)$

75. $y < -0.5x^2 + 150x - 11226.6$
$y > -0.11x + 38$

76. $y < x^2$
$y > 2^x$
$y < 5x - 6$

Write a system of inequalities that describes the possible solutions to each problem and graph the solution set to the system.

77. *Size Restrictions* United Parcel Service defines the girth of a box as the sum of the length, twice the width, and twice the height. The maximum girth that UPS will accept is 130 in. If the length of a box is 50 in., then what inequality must be satisfied by the width and height? Draw a graph showing the acceptable widths and heights for a length of 50 in.

78. *More Restrictions* United Parcel Service defines the girth of a box as the sum of the length, twice the width, and twice the height. The maximum girth that UPS will accept is 130 in. A shipping clerk wants to ship parts in a box that has a height of 24 in. For easy handling, he wants the box to have a width that is less than or equal to two-thirds of the length. Write a system of inequalities that the box must satisfy and draw a graph showing the possible lengths and widths.

79. *Inventory Control* A car dealer stocks mid-size and full-size cars on her lot, which cannot hold more than 110 cars. On the average, she borrows \$10,000 to purchase a mid-size car and \$15,000 to purchase a full-size car. How many cars of each type could she stock if her total debt cannot exceed \$1.5 million?

80. *Delicate Balance* A fast food restaurant must have a minimum of 30 employees and a maximum of 50. To avoid charges of sexual bias, the company has a policy that the number of employees of one sex must never exceed the number of employees of the other sex by more than six. How many persons of each sex could be employed at this restaurant?

81. *Political Correctness* A political party is selling \$50 tickets and \$100 tickets for a fund-raising banquet in a hall that cannot hold more than 500 people. To show that the party represents the people, the number of \$100 tickets must not be greater than 20% of the total number of tickets sold. How many tickets of each type can be sold?

82. *Mixing Alloys* A metallurgist has two alloys available. Alloy A is 20% zinc and 80% copper, while alloy B is 60% zinc and 40% copper. He wants to melt down and mix x ounces of alloy A with y ounces of alloy B to get a metal that is at most 50% zinc and at most 60% copper. How many ounces of each alloy should he use if the new piece of metal must not weigh more than 20 ounces?

Linking Concepts

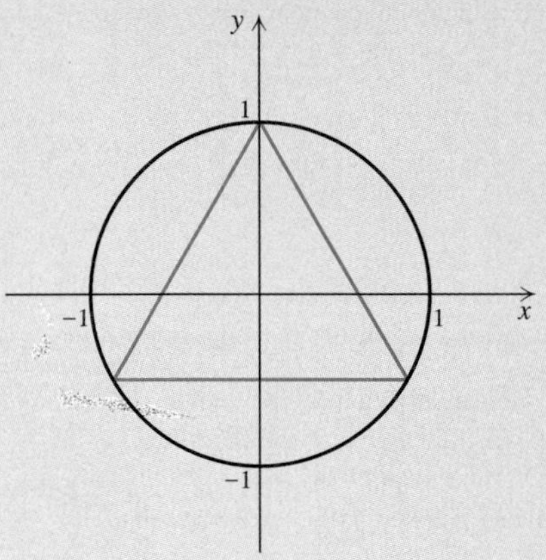

For Individual or Group Explorations

Regular Polygons The equilateral triangle in the figure is inscribed in the circle $x^2 + y^2 = 1$.

a) Write a system of inequalities whose graph consists of all points on and inside the equilateral triangle.

b) Write a system of inequalities whose graph consists of all points on and inside a square inscribed in $x^2 + y^2 = 1$. Position the square in any manner that you choose.

c) Write a system of inequalities whose graph consists of all points on and inside a regular pentagon inscribed in $x^2 + y^2 = 1$. You will need trigonometry for this one.

d) Write a system of inequalities whose graph consists of all points on and inside a regular hexagon inscribed in $x^2 + y^2 = 1$.

5.6 Linear Programming

In this section we apply our knowledge of systems of linear inequalities to solving linear programming problems. Linear programming is a method that can be used to solve many practical business problems. Linear programming can tell us how to allocate resources to achieve a maximum profit, minimum labor cost, or a most nutritious meal.

Graphing the Constraints

In the simplest linear programming applications we have two variables that must satisfy several linear inequalities. These inequalities are called the **constraints** because they restrict the variables to only certain values. The graph of the solution set to the system is used to indicate the points that satisfy all of the constraints. Any point that satisfies all of the constraints is called a **feasible solution** to the problem.

EXAMPLE 1 Graphing the constraints

Graph the solution set to the system of inequalities and identify each vertex of the region.

$$x \geq 0, \quad y \geq 0$$
$$2x + y \leq 6$$
$$x + y \leq 4$$

Solution

The points on or to the right of the y-axis satisfy $x \geq 0$. The points on or above the x-axis satisfy $y \geq 0$. The points on or below the line $2x + y = 6$ satisfy $2x + y \leq 6$. The points on or below the line $x + y = 4$ satisfy $x + y \leq 4$. Graph each straight line and shade the region that satisfies all four inequalities, as shown in Fig. 5.36. Three of the vertices are easily identified as (0, 0), (0, 4), and (3, 0). The fourth vertex is at the intersection of $x + y = 4$ and $2x + y = 6$. Multiply $x + y = 4$ by -1 and add the result to $2x + y = 6$:

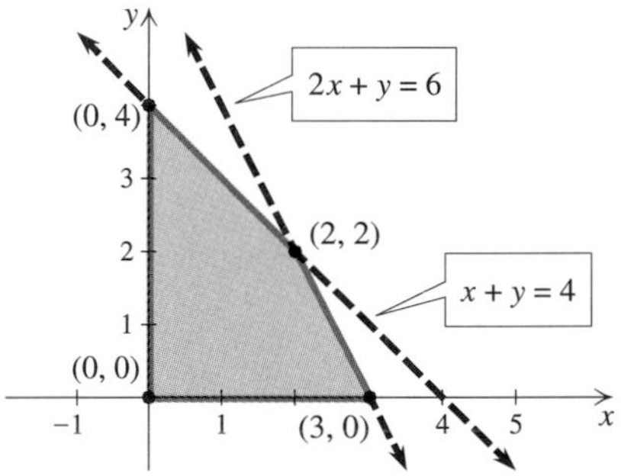

Figure 5.36

$$\begin{aligned} -x - y &= -4 \\ 2x + y &= 6 \\ \hline x \quad &= 2 \end{aligned}$$

If $x = 2$ and $x + y = 4$, then $y = 2$. So the fourth vertex is (2, 2).

In linear programming, the constraints usually come from physical limitations of resources described within the specific problem. Constraints that are always satisfied are called **natural constraints**. For example, the requirement that the number of employees be greater than or equal to zero is a natural constraint. In the next example we write the constraints and then graph the points in the coordinate plane that satisfy all of the constraints.

EXAMPLE 2 Finding and graphing the constraints

Bruce builds portable storage buildings. He uses 10 sheets of plywood and 15 studs in a small building, and he uses 15 sheets of plywood and 45 studs in a large building. Bruce has available only 60 sheets of plywood and 135 studs. Write the constraints on the number of small and large portable buildings that he can build with the available supplies, and graph the solution set to the system of constraints.

Solution

Let x represent the number of small buildings and y represent the number of large buildings. The natural constraints are

$$x \geq 0 \quad \text{and} \quad y \geq 0$$

because he cannot build a negative number of buildings. Since he has only 60 sheets of plywood available, we must have

$$10x + 15y \leq 60$$

$$2x + 3y \leq 12 \qquad \text{Divide each side by 5.}$$

Since he has only 135 studs available, we must have

$$15x + 45y \leq 135$$

$$x + 3y \leq 9 \qquad \text{Divide each side by 15.}$$

The conditions stated lead to the following system of constraints:

$$x \geq 0, \quad y \geq 0 \qquad \text{Natural constraints}$$

$$2x + 3y \leq 12 \qquad \text{Constraint on amount of plywood}$$

$$x + 3y \leq 9 \qquad \text{Constraint on number of studs}$$

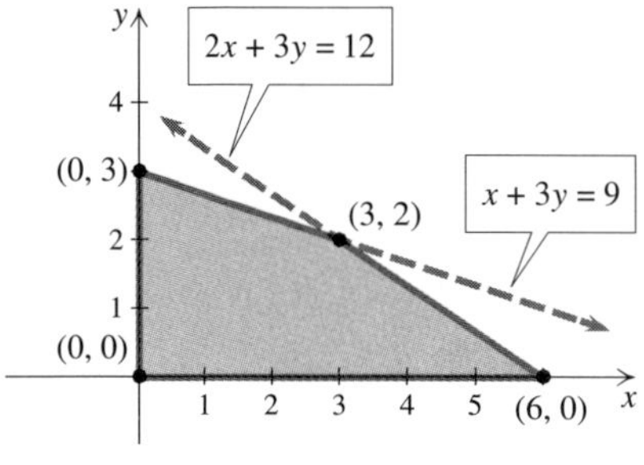

Figure 5.37

The graph of the solution set to this system is shown in Fig. 5.37.

Maximizing or Minimizing a Linear Function

In Example 2, any ordered pair within the shaded region of Fig. 5.37 is a feasible solution to the problem of deciding how many buildings of each type could be built. For each feasible solution within the shaded region Bruce makes some amount of profit. Of course, Bruce wants to find a feasible solution that will yield the maximum possible profit. In general, the function that we wish to maximize or minimize, subject to the constraints, is called the **objective function.**

If Bruce makes a profit of \$400 on a small building and \$500 on a large building, then the total profit from x small and y large buildings is

$$P = 400x + 500y.$$

Since the profit is a function of x and y, we write the objective function as

$$P(x, y) = 400x + 500y.$$

The function P is a linear function of x and y. The domain of P is the region graphed in Fig. 5.37.

Definition: Linear Function in Two Variables

A **linear function in two variables** is a function of the form

$$f(x, y) = Ax + By + C,$$

where A, B, and C are real numbers such that A and B are not both zero.

Bruce is interested in the maximum profit, subject to the constraints on x and y. Suppose $x = 1$ and $y = 1$; then the profit is

$$P(1, 1) = 400(1) + 500(1) = \$900.$$

In fact, the profit is \$900 for any x and y that satisfy

$$400x + 500y = 900.$$

If he wants \$1300 profit, then x and y must satisfy

$$400x + 500y = 1300,$$

and the profit is \$1800 if x and y satisfy

$$400x + 500y = 1800.$$

The graphs of these lines are shown in Fig. 5.38. Notice that the larger profit is found on the higher *profit line* and all of the profit lines are parallel. Bruce wants the highest profit line that intersects the region of feasible solutions. You can see in Fig. 5.38 that the highest line that intersects the region and is parallel to the other profit lines will intersect the region at the vertex (6, 0). So he should build six small buildings and no large buildings to maximize the profit. The maximum profit is $P(6, 0) = 400(6) + 500(0) = \$2400$.

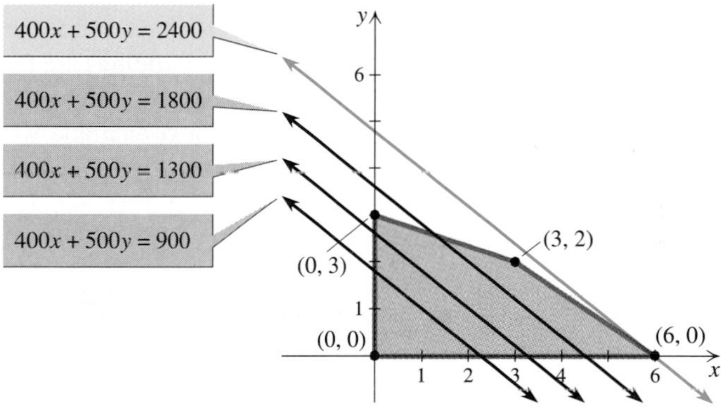

Figure 5.38

In another case we may be looking for the point that would *minimize* a linear function of the two variables. *In general, if the maximum or minimum value exists, then the maximum or minimum value of a linear function subject to linear constraints occurs at a vertex of the region determined by the constraints.* The minimum profit for the portable buildings occurs when no buildings are built, at the vertex (0, 0). It is possible that the maximum or minimum value occurs at two adjacent vertices and at every point along the line segment joining them.

It is a bit cumbersome and possibly inaccurate to graph parallel lines and then find the highest (or lowest) one that intersects the region determined by the constraints. Instead, we can use the following procedure for linear programming, which does not depend as much on graphing.

PROCEDURE

Linear Programming

Use the following steps to find the maximum or minimum value of a linear function subject to linear constraints.

1. Graph the region that satisfies all of the constraints.
2. Determine the coordinates of each vertex of the region.
3. Evaluate the function at each vertex of the region.
4. Identify which vertex gives the maximum or minimum value of the function.

To use the new procedure on Bruce's buildings, note that in Fig. 5.38 the vertices are (0, 0), (6, 0), (0, 3), and (3, 2). Compute the profit at each vertex:

$$P(0, 0) = 400(0) + 500(0) = \$0 \qquad \text{Minimum profit}$$
$$P(6, 0) = 400(6) + 500(0) = \$2400 \qquad \text{Maximum profit}$$
$$P(0, 3) = 400(0) + 500(3) = \$1500$$
$$P(3, 2) = 400(3) + 500(2) = \$2200$$

From this list, we see that the maximum profit is \$2400, when six small buildings and no large buildings are built, and the minimum profit is \$0, when no buildings of either type are built.

In the next example we use the linear programming technique to find the minimum value of a linear function subject to a system of constraints.

EXAMPLE 3 Finding the minimum value of a linear function

One serving of Muesli breakfast cereal contains 4 grams of protein and 30 grams of carbohydrates. One serving of Multi Bran Chex contains 2 grams of protein and 25 grams of carbohydrates. A dietitian wants to mix these two cereals to make a batch that contains at least 44 grams of protein and at least 450 grams of carbohydrates. If the cost of Muesli is 21 cents per serving and the cost of Multi Bran Chex is 14 cents per serving, then how many servings of each cereal would minimize the cost and satisfy the constraints?

Solution

Let x = the number of servings of Muesli and y = the number of servings of Multi Bran Chex. If the batch is to contain at least 44 grams of protein, then

$$4x + 2y \geq 44.$$

If the batch is to contain at least 450 grams of carbohydrates, then

$$30x + 25y \geq 450.$$

Simplify each inequality and use the two natural constraints to get the following system:

$$x \geq 0, \quad y \geq 0$$
$$2x + y \geq 22$$
$$6x + 5y \geq 90$$

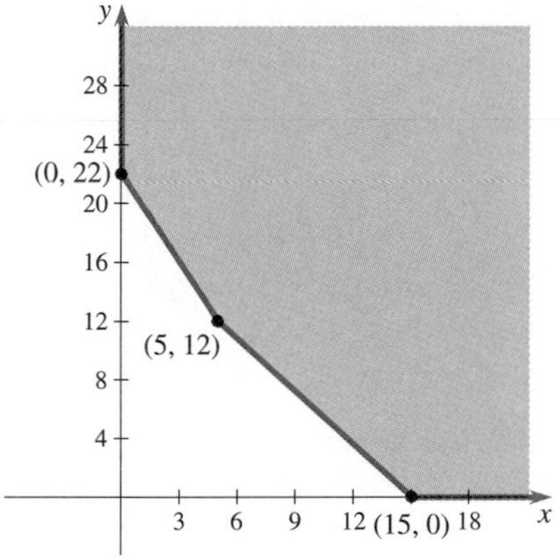

Figure 5.39

The graph of the constraints is shown in Fig. 5.39. The vertices are (0, 22), (5, 12), and (15, 0). The cost in dollars for x servings of Muesli and y servings of Multi Bran Chex is $C(x, y) = 0.21x + 0.14y$. Evaluate the cost at each vertex.

$$C(0, 22) = 0.21(0) + 0.14(22) = \$3.08$$
$$C(5, 12) = 0.21(5) + 0.14(12) = \$2.73 \qquad \text{Minimum cost}$$
$$C(15, 0) = 0.21(15) + 0.14(0) = \$3.15$$

The minimum cost of \$2.73 is attained by using 5 servings of Muesli and 12 servings of Multi Bran Chex. Note that $C(x, y)$ does not have a maximum value on this region. The cost increases without bound as x and y increase.

The examples of linear programming given in this text are simple examples. Problems in linear programming in business can involve a hundred or more variables subject to as many inequalities. These problems are solved by computers using matrix methods, but the basic idea is the same as we have seen in this section.

FOR THOUGHT True or False? Explain.

1. The graph of $x \geq 0$ in the coordinate plane consists only of points on the x-axis that are at or to the right of the origin.
2. The graph of $y \geq 2$ in the coordinate plane consists of the points on or to the right of the line $y = 2$.
3. The graph of $x + y \leq 5$ does not include the origin.
4. The graph of $2x + 3y = 12$ has x-intercept (0, 4) and y-intercept (6, 0).
5. The graph of a system of inequalities is the intersection of their individual solution sets.
6. In linear programming, constraints are inequalities that restrict the values of the variables.
7. The function $f(x, y) = 4x^2 + 9y^2 + 36$ is a linear function of x and y.
8. The value of $R(x, y) = 30x + 15y$ at the point (1, 3) is 75.
9. If $C(x, y) = 7x + 9y + 3$, then $C(0, 5) = 45$.
10. To solve a linear programming problem, we evaluate the objective function at the vertices of the region determined by the constraints.

5.6 EXERCISES

Tape 17 Disk

Graph the solution set to each system of inequalities and identify each vertex of the region.

1. $x \geq 0, y \geq 0$
$x + y \leq 4$

2. $x \geq 0, y \geq 0$
$2x + y \leq 4$

3. $x \geq 0, y \geq 0$
$x \leq 1, y \leq 3$

4. $x \geq 0, y \geq 0$
$y \leq x, x \leq 3$

5. $x \geq 0, y \geq 0$
$x + y \leq 4$
$2x + y \leq 6$

6. $x \geq 0, y \geq 0$
$50x + 40y \leq 200$
$10x + 20y \leq 60$

7. $x \geq 0, y \geq 0$
$2x + y \geq 4$
$x + y \geq 3$

8. $x \geq 0, y \geq 0$
$20x + 10y \geq 40$
$5x + 5y \geq 15$

9. $x \geq 0, y \geq 0$
$3x + y \geq 6$
$x + y \geq 4$

10. $x \geq 0, y \geq 0$
$2x + y \geq 6$
$x + 2y \geq 6$

11. $x \geq 0, y \geq 0$
$3x + y \geq 8$
$x + y \geq 6$

12. $x \geq 0, y \geq 0$
$x + 4y \leq 20$
$4x + y \leq 64$

Find the maximum value of the objective function $T(x, y) = 2x + 3y$ on each given region.

13.

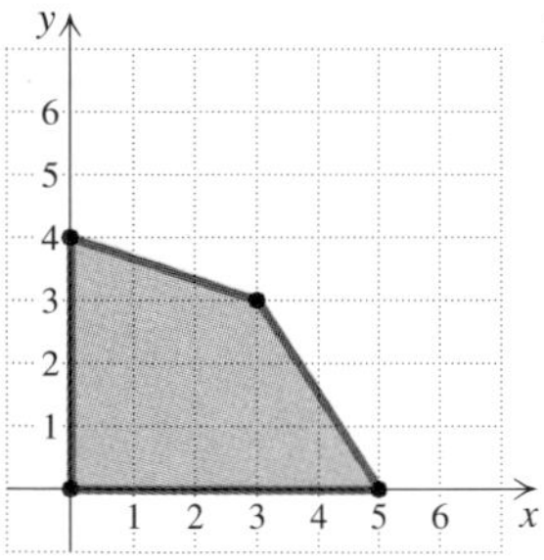

14.

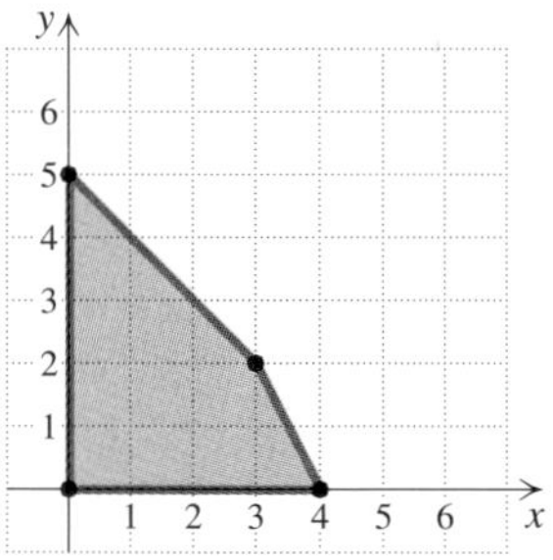

Find the minimum value of the objective function $H(x, y) = 2x + 2y$ on each given region.

15.

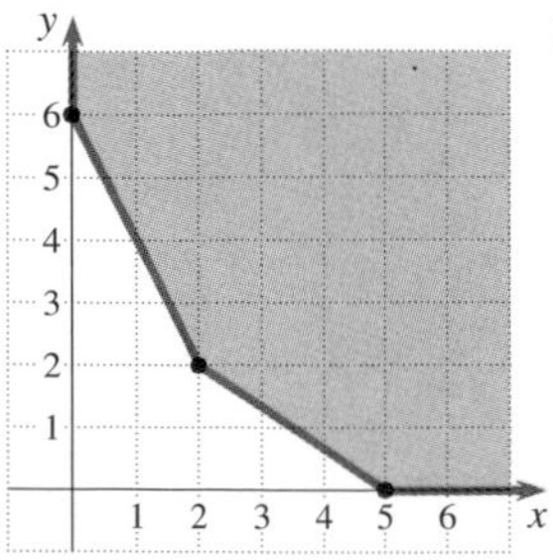

16.

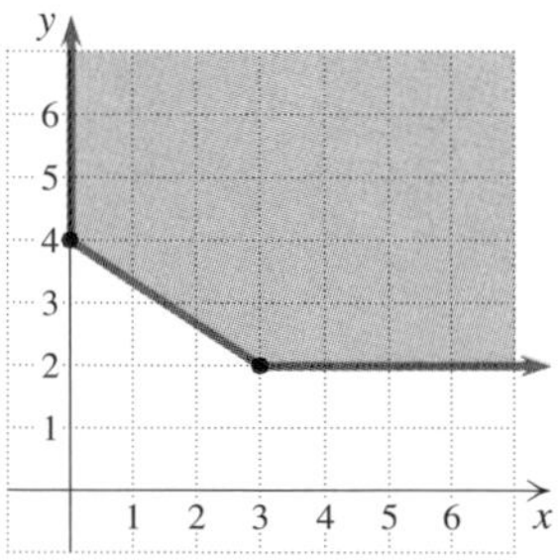

Find the maximum or minimum value of each objective function subject to the given constraints.

17. Maximize $P(x, y) = 5x + 9y$ subject to $x \geq 0$, $y \geq 0$, and $x + 2y \leq 6$.

18. Maximize $P(x, y) = 25x + 31y$ subject to $x \geq 0$, $y \geq 0$, and $5x + 6y \leq 30$.

19. Minimize $C(x, y) = 10x + 20y$ subject to $x \geq 0$, $y \geq 0$, $x + y \geq 8$, and $3x + 5y \geq 30$.

20. Maximize $R(x, y) = 50x + 20y$ subject to $x \geq 0$, $y \geq 0$, $3x + y \leq 18$, and $2x + y \leq 14$.

Solve each problem by linear programming.

21. *Maximizing Revenue* Bob and Betty make birdhouses and mailboxes in their craft shop near Gatlinburg. Each bird house requires 3 hr work from Bob and 1 hr from Betty. Each mailbox requires 4 hr work from Bob and 2 hr work from Betty. Bob cannot work more than 48 hr per week and Betty cannot work more than 20 hr per week. If each bird house sells for \$12 and each mailbox sells for \$20, then how many of each should they make to maximize their revenue?

22. *Maximizing Revenue* At Taco Town a taco contains 2 oz of ground beef and 1 oz of chopped tomatoes. A burrito contains 1 oz of ground beef and 3 oz of chopped tomatoes. Near closing time the cook discovers that they have only 22 oz of ground beef and 36 oz of tomatoes left. The manager directs the cook to use the available resources to maximize their revenue for the remainder of the shift. If a taco sells for 40 cents and a burrito sells for 65 cents, then how many of each should they make to maximize their revenue?

23. *Bird Houses and Mailboxes* If a bird house sells for \$18 and a mailbox for \$20, then how many of each should Bob and Betty build to maximize their revenue, subject to the constraints of Exercise 21?

24. *Tacos and Burritos* If a taco sells for 20 cents and a burrito for 65 cents, then how many of each should be made to maximize the revenue, subject to the constraints of Exercise 22?

25. *Minimizing Operating Costs* Kimo's Material Company hauls gravel to a construction site, using a small truck and a large truck. The carrying capacity and operating cost per load are given in the accompanying table. Kimo must deliver a minimum of 120 yd^3 per day to satisfy his contract with the builder. The union contract with his drivers requires that the total number of loads per day be a minimum of 8. How many loads should be made in each truck per day to minimize the total cost?

Table for Exercise 25

	Small truck	Large truck
Capacity (yd^3)	12	20
Cost/load	\$70	\$60

26. *Minimizing Labor Costs* Tina's Telemarketing employs part-time and full-time workers. The number of hours worked per week and the pay per hour for each is given in the accompanying table. Tina needs at least 1200 hr of work done per week. To qualify for certain tax breaks, she must have at least 45 employees. How many part-time and full-time employees should be hired to minimize Tina's weekly labor cost?

Table for Exercise 26

	Part-time	Full-time
Hr/wk	20	40
Pay/hr	\$6	\$8

27. *Small Trucks and Large Trucks* If it costs \$70 per load to operate the small truck and \$75 per load to operate the large truck, then how many loads should be made in each truck per day to minimize the total cost, subject to the constraints of Exercise 25?

28. *Part-Time and Full-Time Workers* If the labor cost for a part-timer is \$9/hr and the labor cost for a full-timer is \$8/hr, then how many of each should be employed to minimize the weekly labor cost, subject to the constraints of Exercise 26?

LINKING CONCEPTS

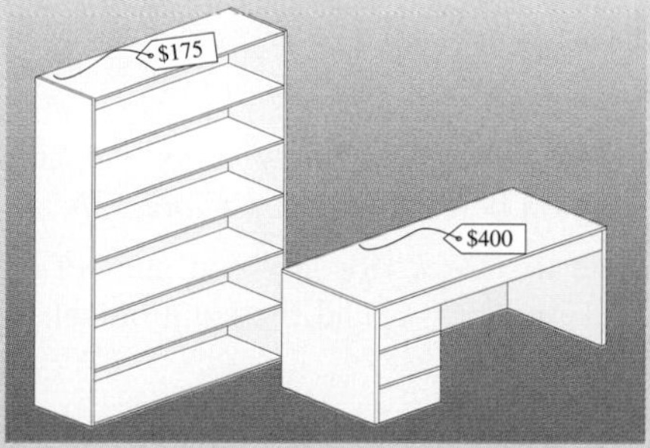

For Individual or Group Explorations

Numerous Constraints Lucy's Woodworks makes desks and bookcases for furniture stores. Lucy sells the desks for \$400 and the bookcases for \$175. Each desk takes 30 ft^2 of oak plywood, 12 ft of molding, 1 quart of stain, 3 pints of lacquer, 9 drawer pulls, 7 drawer glides, and 20 hours of labor. Each bookcase requires 20 ft^2 of oak plywood, 15 ft of molding, 1 pint of stain, 1 quart of lacquer, 20 shelf pins, and 12 hours of labor. Lucy has available only 960 ft^2 of oak plywood, 480 ft of molding, 8 gallons of stain, 15 gallons of lacquer, 270 drawer pulls, 350 drawer glides, 580 shelf pins, and 720 hours of labor.

a) How many desks and how many bookcases should Lucy make to maximize her revenue?

b) Suppose Lucy could increase her supply of one item. Which item would have the most significant impact on her revenue? Explain your answer.

HIGHLIGHTS

- **Section 5.1**
Systems of Linear Equations in Two Variables

1. A linear equation in two variables has the form $Ax + By = C$, where A and B are not both 0.
2. The solution set to a system of equations in two variables is the intersection of the graphs of the equations, or the set of ordered pairs that satisfy all of the equations.
3. The methods of addition and substitution can be used on linear systems to eliminate variables.
4. A system of linear equations in two variables may have one solution (independent system), no solution (inconsistent system), or infinitely many solutions (dependent system).

- **Section 5.2**
Systems of Linear Equations in Three Variables

1. A linear equation in three variables has the form $Ax + By + Cz = D$, where A, B, and C are not all 0.
2. A system of linear equations in three variables can have a single ordered triple as a solution, no solution, or infinitely many solutions.

- **Section 5.3**
Nonlinear Systems of Equations

1. A system with at least one nonlinear equation is a nonlinear system.
2. The graphs of a nonlinear system may intersect at more than one point.
3. The methods of addition and substitution can be used on nonlinear systems to eliminate variables.

- **Section 5.4**
Partial Fractions

1. In partial fraction decomposition we reverse the process of addition of rational expressions.
2. The numerators of the partial fractions are found by solving a system of linear equations.

- **Section 5.5**
Inequalities and Systems of Inequalities in Two Variables

1. The solution set to $y > mx + b$ is the region above the line $y = mx + b$, and the solution set to $y < mx + b$ is the region below the line $y = mx + b$.
2. The solution set in the coordinate plane to $x > k$ (for any real number k) is the region to the right of the vertical line $x = k$, and the solution set to $x < k$ is the region to the left of $x = k$.
3. The solution set to a $\leq$ or $\geq$ inequality includes the boundary (drawn solid), while the solution set to a $<$ or $>$ inequality excludes the boundary (drawn dashed).
4. A system of inequalities is solved by testing a point from each region formed by the boundary lines or curves.

- **Section 5.6 Linear Programming**

1. Inequalities concerning the variables in linear programming are called constraints.
2. A system of constraints determines a region of possible values for the variables in a linear programming problem.
3. To find the maximum or minimum value of a linear objective function subject to linear constraints, evaluate the function at each vertex of the region determined by the constraints.

CHAPTER 5 REVIEW EXERCISES

Solve each system by graphing.

1. $2x - 3y = -9$, $3x + y = 14$

2. $3x - 2y = 0$, $y = -2x - 7$

3. $x + y = 2$, $2y - 3x = 9$

4. $x - y = 30$, $2x + 3y = 10$

Solve each system by the method of your choice. Indicate whether each system is independent, dependent, or inconsistent.

5. $3x - 5y = 19$, $y = x$

6. $x + y = 9$, $y = x - 3$

7. $4x - 3y = 6$, $3x + 2y = 9$

8. $3x - 2y = 4$, $5x + 7y = 1$

9. $6x + 2y = 2$, $y = -3x + 1$

10. $x - y = 9$, $2y - 2x = -18$

11. $3x - 4y = 12$, $8y - 6x = 9$

12. $y = -5x + 3$, $5x + y = 6$

Solve each system.

13. $x + y - z = 8$, $2x + y + z = 1$, $x + 2y + 3z = -5$

14. $2x + 3y - 2z = 8$, $3x - y + 4z = -20$, $x + y - z = 3$

15. $x + y + z = 1$, $2x - y + 2z = 2$, $2x + 2y + 2z = 2$

16. $x - y - z = 9$, $x + y + 2z = -9$, $-2x + 2y + 2z = -18$

17. $x + y + z = 1$, $2x - y + 3z = 5$, $x + y + z = 4$

18. $2x - y + z = 4$, $x - y + z = -1$, $-x + y - z = 0$

Solve each nonlinear system of equations. Find real solutions only.

19. $x^2 + y^2 = 4$, $x = y^2$

20. $x^2 - y^2 = 9$, $x^2 + y^2 = 7$

21. $y = |x|$, $y = x^2$

22. $y = 2x^2 + x - 3$, $6x + y = 12$

Find the partial fraction decomposition for each rational expression.

23. $\dfrac{7x - 7}{(x - 3)(x + 4)}$

24. $\dfrac{x - 13}{x^2 - 6x + 5}$

25. $\dfrac{7x^2 - 7x + 23}{x^3 - 3x^2 + 4x - 12}$

26. $\dfrac{10x^2 - 6x + 2}{(x - 1)^2(x + 2)}$

Graph the solution set to each inequality.

27. $x^2 + (y - 3)^2 < 9$

28. $2x - 9y \le 18$

29. $x \le (y - 1)^2$

30. $y < 6 - 2x^2$

Graph the solution set to each system of inequalities.

31. $2x - 3y \ge 6$, $x \le 2$

32. $x \le 3$, $y \ge 1$, $x - y \ge -5$

33. $y \ge 2x^2 - 6$, $x^2 + y^2 \le 9$

34. $x^2 + y^2 \ge 16$, $2y \ge x^2 - 16$

35. $x \ge 0$, $y \ge 1$, $x + 2y \le 10$, $3x + 4y \le 24$

36. $x \ge 0$, $y \ge 0$, $30x + 60y \le 1200$, $x + y \le 30$

37. $x \ge 0$, $y \ge 0$, $x + 6y \ge 60$, $x + y \ge 35$

38. $x \ge 0$, $y \ge x + 1$, $x + y \le 5$

Solve each problem, using a system of equations.

39. Find the equation of the line through $(-2, 3)$ and $(4, -1)$.

40. Find the equation of the line through $(4, 7)$ and $(-2, -3)$.

41. Find the equation of the parabola through $(1, 4)$, $(3, 20)$, and $(-2, 25)$.

42. Find the equation of the parabola through $(-1, 10)$, $(2, -5)$, and $(3, -18)$.

43. *Tacos and Burritos* At Taco Town a taco contains 1 oz of meat and 2 oz of cheese, while a burrito contains 2 oz of meat and 3 oz of cheese. In one hour the cook used 181 oz of meat and 300 oz of cheese making tacos and burritos. How many of each were made?

44. *Imported and Domestic Cars* Nicholas had 10% imports on his used car lot, and Seymour had 30% imports on his used car lot. After Nicholas bought Seymour's entire stock, Nicholas had 300 cars, of which 22% were imports. How many cars were on each lot originally?

45. *Daisies, Carnations, and Roses* Esther's Flower Shop sells a bouquet containing five daisies, three carnations, and two roses for \$3.05. Esther also sells a bouquet containing three daisies, one carnation, and four roses for \$2.75. Her Valentine's Day Special contains four daisies, two carnations, and one rose for \$2.10. How much should she charge for her Economy Special, which contains one daisy, one carnation, and one rose?

46. *Peppers, Tomatoes, and Eggplants* Ngan planted 81 plants in his garden at a total cost of \$23.85. The 81 plants consisted of peppers at 20 cents each, tomatoes at 35 cents each, and eggplants at 30 cents each. If the total number of peppers and tomatoes was only half the number of eggplants, then how many of each did he plant?

Solve each linear programming problem.

47. *Minimum* Find the minimum value of the function $C(x, y) = 0.42x + 0.84y$ subject to the constraints $x \geq 0$, $y \geq 0$, $x + 6y \geq 60$, and $x + y \geq 35$.

48. *Maximum* Find the maximum value of the function $P(x, y) = 1.23x + 1.64y$ subject to the constraints $x \geq 0$, $y \geq 0$, $x + 2y \leq 40$, and $x + y \leq 30$.

49. *Pipeline or Barge* A refinery gets its oil either from a pipeline or from barges. The refinery needs at least 12 million barrels per day. The maximum capacity of the pipeline is 12 million barrels per day, and the maximum that can be delivered by barge is 8 million barrels per day. The refinery has a contract to buy at least 6 million barrels per day through the pipeline. If the cost of oil by barge is \$18 per barrel and the cost of oil from the pipeline is \$20 per barrel, then how much oil should be purchased from each source to minimize the cost and satisfy the constraints?

50. *Fluctuating Costs* If the cost of oil by barge goes up to \$21 per barrel while the cost of oil by pipeline stays at \$20 per barrel, then how much oil should be purchased from each source to minimize the cost and satisfy the constraints of Exercise 49?

CHAPTER 5 TEST

Solve the system by the indicated method.

1. Graphing:
$$2x + 3y = 6$$
$$y = \frac{1}{3}x + 5$$

2. Substitution:
$$2x + y = 4$$
$$3x - 4y = 9$$

3. Addition:
$$10x - 3y = 22$$
$$7x + 2y = 40$$

Determine whether each of the following systems is independent, inconsistent, or dependent.

4. $x = 6 - y$
$3x + 3y = 4$

5. $y = \frac{1}{2}x + 3$
$x - 2y = -6$

6. $y = 2x - 1$
$y = 3x + 20$

7. $y = -x + 2$
$y = -x + 5$

Find the solution set to each system of equations in three variables.

8. $2x - y + z = 4$
$-x + 2y - z = 6$

9. $x - 2y - z = 2$
$2x + 3y + z = -1$
$3x - y - 3z = -4$

10. $x + y + z = 1$
$x + y - z = 4$
$-x - y + z = 2$

Solve each system.

11. $x^2 + y^2 = 16$
$x^2 - 4y^2 = 16$

12. $x + y = -2$
$y = x^2 - 5x$

Find the partial fraction decomposition for each rational expression.

13. $\dfrac{2x + 10}{x^2 - 2x - 8}$

14. $\dfrac{4x^2 + x - 2}{x^3 - x^2}$

Graph the solution set to each inequality or system of inequalities.

15. $2x - y < 8$

16. $x + y \leq 5$
$x - y < 0$

17. $x^2 + y^2 \leq 9$
$y \leq 1 - x^2$

Solve Each Problem

18. In a survey of 52 students in the cafeteria, it was found that 15 were commuters. If one-quarter of the female students and one-third of the male students in the survey were commuters, then how many of each sex were surveyed?

19. General Hospital is planning an aggressive advertising campaign to bolster the hospital's image in the community. Each television commercial reaches 14,000 people and costs \$9000, while each newspaper ad reaches 6000 people and costs \$3000. The advertising budget for the campaign is limited to \$99,000 and the advertising agency has the capability of producing a maximum of 23 ads and/or commercials during the time allotted for the campaign. What mix of television commercials and newspaper ads will maximize the audience exposure, subject to the given constraints?

TYING IT ALL TOGETHER

Chapters P–5

Solve each equation.

1. $\frac{x-2}{x+5} = \frac{11}{24}$

2. $\frac{1}{x} + \frac{x-2}{x+5} = \frac{11}{24}$

3. $5 - 3(x + 2) - 2(x - 2) = 7$

4. $|3 - 2x| = 5$

5. $\sqrt{3 - 2x} = 5$

6. $3x^2 - 4 = 0$

7. $\frac{(x-2)^2}{x^2} = 1$

8. $2^{x-1} = 9$

9. $\log(x + 1) + \log(x + 4) = 1$

10. $x^{-2/3} = 0.25$

11. $x^2 - 3x = 6$

12. $2(x - 3)^2 - 1 = 0$

Solve each inequality and graph the solution set on the number line.

13. $3 - 2x > 0$

14. $|3 - 2x| > 0$

15. $x^2 \geq 9$

16. $(x - 2)(x + 4) \leq 27$

Graph the solution set to each inequality in the coordinate plane.

17. $3 - 2x > y$

18. $|3 - 2x| > y$

19. $x^2 \geq 9$

20. $(x - 2)(x + 4) \leq y$

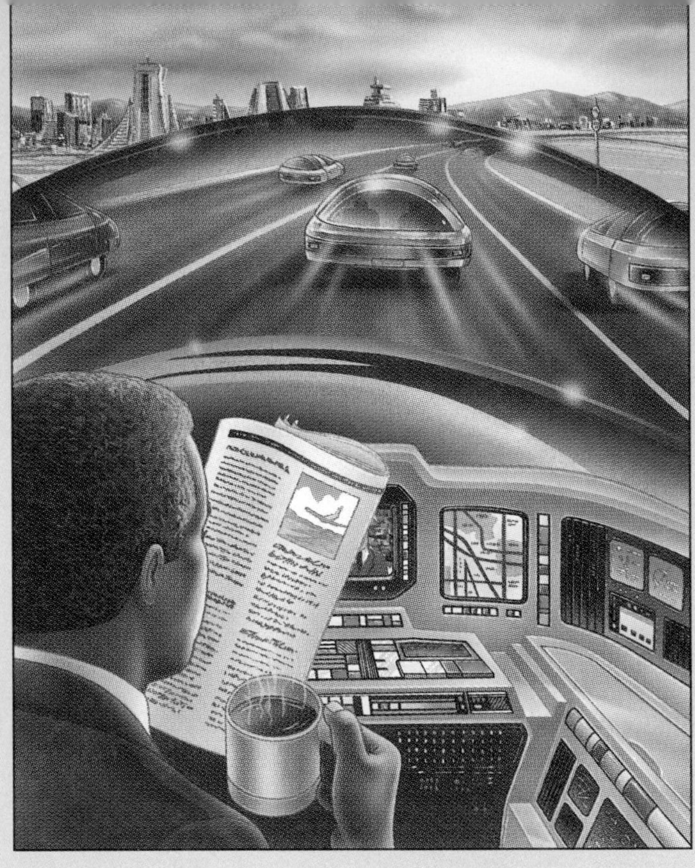

In the year 2010, traffic gridlocks on Los Angeles freeways will be history. In response to electronic commands, great platoons of linked vehicles will cruise the once congested lanes at an unvarying 75 miles per hour. Safely locked inside their "smart cars," commuters will read, work on their laptop computers, even catch a few extra winks on the way to work. It seems they'll do almost anything except keep their eyes on the road.

This is the vision of Intelligent Vehicle/Highway Systems (IVHS), a nonprofit group whose members foresee a cluster of emerging technologies transforming modern highways. The organization is already testing on-board navigation systems along a 13-mile "smart corridor" of southern California's Santa Monica Freeway. In the new interactive systems, powerful computers will pinpoint drivers' locations, select routes, display maps, warn of impending collisions, and even issue weather reports. At last, driving will come up to speed.

To perform such feats, IVHS systems must instantly solve huge linear programming problems involving hundreds of thousands of variables. Problems of similar complexity—impossible to solve even with the help of super computers—are echoed in AT&T's domestic long-distance communication network, which manipulates 800,000 variables! However, in 1984, an employee of Bell Lab discovered a new matrix technique that allowed computers to solve highly complex problems within a short period of time. By implementing this method, AT&T was able to solve one of its unmanageable linear programming problems in less than half an hour and obtain another 10 percent capacity out of its $15 billion system.

In Chapter 5, we covered basic linear programming. But as AT&T discovered, huge problems require more efficient techniques. In this chapter we'll study matrices and several methods of using them to solve systems of linear equations. These methods lie at the core of the large-scale computer programs being used today to manage traffic flow, schedule airline flights, and track data in inventory control, accounting, and purchasing.

6 MATRICES AND DETERMINANTS

6.1 Solving Linear Systems Using Matrices
6.2 Operations with Matrices
6.3 Multiplication of Matrices
6.4 Inverses of Matrices
6.5 Solution of Linear Systems in Two Variables Using Determinants
6.6 Solution of Linear Systems in Three Variables Using Determinants

6.1 Solving Linear Systems Using Matrices

In this section we learn a method for solving systems of linear equations that is an improved version of the addition method of Section 5.1. The new method requires some new terminology.

Matrices

Twenty-six female students and twenty-four male students responded to a survey on income in a college algebra class. Among the female students, 5 classified themselves as low-income, 10 as middle-income, and 11 as high-income. Among the male students, 9 were low-income, 2 were middle-income, and 13 were high-income. Each student is classified in two ways, according to sex and income. This information can be written in a *matrix*:

$$\begin{array}{l} \\ \text{Female} \\ \text{Male} \end{array} \begin{array}{c} \begin{array}{ccc} \text{L} & \text{M} & \text{H} \end{array} \\ \begin{bmatrix} 5 & 10 & 11 \\ 9 & 2 & 13 \end{bmatrix} \end{array}$$

In this matrix we can see the class makeup according to sex and income. A matrix provides a convenient way to organize a two-way classification of data.

A **matrix** is a rectangular array of real numbers. The **rows** of a matrix run horizontally, and the **columns** run vertically. A **row matrix** is a matrix with only one row, and a **column matrix** is a matrix with only one column. A matrix with m rows and n columns has **order** $m \times n$ (read "m by n") or **dimension** $m \times n$. The number of rows is always given first. For example, the matrix used to classify the students is a 2×3 matrix.

EXAMPLE 1 Finding the order of a matrix

Determine the order of each matrix.

a) $\begin{bmatrix} -4 & 3 & -2 \end{bmatrix}$ **b)** $\begin{bmatrix} 3 & -1 \\ 4 & 2 \end{bmatrix}$ **c)** $\begin{bmatrix} -5 & 19 \\ 14 & 2 \\ 0 & -1 \end{bmatrix}$ **d)** $\begin{bmatrix} -4 & 46 & 8 \\ 1 & 0 & 13 \\ -12 & -5 & 2 \end{bmatrix}$

Solution

Matrix (a) is a row matrix with order 1×3. Matrix (b) has order 2×2, matrix (c) has order 3×2, and matrix (d) has order 3×3.

A **square** matrix has an equal number of rows and columns. Matrices (b) and (d) of Example 1 are square matrices. Each number in a matrix is called an **entry** or an **element**. The matrix [5] is a 1×1 matrix with only one entry, 5.

The Augmented Matrix

We now see how matrices are used to represent systems of linear equations. The solution to a system of linear equations such as

$$\begin{aligned} x - 3y &= 11 \\ 2x + y &= 1 \end{aligned}$$

depends on the coefficients of x and y and the constants on the right-hand side of the equation. The **coefficient matrix** for this system is the matrix

$$\begin{bmatrix} 1 & -3 \\ 2 & 1 \end{bmatrix},$$

whose entries are the coefficients of the variables. (The coefficient of y in $x - 3y = 11$ is -3.) The constants from the right-hand side of the system are attached to the matrix of coefficients, to form the **augmented matrix** of the system:

$$\left[\begin{array}{cc|c} 1 & -3 & 11 \\ 2 & 1 & 1 \end{array}\right]$$

Each row of the augmented matrix represents an equation of the system, while the columns represent the coefficients of x, the coefficients of y, and the constants, respectively. The vertical line represents the equal signs. Two systems of linear equations are **equivalent** if they have the same solution set, while two augmented matrices are **equivalent** if the systems they represent are equivalent.

EXAMPLE 2 Determining the augmented matrix

Write the augmented matrix for each system of equations.

a) $\begin{aligned} x &= y + 3 \\ y &= 4 - x \end{aligned}$ **b)** $\begin{aligned} x + 2y - z &= 1 \\ 2x \quad\quad + 3z &= 5 \\ 3x - 2y + z &= 0 \end{aligned}$ **c)** $\begin{aligned} x - y &= 2 \\ y - z &= 3 \end{aligned}$

Solution

a) To write the augmented matrix, the equations must be in standard form with the variables on the left-hand side and constants on the right-hand side:

$$x - y = 3$$
$$x + y = 4$$

We write the augmented matrix using the coefficients of the variables and the constants:

$$\left[\begin{array}{cc|c} 1 & -1 & 3 \\ 1 & 1 & 4 \end{array}\right]$$

b) Use the coefficient 0 for each variable that is missing:

$$\left[\begin{array}{ccc|c} 1 & 2 & -1 & 1 \\ 2 & 0 & 3 & 5 \\ 3 & -2 & 1 & 0 \end{array}\right]$$

c) The augmented matrix for this system is a 2×4 matrix:

$$\left[\begin{array}{ccc|c} 1 & -1 & 0 & 2 \\ 0 & 1 & -1 & 3 \end{array}\right]$$

EXAMPLE 3 Writing a system for an augmented matrix

Write the system of equations represented by each augmented matrix.

a) $\left[\begin{array}{cc|c} 1 & 3 & -5 \\ 2 & -3 & 1 \end{array}\right]$ **b)** $\left[\begin{array}{cc|c} 1 & 0 & 7 \\ 0 & 1 & 3 \end{array}\right]$ **c)** $\left[\begin{array}{ccc|c} 2 & 5 & 1 & 2 \\ -3 & 0 & 4 & -1 \\ 4 & -5 & 2 & 3 \end{array}\right]$

Solution

a) Use the first two numbers in each row as the coefficients of x and y and the last number as the constant to get the following system:

$$x + 3y = -5$$
$$2x - 3y = 1$$

b) The augmented matrix represents the following system:

$$\begin{aligned} x &= 7 \\ y &= 3 \end{aligned}$$

c) Use the first three numbers in each row as the coefficients of x, y, and z and the last number as the constant to get the following system:

$$\begin{aligned} 2x + 5y + z &= 2 \\ -3x \qquad + 4z &= -1 \\ 4x - 5y + 2z &= 3 \end{aligned}$$

Recall that to solve a single equation, we write simpler and simpler equivalent equations to get an equation whose solution is obvious. Similarly, to solve a system of equations, we write simpler and simpler equivalent systems to get a system whose solution is obvious. We now look at operations that can be performed on augmented matrices to obtain simpler equivalent augmented matrices.

The Gaussian Elimination Method

The rows of an augmented matrix represent the equations of a system. Since the equations of a system can be written in any order, two rows of an augmented matrix can be interchanged if necessary. Since multiplication of both sides of an equation by the same nonzero number produces an equivalent equation, multiplying each entry in a row of the augmented matrix by a nonzero number produces an equivalent augmented matrix. In Section 5.1, two equations were added to eliminate a variable. In the augmented matrix, elimination of variables is accomplished by adding the entries in one row to the corresponding entries in another row. These two row operations can be combined to add a multiple of one row to another, just as was done in solving systems by addition. The three **row operations** for an augmented matrix are summarized as follows.

SUMMARY

Row Operations

Any of the following row operations on an augmented matrix gives an equivalent augmented matrix:

1. Interchanging two rows of the matrix.
2. Multiplying every entry in a row by the same nonzero real number.
3. Adding to a row a multiple of another row.

To solve a system of two linear equations in two variables using the **Gaussian elimination method**, we use row operations to obtain simpler and simpler augmented matrices. We want to get an augmented matrix that corresponds to a system whose solution is obvious. An augmented matrix of the following form is the simplest:

$$\left[\begin{array}{cc|c} 1 & 0 & a \\ 0 & 1 & b \end{array}\right]$$

Notice that this augmented matrix corresponds to the system $x = a$ and $y = b$, for which the solution set is $\{(a, b)\}$.

The **diagonal** of a matrix consists of the entries in the first row first column, second row second column, third row third column, and so on. The goal of the Gaussian elimination method is to convert the original augmented matrix into an equivalent augmented matrix that has 1's on its diagonal and 0's above and below the 1's. If the system has a unique solution, then it will appear in the right-most column of the final augmented matrix.

EXAMPLE 4 Using the Gaussian elimination method

Use row operations to solve the system.

$$\begin{aligned} 2x - 4y &= 16 \\ 3x + y &= 3 \end{aligned}$$

Solution

Start with the augmented matrix:

$$\left[\begin{array}{cc|c} 2 & -4 & 16 \\ 3 & 1 & 3 \end{array}\right]$$

The first step is to multiply the first row R_1 by $\frac{1}{2}$ to get a 1 in the first position on the diagonal. Think of this step as replacing R_1 by $\frac{1}{2}R_1$. We show this in symbols as $\frac{1}{2}R_1 \to R_1$.

$$\left[\begin{array}{cc|c} 1 & -2 & 8 \\ 3 & 1 & 3 \end{array}\right] \qquad \frac{1}{2}R_1 \to R_1$$

To get a 0 in the first position of R_2, multiply R_1 by -3 and add the result to R_2. Since $-3R_1 = [-3, 6, -24]$ and $R_2 = [3, 1, 3]$, $-3R_1 + R_2 = [0, 7, -21]$. We are replacing R_2 by $-3R_1 + R_2$:

$$\left[\begin{array}{cc|c} 1 & -2 & 8 \\ 0 & 7 & -21 \end{array}\right] \qquad -3R_1 + R_2 \to R_2$$

To get a 1 in the second position on the diagonal, multiply R_2 by $\frac{1}{7}$:

$$\left[\begin{array}{cc|c} 1 & -2 & 8 \\ 0 & 1 & -3 \end{array}\right] \qquad \frac{1}{7}R_2 \to R_2$$

Now row 2 is in the form needed to solve the system. We next get a 0 as the second entry in R_1. Multiply row 2 by 2 and add the result to row 1. Since

$2R_2 = [0, 2, -6]$ and $R_1 = [1, -2, 8]$, $2R_2 + R_1 = [1, 0, 2]$. We get the following matrix.

$$\left[\begin{array}{cc|c} 1 & 0 & 2 \\ 0 & 1 & -3 \end{array}\right] \qquad 2R_2 + R_1 \to R_1$$

Note that the coefficient of y in the first equation is now 0. The system associated with the last augmented matrix is $x = 2$ and $y = -3$. So the solution set to the system is $\{(2, -3)\}$. Check in the original system.

The procedure used in Example 4 to solve a system that has a unique solution is summarized below. For inconsistent and dependent systems see Examples 6, 7, and 8.

PROCEDURE

The Gaussian Elimination Method for an Independent System of Two Equations

To solve a system of two linear equations in two variables using Gaussian elimination, perform the following row operations on the augmented matrix.

1. If necessary, interchange R_1 and R_2 so that R_1 begins with a nonzero entry.
2. Get a 1 in the first position on the diagonal by multiplying R_1 by the reciprocal of the first entry in R_1.
3. Add an appropriate multiple of R_1 to R_2 to get 0 below the first 1.
4. Get a 1 in the second position on the diagonal by multiplying R_2 by the reciprocal of the second entry in R_2.
5. Add an appropriate multiple of R_2 to R_1 to get 0 above the second 1.
6. Read the unique solution from the last column of the final augmented matrix.

In the next example, Gaussian elimination is used on a system involving three variables. For three linear equations in three variables, x, y, and z, we try to get the augmented matrix into the form

$$\left[\begin{array}{ccc|c} 1 & 0 & 0 & a \\ 0 & 1 & 0 & b \\ 0 & 0 & 1 & c \end{array}\right],$$

from which we conclude that $x = a$, $y = b$, and $z = c$.

EXAMPLE 5 Gaussian elimination with three variables

Use the Gaussian elimination method to solve the following system:

$$\begin{aligned} 2x - y + z &= 1 \\ x + y - 2z &= 5 \\ 3x - y - z &= 8 \end{aligned}$$

Solution

Write the augmented matrix:

$$\left[\begin{array}{rrr|r} 2 & -1 & 1 & 1 \\ 1 & 1 & -2 & 5 \\ 3 & -1 & -1 & 8 \end{array}\right]$$

To get the first 1 on the diagonal, we could multiply R_1 by $\frac{1}{2}$ or interchange R_1 and R_2. Interchanging R_1 and R_2 is simpler because it avoids getting fractions in the entries:

$$\left[\begin{array}{rrr|r} 1 & 1 & -2 & 5 \\ 2 & -1 & 1 & 1 \\ 3 & -1 & -1 & 8 \end{array}\right] \qquad R_1 \leftrightarrow R_2$$

Now use the first row to get 0's below the first 1 on the diagonal. First, to get a 0 in the first position of the second row, multiply the first row by -2 and add the result to the second row. Then, to get a 0 in the first position of the third row, multiply the first row by -3 and add the result to the third row. These two steps eliminate the variable x from the second and third rows.

$$\left[\begin{array}{rrr|r} 1 & 1 & -2 & 5 \\ 0 & -3 & 5 & -9 \\ 0 & -4 & 5 & -7 \end{array}\right] \qquad \begin{array}{l} -2R_1 + R_2 \to R_2 \\ -3R_1 + R_3 \to R_3 \end{array}$$

To get a 1 in the second position on the diagonal, multiply the second row by $-\frac{1}{3}$:

$$\left[\begin{array}{rrr|r} 1 & 1 & -2 & 5 \\ 0 & 1 & -\frac{5}{3} & 3 \\ 0 & -4 & 5 & -7 \end{array}\right] \qquad -\frac{1}{3}R_2 \to R_2$$

To get a 0 above the second 1 on the diagonal, multiply R_2 by -1 and add the result to R_1. To get a 0 below the second 1 on the diagonal, multiply R_2 by 4 and add the result to the third row:

$$\left[\begin{array}{rrr|r} 1 & 0 & -\frac{1}{3} & 2 \\ 0 & 1 & -\frac{5}{3} & 3 \\ 0 & 0 & -\frac{5}{3} & 5 \end{array}\right] \qquad \begin{array}{l} -1R_2 + R_1 \to R_1 \\ \\ 4R_2 + R_3 \to R_3 \end{array}$$

To get a 1 in the third position on the diagonal, multiply R_3 by $-\frac{3}{5}$:

$$\left[\begin{array}{rrr|r} 1 & 0 & -\frac{1}{3} & 2 \\ 0 & 1 & -\frac{5}{3} & 3 \\ 0 & 0 & 1 & -3 \end{array}\right] \qquad -\frac{3}{5}R_3 \to R_3$$

Use the third row to get 0's above the third 1 on the diagonal:

$$\left[\begin{array}{rrr|r} 1 & 0 & 0 & 1 \\ 0 & 1 & 0 & -2 \\ 0 & 0 & 1 & -3 \end{array}\right] \qquad \begin{array}{l} \frac{1}{3}R_3 + R_1 \to R_1 \\ \frac{5}{3}R_3 + R_2 \to R_2 \end{array}$$

This last matrix corresponds to $x = 1$, $y = -2$, and $z = -3$. So the solution set to the system is $\{(1, -2, -3)\}$. Check in the original system.

In Example 5, there were two different row operations that would produce the first 1 on the diagonal. When doing Gaussian elimination by hand, you should choose the row operations that make the computations the simplest. When this method is performed by a computer, the same sequence of steps is used on every system. Any system of three equations that has a unique solution can be solved with the following sequence of steps.

PROCEDURE

The Gaussian Elimination Method for an Independent System of Three Equations

To solve a system of three linear equations in three variables using Gaussian elimination, perform the following row operations on the augmented matrix.

1. Get a 1 in the first position on the diagonal by multiplying R_1 by the reciprocal of the first entry in R_1. (First interchange rows if necessary.)
2. Add appropriate multiples of R_1 to R_2 and R_3 to get 0's below the first 1.
3. Get a 1 in the second position on the diagonal by multiplying R_2 by the reciprocal of the second entry in R_2. (First interchange rows if necessary.)
4. Add appropriate multiples of R_2 to R_1 and R_3 to get 0's above and below the second 1.
5. Get a 1 in the third position on the diagonal by multiplying R_3 by the reciprocal of the third entry in R_3.
6. Add appropriate multiples of R_3 to R_1 and R_2 to get 0's above the third 1.
7. Read the unique solution from the last column of the final augmented matrix.

Inconsistent and Dependent Equations

Inconsistent and dependent systems can also be solved by using Gaussian elimination, but in these cases it is impossible to get the augmented matrix into the desired form.

EXAMPLE 6 An inconsistent system in two variables

Solve the system.

$$\begin{aligned} x &= y + 3 \\ 2y &= 2x + 5 \end{aligned}$$

Solution

Write both equations in the form $Ax + By = C$:

$$\begin{aligned} x - y &= 3 \\ -2x + 2y &= 5 \end{aligned}$$

Start with the augmented matrix:

$$\left[\begin{array}{rr|r} 1 & -1 & 3 \\ -2 & 2 & 5 \end{array}\right]$$

To get a 0 in the first position of R_2, multiply R_1 by 2 and add the result to R_2:

$$\left[\begin{array}{rr|r} 1 & -1 & 3 \\ 0 & 0 & 11 \end{array}\right] \qquad 2R_1 + R_2 \to R_2$$

It is impossible to convert this augmented matrix to the desired form of the Gaussian elimination method. However, we can obtain the solution by observing that the second row corresponds to the equation $0 = 11$. So the system is inconsistent, and there is no solution.

Applying Gaussian elimination to an inconsistent system (as in Example 6) causes a row to appear with 0 as the entry for each coefficient but a nonzero entry for the constant. For a dependent system of two equations in two variables (as in the next example), a 0 will appear in every entry of some row.

EXAMPLE 7 **A dependent system in two variables**

Solve the system.

$$\begin{aligned} 2x + y &= 4 \\ 4x + 2y &= 8 \end{aligned}$$

Solution

Start with the augmented matrix:

$$\left[\begin{array}{rr|r} 2 & 1 & 4 \\ 4 & 2 & 8 \end{array}\right]$$

To get a 0 in the first position of R_2, multiply R_1 by -2 and add the result to R_2:

$$\left[\begin{array}{rr|r} 2 & 1 & 4 \\ 0 & 0 & 0 \end{array}\right] \qquad -2R_1 + R_2 \to R_2$$

The second row of this augmented matrix gives us the equation $0 = 0$. So the system is dependent. Every ordered pair that satisfies the first equation satisfies both equations. The solution set is $\{(x, y) | 2x + y = 4\}$.

Solving a dependent system that has fewer equations than variables does not necessarily result in a row in which all entries are 0. The next example shows a dependent system of two equations in three variables solved by Gaussian elimination.

EXAMPLE 8 A dependent system in three variables

Solve the system.

$$\begin{aligned} x - y + z &= 2 \\ -2x + y + 2z &= 5 \end{aligned}$$

Solution

Start with the augmented matrix:

$$\left[\begin{array}{rrr|r} 1 & -1 & 1 & 2 \\ -2 & 1 & 2 & 5 \end{array}\right]$$

Perform the following row operations to get 1's and 0's in the first two columns as you would for a 2×2 matrix:

$$\left[\begin{array}{rrr|r} 1 & -1 & 1 & 2 \\ 0 & -1 & 4 & 9 \end{array}\right] \qquad 2R_1 + R_2 \to R_2$$

$$\left[\begin{array}{rrr|r} 1 & -1 & 1 & 2 \\ 0 & 1 & -4 & -9 \end{array}\right] \qquad -1R_2 \to R_2$$

$$\left[\begin{array}{rrr|r} 1 & 0 & -3 & -7 \\ 0 & 1 & -4 & -9 \end{array}\right] \qquad R_2 + R_1 \to R_1$$

The last matrix corresponds to the system

$$\begin{aligned} x - 3z &= -7 \\ y - 4z &= -9 \end{aligned}$$

or $x = 3z - 7$ and $y = 4z - 9$. The system is dependent, and the solution set is

$$\{(3z - 7, 4z - 9, z) \mid z \text{ is any real number}\}.$$

Replace x by $3z - 7$ and y by $4z - 9$ in the original equations to check.

The Gaussian elimination method can be applied to a system of n linear equations in m unknowns. However, it is a rather tedious method to perform when n and m are greater than 2, especially when fractions are involved. Computers can be programmed to work with matrices, and Gaussian elimination is frequently used for computer solutions.

You can enter matrices into a graphing calculator and perform row operations with the calculator. However, performing row operations with a calculator is still rather tedious. In Section 6.4 we will see a much simpler method for solving systems with a calculator.

Applications

There is much discussion among city planners and transportation experts about Intelligent Vehicle/Highway Systems. (See *Omni*, January 1992.) Such a system would use a central computer and computers within vehicles to manage traffic flow by controlling traffic lights, rerouting traffic away from congested areas, and perhaps even driving the vehicle for you. It is easy to say that computers will control the complex traffic system of the future, but a computer does only what it is programmed to do. The next example shows one type of problem that a computer would be continually solving in order to control traffic flow.

EXAMPLE 9 Traffic control in the future

Figure 6.1 shows the intersections of four one-way streets. The numbers on the arrows represent the number of cars per hour that desire to enter each intersection and leave each intersection. For example, 400 cars per hour want to enter intersection P from the north on First Ave. while 300 cars per hour want to head east from intersection Q on Elm Street. The letters w, x, y, and z represent the number of cars per hour passing the four points between these four intersections, as shown in Fig. 6.1.

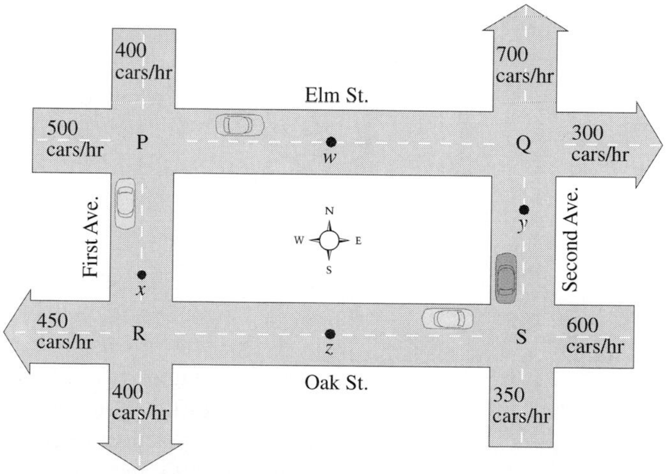

Figure 6.1

a) Find values for w, x, y, and z that would realize this desired traffic flow.

b) If construction on Oak Street limits z to 300 cars per hour, then how many cars per hour would have to pass w, x, and y?

Solution

a) The solution to the problem is based on the fact that the number of cars entering an intersection per hour must equal the number leaving that intersection per hour, if the traffic is to keep flowing. Since 900 cars (400 + 500) enter intersection P, 900 must leave, $x + w = 900$. Writing a similar equation for each intersection yields the system shown next.

$$\begin{aligned} w + x &= 900 \\ w + y &= 1000 \\ x + z &= 850 \\ y + z &= 950 \end{aligned}$$

For this system, the augmented matrix is

$$\left[\begin{array}{cccc|c} 1 & 1 & 0 & 0 & 900 \\ 1 & 0 & 1 & 0 & 1000 \\ 0 & 1 & 0 & 1 & 850 \\ 0 & 0 & 1 & 1 & 950 \end{array}\right].$$

Use row operations to get the equivalent matrix:

$$\left[\begin{array}{cccc|c} 1 & 1 & 0 & 0 & 900 \\ 0 & 1 & 0 & 1 & 850 \\ 0 & 0 & 1 & 1 & 950 \\ 0 & 0 & 0 & 0 & 0 \end{array}\right].$$

The system is a dependent system and does not have a unique solution. From this matrix we get $w = 50 + z$, $x = 850 - z$, and $y = 950 - z$. The problem is solved by any ordered 4-tuple of the form $(50 + z, 850 - z, 950 - z, z)$ where z is a nonnegative integer.

b) If z is limited to 300 because of construction, then the solution is (350, 550, 650, 300). To keep traffic flowing when $z = 300$, the system must route 350 cars past w, 550 past x, and 650 past y.

One-way streets were used in Example 9 to keep the problem simple, but you can imagine two-way streets where at each intersection there are three ways to route traffic. You can also imagine many streets and many intersections all subject to systems of equations, with computers continually controlling flow to keep all traffic moving. Computerized traffic control may be a few years away, but similar problems are being solved today at AT&T to route long-distance calls and at American Airlines to schedule flight crews and equipment.

For Thought True or False? Explain.

1. The augmented matrix for a system of two linear equations in two unknowns is a 2×2 matrix.

2. The augmented matrix for $\begin{aligned} x - y &= 4 \\ 3x + y &= 5 \end{aligned}$ is $\left[\begin{array}{cc|c} x & -y & 4 \\ 3x & y & 5 \end{array}\right]$.

3. The augmented matrix for $\begin{aligned} 3x - 2y &= 4 \\ x + y &= 6 \end{aligned}$ is $\left[\begin{array}{cc|c} 3 & -2 & 4 \\ 1 & 1 & 6 \end{array}\right]$.

4. The matrix $\left[\begin{array}{cc|c} 1 & 0 & 2 \\ 1 & -1 & -3 \end{array}\right]$ corresponds to the system $\begin{aligned} x &= 2 \\ x - y &= -3. \end{aligned}$

5. The matrix $\left[\begin{array}{cc|c} 1 & 3 & -2 \\ -1 & -5 & 4 \end{array}\right]$ is equivalent to $\left[\begin{array}{cc|c} 1 & 3 & -2 \\ 0 & -2 & 2 \end{array}\right]$.

6. The matrix $\left[\begin{array}{cc|c} 1 & 2 & 3 \\ 1 & -3 & 2 \end{array}\right]$ is equivalent to $\left[\begin{array}{cc|c} 1 & 2 & 3 \\ 0 & -5 & -1 \end{array}\right]$.

7. The matrix $\left[\begin{array}{cc|c} 1 & 0 & 2 \\ 0 & 1 & 7 \end{array}\right]$ corresponds to the system $\begin{aligned} x + y &= 2 \\ x + y &= 7. \end{aligned}$

8. The system corresponding to $\left[\begin{array}{cc|c} 1 & 3 & 5 \\ 0 & 0 & 7 \end{array}\right]$ is inconsistent.

9. The system corresponding to $\left[\begin{array}{cc|c} -1 & 2 & -3 \\ 0 & 0 & 0 \end{array}\right]$ is inconsistent.

10. The notation $2R_1 + R_3 \to R_3$ means to replace R_3 by $2R_1 + R_3$.

6.1 Exercises

Tape 18 Disk

Determine the order of each matrix.

1. $[1 \quad 5 \quad 8]$

2. $\begin{bmatrix} -3 \\ y \\ 5 \end{bmatrix}$

3. $[7]$

4. $\begin{bmatrix} x & y \\ z & w \end{bmatrix}$

5. $\begin{bmatrix} -5 & 12 \\ 99 & 6 \\ 0 & 0 \end{bmatrix}$

6. $\begin{bmatrix} 1 & 5 & 7 \\ 3 & 0 & 5 \\ 2 & -6 & -3 \end{bmatrix}$

Write the augmented matrix for each system of equations.

7. $\begin{aligned} x - 2y &= 4 \\ 3x + 2y &= -5 \end{aligned}$

8. $\begin{aligned} 4x - y &= 1 \\ x + 3y &= 5 \end{aligned}$

9. $\begin{aligned} x - y - z &= 4 \\ x + 3y - z &= 1 \\ 2y - 5z &= -6 \end{aligned}$

10. $\begin{aligned} x + 3y \phantom{{}- 4z} &= 5 \\ y - 4z &= 8 \\ -2x \phantom{{}+ 3y} + 5z &= 7 \end{aligned}$

Write the system of equations represented by each augmented matrix.

11. $\left[\begin{array}{cc|c} 3 & 4 & -2 \\ 3 & -5 & 0 \end{array}\right]$

12. $\left[\begin{array}{cc|c} 1 & 0 & -7 \\ 0 & 1 & 5 \end{array}\right]$

13. $\left[\begin{array}{ccc|c} 5 & 0 & 0 & 6 \\ -4 & 0 & 2 & -1 \\ 4 & 4 & 0 & 7 \end{array}\right]$

14. $\left[\begin{array}{ccc|c} 1 & 0 & 1 & 2 \\ 0 & 1 & -1 & -6 \\ 1 & -1 & 1 & 5 \end{array}\right]$

Describe in words the row operation(s) that will convert the first augmented matrix into the second.

15. $\left[\begin{array}{cc|c} 2 & 4 & 14 \\ 5 & 4 & 5 \end{array}\right], \left[\begin{array}{cc|c} 1 & 2 & 7 \\ 5 & 4 & 5 \end{array}\right]$

16. $\left[\begin{array}{cc|c} 1 & 2 & 7 \\ 5 & 4 & 5 \end{array}\right], \left[\begin{array}{cc|c} 1 & 2 & 7 \\ 0 & -6 & -30 \end{array}\right]$

17. $\left[\begin{array}{cc|c} 1 & 2 & 7 \\ 0 & -6 & -30 \end{array}\right], \left[\begin{array}{cc|c} 1 & 2 & 7 \\ 0 & 1 & 5 \end{array}\right]$

18. $\left[\begin{array}{cc|c} 1 & 2 & 7 \\ 0 & 1 & 5 \end{array}\right], \left[\begin{array}{cc|c} 1 & 0 & -3 \\ 0 & 1 & 5 \end{array}\right]$

Solve each system using Gaussian elimination.

19. $x + y = 5$
$-2x + y = -1$

20. $x - y = 2$
$3x - y = 12$

21. $y = 4 - 2x$
$x = 8 + y$

22. $3x = 1 + 2y$
$y = 2 - x$

23. $2x - y = 3$
$3x + 2y = 15$

24. $2x - 3y = -1$
$3x - 2y = 1$

25. $0.4x - 0.2y = 0$
$x + 1.5y = 2$

26. $0.2x + 0.6y = 0.7$
$0.5x - y = 0.5$

27. $3a - 5b = 7$
$-3a + 5b = 4$

28. $2s - 3t = 9$
$4s - 6t = 1$

29. $0.5u + 1.5v = 2$
$3u + 9v = 12$

30. $m - 2.5n = 0.5$
$-4m + 10n = -2$

31. $x + y + z = 6$
$x - y - z = 0$
$2y - z = 3$

32. $x - y + z = 2$
$-x + y + z = 4$
$-x + z = 2$

33. $2x + y = 2 + z$
$x + 2y = 2 + z$
$x + 2z = 2 + y$

34. $3x = 4 + y$
$x + y = z - 1$
$2z = 3 - x$

35. $2a - 2b + c = -2$
$a + b - 3c = 3$
$a - 3b + c = -5$

36. $r - 3s - t = -3$
$-r - s + 2t = 1$
$-r + 2s - t = -2$

37. $3y = x + z$
$x - y - 3z = 4$
$x + y + 2z = -1$

38. $z = 2 + x$
$2x - y = 1$
$y + 3z = 15$

39. $x - 2y + 3z = 1$
$2x - 4y + 6z = 2$
$-3x + 6y - 9z = -3$

40. $4x - 2y + 6z = 4$
$2x - y + 3z = 2$
$-2x + y - 3z = -2$

41. $x - y + z = 2$
$2x + y - z = 1$
$2x - 2y + 2z = 5$

42. $x - y + z = 4$
$x + y - z = 1$
$x + y - z = 3$

43. $x + y - z = 3$
$3x + y + z = 7$
$x - y + 3z = 1$

44. $x + 2y + 2z = 4$
$2x + y + z = 1$
$-x + y + z = 3$

45. $2x - y + 3z = 1$
$x + y - z = 4$

46. $x + 3y + z = 6$
$-x + y - z = 2$

47. $x - y + z - w = 2$
$-x + 2y - z - w = -1$
$2x - y - z + w = 4$
$x + 3y - 2z - 3w = 6$

48. $3a - 2b + c + d = 0$
$a - b + c - d = -4$
$-2a + b + 3c - 2d = 5$
$2a + 3b - c - d = -3$

Write a system of linear equations for each problem and solve the system using Gaussian elimination.

49. *Wages from Two Jobs* Mike works a total of 60 hr per week at his two jobs. He makes \$5 per hour at Burgers-R-Us and \$5.50 per hour at the Soap Opera Laundromat. If his total pay for one week is \$311 before taxes, then how many hours does he work at each job?

50. *Postal Rates* Noriko spent \$16.80 on postage inviting a total of 60 guests to her promotion party. Each woman was sent a picture postcard showing the company headquarters in Tokyo while each man was invited with a letter. If she put a 20-cent stamp on each postcard and a 32-cent stamp on each letter, then how many guests of each sex were invited?

51. *Investment Portfolio* Petula invested a total of \$40,000 in a no-load mutual fund, treasury bills, and municipal bonds. Her total return of \$3660 came from an 8% return on her investment in the no-load mutual fund, a 9% return on the treasury bills, and 12% return on the municipal bonds. If her total investment in treasury bills and municipal bonds was equal to her investment in the no-load mutual fund, then how much did she invest in each?

52. *Nutrition* The table shows the percentage of U.S. Recommended Daily Allowances (RDA) for phosphorus, magnesium, and calcium in one ounce of three breakfast cereals (without milk). If Hulk Hogan got 98% of the RDA of phosphorus, 84% of the RDA of magnesium, and 38% of the RDA of calcium by eating a large bowl of each (without milk), then how many ounces of each cereal did he eat?

Table for Exercise 52

	Percentage of U.S. RDA		
	Phosphorus	**Magnesium**	**Calcium**
Kix	4%	2%	4%
Quick Oats	10%	10%	0%
Muesli	8%	8%	2%

53. *Cubic Curve Fitting* Find a, b, and c such that the graph of $y = ax^3 + bx + c$ goes through the points $(-1, 4)$, $(1, 2)$, and $(2, 7)$.

54. *Quadratic Curve Fitting* Find a, b, and c such that the graph of $y = ax^2 + bx + c$ goes through the points $(-1, 0)$, $(1, 0)$, and $(3, 0)$.

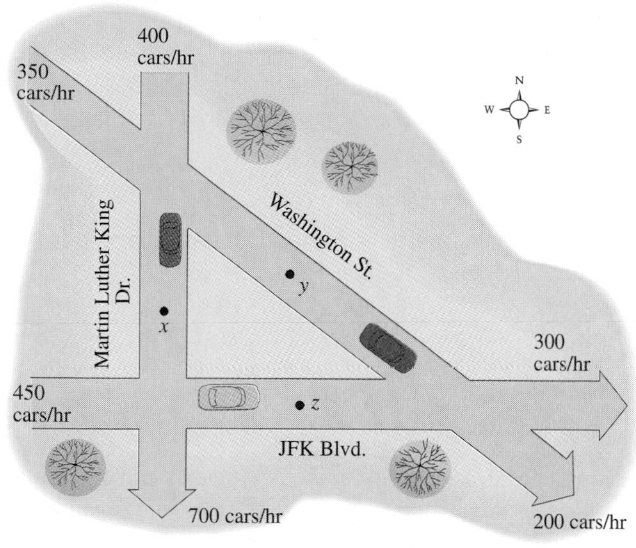

Figure for Exercise 55

55. *Traffic Control I* The diagram shows the number of cars that desire to enter and leave each of three intersections on three one-way streets in a 30-minute period. The letters x, y, and z represent the number of cars passing the three points between these three intersections as shown in the diagram. Find values for x, y, and z that would realize this desired traffic flow. If construction on JFK Boulevard limits the value of z to 50, then what values for x and y would keep the traffic flowing?

56. *Traffic Control II* Southbound, M. L. King Drive in the figure for Exercise 55 leads to another intersection that cannot always handle 700 cars in a 30-minute period. Change 700 to 600 in the figure and write a system of three equations in x, y, and z. What is the solution to this system? If you had control over the 400 cars coming from the north into the first intersection on M. L. King Drive, what number would you use in place of 400 to get the system flowing again?

Linking Concepts

For Individual or Group Explorations

More Congestion *The accompanying figure shows two one-way streets and two two-way streets. The numbers and arrows represent the number of cars per hour that desire to enter or leave each intersection. The variables a, b, c, d, e, and f*

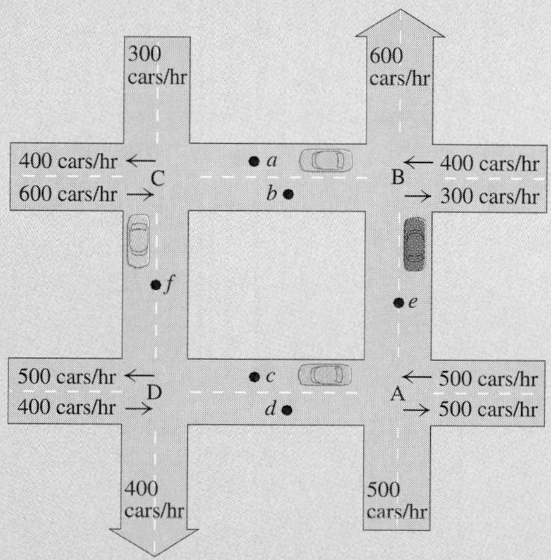

represent the number of cars per hour that pass the six points between the four intersections. To keep traffic flowing, the number of cars per hour that enter an intersection must equal the number per hour that leave an intersection.

a) Write four equations (one for each intersection) with the six variables.

b) Does the system have a unique solution?

c) If $a = 300$, $c = 400$, and $f = 200$ cars/hr, find b, d, and e.

d) If the number of cars entering intersection A from the south is changed from 500 to 800 cars/hr, then what happens to the system and why?

6.2 Operations with Matrices

In Section 6.1, matrices were used to keep track of the coefficients and constants in systems of equations. Matrices are also useful for simplifying and organizing information ranging from inventories to win-loss records of sports teams. In this section we study matrices in more detail and learn how operations with matrices are used in applications.

Notation

In Section 6.1 a matrix was defined as a rectangular array of numbers. Capital letters are used to name matrices and lowercase letters to name their entries. A general $m \times n$ matrix with m rows and n columns is given as follows:

$$A = \begin{bmatrix} a_{11} & a_{12} & a_{13} & \cdots & a_{1n} \\ a_{21} & a_{22} & a_{23} & \cdots & a_{2n} \\ a_{31} & a_{32} & a_{33} & \cdots & a_{3n} \\ \vdots & \vdots & \vdots & & \vdots \\ a_{m1} & a_{m2} & a_{m3} & \cdots & a_{mn} \end{bmatrix}$$

The subscripts indicate the position of each entry. For example, a_{32} is the entry in the third row and second column. The entry in the ith row and jth column is denoted by a_{ij}.

Two matrices are **equal** if they have the same order and the corresponding entries are equal. We write

$$\begin{bmatrix} 0.5 \\ 0.25 \end{bmatrix} = \begin{bmatrix} \frac{1}{2} \\ \frac{1}{4} \end{bmatrix}$$

because these matrices have the same order and their corresponding entries are equal. The matrices

$$\begin{bmatrix} 4 & 3 \\ 2 & 1 \end{bmatrix} \quad \text{and} \quad \begin{bmatrix} 1 & 3 \\ 2 & 1 \end{bmatrix}$$

have the same order, but they are not equal because the entries in the first row and first column are not equal. The matrices

$$[3 \quad 5] \quad \text{and} \quad \begin{bmatrix} 3 \\ 5 \end{bmatrix}$$

are not equal because the first has order 1×2 (a row matrix) and the second has order 2×1 (a column matrix).

EXAMPLE 1 Equal matrices

Determine the values of x, y, and z that make the following matrix equation true:

$$\begin{bmatrix} x & y - 1 \\ z & 7 \end{bmatrix} = \begin{bmatrix} 1 & 3 \\ 2 & 7 \end{bmatrix}$$

Solution

If these matrices are equal, then the corresponding entries are equal. So $x = 1$, $y = 4$, and $z = 2$.

Addition and Subtraction of Matrices

Matrices are rectangular arrays of real numbers, and in many ways they behave like real numbers. We can define matrix operations that have many properties similar to the properties of the real numbers.

Definition: Matrix Addition

The sum of two $m \times n$ matrices A and B is the $m \times n$ matrix denoted $A + B$ whose entries are the sums of the corresponding entries of A and B.

Note that only matrices that have the same order can be added. There is no definition for the sum of matrices of different orders.

EXAMPLE 2 Sum of matrices

Find $A + B$ given that $A = \begin{bmatrix} -4 & 3 \\ 5 & -2 \end{bmatrix}$ and $B = \begin{bmatrix} 7 & -3 \\ 2 & -5 \end{bmatrix}$.

Solution

To find $A + B$, add the corresponding entries of A and B:

$$A + B = \begin{bmatrix} -4 & 3 \\ 5 & -2 \end{bmatrix} + \begin{bmatrix} 7 & -3 \\ 2 & -5 \end{bmatrix} = \begin{bmatrix} -4 + 7 & 3 + (-3) \\ 5 + 2 & -2 + (-5) \end{bmatrix} = \begin{bmatrix} 3 & 0 \\ 7 & -7 \end{bmatrix}$$

To check, define matrices A and B on your graphing calculator using the matrix edit feature. Then use matrix names feature to display $A + B$ and find the sum as in Fig. 6.2(a), (b), and (c).

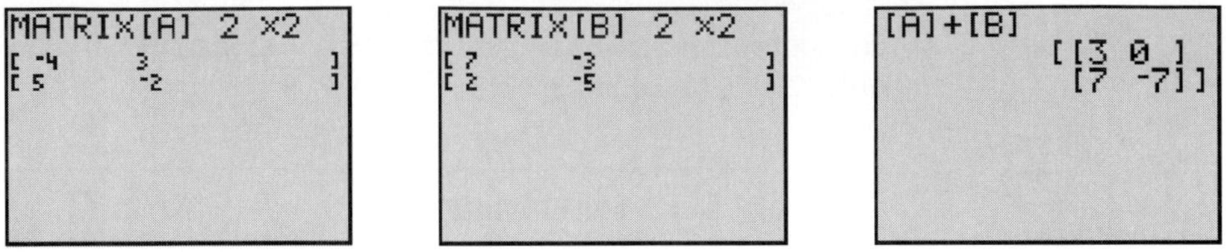

(a) (b) (c)

Figure 6.2

If all of the entries of a matrix are zero, the matrix is called a **zero matrix**. There is a zero matrix for every order. In matrix addition, the zero matrix behaves just like the additive identity 0 in the set of real numbers. For example,

$$\begin{bmatrix} 5 & -2 \\ 3 & -4 \end{bmatrix} + \begin{bmatrix} 0 & 0 \\ 0 & 0 \end{bmatrix} = \begin{bmatrix} 5 & -2 \\ 3 & -4 \end{bmatrix}.$$

In general, an $n \times n$ zero matrix is called the **additive identity** for $n \times n$ matrices.

For any matrix A, the **additive inverse** of A, denoted $-A$, is the matrix of the same order as A such that each entry of $-A$ is the opposite of the corresponding entry of A. Since corresponding entries are added in matrix addition, $A + (-A)$ is a zero matrix.

EXAMPLE 3 Additive inverses of matrices

Find $-A$ and $A + (-A)$ for $A = \begin{bmatrix} -1 & 2 & 0 \\ 4 & -3 & 5 \\ 2 & 0 & -9 \end{bmatrix}$.

Solution

To find $-A$, find the opposite of every entry of A:

$$-A = \begin{bmatrix} 1 & -2 & 0 \\ -4 & 3 & -5 \\ -2 & 0 & 9 \end{bmatrix}$$

$$A + (-A) = \begin{bmatrix} -1 & 2 & 0 \\ 4 & -3 & 5 \\ 2 & 0 & -9 \end{bmatrix} + \begin{bmatrix} 1 & -2 & 0 \\ -4 & 3 & -5 \\ -2 & 0 & 9 \end{bmatrix} = \begin{bmatrix} 0 & 0 & 0 \\ 0 & 0 & 0 \\ 0 & 0 & 0 \end{bmatrix}$$

Therefore, the sum of A and $-A$ is the additive identity for 3×3 matrices.

The difference of two real numbers a and b is defined by $a - b = a + (-b)$. The difference of two matrices of the same order is defined similarly.

Definition: Matrix Subtraction

The difference of two $m \times n$ matrices A and B is the $m \times n$ matrix denoted $A - B$, where $A - B = A + (-B)$.

Even though subtraction is defined as addition of the additive inverse, we can certainly find the difference for two matrices by subtracting their corresponding entries. Note that we can subtract corresponding entries only if the matrices have the same order.

EXAMPLE 4 Subtraction of matrices

Let $A = [3 \quad 5 \quad 8]$, $B = [3 \quad -1 \quad 6]$, $C = \begin{bmatrix} -3 \\ 5 \\ 6 \end{bmatrix}$, and $D = \begin{bmatrix} 4 \\ 7 \\ 2 \end{bmatrix}$. Find the following matrices.

a) $A - B$ **b)** $C - D$ **c)** $A - C$

Solution

a) To find $A - B$, subtract the corresponding entries of the matrices:

$$A - B = [3 \quad 5 \quad 8] - [3 \quad -1 \quad 6] = [0 \quad 6 \quad 2]$$

b) To find $C - D$, subtract the corresponding entries:

$$C - D = \begin{bmatrix} -3 \\ 5 \\ 6 \end{bmatrix} - \begin{bmatrix} 4 \\ 7 \\ 2 \end{bmatrix} = \begin{bmatrix} -7 \\ -2 \\ 4 \end{bmatrix}$$

c) Since A and C do not have the same order, $A - C$ is not defined.

Scalar Multiplication

A matrix of order 1×1 is a matrix with only one entry. To distinguish a 1×1 matrix from a real number, a real number is called a **scalar** when we are dealing with matrices. We define multiplication of a matrix by a scalar as follows.

Definition: Scalar Multiplication

If A is an $m \times n$ matrix and b is a scalar, then the matrix bA is the $m \times n$ matrix obtained by multiplying each entry of A by the real number b.

EXAMPLE 5 Scalar multiplication

Given that $A = [-2 \quad 3 \quad 5]$ and $B = \begin{bmatrix} -3 & 4 \\ 2 & -6 \end{bmatrix}$, find the following matrices.

a) $3A$ **b)** $-2B$ **c)** $-1A$

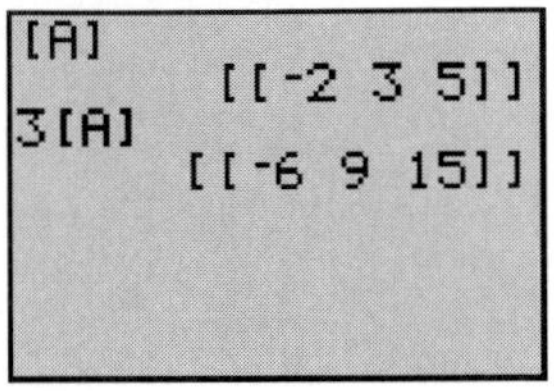

Figure 6.3

Solution

a) The matrix $3A$ is the 1×3 matrix formed by multiplying each entry of A by 3:

$$3A = [-6 \quad 9 \quad 15]$$

To check, define A on your calculator and perform scalar multiplication as in Fig. 6.3.

b) Multiply each entry of B by -2:

$$-2B = \begin{bmatrix} 6 & -8 \\ -4 & 12 \end{bmatrix}$$

c) Multiply each entry of A by -1:

$$-1A = [2 \quad -3 \quad -5]$$

The scalar product of A and -1 is the additive inverse of A, $-1A = -A$.

In Section 6.3 multiplication of matrices will be defined in a manner that is very different from scalar multiplication.

Applications

In Section 6.1 we saw how a matrix is used to represent a system of equations. Just the essential parts of the system, the coefficients, are listed in the matrix. Matrices are also very useful in two-way classifications of data. The matrix just contains the essentials, and we must remember what the entries represent. For example, an electronics manufacturer buys transistors and resistors from suppliers A and B. The following table shows the number of transistors and resistors supplied by each for the first week of January.

	A	B
Transistors	400	800
Resistors	600	500

This supply information can be listed in a 2 × 2 matrix S_1:

$$S_1 = \begin{bmatrix} 400 & 800 \\ 600 & 500 \end{bmatrix}$$

If the manufacturer wants to increase the orders by 10% for the second week, the supply matrix for the second week is $S_1 + 0.1S_1$ or $1.1S_1$, which is found by scalar multiplication:

$$S_2 = 1.1S_1 = (1.1)\begin{bmatrix} 400 & 800 \\ 600 & 500 \end{bmatrix} = \begin{bmatrix} 440 & 880 \\ 660 & 550 \end{bmatrix}$$

The matrix $T = S_1 + S_2$ gives the total supplies for the first two weeks.

$$T = S_1 + S_2 = \begin{bmatrix} 400 & 800 \\ 600 & 500 \end{bmatrix} + \begin{bmatrix} 440 & 880 \\ 660 & 550 \end{bmatrix} = \begin{bmatrix} 840 & 1680 \\ 1260 & 1050 \end{bmatrix}$$

This simple example illustrates the operations of addition and scalar multiplication, but you can imagine the same operations with matrices showing amounts of many different items ordered from many different suppliers. In applications involving large matrices, computers are used to store the matrices and perform the matrix operations.

FOR THOUGHT True or False? Explain.

The following statements refer to the matrices

$$A = \begin{bmatrix} 1 \\ 3 \end{bmatrix}, \quad B = \begin{bmatrix} 1 \\ 3 \end{bmatrix}, \quad C = \begin{bmatrix} 1 & 1 \\ 3 & 3 \end{bmatrix},$$

$$D = \begin{bmatrix} -3 & 5 \\ 1 & -2 \end{bmatrix}, \quad \text{and} \quad E = \begin{bmatrix} -2 & 6 \\ 4 & 1 \end{bmatrix}.$$

1. $A = B$ **2.** $A = C$ **3.** $A + B = C$

4. $C + D = E$

5. $A - B = \begin{bmatrix} 0 \\ 0 \end{bmatrix}$ **6.** $3B = \begin{bmatrix} 3 \\ 3 \end{bmatrix}$ **7.** $-A = \begin{bmatrix} 3 \\ 1 \end{bmatrix}$

8. $A + C = \begin{bmatrix} 2 & 1 \\ 6 & 3 \end{bmatrix}$ **9.** $C - A = \begin{bmatrix} 1 & 0 \\ 3 & 0 \end{bmatrix}$

10. $C + 2D = \begin{bmatrix} -5 & 11 \\ 4 & 1 \end{bmatrix}$

6.2 EXERCISES

Tape 18 Disk

Determine the values of x, y, and z that make each matrix equation true.

1. $\begin{bmatrix} x \\ 5 \end{bmatrix} = \begin{bmatrix} 2 \\ y \end{bmatrix}$

2. $\begin{bmatrix} 2z \\ 5x \end{bmatrix} = \begin{bmatrix} -1 \\ 2 \end{bmatrix}$

3. $\begin{bmatrix} 2x & 4y \\ 3z & 8 \end{bmatrix} = \begin{bmatrix} 6 & 10 \\ z + y & 8 \end{bmatrix}$

4. $\begin{bmatrix} -x & 2y \\ 3 & x + y \end{bmatrix} = \begin{bmatrix} 3 & -6 \\ 3 & 4z \end{bmatrix}$

Perform the following operations. If it is not possible to perform an operation, explain.

5. $\begin{bmatrix} 3 \\ 5 \end{bmatrix} + \begin{bmatrix} 2 \\ 1 \end{bmatrix}$

6. $\begin{bmatrix} -1 \\ 2 \end{bmatrix} + \begin{bmatrix} 0 \\ 3 \end{bmatrix}$

7. $\begin{bmatrix} 0.2 & 0.1 \\ 0.4 & 0.3 \end{bmatrix} + \begin{bmatrix} 0.2 & 0.05 \\ 0.3 & 0.8 \end{bmatrix}$

8. $\begin{bmatrix} \frac{1}{2} & \frac{1}{3} \\ \frac{1}{4} & 1 \end{bmatrix} - \begin{bmatrix} -\frac{1}{2} & \frac{1}{6} \\ 1 & \frac{1}{3} \end{bmatrix}$

9. $3\begin{bmatrix} \frac{1}{6} & \frac{1}{2} \\ 1 & -4 \end{bmatrix}$

10. $-2\begin{bmatrix} -1 & 4 \\ 3 & 1 \end{bmatrix}$

11. $\begin{bmatrix} -2 & 4 \\ 6 & 8 \end{bmatrix} - 4\begin{bmatrix} -3 & 1 \\ 2 & -2 \end{bmatrix}$

12. $2\begin{bmatrix} \frac{1}{4} \\ \frac{1}{3} \end{bmatrix} + 3\begin{bmatrix} \frac{1}{4} \\ \frac{1}{6} \end{bmatrix}$

13. $\begin{bmatrix} -1 & 3 \\ 2 & 5 \end{bmatrix} + 5\begin{bmatrix} -2 \\ 4 \end{bmatrix}$

14. $[\sqrt{2} \quad 4 \quad \sqrt{12}] + [\sqrt{8} \quad -2 \quad \sqrt{3}]$

15. $2\begin{bmatrix} -\sqrt{2} \\ \sqrt{5} \\ \sqrt{27} \end{bmatrix} - \begin{bmatrix} -\sqrt{8} \\ \sqrt{20} \\ -\sqrt{3} \end{bmatrix}$

16. $[1 \quad 3 \quad 7] + \begin{bmatrix} -1 \\ 2 \\ 3 \end{bmatrix}$

17. $2\begin{bmatrix} a \\ b \end{bmatrix} + 3\begin{bmatrix} 2a \\ 4b \end{bmatrix} - 5\begin{bmatrix} -a \\ 3b \end{bmatrix}$

18. $2\begin{bmatrix} -a \\ b \\ c \end{bmatrix} - \begin{bmatrix} -3a \\ 4b \\ -c \end{bmatrix} - 6\begin{bmatrix} a \\ -b \\ 2c \end{bmatrix}$

19. $0.4\begin{bmatrix} -x & y \\ 2x & 8y \end{bmatrix} - 0.3\begin{bmatrix} 2x & 3y \\ 5x & -y \end{bmatrix}$

20. $a\begin{bmatrix} a & b \\ c & 2b \end{bmatrix} - b\begin{bmatrix} b & a \\ 0 & 3a \end{bmatrix}$

21. $\begin{bmatrix} -5 & 4 \\ -2 & 3 \\ 0 & 1 \end{bmatrix} - \begin{bmatrix} -4 & -9 \\ 7 & 0 \\ -6 & 3 \end{bmatrix}$

22. $\begin{bmatrix} 2 & 3 & 4 \\ 4 & 6 & 8 \\ 6 & 3 & 1 \end{bmatrix} + \begin{bmatrix} 1 & 0 & 0 \\ 0 & 1 & 0 \\ 0 & 0 & 1 \end{bmatrix}$

23. $2\begin{bmatrix} x & y & z \\ -x & 2y & 3z \\ x & -y & -3z \end{bmatrix} - \begin{bmatrix} -x & 0 & 3z \\ 4x & y & -z \\ 2x & 5y & z \end{bmatrix}$

24. $\frac{1}{2}\begin{bmatrix} 2 & -8 & -4 \\ 14 & 16 & 10 \\ 4 & -6 & 2 \end{bmatrix} + \frac{1}{3}\begin{bmatrix} 6 & -9 & 0 \\ 0 & 12 & -6 \\ 21 & -18 & -3 \end{bmatrix}$

Let $A = \begin{bmatrix} -4 & 1 \\ 3 & 0 \end{bmatrix}$, $B = \begin{bmatrix} -1 & -2 \\ 7 & 4 \end{bmatrix}$, $C = \begin{bmatrix} -3 & -4 \\ 2 & -5 \end{bmatrix}$, $D = \begin{bmatrix} -4 \\ 5 \end{bmatrix}$, and $E = \begin{bmatrix} -1 \\ 2 \end{bmatrix}$. Find each of the following matrices, if possible.

25. $A + B$ **26.** $A - B$ **27.** $A - D$

28. $E - D$ **29.** $(A + B) + C$ **30.** $3A + 3C$

31. $A + (B + C)$ **32.** $3(A + C)$ **33.** $D + E$

34. $D + A$ **35.** $E + D$ **36.** $B + A$

37. $2A - B$ **38.** $-C - 3B$ **39.** $2D - 3E$

40. $4D + 0E$

Each of the following matrix equations corresponds to a system of linear equations. Write the system of equations and solve it by the method of your choice.

41. $\begin{bmatrix} x + y \\ x - y \end{bmatrix} = \begin{bmatrix} 5 \\ 1 \end{bmatrix}$

42. $\begin{bmatrix} x - y \\ 2x + y \end{bmatrix} = \begin{bmatrix} -1 \\ 4 \end{bmatrix}$

43. $\begin{bmatrix} 2x + 3y \\ x - 4y \end{bmatrix} = \begin{bmatrix} 7 \\ -13 \end{bmatrix}$

44. $\begin{bmatrix} x - 3y \\ 2x + y \end{bmatrix} = \begin{bmatrix} 1 \\ -5 \end{bmatrix}$

45. $\begin{bmatrix} x + y + z \\ x - y - z \\ x - y + z \end{bmatrix} = \begin{bmatrix} 8 \\ -7 \\ 2 \end{bmatrix}$

46. $\begin{bmatrix} 2x + y + z \\ x - 2y - z \\ x - y + z \end{bmatrix} = \begin{bmatrix} 7 \\ -6 \\ 2 \end{bmatrix}$

Solve each problem.

47. *Budgeting* In January, Terry spent \$120 on food, \$30 on clothing, and \$40 on utilities. In February she spent \$130 on food, \$70 on clothing, and \$50 on utilities. In March she spent \$140 on food, \$60 on clothing, and \$45 on utilities. Write a 3×1 matrix for each month's expenditures and find the sum of the three matrices. What do the entries in the sum represent?

48. *Nutritional Content* According to manufacturers' labels, one serving of Kix contains 110 calories, 2 g of protein, and 40 mg of potassium. One serving of Quick Oats contains 100 calories, 4 g of protein, and 100 mg of potassium. One serving of Muesli contains 120 calories, 3 g of protein, and 115 mg of potassium. Write 1×3 matrices K, Q, and M, which express the nutritional content of each cereal. Find $2K + 2Q + 3M$ and indicate what the entries represent.

49. *Arming the Villagers* In preparation for an attack by rampaging warlords, Xena accumulated 40 swords, 30 longbows, and 80 arrows to arm the villagers. Her companion Gabrielle, obtained 80 swords, 90 longbows, and 200 arrows from a passing arms dealer. Write this information as a 3×2 matrix. One week later both Xena and Gabrielle managed to increase their supplies in each category by 50%. Write a 3×2 matrix for the supply of armaments in the second week.

50. *Recommended Daily Allowances* The percentages of the U.S. Recommended Daily Allowances for phosphorus, magnesium, and calcium for a 1-oz serving of Kix are 4, 2, and 4, respectively. The percentages of the U.S. Recommended Daily Allowances for phosphorus, magnesium, and calcium in 1/2 cup of milk are 11, 4, and 16, respectively. Write a 1×3 matrix K giving the percentages for 1 oz of Kix and a 1×3 matrix M giving the percentages for 1/2 cup of milk. Find the matrix $K + 2M$; indicate what its entries represent.

For Writing/Discussion

Let $A = \begin{bmatrix} a_{11} & a_{12} \\ a_{21} & a_{22} \end{bmatrix}$, $B = \begin{bmatrix} b_{11} & b_{12} \\ b_{21} & b_{22} \end{bmatrix}$, and $C = \begin{bmatrix} c_{11} & c_{12} \\ c_{21} & c_{22} \end{bmatrix}$ for the following problems.

51. Is $A + B = B + A$? Is addition of 2×2 matrices commutative? Explain.

52. Is addition of 3×3 matrices commutative? Explain.

53. Is $(A + B) + C = A + (B + C)$? Is addition of 2×2 matrices associative? Explain.

54. Is addition of 3×3 matrices associative? Explain.

55. Is $k(A + B) = kA + kB$ for any constant k? Is scalar multiplication distributive over addition of 2×2 matrices?

56. Is scalar multiplication distributive over addition of $n \times n$ matrices for each natural number n?

57. In the set of 2×2 matrices, which matrix is the additive identity?

58. Does every 2×2 matrix have an additive inverse with respect to the appropriate additive identity?

6.3 Multiplication of Matrices

In Sections 6.1 and 6.2 we saw how matrices are used for solving equations and for representing two-way classifications of data. We saw how addition and scalar multiplication of matrices could be useful in applications. In this section you will learn to find the product of two matrices. Matrix multiplication is more complicated than addition or subtraction, but it is also very useful in applications.

An Application

Before presenting the general definition of multiplication, let us look at an example where multiplication of matrices is useful. Table 6.1 shows the number of economy, mid-size, and large cars rented by individuals and corporations at a rental agency in a single day. Table 6.2 shows the number of bonus points and free miles given in a promotional program for each of the three car types.

Table 6.1

	Econo	Mid	Large
Individuals	3	2	6
Corporations	5	2	4

Table 6.2

	Bonus points	Free miles
Econo	20	50
Mid	30	100
Large	40	150

The 1×3 row matrix $[3 \quad 2 \quad 6]$ from Table 6.1 represents the number of economy, mid-size, and large cars rented by individuals. The 3×1 column matrix $\begin{bmatrix} 20 \\ 30 \\ 40 \end{bmatrix}$ from Table 6.2 represents the bonus points given for each economy, mid-size, and large car that is rented. The product of these two matrices is a 1×1 matrix whose entry is the sum of the products of the corresponding entries:

$$[3 \quad 2 \quad 6]\begin{bmatrix} 20 \\ 30 \\ 40 \end{bmatrix} = [3(20) + 2(30) + 6(40)] = [360]$$

The product of this row matrix and this column matrix gives the total number of bonus points given to individuals on the rental of the 11 cars.

Now write Table 6.1 as a 2×3 matrix giving the number of cars of each type rented by individuals and corporations and write Table 6.2 as a 3×2 matrix giving the number of bonus points and free miles for each type of car rented.

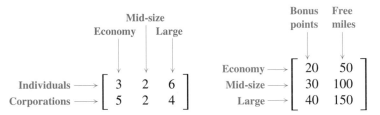

Columns: **Economy**, **Mid-size**, **Large**; rows: **Individuals**, **Corporations**

$$\begin{bmatrix} 3 & 2 & 6 \\ 5 & 2 & 4 \end{bmatrix}$$

Columns: **Bonus points**, **Free miles**; rows: **Economy**, **Mid-size**, **Large**

$$\begin{bmatrix} 20 & 50 \\ 30 & 100 \\ 40 & 150 \end{bmatrix}$$

The product of these two matrices is a 2×2 matrix that gives the total bonus points and free miles both for individuals and corporations.

Columns of the product: **Bonus points**, **Free miles**; rows: **Individuals**, **Corporations**

$$\begin{bmatrix} 3 & 2 & 6 \\ 5 & 2 & 4 \end{bmatrix}\begin{bmatrix} 20 & 50 \\ 30 & 100 \\ 40 & 150 \end{bmatrix} = \begin{bmatrix} 360 & 1250 \\ 320 & 1050 \end{bmatrix}$$

The product of the 2×3 matrix and the 3×2 matrix is a 2×2 matrix. Take a careful look at where the entries in the 2×2 matrix come from:

$$3(20) + 2(30) + 6(40) = 360 \qquad \textbf{Total bonus points for individuals}$$
$$3(50) + 2(100) + 6(150) = 1250 \qquad \textbf{Total free miles for individuals}$$
$$5(20) + 2(30) + 4(40) = 320 \qquad \textbf{Total bonus points for corporations}$$
$$5(50) + 2(100) + 4(150) = 1050 \qquad \textbf{Total free miles for corporations}$$

Each entry in the 2×2 matrix is found by multiplying the entries of a *row* of the first matrix by the corresponding entries of a *column* of the second matrix and adding the results. To multiply any matrices, the number of entries in a row of the first matrix must equal the number of entries in a column of the second matrix.

Matrix Multiplication

The product of two matrices is illustrated by the previous example of the rental cars. The general definition of matrix multiplication follows.

Definition: Matrix Multiplication

The product of an $m \times n$ matrix A and an $n \times p$ matrix B is an $m \times p$ matrix AB whose entries are found as follows. The entry in the ith row and jth column of AB is found by multiplying each entry in the ith row of A by the corresponding entry in the jth column of B and adding the results.

Note that by the definition we can multiply only an $\boldsymbol{m} \times n$ matrix and an $n \times \boldsymbol{p}$ matrix to get an $\boldsymbol{m} \times \boldsymbol{p}$ matrix. To find a product AB, *each row of A must have the same number of entries as each column of B*. The entry c_{ij} in AB comes from the ith row of A and the jth column of B as shown below.

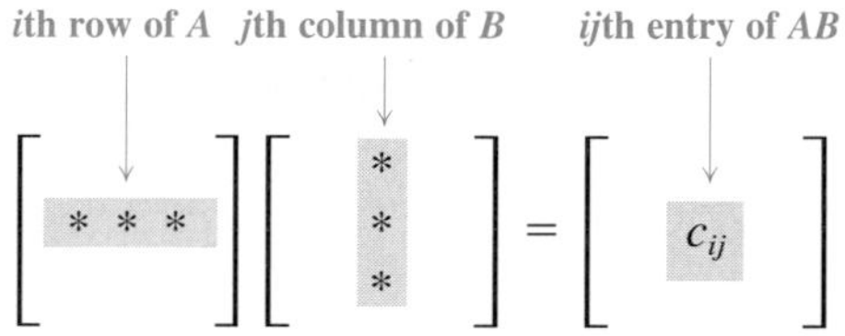

EXAMPLE 1 Multiplying matrices

Find the following products.

a) $\begin{bmatrix} 2 & 3 \end{bmatrix}\begin{bmatrix} -3 \\ 4 \end{bmatrix}$ **b)** $\begin{bmatrix} 1 & 2 \\ 3 & 4 \end{bmatrix}\begin{bmatrix} -2 \\ 5 \end{bmatrix}$ **c)** $\begin{bmatrix} 1 & 3 \\ 5 & 7 \end{bmatrix}\begin{bmatrix} 2 & 4 \\ 6 & 8 \end{bmatrix}$

Solution

a) The product of a $\mathbf{1} \times 2$ matrix and a $2 \times \mathbf{1}$ matrix is a $\mathbf{1} \times \mathbf{1}$ matrix. The only entry in the product is found by multiplying the first row of the first matrix by the first column of the second matrix:

$$\begin{bmatrix} 2 & 3 \end{bmatrix}\begin{bmatrix} -3 \\ 4 \end{bmatrix} = [2(-3) + 3(4)] = [6]$$

b) The product of a $\mathbf{2} \times 2$ matrix and a $2 \times \mathbf{1}$ matrix is a $\mathbf{2} \times \mathbf{1}$ matrix. Multiply the corresponding entries in each row of the first matrix and the only column of the second matrix:

$$\begin{bmatrix} 1 & 2 \\ 3 & 4 \end{bmatrix}\begin{bmatrix} -2 \\ 5 \end{bmatrix} = \begin{bmatrix} 8 \\ 14 \end{bmatrix} \qquad \begin{aligned} 1(-2) + 2(5) &= 8 \\ 3(-2) + 4(5) &= 14 \end{aligned}$$

c) The product of a $\mathbf{2} \times 2$ matrix and a $2 \times \mathbf{2}$ matrix is a $\mathbf{2} \times \mathbf{2}$ matrix. Multiply the corresponding entries in each row of the first matrix and each column of the second matrix.

$$\begin{bmatrix} 1 & 3 \\ 5 & 7 \end{bmatrix}\begin{bmatrix} 2 & 4 \\ 6 & 8 \end{bmatrix} = \begin{bmatrix} 20 & 28 \\ 52 & 76 \end{bmatrix} \qquad \begin{aligned} 1 \cdot 2 + 3 \cdot 6 &= 20 \\ 1 \cdot 4 + 3 \cdot 8 &= 28 \\ 5 \cdot 2 + 7 \cdot 6 &= 52 \\ 5 \cdot 4 + 7 \cdot 8 &= 76 \end{aligned}$$

EXAMPLE 2 **Multiplying matrices**

Find AB and BA in each case.

a) $A = \begin{bmatrix} 1 & 3 \\ 5 & 7 \\ 8 & 2 \end{bmatrix}$, $B = \begin{bmatrix} 2 & 4 & -1 \\ 6 & -3 & 2 \end{bmatrix}$

b) $A = \begin{bmatrix} 1 & 3 & 4 \\ 2 & 5 & 6 \\ 7 & 8 & 9 \end{bmatrix}$, $B = \begin{bmatrix} 1 & 0 & 1 \\ 0 & 1 & 0 \\ 0 & 1 & 1 \end{bmatrix}$

Solution

a) The product of 3×2 matrix A and 2×3 matrix B is the 3×3 matrix AB. The first row of AB is found by multiplying the corresponding entries in the first row of A and each column of B:

$$1 \cdot 2 + 3 \cdot 6 = 20, \quad 1 \cdot 4 + 3(-3) = -5, \quad \text{and} \quad 1(-1) + 3 \cdot 2 = 5$$

So 20, -5, and 5 form the first row of AB. The second row of AB is formed from multiplying the second row of A and each column from B:

$$5 \cdot 2 + 7 \cdot 6 = 52, \quad 5 \cdot 4 + 7(-3) = -1, \quad \text{and} \quad 5(-1) + 7 \cdot 2 = 9$$

So 52, -1, and 9 form the second row of AB. The third row of AB is formed by multiplying corresponding entries in the third row of A and each column of B.

$$AB = \begin{bmatrix} 1 & 3 \\ 5 & 7 \\ 8 & 2 \end{bmatrix} \begin{bmatrix} 2 & 4 & -1 \\ 6 & -3 & 2 \end{bmatrix} = \begin{bmatrix} 20 & -5 & 5 \\ 52 & -1 & 9 \\ 28 & 26 & -4 \end{bmatrix}$$

The product of 2×3 matrix B and 3×2 matrix A is the 2×2 matrix BA:

$$BA = \begin{bmatrix} 2 & 4 & -1 \\ 6 & -3 & 2 \end{bmatrix} \begin{bmatrix} 1 & 3 \\ 5 & 7 \\ 8 & 2 \end{bmatrix}$$

$$= \begin{bmatrix} 14 & 32 \\ 7 & 1 \end{bmatrix} \quad \begin{aligned} 2(1) + 4(5) + (-1)(8) &= 14 \\ 2(3) + 4(7) + (-1)(2) &= 32 \\ 6(1) + (-3)(5) + 2(8) &= 7 \\ 6(3) + (-3)(7) + 2(2) &= 1 \end{aligned}$$

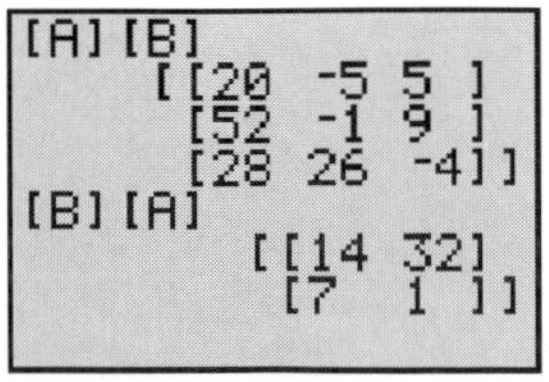

Figure 6.4

To check, enter A and B into a graphing calculator and find the products as shown in Fig. 6.4. □

b) The product of 3×3 matrix A and 3×3 matrix B is the 3×3 matrix AB:

$$AB = \begin{bmatrix} 1 & 3 & 4 \\ 2 & 5 & 6 \\ 7 & 8 & 9 \end{bmatrix} \begin{bmatrix} 1 & 0 & 1 \\ 0 & 1 & 0 \\ 0 & 1 & 1 \end{bmatrix} = \begin{bmatrix} 1 & 7 & 5 \\ 2 & 11 & 8 \\ 7 & 17 & 16 \end{bmatrix}$$

The product of 3×3 matrix B and 3×3 matrix A is the 3×3 matrix BA:

$$BA = \begin{bmatrix} 1 & 0 & 1 \\ 0 & 1 & 0 \\ 0 & 1 & 1 \end{bmatrix} \begin{bmatrix} 1 & 3 & 4 \\ 2 & 5 & 6 \\ 7 & 8 & 9 \end{bmatrix} = \begin{bmatrix} 8 & 11 & 13 \\ 2 & 5 & 6 \\ 9 & 13 & 15 \end{bmatrix}$$

Operations with matrices have some properties similar to the properties of operations with real numbers. For example, multiplication of matrices is associative, multiplicative inverses exist for some matrices, and there is a multiplicative identity. However, multiplication of matrices is generally not commutative. Example 2 shows matrices A and B where $AB \neq BA$. In Example 2(a), AB and BA are not even the same order. It can also happen that AB is defined but BA is undefined because of the orders of A and B. We will not study all of the properties of the operations with matrices in detail in this text, but some properties for 2×2 matrices are discussed in the exercises.

Matrix Equations

Recall that two matrices are equal provided they are the same order and their corresponding entries are equal. This definition is used to solve matrix equations.

EXAMPLE 3 Solving a matrix equation

Find the values of x and y that satisfy the matrix equation

$$\begin{bmatrix} 3 & 2 \\ 4 & -1 \end{bmatrix} \begin{bmatrix} x \\ y \end{bmatrix} = \begin{bmatrix} 1 \\ -6 \end{bmatrix}.$$

Solution

Multiply the matrices on the left-hand side to get the matrix equation

$$\begin{bmatrix} 3x + 2y \\ 4x - y \end{bmatrix} = \begin{bmatrix} 1 \\ -6 \end{bmatrix}.$$

Since matrices of the same order are equal only when all of their corresponding entries are equal, we have the following system of equations:

(1) $\quad 3x + 2y = 1$

(2) $\quad 4x - y = -6$

Solve by eliminating y:

$$\begin{aligned} 3x + 2y &= 1 \\ 8x - 2y &= -12 \qquad \text{Eq. (2) multiplied by 2} \\ \hline 11x \qquad &= -11 \\ x &= -1 \end{aligned}$$

Now use $x = -1$ in Eq. (1):

$$\begin{aligned} 3(-1) + 2y &= 1 \\ 2y &= 4 \\ y &= 2 \end{aligned}$$

You should check that the matrix equation is satisfied if $x = -1$ and $y = 2$.

In Example 3, we rewrote a matrix equation as a system of equations. We will now write a system of equations as a matrix equation. In fact, any system of linear equations can be written as a matrix equation in the form $AX = B$, where A is a matrix of coefficients, X is a column matrix of variables, and B is a column matrix of constants. In the next example we write a system of two equations as a matrix equation. In Section 6.4 we will solve systems of equations by using matrix operations on the corresponding matrix equations.

EXAMPLE 4 Writing a matrix equation

Write the following system as an equivalent matrix equation in the form $AX = B$.

$$\begin{aligned} 2x + y &= 5 \\ x - y &= 4 \end{aligned}$$

Solution

Let $A = \begin{bmatrix} 2 & 1 \\ 1 & -1 \end{bmatrix}$, $X = \begin{bmatrix} x \\ y \end{bmatrix}$, and $B = \begin{bmatrix} 5 \\ 4 \end{bmatrix}$. The system of equations is equivalent to the matrix equation

$$\begin{bmatrix} 2 & 1 \\ 1 & -1 \end{bmatrix} \begin{bmatrix} x \\ y \end{bmatrix} = \begin{bmatrix} 5 \\ 4 \end{bmatrix},$$

which is of the form $AX = B$. Check by multiplying the two matrices on the left-hand side to get

$$\begin{bmatrix} 2x + y \\ x - y \end{bmatrix} = \begin{bmatrix} 5 \\ 4 \end{bmatrix}.$$

By the definition of equal matrices, this matrix equation is correct provided that $2x + y = 5$ and $x - y = 4$, which is the original system.

FOR THOUGHT True or False? Explain.

The following statements refer to the matrices

$$A = \begin{bmatrix} 1 \\ 6 \end{bmatrix},\ B = [7 \quad 9],\ C = \begin{bmatrix} 2 & 3 \\ 4 & 5 \end{bmatrix},$$

$$D = \begin{bmatrix} 1 & 0 \\ 0 & 1 \end{bmatrix},\ \text{and } E = \begin{bmatrix} 2 & -1 \\ 0 & 3 \end{bmatrix}.$$

1. The order of AB is 2×2.
2. The order of BA is 1×1.
3. The order of AC is 1×2.
4. The order of CA is 2×2.
5. $DC = C$ and $CD = C$.
6. $BC = [50 \quad 66]$
7. $AB = \begin{bmatrix} 7 & 9 \\ 42 & 54 \end{bmatrix}$
8. $CE = \begin{bmatrix} 4 & 7 \\ 8 & 11 \end{bmatrix}$
9. $BA = [61]$
10. $CE = EC$

6.3 EXERCISES

Tape 18 Disk

Find the order of AB in each case if the matrices can be multiplied.

1. A has order 3×2, B has order 2×5
2. A has order 3×1, B has order 1×3
3. A has order 1×4, B has order 4×1
4. A has order 4×2, B has order 2×5
5. A has order 5×1, B has order 1×5
6. A has order 2×1, B has order 1×6
7. A has order 3×3, B has order 3×3
8. A has order 4×4, B has order 4×1
9. A has order 3×4, B has order 3×4
10. A has order 4×2, B has order 3×4

Let $A = \begin{bmatrix} 2 \\ -3 \\ 1 \end{bmatrix}$, $B = [2 \quad 3 \quad 4]$, $C = \begin{bmatrix} 2 & 3 \\ 4 & 5 \\ 1 & 0 \end{bmatrix}$,

$D = \begin{bmatrix} 2 & -1 & 1 \\ 0 & 3 & 2 \end{bmatrix}$, and $E = \begin{bmatrix} 1 & 1 & 1 \\ 0 & 1 & 1 \\ 0 & 0 & 1 \end{bmatrix}$.

Find each product if possible.

11. AB	**12.** BC	**13.** BE	**14.** BA
15. CE	**16.** DE	**17.** EC	**18.** BD
19. DC	**20.** CD	**21.** ED	**22.** EE
23. EA	**24.** DA	**25.** AE	**26.** EB

Find each product if possible.

27. $\begin{bmatrix} 2 & 0 \\ 3 & 1 \end{bmatrix} \begin{bmatrix} 1 & 1 \\ 0 & 1 \end{bmatrix}$

28. $\begin{bmatrix} -1 & 2 \\ 3 & 4 \end{bmatrix} \begin{bmatrix} 2 & -1 \\ 5 & 2 \end{bmatrix}$

29. $\begin{bmatrix} 7 & 4 \\ 5 & 3 \end{bmatrix} \begin{bmatrix} 3 & -4 \\ -5 & 7 \end{bmatrix}$

30. $\begin{bmatrix} -2 & -3 \\ 5 & 8 \end{bmatrix} \begin{bmatrix} -8 & -3 \\ 5 & 2 \end{bmatrix}$

31. $\begin{bmatrix} -0.5 & 4 \\ 9 & 0.7 \end{bmatrix} \begin{bmatrix} 1 & 0 \\ 0 & 1 \end{bmatrix}$

32. $\begin{bmatrix} 1 & 0 \\ 0 & 1 \end{bmatrix} \begin{bmatrix} -0.7 & 1.2 \\ 3 & 1.1 \end{bmatrix}$

33. $[-2 \quad 3] \begin{bmatrix} a & 3b \\ 2a & b \end{bmatrix}$

34. $\begin{bmatrix} -2 & 5 \\ 6 & 4 \end{bmatrix} \begin{bmatrix} x \\ y \end{bmatrix}$

35. $\begin{bmatrix} a & 0 \\ 0 & b \end{bmatrix} \begin{bmatrix} -2 & 5 & 3 \\ 1 & 4 & 6 \end{bmatrix}$

36. $\begin{bmatrix} 1 & 1 \\ 1 & -1 \end{bmatrix} \begin{bmatrix} x \\ y \end{bmatrix}$

37. $[1 \quad 2 \quad 3] \begin{bmatrix} 1 & 0 & 1 \\ 0 & 1 & 1 \\ 1 & 0 & 1 \end{bmatrix}$

38. $\begin{bmatrix} 1 & 1 & 0 \\ 1 & 0 & 1 \\ 0 & 1 & 1 \end{bmatrix} \begin{bmatrix} -2 \\ 3 \\ 5 \end{bmatrix}$

39. $[-1 \quad 0 \quad 3] \begin{bmatrix} -5 \\ 1 \\ 4 \end{bmatrix}$

40. $\begin{bmatrix} -5 \\ 1 \\ 4 \end{bmatrix} [-1 \quad 0 \quad 3]$

41. $\begin{bmatrix} x \\ y \end{bmatrix} [x \quad y]$

42. $[x \quad y] \begin{bmatrix} x \\ y \end{bmatrix}$

43. $\begin{bmatrix} -1 & 2 & 3 \\ 3 & 4 & 4 \end{bmatrix} \begin{bmatrix} \sqrt{2} \\ 0 \\ \sqrt{2} \end{bmatrix}$

44. $[2 \quad 3] \begin{bmatrix} 0 & \sqrt{2} & 5 \\ -1 & \sqrt{8} & 0 \end{bmatrix}$

45. $\begin{bmatrix} \frac{1}{2} & \frac{1}{3} \\ \frac{1}{4} & \frac{1}{5} \end{bmatrix} \begin{bmatrix} -8 & 12 \\ -5 & 15 \end{bmatrix}$

46. $\begin{bmatrix} \frac{1}{4} & \frac{1}{2} \\ \frac{1}{8} & -\frac{1}{2} \end{bmatrix} \begin{bmatrix} -\frac{1}{2} & \frac{1}{4} \\ \frac{1}{2} & \frac{1}{4} \end{bmatrix}$

47. $[3 \quad 0 \quad 3][2 \quad 4 \quad 6]$

48. $\begin{bmatrix} 2 \\ 5 \end{bmatrix} \begin{bmatrix} 7 & 2 \\ 3 & 1 \end{bmatrix}$

49. $\begin{bmatrix} 1 & 0 & -1 \\ 0 & 1 & 0 \\ 1 & 1 & 1 \end{bmatrix} \begin{bmatrix} 9 & 8 & 10 \\ 3 & 5 & 2 \\ 7 & 8 & 4 \end{bmatrix}$

50. $\begin{bmatrix} 1 & 1 & 1 \\ 0 & 1 & 1 \\ 0 & 0 & 1 \end{bmatrix} \begin{bmatrix} -2 & 3 & -4 \\ 2 & 5 & 7 \\ -3 & 0 & -6 \end{bmatrix}$

51. $\begin{bmatrix} 5 & -3 & -2 \\ 4 & 2 & 6 \\ 2 & 3 & -8 \end{bmatrix} \begin{bmatrix} 0.2 & 0.3 \\ 0.2 & -0.4 \\ 0.3 & 0.5 \end{bmatrix}$

52. $\begin{bmatrix} 0.2 & 0.1 & 0.7 \\ 0.3 & 0.3 & 0.4 \end{bmatrix} \begin{bmatrix} 20 & 30 & 40 \\ 10 & 20 & 50 \\ 60 & 50 & 40 \end{bmatrix}$

Write each matrix equation as a system of equations and solve the system by the method of your choice.

53. $\begin{bmatrix} 2 & -3 \\ 1 & 2 \end{bmatrix} \begin{bmatrix} x \\ y \end{bmatrix} = \begin{bmatrix} 0 \\ 7 \end{bmatrix}$

54. $\begin{bmatrix} 1 & 5 \\ -2 & 4 \end{bmatrix} \begin{bmatrix} x \\ y \end{bmatrix} = \begin{bmatrix} 2 \\ 10 \end{bmatrix}$

55. $\begin{bmatrix} 2 & 3 \\ 4 & 6 \end{bmatrix} \begin{bmatrix} x \\ y \end{bmatrix} = \begin{bmatrix} 5 \\ 9 \end{bmatrix}$

56. $\begin{bmatrix} 1 & -3 \\ -2 & 6 \end{bmatrix} \begin{bmatrix} x \\ y \end{bmatrix} = \begin{bmatrix} 1 \\ -2 \end{bmatrix}$

57. $\begin{bmatrix} 1 & 1 & 1 \\ 0 & 1 & 1 \\ 0 & 0 & 1 \end{bmatrix} \begin{bmatrix} x \\ y \\ z \end{bmatrix} = \begin{bmatrix} 4 \\ 5 \\ 6 \end{bmatrix}$

58. $\begin{bmatrix} 2 & 3 & 1 \\ 0 & 1 & 4 \\ 0 & 0 & 2 \end{bmatrix} \begin{bmatrix} x \\ y \\ z \end{bmatrix} = \begin{bmatrix} 0 \\ 3 \\ 6 \end{bmatrix}$

Write a matrix equation of the form $AX = B$ that corresponds to each system of equations.

59. $2x + 3y = 9$
$4x - y = 6$

60. $x - y = -7$
$x + 2y = 8$

61. $x + 2y - z = 3$
$3x - y + 3z = 1$
$2x + y - 4z = 0$

62. $x + y + z = 1$
$2x + y - z = 4$
$x - y - 3z = 2$

Solve each problem.

63. *Building Costs* A contractor builds two types of houses. The costs for labor and materials for the economy model and the deluxe model are shown in the table. Write the information in the table as a matrix A. Suppose that the contractor built four economy models and seven deluxe models. Write a matrix Q of the appropriate order containing the quantity of each type. Find the product matrix AQ. What do the entries of AQ represent?

Table for Exercise 63

	Economy	Deluxe
Labor	\$24,000	\$40,000
Materials	\$38,000	\$70,000

64. *Nutritional Information* According to the manufacturers, the breakfast cereals Almond Delight and Basic 4 contain the numbers of grams of protein, carbohydrates, and fat per serving listed in the table. Write the information in the table as a matrix A. In one week Julia ate four servings of Almond Delight and three servings of Basic 4. Write a matrix Q of the appropriate order expressing the quantity of each type that Julia ate. Find the product AQ. What do the entries of AQ represent?

Table for Exercise 64

	Almond Delight	Basic 4
Protein	2	3
Carbohydrates	23	28
Fat	2	2

For Writing/Discussion

Let $A = \begin{bmatrix} a_{11} & a_{12} \\ a_{21} & a_{22} \end{bmatrix}$, $B = \begin{bmatrix} b_{11} & b_{12} \\ b_{21} & b_{22} \end{bmatrix}$, and $C = \begin{bmatrix} c_{11} & c_{12} \\ c_{21} & c_{22} \end{bmatrix}$. Determine whether each of the following statements is true, and explain your answer.

65. $AB = BA$ (commutative)

66. $(AB)C = A(BC)$ (associative)

67. For any real number k, $k(A + B) = kA + kB$.

68. $A(B + C) = AB + AC$ (distributive)

69. For any real numbers s and t, $sA + tA = (s + t)A$.

70. Multiplication of 1×1 matrices is commutative.

LINKING CONCEPTS

	A	B	C	D
A	0	1	1	0
B	0	0	1	1
C	0	0	0	1
D	1	0	0	0

For Individual or Group Explorations

Ranking Soccer Teams *The table shown here gives the records of all four teams in a soccer league. An entry of 1 indicates that the row team has defeated the column team. (There are no ties.) The teams are ranked, not by their percentage of victories, but by the number of points received under a ranking scheme that gives a team credit for the quality of the team it defeats. Since team A defeated B and C, A gets two points. Since B defeated C and D, and C defeated D, A gets 3 more secondary points for a total of 5 points. Since D's only victory is over A, D gets 1 point for that victory plus 2 secondary points for A's defeats of B and C, giving D a total of 3 points.*

a) Write the table as a 4×4 matrix M and find M^2.

b) Explain what the entries of M^2 represent.

c) Now let T be a 4×1 matrix with a 1 in every entry. Find $(M + M^2)T$.

d) Explain what the entries of $(M + M^2)T$ represent.

e) Is it possible for one team to have a better win-loss record than another, but end up ranked lower than the other because of this point scheme? Give an example to support your answer.

f) Make up a win-loss table (with no ties) for a six-team soccer league like the given table. Use a graphing calculator to find $(M + M^2)T$. Compare the percentage of games won by each team with its ranking by this scheme. Is it possible for a team to have a higher percentage of wins but still be ranked lower than another team?

g) Find $(2M + M^2)T$ for the matrix M from part (f) and explain its entries. What is the significance of the number 2?

h) Compare the ranking of the six teams using $(2M + M^2)T$ and $(M + M^2)T$. Is it possible that $(2M + M^2)T$ could change the order of the teams?

6.4 Inverses of Matrices

In previous sections we learned to add, subtract, and multiply matrices. There is no definition for division of matrices. However, there is an identity matrix for multiplication that behaves like the multiplicative identity 1 in the real number system. (The number 1 is called the multiplicative identity because $1 \cdot a = a$ and $a \cdot 1 = a$ for any real number a.) In this section we will see that for certain matrices there are inverse matrices such that the product of a matrix and its inverse matrix is the identity matrix.

The Identity Matrix

Consider the product of a 2×2 matrix $A = \begin{bmatrix} a_{11} & a_{12} \\ a_{21} & a_{22} \end{bmatrix}$ and $I = \begin{bmatrix} 1 & 0 \\ 0 & 1 \end{bmatrix}$:

$$\begin{bmatrix} a_{11} & a_{12} \\ a_{21} & a_{22} \end{bmatrix}\begin{bmatrix} 1 & 0 \\ 0 & 1 \end{bmatrix} = \begin{bmatrix} a_{11} & a_{12} \\ a_{21} & a_{22} \end{bmatrix} \quad \text{and} \quad \begin{bmatrix} 1 & 0 \\ 0 & 1 \end{bmatrix}\begin{bmatrix} a_{11} & a_{12} \\ a_{21} & a_{22} \end{bmatrix} = \begin{bmatrix} a_{11} & a_{12} \\ a_{21} & a_{22} \end{bmatrix}$$

Since $AI = A$ and $IA = A$, the matrix I is called the 2×2 **identity matrix**. The matrix I has 1's on its diagonal and 0's elsewhere. The 2×2 identity matrix is not an identity matrix for 3×3 matrices, but a 3×3 matrix with 1's on the diagonal and 0's elsewhere is the identity matrix for 3×3 matrices.

Definition: Identity Matrix

For each positive integer n, the $\boldsymbol{n \times n}$ **identity matrix** $\boldsymbol{I}$ is an $n \times n$ matrix with 1's on the diagonal and 0's elsewhere. In symbols,

$$I = \begin{bmatrix} 1 & 0 & 0 & \dots & 0 \\ 0 & 1 & 0 & \dots & 0 \\ 0 & 0 & 1 & \dots & 0 \\ \vdots & \vdots & \vdots & & \vdots \\ 0 & 0 & 0 & \dots & 1 \end{bmatrix}.$$

We use the letter I for the identity matrix for any order, but the order of I should be clear from the context.

EXAMPLE 1 Using an identity matrix

Find a matrix I such that $BI = B$ and $IB = B$ for

$$B = \begin{bmatrix} 2 & 3 & 5 \\ 1 & 0 & 4 \\ 5 & 7 & 2 \end{bmatrix}.$$

Solution

The matrix I must be the 3×3 identity matrix

$$I = \begin{bmatrix} 1 & 0 & 0 \\ 0 & 1 & 0 \\ 0 & 0 & 1 \end{bmatrix}.$$

Check that $BI = B$:

$$\begin{bmatrix} 2 & 3 & 5 \\ 1 & 0 & 4 \\ 5 & 7 & 2 \end{bmatrix}\begin{bmatrix} 1 & 0 & 0 \\ 0 & 1 & 0 \\ 0 & 0 & 1 \end{bmatrix} = \begin{bmatrix} 2 & 3 & 5 \\ 1 & 0 & 4 \\ 5 & 7 & 2 \end{bmatrix}$$

$2 \cdot 1 + 3 \cdot 0 + 5 \cdot 0 = 2$
$2 \cdot 0 + 3 \cdot 1 + 5 \cdot 0 = 3$
$2 \cdot 0 + 3 \cdot 0 + 5 \cdot 1 = 5$

The computations at the right show that 2, 3, and 5 form the first row of BI. The second and third rows are found similarly. Now check that $IB = B$:

$$\begin{bmatrix} 1 & 0 & 0 \\ 0 & 1 & 0 \\ 0 & 0 & 1 \end{bmatrix} \begin{bmatrix} 2 & 3 & 5 \\ 1 & 0 & 4 \\ 5 & 7 & 2 \end{bmatrix} = \begin{bmatrix} 2 & 3 & 5 \\ 1 & 0 & 4 \\ 5 & 7 & 2 \end{bmatrix}$$

The Inverse of a Matrix

Every nonzero real number a has a multiplicative inverse $1/a$ such that $a \cdot (1/a) = 1$. The product of a real number and its multiplicative inverse is the multiplicative identity. The situation is similar for matrices.

Definition: Inverse of a Matrix

The **inverse** of an $n \times n$ matrix A is an $n \times n$ matrix A^{-1} (if it exists) such that $AA^{-1} = I$ and $A^{-1}A = I$. (Read A^{-1} as "A inverse.")

If A has an inverse, then A is **invertible**. Before we learn how to find the inverse of a matrix, we use the definition to determine whether two given matrices are inverses.

EXAMPLE 2 Using the definition of inverse matrices

Determine whether $A = \begin{bmatrix} 3 & 4 \\ 5 & 7 \end{bmatrix}$ and $B = \begin{bmatrix} 7 & -4 \\ -5 & 3 \end{bmatrix}$ are inverses of each other.

Solution

Find the products AB and BA:

$$AB = \begin{bmatrix} 3 & 4 \\ 5 & 7 \end{bmatrix} \begin{bmatrix} 7 & -4 \\ -5 & 3 \end{bmatrix} = \begin{bmatrix} 1 & 0 \\ 0 & 1 \end{bmatrix}$$

$$BA = \begin{bmatrix} 7 & -4 \\ -5 & 3 \end{bmatrix} \begin{bmatrix} 3 & 4 \\ 5 & 7 \end{bmatrix} = \begin{bmatrix} 1 & 0 \\ 0 & 1 \end{bmatrix}$$

A and B are inverses because $AB = BA = I$, where I is the 2×2 identity matrix.

If we are given the matrix

$$A = \begin{bmatrix} 3 & 4 \\ 5 & 7 \end{bmatrix}$$

from Example 2, how do we find its inverse if it is not already known? According to the definition, A^{-1} is a 2×2 matrix such that $AA^{-1} = I$ and $A^{-1}A = I$. So if

$$A^{-1} = \begin{bmatrix} x & y \\ z & w \end{bmatrix},$$

then

$$\begin{bmatrix} 3 & 4 \\ 5 & 7 \end{bmatrix}\begin{bmatrix} x & y \\ z & w \end{bmatrix} = \begin{bmatrix} 1 & 0 \\ 0 & 1 \end{bmatrix} \quad \text{and} \quad \begin{bmatrix} x & y \\ z & w \end{bmatrix}\begin{bmatrix} 3 & 4 \\ 5 & 7 \end{bmatrix} = \begin{bmatrix} 1 & 0 \\ 0 & 1 \end{bmatrix}.$$

To find A^{-1} we solve these matrix equations. If A is invertible, both equations will have the same solution. We will work with the first one. Multiply the two matrices on the left-hand side of the first equation to get the following equation:

$$\begin{bmatrix} 3x + 4z & 3y + 4w \\ 5x + 7z & 5y + 7w \end{bmatrix} = \begin{bmatrix} 1 & 0 \\ 0 & 1 \end{bmatrix}$$

Equate the corresponding terms from these equal matrices to get the following two systems:

$$\begin{aligned} 3x + 4z &= 1 \\ 5x + 7z &= 0 \end{aligned} \qquad\qquad \begin{aligned} 3y + 4w &= 0 \\ 5y + 7w &= 1 \end{aligned}$$

We can solve these two systems by using the Gaussian elimination method from Section 6.1. The augmented matrices for these systems are

$$\left[\begin{array}{cc|c} 3 & 4 & 1 \\ 5 & 7 & 0 \end{array}\right] \quad \text{and} \quad \left[\begin{array}{cc|c} 3 & 4 & 0 \\ 5 & 7 & 1 \end{array}\right].$$

Note that the two augmented matrices have the same coefficient matrix. Since we would use the same row operations on each of them, we can solve the systems simultaneously by combining the two systems into one augmented matrix denoted $[A|I]$:

$$[A|I] = \left[\begin{array}{cc|cc} 3 & 4 & 1 & 0 \\ 5 & 7 & 0 & 1 \end{array}\right]$$

So the problem of finding A^{-1} is equivalent to the problem of solving two systems by Gaussian elimination. A is invertible if and only if these systems have a solution. Multiply the first row of $[A|I]$ by $\frac{1}{3}$ to get a 1 in the first row, first column:

$$\left[\begin{array}{cc|cc} 1 & \frac{4}{3} & \frac{1}{3} & 0 \\ 5 & 7 & 0 & 1 \end{array}\right] \qquad \frac{1}{3}R_1 \to R_1$$

Now multiply row 1 by -5 and add the result to row 2:

$$\left[\begin{array}{cc|cc} 1 & \frac{4}{3} & \frac{1}{3} & 0 \\ 0 & \frac{1}{3} & -\frac{5}{3} & 1 \end{array}\right] \qquad -5R_1 + R_2 \to R_2$$

Multiply row 2 by 3:

$$\left[\begin{array}{cc|cc} 1 & \frac{4}{3} & \frac{1}{3} & 0 \\ 0 & 1 & -5 & 3 \end{array}\right] \qquad 3R_2 \to R_2$$

Multiply row 2 by $-\frac{4}{3}$ and add the result to row 1:

$$\left[\begin{array}{cc|cc} 1 & 0 & 7 & -4 \\ 0 & 1 & -5 & 3 \end{array}\right] \qquad -\frac{4}{3}R_2 + R_1 \to R_1$$

The numbers in the first column to the right of the bar give the values of x and z, while the numbers in the second column give the values of y and w. So $x = 7$, $y = -4$, $z = -5$, and $w = 3$ give the solutions to the two systems, and

$$A^{-1} = \begin{bmatrix} 7 & -4 \\ -5 & 3 \end{bmatrix}.$$

Since A^{-1} is the same matrix that was called B in Example 2, we can be sure that $AA^{-1} = I$ and $A^{-1}A = I$. Note that the matrix A^{-1} actually appeared on the right-hand side of the final augmented matrix, while the 2×2 identity matrix I appeared on the left-hand side. So A^{-1} is found by simply using row operations to convert the matrix $[A|I]$ into the matrix $[I|A^{-1}]$.

The essential steps for finding the inverse of a matrix are listed as follows.

PROCEDURE

Finding A^{-1}

Use the following steps to find the inverse of a square matrix A.

1. Write the augmented matrix $[A|I]$, where I is the identity matrix of the same order as A.
2. Use row operations (the Gaussian elimination method) to convert the left-hand side of the augmented matrix into I.
3. If the left-hand side can be converted to I, then $[A|I]$ becomes $[I|A^{-1}]$, and A^{-1} appears on the right-hand side of the augmented matrix.
4. If the left-hand side cannot be converted to I, then A is not invertible.

EXAMPLE 3 Finding the inverse of a matrix

Find the inverse of the matrix

$$A = \begin{bmatrix} 2 & -3 \\ 1 & 1 \end{bmatrix}.$$

Solution

Write the augmented matrix $[A|I]$:

$$\left[\begin{array}{cc|cc} 2 & -3 & 1 & 0 \\ 1 & 1 & 0 & 1 \end{array}\right]$$

Use row operations to convert $[A|I]$ into $[I|A^{-1}]$:

$$\left[\begin{array}{rr|rr} 1 & 1 & 0 & 1 \\ 2 & -3 & 1 & 0 \end{array}\right] \qquad R_1 \leftrightarrow R_2$$

$$\left[\begin{array}{rr|rr} 1 & 1 & 0 & 1 \\ 0 & -5 & 1 & -2 \end{array}\right] \qquad -2R_1 + R_2 \rightarrow R_2$$

$$\left[\begin{array}{rr|rr} 1 & 1 & 0 & 1 \\ 0 & 1 & -\frac{1}{5} & \frac{2}{5} \end{array}\right] \qquad -\frac{1}{5}R_2 \rightarrow R_2$$

$$\left[\begin{array}{rr|rr} 1 & 0 & \frac{1}{5} & \frac{3}{5} \\ 0 & 1 & -\frac{1}{5} & \frac{2}{5} \end{array}\right] \qquad -R_2 + R_1 \rightarrow R_1$$

Since the last matrix is in the form $[I|A^{-1}]$, we get

$$A^{-1} = \begin{bmatrix} \frac{1}{5} & \frac{3}{5} \\ -\frac{1}{5} & \frac{2}{5} \end{bmatrix}.$$

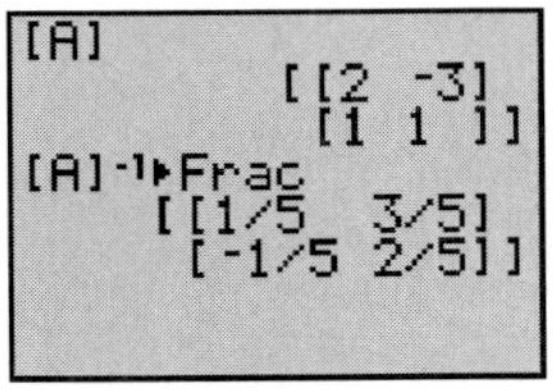

Figure 6.5

Check that $AA^{-1} = A^{-1}A = I$.

The inverse of a matrix can be found with a graphing calculator. Enter the matrix A from Example 3 into a calculator. Then use the x^{-1} key to find A^{-1} as shown in Fig. 6.5.

EXAMPLE 4 A noninvertible matrix

Find the inverse of the matrix

$$A = \begin{bmatrix} 1 & -3 \\ -1 & 3 \end{bmatrix}.$$

Solution

Use row operations on the augmented matrix $[A|I]$:

$$\left[\begin{array}{rr|rr} 1 & -3 & 1 & 0 \\ -1 & 3 & 0 & 1 \end{array}\right]$$

$$\left[\begin{array}{rr|rr} 1 & -3 & 1 & 0 \\ 0 & 0 & 1 & 1 \end{array}\right] \qquad R_1 + R_2 \rightarrow R_2$$

The row containing all zeros on the left-hand side of the augmented matrix indicates that the left-hand side (the matrix A) cannot be converted to I using row operations. So A is not invertible. Note that the row of 0's in the augmented matrix means that there is no solution to the systems that must be solved to find A^{-1}.

EXAMPLE 5 **The inverse of a 3 × 3 matrix**

Find the inverse of the matrix

$$A = \begin{bmatrix} 0 & 1 & 2 \\ 1 & 0 & 3 \\ 0 & 1 & 4 \end{bmatrix}.$$

Solution

Perform row operations to convert $[A|I]$ into $[I|A^{-1}]$, where I is the 3×3 identity matrix.

$$\left[\begin{array}{ccc|ccc} 0 & 1 & 2 & 1 & 0 & 0 \\ 1 & 0 & 3 & 0 & 1 & 0 \\ 0 & 1 & 4 & 0 & 0 & 1 \end{array}\right] \qquad \text{The augmented matrix}$$

Interchange the first and second rows to get the first 1 on the diagonal:

$$\left[\begin{array}{ccc|ccc} 1 & 0 & 3 & 0 & 1 & 0 \\ 0 & 1 & 2 & 1 & 0 & 0 \\ 0 & 1 & 4 & 0 & 0 & 1 \end{array}\right] \qquad R_1 \leftrightarrow R_2$$

Since the first column is now in the desired form, we work on the second column:

$$\left[\begin{array}{ccc|ccc} 1 & 0 & 3 & 0 & 1 & 0 \\ 0 & 1 & 2 & 1 & 0 & 0 \\ 0 & 0 & 2 & -1 & 0 & 1 \end{array}\right] \qquad -R_2 + R_3 \rightarrow R_3$$

We now get a 1 in the last position on the diagonal and 0's above it:

$$\left[\begin{array}{ccc|ccc} 1 & 0 & 3 & 0 & 1 & 0 \\ 0 & 1 & 2 & 1 & 0 & 0 \\ 0 & 0 & 1 & -0.5 & 0 & 0.5 \end{array}\right] \qquad \frac{1}{2}R_3 \rightarrow R_3$$

$$\left[\begin{array}{ccc|ccc} 1 & 0 & 0 & 1.5 & 1 & -1.5 \\ 0 & 1 & 0 & 2 & 0 & -1 \\ 0 & 0 & 1 & -0.5 & 0 & 0.5 \end{array}\right] \qquad -3R_3 + R_1 \rightarrow R_1 \text{ and } -2R_3 + R_2 \rightarrow R_2$$

So

$$A^{-1} = \begin{bmatrix} 1.5 & 1 & -1.5 \\ 2 & 0 & -1 \\ -0.5 & 0 & 0.5 \end{bmatrix}.$$

Check that $AA^{-1} = I$ and $A^{-1}A = I$.

Of course it is a lot simpler to find A^{-1} with a calculator as shown in Fig. 6.6.

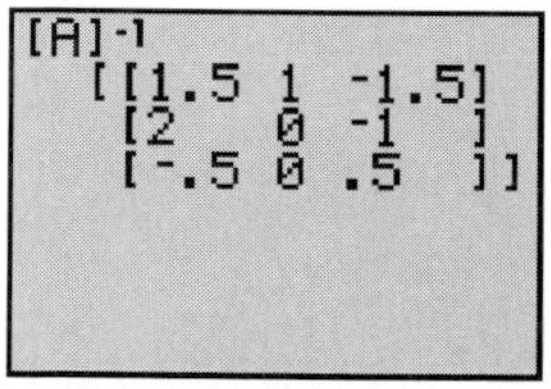

Figure 6.6

Finding the inverse of matrices larger than 2×2 is rather tedious, but technology can be used to great advantage here. We will now see why inverse matrices are so important.

Solving Systems of Equations Using Matrix Inverses

In Section 6.3 we saw that a system of n linear equations in n unknowns could be written as a matrix equation of the form $AX = B$, where A is a matrix of coefficients, X is a matrix of variables, and B is a matrix of constants. If A^{-1} exists, then we can multiply each side of this equation by A^{-1}. Since matrix multiplication is not commutative in general, A^{-1} is placed to the left of the matrices on each side:

$$
\begin{aligned}
AX &= B && \\
A^{-1}(AX) &= A^{-1}B && \text{Multiply each side by } A^{-1}. \\
(A^{-1}A)X &= A^{-1}B && \text{Matrix multiplication is associative.} \\
IX &= A^{-1}B && \text{Since } A^{-1}A = I \\
X &= A^{-1}B && \text{Since } I \text{ is the identity matrix}
\end{aligned}
$$

The last equation indicates that the values of the variables in the matrix X are equal to the entries in the matrix $A^{-1}B$. So solving the system is equivalent to finding $A^{-1}B$. This result is summarized in the following theorem.

Theorem: Solving a System Using A^{-1}

If a system of n linear equations in n variables has a unique solution, then the solution is given by

$$X = A^{-1}B,$$

where A is the matrix of coefficients, B is the matrix of constants, and X is the matrix of variables.

If there is no solution or there are infinitely many solutions, the matrix of coefficients is not invertible.

EXAMPLE 6 Using the inverse of a matrix to solve a system

Solve the system by using A^{-1}.

$$
\begin{aligned}
2x - 3y &= 1 \\
x + y &= 8
\end{aligned}
$$

Solution

For this system,

$$A = \begin{bmatrix} 2 & -3 \\ 1 & 1 \end{bmatrix}, \quad X = \begin{bmatrix} x \\ y \end{bmatrix}, \quad \text{and} \quad B = \begin{bmatrix} 1 \\ 8 \end{bmatrix}.$$

Since the matrix A is the same as in Example 3, use A^{-1} from Example 3. Multiply A^{-1} and B to obtain

$$\begin{bmatrix} \frac{1}{5} & \frac{3}{5} \\ -\frac{1}{5} & \frac{2}{5} \end{bmatrix} \begin{bmatrix} 1 \\ 8 \end{bmatrix} = \begin{bmatrix} 5 \\ 3 \end{bmatrix}.$$

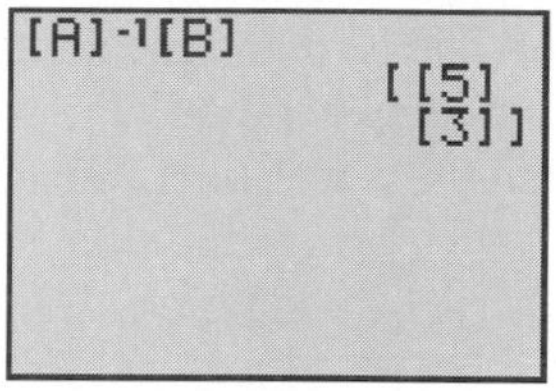

Figure 6.7

Since $X = A^{-1}B$, we have

$$X = \begin{bmatrix} x \\ y \end{bmatrix} = \begin{bmatrix} 5 \\ 3 \end{bmatrix}.$$

So $x = 5$ and $y = 3$. Check this solution in the original system.

With a calculator you can enter A and B, then find $A^{-1}B$ as shown in Fig. 6.7.

Since computers and even pocket calculators can find inverses of matrices, the method of solving an equation by first finding A^{-1} is a popular one for use with machines. By hand, this method may seem somewhat tedious. However, it is useful when there are many systems to solve with the same coefficients. The following system has the same coefficients as the system of Example 6:

$$\begin{aligned} 2x - 3y &= -1 \\ x + y &= -3 \end{aligned}$$

So

$$X = A^{-1}B = \begin{bmatrix} \frac{1}{5} & \frac{3}{5} \\ -\frac{1}{5} & \frac{2}{5} \end{bmatrix} \begin{bmatrix} -1 \\ -3 \end{bmatrix} = \begin{bmatrix} -2 \\ -1 \end{bmatrix}$$

or $x = -2$ and $y = -1$. As long as the coefficients of x and y are unchanged, the same A^{-1} is used to solve the system.

Application of Matrices to Secret Codes

There are many ways of encoding a message so that no one other than the intended recipient can understand the message. One way is to use a matrix to encode the message and the inverse matrix to decode the message. Assume that a space is 0, A is 1, B is 2, C is 3, and so on. The numerical equivalent of the word HELP is **8**, **5**, **12**, **16**. List these numbers in two 2×1 matrices. Then multiply the matrices by a coding matrix. We will use the 2×2 matrix A from Example 6:

$$\begin{bmatrix} 2 & -3 \\ 1 & 1 \end{bmatrix} \begin{bmatrix} 8 \\ 5 \end{bmatrix} = \begin{bmatrix} 1 \\ 13 \end{bmatrix} \quad \begin{bmatrix} 2 & -3 \\ 1 & 1 \end{bmatrix} \begin{bmatrix} 12 \\ 16 \end{bmatrix} = \begin{bmatrix} -24 \\ 28 \end{bmatrix}$$

So the encoded message sent is 1, 13, -24, 28. The person receiving the message must know that

$$A^{-1} = \begin{bmatrix} \frac{1}{5} & \frac{3}{5} \\ -\frac{1}{5} & \frac{2}{5} \end{bmatrix}.$$

To decode the message, we find

$$\begin{bmatrix} \frac{1}{5} & \frac{3}{5} \\ -\frac{1}{5} & \frac{2}{5} \end{bmatrix}\begin{bmatrix} 1 \\ 13 \end{bmatrix} = \begin{bmatrix} 8 \\ 5 \end{bmatrix} \quad \text{and} \quad \begin{bmatrix} \frac{1}{5} & \frac{3}{5} \\ -\frac{1}{5} & \frac{2}{5} \end{bmatrix}\begin{bmatrix} -24 \\ 28 \end{bmatrix} = \begin{bmatrix} 12 \\ 16 \end{bmatrix}.$$

So the message is 8, 5, 12, 16, or HELP. Of course, any invertible matrix of any order and its inverse could be used for this coding scheme, and all of the encoding and decoding could be done by a computer. If a person or computer didn't know A or A^{-1} or even their order, then how hard do you think it would be to break this code?

FOR THOUGHT True or False? Explain.

The following statements refer to the matrices

$$A = \begin{bmatrix} 2 & 3 \\ 3 & 5 \end{bmatrix}, \quad B = \begin{bmatrix} 5 & -3 \\ -3 & 2 \end{bmatrix}, \quad C = \begin{bmatrix} 4 & 6 \\ 3 & 5 \\ 2 & 1 \end{bmatrix},$$

$$D = \begin{bmatrix} 11 \\ 19 \end{bmatrix}, \quad \text{and} \quad I = \begin{bmatrix} 1 & 0 \\ 0 & 1 \end{bmatrix}.$$

1. $AB = BA = I$

2. $B = A^{-1}$

3. B is an invertible matrix.

4. $AC = CA$

5. $CI = C$

6. C is an invertible matrix.

7. The system $\begin{aligned} 2x + 3y &= 11 \\ 3x + y &= 19 \end{aligned}$ is equivalent to $A\begin{bmatrix} x \\ y \end{bmatrix} = D$.

8. $A^{-1}D = \begin{bmatrix} -2 \\ 5 \end{bmatrix}$

9. The solution set to the system $\begin{aligned} 2x + 3y &= 11 \\ 3x + y &= 19 \end{aligned}$ is $\{(-2, 5)\}$.

10. The solution set to the system $\begin{aligned} 2x + 3y &= 3 \\ 3x + y &= -7 \end{aligned}$ is obtained from $A^{-1}\begin{bmatrix} 3 \\ -7 \end{bmatrix}$.

6.4 EXERCISES

Tape 19 Disk

Find the following products.

1. $\begin{bmatrix} 1 & 0 \\ 0 & 1 \end{bmatrix}\begin{bmatrix} -3 & 5 \\ 12 & 6 \end{bmatrix}$

2. $\begin{bmatrix} 4 & 8 \\ 9 & -2 \end{bmatrix}\begin{bmatrix} 1 & 0 \\ 0 & 1 \end{bmatrix}$

3.

4. $\begin{bmatrix} 3 & 2 \\ 3 & 3 \end{bmatrix}\begin{bmatrix} 1 & -\frac{2}{3} \\ -1 & 1 \end{bmatrix}$

5. $\begin{bmatrix} 3 & 4 \\ 3 & 5 \end{bmatrix}\begin{bmatrix} \frac{5}{3} & -\frac{4}{3} \\ -1 & 1 \end{bmatrix}$

6. $\begin{bmatrix} 1 & 2 \\ 4 & 5 \end{bmatrix}\begin{bmatrix} -\frac{5}{3} & \frac{2}{3} \\ \frac{4}{3} & -\frac{1}{3} \end{bmatrix}$

7. $\begin{bmatrix} 3 & 5 & 1 \\ 4 & 5 & 7 \\ 4 & 9 & 2 \end{bmatrix}\begin{bmatrix} 1 & 0 & 0 \\ 0 & 1 & 0 \\ 0 & 0 & 1 \end{bmatrix}$

8. $\begin{bmatrix} 1 & 0 & 0 \\ 0 & 1 & 0 \\ 0 & 0 & 1 \end{bmatrix}\begin{bmatrix} 4 & 0 & 5 \\ 0 & 7 & 9 \\ 3 & 1 & 2 \end{bmatrix}$

9. $\begin{bmatrix} 1 & 0 & 2 \\ 1 & 3 & 0 \\ 0 & 1 & 0 \end{bmatrix}\begin{bmatrix} 0 & 1 & -3 \\ 0 & 0 & 1 \\ 0.5 & -0.5 & 1.5 \end{bmatrix}$

10. $\begin{bmatrix} 2 & 1 & 0 \\ 1 & 0 & 2 \\ 0 & 1 & 1 \end{bmatrix}\begin{bmatrix} 0.4 & 0.2 & -0.4 \\ 0.2 & -0.4 & 0.8 \\ -0.2 & 0.4 & 0.2 \end{bmatrix}$

11. $\begin{bmatrix} 1 & 1 & 0 \\ 0 & 1 & 1 \\ 1 & 0 & 1 \end{bmatrix}\begin{bmatrix} 0.5 & -0.5 & 0.5 \\ 0.5 & 0.5 & -0.5 \\ -0.5 & 0.5 & 0.5 \end{bmatrix}$

12. $\begin{bmatrix} 0.4 & 0.2 & -0.4 \\ 0.2 & -0.4 & 0.8 \\ -0.2 & 0.4 & 0.2 \end{bmatrix}\begin{bmatrix} 2 & 1 & 0 \\ 1 & 0 & 2 \\ 0 & 1 & 1 \end{bmatrix}$

Determine whether the matrices in each pair are inverses of each other.

13. $\begin{bmatrix} 3 & 1 \\ 11 & 4 \end{bmatrix}, \begin{bmatrix} 4 & -1 \\ -11 & 3 \end{bmatrix}$

14. $\begin{bmatrix} \frac{1}{2} & 0 \\ 0 & \frac{1}{2} \end{bmatrix}, \begin{bmatrix} 2 & 0 \\ 0 & 2 \end{bmatrix}$

15. $\begin{bmatrix} \frac{1}{2} & -1 \\ 3 & -12 \end{bmatrix}, \begin{bmatrix} 4 & 2 \\ 1 & 1 \end{bmatrix}$

16. $\begin{bmatrix} 1 & 2 & 3 \\ 0 & 1 & 2 \\ 0 & 0 & 1 \end{bmatrix}, \begin{bmatrix} 1 & -2 & 1 \\ 0 & 1 & -2 \\ 0 & 0 & 1 \end{bmatrix}$

17. $\begin{bmatrix} 1 & 0 & 0 \\ 0 & \frac{1}{2} & 0 \end{bmatrix}, \begin{bmatrix} 1 & 0 \\ 0 & 2 \\ 3 & 4 \end{bmatrix}$

18. $\begin{bmatrix} 1 & 2 \\ 3 & 4 \\ 5 & 6 \end{bmatrix}, \begin{bmatrix} 1 & \frac{1}{2} \\ \frac{1}{3} & \frac{1}{4} \\ \frac{1}{5} & \frac{1}{6} \end{bmatrix}$

Find the inverse of each matrix A if possible. Check that $AA^{-1} = I$ and $A^{-1}A = I$.

19. $\begin{bmatrix} 1 & 4 \\ 0 & 2 \end{bmatrix}$

20. $\begin{bmatrix} 1 & 3 \\ 0 & -1 \end{bmatrix}$

21. $\begin{bmatrix} 1 & 6 \\ 1 & 9 \end{bmatrix}$

22. $\begin{bmatrix} 1 & 4 \\ 3 & 8 \end{bmatrix}$

23. $\begin{bmatrix} -2 & -3 \\ 3 & 4 \end{bmatrix}$

24. $\begin{bmatrix} 3 & 4 \\ 4 & 5 \end{bmatrix}$

25. $\begin{bmatrix} 1 & -5 \\ -1 & 3 \end{bmatrix}$

26. $\begin{bmatrix} 4 & 3 \\ -3 & -2 \end{bmatrix}$

27. $\begin{bmatrix} -1 & 5 \\ 2 & -10 \end{bmatrix}$

28. $\begin{bmatrix} 2 & 6 \\ 1 & 3 \end{bmatrix}$

29. $\begin{bmatrix} 1 & 1 & 0 \\ 0 & -1 & -1 \\ 1 & 0 & -1 \end{bmatrix}$

30. $\begin{bmatrix} 1 & -1 & 2 \\ 1 & 2 & 3 \\ 2 & 1 & 5 \end{bmatrix}$

31. $\begin{bmatrix} 1 & 1 & 1 \\ 1 & -1 & -1 \\ 1 & -1 & 1 \end{bmatrix}$

32. $\begin{bmatrix} 1 & 0 & 2 \\ 0 & 2 & 0 \\ 1 & 3 & 0 \end{bmatrix}$

33. $\begin{bmatrix} 0 & 2 & 0 \\ 3 & 3 & 2 \\ 2 & 5 & 1 \end{bmatrix}$

34. $\begin{bmatrix} 4 & 1 & -3 \\ 0 & 1 & 0 \\ -3 & 1 & 2 \end{bmatrix}$

Solve each system of equations by using A^{-1}. Note that the matrix of coefficients in each system is a matrix from Exercises 19–34.

35. $\begin{aligned} x + 6y &= -3 \\ x + 9y &= -6 \end{aligned}$

36. $\begin{aligned} x + 4y &= 5 \\ 3x + 8y &= 7 \end{aligned}$

37. $\begin{aligned} x + 6y &= 4 \\ x + 9y &= 5 \end{aligned}$

38. $\begin{aligned} x + 4y &= 1 \\ 3x + 8y &= 5 \end{aligned}$

39. $\begin{aligned} -2x - 3y &= 1 \\ 3x + 4y &= -1 \end{aligned}$

40. $\begin{aligned} 3x + 4y &= 1 \\ 4x + 5y &= 2 \end{aligned}$

41. $\begin{aligned} x - 5y &= -5 \\ -x + 3y &= 1 \end{aligned}$

42. $\begin{aligned} 4x + 3y &= 2 \\ -3x - 2y &= -1 \end{aligned}$

43. $\begin{aligned} x + y + z &= 3 \\ x - y - z &= -1 \\ x - y + z &= 5 \end{aligned}$

44. $\begin{aligned} x + 2z &= -4 \\ 2y &= 6 \\ x + 3y &= 7 \end{aligned}$

45. $\begin{aligned} 2y &= 6 \\ 3x + 3y + 2z &= 16 \\ 2x + 5y + z &= 19 \end{aligned}$

46. $\begin{aligned} 4x + y - 3z &= 3 \\ y &= -2 \\ -3x + y + 2z &= -5 \end{aligned}$

Solve each system of equations by using A^{-1} if possible.

47. $\begin{aligned} 0.3x &= 3 - 0.1y \\ 4y &= 7 - 2x \end{aligned}$

48. $\begin{aligned} 2x &= 3y - 7 \\ y &= x + 4 \end{aligned}$

49. $\begin{aligned} x - y + z &= 5 \\ 2x - y + 3z &= 1 \\ y + z &= -9 \end{aligned}$

50. $\begin{aligned} x + y - z &= 4 \\ 2x - 3y + z &= 2 \\ 4x - y - z &= 6 \end{aligned}$

51. $\begin{aligned} x + y + z &= 1 \\ 2x + 4y + z &= 2 \\ x + 3y + 6z &= 3 \end{aligned}$

52. $\begin{aligned} 0.5x - 0.25y + 0.1z &= 3 \\ 0.2x - 0.5y + 0.2z &= -2 \\ 0.1x + 0.3y - 0.5z &= -8 \end{aligned}$

Determine whether each matrix is invertible. If so, state the inverse.

53. $\begin{bmatrix} 1 & 0 & 1 \\ 0 & 2 & 2 \\ 2 & 1 & 0 \end{bmatrix}$

54. $\begin{bmatrix} 1 & 3 & 0 \\ 0 & 3 & -2 \\ 0 & -5 & 3 \end{bmatrix}$

55. $\begin{bmatrix} 0 & 4 & 2 \\ 0 & 3 & 2 \\ 1 & -1 & 1 \end{bmatrix}$

56. $\begin{bmatrix} 1 & 2 & 0 \\ 2 & 3 & -1 \\ 0 & 1 & 2 \end{bmatrix}$

57. $\begin{bmatrix} 1 & 2 & 3 & 4 \\ 0 & 1 & 2 & 3 \\ 0 & 0 & 1 & 2 \\ 0 & 0 & 0 & 1 \end{bmatrix}$

58. $\begin{bmatrix} 1 & 0 & 0 & 0 \\ -2 & 1 & 0 & 0 \\ 3 & -2 & 1 & 0 \\ 5 & 3 & -2 & 1 \end{bmatrix}$

Most graphing calculators can perform operations with matrices, including matrix inversion and multiplication. Solve the following systems, using a graphing calculator to find A^{-1} and the product $A^{-1}B$.

59. $\begin{aligned} 0.1x + 0.2y + 0.1z &= 27 \\ 0.5x + 0.2y + 0.3z &= 9 \\ 0.4x + 0.8y + 0.1z &= 36 \end{aligned}$

60. $\begin{aligned} 3x + 6y + 4z &= 9 \\ x + 2y - 2z &= -18 \\ -x + 4y + 3z &= 54 \end{aligned}$

61. $\begin{aligned} 1.5x - 5y + 3z &= 16 \\ 2.25x - 4y + z &= 24 \\ 2x + 3.5y - 3z &= -8 \end{aligned}$

62. $\begin{aligned} 2.1x - 3.4y + 5z &= 100 \\ 1.3x + 2y - 8z &= 250 \\ 2.5x + 3y - 9.1z &= 300 \end{aligned}$

Solve each problem.

63. Find all matrices A such that $A = \begin{bmatrix} a & 7 \\ 3 & b \end{bmatrix}$, $A^{-1} = \begin{bmatrix} -b & 7 \\ 3 & -a \end{bmatrix}$, and a and b are positive integers.

64. Find all matrices of the form $A = \begin{bmatrix} a & a \\ 0 & c \end{bmatrix}$ such that $A^2 = I$.

Write a system of equations for each problem. Solve the system using an inverse matrix.

65. *Eggs and Magazines* Stephanie bought a dozen eggs and a magazine at the Handy Mart. Her bill including tax was \$2.70. If groceries are taxed at 8% and magazines at 5% and she paid 15 cents in tax on the purchase, then what was the price of each item?

66. *Dogs and Suds* The French Club sold 48 hot dogs and 120 soft drinks at the game on Saturday for a total of \$103.20. If the price of a hot dog was 40 cents more than the price of a soft drink, then what was the price of each item?

67. *Plywood and Insulation* A contractor purchased four loads of plywood and six loads of insulation on Monday for \$2500, and three loads of plywood and five loads of insulation on Tuesday for \$1950. Find the cost of one load of plywood and the cost of one load of insulation.

68. *Virtual Pets* One Monday KZ Toys received a shipment of 12 Nano Puppies and 6 Giga Pets for a total cost of \$138. Tuesday's shipment contained 4 Nano Puppies and 8 Giga Pets for a total cost of \$76. Wednesday's shipment of 25 Nano Puppies and 33 Giga Pets did not include an invoice. What was the cost of Wednesday's shipment?

The following messages were encoded by using the matrix $A = \begin{bmatrix} 3 & 1 \\ 5 & 2 \end{bmatrix}$ and the coding scheme described in this section. Find A^{-1} and use it to decode the messages.

69. 36, 65, 49, 83, 12, 24, 66, 111, 33, 55

70. 15, 29, 26, 45, 24, 46, 3, 5, 46, 83, 6, 12, 77, 133

Write a system of equations for each of the following problems and solve the system using matrix inversion and matrix multiplication on a graphing calculator.

71. *Mixing Investments* The Asset Manager fund keeps 76% of its money in stocks while the Magellan fund keeps 90% of its money in stocks. How should an investor divide \$60,000 between these two mutual funds so that 86% of the money is in stocks?

72. *Mixing Investments* The Asset Manager mutual fund investment mix is 76% stocks, 20% bonds, and 4% cash. The Magellan fund mix is 90% stocks, 9% bonds, and 1%

Table for Exercise 74

	Jambalaya	Crawfish pie	Filé gumbo	Iced tea	Dessert	Cost
Boudreaux	36	28	35	90	68	\$344.35
Thibodeaux	37	19	56	84	75	\$369.10
Fontenot	49	55	70	150	125	\$588.90
Arceneaux	58	34	52	122	132	\$529.50
Gautreaux	44	65	39	133	120	\$521.65

cash. The Puritan fund mix is 60% stocks, 33% bonds, and 7% cash. How should an investor divide $50,000 between these three funds so that 74% of the money is in stocks, 21.7% is in bonds, and 4.3% is in cash?

73. *Stocking Supplies* Fernando purchases supplies for an import store. His first shipment on Monday was for 24 animal totems, 33 trade-bead necklaces, and 12 tribal masks for a total price of $202.23. His second shipment was for 19 animal totems, 40 trade-bead necklaces, and 22 tribal masks for a total price of $209.38. His third shipment was for 30 animal totems, 9 trade-bead necklaces, and 19 tribal masks for a total price of $167.66. For the fourth shipment the computer was down, and Fernando did not know the price of each item. What is the price of each item?

74. *On the Bayou* A-Bear's Catering Service charges its customers according to the number of servings of each item that is supplied at the party. The table shows the number of servings of jambalaya, crawfish pie, filé gumbo, iced tea, and dessert for the last five customers, along with the total cost of each party. What amount does A-Bear's charge per serving of each item?

LINKING CONCEPTS

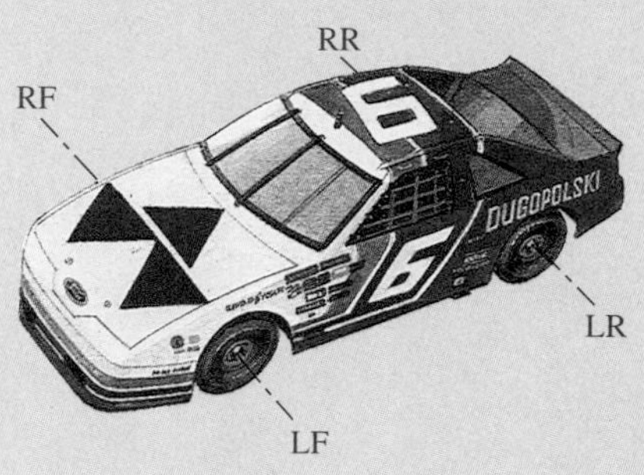

For Individual or Group Explorations

Racing Rules Race car drivers adjust the weight distribution of their cars according to track conditions. However, according to a NASCAR rule, no more than 52% of a car's weight can be on any pair of tires.

a) A driver of a 1250-pound car wants to have 50% of its weight on the left rear and left front tires and 48% of its weight on the left rear and right front tires. If the right front weight is fixed at 288 pounds, then what amount of weight should be on the other three tires?

b) Is the NASCAR rule satisfied with the weight distribution found in part (a)?

c) A driver of a 1300-pound car wants to have 50% of the car's weight on the left front and left rear tires, 48% on the left rear and right front tires, and 51% on the left rear and right rear tires. How much weight should be on each of the four tires?

d) The driver of the 1300-pound car wants to satisfy the NASCAR 52% rule and have as much weight as possible on the left front tire. What is the maximum amount of weight that can be on that tire?

6.5 Solution of Linear Systems in Two Variables Using Determinants

We have solved linear systems of equations by graphing, substitution, addition, Gaussian elimination, and inverse matrices. Graphing, substitution, and addition are feasible only with relatively simple systems. By contrast, the Gaussian elimination and inverse matrix methods are readily performed by computers or even hand-held calculators. With a machine doing the work, they can be applied to complicated systems such as those in Exercises 71 and 72 of Section 6.4. Determinants, which we now discuss, can also be used by computers and calculators, and give us another method that is not limited to simple systems.

The Determinant of a 2 × 2 Matrix

Before we can solve a system of equations by using determinants, we need to learn what a determinant is and how to find it. The **determinant** of a square matrix is a real number associated with the matrix. Every square matrix has a determinant. The determinant of a 1 × 1 matrix is the single entry of the matrix. For a 2 × 2 matrix the determinant is defined as follows.

Definition: Determinant of a 2 × 2 Matrix

The **determinant** of the matrix $\begin{bmatrix} a_{11} & a_{12} \\ a_{21} & a_{22} \end{bmatrix}$ is the real number $a_{11}a_{22} - a_{21}a_{12}$. In symbols,

$$\begin{vmatrix} a_{11} & a_{12} \\ a_{21} & a_{22} \end{vmatrix} = a_{11}a_{22} - a_{21}a_{12}.$$

If a matrix is named A, then the determinant of that matrix is denoted as $|A|$ or $\det(A)$. Even though the symbol for determinant looks like the absolute value symbol, the value of a determinant may be any real number. For a 2 × 2 matrix, that number is found by subtracting the products of the diagonal entries:

$$\begin{vmatrix} a_{11} & a_{12} \\ a_{21} & a_{22} \end{vmatrix} = a_{11}a_{22} - a_{21}a_{12}$$

EXAMPLE 1 The determinant of a 2 × 2 matrix

Find the determinant of each matrix.

a) $\begin{bmatrix} 3 & -1 \\ 4 & -5 \end{bmatrix}$ **b)** $\begin{bmatrix} 4 & -6 \\ 2 & -3 \end{bmatrix}$

Solution

a) $\begin{vmatrix} 3 & -1 \\ 4 & -5 \end{vmatrix} = 3(-5) - (4)(-1) = -15 + 4 = -11$

b) $\begin{vmatrix} 4 & -6 \\ 2 & -3 \end{vmatrix} = 4(-3) - (2)(-6) = -12 + 12 = 0$

To find these determinants with a calculator enter the matrices A and B, then use the determinant function as in Fig. 6.8.

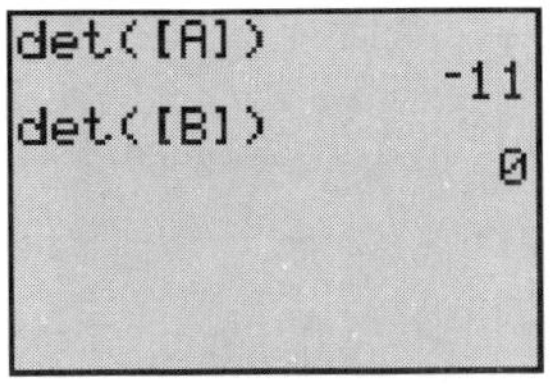

Figure 6.8

Cramer's Rule for Systems in Two Variables

We will now see how determinants arise in the solution of a system of two linear equations in two unknowns. Consider a general system of two linear equations in two unknowns, x and y,

$$(1) \qquad a_1x + b_1y = c_1$$

$$(2) \qquad a_2x + b_2y = c_2$$

where a_1, a_2, b_1, b_2, c_1, and c_2 are real numbers.

To eliminate y, multiply Eq. (1) by b_2 and Eq. (2) by $-b_1$:

$$
\begin{aligned}
b_2a_1x + b_2b_1y &= b_2c_1 && \text{Eq. (1) multiplied by } b_2\\
-b_1a_2x - b_1b_2y &= -b_1c_2 && \text{Eq. (2) multiplied by } -b_1\\
a_1b_2x - a_2b_1x &= c_1b_2 - c_2b_1 && \text{Add.}\\
(a_1b_2 - a_2b_1)x &= c_1b_2 - c_2b_1 && \text{Factor out } x.\\
x &= \frac{c_1b_2 - c_2b_1}{a_1b_2 - a_2b_1} && \text{Provided that } a_1b_2 - a_2b_1 \neq 0
\end{aligned}
$$

This formula for x can be written using determinants as

$$x = \frac{\begin{vmatrix} c_1 & b_1 \\ c_2 & b_2 \end{vmatrix}}{\begin{vmatrix} a_1 & b_1 \\ a_2 & b_2 \end{vmatrix}}, \quad \text{provided } a_1b_2 - a_2b_1 \neq 0.$$

The same procedure is used to eliminate x and get the following formula for y in terms of determinants:

$$y = \frac{\begin{vmatrix} a_1 & c_1 \\ a_2 & c_2 \end{vmatrix}}{\begin{vmatrix} a_1 & b_1 \\ a_2 & b_2 \end{vmatrix}}, \quad \text{provided } a_1b_2 - a_2b_1 \neq 0$$

Notice that there are three determinants involved in solving for x and y. Let

$$D = \begin{vmatrix} a_1 & b_1 \\ a_2 & b_2 \end{vmatrix}, \quad D_x = \begin{vmatrix} c_1 & b_1 \\ c_2 & b_2 \end{vmatrix}, \quad \text{and} \quad D_y = \begin{vmatrix} a_1 & c_1 \\ a_2 & c_2 \end{vmatrix}.$$

Note that D is the determinant of the original matrix of coefficients of x and y. D appears in the denominator for both x and y. D_x is the determinant D with the constants c_1 and c_2 replacing the first column of D. D_y is the determinant D with the constants c_1 and c_2 replacing the second column of D. These formulas for solving a system of two linear equations in two variables are known as **Cramer's rule**.

Cramer's Rule for Systems in Two Variables

The solution to the system

$$a_1x + b_1y = c_1$$
$$a_2x + b_2y = c_2$$

is given by $x = \dfrac{D_x}{D}$ and $y = \dfrac{D_y}{D}$, where

$$D = \begin{vmatrix} a_1 & b_1 \\ a_2 & b_2 \end{vmatrix}, \quad D_x = \begin{vmatrix} c_1 & b_1 \\ c_2 & b_2 \end{vmatrix}, \quad \text{and} \quad D_y = \begin{vmatrix} a_1 & c_1 \\ a_2 & c_2 \end{vmatrix},$$

provided that $D \neq 0$.

EXAMPLE 2 Applying Cramer's rule

Use Cramer's rule to solve the system.

$$3x = 2y + 9$$
$$3y = x + 3$$

Solution

To apply Cramer's rule, rewrite both equations in the form $Ax + By = C$:

$$3x - 2y = 9$$
$$-x + 3y = 3$$

First find the determinant of the coefficient matrix using the coefficients of x and y:

$$D = \begin{vmatrix} 3 & -2 \\ -1 & 3 \end{vmatrix} = 3(3) - (-1)(-2) = 7$$

Next we find the determinants D_x and D_y. For D_x, use 9 and 3 in the x-column, and for D_y, use 9 and 3 in the y-column:

$$D_x = \begin{vmatrix} 9 & -2 \\ 3 & 3 \end{vmatrix} = 33 \quad \text{and} \quad D_y = \begin{vmatrix} 3 & 9 \\ -1 & 3 \end{vmatrix} = 18.$$

By Cramer's rule,

$$x = \frac{D_x}{D} = \frac{33}{7} \quad \text{and} \quad y = \frac{D_y}{D} = \frac{18}{7}.$$

Check that $x = 33/7$ and $y = 18/7$ satisfy the original system.

To check this result with a calculator, define matrices D and A as in Fig. 6.9(a). Find x with Cramer's rule as in Fig. 6.9(b). You can find y in a similar manner.

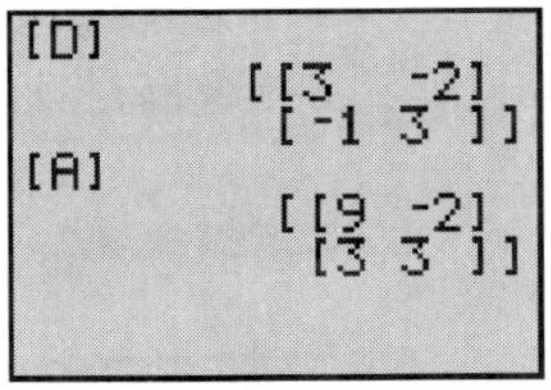

(a)

(b)

Figure 6.9

A system of two linear equations in two unknowns may have a unique solution, no solution, or infinitely many solutions. Cramer's rule works only on systems that have a unique solution. For inconsistent or dependent systems, $D = 0$ and Cramer's rule will not give the solution. If $D = 0$, then another method must be used to determine the solution set.

EXAMPLE 3 Inconsistent and dependent systems

Use Cramer's rule to solve each system if possible.

a) $\begin{aligned} 2x - 4y &= 8 \\ -x + 2y &= -4 \end{aligned}$ **b)** $\begin{aligned} 2x - 4y &= 8 \\ -x + 2y &= 6 \end{aligned}$

Solution

The matrix D is the same for both systems:

$$D = \begin{vmatrix} 2 & -4 \\ -1 & 2 \end{vmatrix} = 0$$

So Cramer's rule does not apply to either system. Multiply the second equation in each system by 2 and add the equations:

a) $\begin{aligned} 2x - 4y &= 8 \\ -2x + 4y &= -8 \\ \hline 0 &= 0 \end{aligned}$ **b)** $\begin{aligned} 2x - 4y &= 8 \\ -2x + 4y &= 12 \\ \hline 0 &= 20 \end{aligned}$

System (a) is dependent, and the solution set is $\{(x, y) \mid -x + 2y = -4\}$. System (b) is inconsistent and has no solution.

A system of two linear equations in two variables has a unique solution if and only if the determinant of the matrix of coefficients is nonzero. We have not yet seen how to find a determinant of a larger matrix, but the same result is true for a system of n linear equations in n variables. In Section 6.4 we learned that a system of n linear equations in n variables has a unique solution if and only if the matrix of coefficients is invertible. These two results are combined in the following theorem to give a means of identifying whether a matrix is invertible.

Theorem: Invertible Matrices

A matrix is invertible if and only if it has a nonzero determinant.

EXAMPLE 4 Determinants and inverse matrices

Are the matrices $A = \begin{bmatrix} 2 & -3 \\ 4 & 5 \end{bmatrix}$ and $B = \begin{bmatrix} 3 & 5 \\ 6 & 10 \end{bmatrix}$ invertible?

Solution

Since $|A| = 10 - (-12) = 22$, A is an invertible matrix. However, because $|B| = 30 - 30 = 0$, B is not an invertible matrix.

FOR THOUGHT True or False? Explain.

The following statements refer to the matrices

$$A = \begin{bmatrix} 3 & -5 \\ 1 & 4 \end{bmatrix}, B = \begin{bmatrix} 4 & -2 \\ -10 & 5 \end{bmatrix},$$
$$C = \begin{bmatrix} 2 & -5 \\ 6 & 4 \end{bmatrix}, \text{ and } E = \begin{bmatrix} 3 & 2 \\ 1 & 6 \end{bmatrix}.$$

1. $|A| = 7$
2. A is invertible.
3. $|B| = 0$
4. B is invertible.
5. The system $\begin{aligned} 3x - 5y &= 2 \\ x + 4y &= 6 \end{aligned}$ is independent.
6. $|CE| = |C| \cdot |E|$
7. The solution to the system $\begin{aligned} 3x^2 - 5y^2 &= 2 \\ x^2 + 4y &= 6 \end{aligned}$ is $x = \frac{|C|}{|A|}$ and $y = \frac{|E|}{|A|}$.
8. The determinant of the 2×2 identity matrix I is 1.
9. The matrix $\begin{bmatrix} 2 & 0.1 \\ 100 & 5 \end{bmatrix}$ is invertible.
10. $\begin{bmatrix} 5 & 3 \\ 1 & 6 \end{bmatrix} = 27$

6.5 EXERCISES

Tape 19 Disk

Find the determinant of each matrix.

1. $\begin{bmatrix} 1 & 3 \\ 0 & 2 \end{bmatrix}$
2. $\begin{bmatrix} 0 & 4 \\ 2 & -1 \end{bmatrix}$
3. $\begin{bmatrix} 3 & 4 \\ 2 & 9 \end{bmatrix}$
4. $\begin{bmatrix} 7 & 2 \\ 3 & -2 \end{bmatrix}$
5. $\begin{bmatrix} -0.3 & -0.5 \\ -0.7 & 0.2 \end{bmatrix}$
6. $\begin{bmatrix} -\frac{1}{3} & \frac{4}{3} \\ -3 & \frac{2}{3} \end{bmatrix}$
7. $\begin{bmatrix} \frac{1}{8} & -\frac{3}{8} \\ 2 & -\frac{1}{4} \end{bmatrix}$
8. $\begin{bmatrix} -1 & -3 \\ -5 & -8 \end{bmatrix}$
9. $\begin{bmatrix} 0.02 & 0.4 \\ 1 & 20 \end{bmatrix}$
10. $\begin{bmatrix} -0.3 & 0.4 \\ 3 & -4 \end{bmatrix}$
11. $\begin{bmatrix} 3 & -5 \\ -9 & 15 \end{bmatrix}$
12. $\begin{bmatrix} -6 & 2 \\ 3 & -1 \end{bmatrix}$

Solve each system, using Cramer's rule when possible.

13. $\begin{aligned} 2x - y &= -11 \\ x + 3y &= 12 \end{aligned}$
14. $\begin{aligned} 3x - 2y &= -4 \\ -5x + 4y &= -1 \end{aligned}$
15. $\begin{aligned} x &= y + 6 \\ x + y &= 5 \end{aligned}$
16. $\begin{aligned} 3x + y &= 7 \\ 4x &= y - 4 \end{aligned}$
17. $\begin{aligned} \frac{1}{2}x - \frac{1}{3}y &= 4 \\ \frac{1}{4}x + \frac{1}{2}y &= 6 \end{aligned}$
18. $\begin{aligned} \frac{1}{4}x + \frac{2}{3}y &= 25 \\ \frac{3}{5}x - \frac{1}{10}y &= 12 \end{aligned}$
19. $\begin{aligned} 0.2x + 0.12y &= 148 \\ x + y &= 900 \end{aligned}$
20. $\begin{aligned} 0.08x + 0.05y &= 72 \\ 2x - y &= 0 \end{aligned}$
21. $\begin{aligned} 3x + y &= 6 \\ -6x - 2y &= -12 \end{aligned}$
22. $\begin{aligned} 8x - 4y &= 2 \\ 4x - 2y &= 1 \end{aligned}$
23. $\begin{aligned} 8x - y &= 9 \\ -8x + y &= 10 \end{aligned}$
24. $\begin{aligned} 12x + 3y &= 9 \\ 4x + y &= 6 \end{aligned}$
25. $\begin{aligned} y &= x - 3 \\ y &= 3x + 9 \end{aligned}$
26. $\begin{aligned} y &= \frac{x - 3}{2} \\ x + 2y &= 15 \end{aligned}$
27. $\begin{aligned} \sqrt{2}\,x + \sqrt{3}\,y &= 4 \\ \sqrt{18x} - \sqrt{12y} &= -3 \end{aligned}$
28. $\begin{aligned} \frac{\sqrt{3x}}{3} + y &= 1 \\ x - \sqrt{3y} &= 0 \end{aligned}$
29. $\begin{aligned} x^2 + y^2 &= 25 \\ x^2 - y &= 5 \end{aligned}$
30. $\begin{aligned} x^2 + y &= 8 \\ x^2 - y &= 4 \end{aligned}$
31. $\begin{aligned} x - 2y &= y^2 \\ \frac{1}{2}x - y &= 2 \end{aligned}$
32. $\begin{aligned} y &= 0.15x \\ x + y &= 736 \end{aligned}$

Determine whether each matrix is invertible by finding the determinant of the matrix.

33. $\begin{bmatrix} 4 & 0.5 \\ 2 & 3 \end{bmatrix}$
34. $\begin{bmatrix} -5 & 2 \\ 4 & -1 \end{bmatrix}$

35. $\begin{bmatrix} 3 & -4 \\ 9 & -12 \end{bmatrix}$ **36.** $\begin{bmatrix} \frac{1}{2} & 12 \\ \frac{1}{3} & 8 \end{bmatrix}$

Solve the following systems using Cramer's rule and a graphing calculator.

37. $3.47x + 23.09y = 5978.95$
$12.48x + 3.98y = 2765.34$

38. $0.0875x + 0.1625y = 564.40$
$x + y = 4232$

Solve each problem, using two linear equations in two variables and Cramer's rule.

39. *The Survey Says* A survey of 615 teenagers found that 44% of the boys and 35% of the girls would like to be taller. If altogether 231 teenagers in the survey wished they were taller, how many boys and how many girls were in the survey?

40. *Average Weight* The average weight for the Packer's starting quarterback Brett Favre and his backup Steve Bono in 1997 was 219 lb. If Favre was 14 lb heavier than Bono, then how much did each weigh?

41. *Modern Maturity* One morning Sarah awoke to find that the digits in her age had reversed, and she was 72 years older than she was when she went to bed. If the sum of the digits in her age is 10, then how old was she when she went to bed?

42. *Acute Angles* One acute angle of a right triangle is 1° larger than twice the other acute angle. What are the measures of the acute angles?

43. *An Isosceles Triangle* If the smallest angle of an isosceles triangle is 2° smaller than any other angle, then what is the measure of each angle?

44. *A Losing Situation* Morton Motor Express lost a full truckload of TVs and VCRs. The truck carrying the shipment had a capacity of 2350 ft^3. On the insurance claim the TVs were valued at \$400 each and the VCRs were valued at \$225 each, for a total value of \$147,500. If each TV was in a carton with a volume of 8 ft^3 and each VCR was in a carton of 2.5 ft^3, then how many TVs and VCRs were in the shipment?

For Writing/Discussion

The following exercises investigate some of the properties of determinants. For these exercises let $M = \begin{bmatrix} 3 & 2 \\ 5 & 4 \end{bmatrix}$ and $N = \begin{bmatrix} 2 & 7 \\ 1 & 5 \end{bmatrix}$.

45. Find $|M|$, $|N|$, and $|MN|$. Is $|MN| = |M| \cdot |N|$?

46. Find M^{-1} and $|M^{-1}|$. Is $|M^{-1}| = 1/|M|$?

47. Prove that the determinant of a product of two 2×2 matrices is equal to the product of their determinants.

48. Prove that if A is any 2×2 invertible matrix, then the determinant of A^{-1} is the reciprocal of the determinant of A.

49. Find $|-2M|$. Is $|-2M| = -2 \cdot |M|$?

50. Prove that if k is any scalar and A is any 2×2 matrix, then $|kA| = k^2 \cdot |A|$.

6.6 Solution of Linear Systems in Three Variables Using Determinants

The determinant can be defined for any square matrix. In this section we define the determinant of a 3×3 matrix by extending the definition of the determinant for 2×2 matrices. We can then solve linear systems of three equations in three unknowns, using an extended version of Cramer's rule. The first step is to define a determinant of a certain part of a matrix, a *minor*.

Minors

To each entry of a 3×3 matrix there corresponds a 2×2 matrix, which is obtained by deleting the row and column in which that entry appears. The determinant of this 2×2 matrix is called the **minor** of that entry.

EXAMPLE 1 Finding the minor of an entry

Find the minors for the entries -2, 5, and 6 in the 3×3 matrix $\begin{bmatrix} -2 & -3 & -1 \\ -7 & 4 & 5 \\ 0 & 6 & 1 \end{bmatrix}$.

Solution

To find the minor for the entry -2, delete the first row and first column.

$$\begin{bmatrix} \not{-2} & \not{-3} & \not{-1} \\ \not{-7} & 4 & 5 \\ \not{0} & 6 & 1 \end{bmatrix}$$

The minor for -2 is $\begin{vmatrix} 4 & 5 \\ 6 & 1 \end{vmatrix} = 4 - (30) = -26$. To find the minor for the entry 5, delete the second row and third column.

$$\begin{bmatrix} -2 & -3 & \not{-1} \\ \not{-7} & \not{4} & \not{5} \\ 0 & 6 & \not{1} \end{bmatrix}$$

The minor for 5 is $\begin{vmatrix} -2 & -3 \\ 0 & 6 \end{vmatrix} = -12 - (0) = -12$. To find the minor for the entry 6, delete the third row and second column.

$$\begin{bmatrix} -2 & \not{-3} & -1 \\ -7 & \not{4} & 5 \\ \not{0} & \not{6} & \not{1} \end{bmatrix}$$

The minor for 6 is $\begin{vmatrix} -2 & -1 \\ -7 & 5 \end{vmatrix} = -10 - (7) = -17$.

The Determinant of a 3 × 3 Matrix

The determinant of a 3×3 matrix is defined in terms of minors. Let M_{ij} be the 2×2 matrix obtained from M by deleting the ith row and jth column. The determinant of M_{ij}, $|M_{ij}|$, is the minor for a_{ij}.

Definition: Determinant of a 3 × 3 Matrix

If $A = \begin{bmatrix} a_{11} & a_{12} & a_{13} \\ a_{21} & a_{22} & a_{23} \\ a_{31} & a_{32} & a_{33} \end{bmatrix}$, then the determinant of A, $|A|$, is defined as

$$|A| = a_{11}|M_{11}| - a_{21}|M_{21}| + a_{31}|M_{31}|.$$

To find the determinant of A, each entry in the first column of A is multiplied by its minor. This process is referred to as **expansion by minors** about the first column. Note the sign change on the middle term in the expansion.

EXAMPLE 2 The determinant of a 3 × 3 matrix

Find $|A|$, given that $A = \begin{bmatrix} -2 & -3 & -1 \\ -7 & 4 & 5 \\ 0 & 6 & 1 \end{bmatrix}$.

Solution

Use the definition to expand by minors about the first column.

$$\begin{aligned} |A| &= -2 \cdot \begin{vmatrix} 4 & 5 \\ 6 & 1 \end{vmatrix} - (-7) \cdot \begin{vmatrix} -3 & -1 \\ 6 & 1 \end{vmatrix} + 0 \cdot \begin{vmatrix} -3 & -1 \\ 4 & 5 \end{vmatrix} \\ &= -2(-26) + 7(3) + 0(-11) \\ &= 73 \end{aligned}$$

The value of the determinant of a 3 × 3 matrix can be found by expansion by minors about any row or column. However, you must use alternating plus and minus signs to precede the coefficients of the minors according to the following **sign array**:

$$\begin{bmatrix} + & - & + \\ - & + & - \\ + & - & + \end{bmatrix}$$

The signs in the sign array are used for the determinant of *any* 3 × 3 matrix and they are independent of the signs of the entries in the matrix. Notice that in Example 2, when we expanded by minors about the first column, we used the signs "+ − +" from the first column of the sign array. These signs were used in addition to the signs that appear on the entries themselves (−2, −7, and 0).

EXAMPLE 3 Expansion by minors about the second column

Expand by minors using the second column to find $|A|$, given that $A = \begin{bmatrix} -2 & -3 & -1 \\ -7 & 4 & 5 \\ 0 & 6 & 1 \end{bmatrix}$.

Solution

Use the signs "− + −" from the second column of the sign array:

$$\begin{bmatrix} + & - & + \\ - & + & - \\ + & - & + \end{bmatrix}$$

The coefficients -3, 4, and 6 from the second column of A are preceded by the signs from the second column of the sign array:

From the sign array

$$|A| = -(-3) \cdot \begin{vmatrix} -7 & 5 \\ 0 & 1 \end{vmatrix} + (4) \cdot \begin{vmatrix} -2 & -1 \\ 0 & 1 \end{vmatrix} - (6) \cdot \begin{vmatrix} -2 & -1 \\ -7 & 5 \end{vmatrix}$$

From second column of A

$$= 3(-7) + 4(-2) - 6(-17)$$
$$= 73$$

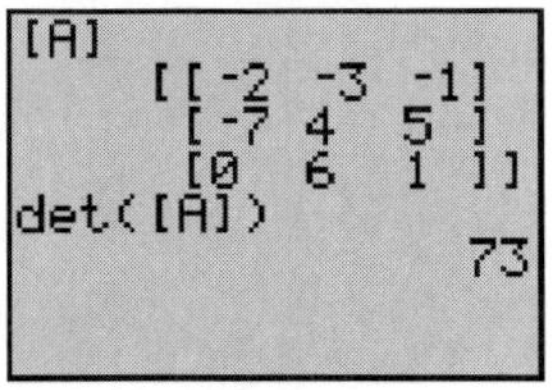

Figure 6.10

To check, define matrix A on a calculator and find the determinant as in Fig. 6.10.

In Examples 2 and 3 we got the same value for $|A|$ by using two different expansions. Expanding about any row or column gives the same result, but the computations can be easier if we examine the matrix and choose the row or column that contains the most 0's. Using 0's for the coefficients of one or more minors simplifies the work, because we do not have to evaluate the minors that are multiplied by zero. If a row or column of a matrix contains all 0's, then the determinant of the matrix is 0.

EXAMPLE 4 **Expansion by minors using the simplest row or column**

Find $|B|$, given that $B = \begin{bmatrix} 4 & 2 & 1 \\ -6 & 3 & 5 \\ 0 & 0 & -7 \end{bmatrix}$.

Solution

Since the third row has two 0's, we expand by minors about the third row. Use the signs "+ − +" from the third row of the sign array and the coefficients 0, 0, and -7 from the third row of B:

$$|B| = 0 \cdot \begin{vmatrix} 2 & 1 \\ 3 & 5 \end{vmatrix} - 0 \cdot \begin{vmatrix} 4 & 1 \\ -6 & 5 \end{vmatrix} + (-7) \cdot \begin{vmatrix} 4 & 2 \\ -6 & 3 \end{vmatrix}$$
$$= -7(24)$$
$$= -168$$

Determinant of a 4 × 4 Matrix

The determinant of a 4 × 4 matrix is also found by expanding by minors about a row or column. The following 4 × 4 sign array of alternating + and − signs (starting with + in the upper-left position) is used for the signs in the expansion:

$$\begin{bmatrix} + & - & + & - \\ - & + & - & + \\ + & - & + & - \\ - & + & - & + \end{bmatrix}$$

The minor for an entry of a 4 × 4 matrix is the determinant of the 3 × 3 matrix found by deleting the row and column of that entry. In general, the determinant of an $n \times n$ matrix is defined in terms of determinants of $(n-1) \times (n-1)$ matrices in the same manner.

EXAMPLE 5 **Determinant of a 4 × 4 matrix**

Find $|A|$, given that $A = \begin{bmatrix} -2 & -3 & 0 & 4 \\ 1 & -6 & 1 & -1 \\ 2 & 0 & 1 & 5 \\ 4 & 0 & 3 & 1 \end{bmatrix}$

Solution

Since the second column has two zeros, we expand by minors about the second column, using the signs "− + − +" from the second column of the sign array:

$$|A| = -(-3)\begin{vmatrix} 1 & 1 & -1 \\ 2 & 1 & 5 \\ 4 & 3 & 1 \end{vmatrix} + (-6)\begin{vmatrix} -2 & 0 & 4 \\ 2 & 1 & 5 \\ 4 & 3 & 1 \end{vmatrix}$$

$$- 0\begin{vmatrix} -2 & 0 & 4 \\ 1 & 1 & -1 \\ 4 & 3 & 1 \end{vmatrix} + 0\begin{vmatrix} -2 & 0 & 4 \\ 1 & 1 & -1 \\ 2 & 1 & 5 \end{vmatrix}$$

Evaluate the determinant of the first two 3 × 3 matrices to get

$$|A| = 3(2) - 6(36) = -210.$$

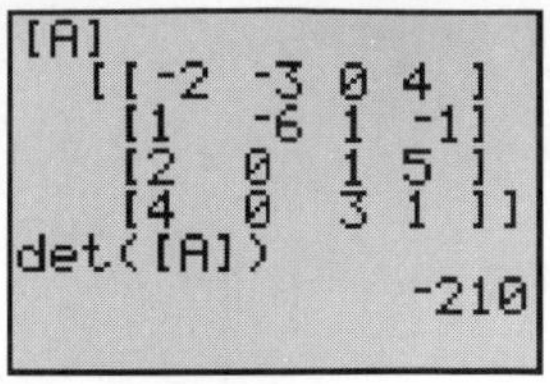

Figure 6.11

To check, enter A into your calculator and find its determinant as in Fig. 6.11.

Cramer's Rule for Systems in Three Variables

Cramer's rule for solving a system of three linear equations in three variables is similar to Cramer's rule for two variables. The rule consists of formulas for x, y, and z in terms of determinants. The development of Cramer's rule for three variables is similar to the development for two variables that was presented in Section 6.5, and so we will omit it.

Cramer's Rule for Systems in Three Variables

The solution to the system

$$\begin{aligned} a_1x + b_1y + c_1z &= d_1 \\ a_2x + b_2y + c_2z &= d_2 \\ a_3x + b_3y + c_3z &= d_3 \end{aligned}$$

is given by $x = \dfrac{D_x}{D}$, $y = \dfrac{D_y}{D}$, and $z = \dfrac{D_z}{D}$, where

$$D = \begin{vmatrix} a_1 & b_1 & c_1 \\ a_2 & b_2 & c_2 \\ a_3 & b_3 & c_3 \end{vmatrix}, \quad D_x = \begin{vmatrix} d_1 & b_1 & c_1 \\ d_2 & b_2 & c_2 \\ d_3 & b_3 & c_3 \end{vmatrix},$$

$$D_y = \begin{vmatrix} a_1 & d_1 & c_1 \\ a_2 & d_2 & c_2 \\ a_3 & d_3 & c_3 \end{vmatrix}, \quad \text{and} \quad D_z = \begin{vmatrix} a_1 & b_1 & d_1 \\ a_2 & b_2 & d_2 \\ a_3 & b_3 & d_3 \end{vmatrix}, \quad \text{for } D \neq 0.$$

Note that D_x, D_y, and D_z are obtained by replacing, respectively, the first, second, and third columns of D by the constants d_1, d_2, and d_3.

EXAMPLE 6 Solving a system using Cramer's rule

Use Cramer's rule to solve the system

$$\begin{aligned} x + y + z &= 0 \\ 2x - y + z &= -1 \\ -x + 3y - z &= -8. \end{aligned}$$

Solution

To use Cramer's rule, we first evaluate D, D_x, D_y, and D_z:

$$D = \begin{vmatrix} 1 & 1 & 1 \\ 2 & -1 & 1 \\ -1 & 3 & -1 \end{vmatrix} = 1 \cdot \begin{vmatrix} -1 & 1 \\ 3 & -1 \end{vmatrix} - 2 \cdot \begin{vmatrix} 1 & 1 \\ 3 & -1 \end{vmatrix} + (-1) \cdot \begin{vmatrix} 1 & 1 \\ -1 & 1 \end{vmatrix}$$

$$= 1(-2) - 2(-4) - 1(2)$$

$$= 4$$

To find D_x, D_y, or D_z, expand by minors about the first row because the first row contains a zero in each case:

$$D_x = \begin{vmatrix} 0 & 1 & 1 \\ -1 & -1 & 1 \\ -8 & 3 & -1 \end{vmatrix} = 0 \cdot \begin{vmatrix} -1 & 1 \\ 3 & -1 \end{vmatrix} - (1) \cdot \begin{vmatrix} -1 & 1 \\ -8 & -1 \end{vmatrix} + (1) \cdot \begin{vmatrix} -1 & -1 \\ -8 & 3 \end{vmatrix}$$

$$= -1(9) + 1(-11)$$

$$= -20$$

$$D_y = \begin{vmatrix} 1 & 0 & 1 \\ 2 & -1 & 1 \\ -1 & -8 & -1 \end{vmatrix} = 1 \cdot \begin{vmatrix} -1 & 1 \\ -8 & -1 \end{vmatrix} - (0) \cdot \begin{vmatrix} 2 & 1 \\ -1 & -1 \end{vmatrix} + (1) \cdot \begin{vmatrix} 2 & -1 \\ -1 & -8 \end{vmatrix}$$

$$= 1(9) + 1(-17)$$

$$= -8$$

$$D_z = \begin{vmatrix} 1 & 1 & 0 \\ 2 & -1 & -1 \\ -1 & 3 & -8 \end{vmatrix} = 1 \cdot \begin{vmatrix} -1 & -1 \\ 3 & -8 \end{vmatrix} - (1) \cdot \begin{vmatrix} 2 & -1 \\ -1 & -8 \end{vmatrix} + (0) \cdot \begin{vmatrix} 2 & -1 \\ 1 & 3 \end{vmatrix}$$

$$= 1(11) - 1(-17)$$

$$= 28$$

Now, by Cramer's rule,

$$x = \frac{D_x}{D} = \frac{-20}{4} = -5, \quad y = \frac{D_y}{D} = \frac{-8}{4} = -2, \quad \text{and} \quad z = \frac{D_z}{D} = \frac{28}{4} = 7.$$

Check that the ordered triple $(-5, -2, 7)$ satisfies all three equations. The solution set to the system is $\{(-5, -2, 7)\}$.

Cramer's rule can provide the solution to any system of three linear equations in three variables that has a unique solution. Its advantage is that it can give the value of any one of the variables without having to solve for the others. If $D = 0$, then Cramer's rule does not give the solution to the system, but it does indicate that the system is either dependent or inconsistent. If $D = 0$, then we must use another method to complete the solution to the system.

EXAMPLE 7 **Solving a system with $D = 0$**

Use Cramer's rule to solve the system.

$$\begin{aligned} &(1) \quad 2x + y - z = 3 \\ &(2) \quad 4x + 2y - 2z = 6 \\ &(3) \quad 6x + 3y - 3z = 9 \end{aligned}$$

Solution

To use Cramer's rule, we first evaluate D:

$$\begin{aligned} D = \begin{vmatrix} 2 & 1 & -1 \\ 4 & 2 & -2 \\ 6 & 3 & -3 \end{vmatrix} &= 2 \cdot \begin{vmatrix} 2 & -2 \\ 3 & -3 \end{vmatrix} - 4 \cdot \begin{vmatrix} 1 & -1 \\ 3 & -3 \end{vmatrix} + 6 \cdot \begin{vmatrix} 1 & -1 \\ 2 & -2 \end{vmatrix} \\ &= 2(0) - 4(0) + 6(0) \\ &= 0 \end{aligned}$$

Because $D = 0$, Cramer's rule cannot be used to solve the system. We could use the Gaussian elimination method, but note that Eqs. (2) and (3) are obtained by multiplying Eq. (1) by 2 and 3, respectively. Since all three equations are equivalent, the solution set to the system is $\{(x, y, z) | 2x + y - z = 3\}$.

In the last two chapters we have discussed several different methods for solving systems of linear equations. Studying different methods increases our understanding of systems of equations. A small system can usually be solved by any of these methods, but for large systems that are solved with computers, the most efficient and popular method is probably the Gaussian elimination method or a variation of it. Since you can find determinants, invert matrices, and perform operations with them on a graphing calculator, you can use either Cramer's rule or inverse matrices with a graphing calculator.

FOR THOUGHT True or False? Explain.

Statements 1–5 reference the matrix $A = \begin{bmatrix} 2 & -3 & 1 \\ 3 & 4 & 2 \\ 0 & 0 & 1 \end{bmatrix}$.

1. The sign array is used to determine whether $|A|$ is positive or negative.

2. $|A| = 2 \cdot \begin{vmatrix} 4 & 2 \\ 0 & 1 \end{vmatrix} - (-3) \cdot \begin{vmatrix} 3 & 2 \\ 0 & 1 \end{vmatrix} + 1 \cdot \begin{vmatrix} 2 & -3 \\ 3 & 4 \end{vmatrix}$

3. We can find $|A|$ by expanding about any row or column.

4. $|A| = \begin{vmatrix} 2 & -3 \\ 3 & 4 \end{vmatrix}$

5. We can find $|A|$ by expanding by minors about the diagonal.

6. A minor is a 2×2 matrix.

7. By Cramer's rule, the value of x is D/D_x.

8. If a matrix has a row in which all entries are zero, then the determinant of the matrix is 0.

9. By Cramer's rule, there is no solution to a system for which $D = 0$.

10. Cramer's rule works on any system of nonlinear equations.

6.6 EXERCISES

Tape 19 Disk

Find the indicated minors, using the matrix $\begin{bmatrix} 2 & -3 & 1 \\ 4 & 5 & -6 \\ 7 & 9 & -8 \end{bmatrix}$.

1. Minor for 2 **2.** Minor for -3 **3.** Minor for 1

4. Minor for 4 **5.** Minor for 5 **6.** Minor for -6

7. Minor for 9 **8.** Minor for -8

Find the determinant of each 3×3 matrix, using expansion by minors about the first column.

9. $\begin{bmatrix} 1 & -4 & 0 \\ -3 & 1 & -2 \\ 3 & -1 & 5 \end{bmatrix}$ **10.** $\begin{bmatrix} 1 & -3 & 2 \\ 3 & 1 & -4 \\ 2 & 3 & 6 \end{bmatrix}$

11. $\begin{bmatrix} 3 & -1 & 2 \\ 0 & 4 & -1 \\ 5 & 1 & -2 \end{bmatrix}$ **12.** $\begin{bmatrix} -1 & 3 & -1 \\ 0 & 2 & -3 \\ 2 & 6 & -9 \end{bmatrix}$

13. $\begin{bmatrix} -2 & 5 & 1 \\ -3 & 0 & -1 \\ 0 & 2 & -7 \end{bmatrix}$ **14.** $\begin{bmatrix} 0 & -6 & 2 \\ -1 & 4 & -2 \\ 5 & 3 & -1 \end{bmatrix}$

15. $\begin{bmatrix} 0.1 & 30 & 1 \\ 0.4 & 20 & 6 \\ 0.7 & 90 & 8 \end{bmatrix}$ **16.** $\begin{bmatrix} 3 & 0.3 & 10 \\ 5 & 0.5 & 30 \\ 8 & 0.1 & 80 \end{bmatrix}$

Evaluate the following determinants, using expansion by minors about the row or column of your choice.

17. $\begin{vmatrix} -1 & 3 & 5 \\ -2 & 0 & 0 \\ 4 & 3 & -4 \end{vmatrix}$ **18.** $\begin{vmatrix} 8 & -9 & 1 \\ 3 & 4 & 0 \\ -2 & 1 & 0 \end{vmatrix}$

19. $\begin{vmatrix} 1 & 1 & 1 \\ 2 & 2 & 2 \\ 4 & 4 & 4 \end{vmatrix}$ **20.** $\begin{vmatrix} 4 & -1 & 3 \\ 4 & -1 & 3 \\ 4 & -1 & 3 \end{vmatrix}$

21. $\begin{vmatrix} 0 & -1 & 0 \\ 3 & 4 & 6 \\ -2 & 3 & -5 \end{vmatrix}$ **22.** $\begin{vmatrix} 2 & 0 & 0 \\ 56 & 3 & -4 \\ 88 & 5 & -2 \end{vmatrix}$

23. $\begin{vmatrix} 2 & 0 & 1 \\ 4 & 0 & 6 \\ -7 & 9 & -8 \end{vmatrix}$ **24.** $\begin{vmatrix} 2 & -3 & 1 \\ -2 & 5 & -6 \\ 0 & 0 & 0 \end{vmatrix}$

25. $\begin{vmatrix} 3 & 0 & 1 & 5 \\ 2 & -3 & 2 & 0 \\ -2 & 3 & 1 & 2 \\ 2 & -4 & 1 & 3 \end{vmatrix}$

26. $\begin{vmatrix} 1 & -4 & 2 & 0 \\ -2 & -1 & 0 & -3 \\ 2 & 2 & 4 & 1 \\ 3 & 0 & -3 & 1 \end{vmatrix}$

27. $\begin{vmatrix} 2 & -3 & 4 & 6 \\ 1 & -5 & 0 & 0 \\ 1 & 3 & 1 & -3 \\ -2 & 0 & 2 & 1 \end{vmatrix}$

28. $\begin{vmatrix} -2 & 4 & 0 & 5 \\ 2 & -1 & 0 & 7 \\ 3 & 2 & 0 & -1 \\ 2 & 2 & -3 & 4 \end{vmatrix}$

Solve each system, using Cramer's rule where possible.

29. $\begin{aligned} x + y + z &= 6 \\ x - y + z &= 2 \\ 2x + y + z &= 7 \end{aligned}$ **30.** $\begin{aligned} 2x - 2y + 3z &= 7 \\ x + y - z &= -2 \\ 3x + y - 2z &= 5 \end{aligned}$

31. $$\begin{aligned} x + 2y \quad\;\; &= 8 \\ x - 3y + z &= -2 \\ 2x - y \quad\;\; &= 1 \end{aligned}$$

32. $$\begin{aligned} 2x + y \quad\;\; &= -4 \\ 3y - z &= -1 \\ x \quad + 3z &= -16 \end{aligned}$$

33. $$\begin{aligned} 2x - 3y + z &= 1 \\ x + 4y - z &= 0 \\ 3x - y + 2z &= 0 \end{aligned}$$

34. $$\begin{aligned} -2x + y - z &= 0 \\ x - y + 3z &= 1 \\ 3x + 3y + 2z &= 0 \end{aligned}$$

35. $$\begin{aligned} x + y + z &= 2 \\ 2x - y + 3z &= 0 \\ 3x + y - z &= 0 \end{aligned}$$

36. $$\begin{aligned} x - 2y - z &= 0 \\ -x + y + 3z &= 0 \\ x + 3y + z &= 3 \end{aligned}$$

37. $$\begin{aligned} x + y - 2z &= 1 \\ x - 2y + z &= 2 \\ 2x - y - z &= 3 \end{aligned}$$

38. $$\begin{aligned} x + y + z &= 4 \\ -2x - y + 3z &= 1 \\ y + 5z &= 9 \end{aligned}$$

39. $$\begin{aligned} x - y + z &= 5 \\ x + 2y + 3z &= 8 \\ 2x - 2y + 2z &= 16 \end{aligned}$$

40. $$\begin{aligned} 3x + 6y + 9z &= 12 \\ x + 2y + 3z &= 0 \\ x - y - 3z &= 0 \end{aligned}$$

Solve each problem, using a system of three equations in three unknowns and Cramer's rule.

41. *Age Disclosure* Jackie, Rochelle, and Alisha will not disclose their ages. However, the average of the ages of Jackie and Rochelle is 33, the average for Rochelle and Alisha is 25, and the average for Jackie and Alisha is 19. How old is each?

42. *Bennie's Coins* Bennie emptied his pocket of 49 coins to pay for his \$5.50 lunch. He used only nickels, dimes, and quarters, and the total number of dimes and quarters was one more than the number of nickels. How many of each type of coin did he use?

43. *What a Difference a Weight Makes* A sociology professor gave two one-hour exams and a final exam. Ian was distressed with his average score of 60 for the three tests and went to see the professor. Because of Ian's improvement during the semester, the professor offered to count the final exam as 60% of the grade and the two tests equally, giving Ian a weighted average of 76. Ian countered that since he improved steadily during the semester, the first test should count 10%, the second 20%, and the final 70% of the grade, giving a weighted average of 83. What were Ian's actual scores on the two tests and the final exam?

44. *Cookie Time* Cheryl, of Cheryl's Famous Cookies, set out 18 cups of flour, 14 cups of sugar, and 13 cups of shortening for her employees to use in making some batches of chocolate chip, oatmeal, and peanut butter cookies. She left for the day without telling them how many batches of each type to bake. The table gives the number of cups of each ingredient required for one batch of each type of cookie. How many batches of each were they supposed to bake?

Table for Exercise 44

	Flour	Sugar	Shortening
Chocolate chip	2 cups	2 cups	1 cup
Oatmeal	1 cup	1 cup	2 cups
Peanut butter	4 cups	2 cups	1 cup

Use the determinant feature of a graphing calculator to solve each system by Cramer's rule.

45. $$\begin{aligned} 0.2x - 0.3y + 1.2z &= 13.11 \\ 0.25x + 0.35y - 0.9z &= -1.575 \\ 2.4x - y + 1.25z &= 42.02 \end{aligned}$$

46. $$\begin{aligned} 3.6x + 4.5y + 6.8z &= 45{,}300 \\ 0.09x + 0.05y + 0.04z &= 474 \\ x + y - z &= 0 \end{aligned}$$

Solve each problem, using Cramer's rule and a graphing calculator.

47. *Gasoline Sales* The Runway Deli sells regular unleaded, plus unleaded, and supreme unleaded Chevron gasoline. The number of gallons of each grade and the total receipts for gasoline are shown in the table for the first three weeks of February. What was the price per gallon for each grade?

	Regular	Plus	Supreme	Receipts
Week 1	1270	980	890	\$3728.66
Week 2	1450	1280	1050	\$4496.82
Week 3	1340	1190	1060	\$4279.01

48. *Gasoline Sales* John's Curb Market sells regular unleaded, plus unleaded, and supreme unleaded Citgo gasoline. John was experimenting with prices during the first three weeks of February. The prices that he charged during each week for each grade and the total receipts for each week are given in the table. John unexpectedly found that, regardless of price, the number of gallons sold in each grade was the same for all three weeks. How many gallons of each grade did he sell every week?

	Regular	Plus	Supreme	Receipts
Week 1	\$1.099	\$1.209	\$1.259	\$5457.47
Week 2	\$1.069	\$1.219	\$1.289	\$5455.17
Week 3	\$1.029	\$1.239	\$1.299	\$5422.47

Extend Cramer's rule to four linear equations in four unknowns. Solve each system, using the extended Cramer's rule.

49. $$\begin{aligned} w + x + y + z &= 4 \\ 2w - x + y + 3z &= 13 \\ w + 2x - y + 2z &= -2 \\ w - x - y + 4z &= 8 \end{aligned}$$

50. $$\begin{aligned} 2w + 2x - 2y + z &= 9 \\ w + x + y + z &= 7 \\ 4w - 3x + 2y - 5z &= -13 \\ w + 3x - y + 9z &= 10 \end{aligned}$$

The equation of a line through two points can be expressed as an equation involving a determinant.

51. Show that the following equation is equivalent to the equation of the line through $(3, -5)$ and $(-2, 6)$.

$$\begin{vmatrix} x & y & 1 \\ 3 & -5 & 1 \\ -2 & 6 & 1 \end{vmatrix} = 0$$

52. Show that the following equation is equivalent to the equation of the line through (x_1, y_1) and (x_2, y_2).

$$\begin{vmatrix} x & y & 1 \\ x_1 & y_1 & 1 \\ x_2 & y_2 & 1 \end{vmatrix} = 0$$

For Writing/Discussion

Prove each of the following statements for any 3×3 matrix A.

53. If all entries in any row or column of A are zero, then $|A| = 0$.

54. If A has two identical rows (or columns), then $|A| = 0$.

55. If all entries in a row (or column) of A are multiplied by a constant k, then the determinant of the new matrix is $k \cdot |A|$.

56. If two rows (or columns) of A are interchanged, then the determinant of thc new matrix is $-|A|$.

HIGHLIGHTS

- **Section 6.1 Solving Linear Systems Using Matrices**

1. An $m \times n$ matrix is a rectangular array of numbers with m rows and n columns.
2. For a system of linear equations, the augmented matrix is a matrix whose entries are the coefficients of the variables of the system together with the constants.
3. In the Gaussian elimination method for solving a system of linear equations, there are three row operations that can be used on the augmented matrix to simplify it.
4. Applying the Gaussian elimination method to a dependent system causes an entire row to appear with zero in each entry.
5. Applying the Gaussian elimination method to an inconsistent system causes a row to appear with zero as the entry for each coefficient but a nonzero entry for the constant.

- **Section 6.2 Operations with Matrices**

1. Two matrices are equal if they have the same order and all corresponding entries are equal.
2. Two matrices of the same order can be added or subtracted by adding or subtracting the corresponding entries.
3. Matrices of a given order have an additive identity and each matrix has an additive inverse.
4. The product of a real number (scalar) and a matrix is found by multiplying each entry of the matrix by the real number.

- **Section 6.3 Multiplication of Matrices**

1. The product of an $m \times n$ matrix A and an $n \times p$ matrix B is an $m \times p$ matrix AB, where the ijth entry of AB is the sum of the products of the corresponding entries in the ith row of A and the jth column of B.
2. The product AB is defined only if the number of columns of A is equal to the number of rows of B. AB is not necessarily equal to BA.
3. A linear system of equations can be written as a matrix equation in the form $AX = B$, where A is the matrix of coefficients, X is a column matrix containing the variables, and B is a column matrix containing the constants.

- **Section 6.4 Inverses of Matrices**

1. The $n \times n$ identity matrix has 1's on the diagonal and 0's elsewhere.
2. If A is an $n \times n$ matrix and I is the $n \times n$ identity matrix, then $AI = IA = A$.
3. If A is an $n \times n$ matrix for which there is a matrix A^{-1} such that $AA^{-1} = A^{-1}A = I$, then A is an invertible matrix and A and A^{-1} are inverses of each other.
4. If A is an invertible matrix, then the solution to $AX = B$ is $X = A^{-1}B$.

- **Section 6.5 Solution of Linear Systems in Two Variables Using Determinants**

1. The determinant of a square matrix is a real number associated with the matrix.
2. Cramer's rule gives formulas involving determinants for finding the solution to an independent system of two linear equations in two unknowns or three linear equations in three unknowns.
3. A matrix has an inverse if and only if its determinant is nonzero.

- **Section 6.6 Solution of Linear Systems in Three Variables Using Determinants**

1. The determinant of a 3×3 matrix is defined in terms of determinants of 2×2 matrices, using the idea of minors and the sign array.
2. Cramer's rule will not work if the determinant of the matrix of coefficients is zero.

CHAPTER 6 REVIEW EXERCISES

Let $A = \begin{bmatrix} 2 & -3 \\ -2 & 4 \end{bmatrix}$, $B = \begin{bmatrix} 3 & 7 \\ 1 & 2 \end{bmatrix}$, $C = \begin{bmatrix} -1 \\ 3 \end{bmatrix}$,

$D = \begin{bmatrix} 5 \\ -3 \end{bmatrix}$, $E = \begin{bmatrix} 1 \\ -4 \\ 3 \end{bmatrix}$, $F = [3 \quad 2 \quad -1]$, and

$G = \begin{bmatrix} -1 & 0 & 0 \\ 1 & 1 & 0 \\ -2 & 3 & 1 \end{bmatrix}$. Find each of the following matrices or determinants if possible.

1. $A + B$ **2.** $A - B$ **3.** $2A - B$ **4.** $2A + 3B$

5. AB **6.** BA **7.** $D + E$ **8.** $F + G$

9. AC **10.** BD **11.** EF **12.** FE

13. FG **14.** GE **15.** GF **16.** EG

17. A^{-1} **18.** B^{-1} **19.** G^{-1} **20.** $A^{-1}C$

21. $(AB)^{-1}$ **22.** $A^{-1}B^{-1}$ **23.** AA^{-1} **24.** GG^{-1}

25. $|A|$ **26.** $|B|$ **27.** $|G|$ **28.** $|C|$

Solve each of the following systems by all three methods: Gaussian elimination, matrix inversion, and Cramer's rule.

29. $\begin{aligned} x + y &= 9 \\ 2x - y &= 1 \end{aligned}$

30. $\begin{aligned} x - 2y &= 3 \\ x + 2y &= 2 \end{aligned}$

31. $\begin{aligned} 2x + y &= -1 \\ 3x + 2y &= 0 \end{aligned}$

32. $\begin{aligned} 3x - y &= 1 \\ -2x + y &= 1 \end{aligned}$

33. $\begin{aligned} x - 5y &= 9 \\ -2x + 10y &= -18 \end{aligned}$

34. $\begin{aligned} 3x - y &= 4 \\ 6x - 2y &= 6 \end{aligned}$

35. $\begin{aligned} 0.05x + 0.1y &= 1 \\ 10x + 20y &= 20 \end{aligned}$

36. $\begin{aligned} 0.04x - 0.2y &= 3 \\ 2x - 10y &= 150 \end{aligned}$

37. $\begin{aligned} x + y - 2z &= -3 \\ -x + 2y - z &= 0 \\ -x - y + 3z &= 6 \end{aligned}$

38. $\begin{aligned} x - y + z &= 5 \\ x + y + 3z &= 11 \\ -x + 2y - z &= -5 \end{aligned}$

39. $\begin{aligned} y - 3z &= 1 \\ x + 2y \quad &= 5 \\ x \quad + 4z &= 1 \end{aligned}$

40. $\begin{aligned} 3x + y - 2z &= 0 \\ -2y + z &= 0 \\ y + 3z &= 14 \end{aligned}$

41. $\begin{aligned} x - y + z &= 2 \\ x - 2y - z &= 1 \\ 2x - 3y \quad &= 3 \end{aligned}$

42. $\begin{aligned} x + 2y + z &= 1 \\ 2x + 4y + 2z &= 0 \\ -x - 2y - z &= 2 \end{aligned}$

43. $\begin{aligned} x - 3y - z &= 2 \\ x - 3y - z &= 1 \\ x - 3y - z &= 0 \end{aligned}$

44. $\begin{aligned} 2x - y - z &= 0 \\ x + y + z &= 3 \\ 3x \quad &= 3 \end{aligned}$

Find the values of x, y, and z that make each of the equations true.

45. $\begin{bmatrix} x \\ x + y \end{bmatrix} = \begin{bmatrix} 9 \\ -3 \end{bmatrix}$

46. $\begin{bmatrix} x^2 \\ x - y \end{bmatrix} = \begin{bmatrix} 4 \\ 1 \end{bmatrix}$

47. $\begin{bmatrix} 1 & 1 \\ 2 & 1 \end{bmatrix} \begin{bmatrix} x \\ y \end{bmatrix} = \begin{bmatrix} 6 \\ 8 \end{bmatrix}$

48. $\begin{bmatrix} 0 & 0.5 \\ 1 & 1 \end{bmatrix} \begin{bmatrix} x \\ y \end{bmatrix} = \begin{bmatrix} 7 \\ 9 \end{bmatrix}$

49. $\begin{bmatrix} x \\ y \end{bmatrix} + \begin{bmatrix} y \\ -x \end{bmatrix} = \begin{bmatrix} -3 \\ y \end{bmatrix}$

50. $\begin{bmatrix} y \\ x \end{bmatrix} - \begin{bmatrix} x \\ y \end{bmatrix} = \begin{bmatrix} 4 \\ 5 \end{bmatrix}$

51. $\begin{bmatrix} x + y & 0 & 0 \\ 0 & y + z & 0 \\ 0 & 0 & x + z \end{bmatrix} = \begin{bmatrix} 1 & 0 & 0 \\ 0 & 1 & 0 \\ 0 & 0 & 1 \end{bmatrix}$

52. $\begin{bmatrix} 0 & x - y & 2z \\ 0 & 0 & y - z \\ 0 & 0 & 0 \end{bmatrix} = \begin{bmatrix} 0 & 2 & 5 \\ 0 & 0 & 3 \\ 0 & 0 & 0 \end{bmatrix}$

53. $\begin{bmatrix} 1 & 1 & 0 \\ 0 & 1 & 2 \\ 1 & 0 & 3 \end{bmatrix} \begin{bmatrix} x \\ y \\ z \end{bmatrix} = \begin{bmatrix} -1 \\ 7 \\ 17 \end{bmatrix}$

54. $\begin{bmatrix} 1 & 1 & 1 \\ -1 & 1 & -1 \\ 1 & 0 & 2 \end{bmatrix} \begin{bmatrix} x \\ y \\ z \end{bmatrix} = \begin{bmatrix} 0 \\ 0 \\ 0 \end{bmatrix}$

Solve each problem, using a system of linear equations in two or three variables. Use the method of your choice from this chapter.

55. *Fine for Polluting* A small manufacturing plant must pay a fine of \$10 for each gallon of pollutant A and \$6 for each gallon of pollutant B per day that it discharges into a nearby stream. If the manufacturing process produces three gallons of pollutant A for every four gallons of pollutant B and the daily fine is \$4060, then how many gallons of each are discharged each day?

56. *Friends* Ross, Joey, and Chandler spent a total of \$216 on coffee and pastry last month at the Central Perk coffee shop. Joey and Ross's expenses totaled only half as much as Chandler's. If Joey spent \$12 more than Ross, then how much did each spend?

57. *Utility Bills* Bette's total expense for water, gas, and electricity for one month including tax was \$189.83. There is a 6% state tax on electricity, a 5% city tax on gas, and a 4% county tax on water. If her total expenses included \$9.83 in taxes and her electric bill including tax was twice the gas bill including tax, then how much was each bill including tax?

58. *Predicting Car Sales* In the first three months of the year, West Coast Cadillac sold 38, 42, and 49 new cars, respectively. Find the equation of the parabola of the form $y = ax^2 + bx + c$ that passes through (1, 38), (2, 42), and (3, 49) as shown in the figure. (The fact that each point satisfies $y = ax^2 + bx + c$ gives three linear equations in a, b, and c.) Assuming that the fourth month's sales will fall on that same parabola, what would be the predicted sales for the fourth month?

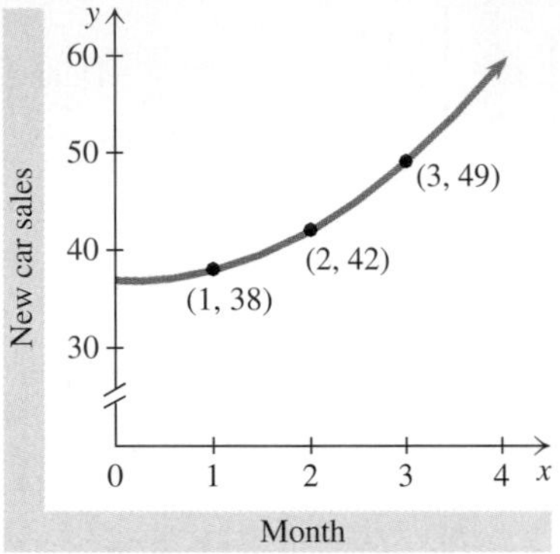

Figure for Exercise 58

CHAPTER 6 TEST

Solve each system, using Gaussian elimination.

1. $2x - 3y = 1$
$x + 9y = 4$

2. $2x - y + z = 5$
$x - 2y - z = -2$
$3x - y - z = 6$

3. $x - y - z = 1$
$2x + y - z = 0$
$5x - 2y - 4z = 3$

Let $A = \begin{bmatrix} 1 & -1 \\ -2 & 4 \end{bmatrix}$, $B = \begin{bmatrix} 2 & -3 \\ -4 & 6 \end{bmatrix}$, $C = \begin{bmatrix} -2 \\ 1 \end{bmatrix}$,

$D = \begin{bmatrix} 3 \\ -2 \end{bmatrix}$, $E = \begin{bmatrix} 2 \\ 3 \\ -1 \end{bmatrix}$, $F = [1 \quad 0 \quad -1]$, and

$G = \begin{bmatrix} -2 & 3 & 1 \\ -3 & 1 & 3 \\ 0 & 2 & -1 \end{bmatrix}$. Find each of the following matrices or determinants if possible.

4. $A + B$

5. $2A - B$

6. AB

7. AC

8. CB

9. FG

10. EF

11. A^{-1}

12. G^{-1}

13. $|A|$

14. $|B|$

15. $|G|$

Solve each system, using Cramer's rule.

16. $x - y = 2$
$-2x + 4y = 2$

17. $2x - 3y = 6$
$-4x + 6y = 1$

18. $-2x + 3y + z = -2$
$-3x + y + 3z = -4$
$2y - z = 0$

Solve each system by using inverse matrices.

19. $x - y = 1$
$-2x + 4y = -8$

20. $-2x + 3y + z = 1$
$-3x + y + 3z = 0$
$2y - z = -1$

Solve by using a method from this chapter.

21. The manager of a computer store bought x copies of the program Math Skillbuilder for \$10 each at the beginning of the year and sold y copies of the program for \$35 each. At the end of the year the program was obsolete and she destroyed 12 unsold copies. If her net profit for the year was \$730, then how many were bought and how many were sold?

22. Find a, b, and c such that the graph of $y = ax^2 + b\sqrt{x} + c$ goes through the points (0, 3), (1, −1/2), and (4, 3).

TYING IT ALL TOGETHER

Chapters P–6

Solve each equation.

1. $2(x + 3) - 5x = 7$

2. $\frac{1}{2}\left(x - \frac{1}{3}\right) + \frac{1}{5} = 1$

3. $\frac{1}{2}(2x - 2)(6x - 8) = 4$

4. $1 - \frac{1}{2}(8x - 4) = 9$

Solve each system of equations by the specified method.

5. Graphing:
$2x + y = 6$
$x - 2y = 8$

6. Substitution:
$y + 2x = 1$
$2x + 6y = 2$

7. Addition:
$2x - 0.06y = 20$
$3x + 0.01y = 20$

8. Gaussian elimination:
$2x - y = 1$
$x + 3y = -11$

9. Matrix inversion:
$3x - 5y = -7$
$-x + y = 1$

10. Cramer's rule:
$4x - 3y = 5$
$3x - 5y = 1$

11. Your choice:
$x^2 + y^2 = 25$
$x - y = -1$

12. Your choice:
$x^2 - y = 1$
$x + y = 1$

Flight controllers on Earth breathed a sigh of relief as the shuttle Atlantis slowly snuggled up to Mir without incident. Shuttle and space station would remain linked for six days while astronaut David Wolf boarded and crucial supplies—including water, food, and a new computer—were swapped.

Both physician and engineer, Wolf had prepared well for his four-month tour aboard the aging Mir. Numerous mishaps, including a life-threatening collision with a supply vessel, a fire, power outages, and several computer failures, were threatening to turn Mir from a science laboratory into a space repair shop. As one U.S. academic remarked, "We Americans have a throw-away mentality. The Russians say, "Let's put duct tape on it and get it fixed. . . ."

Mir, which means "peace" in Russian, is Earth's first orbiting space station. Although originally built by the Russians to last five years, Mir has hosted 23 crews for over a decade (up to six people at a time for periods up to one month). Currently, under an agreement between NASA and Russia, a series of shuttle missions is promoting an astronaut exchange program and setting the groundwork for construction of the first International Space Station.

On *Mir*, astronauts operate as the world's highest-flying laboratory technicians, exploring a broad range of subjects, while scientists whose experiments were chosen wait eagerly on earth for the results. Wolf was scheduled to participate in nearly 35 scientific experiments, as well as assist with routine station maintenance.

How do scientists determine Mir's location as it orbits earth? How do they communicate with deep-space probes? How do telescopes gather light emitted from objects that are light years away? The answers to all these questions involve mathematics, including the mathematics of conic sections, which we study in this chapter. For it is the curves found on cones that scientists use to model orbits of planets, comets, and satellites and to design telescopes and antennae that receive light and radio signals, traveling through space from distant galaxies.

7 The Conic Sections

7.1 The Parabola

7.2 The Ellipse and the Circle

7.3 The Hyperbola

7.1 The Parabola

The parabola, circle, ellipse, and hyperbola can be defined as the four curves that are obtained by intersecting a right circular cone and a plane as shown in Fig. 7.1. That is why these curves are known as **conic sections**. If the plane passes through the vertex of the cone, then the intersection of the cone and the plane is called a **degenerate conic**. The conic sections can also be defined as the graphs of certain equations (as we did for the parabola in Section 3.2). However, the useful properties of the curves are not apparent in either of these approaches. In this chapter we give geometric definitions from which we derive equations for the conic sections. This approach allows us to better understand the properties of the conic sections.

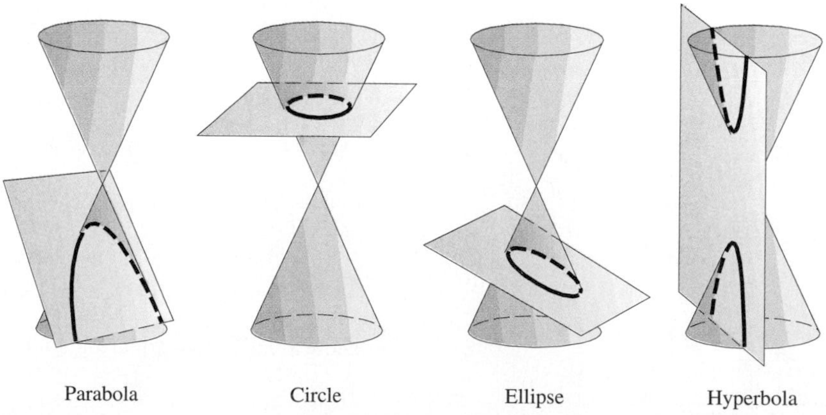

Figure 7.1

Definition

Previously in Section 3.2, we defined a parabola algebraically as the graph of $y = ax^2 + bx + c$ for $a \neq 0$. This equation can be expressed also in the form $y = a(x - h)^2 + k$. No equation is mentioned in our geometric definition of a parabola.

Definition: Parabola

A **parabola** is the set of all points in the plane that are equidistant from a fixed line (the **directrix**) and a fixed point not on the line (the **focus**).

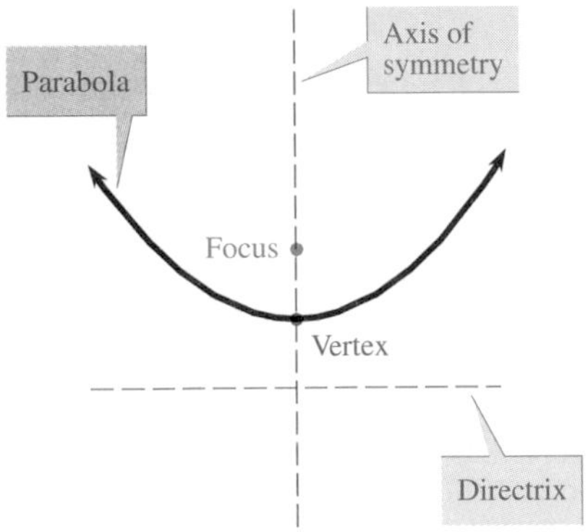

Figure 7.2

Figure 7.2 shows a parabola with its directrix, focus, axis of symmetry, and vertex. In terms of the directrix and focus, the **axis of symmetry** can be described as the line perpendicular to the directrix and containing the focus. The **vertex** is the point on the axis of symmetry that is equidistant from the focus and directrix. If we position the directrix and focus in a coordinate plane with the directrix horizontal, we can find an equation that is satisfied by all points of the parabola.

Developing the Equation

Start with a focus and a horizontal directrix as shown in Fig. 7.3. If we use the coordinates (h, k) for the vertex, then the focus is $(h, k + p)$ and the directrix is $y = k - p$, where p (the **focal length**) is the directed distance from the vertex to the focus. If the focus is above the vertex, then $p > 0$, and if the focus is below the vertex, then $p < 0$. The distance from the vertex to the focus or the vertex to the directrix is $|p|$.

The distance d_1 from an arbitrary point (x, y) on the parabola to the directrix is the distance from (x, y) to $(x, k - p)$, as shown in Fig. 7.3. We use the distance formula from Section 2.1 to find d_1:

$$d_1 = \sqrt{(x - x)^2 + (y - (k - p))^2} = \sqrt{y^2 - 2(k - p)y + (k - p)^2}$$

Now we find the distance d_2 between (x, y) and the focus $(h, k + p)$:

$$d_2 = \sqrt{(x - h)^2 + (y - (k + p))^2} = \sqrt{(x - h)^2 + y^2 - 2(k + p)y + (k + p)^2}$$

Since $d_1 = d_2$ for every point (x, y) on the parabola, we have the following equation.

$$\sqrt{y^2 - 2(k - p)y + (k - p)^2} = \sqrt{(x - h)^2 + y^2 - 2(k + p)y + (k + p)^2}$$

You should verify that squaring each side and simplifying yields

$$y = \frac{1}{4p}(x - h)^2 + k.$$

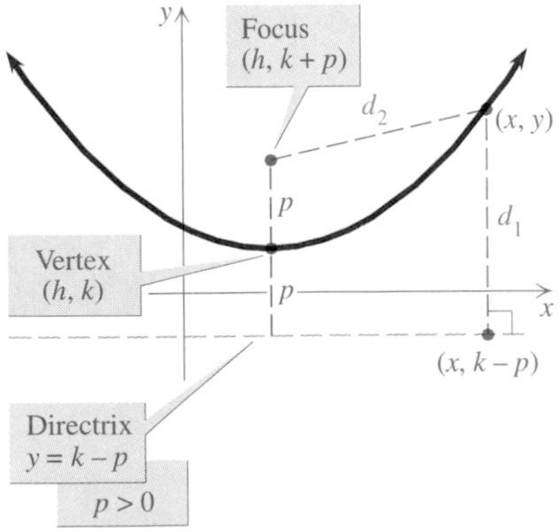

Figure 7.3

This equation is of the form $y = a(x - h)^2 + k$, where $a = 1/(4p)$. So the curve determined by the geometric definition has an equation that is an equation of a parabola according to the algebraic definition. We state these results as follows.

Theorem: The Equation of a Parabola

The equation of a parabola with focus $(h, k + p)$ and directrix $y = k - p$ is

$$y = a(x - h)^2 + k,$$

where $a = 1/(4p)$ and (h, k) is the vertex.

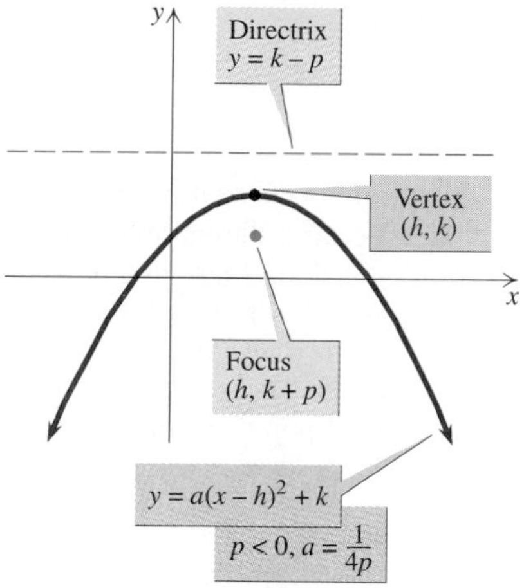

Figure 7.4

The link between the geometric definition and the equation of a parabola is

$$a = \frac{1}{4p}.$$

For any particular parabola, a and p have the same sign. If they are both positive, the parabola opens upward and the focus is above the directrix. If they are both negative, the parabola opens downward and the focus is below the directrix. Figure 7.4 shows the positions of the focus, directrix, and vertex for a parabola with $p < 0$. Since a is inversely proportional to p, smaller values of $|p|$ correspond to larger values of $|a|$ and to "narrower" parabolas.

EXAMPLE 1 Writing the equation from the focus and directrix

Find the equation of the parabola with focus $(-1, 3)$ and directrix $y = 2$.

Solution

The focus is one unit above the directrix as shown in Fig. 7.5. So $p = 1/2$. Therefore $a = 1/(4p) = 1/2$. Since the vertex is halfway between the focus and directrix, the y-coordinate of the vertex is $(3 + 2)/2$ and the vertex is $(-1, 5/2)$. Use $a = 1/2$, $h = -1$, and $k = 5/2$ in the formula $y = a(x - h)^2 + k$ to get the equation

$$y = \frac{1}{2}(x - (-1))^2 + \frac{5}{2}.$$

Simplify to get the equation $y = \frac{1}{2}x^2 + x + 3$.

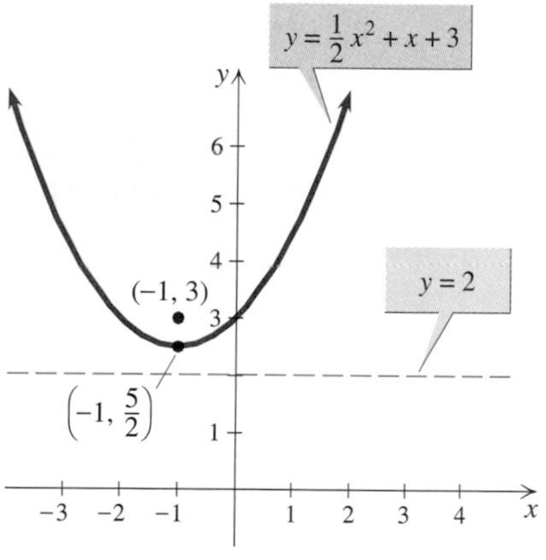

Figure 7.5

The Standard Equation of a Parabola

If we start with the standard equation of a parabola, $y = ax^2 + bx + c$, we can identify the vertex, focus, and directrix by rewriting it in the form $y = a(x - h)^2 + k$.

EXAMPLE 2 Finding the vertex, focus, and directrix

Find the vertex, focus, and directrix of the graph of $y = -3x^2 - 6x + 2$.

Solution

Use completing the square to write the equation in the form $y = a(x - h)^2 + k$:

$$\begin{aligned} y &= -3(x^2 + 2x) + 2 \\ &= -3(x^2 + 2x + 1 - 1) + 2 \\ &= -3(x^2 + 2x + 1) + 2 + 3 \\ &= -3(x + 1)^2 + 5 \end{aligned}$$

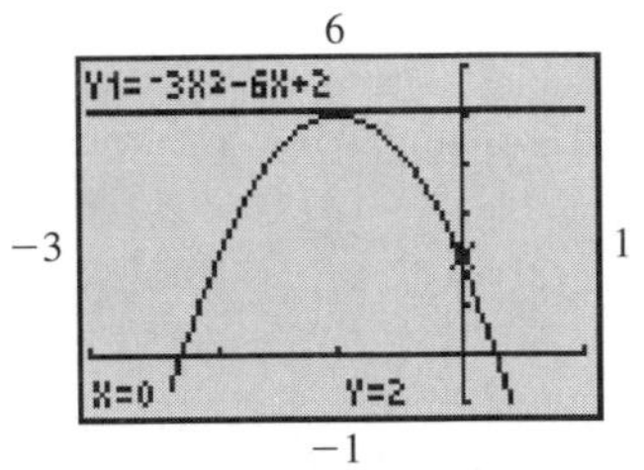

Figure 7.6

The vertex is $(-1, 5)$, and the parabola opens downward because $a = -3$. Since $a = 1/(4p)$, we have $1/(4p) = -3$, or $p = -1/12$. Because the parabola opens downward, the focus is 1/12 unit below the vertex $(-1, 5)$ and the directrix is a horizontal line 1/12 unit above the vertex. The focus is $(-1, 59/12)$, and the directrix is $y = 61/12$.

The graphs of $y_1 = -3x^2 - 6x + 2$ and $y_2 = 61/12$ in Fig. 7.6 show how close the directrix is to the vertex for this parabola.

In Section 3.2 we learned that the x-coordinate of the vertex of the parabola $y = ax^2 + bx + c$ is $-b/(2a)$. We can use $x = -b/(2a)$ and $a = 1/(4p)$ to determine the focus and directrix without completing the square.

EXAMPLE 3 Finding the vertex, focus, and directrix

Find the vertex, focus, and directrix of the parabola $y = 2x^2 + 6x - 7$ without completing the square, and determine whether the parabola opens upward or downward.

Solution

First use $x = -b/(2a)$ to find the x-coordinate of the vertex:

$$x = \frac{-b}{2a} = \frac{-6}{2 \cdot 2} = -\frac{3}{2}$$

To find the y-coordinate of the vertex, let $x = -3/2$ in $y = 2x^2 + 6x - 7$:

$$y = 2\left(-\frac{3}{2}\right)^2 + 6\left(-\frac{3}{2}\right) - 7 = \frac{9}{2} - 9 - 7 = -\frac{23}{2}$$

The vertex is $(-3/2, -23/2)$. Use $2 = 1/(4p)$ to get $p = 1/8$. Since the parabola opens upward, the directrix is 1/8 unit below the vertex and the focus is 1/8 unit above the vertex. The directrix is $y = -93/8$, and the focus is $(-3/2, -91/8)$.

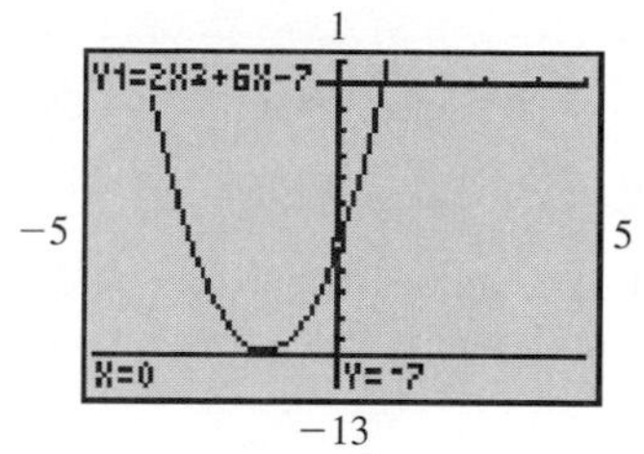

Figure 7.7

You can graph $y_1 = -93/8$ and $y_2 = 2x^2 + 6x - 7$ as in Fig. 7.7 to check the position of the parabola and its directrix.

Graphing a Parabola

According to the geometric definition of a parabola, every point on the parabola is equidistant from its focus and directrix. However, it is not easy to find points satisfying that condition. The German mathematician Johannes Kepler (1571–1630) devised a method for drawing a parabola with a given focus and directrix: A piece of string, with length equal to the length of the T-square, is attached to the end of the T-square and the focus, as shown in Fig. 7.8. A pencil is moved down the edge of the T-square, holding the string against it, while the T-square is moved along the directrix toward the focus. While the pencil is aligning the string with the edge of the T-square, it remains equidistant from the focus and directrix.

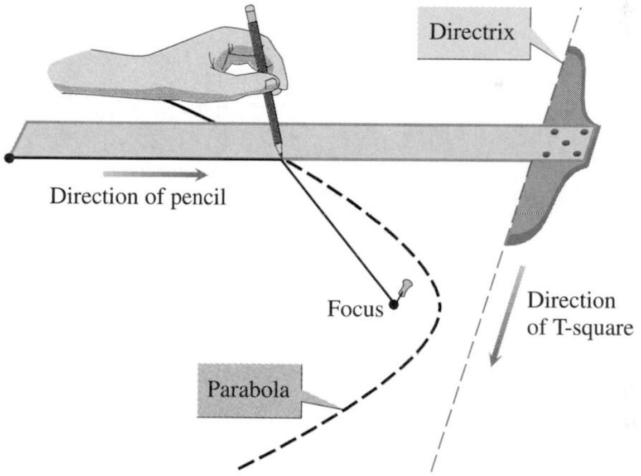

Figure 7.8

Although Kepler's method does give the graph of the parabola from the focus and directrix, you may not want to use it too often. Next, we find the equation of a parabola with a given focus and directrix and then graph it as in Section 3.2.

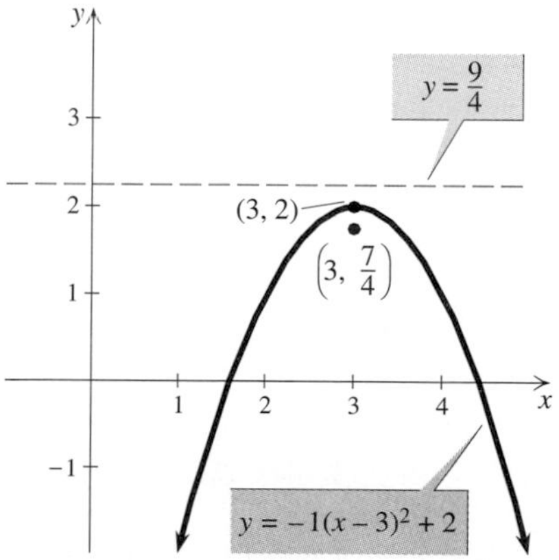

Figure 7.9

EXAMPLE 4 Graphing a parabola given its focus and directrix

Find the vertex, axis of symmetry, x-intercepts, and y-intercept of the parabola that has focus $(3, 7/4)$ and directrix $y = 9/4$. Sketch the graph, showing the focus and directrix.

Solution

First draw the focus and directrix on the graph as shown in Fig. 7.9. Since the vertex is midway between the focus and directrix, the vertex is $(3, 2)$. The distance between the focus and vertex is $1/4$. Since the directrix is above the focus, the parabola opens downward, and we have $p = -1/4$. Use $a = 1/(4p)$ to get $a = -1$. Use $a = -1$ and the vertex (3, 2) in the equation $y = a(x - h)^2 + k$ to get

$$y = -1(x - 3)^2 + 2.$$

The axis of symmetry is the vertical line $x = 3$. If $x = 0$, then $y = -1(0 - 3)^2 + 2 = -7$. So the y-intercept is $(0, -7)$. Find the x-intercepts by setting y equal to 0 in the equation:

$$\begin{aligned} -1(x - 3)^2 + 2 &= 0 \\ (x - 3)^2 &= 2 \\ x - 3 &= \pm\sqrt{2} \\ x &= 3 \pm \sqrt{2} \end{aligned}$$

The x-intercepts are $(3 - \sqrt{2}, 0)$ and $(3 + \sqrt{2}, 0)$. Two additional points that satisfy $y = -1(x - 3)^2 + 2$ are $(1, -2)$ and $(4, 1)$. Using all of this information, we get the graph shown in Fig. 7.9.

Parabolas Opening to the Left or Right

If we interchange x and y in $y = a(x - h)^2 + k$, we get

$$x = a(y - h)^2 + k.$$

The graph of this equation is also a parabola. By interchanging the variables, the roles of the x- and y-axes are interchanged. For example, the graph of $y = 2x^2$ opens upward, while the graph of $x = 2y^2$ opens to the right. For parabolas opening right or left, the directrix is a vertical line. If the focus is to the right of the directrix, then the parabola opens to the right, and if the focus is to the left of the directrix, then the parabola opens to the left. Figure 7.10 shows the relative

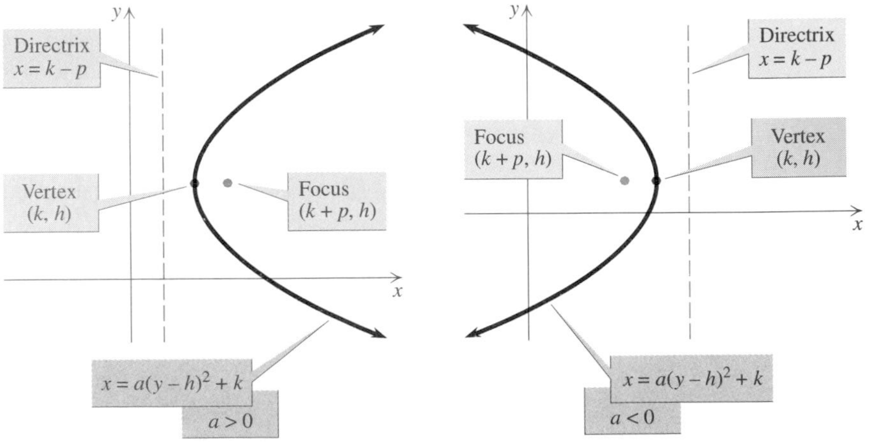

Figure 7.10

locations of the vertex, focus, and directrix for parabolas of the form $x = a(y - h)^2 + k$.

The equation $x = a(y - h)^2 + k$ can be written as $x = ay^2 + by + c$. So the graph of $x = ay^2 + by + c$ is also a parabola opening to the right for $a > 0$ and to the left for $a < 0$. Because the roles of x and y are interchanged, $-b/(2a)$ is now the y-coordinate of the vertex, and the axis of symmetry is the horizontal line $y = -b/(2a)$.

EXAMPLE 5 **Graphing a parabola with a vertical directrix**

Find the vertex, axis of symmetry, y-intercepts, focus, and directrix for the parabola $x = y^2 - 2y$. Find several other points on the parabola and sketch the graph.

Solution

Because $a = 1$, the parabola opens to the right. The y-coordinate of the vertex is

$$y = \frac{-b}{2a} = \frac{-(-2)}{2(1)} = 1.$$

If $y = 1$, then $x = (1)^2 - 2(1) = -1$ and the vertex is $(-1, 1)$. The axis of symmetry is the horizontal line $y = 1$. To find the y-intercepts, we solve $y^2 - 2y = 0$ by factoring:

$$y(y - 2) = 0$$

$$y = 0 \quad \text{or} \quad y = 2$$

The y-intercepts are $(0, 0)$ and $(0, 2)$. Using all of this information and the additional points $(3, 3)$ and $(3, -1)$, we get the graph shown in Fig. 7.11. Because $a = 1$ and $a = 1/(4p)$, we have $p = 1/4$. So the focus is 1/4 unit to the right of the vertex at $(-3/4, 1)$. The directrix is the vertical line 1/4 unit to the left of the vertex, $x = -5/4$.

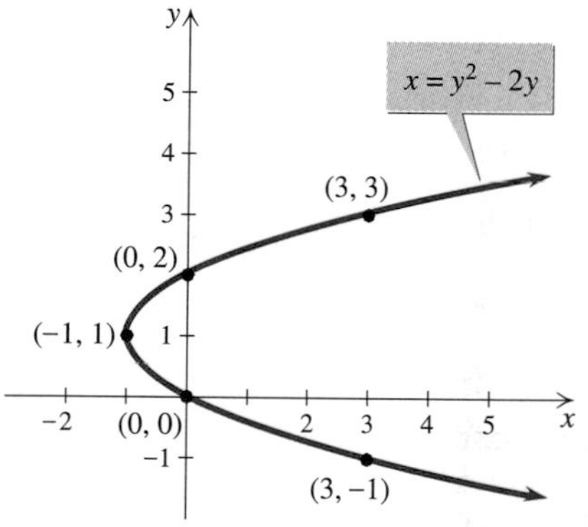

Figure 7.11

Applications

In Section 3.2 we saw an important application of a parabola. Because of the shape of a parabola, a quadratic function has a maximum value or a minimum value at the vertex of the parabola. However, parabolas are important for another totally different reason. When a ray of light, traveling parallel to the axis of symmetry, hits a parabolic reflector, it is reflected toward the focus of the parabola. See Fig. 7.12. This property is used in telescopes to magnify the light from distant stars. For spotlights, in which the light source is at the focus, the reflecting property is used in reverse. Light originating at the focus is reflected off the parabolic reflector and projected outward in a narrow beam. The reflecting property is used also in telephoto camera lenses, radio antennas, satellite dishes, eavesdropping devices, and flashlights.

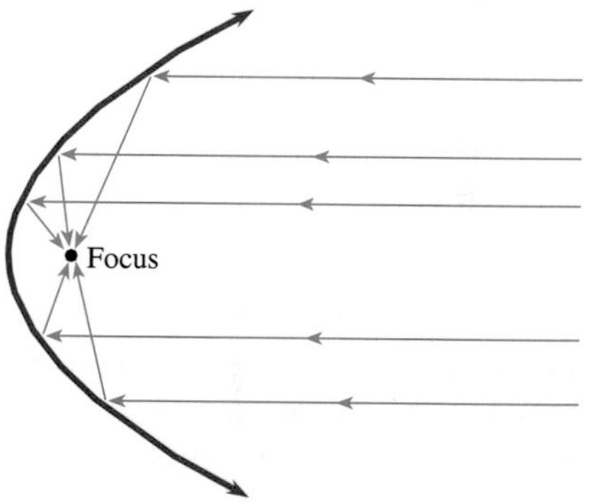

Figure 7.12

EXAMPLE 6 **The Hubble telescope**

The Hubble space telescope uses a glass mirror with a parabolic cross section and a diameter of 2.4 meters, as shown in Fig. 7.13. If the focus of the parabola is 57.6 meters from the vertex, then what is the equation of the parabola used for the mirror? How much thicker is the mirror at the edge than at its center?

Solution

If we position the parabola with the vertex at the origin and opening upward as shown in Fig. 7.13, then its equation is of the form $y = ax^2$. Since the distance between the focus and vertex is 57.6 meters, $p = 57.6$. Using $a = 1/(4p)$, we get $a = 1/(4 \cdot 57.6) \approx 0.004340$. The equation of the parabola is

$$y = 0.004340x^2.$$

To find the difference in thickness at the edge, let $x = 1.2$ in the equation of the parabola:

$$y = 0.004340(1.2)^2 \approx 0.006250 \text{ meter}$$

The mirror is 0.006250 meter thicker at the edge than at the center.

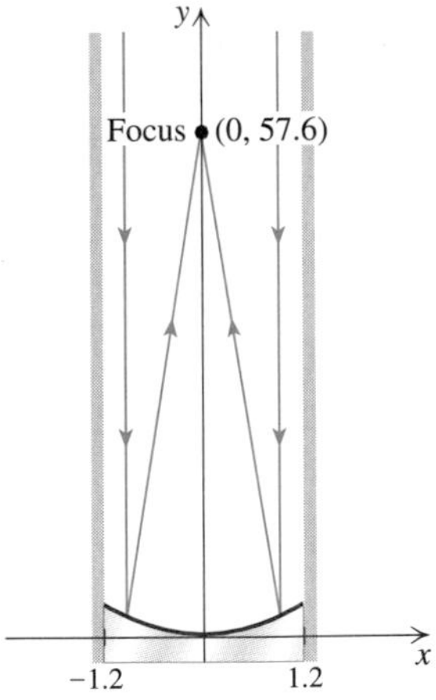

Figure 7.13

For a telescope to work properly, the glass mirror must be ground with great precision. Prior to the 1993 repair, the Hubble space telescope did not work as well as hoped (*Scientific American*, June 1992), because the mirror was actually ground to be only 0.006248 meter thicker at the outside edge, two millionths of a meter smaller than it should have been!

FOR THOUGHT True or False? Explain.

1. A parabola with focus (0, 0) and directrix $y = -1$ has vertex (0, 1).
2. A parabola with focus (3, 0) and directrix $x = -1$ opens to the right.
3. A parabola with focus (4, 5) and directrix $x = 1$ has vertex (5/2, 5).
4. For $y = x^2$, the focus is (0, −1/4).
5. For $x = y^2$, the focus is (1/4, 0).
6. A parabola with focus (2, 3) and directrix $y = -5$ has no x-intercepts.
7. The parabola $y = 4x^2 - 9x$ has no y-intercept.
8. The vertex of the parabola $x = 2(y + 3)^2 - 5$ is (−5, −3).
9. The parabola $y = (x - 5)^2 + 4$ has its focus at (5, 4).
10. The parabola $x = -3y^2 + 7y - 5$ opens downward.

7.1 EXERCISES

Tape 20 Disk

Find the equation of the parabola with the given focus and directrix.

1.

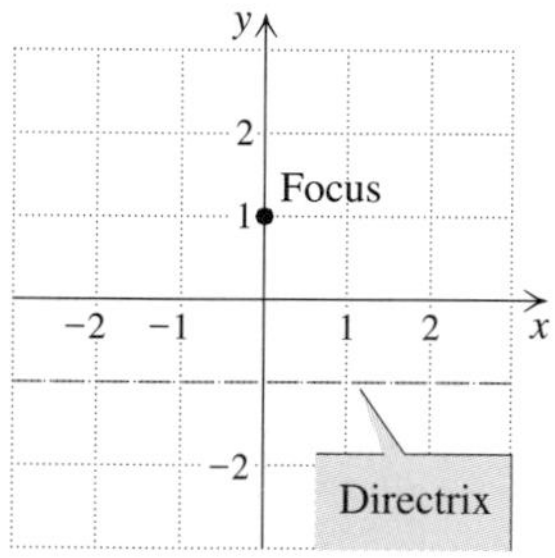

2.

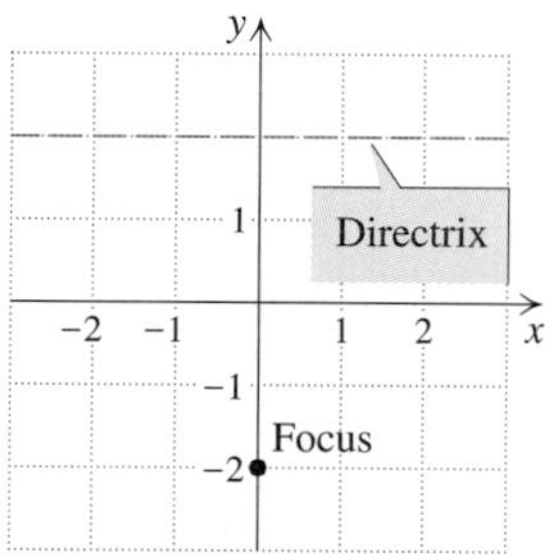

3.

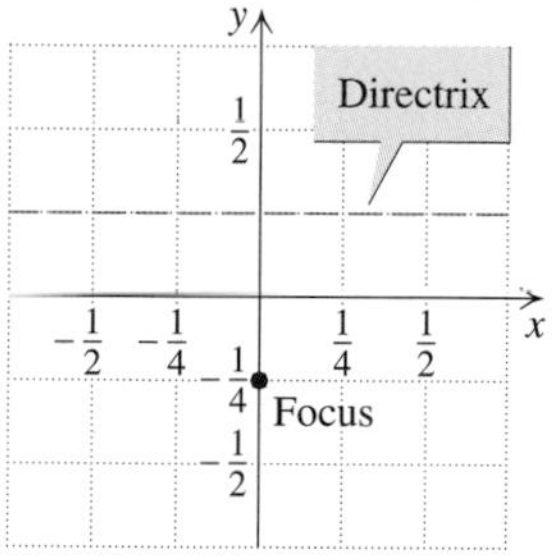

4.

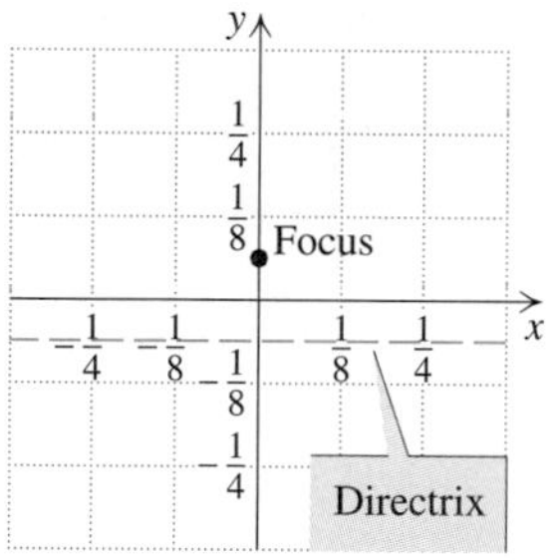

Find the equation of the parabola with the given focus and directrix.

5. Focus (3, 5), directrix $y = 2$
6. Focus (−1, 5), directrix $y = 3$
7. Focus (1, −3), directrix $y = 2$
8. Focus (1, −4), directrix $y = 0$
9. Focus (−2, 1.2), directrix $y = 0.8$
10. Focus (3, 9/8), directrix $y = 7/8$

Use completing the square to write each equation in the form $y = a(x - h)^2 + k$. Identify the vertex, focus, and directrix.

11. $y = x^2 - 8x + 3$
12. $y = x^2 + 2x - 5$
13. $y = 2x^2 + 12x + 5$
14. $y = 3x^2 + 12x + 1$
15. $y = -2x^2 + 6x + 1$
16. $y = -3x^2 - 6x + 5$
17. $y = 5x^2 + 30x$
18. $y = -2x^2 + 12x$
19. $y = \frac{1}{8}x^2 - \frac{1}{2}x + \frac{9}{2}$
20. $y = \frac{1}{4}x^2 + \frac{1}{2}x - \frac{7}{4}$

Find the vertex, focus, and directrix of each parabola without completing the square, and determine whether the parabola opens upward or downward.

21. $y = x^2 - 4x + 3$
22. $y = x^2 - 6x - 7$
23. $y = -x^2 + 2x - 5$
24. $y = -x^2 + 4x + 3$
25. $y = 3x^2 - 6x + 1$
26. $y = 2x^2 + 4x - 1$
27. $y = -\frac{1}{2}x^2 - 3x + 2$
28. $y = -\frac{1}{2}x^2 + 3x - 1$
29. $y = \frac{1}{4}x^2 + 5$
30. $y = -\frac{1}{8}x^2 - 6$

Find the vertex, axis of symmetry, x-intercepts, and y-intercept of the parabola that has the given focus and directrix. Sketch the graph, showing the focus and directrix.

31. Focus (1/2, −2), directrix $y = -5/2$

32. Focus $(1, -35/4)$, directrix $y = -37/4$

33. Focus $(-1/2, 6)$, directrix $y = 13/2$

34. Focus $(-1, 35/4)$, directrix $y = 37/4$

Find the vertex, axis of symmetry, x-intercepts, y-intercept, focus, and directrix for each parabola. Sketch the graph, showing the focus and directrix.

35. $y = \frac{1}{2}(x + 2)^2 + 2$

36. $y = \frac{1}{2}(x - 4)^2 + 1$

37. $y = -\frac{1}{4}(x + 4)^2 + 2$

38. $y = -\frac{1}{4}(x - 2)^2 + 4$

39. $y = \frac{1}{2}x^2 - 2$

40. $y = -\frac{1}{4}x^2 + 4$

41. $y = x^2 - 4x + 4$

42. $y = (x - 4)^2$

43. $y = \frac{1}{3}x^2 - x$

44. $y = \frac{1}{5}x^2 + x$

Find the vertex, axis of symmetry, x-intercept, y-intercepts, focus, and directrix for each parabola. Sketch the graph, showing the focus and directrix.

45. $x = -y^2$

46. $x = y^2 - 2$

47. $x = -\frac{1}{4}y^2 + 1$

48. $x = \frac{1}{2}(y - 1)^2$

49. $x = y^2 + y - 6$

50. $x = y^2 + y - 2$

51. $x = -\frac{1}{2}y^2 - y - 4$

52. $x = -\frac{1}{2}y^2 + 3y + 4$

53. $x = 2(y - 1)^2 + 3$

54. $x = 3(y + 1)^2 - 2$

55. $x = -\frac{1}{2}(y + 2)^2 + 1$

56. $x = -\frac{1}{4}(y - 2)^2 - 1$

Find the equation of the parabola determined by the given information.

57. Focus $(1, 5)$, vertex $(1, 4)$

58. Directrix $y = 5$, vertex $(2, 3)$

59. Vertex $(0, 0)$, directrix $x = -2$

60. Focus $(-2, 3)$, vertex $(-9/4, 3)$

Solve each problem.

61. *The Hale Telescope* The focus of the Hale telescope on Palomar Mountain in California is 55 ft above the mirror (at the vertex). The Pyrex glass mirror is 200 in. in diameter and 23 in. thick at the center. Find the equation for the parabola that was used to shape the glass. How thick is the glass on the outside edge?

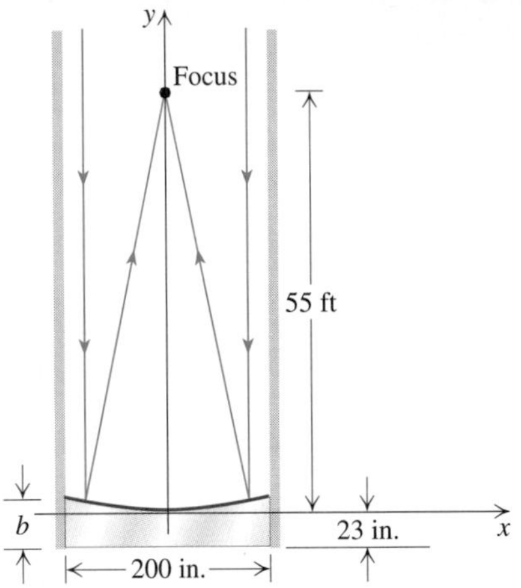

Figure for Exercise 61

62. *Eavesdropping* From the Edmund Scientific catalog you can buy a device that will ''pull in voices up to three-quarters of a mile away with our electronic parabolic microphone.'' The 18.75-in.-diameter plastic shield reflects sound waves to a microphone located at the focus. Given that the microphone is located 6 in. from the vertex of the parabolic shield, find the equation for a cross section of the shield. What is the depth of the shield?

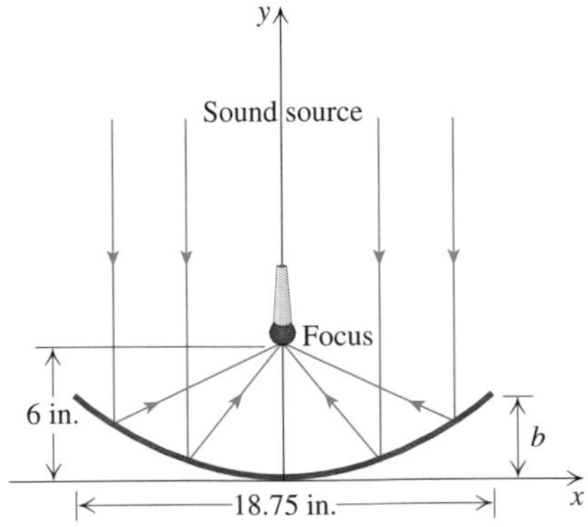

Figure for Exercise 62

Use a graphing calculator to solve each problem.

63. Graph $y = x^2$ using the viewing window with $-1 \leq x \leq 1$ and $0 \leq y \leq 1$. Graph $y = 2x^2 - 4x + 5$ using the viewing window with $-1 \leq x \leq 3$ and $3 \leq y \leq 11$. What can you say about the two graphs?

64. Find two different viewing windows in which the graph of $y = 3x^2 + 30x + 71$ looks just like the graph of $y = x^2$ in the viewing window with $-1 \leq x \leq 1$ and $0 \leq y \leq 1$.

65. The graph of $x = -y^2$ is a parabola opening to the left with vertex at the origin. Find two functions whose graphs will together form this parabola and graph them on your calculator.

66. You can illustrate the reflective property of the parabola $x = -y^2$ on the screen of your calculator. First graph $x = -y^2$ as in the previous exercise. Next, by using the graphs of two line segments, make it appear that a particle coming in horizontally from the left toward $(-1, 1)$ is reflected off the parabola and heads toward the focus $(-1/4, 0)$. Consult your manual to see how to graph line segments.

For Writing/Discussion

67. *Derive the Equation* In this section we derived the equation of a parabola with a given focus and a horizontal directrix. Use the same technique to derive the equation of a parabola with a given focus and a vertical directrix.

68. *Cooperative Learning* Working in groups, write a summary of everything that we have learned about parabolas. Be sure to include parabolas opening up and down and parabolas opening left and right.

Linking Concepts

For Individual or Group Explorations

• CALCULUS •

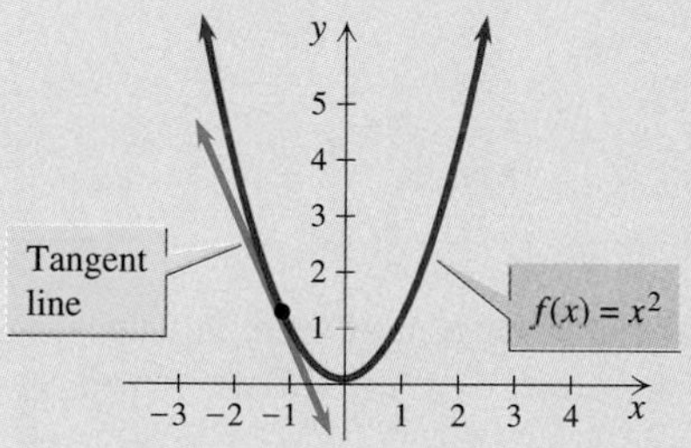

Tangent Lines A line tangent to a parabola must intersect the parabola at exactly one point as shown in the accompanying figure. Of course it is impossible to draw a graph that really looks like there is one point of intersection. However, if we find a tangent line algebraically then we can be sure that it intersects only once.

a) Consider $y = x^2$ and the point $(3, 9)$. If $y - 9 = m(x - 3)$ is tangent to the parabola at $(3, 9)$ then the system

$$y = x^2$$
$$y - 9 = m(x - 3)$$

must have exactly one solution. For what value of m does the system have exactly one solution?

b) Graph the parabola and the tangent line found in part (a) on a graphing calculator and then use the intersect feature of the calculator to find the point of intersection.

c) There is another line that intersects $y = x^2$ at $(3, 9)$ and only at $(3, 9)$. What is it? Why did it not appear in part (a)?

d) Use the same reasoning as used in part (a) to find the slope of the tangent line to $y = x^2$ at (x_1, y_1).

e) Use the result of part (d) to find the equation of the tangent line to $y = x^2$ at $(-2.5, 6.25)$.

7.2 The Ellipse and the Circle

Ellipses and circles can be obtained by intersecting planes and cones as shown in Fig. 7.1. Here we will develop general equations for ellipses, and then circles, by starting with geometric definitions as we did with the parabola.

Definition of Ellipse

An easy way to draw an ellipse is illustrated in Fig. 7.14. A string is attached at two fixed points, and a pencil is used to take up the slack. As the pencil is moved around the paper, the sum of the distances of the pencil point from the two fixed points remains constant, because the length of the string is constant. This method of drawing an ellipse illustrates the geometric definition for an ellipse.

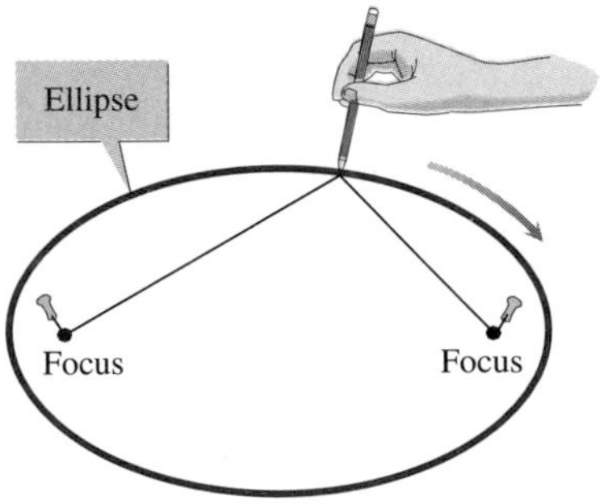

Figure 7.14

Definition: Ellipse

An **ellipse** is the set of points in a plane such that the sum of their distances from two fixed points is a constant. Each fixed point is called a **focus** (plural: foci) of the ellipse.

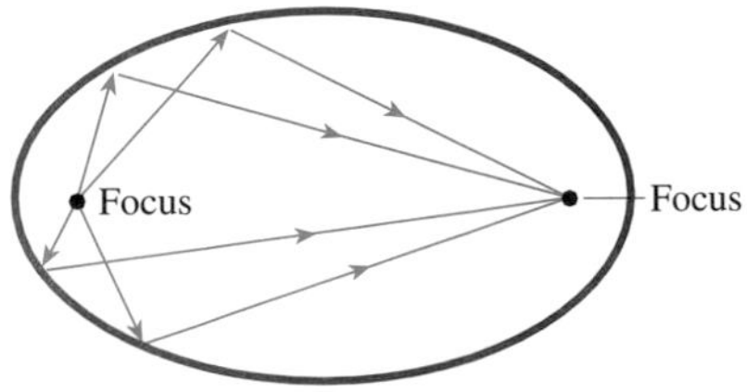

Figure 7.15

The ellipse, like the parabola, has interesting reflecting properties. All light or sound waves emitted from one focus are reflected off the ellipse to concentrate at the other focus as shown in Fig. 7.15. This property is used in light fixtures such as a dentist's light, for which a concentration of light at a point is desired, and in a whispering gallery like the one in the U.S. Capitol Building. In a whispering gallery, a whisper emitted at one focus is reflected off the elliptical ceiling and is amplified so that it can be heard at the other focus, but not anywhere in between.

The orbits of the planets around the sun and satellites around the earth are elliptical. For the orbit of the earth, the sun is at one focus of the elliptical path.

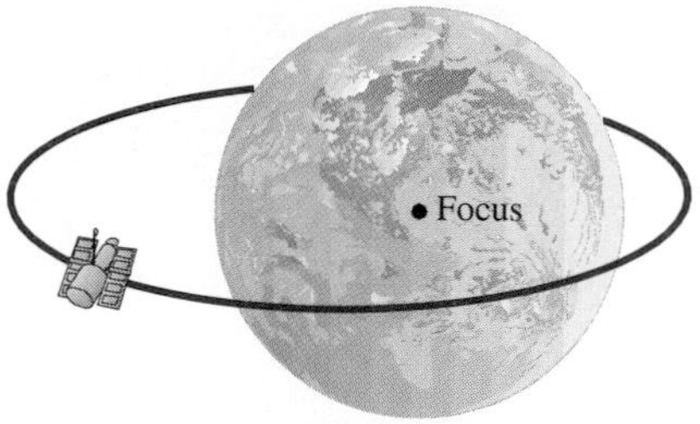

Figure 7.16

For the orbit of a satellite such as the Hubble space telescope, the center of the earth is one focus. See Fig. 7.16.

The Equation of the Ellipse

The ellipse shown in Fig. 7.17 has foci at $(c, 0)$ and $(-c, 0)$, and y-intercepts $(0, b)$ and $(0, -b)$, where $c > 0$ and $b > 0$. The line segment $\overline{V_1V_2}$ is the **major axis**, and the line segment $\overline{B_1B_2}$ is the **minor axis**. For any ellipse, the major axis is longer than the minor axis and the foci are on the major axis. The **center** of an ellipse is the midpoint of the major (or minor) axis. The ellipse in Fig. 7.17 is centered at the origin. The **vertices** of an ellipse are the endpoints of the major axis. The vertices of the ellipse in Fig. 7.17 are the x-intercepts.

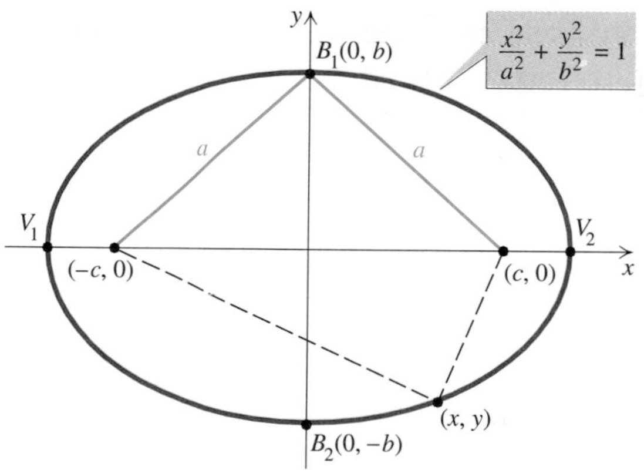

Figure 7.17

Let a be the distance between $(c, 0)$ and the y-intercept $(0, b)$ as shown in Fig. 7.17. The sum of the distances from the two foci to $(0, b)$ is $2a$. So for any point (x, y) on the ellipse, the distance from (x, y) to $(c, 0)$ plus the distance from (x, y) to $(-c, 0)$ is equal to $2a$. Writing this last statement as an equation (using the distance formula) gives the equation of the ellipse shown in Fig. 7.17:

$$\sqrt{(x - c)^2 + (y - 0)^2} + \sqrt{(x - (-c))^2 + (y - 0)^2} = 2a$$

With some effort, this equation can be greatly simplified. We will provide the major steps in simplifying it, and leave the details as an exercise. First simplify inside the radicals and isolate them to get

$$\sqrt{x^2 - 2xc + c^2 + y^2} = 2a - \sqrt{x^2 + 2xc + c^2 + y^2}.$$

Next, square each side and simplify again to get

$$a\sqrt{x^2 + 2xc + c^2 + y^2} = a^2 + xc.$$

Squaring each side again yields

$$a^2x^2 - c^2x^2 + a^2y^2 = a^4 - a^2c^2$$

$$(a^2 - c^2)x^2 + a^2y^2 = a^2(a^2 - c^2) \qquad \text{Factor.}$$

Since $a^2 = b^2 + c^2$, or $a^2 - c^2 = b^2$ (from Fig. 7.17), replace $a^2 - c^2$ by b^2:

$$b^2x^2 + a^2y^2 = a^2b^2$$

$$\frac{x^2}{a^2} + \frac{y^2}{b^2} = 1 \qquad \text{Divide each side by } a^2b^2.$$

We have proved the following theorem.

Theorem: Equation of an Ellipse with Center (0, 0) and Horizontal Major Axis

The equation of an ellipse centered at the origin with foci $(c, 0)$ and $(-c, 0)$ and y intercepts $(0, b)$ and $(0, -b)$ is

$$\frac{x^2}{a^2} + \frac{y^2}{b^2} = 1, \quad \text{where } a^2 = b^2 + c^2.$$

If we start with the foci on the y-axis and x-intercepts $(b, 0)$ and $(-b, 0)$, then we can develop a similar equation, which is stated in the following theorem. If the foci are on the y-axis, then the y-intercepts are the vertices and the major axis is vertical.

Theorem: Equation of an Ellipse with Center (0, 0) and Vertical Major Axis

The equation of an ellipse centered at the origin with foci $(0, c)$ and $(0, -c)$ and x-intercepts $(b, 0)$ and $(-b, 0)$ is

$$\frac{x^2}{b^2} + \frac{y^2}{a^2} = 1, \quad \text{where } a^2 = b^2 + c^2.$$

Consider the ellipse in Fig. 7.18 with foci $(c, 0)$ and $(-c, 0)$ and equation

$$\frac{x^2}{a^2} + \frac{y^2}{b^2} = 1,$$

where $a > b > 0$. If $y = 0$ in this equation, then $x = \pm a$. So the vertices (or x-intercepts) of the ellipse are $(a, 0)$ and $(-a, 0)$, and a is the distance from the center to a vertex. The distance from a focus to an endpoint of the minor axis is a also. *So in any ellipse the distance from the focus to an endpoint of the minor axis is the same as the distance from the center to a vertex.* The graphs in Fig. 7.18 will help you remember the relationship between a, b, and c.

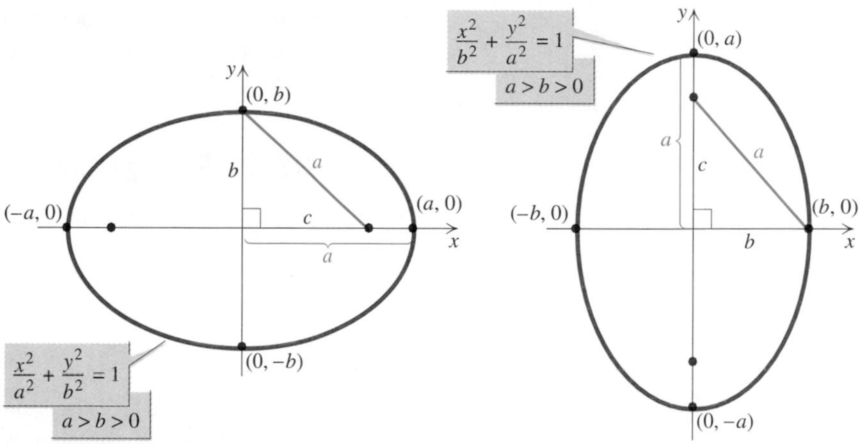

Figure 7.18

EXAMPLE 1 Writing the equation of an ellipse

Sketch an ellipse with foci at (0, 4) and (0, −4) and vertices (0, 6) and (0, −6), and find the equation for this ellipse.

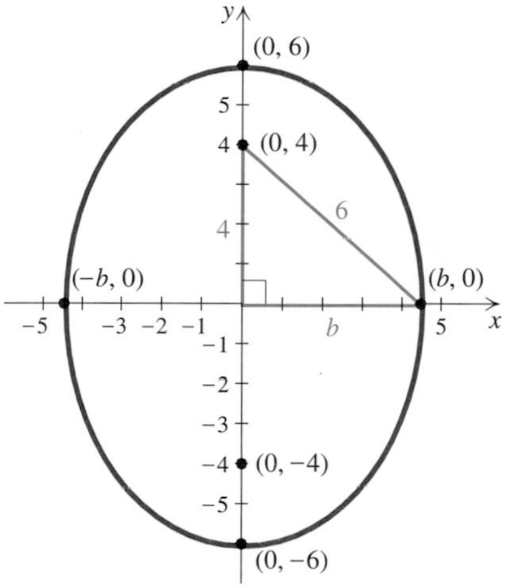

Figure 7.19

Solution

Since the vertices are on the y-axis, the major axis is vertical and the ellipse is elongated in the direction of the y-axis. A sketch of the ellipse appears in Fig. 7.19. Because a is the distance from the center to a vertex, $a = 6$. Because c is the distance from the center to a focus, $c = 4$. To write the equation of the ellipse, we need the value of b^2. Use $a = 6$ and $c = 4$ in $a^2 = b^2 + c^2$ to get

$$6^2 = b^2 + 4^2$$

$$b^2 = 20$$

So the x-intercepts are $(\sqrt{20}, 0)$ and $(-\sqrt{20}, 0)$, and the equation of the ellipse is

$$\frac{x^2}{20} + \frac{y^2}{36} = 1.$$

Graphing an Ellipse Centered at the Origin

For $a > b > 0$, the graph of

$$\frac{x^2}{a^2} + \frac{y^2}{b^2} = 1$$

is an ellipse centered at the origin with a horizontal major axis, x-intercepts $(a, 0)$ and $(-a, 0)$, and y-intercepts $(0, b)$ and $(0, -b)$, as shown in Fig. 7.18. The foci are on the major axis and are determined by $a^2 = b^2 + c^2$ or $c^2 = a^2 -$

b^2. Remember that when the denominator for x^2 is larger than the denominator for y^2, the major axis is horizontal. *To sketch the graph of an ellipse centered at the origin, simply locate the four intercepts and draw an ellipse through them.*

EXAMPLE 2 **Graphing an ellipse with foci on the x-axis**

Sketch the graph and identify the foci of the ellipse

$$\frac{x^2}{9} + \frac{y^2}{4} = 1.$$

Solution

To sketch the ellipse, we find the x-intercepts and the y-intercepts. If $x = 0$, then $y^2 = 4$ or $y = \pm 2$. So the y-intercepts are $(0, 2)$ and $(0, -2)$. If $y = 0$, then $x = \pm 3$. So the x-intercepts are $(3, 0)$ and $(-3, 0)$. To make a rough sketch of an ellipse, plot only the intercepts and draw an ellipse through them, as shown in Fig. 7.20. Since this ellipse is elongated in the direction of the x-axis, the foci are on the x-axis. Use $a = 3$ and $b = 2$ in $c^2 = a^2 - b^2$, to get $c^2 = 9 - 4 = 5$. So $c = \pm\sqrt{5}$, and the foci are $(\sqrt{5}, 0)$ and $(-\sqrt{5}, 0)$.

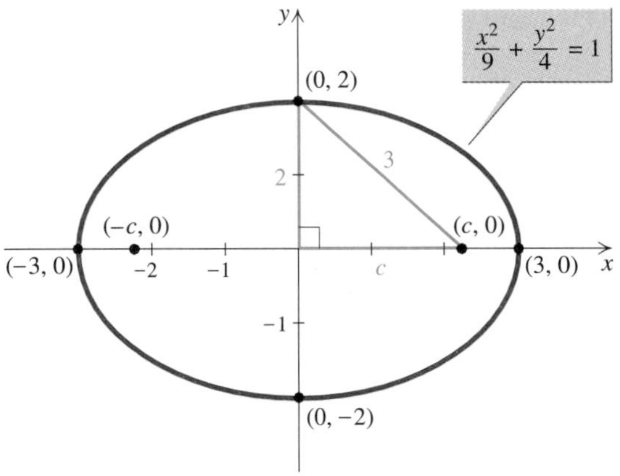

Figure 7.20

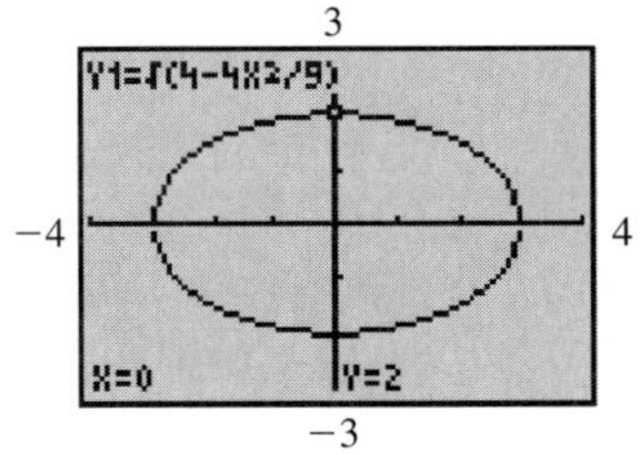

Figure 7.21

To check, graph $y_1 = \sqrt{4 - 4x^2/9}$ and $y_2 = -y_1$ as shown in Fig. 7.21.

For $a > b > 0$, the graph of

$$\frac{x^2}{b^2} + \frac{y^2}{a^2} = 1$$

is an ellipse with a vertical major axis, x-intercepts $(b, 0)$ and $(-b, 0)$, and y-intercepts $(0, a)$ and $(0, -a)$, as shown in Fig. 7.18. When the denominator for y^2 is larger than the denominator for x^2, the major axis is vertical. The foci are always on the major axis and are determined by $a^2 = b^2 + c^2$ or $c^2 = a^2 -$

b^2. Remember that this relationship between a, b, and c is determined by the Pythagorean theorem, because a, b, and c are the lengths of sides of a right triangle, as shown in Fig. 7.18.

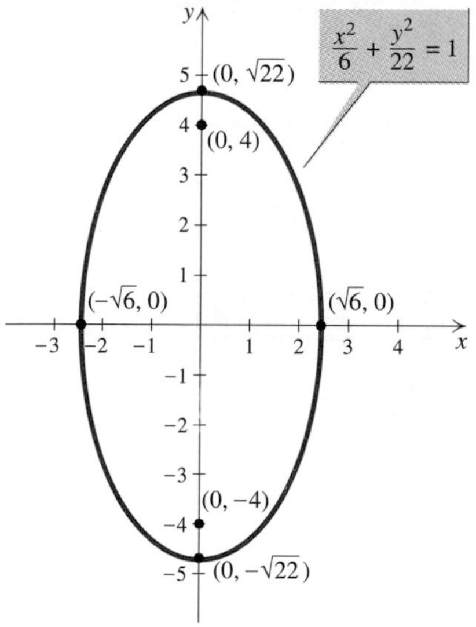

Figure 7.22

EXAMPLE 3 **Graphing an ellipse with foci on the *y*-axis**

Sketch the graph of $11x^2 + 3y^2 = 66$ and identify the foci.

Solution

We first divide each side of the equation by 66 to get the standard equation:

$$\frac{x^2}{6} + \frac{y^2}{22} = 1$$

If $x = 0$, then $y^2 = 22$ or $y = \pm\sqrt{22}$. So the y-intercepts are $(0, \sqrt{22})$ and $(0, -\sqrt{22})$. If $y = 0$, then $x = \pm\sqrt{6}$ and the x-intercepts are $(\sqrt{6}, 0)$ and $(-\sqrt{6}, 0)$. Plot these four points and draw an ellipse through them as shown in Fig. 7.22. Since this ellipse is elongated in the direction of the y-axis, the foci are on the y-axis. Use $a^2 = 22$ and $b^2 = 6$ in $c^2 = a^2 - b^2$ to get $c^2 = 22 - 6 = 16$. So $c = \pm 4$, and the foci are $(0, 4)$ and $(0, -4)$.

Translations of Ellipses

Although an ellipse is not the graph of a function, its graph can be translated in the same manner. Figure 7.23 shows the graphs of

$$\frac{(x-h)^2}{a^2} + \frac{(y-k)^2}{b^2} = 1 \quad \text{and} \quad \frac{x^2}{a^2} + \frac{y^2}{b^2} = 1.$$

They have the same size and shape, but the graph of the first equation is centered at (h, k) rather than at the origin. So the graph of the first equation is obtained by translating the graph of the second horizontally h units and vertically k units.

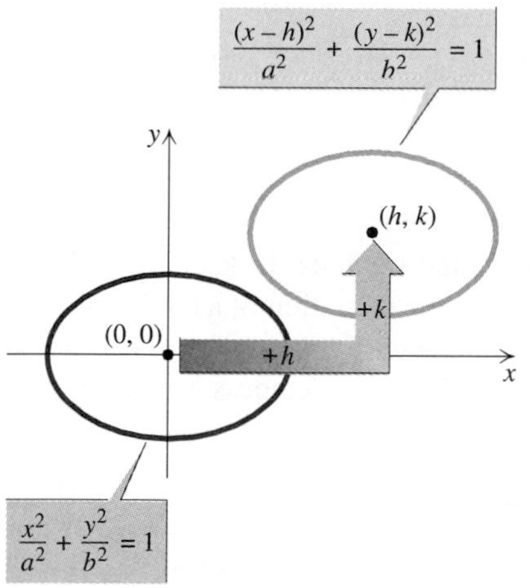

Figure 7.23

EXAMPLE 4 **Graphing an ellipse centered at (h, k)**

Sketch the graph and identify the foci of the ellipse

$$\frac{(x-3)^2}{25} + \frac{(y+1)^2}{9} = 1.$$

Solution

The graph of this equation is a translation of the graph of

$$\frac{x^2}{25} + \frac{y^2}{9} = 1,$$

three units to the right and one unit downward. The center of the ellipse is $(3, -1)$. Since $a^2 = 25$, the vertices lie five units to the right and five units to the left of $(3, -1)$. So the ellipse goes through $(8, -1)$ and $(-2, -1)$. Since $b^2 = 9$, the graph includes points three units above and three units below the center. So the ellipse goes through $(3, 2)$ and $(3, -4)$, as shown in Fig. 7.24. Since $c^2 = 25 - 9 = 16$, $c = \pm 4$. The major axis is on the horizontal line $y = -1$. So the foci are found four units to the right and four units to the left of the center at $(7, -1)$ and $(-1, -1)$.

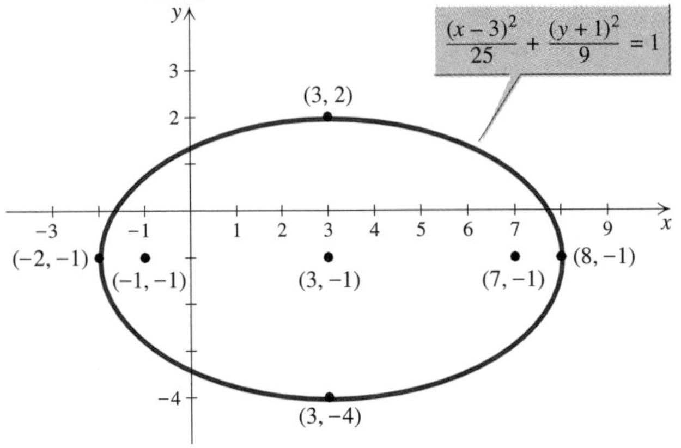

Figure 7.24

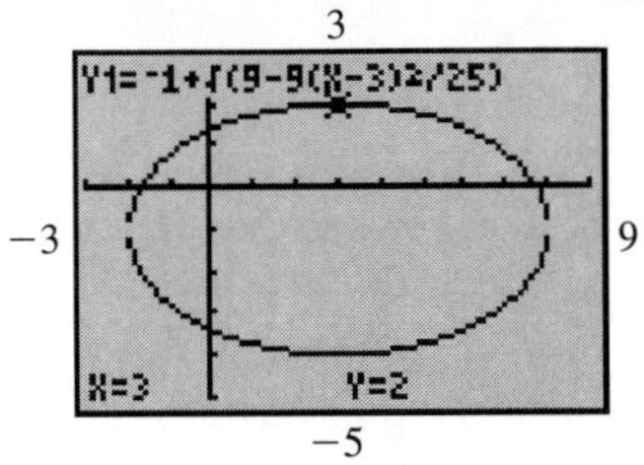

Figure 7.25

To check the location of the ellipse, graph

$$y_1 = -1 + \sqrt{9 - 9(x-3)^2/25} \quad \text{and} \quad y_2 = -1 - \sqrt{9 - 9(x-3)^2/25}$$

on a graphing calculator as in Fig. 7.25.

The Circle

The circle is the simplest curve of the four conic sections. In keeping with our approach to the conic sections, we give a geometric definition of a circle and then use the distance formula to derive the standard equation for a circle. The standard equation was derived in this way in Section 2.3. For completeness, we now repeat the definition and derivation.

Definition: Circle

A **circle** is the set of all points in a plane such that their distance from a fixed point (the **center**) is a constant (the **radius**).

As shown in Fig. 7.26, a point (x, y) is on a circle with center (h, k) and radius r if and only if

$$\sqrt{(x-h)^2 + (y-k)^2} = r.$$

If we square both sides of this equation, we get the standard equation of a circle.

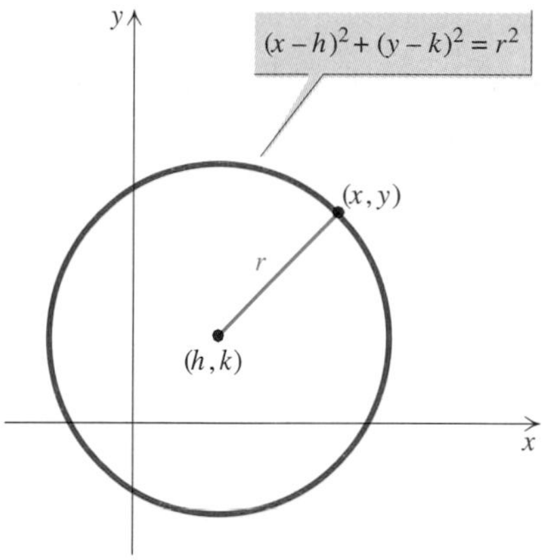

Figure 7.26

Theorem: Standard Equation of a Circle

The standard equation of a circle with center (h, k) and radius $r (r > 0)$ is

$$(x-h)^2+(y-k)^2=r^2.$$

The equation $(x-h)^2+(y-k)^2=a$ is a circle of radius $\sqrt{a}$ if $a > 0$. If $a = 0$, only (h, k) satisfies $(x-h)^2+(y-k)^2=0$ and the point (h, k) is a degenerate circle. If $a < 0$, then no ordered pair satisfies $(x-h)^2+(y-k)^2=a$. If h and k are zero, then we get the standard equation of a circle centered at the origin, $x^2+y^2=r^2$.

A circle can be thought of as an ellipse in which the two foci coincide at the center. If the foci are identical, then $a = b$ and the equation for an ellipse becomes the equation for a circle with radius a.

EXAMPLE 5 Finding the equation for a circle

Write the equation for the circle that has center (4, 5) and passes through (−1, 2).

Solution

The radius is the distance from (4, 5) to (−1, 2):

$$r=\sqrt{(4-(-1))^2+(5-2)^2}=\sqrt{25+9}=\sqrt{34}$$

Use $h = 4$, $k = 5$, and $r=\sqrt{34}$ in $(x-h)^2+(y-k)^2=r^2$ to get the equation

$$(x-4)^2+(y-5)^2=34.$$

To graph a circle, we must know the center and radius. A compass or a string can be used to keep the pencil at a fixed distance from the center.

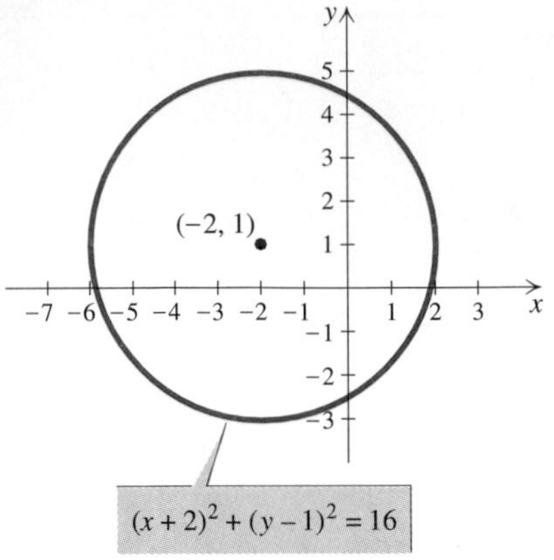

Figure 7.27

EXAMPLE 6 Finding the center and radius

Determine the center and radius, and sketch the graph of $x^2 + 4x + y^2 - 2y = 11$.

Solution

Use completing the square to get the equation into the standard form:

$$x^2 + 4x + 4 + y^2 - 2y + 1 = 11 + 4 + 1$$
$$(x + 2)^2 + (y - 1)^2 = 16$$

From the standard form we recognize that the center is $(-2, 1)$ and the radius is 4. The graph is shown in Fig. 7.27.

Applications

The **eccentricity** e of an ellipse is defined by $e = c/a$, where c is the distance from the center to a focus and a is one-half the length of the major axis. Since $0 < c < a$, we have $0 < e < 1$. For an ellipse that appears circular, the foci are close to the center and c is small compared with a. So its eccentricity is near 0. For an ellipse that is very elongated, the foci are close to the vertices and c is nearly equal to a. So its eccentricity is near 1. See Fig. 7.28. A satellite that orbits the earth has an elliptical orbit that is nearly circular, with eccentricity close to 0. On the other hand, Halley's comet is in a very elongated elliptical orbit around the sun, with eccentricity close to 1.

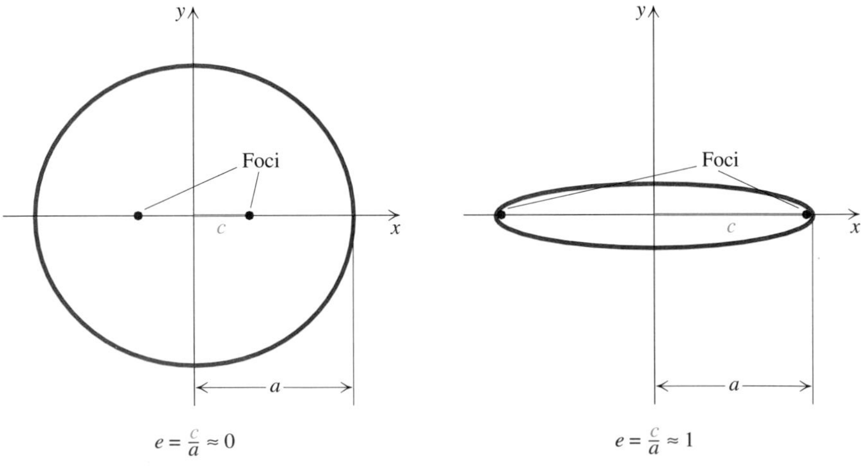

Figure 7.28

EXAMPLE 7 Eccentricity of an orbit

The first artificial satellite to orbit the earth was Sputnik I, launched by the Soviet Union in 1957. The altitude of Sputnik varied from 132 miles to 583 miles above the surface of the earth. If the center of the earth is one focus of its elliptical orbit and the radius of the earth is 3950 miles, then what was the eccentricity of the orbit?

Solution

The center of the earth is one focus F_1 of the elliptical orbit as shown in Fig. 7.29. When Sputnik I was 132 miles above the earth, it was at a vertex V_1, 4082 miles from F_1 ($4082 = 3950 + 132$). When Sputnik I was 583 miles above the earth, it was at the other vertex V_2, 4533 miles from F_1 ($4533 = 3950 + 583$). So the length of the major axis is $4082 + 4533 = 8615$ miles. Since the length of the major axis is $2a$, we get $a = 4307.5$ miles. Since c is the distance from the center of the ellipse C to F_1, we get $c = 4307.5 - 4082 = 225.5$. Use $e = c/a$ to get

$$e = \frac{225.5}{4307.5} \approx 0.052.$$

So the eccentricity of the orbit was approximately 0.052.

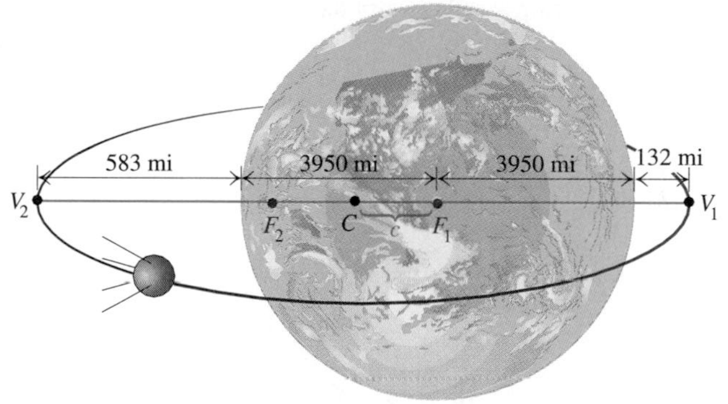

Figure 7.29

FOR THOUGHT True or False? Explain.

1. The x-intercepts for $\frac{x^2}{9} + \frac{y^2}{4} = 1$ are $(9, 0)$ and $(-9, 0)$.
2. The graph of $2x^2 + y^2 = 1$ is an ellipse.
3. The ellipse $\frac{x^2}{16} + \frac{y^2}{25} = 1$ has a major axis of length 10.
4. The x-intercepts for $0.5x^2 + y^2 = 1$ are $(\sqrt{2}, 0)$ and $(-\sqrt{2}, 0)$.
5. The y-intercepts for $x^2 + \frac{y^2}{3} = 1$ are $(0, \sqrt{3})$ and $(0, -\sqrt{3})$.
6. A circle is a set of points, and the center is one of those points.

7. If the foci of an ellipse coincide, then the ellipse is a circle.

8. No ordered pair satisfies the equation $(x - 3)^2 + (y + 1)^2 = 0$.

9. The graph of $(x - 1)^2 + (y + 2)^2 + 9 = 0$ is a circle of radius 3.

10. The radius of the circle $x^2 - 4x + y^2 + y = 9$ is 3.

7.2 EXERCISES

Tape 20 Disk

Find the equation of each ellipse described below and sketch its graph.

1. Foci $(-2, 0)$ and $(2, 0)$, and y-intercepts $(0, -3)$ and $(0, 3)$
2. Foci $(-3, 0)$ and $(3, 0)$, and y-intercepts $(0, -4)$ and $(0, 4)$
3. Foci $(-4, 0)$ and $(4, 0)$, and x-intercepts $(-5, 0)$ and $(5, 0)$
4. Foci $(-1, 0)$ and $(1, 0)$, and x-intercepts $(-3, 0)$ and $(3, 0)$
5. Foci $(0, 2)$ and $(0, -2)$, and x-intercepts $(2, 0)$ and $(-2, 0)$
6. Foci $(0, 6)$ and $(0, -6)$, and x-intercepts $(2, 0)$ and $(-2, 0)$
7. Foci $(0, 4)$ and $(0, -4)$, and y-intercepts $(0, 7)$ and $(0, -7)$
8. Foci $(0, 3)$ and $(0, -3)$, and y-intercepts $(0, 4)$ and $(0, -4)$

Sketch the graph of each ellipse and identify the foci.

9. $\dfrac{x^2}{16} + \dfrac{y^2}{4} = 1$

10. $\dfrac{x^2}{16} + \dfrac{y^2}{9} = 1$

11. $\dfrac{x^2}{9} + \dfrac{y^2}{36} = 1$

12. $x^2 + \dfrac{y^2}{4} = 1$

13. $\dfrac{x^2}{25} + y^2 = 1$

14. $\dfrac{x^2}{6} + \dfrac{y^2}{10} = 1$

15. $\dfrac{y^2}{25} + \dfrac{x^2}{9} = 1$

16. $\dfrac{y^2}{9} + \dfrac{x^2}{4} = 1$

17. $9x^2 + y^2 = 9$

18. $x^2 + 4y^2 = 4$

19. $4x^2 + 9y^2 = 36$

20. $9x^2 + 25y^2 = 225$

Sketch the graph of each ellipse and identify the foci.

21. $\dfrac{(x - 1)^2}{16} + \dfrac{(y + 3)^2}{9} = 1$

22. $\dfrac{(x + 2)^2}{16} + \dfrac{(y + 1)^2}{4} = 1$

23. $\dfrac{(x - 3)^2}{9} + \dfrac{(y + 2)^2}{25} = 1$

24. $(x - 5)^2 + \dfrac{(y - 3)^2}{9} = 1$

25. $(x + 4)^2 + 36(y + 3)^2 = 36$

26. $9(x - 1)^2 + 4(y + 3)^2 = 36$

27. $9x^2 - 18x + 4y^2 + 16y = 11$

28. $4x^2 + 16x + y^2 - 6y = -21$

Find the equation of each ellipse and identify its foci.

29.

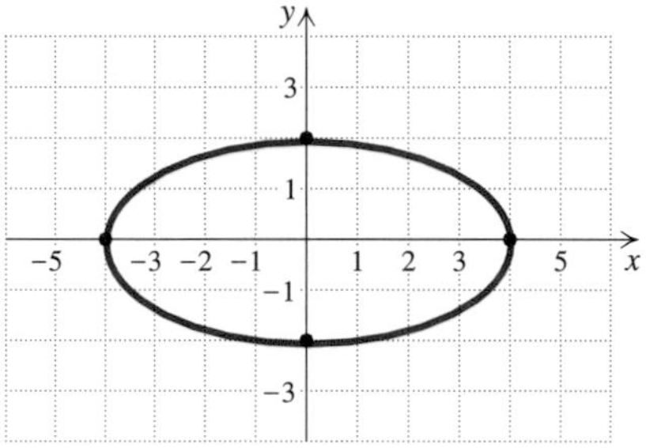

30.

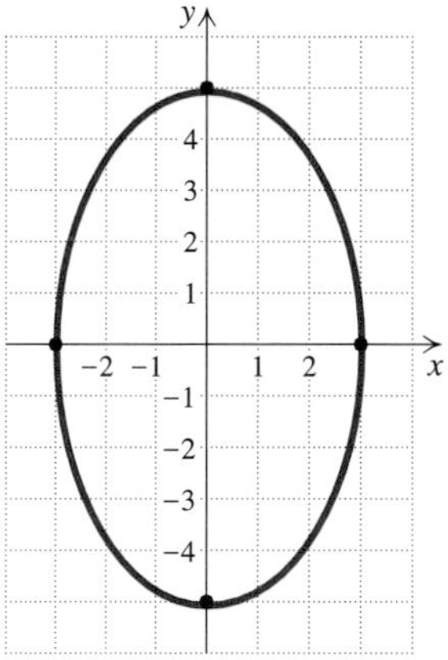

31.

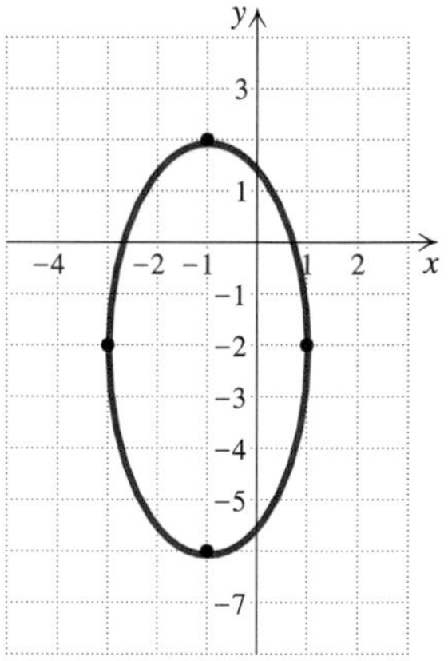

32.

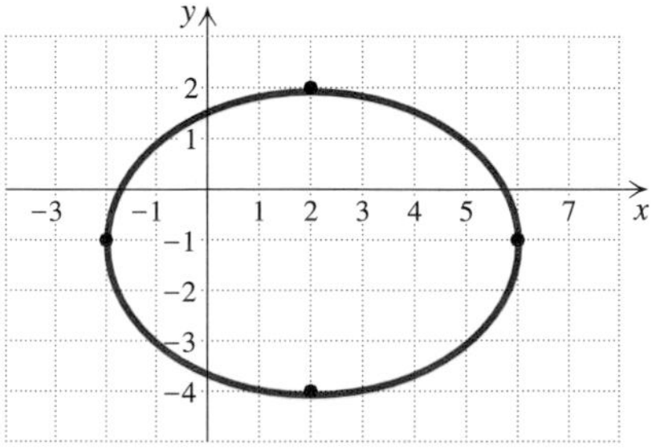

For Exercises 33–40, write the equation for each circle described.

33. Center (0, 0) and radius 2

34. Center (0, 0) and radius $\sqrt{5}$

35. Center (0, 0) and passing through (4, 5)

36. Center (0, 0) and passing through $(-3, -4)$

37. Center $(2, -3)$ and passing through (4, 1)

38. Center $(-2, -4)$ and passing through $(1, -1)$

39. Diameter has endpoints (3, 4) and $(-1, 2)$.

40. Diameter has endpoints $(3, -1)$ and $(-4, 2)$.

Determine the center and radius of each circle and sketch its graph.

41. $x^2 + y^2 = 100$

42. $x^2 + y^2 = 25$

43. $(x - 1)^2 + (y - 2)^2 = 4$

44. $(x + 2)^2 + (y - 3)^2 = 9$

45. $(x + 2)^2 + (y + 2)^2 = 8$

46. $x^2 + (y - 3)^2 = 9$

Find the center and radius of each circle.

47. $x^2 + y^2 + 2y = 8$

48. $x^2 - 6x + y^2 = 1$

49. $x^2 + 8x + y^2 = 10y$

50. $x^2 + y^2 = 12x - 12y$

51. $x^2 + 4x + y^2 = 5$

52. $x^2 + y^2 - 6y = 0$

53. $x^2 - x + y^2 + y = \frac{1}{2}$

54. $x^2 + 5x + y^2 + 3y = \frac{1}{2}$

55. $x^2 + \frac{2}{3}x + y^2 + \frac{1}{3}y = \frac{1}{9}$

56. $x^2 + \frac{1}{2}x + y^2 + \frac{1}{2}y = \frac{1}{8}$

57. $2x^2 + 4x + 2y^2 = 1$

58. $x^2 + y^2 = \frac{3}{2}y$

Write each of the following equations in one of the forms:
$y = a(x - h)^2 + k,\quad x = a(y - h)^2 + k,$
$\frac{(x - h)^2}{a^2} + \frac{(y - k)^2}{b^2} = 1,\quad \text{or} \quad (x - h)^2 + (y - k)^2 = r^2.$
Then identify each equation as the equation of a parabola, an ellipse, or a circle.

59. $y = x^2 + y^2$

60. $4x = x^2 + y^2$

61. $4x^2 + 12y^2 = 4$

62. $2x^2 + 2y^2 = 4 - y$

63. $2x^2 + 4x = 4 - y$

64. $2x^2 + 4y^2 = 4 - y$

65. $2 - x = (2 - y)^2$

66. $3(3 - x)^2 = 9 - y$

67. $2(4 - x)^2 = 4 - y^2$

68. $\frac{x^2}{4} + \frac{y^2}{4} = 1$

69. $9x^2 = 1 - 9y^2$

70. $9x^2 = 1 - 9y$

Solve each problem.

71. *Foci and a Point* Find the equation of the ellipse that goes through the point (2, 3) and has foci (2, 0) and $(-2, 0)$.

72. *Foci and Eccentricity* Find the equation of an ellipse with foci $(\pm 5, 0)$ and eccentricity 1/2.

73. *Focus of Elliptical Reflector* An elliptical reflector is 10 in. in diameter and 3 in. deep. A light source is at one focus (0, 0), 3 in. from the vertex $(-3, 0)$, as shown in the figure. Find the location of the other focus.

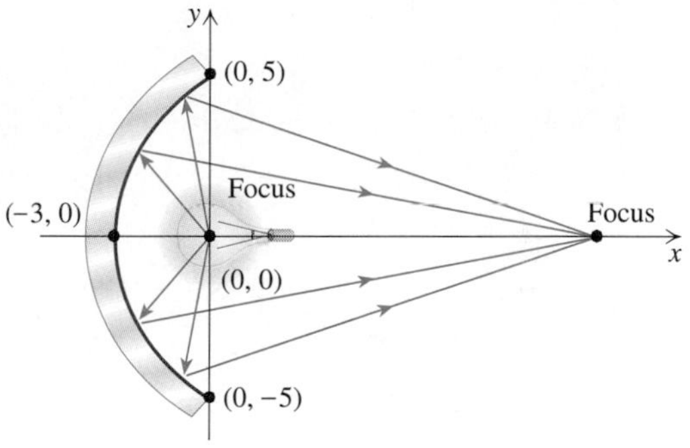

Figure for Exercise 73

74. *Constructing an Elliptical Arch* A mason is constructing an elliptical form for an arch out of a 4-ft by 12-ft sheet of plywood, as shown in the figure. What length string is needed to draw the ellipse on the plywood (by attaching the ends at the foci as discussed at the beginning of this section)? Where are the foci for this ellipse?

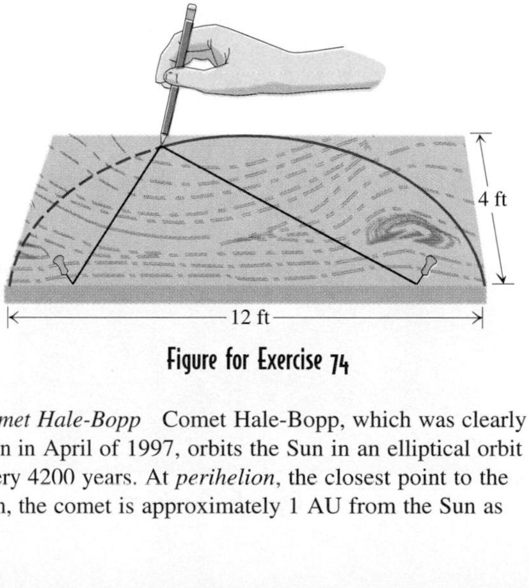

Figure for Exercise 74

75. *Comet Hale-Bopp* Comet Hale-Bopp, which was clearly seen in April of 1997, orbits the Sun in an elliptical orbit every 4200 years. At *perihelion*, the closest point to the Sun, the comet is approximately 1 AU from the Sun as shown in the accompanying figure (*Sky and Telescope*, April 1997, used with permission. Guy Ottewell, Astronomical Calendar 1997, p. 59.) At *aphelion*, the farthest point from the Sun, the comet is 520 AU from the Sun. Find the equation of the ellipse. What is the eccentricity of the orbit?

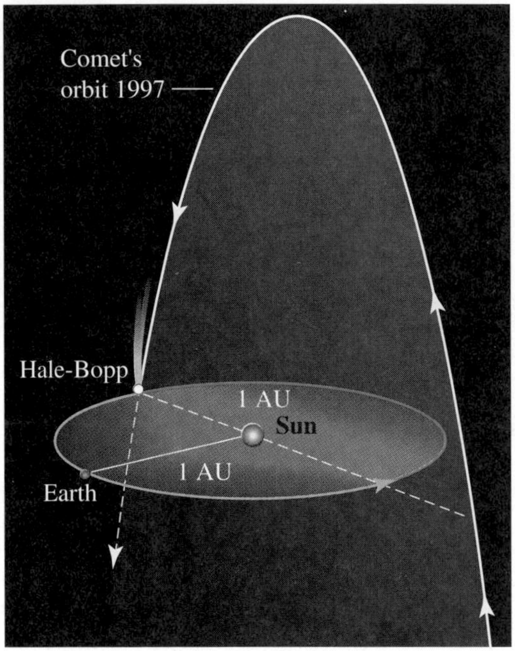

Figure for Exercise 75

76. *Orbit of the Moon* The moon travels on an elliptical path with the earth at one focus. If the maximum distance from the moon to the earth is 405,500 km and the minimum distance is 363,300 km, then what is the eccentricity of the orbit?

77. *Halley's Comet* The comet Halley, last seen in 1986, travels in an elliptical orbit with the sun at one focus. Its minimum distance to the sun is 8×10^7 km, and the eccentricity of its orbit is 0.97. What is its maximum distance from the sun? Comet Halley will next be visible from the earth in 2062.

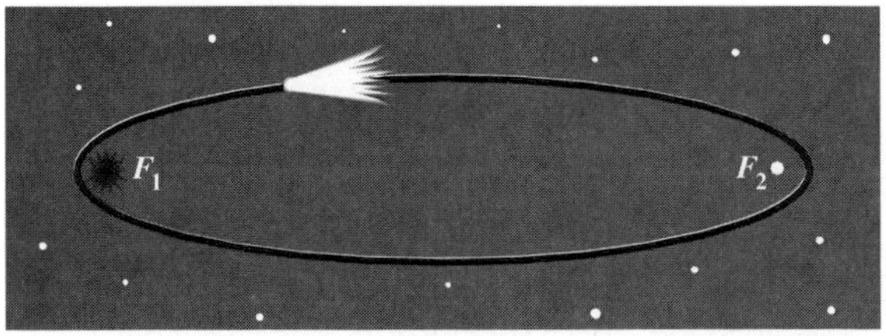

Figure for Exercise 77

78. *Adjacent Circles* A 13-in.-diameter mag wheel and a 16-in.-diameter mag wheel are placed in the first quadrant as shown in the figure. Write an equation for the circular boundary of each wheel.

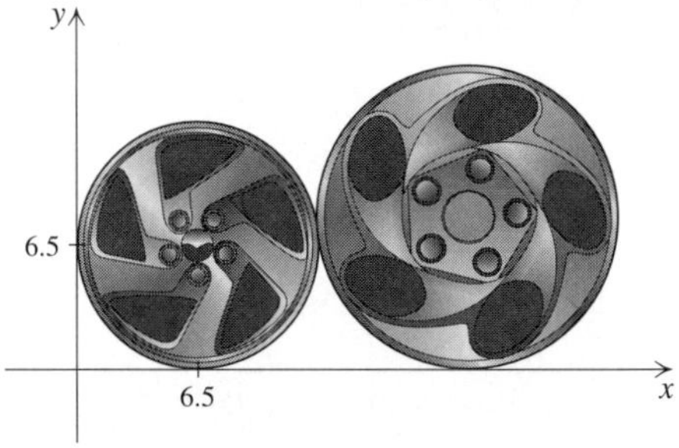

Figure for Exercise 78

79. *Picturing Earth* Solve $x^2 + y^2 = 6360^2$ for y and graph the resulting two equations on a graphing calculator to get a picture of earth (a circle of radius 6360 km) as seen from space. The center of earth is at the origin.

80. *Orbit of Mir* The elliptical orbit of the Mir space station has the equation

$$\frac{(x - 5)^2}{6735^2} + \frac{y^2}{6734.998^2} = 1$$

where one of the foci is at the center of earth. Solve this equation for y and graph the resulting two functions on the same screen as the graph obtained in the previous exercise. The graph of the circle and the ellipse together will give you an idea of the altitude of Mir compared with the size of earth. What is the eccentricity of the orbit of Mir?

81. *Tangent to an Ellipse* It can be shown that the line tangent to the ellipse $x^2/a^2 + y^2/b^2 = 1$ at the point (x_1, y_1) has the equation

$$\frac{x_1 x}{a^2} + \frac{y_1 y}{b^2} = 1.$$

a. Find the equation of the line tangent to the ellipse $x^2/25 + y^2/9 = 1$ at $(-4, 9/5)$.

b. Graph the ellipse and the tangent line on a graphing calculator and use the intersect feature to find the point of intersection.

82. *Points on an Ellipse* Find all points of the ellipse $9x^2 + 25y^2 = 225$ that are twice as far from one focus as they are from the other focus.

Figure for Exercise 80

For Writing/Discussion

83. *Best Reflector* Both parabolic and elliptical reflectors reflect sound waves to a focal point where they are amplified. To eavesdrop on the conversation of the quarterback on a football field, which type of reflector is preferable and why?

84. *Details* Fill in the details in the development of the equation of the ellipse that is outlined in this section.

85. *Development* Develop the equation of the ellipse with foci at $(0, \pm c)$ and x-intercepts $(\pm b, 0)$ from the definition of ellipse.

86. *Cooperative Learning* To make drapes gather properly, they must be made 2.5 times as wide as the window. Discuss with your classmates the problem of making drapes for a semicircular window. Explain in detail how to cut the material. Of course these drapes do not open and close, they just hang there. This problem was given to this author by a friend in the drapery business.

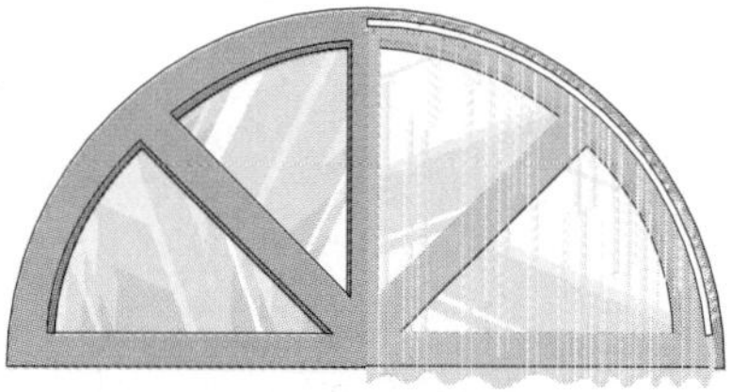

Figure for Exercise 86

Linking Concepts

For Individual or Group Explorations

Apogee, Perigee, and Eccentricity *For a planet or satellite in an elliptical orbit around a focus of the ellipse, perigee (P) is defined to be its closest distance to the focus and apogee (A) is defined to be its greatest distance from the focus.*

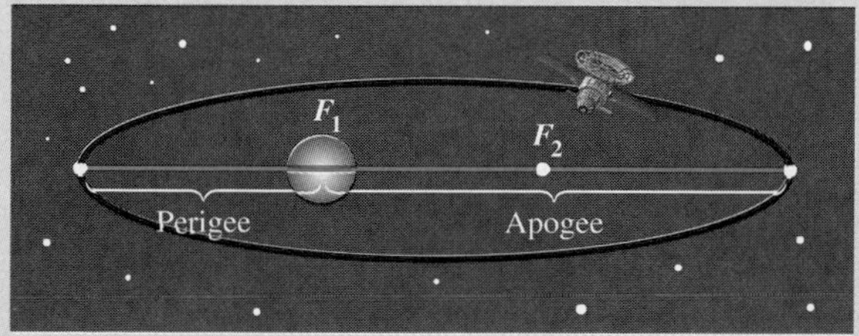

a) Show that $(A - P)/(A + P)$ is equal to the eccentricity of the orbit.

b) Find the apogee and perigee for the satellite Sputnik I discussed in Example 7 of this section.

c) Use the apogee and perigee to find the eccentricity of the orbit of Sputnik I.

d) Pluto orbits the Sun in an elliptical orbit with eccentricity 0.2484. The perigee for Pluto is 29.64 AU (astronomical units). Find the apogee for Pluto.

e) Mars orbits the sun in an elliptical orbit with eccentricity 0.0934. The apogee for Mars is 2.492×10^8 km. Find the perigee for Mars.

7.3 The Hyperbola

The last of the four conic sections, the hyperbola, has two branches and each one has a focus as shown in Fig. 7.30. The hyperbola also has a useful reflecting property. A light ray aimed at one focus is reflected toward the other focus as shown in the figure. This reflecting property is used in telescopes, as shown in Fig. 7.31. Within the telescope, a small hyperbolic mirror with the same focus as the large parabolic mirror reflects the light to a more convenient location for viewing.

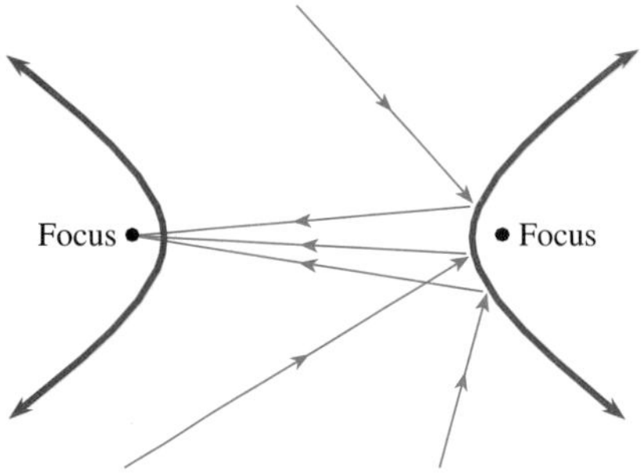

Figure 7.30

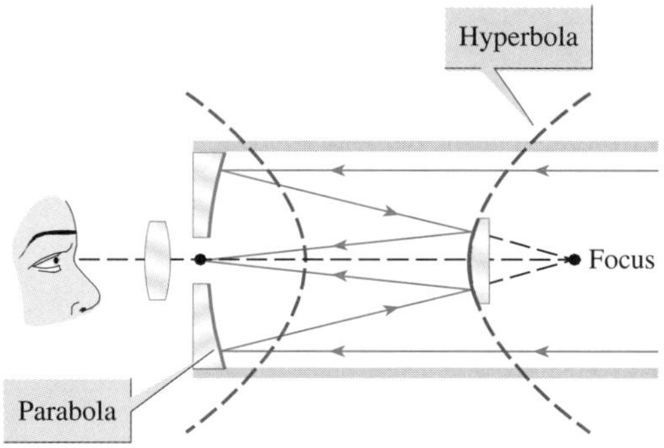

Figure 7.31

Hyperbolas also occur in the context of supersonic noise pollution. The sudden change in air pressure from an aircraft traveling at supersonic speed creates a cone-shaped wave through the air. The plane of the ground and this cone intersect along one branch of a hyperbola as shown in Fig. 7.32. A sonic boom is heard on the ground along this branch of the hyperbola. Since this curve where the sonic boom is heard travels along the ground with the aircraft, supersonic planes such as the Concorde are restricted from flying across the continental United States.

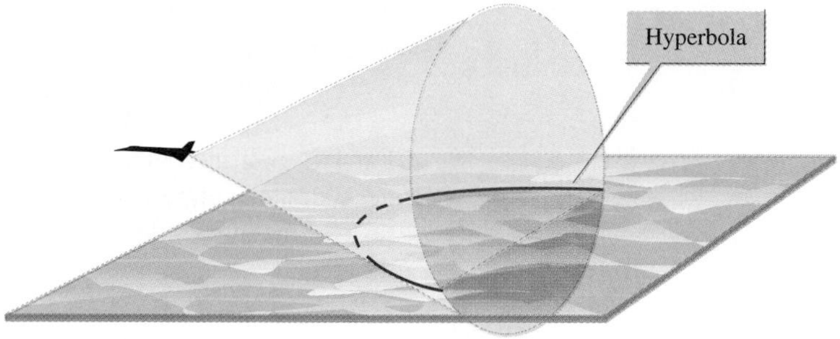

Figure 7.32

A hyperbola may also occur as the path of a moving object such as a spacecraft traveling past the moon on its way toward Venus, a comet passing in the neighborhood of the sun, or an alpha particle passing by the nucleus of an atom.

The Definition

A hyperbola can be defined as the intersection of a cone and a plane, as shown in Fig. 7.1. As we did for the other conic sections, we will give a geometric definition of a hyperbola and use the distance formula to derive its equation.

Definition: Hyperbola

A **hyperbola** is the set of points in a plane such that the difference between the distances from two fixed points (foci) is constant.

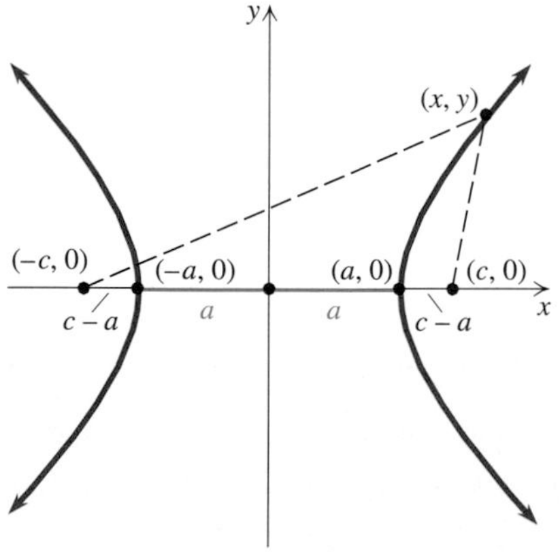

Figure 7.33

For a point on a hyperbola the *difference* between the distances from two fixed points is constant, and for a point on an ellipse the *sum* of the distances from two fixed points is constant. The definitions of a hyperbola and an ellipse are similar, and we will see that their equations are similar also. Their graphs, however, are very different. In the hyperbola shown in Fig. 7.33, the branches look like parabolas, *but they are not parabolas* because they do not satisfy the geometric definition of a parabola. A hyperbola with foci on the x-axis as in Fig. 7.33 is said to open to the left and right.

Developing the Equation

The hyperbola shown in Fig. 7.33 has foci at $(c, 0)$ and $(-c, 0)$ and x-intercepts or **vertices** at $(a, 0)$ and $(-a, 0)$, where $a > 0$ and $c > 0$. The line segment between the vertices is the **transverse axis**. The point $(0, 0)$, halfway between the foci, is the **center**. The point $(a, 0)$ is on the hyperbola. The distance from $(a, 0)$ to the focus $(c, 0)$ is $c - a$. The distance from $(a, 0)$ to $(-c, 0)$ is $c + a$.

So the constant difference is $2a$. For an arbitrary point (x, y), the distance to $(c, 0)$ is subtracted from the distance to $(-c, 0)$ to get the constant $2a$:

$$\sqrt{(x-(-c))^2 + (y-0)^2} - \sqrt{(x-c)^2 + (y-0)^2} = 2a$$

For (x, y) on the other branch we would subtract in the opposite order, but we get the same simplified form. The simplified form for this equation is similar to that for the ellipse. We will provide the major steps for simplifying it and leave the details as an exercise. First simplify inside the radicals and isolate them to get

$$\sqrt{x^2 + 2xc + c^2 + y^2} = 2a + \sqrt{x^2 - 2xc + c^2 + y^2}.$$

Squaring each side and simplifying yields

$$xc - a^2 = a\sqrt{x^2 - 2xc + c^2 + y^2}.$$

Square each side again and simplify to get

$$c^2x^2 - a^2x^2 - a^2y^2 = c^2a^2 - a^4$$

$$(c^2 - a^2)x^2 - a^2y^2 = a^2(c^2 - a^2) \qquad \text{Factor.}$$

Now $c^2 - a^2$ is positive, because $c > a$. So let $b^2 = c^2 - a^2$ and substitute:

$$b^2x^2 - a^2y^2 = a^2b^2$$

$$\frac{x^2}{a^2} - \frac{y^2}{b^2} = 1 \qquad \text{Divide each side by } a^2b^2.$$

We have proved the following theorem.

Theorem: Equation of a Hyperbola Centered at (0, 0) Opening Left and Right

The equation of a hyperbola centered at the origin with foci $(c, 0)$ and $(-c, 0)$ and x-intercepts $(a, 0)$ and $(-a, 0)$ is

$$\frac{x^2}{a^2} - \frac{y^2}{b^2} = 1, \quad \text{where } b^2 = c^2 - a^2.$$

If the foci are positioned on the y-axis, then we say that the hyperbola opens up and down. For hyperbolas that open up and down we have the following theorem.

Theorem: Equation of a Hyperbola Centered at (0, 0) Opening Up and Down

The equation of a hyperbola centered at the origin with foci $(0, c)$ and $(0, -c)$ and y-intercepts $(0, a)$ and $(0, -a)$ is

$$\frac{y^2}{a^2} - \frac{x^2}{b^2} = 1, \quad \text{where } b^2 = c^2 - a^2.$$

Graphing a Hyperbola Centered at (0, 0)

If we solve the equation

$$\frac{x^2}{a^2} - \frac{y^2}{b^2} = 1$$

for y, we get the following:

$$\frac{y^2}{b^2} = \frac{x^2}{a^2} - 1$$

$$y^2 = \frac{b^2x^2}{a^2} - b^2 \qquad \text{Multiply each side by } b^2.$$

$$y^2 = \frac{b^2x^2}{a^2} - \frac{a^2b^2x^2}{a^2x^2} \qquad \text{Write } b^2 \text{ as } \frac{a^2b^2x^2}{a^2x^2}.$$

$$y^2 = \frac{b^2x^2}{a^2}\left(1 - \frac{a^2}{x^2}\right) \qquad \text{Factor out } \frac{b^2x^2}{a^2}.$$

$$y = \pm\frac{b}{a}x\sqrt{1 - \frac{a^2}{x^2}}$$

As $x \to \infty$, the value of $(a^2/x^2) \to 0$ and the value of y can be approximated by $y = \pm(b/a)x$. So the lines

$$y = \frac{b}{a}x \quad \text{and} \quad y = -\frac{b}{a}x$$

are oblique asymptotes for the graph of the hyperbola. The graph of this hyperbola is shown with its asymptotes in Fig. 7.34. The asymptotes are essential for determining the proper shape of the hyperbola. The asymptotes go through the points (a, b), $(a, -b)$, $(-a, b)$, and $(-a, -b)$. The rectangle with these four points as vertices is called the **fundamental rectangle**. The line segment with endpoints

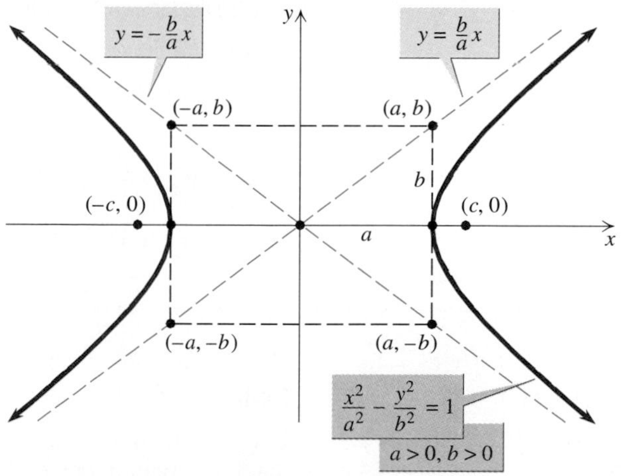

Figure 7.34

$(0, b)$ and $(0, -b)$ is called the **conjugate axis**. The location of the fundamental rectangle is determined by the conjugate axis and the transverse axis.

Note that since $c^2 = a^2 + b^2$, the distance c from the center to a focus is equal to the distance from the center to (a, b), as shown in Fig. 7.35.

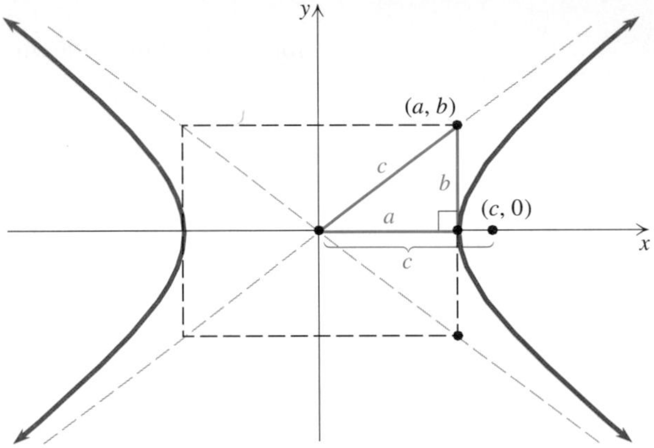

Figure 7.35

The following steps will help you graph hyperbolas opening left and right.

PROCEDURE

Graphing the Hyperbola $\frac{x^2}{a^2} - \frac{y^2}{b^2} = 1$

To graph $\frac{x^2}{a^2} - \frac{y^2}{b^2} = 1$ for $a > 0$ and $b > 0$, do the following:

1. Locate the x-intercepts $(a, 0)$ and $(-a, 0)$.
2. Draw the rectangle through $(\pm a, 0)$ and through $(0, \pm b)$.
3. Extend the diagonals of the rectangle to get the asymptotes.
4. Draw a hyperbola opening to the left and right from the x-intercepts approaching the asymptotes.

EXAMPLE 1 Graphing a hyperbola opening left and right

Determine the foci and the equations of the asymptotes, and sketch the graph of

$$\frac{x^2}{36} - \frac{y^2}{9} = 1.$$

Solution

If $y = 0$, then $x = \pm 6$. The x-intercepts are $(6, 0)$ and $(-6, 0)$. Since $b^2 = 9$, the fundamental rectangle goes through $(0, 3)$ and $(0, -3)$ and through the x-intercepts. Next draw the fundamental rectangle and extend its diagonals to get

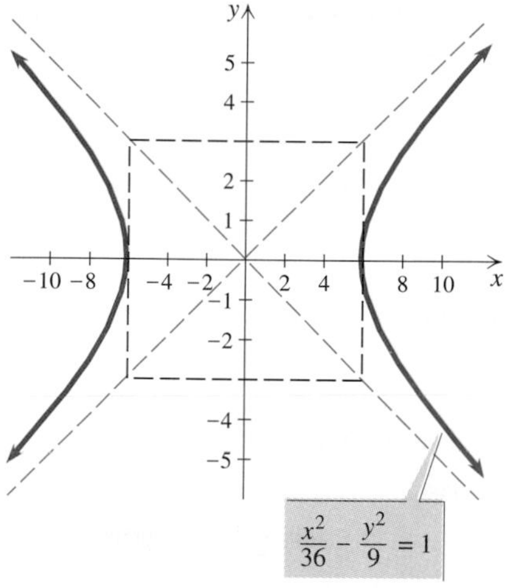

Figure 7.36

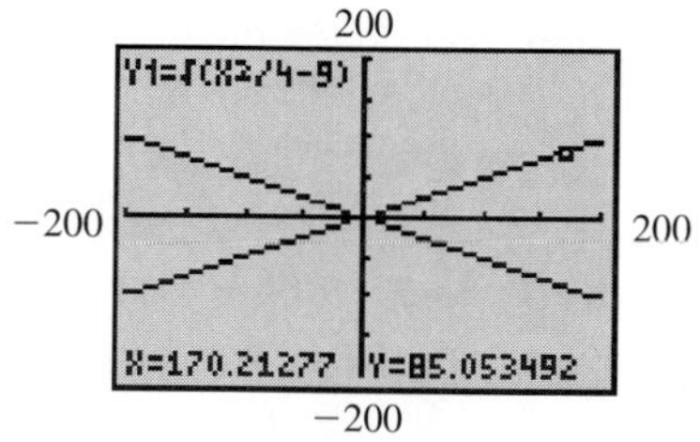

Figure 7.37

the asymptotes. Draw the hyperbola opening to the left and right, as shown in Fig. 7.36. To find the foci, use $c^2 = a^2 + b^2$:

$$c^2 = 36 + 9 = 45$$
$$c = \pm\sqrt{45}$$

So the foci are $(\sqrt{45},\ 0)$ and $(-\sqrt{45},\ 0)$. From the graph we get the asymptotes

$$y = \frac{1}{2}x \quad \text{and} \quad y = -\frac{1}{2}x.$$

The graph of a hyperbola gets closer and closer to its asymptotes. So in a large viewing window you cannot tell the difference between the hyperbola and its asymptotes. For example, the hyperbola of Example 1, $y = \pm\sqrt{x^2/4 - 9}$, looks like its asymptotes $y = \pm(1/2)x$ in Fig. 7.37. □

We can show that hyperbolas opening up and down have asymptotes just as hyperbolas opening right and left have. The lines

$$y = \frac{a}{b}x \quad \text{and} \quad y = -\frac{a}{b}x$$

are asymptotes for the graph of

$$\frac{y^2}{a^2} - \frac{x^2}{b^2} = 1,$$

as shown in Fig. 7.38. Note that the asymptotes are essential for determining the shape of the hyperbola.

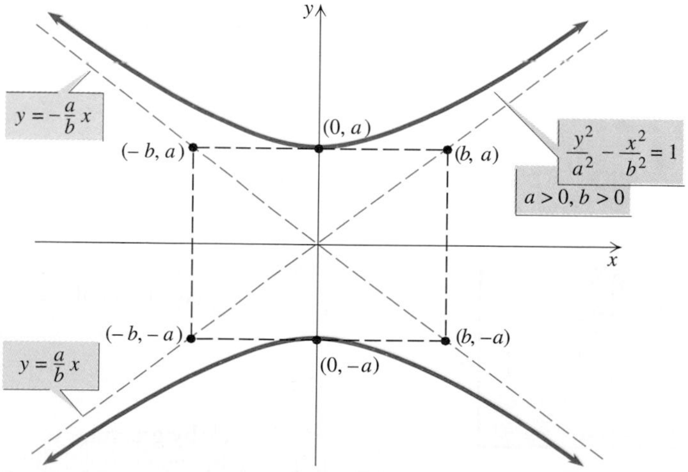

Figure 7.38

The following steps will help you graph hyperbolas opening up and down.

PROCEDURE **Graphing the Hyperbola $\frac{y^2}{a^2} - \frac{x^2}{b^2} = 1$**

To graph $\frac{y^2}{a^2} - \frac{x^2}{b^2} = 1$ for $a > 0$ and $b > 0$, do the following:

1. Locate the y-intercepts $(0, a)$ and $(0, -a)$.
2. Draw the rectangle through $(0, \pm a)$ and through $(b, 0)$ and $(-b, 0)$.
3. Extend the diagonals of the rectangle to get the asymptotes.
4. Draw a hyperbola opening up and down from the y-intercepts approaching the asymptotes.

EXAMPLE 2 **Graphing a hyperbola opening up and down**

Determine the foci and the equations of the asymptotes, and sketch the graph of

$$4y^2 - 9x^2 = 36.$$

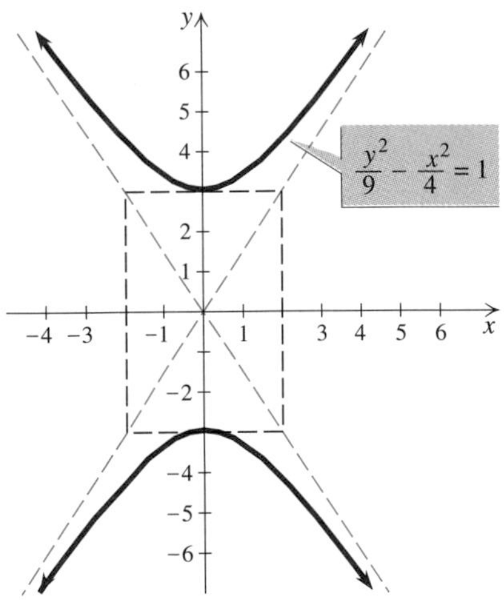

Figure 7.39

Solution

Divide each side of the equation by 36 to get the equation into the standard form for the equation of the hyperbola:

$$\frac{y^2}{9} - \frac{x^2}{4} = 1$$

If $x = 0$, then $y = \pm 3$. The y-intercepts are $(0, 3)$ and $(0, -3)$. Since $b^2 = 4$, the fundamental rectangle goes through the y-intercepts and through $(2, 0)$ and $(-2, 0)$. Draw the fundamental rectangle and extend its diagonals to get the asymptotes. Draw a hyperbola opening up and down from the y-intercepts approaching the asymptotes as in Fig. 7.39. To find the foci, use $c^2 = a^2 + b^2$:

$$c^2 = 9 + 4 = 13$$
$$c = \pm\sqrt{13}$$

The foci are $(0, \sqrt{13})$ and $(0, -\sqrt{13})$. From the fundamental rectangle we can see that the equations of the asymptotes are

$$y = \frac{3}{2}x \quad \text{and} \quad y = -\frac{3}{2}x.$$

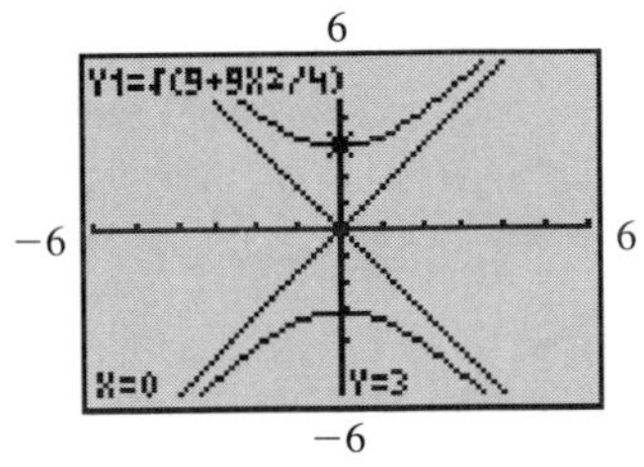

Figure 7.40

Check by graphing $y = \pm\sqrt{9 + 9x^2/4}$ and $y = \pm 1.5x$ with a calculator as in Fig. 7.40.

Hyperbolas Centered at (h, k)

The graph of a hyperbola can be translated horizontally and vertically by replacing x by $x - h$ and y by $y - k$.

Theorem: Hyperbolas Centered at (h, k)

A hyperbola centered at (h, k), opening left and right, has a horizontal transverse axis and equation

$$\frac{(x-h)^2}{a^2} - \frac{(y-k)^2}{b^2} = 1.$$

A hyperbola centered at (h, k), opening up and down, has a vertical transverse axis and equation

$$\frac{(y-k)^2}{a^2} - \frac{(x-h)^2}{b^2} = 1.$$

EXAMPLE 3 Graphing a hyperbola centered at (h, k)

Determine the foci and the equations of the asymptotes, and sketch the graph of

$$\frac{(x-2)^2}{9} - \frac{(y+1)^2}{4} = 1.$$

Solution

The graph that we seek is the graph of

$$\frac{x^2}{9} - \frac{y^2}{4} = 1$$

translated so that its center is $(2, -1)$. Since $a = 3$, the vertices are three units from $(2, -1)$ at $(5, -1)$ and $(-1, -1)$. Because $b = 2$, the fundamental rectangle passes through the vertices and points that are two units above and below $(2, -1)$. Draw the fundamental rectangle through the vertices and through $(2, 1)$ and $(2, -3)$. Extend the diagonals of the fundamental rectangle for the asymptotes, and draw the hyperbola opening to the left and right as shown in Fig. 7.41 on the next page. Use $c^2 = a^2 + b^2$ to get $c = \sqrt{13}$. The foci are on the transverse axis, $\sqrt{13}$ units from the center $(2, -1)$. So the foci are $(2 + \sqrt{13}, -1)$ and $(2 - \sqrt{13}, -1)$. The asymptotes have slopes $\pm 2/3$ and pass through $(2, -1)$. Using the point-slope form for the equation of the line, we get the equations

$$y = \frac{2}{3}x - \frac{7}{3} \quad \text{and} \quad y = -\frac{2}{3}x + \frac{1}{3}$$

as the equations of the asymptotes.

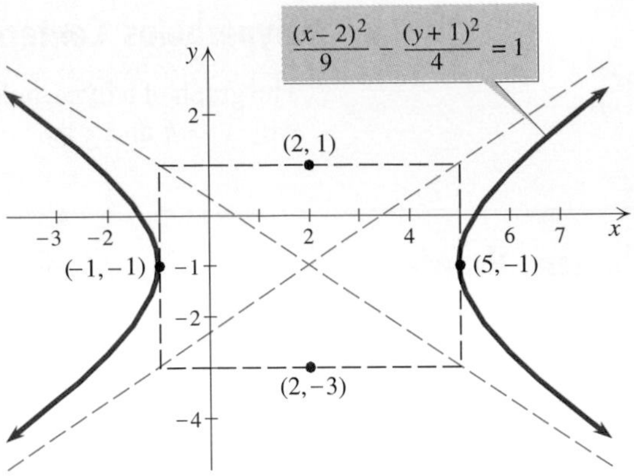

Figure 7.41

Check by graphing

$$y_1 = -1 + \sqrt{4(x - 2)^2/9 - 4},$$
$$y_2 = -1 - \sqrt{4(x - 2)^2/9 - 4},$$
$$y_3 = (2x - 7)/3, \text{ and}$$
$$y_4 = (-2x + 1)/3$$

on a graphing calculator as in Fig. 7.42.

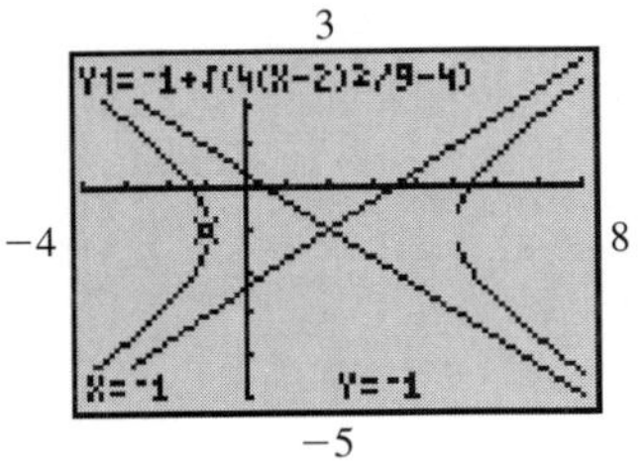

Figure 7.42

Finding the Equation of a Hyperbola

The equation of a hyperbola depends on the location of the foci, center, vertices, transverse axis, conjugate axis, and asymptotes. However, it is not necessary to have all of this information to write the equation. In the next example, we find the equation of a hyperbola given only its transverse axis and conjugate axis.

EXAMPLE 4 Writing the equation of a hyperbola

Find the equation of the hyperbola whose transverse axis has endpoints $(0, \pm 4)$ and whose conjugate axis has endpoints $(\pm 2, 0)$.

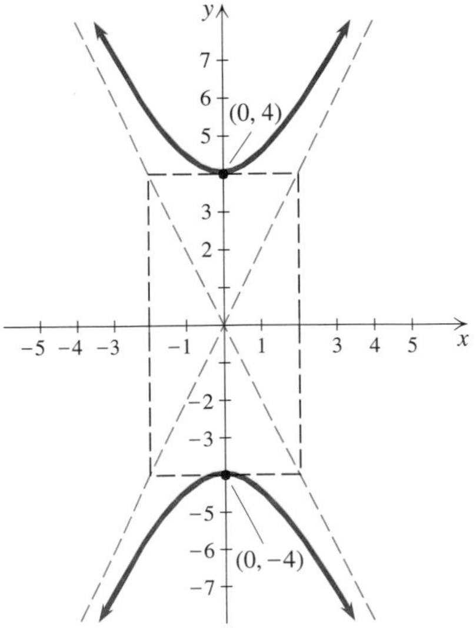

Figure 7.43

Solution

Since the vertices are the endpoints of the transverse axis, the vertices are $(0, \pm 4)$ and the hyperbola opens up and down. Since the fundamental rectangle goes through the endpoints of the transverse axis and the endpoints of the conjugate axis, we can sketch the hyperbola shown in Fig. 7.43. Since the hyperbola opens up and down and is centered at the origin, its equation is of the form

$$\frac{y^2}{a^2} - \frac{x^2}{b^2} = 1,$$

for $a > 0$ and $b > 0$. From the fundamental rectangle we get $a = 4$ and $b = 2$. So the equation of the hyperbola is

$$\frac{y^2}{16} - \frac{x^2}{4} = 1.$$

FOR THOUGHT True or False? Explain.

1. The graph of $\frac{x^2}{4} - \frac{y}{9} = 1$ is a hyperbola.
2. The graph of $\frac{x^2}{16} - \frac{y^2}{9} = 1$ has y-intercepts at $(0, 3)$ and $(0, -3)$.
3. The hyperbola $y^2 - x^2 = 1$ opens up and down.
4. The graph of $y^2 = 4 + 16x^2$ is a hyperbola.
5. Every point that satisfies $y = \frac{b}{a}x$ must satisfy $\frac{x^2}{a^2} - \frac{y^2}{b^2} = 1.$
6. The asymptotes for $x^2 - \frac{y^2}{4} = 1$ are $y = 2x$ and $y = -2x$.
7. The foci for $\frac{x^2}{16} - \frac{y^2}{9} = 1$ are $(5, 0)$ and $(-5, 0)$.
8. The points $(0, \sqrt{8})$ and $(0, -\sqrt{8})$ are the foci for $\frac{y^2}{3} - \frac{x^2}{5} = 1.$
9. The graph of $\frac{x^2}{9} - \frac{y^2}{4} = 1$ intersects the line $y = \frac{2}{3}x.$
10. The graph of $y^2 = 1 - x^2$ is a hyperbola centered at the origin.

7.3 EXERCISES

Tape 20 Disk

Determine the foci and the equations of the asymptotes, and sketch the graph of each hyperbola.

1. $\frac{x^2}{4} - \frac{y^2}{9} = 1$
2. $\frac{x^2}{16} - \frac{y^2}{9} = 1$
3. $\frac{y^2}{4} - \frac{x^2}{25} = 1$
4. $\frac{y^2}{9} - \frac{x^2}{16} = 1$
5. $\frac{x^2}{4} - y^2 = 1$
6. $x^2 - \frac{y^2}{4} = 1$
7. $x^2 - \frac{y^2}{9} = 1$
8. $\frac{x^2}{9} - y^2 = 1$
9. $16x^2 - 9y^2 = 144$
10. $9x^2 - 25y^2 = 225$
11. $x^2 - y^2 = 1$
12. $y^2 - x^2 = 1$

Sketch the graph of each hyperbola. Determine the foci and the equations of the asymptotes.

13. $\frac{(x+1)^2}{4} - \frac{(y-2)^2}{9} = 1$
14. $\frac{(x+3)^2}{16} - \frac{(y+2)^2}{25} = 1$
15. $\frac{(y-1)^2}{4} - (x+2)^2 = 1$
16. $\frac{(y-2)^2}{4} - \frac{(x-1)^2}{9} = 1$
17. $\frac{(x+2)^2}{16} - \frac{(y-3)^2}{9} = 1$
18. $\frac{(x+1)^2}{16} - \frac{(y+2)^2}{25} = 1$
19. $(y-3)^2 - (x-3)^2 = 1$
20. $(y+2)^2 - (x+2)^2 = 1$

Find the equation of each hyperbola described below.

21. Asymptotes $y = \frac{1}{2}x$ and $y = -\frac{1}{2}x$, and x-intercepts $(6, 0)$ and $(-6, 0)$

22. Asymptotes $y = x$ and $y = -x$, and y-intercepts $(0, 2)$ and $(0, -2)$

23. Foci $(5, 0)$ and $(-5, 0)$, and x-intercepts $(3, 0)$ and $(-3, 0)$

24. Foci $(0, 5)$ and $(0, -5)$, and y-intercepts $(0, 4)$ and $(0, -4)$

25. Vertices of the fundamental rectangle $(3, \pm 5)$ and $(-3, \pm 5)$, and opening left and right

26. Vertices of the fundamental rectangle $(1, \pm 7)$ and $(-1, \pm 7)$, and opening up and down

Find the equation of each hyperbola.

27.
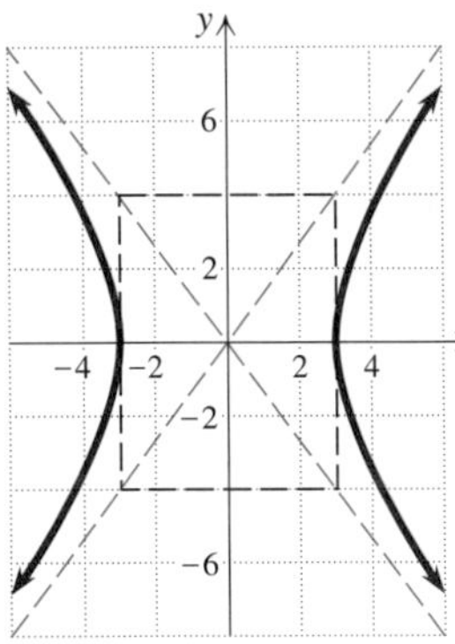

28.
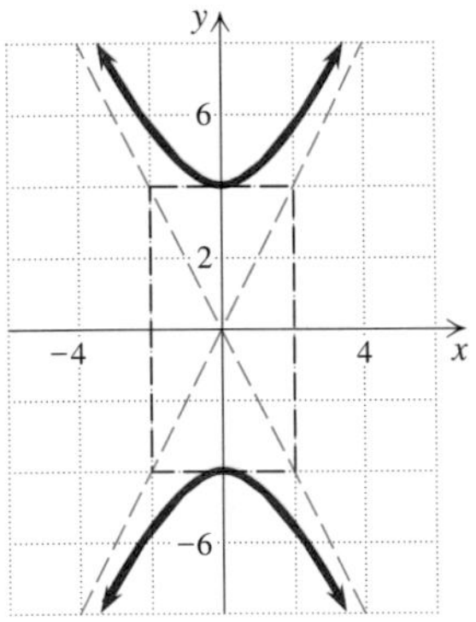

29.
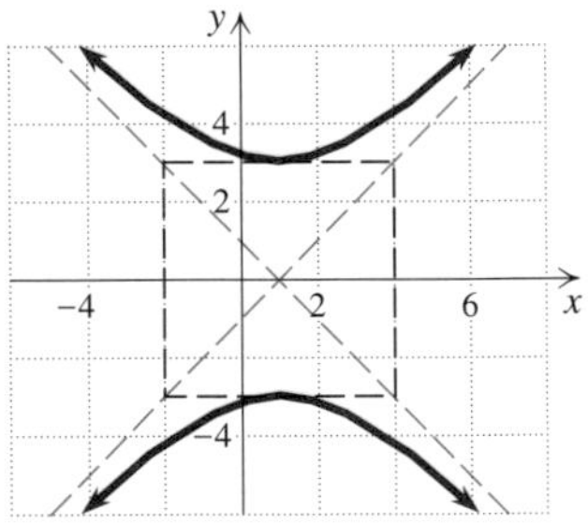

30.
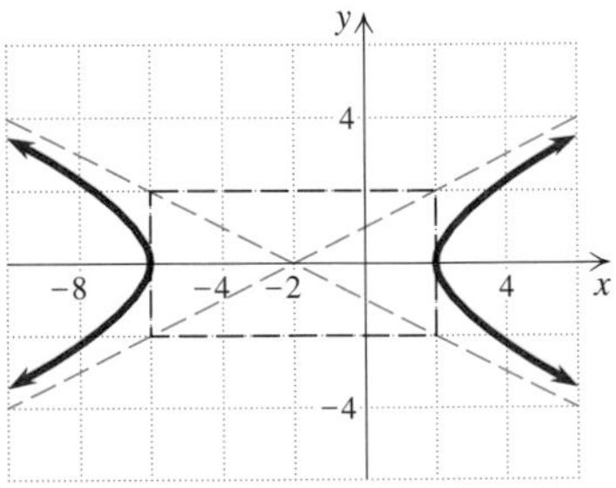

Determine which conic section each equation represents by rewriting each equation in one of the standard forms of the conic sections.

31. $y^2 - x^2 + 2x = 2$

32. $4y^2 + x^2 - 2x = 15$

33. $y - x^2 = 2x$

34. $y^2 + x^2 - 2x = 0$

35. $25x^2 = 2500 - 25y^2$

36. $100x^2 = 25y^2 + 2500$

37. $100y^2 = 2500 - 25x$

38. $100y^2 = 2500 - 25x^2$

39. $2x^2 - 4x + 2y^2 - 8y = -9$

40. $2x^2 + 4x + y = -7$

41. $2x^2 + 4x + y^2 + 6y = -7$

42. $9x^2 - 18x + 4y^2 + 16y = 11$

Solve each problem.

43. *Telephoto Lens* The focus of the main parabolic mirror of a telephoto lens is 10 in. above the vertex as shown in the drawing. A hyperbolic mirror is placed so that its vertex is at $(0, 8)$. If the hyperbola has center $(0, 0)$ and foci $(0, \pm 10)$, then what is the equation of the hyperbola?

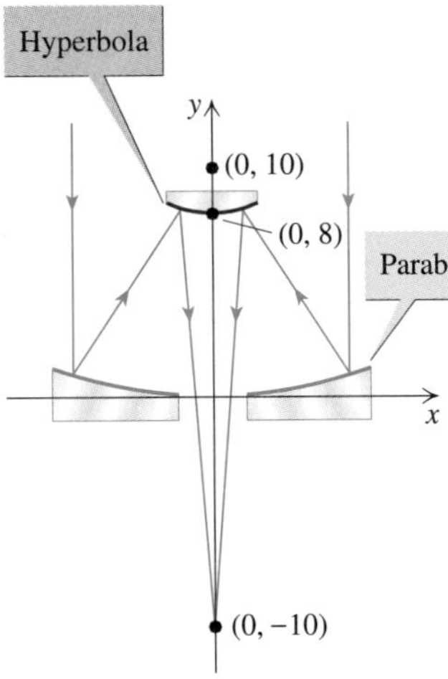

Figure for Exercise 43

44. *Parabolic Mirror* Find the equation of the cross section of the parabolic mirror described in Exercise 43.

45. *Marine Navigation* In 1990 the loran (long range navigation) system had about 500,000 users (*Popular Mechanics*, March 1990). A \$300 loran unit measures the difference in time that it takes for radio signals from pairs of fixed points to reach a ship. The unit then finds the equations of two hyperbolas that pass through the location of the ship and determines the location of the ship. Suppose that the hyperbolas $9x^2 - 4y^2 = 36$ and $16y^2 - x^2 = 16$ pass through the location of a ship in the first quadrant. Find the exact location of the ship.

46. *Air Navigation* A pilot is flying in the coordinate system shown in the figure. Using radio signals emitted from $(4, 0)$

and $(-4, 0)$, he knows that he is four units closer to $(4, 0)$ than he is to $(-4, 0)$. Find the equation of the hyperbola with foci $(\pm 4, 0)$ that passes through his location. He also

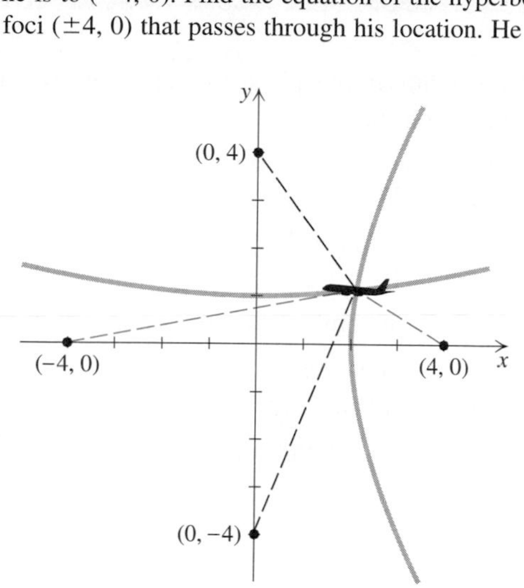

Figure for Exercise 46

knows that he is two units closer to $(0, 4)$ than he is to $(0, -4)$. Find the equation of the hyperbola with foci $(0, \pm 4)$ that passes through his location. If the pilot is in the first quadrant, then what is his exact location?

47. *Points on a Hyperbola* Find all points of the hyperbola $x^2 - y^2 = 1$ that are twice as far from one focus as they are from the other focus.

48. *Perpendicular Asymptotes* For what values of a and b are the asymptotes of the hyperbola $x^2/a^2 - y^2/b^2 = 1$ perpendicular?

Graph each hyperbola on a graphing calculator along with its asymptotes. Observe how close the hyperbola gets to its asymptotes as x gets larger and larger. If $x = 50$, what is the difference between the y-value on the asymptotes and the y-value on the hyperbola?

49. $x^2 - y^2 = 1$

50. $\frac{y^2}{4} - \frac{x^2}{9} = 1$

For Writing/Discussion

51. *Details* Fill in the details in the development of the equation of a hyperbola that is outlined at the beginning of this section.

52. *Cooperative Learning* Work in groups to find conditions on A, B, C, D, and E that will determine whether $Ax^2 + Bx + Cy^2 + Dy = E$ represents a parabola, an ellipse, a circle, or a hyperbola (without rewriting the equation). Test your conditions on the equations of Exercises 31–42.

53. *DeAlwis's Theorem* Let (x_1, y_1) and (x_2, y_2) be any two points on the hyperbola $x^2 - y^2 = 1$. Consider the parallelogram that has these two points as opposite vertices and sides that are parallel to the asymptotes of the hyperbola. Show that the other two opposite vertices of the parallelogram lie on a line through the origin.

Linking Concepts

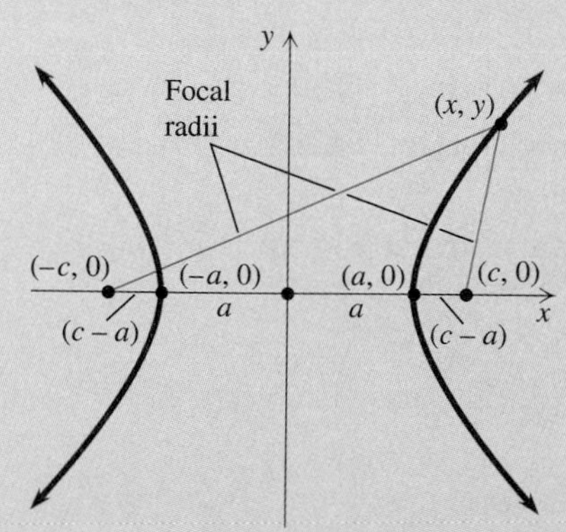

For Individual or Group Explorations

Eccentricity and Focal Radii For the hyperbola $x^2/a^2 - y^2/b^2 = 1$ the eccentricity e is defined as $e = c/a$, where $a^2 + b^2 = c^2$. The line segments joining a point on the hyperbola to the foci as shown in the figure are called the focal radii.

a) What is the eccentricity of a hyperbola if the asymptotes are perpendicular?

b) Let e_1 and e_2 be the eccentricities of the hyperbolas $x^2/a^2 - y^2/b^2 = 1$ and $y^2/b^2 - x^2/a^2 = 1$, respectively. Show that $e_1^2 e_2^2 = e_1^2 + e_2^2$.

c) Show that for a point (x_1, y_1) on the right-hand branch of $x^2/a^2 - y^2/b^2 = 1$ the length of the shorter focal radius is $x_1 e - a$ and the length of the longer focal radius is $x_1 e + a$. (*Hint*: Start with the definition of the hyperbola.)

HIGHLIGHTS

• Section 7.1
The Parabola

1. A parabola is the set of all points in the plane that are equidistant from a fixed line and a fixed point not on the line.
2. The graph of $y = ax^2 + bx + c$ is a parabola opening up if $a > 0$ and down if $a < 0$, and the x-coordinate of the vertex is $x = -b/(2a)$.
3. The graph of $x = ay^2 + by + c$ is a parabola opening right if $a > 0$ and left if $a < 0$, and the y-coordinate of the vertex is $y = -b/(2a)$.
4. The vertex of a parabola is halfway between the focus and directrix.
5. The parabola $y = a(x - h)^2 + k$ has vertex (h, k), focus $(h, k + p)$, and directrix $y = k - p$, where $a = 1/(4p)$.
6. The parabola $x = a(y - h)^2 + k$ has vertex (k, h), focus $(k + p, h)$, and directrix $x = k - p$, where $a = 1/(4p)$.

• Section 7.2
The Ellipse and Circle

1. An ellipse is the set of points in the plane such that the sum of their distances from two fixed points is a constant.
2. If $a > b > 0$, the ellipse $\frac{x^2}{a^2} + \frac{y^2}{b^2} = 1$ has intercepts $(\pm a, 0)$ and $(0, \pm b)$ and foci $(\pm c, 0)$, where $c^2 = a^2 - b^2$.
3. If $a > b > 0$, the ellipse $\frac{x^2}{b^2} + \frac{y^2}{a^2} = 1$ has intercepts $(\pm b, 0)$ and $(0, \pm a)$ and foci $(0, \pm c)$, where $c^2 = a^2 - b^2$.
4. A circle is the set of points in a plane such that their distance from a fixed point is a constant.
5. The graph of $(x - h)^2 + (y - k)^2 = r^2$ for $r > 0$ is a circle with radius r and center (h, k).
6. The graph of $x^2 + y^2 = r^2$ for $r > 0$ is a circle with radius r centered at $(0, 0)$.

• Section 7.3
The Hyperbola

1. A hyperbola is the set of points in a plane such that the difference between the distances from two fixed points is constant.
2. The equation of the hyperbola with foci $(\pm c, 0)$ and x-intercepts $(\pm a, 0)$ is $\frac{x^2}{a^2} - \frac{y^2}{b^2} = 1$, where $b^2 = c^2 - a^2$.
3. The equation of the hyperbola with foci $(0, \pm c)$ and y-intercepts $(0, \pm a)$ is $\frac{y^2}{a^2} - \frac{x^2}{b^2} = 1$, where $b^2 = c^2 - a^2$.

4. Draw the fundamental rectangle and the asymptotes to determine the shape of the hyperbola.

5. By replacing x by $x - h$ and y by $y - k$, an ellipse, circle, or hyperbola with center (0, 0) is translated to one with center at (h, k).

CHAPTER 7 REVIEW EXERCISES

For Exercises 1–6, sketch the graph of each parabola. Determine the x- and y-intercepts, vertex, axis of symmetry, focus, and directrix for each.

1. $y = x^2 + 4x - 12$

2. $y = 4x - x^2$

3. $y = 6x - 2x^2$

4. $y = 2x^2 - 4x + 2$

5. $x = y^2 + 4y - 6$

6. $x = -y^2 + 6y - 9$

Sketch the graph of each ellipse, and determine its foci.

7. $\frac{x^2}{16} + \frac{y^2}{36} = 1$

8. $\frac{x^2}{64} + \frac{y^2}{16} = 1$

9. $\frac{(x-1)^2}{8} + \frac{(y-1)^2}{24} = 1$

10. $\frac{(x+2)^2}{16} + \frac{(y-1)^2}{7} = 1$

11. $5x^2 - 10x + 4y^2 + 24y = -1$

12. $16x^2 - 16x + 4y^2 + 4y = 59$

Determine the center and radius of each circle, and sketch its graph.

13. $x^2 + y^2 = 81$

14. $6x^2 + 6y^2 = 36$

15. $(x + 1)^2 + y^2 = 4$

16. $(x - 2)^2 + (y + 3)^2 = 9$

17. $x^2 + 5x + y^2 + \frac{1}{4} = 0$

18. $x^2 + 3x + y^2 + 5y = \frac{1}{2}$

Write the standard equation for each circle with the given center and radius.

19. Center $(0, -4)$, radius 3

20. Center $(-2, -5)$, radius 1

21. Center $(-2, -7)$, radius $\sqrt{6}$

22. Center $\left(\frac{1}{2}, -\frac{1}{4}\right)$, radius $\frac{\sqrt{2}}{2}$

Sketch the graph of each hyperbola. Determine the foci and the equations of the asymptotes.

23. $\frac{x^2}{64} - \frac{y^2}{36} = 1$

24. $\frac{y^2}{100} - \frac{x^2}{64} = 1$

25. $\frac{(y-2)^2}{64} - \frac{(x-4)^2}{16} = 1$

26. $\frac{(x-5)^2}{100} - \frac{(y-10)^2}{225} = 1$

27. $x^2 - 4x - 4y^2 + 32y = 64$

28. $y^2 - 6y - 4x^2 + 48x = 279$

Identify each equation as the equation of a parabola, ellipse, circle, or hyperbola. Try to do these problems without rewriting the equations.

29. $x^2 = y^2 + 1$

30. $x^2 + y^2 = 1$

31. $x^2 = 1 - 4y^2$

32. $x^2 + 4x + y^2 = 0$

33. $x^2 + y = 1$

34. $y^2 = 1 - x$

35. $x^2 + 4x = y^2$

36. $9x^2 + 7y^2 = 63$

Write each equation in standard form, then sketch the graph of the equation.

37. $x^2 = 4 - y^2$

38. $x^2 = 4y^2 + 4$

39. $x^2 = 4y + 4$

40. $y^2 = 4x - 4$

41. $x^2 = 4 - 4y^2$

42. $x^2 = 4y - y^2$

43. $4y^2 = 4x - x^2$

44. $x^2 - 4x = y^2 + 4y + 4$

Determine the equation of each conic section described below.

45. A parabola with focus (1, 3) and directrix $x = 1/2$

46. A circle centered at the origin and passing through (2, 8)

47. An ellipse with foci $(\pm 4, 0)$ and vertices $(\pm 6, 0)$

48. A hyperbola with asymptotes $y = \pm 3x$ and y-intercepts $(0, \pm 3)$

49. A circle with center (1, 3) and passing through $(-1, -1)$

50. An ellipse with foci (3, 2) and $(-1, 2)$, and vertices (5, 2) and $(-3, 2)$

51. A hyperbola with foci $(\pm 3, 0)$ and x-intercepts $(\pm 2, 0)$

52. A parabola with focus (0, 3) and directrix $y = 1$

Assume that each of the following graphs is the graph of a parabola, ellipse, circle, or hyperbola. Find the equation for each graph.

53.

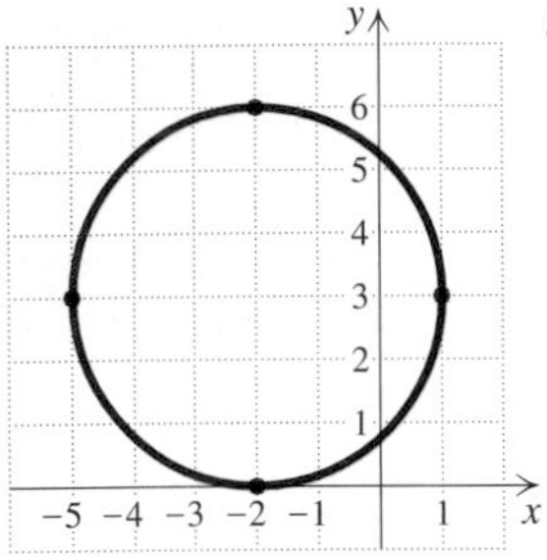

54.

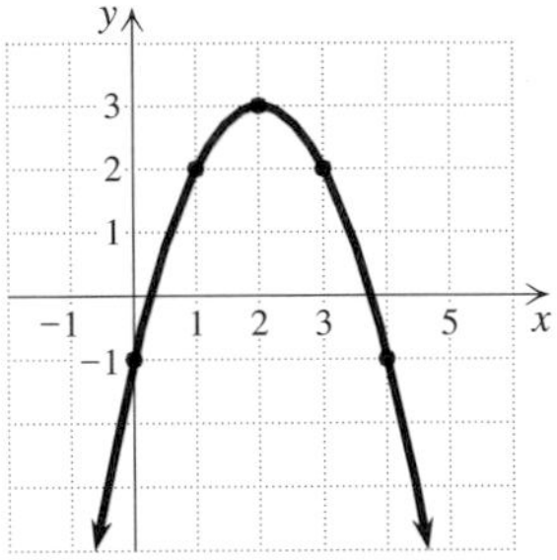

55.

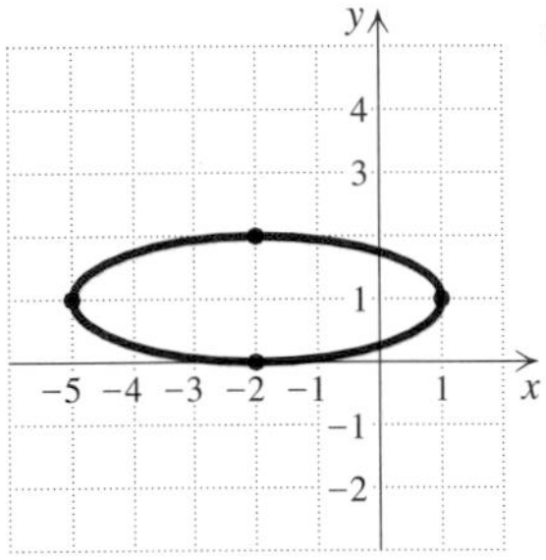

56.

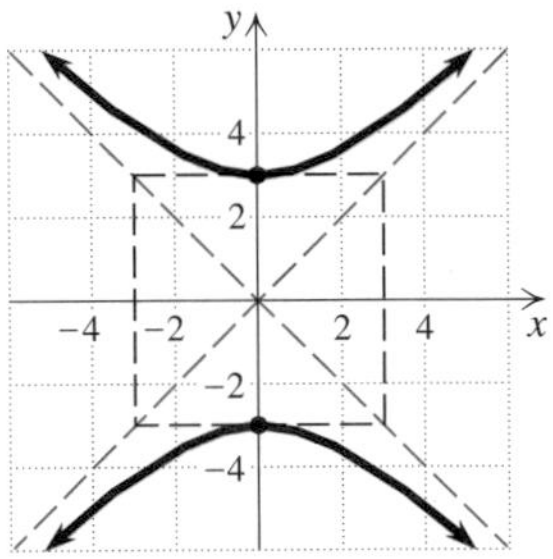

57.

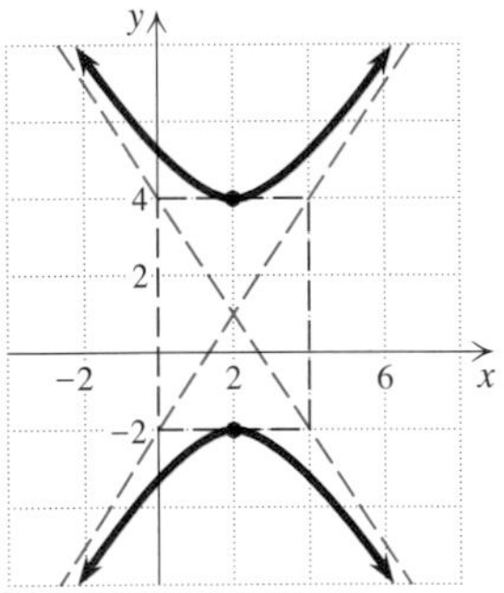

58.

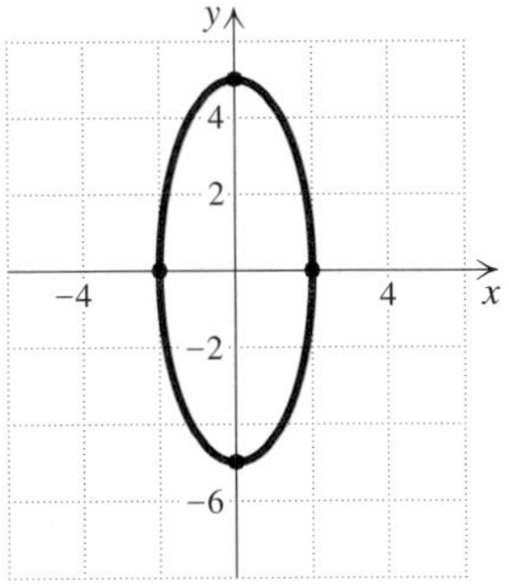

Solve each problem.

59. *Nuclear Power* A cooling tower for a nuclear power plant has a hyperbolic cross section as shown in the figure. The diameter of the tower at the top and bottom is 240 ft, while the diameter at the middle is 200 ft. The height of the tower is $48\sqrt{11}$ ft. Find the equation of the hyperbola, using the coordinate system shown in the figure.

60. *Searchlight* The bulb in a searchlight is positioned 10 in. above the vertex of its parabolic reflector, as shown in the figure. The width of the reflector is 30 in. Using the coordinate system given in the figure, find the equation of the parabola and the thickness t of the reflector at its outside edge.

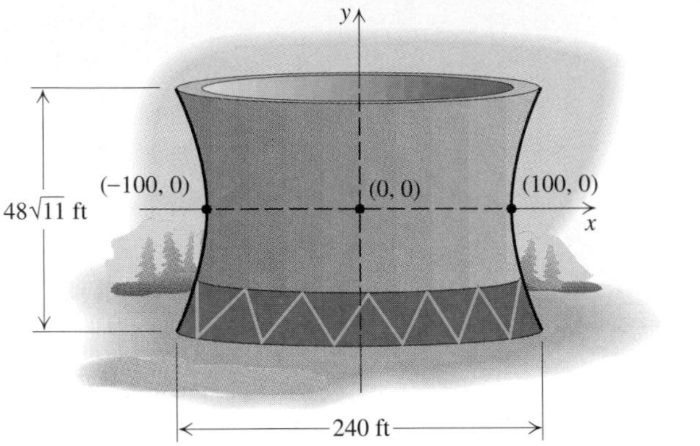

Figure for Exercise 59

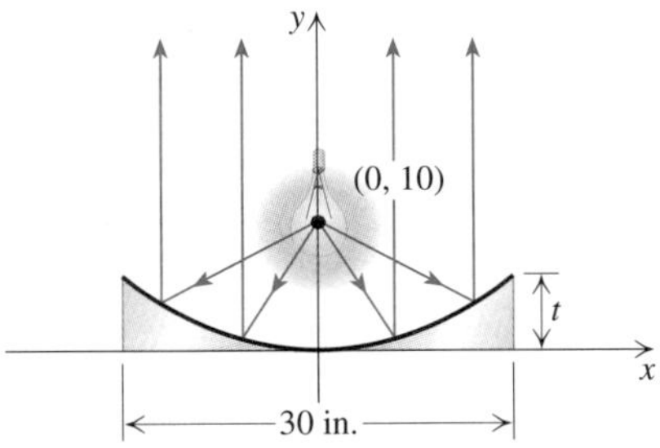

Figure for Exercise 60

61. *Whispering Gallery* In the whispering gallery shown in the figure, the foci of the ellipse are 60 ft apart. Each focus is 4 ft from the vertex of an elliptical reflector. Using the coordinate system given in the figure, find the equation of the ellipse that is used to make the elliptical reflectors and the dimension marked h in the figure.

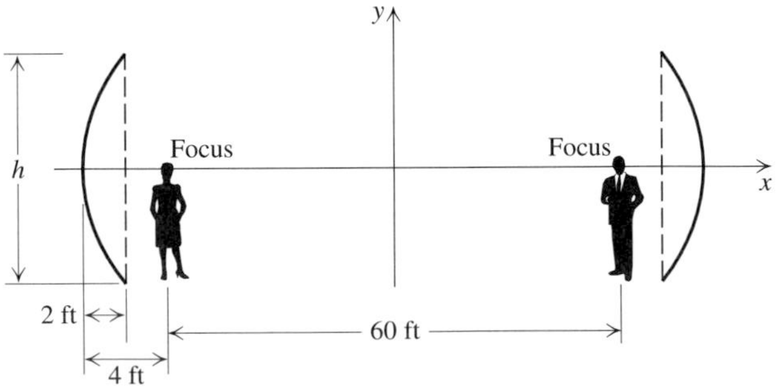

Figure for Exercise 61

CHAPTER 7 TEST

Sketch the graph of each equation.

1. $x^2 + y^2 = 8$

2. $100x^2 + 9y^2 = 900$

3. $x^2 + 6x - y = -8$

4. $\dfrac{y^2}{25} - \dfrac{x^2}{9} = 1$

5. $x^2 + 6x + y^2 - 2y = 0$

6. $\dfrac{(x-2)^2}{9} - \dfrac{(y+3)^2}{4} = 1$

Determine whether each equation is the equation of a parabola, an ellipse, a circle, or a hyperbola.

7. $x^2 - 8x = y^2$

8. $x^2 - 8x = y$

9. $8x - x^2 = y^2$

10. $8x - x^2 = 8y^2$

Determine the equation of each conic section.

11. A circle with center $(-3, 4)$ and radius $2\sqrt{3}$

12. A parabola with focus $(2, 0)$ and directrix $x = -2$

13. An ellipse with focus $(0, \pm\sqrt{6})$ and x-intercepts $(\pm 2, 0)$

14. A hyperbola with foci $(\pm 8, 0)$ and vertices $(\pm 6, 0)$

Solve each problem.

15. Find the focus, directrix, vertex, and axis of symmetry for the parabola $y = x^2 - 4x$.

16. Find the foci, length of the major axis, and length of the minor axis for the ellipse $x^2 + 4y^2 = 16$.

17. Find the foci, vertices, equations of the asymptotes, length of the transverse axis, and length of the conjugate axis for the hyperbola $16y^2 - x^2 = 16$.

18. Find the center and radius of the circle

$$x^2 + x + y^2 - 3y = -\frac{1}{4}.$$

19. A lithotripter is used to disintegrate kidney stones by bombarding them with high-energy shock waves generated at one focus of an elliptical reflector. The lithotripter is positioned so that the kidney stone is at the other focus of the reflector. If the equation $81x^2 + 225y^2 = 18{,}225$ is used for the cross section of the elliptical reflector, with centimeters as the unit of measurement, then how far from the point of generation will the waves be focused?

TYING IT ALL TOGETHER

Chapters P–7

Sketch the graph of each equation.

1. $y = 6x - x^2$

2. $y = 6x$

3. $y = 6 - x^2$

4. $y^2 = 6 - x^2$

5. $y = 6 + x$

6. $y^2 = 6 - x$

7. $y = (6 - x)^2$

8. $y = |x + 6|$

9. $y = 6^x$

10. $y = \log_6(x)$

11. $y = \dfrac{1}{x^2 - 6}$

12. $4x^2 + 9y^2 = 36$

13. $4x^2 - 9y^2 = 36$

14. $y = 6x - x^3$

15. $2x + 3y = 6$

16. $x^2 - 6x = y^2 - 6$

Solve each equation.

17. $3(x - 3) + 5 = -9$

18. $5x - 4(2x - 3) = 17$

19. $2\left(x - \dfrac{1}{3}\right) - 3\left(\dfrac{1}{2} - x\right) = \dfrac{3}{2}$

20. $-4\left(x - \dfrac{1}{2}\right) = 3\left(x + \dfrac{2}{3}\right)$

21. $\dfrac{1}{2}x - \dfrac{1}{3} = \dfrac{1}{4}x + \dfrac{3}{2}$

22. $0.05(x - 20) + 0.02(x + 10) = 2.7$

23. $2x^2 + 31x = 51$

24. $2x^2 + 31x = 0$

25. $x^2 - 34x + 286 = 0$

26. $x^2 - 34x + 290 = 0$

Across the eastern United States, people watched the spectacular meteor shower light up the night sky. Fiery trails cast an eerie glow over the spectators gathering below, yet few onlookers displayed any signs of fear. After all, what was the chance of a direct hit? Then a giant fireball suddenly loomed over New York City and began descending rapidly. Crowds panicked as a mighty blast tore apart skyscrapers and rocked the concrete canyons. The asteroid's impact, which resembled a several-megaton nuclear blast, triggered an earthquake that, in turn, leveled more buildings and plunged the streets in darkness. In a matter of minutes, the great city had been reduced to rubble.

Science fiction? Perhaps, but chunks of rock and ice have been tumbling about in space since the creation of the solar system. And small meteoroids, packing the wallop of a thousand tons of TNT, do enter the Earth's orbit every year or so. Usually they disintegrate upon contact with the atmosphere. However, in 1908, a comet possibly the size of a football field struck Siberia. And in July 1994, the Hubble Telescope captured images of collisions between the planet Jupiter and fragments of the comet Shoemaker-Levy 9. Apparently the impact of only one of the smaller fragments released enough explosive energy to vaporize the city of Los Angeles. Said one planetary scientist, "We can be very glad that this comet was heading for Jupiter and not Earth."

Today scientists are still debating the probability that a "doomsday rock" will collide with Earth. In 1992 a NASA team called for an international watch for these unwelcome visitors.

In this chapter we'll see how discrete mathematics can help us calculate accumulated monthly savings, various ways to select lottery numbers, and the economic impact of a $2,000,000 company payroll on a community. We'll also see how, based on estimates of past collisions, we can determine the probability of an asteroid strike during the next 100 years—an event that could have a lasting impact on your future!

8 Sequences, Series, and Probability

8.1 Sequences
8.2 Series
8.3 Geometric Sequences and Series
8.4 Counting and Permutations
8.5 Combinations, Labeling, and the Binomial Theorem
8.6 Probability
8.7 Mathematical Induction

8.1 Sequences

In everyday life, we hear the term *sequence* in many contexts. We describe a sequence of events, we make a sequence of car payments, or we get something out of sequence. In this section we will give a mathematical definition of the term and explore several applications.

Definition

We can think of a sequence of numbers as *an ordered list* of numbers. For example, your grades on the first four algebra tests can be listed to form a finite sequence. The sequence

$$10, 20, 30, 40, 50 \ldots$$

is an infinite sequence that lists the positive multiples of 10.

We can think of a sequence as a list, but saying that a sequence is a list is too vague for mathematics. The definition of sequence can be clearly stated by using the terminology of functions.

Definition: Sequence

A **finite sequence** is a function whose domain is $\{1, 2, 3, \ldots, n\}$, the positive integers less than or equal to a fixed positive integer n.

An **infinite sequence** is a function whose domain is the set of all positive integers.

When the domain is apparent, we will refer to either a finite or an infinite sequence as a sequence.

The function $f(n) = n^2$ with domain $\{1, 2, 3, 4, 5\}$ is a finite sequence. For the independent variable of a sequence, we usually use n (for natural number) rather than x, and assume that only natural numbers can be used in place of n. For the dependent variable $f(n)$, we generally write a_n (read "a sub n"). So this finite sequence is also defined by

$$a_n = n^2 \quad \text{for } 1 \leq n \leq 5.$$

The **terms** of the sequence are the values of the dependent variable a_n. We call a_n the ***n*th term** or the **general term** of the sequence. The equation $a_n = n^2$ provides a formula for finding the nth term. The five terms of this sequence are $a_1 = 1$, $a_2 = 4$, $a_3 = 9$, $a_4 = 16$, and $a_5 = 25$. We refer to a listing of the terms as the sequence. So 1, 4, 9, 16, 25 is a finite sequence with five terms.

EXAMPLE 1 Listing terms of a finite sequence

Find all terms of the sequence

$$a_n = n^2 - n + 2 \quad \text{for } 1 \leq n \leq 4.$$

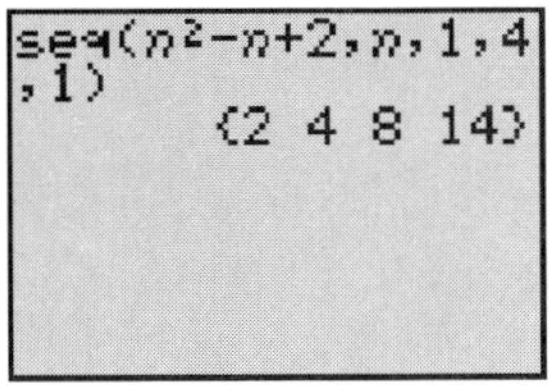

Figure 8.1

Solution

Replace n in the formula by each integer from 1 through 4:

$$a_1 = 1^2 - 1 + 2 = 2, \quad a_2 = 2^2 - 2 + 2 = 4, \quad a_3 = 8, \quad a_4 = 14$$

The four terms of the sequence are 2, 4, 8, and 14.

On a calculator the terms of the sequence $n^2 - n + 2$ are found for $n = 1$ through $n = 4$ in increments of 1 as shown in Fig. 8.1.

EXAMPLE 2 Listing terms of an infinite sequence

Find the first four terms of the infinite sequence

$$a_n = \frac{(-1)^{n-1}2^n}{n}.$$

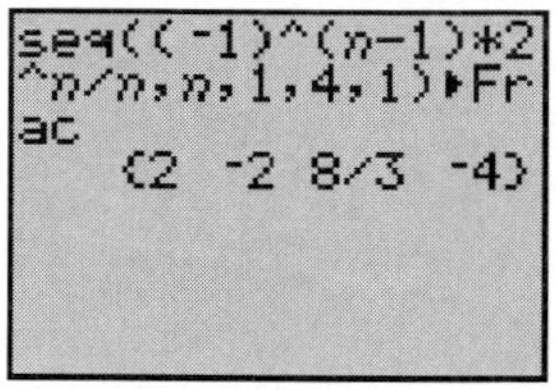

Figure 8.2

Solution

Replace n in the formula by each integer from 1 through 4:

$$a_1 = \frac{(-1)^{1-1}2^1}{1} = 2 \qquad a_2 = \frac{(-1)^1 2^2}{2} = -2$$

$$a_3 = \frac{(-1)^2 2^3}{3} = \frac{8}{3} \qquad a_4 = \frac{(-1)^3 2^4}{4} = -4$$

The first four terms of the infinite sequence are 2, -2, 8/3, and -4.

This sequence is defined on a calculator and the first four terms are listed in Fig. 8.2.

Factorial Notation

Products of consecutive positive integers occur often in sequences and other functions discussed in this chapter. For example, consider the product

$$5 \cdot 4 \cdot 3 \cdot 2 \cdot 1 = 120.$$

The notation 5! (read "five factorial") is used to represent the product of the positive integers from 1 through 5. So $5! = 5 \cdot 4 \cdot 3 \cdot 2 \cdot 1 = 120$. In general, $n!$ is the product of the positive integers from 1 through n. We will find it convenient when writing formulas to have a meaning for 0! even though it does not represent a product of positive integers. The value given to 0! is 1.

Definition: Factorial Notation

For any positive integer n, the notation $n!$ (read **"n factorial"**) is defined by

$$n! = n \cdot (n-1) \cdot \cdots \cdot 3 \cdot 2 \cdot 1$$

The symbol 0! is defined to be 1, $0! = 1$.

EXAMPLE 3 A sequence involving factorial notation

Find the first five terms of the infinite sequence whose nth term is

$$a_n = \frac{(-1)^n}{(n-1)!}.$$

Solution

Replace n in the formula by each integer from 1 through 5. Use $0! = 1$, $1! = 1$, $2! = 2 \cdot 1 = 2$, $3! = 3 \cdot 2 \cdot 1 = 6$, and $4! = 4 \cdot 3 \cdot 2 \cdot 1 = 24$.

$$a_1 = \frac{(-1)^1}{(1-1)!} = \frac{-1}{0!} = -1 \qquad a_2 = \frac{(-1)^2}{1!} = 1$$

$$a_3 = \frac{(-1)^3}{2!} = -\frac{1}{2} \qquad a_4 = \frac{(-1)^4}{3!} = \frac{1}{6} \qquad a_5 = \frac{(-1)^5}{4!} = -\frac{1}{24}$$

Figure 8.3

Using these five terms, we write the infinite sequence as follows:

$$-1, 1, -\frac{1}{2}, \frac{1}{6}, -\frac{1}{24}, \ldots$$

You can use the factorial function and the fraction feature to find a_3, a_4, and a_5 with a calculator as in Fig. 8.3.

Most scientific calculators have a factorial key, which might be labeled $x!$ or $n!$. The factorial key is used to find values such as 12!, which is 479,001,600. The value of 69! is the largest factorial that many calculators can calculate, because 70! is larger than 10^{100}. Try your calculator to find the largest factorial that it can calculate.

Finding a Formula for the *n*th Term

We often have the terms of a sequence and want to write a formula for the sequence. For example, consider your two parents, four grandparents, eight great-grandparents, and so on. The sequence 2, 4, 8, . . . lists the number of ancestors you have in each generation going backward in time. Continuing the sequence, the number of great-great-grandparents is 16. Therefore, we want a formula in which each term is twice the one preceding it. The formula

$$a_n = 2^n$$

is the correct formula for the sequence of ancestors. Note that the formula of Example 1,

$$a_n = n^2 - n + 2,$$

gives the same first three terms as this formula but does not have 16 as the fourth term. So if a sequence such as 2, 4, 8, . . . is given out of context, many formulas might be found that will produce the same given terms. When attempting to write a formula for a given sequence of terms, look for an "obvious" pattern. Of course, some sequences do not have an obvious pattern. For example, there is no known pattern to the infinite sequence 2, 3, 5, 7, 11, 13, 17, . . . , the sequence of prime numbers.

EXAMPLE 4 Finding a formula for a sequence

Write a formula for the general term of each infinite sequence.

a) 6, 8, 10, 12, . . . **b)** 3, 5, 7, 9, . . . **c)** $1, -\frac{1}{4}, \frac{1}{9}, -\frac{1}{16}, \ldots$

Solution

a) Assuming that all terms of the sequence are multiples of 2, we might try the expression $2n$ for the general term. Since the domain of a sequence is the set of positive integers, the expression $2n$ gives the values 2, 4, 6, 8, and so on.

But 2 and 4 are not part of this sequence. To get the desired sequence, use the formula $a_n = 2n + 4$. Check by using $a_n = 2n + 4$ to find

$$a_1 = 2(1) + 4 = 6, \qquad a_2 = 2(2) + 4 = 8,$$
$$a_3 = 2(3) + 4 = 10, \qquad a_4 = 2(4) + 4 = 12.$$

b) Since every odd number is an even number plus 1, let $a_n = 2n + 1$. Using $a_n = 2n + 1$, we get $a_1 = 3$, $a_2 = 5$, $a_3 = 7$, and so on. So the general term of the given sequence is $a_n = 2n + 1$.

c) To obtain alternating signs, use a power of -1. The expression $(-1)^{n+1}$ for the numerator will give a value of 1 in the numerator when n is odd and a value of -1 in the numerator when n is even. Since the denominators are the squares of the positive integers, we use n^2 to produce the denominators. So the nth term of the sequence is

$$a_n = \frac{(-1)^{n+1}}{n^2}.$$

Figure 8.4

Check the formula by finding a_1, a_2, a_3, and a_4 with a calculator as in Fig. 8.4.

Recursion Formulas

So far, the formulas used for the nth term of a sequence have expressed the nth term as a function of n, the number of the term. In another approach, a **recursion formula** gives the nth term as a function *of the previous term*. If the first term is known, then a recursion formula determines the remaining terms of the sequence.

EXAMPLE 5 A recursion formula

Find the first four terms of the infinite sequence in which $a_1 = 3$ and $a_n = (a_{n-1})^2 - 5$ for $n \geq 2$.

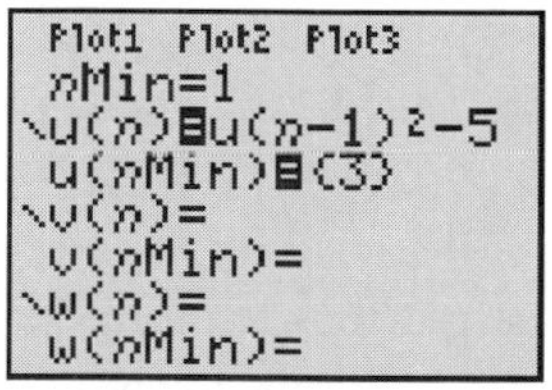

Figure 8.5

Solution

We are given $a_1 = 3$. If $n = 2$, the recursion formula is $a_2 = (a_1)^2 - 5$. Since $a_1 = 3$, we get $a_2 = 3^2 - 5 = 4$. To find the next two terms, we let $n = 3$ and $n = 4$ in the recursion formula:

$$a_3 = (a_2)^2 - 5 = 4^2 - 5 = 11 \qquad \text{Since } a_2 = 4$$
$$a_4 = (a_3)^2 - 5 = 11^2 - 5 = 116 \qquad \text{Since } a_3 = 11$$

Figure 8.6

So the first four terms of the infinite sequence are 3, 4, 11, and 116.

On a calculator a recursion formula is defined in the sequence mode using the Y= key as in Fig. 8.5. The first four terms are shown in Fig. 8.6.

Arithmetic Sequences

In the sequence

$$4, 9, 14, 19, 24, \ldots$$

each term is 5 larger than the previous term. So the sequence could be written as follows:

$$4, \quad 4 + 5, \quad 4 + 2 \cdot 5, \quad 4 + 3 \cdot 5, \quad 4 + 4 \cdot 5, \ldots$$

A formula for the general term of this sequence is

$$a_n = 4 + (n - 1)5.$$

This sequence is an example of an arithmetic sequence.

Definition: Arithmetic Sequence

A sequence that has an nth term of the form

$$a_n = a_1 + (n - 1)d,$$

where a_1 and d are any real numbers, is called an **arithmetic sequence**.

In an arithmetic sequence, each term after the first is obtained by adding a constant to the previous term. The first term is a_1, the second term is $a_1 + d$, the third term is $a_1 + 2d$, the fourth term is $a_1 + 3d$, and so on. The number d is called the **common difference**. A sequence is an arithmetic sequence if and only if there is a common difference between consecutive terms.

EXAMPLE 6 Finding a formula for an arithmetic sequence

Determine whether each sequence is arithmetic. If it is, then write a formula for the general term of the sequence.

a) 3, 7, 11, 15, 19, . . . **b)** 5, 2, −1, −4, . . . **c)** 2, 6, 18, 54, . . .

Solution

a) Since each term of the sequence is 4 larger than the previous term, the sequence is arithmetic and $d = 4$. Since $a_1 = 3$, the formula for the nth term is

$$a_n = 3 + (n - 1)4.$$

This formula can be simplified to

$$a_n = 4n - 1.$$

Points that satisfy $u_n = 4n - 1$ lie in a straight line in Fig. 8.7 because an arithmetic sequence has the same form as a linear function. □

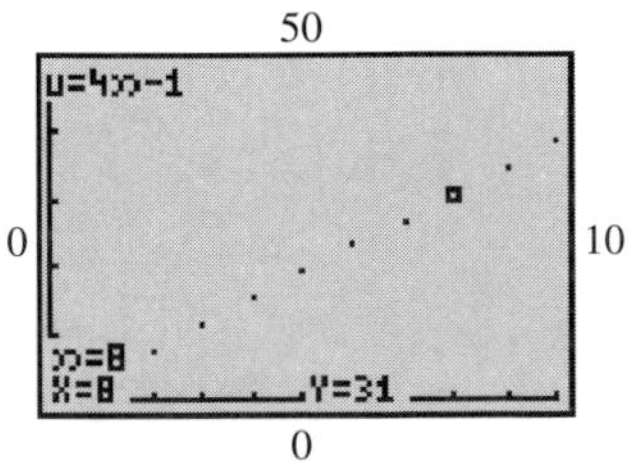

Figure 8.7

b) Since each term of this sequence is 3 smaller than the previous term, the sequence is arithmetic and $d = -3$. Since $a_1 = 5$, we have

$$a_n = 5 + (n - 1)(-3).$$

This formula can be simplified to

$$a_n = -3n + 8.$$

c) For 2, 6, 18, 54, . . . , we get $6 - 2 = 4$ and $18 - 6 = 12$. Since the difference between consecutive terms is not constant, the sequence is not arithmetic.

Note that when the general term of an arithmetic sequence is simplified, it has the form of a linear function.

EXAMPLE 7 Finding the terms of an arithmetic sequence

Find the first four terms and the 30th term of each arithmetic sequence.

a) $a_n = -7 + (n - 1)6$ **b)** $a_n = -\frac{1}{2}n + 8$

Solution

a) Let n take the values from 1 through 4:

$$a_1 = -7 + (1 - 1)6 = -7$$
$$a_2 = -7 + (2 - 1)6 = -1$$
$$a_3 = -7 + (3 - 1)6 = 5$$
$$a_4 = -7 + (4 - 1)6 = 11$$

The first four terms of the sequence are $-7, -1, 5, 11$. The 30th term is

$$a_{30} = -7 + (30 - 1)6 = 167.$$

b) Let n take the values from 1 through 4:

$$a_1 = -\frac{1}{2}(1) + 8 = \frac{15}{2}$$
$$a_2 = -\frac{1}{2}(2) + 8 - 7$$
$$a_3 = -\frac{1}{2}(3) + 8 = \frac{13}{2}$$
$$a_4 = -\frac{1}{2}(4) + 8 = 6$$

The first four terms of the sequence are 15/2, 7, 13/2, 6. The 30th term is

$$a_{30} = -\frac{1}{2}(30) + 8 = -7.$$

The formula $a_n = a_1 + (n - 1)d$ involves four quantities, a_n, a_1, n, and d. If any three of them are known, the fourth can be found.

EXAMPLE 8 Finding a term of an arithmetic sequence

An insurance representative made \$30,000 her first year and \$60,000 her seventh year. Assume that her annual salary figures form an arithmetic sequence and predict what she will make in her tenth year.

Solution

The seventh term is $a_7 = a_1 + (7 - 1)d$. Use $a_7 = 60{,}000$ and $a_1 = 30{,}000$ in this equation to find d:

$$\begin{aligned} 60{,}000 &= 30{,}000 + (7 - 1)d \\ 30{,}000 &= 6d \\ 5000 &= d \end{aligned}$$

Now use the formula $a_n = a_1 + (n - 1)(5000)$ to find a_{10}:

$$\begin{aligned} a_{10} &= 30{,}000 + (10 - 1)(5000) \\ a_{10} &= 75{,}000 \end{aligned}$$

So the predicted salary for her tenth year is \$75,000.

Using a recursion formula, an arithmetic sequence with first term a_1 and constant difference d is defined by $a_n = a_{n-1} + d$ for $n \geq 2$.

EXAMPLE 9 A recursion formula for an arithmetic sequence

Find the first four terms of the sequence in which $a_1 = -7$ and $a_n = a_{n-1} + 6$ for $n \geq 2$.

Solution

The recursion formula indicates that each term after the first is obtained by adding 6 to the previous term:

$$\begin{aligned} a_1 &= -7 \\ a_2 &= a_1 + 6 = -7 + 6 = -1 \\ a_3 &= a_2 + 6 = -1 + 6 = 5 \\ a_4 &= a_3 + 6 = 5 + 6 = 11 \end{aligned}$$

The first four terms are -7, -1, 5, and 11. Note that this recursion formula produces the same sequence as the formula in Example 7(a).

On a calculator define u_n to be $u_{n-1} + 6$ and get the first four terms as shown in Fig. 8.8.

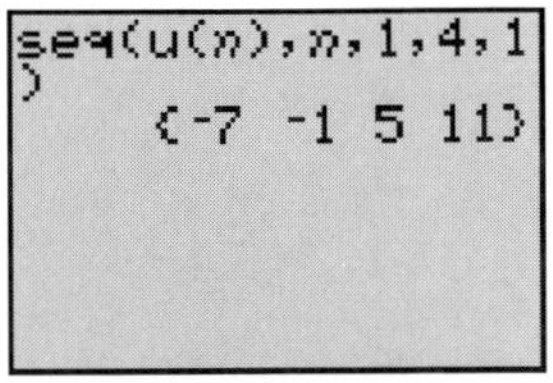

Figure 8.8

FOR THOUGHT True or False? Explain.

1. The equation $a_n = e^n$ for n a natural number defines a sequence.
2. The domain of a finite sequence is the set of positive integers.
3. We can think of a sequence as a list of the values of the dependent variable.
4. The letter n is used to represent the dependent variable.
5. The first four terms of $a_n = (-1)^{n-1}n^3$ are $-1, 8, -27, 81$.
6. The fifth term of $a_n = -3 + (n - 1)6$ is 5.
7. The common difference in the arithmetic sequence 7, 4, 1, −2, . . . is 3.
8. The sequence 1, 4, 9, 16, 25, 36, . . . is an arithmetic sequence.
9. If the first term of an arithmetic sequence is 4 and the third term is 14, then the fourth term is 24.
10. The sequence $a_n = 5 + 2n$ is an arithmetic sequence.

8.1 EXERCISES

Tape 21 Disk

Find all terms of each finite sequence.

1. $a_n = n^2,\ 1 \le n \le 7$
2. $a_n = (n - 1)^2,\ 1 \le n \le 5$
3. $b_n = \dfrac{(-1)^{n+1}}{n + 1},\ 1 \le n \le 8$
4. $b_n = (-1)^n 3n,\ 1 \le n \le 4$
5. $c_n = (-2)^{n-1},\ 1 \le n \le 6$
6. $c_n = (-3)^{n-2},\ 1 \le n \le 6$
7. $a_n = 2^{2-n},\ 1 \le n \le 5$
8. $a_n = \left(\dfrac{1}{2}\right)^{3-n},\ 1 \le n \le 7$
9. $a_n = -6 + (n - 1)(-4),\ 1 \le n \le 5$
10. $a_n = -2 + (n - 1)4,\ 1 \le n \le 8$
11. $b_n = 5 + (n - 1)(0.5),\ 1 \le n \le 7$
12. $b_n = \dfrac{1}{4} + (n - 1)\left(-\dfrac{1}{2}\right),\ 1 \le n \le 5$
13. $c_n = \dfrac{n^2}{n!},\ 1 \le n \le 5$
14. $c_n = \dfrac{n!}{(n - 2)!},\ 2 \le n \le 9$
15. $a_n = (n - 1)!,\ 1 \le n \le 7$
16. $t_n = (2n)!,\ 1 \le n \le 4$

Find the first four terms and the 10th term of each infinite sequence whose nth term is given.

17. $a_n = 8 + (n - 1)(-3)$
18. $a_n = -7 + (n - 1)(0.5)$
19. $a_n = \dfrac{4}{2n + 1}$
20. $a_n = \dfrac{2}{n^2 + 1}$
21. $a_n = \dfrac{(-1)^n}{(n + 1)(n + 2)}$
22. $a_n = \dfrac{(-1)^{n+1}}{(n + 1)^2}$
23. $a_n = \dfrac{2^n}{n!}$
24. $a_n = \dfrac{(-1)^n}{(n + 1)!}$
25. $a_n = \dfrac{(-2)^{2n-1}}{(n - 1)!}$
26. $a_n = \dfrac{e^{-n}}{(n + 2)!}$
27. $a_n = -0.1n + 9$
28. $a_n = 0.3n - 0.4$

Find the first four terms and the eighth term of each infinite sequence given by a recursion formula.

29. $a_n = 3a_{n-1} + 2,\ a_1 = -4$
30. $a_n = 1 - \dfrac{1}{a_{n-1}},\ a_1 = 2$
31. $a_n = (a_{n-1})^2 - 3,\ a_1 = 2$
32. $a_n = (a_{n-1})^2 - 2,\ a_1 = -2$
33. $a_n = a_{n-1} + 7,\ a_1 = -15$
34. $a_n = \dfrac{1}{2}a_{n-1},\ a_1 = 8$

For Exercises 35–46, write a formula for the nth term of each infinite sequence. Do not use a recursion formula.

35. 2, 4, 6, 8, . . .
36. 1, 3, 5, 7, . . .
37. 9, 11, 13, 15, . . .
38. 14, 16, 18, 20, . . .
39. 1, −1, 1, −1, . . .
40. $-\dfrac{1}{2}, \dfrac{1}{2}, -\dfrac{1}{2}, \dfrac{1}{2}, \ldots$

41. 1, 8, 27, 64, . . .

42. $1, -\frac{1}{8}, \frac{1}{27}, -\frac{1}{64}, \ldots$

43. $e, e^2, e^3, e^4, \ldots$

44. $\pi, 4\pi, 9\pi, 16\pi, \ldots$

45. $1, \frac{1}{2}, \frac{1}{4}, \frac{1}{8}, \ldots$

46. $1, -3, 9, -27, \ldots$

Determine whether each given sequence could be an arithmetic sequence.

47. 2, 3, 4, 5, . . .

48. $-7, -4, -2, 0, \ldots$

49. 1, 0.5, 1, 0.5, . . .

50. $3, 0, -3, -6, \ldots$

51. 2, 4, 8, 16, . . .

52. $1, \frac{5}{4}, \frac{3}{2}, \frac{7}{4}, \ldots$

53. $\frac{\pi}{4}, \frac{\pi}{2}, \frac{3\pi}{4}, \pi, \ldots$

54. 1, 2, 3, 2, 3, . . .

Write a formula for the *n*th term of each arithmetic sequence. Do not use a recursion formula.

55. 1, 6, 11, 16, . . .

56. 2, 5, 8, 11, . . .

57. 0, 2, 4, 6, . . .

58. $-3, 3, 9, 15, \ldots$

59. $5, 1, -3, -7, \ldots$

60. $1, -1, -3, -5, \ldots$

61. 1, 1.1, 1.2, 1.3, . . .

62. 2, 2.75, 3.5, 4.25, . . .

63. $\frac{\pi}{6}, \frac{\pi}{3}, \frac{\pi}{2}, \frac{2\pi}{3}, \ldots$

64. $\frac{\pi}{12}, \frac{\pi}{6}, \frac{\pi}{4}, \frac{\pi}{3}, \ldots$

65. 20, 35, 50, 65, . . .

66. 70, 60, 50, 40, . . .

Find the first four terms and the 10th term of each arithmetic sequence.

67. $a_n = 6 + (n - 1)(-3)$

68. $b_n = -12 + (n - 1)4$

69. $c_n = 1 + (n - 1)(-0.1)$

70. $q_n = 10 - 5n$

71. $w_n = -\frac{1}{3}n + 5$

72. $t_n = \frac{1}{2}n + \frac{1}{2}$

Find the indicated part of each arithmetic sequence.

73. Find the eighth term of the sequence that has a first term of -3 and a common difference of 5.

74. Find the 11th term of the sequence that has a first term of 4 and a common difference of -0.8.

75. Find the 10th term of the sequence whose third term is 6 and whose seventh term is 18.

76. Find the eighth term of the sequence whose second term is 20 and whose fifth term is 10.

77. Find the common difference of the sequence in which the first term is 12 and the 21st term is 96.

78. Find the common difference of the sequence in which the first term is 5 and the 11th term is -10.

79. Find a formula for a_n, given that $a_3 = 10$ and $a_7 = 20$.

80. Find a formula for a_n, given that $a_5 = 30$ and $a_{10} = -5$.

Write a recursion formula for each sequence.

81. 3, 12, 21, 30, . . .

82. 30, 25, 20, 15, . . .

83. $\frac{1}{3}, 1, 3, 9, \ldots$

84. $4, -1, \frac{1}{4}, -\frac{1}{16}, \ldots$

85. $16, 4, 2, \sqrt{2}, \ldots$

86. $t^2, t^4, t^8, t^{16}, \ldots$

Solve each problem.

87. *Recursive Pricing* The MSRP for a 1998 Jeep Cherokee Classic is $20,480 (Edmund's 1998 New Truck Prices). Analysts estimate that the MSRP will increase 6% per year for the next five years. Find the price to the nearest dollar for the Jeep Cherokee Classic from 1999 through 2003. Write a formula for this sequence.

88. *Rising Salary* Suppose that you made $20,480 in 1998 and your boss promised that you will get a $1229 raise each year for the next five years. Find your salary for the years 1999 through 2003. Write a formula for this sequence. Is your salary in 2003 equal to the price of the 2003 Jeep Cherokee Classic from the previous exercise?

89. *Reading Marathon* On November 1 an English teacher had his class read five pages of a long novel. He then told them to increase their daily reading by three pages each day. For example, on November 2 they should read eight pages. Write a formula for the number of pages that they will read on the *n*th day of November. If they follow the teacher's instructions, then how many pages will they be reading on the last day of November?

90. *Stiff Penalty* If a contractor does not complete a multimillion-dollar construction project on time, he must pay a penalty of $500 for the first day that he is late, $700 for the second day, $900 for the third day, and so on. Each day the penalty is $200 larger than the previous day. Write a formula for the penalty on the *n*th day. What is the penalty for the 10th day?

91. *Nursing Home Care* The estimated annual cost to reside in a nursing home in 1996 was $35,000 (*Fortune*, October 14, 1996). If the average increase in the annual cost is $1800 each year, then what will the annual cost be in 2010?

92. *Good Planning* Sam's retirement plan gives her a fixed raise of *d* dollars each year. If her retirement income was

\$24,500 in her fifth year of retirement and \$25,700 in her ninth year, then what was her income her first year? What will her income be in her 13th year of retirement?

93. *Countertops* It takes C_n corner tiles, E_n edge tiles, and I_n interior tiles to cover an n ft by n ft island as shown in the accompanying figure. Assuming all tiles are 6 in. by 6 in., write expressions for C_n, E_n, and I_n.

94. *Countertop Pricing* In the previous exercise, corner tiles cost \$0.89 each, edge tiles cost \$0.79 each, and interior tiles cost \$0.69 each. Write an expression for K_n, the cost of the tiles to cover an n ft by n ft countertop.

For Writing/Discussion

95. Explain the difference between a function and a sequence.

96. *Cooperative Learning* Write your own formula for the nth term of a sequence on a piece of paper and list the first five terms. Disclose the terms one at a time to your classmates, giving them the opportunity to guess the formula after each disclosed term.

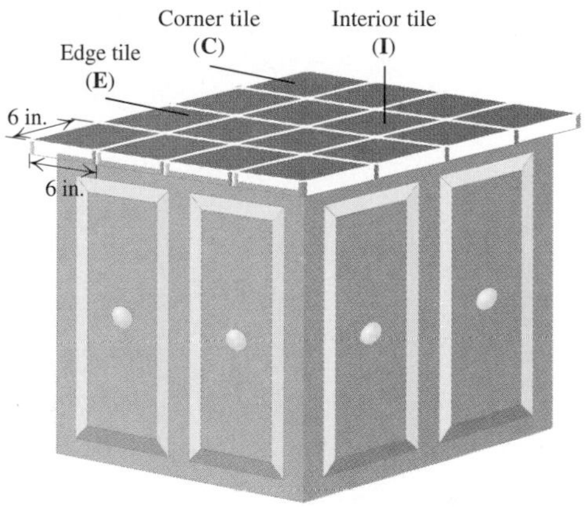

Figure for Exercise 93

LINKING CONCEPTS

For Individual or Group Explorations

Converging Sequences *If the terms of a sequence a_n get closer and closer to some number L as n gets larger and larger without bound, then we say that the sequence converges to L and L is the limit of the sequence. If the sequence does not converge then it diverges. These terms are defined with greater precision and studied extensively in calculus, but with a graphing calculator we can gain a good understanding of these ideas.*

a) Let $a_n = (0.99999)^n$ and find a_n for $n = 100$, 1000, and 1,000,000. What do you think is the limit of this sequence?

b) Let $a_n = (1.00001)^n$ and find a_n for $n = 100$, 1000, and 1,000,000. What do you think is the limit of this sequence?

c) Let $a_n = (1 + 1/n)^n$ and find a_n for $n = 1000$, 10,000, and 100,000. This sequence converges to a number that you have seen earlier in this course. What is the limit? What is the difference between this sequence and the one in part (b)?

d) Graph the functions $f(x) = (0.99999)^x$, $g(x) = (1.00001)^x$, and $h(x) = (1 + 1/x)^x$. Identify any horizontal asymptotes for these graphs. What is the relationship between horizontal asymptotes and limits of sequences?

e) Consider the sequence $a_n = k^n$ where k is a fixed real number. For what values of k do you think the sequence converges and for what values do you think it diverges?

8.2 Series

In this section we continue the study of sequences, but here we concentrate on finding the sum of the terms of a sequence. For example, if an employee starts at \$20,000 per year and gets a \$1200 raise each year for the next 39 years, then the total pay for 40 years of work is the sum of 40 terms of an arithmetic sequence:

$$20{,}000 + 21{,}200 + 22{,}400 + \cdots + 46{,}800$$

This sum is an example of a *series*. We could add the 40 terms to find the total pay, but we will soon discover a formula for this sum that involves only the first term, the last term, and the number of terms. (The total investment in an employee over a lifetime of work is often cited as a reason for selecting personnel carefully.)

Summation Notation

As a shorthand way to indicate the sum of the terms of a sequence, we adopt a new notation called **summation notation**. We use the Greek letter Σ (sigma) in summation notation. For example, the sum of the annual salaries for 40 years of work is written as

$$\sum_{n=1}^{40} [20{,}000 + (n - 1)(1200)].$$

Following the letter sigma is the formula for the nth term of an arithmetic sequence in which the first term is 20,000 and the common difference is 1200. The numbers below and above the letter sigma indicate that this is an expression for the sum of the first through fortieth terms of this sequence.

As another example, the sum of the squares of the first five positive integers can be written as

$$\sum_{i=1}^{5} i^2.$$

To evaluate this sum, let i take the integral values from 1 through 5 in the expression i^2. Thus

$$\sum_{i=1}^{5} i^2 = 1^2 + 2^2 + 3^2 + 4^2 + 5^2 = 1 + 4 + 9 + 16 + 25 = 55.$$

The letter i in the summation notation is called the **index of summation**. Although we usually use i or n for the index of summation, any letter may be used. For example, the expressions

$$\sum_{n=1}^{5} n^2, \quad \sum_{j=1}^{5} j^2, \quad \text{and} \quad \sum_{i=1}^{5} i^2$$

all have the same value.

In the summation notation, the expression following the letter sigma is the *general term* of a sequence. The numbers below and above sigma indicate which terms of the sequence are to be added.

EXAMPLE 1 Evaluating summations

Find the sum in each case.

a) $\sum_{i=1}^{6} (-1)^i 2^{i-1}$ **b)** $\sum_{n=3}^{7} (2n - 1)$ **c)** $\sum_{i=1}^{5} 4$

Solution

a) Evaluate $(-1)^i 2^{i-1}$ for $i = 1$ through 6 and add the resulting terms:

$$\sum_{i=1}^{6} (-1)^i 2^{i-1} = (-1)^1 2^0 + (-1)^2 2^1 + (-1)^3 2^2 + (-1)^4 2^3 + (-1)^5 2^4 + (-1)^6 2^5$$
$$= -1 + 2 - 4 + 8 - 16 + 32 = 21$$

b) Find the third through seventh terms of the sequence whose general term is $2n - 1$ and add the results:

$$\sum_{n=3}^{7} (2n - 1) = 5 + 7 + 9 + 11 + 13 = 45$$

c) Every term of this series is 4. The notation $i = 1$ through 5 means that we add the first five 4's from a sequence in which every term is 4:

$$\sum_{i=1}^{5} 4 = 4 + 4 + 4 + 4 + 4 = 20$$

An expression in summation notation or an expression such as $1 + 2 + 3$, in which we have not actually performed the addition, is called an **indicated sum**.

Definition: Series

The indicated sum of the terms of a sequence is called a **series**.

Just as a sequence may be finite or infinite, a series may be finite or infinite. In this section we will discuss only finite series. In Section 8.3 we will discuss one type of infinite series.

To write a series in summation notation, a formula must be found for the nth term of the corresponding sequence.

EXAMPLE 2 Writing a series in summation notation

Write each series using summation notation.

a) $2 + 4 + 6 + 8 + 10$ **b)** $\frac{1}{5} - \frac{1}{7} + \frac{1}{9} - \frac{1}{11} + \frac{1}{13}$

Solution

a) The series consists of a sequence of even integers. The nth term for this sequence is $a_n = 2n$. This series consists of five terms of this sequence.

$$2 + 4 + 6 + 8 + 10 = \sum_{i=1}^{5} 2i$$

b) This series has two features of interest: The denominators of the fractions are odd integers and the signs alternate. If we use $a_n = 2n + 1$ as the general term for odd integers, we get $a_1 = 3$ and $a_2 = 5$. We do not always have to use $i = 1$ in the summation notation. In this series it is easier to use $i = 2$ through 6. Use $(-1)^i$ to get the alternating signs:

$$\frac{1}{5} - \frac{1}{7} + \frac{1}{9} - \frac{1}{11} + \frac{1}{13} = \sum_{i=2}^{6} \frac{(-1)^i}{2i + 1}$$

Changing the Index of Summation

In Example 2(b) the index of summation i ranged from 2 through 6, but the starting point of the index is arbitrary. The notation

$$\sum_{j=3}^{7} \frac{(-1)^{j-1}}{2j - 1}$$

is another notation for the same sum. The summation notation for a given series can be written so that the index begins at any given number.

EXAMPLE 3 Adjusting the index of summation

Rewrite the series so that instead of index i, it has index j, where j starts at 1.

$$\sum_{i=2}^{6} \frac{(-1)^i}{2i + 1}$$

Solution

In this series, i takes the values 2 through 6. If j starts at 1, then $j = i - 1$, and j takes the values 1 through 5. If $j = i - 1$, then $i = j + 1$. To change the formula for the general term of the series, replace i by $j + 1$:

$$\sum_{i=2}^{6} \frac{(-1)^i}{2i + 1} = \sum_{j=1}^{5} \frac{(-1)^{j+1}}{2(j + 1) + 1} = \sum_{j=1}^{5} \frac{(-1)^{j+1}}{2j + 3}$$

Check that these two series have exactly the same five terms.

The Mean

If you take a sequence of three tests, then your ''average'' is the sum of the three test scores divided by 3. What is commonly called the ''average'' is called the *mean* or *arithmetic mean* in mathematics. The mean can be defined using summation notation.

Definition: Mean

The **mean** of the numbers $x_1, x_2, x_3, \ldots, x_n$ is the number $\bar{x}$, given by

$$\bar{x} = \frac{\sum_{i=1}^{n} x_i}{n}.$$

EXAMPLE 4 Finding the mean of a sequence of numbers

Find the mean of the numbers $-12, 3, 0, 5, -2, 9$.

Solution

To find the mean, divide the total of the six numbers by 6:

$$\bar{x} = \frac{-12 + 3 + 0 + 5 + (-2) + 9}{6} = \frac{1}{2}$$

The mean is 1/2.

Arithmetic Series

The indicated sum of an arithmetic sequence is an **arithmetic series**. The sum of a finite arithmetic series can be found without actually adding up all of the terms. Let S represent the sum of the even integers from 2 through 50. We can find S by using the following procedure:

$$S = 2 + 4 + 6 + 8 + \cdots + 48 + 50$$

$$S = 50 + 48 + 46 + 44 + \cdots + 4 + 2 \quad \text{Write terms in reverse order.}$$

$$2S = 52 + 52 + 52 + 52 + \cdots + 52 + 52 \quad \text{Add corresponding terms.}$$

Since there are 25 numbers in the series of even integers from 2 through 50, the number 52 appears 25 times on the right-hand side of the last equation.

$$2S = 25(52) = 1300$$

$$S = 650$$

So the sum of the even integers from 2 through 50 is 650.

We can use the same idea to develop a formula for the sum of n terms of any arithmetic series. Let $S_n = a_1 + a_2 + a_3 + \cdots + a_n$ be a finite arithmetic series. Since there is a constant difference between the terms, S_n can be written forwards and backwards as follows:

$$\begin{array}{rl} S_n = a_1 & + (a_1 + d) + (a_1 + 2d) + \cdots + a_n \\ S_n = a_n & + (a_n - d) + (a_n - 2d) + \cdots + a_1 \\ \hline 2S_n = (a_1 + a_n) & + (a_1 + a_n) + (a_1 + a_n) + \cdots + (a_1 + a_n) \end{array} \quad \text{Add.}$$

Now, there are n terms of the type $a_1 + a_n$ on the right-hand side of the last equation, so the right-hand side can be simplified:

$$2S_n = n(a_1 + a_n)$$

$$S_n = \frac{n}{2}(a_1 + a_n)$$

This result is summarized as follows.

Theorem: Sum of an Arithmetic Series

The sum S_n of the first n terms of an arithmetic series with first term a_1 and nth term a_n is given by the formula

$$S_n = \frac{n}{2}(a_1 + a_n).$$

So, to find the sum of any arithmetic series, all we need to know is the first term, the last term, and the number of terms.

EXAMPLE 5 Finding the sum of an arithmetic series

Find the sum of each arithmetic series.

a) $\sum_{i=1}^{15} (3i - 5)$ **b)** $36 + 41 + 46 + 51 + \cdots + 91$

Solution

a) To find the sum, we need to know the first term, the last term, and the number of terms. For this series, $a_1 = -2$, $a_{15} = 40$, and $n = 15$. So

$$\sum_{i=1}^{15} (3i - 5) = \frac{15}{2}(-2 + 40) = 285.$$

b) For this series, $a_1 = 36$ and $a_n = 91$, but to find S_n, the number of terms n must be known. We can find n from the formula for the general term of the arithmetic sequence $a_n = a_1 + (n - 1)d$ by using $d = 5$:

$$91 = 36 + (n - 1)5$$
$$55 = (n - 1)5$$
$$11 = n - 1$$
$$12 = n$$

We can now use $n = 12$, $a_1 = 36$, and $a_{12} = 91$ in the formula to get the sum of the 12 terms of this arithmetic series:

$$S_{12} = \frac{12}{2}(36 + 91) = 762$$

In the next example we return to the problem of finding the sum of the 40 annual salaries presented at the beginning of this section.

EXAMPLE 6 Total salary for 40 years of work

Find the total salary for an employee who is paid \$20,000 for the first year and receives a \$1200 raise each year for the next 39 years. Find the mean of the 40 annual salaries.

Solution

The salaries form an arithmetic sequence whose nth term is $20{,}000 + (n - 1)1200$. The total of the first 40 terms of the sequence of salaries is given by

$$S_{40} = \sum_{n=1}^{40} [20{,}000 + (n - 1)1200] = \frac{40}{2}(20{,}000 + 66{,}800) = 1{,}736{,}000.$$

The total salary for 40 years of work is \$1,736,000. Divide the total salary by 40 to get a mean salary of \$43,400.

FOR THOUGHT True or False? Explain.

1. $\sum_{i=1}^{3} (-2)^i = -6$
2. $\sum_{i=1}^{6} (0 \cdot i + 5) = 30$
3. $\sum_{i=1}^{k} 5i = 5\left(\sum_{i=1}^{k} i\right)$
4. $\sum_{i=1}^{k} (i^2 + 1) = \left(\sum_{i=1}^{k} i^2\right) + k$

5. There are nine terms in the series $\sum_{i=5}^{14} 3i^2$.

6. $\sum_{i=2}^{8} (-1)^i 3i^2 = \sum_{j=1}^{7} (-1)^{j-1} 3(j+1)^2$

7. The series $\sum_{n=1}^{9} (3n-5)$ is an arithmetic series.

8. The sum of the first n counting numbers is $\frac{n(n+1)}{2}$.

9. The sum of the even integers from 8 through 68 inclusive is $\frac{60}{2}(8+68)$.

10. $\sum_{i=1}^{10} i^2 = \frac{10}{2}(1+100)$

8.2 EXERCISES

Tape 21 Disk

Find the sum of each series.

1. $\sum_{i=1}^{5} i^2$
2. $\sum_{j=0}^{4} (j+1)^2$
3. $\sum_{j=1}^{6} (2j-1)$
4. $\sum_{i=1}^{7} (2i+5)$
5. $\sum_{n=2}^{5} 2^{-n}$
6. $\sum_{i=1}^{4} (-1)^n i^2$
7. $\sum_{i=4}^{100} 5i^0$
8. $\sum_{i=7}^{36} (10+i^0)$
9. $\sum_{i=3}^{47} (-1)^{i+1}$
10. $\sum_{i=1}^{10} 3$
11. $\sum_{j=7}^{44} (-1)^j$
12. $\sum_{i=0}^{5} i(i-1)(i-2)$

Write each series in summation notation. Use the index i and let i begin at 1 in each summation.

13. $1+2+3+4+5+6$
14. $2+4+6+8+10+12$
15. $-1+3-5+7-9$
16. $3-6+9-12+15-18$
17. $1+4+9+16+25$
18. $1+3+9+27+81$
19. $1-\frac{1}{2}+\frac{1}{4}-\frac{1}{8}+\frac{1}{16}$
20. $-1+\frac{1}{2}-\frac{1}{3}+\frac{1}{4}$
21. $\ln(x_1)+\ln(x_2)+\ln(x_3)$
22. $x^3+x^4+x^5+x^6+x^7$
23. $a+ar+ar^2+\cdots+ar^{10}$
24. $b^2+b^3+b^4+\cdots+b^{12}$

Rewrite each series using the new index j as indicated.

25. $\sum_{i=1}^{32} (-1)^i = \sum_{j=0}$
26. $\sum_{i=1}^{10} 2^i = \sum_{j=0}$
27. $\sum_{i=4}^{13} (2i+1) = \sum_{j=1}$
28. $\sum_{i=7}^{12} (3i-4) = \sum_{j=1}$
29. $\sum_{x=2}^{10} \frac{10!}{x!(10-x)!} = \sum_{j=0}$
30. $\sum_{i=2}^{9} \frac{x^i}{i!} = \sum_{j=0}$
31. $\sum_{n=2}^{6} \frac{5^n e^{-5}}{n!} = \sum_{j=5}$
32. $\sum_{i=0}^{5} 3^{2i-1} = \sum_{j=3}$

Write out all of the terms of each series.

33. $\sum_{i=0}^{5} 0.5r^i$
34. $\sum_{i=1}^{6} i^n$
35. $\sum_{j=0}^{4} a^{4-j}b^j$
36. $\sum_{j=0}^{3} (-1)^j x^{3-j} y^j$
37. $\sum_{i=0}^{2} \frac{2}{i!(2-i)!} a^{2-i}b^i$
38. $\sum_{j=0}^{3} \frac{6}{j!(3-j)!} a^{3-j}b^j$

Find the mean of each sequence of numbers.

39. 6, 23, 45
40. 33, 42, 78, 19
41. −6, 0, 3, 4, 3, 92
42. 12, 20, 12, 30, 28, 28, 10
43. $\sqrt{2}$, π, 33.6, −19.4, 52
44. $\sqrt{5}$, $-3\sqrt{3}$, $\pi/2$, e, 98.6

Find the sum of each arithmetic series.

45. $1+2+3+\cdots+47$
46. $2+4+6+\cdots+88$
47. $8+5+2+(-1)+\cdots+(-16)$
48. $5+1+(-3)+(-7)+\cdots+(-27)$
49. $3+7+11+15+\cdots+55$
50. $-6+1+8+15+\cdots+50$
51. $\frac{1}{2}+\frac{3}{4}+1+\frac{5}{4}+\cdots+5$

52. $1 + \frac{4}{3} + \frac{5}{3} + 2 + \cdots + \frac{22}{3}$

53. $\sum_{i=1}^{12} (6i - 9)$

54. $\sum_{i=1}^{11} (0.5i + 4)$

55. $\sum_{n=3}^{15} (-0.1n + 1)$

56. $\sum_{i=4}^{20} (-0.3n + 2)$

Solve each problem using the ideas of series.

57. *Total Salary* If a graphic artist makes \$30,000 his first year and gets a \$1000 raise each year, then what will be his total salary for 30 years of work? What is his mean annual salary for 30 years of work?

58. *Assigned Reading* An English teacher with a minor in mathematics told her students that if they read $2n + 1$ pages of a long novel on the nth day of October, for each day of October, then they will exactly finish the novel in October. How many pages are there in this novel? What is the mean number of pages read per day?

59. *Mount of Cans I* A grocer wants to build a ''mountain'' out of cans of mountain-grown coffee as shown in the figure. The first level is to be a rectangle containing 9 rows of 12 cans in each row. Each level after the first is to contain one less row with 12 cans in each row. Finally, the top level is to contain one row of 12 cans. Write a sequence whose terms are the number of cans at each level. Write the sum of the terms of this sequence in summation notation and find the number of cans in the mountain.

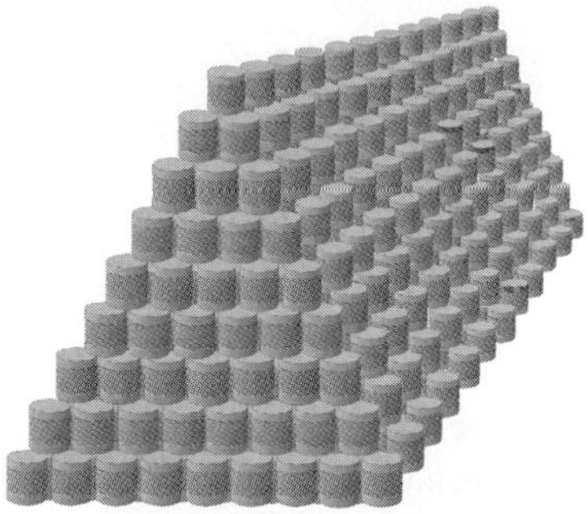

Figure for Exercise 59

60. *Mount of Cans II* Suppose that the grocer in Exercise 59 builds the ''mountain'' so that each level after the first contains one less row and one less can in each row. Write a sequence whose terms are the number of cans at each level. Write the sum of the terms of this sequence in summation notation and find the number of cans in the mountain.

61. *Annual Payments* Wilma will deposit \$1000 into an account paying 5% compounded annually each January 1 for 10 consecutive years. Given that the first deposit was made January 1, 1990, and the last will be made January 1, 1999, write a series in summation notation whose sum is the amount in the account on January 1, 2000.

62. *Compounded Quarterly* Duane deposited \$100 into an account paying 4% compounded quarterly each January 1 for eight consecutive years. Given that the first deposit was made January 1, 1990, and the last was made January 1, 1997, write a series in summation notation whose sum is the amount in the account on January 1, 2000.

63. *Drug Therapy* A doctor instructed a patient to start taking 200 mg of Dilantin every 8 hours to control seizures. If the half-life of Dilantin for this patient is 12 hours, then he still has 63% of the last dose in his body when he takes the next dose. Immediately after taking the fourth pill, the amount of Dilantin in the patient's body is

$$200 + 200(0.63) + 200(0.63)^2 + 200(0.63)^3.$$

Find the sum of this series.

64. *Extended Drug Therapy* Use summation notation to write a series for the amount of Dilantin in the body of the patient in the previous exercise after one week on Dilantin. Find the sum of the series.

Find the indicated mean.

65. Find the mean of the 9th through the 60th terms inclusive of the sequence $a_n = 5n + 56$.

66. Find the mean of the 15th through the 55th terms inclusive of the sequence $a_n = 7 - 4n$.

67. Find the mean of the seventh through 10th terms inclusive of the sequence in which $a_1 = -2$ and $a_n = (a_{n-1})^2 - 3$ for $n \geq 2$.

68. Find the mean of the fifth through the eighth terms inclusive of the sequence in which $a_n = (-1/2)^n$.

For Writing/Discussion

69. Explain the difference between a sequence and a series.

70. *Cooperative Learning* Write your own formula for the nth term a_n of a sequence on a piece of paper. Find the *partial sums* $S_1 = a_1$, $S_2 = a_1 + a_2$, $S_3 = a_1 + a_2 + a_3$, and so on. Disclose the first five partial sums one at a time to your classmates, giving them the opportunity to guess the formula for a_n after each disclosed partial sum.

Linking Concepts

Company/CEO	5-year total compensation ($ millions)
Conseco/ S. C. Hilbert	277
Travelers Group/ S. I. Weill	274
Walt Disney/ M. D. Eisner	236
Green Tree Financial/ L. M. Coss	216
Gateway 2000/ T. W. Waitt	145

For Individual or Group Explorations

Other Means *When we find the mean of a set of numbers we are attempting to find the "middle" or "center" of the set of numbers. The arithmetic mean that we defined in this section is used so frequently that it is hard to believe that there could be any other way to find the middle. However, there are several other ways to find the middle that produce approximately the same results. Consider the salaries of the five highest paid CEOs shown in the table (Forbes, May 19, 1997).*

a) Find the *median*, the score for which approximately one-half of the scores are lower and one-half of the scores are higher.

b) Find the *geometric mean*, the nth root of the product of the n scores:

$$GM = \sqrt[n]{x_1 \cdot x_2 \cdot x_3 \cdot \cdots \cdot x_n}$$

c) Find the *harmonic mean*, the number of scores divided by the sum of the reciprocals of the scores:

$$HM = \frac{n}{\sum \frac{1}{x}}$$

d) Find the *quadratic mean* by finding the sum of the squares of the scores, divide by n, then take the square root:

$$QM = \sqrt{\frac{\sum x^2}{n}}$$

8.3 Geometric Sequences and Series

We saw that in an arithmetic sequence there is a constant difference between consecutive terms. This simple relationship allowed us to find a formula for the sum of n terms of an arithmetic sequence. Another type of sequence that has a simple pattern is a *geometric sequence*. In a geometric sequence consecutive terms have a *constant ratio*. In this section we will make use of the constant ratio to find a formula for the sum of n terms of a geometric sequence. We study arithmetic and geometric sequences in detail because they occur in many applications and because they are two of the very few sequences for which we can find formulas for the sum of n terms.

Geometric Sequences

In the sequence

$$100, 50, 25, 12.5, \ldots$$

each term after the first is half of the term preceding it. This sequence can be written as

$$100, \quad 100\left(\frac{1}{2}\right), \quad 100\left(\frac{1}{2}\right)^2, \quad 100\left(\frac{1}{2}\right)^3, \ldots$$

and its nth term is given by

$$a_n = 100\left(\frac{1}{2}\right)^{n-1}.$$

Any sequence in which each term after the first is a constant multiple of the preceding term is called a geometric sequence.

Definition: Geometric Sequence

A sequence with general term $a_n = ar^{n-1}$ is called a **geometric sequence** with **common ratio r**, where $r \neq 1$ and $r \neq 0$.

According to the definition, every geometric sequence is of the form

$$a, \quad ar, \quad ar^2, \quad ar^3, \quad ar^4, \ldots$$

Notice also that in a geometric sequence the ratio of any term (after the first) and the term preceding it is r. To write a formula for the nth term of a geometric sequence, all we need to know is the first term and the constant ratio.

EXAMPLE 1 Finding a formula for the *n*th term

Write a formula for the nth term of each geometric sequence.

a) 0.3, 0.03, 0.003, 0.0003, . . . **b)** 2, −6, 18, −54, . . .

Solution

a) Since each term after the first is one-tenth of the term preceding it, we use $r = 0.1$ and $a = 0.3$ in the formula $a_n = ar^{n-1}$:

$$a_n = 0.3(0.1)^{n-1}$$

b) Choose any two consecutive terms and divide the second by the first to obtain the common ratio. So $r = -6/2 = -3$. Since the first term is 2,

$$a_n = 2(-3)^{n-1}.$$

In Section 8.2 we learned some interesting facts about arithmetic sequences. However, this information is useful only if we can determine whether a given sequence is arithmetic. In this section we will learn some facts about geometric sequences. Likewise, we must be able to determine whether a given sequence is geometric even when it is not given in exactly the same form as the definition.

EXAMPLE 2 Identifying a geometric sequence

Find the first four terms of each sequence and determine whether the sequence is geometric.

a) $b_n = (-2)^{3n}$ **b)** $a_1 = 1.25$ and $a_n = -2a_{n-1}$ for $n \geq 2$ **c)** $c_n = 3n$

Solution

a) Use $n = 1, 2, 3,$ and 4 in the formula $b_n = (-2)^{3n}$ to find the first four terms:

$$-8, \quad 64, \quad -512, \quad 4096, \ldots$$

The ratio of any term and the preceding term can be found from the formula

$$\frac{b_n}{b_{n-1}} = \frac{(-2)^{3n}}{(-2)^{3(n-1)}} = (-2)^3 = -8.$$

Since there is a common ratio of -8, the sequence is geometric.

b) Use the recursion formula $a_n = -2a_{n-1}$ to obtain each term after the first:

$$a_1 = 1.25$$
$$a_2 = -2a_1 = -2.5$$
$$a_3 = -2a_2 = 5$$
$$a_4 = -2a_3 = -10$$

This recursion formula defines the sequence

$$1.25, \ -2.5, \ 5, \ -10, \ldots$$

The formula $a_n = -2a_{n-1}$ means that each term is a constant multiple of the term preceding it. The constant ratio is -2 and the sequence is geometric.

c) Use $c_n = 3n$ to find $c_1 = 3$, $c_2 = 6$, $c_3 = 9$, and $c_4 = 12$. Since $6/3 = 2$ and $9/6 = 1.5$, there is no common ratio for consecutive terms. The sequence is not geometric. Because each term is 3 larger than the preceding term, the sequence is arithmetic.

In the next example we see that a geometric sequence occurs in an investment earning compound interest. If we have an initial deposit earning compound interest, then the amounts in the account at the end of consecutive years form a geometric sequence.

EXAMPLE 3 A geometric sequence in investment

The parents of a newborn decide to start saving early for her college education. On the day of her birth, they invest \$3000 at 6% compounded annually. Find the amount of the investment at the end of each of the first four years and find a formula for the amount at the end of the nth year. Find the amount at the end of the 18th year.

Solution

At the end of the first year the amount is \$3000(1.06). At the end of the second year the amount is $\$3000(1.06)^2$. Use a calculator to find the amounts at the ends of the first four years.

$$\begin{aligned} a_1 &= 3000(1.06)^1 = \$3180 \\ a_2 &= 3000(1.06)^2 = \$3370.80 \\ a_3 &= 3000(1.06)^3 = \$3573.05 \\ a_4 &= 3000(1.06)^4 = \$3787.43 \end{aligned}$$

A formula for the nth term of this geometric sequence is $a_n = 3000(1.06)^n$. The amount at the end of the 18th year is

$$a_{18} = 3000(1.06)^{18} = \$8563.02.$$

The formula for the general term of a geometric sequence involves a_n, a, n, and r. If we know the value of any three of these quantities, then we can find the value of the fourth.

EXAMPLE 4 Find the number of terms in a geometric sequence

A certain ball always rebounds 2/3 of the distance from which it falls. If the ball is dropped from a height of 9 feet, and later it is observed rebounding to a height of 64/81 feet, then how many times did it bounce?

Solution

After the first bounce the ball rebounds to $9(2/3) = 6$ feet. After the second bounce the ball rebounds to a height of $9(2/3)^2 = 4$ feet. The first term of this sequence is 6 and the common ratio is 2/3. So after the nth bounce the ball rebounds to a height h_n given by

$$h_n = 6\left(\frac{2}{3}\right)^{n-1}.$$

To find the number of bounces, solve the following equation, which states that the nth bounce rebounds to 64/81 feet.

$$6\left(\frac{2}{3}\right)^{n-1} = \frac{64}{81}$$

$$\left(\frac{2}{3}\right)^{n-1} = \frac{32}{243} = \left(\frac{2}{3}\right)^{5}$$

$$n - 1 = 5$$

$$n = 6$$

The ball bounced six times.

Geometric Series

Geometric sequences occur in applications ranging from bouncing balls to investing money. The sum of n terms of a geometric sequence can give the total distance traveled by a bouncing ball or the total value of some periodic deposits earning compound interest. The indicated sum of the terms of a geometric sequence is called a **geometric series**. To find the actual sum of a geometric series, we can use a procedure similar to that used for finding the sum of an arithmetic series.

Consider the geometric sequence $a_n = 2^{n-1}$. Let S_{10} represent the sum of the first ten terms of this geometric sequence:

$$S_{10} = 1 + 2 + 4 + 8 + \cdots + 512$$

If we multiply each side of this equation by -2, the opposite of the common ratio, we get

$$-2S_{10} = -2 - 4 - 8 - 16 - \cdots - 512 - 1024.$$

Adding S_{10} and $-2S_{10}$ eliminates most of the terms:

$$\begin{aligned} S_{10} &= 1 + 2 + 4 + 8 + 16 + \cdots + 512 \\ -2S_{10} &= \quad\; - 2 - 4 - 8 - 16 - \cdots - 512 - 1024 \\ \hline -S_{10} &= 1 \qquad\qquad\qquad\qquad\qquad\qquad\qquad - 1024 \qquad \text{Add.} \\ -S_{10} &= -1023 \\ S_{10} &= 1023 \end{aligned}$$

The "trick" to finding the sum of n terms of a geometric series is to change the signs and shift the terms so that most terms "cancel out" when the two equations are added. This method can also be used to find a general formula for the sum of a geometric series.

Let S_n represent the sum of the first n terms of the geometric sequence $a_n = ar^{n-1}$.

$$S_n = a + ar + ar^2 + \cdots + ar^{n-1}$$

Adding S_n and $-rS_n$ eliminates most of the terms:

$$\begin{aligned} S_n &= a + ar + ar^2 + \cdots + ar^{n-1} \\ -rS_n &= \quad - ar - ar^2 - ar^3 - \cdots - ar^{n-1} - ar^n \\ \hline S_n - rS_n &= a - ar^n \qquad \text{Add.} \\ (1 - r)S_n &= a(1 - r^n) \\ S_n &= \frac{a(1 - r^n)}{1 - r} \qquad \text{Provided that } r \neq 1. \end{aligned}$$

So the sum of a geometric series can be found if we know the first term, the constant ratio, and the number of terms. This result is summarized in the following theorem.

Theorem: Sum of a Finite Geometric Series

If S_n represents the sum of the first n terms of a geometric series with first term a and common ratio r ($r \neq 1$), then

$$S_n = \frac{a(1 - r^n)}{1 - r}.$$

EXAMPLE 5 **Finding the sum of a geometric series**

Find the sum of each geometric series.

a) $1 + \frac{1}{3} + \frac{1}{9} + \cdots + \frac{1}{243}$ **b)** $\sum_{j=0}^{10} 100(1.05)^j$

Solution

a) To find the sum of a finite geometric series, we need the first term a, the ratio r, and the number of terms n. To find n, use $a = 1$, $a_n = 1/243$, and $r = 1/3$ in the formula $a_n = ar^{n-1}$:

$$\begin{aligned} 1\left(\frac{1}{3}\right)^{n-1} &= \frac{1}{243} \\ n - 1 &= 5 \qquad \text{Because } \left(\frac{1}{3}\right)^5 = \frac{1}{243} \\ n &= 6 \end{aligned}$$

Now use $n = 6$, $a = 1$, $r = 1/3$ in the formula $S_n = \frac{a(1 - r^n)}{1 - r}$:

$$S_6 = \frac{1\left(1 - \left(\frac{1}{3}\right)^6\right)}{1 - \frac{1}{3}} = \frac{\frac{728}{729}}{\frac{2}{3}} = \frac{364}{243}$$

b) First write out some terms of the series:

$$\sum_{j=0}^{10} 100(1.05)^j = 100 + 100(1.05) + 100(1.05)^2 + \cdots + 100(1.05)^{10}$$

In this geometric series, $a = 100$, $r = 1.05$, and $n = 11$:

$$\sum_{j=0}^{10} 100(1.05)^j = S_{11} = \frac{100(1 - (1.05)^{11})}{1 - 1.05} \approx 1420.68$$

One of the most important applications of the sum of a finite geometric series is in projecting the value of an annuity. An **annuity** is a sequence of equal periodic payments. If each payment earns the same rate of compound interest, then the total value of the annuity can be found by using the formula for the sum of a finite geometric series.

EXAMPLE 6 Finding the value of an annuity

To maintain his customary style of living after retirement, a single man earning \$50,000 per year at age 65 must have saved a minimum of \$250,000 (Fidelity Investments, Boston). If Chad invests \$1000 at the beginning of each year for 40 years in an investment paying 6% compounded annually, then what is the value of this annuity at the end of the 40th year?

Solution

The last deposit earns interest for only one year and amounts to 1000(1.06). The second to last deposit earns interest for two years and amounts to $1000(1.06)^2$. This pattern continues down to the first deposit, which earns interest for 40 years and amounts to $1000(1.06)^{40}$. The value of the annuity is the sum of a finite geometric series:

$$S_{40} = 1000(1.06) + 1000(1.06)^2 + \cdots + 1000(1.06)^{40}$$

Since the first term is 1000(1.06), use $a = 1000(1.06)$, $r = 1.06$, and $n = 40$:

$$S_{40} = \frac{1000(1.06)(1 - (1.06)^{40})}{1 - 1.06} = \$164{,}047.68$$

Infinite Geometric Series

In the geometric series

$$2 + 4 + 8 + 16 + \cdots,$$

in which $r = 2$, the terms get larger and larger. So the sum of the first n terms increases without bound as n increases. In the geometric series

$$\frac{1}{2} + \frac{1}{4} + \frac{1}{8} + \frac{1}{16} + \cdots,$$

in which $r = 1/2$, the terms get smaller and smaller. The sum of n terms of this series is less than 1 no matter how large n is. (To see this, add some terms on your calculator.) We can explain the different behavior of these series by examining the term r^n in the formula

$$S_n = \frac{a(1 - r^n)}{1 - r}.$$

If $r = 2$, the values of r^n increase without bound as n gets larger, causing S_n to increase without bound. If $r = 1/2$, the values of $(1/2)^n$ approach 0 as n gets larger. In symbols, $(1/2)^n \to 0$ as $n \to \infty$. If $r^n \to 0$, then $1 - r^n$ is approximately 1. If we replace $1 - r^n$ by 1 in the formula for S_n, we get

$$S_n \approx \frac{a}{1 - r} \quad \text{for large values of } n.$$

In the above series $r = 1/2$ and $a = 1/2$. So if n is large we have

$$S_n \approx \frac{\frac{1}{2}}{1 - \frac{1}{2}} = 1.$$

Figure 8.9

In general, it can be proved that $r^n \to 0$ as $n \to \infty$ provided that $|r| < 1$, and r^n does not get close to 0 as $n \to \infty$ for $|r| \geq 1$. You will better understand these ideas if you use a calculator to find some large powers of r for various values of r as in Fig. 8.9. So, if $|r| < 1$ and n is large, then S_n is approximately $a/(1 - r)$. Furthermore, by using more terms in the sum we can get a sum that is arbitrarily close to the number $a/(1 - r)$. In this sense we say that the sum of all terms of the infinite geometric series is $a/(1 - r)$.

Theorem: Sum of an Infinite Geometric Series

If $a + ar + ar^2 + \cdots$ is an infinite geometric series with $|r| < 1$, then the sum S of all of the terms is given by

$$S = \frac{a}{1 - r}.$$

We can use the infinity symbol ∞ to indicate the sum of infinitely many terms of an infinite geometric series as follows:

$$a + ar + ar^2 + \cdots = \sum_{i=1}^{\infty} ar^{i-1}$$

EXAMPLE 7 **Finding the sum of an infinite geometric series**

Find the sum of each infinite geometric series.

a) $1 + \frac{1}{3} + \frac{1}{9} + \cdots$ **b)** $\sum_{j=1}^{\infty} 100(-0.99)^j$ **c)** $\sum_{i=0}^{\infty} 3(1.01)^i$

Solution

a) The first term is 1, and the common ratio is 1/3. So the sum of the infinite series is

$$S = \frac{1}{1 - \frac{1}{3}} = \frac{3}{2}.$$

b) The first term is -99, and the common ratio is -0.99. So the sum of the infinite geometric series is

$$S = \frac{-99}{1 - (-0.99)} = \frac{-99}{1.99} = -\frac{9900}{199}.$$

c) The first term is 3, and the common ratio is 1.01. Since the absolute value of the ratio is greater than 1, this infinite geometric series has no sum.

In the next example we use the formula for the sum of an infinite geometric series in a physical situation. Since physical processes do not continue infinitely, the formula for the sum of infinitely many terms is used as an approximation for the sum of a large finite number of terms of a geometric series in which $|r| < 1$.

EXAMPLE 8 **Total distance traveled in bungee jumping**

A man jumping from a bridge with a bungee cord tied to his legs falls 120 feet before being pulled back upward by the bungee cord. If he always rebounds 1/3 of the distance that he has fallen and then falls 2/3 of the distance of his last rebound, then approximately how far does the man travel before coming to rest?

Solution

In actual practice, the man does not go up and down infinitely on the bungee cord. However, to get an approximate answer, we can model this situation shown in Fig. 8.10 with two infinite geometric sequences, the sequence of falls and the sequence of rises. The man falls 120 feet, then rises 40 feet. He falls 80/3 feet, then rises 80/9 feet. He falls 160/27 feet, then rises 160/81 feet. The total distance the man falls is given by the following series:

$$F = 120 + \frac{80}{3} + \frac{160}{27} + \cdots = \frac{120}{1 - \frac{2}{9}} = \frac{1080}{7} \text{ feet}$$

The total distance the man rises is given by the following series:

$$R = 40 + \frac{80}{9} + \frac{160}{81} + \cdots = \frac{40}{1 - \frac{2}{9}} = \frac{360}{7} \text{ feet}$$

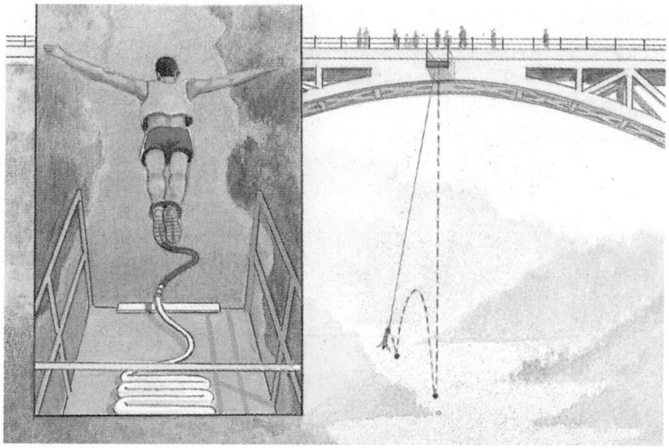

Figure 8.10

The total distance he travels before coming to rest is the sum of these distances, 1440/7 feet, or approximately 205.7 feet.

Repeating Decimals

Another application of geometric series occurs in rational numbers that are infinite repeating decimals. For example, the fraction 1/3 is a repeating decimal. This repeating decimal can be viewed as the sum of a geometric series:

$$\begin{aligned}\frac{1}{3} &= 0.3333\ldots \\ &= 0.3 + 0.3 \cdot 10^{-1} + 0.3 \cdot 10^{-2} + 0.3 \cdot 10^{-3} + \cdots\end{aligned}$$

The first term is 0.3, and the ratio is 10^{-1}. The formula for the sum of an infinite geometric series can be used to convert the repeating decimal number back into a fraction.

$$\begin{aligned}0.3333\ldots &= \frac{0.3}{1 - 10^{-1}} \\ &= \frac{0.3}{0.9} = \frac{1}{3}\end{aligned}$$

The next example uses this same idea on a decimal number in which some of the digits do not repeat.

EXAMPLE 9 Converting a repeating decimal to a fraction

Use the formula for the sum of an infinite geometric series to convert the repeating decimal number 1.2417417417 . . . into a fraction.

Solution

Separate the repeating part of the number from the nonrepeating part and convert the repeating part into a fraction, using the formula for the sum of an infinite geometric series:

$$\begin{aligned}1.2417417417417\ldots &= 1.2 + 0.0417417417417\ldots\\ &= 1.2 + 417\cdot 10^{-4} + 417\cdot 10^{-7} + 417\cdot 10^{-10} + \cdots\\ &= \frac{12}{10} + \frac{417\cdot 10^{-4}}{1 - 10^{-3}} \qquad a = 417\cdot 10^{-4} \text{ and } r = 10^{-3}\\ &= \frac{12}{10} + \frac{417}{10{,}000 - 10}\\ &= \frac{12}{10} + \frac{417}{9990}\\ &= \frac{12405}{9990}\end{aligned}$$

FOR THOUGHT True or False? Explain.

1. The sequence 2, 6, 24, . . . is a geometric sequence.
2. The sequence $a_n = 3(2)^{3-n}$ is a geometric sequence.
3. The firest term of the geometric sequence $a_n = 5(0.3)^n$ is 1.5.
4. The common ratio in the geometric sequence $a_n = 5^{-n}$ is 5.
5. A geometric series is the indicated sum of a geometric sequence.
6. $3 + 6 + 12 + 24 + \cdots = \dfrac{3}{1 - 2}$
7. $\displaystyle\sum_{i=1}^{9} 3(0.6)^i = \frac{1.8(1 - 0.6)^9}{1 - 0.6}$
8. $\displaystyle\sum_{i=0}^{4} 2(10)^i = 22{,}222$
9. $\displaystyle\sum_{i=1}^{\infty} 3(0.1)^i = \frac{1}{3}$
10. $\displaystyle\sum_{i=1}^{\infty} \left(\frac{1}{2}\right)^i = 1$

8.3 EXERCISES

Tape 21 Disk

Write a formula for the nth term of each geometric sequence. Do not use a recursion formula.

1. $\frac{1}{6}, \frac{1}{3}, \frac{2}{3}, \frac{4}{3}, \ldots$
2. $\frac{1}{6}$, 0.5, 1.5, 4.5, . . .
3. 0.9, 0.09, 0.009, 0.0009, . . .
4. 3, 9, 27, 81, . . .
5. 4, −12, 36, −108, . . .
6. $5, -1, \frac{1}{5}, -\frac{1}{25}, \ldots$

Identify each sequence as arithmetic, geometric, or neither.

7. $\frac{1}{6}, \frac{1}{3}, \frac{1}{2}, \frac{2}{3}, \ldots$
8. $\frac{1}{6}, \frac{1}{3}, 1, 4, \ldots$
9. $\frac{1}{6}, \frac{1}{3}, \frac{2}{3}, \frac{4}{3}, \ldots$
10. 3, 2, 1, 0, −1, . . .
11. 1, 4, 9, 16, 25, . . .
12. $5, 1, \frac{1}{5}, \frac{1}{25}, \ldots$

Find the first four terms of each sequence and identify each sequence as arithmetic, geometric, or neither.

13. $a_n = 2n$

14. $a_n = 2^n$

15. $a_n = n^2$

16. $a_n = n!$

17. $a_n = 2^{-n}$

18. $a_n = n + 2$

19. $b_n = 2^{2n+1}$

20. $d_n = \dfrac{1}{3^{n/2}}$

21. $c_1 = 3$, $c_n = -3c_{n-1}$ for $n \geq 2$

22. $h_1 = \sqrt{2}$, $h_n = \sqrt{3}h_{n-1}$ for $n \geq 2$

List the first three terms and the 10th term of the geometric sequence whose nth term is given.

23. $a_n = 3(-2)^{n-1}$

24. $a_n = 2(-0.5)^{n-1}$

25. $a_n = 4(0.1)^{n-1}$

26. $a_n = 0.01(5)^{n-1}$

Find the required part of each geometric sequence.

27. Find the number of terms of a geometric sequence with first term 3, common ratio 1/2, and last term 3/1024.

28. Find the number of terms of a geometric sequence with first term 1/64, common ratio -2, and last term -512.

29. Find the first term of a geometric sequence with sixth term 1/81 and common ratio of 1/3.

30. Find the first term of a geometric sequence with seventh term 1/16 and common ratio 1/2.

31. Find the common ratio for a geometric sequence with first term 2/3 and third term 6.

32. Find the common ratio for a geometric sequence with second term -1 and fifth term -27.

33. Find a formula for the nth term of a geometric sequence with third term -12 and sixth term 96.

34. Find a formula for the nth term of a geometric sequence with second term -40 and fifth term 0.04.

Find the sum of each finite geometric series by using the formula for S_n.

35. $6 + 2 + \dfrac{2}{3} + \dfrac{2}{9} + \dfrac{2}{27}$

36. $2 + 10 + 50 + 250 + 1250$

37. $1.5 - 3 + 6 - 12 + 24 - 48 + 96 - 192$

38. $1 - \dfrac{1}{2} + \dfrac{1}{4} - \dfrac{1}{8} + \dfrac{1}{16} - \dfrac{1}{32} + \dfrac{1}{64}$

39. $\sum_{i=1}^{12} 2(1.05)^{i-1}$

40. $\sum_{i=0}^{30} 300(1.08)^i$

41. $\sum_{i=0}^{7} 200(1.01)^i$

42. $\sum_{i=1}^{20} 421(1.09)^{i-1}$

Write each geometric series in summation notation.

43. $3 - 1 + \dfrac{1}{3} - \dfrac{1}{9} + \dfrac{1}{27}$

44. $2 + 1 + \dfrac{1}{2} + \dfrac{1}{4} + \dfrac{1}{8} + \dfrac{1}{16}$

45. $0.6 + 0.06 + 0.006 + \cdots$

46. $4 - 1 + \dfrac{1}{4} - \dfrac{1}{16} + \cdots$

47. $-4.5 + 1.5 - 0.5 + \dfrac{1}{6} - \cdots$

48. $a + ab + ab^2 + \cdots + ab^{37}$

Find the sum of each infinite geometric series where possible.

49. $3 - 1 + \dfrac{1}{3} - \dfrac{1}{9} + \dfrac{1}{27} - \cdots$

50. $1 + \dfrac{1}{2} + \dfrac{1}{4} + \dfrac{1}{8} + \dfrac{1}{16} + \cdots$

51. $0.9 + 0.09 + 0.009 + \cdots$

52. $-1 + \dfrac{1}{4} - \dfrac{1}{16} + \dfrac{1}{64} - \cdots$

53. $-9.9 + 3.3 - 1.1 + \cdots$

54. $1.2 - 2.4 + 4.8 - 9.6 + \cdots$

55. $\sum_{i=1}^{\infty} 34(0.01)^i$

56. $\sum_{i=0}^{\infty} 300(0.99)^i$

57. $\sum_{i=0}^{\infty} 300(-1.06)^i$

58. $\sum_{i=0}^{\infty} (0.98)^i$

59. $\sum_{i=1}^{\infty} 6(0.1)^i$

60. $\sum_{i=0}^{\infty} (0.1)^i$

61. $\sum_{i=0}^{\infty} 34(-0.7)^i$

62. $\sum_{i=1}^{\infty} 123(0.001)^i$

Use the formula for the sum of an infinite geometric series to write each repeating decimal number as a fraction.

63. 0.04444 . . .

64. 0.0121212 . . .

65. 8.2545454 . . .

66. 3.65176176176 . . .

Use the ideas of geometric series to solve each problem.

67. *Drug Therapy* If a patient starts taking 100 mg of Lamictal every 8 hours to control seizures, then the amount of Lamictal in the patient's body after taking the twenty-fifth pill is given by the finite geometric series

$$\sum_{n=1}^{25} 100(0.69)^{n-1}.$$

Use the formula for S_n to find this sum.

68. *Long-Range Therapy* The long-range build-up of Lamictal in the body of a patient taking 100 mg three times a day is given by the infinite geometric series

$$\sum_{n=1}^{\infty} 100(0.69)^{n-1}.$$

Find the sum of this series. What is the difference between the long-range build-up and the amount in the person after 25 pills?

69. *Sales Goals* A group of college students selling magazine subscriptions during the summer sold one subscription on June 1. The sales manager was encouraged by this and said that their daily goal for each day of June is to double the sales of the previous day. If the students work every day during June and meet this goal, then what is the total number of magazine subscriptions that they will sell during June?

70. *Family Tree* Consider yourself, your parents, your grandparents, your great-grandparents, your great-great-grandparents, and so on, back to your grandparents with the word ''great'' used in front 40 times. What is the total number of people that you are considering?

71. *Compound Interest* Given that \$4000 is deposited at the beginning of a quarter into an account earning 8% annual interest compounded quarterly, write a formula for the amount in the account at the end of the nth quarter. How much is in the account at the end of 37 quarters?

72. *Compound Interest* Given that \$8000 is deposited at the beginning of a month into an account earning 6% annual interest compounded monthly, write a formula for the amount in the account at the end of the nth month. How much is in the account at the end of the 56th month?

73. *Value of an Annuity* If you deposit \$200 on the first of each month for 12 months into an account paying 12% annual interest compounded monthly, then how much is in the account at the end of the 12th month? Note that each deposit earns 1% per month for a different number of months.

74. *Saving for Retirement* If you deposit \$9000 on the first of each year for 40 years into an account paying 8% compounded annually, then how much is in the account at the end of the 40th year? Use the formula for the sum of a finite geometric series.

75. *Saving for Retirement* If \$100 is deposited at the end of each month for 30 years in a retirement account earning 9% compounded monthly, then what is the value of this annuity immediately after the last payment?

76. *Down Payment* To get a down payment for a house, a couple plan to deposit \$800 at the end of each quarter for 36 quarters in an account paying 6% compounded quarterly. What is the value of this annuity immediately after the last deposit?

77. *Bouncing Ball* Suppose that a ball always rebounds 2/3 of the distance from which it falls. If this ball is dropped from a height of 9 ft, then approximately how far does it travel before coming to rest?

78. *Saturating the Market* A sales manager has set a team goal of \$1 million in sales. The team plans to get 1/2 of the goal the first week of the campaign. Every week after the first they will sell only 1/2 as much as the previous week because the market will become saturated. How close will the team be to the sales goal after 15 weeks of selling?

79. *Economic Impact* The anticipated payroll of the new General Dynamics plant in Hammond, Louisiana, is \$2 million annually. It is estimated that 75% of this money is spent in Hammond by its recipients. The people who receive that money again spend 75% of it in Hammond, and so on. The total of all this spending is the *total economic impact* of the plant on Hammond. What is the total economic impact of the plant?

80. *Disaster Relief* If the federal government provides \$300 million in disaster relief to the people of Louisiana after a hurricane, then 80% of that money is spent in Louisiana. If the money is respent over and over at a rate of 80% in Louisiana, then what is the total economic impact of the \$300 million on the state?

For Writing/Discussion

81. Can an arithmetic sequence and a geometric sequence have the same first three terms? Explain your answer.

82. *Cooperative Learning* Get a ''super ball'' and work in a small group to measure the distance that it rebounds from falls of 8 ft, 6 ft, 4 ft, and 2 ft. Is it reasonable to assume that the rebound distance is a constant percentage of the fall distance? Use an infinite series to find the total distance that your ball travels vertically before coming to rest when it is dropped from a height of 8 ft. Discuss the applicability of the infinite series model to this physical experiment.

83. Consider the functions $y = r^x$ and $y = x^r$ for $x > 0$. Graph these functions for some values of r with $|r| < 1$ and for some values of r with $|r| \geq 1$. Make a conjecture about the relationship between the value of r and the values of r^x and x^r as $x \to \infty$.

84. Consider the function $y = 5(1 - r^x)/(1 - r)$ for $x > 0$. Graph this function for some values of r with $|r| < 1$ and for some values of r with $|r| > 1$. Make a conjecture about the relationship between the value of r and the value of y as $x \to \infty$. Explain how these graphs are related to series.

LINKING CONCEPTS

For Individual or Group Explorations

Annuities *An annuity consists of periodic payments into an account paying compound interest. Suppose that R dollars is deposited at the beginning of each period for n periods and the compound interest rate is i per period.*

a) What is the amount of the first deposit at the time of the last deposit?

b) Write a series for the total amount of all n deposits at the time of the last deposit.

c) Use the formula for the sum of a finite geometric series to show that the sum of this series is $R((1 + i)^n - 1)/i$.

d) Find the amount at the time of the last deposit for an annuity of \$2000 per year for 30 years at 12% compounded annually.

e) Find the amount 7 months after the last deposit for an annuity of \$800 per month for 13 years with a rate of 9% compounded monthly.

8.4 Counting and Permutations

In the first three sections of this chapter we studied sequences and series. In this section we will look at sequences of events and count the number of ways that a sequence of events can occur. But in this instance, we actually try to avoid counting in the usual sense because this method of counting would probably be too cumbersome or tedious. In this new context, counting means finding the number of ways in which something can be done without actually listing all of the ways and counting them.

The Fundamental Counting Principle

Let's say that the cafeteria lunch special includes a choice of sandwich and dessert. In the terminology of counting, choosing a sandwich is an **event** and choosing a dessert is another event. Suppose that there are three outcomes, ham, salami, or tuna, for the first event and two outcomes, pie or cake, for the second event.

How many different lunches are available using these choices? We can make a diagram showing all of the possibilities as in Fig. 8.11. This diagram is called a **tree diagram**. Considering only the types of sandwich and dessert, the tree diagram shows six different lunches. Of course, 6 can be obtained by multiplying 3 and 2. This example illustrates the fundamental counting principle.

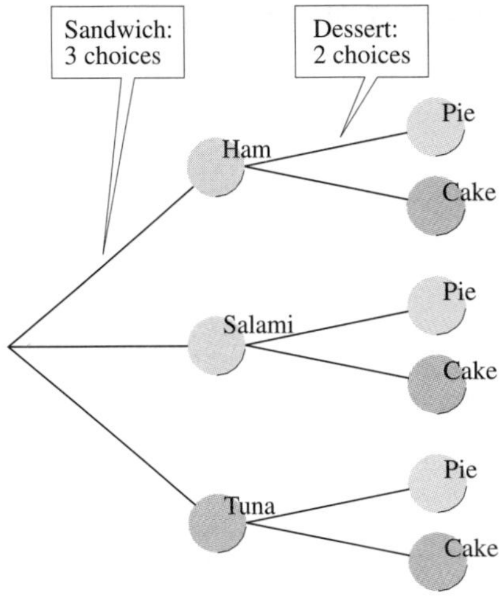

Figure 8.11

Fundamental Counting Principle

If event A has m different outcomes and event B has n different outcomes, then there are mn different ways for events A and B to occur together.

The fundamental counting principle can also be used for more than two events, as is illustrated in the next example.

EXAMPLE 1 Applying the fundamental counting principle

When ordering a new car, you are given a choice of three engines, two transmissions, six colors, three interior designs, and whether or not to get air conditioning. How many different cars can be ordered considering these choices?

Solution

There are three outcomes to the event of choosing the engine, two outcomes to choosing the transmission, six outcomes to choosing the color, three outcomes to choosing the interior, and two outcomes to choosing the air conditioning (to have it or not). So the number of different cars available is $3 \cdot 2 \cdot 6 \cdot 3 \cdot 2 = 216$.

EXAMPLE 2 **Applying the fundamental counting principle**

How many different license plates are possible if each plate consists of three letters followed by a three-digit number? Assume that repetitions in the letters or numbers are allowed and that any of the ten digits may be used.

Solution

Since there are 26 choices for each of the three letters and ten choices for each of the three numbers, by the fundamental counting principle, the number of license plates is $26 \cdot 26 \cdot 26 \cdot 10 \cdot 10 \cdot 10 = 26^3 \cdot 10^3 = 17{,}576{,}000$.

Permutations

In Examples 1 and 2, each choice was independent of the previous choices. But the number of ways in which an event can occur often depends on what has already occurred. For example, in arranging three students in a row, the choice of the student for the second seat depends on which student was placed first. Consider the following six different sequential arrangements (or permutations) of three students, Ann, Bob, and Carol:

Ann, Bob, Carol	Bob, Ann, Carol	Carol, Ann, Bob
Ann, Carol, Bob	Bob, Carol, Ann	Carol, Bob, Ann

These six permutations can also be shown in a tree diagram as we did in Fig. 8.12. Since the students are distinct objects, there can be no repetition of students. An arrangement such as Ann, Ann, Bob is not allowed. A **permutation** is

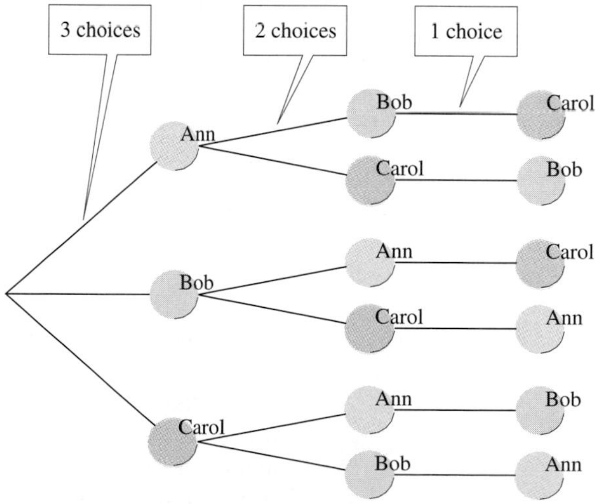

Figure 8.12

an ordering or arrangement of distinct objects in a sequential manner. There is a first, a second, a third, and so on. High-level diplomats are usually *not* seated so that they are arranged in a sequential order. They are seated at round tables so that no one is first, or second, or third.

EXAMPLE 3 Finding permutations

A Federal Express driver must make ten deliveries to ten different addresses. In how many ways can she make those deliveries?

Solution

For the event of choosing the first address there are ten outcomes. For the event of choosing the second address, there are nine outcomes (since one has already been chosen). For the third address, there are eight outcomes, and so on. So, according to the fundamental counting principle, the number of permutations of the ten addresses is

$$10 \cdot 9 \cdot 8 \cdot 7 \cdot 6 \cdot 5 \cdot 4 \cdot 3 \cdot 2 \cdot 1 = 10! = 3{,}628{,}800.$$

In general, the number of arrangements of any n distinct objects in a sequential manner is referred to as the number of permutations of n things taken n at a time and the notation $P(n, n)$ is used to represent this number. The phrase "taken n at a time" indicates that all of the n objects are used in the arrangements. (We will soon discuss permutations in which not all of the n objects are used.) Since there are n choices for the first object, $n - 1$ choices for the second object, $n - 2$ choices for the third object, and so on, we have $P(n, n) = n!$.

Theorem: Permutations of n Things Taken n at a Time

The notation $P(n, n)$ represents the number of permutations of n things taken n at a time, and $P(n, n) = n!$.

Sometimes we are interested in permutations in which not all of the objects are used. For example, if the Federal Express driver wants to deliver three of the ten packages before lunch, then how many ways are there to make those three deliveries? There are ten possibilities for the first delivery, nine for the second, and eight for the third. So the number of ways to make three stops before lunch, or the number of permutations of 10 things taken 3 at a time, is $10 \cdot 9 \cdot 8 = 720$.

The notation $P(10, 3)$ is used to represent the number of permutations of 10 things taken 3 at a time. Notice that

$$P(10, 3) = \frac{10!}{7!} = \frac{10 \cdot 9 \cdot 8 \cdot 7 \cdot 6 \cdot 5 \cdot 4 \cdot 3 \cdot 2 \cdot 1}{7 \cdot 6 \cdot 5 \cdot 4 \cdot 3 \cdot 2 \cdot 1} = 10 \cdot 9 \cdot 8 = 720.$$

In general, we have the following theorem.

Theorem: Permutations of *n* Things Taken *r* at a Time

The notation $P(n, r)$ represents the number of permutations of n things taken r at a time, and

$$P(n, r) = \frac{n!}{(n - r)!} \quad \text{for } 0 \leq r \leq n.$$

Even though n things are usually not taken 0 at a time, 0 is allowed in the formula. Recall that by definition, $0! = 1$. For example, $P(8, 0) = 8!/8! = 1$ and $P(0, 0) = 0!/0! = 1/1 = 1$. The one way to choose 0 objects from 8 objects (or no objects) is to do nothing.

Many calculators have a key that gives the value of $P(n, r)$ when given n and r. The notation nPr is often used on calculators and elsewhere for $P(n, r)$.

EXAMPLE 4 Finding permutations of *n* things taken *r* at a time

The Beau Chene Garden Club has 12 members. They plan to elect a president, a vice-president, and a treasurer. How many different outcomes are possible for the election if each member is eligible for each office and no one can hold two offices?

Solution

The number of ways in which these offices can be filled is precisely the number of permutations of 12 things taken 3 at a time:

$$P(12, 3) = \frac{12!}{(12 - 3)!} = \frac{12!}{9!} = 12 \cdot 11 \cdot 10 = 1320$$

Use the permutation function or factorials on a calculator to check as in Fig. 8.13.

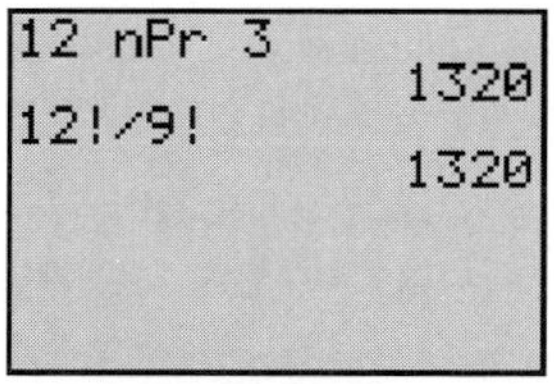

Figure 8.13

FOR THOUGHT True or False? Explain.

1. If a product code consists of a single letter followed by a two-digit number from 10 through 99, then $89 \cdot 26$ different codes are available.
2. If a fraternity name consists of three Greek letters chosen from the 24 letters in the Greek alphabet with repetitions allowed, then $23 \cdot 22 \cdot 21$ different fraternity names are possible.
3. If an outfit consists of a tie, a shirt, a pair of pants, and a coat, and John has three ties, five shirts, and three coats that all match his only pair of pants, then John has 11 outfits available to wear.
4. The number of ways in which five people can line up to buy tickets is 120.
5. The number of permutations of 10 things taken 2 at a time is 90.
6. The number of different ways to mark the answers to a 20-question multiple-choice test with each question having four choices is $P(20, 4)$.
7. The number of different ways to mark the answers to this sequence of 10 ''For Thought'' questions is 2^{10}.
8. $\dfrac{1000!}{998!} = 999{,}000$
9. $P(10, 9) = P(10, 1)$
10. $P(29, 1) = 1$

8.4 EXERCISES

Tape 22 Disk

Solve each problem using the fundamental counting principle.

1. *Traveling Sales Representative* A sales representative can take either of two different routes from Sacramento to Stockton and any one of four different routes from Stockton to San Francisco. How many different routes can she take from Sacramento to San Francisco, going through Stockton?
2. *Optional Equipment* A new car can be ordered in any one of nine different colors, with three different engines, two different transmissions, three different body styles, two different interior designs, and four different stereo systems. How many different cars are available?
3. *Sleepless Night* The ghosts of Christmas Past, Present, and Future plan on visiting Scrooge at 1:00, 2:00, and 3:00 in the morning. All three are available for haunting at all three times. Make a tree diagram showing all of the different orders in which the three ghosts can visit Scrooge. How many different arrangements are possible for this haunting schedule?
4. *Track Competition* Juan, Felix, Ronnie, and Ted are in a 100-m race. Make a tree diagram showing all possible orders in which they can finish the race, assuming that there are no ties. In how many ways can they finish the race?
5. *Poker Hands* A poker hand consists of five cards drawn from a deck of 52. How many different poker hands are there consisting of an ace, king, queen, jack, and ten?
6. *Drawing Cards from a Deck* In a certain card game, four cards are drawn from a deck of 52. How many different hands are there containing one heart, one spade, one club, and one diamond?
7. *Have It Your Way* Wendy's Old Fashioned Hamburgers once advertised that 256 different hamburgers were available at Wendy's. This number was obtained by using the fundamental counting principle and considering whether or not to include each one of several different options on the burger. How many different optional items were used to get this number?
8. *Choosing a Pizza* A pizza can be ordered in three sizes with either thick crust or thin crust. You have to decide whether to include each of four different meats on your pizza. You must also decide whether to include green peppers, onions, mushrooms, anchovies, and/or black olives. How many different pizzas can you order?

Evaluate each expression.

9. $P(7, 3)$
10. $P(16, 4)$
11. $P(99, 0)$
12. $P(55, 1)$

13. $\dfrac{P(9, 5)}{5!}$

14. $\dfrac{P(7, 2)}{2!}$

15. $\dfrac{P(11, 3)}{3!}$

16. $\dfrac{P(15, 5)}{5!}$

17. $\dfrac{16!}{4!12!}$

18. $\dfrac{19!}{17!2!}$

19. $\dfrac{88!}{85!3!}$

20. $\dfrac{102!}{100!2!}$

Solve each problem using the idea of permutations.

21. *Atonement of Hercules* The King of Tiryens ordered the Greek hero Hercules to atone for murdering his own family by performing 12 difficult and dangerous tasks. How many different orders are there for Hercules to perform the 12 tasks?

22. *Parading in Order* A small Mardi Gras parade consists of eight floats and three marching bands. In how many different orders can they line up to parade?

23. *Inspecting Restaurants* A health inspector must visit 3 of 15 restaurants on Monday. In how many ways can she pick a first, second, and third restaurant to visit?

24. *Assigned Reading* In how many ways can an English professor randomly give out one copy each of *War and Peace*, *The Grapes of Wrath*, *Moby Dick*, and *Gone with the Wind*, and four copies of *Jurassic Park* to a class of eight students?

25. *Scheduling Radio Shows* The program director for a public radio station has 26 half-hour shows available for Sunday evening. How many different schedules are possible for the 6:00 to 10:00 P.M. time period?

26. *Choosing Songs* A disc jockey must choose eight songs from the top 20 to play in the next 30-minute segment of her show. How many different arrangements are possible for this segment?

Solve each counting problem.

27. *Multiple-Choice Test* How many different ways are there to mark the answers to a six-question multiple-choice test in which each question has four possible answers?

28. *Choosing a Name* A novelist has decided on four possible first names and three possible last names for the main character in his next book. In how many ways can he name the main character?

29. *Choosing a Prize* A committee has four different VCRs, three different CD players, and six different CDs available. The person chosen as Outstanding Freshman will receive one item from each category. How many different prizes are possible?

30. *Phone Extensions* How many different four-digit extensions are available for a company phone system if the first digit cannot be 0?

31. *Computer Passwords* How many different three-letter computer passwords are available if any letters can be used but repetition of letters is not allowed?

32. *Company Cars* A new Cadillac, a new Dodge, and a used Taurus are to be assigned randomly to three of ten real estate salespersons. In how many ways can the assignment be made?

33. *Phone Numbers* How many different seven-digit phone numbers are available in Jamestown if the first three digits are either 345, 286, or 329?

34. *Electronic Mail* Bob Smith's e-mail address at International Plumbing Supply is bobs@ips.com. If all e-mail addresses at IPS consist of four letters followed by @ips.com, then how many possible addresses are there?

35. *Clark to the Rescue* The archvillain Lex Luther fires a nuclear missile into California's fault line, causing a tremendous earthquake. During the resulting chaos, Superman has to save Lois from a rock slide, rescue Jimmy from a bursting dam, stop a train from derailing, and catch a school bus that is plummeting from the Golden Gate Bridge. How many different ways are there for the Man of Steel to arrange these four rescues?

36. *Bus Routes* A bus picks up passengers from five hotels in downtown Seattle before heading to the Seattle-Tacoma Airport. How many different ways are there to arrange these stops?

37. *Taking a Test* How many ways are there to choose the answers to a test that consists of five true-false questions followed by six multiple-choice questions with four options each?

38. *Granting Tenure* A faculty committee votes on whether or not to grant tenure to each of four candidates. How many different possible outcomes are there to the vote?

39. *Possible Words* Ciara is entering a contest (sponsored by a detergent maker) that requires finding all three-letter words that can be made from the word WASHING. No letter may be used more than once. To make sure that she does not miss any, Ciara plans to write down all possible three-letter words and then look up each one in a dictionary. How many possible three-letter words will be on her list?

40. *Listing Permutations* Make a list of all of the permutations of the letters A, B, C, D, and E taken three at a time. How many permutations should be in your list?

For Writing/Discussion

41. *Listing Subsets* List all of the subsets of each of the sets $\{A\}$, $\{A, B\}$, $\{A, B, C\}$, and $\{A, B, C, D\}$. Find a formula for the number of subsets of a set of n elements.

42. *Number of Subsets* Explain how the fundamental counting principle can be used to find the number of subsets of a set of n elements.

Linking Concepts

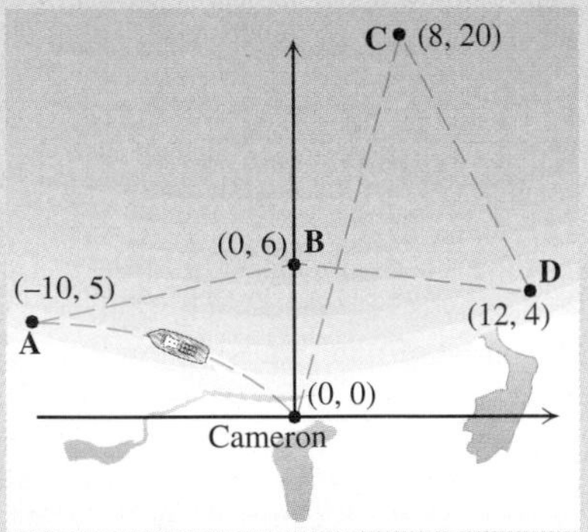

For Individual or Group Explorations

The Shortest Route A supply boat must travel from Cameron, Louisiana to four oil rigs in the Gulf of Mexico and return to Cameron. In a rectangular coordinate system Cameron is at (0, 0). The coordinates of the rigs are (−10, 5), (0, 6), (8, 20), and (12, 4) where the units are miles.

a) How many routes are possible?

b) List all of the possible routes.

c) Find the distance to the nearest tenth of a mile from each location to every other location.

d) Find the length of each of the listed routes. Are there any duplications?

e) What is the shortest possible route?

f) If the supply boat had to stop at 40 oil rigs, then how many routes are possible? How long would it take you to find the shortest route?

8.5 Combinations, Labeling, and the Binomial Theorem

In Section 8.4 we learned the fundamental counting principle, and we applied it to finding the number of permutations of n objects taken r at a time. In permutations, the r objects are arranged in a sequential manner. In this section, we will find the number of ways to choose r objects from n distinct objects when the order in which the objects are chosen or placed is unimportant.

Combinations of n Things Taken r at a Time

In how many ways can two students be selected to go to the board from a class of four students: Adams, Baird, Campbell, and Dalton? Assuming that the two selected are treated identically, the choice of Adams and Baird is no different from the choice of Baird and Adams. Set notation provides a convenient way of listing all possible choices of two students from the four available students because in set notation $\{A, B\}$ is the same as $\{B, A\}$. We can easily list all subsets or **combinations** of two elements taken from the set $\{A, B, C, D\}$:

$$\{A, B\} \quad \{A, C\} \quad \{A, D\} \quad \{B, C\} \quad \{B, D\} \quad \{C, D\}$$

The number of these subsets is the number of combinations of four things taken two at a time, denoted by $C(4, 2)$. Since there are six subsets, $C(4, 2) = 6$.

If we had a first and a second prize to give to two of the four students, then $P(4, 2) = 4 \cdot 3 = 12$ is the number of ways to award the prizes. Since the prizes are different, AB is different from BA, AC is different from CA, and so on. The 12 permutations are listed here:

$$\begin{matrix} AB & AC & AD & BC & BD & CD \\ BA & CA & DA & CB & DB & DC \end{matrix}$$

From the list of combinations of four things taken two at a time, we made the list of permutations of four things two at a time by rearranging each combination. So $P(4, 2) = 2 \cdot C(4, 2)$.

In general, we can list all combinations of n things taken r at a time, then rearrange each of those subsets of r things in $r!$ ways to obtain all of the permutations of n things taken r at a time. So $P(n, r) = r!\, C(n, r)$, or

$$C(n, r) = \frac{P(n, r)}{r!}.$$

Since $P(n, r) = n!/(n - r)!$, we have

$$C(n, r) = \frac{n!}{(n - r)!\, r!}.$$

These results are summarized in the following theorem.

Theorem: Combinations of n Things Taken r at a Time

The number of combinations of n things taken r at a time (or the number of subsets of size r from a set of n elements) is given by the formula

$$C(n, r) = \frac{n!}{(n - r)!\, r!} \quad \text{for } 0 \le r \le n.$$

Many calculators can calculate the value of $C(n, r)$ when given the value of n and r. The notations $\binom{n}{r}$ or nCr may be used on your calculator or elsewhere for $C(n, r)$. Note that

$$C(n, n) = \frac{n!}{0!\, n!} = 1 \quad \text{and} \quad C(n, 0) = \frac{n!}{n!\, 0!} = 1.$$

There is only one way to choose all n objects from a group of n objects if the order does not matter, and there is only one way to choose no object ($r = 0$) from a group of n objects.

EXAMPLE 1 Combinations

To raise money, an alumni association prints lottery tickets with the numbers 1 through 11 printed on each ticket, as shown in Fig. 8.14. To play the lottery, one must circle three numbers on the ticket. In how many ways can three numbers be chosen out of 11 numbers on the ticket?

Figure 8.14

Solution

Choosing three numbers from a list of 11 numbers is the same as choosing a subset of size 3 from a set of 11 elements. So the number of ways to choose the three numbers in playing the lottery is the number of combinations of 11 things taken 3 at a time:

$$C(11, 3) = \frac{11!}{8!\,3!}$$

$$= \frac{11 \cdot 10 \cdot 9 \cdot 8 \cdot 7 \cdot 6 \cdot 5 \cdot 4 \cdot 3 \cdot 2 \cdot 1}{8 \cdot 7 \cdot 6 \cdot 5 \cdot 4 \cdot 3 \cdot 2 \cdot 1 \cdot 3 \cdot 2 \cdot 1}$$

$$= \frac{11 \cdot 10 \cdot 9}{3 \cdot 2 \cdot 1} \quad \text{Divide numerator and denominator by } 8!.$$

$$= 165$$

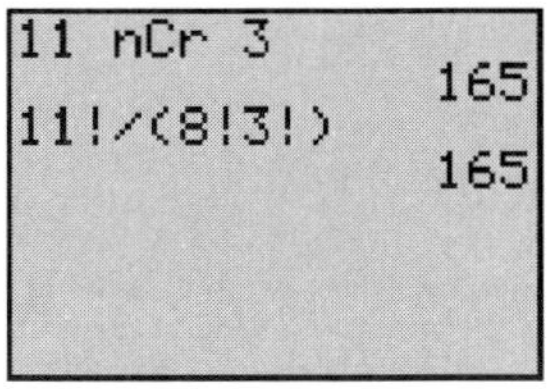

Figure 8.15

Use the combination function or factorials on a calculator to check as in Fig. 8.15.

Note that the combination formula counts the number of subsets of n objects taken r at a time. The objects are not necessarily placed in a subset, but they are all treated alike as in Example 1. In counting the combinations in Example 1, the only important thing is *which* numbers are circled; the order in which the numbers are circled is not important. By contrast, the permutation formula counts the number of ways to select r objects from n, *in order*. For example, the number of ways to award a first, second, and third prize to three of five people is $P(5, 3) = 60$, because the order matters. The number of ways to give three identical prizes to three of five people is $C(5, 3) = 10$, because the order of the awards doesn't matter.

Do not forget that we also have the fundamental counting principle to count the number of ways in which a sequence of events can occur. In the next example we use both the fundamental counting principle and the combination formula.

EXAMPLE 2 Combinations and the counting principle

A company employs nine male welders and seven female welders. A committee of three male welders and two female welders is to be chosen to represent the welders in negotiations with management. How many different committees can be chosen?

Solution

First observe that three male welders can be chosen from the nine available in $C(9, 3) = 84$ ways. Next observe that two female welders can be chosen from seven available in $C(7, 2) = 21$ ways. Now use the fundamental counting principle to get $84 \cdot 21 = 1764$ ways to choose the males and then the females.

Labeling

In a **labeling problem**, n distinct objects are to be given labels, each object getting exactly one label. For example, each person in a class is "labeled" with a letter grade at the end of the semester. Each student living on campus is "labeled" with the name of the dormitory in which the student resides. In a labeling problem, each distinct object gets one label, but there may be several types of labels and many labels of each type.

EXAMPLE 3 A labeling problem

Twelve students have volunteered to help with a political campaign. The campaign director needs three telephone solicitors, four door-to-door solicitors, and five envelope stuffers. In how many ways can these jobs (labels) be assigned to these 12 students?

Solution

Since the three telephone solicitors all get the same type of label, the number of ways to select the three students is $C(12, 3)$. The number of ways to select four door-to-door solicitors from the remaining nine students is $C(9, 4)$. The number of ways to select the five envelope stuffers from the remaining five students is

$C(5, 5)$. By the fundamental counting principle the number of ways to make all three selections is

$$\begin{aligned} C(12, 3) \cdot C(9, 4) \cdot C(5, 5) &= \frac{12!}{9!\,3!} \cdot \frac{9!}{5!\,4!} \cdot \frac{5!}{0!\,5!} \\ &= \frac{12!}{3!\,4!\,5!} \\ &= 27{,}720. \end{aligned}$$

Note that in Example 3 there were 12 distinct objects to be labeled with three labels of one type, four labels of another type, and five labels of a third type, and the number of ways to assign those labels was found to be $12!/(3!\,4!\,5!)$. Instead of using combinations and the fundamental counting principle as in Example 3, we can use the following theorem.

Labeling Theorem

If each of n distinct objects is to be assigned one label and there are r_1 labels of the first type, r_2 labels of the second type, . . . , and r_k labels of the kth type, where $r_1 + r_2 + \cdots + r_k = n$, then the number of ways to assign the labels is

$$\frac{n!}{r_1!\, r_2! \cdot \cdots \cdot r_k!}.$$

EXAMPLE 4 Rearrangements of letters in a word

How many different arrangements are there for the 11 letters in the word MISSISSIPPI?

Solution

This problem is a labeling problem if we think of the 11 positions for the letters as 11 distinct objects to be labeled. There is one M-label, and there are four S-labels, four I-labels, and two P-labels. So the number of ways to arrange the letters in MISSISSIPPI is

$$\frac{11!}{1!\,4!\,4!\,2!} = 34{,}650.$$

To calculate this value, either enclose the denominator in parentheses or divide by each factorial in the denominator as in Fig. 8.16.

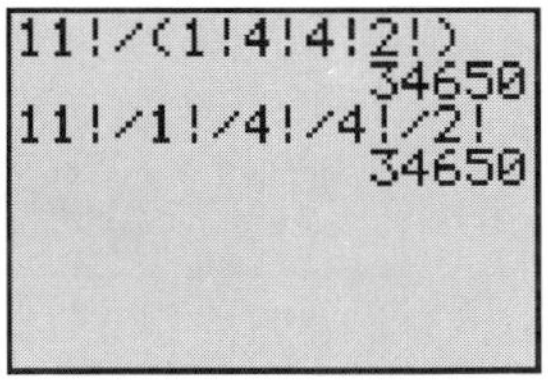

Figure 8.16

You may think of permutation and combination problems as being very different, but they both are labeling problems in actuality. For example, to find the number of subsets of size 3 from a set of size 5, we are assigning three I-labels and two N-labels (I for "in the subset" and N for "not in the subset") to the five distinct objects of the set. Note that

$$\frac{5!}{3!\ 2!} = C(5, 3).$$

To find the number of ways to give a first, second, and third prize to three of ten people, we are assigning one F-label, one S-label, one T-label, and seven N-labels (N for "no prize") to the ten distinct people. Note that

$$\frac{10!}{1!\ 1!\ 1!\ 7!} = P(10, 3).$$

Binomial Expansion

One of the first facts we learn in algebra is

$$(a + b)^2 = a^2 + 2ab + b^2.$$

Our new labeling technique can be used to find the *coefficients* for the square of $a + b$ and any higher power of $a + b$. But before using counting techniques, let's look for patterns in the higher powers of $a + b$ and learn some shortcuts that can be applied in the simple cases.

We can find $(a + b)^3$ by multiplying $(a + b)^2$ by $a + b$. The powers of $a + b$ from the zero power to the fourth power are listed here.

$$\begin{aligned}
(a + b)^0 &= 1 \\
(a + b)^1 &= a + b \\
(a + b)^2 &= a^2 + 2ab + b^2 \\
(a + b)^3 &= a^3 + 3a^2b + 3ab^2 + b^3 \\
(a + b)^4 &= a^4 + 4a^3b + 6a^2b^2 + 4ab^3 + b^4
\end{aligned}$$

The right-hand sides of these equations are referred to as **binomial expansions**. Each one after the first one is obtained by multiplying the previous expansion by $a + b$. To find the expansion for $(a + b)^5$, multiply the expansion for $(a + b)^4$ by $a + b$:

$$\begin{array}{rrrrrr}
 & a^4 & +\ 4a^3b & +\ 6a^2b^2 & +\ 4ab^3 & +\ b^4 \\
 & & & & a & +\ b \\
\hline
 & a^4b & +\ 4a^3b^2 & +\ 6a^2b^3 & +\ 4ab^4 & +\ b^5 \\
a^5 & +\ 4a^4b & +\ 6a^3b^2 & +\ 4a^2b^3 & +\ ab^4 & \\
\hline
a^5 & +\ 5a^4b & +\ 10a^3b^2 & +\ 10a^2b^3 & +\ 5ab^4 & +\ b^5
\end{array}$$

Note that there are six terms in the expansion of $(a + b)^5$ and that the exponents in each term have a sum of 5. The powers of a decrease from left to right, while the powers of b increase. The **binomial coefficients** 1, 5, 10, 10, 5, 1 are obtained by adding the coefficients from consecutive terms in the expansion of $(a + b)^4$ as follows:

$$(a + b)^4 = \mathbf{1}a^4 + \mathbf{4}a^3b + 6a^2b^2 + 4ab^3 + 1b^4$$

$$(a + b)^5 = a^5 + \mathbf{5}a^4b + 10a^3b^2 + 10a^2b^3 + 5ab^4 + b^5$$

It is easy to remember the coefficients for the first few powers of $a + b$ by using **Pascal's triangle**. Pascal's triangle is a triangular array of binomial coefficients:

1	Coefficient of $(a + b)^0$
1 1	Coefficients of $(a + b)^1$
1 2 1	Coefficients of $(a + b)^2$
1 3 3 1	Coefficients of $(a + b)^3$
1 4 6 4 1	Coefficients of $(a + b)^4$
1 5 10 10 5 1	Coefficients of $(a + b)^5$

Pascal's triangle

Each row in Pascal's triangle starts and ends with 1 and the coefficients between the 1's are obtained from the previous row by adding consecutive entries.

EXAMPLE 5 Using Pascal's triangle

Use Pascal's triangle to find each binomial expansion.

a) $(2x - 3y)^4$ **b)** $(a + b)^6$

Solution

a) Think of $(2x - 3y)^4$ as $(2x + (-3y))^4$ and use the coefficients 1, 4, 6, 4, 1 from Pascal's triangle:

$$\begin{aligned}(2x - 3y)^4 &= 1(2x)^4(-3y)^0 + 4(2x)^3(-3y)^1 + 6(2x)^2(-3y)^2 \\ &\quad + 4(2x)^1(-3y)^3 + 1(2x)^0(-3y)^4 \\ &= 16x^4 - 96x^3y + 216x^2y^2 - 216xy^3 + 81y^4\end{aligned}$$

b) Use the coefficients 1, 5, 10, 10, 5, 1 from Pascal's triangle to obtain the coefficients 1, 6, 15, 20, 15, 6, 1. Use these coefficients with decreasing powers of a and increasing powers of b to get

$$(a + b)^6 = a^6 + 6a^5b + 15a^4b^2 + 20a^3b^3 + 15a^2b^4 + 6ab^5 + b^6.$$

The Binomial Theorem

Pascal's triangle provides an easy way to find a binomial expansion for small powers of a binomial, but it is not practical for large powers. To get the binomial coefficients for larger powers, we count the number of like terms of each type using the idea of labeling. For example, consider

$$(a + b)^3 = (a + b)(a + b)(a + b) = a^3 + 3a^2b + 3ab^2 + b^3.$$

The terms of the product come from all of the different ways there are to select either a or b from each of the three factors and multiply the selections. The coefficient of a^3 is 1 because we get a^3 only from aaa, where a is chosen from each factor. The coefficient of a^2b is 3 because a^2b is obtained from aab, aba, and baa. From the labeling theorem, the number of ways to rearrange the letters aab is $3!/(2!\,1!) = 3$. So the coefficient of a^2b is the number of ways to label the three factors with two a-labels and one b-label.

In general, the expansion of $(a + b)^n$ contains $n + 1$ terms in which the exponents have a sum of n. The coefficient of $a^{n-r}b^r$ is the number of ways to label n factors with $(n - r)$ of the a-labels and r of the b-labels:

$$\frac{n!}{r!(n - r)!}$$

The *binomial theorem* expresses this result using summation notation.

The Binomial Theorem

If n is a positive integer, then for any real numbers a and b,

$$(a + b)^n = \sum_{r=0}^{n} \binom{n}{r} a^{n-r}b^r, \quad \text{where } \binom{n}{r} = \frac{n!}{r!(n - r)!}.$$

EXAMPLE 6 Using the binomial theorem

What is the coefficient of a^7b^2 in the binomial expansion of $(a + b)^9$?

Solution

Compare a^7b^2 to $a^{n-r}b^r$ to see that $r = 2$ and $n = 9$. According to the binomial theorem, the coefficient of a^7b^2 is

$$\binom{9}{2} = \frac{9!}{2!\,7!} = \frac{9 \cdot 8}{2} = 36.$$

The term $36a^7b^2$ occurs in the expansion of $(a + b)^9$.

EXAMPLE 7 Using the binomial theorem

Write out the first three terms in the expansion of $(x - 2y)^{10}$.

Solution

Write $(x - 2y)^{10}$ as $(x + (-2y))^{10}$. According to the binomial theorem,

$$(x + (-2y))^{10} = \sum_{r=0}^{10} \binom{10}{r} x^{10-r}(-2y)^r.$$

To find the first three terms, let $r = 0$, 1, and 2:

$$\begin{aligned}(x - 2y)^{10} &= \binom{10}{0} x^{10}(-2y)^0 + \binom{10}{1} x^9(-2y)^1 + \binom{10}{2} x^8(-2y)^2 + \cdots \\ &= x^{10} - 20x^9y + 180x^8y^2 + \cdots\end{aligned}$$

In the next example, labeling is used to find the coefficient of a term in a power of a trinomial.

EXAMPLE 8 Trinomial coefficients

What is the coefficient of a^3b^2c in the expansion of $(a + b + c)^6$?

Solution

The terms of the product $(a + b + c)^6$ come from all of the different ways there are to select a, b, or c from each of the six distinct factors and multiply the selections. The number of times that a^3b^2c occurs is the same as the number of rearrangements of *aaabbc*, which is a labeling problem. The number of rearrangements of *aaabbc* is

$$\frac{6!}{3!\,2!\,1!} = \frac{6 \cdot 5 \cdot 4 \cdot 3 \cdot 2 \cdot 1}{3 \cdot 2 \cdot 1 \cdot 2 \cdot 1 \cdot 1} = 60.$$

So the term $60a^3b^2c$ occurs in the expansion of $(a + b + c)^6$.

FOR THOUGHT True or False? Explain.

1. The number of ways to choose three questions to answer out of five questions on an essay test is $C(5, 2)$.
2. The number of ways to answer a five-question multiple-choice test in which each question has three choices is $P(5, 3)$.
3. The number of ways to pick a Miss America and the first runner-up from the five finalists is 20.
4. The binomial expansion for $(x + y)^n$ contains n terms.
5. For any real numbers a and b, $(a + b)^5 = \sum_{i=0}^{5} \binom{5}{i} a^i b^{5-i}$.
6. The sum of the binomial coefficients in $(a + b)^n$ is 2^n.
7. $P(8, 3) < C(8, 3)$
8. $P(7, 3) = (3!) \cdot C(7, 3)$
9. $P(8, 3) = P(8, 5)$
10. $C(1, 1) = 1$

8.5 EXERCISES

Tape 22 Disk

Solve each problem.

1. *Poker Hands* How many five-card poker hands are there if you draw five cards from a deck of 52?
2. *Bridge Hands* How many 13-card bridge hands are there if you draw 13 cards from a deck of 52?
3. *Playing a Lottery* In a certain lottery the player chooses six numbers from the numbers 1 through 49. In how many ways can the six numbers be chosen?
4. *Fantasy Five* In a different lottery the player chooses five numbers from the numbers 1 through 39. In how many ways can the five numbers be chosen?
5. *Job Candidates* The search committee has narrowed the applicants to five unranked candidates. In how many ways can three be chosen for an in-depth interview?
6. *Spreading the Flu* In how many ways can nature select five students out of a class of 20 students to get the flu?

The following problems may involve combinations, permutations, or the fundamental counting principle.

7. *Television Schedule* Arnold plans to spend the evening watching television. Each of the three networks runs one-hour shows starting at 7:00, 8:00, and 9:00 P.M. In how many ways can he watch three complete shows from 7:00 to 10:00 P.M.?
8. *Cafeteria Meals* For \$3.98 you can get a salad, main course, and dessert at the cafeteria. If you have a choice of five different salads, six different main courses, and four different desserts, then how many different \$3.98 meals are there?
9. *Fire Code Inspections* The fire inspector in Cincinnati must select three night clubs from a list of eight for an inspection of their compliance with the fire code. In how many ways can she select the three night clubs?
10. *Prize-Winning Pigs* In how many ways can a red ribbon, a blue ribbon, and a green ribbon be awarded to three of six pigs at the county fair?
11. *Choosing a Team* From the nine male and six female sales representatives for an insurance company, a team of three men and two women will be selected to attend a national conference on insurance fraud. In how many ways can the team of five be selected?
12. *Poker Hands* How many five-card poker hands are there containing three hearts and two spades?
13. *Returning Exam Papers* In how many different orders can Professor Pereira return 12 exam papers to 12 students?
14. *Saving the Best Till Last* In how many ways can Professor Yang return examination papers to 12 students if she always returns the worst paper first and the best paper last?
15. *Rolling Dice* A die is to be rolled twice, and each time the number of dots showing on the top face is to be recorded. If we think of the outcome of two rolls as an ordered pair of numbers, then how many outcomes are there?

16. *Having Children* A couple plans to have three children. Considering the sex and order of birth of each of the three children, how many outcomes are possible to this plan?

17. *Marching Bands* In how many ways can four marching bands and three floats line up for a parade if a marching band must lead the parade and two bands cannot march next to one another?

18. *Marching Bands* In how many ways can four marching bands and four floats line up for a parade if a marching band must lead the parade and two bands cannot march next to one another?

Solve each problem using the idea of labeling.

19. *Rearranging Letters* How many permutations are possible using the letters in the word ALABAMA?

20. *Spelling Mistakes* How many incorrect spellings of the word FLORIDA are there, using all of the correct letters?

21. *Assigning Topics* An instructor in a history class of ten students wants term papers written on World War II, World War I, and the Civil War. If he randomly assigns World War II to five students, World War I to three students, and the Civil War to two students, then in how many ways can these assignments be made?

22. *Parking Tickets* Officer O'Reilly is in charge of 12 officers working a sporting event. In how many ways can she choose six to control parking, four to work security, and two to handle emergencies?

23. *Determining Chords* How many distinct chords (line segments with endpoints on the circle) are determined by three points lying on a circle? By four points? By five points? By n points?

24. *Determining Triangles* How many distinct triangles are determined by six points lying on a circle, where the vertices of each triangle are chosen from the six points?

25. *Assigning Volunteers* Ten students volunteered to work in the governor's reelection campaign. Three will be assigned to making phone calls, two will be assigned to stuffing envelopes, and five will be assigned to making signs. In how many ways can the assignments be made?

26. *Assigning Vehicles* Three identical Buicks, four identical Fords, and three identical Toyotas are to be assigned to ten traveling salespeople. In how many ways can the assignments be made?

Write the complete binomial expansion for each of the following powers of a binomial.

27. $(a - 2)^3$
28. $(b^2 - 3)^3$
29. $(2a + b^2)^3$
30. $(x + 3)^3$
31. $(x - 2y)^4$
32. $(y - 2)^4$
33. $(x^2 + 1)^4$
34. $(2r + 3t^2)^4$

Write the first three terms of each binomial expansion.

35. $(x + y)^9$
36. $(a - b)^{10}$
37. $(2x - y)^{12}$
38. $(a + 2b)^{11}$
39. $(2s - 0.5t)^8$
40. $(3y^2 + a)^{10}$
41. $(m^2 - 2w^3)^9$
42. $(ab^2 - 5c)^8$

Solve each problem.

43. What is the coefficient of w^3y^5 in the expansion of $(w + y)^8$?

44. What is the coefficient of a^3z^9 in the expansion of $(a + z)^{12}$?

45. What is the coefficient of a^5b^8 in the expansion of $(b - 2a)^{13}$?

46. What is the coefficient of x^6y^5 in the expansion of $(0.5x - y)^{11}$?

47. What is the coefficient of $a^2b^4c^6$ in the expansion of $(a + b + c)^{12}$?

48. What is the coefficient of $x^3y^2z^3$ in the expansion of $(x - y - 2z)^8$?

49. What is the coefficient of a^3b^7 in the expansion of $(a + b + 2c)^{10}$?

50. What is the coefficient of $w^2xy^3z^9$ in the expansion of $(w + x + y + z)^{15}$?

For Writing/Discussion

51. Explain why $C(n, r) = C(n, n - r)$.

52. Is $P(n, r) = P(n, n - r)$? Explain your answer.

53. Explain the difference between permutations and combinations.

54. Prove that $\sum_{i=0}^{n} \binom{n}{i} = 2^n$ for any positive integer n.

55. *Cooperative Learning* Make up three counting problems: a combination problem, a permutation problem, and a labeling problem. Give them to a classmate to solve.

Linking Concepts

For Individual or Group Explorations

Poker Hands *In the game of poker, a hand consists of five cards drawn from a deck of 52. The best hand is the royal flush. It consists of the ace, king, queen, jack, and ten of one suit. There are only four possible royal flushes. Find the number of possible hands of each of the following types. Explain your reasoning. The hands are listed in order from the least to the most common. Of course, a hand in this list beats any hand that is below it in this list.*

a) Straight flush (five cards in a sequence in the same suit, but not a royal flush)

b) Four of a kind (for example, four kings or four twos)

c) Full house (three of a kind together with a pair)

d) Flush (five cards in a single suit, but not a straight)

e) Three of a kind

f) Two pairs

g) One pair

8.6 Probability

We often hear statements such as ''The probability of rain today is 70%,'' ''I have a 50-50 chance of passing English,'' or ''I have one chance in a million of winning the lottery.'' It is clear that we have a better chance of getting rained on than we have of winning a lottery, but what exactly is probability? In this section we will study probability and use the counting techniques of Sections 8.4 and 8.5 to make precise statements concerning probabilities.

The Probability of an Event

An **experiment** is any process for which the outcome is uncertain. For example, if we toss a coin, roll a die, draw a poker hand from a deck, or arrange people in a line in a manner that makes the outcome uncertain, then these processes are experiments. A **sample space** is the set of all possible outcomes to an experiment. If each outcome occurs with about the same frequency when the experiment is repeated many times, then the outcomes are called **equally likely**.

The simplest experiments for determining probabilities are ones in which the outcomes are equally likely. For example, if a fair coin is tossed, then the sample space S consists of two equally likely outcomes, heads and tails:

$$S = \{H, T\}$$

An **event** is a subset of a sample space. The subset $E = \{H\}$ is the event of getting heads when the coin is tossed. If $n(S)$ is the number of equally likely outcomes in S, then $n(S) = 2$. If $n(E)$ is the number of equally likely outcomes in E, then $n(E) = 1$. The probability of the event E, $P(E)$, is the ratio of $n(E)$ to $n(S)$:

$$P(E) = \frac{n(E)}{n(S)} = \frac{1}{2}$$

Definition: Probability of an Event

If S is a sample space of equally likely outcomes to an experiment and the event E is a subset of S, then the **probability of E**, $P(E)$, is defined by

$$P(E) = \frac{n(E)}{n(S)}.$$

Since E is a subset of S, $0 \le n(E) \le n(S)$. So $P(E)$ is a number between 0 and 1, inclusive. If there are no outcomes in the event E, then $P(E) = 0$ and it is impossible for E to occur. If $n(E) = n(S)$ then $P(E) = 1$ and the event E is certain to occur. For example, if E is the event of getting two heads on a single toss of a coin, then $n(E) = 0$ and $P(E) = 0/2 = 0$. If E is the event of getting fewer than two heads on a single toss of a coin, then for either outcome H or T there are fewer than two heads. So $E = \{H, T\}$, $n(E) = 2$, and $P(E) = 2/2 = 1$.

Probability predicts the future, but not exactly. The probability that heads will occur when a coin is tossed is 1/2, but there is no way to know whether heads or tails will occur on any future toss. However, the probability 1/2 indicates that if the coin is tossed many times, then *about* 1/2 of the tosses will be heads.

EXAMPLE 1 Rolling a single die

What is the probability of getting a number larger than 4 when a single die is rolled?

Solution

When a die is rolled, the number of dots showing on the upper face of the die is counted. So the sample space of equally likely outcomes is

$$S = \{1, 2, 3, 4, 5, 6\} \quad \text{and} \quad n(S) = 6.$$

Since only 5 and 6 are larger than 4,

$$E = \{5, 6\} \quad \text{and} \quad n(E) = 2.$$

According to the definition of probability,

$$P(E) = \frac{n(E)}{n(S)} = \frac{2}{6} = \frac{1}{3}.$$

EXAMPLE 2 Tossing a pair of coins

What is the probability of getting at least one head when a pair of coins is tossed?

Solution

Since there are two equally likely outcomes for the first coin and two equally likely outcomes for the second coin, by the fundamental counting principle there are four equally likely outcomes to the experiment of tossing a pair of coins. We can list the outcomes as ordered pairs:

$$S = \{(H, H), (H, T), (T, H), (T, T)\}.$$

The ordered pairs (H, T) and (T, H) must be listed as different outcomes, because the first coordinate is the result from the first coin and the second coordinate is the result from the second coin. Since three of these outcomes result in at least one head,

$$E = \{(H, H), (H, T), (T, H)\}$$

and $n(E) = 3$. So

$$P(E) = \frac{n(E)}{n(S)} = \frac{3}{4}.$$

EXAMPLE 3 Rolling a pair of dice

What is the probability of getting a sum of 5 when a pair of dice is rolled?

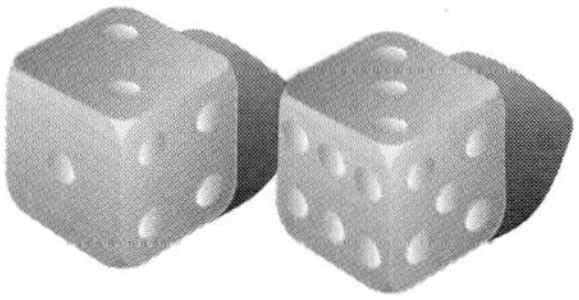

Figure 8.17

Solution

Since there are six equally likely outcomes for each die, by the fundamental counting principle there are 36 equally likely outcomes to rolling the pair. See Fig. 8.17. We can list the 36 outcomes as ordered pairs as shown.

$$\begin{aligned} S = \{&(1, 1), (1, 2), (1, 3), (1, 4), (1, 5), (1, 6),\\ &(2, 1), (2, 2), (2, 3), (2, 4), (2, 5), (2, 6),\\ &(3, 1), (3, 2), (3, 3), (3, 4), (3, 5), (3, 6),\\ &(4, 1), (4, 2), (4, 3), (4, 4), (4, 5), (4, 6),\\ &(5, 1), (5, 2), (5, 3), (5, 4), (5, 5), (5, 6),\\ &(6, 1), (6, 2), (6, 3), (6, 4), (6, 5), (6, 6)\} \end{aligned}$$

The ordered pairs in color in the sample space are the ones in which the sum of the entries is 5. So the phrase "sum of the numbers is 5" describes the event

$$E = \{(4, 1), (3, 2), (2, 3), (1, 4)\},$$

and

$$P(E) = \frac{n(E)}{n(S)} = \frac{4}{36} = \frac{1}{9}.$$

EXAMPLE 4 Probability of winning a lottery

To play the Alumni Association Lottery, you pick three numbers from the integers from 1 through 11 inclusive. What is the probability that you win the $50 prize by picking the same three numbers as the ones chosen by the Alumni Association?

Solution

The number of ways to select three numbers out of 11 numbers is $C(11, 3) = 165$. Since only one of the 165 equally likely outcomes is the winning combination, the probability of winning is 1/165.

Complementary Events

If the probability of rain today is 60%, then the probability that it does not rain is 40%. Rain and not rain are called complementary events. For complementary events there is no possibility that both occur, and one of them must occur. If A is an event, then A' (read "A prime" or "A complement") represents the complement of the event A.

Definition: Complementary Events

Two events A and A' are called **complementary events** if $A \cap A' = \varnothing$ and $P(A) + P(A') = 1$.

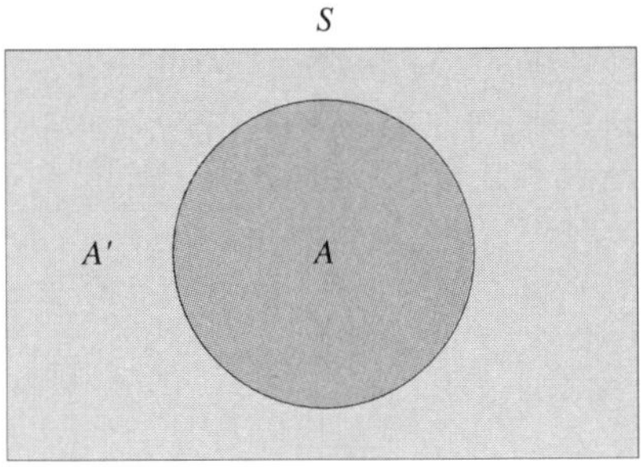

Figure 8.18

The diagram in Fig. 8.18 illustrates complementary events. The region inside the rectangle represents the sample space S, while the region inside the circle represents the event A. The event A' is the region inside the rectangle, but outside the circle. Note that $A \cup A' = S$ and $A \cap A' = \varnothing$.

EXAMPLE 5 The probability of complementary events

What is the probability of getting a sum that is not equal to 5 when rolling a pair of dice?

Solution

From Example 3, the probability that the sum is 5 is 1/9. Getting a sum that is not equal to 5 is the complement of getting a sum that is equal to 5. So

$$P(\text{sum is } 5) + P(\text{sum is not } 5) = 1$$
$$P(\text{sum is not } 5) = 1 - P(\text{sum is } 5).$$

So the probability that the sum is not 5 is $1 - 1/9$, or 8/9.

Scientists estimate that a "killer asteroid," an asteroid with diameter greater than one kilometer, hits the earth on the average of once in a million years. Simply knowing the average number of hits in a million years allows them to find the probability of any number of hits in a million years. The main concern is the probability of no hits. Probabilities of this type involve e, the base of the natural logarithm. In general, the expression $e^{-\alpha}$ gives an estimate of the probability that there are no occurrences of an event in a period of time. Here α is the average number of occurrences in that time period. For the killer asteroid, the average is one hit per one million years, $\alpha = 1$. So the probability that there are no asteroid hits in the next one million years is

$$e^{-1} = 0.3678 \approx 37\%.$$

The probability that there is at least one hit (the complement of no hits) in the next one million years is about 63%, quite high, but a million years is a long time. In the next example, probabilities are calculated for a time period that is more meaningful to humans.

EXAMPLE 6 **The probability of collision with an asteroid**

Find the probability that the earth will survive the next 100 years without a hit by a killer asteroid, and find the probability that during the next 100 years the earth will be hit at least once by a killer asteroid. Use the estimate of an average of one hit per one million years.

Solution

To find the required probability, we need the average number of hits in 100 years. Convert the ratio of one hit per one million years into the equivalent ratio of 0.0001 hits per 100 years:

$$\frac{1 \text{ hit}}{1{,}000{,}000 \text{ years}} = \frac{0.0001 \text{ hits}}{100 \text{ years}}$$

Since 0.0001 is the average number of hits per 100 years, use $\alpha = 0.0001$ in the expression $e^{-\alpha}$. The probability of no hits in the next 100 years is $e^{-0.0001} \approx 0.999900005$. The probability of at least one hit (the complement of no hits) in the next 100 years is

$$1 - 0.999900005 = 0.000099995,$$

about 1/10,000. Nothing to worry about.

Odds

If the probability is 4/5 that the Braves will win the World Series and 1/5 that they will lose, then they are four times as likely to win as they are to lose. We say that the *odds* in favor of the Braves winning the World Series are 4 to 1.

Notice that odds are *not* probabilities. Odds are ratios of probabilities. Odds are usually written as ratios of whole numbers.

Definition: Odds

If A is any event, then the **odds in favor of A** are defined as the ratio $P(A)$ to $P(A')$ and the **odds against A** are defined as the ratio of $P(A')$ to $P(A)$.

EXAMPLE 7 Odds for and against an event

What are the odds in favor of getting a sum of 5 when rolling a pair of dice? What are the odds against a sum of 5?

Solution

From Examples 3 and 5, $P(\text{sum is } 5) = 1/9$ and $P(\text{sum is not } 5) = 8/9$. The odds in favor of getting a sum of 5 are 1/9 to 8/9. Multiply each fraction by 9 to get the odds 1 to 8. The odds against a sum of 5 are 8 to 1.

The odds in favor of A are found from the ratio of $P(A)$ to $P(A')$, but the equation $P(A) + P(A') = 1$ can be used to find $P(A)$ and $P(A')$.

EXAMPLE 8 Finding probabilities from odds

If the odds in favor of the Rams going to the Super Bowl are 1 to 5, then what is the probability that the Rams will go to the Super Bowl?

Solution

Since 1 to 5 is the ratio of the probability of going to that of not going, the probability of not going is 5 times as large as the probability of going. Let $P(G) = x$ and $P(G') = 5x$. Since

$$P(G) + P(G') = 1,$$

we have

$$\begin{aligned} x + 5x &= 1 \\ 6x &= 1 \\ x &= \frac{1}{6} \end{aligned}$$

So the probability that the Rams will go to the Super Bowl is 1/6.

We can express the idea found in Example 8 as a theorem relating odds to probabilities.

Theorem: Converting from Odds to Probability

If the odds in favor of event E are a to b, then

$$P(E) = \frac{a}{a+b} \quad \text{and} \quad P(E') = \frac{b}{a+b}.$$

The Addition Rule

One way to find the probability of an event is to list all of the equally likely outcomes to the experiment and count those outcomes that make up the event. The relationship between the probabilities of complementary events allows us to find the probability of the complement of an event by subtraction rather than counting. The addition rule provides another relationship between probabilities of events that can be used to find probabilities.

In rolling a pair of dice, let A be the event that doubles occur and let B be the event that the sum is 4. We have

$$A = \{(1, 1), (2, 2), (3, 3), (4, 4), (5, 5), (6, 6)\}$$

and

$$B = \{(3, 1), (2, 2), (1, 3)\}.$$

The event that doubles *and* a sum of 4 occur is the event $A \cap B$:

$$A \cap B = \{(2, 2)\}$$

Count the ordered pairs to get $P(A) = 6/36$, $P(B) = 3/36$, and $P(A \cap B) = 1/36$. The event that either doubles *or* a sum of 4 occurs is the event $A \cup B$:

$$A \cup B = \{(1, 1), (2, 2), (3, 3), (4, 4), (5, 5), (6, 6), (3, 1), (1, 3)\}$$

Even though (2, 2) belongs to both A and B, it is listed only once in $A \cup B$. Note that $P(A \cup B)$ is 8/36 and

$$\frac{8}{36} = \frac{6}{36} + \frac{3}{36} - \frac{1}{36}.$$

Since (2, 2) is counted in both A and B, $P(A \cap B)$ is subtracted from $P(A) + P(B)$. This example illustrates the addition rule.

The Addition Rule

If A and B are any events in a sample space, then

$$P(A \cup B) = P(A) + P(B) - P(A \cap B).$$

If $P(A \cap B) = 0$, then A and B are called **mutually exclusive** events and

$$P(A \cup B) = P(A) + P(B).$$

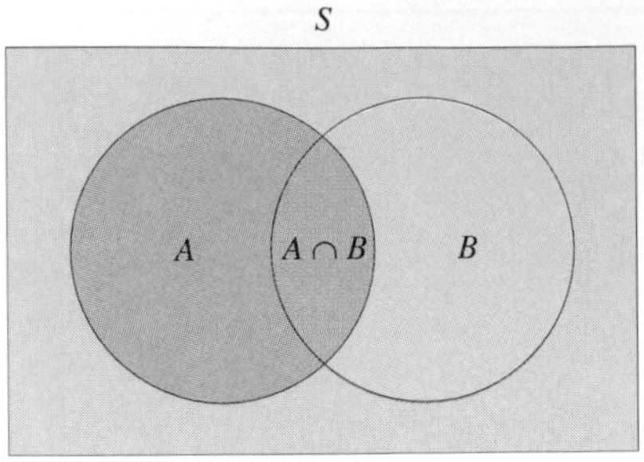

Figure 8.19

Two events are mutually exclusive if it is impossible for both to occur. For example, when a pair of dice is rolled once, getting a sum of 4 and getting a sum of 5 are mutually exclusive events.

The diagram in Fig. 8.19 illustrates the addition rule for two sets A and B whose intersection is not empty. To find $P(A \cup B)$ we add $P(A)$ and $P(B)$, but we must subtract $P(A \cap B)$ because it is counted in both $P(A)$ and $P(B)$. Since $P(A \cap B) = 0$ for mutually exclusive events, the addition rule for mutually exclusive events is a special case of the general addition rule. Note that complementary events are mutually exclusive, but mutually exclusive events are not necessarily complementary.

EXAMPLE 9 Using the addition rule

At Hillside Community College, 60% of the students are commuters (C), 50% are female (F), and 30% are female commuters. If a student is selected at random, what is the probability that the student is either a female or a commuter?

Solution

By the addition rule, the probability of selecting either a female or a commuter is

$$\begin{aligned} P(F \cup C) &= P(F) + P(C) - P(F \cap C) \\ &= 0.50 + 0.60 - 0.30 \\ &= 0.80 \end{aligned}$$

EXAMPLE 10 The addition rule with mutually exclusive events

In rolling a pair of dice, what is the probability that the sum is 12 or at least one die shows a 2?

Solution

Let A be the event that the sum is 12 and B be the event that at least one die shows a 2. Since A occurs on only one of the 36 equally likely outcomes, (6, 6) (see Example 3), $P(A) = 1/36$. B occurs on 11 of the equally likely outcomes, so $P(B) = 11/36$. Since A and B are mutually exclusive, we have

$$P(A \cup B) = P(A) + P(B) = \frac{1}{36} + \frac{11}{36} = \frac{12}{36} = \frac{1}{3}.$$

FOR THOUGHT True or False? Explain.

1. If S is a sample space of equally likely outcomes and E is a subset of S, then $P(E) = n(E)$.
2. If an experiment consists of tossing four coins, then the sample space consists of eight equally likely outcomes.
3. If a single coin is tossed twice, then P(at least one tail) $= 0.75$.
4. If a pair of dice is rolled, then P(at least one 4) $= 11/36$.
5. If four coins are tossed, then P(at least one head) $= 4/16$.
6. If two coins are tossed, then the complement of getting exactly two heads is getting exactly two tails.
7. If the probability of getting exactly three tails in a toss of three coins is 1/8, then the probability of getting at least one head is 7/8.
8. If P(snow today) $= 0.7$, then the odds in favor of snow are 7 to 10.
9. If the odds in favor of an event E are 3 to 4, then $P(E) = 3/4$.
10. The ratio of 1/5 to 4/5 is equivalent to the ratio of 4 to 1.

8.6 EXERCISES

Tape 22 Disk

Solve each probability problem.

1. *Business Expansion* The board of directors for a major corporation cannot decide whether to build its new assembly plant in Dallas, Memphis, or Chicago. If one of these cities is chosen at random, then what is the probability that
   a. Dallas is chosen?
   b. Memphis is not chosen?
   c. Topeka is chosen?
2. *Scratch and Win* A batch of 100,000 scratch-and-win tickets contains 10,000 that are redeemable for a free order of French fries. If you randomly select one of these tickets, then what is the probability that
   a. you win an order of French fries?
   b. you do not win an order of French fries?
3. *Rolling a Die* If a single die is rolled, then what is the probability of getting
   a. a number larger than 2?
   b. a number less than or equal to 6?
   c. a number other than 4?
   d. a number larger than 8?
   e. a number smaller than 2?
4. *Tossing a Coin* If a single coin is tossed, then what is the probability of getting
   a. heads?
   b. fewer than two tails?
   c. exactly three tails?
5. *Tossing Two Coins Once* If a pair of coins is tossed, then what is the probability of getting
   a. exactly two tails?
   b. at least one head?
   c. exactly two heads?
   d. at most one head?
6. *Tossing One Coin Twice* If a single coin is tossed twice, then what is the probability of getting
   a. heads followed by tails?
   b. two tails in a row?
   c. heads on the second toss?
   d. exactly one head?
7. *Rolling a Pair of Dice* If a pair of dice is rolled, then what is the probability of getting
   a. a pair of 3's?
   b. at least one 3?
   c. a sum of 6?
   d. a sum greater than 2?
   e. a sum less than 3?
8. *Rolling a Die Twice* If a single die is rolled twice, then what is the probability of getting
   a. a 1 followed by a 6?
   b. a sum of 4?
   c. a 5 on the second roll?
   d. no more than two 4's?
   e. an even number followed by an odd number?

9. *Colored Marbles* A marble is selected at random from a jar containing three red marbles, four yellow marbles, and six green marbles. What is the probability that
a. the marble is red?
b. the marble is not yellow?
c. the marble is either red or green?
d. the marble is neither red nor green?

10. *Choosing a Chairperson* A committee consists of one Democrat, six Republicans, and seven Independents. If one person is randomly selected from the committee to be the chairperson, then what is the probability that
a. the person is a Democrat?
b. the person is either a Democrat or a Republican?
c. the person is not a Republican?

11. *Numbered Marbles* A jar contains nine marbles numbered 1 through 9. Two marbles are randomly selected one at a time without replacement. What is the probability that
a. 1 is selected first and 9 is selected second?
b. the sum of the numbers selected is 4?
c. the sum of the numbers selected is 5?

12. *Foul Play* A company consists of a president, a vice-president, and 10 salespeople. If 2 of the 12 people are randomly selected to win a Hawaiian vacation, then what is the probability that none of the salespeople is a winner?

13. *Poker Hands* If a five-card poker hand is drawn from a deck of 52, then what is the probability that
a. the hand contains the ace, king, queen, jack, and 10 of hearts?
b. the hand contains one 3, one 4, one 5, one 6, and one 7?

14. *Lineup* If four people with different names and different weights randomly line up to buy concert tickets, then what is the probability that
a. they line up in alphabetical order?
b. they line up in order of increasing weight?

Solve each problem.

15. *Drive Defensively* If the probability of surviving a head-on car accident at 55 mph is 0.001, then what is the probability of not surviving?

16. *Tax Time* If the probability of a tax return not being audited by the IRS is 0.91, then what is the probability of a tax return being audited?

17. *Rolling Fours* A pair of dice is rolled. What is the probability of
a. getting a pair of 4's?
b. not getting a pair of 4's?
c. getting at least one number that is not a 4?

18. *Tossing Triplets* Three coins are tossed. What is the probability of
a. getting three heads?
b. not getting three heads?
c. getting at least one head?

19. *Killer Asteroids* Some scientists estimate that the earth takes an average of three hits per one million years by killer asteroids. In this case, what is the probability that there will be no hits in the next one million years? What is the probability that there will be at least one hit in the next million years?

Figure for Exercise 19

20. *Safe at Last* Use the estimate of an average of three hits in one million years to find the probability that there will be no hits by a killer asteroid in the next 200 years. Find the probability that there will be at least one hit in the next 200 years.

Solve each problem.

21. *Hurricane Alley* If the probability is 80% that the eye of hurricane Zelda comes ashore within 30 mi of Biloxi, then what are the odds in favor of the eye coming ashore within 30 mi of Biloxi?

22. *On Target* If the probability that an arrow hits its target is 7/9, then what are the odds
a. in favor of the arrow hitting its target?
b. against the arrow hitting its target?

23. *Stock Market Rally* If the probability that the stock market goes up tomorrow is 1/4, then what are the odds
a. in favor of the stock market going up tomorrow?
b. against the stock market going up tomorrow?

24. *Read My Lips* If the probability of new taxes this year is 4/5, then what are the odds
a. in favor of new taxes?
b. against new taxes?

25. *Weather Forecast* If the odds in favor of rain today are 4 to 1, then what is the probability of rain today?

26. *Checkmate* If the odds are 2 to 1 in favor of Big Blue (the IBM computer) beating the Russian grand master, then what is the probability that Big Blue wins?

27. *Morning Line* If the Las Vegas odds makers set the odds at 9 to 1 in favor of the Tigers winning their next game, then
a. what are the odds against the Tigers winning their next game?
b. what is the probability that the Tigers win their next game?

28. *Public Opinion* If a pollster says that the odds are 7 to 1 against the reelection of the mayor, then
a. what are the odds in favor of the reelection of the mayor?
b. what is the probability that the mayor will be reelected?

29. *Two Out of Four* What are the odds in favor of getting exactly two heads in four tosses of a coin?

30. *Rolling Once* What are the odds in favor of getting a 5 in a single roll of a die?

31. *Lucky Seven* What are the odds in favor of getting a sum of 7 when rolling a pair of dice?

32. *Four Out of Two* What are the odds in favor of getting at least one 4 when rolling a pair of dice?

33. *Only One Winner* If 2 million lottery tickets are sold and only one of them is the winning ticket, then what are the odds in favor of winning if you hold a single ticket?

34. *Pick Six* What are the odds in favor of winning a lottery in which you must choose six numbers from the numbers 1 through 49?

35. *Five in Five* If the odds in favor of getting five heads in five tosses of a coin are 1 to 31, then what is the probability of getting five heads in five tosses of a coin?

36. *Electing Jones* If the odds against Jones winning the election are 3 to 5, then what is the probability that Jones will win the election?

Use the addition rule to solve each problem.

37. *Insurance Categories* Among the drivers insured by American Insurance, 64% are women, 38% of the drivers are in a high-risk category, and 24% of the drivers are high-risk women. If a driver is randomly selected from that company, what is the probability that the driver is either high-risk or a woman?

38. *Six or Four* What is the probability of getting either a sum of 6 or at least one 4 in the roll of a pair of dice?

39. *Family of Five* A couple plan to have three children. Assuming that males and females are equally likely, what is the probability that they have either three boys or three girls?

40. *Ten or Four* What is the probability of getting either a sum of 10 or a sum of 4 in the roll of a pair of dice?

41. *Pick a Card* What is the probability of getting either a heart or a king when drawing a single card from a deck of 52 cards?

42. *Any Card* What is the probability of getting either a heart or a diamond when drawing a single card from a deck of 52 cards?

43. *Selecting Students* The accompanying pie chart shows the percentages of freshmen, sophomores, juniors, and seniors at Washington High School. Find the probability that a randomly selected student is
a. a freshman.
b. a junior or a senior.
c. not a sophomore.

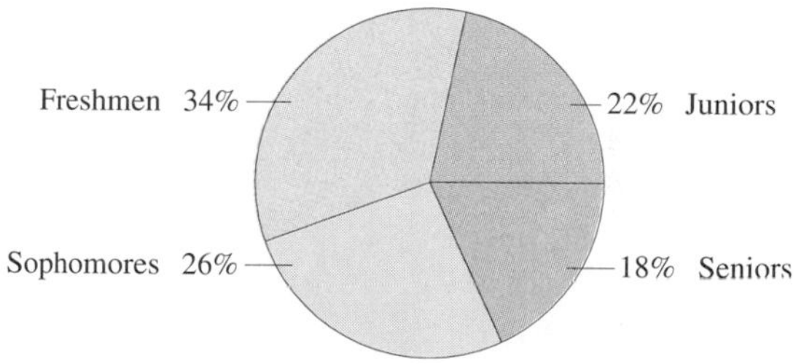

Figure for Exercise 43

44. *Earth-Crossing Asteroids* To reduce the risk of being hit by a killer asteroid, scientists want to locate and track all of the asteroids that are visible through telescopes. Each asteroid would be classified according to whether its diameter D is less than 1 km and whether its orbit crossed the orbit of earth (an earth-crossing orbit). The table with the figure shows a hypothetical classification of 900 asteroids visible to an amateur astronomer. If an amateur astronomer randomly spots an asteroid one night, then what is the probability that
a. it is an earth-crossing asteroid and its diameter is greater than or equal to 1 km?
b. it is either an earth-crossing asteroid or its diameter is greater than or equal to 1 km?

Asteroid classification	$D < 1$ km	$D \geq 1$ km
Earth-crossing	210	90
Not Earth-crossing	340	260

Figure for Exercise 44

For Writing/Discussion

45. Explain the difference between mutually exclusive events and complementary events.

46. Explain the difference between probability and odds.

47. *Cooperative Learning* Put three pennies into a can. Shake and toss the pennies 100 times. After each toss have your helper record whether 0, 1, 2, or 3 heads are showing. On what percent of the tosses did 0, 1, 2, and 3 heads occur? What are the theoretical probabilities of obtaining 0, 1, 2, and 3 heads in a toss of three coins? How well does the theoretical model fit the actual coin toss?

48. Use a calculator to generate 100 random numbers between 0 and 1 to simulate tossing three pennies 100 times. Record 0 heads if the random number x satisfies $0 < x < 0.125$, record 1 head for $0.125 < x < 0.5$, record 2 heads for $0.5 < x < 0.875$, and record 3 heads for $0.875 < x < 1$. On what percent of the simulated tosses did 0, 1, 2, and 3 heads occur? Compare your results with those of Exercise 47.

49. Explain how you could simulate tossing three coins 100 times using a random number generator, knowing only that heads and tails are equally likely on each toss. Use your method and compare the results with those of the previous exercise.

Linking Concepts

For Individual or Group Explorations

Dumping Pennies *Place 100 pennies into a can, shake well, then dump them onto the floor. Set aside all of the pennies that show heads and place the remaining pennies into the can and repeat this process until there are no pennies left to be placed back into the can. Let a_n be the number of pennies that are in the can on the nth dump.*

a) Enter the ordered pairs (n, a_n) into a graphing calculator and use regression to find the equations of the best line and the best exponential curve that fits the data.

b) Graph the line and the exponential curve along with the ordered pairs (n, a_n). Which model looks best with the data?

c) Based on probability, what fraction of the pennies dumped would you expect to place back into the can after each dump? Does a number close to this number appear in your exponential function?

d) What does the exponential curve $y = 200(1/2)^x$ have to do with all of this?

e) Based on probability, about how many dumps would you expect to make before you have no pennies left?

8.7 Mathematical Induction

A **statement** is a sentence or equation that is either true or false. Statements involving positive integers are common in the study of sequences and series and in other areas of mathematics. For example, the statement

$$1^3 + 2^3 + 3^3 + \cdots + n^3 = \frac{n^2(n + 1)^2}{4}$$

is true for every positive integer n. This statement claims that

$$1^3 = \frac{1^2(1 + 1)^2}{4}, \quad 1^3 + 2^3 = \frac{2^2(2 + 1)^2}{4}, \quad 1^3 + 2^3 + 3^3 = \frac{3^2(3 + 1)^2}{4},$$

and so on. You can easily verify that the statement is true in these first three cases, but that does not prove that the statement is true for *every* positive integer n. In this section we will learn how to prove that a statement is true for every positive integer without performing infinitely many verifications.

Statements Involving Positive Integers

Suppose that S_n is used to denote the statement that $2(n + 3)$ is equal to $2n + 6$ for any positive integer n. This statement is written in symbols as follows:

$$S_n\text{:} \quad 2(n + 3) = 2n + 6$$

Is S_n true for every positive integer n? We can check that $2(\mathbf{1} + 3) = 2(\mathbf{1}) + 6$, $2(\mathbf{2} + 3) = 2(\mathbf{2}) + 6$, and $2(\mathbf{3} + 3) = 2(\mathbf{3}) + 6$ are all correct, but no matter how many of these equations are checked, it will not prove that S_n is true for every positive integer n. However, we do know that $a(b + c) = ab + ac$ for any real numbers a, b, and c. Since positive integers are real numbers, S_n is true for every positive integer n because of the distributive property of the real numbers.

In Sections 8.2 and 8.3, statements involving positive integers occurred in the study of series. Consider the statement that the sum of the first n positive odd integers is n^2. In symbols,

$$S_n\text{:} \quad 1 + 3 + 5 + 7 + \cdots + (2n - 1) = n^2.$$

Note that S_n is a statement about the sum of an arithmetic series. However, we do know that the sum of n terms of an arithmetic series is the sum of the first and last terms multiplied by $n/2$:

$$S = \frac{n}{2}(a_1 + a_n) = \frac{n}{2}(1 + 2n - 1) = n^2$$

So S_n is true for every positive integer n.

The Principle of Mathematical Induction

So far we have proved two statements true for every positive integer n. One statement was proved by using a property of the real numbers, and the other followed from the formula for the sum of an arithmetic series. The techniques used so far will not always work. For this reason, mathematicians have developed another technique, which is called the **principle of mathematical induction** or simply **mathematical induction**.

Suppose you want to prove that you can make a very long (possibly infinite) journey on foot. You could argue that you can take the first step. Next you could argue that after every step you will use your forward momentum to take another step. Therefore, you can make the journey. Note that both parts of the argument are necessary. If every step taken leads to another step, but it is not possible to make the first step, then the journey can't be made. Likewise, if all you can make is the first step, then you will certainly not make the journey. A proof by mathematical induction is like this example of proving you can make a long journey.

The first step in mathematical induction is to prove that a statement S_n is true in the very first case. S_1 must be proved true. The second step is to prove that the truth of S_k implies the truth of S_{k+1} for any positive integer k. Note that S_k is not proved true. We assume that S_k is true and show that this assumption leads to the truth of S_{k+1}. Mathematicians agree that these two steps are necessary and sufficient to prove that S_n is true for all positive integers n, and so the principle of mathematical induction is accepted as an axiom in mathematics.

Principle of Mathematical Induction

Let S_n be a statement for every positive integer n. If

1. S_1 is true, and

2. the truth of S_k implies the truth of S_{k+1} for every positive integer k,

then S_n is true for every positive integer n.

To prove that a statement S_n is true for every positive integer n, the only statements that we work with are S_1, S_k, and S_{k+1}. So in the first example, we will practice simply writing those statements (correctly).

EXAMPLE 1 Writing S_1, S_k, and S_{k+1}

For the given statement S_n, write the three statements S_1, S_k, and S_{k+1}.

$$S_n\text{:}\quad 1^2 + 2^2 + 3^2 + \cdots + n^2 = \frac{n(n+1)(2n+1)}{6}$$

Solution

Write S_1 by replacing n by 1 in the statement S_n. For $n = 1$ the left-hand side contains only one term:

$$1^2 = \frac{1(1+1)(2 \cdot 1 + 1)}{6}$$

S_k can be written by replacing n by k in the statement S_n:

$$1^2 + 2^2 + 3^2 + \cdots + k^2 = \frac{k(k+1)(2k+1)}{6}$$

S_{k+1} can be written by replacing n by $k + 1$ in the statement S_n:

$$1^2 + 2^2 + 3^2 + \cdots + (k+1)^2 = \frac{(k+1)(k+1+1)(2(k+1)+1)}{6}$$

$$1^2 + 2^2 + 3^2 + \cdots + (k+1)^2 = \frac{(k+1)(k+2)(2k+3)}{6}$$

Earlier we proved that the sum of the first n odd integers was n^2 by using the formula for the sum of an arithmetic series. In the next example we prove the statement again, this time using mathematical induction. Note that we name the statements in this example as T_n to distinguish them from the statements S_n in the last example.

EXAMPLE 2 Proof by mathematical induction

Use mathematical induction to prove that the statement

$$T_n: \quad 1 + 3 + 5 + \cdots + (2n - 1) = n^2$$

is true for every positive integer n.

Solution

Step 1: Write T_1 by replacing n by 1 in the given equation.

$$T_1: \quad 1 = 1^2$$

Since $1 = 1^2$ is correct, T_1 is true.

Step 2: Assume that T_k is true. That is, we assume that the equation

$$1 + 3 + 5 + \cdots + (2k - 1) = k^2$$

is correct. Now we show that the truth of T_k implies that T_{k+1} is true. The next odd integer after $2k - 1$ is $2(k + 1) - 1$. Since T_k is true, we can add $2(k + 1) - 1$ to each side of the equation:

$$1 + 3 + 5 + \cdots + (2k - 1) + (2(k+1) - 1) = k^2 + 2(k+1) - 1$$

It is not necessary to write $2k - 1$ on the left-hand side because $2k - 1$ is the odd integer that precedes the odd integer $2(k + 1) - 1$:

$$1 + 3 + 5 + \cdots + 2(k+1) - 1 = k^2 + 2k + 1 \qquad \text{Simplify.}$$

$$1 + 3 + 5 + \cdots + 2(k+1) - 1 = (k+1)^2 \qquad \text{This statement is } T_{k+1}.$$

We have shown that the truth of T_k implies that T_{k+1} is true. So by the principle of mathematical induction, the statement T_n is true for every positive integer n.

Dominoes can be used to illustrate mathematical induction. Imagine an infinite sequence of dominoes, arranged so that when one falls over, it knocks over the one next to it, which knocks over the next one, and so on. If we can topple the first domino, then all dominoes in the infinite sequence will (theoretically) fall over.

In the next example, mathematical induction is used to prove a statement that was presented earlier in this section and would be difficult to prove without it.

EXAMPLE 3 Proof by mathematical induction

Use mathematical induction to prove that the statement

$$T_n: \quad 1^3 + 2^3 + 3^3 + \cdots + n^3 = \frac{n^2(n+1)^2}{4}$$

is true for every positive integer n.

Solution

Step 1: If $n = 1$, then the statement T_1 is

$$1^3 = \frac{1^2(1+1)^2}{4}.$$

This equation is correct by arithmetic, so T_1 is true.
Step 2: If we assume that T_k is true, then

$$1^3 + 2^3 + 3^3 + \cdots + k^3 = \frac{k^2(k+1)^2}{4}.$$

Add $(k+1)^3$ to each side of the equation:

$$1^3 + 2^3 + 3^3 + \cdots + k^3 + (k+1)^3 = \frac{k^2(k+1)^2}{4} + (k+1)^3$$

$$1^3 + 2^3 + 3^3 + \cdots + (k+1)^3 = \frac{k^2(k+1)^2}{4} + \frac{4(k+1)^3}{4} \qquad \text{Find the LCD.}$$

$$1^3 + 2^3 + 3^3 + \cdots + (k+1)^3 = \frac{(k+1)^2(k^2 + 4(k+1))}{4} \qquad \text{Factor out } (k+1)^2.$$

$$1^3 + 2^3 + 3^3 + \cdots + (k+1)^3 = \frac{(k+1)^2(k+2)^2}{4}$$

Since the last equation is T_{k+1}, we have shown that the truth of T_k implies the truth of T_{k+1} for every positive integer k. By the principle of mathematical induction, T_n is true for every positive integer n.

Mathematical induction can be applied to the situation of a new college graduate who seeks a lifelong career. The graduate believes he or she can get a first job. The graduate also believes that every job will provide some experience that will guarantee getting a next job. If these two beliefs are really correct, then the graduate will have a lifelong career.

In the next example, we prove a statement involving inequality.

EXAMPLE 4 Proof by mathematical induction

Use mathematical induction to prove that the statement

$$W_n: \quad 3^n < (n + 2)!$$

is true for every positive integer n.

Solution

Step 1: If $n = 1$, then the statement W_1 is

$$3^1 < (1 + 2)!$$

This inequality is correct and W_1 is true because $(1 + 2)! = 3! = 6$.

Step 2: If W_k is assumed to be true for a positive integer k, then

$$3^k < (k + 2)!$$

Multiply each side by 3 to get

$$3^{k+1} < 3 \cdot (k + 2)!$$

Since $3 < k + 3$ for $k \geq 1$, we have

$$3^{k+1} < (k + 3) \cdot (k + 2)!$$
$$3^{k+1} < (k + 3)!$$
$$3^{k+1} < ((k + 1) + 2)!$$

Since the last inequality is W_{k+1}, we have shown that the truth of W_k implies that W_{k+1} is true. By the principle of mathematical induction, W_n is true for all positive integers n.

When writing a mathematical proof we try to convince the reader of the truth of a statement. Most proofs that are included in this text are fairly "mechanical," in that an obvious calculation or simplification gives the desired result. Mathematical induction provides a framework for a certain kind of proof, but within that framework we may still need some human ingenuity (as in the second step of Example 4). How much ingenuity is required in a proof is not necessarily related to the complexity of the statement. The story of Fermat's last theorem provides a classic example of how difficult it can be to prove a simple statement. The French mathematician Pierre de Fermat studied the problem of finding all positive integers that satisfy $a^n + b^n = c^n$. Of course, if $n = 2$ there are solutions such as $3^2 + 4^2 = 5^2$ and $5^2 + 12^2 = 13^2$. In 1637, Fermat stated that there are no solutions with $n \geq 3$. The proof of this simple statement (Fermat's last theorem) stumped mathematicians for over 350 years. Finally, in 1993 Dr. Andrew Wiles of Princeton University announced that he had proved it (*New York Times*, June 24, 1993). Dr. Wiles estimated that the written details of his proof, which he worked on for seven years, will take over 200 pages!

FOR THOUGHT True or False? Explain.

1. The equation $\sum_{i=1}^{n} (4i - 2) = 2n^2$ is true if $n = 1$.
2. The inequality $n^3 < 4n + 15$ is true for $n = 1, 2,$ and 3.
3. For each positive integer n, $\dfrac{n-1}{n+1} < 0.9$.
4. Mathematical induction can be used to prove that $3(x + 1) = 3x + 3$ for every real number x.
5. If S_0 is true and the truth of S_{k-1} implies the truth of S_k for every positive integer k, then S_n is true for every nonnegative integer n.
6. If $n = k + 1$, then $\sum_{i=1}^{n} \dfrac{1}{i(i+1)} = \dfrac{n}{n+1}$ becomes $\sum_{i=1}^{k} \dfrac{1}{i(i+1)} = \dfrac{k+1}{k+2}$.
7. Mathematical induction can be used to prove that $\sum_{i=1}^{\infty} 2^{-i} = 1$.
8. The statement $n^2 - n > 0$ is true for $n = 1$.
9. The statement $n^2 - n > 0$ is true for $n > 1$.
10. The statement $n^2 - n > 0$ is true for every positive integer n.

8.7 EXERCISES

Tape 22 Disk

Determine whether each statement is true for $n = 1, 2,$ and 3.

1. $\sum_{i=1}^{n} (3i - 1) = \dfrac{3n^2 + n}{2}$
2. $\sum_{i=1}^{n} \left(\dfrac{1}{2}\right)^i = 1 - 2^{-n}$
3. $\sum_{i=1}^{n} \dfrac{1}{i(i+1)} = \dfrac{n}{n+1}$
4. $\sum_{i=1}^{n} 3^i = \dfrac{3(3^n - 1)}{2}$
5. $\sum_{i=1}^{n} i^2 = 4n - 3$
6. $\sum_{i=1}^{n} 4i = 8n - 4$
7. $n^2 < n^3$
8. $(0.5)^{n-1} > 0.5$

For each given statement S_n, write the statements S_1, S_k, and S_{k+1}.

9. S_n: $\sum_{i=1}^{n} 2i = n(n + 1)$
10. S_n: $\sum_{i=1}^{n} 5i = \dfrac{5n(n+1)}{2}$
11. S_n: $2 + 6 + 10 + \cdots + (4n - 2) = 2n^2$
12. S_n: $3 + 8 + 13 + \cdots + (5n - 2) = \dfrac{n(5n+1)}{2}$
13. S_n: $\sum_{i=1}^{n} 2^i = 2^{n+1} - 2$
14. S_n: $\sum_{i=1}^{n} 5^{i+1} = \dfrac{5^{n+2} - 25}{4}$
15. S_n: $(ab)^n = a^n b^n$
16. S_n: $(a + b)^n = a^n + b^n$
17. S_n: If $0 < a < 1$, then $0 < a^n < 1$.
18. S_n: If $a > 1$, then $a^n > 1$.

Use mathematical induction to prove that each statement is true for each positive integer n.

19. $1 + 2 + 3 + \cdots + n = \dfrac{n(n+1)}{2}$
20. $2 + 4 + 6 + \cdots + 2n = n(n + 1)$
21. $3 + 7 + 11 + \cdots + (4n - 1) = n(2n + 1)$
22. $2 + 7 + 12 + \cdots + (5n - 3) = \dfrac{n(5n-1)}{2}$
23. $\sum_{i=1}^{n} 2^i = 2^{n+1} - 2$
24. $\sum_{i=1}^{n} 5^{i+1} = \dfrac{5^{n+2} - 25}{4}$
25. $\sum_{i=1}^{n} (3i - 1) = \dfrac{3n^2 + n}{2}$
26. $\sum_{i=1}^{n} \left(\dfrac{1}{2}\right)^i = 1 - 2^{-n}$

27. $1^2 + 2^2 + 3^2 + \cdots + n^2 = \frac{n(n+1)(2n+1)}{6}$

28. $\frac{1}{1\cdot 2} + \frac{1}{2\cdot 3} + \frac{1}{3\cdot 4} + \cdots + \frac{1}{n(n+1)} = \frac{n}{n+1}$

29. $1\cdot 3 + 2\cdot 4 + 3\cdot 5 + \cdots + n(n+2) = \frac{n}{6}(n+1)(2n+7)$

30. $\frac{1}{1\cdot 4} + \frac{1}{4\cdot 7} + \frac{1}{7\cdot 10} + \cdots + \frac{1}{(3n-2)(3n+1)} = \frac{n}{3n+1}$

31. $\frac{1}{1\cdot 3} + \frac{1}{3\cdot 5} + \frac{1}{5\cdot 7} + \cdots + \frac{1}{(2n-1)(2n+1)} = \frac{n}{2n+1}$

32. $\frac{1}{1\cdot 2\cdot 3} + \frac{1}{2\cdot 3\cdot 4} + \frac{1}{3\cdot 4\cdot 5} + \cdots + \frac{1}{n(n+1)(n+2)} = \frac{n(n+3)}{4(n+1)(n+2)}$

33. If $0 < a < 1$, then $0 < a^n < 1$.

34. If $a > 1$, then $a^n > 1$.

35. $n < 2^n$

36. $2^{n-1} \leq n!$

37. The integer $5^n - 1$ is divisible by 4 for every positive integer n.

38. The integer $7^n - 1$ is divisible by 6 for every positive integer n.

39. If x is any real number with $x \neq 1$, then $1 + x + x^2 + x^3 + \cdots + x^n = \frac{x^{n+1} - 1}{x - 1}$.

40. If a and b are constants, then $(ab)^n = a^n b^n$.

41. If a and m are constants, then $(a^m)^n = a^{mn}$.

For Writing/Discussion

42. Write a paragraph describing mathematical induction in your own words.

43. Mathematical induction is used to prove that a statement is true for every positive integer. What steps do you think are necessary to prove that a statement is true for every integer?

HIGHLIGHTS

- **Section 8.1 Sequences**

1. A finite sequence is a function whose domain is the set of positive integers less than or equal to some fixed positive integer.
2. An infinite sequence is a function whose domain is the set of all positive integers.
3. The nth term of an arithmetic sequence is $a_n = a_1 + (n-1)d$, where a_1 is the first term and d is the common difference.

- **Section 8.2 Series**

1. A series is the indicated sum of a sequence.
2. An arithmetic series is the indicated sum of an arithmetic sequence.
3. The sum of the terms a_1 through a_n of an arithmetic series is given by $S_n = \frac{n}{2}(a_1 + a_n)$.

- **Section 8.3 Geometric Sequences and Series**

1. The nth term of a geometric sequence is $a_n = ar^{n-1}$, where $r \neq 1$ and $r \neq 0$, and r is the common ratio.
2. A geometric series is an indicated sum of a geometric sequence.
3. The sum of the first n terms of a geometric series is given by $S_n = \frac{a(1 - r^n)}{1 - r}$.

4. The sum of all terms of an infinite geometric series in which $|r| < 1$ is given by $S = \dfrac{a}{1 - r}$.

- **Section 8.4**
  Counting and Permutations

1. If event A has m outcomes and event B has n outcomes, then there are mn ways for A and B to occur.
2. The number of permutations of n things taken r at a time is denoted by $P(n, r)$ and given by the formula $P(n, r) = \dfrac{n!}{(n - r)!}$ for $0 \leq r \leq n$.
3. $P(n, r)$ is the number of ordered arrangements of r things taken from a group of n things.

- **Section 8.5**
  Combinations, Labeling, and the Binomial Theorem

1. The number of combinations of n things taken r at a time is denoted by $C(n, r)$ or $\binom{n}{r}$ and is given by the formula $C(n, r) = \dfrac{n!}{(n - r)!\, r!}$ for $0 \leq r \leq n$.
2. $C(n, r)$ is the number of unordered choices of r things taken from a group of n things.
3. The number of ways to label n objects with r_1 labels of type 1, r_2 labels of type 2, . . . , and r_k labels of type k, where $r_1 + r_2 + \cdots + r_k = n$, is $\dfrac{n!}{r_1!\, r_2! \cdot \cdots \cdot r_k!}$.
4. The binomial theorem says that $(a + b)^n = \sum_{r=0}^{n} \binom{n}{r} a^{n-r}b^r$ for every positive integer n.

- **Section 8.6**
  Probability

1. If S is a sample space of equally likely outcomes to an experiment and E is a subset of S, then $P(E) = n(E)/n(S)$.
2. If A and B are any events in a sample space, then $P(A \cup B) = P(A) + P(B) - P(A \cap B)$.
3. If A and B are mutually exclusive events, then $P(A \cup B) = P(A) + P(B)$.
4. Two events A and A' are called complementary events if $A \cap A' = \varnothing$ and $P(A) + P(A') = 1$.
5. The odds in favor of event A are defined as the ratio $P(A)$ to $P(A')$, and the odds against A are defined as the ratio $P(A')$ to $P(A)$.
6. If the odds in favor of E are a to b, then $P(E) = a/(a + b)$ and $P(E') = b/(a + b)$.

- **Section 8.7**
  Mathematical Induction

1. To prove that the statement S_n is true for every positive integer n, prove that S_1 is true and prove that for any positive integer k, assuming that S_k is true implies that S_{k+1} is true.

CHAPTER 8 REVIEW EXERCISES

List all terms of each finite sequence.

1. $a_n = 2^{n-1}$ for $1 \le n \le 5$

2. $a_n = 3n - 2$ for $1 \le n \le 4$

3. $a_n = \dfrac{(-1)^n}{n!}$ for $1 \le n \le 4$

4. $a_n = (n - 2)^2$ for $1 \le n \le 6$

List the first three terms of each infinite sequence.

5. $a_n = 3(0.5)^{n-1}$

6. $b_n = \dfrac{1}{n(n + 1)}$

7. $c_n = -3n + 6$

8. $d_n = \dfrac{(-1)^n}{n^2}$

Find the sum of each series.

9. $\sum_{i=1}^{4} (0.5)^i$

10. $\sum_{i=1}^{3} (5i - 1)$

11. $\sum_{i=1}^{50} (4i + 7)$

12. $\sum_{i=1}^{4} 6$

13. $\sum_{i=1}^{\infty} 0.3(0.1)^{n-1}$

14. $\sum_{i=1}^{\infty} 5(-0.8)^n$

15. $\sum_{i=1}^{20} 1000(1.05)^{i-1}$

16. $\sum_{i=1}^{10} 6\left(\frac{1}{3}\right)^i$

Write a formula for the *n*th term of each sequence.

17. $-\frac{1}{3}, \frac{1}{4}, -\frac{1}{5}, \ldots$

18. $20, 17, 14, \ldots$

19. $6, 1, \frac{1}{6}, \ldots$

20. $1, -4, 9, -16, \ldots$

Write each series in summation notation. Use the index i and let i begin at 1.

21. $\frac{1}{2} - \frac{1}{3} + \frac{1}{4} - \cdots$

22. $5 + \frac{5}{2} + \frac{5}{4} + \frac{5}{8} + \cdots$

23. $2 + 4 + 6 + \cdots + 28$

24. $1 + 4 + 9 + \cdots + n^2$

Solve each problem.

25. *Common Ratio* Find the common ratio for a geometric sequence that has a first term of 4 and a seventh term of 256.

26. *Common Difference* Find the common difference for an arithmetic sequence that has a first term of 5 and a seventh term of 29.

27. *Compounded Quarterly* If \$100 is deposited in an account paying 9% compounded quarterly, then how much will be in the account at the end of 10 years?

28. *Compounded Monthly* If \$40,000 is deposited in an account paying 6% compounded monthly, then how much will be in the account at the end of nine years?

29. *Annual Payments* If \$1000 is deposited at the beginning of each year for 10 years in an account paying 6% compounded annually, then what is the total value of the 10 deposits at the end of the 10th year?

30. *Monthly Payments* If \$50 is deposited at the beginning of each month for 20 years in an account paying 6% compounded monthly, then what is the total value of the 240 deposits at the end of the 20th year?

31. *Tummy Masters* TV Specialties sold 100,000 plastic Tummy Masters for \$99.95 each during the first month that the product was advertised. Each month thereafter, sales levels were about 90% of the sales in the previous month. If this pattern continues, then what is the approximate total number of Tummy Masters that could be sold?

32. *Bouncing Ball* A ball made from a new synthetic rubber will rebound 97% of the distance from which it is dropped. If the ball is dropped from a height of 6 ft, then approximately how far will it travel before coming to rest?

Write the complete binomial expansion for each of the following powers of a binomial.

33. $(a + 2b)^4$

34. $(x - 5)^3$

35. $(2a - b)^5$

36. $(w + 2)^6$

Write the first three terms of each binomial expansion.

37. $(a + b)^{10}$

38. $(x - 2y)^9$

39. $\left(2x + \frac{y}{2}\right)^8$

40. $(2a - 3b)^7$

Solve each problem.

41. What is the coefficient of a^4b^9 in the expansion of $(a + b)^{13}$?

42. What is the coefficient of x^8y^7 in the expansion of $(2x - y)^{15}$?

43. What is the coefficient of $w^2x^3y^6$ in the expansion of $(w + 2x + y)^{11}$?

44. What is the coefficient of x^5y^7 in the expansion of $(2w + x + y)^{12}$?

45. How many terms are there in the expansion of $(a + b)^{23}$?

46. Write out the first seven rows of Pascal's triangle.

Solve each counting problem.

47. *Multiple-Choice Test* How many different ways are there to mark the answers to a nine-question multiple-choice test in which each question has five possible answers?

48. *Scheduling Departures* Six airplanes are scheduled to depart at 2:00 on the same runway. In how many ways can they line up for departure?

49. *Three-Letter Words* John is trying to find all three-letter English words that can be formed without repetition using the letters in the word FLORIDA. He plans to write down all possible three-letter "words" and then check each one with a dictionary to see if it is actually a word in the English language. How many possible three-letter "words" are there?

50. *Selecting a Team* The sales manager for an insurance company must select a team of two agents to give a presentation. If the manager has five male agents and six female agents available, and the team must consist of one man and one woman, then how many different teams are possible?

51. *Placing Advertisements* A candidate for city council is going to place one newspaper advertisement, one radio advertisement, and one television advertisement. If there are three newspapers, five radio stations, and four television stations available, then in how many ways can the advertisements be placed?

52. *Signal Flags* A ship has nine different flags available. A signal consists of three flags displayed on a vertical pole. How many different signals are possible?

53. *Choosing a Vacation* A travel agent offers a vacation in which you can visit any five cities, chosen from Paris, Rome, London, Istanbul, Monte Carlo, Vienna, Madrid, and Berlin. How many different vacations are possible, not counting the order in which the cities are visited?

54. *Counting Subsets* How many four-element subsets are there for the set $\{a, b, c, d, e, f, g\}$?

55. *Choosing a Committee* A city council consists of five Democrats and three Republicans. In how many ways can four council members be selected by the mayor to go to a convention in San Francisco if the mayor

a. may choose any four?

b. must choose four Democrats?

c. must choose two Democrats and two Republicans?

56. *Full House* In a five-card poker hand, a full house is three cards of one kind and two cards of another. How many full houses are there consisting of three queens and two 10's?

57. *Arranging Letters* How many different arrangements are there for the letters in the word KANSAS? In TEXAS?

58. *Marking Pickup Trucks* Eight Mazda pickups of different colors are on sale through Saturday only. How many ways are there for the dealer to mark two of them \$7000, two of them \$8000, and four of them \$9000?

59. *Possible Families* A couple plan to have seven children. How many different families are possible, considering the sex of each child and the order of birth?

60. *Triple Feature* The Galaxy Theater has three horror films to show on Saturday night. In how many different ways can the program for a triple feature be arranged?

Solve each probability problem.

61. *Just Guessing* If Miriam randomly marks the answers to a ten-question true-false test, then what is the probability that she gets all ten correct? What is the probability that she gets all ten wrong?

62. *Large Family* If a couple plan to have six children, then what is the probability that they get six girls?

63. *Jelly Beans* There are six red, five green, and two yellow jelly beans in a jar. If a jelly bean is selected at random, then what is the probability that

a. the selected bean is green?

b. the selected bean is either yellow or red?

c. the selected bean is blue?

d. the selected bean is not purple?

64. *Rolling Dice* A pair of dice is rolled. What is the probability that

a. at least one die shows an even number?

b. the sum is an even number?

c. the sum is 4?

d. the sum is 4 and at least one die shows an even number?

e. the sum is 4 or at least one die shows an even number?

65. *My Three Sons* Suppose a couple plan to have three children. What are the odds in favor of getting three boys?

66. *Sum of Six* Suppose a pair of dice is rolled. What are the odds in favor of the sum being 6?

67. *Gone Fishing* If 90% of the fish in Lake Louise are perch and a single fish is caught at random, then what are the odds in favor of catching a perch?

68. *Future Plans* In a survey of high school students, 80% said they planned to attend college. If a student is selected at random from this group, then what are the odds against getting one who plans to attend college?

69. *Enrollment Data* At MSU, 60% of the students are enrolled in an English class, 70% are enrolled in a mathematics class, and 40% are enrolled in both. If a student is selected at random, then what is the probability that the student is enrolled in either mathematics or English?

70. *Rolling Again* If a pair of dice is rolled, then what is the probability that the sum is either 5 or 6?

Evaluate each expression.

71. $8!$

72. $1!$

73. $\frac{5!}{3!}$

74. $\frac{7!}{(7-3)!}$

75. $\frac{9!}{3!\,3!\,3!}$

76. $\frac{10!}{0!\,2!\,3!\,5!}$

77. $C(8, 6)$

78. $\binom{12}{3}$

79. $P(8, 4)$

80. $P(4, 4)$

81. $C(8, 1)$

82. $C(12, 0)$

Use mathematical induction to prove that each statement is true for each positive integer n.

83. $3 + 6 + 9 + \cdots + 3n = \frac{3}{2}(n^2 + n)$

84. $\sum_{i=1}^{n} 2(3)^{i-1} = 3^n - 1$

CHAPTER 8 TEST

List all terms of each finite sequence.

1. $a_n = 2.3 + (n - 1)(0.5)$, $1 \le n \le 4$

2. $c_1 = 20$ and $c_n = \frac{1}{2}c_{n-1}$ for $2 \le n \le 4$

Write a formula for the nth term of each infinite sequence. Do not use a recursion formula.

3. $0, -1, 4, -9, 16, \ldots$

4. $7, 10, 13, 16, \ldots$

5. $\frac{1}{3}, -\frac{1}{6}, \frac{1}{12}, -\frac{1}{24}, \ldots$

Find the sum of each series.

6. $\sum_{j=0}^{53} (3j - 5)$

7. $\sum_{i=1}^{23} 300(1.05)^i$

8. $\sum_{n=1}^{\infty} (0.98)^n$

Solve each problem.

9. Find a formula for the nth term of an arithmetic sequence whose first term is -3 and whose ninth term is 9.

10. How many different nine-letter ''words'' can be made from the nine letters in TENNESSEE?

11. On the first day of April, \$300 worth of lottery tickets were sold at the Quick Mart. If sales increased by \$10 each day, then what was the mean of the daily sales amounts for April?

12. A customer at an automatic bank teller must enter a four-digit secret number to use the machine. If any of the integers from 0 through 9 can be used for each of the four digits, then how many secret numbers are there? If the machine gives a customer three tries to enter the secret number, then what is the probability that an unauthorized person can guess the secret number and gain access to the account?

13. Desmond and Molly Jones can save \$700 per month toward their \$120,000 dream house by living with Desmond's parents. If \$700 is deposited at the beginning of each month for 300 months into an account paying 6% compounded monthly, then what is the value of this annuity at the end of the 300th month? If the price of a \$120,000 house increases 6% each year, then what will it cost at the end of the 25th year? After 25 years of saving, will they have enough to buy the house and move out of his parents' house?

14. Write all of the terms of the binomial expansion for $(a - 2x)^5$.

15. Write the first three terms of the binomial expansion for $(x + y^2)^{24}$.

16. Write the binomial expansion for $(m + y)^{30}$ using summation notation.

17. If a pair of fair dice is rolled, then what is the probability that the sum of the numbers showing is 7? What are the odds in favor of rolling a 7?

18. An employee randomly selects three of the 12 months of the year in which to take a vacation. In how many ways can this selection be made?

19. In a seventh grade class of 12 boys and 10 girls, the teacher randomly selects two boys and two girls to be crossing guards. How many outcomes are there to this process?

20. In a race of eight horses, a bettor randomly selects three horses for the categories of win, place, and show. What is the probability that the bettor gets the horses and the order of finish correct?

21. Use mathematical induction to prove that $\sum_{i=1}^{n} \left(\frac{1}{2}\right)^i < 1$ for every positive integer n.

ANSWERS TO SELECTED EXERCISES

CHAPTER P

Section P.1

1. T **3.** T **5.** T **7.** T **9.** All
11. $\{-\sqrt{2}, \sqrt{3}, \pi, 5.090090009\ldots\}$ **13.** $\{0, 1\}$ **15.** $x + 7$
17. $5x + 15$ **19.** $5(x + 1)$ **21.** $(-13 + 4) + x$ **23.** 8
25. $\sqrt{3}$ **27.** $y^2 - x^2$ **29.** 7.2 **31.** $\sqrt{5}$ **33.** 5.5 **35.** 22
37. 72/5 **39.** -38 **41.** 52 **43.** 490 **45.** 52 **47.** 1
49. -3.4 **51.** 143 **53.** -61 **55.** -14 **57.** 1 **59.** 0
61. $-2x$ **63.** $0.85x$ **65.** $-6xy$ **67.** $-24wx$ **69.** $-3x - 6$
71. $0.97x - 6$ **73.** $2zy - 2$ **75.** $7x/12$ **77.** $3xy/4$ **79.** $3 - 2x$
81. $3x - y$ **83.** $-\frac{5}{4}y + 3$ **85.** $-\frac{12}{5}$ **87.** 0
89. $-1/2, -5/12, -1/3, 0, 1/3, 5/12, 1/2$
91. Approximately 1945 through 1985, approximately 1964
93. $132 - 0.6a + 0.4r$ **95.** 3 and 4 **97.** Big gray elephant
99. Same percent

Section P.2

1. 64 **3.** -16 **5.** 1/8 **7.** 13 **9.** 50 **11.** -5 **13.** -5
15. -35 **17.** -72 **19.** 1/81 **21.** 25 **23.** 11/30 **25.** 24
27. 8 **29.** 16 **31.** 5360 **33.** $-6x^{11}y^{11}$ **35.** 225 **37.** $-4x^6$
39. $-\frac{8}{27}x^6$ **41.** $3x^4$ **43.** $\frac{25}{y^4}$ **45.** $\frac{1}{6y^3}$ **47.** $\frac{n^2}{2}$ **49.** $17a^6$
51. $-\frac{x^2}{3y^6}$ **53.** $\frac{16x^{20}}{81y^8}$ **55.** x^{b+5} **57.** $\frac{-125a^{6t}}{b^{9t}}$ **59.** $\frac{-3y^{6v}}{2x^{5w}}$
61. $a^{-3s+15}b^{6t-27}$ **63.** 43,000 **65.** -0.0000356 **67.** 5×10^6
69. -6.72×10^{-5} **71.** 0.000000007 **73.** -2×10^{10}
75. 2×10^{16} **77.** 2×10^{-3} **79.** 4×10^{-29} **81.** 1×10^{23}
83. -9.936×10^{-5} **85.** 4.78×10^{-41} **87.** 2.7×10^9 **89.** 29
91. 367 **93.** \$5138.34 **95.** 1.577×10^{24} tons **97.** 6379 km
99. 3.3×10^5 **101.** 8.45×10^{16} Btu

Section P.3

1. -3 **3.** 8 **5.** -4 **7.** 81 **9.** 1/16 **11.** $|x|$ **13.** a^3
15. a^2 **17.** xy^2 **19.** $|a|b^2$ **21.** $x^2y^{1/2}$ **23.** $6a^{3/2}$ **25.** $3a^{1/6}$
27. $a^{7/3}$ **29.** $a^{2m}b^{4n}$ **31.** $\frac{x^2y}{z^3}$ **33.** 30 **35.** -2 **37.** -2
39. $-1/5$ **41.** 8 **43.** $\sqrt[3]{10^2}$ **45.** $\frac{3}{\sqrt[5]{y^3}}$ **47.** $x^{-1/2}$ **49.** $x^{3/5}$
51. $4x$ **53.** $2y^3$ **55.** $\frac{\sqrt{xy}}{10}$ **57.** $\frac{-2a}{b^5}$ **59.** $2\sqrt{7}$ **61.** $\frac{\sqrt{5}}{5}$
63. $\frac{\sqrt{2x}}{4}$ **65.** $2\sqrt[3]{5}$ **67.** $-5x\sqrt[3]{2x}$ **69.** $\frac{\sqrt[3]{4}}{2}$
71. $2\sqrt{2} + 2\sqrt{5} - 2\sqrt{3}$ **73.** $-30\sqrt{2}$ **75.** $60a$ **77.** 75
79. $\frac{3\sqrt{a}}{a^2}$ **81.** $\frac{5\sqrt{x}}{x}$ **83.** $5x\sqrt{5x}$ **85.** $\sqrt[6]{72}$ **87.** $\sqrt[12]{81x^3}$
89. $\sqrt[6]{4x^5y^5}$ **91.** $\sqrt[6]{7}$ **93.** 46 **95.** 25.4 **97.** 47.5%
99. 14 in. **101.** $6\sqrt{10}$ ft² **103.** b **105.** no

Section P.4

1. 3, 1 **3.** Not a polynomial **5.** 0, 79 **7.** $8x^2 + 3x - 1$
9. $-5x^2 + x - 3$ **11.** $(-5a^2 + 4a)x^3 + 2a^2x - 3$
13. $2x - 1$ **15.** $3x^2 - 3x - 6$ **17.** $-18a^5 + 15a^4 - 6a^3$
19. $3b^3 - 14b^2 + 17b - 6$ **21.** $8x^3 - 1$ **23.** $xz - 4z + 3x - 12$
25. $a^3 - b^3$ **27.** $a^2 + 7a - 18$ **29.** $2y^2 + 15y - 27$
31. $4x^2 - 81$ **33.** $4x^2 + 20x + 25$ **35.** $6x^4 + 22x^2 + 20$
37. $5 + 4\sqrt{2}$ **39.** $14 - 11\sqrt{2}$ **41.** $9 - \sqrt{6}$ **43.** $8 + 2\sqrt{15}$
45. $4 + 4\sqrt{x} + x$ **47.** $9x^2 + 30x + 25$ **49.** $x^{2n} - 9$ **51.** -23
53. $55 - 6\sqrt{6}$ **55.** $9x^6 - 24x^3 + 16$ **57.** $5\sqrt{2} + 2\sqrt{10}$
59. $\frac{2\sqrt{6} - \sqrt{2}}{11}$ **61.** $-9x^3$ **63.** $-x + 2$ **65.** $x + 3$
67. $a^2 + a + 1$ **69.** $x + 5, 13$ **71.** $2x - 6, 13$
73. $x^2 + x + 1, 0$ **75.** $ab + b, 0$ **77.** $x + 1 + \frac{-1}{x - 1}$
79. $2x - 3 + \frac{1}{x}$ **81.** $x + 1 + \frac{1}{x}$ **83.** $x + \frac{1}{x - 2}$ **85.** 12
87. 72 **89.** 77 **91.** 59/27 **93.** $x^2 + 2x - 24$
95. $2a^{10} - 3a^5 - 27$ **97.** $-y - 9$ **99.** $6a - 6$
101. $w^2 + 8w + 16$ **103.** $16x^2 - 81$ **105.** $3y^5 - 9xy^2$
107. $2b - 1$ **109.** $x^6 - 64$ **111.** $9w^4 - 12w^2n + 4n^2$ **113.** 1
115. $2x^2 + 5x - 3$ **117.** $x^2 + 5x + 6$ **119.** $4x^2 - 20x + 24$
121. \$0.45 **123.** \$5500 **125.** $\$1.788 \times 10^9$, 4.8%
127. 100 ft² less

Section P.5

1. $6x^2(x - 2), -6x^2(-x + 2)$ **3.** $4a(1 - 2b), -4a(-1 + 2b)$
5. $ax(-x^2 + 5x - 5), -ax(x^2 - 5x + 5)$
7. $1(m - n), -1(-m + n)$ **9.** $(x^2 + 5)(x + 2)$
11. $(y^2 - 3)(y - 1)$ **13.** $(d - w)(ay + 1)$ **15.** $(y^2 - b)(x^2 - a)$
17. $(x + 2)(x + 8)$ **19.** $(x - 6)(x + 2)$ **21.** $(m - 2)(m - 10)$
23. $(t - 7)(t + 12)$ **25.** $(2x + 1)(x - 4)$ **27.** $(4x + 1)(2x - 3)$
29. $(3y + 5)(2y - 1)$ **31.** $(t - u)(t + u)$ **33.** $(t + 1)^2$
35. $(2w - 1)^2$ **37.** $(y^{2t} - 5)(y^{2t} + 5)$ **39.** $(3zx + 4)^2$
41. $(t - u)(t^2 + tu + u^2)$ **43.** $(a - 2)(a^2 + 2a + 4)$

45. $(3y + 2)(9y^2 - 6y + 4)$
47. $(3xy^2 - 2z^3)(9x^2y^4 + 6xy^2z^3 + 4z^6)$ **49.** $(y^3 + 5)^2$
51. $(2a^2b^4 + 1)(2a^2b^4 - 5)$ **53.** $(2a + 7)(2a - 3)$
55. $(b^2 - 2)(b^2 + 1)$ **57.** $-3x(x - 3)(x + 3)$
59. $2t(2t + 3w)(4t^2 - 6tw + 9w^2)$ **61.** $(a - 2)(a + 2)(a + 1)$
63. $(x - 2)^2(x^2 + 2x + 4)$ **65.** $-2x(6x + 1)(3x - 2)$
67. $(a - 2)(a^2 + 2a + 4)(a + 2)(a^2 - 2a + 4)(a - 1)$
69. $-(3x + 5)(2x - 3)$ **71.** $(a - 1)(a + 1)(a^2 + 1)$ **73.** Yes
75. Yes **77.** No **79.** $(x - 1)(x + 2)(x + 3)$
81. $(x - 3)(x^2 + 2x + 2)$ **83.** $(x + 2)(x + 3)(x - 1)(x + 1)$
85. $(x^m - 1)(x^m + 1)$ **87.** $(x^q - 2)(x^q - 3)$
89. $(x^n - 1)(x^{2n} + x^n + 1)$ **91.** $(x^{w+1} + 2)(x^w + 1)$
93. $x^2 + x + 1$ **95.** 15 in.^{3}, 20 in.^{3}, 12 in.^{3}, 1 in. **97.** b **99.** No

Section P.6

1. $\{x \mid x \neq -2\}$ **3.** $\{x \mid x \neq 4 \text{ and } x \neq -2\}$
5. $\{x \mid x \neq 3 \text{ and } x \neq -3\}$ **7.** All real numbers **9.** $\frac{3}{x + 2}$
11. $-\frac{2}{3}$ **13.** $\frac{ab^4}{b - a^2}$ **15.** $\frac{y^2z}{x^3}$ **17.** $\frac{a^2 + ab + b^2}{a + b}$ **19.** $\frac{3}{7ab}$
21. $\frac{42}{a^2}$ **23.** $\frac{a + 3}{3}$ **25.** $\frac{2x - 2y}{x + y}$ **27.** $x^2y^2 + xy^3$ **29.** $\frac{16a}{12a^2}$
31. $\frac{x^2 - 8x + 15}{x^2 - 9}$ **33.** $\frac{x^2 + x}{x^2 + 6x + 5}$ **35.** $12a^2b^3$ **37.** $6(a + b)$
39. $(x + 2)(x + 3)(x - 3)$ **41.** $\frac{9 + x}{6x}$ **43.** $\frac{x + 7}{(x - 1)(x + 1)}$
45. $\frac{3a + 1}{a}$ **47.** $\frac{t^2 - 2}{t + 1}$ **49.** $\frac{2x^2 + 3x - 1}{(x + 1)(x + 2)(x + 3)}$
51. $\frac{7}{2x - 6}$ **53.** $\frac{x^2}{x^3 - y^3}$ **55.** $\frac{x^2 + 2x - 1}{x(x^2 - 1)}$ **57.** $\frac{4b^2 - 3ab}{b + 2a}$
59. $\frac{a^2 - a}{3b^2 + b}$ **61.** $\frac{a + 2}{a - 2}$ **63.** $\frac{6t - 3}{2t^2 + t - 4}$ **65.** $\frac{1 + x}{1 - x}$
67. $\frac{a^3b^3 + 1}{ab^3}$ **69.** $-x^2y - xy^2$ **71.** $\frac{m^2n^2}{n^2 - 2mn + m^2}$ **73.** $\frac{3}{7}$
75. $\frac{3}{506}$ **77.** $\frac{1}{5}$ **79.** $\frac{1195}{1191}$ **81.** 3.1087 **83.** 3.00001
85. Decreasing, \$27.14, \$24.17, \$22.27
87. \$6,000,000, \$18,000,000, \$594,000,000, $p \geq 0$ and $p < 100$
89. $\frac{5}{12}$ **91.** 272.7 mph **95.** $\frac{8}{13}, \frac{3}{2}$

Chapter P Review Exercises

1. F **3.** F **5.** F **7.** F **9.** F **11.** F **13.** $17x - 12$
15. $\frac{3x}{10}$ **17.** $\frac{x - 2}{3}$ **19.** $-\frac{3}{4}$ **21.** -2 **23.** 4 **25.** 7
27. 28.5 **29.** 625 **31.** 44 **33.** 3/2 **35.** $-16/3$ **37.** 1/4
39. $5x^2$ **41.** 11 **43.** $2s\sqrt{7s}$ **45.** $-10\sqrt[3]{2}$ **47.** $\frac{\sqrt{10a}}{2a}$
49. $\frac{\sqrt[3]{50}}{5}$ **51.** $8n\sqrt{2n}$ **53.** $3 + \sqrt{3}$ **55.** $\frac{\sqrt{3}}{5}$
57. 320,000,000 **59.** -0.000185 **61.** 5.6×10^{-5}
63. -2.34×10^6 **65.** 1.25×10^{20} **67.** 4×10^6
69. $2x^2 + x - 7$ **71.** $-5x^4 + 3x^3 + 3x$
73. $3a^3 - 8a^2 + 9a - 10$ **75.** $b^2 - 6by + 9y^2$ **77.** $3t^2 - 7t - 6$
79. $-5y^3$ **81.** 7 **83.** $23 + 4\sqrt{15}$ **85.** $2x + 2\sqrt{2x} - 1$
87. $x^2 + 4x - 1$, 1 **89.** $3x + 2$, 4 **91.** $x - 2 + \frac{1}{x + 2}$
93. $2 + \frac{13}{x - 5}$ **95.** $6x(x - 1)(x + 1)$ **97.** $(3h + 4t)^2$
99. $(t + y)(t^2 - ty + y^2)$ **101.** $(x - 3)(x + 3)^2$
103. $(t - 1)(t^2 + t + 1)(t + 1)(t^2 - t + 1)$
105. $(6x + 5)(3x - 4)$ **107.** $ab(a + 6)(a - 3)$
109. $(x - 1)(x + 1)(2x + y)$ **111.** 2 **113.** $\frac{2x + 2}{(x - 2)(x + 4)}$
115. $-\frac{1}{2}$ **117.** $\frac{bc^{17}}{a^8}$ **119.** $\frac{3x + 7}{x^2 - 4}$ **121.** $\frac{5x - 21}{30x^2}$ **123.** $\frac{a - 1}{2}$
125. $\frac{-x^2 + x - 1}{2}$ **127.** $\frac{3a^2 + 8a + 7}{(a + 1)(a - 1)(a + 5)}$ **129.** $\frac{7}{2x - 8}$
131. $\frac{-3y^2 + 7}{4y^2 - 3}$ **133.** $\frac{b^3 - a^2}{ab^2}$ **135.** $\frac{q^3 + p^2}{pq^3}$ **137.** -11
139. -9 **141.** 4/11 **143.** 149/91 **145.** 496.125 ft
147. 5.9×10^{26} **149.** 11.12 **151.** 5/6

Chapter P Test

1. All **2.** $\{-1.22, -1, 0, 2, 10/3\}$
3. $\{-\pi, -\sqrt{3}, \sqrt{5}, 6.020020002\ldots\}$ **4.** $\{0, 2\}$ **5.** 13 **6.** 2
7. $-1/9$ **8.** -1 **9.** $6x^5y^7$ **10.** 0 **11.** $a^2 + 2ab + b^2$
12. $-32a^3b^{10}$ **13.** $3\sqrt{3} + 2\sqrt{2}$ **14.** $\sqrt{3} + 1$ **15.** $\frac{\sqrt[3]{2x^2}}{2x^2}$
16. $2xy^4\sqrt{3xy}$ **17.** $3x^3 + 3x^2 - 12x$ **18.** $-5x^2 + 9x - 13$
19. $x^3 + x^2 - 7x - 3$ **20.** $4h^2 + 2h + 1$ **21.** $x^2 - 6xy - 27y^2$
22. $x^2 + 2x + 2$ **23.** $9x^2 - 48x + 64$ **24.** $4t^8 - 1$
25. $\frac{x^2 - 2x + 4}{x}$ **26.** $\frac{2x^2 + 7x + 21}{(x + 4)(x - 3)(x - 1)}$ **27.** $\frac{2a^2 - 7}{4a^2 - 9}$
28. $\frac{6b^2 - 24a^3b^3}{3a + 4a^2b^2}$ **29.** $a(x - 9)(x - 2)$
30. $m(m - 1)(m + 1)(m^2 + 1)$ **31.** $(3x - 1)(x + 5)$
32. $(bx + w)(x - 3)$ **33.** \$379.31 **34.** 5.9×10^{12} mi
35. 176 ft

CHAPTER 1

Section 1.1

1. No **3.** Yes **5.** $\{5/3\}$ **7.** $\{-2\}$ **9.** $\{1/2\}$ **11.** $\{11\}$
13. $\{-24\}$ **15.** $\{-6\}$ **17.** $\{\sqrt{2} - 2\}$ **19.** $\left\{\frac{5\sqrt{3} + 3\sqrt{2}}{3}\right\}$
21. $\left\{\frac{-1}{\pi}\right\}$ **23.** $\{3\sqrt[3]{4}\}$ **25.** R, identity **27.** $\varnothing$, inconsistent
29. $\{x \mid x \neq 0\}$, identity **31.** $\{w \mid w \neq 1\}$, identity
33. $\{x \mid x \neq 0\}$, identity **35.** $\{9/8\}$, conditional
37. $\{6\}$, conditional **39.** $\varnothing$, inconsistent **41.** $\{-2\}$, conditional
43. $\{-19.952\}$ **45.** $\{2.957\}$ **47.** $\{-2.562\}$ **49.** $\{0.333\}$

51. {0.199} **53.** {0.425} **55.** {−0.380} **57.** {−9, 9} **59.** {−2, 5} **61.** {−10, 0} **63.** {2/3} **65.** ∅ **67.** {200} **69.** {4} **71.** {0} **73.** {−10} **75.** {5} **77.** {−8, −4} **79.** $\{x \mid x \neq 2 \text{ and } x \neq -2\}$, identity **81.** $\{x \mid x \neq -3 \text{ and } x \neq 2\}$, identity **83.** ∅, inconsistent **85.** \$212.5 billion, 1995 **87.** \$18,260.87 **89.** 250,000

Section 1.2

1. $r = \frac{I}{Pt}$ **3.** $C = \frac{5}{9}(F - 32)$ **5.** $r = \frac{C}{2\pi}$ **7.** $y = \frac{C - Ax}{B}$ **9.** $R_2 = \frac{RR_1R_3}{R_1R_3 - RR_3 - RR_1}$ **11.** $d = \frac{a_n - a_1}{n - 1}$ **13.** $D = \frac{5.688 - L + F\sqrt{S}}{2}$ **15.** 5.4% **17.** 2.5 hr **19.** 3.5 in. **21.** $p = -6.9A + 40.3$, 12.7% **23.** 2.0616 **25.** 0.3630 **27.** −6.2710 **29.** \$26,600 **31.** 33.9 in. **33.** \$60,000 **35.** 16 ft, 7 ft, 7 ft **37.** 6400 ft^2 **39.** 112.5 mph **41.** 48 mph **43.** \$106,000 **45.** North side 600, South side 900 **47.** 28.8 hr **49.** 2 P.M. **51.** $y = -\frac{3}{2}x - 3$ **53.** $y = \frac{3}{2}x - 9$ **55.** $y = \frac{3}{4}x + \frac{9}{4}$ **57.** $y = mx - mx_1 + y_1$ **59.** $y = \frac{3}{x + 1}$ **61.** 64.85 acres **63.** 225 ft **65.** 4.15 ft **67.** 1.998 hectares **69.** 34, 35, 36 **71.** \$76,147.10 **73.** 8/3 liters **75.** 28 yr **77.** 4 lb apples, 16 lb apricots **79.** 3 dimes, 5 nickels **81.** 1992, never

Section 1.3

1. Imaginary, $0 + 6i$ **3.** Imaginary, $\frac{1}{3} + \frac{1}{3}i$ **5.** Real, $\sqrt{7} + 0i$ **7.** Real, $\frac{\pi}{2} + 0i$ **9.** $7 + 2i$ **11.** $-2 - 3i$ **13.** $-12 - 18i$ **15.** 26 **17.** 29 **19.** 4 **21.** $-7 + 24i$ **23.** $1 - 4\sqrt{5}i$ **25.** i **27.** −1 **29.** 1 **31.** $-i$ **33.** 90 **35.** 17/4 **37.** 1 **39.** 12 **41.** $\frac{2}{5} + \frac{1}{5}i$ **43.** $\frac{3}{2} - \frac{3}{2}i$ **45.** $-1 + 3i$ **47.** $3 + 3i$ **49.** $\frac{1}{13} - \frac{5}{13}i$ **51.** $-i$ **53.** $-4 + 2i$ **55.** −6 **57.** −10 **59.** $-1 + i\sqrt{5}$ **61.** $-3 + i\sqrt{11}$ **63.** $-4 + 8i$ **65.** $-1 + 2i$ **67.** $\frac{-2 + i\sqrt{2}}{2}$ **69.** $-3 - 2i\sqrt{2}$ **71.** $\frac{3 + \sqrt{21}}{2}$ **73.** 0 **75.** 0 **77.** $9 + 6i$ **79.** 0 **81.** 21 **83.** Yes **85.** No **87.** Yes **89.** Yes **91.** $x^2 + 1$ **93.** $x^2 - 2x + 2$ **95.** $x^2 - 4x + 13$

Section 1.4

1. $\{\pm\sqrt{5}\}$ **3.** $\left\{\pm i\frac{\sqrt{6}}{3}\right\}$ **5.** {0, 6} **7.** {−2, 3} **9.** $\{-2 \pm 2i\}$ **11.** $\left\{\frac{2}{3} \pm \frac{2}{3}i\right\}$ **13.** {−4, 5} **15.** {−2, −1} **17.** $\left\{-\frac{1}{2}, 3\right\}$ **19.** $\left\{\frac{2}{3}, \frac{1}{2}\right\}$ **21.** {−7, 6} **23.** $\left\{-\frac{9}{2}, 2\right\}$ **25.** $x^2 - 12x + 36$ **27.** $r^2 + 3r + \frac{9}{4}$ **29.** $w^2 + \frac{1}{2}w + \frac{1}{16}$ **31.** $\{-3 \pm 2\sqrt{2}\}$ **33.** $\{1 \pm \sqrt{2}\}$ **35.** $\left\{\frac{-3 \pm \sqrt{13}}{2}\right\}$ **37.** $\{-1 \pm 2i\}$ **39.** $\left\{-4, \frac{3}{2}\right\}$ **41.** $\left\{-\frac{1}{3} \pm i\frac{\sqrt{2}}{3}\right\}$ **43.** {−4, 1} **45.** $\left\{-\frac{1}{2}, 3\right\}$ **47.** $\left\{-\frac{1}{3}\right\}$ **49.** $\left\{\pm\frac{\sqrt{6}}{2}\right\}$ **51.** $\{2 \pm i\}$ **53.** $\left\{\frac{-1 \pm \sqrt{2}}{3}\right\}$ **55.** {−3.24, 0.87} **57.** $\{0.71 \pm 2.38i\}$ **59.** 0, 1, real **61.** −4, 2, imaginary **63.** 172, 2, real **65.** $\left\{\frac{2 \pm i}{3}\right\}$ **67.** $\{\pm i\sqrt[4]{2}\}$ **69.** $\left\{-\frac{\sqrt{6}}{6}, \frac{\sqrt{6}}{12}\right\}$ **71.** {−12, 6} **73.** $\left\{\frac{1 \pm \sqrt{5}}{2}\right\}$ **75.** {−6, 8} **77.** $r = \pm\sqrt{\frac{A}{\pi}}$ **79.** $x = -k \pm \sqrt{k^2 - 3}$ **81.** $y = x\left(-1 \pm \frac{\sqrt{6}}{2}\right)$ **83.** 5000 or 35,000 **85.** 2.5 sec **87.** 340 ft **89.** $2\sqrt{205} \approx 28.6$ yd **91.** 18,503.4 pounds **93.** $\frac{-7 + \sqrt{101}}{2} \approx 1.52$ ft **95.** 10 ft/hr **97.** $\frac{9 + \sqrt{53}}{2} \approx 8.14$ days **99.** 40 lb or 20 lb **101.** 4.58 m/sec, 0.93 sec

Section 1.5

1. $x < 12$ **3.** $[-8, \infty)$ **5.** $(-\infty, \pi/2)$ **7.** $x \geq -7$

9. $(5, \infty)$

11. $[2, \infty)$

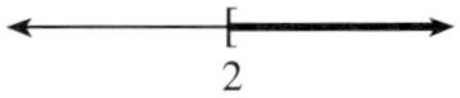

13. $(-\infty, 54)$

15. $(-\infty, 13/3]$

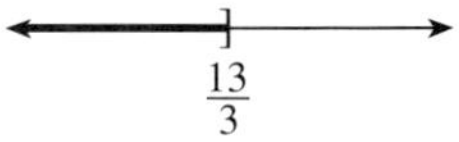

17. $(-\infty, 0]$

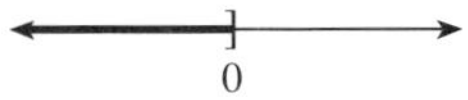

19. $[88, \infty)$

21. $(-3, \infty)$ **23.** ∅ **25.** $(-\infty, 5]$ **27.** $(-3, \infty)$ **29.** [3, 7] **31.** [5, 7) **33.** No **35.** Yes **37.** Yes

39. (3, 6)

41. $(1/2, \infty)$

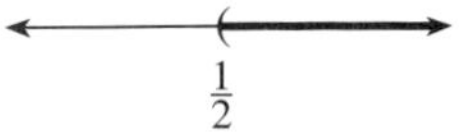

43. $(-\infty, -3) \cup (2, \infty)$ **45.** $(-3, \infty)$

47. $(-\infty, \infty)$ **49.** $\varnothing$

51. $(-1/3, 1)$ **53.** $[1, 3/2]$

55. $(-\infty, -1) \cup (2, \infty)$ **57.** $\varnothing$

59. $\{4\}$ **61.** $\varnothing$

63. $(-\infty, -1) \cup (5, \infty)$ **65.** $(-\infty, 1) \cup (5, \infty)$

67. $|x - 4| > 1$ **69.** $|x - 6| < 2$ **71.** $|x - 4| > 0$
73. $|x - 2| < 5$ **75.** $|x| \geq 9$ **77.** $|x - 7| \leq 4$
79. $|x - 5| > 2$ **81.** $[2, \infty)$ **83.** $(-\infty, 2)$
85. $(-\infty, -3] \cup [3, \infty)$ **87.** [\$0, \$7000] **89.** (93, 115)
91. (86, 102.5) **93.** (0 in., 15 in.] **95.** 96, 79, 68, 56, 47, yes
97. $|66{,}070 - x| > 25{,}000$, $x > 91{,}070$ or $x < 41{,}070$
99. $\dfrac{|x - 35|}{35} < 0.01$, (34.65°, 35.35°) **101.** [2.26 cm, 2.32 cm]
103. All except Saudi Arabia and Norway, all except the United Kingdom

Section 1.6

1. (1.5, ∞) **3.** $(-\infty, -1)$

5. $(5, \infty)$ **7.** $(-\infty, -3)$

9. $(-1, 3/2)$ **11.** $(-\infty, -3) \cup (5, \infty)$

13. $(-\infty, -2] \cup [6, \infty)$ **15.** $[-4, 4]$

17. $\{-3\}$ **19.** $(-\infty, 3/2) \cup (3/2, \infty)$

21. 1 **23.** −1 **25.** 12 **27.** −6.3

29. $(-2, 4]$ **31.** $(-7, 7)$

33. $(-\infty, 0)$ **35.** $[-2, 3] \cup (6, \infty)$

37. $(-\infty, -1/2) \cup (3, 5)$ **39.** $(-\infty, -8) \cup (-3, \infty)$

41. $(-2, 3)$ **43.** $(-\infty, -2) \cup (4, 5)$

45. $[-1, 3] \cup (5, \infty)$

47. $(-\infty, -5] \cup (-3, 3) \cup [5, \infty)$

49. $(-\infty, -4) \cup (6, \infty)$ **51.** $(-3/2, 3/2)$

53. $(-\infty, 5 - \sqrt{2}) \cup (5 + \sqrt{2}, \infty)$

55. $(-\infty, \infty)$ **57.** $(1, \infty)$

59. $(3, \infty)$ **61.** $(-\infty, -2/3) \cup (1, 5)$

63. $(-\infty, 1) \cup (3, 5) \cup (7, \infty)$ **65.** $[-3, 0] \cup [3, \infty)$

67. $\left(-\infty, \frac{3-\sqrt{409}}{10}\right) \cup \left(\frac{3+\sqrt{409}}{10}, \infty\right)$

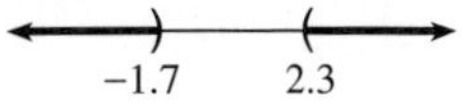

69. $[2, 3) \cup [6, \infty)$

71. $(-\infty, -2) \cup (-1/5, 4)$

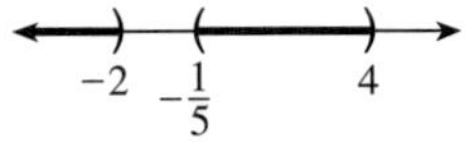

73. $[1, 2] \cup (3, 4)$

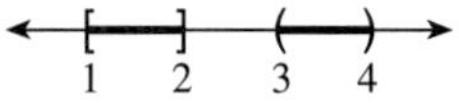

75. $\left(-\infty, \frac{5-\sqrt{17}}{2}\right] \cup (1, 2) \cup \left[\frac{5+\sqrt{17}}{2}, \infty\right)$

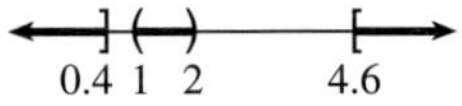

77. $[1/2, \infty)$ **79.** $[-3, 3]$ **81.** $(-\infty, -2) \cup (3, \infty)$
83. $(-\infty, -2) \cup [1, \infty)$ **85.** $(0, 10] \cup [40, 50)$
87. $\sqrt{6}/2 \approx 1.2$ sec **89.** $[6-\sqrt{26}, 6+\sqrt{26}]$ **91.** $r > 6.38\%$
93. 4.788 m/sec **95.** Above 17,982.5 ft **97.** Less than 75%

Chapter 1 Review Exercises

1. $\left\{\frac{2}{3}\right\}$ **3.** $\left\{\pm\frac{\sqrt{6}}{3}\right\}$ **5.** $\left\{\frac{32}{15}\right\}$ **7.** $\{-2\}$

9. $\left\{-\frac{1}{3}\right\}$ **11.** $\{0\}$ **13.** $\left\{\frac{1}{3}, 2\right\}$ **15.** $\{3 \pm i\}$

17. $\{2 \pm \sqrt{3}\}$ **19.** $\left\{\frac{2}{3}, 2\right\}$ **21.** $\left\{\frac{3}{2}\right\}$

23. $(3, \infty)$

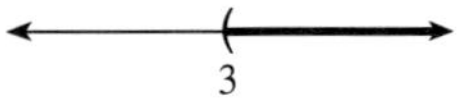

25. $(-\infty, 4)$

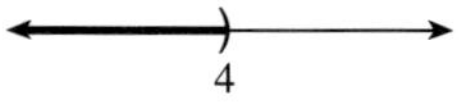

27. $(-\infty, -14/3)$

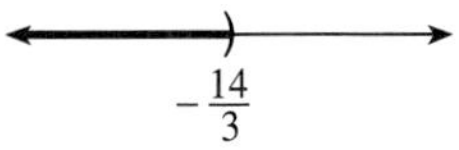

29. $(-1, 13]$

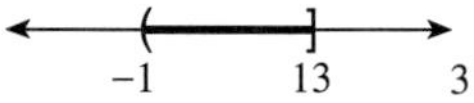

31.

(1/2, 1)

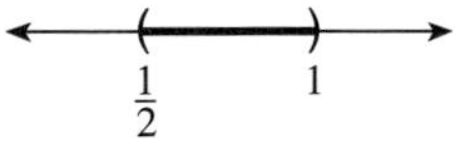

(−4, ∞)

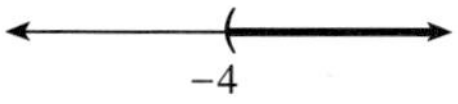

33.

35.

(−∞, 1) ∪ (5, ∞)

{7/2}

37.

39. $(-\infty, \infty)$ **41.** $(1/4, 1/2)$ **43.** $[-5, 3]$ **45.** $[3, 5)$
47. $(-\infty, -2) \cup (0, \infty)$ **49.** $(-\infty, 0) \cup (0, 3) \cup (4, \infty)$
51. $[1-\sqrt{5}, 1+\sqrt{5}]$

53. $(-\infty, 1] \cup [2, 3) \cup (4, \infty)$ **55.** $\varnothing$

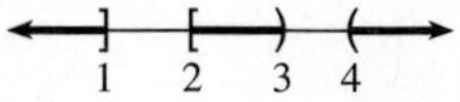

57. $(-\infty, 1) \cup (1, \infty)$

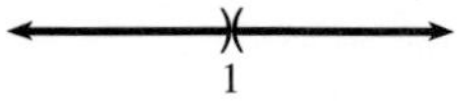

59. $y = \frac{2}{3}x - 2$

61. $y = \frac{1}{x-3}$ **63.** $y = -\frac{a}{b}x + \frac{c}{b}$ **65.** $y = \frac{2x}{x+2}$ **67.** $-1 - i$

69. $-9 - 40i$ **71.** 20 **73.** $-3 - 2i$ **75.** $\frac{1}{5} - \frac{3}{5}i$

77. $-\frac{1}{13} + \frac{5}{13}i$ **79.** $3 + i\sqrt{2}$ **81.** $\frac{3}{4} - \frac{\sqrt{5}}{4}i$ **83.** $-1 - i$

85. $[4/3, \infty)$ **87.** $(-\infty, -5] \cup [5, \infty)$ **89.** 8, two real

91. 0, one real **93.** $\frac{19-\sqrt{209}}{4} \approx 1.14$ in. **95.** 136.4 mi

97. 1600 **99.** 20 mi **101.** [$0, $12.50) **103.** (7 in., 11.5 in.)
105. 2.2×10^{10} gal, 31.8 mpg

Chapter 1 Test

1. $\{1\}$ **2.** $\left\{\pm\frac{\sqrt{6}}{3}\right\}$ **3.** $\{3 \pm 2\sqrt{2}\}$ **4.** $\{2, 7\}$

5. $\{0\}$ **6.** $\{1 \pm 2i\}$

7. $(-\infty, -2)$

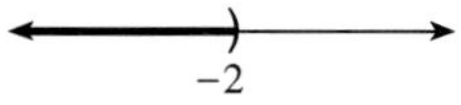

8. $(6, \infty)$

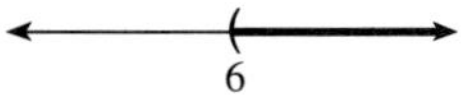

9. $[-1, 2]$

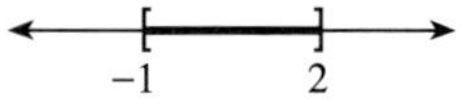

10. $(-\infty, 1) \cup (5, \infty)$

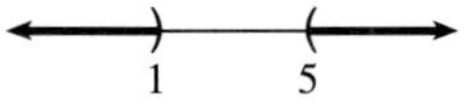

11. $7 - 24i$ **12.** $\frac{1}{2} - \frac{1}{2}i$ **13.** $-1 + i$ **14.** $-4 + 4i\sqrt{3}$

15. $(-2, 4)$ **16.** $(-\infty, 1/2) \cup (3, \infty)$ **17.** $(-\infty, -3] \cup (-1, 4)$

18. -11 **19.** 53 **20.** $y = \frac{1}{3x+2}$ **21.** $[5, \infty)$ **22.** 289 ft^2

23. 20 gal

Tying It All Together Chapters P–1

1. $7x$ **2.** $30x^2$ **3.** $\frac{3}{2x}$ **4.** $x^2 + 6x + 9$ **5.** $6x^2 + x - 2$

6. $2x + h$ **7.** $\frac{2x}{x^2-1}$ **8.** $x^2 + 3x + \frac{9}{4}$ **9.** R **10.** $\left\{0, \frac{11}{30}\right\}$

11. $(-\infty, 0) \cup (0, \infty)$ **12.** $\{0\}$ **13.** $\left\{-\frac{2}{3}, \frac{1}{2}\right\}$

14. $\left\{\frac{8 \pm \sqrt{89}}{5}\right\}$ **15.** $\{0, 1\}$ **16.** $\{1\}$ **17.** 0 **18.** -2

19. $\frac{9}{8}$ **20.** $\frac{44}{27}$ **21.** -8 **22.** -2 **23.** -4 **24.** -2.75

CHAPTER 2

Section 2.1

1. (4, 1), I **3.** (1, 0), x-axis **5.** (5, −1), IV
7. (−4, −2), III **9.** (−2, 4), II

11.

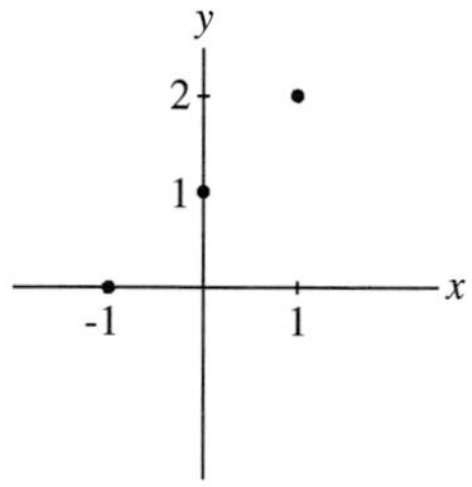

13.

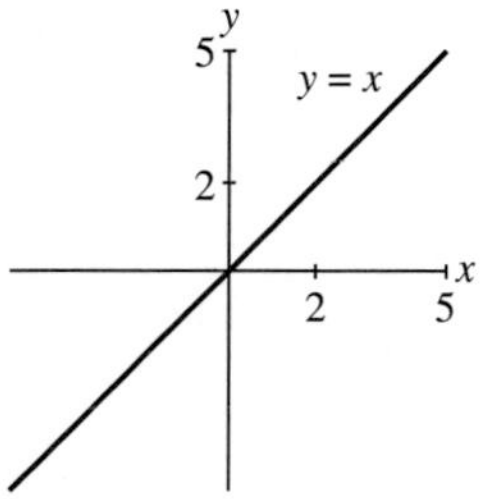

15.

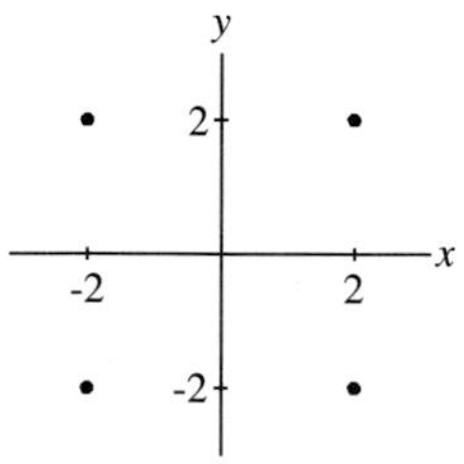

17.

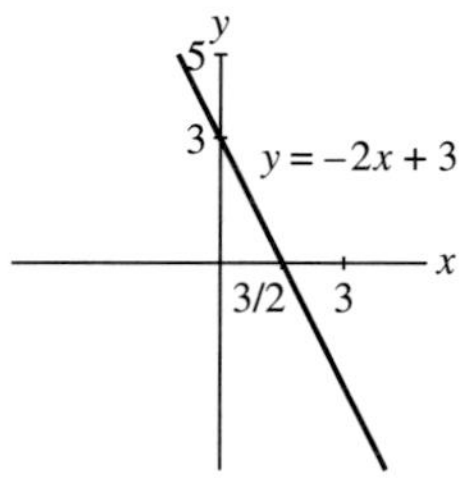

19.

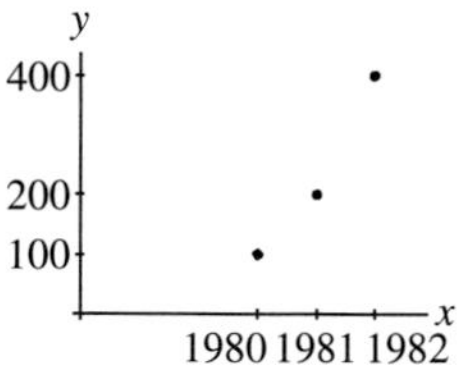

21.

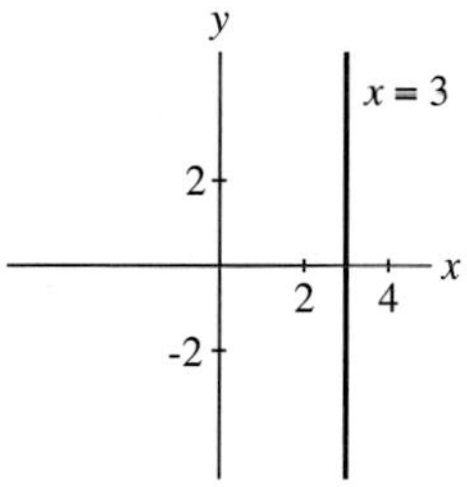

23.

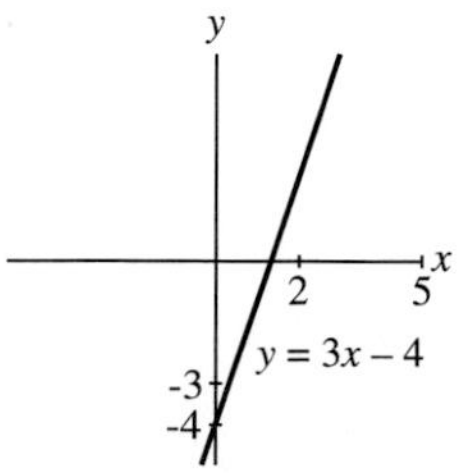

25.

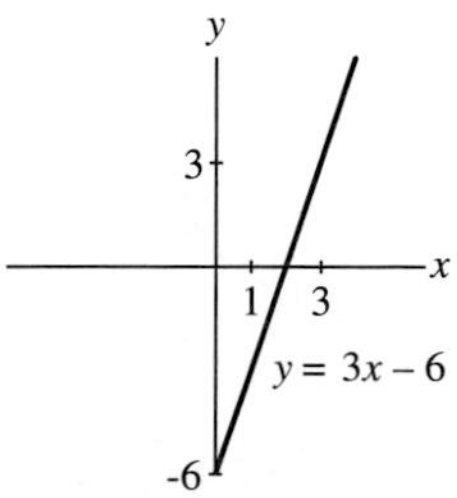

27.

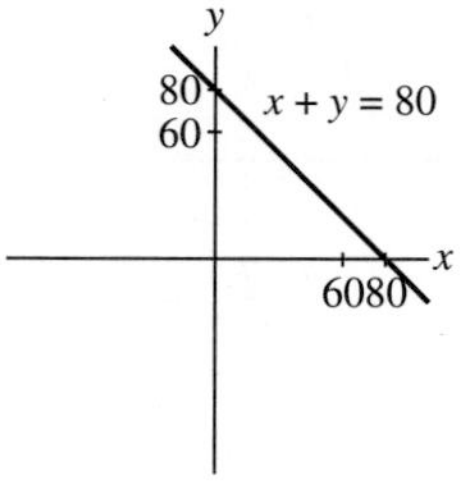

29.

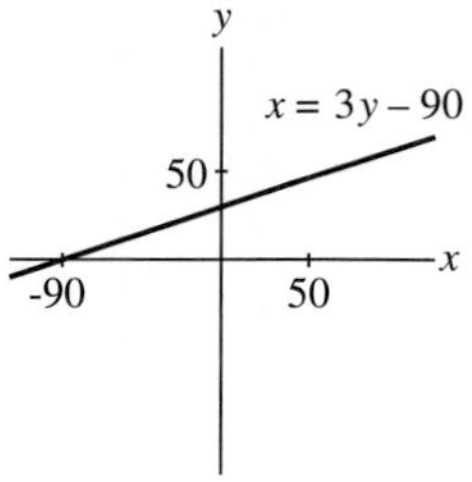

31.

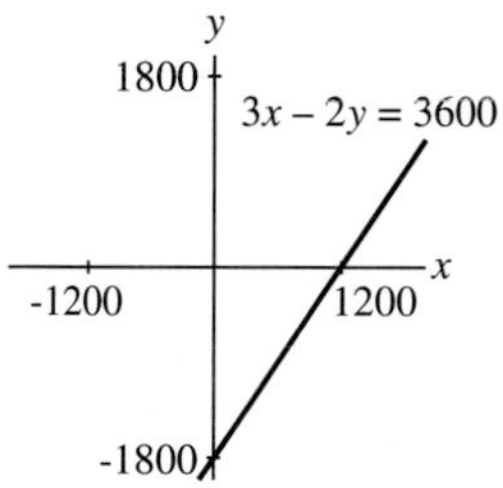

33.

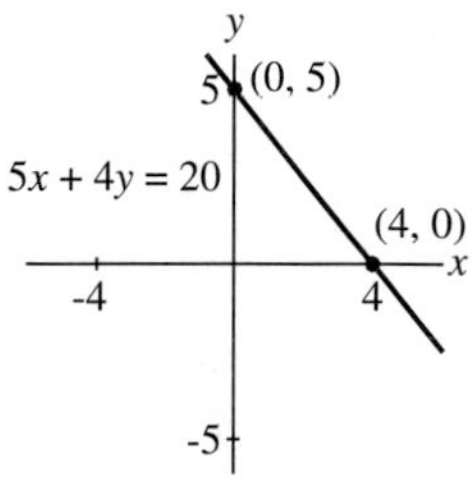

35.

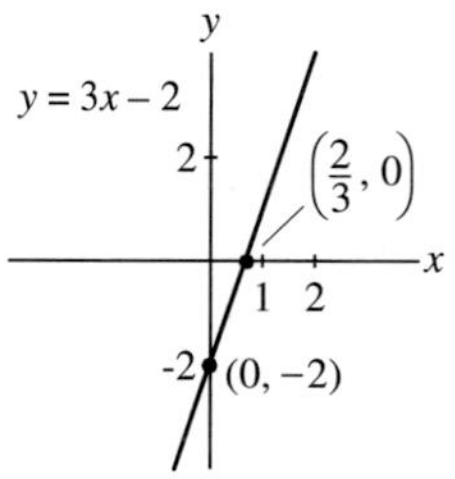

37.

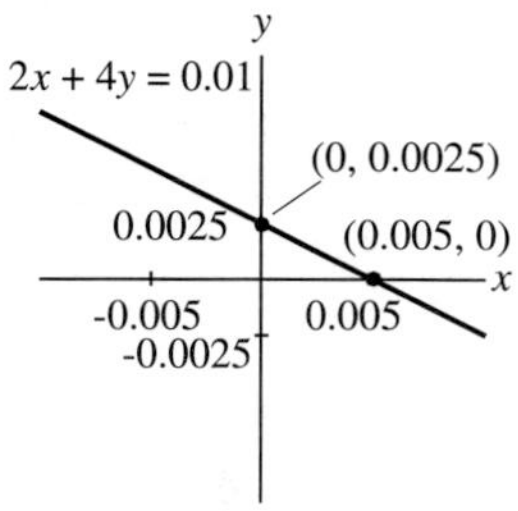

39.

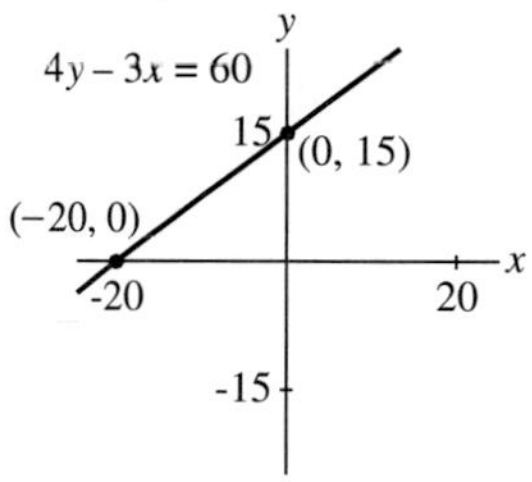

41.

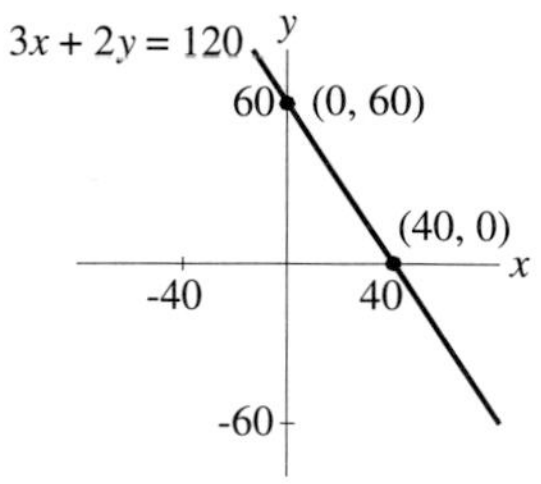

43.

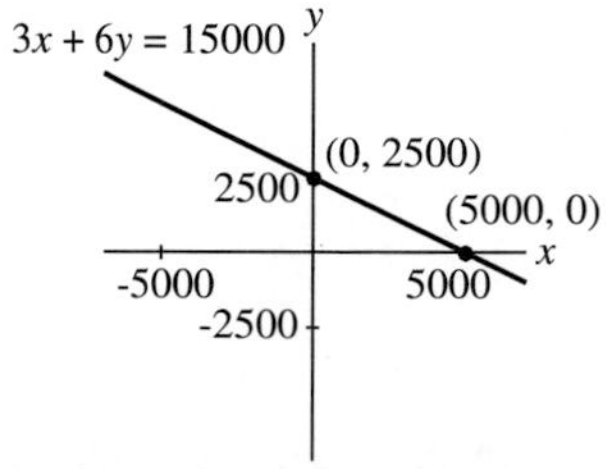

45. {−2.83} **47.** {558.54} **49.** {116,566.67} **51.** {4.91}

53.

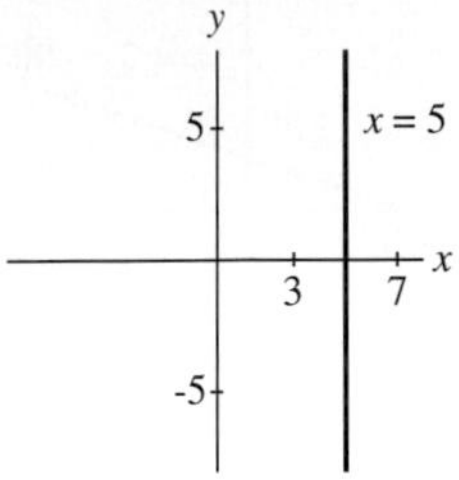

55.

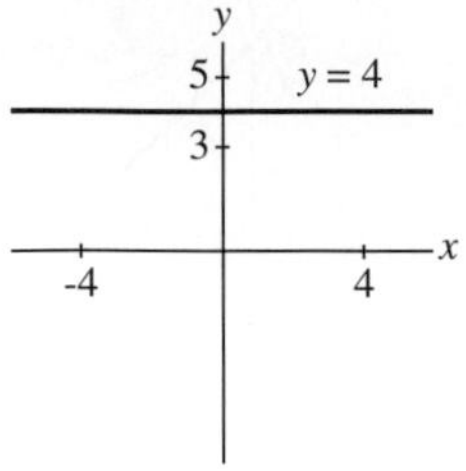

57.

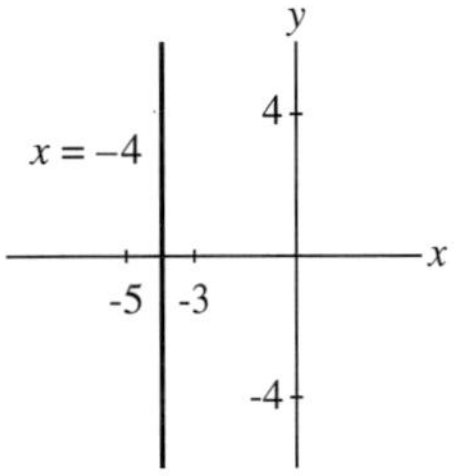

59.

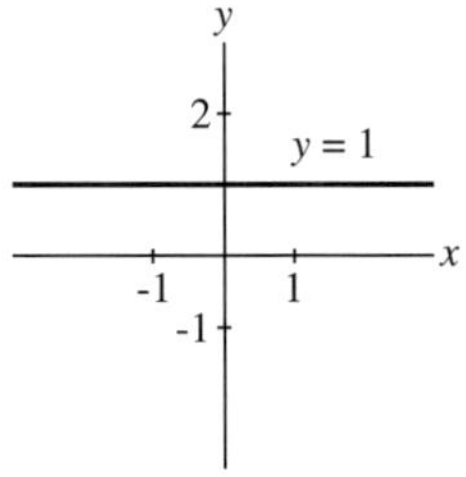

61. (a)

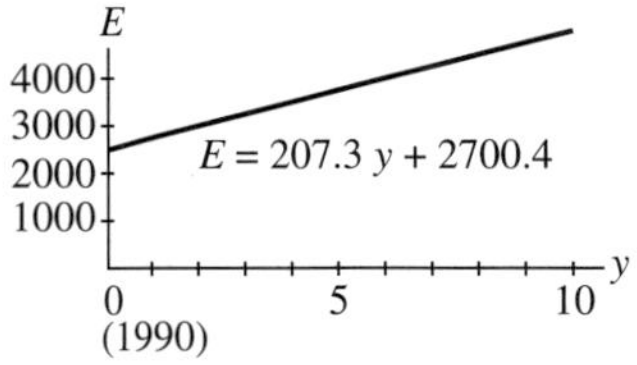

(b) \$3737 (d) No

63. 5, (2.5, 5) **65.** $2\sqrt{2}$, (0, −1)

67. 6, $\left(\dfrac{-2 + 3\sqrt{3}}{2}, \dfrac{5}{2}\right)$ **69.** $\sqrt{74}$, (−1.3, 1.3)

71. $|a - b|$, $\left(\dfrac{a + b}{2}, 0\right)$ **73.** $\dfrac{\sqrt{\pi^2 + 4}}{2}$, $\left(\dfrac{3\pi}{4}, \dfrac{1}{2}\right)$ **75.** No

77. Yes **79.** $-2 \pm 4\sqrt{2}$ **81.** (5, 7) **85.** (1983, 48.75%)

87. C = 1.8

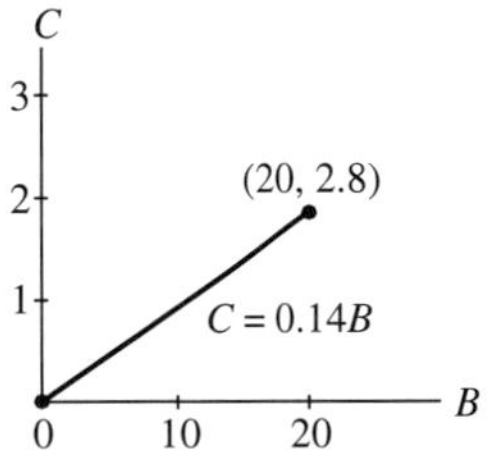

Section 2.2

1. Yes **3.** No **5.** No **7.** Yes **9.** No **11.** Yes **13.** Yes **15.** Yes **17.** Yes **19.** No **21.** Yes **23.** Yes **25.** Both

27. a is a function of b **29.** b is a function of a **31.** Neither

33. Both **35.** $\{-3, \pi, 5\}$, $\{1, \sqrt{2}, 6\}$

37. Domain $(-\infty, \infty)$, range $[5, \infty)$

39. Domain $(-\infty, 0]$, range $(-\infty, \infty)$

41. Domain $[0, \infty)$, range $[5, \infty)$

43. Domain $[1/2, \infty)$, range $[0, \infty)$

45. Domain $(-\infty, \infty)$, range $(-\infty, \infty)$

47. Domain (0, 3), range (2, 5)

49. 10 **51.** −6 **53.** 8 **55.** 0.3808 **57.** −2 **59.** −24

61. $3a^2 - a$ **63.** $3a^2 + 17a + 24$ **65.** $3x^2 + 6xh + 3h^2 - x - h$

67. $6xh + 3h^2 - h$ **69.** $3x^2 + 3x - 2$ **71.** $12x^3 - 10x^2 + 2x$

73. 0, 1/3 **75.** 1, −7 **77.** 114 **79.** 2 **81.** 3 **83.** $6x + 3h$

85. $-2x - h + 1$ **87.** $\dfrac{1}{\sqrt{x + h + 2} + \sqrt{x + 2}}$ **89.** $\dfrac{-1}{x(x + h)}$

91. $A = s^2$ **93.** $s = \dfrac{d\sqrt{2}}{2}$ **95.** $P = 4s$ **97.** $A = \dfrac{P^2}{16}$

99. $C = 353n$ **101.** $C = 50 + 35n$

103. \$434.9 billion, total for 1999, 2003 **105.** −\$2,400/yr

107. 0.02, −0.001, and 0 micrograms/m³/\$

109. −6.2 million hectares/yr

111. At \$18/ticket revenue is increasing at \$1950 per dollar change in ticket price. At \$22/ticket revenue is decreasing at \$2050 per dollar change in ticket price. **113.** $MC(x) = 0.06x + 40.03$, \$46.03

Section 2.3

1. (0, 0), 4

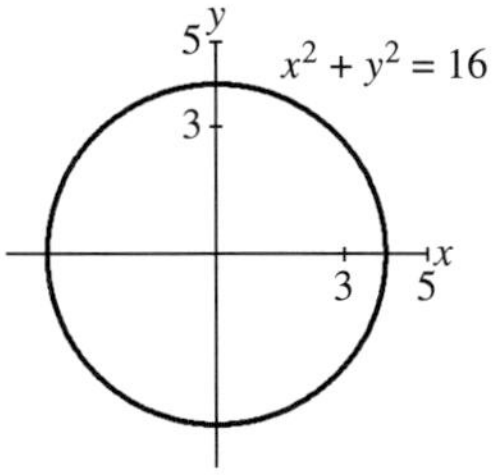

3. (−6, 0), 6

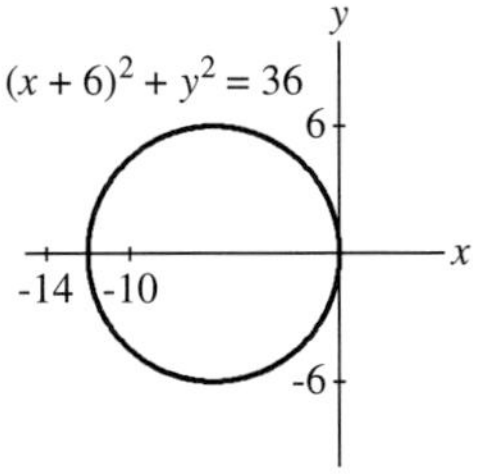

5. (2, −2), $2\sqrt{2}$

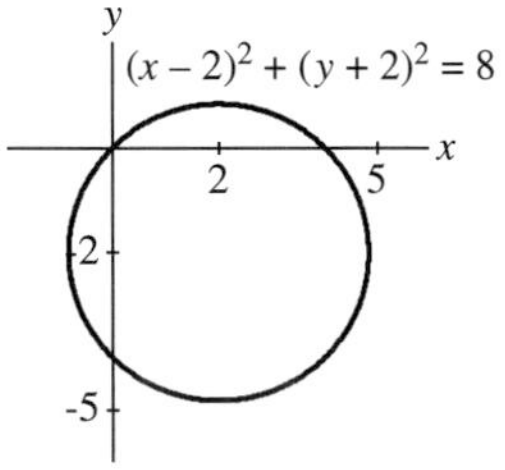

7. $x^2 + y^2 = 7$ **9.** $(x + 2)^2 + (y - 5)^2 = 1/4$

11. $(x - 3)^2 + (y - 5)^2 = 34$ **13.** $(x - 5)^2 + (y + 1)^2 = 32$

15. (0, −3), 3

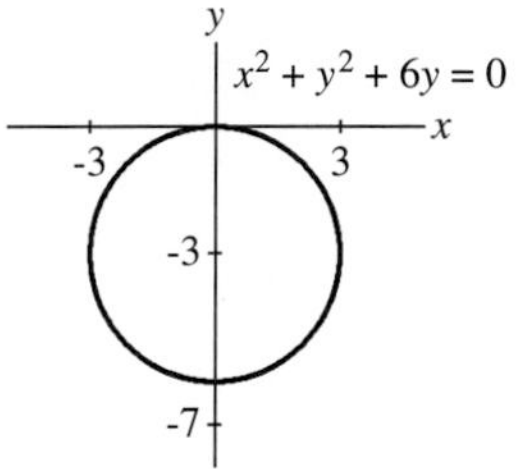

17. (3, 4), 5

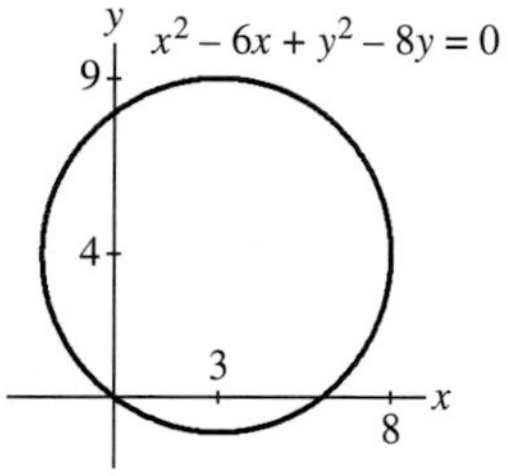

19. (2, 3/2), 5/2

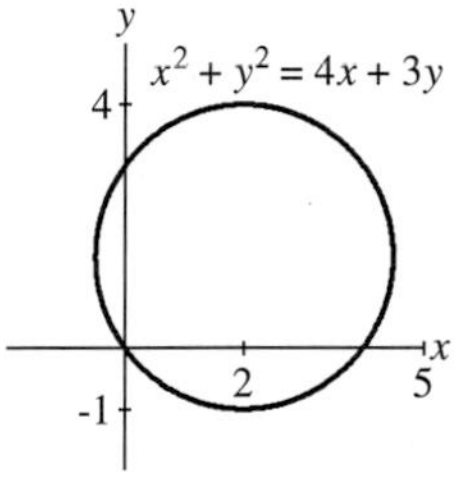

21. (1/4, −1/6), 1/6

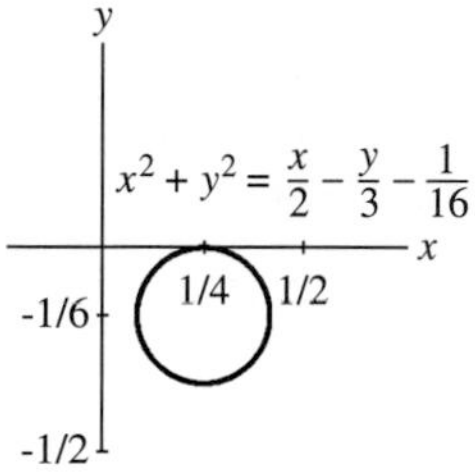

23. Domain $[0, \infty)$, range $[0, \infty)$, yes

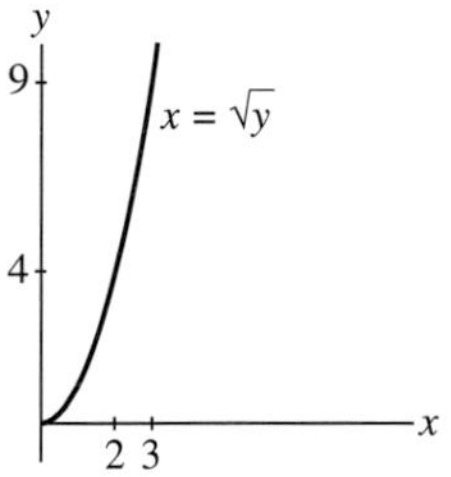

25. Domain $(-\infty, \infty)$, range $[-1, \infty)$, yes

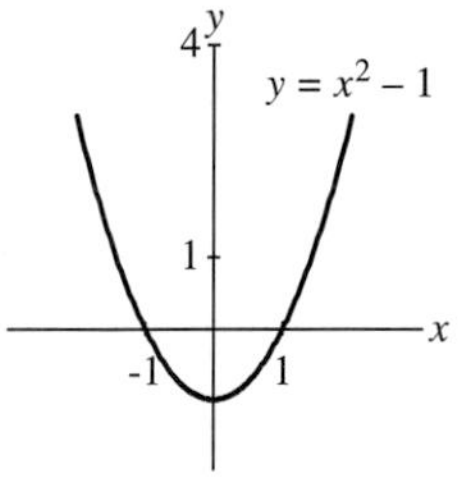

27. Domain $(-\infty, \infty)$, range $\{5\}$, yes

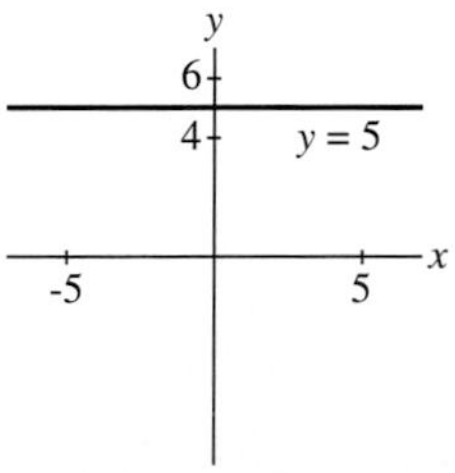

29. Domain $(-\infty, \infty)$, range $(-\infty, \infty)$, yes

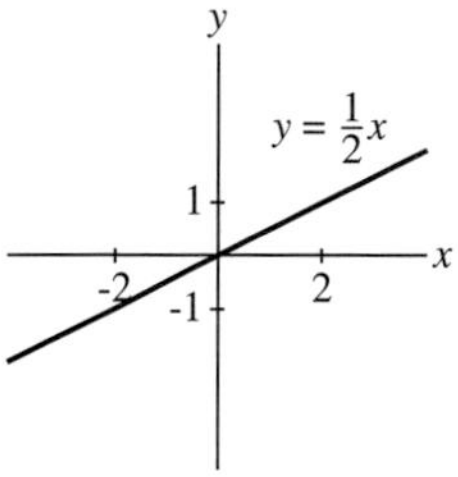

31. Domain $(-\infty, \infty)$, range $(-\infty, \infty)$, yes

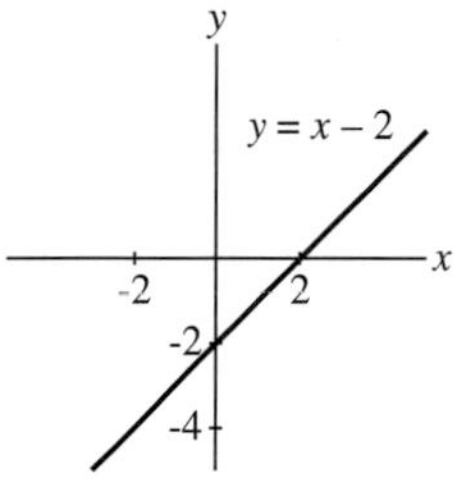

33. Domain $(-\infty, \infty)$, range $[0, \infty)$, yes

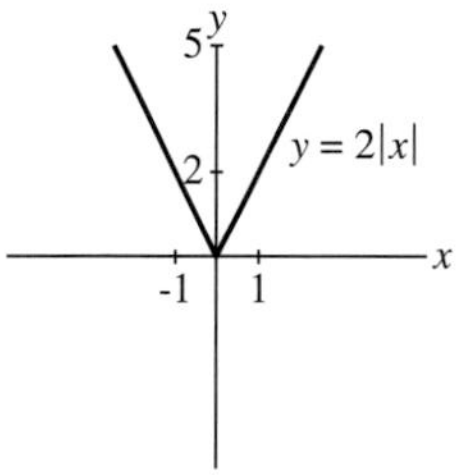

35. Domain $[1, \infty)$, range $(-\infty, \infty)$, no

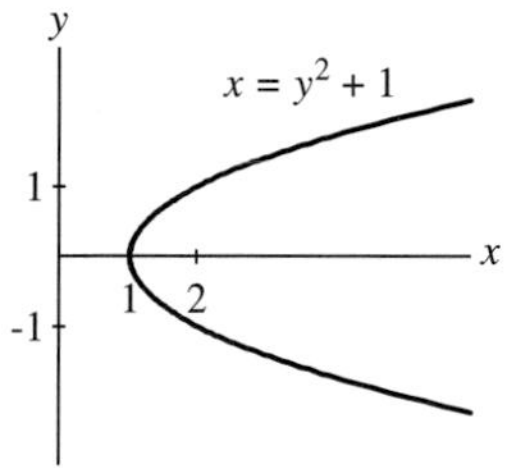

37. Domain $(-\infty, \infty)$, range $[0, \infty)$, yes

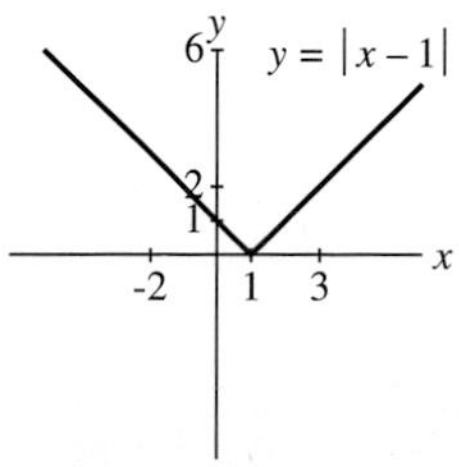

39. Domain $[0, \infty)$, range $(-\infty, \infty)$, no

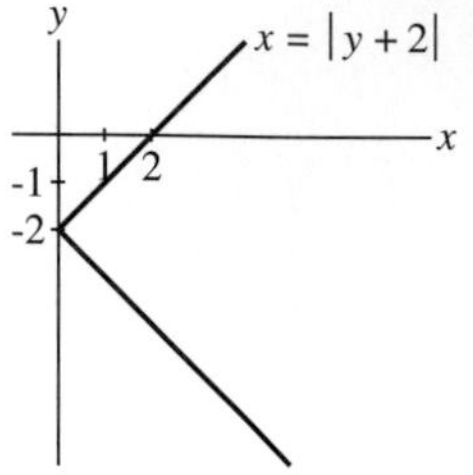

41. Domain $[-1, 1]$, range $[0, 1]$, yes

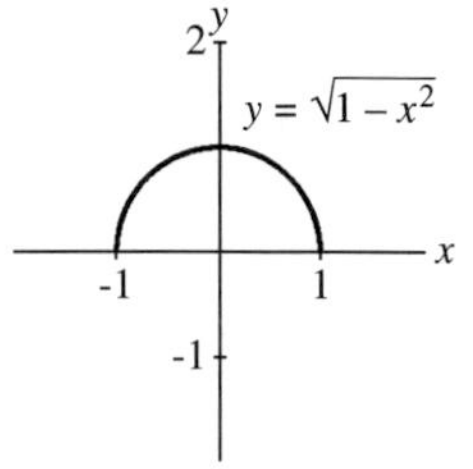

43. Domain $[-1, 1]$, range $[-1, 1]$, no

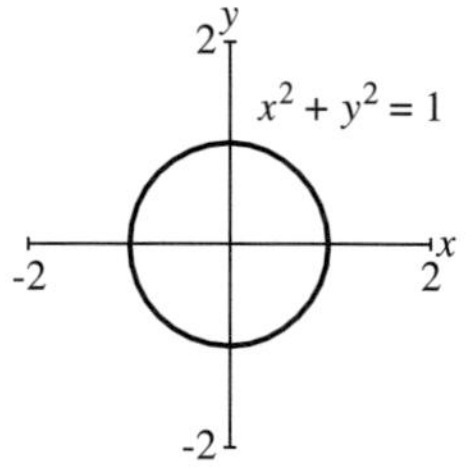

45. Domain $(-\infty, 0]$, range $(-\infty, \infty)$, no

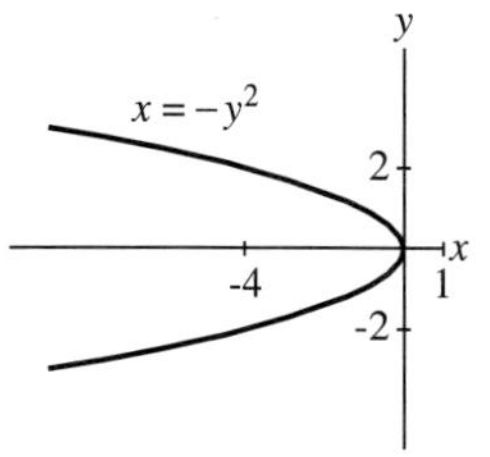

47. Domain $(-\infty, \infty)$, range $(-\infty, 1]$, yes

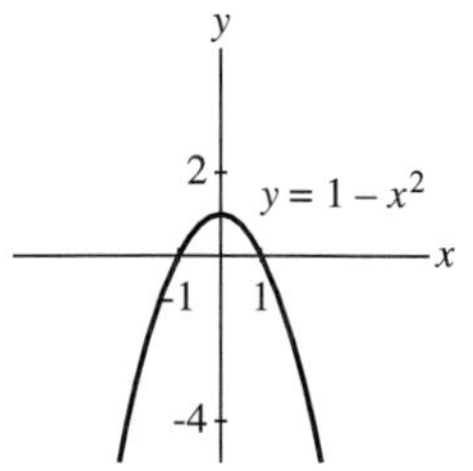

49. Domain $(-\infty, \infty)$, range $(-\infty, 0]$, yes

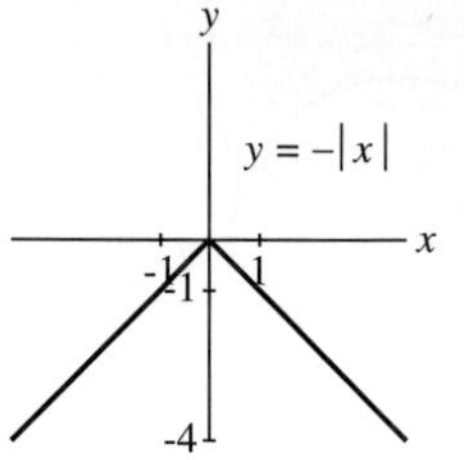

51. No **53.** Yes **55.** Yes

57. Domain $(-\infty, \infty)$, range $\{-2, 2\}$

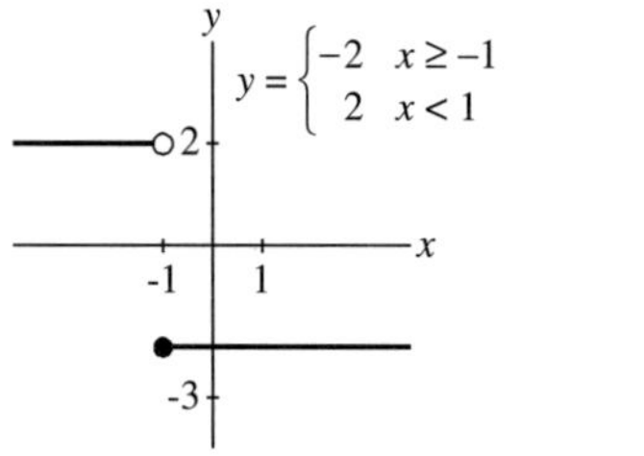

59. Domain $[-2, \infty)$, range $(-\infty, 2]$

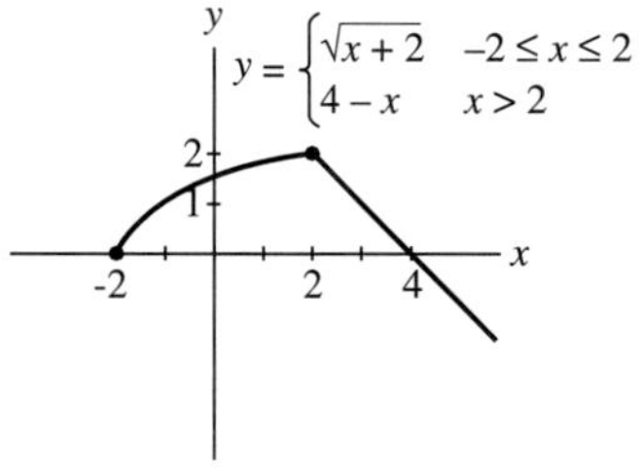

61. Domain $(-\infty, \infty)$, range $[0, \infty)$

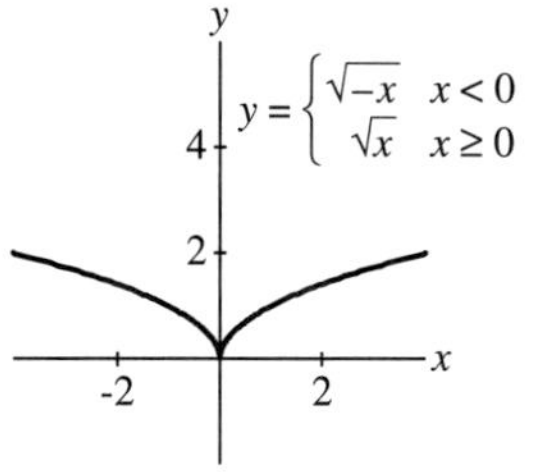

63. Domain $[-2, \infty)$, range $[0, \infty)$

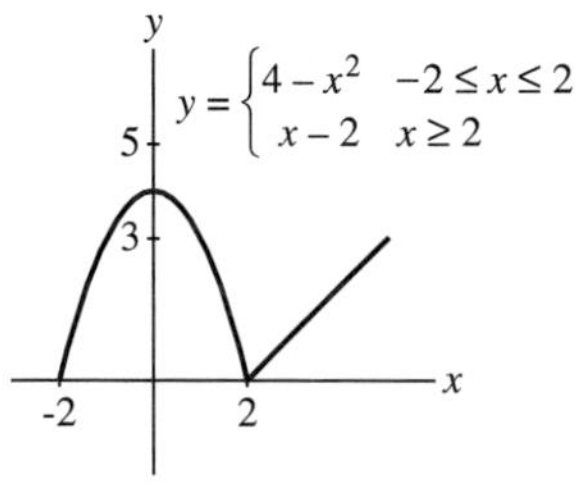

65. Domain $(-\infty, \infty)$, range integers

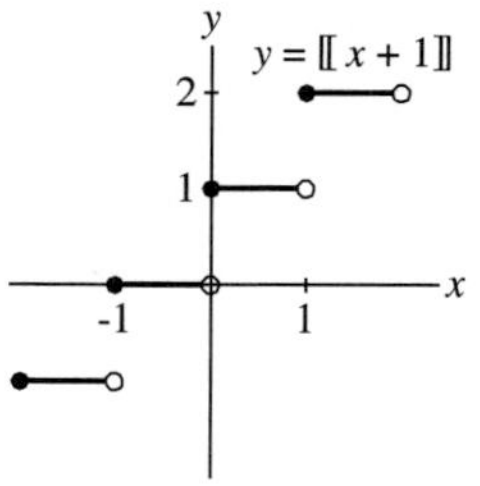

67. Domain $[0, 4)$, range $\{2, 3, 4, 5\}$

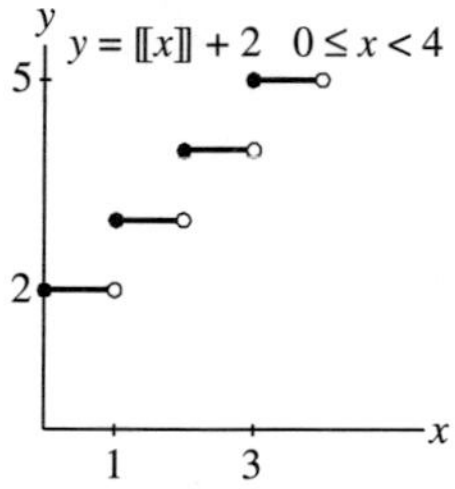

69. Domain $(-\infty, \infty)$, range $(-\infty, 4]$, inc on $(-\infty, 0)$, dec on $(0, \infty)$
71. Domain $(-\infty, 2]$, range $(-\infty, 2]$, inc on $(-\infty, -2)$, constant on $(-2, 2)$
73. Domain $[-4, 4]$, range $[0, 4]$, inc on $(-4, 0)$, dec on $(0, 4)$
75. c **77.** d
79. d inc on $(0, 3)$ and $(6, 15)$, dec on $(3, 6)$ and $(30, 39)$, constant on $(15, 30)$

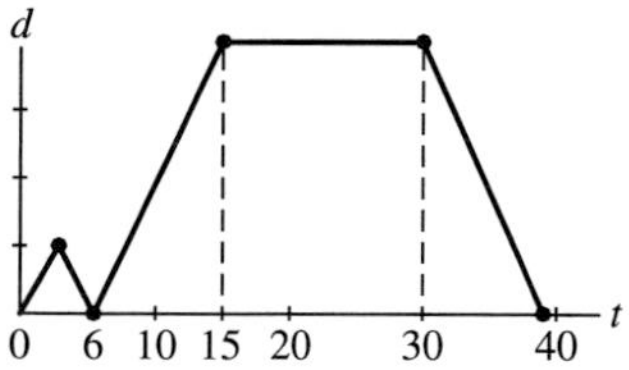

81. a inc on $(0, 2)$, dec on $(2.5, 3.5)$, constant on $(2, 2.5)$

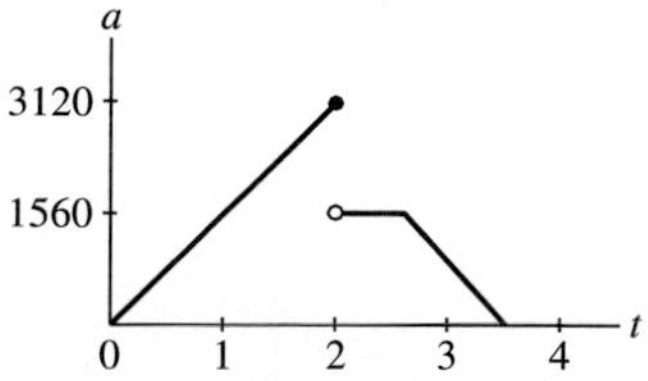

83. Domain $(-\infty, \infty)$, range $(-\infty, \infty)$, inc on $(-\infty, \infty)$

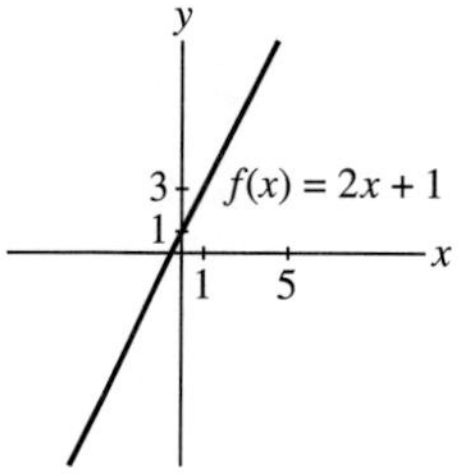

85. Domain $(-\infty, \infty)$, range $[0, \infty)$, dec on $(-\infty, 1)$, inc on $(1, \infty)$

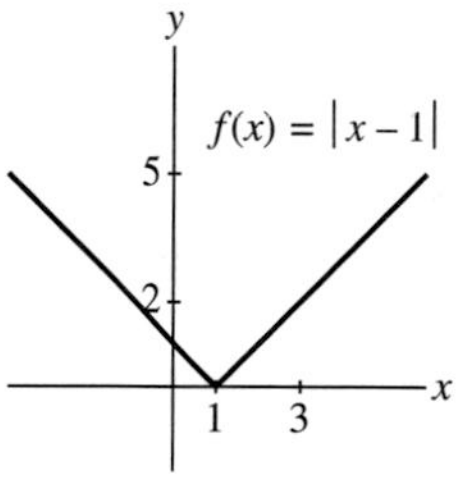

87. Domain $(-\infty, 0) \cup (0, \infty)$, range $\{-2, 2\}$, constant on $(-\infty, 0)$ and $(0, \infty)$

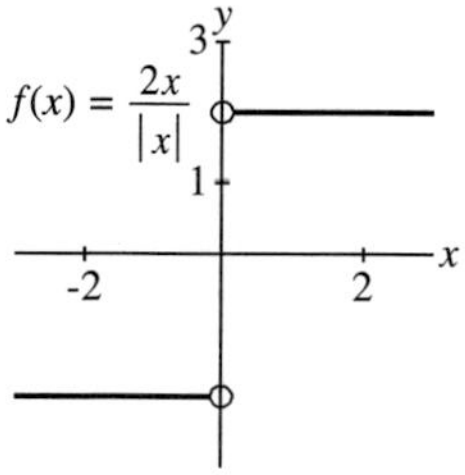

89. Domain $[-1, 1]$, range $[-1, 0]$, dec on $(-1, 0)$, inc on $(0, 1)$

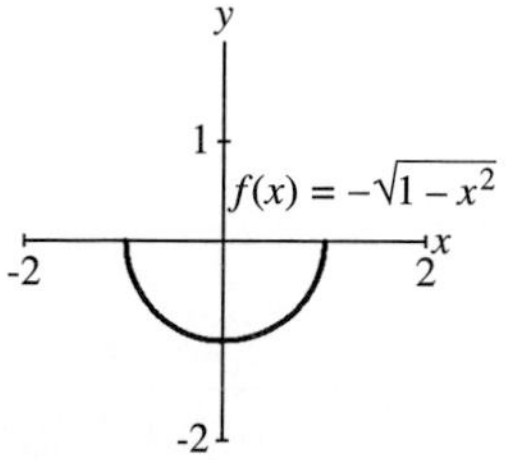

91. Domain $(-\infty, \infty)$, range $(-\infty, \infty)$, inc on $(-\infty, 3)$ and $(3, \infty)$

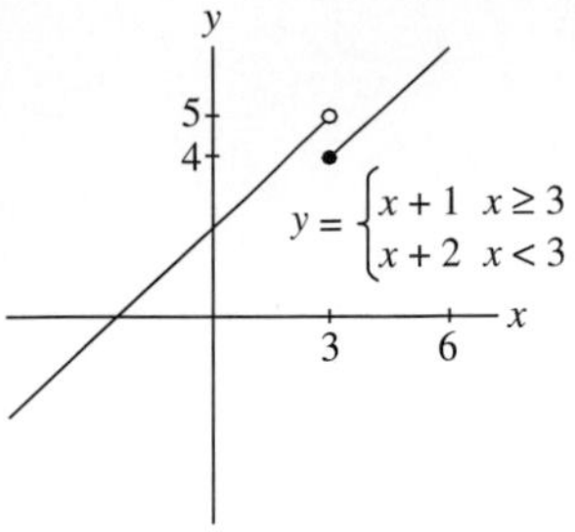

93. Domain $(-\infty, \infty)$, range $(-\infty, 2]$, inc on $(-\infty, -2)$ and $(-2, 0)$, dec on $(0, 2)$ and $(2, \infty)$

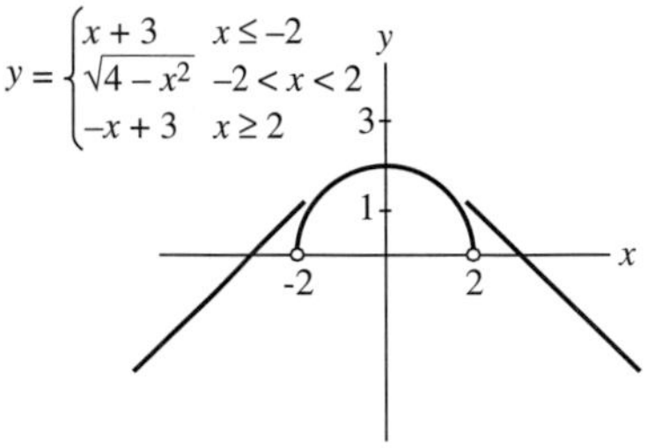

95. dec on $(-\infty, 0.83)$, inc on $(0.83, \infty)$
97. inc on $(-\infty, -1)$ and $(1, \infty)$, dec on $(-1, 1)$
99. dec on $(-\infty, -1.73)$ and $(0, 1.73)$, inc on $(-1.73, 0)$ and $(1.73, \infty)$
101. $(x + 1)^2 + (y - 3)^2 = 29$
103. $f(x) = \begin{cases} -4[\![-x]\!] & 0 < x \leq 3 \\ 15 & 3 < x \leq 8 \end{cases}$

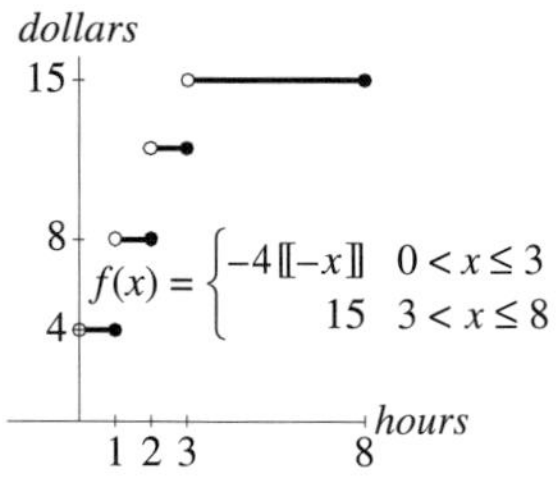

105. $(0, 10^4)$, $(10^4, \infty)$
107. $[5, \infty)$

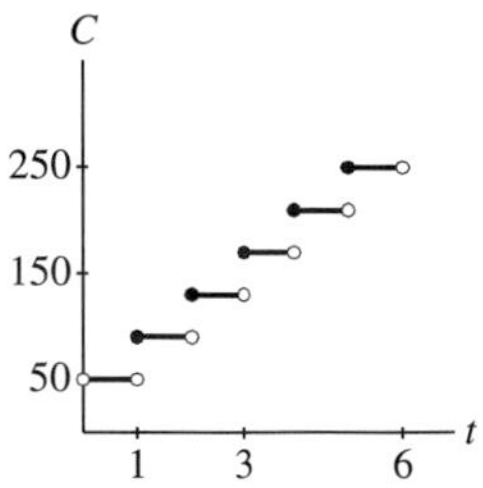

109. 565 million, 720 million, 14.5 million/yr

Section 2.4

1.

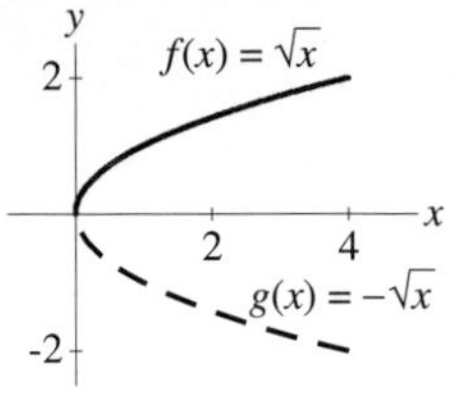

3.

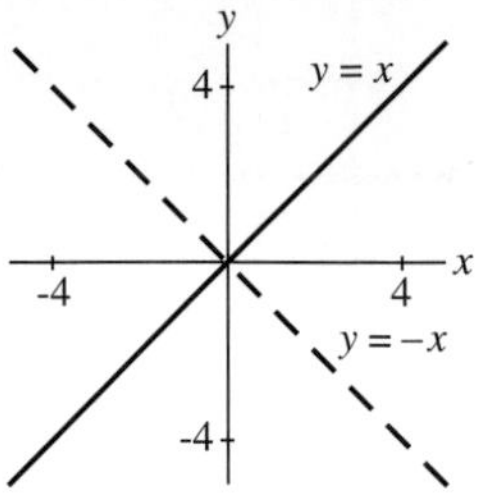

5.

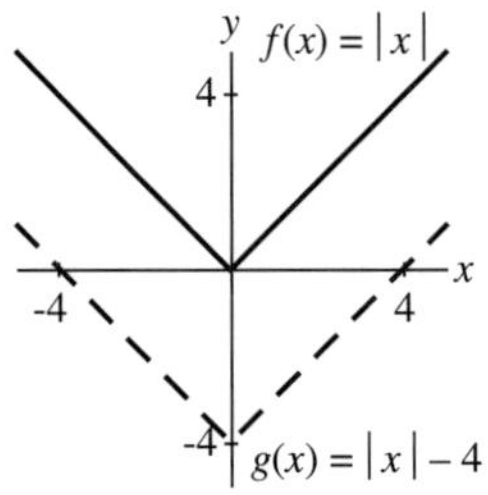

7.

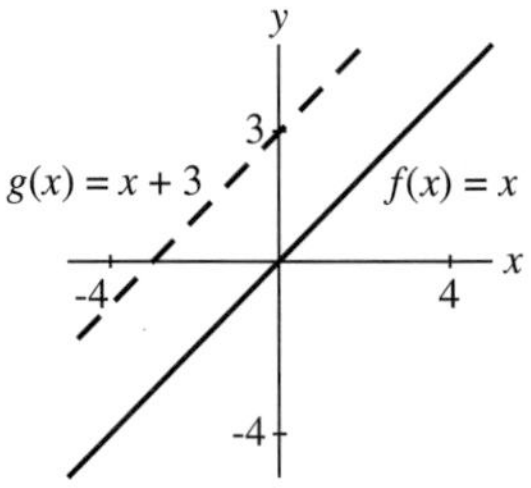

9.

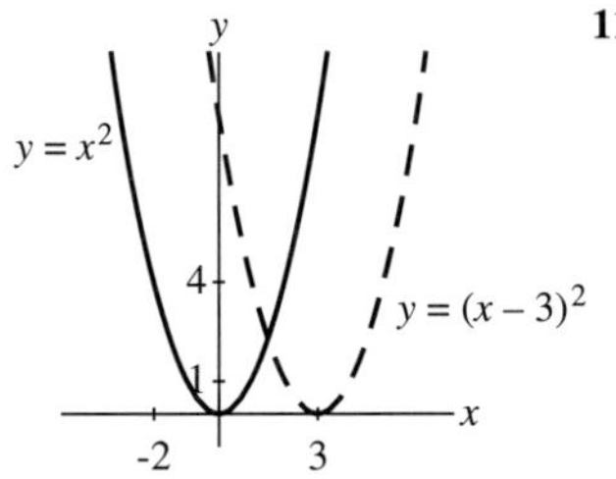

11.

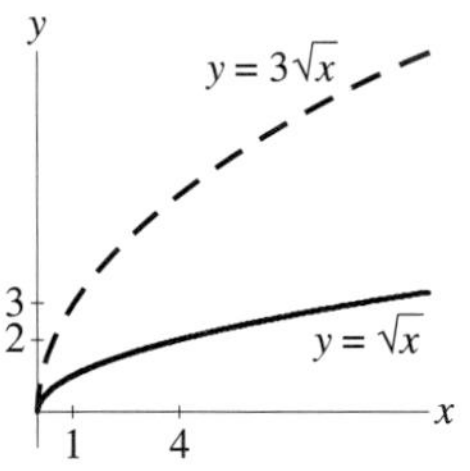

13.

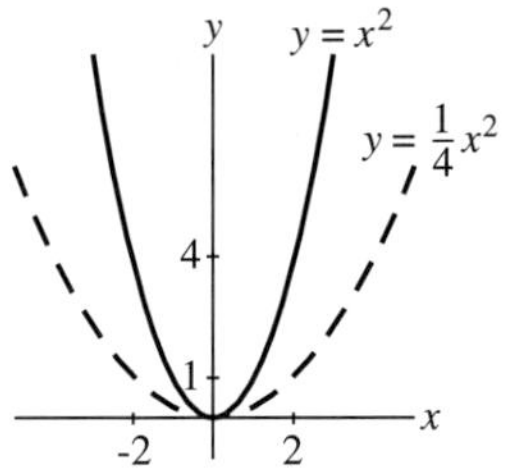

15. (g) **17.** (b) **19.** (c) **21.** (f) **23.** $y = (x - 10)^2 + 4$
25. $y = -3|x - 7| + 9$ **27.** $y = -3\sqrt{x} - 5$
29.

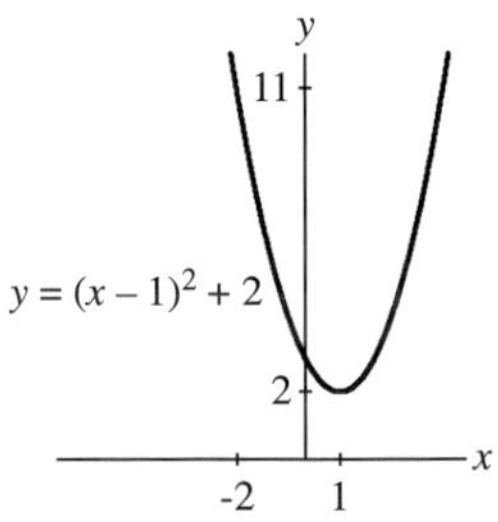

31.

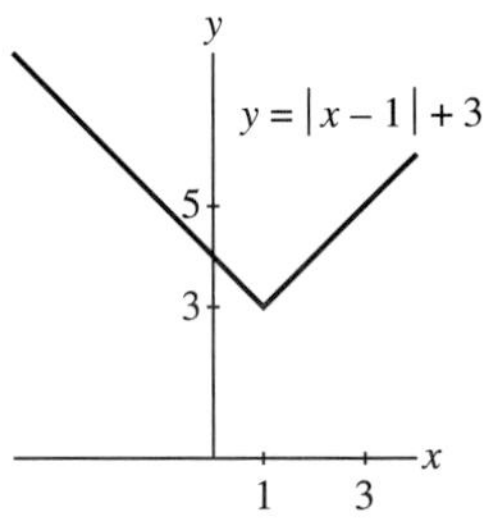

33.
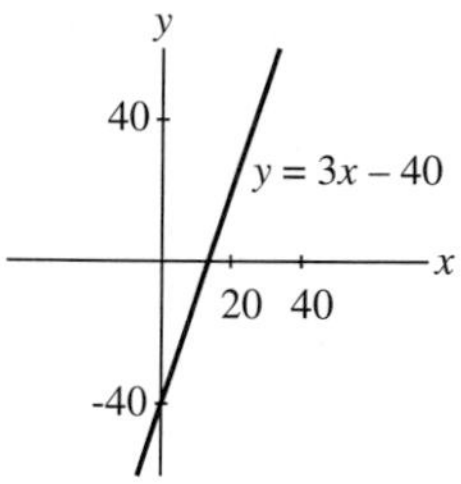

35.
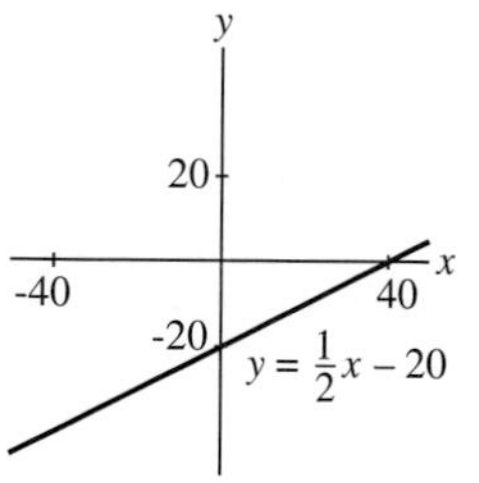

37.
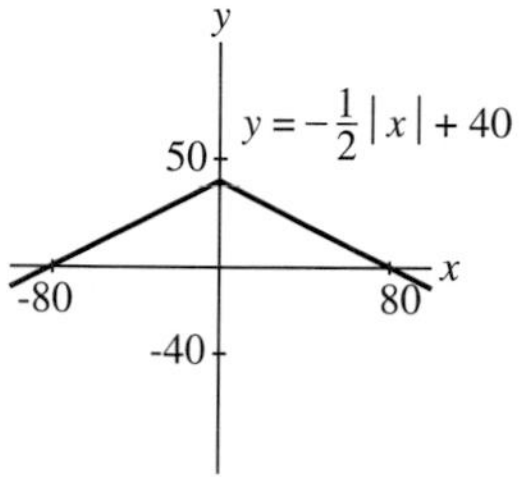

39.
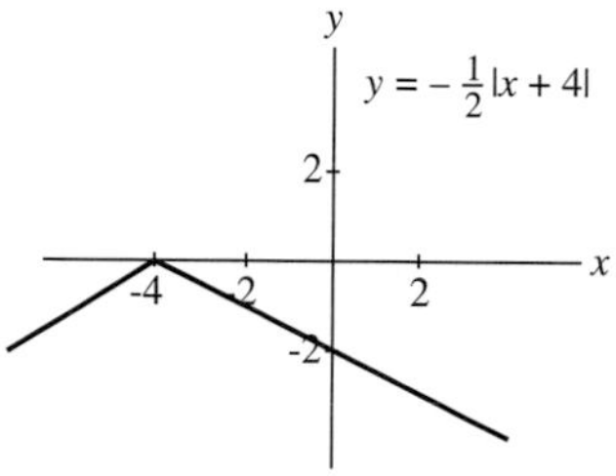

41.
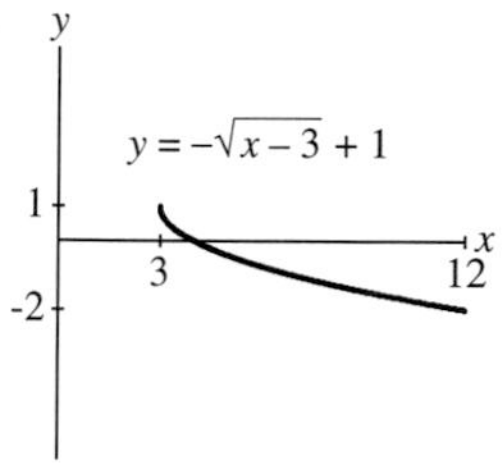

43.
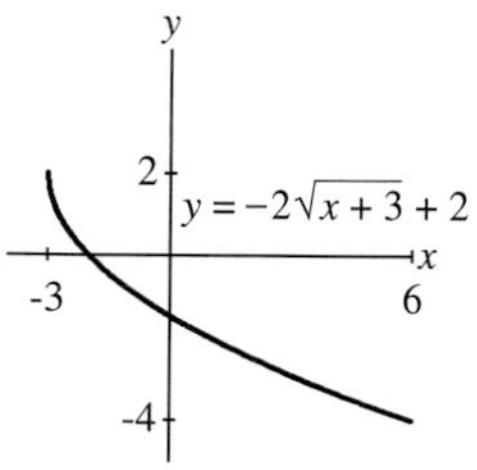

45. Symmetric about y-axis, even **47.** No symmetry, neither
49. Symmetric about $x = -3$, neither
51. No symmetry, neither **53.** Symmetric about origin, odd
55. No symmetry, neither **57.** No symmetry, neither
59. Symmetric about $x = 2$, neither
61. Symmetric about y-axis, even
63. Symmetric about y-axis, even **65.** (e) **67.** (g) **69.** (b)
71. (c) **73.** $(-\infty, -1] \cup [1, \infty)$ **75.** $(-\infty, -1) \cup (5, \infty)$
77. $(-2, 4)$ **79.** $[0, 25]$ **81.** $(-\infty, 2 - \sqrt{3}) \cup (2 + \sqrt{3}, \infty)$
83. $(-5, 5)$ **85.** $(-3.36, 1.55)$ **87.** $N(x) = x + 2000$
89. $x \geq 25\%$ **91.** 1893

Section 2.5

1. 1 **3.** -11 **5.** -8 **7.** 1/12 **9.** $a^2 - 3$
11. $a^3 - 4a^2 + 3a$ **13.** $y = 6x - 1$ **15.** $y = x^2 + 6x + 7$
17. $y = x$ **19.** $y = x$ **21.** $\{(-3, 3), (2, 6)\}$ **23.** $\{(-3, 2)\}$
25. $\{(-3, 2), (2, 0)\}$ **27.** $\{(-3, 0), (1, 0), (4, 4)\}$ **29.** $\{(1, 4)\}$
31. $\{(-3, 4), (1, 4)\}$
33. $(f + g)(x) = \sqrt{x} + x - 4$, $[0, \infty)$
35. $(g \cdot h)(x) = \dfrac{x - 4}{x - 2}$, $(-\infty, 2) \cup (2, \infty)$
37. $\left(\dfrac{f}{g}\right)(x) = \dfrac{\sqrt{x}}{x - 4}$, $[0, 4) \cup (4, \infty)$
39. $\left(\dfrac{g}{h}\right)(x) = x^2 - 6x + 8$, $(-\infty, 2) \cup (2, \infty)$
41. 5 **43.** 5 **45.** 59.8163 **47.** $3x^2 + 2$ **49.** $9x^2 - 6x + 2$
51. $\dfrac{x^2 + 2}{3}$ **53.** $9x - 4$ **55.** $3x^2 - 2x + 1$
57. $(f \circ g)(x) = \sqrt{2x - 1}$, $\left[\dfrac{1}{2}, \infty\right)$
59. $(h \circ f)(x) = \dfrac{1}{\sqrt{x} - 3}$, $[0, 9) \cup (9, \infty)$
61. $(h \circ h)(x) = \dfrac{x - 3}{10 - 3x}$, $(-\infty, 3) \cup \left(3, \dfrac{10}{3}\right) \cup \left(\dfrac{10}{3}, \infty\right)$
63. $(f \circ f)(x) = x^{1/4}$, $[0, \infty)$ **71.** $y = -x$, no
73. $[-1, \infty)$, $[-7, \infty)$ **75.** $[1, \infty)$, $[0, \infty)$ **77.** $[0, \infty)$, $[4, \infty)$
79. $y = (n - 4)^2$ **81.** $y = \dfrac{\sqrt{x + 16}}{8}$ **83.** $F = g \circ h$
85. $H = f \circ g \circ h$ **87.** $N = h \circ g \circ f$ **89.** $P = g \circ f \circ g$
91. $A = d^2/2$ **93.** $(f + g)(x) = 0.55x$
95. $F(x) = \dfrac{31}{15}x + 238$, 279.3 exajoules **97.** $D = \dfrac{1.16 \times 10^7}{L^3}$
99. $W = \dfrac{(8 + \pi)s^2}{8}$ **101.** $s = \dfrac{d\sqrt{2}}{2}$
103. $s = 0.7x$, $c = 0.75s$, $c = 0.525x$, no

Section 2.6

1. $\{(1, 2), (5, 3)\}$, 3, 2 **3.** $\{(-3, -3), (5, 0), (-7, 2)\}$, 0, 2 **5.** No
7. Yes, $\{(0, 3), (5, 2), (6, 4), (9, 7)\}$
9. Yes, $\{(1, 1), (2, 2), (4.5, 4.5)\}$
11. Yes, $\{(x, y) \mid y = x - 2\}$ **13.** Yes, $\left\{(x, y) \mid y = \dfrac{x - 7}{2}\right\}$
15. No **17.** No **19.** Yes **21.** No **23.** No **25.** No
27. Yes **29.** No **31.** No **33.** No **35.** Yes **37.** Yes
39. No **41.** No **43.** $f^{-1}(x) = \dfrac{x + 7}{3}$
45. $f^{-1}(x) = (x - 2)^2 + 3$ for $x \geq 2$ **47.** $f^{-1}(x) = -x - 9$
49. $f^{-1}(x) = \dfrac{5x + 3}{x - 1}$ **51.** $f^{-1}(x) = -\dfrac{1}{x}$
53. $f^{-1}(x) = (x - 5)^3 + 9$ **55.** $f^{-1}(x) = \sqrt{x} + 2$
57. $(g \circ f)(x) = x$, yes **59.** $(g \circ f)(x) = |x|$, no
61. $(g \circ f)(x) = x$, yes **63.** $(g \circ f)(x) = x$, yes
65. y_1 and y_2 are inverse functions. **67.** $f^{-1}(x) = \dfrac{x}{5}$
69. $f^{-1}(x) = x + 88$ **71.** $f^{-1}(x) = \dfrac{x + 7}{3}$ **73.** $f^{-1}(x) = \dfrac{x - 4}{-3}$
75. $f^{-1}(x) = 2x + 18$ **77.** $f^{-1}(x) = -x$ **79.** $f^{-1}(x) = (x + 9)^3$
81. $f^{-1}(x) = \sqrt[3]{\dfrac{x + 7}{2}}$ **83.** No **85.** Yes

87. $f^{-1}(x) = \frac{x-2}{3}$

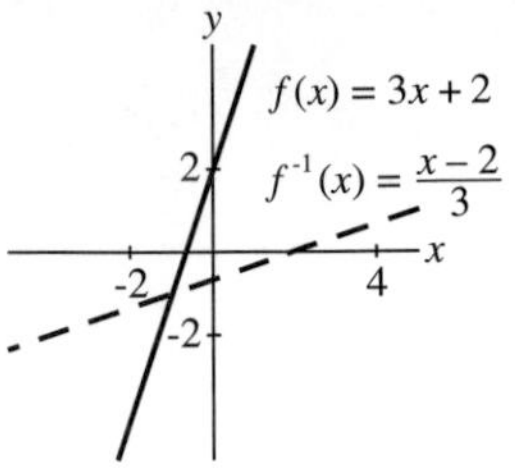

89. $f^{-1}(x) = \sqrt{x+4}$

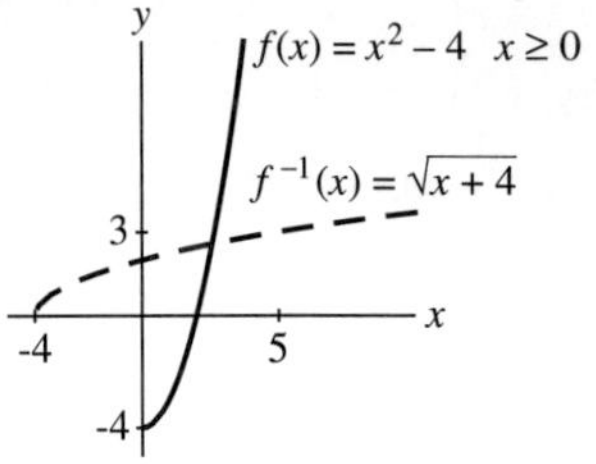

91. $f^{-1}(x) = \sqrt[3]{x}$

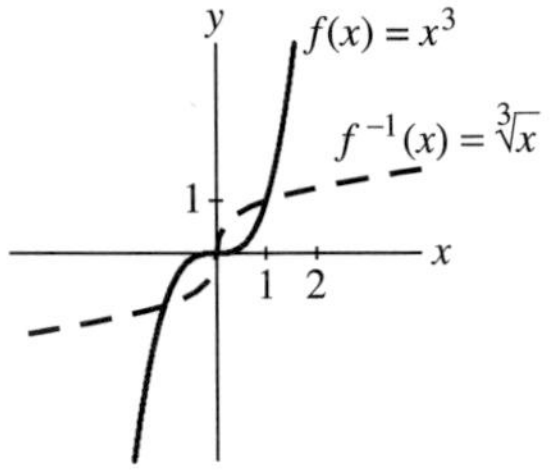

93. $f^{-1}(x) = (x+3)^2$ for $x \geq -3$

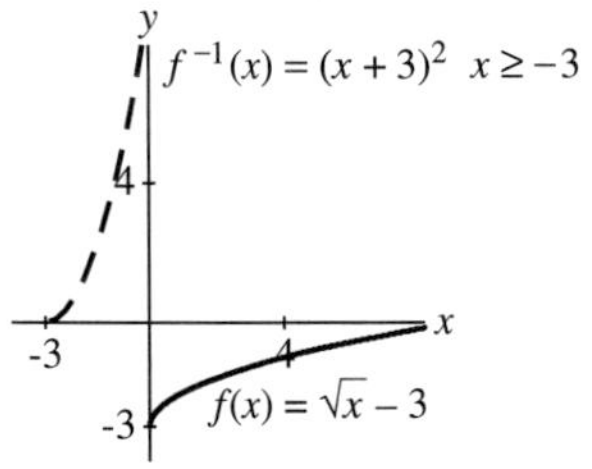

95. $C = 1.08P$, $P = \frac{C}{1.08}$ **97.** Yes, $r = \frac{7.89 - t}{0.39}$, 6

99. $w = \frac{V^2}{1.496}$, 8840 pounds **101.** 10.9%, $V = 50{,}000(1-r)^5$

105. $1 + \frac{-5}{x+2}$

Section 2.7

1. $G = kn$ **3.** $m_1 = \frac{k}{m_2}$ **5.** $C = khr$ **7.** $Y = \frac{kx}{\sqrt{z}}$

9. A varies directly as the square of r.
11. The variable a varies jointly as z and w.
13. No variation
15. The variable y is inversely proportional to x.
17. H varies directly as the square root of t and inversely as s.
19. D varies jointly as L and J and inversely as W.

21. $y = \frac{5}{9}x$ **23.** $T = \frac{-150}{y}$ **25.** $m = 3t^2$ **27.** $y = \frac{1.37x}{\sqrt{z}}$

29. $-\frac{27}{2}$ **31.** 1 **33.** $\sqrt{6}$ **35.** 7/4 **37.** Direct, $L_i = 12L_f$
39. Inverse, $P = 20/n$ **41.** Direct, $S_m = 0.6S_k$ **43.** Neither
45. Direct, $A = 30W$ **47.** Inverse, $n = 5/p$ **49.** 2604 pounds
51. 12.8 hr **53.** \$50.70 **55.** \$19.84 **57.** 18.125 oz
59. 8 ft/yr **61.** No **63.** 38 **65.** 36 ft/sec

Chapter 2 Review Exercises

1. Domain {0, 1, −2}, range {0, 1, −2}, yes

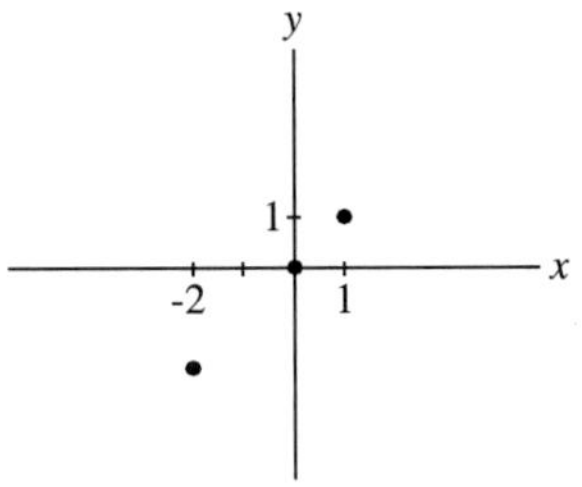

3. Domain $(-\infty, \infty)$, range $(-\infty, \infty)$, yes

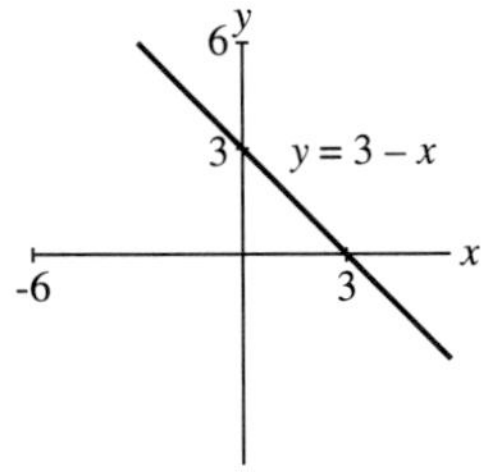

5. Domain {2}, range $(-\infty, \infty)$, no

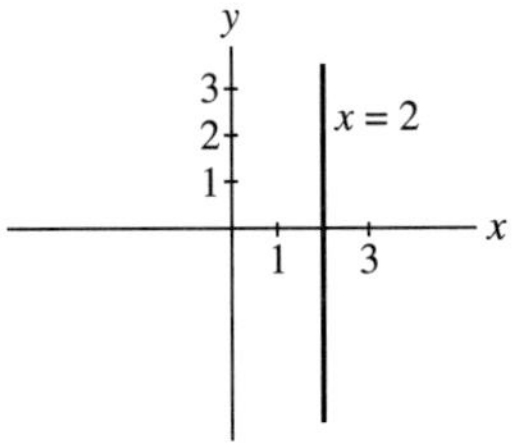

7. Domain [−0.1, 0.1], range [−0.1, 0.1], no

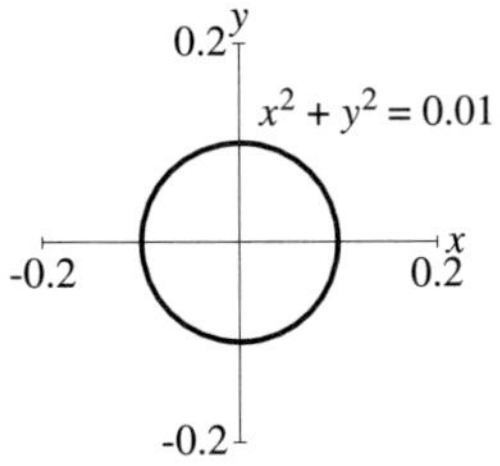

9. Domain $[1, \infty)$, range $(-\infty, \infty)$, no

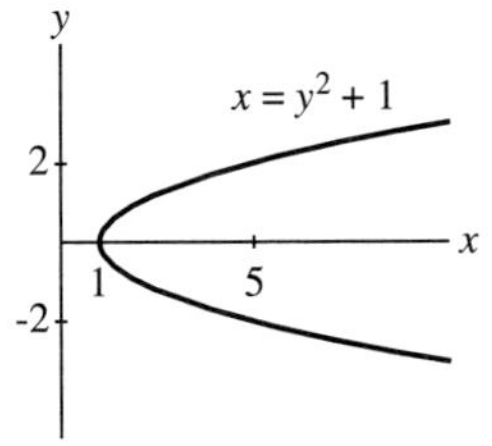

11. Domain $[0, \infty)$, range $[-3, \infty)$, yes

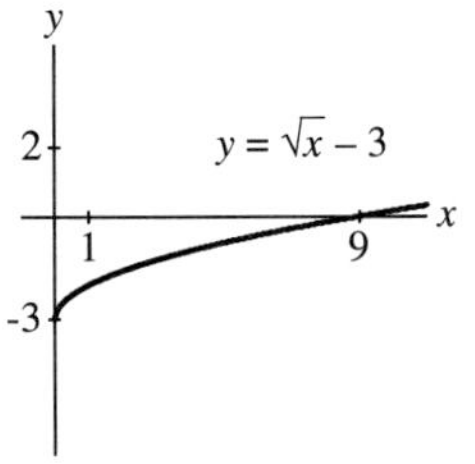

13. 12 **15.** 17 **17.** ±4 **19.** 17 **21.** 4 **23.** −36 **25.** 12
27. $4x^2 - 28x + 52$ **29.** $x^4 + 6x^2 + 12$ **31.** $a^2 + 2a + 4$
33. $6 + h$ **35.** $2x + h$ **37.** x **39.** $\frac{x+7}{2}$

41.

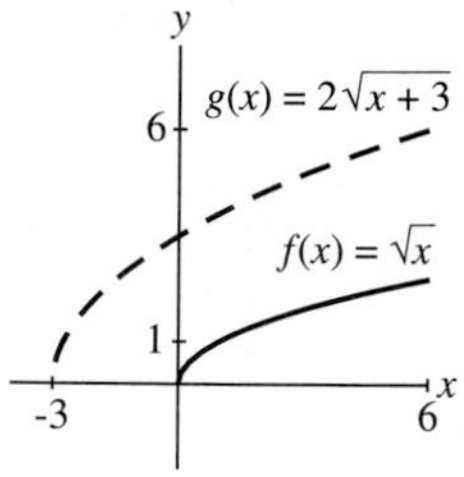

43.

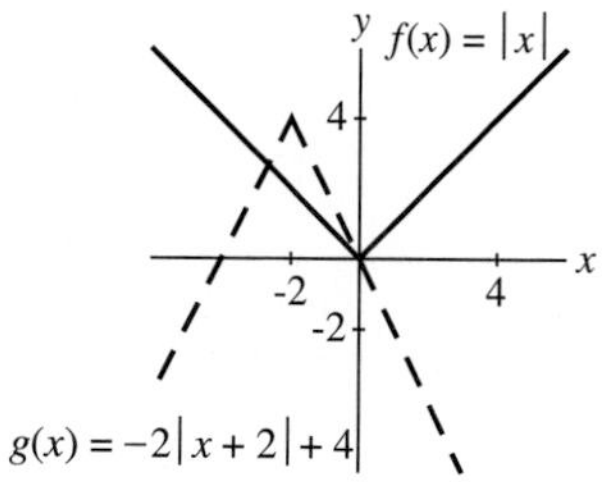

45.

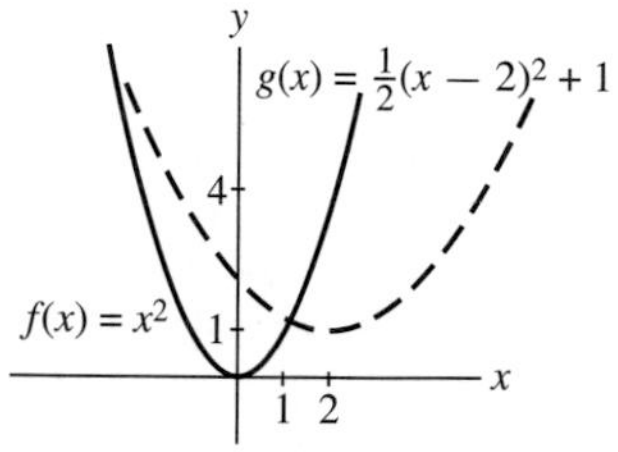

47. $F = f \circ g$ **49.** $H = f \circ h \circ g \circ j$ **51.** $N = h \circ f \circ j$

53. $R = g \circ h \circ j$ **55.** $\sqrt{146}$, $(-1/2, -1/2)$ **57.** $\frac{\sqrt{73}}{12}$, $\left(\frac{3}{8}, \frac{2}{3}\right)$

59. -5 **61.** $\frac{-1}{2x(x+h)}$

63. Domain $[-10, 10]$, range $[0, 10]$, inc on $(-10, 0)$, dec on $(0, 10)$

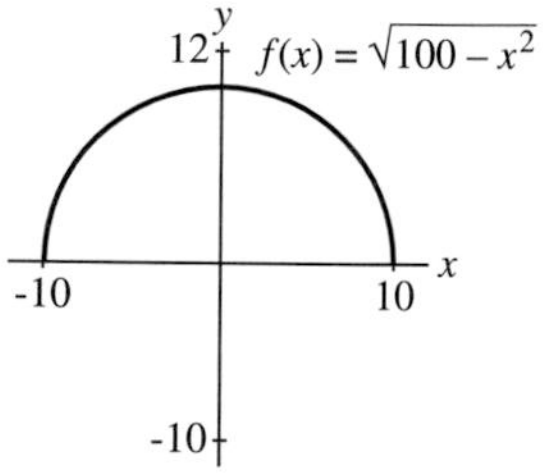

65. Domain $(-\infty, \infty)$, range $(-\infty, \infty)$, inc on $(-\infty, \infty)$

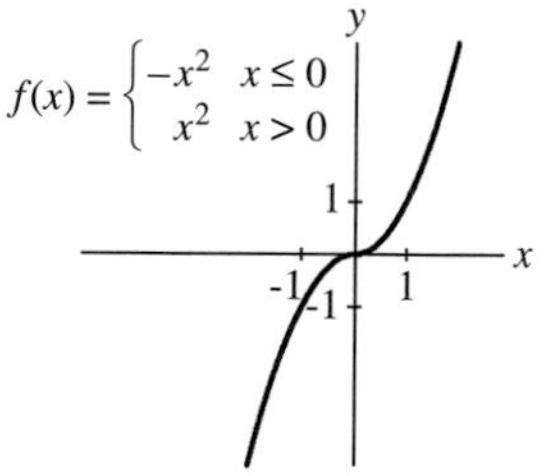

67. Domain $(-\infty, \infty)$, range $[-2, \infty)$, inc on $(-2, 0)$ and $(2, \infty)$, dec on $(-\infty, -2)$ and $(0, 2)$

$$f(x) = \begin{cases} -x-4 & x \le -2 \\ -|x| & -2 < x < 2 \\ x-4 & x \ge 2 \end{cases}$$

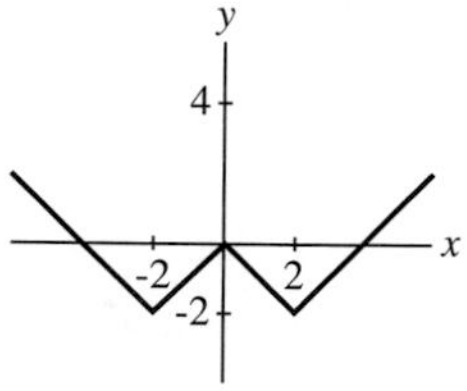

69. $y = |x| - 3$, $(-\infty, \infty)$, $[-3, \infty)$

71. $y = -2|x| + 4$, $(-\infty, \infty)$, $(-\infty, 4]$

73. $y = |x+2| + 1$, $(-\infty, \infty)$, $[1, \infty)$

75. Symmetric about y-axis

77. Symmetric about origin

79. Neither symmetry

81. Symmetric about y-axis

83. $(-\infty, 2] \cup [4, \infty)$ **85.** $(-\sqrt{2}, \sqrt{2})$ **87.** $\varnothing$

89.

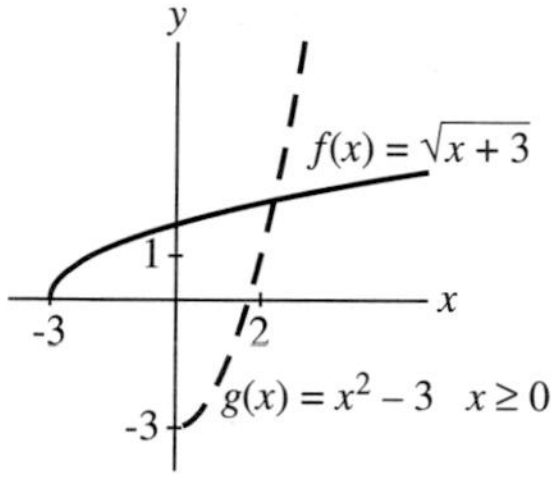

91.

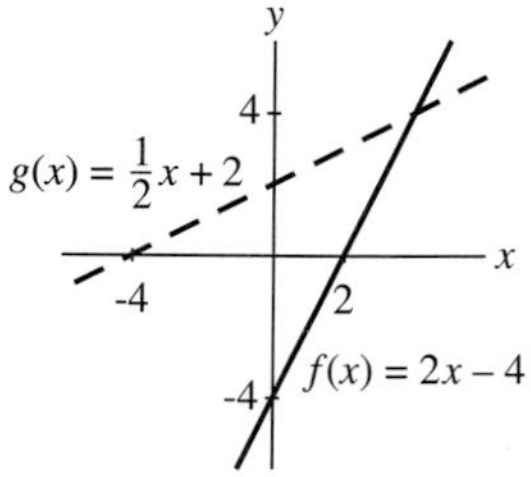

93. Not invertible **95.** $f^{-1}(x) = \frac{x+21}{3}$, $(-\infty, \infty)$, $(-\infty, \infty)$

97. Not invertible **99.** $f^{-1}(x) = x^2 + 9$ for $x \ge 0$, $[0, \infty)$, $[9, \infty)$

101. $f^{-1}(x) = \frac{5x+7}{1-x}$, $(-\infty, 1) \cup (1, \infty)$, $(-\infty, -5) \cup (-5, \infty)$

103. $f^{-1}(x) = -\sqrt{x-1}$, $[1, \infty)$, $(-\infty, 0]$ **105.** $(3, 0)$, $(0, -9/4)$

107. $19/2$

109. $\left(x - \frac{1}{2}\right)^2 + (y+1)^2 = \frac{9}{4}$, center $\left(\frac{1}{2}, -1\right)$, radius $\frac{3}{2}$

111. $A = 4r^2$ **113.** 30 mph/sec **115.** 22.5 **117.** $F = \frac{km_1m_2}{d^2}$

Chapter 2 Test

1. No **2.** Yes **3.** No **4.** Yes **5.** $\{2, 5\}$, $\{-3, -4, 7\}$

6. $[9, \infty)$, $[0, \infty)$ **7.** $[0, \infty)$, $(-\infty, \infty)$

8.

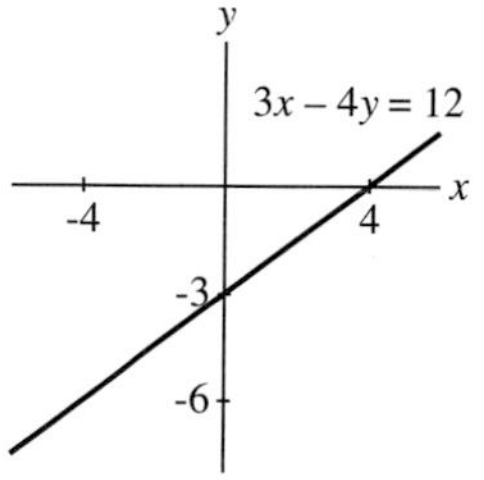

9.

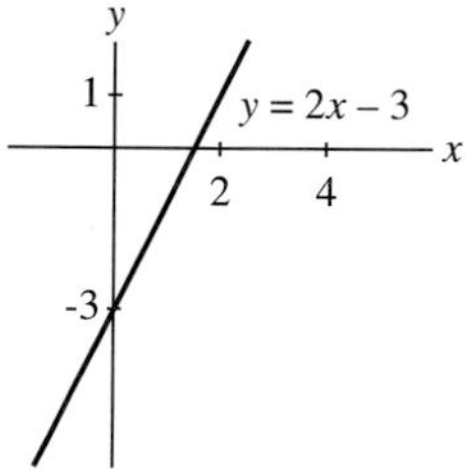

10.

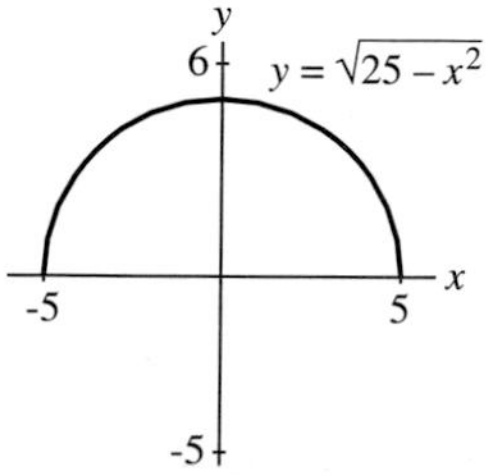

11.

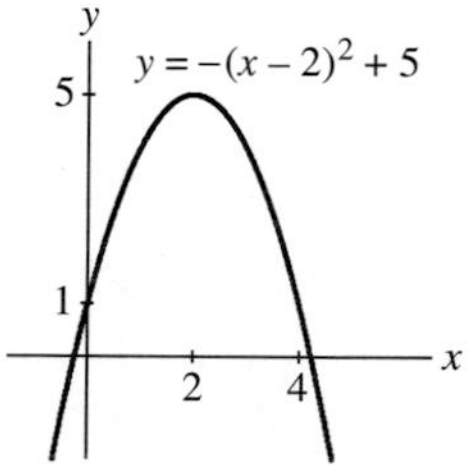

12.
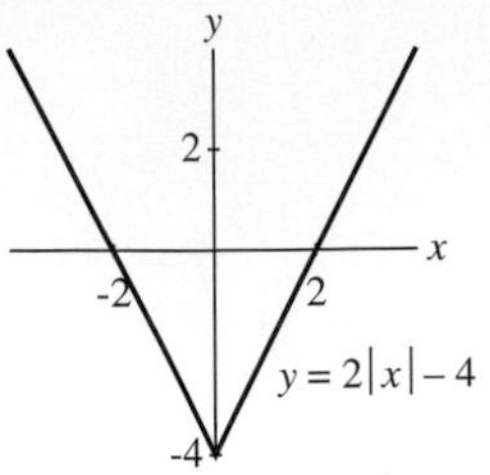

13.
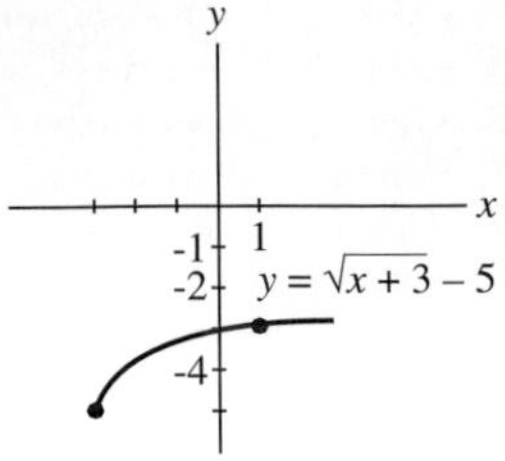

14.
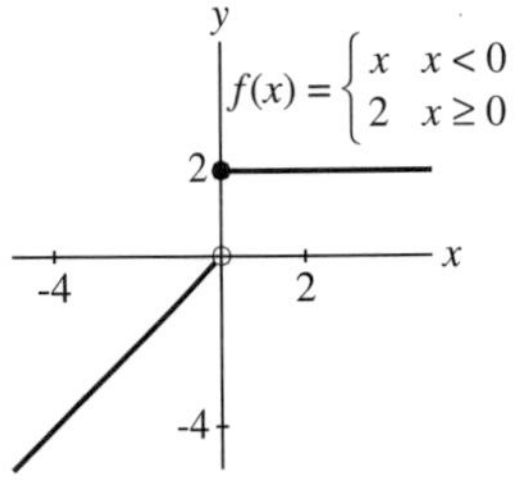

15. 3 **16.** $\sqrt{7}$ **17.** $\sqrt{3x+1}$ **18.** $\dfrac{x+1}{3}$ **19.** 45 **20.** 3
21. (0, 5.5) **22.** $5\sqrt{2}$ **23.** $(x+4)^2 + (y-1)^2 = 7$
24. Dec on $(-\infty, 3)$, inc on $(3, \infty)$ **25.** Symmetric about y-axis
26. $(-2, 4)$ **27.** $g^{-1}(x) = (x-3)^3 + 2$ **28.** \$0.125 per envelope
29. 12 candlepower **30.** $V = \dfrac{\sqrt{2}d^3}{4}$

Tying It All Together Chapters P–2

1. $5x$ **2.** $x^2 - 6x + 9$ **3.** $|x|$ **4.** $x^2 - 4$ **5.** $x^4 + x^3 - 6x^2$
6. 13 **7.** $-2 \pm \sqrt{2}$ **8.** $\dfrac{-3 \pm i\sqrt{6}}{3}$ **9.** $(-\infty, \infty)$
10. $\{\pm\sqrt{5}\}$ **11.** $\{-1, 1\}$ **12.** $\{-2, 2\}$ **13.** $\{2 \pm i\}$
14. $\left\{\dfrac{1 \pm \sqrt{2}}{2}\right\}$ **15.** $\{0, 9\}$ **16.** $\{2\}$

17.
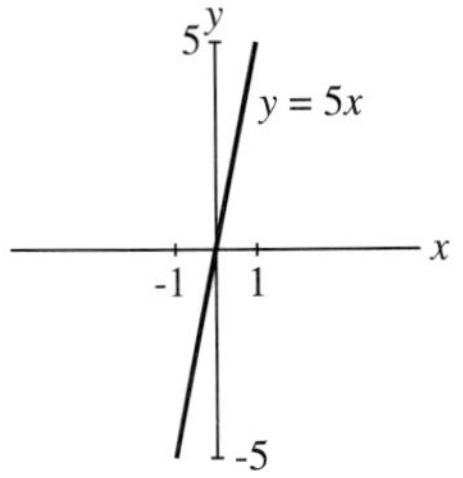

18.
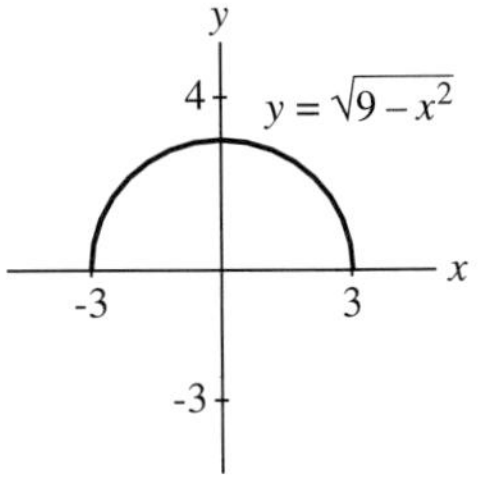

19.
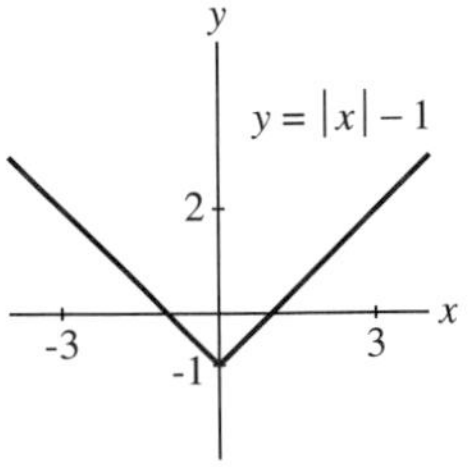

20.
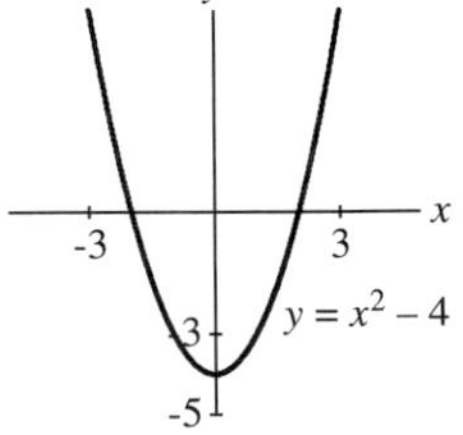

21.
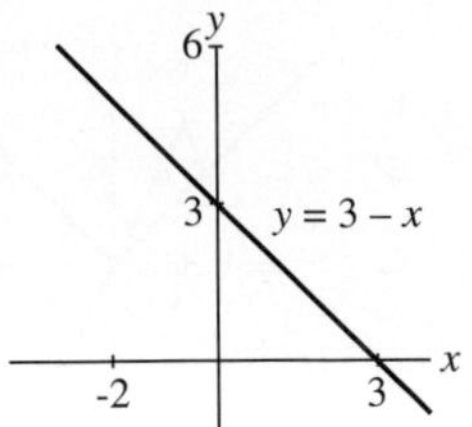

22.
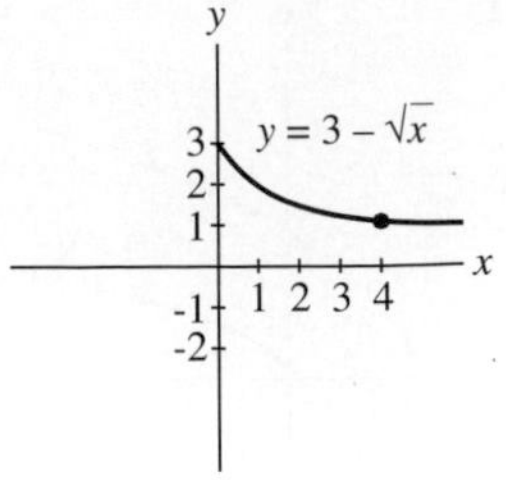

23.
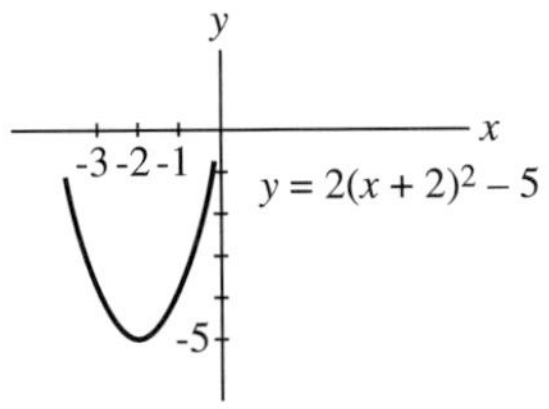

24.
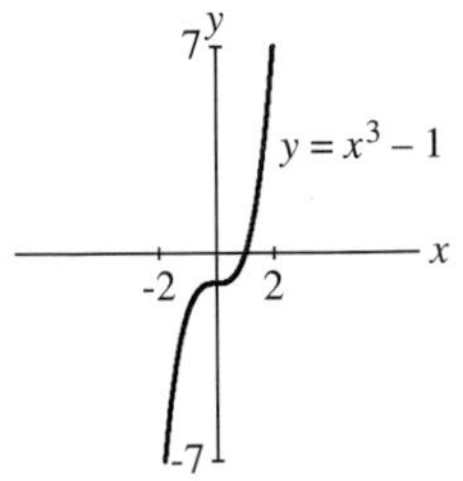

25. $\dfrac{x^2+1}{x}$ **26.** $\dfrac{x^2}{x-1}$ **27.** $\dfrac{3x-10}{x-4}$ **28.** $\dfrac{-2x-7}{x+3}$

CHAPTER 3

Section 3.1

1. $\dfrac{1}{3}$ **3.** -4 **5.** 0 **7.** $-\sqrt{2}$ **9.** $\dfrac{\pi}{8}$
11. $y = \dfrac{3}{5}x - 2, \dfrac{3}{5}, (0, -2)$ **13.** $y = 2x - 5, 2, (0, -5)$
15. $y = \dfrac{1}{2}x + \dfrac{1}{2}, \dfrac{1}{2}, \left(0, \dfrac{1}{2}\right)$ **17.** $y = 4, 0, (0, 4)$
19. $y = 0.03x - 2.6, 0.03, (0, -2.6)$

21.
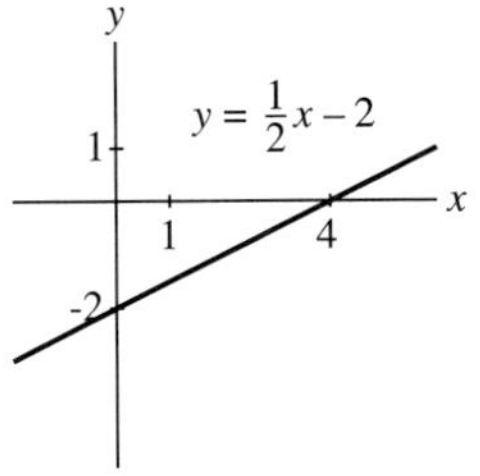

23.
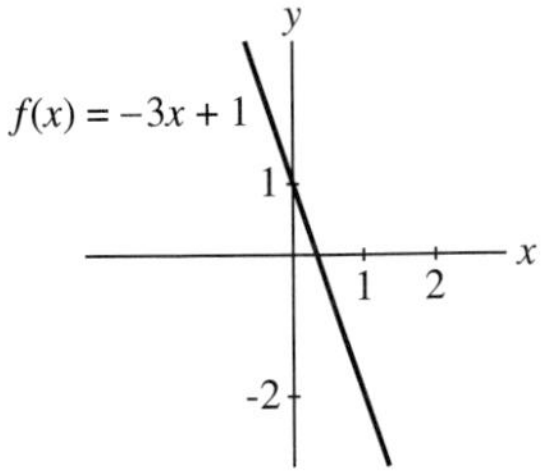

25.
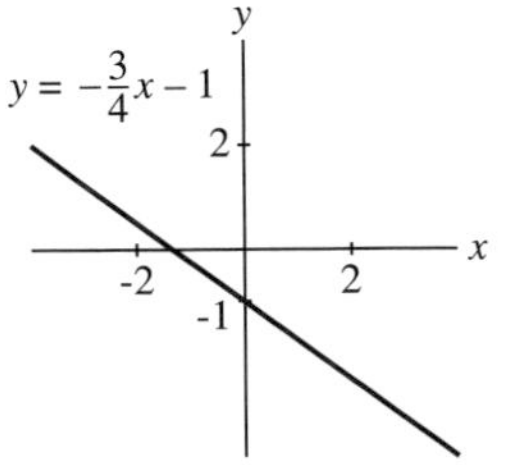

27.
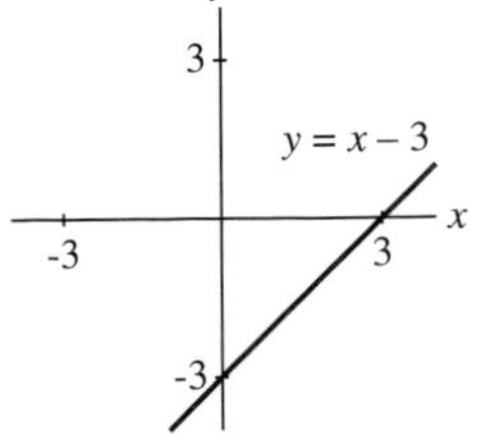

29.

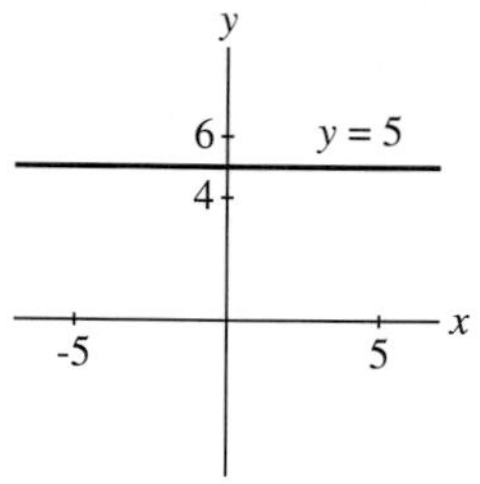

31.

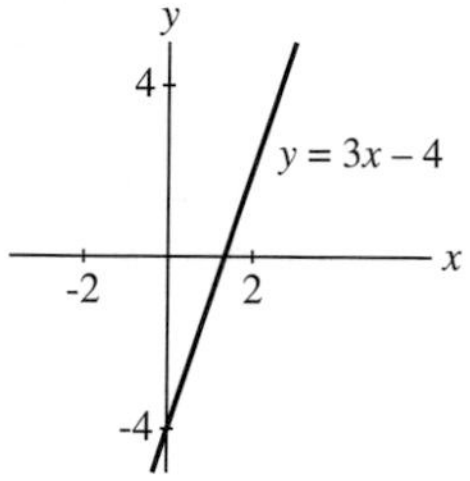

33.

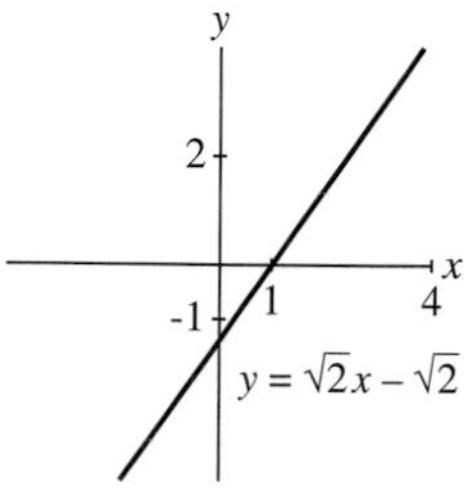

35. $y = \frac{2}{3}x - 1$ **37.** $y = \frac{5}{2}x + \frac{3}{2}$ **39.** $y = -2x + 4$
41. $4x - 3y = 12$ **43.** $4x - 5y = -7$ **45.** $x = -4$
47. $y = -6$ **49.** 0.5 **51.** −1 **53.** 0 **55.** $2x - y = 4$
57. $3x + y = 7$ **59.** $2x + y = -5$ **61.** $y = 5$
63. $3x - 5y = -15$ **65.** 20/3 **67.** −5 **69.** T **71.** T **73.** F
75. T **77.** Yes, no
79. $y = x - 2$, $x^3 - 8 = (x - 2)(x^2 + 2x + 4)$
81. $F = \frac{9}{5}C + 32$, 302°F **83.** $c = 50 - n$, \$400
85. $y = \begin{cases} 0.1a - 3 & 35 \le a \le 50 \\ 0.2a - 8 & 50 < a \le 65 \end{cases}$ 1.7 yr, 4.6 yr
87. $S = -0.005D + 95$
89. $c = -\frac{3}{4}p + 30$, $-\frac{3}{4}$, if the number of printers is increased by 4, then the number of computers must be decreased by 3.
91. $p = -1645a + 16203$, slope −1645, value goes down \$1645/yr., \$4688, 10 yr
93. $C(x) = 50x + 8000$, $MC(x) = 50$, $MC(x)$ is constant on $(0, \infty)$

Section 3.2

1. $y = \left(x - \frac{3}{2}\right)^2 - \frac{9}{4}$

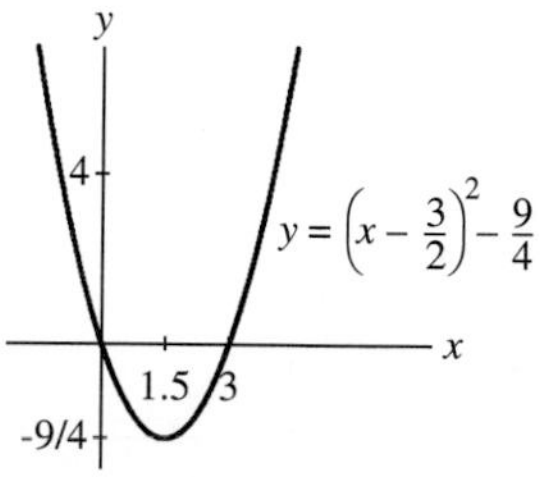

3. $y = 2(x - 3)^2 + 4$

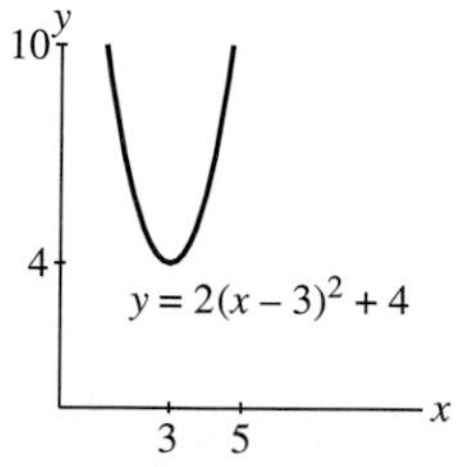

5. $y = -\frac{1}{2}(x - 1)^2 + 3$

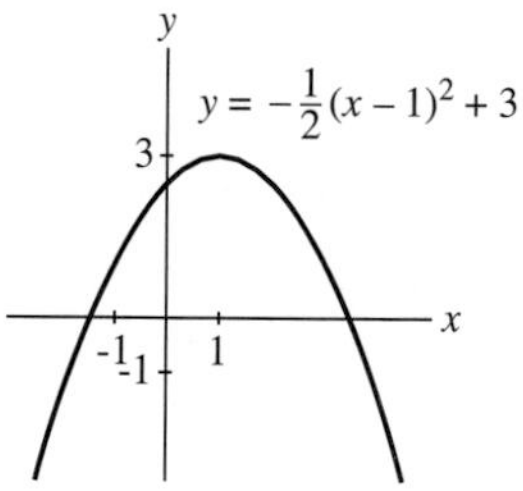

7. $y = \left(x + \frac{3}{2}\right)^2 + \frac{1}{4}$

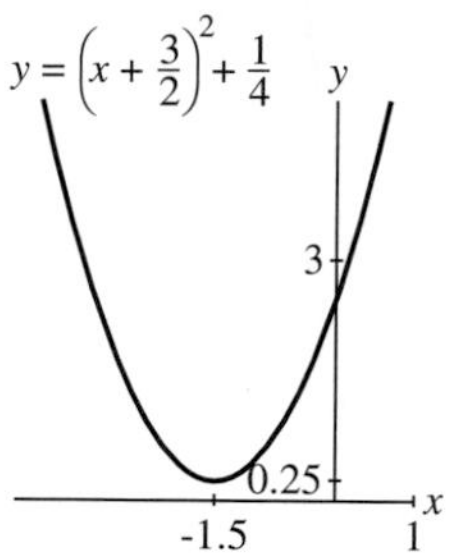

9. $y = -2\left(x - \frac{3}{4}\right)^2 + \frac{1}{8}$

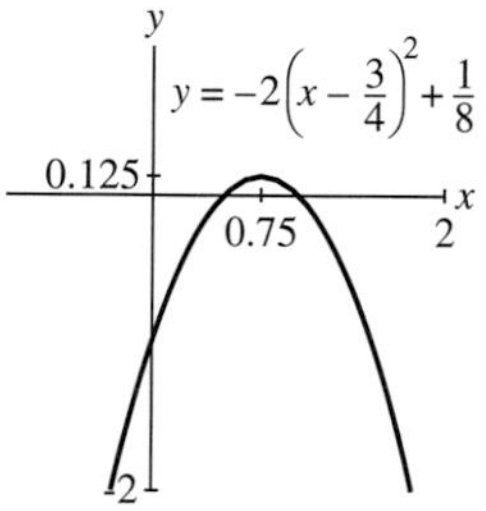

11. $y = -3\left(x - \frac{1}{3}\right)^2 + \frac{1}{3}$

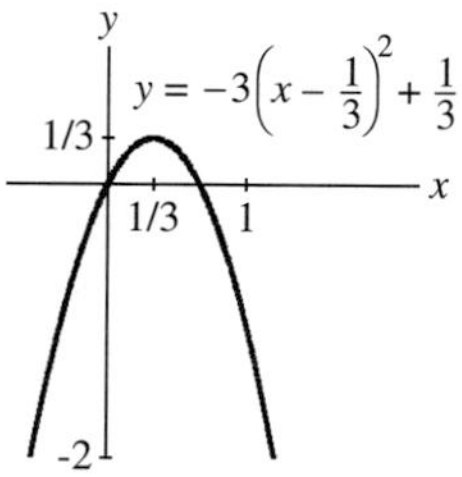

13. (2, −11) **15.** (4, 1) **17.** (−1/3, 1/18)
19. Up, (1, −4), $x = 1$, $[-4, \infty)$, min −4, dec on $(-\infty, 1)$, inc on $(1, \infty)$
21. Range $[-1, \infty)$, min value −1, dec $(-\infty, 1)$, inc $(1, \infty)$
23. Range $(-\infty, \sqrt{3}]$, max value $\sqrt{3}$, inc $(-\infty, 0)$, dec $(0, \infty)$
25. Range $[4, \infty)$, min value 4, dec $(-\infty, 3)$, inc $(3, \infty)$
27. Range $(-\infty, 9]$, max value 9, inc $(-\infty, 1/2)$, dec $(1/2, \infty)$
29. Range $(-\infty, 27/2]$, max value 27/2, inc $(-\infty, 3/2)$, dec $(3/2, \infty)$
31. Range $[-2 - \sqrt{2}, \infty)$, min value $-2 - \sqrt{2}$, dec $(-\infty, -1)$, inc $(-1, \infty)$

33. Vertex (0, −3), axis $x = 0$

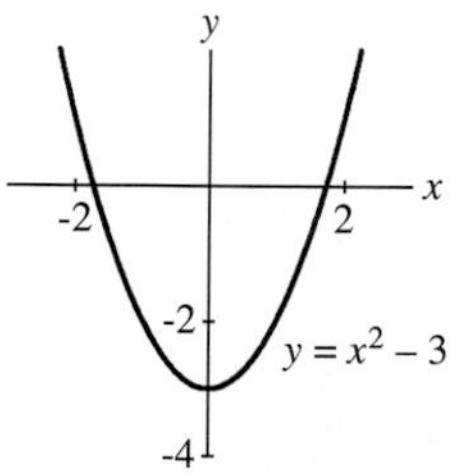

35. Vertex $\left(\frac{\pi}{2}, -\frac{\pi^2}{4}\right)$, axis $x = \frac{\pi}{2}$

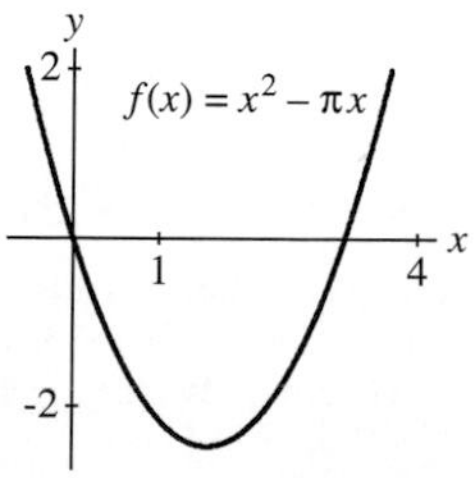

37. Vertex $(-3, 0)$, axis $x = -3$

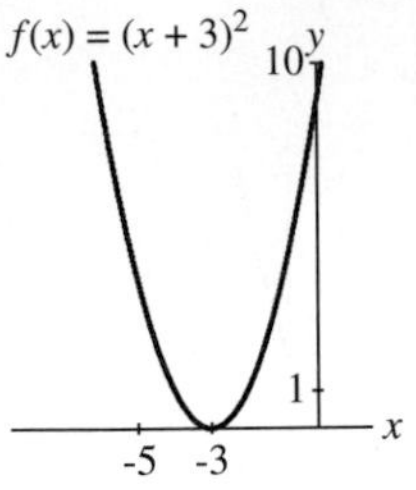

39. Vertex $(3, -4)$, axis $x = 3$

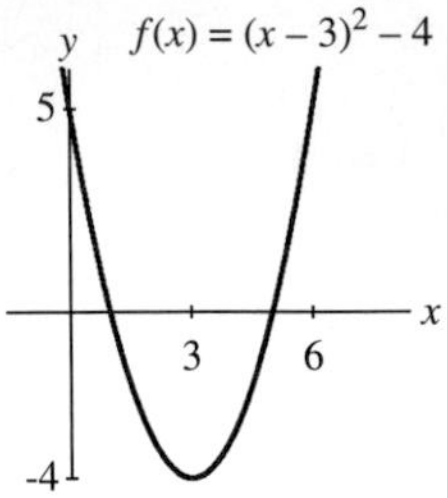

41. Vertex $(2, 12)$, axis $x = 2$

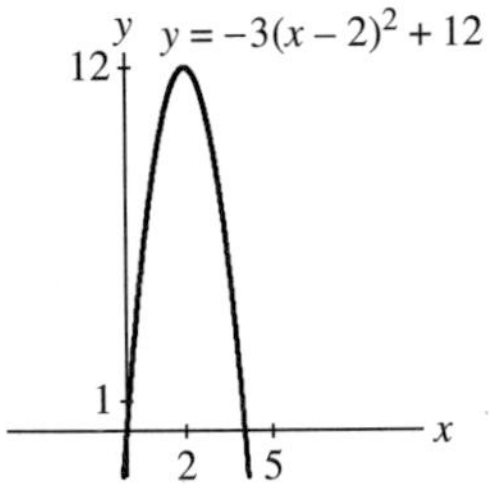

43. Vertex $(1, 3)$, axis $x = 1$

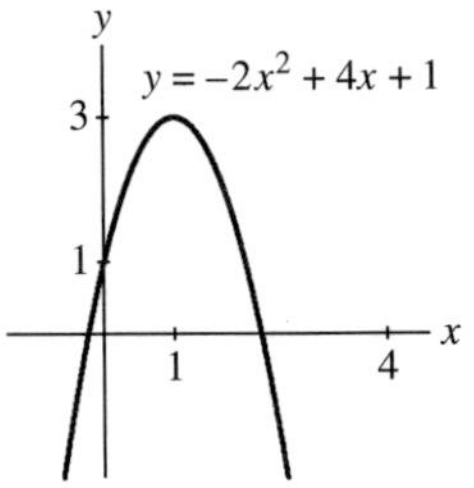

45. $(-\infty, -1] \cup [3, \infty)$ **47.** $(-3, 1)$ **49.** $[-3, 1]$ **51.** $[2, 3]$
53. $(-\infty, 3 - \sqrt{2}) \cup (3 + \sqrt{2}, \infty)$ **55.** $(-\infty, \infty)$
57. $(-0.20, 0.10)$ **59.** $(x - 72)(x + 48)$ **61.** $(x + 30)(2x - 99)$
63. $x = 0, (-2, 4)$ **65.** $x = 1, (0, 8)$ **67.** $x = 1, (3, 16)$
69. $x = -1/2, (1, -13)$ **71.** -12
73. $y = 5(x - 1)^2 - 3$ **75.** $y = -\frac{1}{2}(x + 3)^2 + 2$ **77.** 7/2, 7/2
79. 5 in. wide, 2.5 in. high **81.** 97.24 mph, 13.2 gal/hr **83.** 261 ft
85. $p = 50 - n$, $R = 50n - n^2$, 25 people, \$625 **87.** 1/2
89. a)

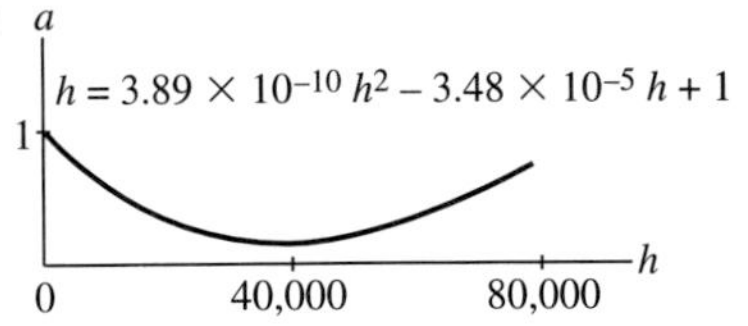

b) decreasing c) dec on (0, 44,730), inc on (44,730, ∞) d) no
e) (0, 30,000)

91. 0.9
93. $t = 14.6n + 110.9$ where n is the number of years after 1985, 300.7, \$1792

Section 3.3

1. $x - 3, 1$ **3.** $-2x^2 + 6x - 14, 33$ **5.** $s^2 + 2, 16$
7. $x + 6, 13$ **9.** $-x + 7, -12$ **11.** $4x^2 + 2x - 4, 0$
13. $2a^2 - 4a + 6, 0$ **15.** $x^3 + x^2 + x + 1, -2$ **17.** $x - \frac{5}{2}, -\frac{1}{4}$
19. 0 **21.** -33 **23.** 5 **25.** 55/8 **27.** 0 **29.** 8
31. $(x + 3)(x + 2)(x - 1)$ **33.** $(x - 4)(x + 3)(x + 5)$
35. Yes **37.** No **39.** Yes **41.** No
43. $\pm(1, 2, 3, 4, 6, 8, 12, 24)$ **45.** $\pm(1, 3, 5, 15)$
47. $\pm\left(1, 3, 5, 15, \frac{1}{2}, \frac{1}{4}, \frac{1}{8}, \frac{3}{2}, \frac{3}{4}, \frac{3}{8}, \frac{5}{2}, \frac{5}{4}, \frac{5}{8}, \frac{15}{2}, \frac{15}{4}, \frac{15}{8}\right)$
49. $\pm\left(1, 2, \frac{1}{2}, \frac{1}{3}, \frac{2}{3}, \frac{1}{6}, \frac{1}{9}, \frac{2}{9}, \frac{1}{18}\right)$ **51.** 2, 3, 4 **53.** $-3, 2 \pm i$
55. $\frac{1}{2}, \frac{3}{2}, \frac{5}{2}$ **57.** $\frac{1}{2}, \frac{1 \pm i}{3}$ **59.** $\pm i, 1, -2$ **61.** $-1, -1, \pm\sqrt{2}$
63. 1/4, 1/3, 1/2 **65.** $1/16, 1 \pm 2i$ **67.** $-6/7, 7/3, \pm i$
69. $2 + \frac{5}{x - 2}$ **71.** $a + \frac{5}{a - 3}$ **73.** $1 + \frac{-3c}{c^2 - 4}$
75. $2 + \frac{-7}{2t + 1}$ **77.** 6 hr **79.** 5 in. by 9 in. by 14 in. **81.** 10

Section 3.4

1. Degree 2, 5 with multiplicity 2
3. Degree 5, ± 3, 0 with multiplicity 3
5. Degree 4, 0, 1 each with multiplicity 2
7. Degree 4, $-4/3$, $3/2$ each with multiplicity 2
9. Degree 3, $0, 2 \pm \sqrt{10}$ **11.** $x^2 + 9$ **13.** $x^2 - 2x - 1$
15. $x^2 - 6x + 13$ **17.** $x^3 - 8x^2 + 37x - 50$
19. $x^2 - 2x - 15 = 0$ **21.** $x^2 + 16 = 0$ **23.** $x^2 - 6x + 10 = 0$
25. $x^3 + 2x^2 + x + 2 = 0$ **27.** $x^3 + 3x = 0$
29. $x^3 - 5x^2 + 8x - 6 = 0$ **31.** $x^3 - 6x^2 + 11x - 6 = 0$
33. $x^3 - 5x^2 + 17x - 13 = 0$ **35.** $24x^3 - 26x^2 + 9x - 1 = 0$
37. $x^4 - 2x^3 + 3x^2 - 2x + 2 = 0$ **39.** 3 neg; 1 neg, 2 imag
41. 1 pos, 2 neg; 1 pos, 2 imag **43.** 4 imag
45. 4 pos; 2 pos, 2 imag; 4 imag **47.** 4 imag and 0
49. $-1 < x < 3$ **51.** $-3 < x < 2$ **53.** $-1 < x < 5$
55. $-1 < x < 3$ **57.** $-2, 1, 5$ **59.** $-3, \frac{3 \pm \sqrt{13}}{2}$
61. $\pm i, 2, -4$ **63.** $-5, \frac{1}{3}, \frac{1}{2}$ **65.** 1, -2 each with multiplicity 2
67. 0, 2 with multiplicity 3 **69.** $0, 1, \pm 2, \pm i\sqrt{3}$
71. $-5 < x < 6$, $-5 < x < 6$ **73.** $-6 < x < 6$, $-5 < x < 5$
75. $-1 < x < 23$, $-1 < x < 23$ **77.** 4 sec and 5 sec **79.** 3 in.
85. $f(x) = -\frac{1}{2}x^3 + 3x^2 - \frac{11}{2}x + 3$

Section 3.5

1. $\{\pm 2, -3\}$ **3.** $\left\{-500, \pm\frac{\sqrt{2}}{2}\right\}$ **5.** $\left\{0, \frac{15 \pm \sqrt{205}}{2}\right\}$
7. $\{0, \pm 2\}$ **9.** $\{\pm 2, \pm 2i\}$ **11.** $\{8\}$ **13.** $\{25\}$ **15.** $\left\{\frac{1}{4}\right\}$
17. $\left\{\frac{2 + \sqrt{13}}{9}\right\}$ **19.** $\{-4, 6\}$ **21.** $\{9\}$ **23.** $\{5\}$ **25.** $\{10\}$
27. $\{\pm 2\sqrt{2}\}$ **29.** $\left\{\pm\frac{1}{8}\right\}$ **31.** $\left\{\frac{1}{49}\right\}$ **33.** $\left\{\frac{5}{4}\right\}$
35. $\{\pm 3, \pm\sqrt{3}\}$ **37.** $\left\{-\frac{17}{2}, \frac{13}{2}\right\}$ **39.** $\left\{\frac{3}{20}, \frac{4}{15}\right\}$

41. $\{-2, -1, 5, 6\}$ **43.** $\{1, 9\}$ **45.** $\{9, 16\}$ **47.** $\{8, 125\}$
49. $\{\pm\sqrt{7}, \pm 1\}$ **51.** $\{0, 8\}$ **53.** $\{-3, 0, 1, 4\}$ **55.** $\{-2, 4\}$
57. $\left\{\frac{1}{2}\right\}$ **59.** $\{\pm 2, 1 \pm i\sqrt{3}, -1 \pm i\sqrt{3}\}$ **61.** $\{\sqrt{3}, 2\}$
63. $\{-2, \pm 1\}$ **65.** $\{5 \pm 9i\}$ **67.** $\left\{\frac{1 \pm 4\sqrt{2}}{3}\right\}$
69. $\{\pm 2\sqrt{6}, \pm\sqrt{35}\}$ **71.** $\{-3, 2, 3\}$ **73.** $\{-11\}$ **75.** $\{2\}$
77. 279.56 m^2 **79.** 23 **81.** $\frac{25}{4}$ and $\frac{49}{4}$ **83.** 5 in. **85.** 1600 ft^2
87. 17,419 pounds **89.** 462.89 in^3 **91.** 1 P.M.
93. \$1.36 billion, \$2.63 billion **95.** 27.4 m

Section 3.6

1. Neither symmetry; crosses at $(-2, 0)$; does not cross at $(1, 0)$; $y \to \infty$ as $x \to \infty$; $y \to -\infty$ as $x \to -\infty$
3. Symmetric about y-axis; no x-intercepts; $y \to \infty$ as $x \to \infty$; $y \to \infty$ as $x \to -\infty$
5. Symmetric about y-axis **7.** Symmetric about $x = 3/2$
9. Neither symmetry **11.** Symmetric about origin
13. Symmetric about $x = 5$ **15.** Symmetric about origin
17. Does not cross at (4, 0) **19.** Crosses at (1/2, 0)
21. Crosses at (1/4, 0) **23.** No x-intercepts
25. Does not cross at (0, 0), crosses at (3, 0)
27. Crosses at (1/2, 0), does not cross at (1, 0)
29. Does not cross at $(-3, 0)$, crosses at (2, 0) **31.** $y \to \infty$
33. $y \to -\infty$ **35.** $y \to -\infty$ **37.** $y \to \infty$ **39.** $y \to \infty$
41. (e) **43.** (g) **45.** (b) **47.** (c)

49.

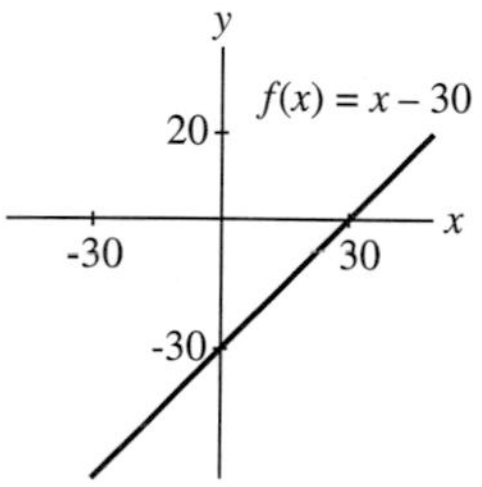

51.

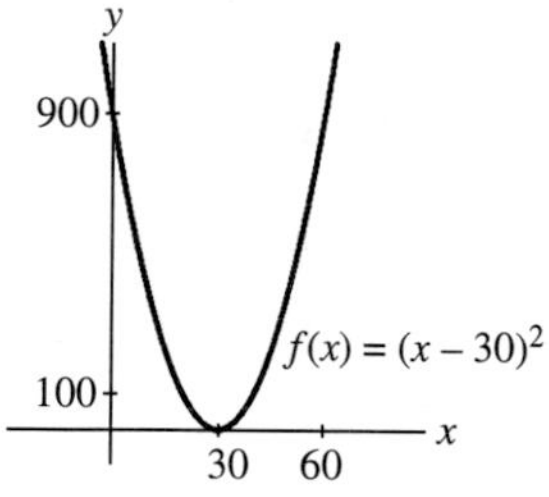

53.

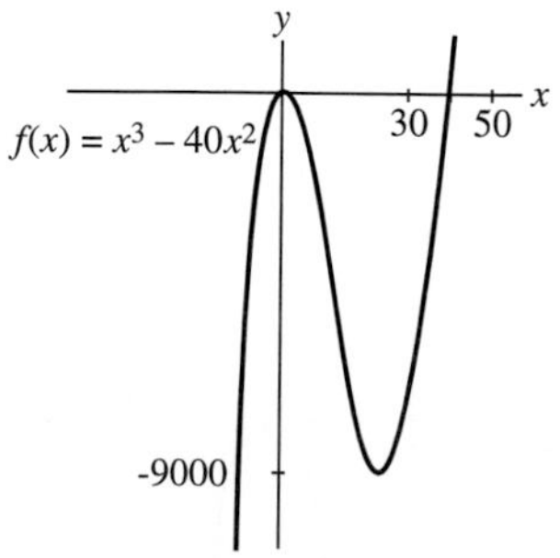

55.

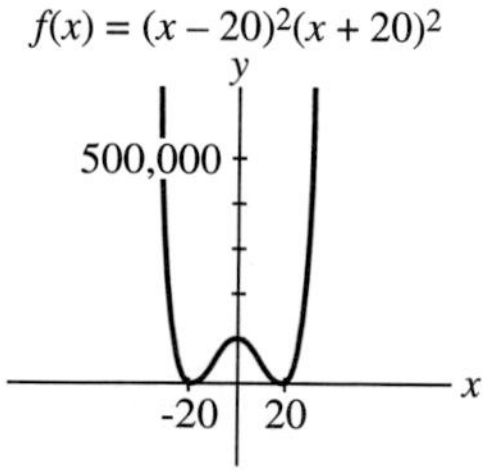

57.

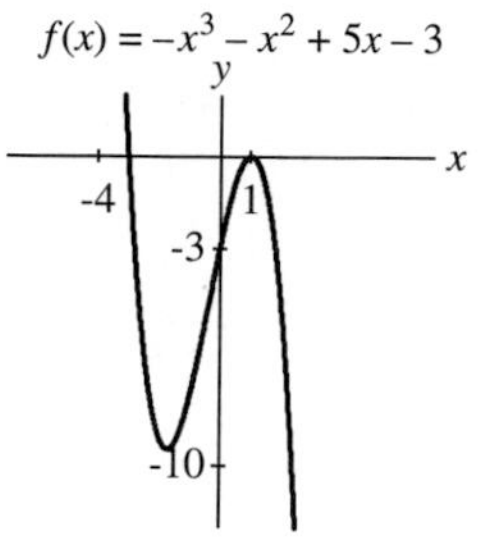

59.

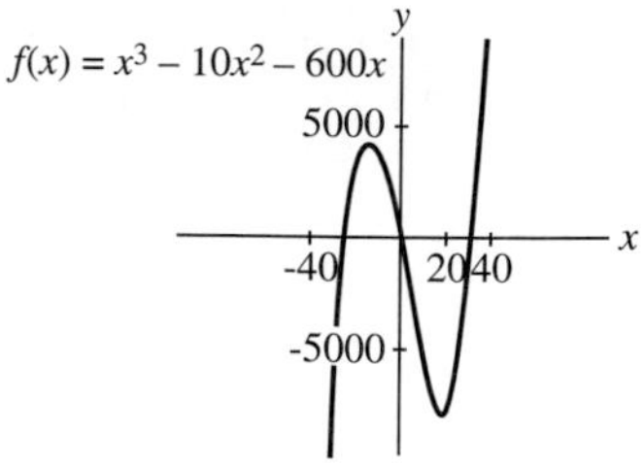

61.

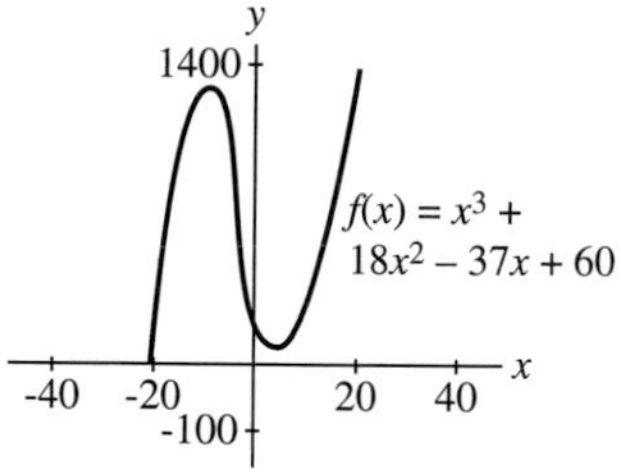

63.

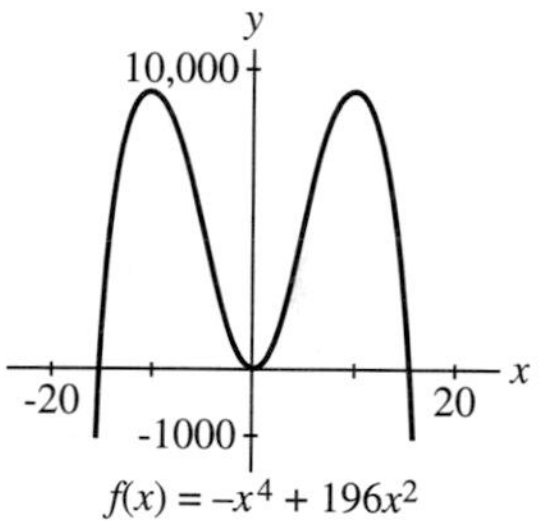

65.

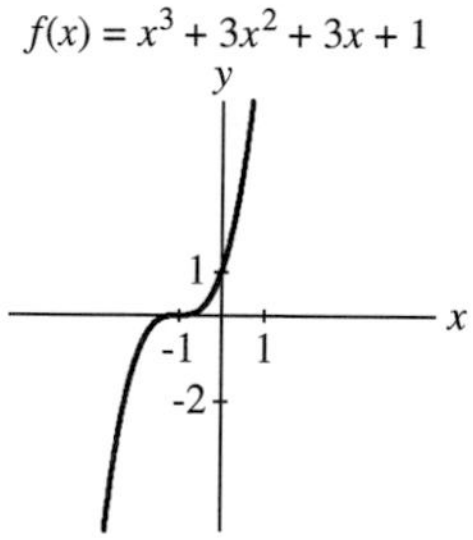

67. $(-\sqrt{3}, 0) \cup (\sqrt{3}, \infty)$ **69.** $(-\infty, -\sqrt{2}] \cup \{0\} \cup [\sqrt{2}, \infty)$
71. $(-4, -1) \cup (1, \infty)$
73. Loc max value 3.11, loc min value 0.37
75. Loc max value 23.74, loc min value -163.74
77. Loc max value 21.01, loc min value 13.99
79. \$3,400, \$2,600, $x >$ \$2,200
81. Decreases to 0 at 30 stores, then increases. Increasing for x in (0, 10) and $(30, \infty)$.
83. $V = 3x^3 - 24x^2 + 48x$, 4/3 in. by 4 in. by 16/3 in.
85. All of the paint would be used to coat the inside of the can, $r = 6.67 \times 10^{-4}$ ft and $h = 11{,}924.69$ ft, $r = 2.82$ ft and $h = 6.67 \times 10^{-4}$ ft

Section 3.7

1. $(-\infty, -2) \cup (-2, \infty)$ **3.** $(-\infty, -2) \cup (-2, 2) \cup (2, \infty)$
5. $(-\infty, 3) \cup (3, \infty)$ **7.** $(-\infty, 0) \cup (0, \infty)$
9. $(-\infty, -1) \cup (-1, 0) \cup (0, 1) \cup (1, \infty)$

11. $(-\infty, -3) \cup (-3, -2) \cup (-2, \infty)$
13. $(-\infty, 2) \cup (2, \infty)$, $y = 0$, $x = 2$
15. $(-\infty, 0) \cup (0, \infty)$, $y = x$, $x = 0$ **17.** $x = 2$, $y = 0$
19. $x = \pm 3$, $y = 0$ **21.** $x = 1$, $y = 2$ **23.** $x = 0$, $y = x - 2$
25. $x = -1$, $y = 3x - 3$ **27.** $x = -2$, $y = -x + 6$

29.

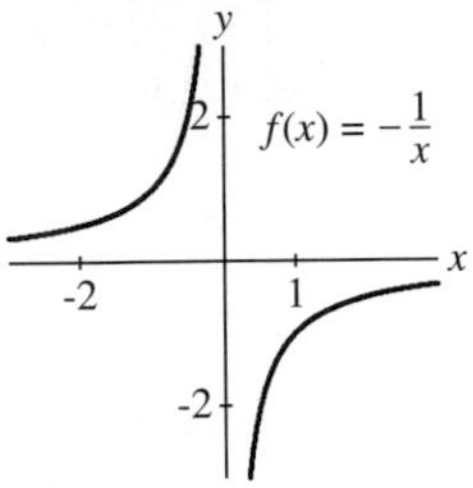

31. (0, −1/2)

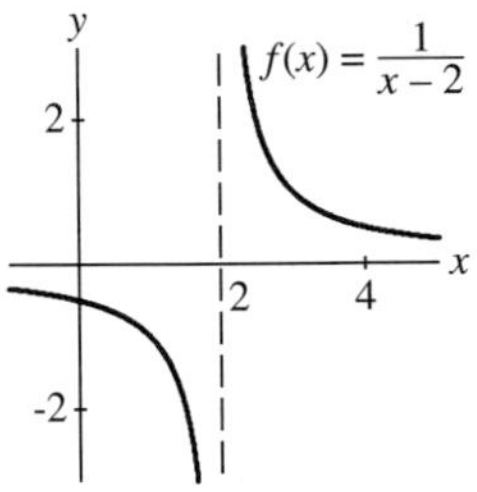

33. (0, −1/4)

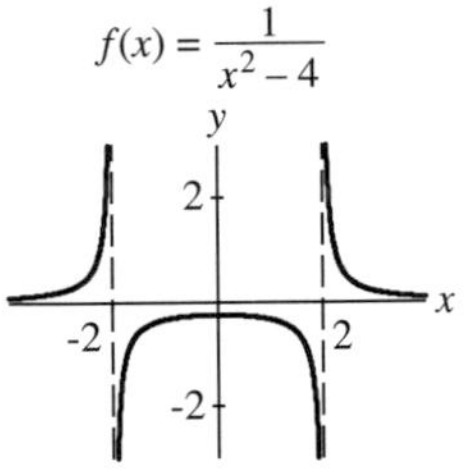

35. (0, −1)

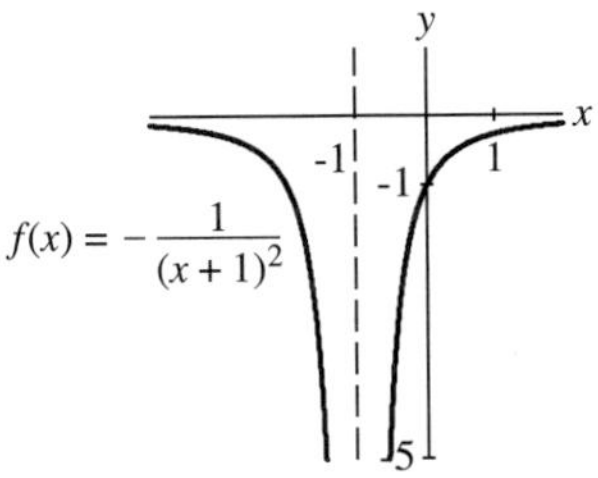

37. (0, −1), (−1/2, 0)

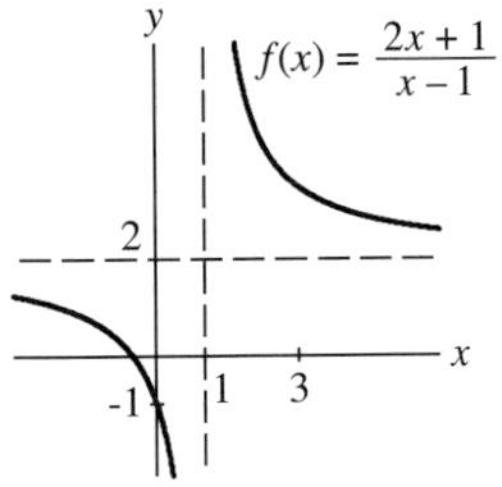

39. (3, 0), (0, −3/2)

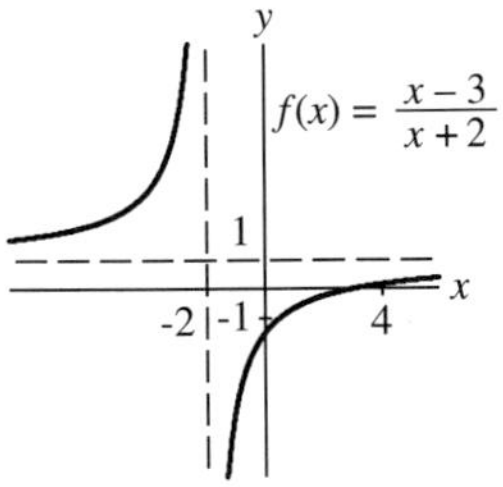

41. (0, 0)

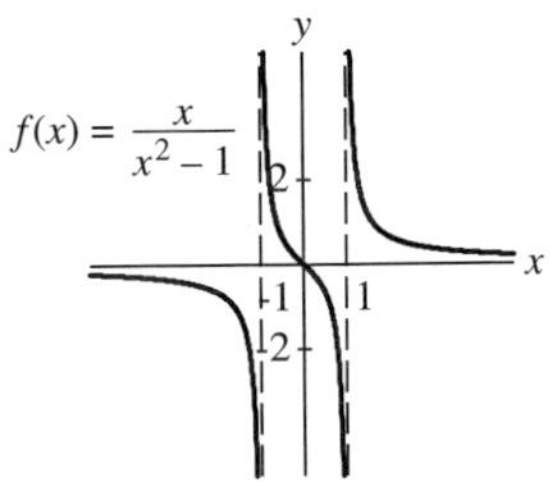

43. (0, 0)

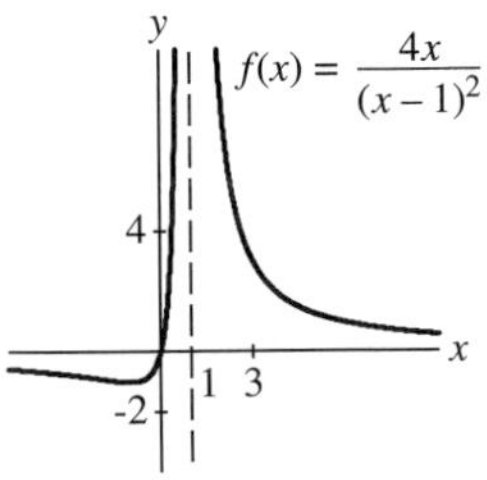

45. $(0, -8/9)$, $(\pm\sqrt{8}, 0)$

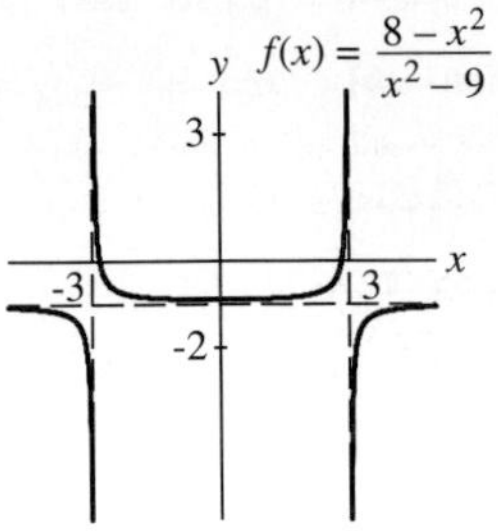

47. $(0, 2)$, $(-2 \pm \sqrt{3}, 0)$

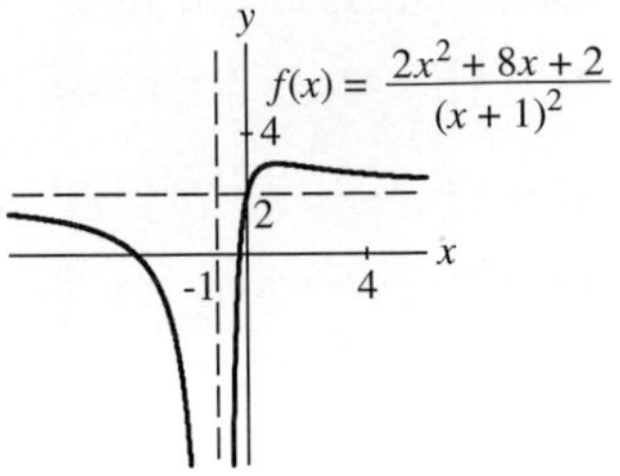

49.

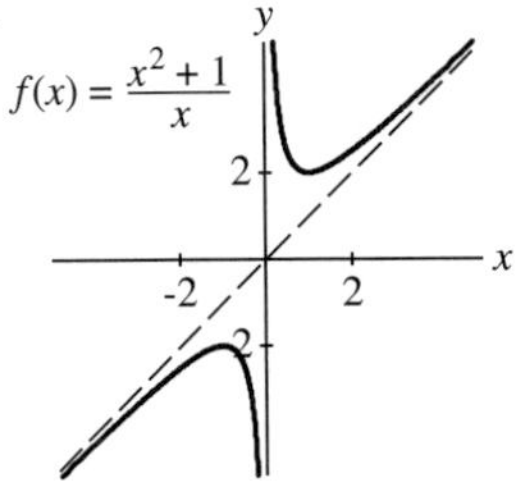

51.

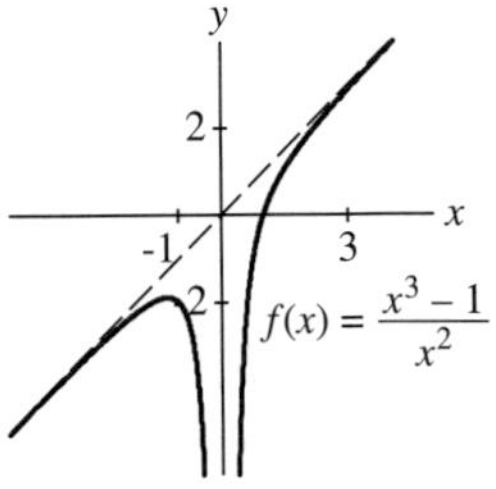

53.

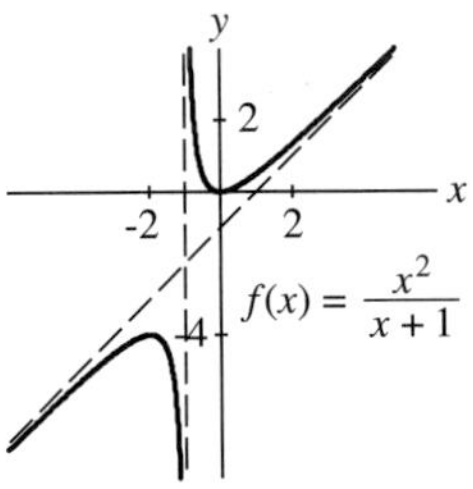

55.

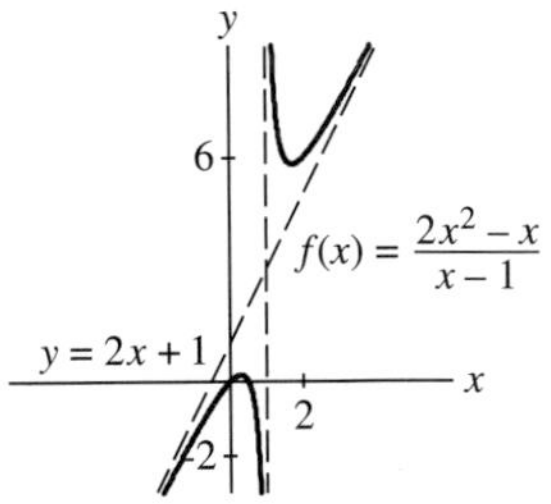

57. (e) **59.** (a) **61.** (b) **63.** (c)

65.

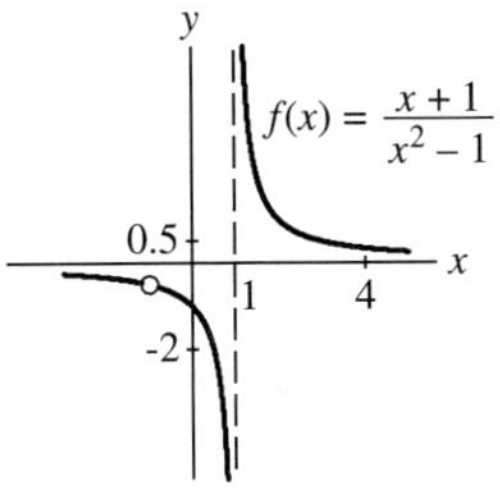

67.

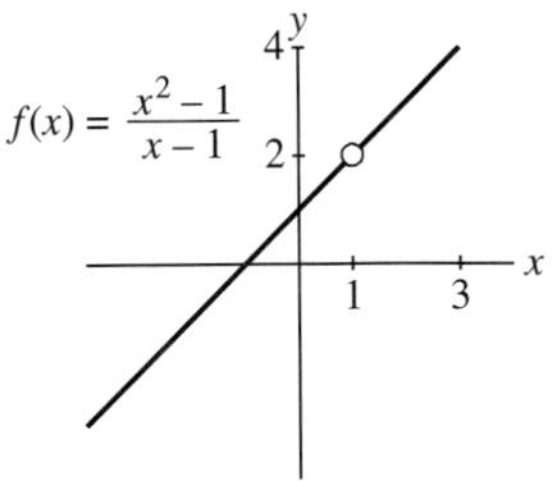

69.

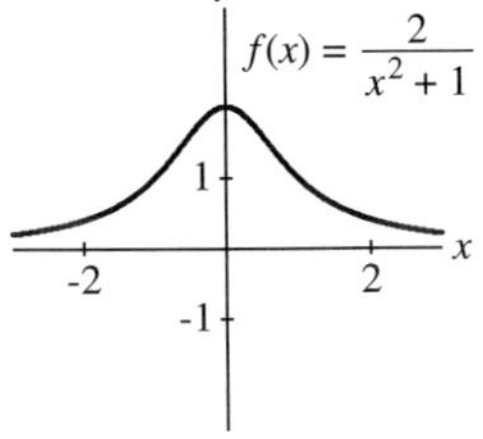

71.

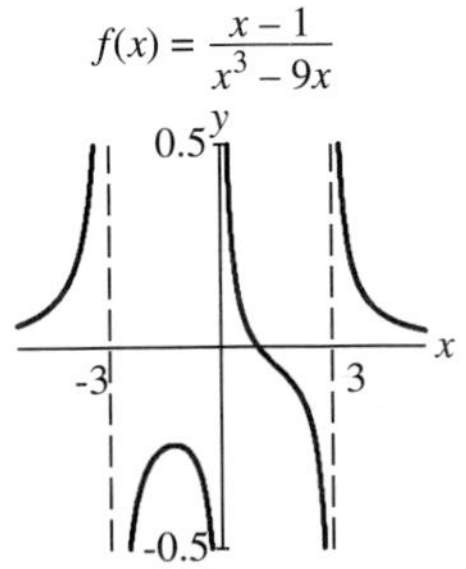

73.

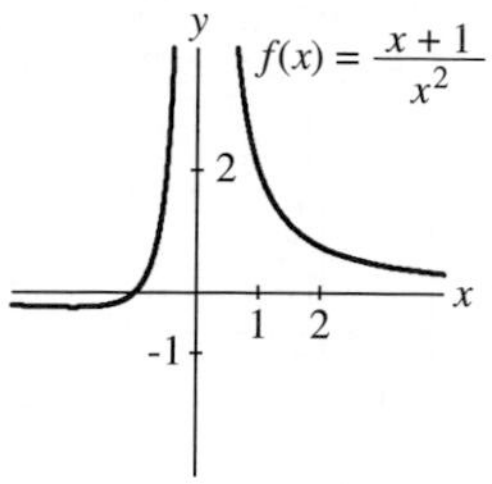

75. 1.5, 0.001 **77.** 0.0, 2.7×10^{-4} **79.** 3.0, 6.7×10^{-8}
81. -0.0, -6.0×10^{-8}
83. Domain $(-\infty, \infty)$, range $(0, 100]$, $y = 0$
85. Domain $(-\infty, \infty)$, range $(0, 1]$, $y = 0$
87. Domain $(-\infty, \infty)$, range $[-5.55, 0.24]$, $y = 0$
89. $(-\infty, -2/9)$ **91.** $(-\sqrt{5}, 5/6) \cup (\sqrt{5}, \infty)$
93. $(-2, 3 - \sqrt{15}] \cup (3, 3 + \sqrt{15}]$
95. $C = \frac{100 + x}{x}$, \$2, $C \to \$1$ **97.** $S = \frac{100}{4 - x}$, $S \to \infty$ as $x \to 4$
99. Approximately 25 min, less than 10 min, coma and permanent brain damage, vertical asymptote is $t = 0$, horizontal is $PPM = 0$, low concentration for a long time or high concentration for a short time can kill you
101. $h = 500/(\pi r^2)$, $S = 2\pi r^2 + 1000/r$, 4.3 ft, \$2789.88

89. (0, 4), (8, 0)

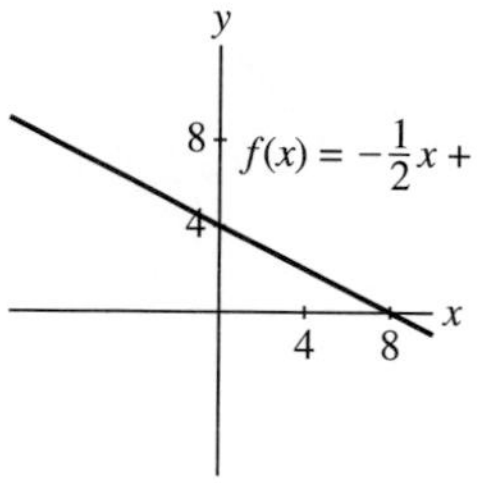

91. $(-1, 0)$, $(2, 0)$, $(0, -2)$

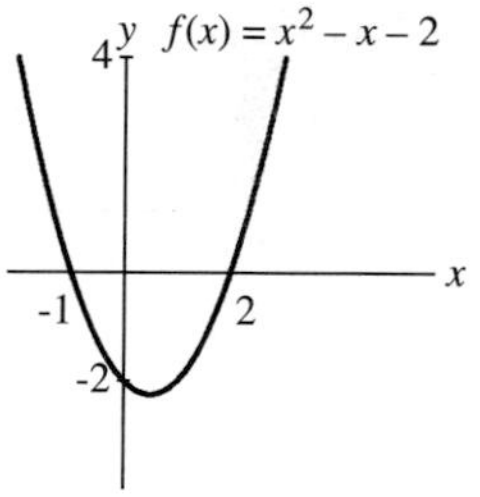

93. $(-1, 0)$, $(2, 0)$, $(0, -2)$

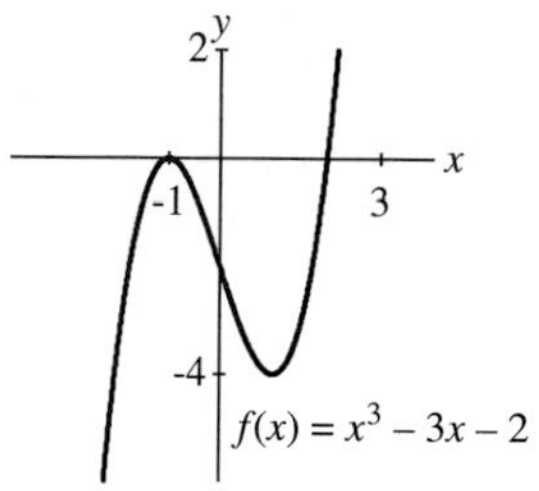

95. $(-2, 0)$, $(1, 0)$, $(2, 0)$, $(0, 2)$

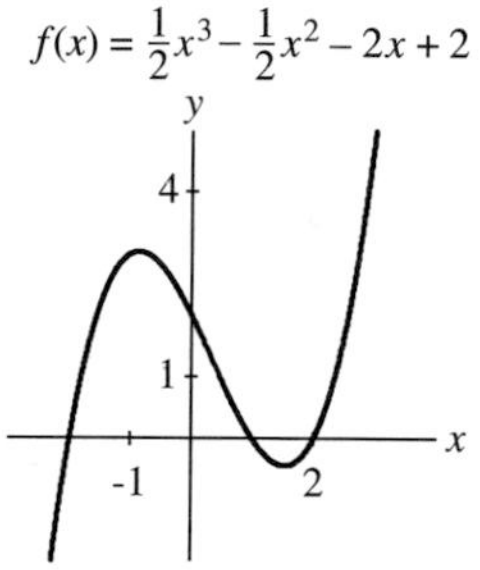

Chapter 3 Review Exercises

1. 3/5 **3.** $y = -\frac{3}{2}x - \frac{1}{2}$ **5.** $f(x) = 3\left(x - \frac{1}{3}\right)^2 + \frac{2}{3}$
7. $(1, -3)$, $x = 1$, $\left(\frac{2 \pm \sqrt{6}}{2}, 0\right)$, $(0, -1)$ **9.** $y = -2x^2 + 4x + 6$
11. $(-3, 1/2)$ **13.** 1/3 **15.** $\pm 2\sqrt{2}$ **17.** $\frac{1}{2}, \frac{-1 \pm i\sqrt{3}}{4}$
19. $\pm\sqrt{10}$, $\pm i\sqrt{10}$ **21.** $-\frac{1}{2}$ and $\frac{1}{2}$ with multiplicity 2
23. 0, $-1 \pm \sqrt{7}$ **25.** 83 **27.** 5 **29.** $\pm\left(1, 2, \frac{1}{3}, \frac{2}{3}\right)$
31. $\pm\left(1, 3, \frac{1}{2}, \frac{1}{3}, \frac{1}{6}, \frac{3}{2}\right)$ **33.** $2x^2 - 5x - 3 = 0$
35. $x^2 - 6x + 13 = 0$ **37.** $x^3 - 4x^2 + 9x - 10 = 0$
39. $x^2 - 4x + 1 = 0$ **41.** 0 with multiplicity 2, 6 imag
43. 1 pos, 2 imag; 3 pos **45.** 3 neg; 1 neg, 2 imag
47. $-4 < x < 3$ **49.** $-1 < x < 8$ **51.** $-1 < x < 1$
53. 1, 2, 3 **55.** $\frac{1}{2}, \frac{1}{3}, \pm i$ **57.** 3, $3 \pm i$ **59.** 2, $1 \pm i\sqrt{2}$
61. 0, $\frac{1}{2}$, $1 \pm \sqrt{3}$ **63.** 1/5 **65.** $\pm\sqrt{2}$ **67.** 30 **69.** 16
71. ± 2 **73.** -7, 9 **75.** $\varnothing$ **77.** 11/4
79. Symmetric about $x = 3/4$ **81.** Symmetric about y-axis
83. Symmetric about origin **85.** $(-\infty, -2.5) \cup (-2.5, \infty)$
87. $(-\infty, \infty)$

97. $(-2, 0)$, $(2, 0)$, $(0, 4)$

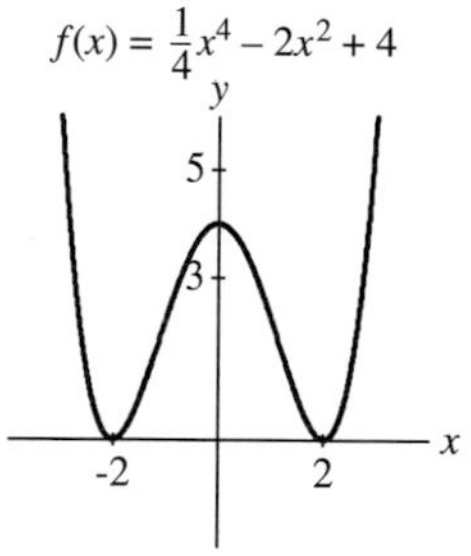

99. (0, 2/3), $x = -3$, $y = 0$

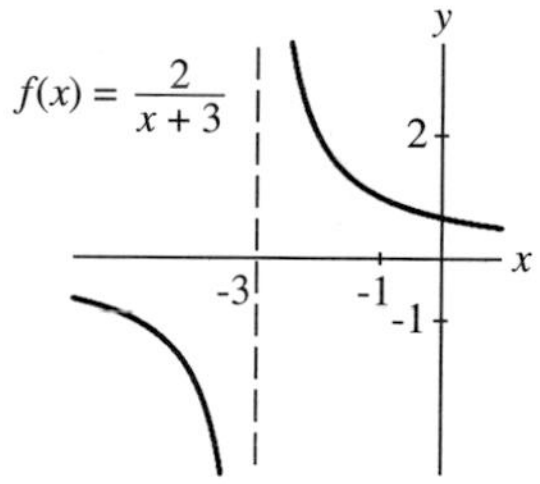

101. (0, 0), $x = \pm 2$, $y = 0$

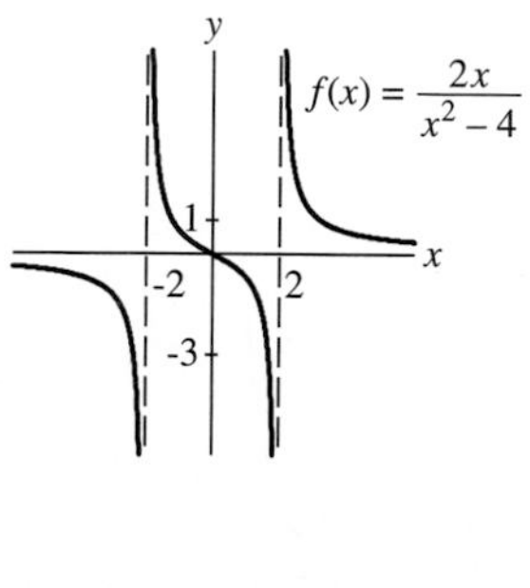

103. (1, 0), $(0, -1/2)$, $x = 2$, $y = x$

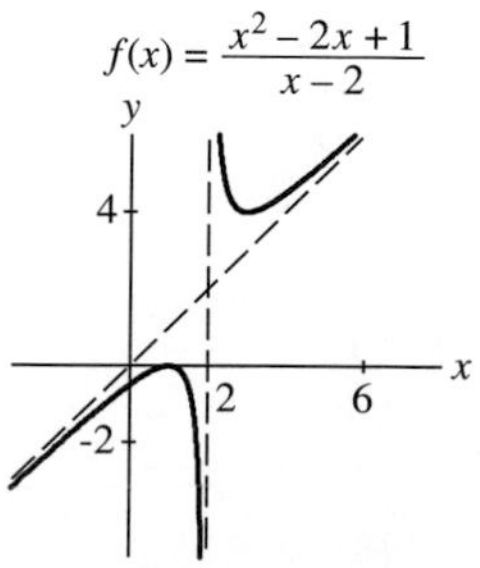

105. (1/2, 0), (0, −1/2), $x = 2$, $y = -2$

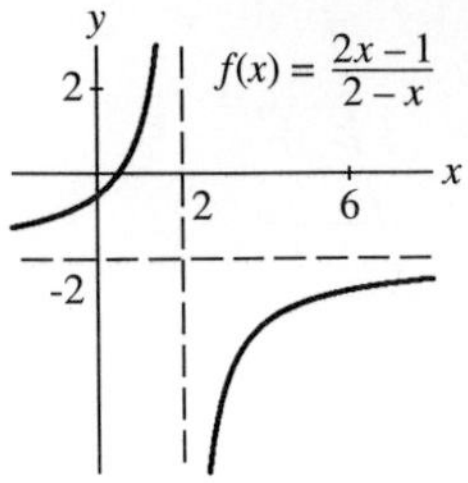

107. (−2, 0), (0, 2)

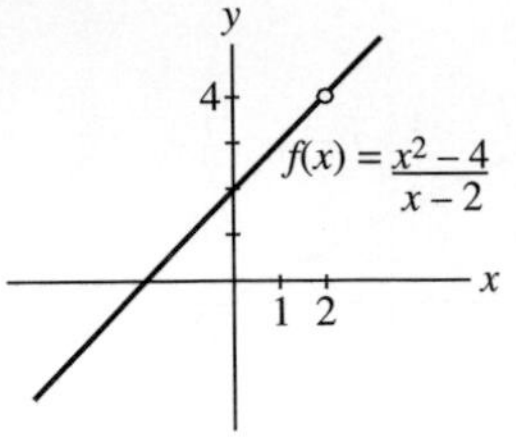

109. [−1/2, 1/2] ∪ [100, ∞) **111.** (−∞, −2) ∪ (0, ∞)
113. (−∞, 0) ∪ (0, 3) ∪ (4, ∞) **115.** (−∞, 1] ∪ [2, 3) ∪ (4, ∞)
117. ∅ **119.** (−∞, 1) ∪ (1, ∞) **121.** $x^2 - 3x$, -15
123. $y = 0.34x + 12.2$ where $x = 0$ corresponds to 1985, 13.22 million
125. $y = 0.8x + 7$ where x is age, 59% **127.** 380.25 ft **129.** No
131. 24 ft wide and 5 ft high **133.** 172.4 ft/sec, $V = 200$, 200 ft/sec

Chapter 3 Test

1. −2 **2.** $\frac{3}{4}$ **3.** $y = \frac{1}{3}x - \frac{14}{3}$ **4.** $y = 3(x-2)^2 - 11$
5. (2, −11), $x = 2$, (0, 1), $\left(\frac{6 \pm \sqrt{33}}{3}, 0\right)$, [−11, ∞) **6.** −11
7. $2x^2 - 6x + 14$, -37 **8.** −14 **9.** $\pm\left(1, 2, 3, 6, \frac{1}{3}, \frac{2}{3}\right)$
10. $x^3 + 3x^2 + 16x + 48 = 0$ **11.** 2 pos, 1 neg; 1 neg, 2 imag
12. 256 ft
13. $P = 1600n + 88{,}000$ where n is the number of years since 1994, \$102,400 **14.** ±3 **15.** ±2, ±2i **16.** $1 \pm \sqrt{6}$, 2
17. ±i each with multiplicity 2
18. 2, 0 with multiplicity 2, $-\frac{3}{2}$ with multiplicity 3
19. $2 \pm i, \frac{1}{2}$

20.

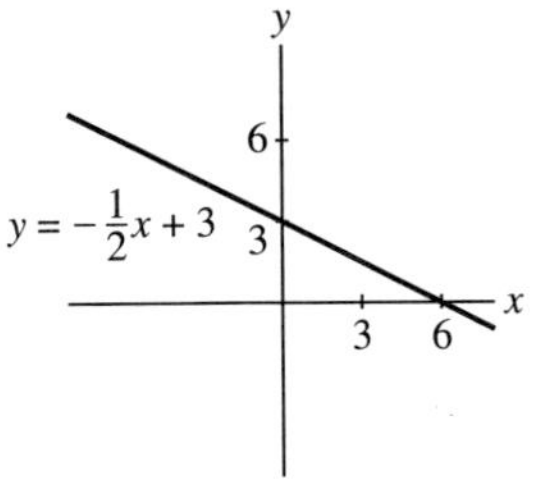

21.

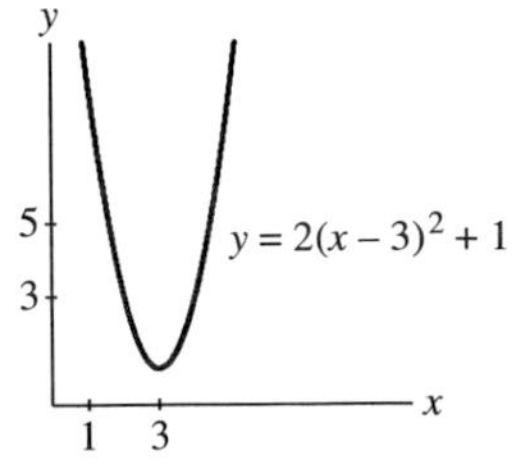

22.

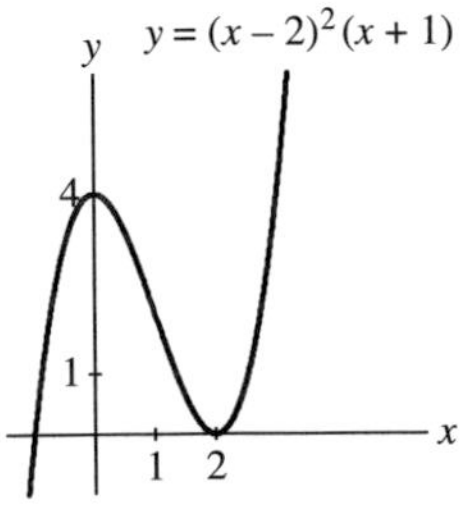

23.

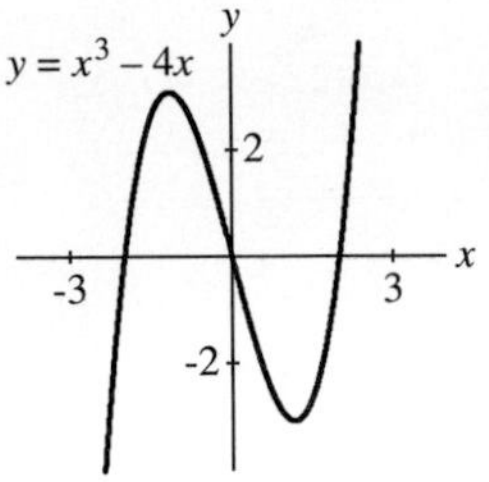

24. $x = 2$, $y = 0$

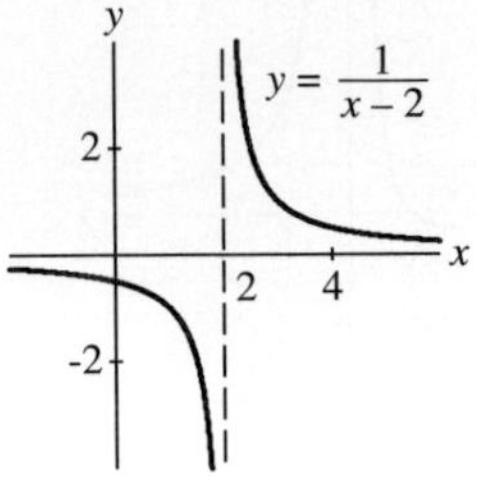

25. $x = 2$, $y = 2$

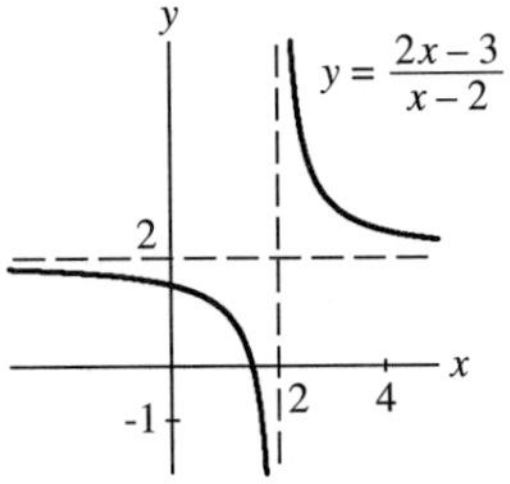

26. $x = 0$, $y = x$

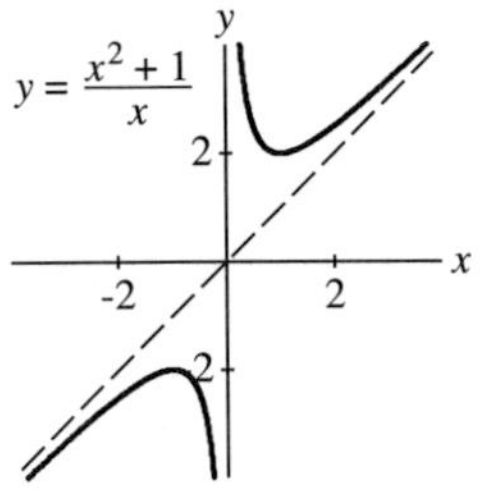

27. $x = \pm 2$, $y = 0$

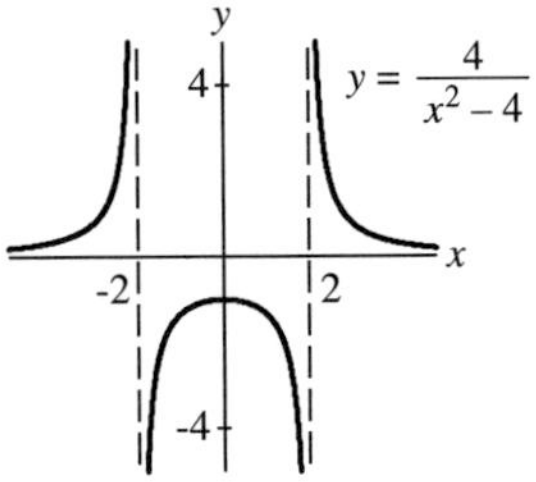

28.

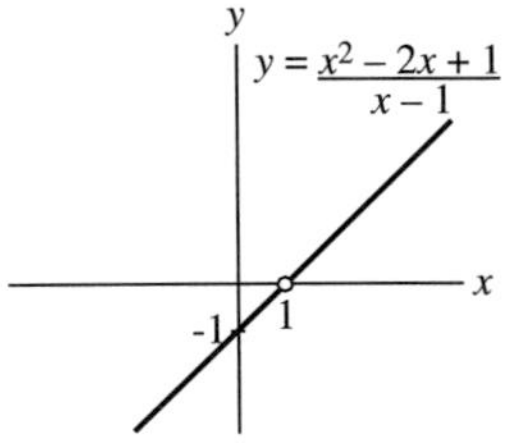

29. (−2, 4) **30.** (−∞, 1/2) ∪ (3, ∞) **31.** (−∞, −3] ∪ (−1, 4)
32. $3 \pm 3\sqrt{3}$ **33.** 16

Tying It All Together Chapters P–3

1. 2 **2.** $-\frac{1}{2}, \frac{1}{3}$ **3.** $-\frac{1}{2}$ **4.** 2 **5.** $1, -\frac{13}{6}$ **6.** $-\frac{1}{2}, \frac{1}{3}, \pm i$
7. 2 **8.** $\pm i\sqrt{3}$ **9.** 0, 4 **10.** 1, −2 **11.** −14, 13 **12.** 0, 2
13. (−∞, ∞), (−∞, ∞) **14.** (−∞, ∞), [0, ∞)

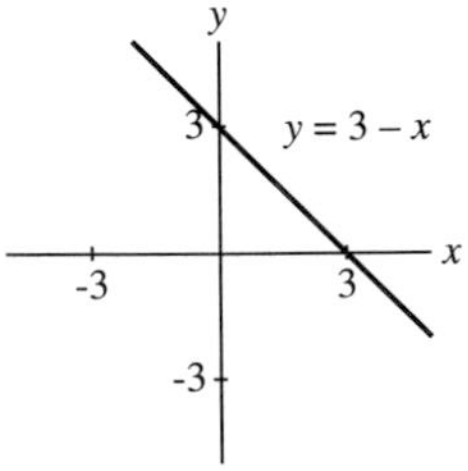

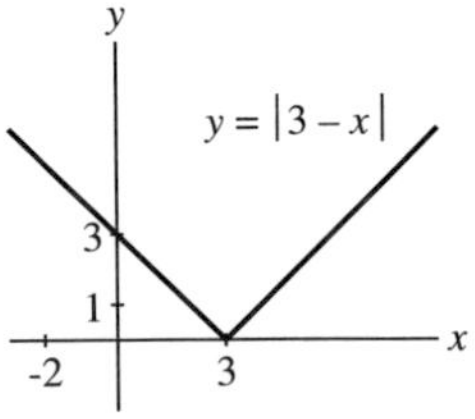

15. $(-\infty, \infty), (-\infty, 3]$

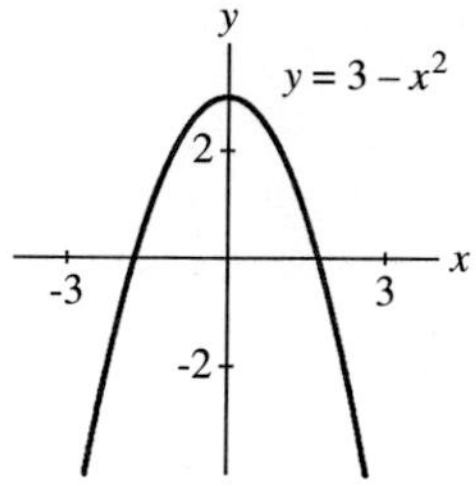

16. $(-\infty, 3) \cup (3, \infty), \{\pm 1\}$

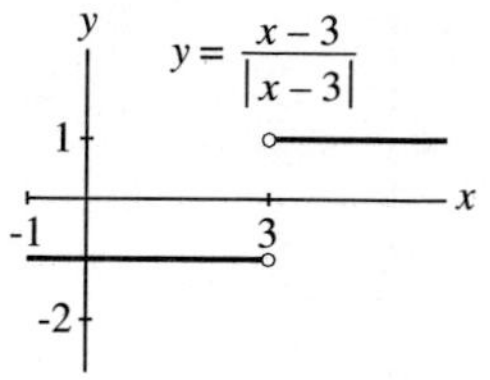

17. $[0, \infty), (-\infty, 3]$

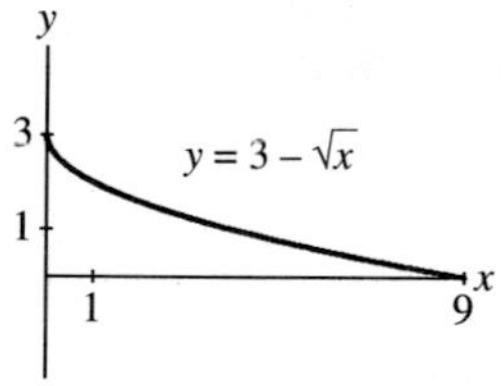

18. $(-\infty, 3) \cup (3, \infty)$, $(-\infty, 0) \cup (0, \infty)$

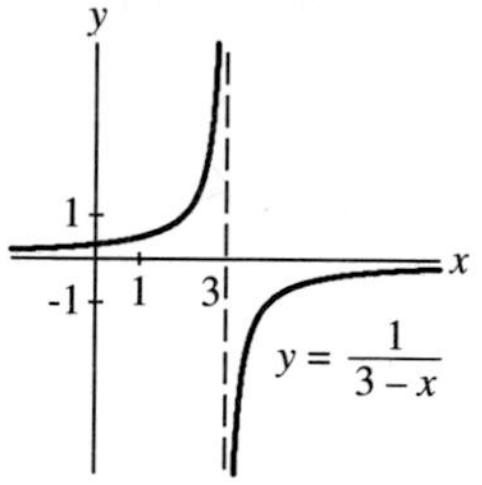

19. $(-\infty, \infty), [0, \infty)$

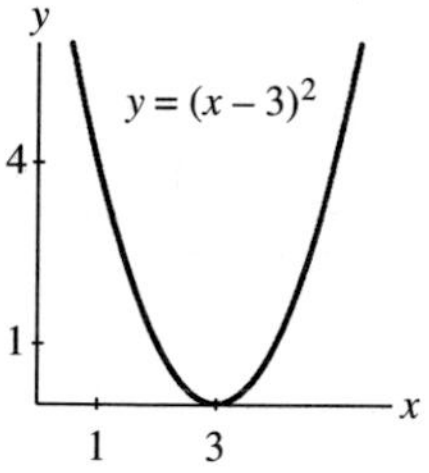

20. $(-\infty, 3) \cup (3, \infty), (0, \infty)$

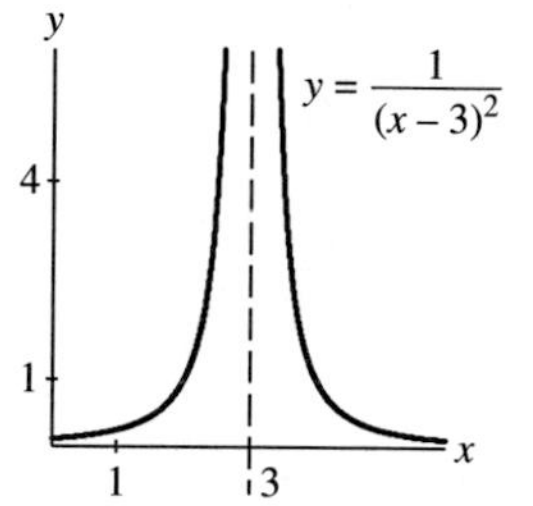

21. $(-\infty, 0) \cup (0, \infty), (-\infty, \infty)$

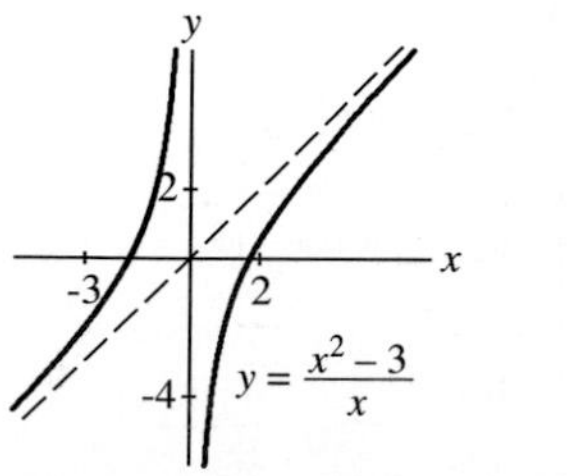

22. $[-\sqrt{3}, \sqrt{3}], [0, \sqrt{3}]$

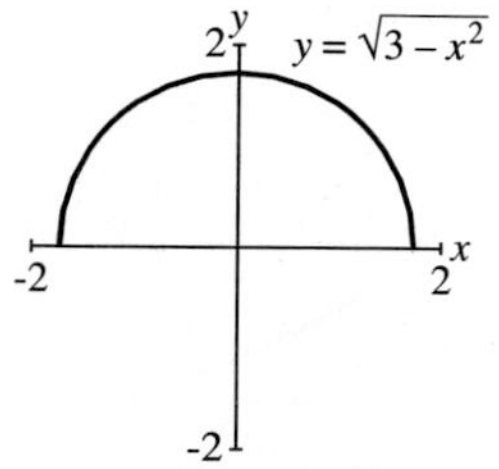

23. $(-\infty, 3], [0, \infty)$

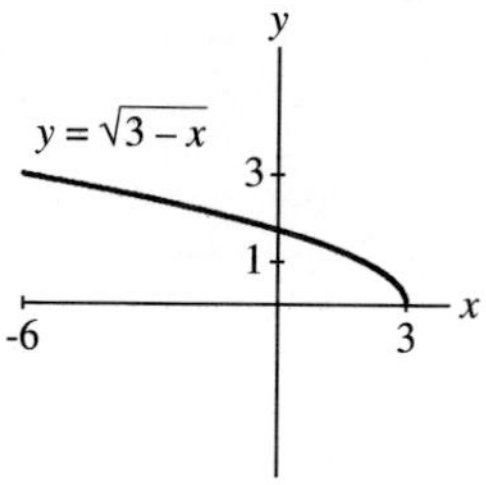

24. $(-\infty, \infty), (-\infty, \infty)$

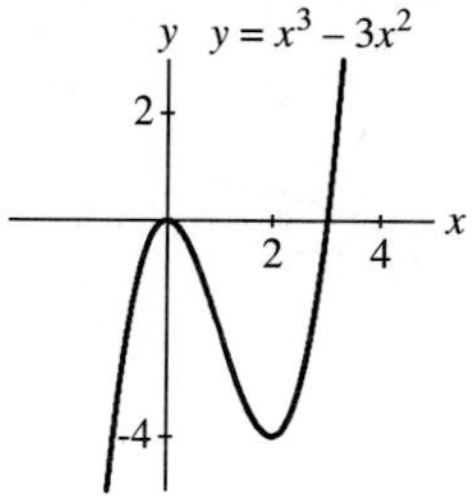

25. 1000 **26.** 1 **27.** $\frac{1}{2}$ **28.** 0.001 **29.** $\frac{1}{9}$ **30.** 3

31. $\frac{1}{9}$ **32.** 8

CHAPTER 4

Section 4.1

1. 9 **3.** 1/9 **5.** 1/2 **7.** 8 **9.** 4 **11.** 2

13. Domain $(-\infty, \infty)$, range $(0, \infty)$, inc

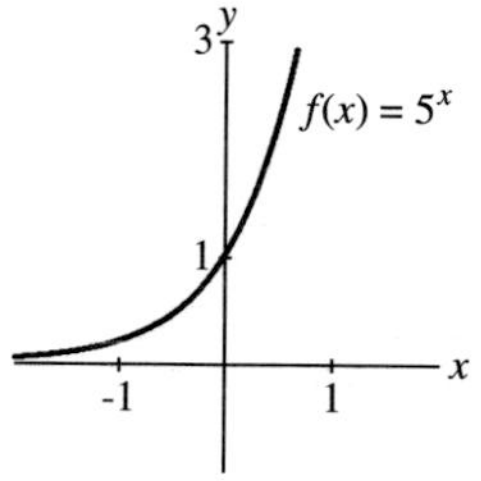

15. Domain $(-\infty, \infty)$, range $(0, \infty)$, dec

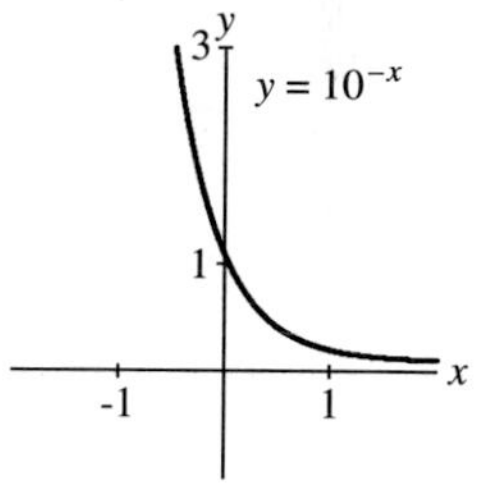

17. Domain $(-\infty, \infty)$, range $(0, \infty)$, dec

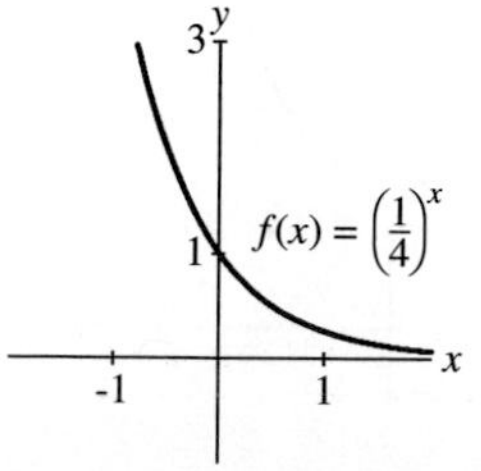

19. Domain $(-\infty, \infty)$, range $(-3, \infty)$, inc.

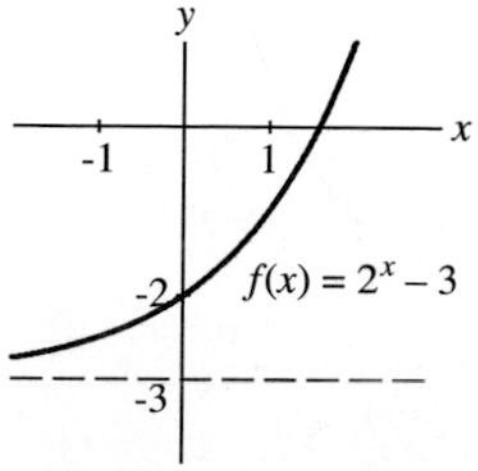

21. Domain $(-\infty, \infty)$, range $(-\infty, 0)$, inc

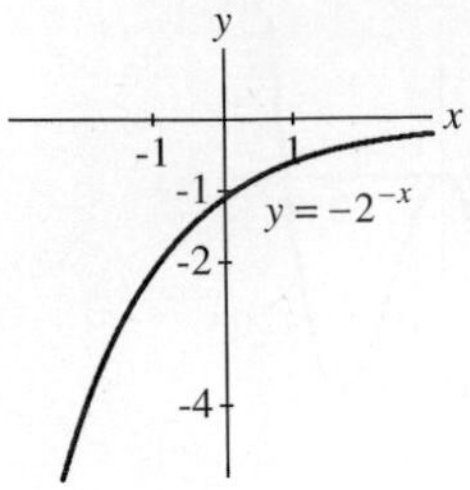

23. Domain $(-\infty, \infty)$, range $(-\infty, 1)$, dec

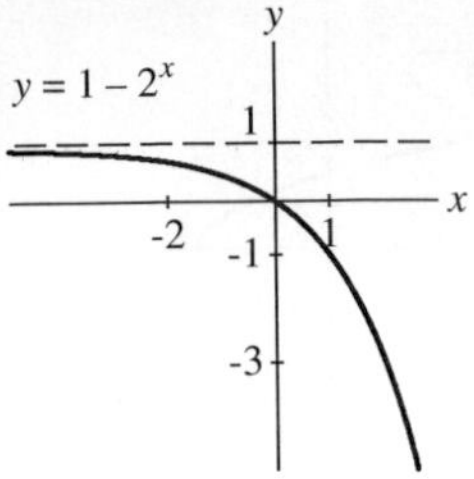

25. Domain $(-\infty, \infty)$, range $(0, \infty)$, inc

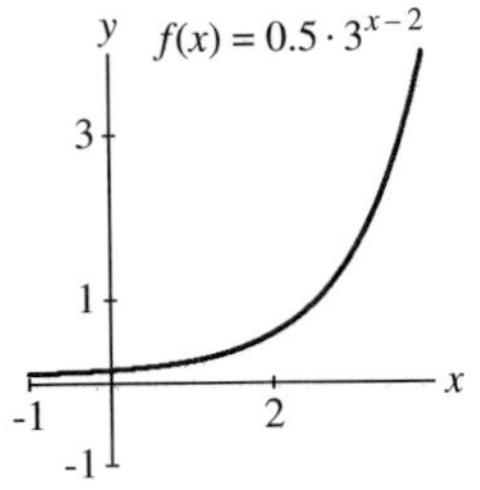

27. Domain $(-\infty, \infty)$, range $(0, \infty)$, dec

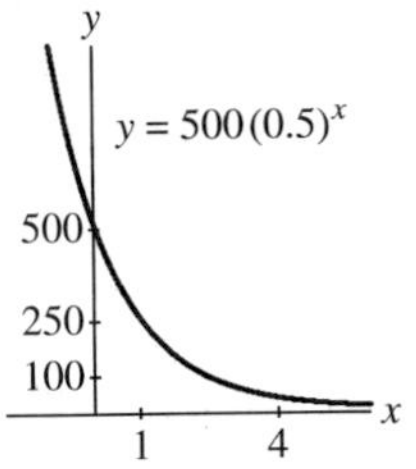

29. Domain $(-\infty, \infty)$, range $(-5, \infty)$, inc

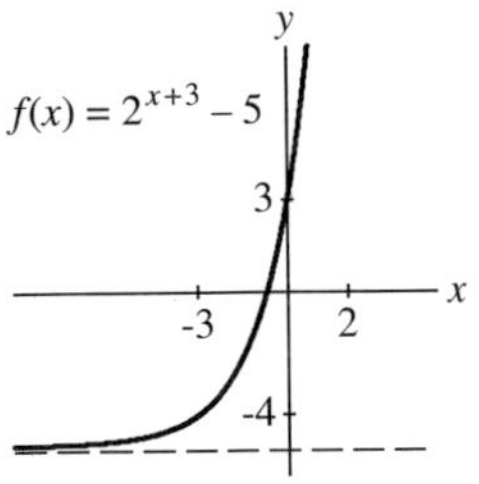

31. $y = 2^{x-5} - 2$ **33.** $y = -(1/4)^{x-1} - 2$ **35.** $\{6\}$ **37.** $\{-1\}$ **39.** $\{3\}$ **41.** $\{-2\}$ **43.** $\{1/3\}$ **45.** $\{-2\}$ **47.** $\{-3\}$ **49.** $\{-1\}$ **51.** 2 **53.** -1 **55.** 0 **57.** -3 **59.** 3 **61.** -1 **63.** 1 **65.** -1 **67.** 9, 1, 1/3, -2 **69.** 1, -2, 5, 1 **71.** -16, -2, $-1/2$, 5 **73.** $\{1.46\}$ **75.** $\{-2.71\}$ **77.** $\{1.20\}$ **79.** $\{11.55\}$

81. $(-\infty, \infty)$, $[1, \infty)$

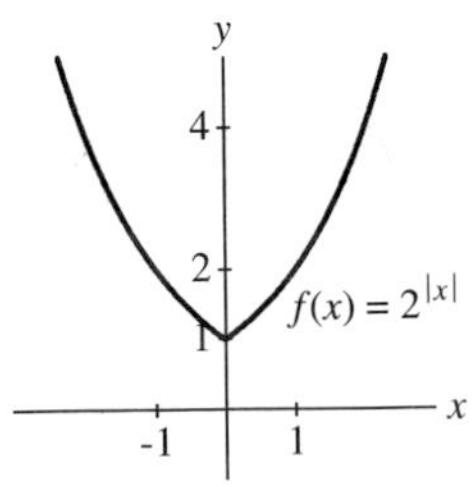

83. $(-\infty, \infty)$, $[2, \infty)$

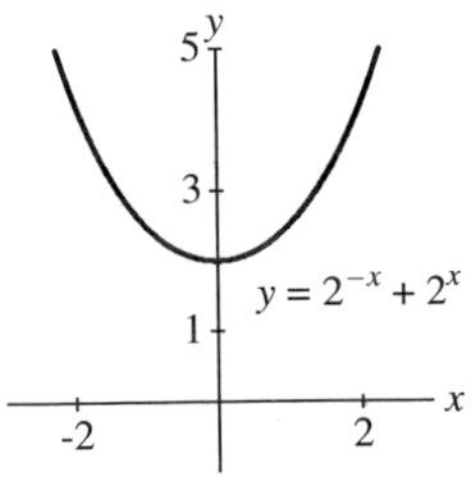

85. $(-\infty, \infty)$, $(0, 1]$

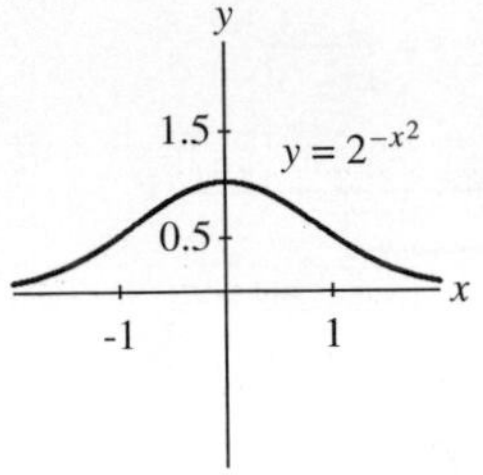

87. \$7237.97, \$2737.97 **89.** \$3251.93 **91.** \$129,196.51, \$129,194.24 **93.** 2,400,000, 3,338,324 **95.** \$9374.46, \$144,003.51 **97.** Decreasing, 89.7%, 71.1%, 67.9%, 59.8% **99.** $P = 10\left(\frac{1}{2}\right)^n$ **101.** 6.8×10^{-5} **103.** 6 **105.** 2

Section 4.2

1. 5, 1 **3.** 1/5, -1 **5.** 3 **7.** 6 **9.** -4 **11.** 1/4 **13.** -3 **15.** -1 **17.** 0 **19.** 1 **21.** -5

23. Domain $(0, \infty)$, range $(-\infty, \infty)$

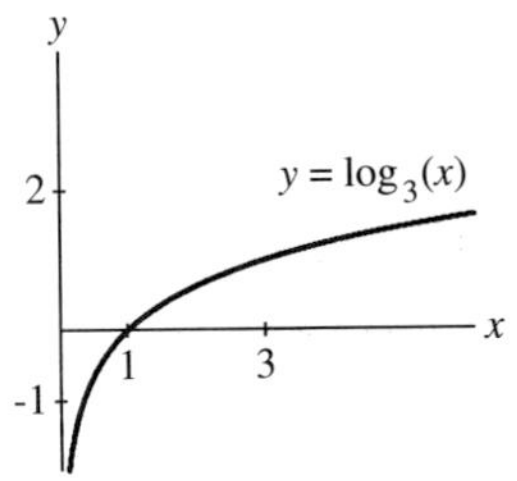

25. Domain $(0, \infty)$, range $(-\infty, \infty)$

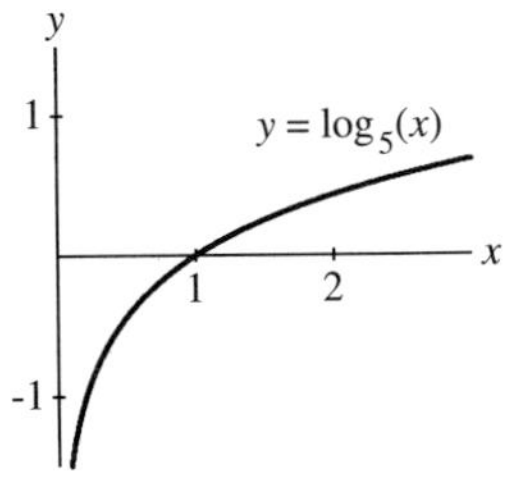

27. Domain $(0, \infty)$, range $(-\infty, \infty)$

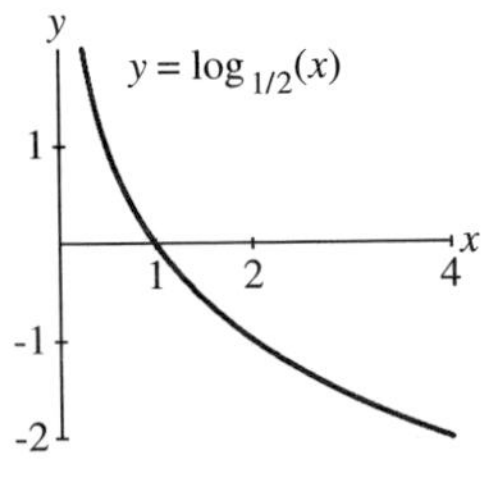

29. Domain $(0, \infty)$, range $(-\infty, \infty)$

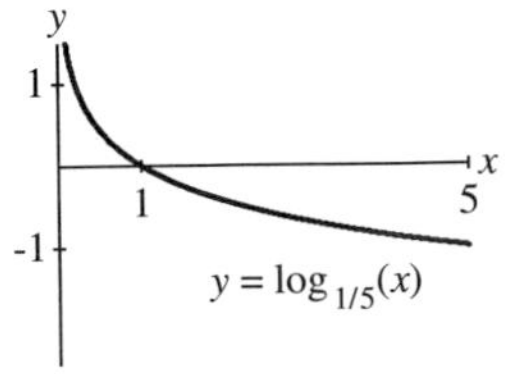

31. Domain $(1, \infty)$, range $(-\infty, \infty)$

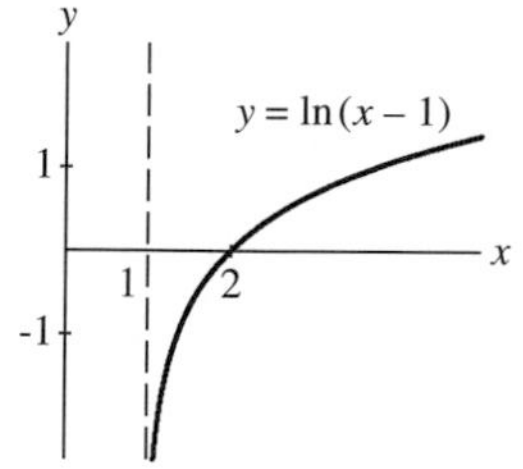

33. Domain $(-2, \infty)$, range $(-\infty, \infty)$

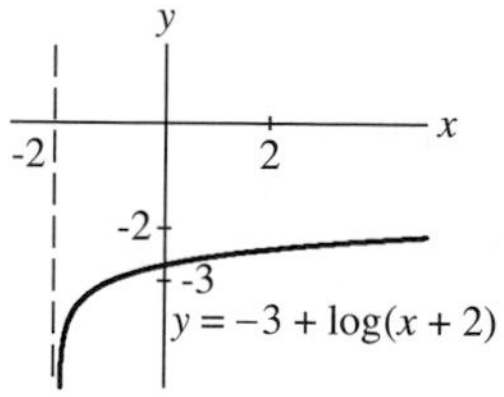

35. Domain $(1, \infty)$, range $(-\infty, \infty)$

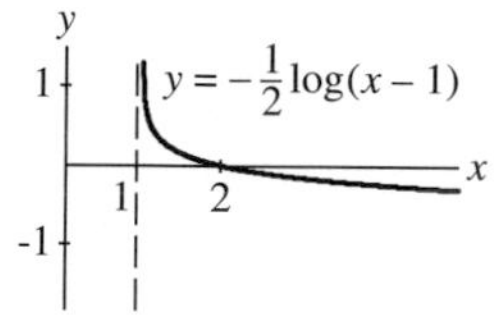

37. $y = \ln(x - 3) - 4$ **39.** $y = -\log_2(x - 5) - 1$ **41.** $10^y = 30$
43. $\log_5(7) = y$ **45.** $2^0 = 1$ **47.** $\ln(2) = 0.09t$ **49.** $b^3 = N/M$
51. $\log_b(y) = 3x$ **53.** 256 **55.** $\sqrt{3}$ **57.** 4 **59.** $\log_3(77)$
61. 6 **63.** $3\sqrt{2}$ **65.** $-1 + \log_3(7)$ **67.** 2 **69.** 0.001
71. 27 **73.** 1/4 **75.** 1.3979 **77.** 1.7712 **79.** 3.8451
81. -12.3315 **83.** 4.7381 **85.** 26.53 **87.** -0.56
89. $f^{-1}(x) = \log_2(x)$ **91.** $f^{-1}(x) = 7^x$ **93.** $f^{-1}(x) = e^x + 1$
95. $f^{-1}(x) = \log_3(x) - 2$
97. $(-\infty, 0) \cup (0, \infty), (-\infty, \infty)$ **99.** $(0, \infty), (-\infty, \infty)$

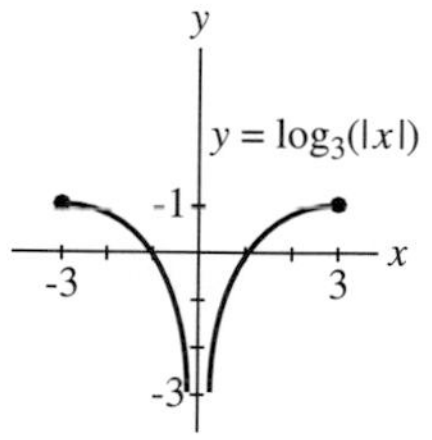

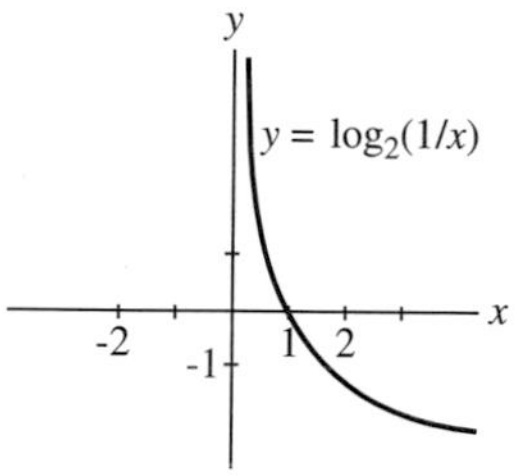

101. 49 yr 125 days **103.** 23.1% **105.** 6.9 yr **107.** 3.5%
109. 9.8 yr **111.** 1.87%, 6.2 billion
113. 2.5 billion gigabits, 2005
115. $p = -45 \log(I) + 190$, log scale is more compact, 0%
117. 4.1 **119.** 3.7 **121.** $c = \ln(b)$, 3.54%

Section 4.3

1. $\log(15)$ **3.** $\log_2(x^2 - x)$ **5.** $\log_4(6)$ **7.** $\ln(x^5)$ **9.** $2 \cdot \log(5)$

11. $-3 \cdot \log(5)$ **13.** $\frac{1}{3}\log(5)$ **15.** $\sqrt{y}$ **17.** $y + 1$ **19.** 999

21. $\log(2) + \log(5)$ **23.** $\log(5) - \log(2)$ **25.** $\log(2) + \frac{1}{2}\log(5)$

27. $2 \cdot \log(2) - 2 \cdot \log(5)$ **29.** $\log_3(5) + \log_3(x)$

31. $\log_2(5) - \log_2(2) - \log_2(y)$ **33.** $\log(3) + \frac{1}{2}\log(x)$

35. $\log(3) + (x - 1)\log(2)$ **37.** $\frac{1}{3}\ln(x) + \frac{1}{3}\ln(y) - \frac{4}{3}\ln(t)$

39. $\ln(6) + \frac{1}{2}\ln(x - 1) - \ln(5) - 3 \cdot \ln(x)$ **41.** $\log_2(5x^3)$

43. $\log_7(x^{-3})$ **45.** $\log\left(\frac{2xy}{z}\right)$ **47.** $\log\left(\frac{z\sqrt{x}}{y \cdot \sqrt[3]{w}}\right)$ **49.** $\log_4(x^{20})$

51. 3.1699 **53.** -3.5864 **55.** 11.8957 **57.** 2.2025
59. 1.5850 **61.** 0.3772 **63.** 13.8695 **65.** 34.3240 **67.** 0.3200
69. 0.0479 **71.** 2.0172 **73.** 1.5928 **75.** T **77.** F **79.** T
81. F **83.** F **85.** F **87.** T **89.** 11 yr 166 days
91. 44 quarters **93.** 5.7% **95.** 3.58% **97.** $\log(I/I_0)$, 3

99. $t = \frac{1}{r}\ln(P) - \frac{1}{r}\ln(P_0)$ **101.** Decreasing, $n \le 4{,}892{,}961$

103. $MR(x) = \log\left[\left(\frac{x + 2}{x + 1}\right)^{500}\right]$, $MR(x) \to 0$

105. $y = 0.201(1.877)^x$ where x is the number of years since 1990, $y = 0.201\, e^{0.630x}$, 63%, 384 million, 2003, never

Section 4.4

1. 80 **3.** 3 **5.** 6 **7.** 20 **9.** 2 **11.** $\frac{\sqrt{5}}{2}$ **13.** $\frac{1}{2}$

15. $\frac{2}{1 + \ln(3)}$ **17.** 3.8074 **19.** 3.5502 **21.** -3.0959
23. No solution **25.** 1.5850 **27.** 0.7677 **29.** -0.2
31. 2.5850 **33.** 1/3 **35.** 1, 100 **37.** 29.4872 **39.** 2
41. $-5/4$ **43.** 0.194, 2.70 **45.** -49.73 **47.** -0.767, 2, 4
49. 19,035 yr ago **51.** 1507 yr **53.** 24,850 yr **55.** 30.5%
57. 34 A.D. **59.** 1 hr 11 min, forever **61.** 5:20 A.M. **63.** 3.4 yr
65. 19,328 yr 307 days **67.** 12,300, 13.9 yr **69.** 1.32 parsecs
71. $P = 10^{-x/6+4.5}$, \$1000, 2003 **73.** 10^{-3} watts/m² **75.** \$6791.91
77. 25°C, 63,491 Ohms **79.** 0.33698, $\ln(1.4) \approx 0.33647$

Chapter 4 Review Exercises

1. 64 **3.** 6 **5.** 0 **7.** 17 **9.** 32 **11.** 3 **13.** 9 **15.** 3
17. -3 **19.** 3 **21.** $\log(x^2 - 3x)$ **23.** $\ln(3x^2y)$

25. $\log(3) + 4 \cdot \log(x)$ **27.** $\log_3(5) + \frac{1}{2}\log_3(x) - 4 \cdot \log_3(y)$

29. $\ln(2) + \ln(5)$ **31.** $\ln(2) + 2 \cdot \ln(5)$ **33.** 10^{10}
35. 3 **37.** -5 **39.** -4 **41.** $2 + \ln(9)$ **43.** -2
45. $100\sqrt{5}$ **47.** 8 **49.** 6 **51.** 3 **53.** 3, -3, $\sqrt{3}$, 0
55. (c) **57.** (b) **59.** (d) **61.** (e)

63. Domain $(-\infty, \infty)$, range $(0, \infty)$, inc, $y = 0$

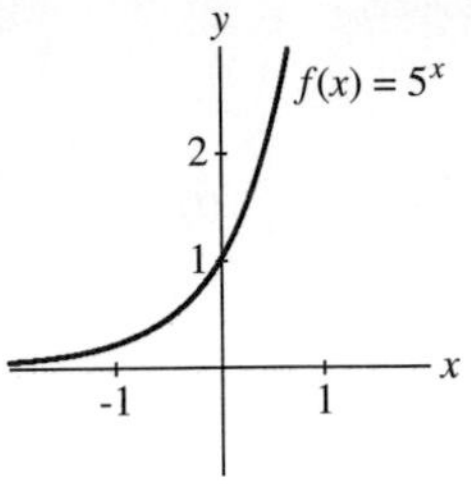

65. Domain $(-\infty, \infty)$, range $(0, \infty)$, dec, $y = 0$

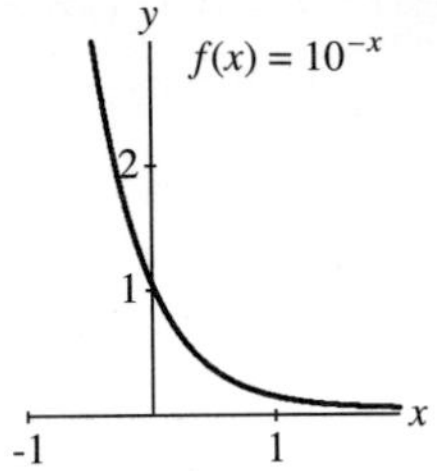

67. Domain $(0, \infty)$, range $(-\infty, \infty)$, inc, $x = 0$

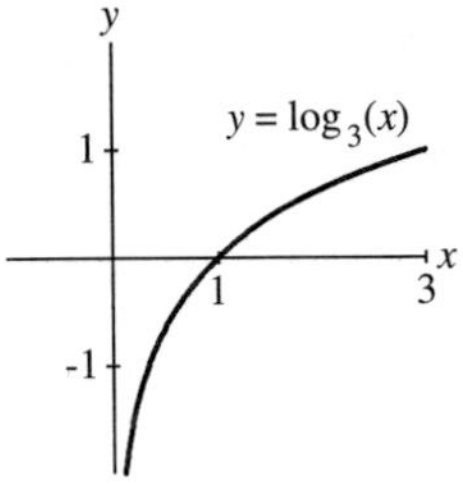

69. Domain $(-3, \infty)$, range $(-\infty, \infty)$, inc, $x = -3$

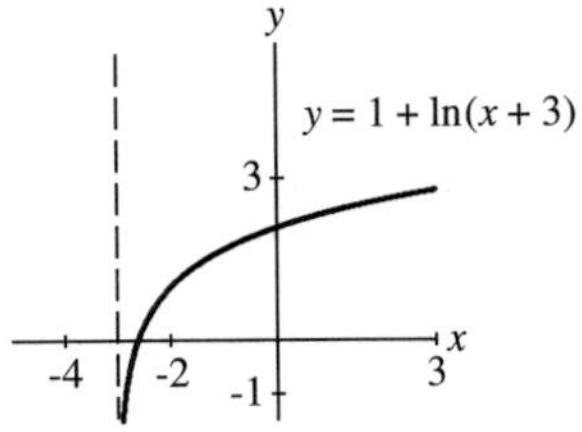

71. Domain $(-\infty, \infty)$, range $(1, \infty)$, inc, $y = 1$

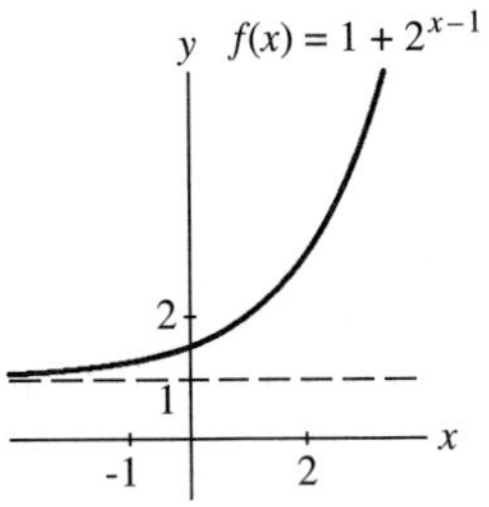

73. Domain $(-\infty, 2)$, range $(-\infty, \infty)$, dec, $x = 2$

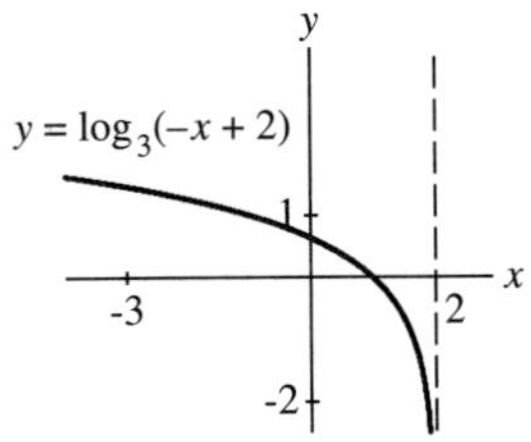

75. $f^{-1}(x) = \log_7(x)$ **77.** $f^{-1}(x) = 5^x$ **79.** 2.0959 **81.** 7.8538 **83.** -4.4243 **85.** The pH of A is one less than the pH of B. **87.** \$122,296.01 **89.** 56 quarters **91.** 25 g, 18.15 g, 2166 yr **93.** 2877 hr **95.** A win will put him \$2 ahead, $n = 20$. **97.** -3

Chapter 4 Test

1. 3 **2.** -2 **3.** 6.47 **4.** $\sqrt{2}$ **5.** $f^{-1}(x) = e^x$ **6.** $f^{-1}(x) = \log_8(x)$ **7.** $\log(xy^3)$ **8.** $\ln\left(\frac{\sqrt{x-1}}{33}\right)$ **9.** $2 \cdot \log_a(2) + \log_a(7)$ **10.** $\log_a(7) - \log_a(2)$ **11.** 4 **12.** 18 **13.** $\frac{\ln(5)}{\ln(5) - \ln(3)} \approx 3.1507$ **14.** $1 + 3^{5.46} \approx 403.7931$

15. Domain $(-\infty, \infty)$, range $(1, \infty)$, inc, $y = 1$

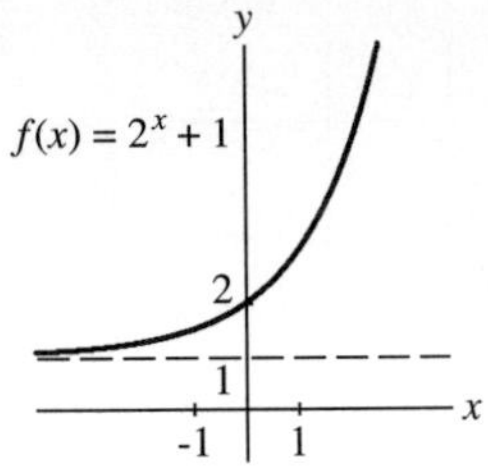

16. Domain $(1, \infty)$, range $(-\infty, \infty)$, dec, $x = 1$

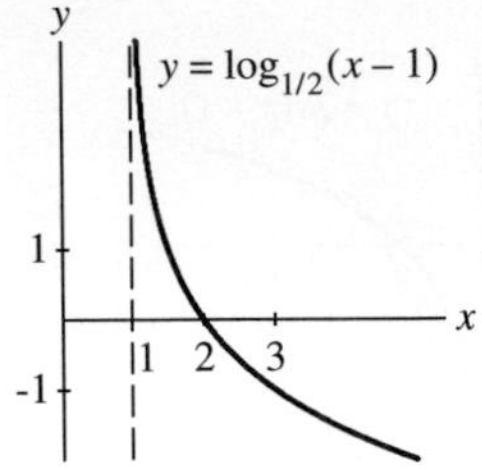

17. (1, 0) **18.** \$9750.88, \$9906.06 **19.** 22.5 watts, 173.3 days, 428.7 days **20.** 61.5 quarters **21.** $p = 0.86$, no

Tying It All Together Chapters P–4

1. 5, 1 **2.** 5 **3.** 19 **4.** 5 **5.** 19 **6.** 7, -1 **7.** $2 \pm \sqrt{2}$ **8.** -3 **9.** 9 **10.** $\log_2(3)$ **11.** 4 **12.** -1, 2, 3

13.

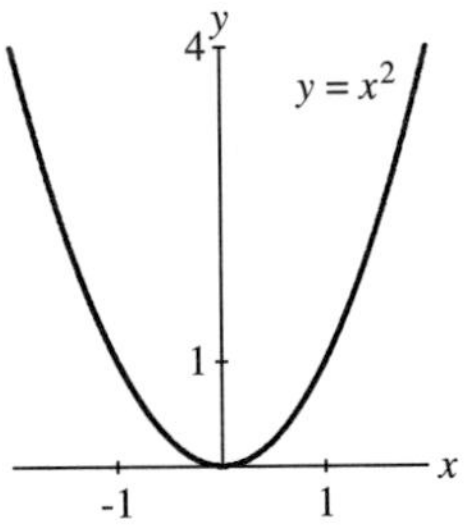

14.

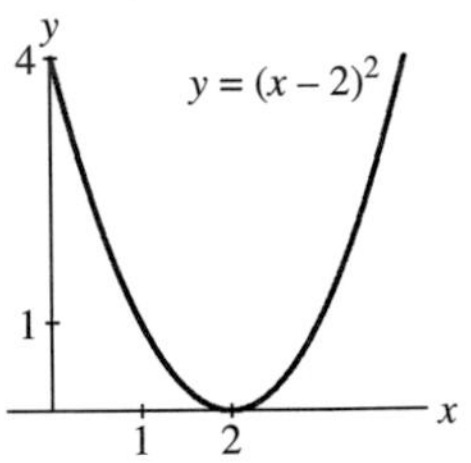

15.

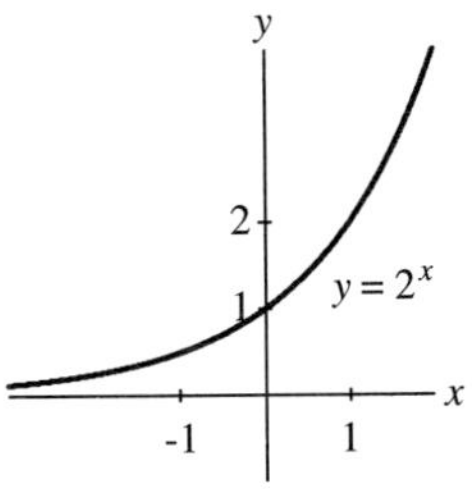

16.

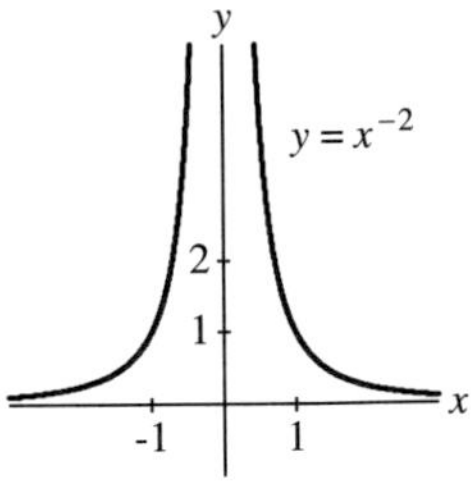

17.

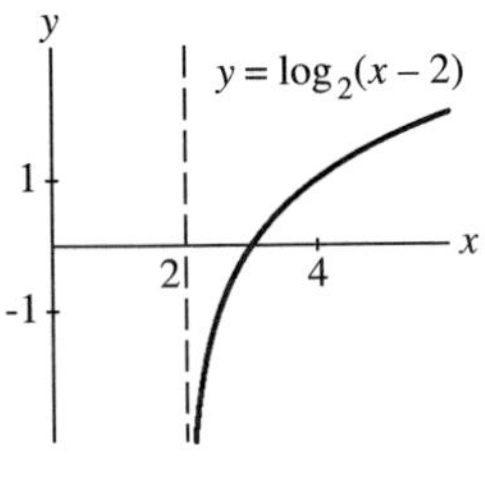

18.

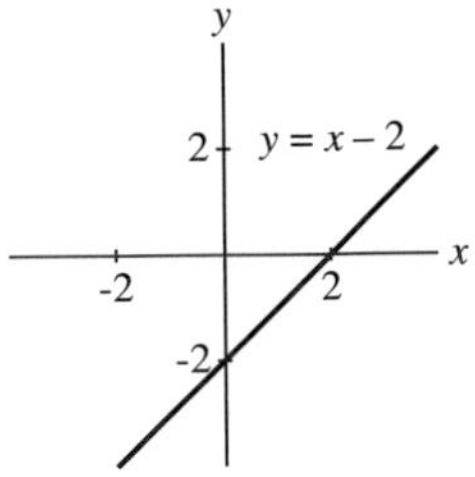

19.

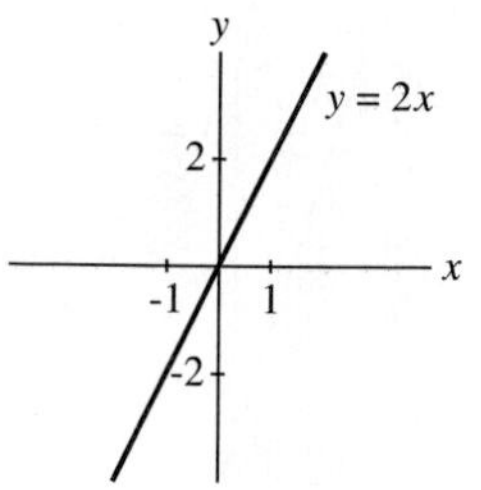

20.

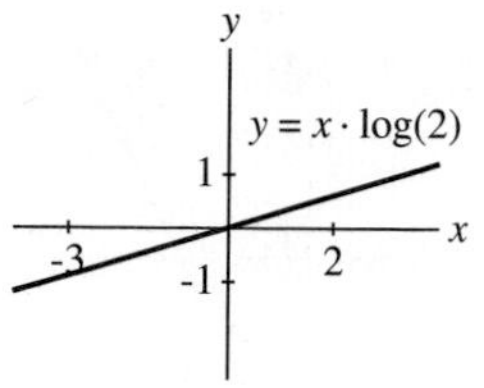

21.

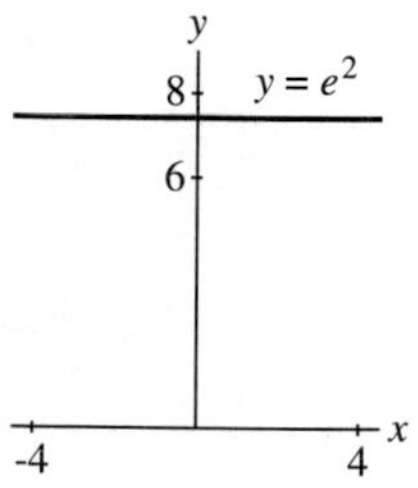

22.

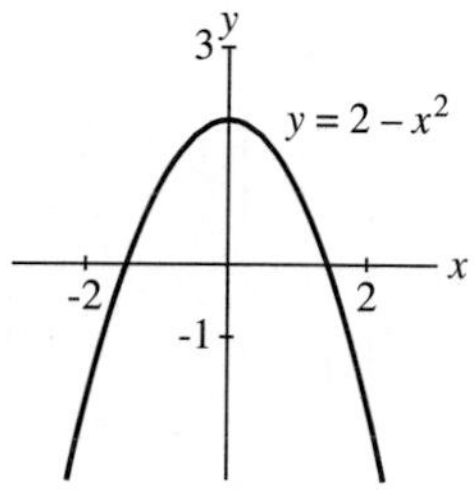

23.

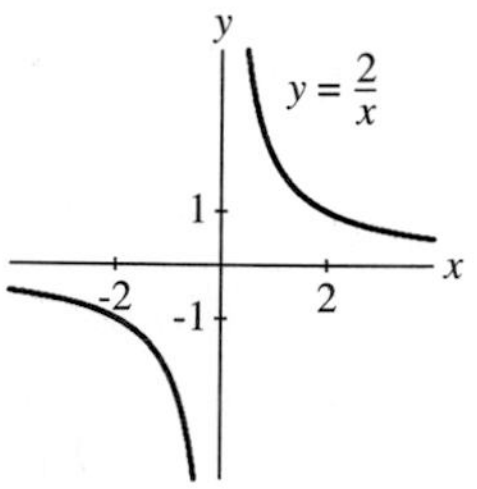

24.

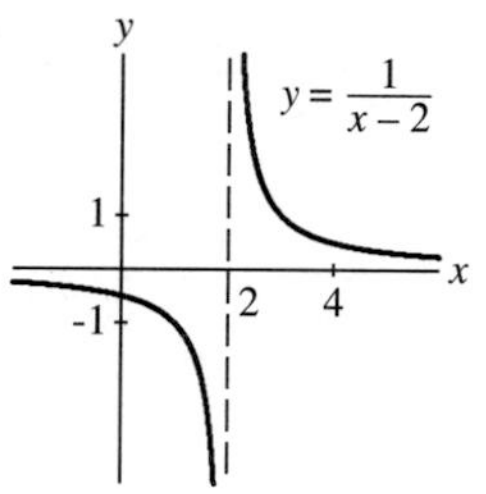

25. $f^{-1}(x) = 3x$ **26.** $f^{-1}(x) = -\log_3(x)$
27. $f^{-1}(x) = x^2 + 2$ for $x \geq 0$ **28.** $f^{-1}(x) = \sqrt[3]{x - 2} + 5$
29. $f^{-1}(x) = (10^x + 3)^2$ **30.** $f^{-1} = \{(1, 3), (4, 5)\}$
31. $f^{-1}(x) = \dfrac{1}{x - 3} + 5$ **32.** $f^{-1}(x) = (\ln(3 - x))^2$ for $x \leq 2$
33. $(p \circ m)(x) = e^{x+5}, (-\infty, \infty), (0, \infty)$
34. $(p \circ q)(x) = e^{\sqrt{x}}, [0, \infty), [1, \infty)$
35. $(q \circ p \circ m)(x) = \sqrt{e^{x+5}}, (-\infty, \infty), (0, \infty)$
36. $(m \circ r \circ q)(x) = \ln(\sqrt{x}) + 5, (0, \infty), (-\infty, \infty)$
37. $(p \circ r \circ m)(x) = e^{\ln(x+5)}, (-5, \infty), (0, \infty)$
38. $(r \circ q \circ p)(x) = \ln(\sqrt{e^x}), (-\infty, \infty), (-\infty, \infty)$ **39.** $F = f \circ g \circ h$
40. $H = g \circ h \circ f$ **41.** $G = h \circ g \circ f$ **42.** $M = h \circ f \circ g$

CHAPTER 5

Section 5.1

1. $\{(1, 2)\}$ **3.** $\varnothing$ **5.** $\{(3, 2)\}$ **7.** $\{(3, 1)\}$ **9.** $\varnothing$
11. $\{(x, y) | x - 2y = 6\}$ **13.** $\{(-1, -1)\}$, independent
15. $\{(11/5, -6/5)\}$, independent
17. $\{(x, y) | y = 3x + 5\}$, dependent **19.** $\varnothing$, inconsistent
21. $\{(150, 50)\}$, independent **23.** $\varnothing$, inconsistent
25. $\{(34, 15)\}$, independent **27.** $\{(13, 7)\}$, independent
29. $\{(4, -1)\}$, independent **31.** $\varnothing$, inconsistent
33. $\{(-1, 1)\}$, independent **35.** $\{(x, y) | x + 2y = 12\}$, dependent
37. $\{(4, 6)\}$, independent **39.** $\{(1.5, 3.48)\}$, independent
41. Independent **43.** Dependent **45.** $(-1000, -497)$
47. $(6.18, -0.54)$ **49.** \$10,000 at 10%, \$15,000 at 8%
51. \$6.50 adult, \$4 child
53. Dependent system,
0 through 12 male memberships, $12 - m$ female memberships
55. Inconsistent system, no solution
57. 600 **59.** 22 nickels, 65 pennies
61. $x = 6$ oz, $y = 10$ oz **63.** Plan A, 12 **65.** 1.1 yr
67. $y = -2x + 3$ **69.** $y = -\dfrac{5}{3}x - \dfrac{1}{3}$

Section 5.2

1.

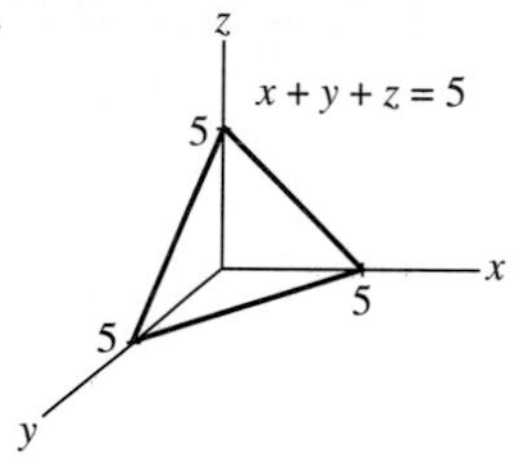

3.

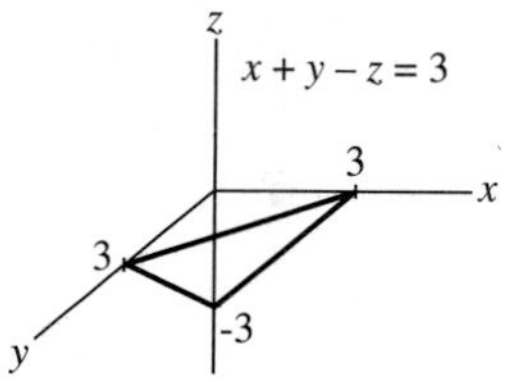

5. $\{(1, 2, 3)\}$ **7.** $\{(-1, 1, 2)\}$ **9.** $\{(0, -5, 5)\}$ **11.** $\varnothing$
13. $\{(x, x - 1, x + 5) | x$ is any real number$\}$
15. $\{(x, y, z) | x + 2y - 3z = 5\}$
17. $\{(2, y, y) | y$ is any real number$\}$
19. $\{(x, 5 - x, 3 - x) | x$ is any real number$\}$ **21.** $\{(5, 7, 9)\}$
23. $\{(1.2, 1.5, 2.4)\}$ **25.** $\{(1000, 2000, 6000)\}$
27. $\{(15/11, 13/11, -3/11)\}$ **29.** $y = x^2 - 3$ **31.** $y = -2x^2 + 5x$
33. $y = x^2 + 4x + 4$ **35.** $x + y + z = 1$ **37.** $2x + y - z = 2$
39. $E = -0.0009s^2 + 0.0357s - 0.1183$, 19.8 mph
41. \$4000 in stocks, \$7000 in bonds, \$14,000 in mutual fund
43. \$0.60, \$0.80, \$0.50
45. Weight distribution (pounds)

LR	LF	RR	RF
280	332	296	292
285	327	291	297
290	322	286	302

47. 116 pennies, 48 nickels, 68 dimes **49.** \$3.95
51. $x = 36$ lb, $y = 24$ lb, $z = 51$ lb

Section 5.3

1. $\{(2, 4), (3, 9)\}$ **3.** $\{(-1, 0), (3, 2)\}$ **5.** $\left\{(0, 0), \left(\dfrac{1}{4}, \dfrac{1}{2}\right)\right\}$
7. $\{(0, 0), (2, 8), (-2, -8)\}$ **9.** $\{(0, 0), (2, 2), (-2, 2)\}$
11. $\{(1, 0)\}$ **13.** $\{(0, 0), (\sqrt{2}, 2), (-\sqrt{2}, 2)\}$
15. $\{(-2 + \sqrt{3}, -2 - \sqrt{3}), (-2 - \sqrt{3}, -2 + \sqrt{3})\}$

17. $\{(1, 1), (1, -1), (-1, 1), (-1, -1)\}$ **19.** $\{(2, 5)\}$
21. $\{(4, -3), (-1, 2)\}$ **23.** $\{(1, -2), (-1, 2)\}$
25. $\left\{\left(-\frac{1}{3}, 2^{2/3}\right)\right\}$ **27.** $\{(2, 1)\}$ **29.** $\{(2, 2)\}$ **31.** $\{(0, 1)\}$
33. $\{(2, 1), (0.3, -1.8)\}$ **35.** $\{(1.9, 0.6), (0.1, -2.0)\}$
37. $(-0.77, 0.59), (2, 2), (4, 16)$ **39.** 9 m and 12 m
41. $6 - 2\sqrt{3}$ ft, $6\sqrt{3} - 6$ ft, $12 - 4\sqrt{3}$ ft
43. $x = 8$ in. and $y = 8$ oz **45.** A 9.6 min, B 48 min
47. $3 + i$ and $3 - i$ **49.** 60 ft and 30 ft
51. 0 and 9.66 yr, 29.5 yr **53.** 6 A.M.

Section 5.4

1. $\dfrac{7x - 5}{(x - 2)(x + 1)}$ **3.** $\dfrac{x^2 - 3x + 5}{(x - 1)(x^2 + 2)}$ **5.** $\dfrac{3x^3 + x^2 + 8x + 5}{(x^2 + 3)^2}$
7. $\dfrac{4x^2 - x - 1}{(x - 1)^3}$ **9.** $A = 2, B = -2$ **11.** $\dfrac{2}{x + 1} + \dfrac{3}{x - 2}$
13. $\dfrac{1/2}{x + 2} + \dfrac{3/2}{x + 4}$ **15.** $\dfrac{1/3}{x - 3} + \dfrac{-1/3}{x + 3}$ **17.** $\dfrac{-1}{x} + \dfrac{1}{x - 1}$
19. $A = 2, B = 5, C = -1$ **21.** $\dfrac{-3}{(x + 2)^2} + \dfrac{-2}{x + 2}$
23. $\dfrac{2x - 3}{x^2 + 1} + \dfrac{4}{x + 1}$ **25.** $\dfrac{-8x}{(x^2 + 9)^2} + \dfrac{3x - 1}{x^2 + 9}$
27. $\dfrac{5}{x - 2} + \dfrac{-2x + 3}{x^2 + 2x + 4}$ **29.** $2x + 1 + \dfrac{2}{x - 1} + \dfrac{3}{x + 1}$
31. $\dfrac{3x + 3}{(x^2 + x + 1)^2} + \dfrac{3x - 5}{x^2 + x + 1}$ **33.** $\dfrac{3x}{x^2 + 4} + \dfrac{1}{x - 2} + \dfrac{-1}{x + 2}$
35. $\dfrac{1}{x} + \dfrac{2}{x^2} + \dfrac{3}{x^3} + \dfrac{4}{x - 1}$ **37.** $\dfrac{3}{x - 2} + \dfrac{-1}{(x - 2)^2} + \dfrac{3}{x - 3}$
39. $\dfrac{4}{x + 5} + \dfrac{2}{x + 2} + \dfrac{3}{x - 3}$ **41.** $\dfrac{-1}{(x - 1)^3} + \dfrac{2}{(x - 1)^2} + \dfrac{1}{x - 1}$
43. $\dfrac{-b/a}{(ax + b)^2} + \dfrac{1/a}{ax + b}$ **45.** $\dfrac{c/b}{x} + \dfrac{1 - ac/b}{ax + b}$
47. $\dfrac{1/b}{x^2} + \dfrac{-a/b^2}{x} + \dfrac{a^2/b^2}{ax + b}$

Section 5.5

1. (c) **3.** (d)

5.

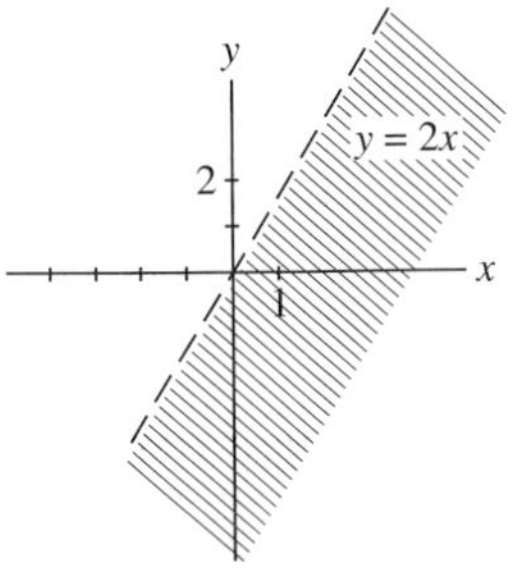

7.

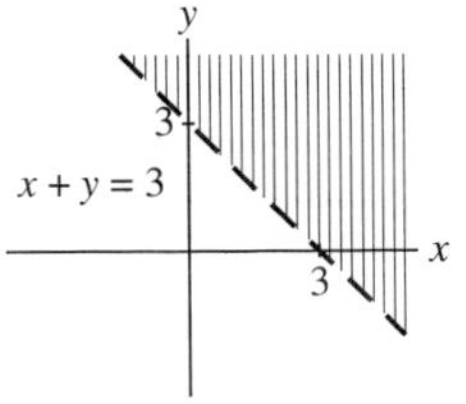

9.

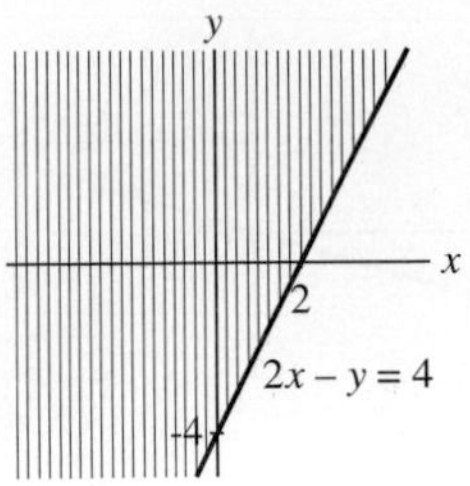

11.

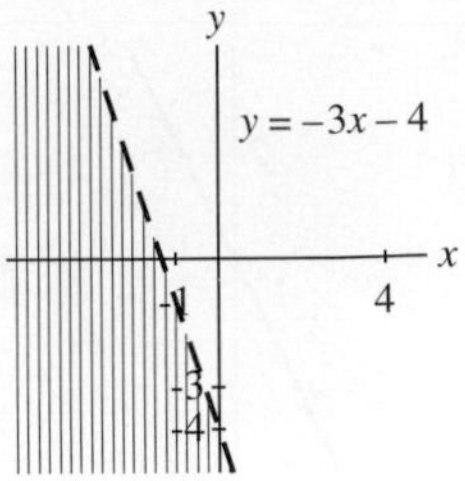

13.

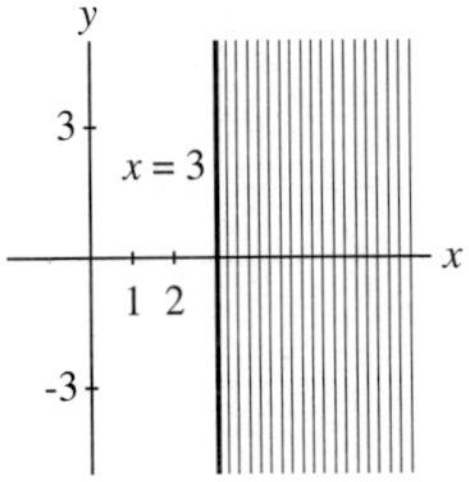

15.

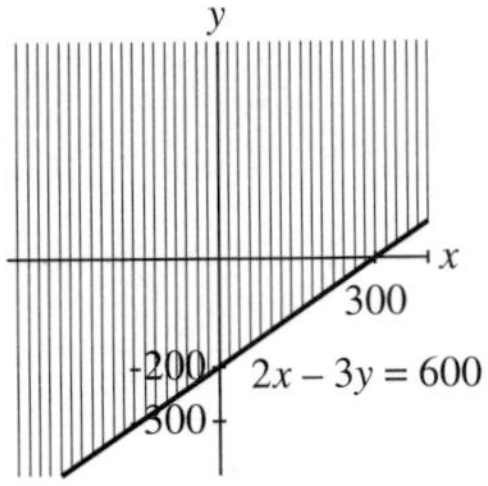

17.

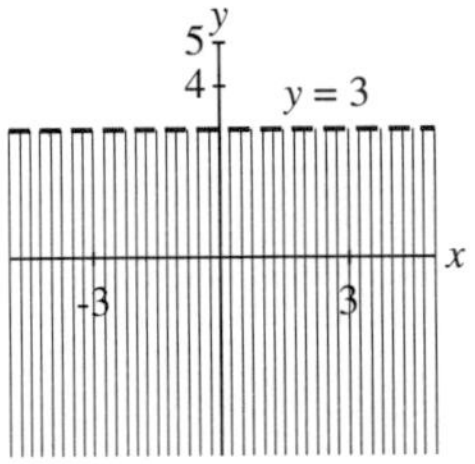

19.

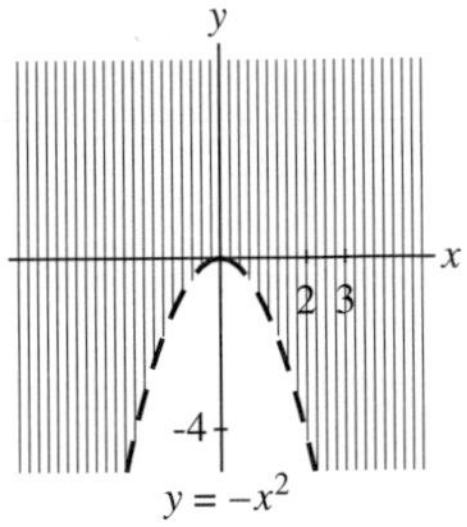

21.

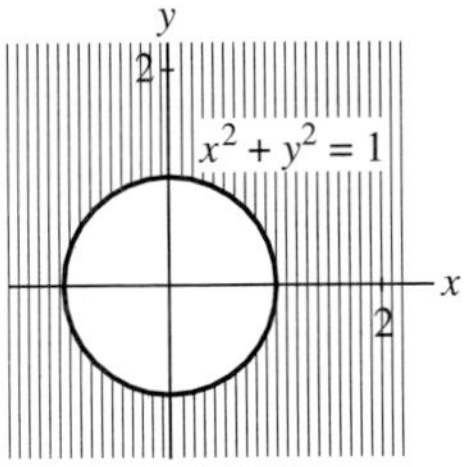

23.

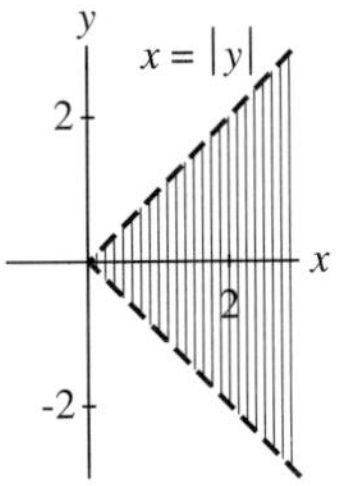

25.

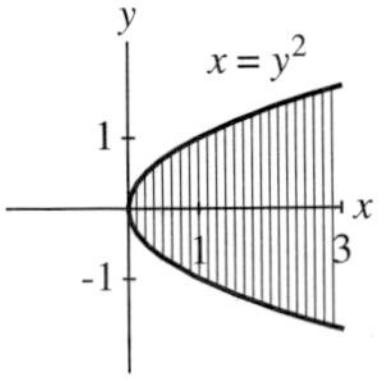

27.

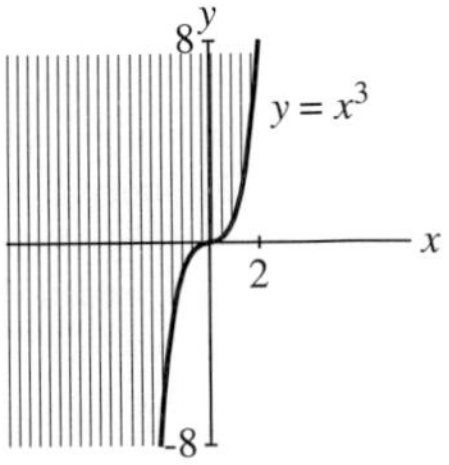

29.

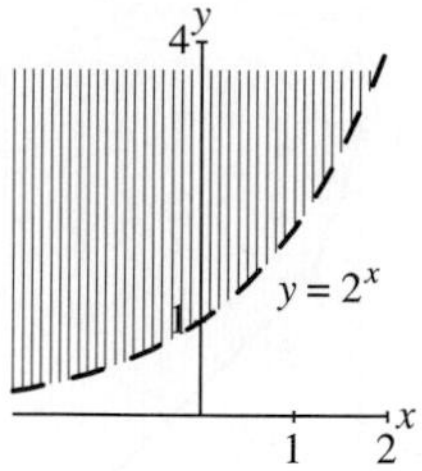

31.

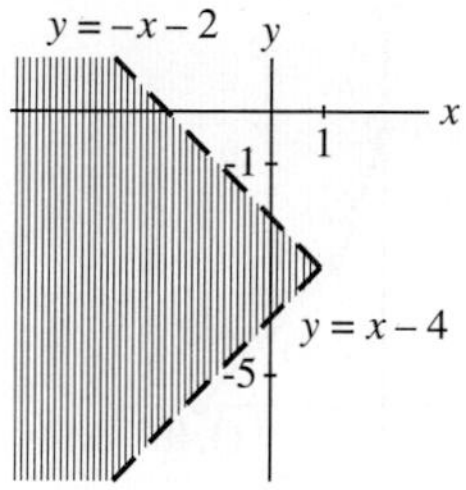

33.

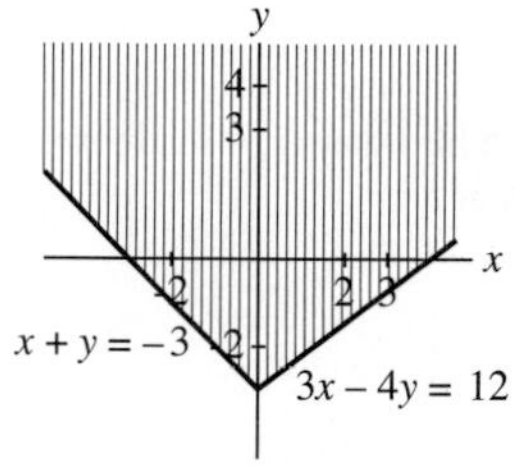

35.

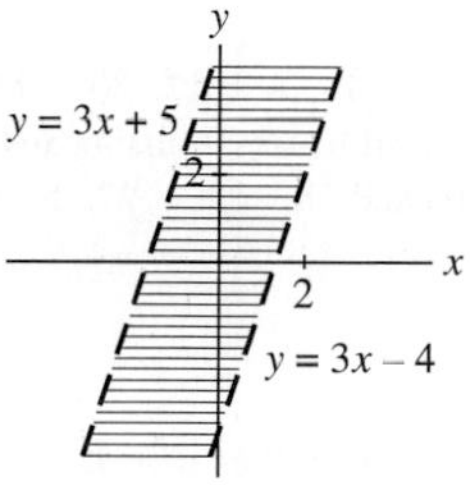

37. No solution

39.

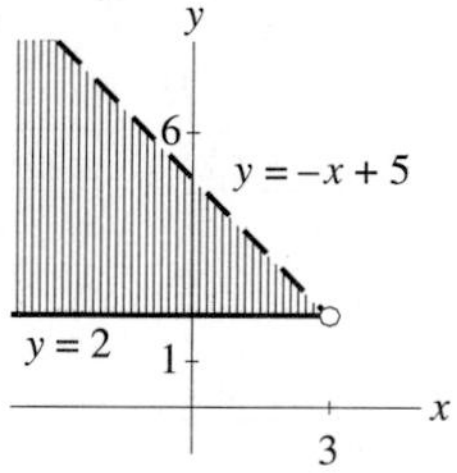

41.

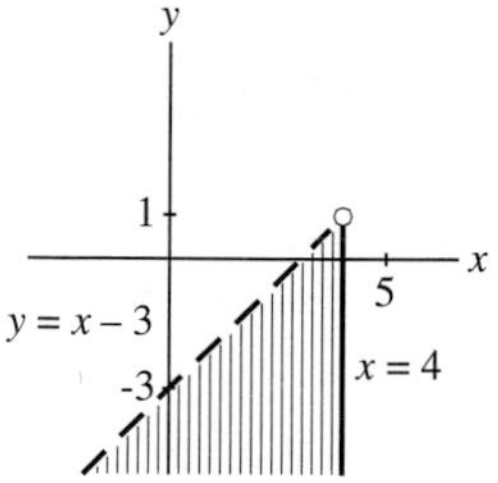

43.

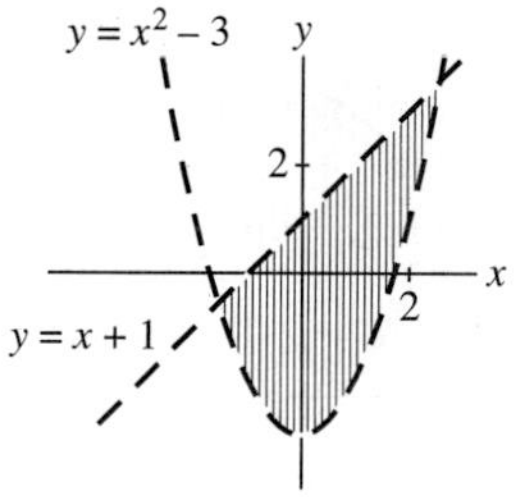

45.

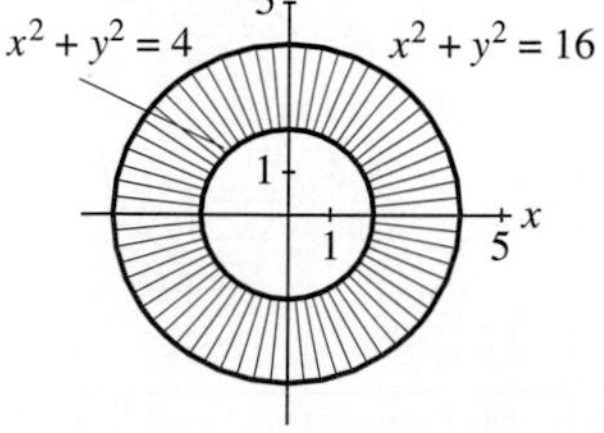

47.

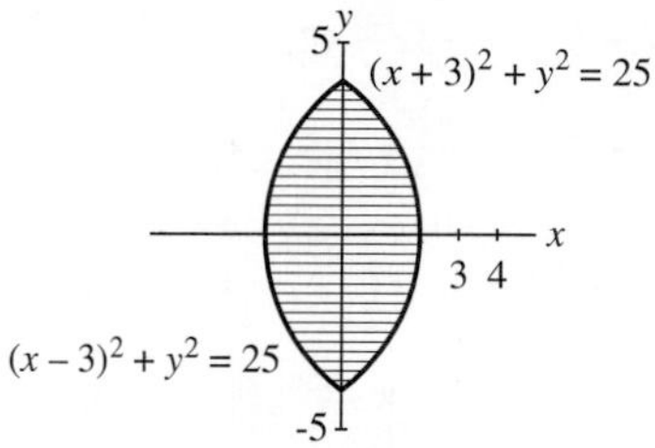

49.

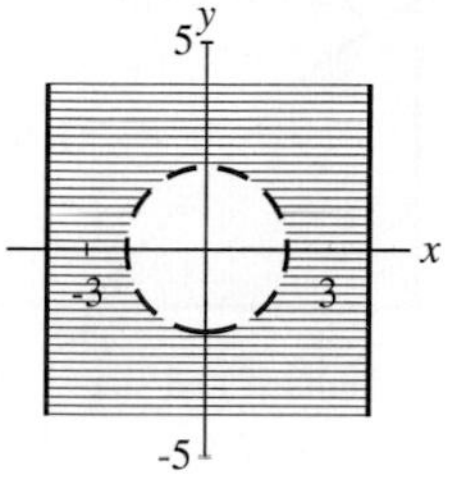

51.

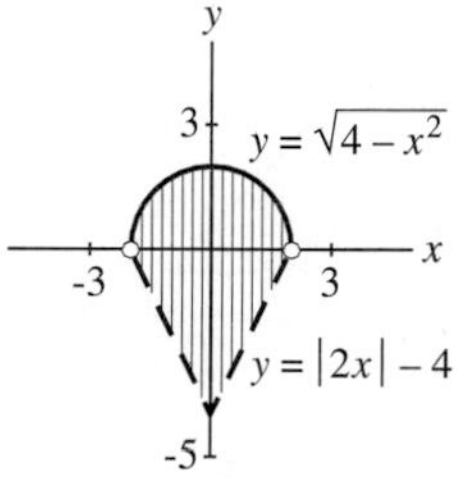

53.

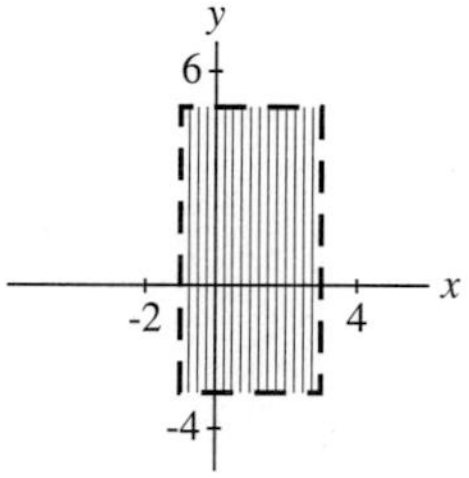

55.

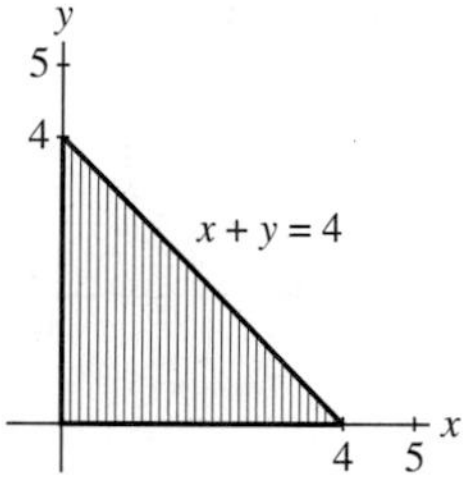

57.

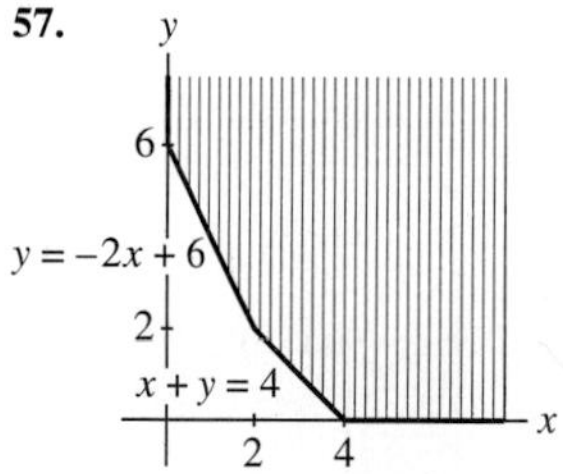

59.

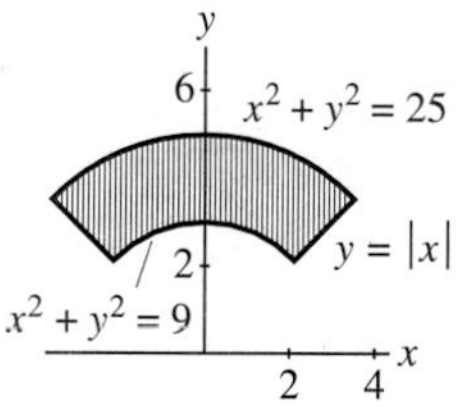

61.

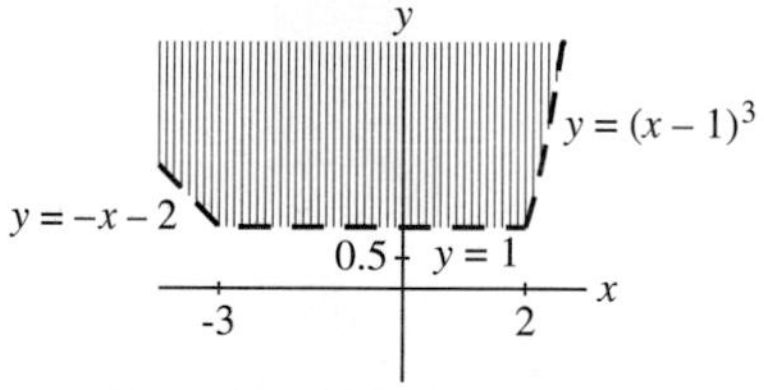

63.

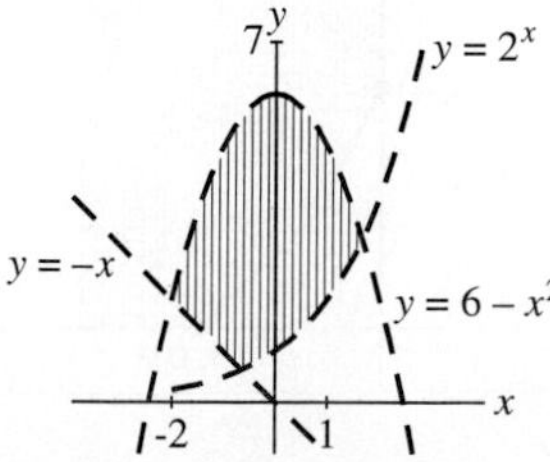

65. $x \geq 0, y \geq 0, y \leq -\frac{2}{3}x + 5, y \leq -3x + 12$

67. $y \geq 0, x \geq 0, y \geq -\frac{1}{2}x + 3, y \geq -\frac{3}{2}x + 5$

69. $|x| < 2, |y| < 2$ **71.** $x > 0, y > 0, x^2 + y^2 < 81$

73. $(-1.17, 1.84)$ **75.** $(150, 22.4)$

77. $w + h \leq 40$

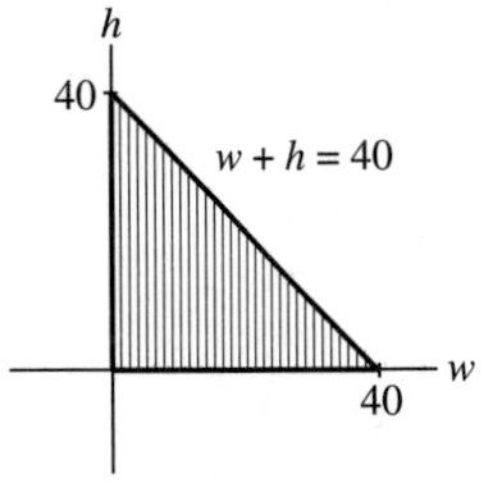

79.

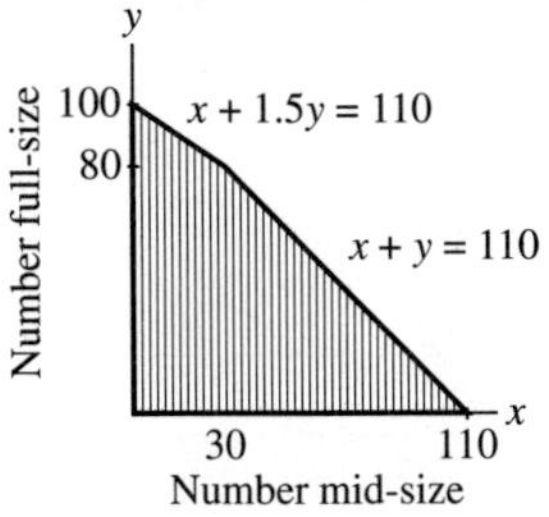

81.

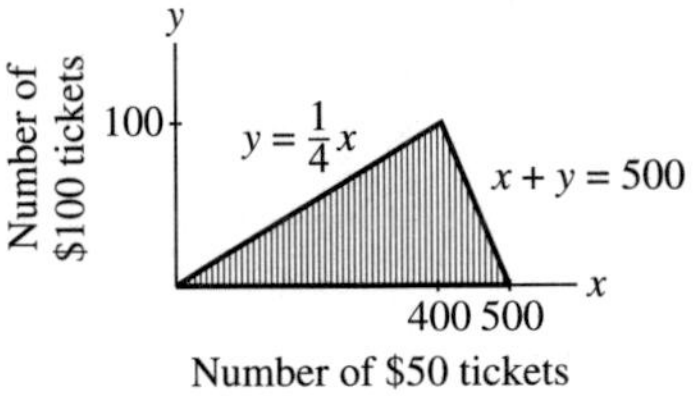

Section 5.6

1.

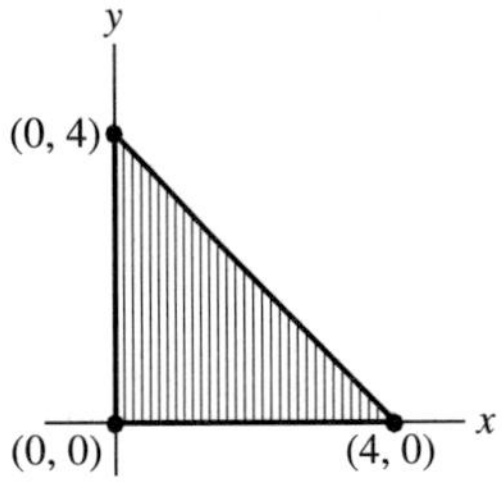

3.

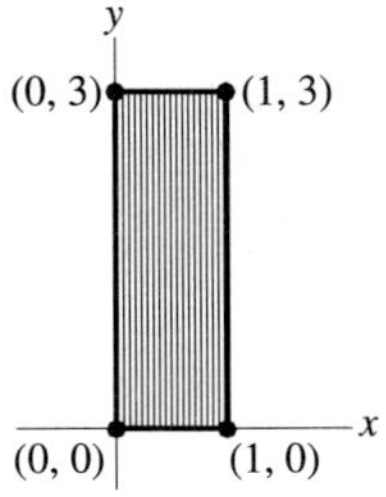

5.

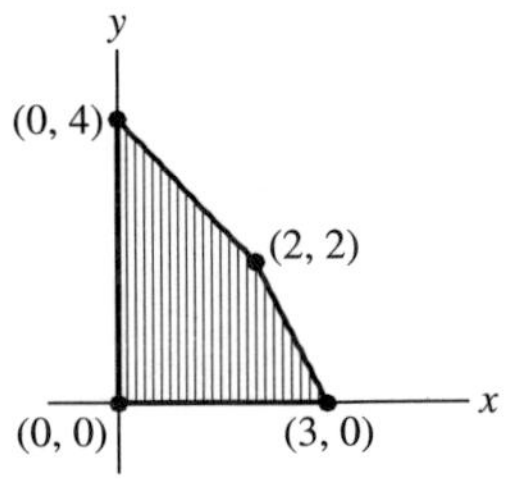

7.

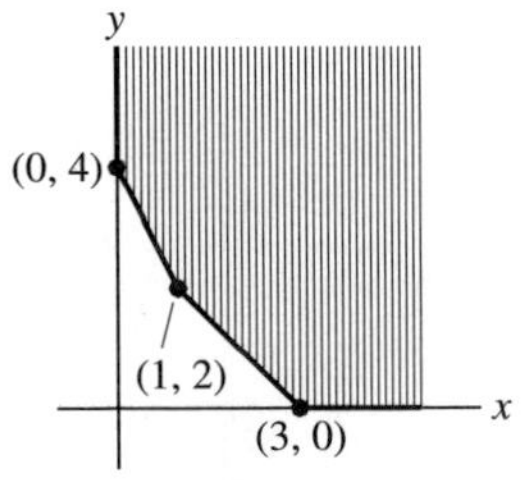

9.

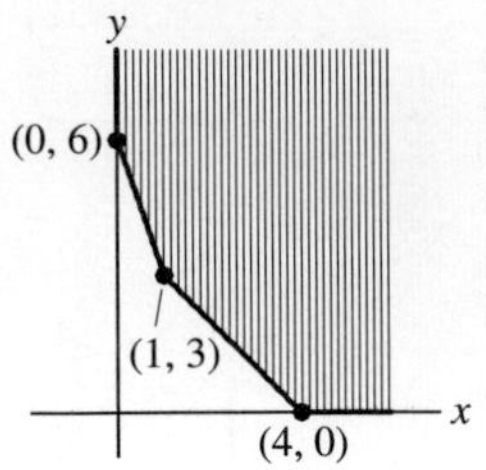

11.

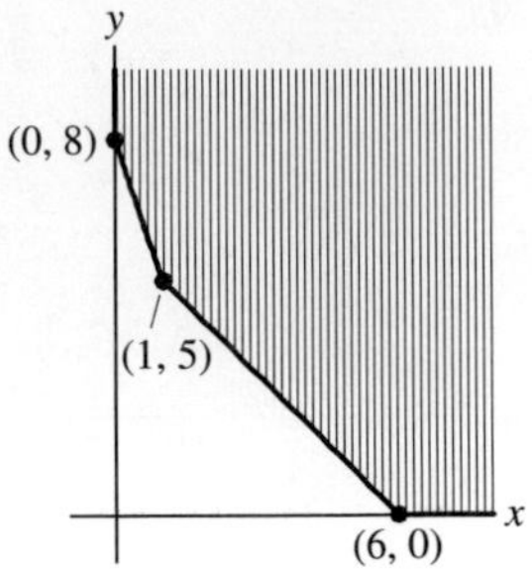

13. 15 **15.** 8 **17.** 30 **19.** 100

21. 8 bird houses, 6 mailboxes **23.** 16 bird houses, 0 mailboxes

25. 0 small, 8 large **27.** 5 small, 3 large

Chapter 5 Review Exercises

1. $\{(3, 5)\}$ **3.** $\{(-1, 3)\}$ **5.** $\{(-19/2, -19/2)\}$, independent

7. $\{(39/17, 18/17)\}$, independent

9. $\{(x, y) | y = -3x + 1\}$, dependent **11.** $\varnothing$, inconsistent

13. $\{(1, 3, -4)\}$ **15.** $\{(x, 0, 1 - x) | x \text{ is any real number}\}$ **17.** $\varnothing$

19. $\left\{\left(\frac{-1 + \sqrt{17}}{2}, \pm\sqrt{\frac{-1 + \sqrt{17}}{2}}\right)\right\}$

21. $\{(0, 0), (1, 1), (-1, 1)\}$ **23.** $\frac{2}{x - 3} + \frac{5}{x + 4}$

25. $\frac{2x - 1}{x^2 + 4} + \frac{5}{x - 3}$

27.

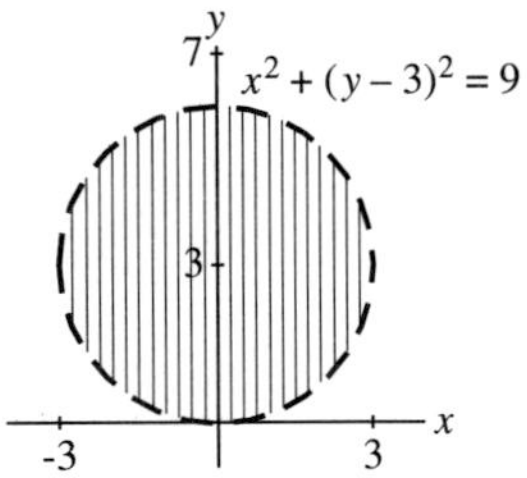

29.

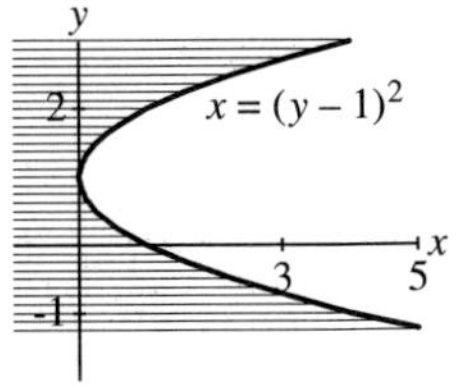

31.

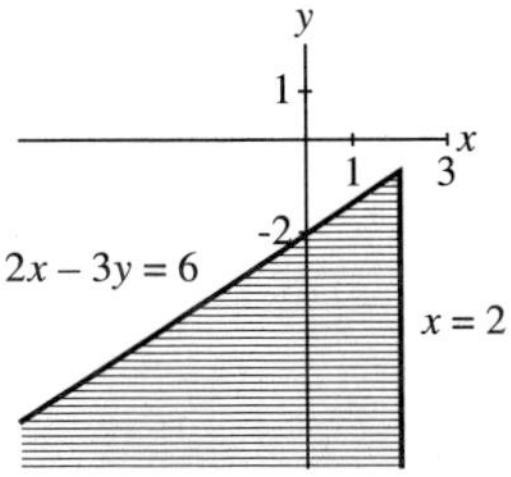

33.

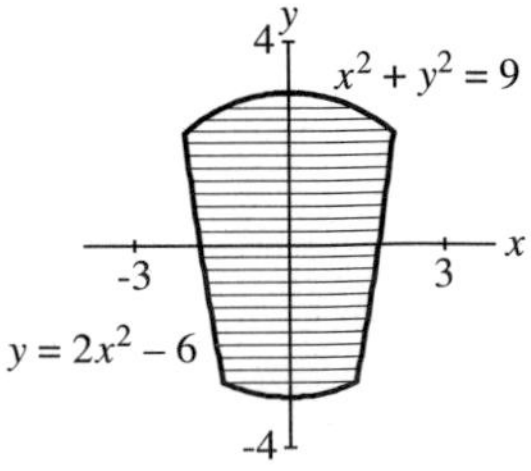

35.

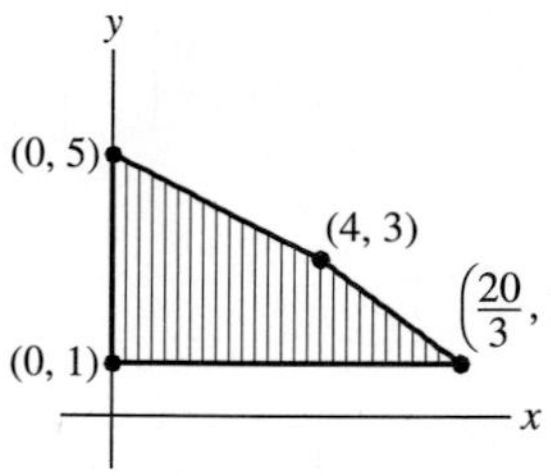

37.

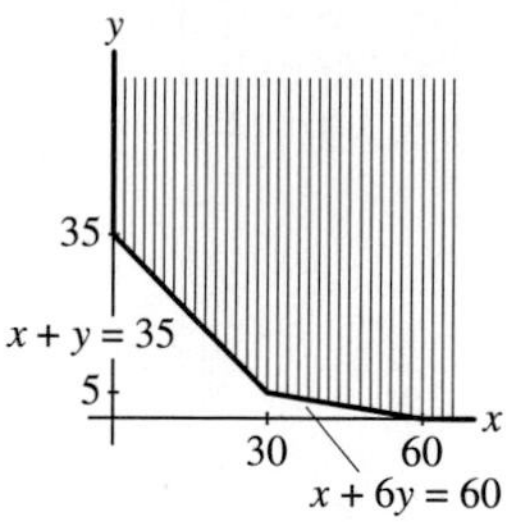

39. $y = -\frac{2}{3}x + \frac{5}{3}$ **41.** $y = 3x^2 - 4x + 5$

43. 57 tacos, 62 burritos **45.** \$0.95 **47.** 16.8

49. 6 million barrels per day from each source

Chapter 5 Test

1. $\{(-3, 4)\}$ **2.** $\{(25/11, -6/11)\}$ **3.** $\{(4, 6)\}$ **4.** Inconsistent

5. Dependent **6.** Independent **7.** Inconsistent

8. $\{(x, 10 - x, 14 - 3x) \mid x \text{ is any real number}\}$ **9.** $\{(1, -2, 3)\}$

10. $\varnothing$ **11.** $\{(4, 0), (-4, 0)\}$

12. $\{(2 + \sqrt{2}, -4 - \sqrt{2}), (2 - \sqrt{2}, -4 + \sqrt{2})\}$

13. $\frac{3}{x - 4} + \frac{-1}{x + 2}$ **14.** $\frac{2}{x^2} + \frac{1}{x} + \frac{3}{x - 1}$

15.

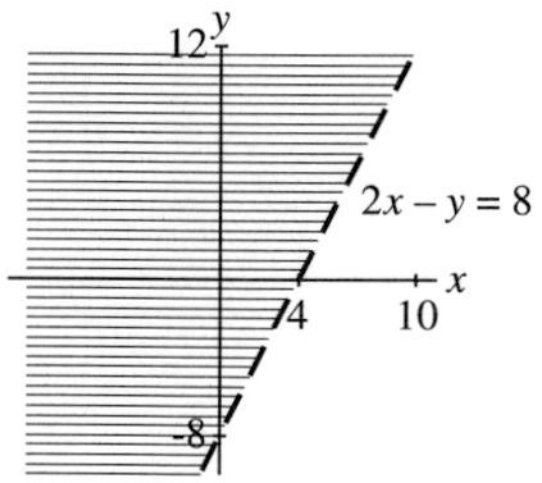

16.

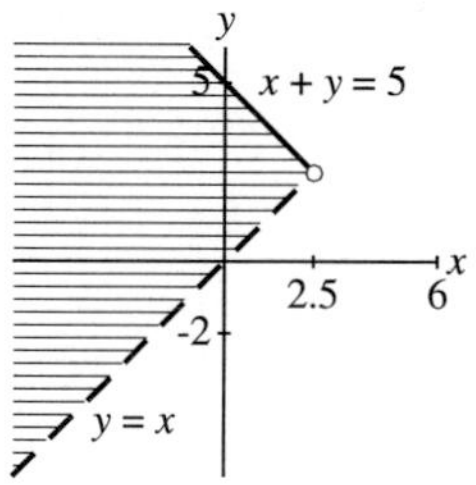

17.

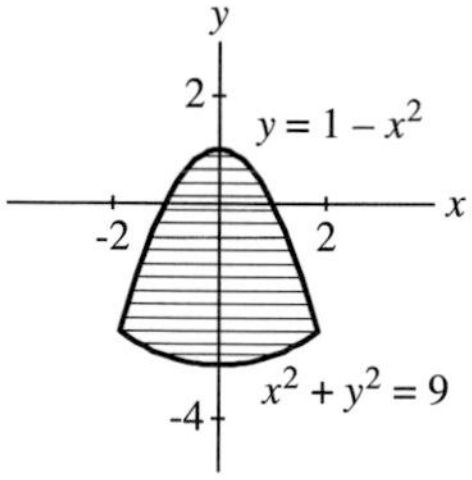

18. 24 males, 28 females **19.** 5 television, 18 newspaper

Tying It All Together Chapters P–5

1. $\left\{\frac{103}{13}\right\}$ **2.** $\left\{\frac{40}{13}, 3\right\}$ **3.** $\left\{-\frac{4}{5}\right\}$ **4.** $\{-1, 4\}$ **5.** $\{-11\}$

6. $\left\{\pm\frac{2\sqrt{3}}{3}\right\}$ **7.** $\{1\}$ **8.** $\{1 + \log_2(9)\}$ **9.** $\{1\}$ **10.** $\{\pm 8\}$

11. $\left\{\frac{3 \pm \sqrt{33}}{2}\right\}$ **12.** $\left\{\frac{6 \pm \sqrt{2}}{2}\right\}$

13. $(-\infty, 3/2)$

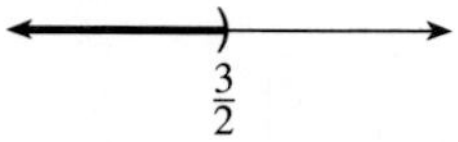

14. $(-\infty, 1.5) \cup (1.5, \infty)$

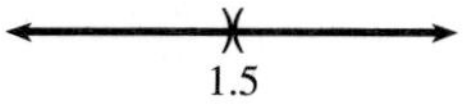

15. $(-\infty, -3] \cup [3, \infty)$

16. $[-7, 5]$

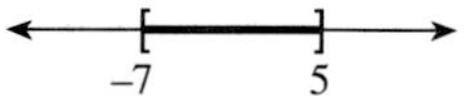

17.

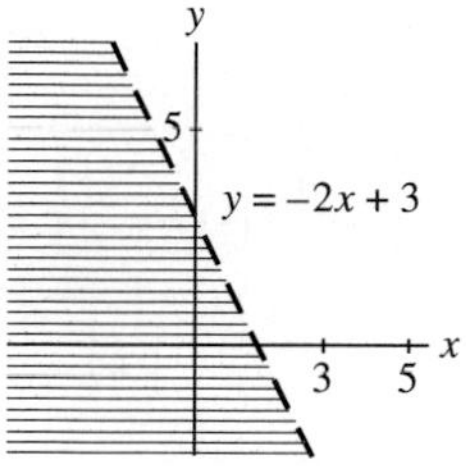

18.

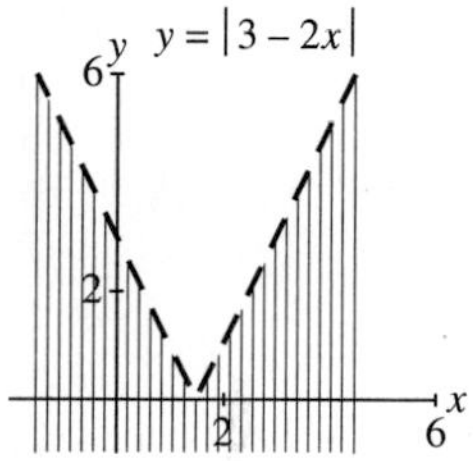

19.

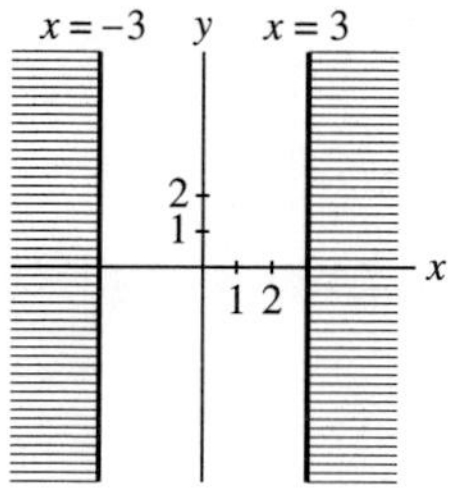

20.

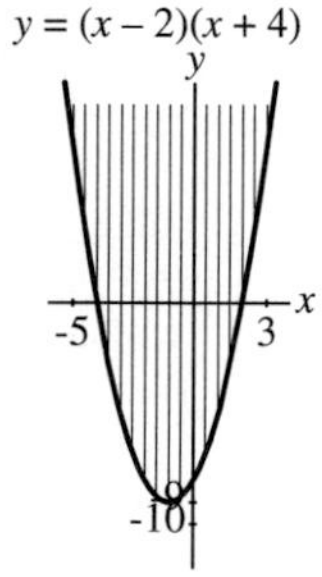

CHAPTER 6

Section 6.1

1. 1×3 **3.** 1×1 **5.** 3×2

7. $\left[\begin{array}{cc|c} 1 & -2 & 4 \\ 3 & 2 & -5 \end{array}\right]$ **9.** $\left[\begin{array}{ccc|c} 1 & -1 & -1 & 4 \\ 1 & 3 & -1 & 1 \\ 0 & 2 & -5 & -6 \end{array}\right]$

11. $\begin{aligned} 3x + 4y &= -2 \\ 3x - 5y &= 0 \end{aligned}$ **13.** $\begin{aligned} 5x \qquad\quad &= 6 \\ -4x \quad + 2z &= -1 \\ 4x + 4y \qquad &= 7 \end{aligned}$

15. Multiply R_1 by 1/2. **17.** Multiply R_2 by $-1/6$. **19.** $\{(2, 3)\}$

21. $\{(4, -4)\}$ **23.** $\{(3, 3)\}$ **25.** $\{(0.5, 1)\}$ **27.** $\varnothing$

29. $\{(u, v) \mid u + 3v = 4\}$ **31.** $\{(3, 2, 1)\}$ **33.** $\{(1, 1, 1)\}$

35. $\{(1, 2, 0)\}$ **37.** $\{(1, 0, -1)\}$ **39.** $\{(x, y, z) \mid x - 2y + 3z = 1\}$

41. $\varnothing$ **43.** $\{(x, 5 - 2x, 2 - x) \mid x \text{ is any real number}\}$

45. $\left\{\left(x, \frac{13 - 5x}{2}, \frac{5 - 3x}{2}\right) \middle| \, x \text{ is any real number}\right\}$

47. $\{(4, 3, 2, 1)\}$ **49.** 38 hr at Burgers-R-Us, 22 hr at Soap Opera

51. \$20,000 mutual fund, \$11,333.33 treasury bills, \$8666.67 bonds

53. $y = x^3 - 2x + 3$

55. $y = 750 - x$, $z = x - 250$, and $250 \le x \le 750$. If $z = 50$, then $x = 300$ and $y = 450$.

Section 6.2

1. $x = 2, y = 5$ **3.** $x = 3, y = 5/2, z = 5/4$ **5.** $\begin{bmatrix} 5 \\ 6 \end{bmatrix}$

7. $\begin{bmatrix} 0.4 & 0.15 \\ 0.7 & 1.1 \end{bmatrix}$ **9.** $\begin{bmatrix} 1/2 & 3/2 \\ 3 & -12 \end{bmatrix}$ **11.** $\begin{bmatrix} 10 & 0 \\ -2 & 16 \end{bmatrix}$

13. Undefined **15.** $\begin{bmatrix} 0 \\ 0 \\ 7\sqrt{3} \end{bmatrix}$ **17.** $\begin{bmatrix} 13a \\ -b \end{bmatrix}$ **19.** $\begin{bmatrix} -x & -0.5y \\ -0.7x & 3.5y \end{bmatrix}$

21. $\begin{bmatrix} -1 & 13 \\ -9 & 3 \\ 6 & -2 \end{bmatrix}$ **23.** $\begin{bmatrix} 3x & 2y & -z \\ -6x & 3y & 7z \\ 0 & -7y & -7z \end{bmatrix}$ **25.** $\begin{bmatrix} -5 & -1 \\ 10 & 4 \end{bmatrix}$

27. Undefined **29.** $\begin{bmatrix} -8 & -5 \\ 12 & -1 \end{bmatrix}$ **31.** $\begin{bmatrix} -8 & -5 \\ 12 & -1 \end{bmatrix}$ **33.** $\begin{bmatrix} -5 \\ 7 \end{bmatrix}$

35. $\begin{bmatrix} -5 \\ 7 \end{bmatrix}$ **37.** $\begin{bmatrix} -7 & 4 \\ -1 & -4 \end{bmatrix}$ **39.** $\begin{bmatrix} -5 \\ 4 \end{bmatrix}$ **41.** $\{(3, 2)\}$

43. $\{(-1, 3)\}$ **45.** $\{(0.5, 3, 4.5)\}$ **47.** $\begin{bmatrix} \$390 \\ \$160 \\ \$135 \end{bmatrix}$

49. $\begin{bmatrix} 40 & 80 \\ 30 & 90 \\ 80 & 200 \end{bmatrix}, \begin{bmatrix} 60 & 120 \\ 45 & 135 \\ 120 & 300 \end{bmatrix}$ **51.** Yes, yes **53.** Yes, yes

55. Yes, yes **57.** $\begin{bmatrix} 0 & 0 \\ 0 & 0 \end{bmatrix}$

Section 6.3

1. 3×5 **3.** 1×1 **5.** 5×5 **7.** 3×3 **9.** AB is undefined.

11. $\begin{bmatrix} 4 & 6 & 8 \\ -6 & -9 & -12 \\ 2 & 3 & 4 \end{bmatrix}$ **13.** $[2 \quad 5 \quad 9]$ **15.** Undefined

17. $\begin{bmatrix} 7 & 8 \\ 5 & 5 \\ 1 & 0 \end{bmatrix}$ **19.** $\begin{bmatrix} 1 & 1 \\ 14 & 15 \end{bmatrix}$ **21.** Undefined **23.** $\begin{bmatrix} 0 \\ -2 \\ 1 \end{bmatrix}$

25. Undefined **27.** $\begin{bmatrix} 2 & 2 \\ 3 & 4 \end{bmatrix}$ **29.** $\begin{bmatrix} 1 & 0 \\ 0 & 1 \end{bmatrix}$ **31.** $\begin{bmatrix} -0.5 & 4 \\ 9 & 0.7 \end{bmatrix}$

33. $[4a \quad -3b]$ **35.** $\begin{bmatrix} -2a & 5a & 3a \\ b & 4b & 6b \end{bmatrix}$ **37.** $[4 \quad 2 \quad 6]$ **39.** $[17]$

41. $\begin{bmatrix} x^2 & xy \\ xy & y^2 \end{bmatrix}$ **43.** $\begin{bmatrix} 2\sqrt{2} \\ 7\sqrt{2} \end{bmatrix}$ **45.** $\begin{bmatrix} -17/3 & 11 \\ -3 & 6 \end{bmatrix}$ **47.** Undefined

49. $\begin{bmatrix} 2 & 0 & 6 \\ 3 & 5 & 2 \\ 19 & 21 & 16 \end{bmatrix}$ **51.** $\begin{bmatrix} -0.2 & 1.7 \\ 3 & 3.4 \\ -1.4 & -4.6 \end{bmatrix}$ **53.** $\{(3, 2)\}$ **55.** $\varnothing$

57. $\{(-1, -1, 6)\}$ **59.** $\begin{bmatrix} 2 & 3 \\ 4 & -1 \end{bmatrix}\begin{bmatrix} x \\ y \end{bmatrix} = \begin{bmatrix} 9 \\ 6 \end{bmatrix}$

61. $\begin{bmatrix} 1 & 2 & -1 \\ 3 & -1 & 3 \\ 2 & 1 & -4 \end{bmatrix}\begin{bmatrix} x \\ y \\ z \end{bmatrix} = \begin{bmatrix} 3 \\ 1 \\ 0 \end{bmatrix}$ **63.** $AQ = \begin{bmatrix} \$376{,}000 \\ \$642{,}000 \end{bmatrix}$

65. F **67.** T **69.** T

Section 6.4

1. $\begin{bmatrix} -3 & 5 \\ 12 & 6 \end{bmatrix}$ **3.** $\begin{bmatrix} 1 & 0 \\ 0 & 1 \end{bmatrix}$ **5.** $\begin{bmatrix} 1 & 0 \\ 0 & 1 \end{bmatrix}$ **7.** $\begin{bmatrix} 3 & 5 & 1 \\ 4 & 5 & 7 \\ 4 & 9 & 2 \end{bmatrix}$

9. $\begin{bmatrix} 1 & 0 & 0 \\ 0 & 1 & 0 \\ 0 & 0 & 1 \end{bmatrix}$ **11.** $\begin{bmatrix} 1 & 0 & 0 \\ 0 & 1 & 0 \\ 0 & 0 & 1 \end{bmatrix}$ **13.** Yes **15.** No **17.** No

19. $\begin{bmatrix} 1 & -2 \\ 0 & 0.5 \end{bmatrix}$ **21.** $\begin{bmatrix} 3 & -2 \\ -1/3 & 1/3 \end{bmatrix}$ **23.** $\begin{bmatrix} 4 & 3 \\ -3 & -2 \end{bmatrix}$

25. $\begin{bmatrix} -1.5 & -2.5 \\ -0.5 & -0.5 \end{bmatrix}$ **27.** No inverse **29.** No inverse

31. $\begin{bmatrix} 0.5 & 0.5 & 0 \\ 0.5 & 0 & -0.5 \\ 0 & -0.5 & 0.5 \end{bmatrix}$ **33.** $\begin{bmatrix} -3.5 & -1 & 2 \\ 0.5 & 0 & 0 \\ 4.5 & 2 & -3 \end{bmatrix}$ **35.** $\{(3, -1)\}$

37. $\{(2, 1/3)\}$ **39.** $\{(1, -1)\}$ **41.** $\{(5, 2)\}$ **43.** $\{(1, -1, 3)\}$
45. $\{(1, 3, 2)\}$ **47.** $\{(11.3, -3.9)\}$
49. $\{(-2z - 4, -z - 9, z) \mid z \text{ is any real number}\}$

51. $\{(1/2, 1/6, 1/3)\}$ **53.** $\begin{bmatrix} 1/3 & -1/6 & 1/3 \\ -2/3 & 1/3 & 1/3 \\ 2/3 & 1/6 & -1/3 \end{bmatrix}$

55. $\begin{bmatrix} 2.5 & -3 & 1 \\ 1 & -1 & 0 \\ -1.5 & 2 & 0 \end{bmatrix}$ **57.** $\begin{bmatrix} 1 & -2 & 1 & 0 \\ 0 & 1 & -2 & 1 \\ 0 & 0 & 1 & -2 \\ 0 & 0 & 0 & 1 \end{bmatrix}$

59. $\{(-165, 97.5, 240)\}$ **61.** $\{(-1.6842, -9.2632, -9.2632)\}$

63. $\begin{bmatrix} 1 & 7 \\ 3 & 20 \end{bmatrix}, \begin{bmatrix} 2 & 7 \\ 3 & 10 \end{bmatrix}, \begin{bmatrix} 4 & 7 \\ 3 & 5 \end{bmatrix}, \begin{bmatrix} 20 & 7 \\ 3 & 1 \end{bmatrix}, \begin{bmatrix} 10 & 7 \\ 3 & 2 \end{bmatrix}, \begin{bmatrix} 5 & 7 \\ 3 & 4 \end{bmatrix}$

65. \$0.75 eggs, \$1.80 magazine **67.** \$400 plywood, \$150 insulation
69. Good luck
71. \$17,142.86 Asset Manager, \$42,857.14 Magellan
73. \$4.20 animal totems, \$2.75 necklaces, \$0.89 tribal masks

Section 6.5

1. 2 **3.** 19 **5.** −0.41 **7.** 23/32 **9.** 0 **11.** 0
13. $\{(-3, 5)\}$ **15.** $\{(11/2, -1/2)\}$ **17.** $\{(12, 6)\}$
19. $\{(500, 400)\}$ **21.** $\{(x, y) \mid 3x + y = 6\}$ **23.** $\varnothing$
25. $\{(-6, -9)\}$ **27.** $\{(\sqrt{2}/2, \sqrt{3})\}$
29. $\{(0, -5), (3, 4), (-3, 4)\}$ **31.** $\{(8, 2), (0, -2)\}$ **33.** Invertible
35. Not invertible **37.** $\{(146, 237)\}$ **39.** 175 boys, 440 girls
41. 19 **43.** 182/3 degrees, 182/3 degrees, 176/3 degrees
45. 2, 3, 6, yes **49.** 8, no

Section 6.6

1. 14 **3.** 1 **5.** −23 **7.** −16 **9.** −33 **11.** −56
13. −115 **15.** 14 **17.** −54 **19.** 0 **21.** −3 **23.** −72
25. −137 **27.** −293 **29.** {(1, 2, 3)} **31.** {(2, 3, 5)}
33. {(7/16, −5/16, −13/16)} **35.** {(−2/7, 11/7, 5/7)}
37. $\{(x, x - 5/3, x - 4/3) \mid x \text{ is any real number}\}$ **39.** ∅
41. Jackie 27, Alisha 11, Rochelle 39 **43.** 30, 50, 100
45. {(16.8, 12.3, 11.2)}
47. \$1.099 regular, \$1.219 plus, \$1.279 supreme **49.** {(1, −2, 3, 2)}

Chapter 6 Review Exercises

1. $\begin{bmatrix} 5 & 4 \\ -1 & 6 \end{bmatrix}$ **3.** $\begin{bmatrix} 1 & -13 \\ -5 & 6 \end{bmatrix}$ **5.** $\begin{bmatrix} 3 & 8 \\ -2 & -6 \end{bmatrix}$ **7.** Undefined

9. $\begin{bmatrix} -11 \\ 14 \end{bmatrix}$ **11.** $\begin{bmatrix} 3 & 2 & -1 \\ -12 & -8 & 4 \\ 9 & 6 & -3 \end{bmatrix}$ **13.** $[1 \quad -1 \quad -1]$

15. Undefined **17.** $\begin{bmatrix} 2 & 1.5 \\ 1 & 1 \end{bmatrix}$ **19.** $\begin{bmatrix} -1 & 0 & 0 \\ 1 & 1 & 0 \\ -5 & -3 & 1 \end{bmatrix}$

21. $\begin{bmatrix} 3 & 4 \\ -1 & -1.5 \end{bmatrix}$ **23.** $\begin{bmatrix} 1 & 0 \\ 0 & 1 \end{bmatrix}$ **25.** 2 **27.** −1

29. {(10/3, 17/3)} **31.** {(−2, 3)} **33.** $\{(x, y) \mid x - 5y = 9\}$
35. ∅ **37.** {(1, 2, 3)} **39.** {(−3, 4, 1)}
41. $\left\{\left(\dfrac{3y+3}{2}, y, \dfrac{1-y}{2}\right) \middle| \; y \text{ is any real number}\right\}$ **43.** ∅
45. {(9, −12)} **47.** {(2, 4)} **49.** {(0, −3)}
51. {(0.5, 0.5, 0.5)} **53.** {(2, −3, 5)}
55. 225.56 gal A, 300.74 gal B
57. \$22.88 water, \$55.65 gas, \$111.30 electric

Chapter 6 Test

1. {(1, 1/3)} **2.** {(3, 2, 1)}

3. $\left\{\left(x, \dfrac{-x-1}{2}, \dfrac{3x-1}{2}\right) \middle| \; x \text{ is any real number}\right\}$ **4.** $\begin{bmatrix} 3 & -4 \\ -6 & 10 \end{bmatrix}$

5. $\begin{bmatrix} 0 & 1 \\ 0 & 2 \end{bmatrix}$ **6.** $\begin{bmatrix} 6 & -9 \\ -20 & 30 \end{bmatrix}$ **7.** $\begin{bmatrix} -3 \\ 8 \end{bmatrix}$ **8.** Undefined

9. $[-2 \quad 1 \quad 2]$ **10.** $\begin{bmatrix} 2 & 0 & -2 \\ 3 & 0 & -3 \\ -1 & 0 & 1 \end{bmatrix}$ **11.** $\begin{bmatrix} 2 & 0.5 \\ 1 & 0.5 \end{bmatrix}$

12. $\begin{bmatrix} 7 & -5 & -8 \\ 3 & -2 & -3 \\ 6 & -4 & -7 \end{bmatrix}$ **13.** 2 **14.** 0 **15.** −1 **16.** {(5, 3)}

17. ∅ **18.** {(6, 2, 4)} **19.** {(−2, −3)} **20.** {(15, 6, 13)}
21. Bought 46, sold 34 **22.** $y = 0.5x^2 - 4\sqrt{x} + 3$

Tying It All Together Chapters P–6

1. {−1/3} **2.** {29/15} **3.** {1/3, 2} **4.** {−3/2} **5.** {(4, −2)}
6. {(2/5, 1/5)} **7.** {(7, −100)} **8.** {(−2, −3)} **9.** {(1, 2)}
10. {(2, 1)} **11.** {(−4, −3), (3, 4)} **12.** {(−2, 3), (1, 0)}

CHAPTER 7

Section 7.1

1. $y = \frac{1}{4}x^2$ **3.** $y = -x^2$ **5.** $y = \frac{1}{6}(x - 3)^2 + \frac{7}{2}$

7. $y = -\frac{1}{10}(x - 1)^2 - \frac{1}{2}$ **9.** $y = 1.25(x + 2)^2 + 1$

11. $y = (x - 4)^2 - 13$, (4, −13), (4, −12.75), $y = -13.125$
13. $y = 2(x + 3)^2 - 13$, (−3, −13), (−3, −103/8), $y = -105/8$
15. $y = -2(x - 1.5)^2 + 5.5$, (1.5, 5.5), (1.5, 5.375), $y = 5.625$
17. $y = 5(x + 3)^2 - 45$, (−3, −45), (−3, −44.95), $y = -45.05$

19. $y = \frac{1}{8}(x - 2)^2 + 4$, (2, 4), (2, 6), $y = 2$

21. (2, −1), (2, −3/4), $y = -5/4$, up
23. (1, −4), (1, −17/4), $y = -15/4$, down
25. (1, −2), (1, −23/12), $y = -25/12$, up
27. (−3, 13/2), (−3, 6), $y = 7$, down
29. (0, 5), (0, 6), $y = 4$, up
31. (1/2, −9/4), $x = 1/2$, (−1, 0), (2, 0), (0, −2)

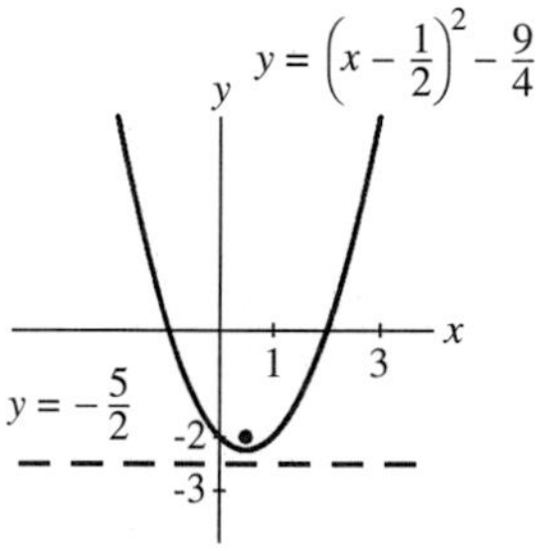

33. (−1/2, 25/4), $x = -1/2$, (−3, 0), (2, 0), (0, 6)

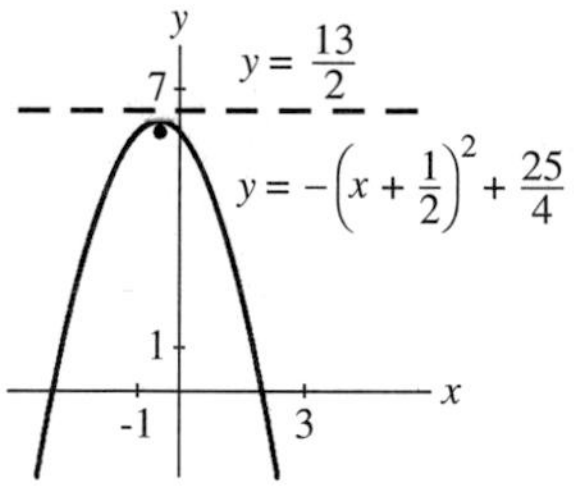

35. (−2, 2), $x = -2$, (0, 4), (−2, 5/2), $y = 3/2$

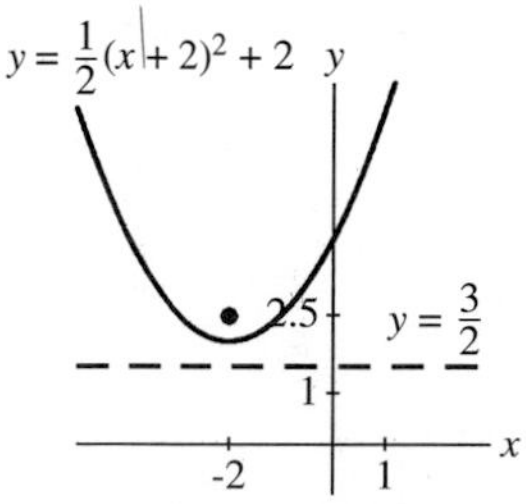

37. $(-4, 2)$, $x = -4$, $(-4 \pm 2\sqrt{2}, 0)$, $(0, -2)$, $(-4, 1)$, $y = 3$

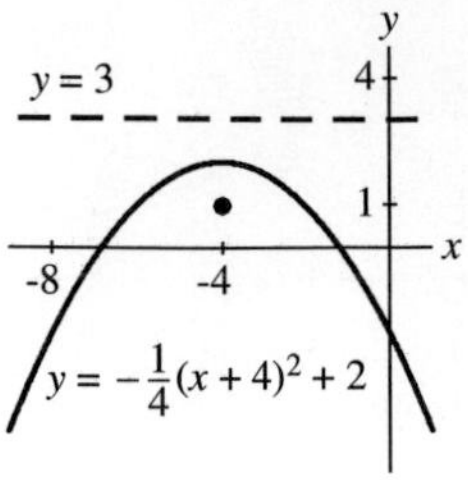

39. $(0, -2)$, $x = 0$, $(\pm 2, 0)$, $(0, -2)$, $(0, -3/2)$, $y = -5/2$

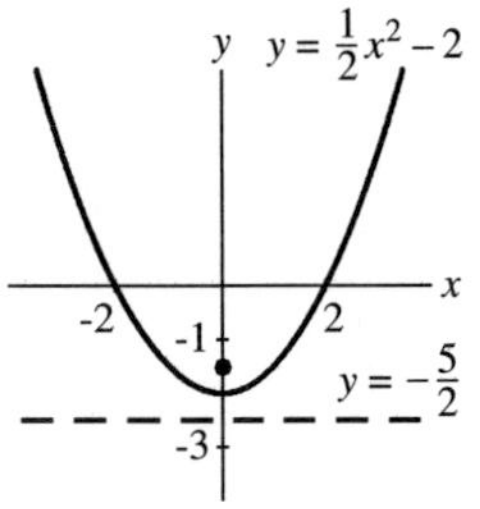

41. $(2, 0)$, $x = 2$, $(2, 0)$, $(0, 4)$, $(2, 1/4)$, $y = -1/4$

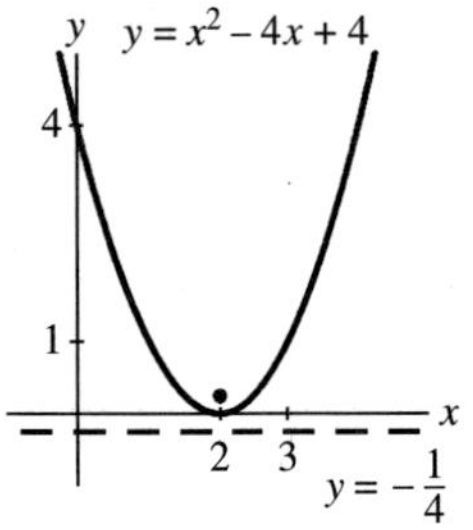

43. $(3/2, -3/4)$, $x = 3/2$, $(0, 0)$, $(3, 0)$, $(0, 0)$, $(3/2, 0)$, $y = -3/2$

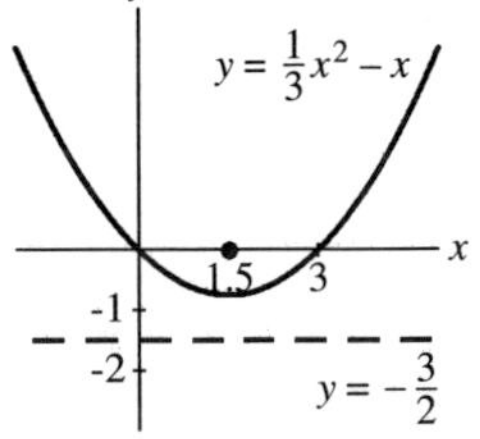

45. $(0, 0)$, $y = 0$, $(0, 0)$, $(0, 0)$, $(-1/4, 0)$, $x = 1/4$

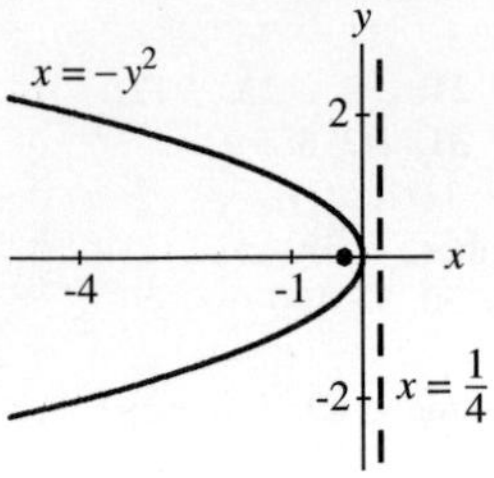

47. $(1, 0)$, $y = 0$, $(1, 0)$, $(0, \pm 2)$, $(0, 0)$, $x = 2$

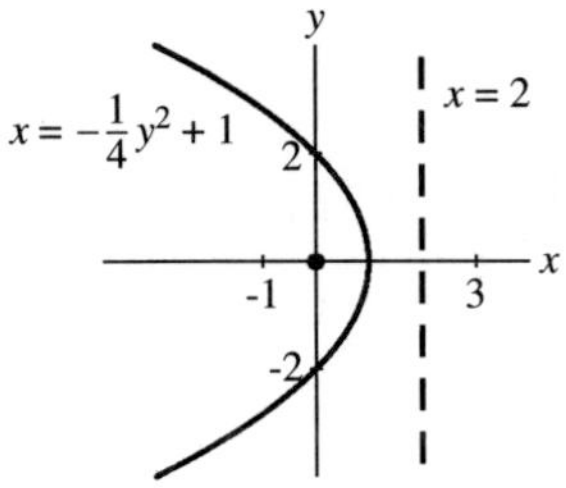

49. $(-25/4, -1/2)$, $y = -1/2$, $(-6, 0)$, $(0, -3)$, $(0, 2)$, $(-6, -1/2)$, $x = -13/2$

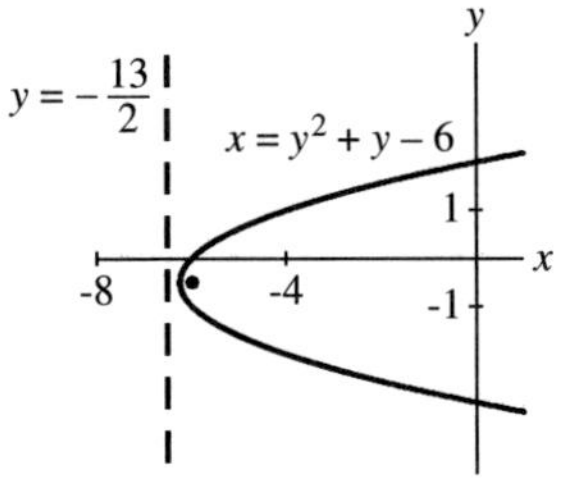

51. $(-7/2, -1)$, $y = -1$, $(-4, 0)$, $(-4, -1)$, $x = -3$

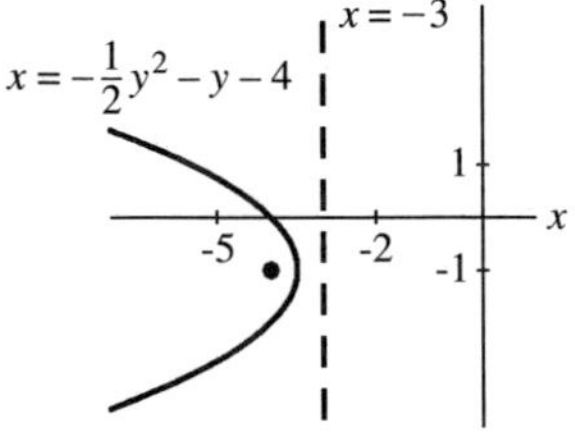

53. (3, 1), $y = 1$, (5, 0), (25/8, 1), $x = 23/8$

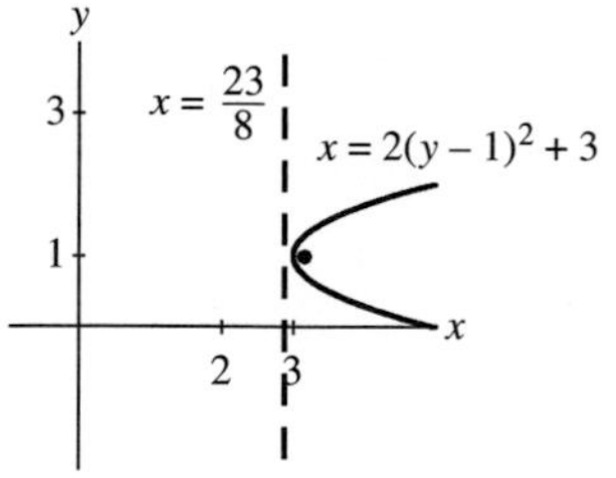

55. (1, −2), $y = -2$, (−1, 0), $\left(0, -2 \pm \sqrt{2}\right)$, (1/2, −2), $x = 3/2$

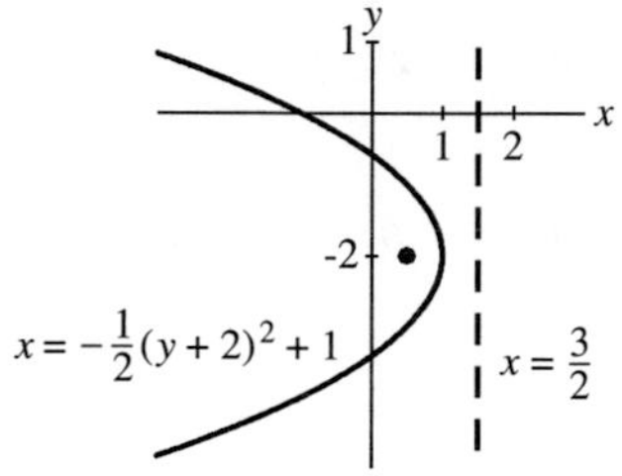

57. $y = \frac{1}{4}(x-1)^2 + 4$ **59.** $x = \frac{1}{8}y^2$ **61.** $y = \frac{1}{2640}x^2$, 26.8 in.
63. They look alike. **65.** $y = \pm\sqrt{-x}$

Section 7.2

1.

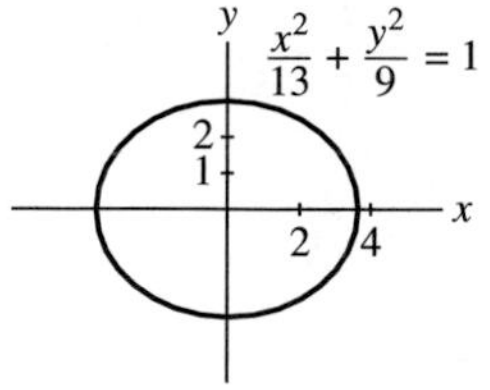

3.

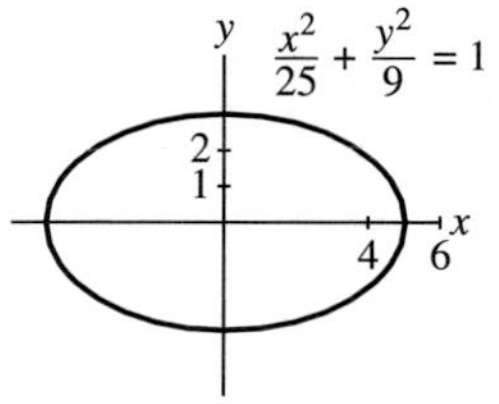

5.

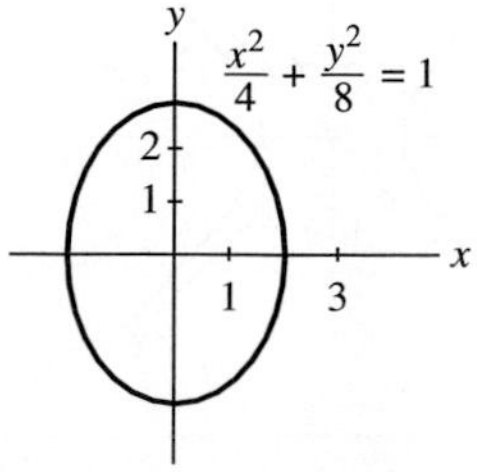

7.

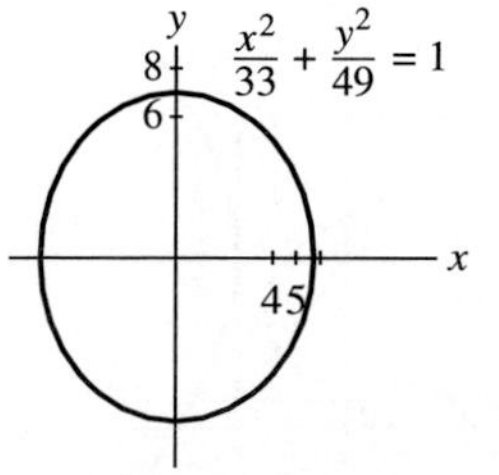

9. $(\pm 2\sqrt{3}, 0)$

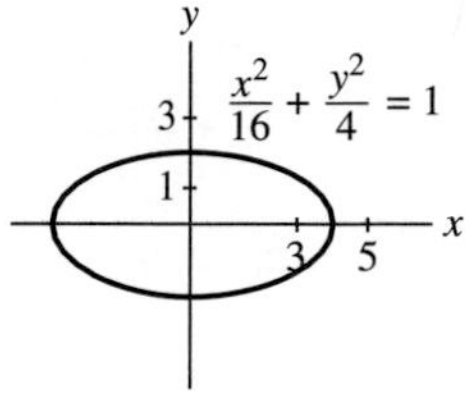

11. $(0, \pm 3\sqrt{3})$

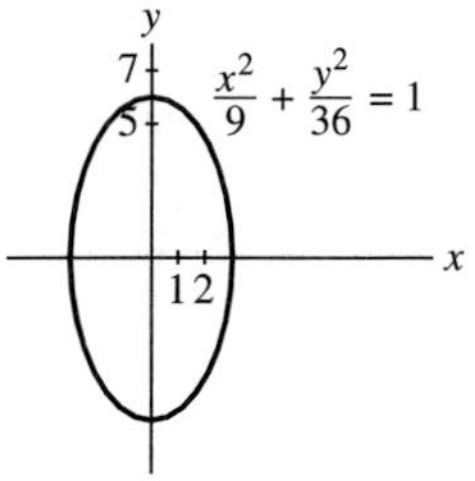

13. $(\pm 2\sqrt{6}, 0)$

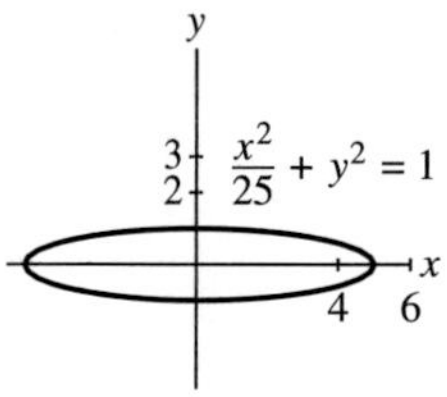

15. $(0, \pm 4)$

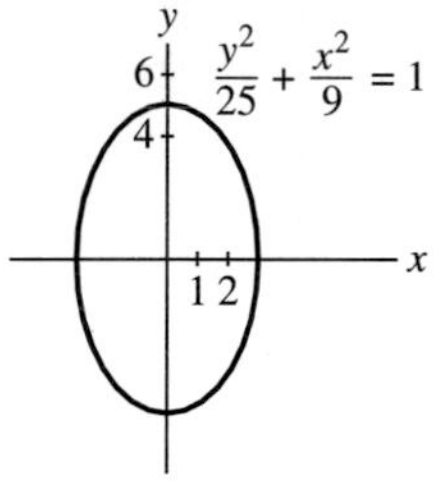

17. $(0, \pm 2\sqrt{2})$

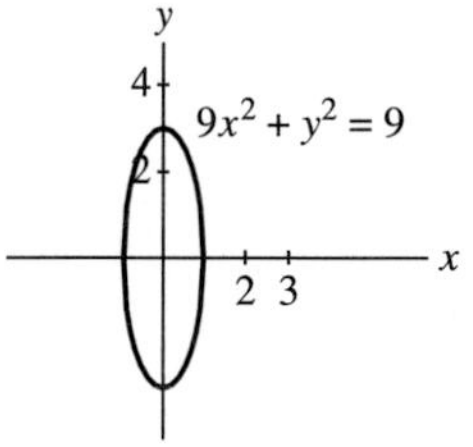

19. $(\pm\sqrt{5}, 0)$

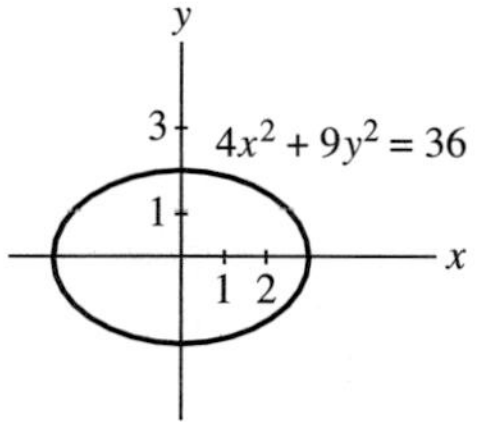

21. $(1 \pm \sqrt{7}, -3)$

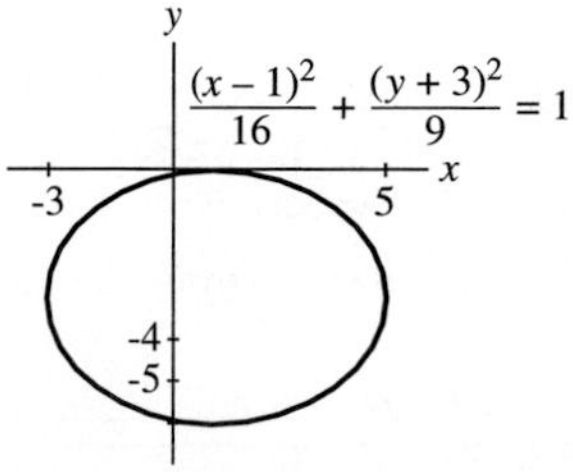

23. (3, 2), (3, −6)

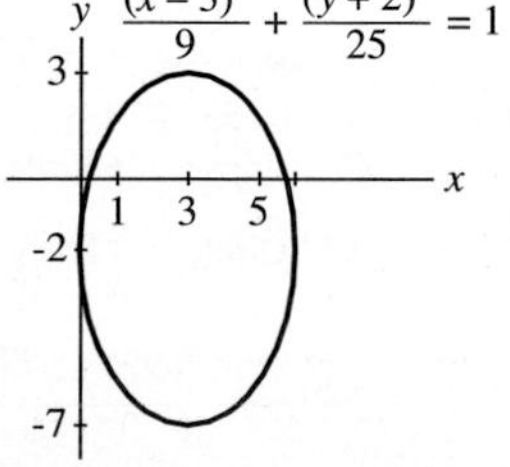

25. $(-4 \pm \sqrt{35}, -3)$

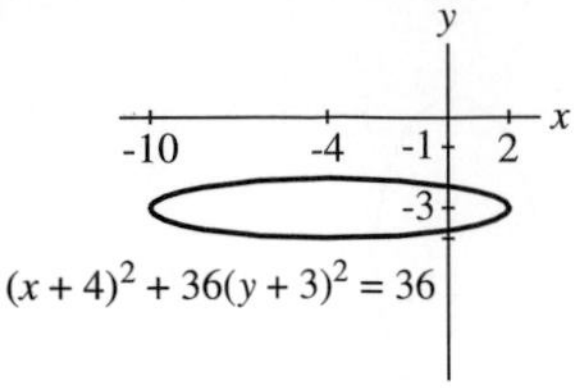

27. $(1, -2 + \sqrt{5})$, $(1, -2 - \sqrt{5})$

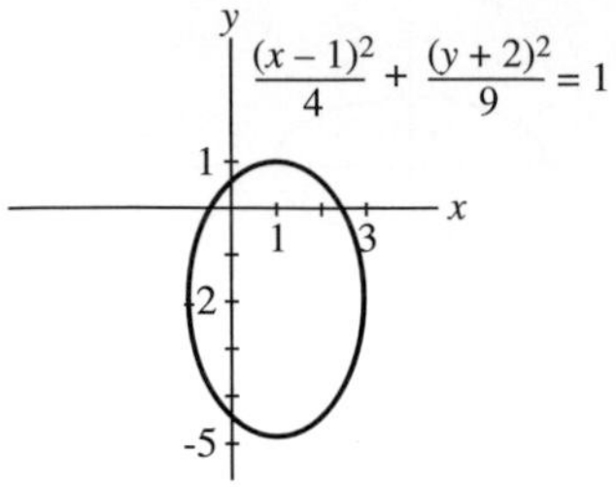

29. $\dfrac{x^2}{16} + \dfrac{y^2}{4} = 1$, $(\pm 2\sqrt{3}, 0)$

31. $\dfrac{(x+1)^2}{4} + \dfrac{(y+2)^2}{16} = 1$, $(-1, -2 \pm 2\sqrt{3})$

33. $x^2 + y^2 = 4$ **35.** $x^2 + y^2 = 41$

37. $(x-2)^2 + (y+3)^2 = 20$ **39.** $(x-1)^2 + (y-3)^2 = 5$

41. $(0, 0)$, 10 **43.** $(1, 2)$, 2

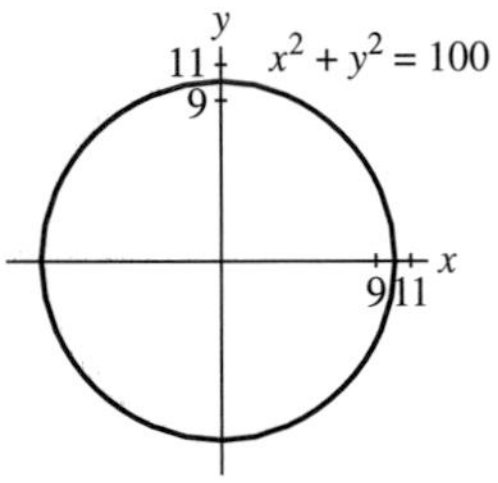

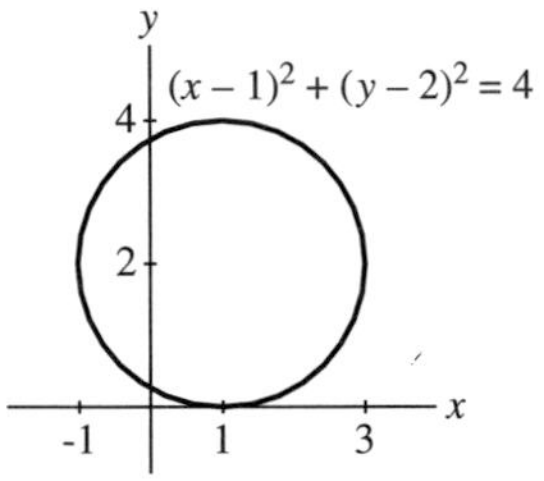

45. $(-2, -2)$, $2\sqrt{2}$

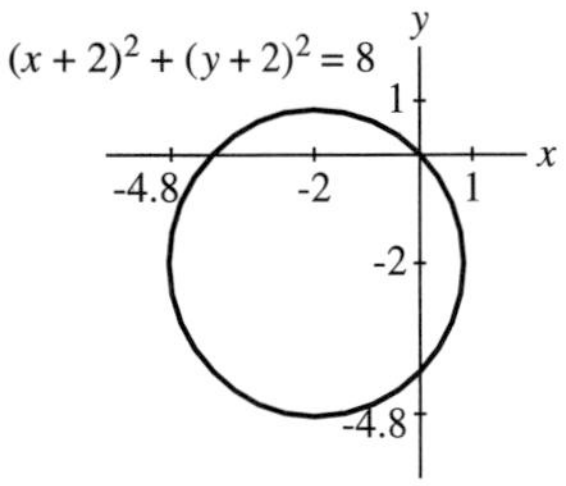

47. $(0, -1)$, 3 **49.** $(-4, 5)$, $\sqrt{41}$ **51.** $(-2, 0)$, 3

53. $(0.5, -0.5)$, 1 **55.** $(-1/3, -1/6)$, 1/2 **57.** $(-1, 0)$, $\sqrt{6}/2$

59. Circle **61.** Ellipse **63.** Parabola **65.** Parabola

67. Ellipse **69.** Circle **71.** $\dfrac{x^2}{16} + \dfrac{y^2}{12} = 1$ **73.** $(12, 0)$

75. $\dfrac{x^2}{260.5^2} + \dfrac{y^2}{520} = 1$, 0.996 **77.** 5.25×10^9 km

79. $y = \pm\sqrt{6360^2 - x^2}$

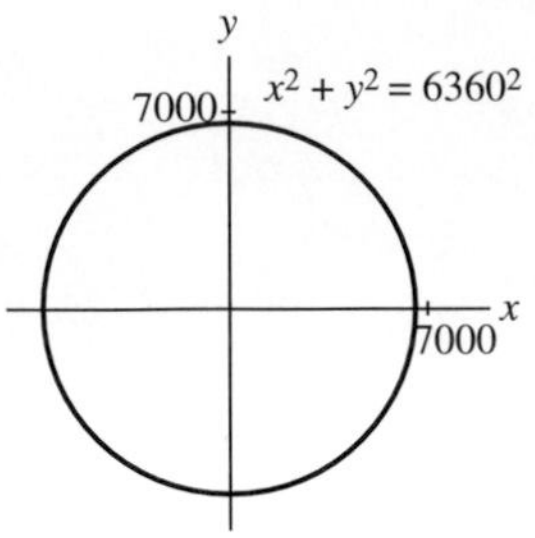

81. $y = \dfrac{4}{5}x + 5$, $(-4, 9/5)$ **83.** Parabolic

Section 7.3

1. $(\pm\sqrt{13}, 0)$, $y = \pm\dfrac{3}{2}x$

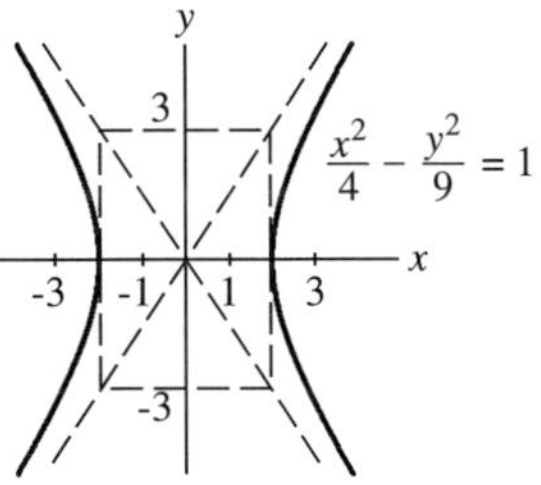

3. $(0, \pm\sqrt{29})$, $y = \pm\dfrac{2}{5}x$

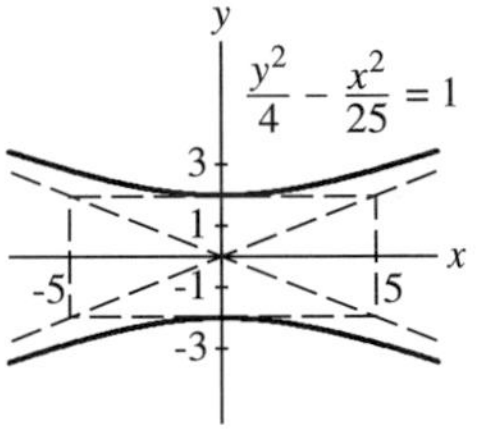

5. $(\pm\sqrt{5}, 0)$, $y = \pm\dfrac{1}{2}x$

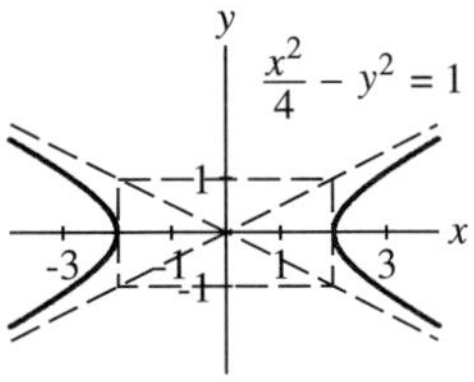

7. $(\pm\sqrt{10}, 0)$, $y = \pm 3x$

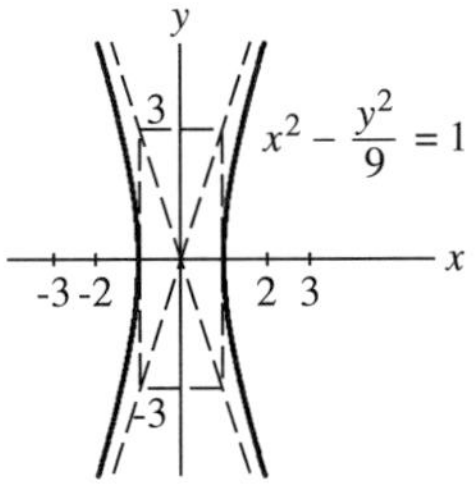

9. $(\pm 5, 0)$, $y = \pm\dfrac{4}{3}x$

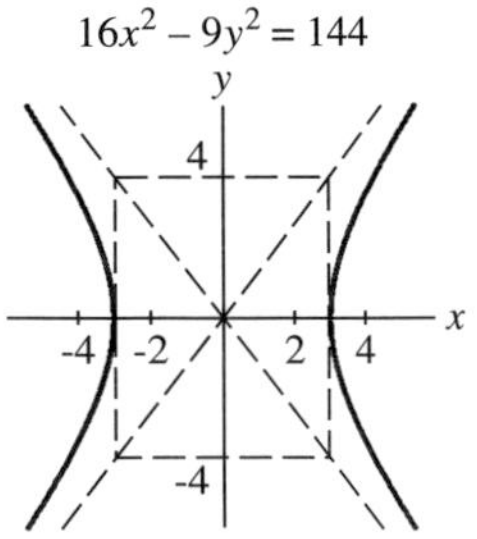

11. $(\pm\sqrt{2}, 0)$, $y = \pm x$

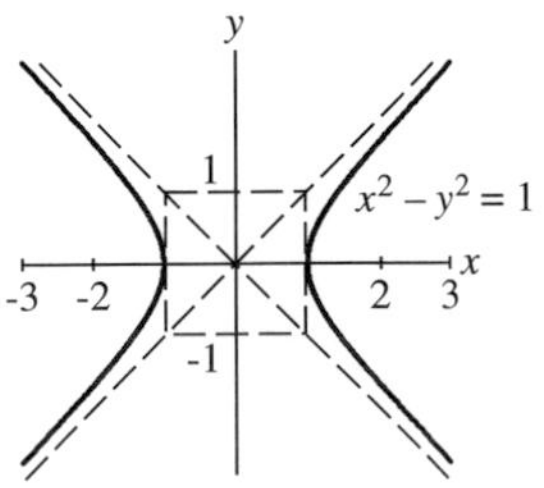

13. $\left(-1 \pm \sqrt{13}, 2\right), y = \frac{3}{2}x + \frac{7}{2}, y = -\frac{3}{2}x + \frac{1}{2}$

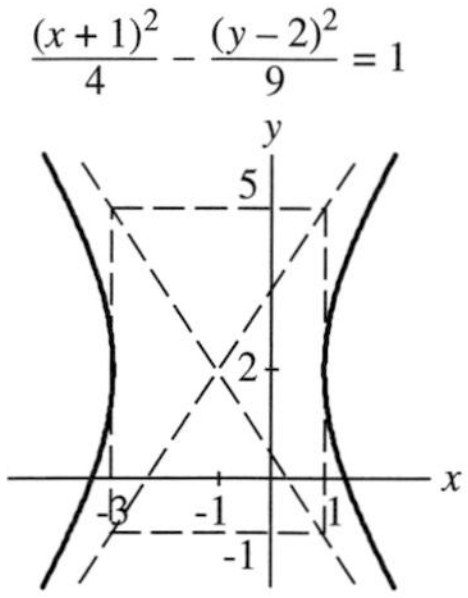

15. $\left(-2, 1 \pm \sqrt{5}\right), y = 2x + 5, y = -2x - 3$

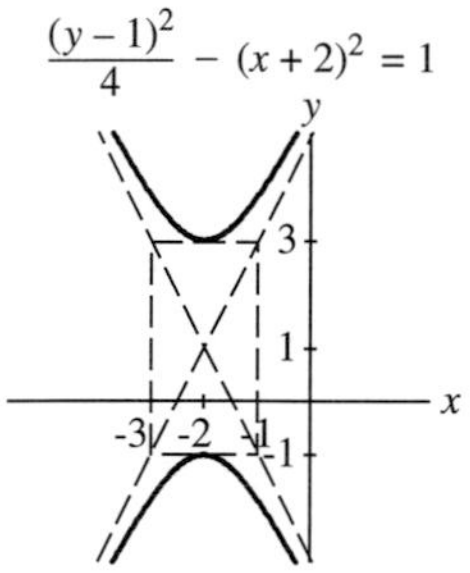

17. $(3, 3), (-7, 3), y = \frac{3}{4}x + \frac{9}{2}, y = -\frac{3}{4}x + \frac{3}{2}$

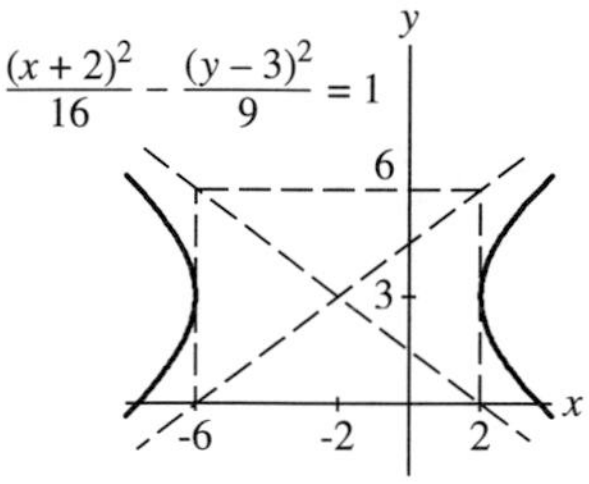

19. $\left(3, 3 \pm \sqrt{2}\right), y = x, y = -x + 6$

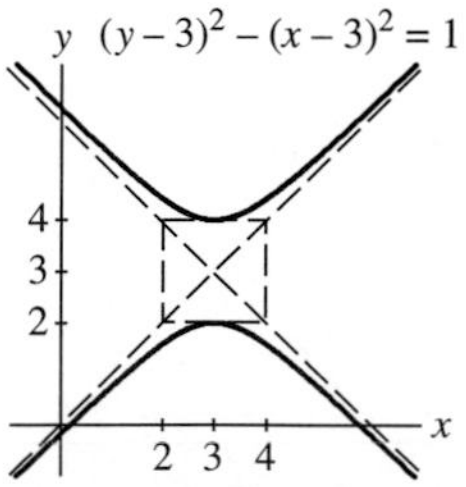

21. $\frac{x^2}{36} - \frac{y^2}{9} = 1$ **23.** $\frac{x^2}{9} - \frac{y^2}{16} = 1$ **25.** $\frac{x^2}{9} - \frac{y^2}{25} = 1$

27. $\frac{x^2}{9} - \frac{y^2}{16} = 1$ **29.** $\frac{y^2}{9} - \frac{(x-1)^2}{9} = 1$

31. $y^2 - (x-1)^2 = 1$, hyperbola **33.** $y = x^2 + 2x$, parabola

35. $x^2 + y^2 = 100$, circle **37.** $x = -4y^2 + 100$, parabola

39. $(x-1)^2 + (y-2)^2 = \frac{1}{2}$, circle

41. $\frac{(x+1)^2}{2} + \frac{(y+3)^2}{4} = 1$, ellipse **43.** $\frac{y^2}{64} - \frac{x^2}{36} = 1$

45. $\left(\frac{4\sqrt{14}}{7}, \frac{3\sqrt{7}}{7}\right)$ **47.** Four points: $\left(\pm\frac{3\sqrt{2}}{2}, \pm\frac{\sqrt{14}}{2}\right)$

49. 0.01

Chapter 7 Review Exercises

1. $(-2, -16), x = -2$
$(-2, -63/4), y = -65/4$

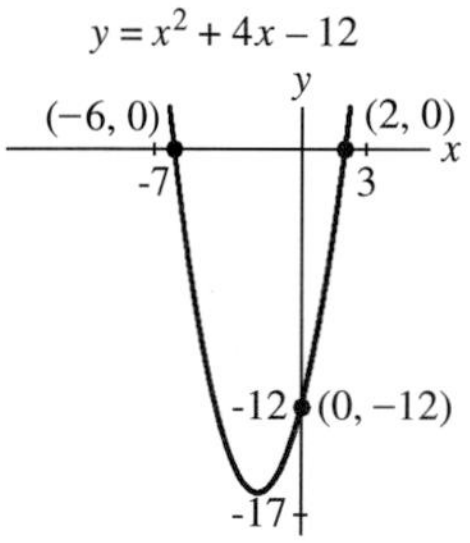

3. $(3/2, 9/2), x = 3/2$
$(3/2, 35/8), y = 37/8$

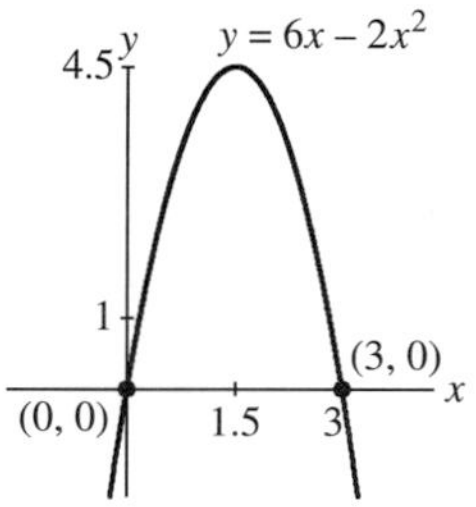

5. $(-10, -2), y = -2$
$(-39/4, -2), x = -41/4$

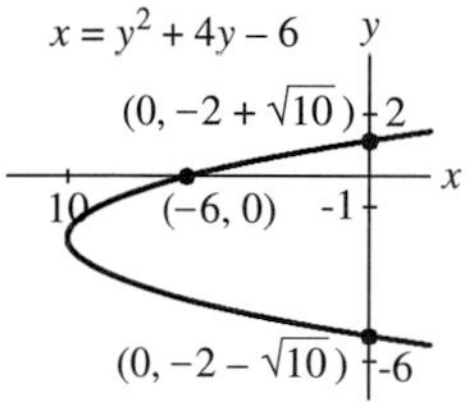

7. $\left(0, \pm 2\sqrt{5}\right)$

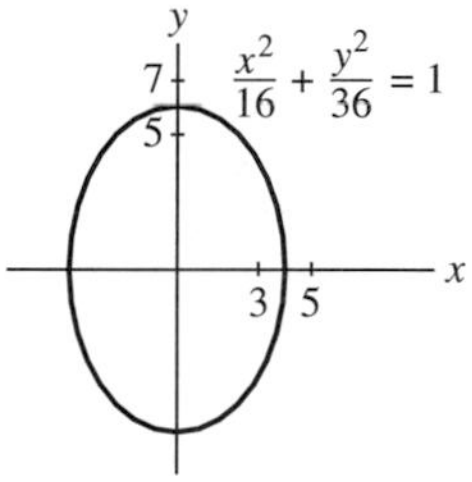

9. $(1, 5), (1, -3)$

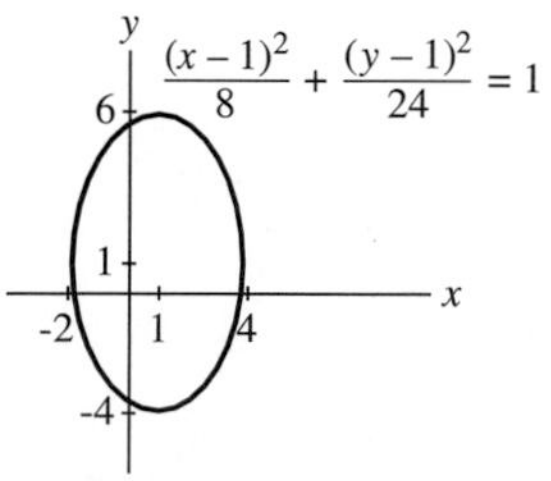

11. $\left(1, -3 \pm \sqrt{2}\right)$

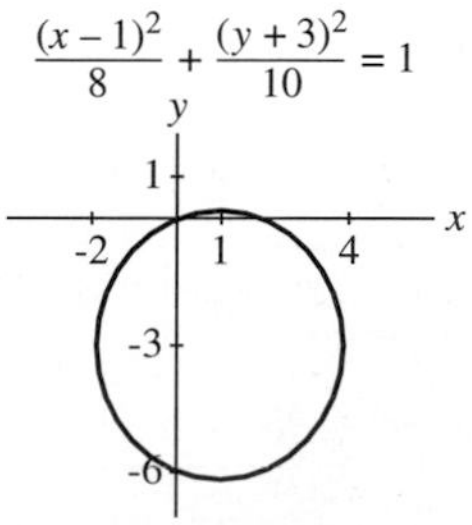

13. (0, 0), 9

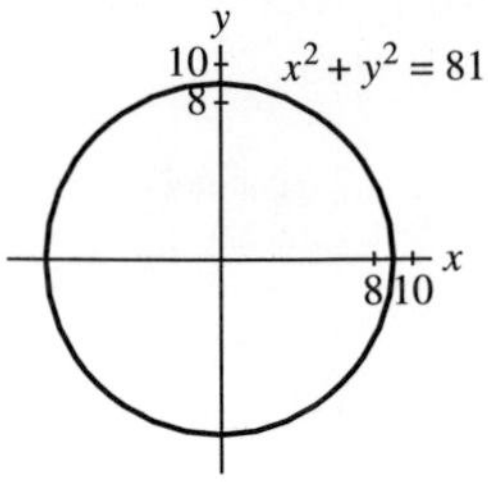

15. (−1, 0), 2

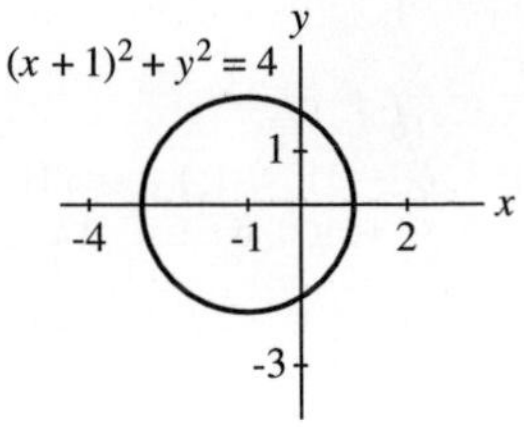

17. $(-5/2, 0), \sqrt{6}$

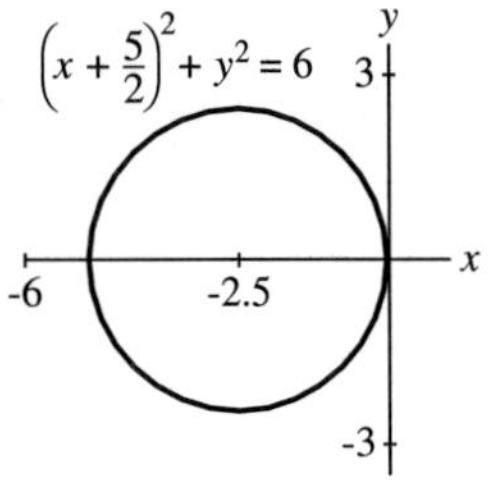

19. $x^2 + (y + 4)^2 = 9$ **21.** $(x + 2)^2 + (y + 7)^2 = 6$

23. $(\pm 10, 0), y = \pm\frac{3}{4}x$

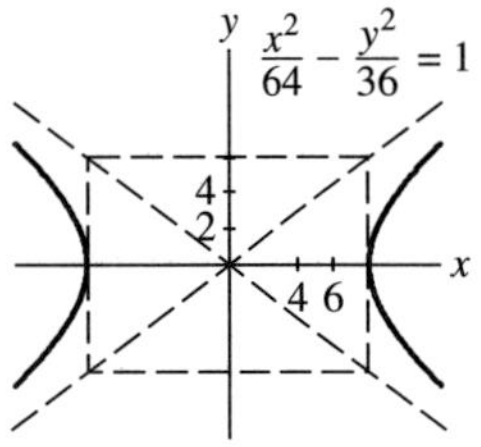

25. $(4, 2 \pm 4\sqrt{5}), y = 2x - 6, y = -2x + 10$

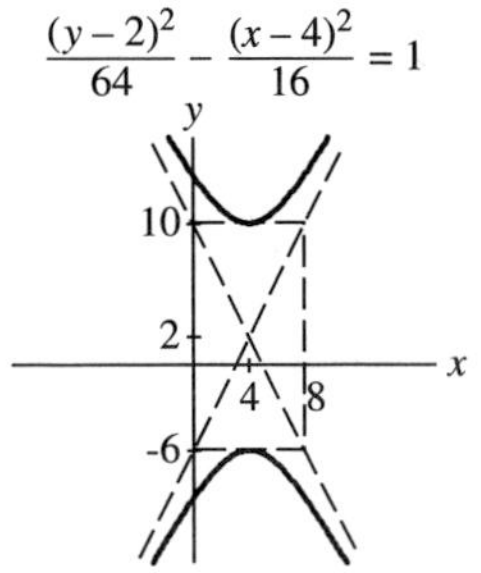

27. $(2 \pm \sqrt{5}, 4), y = \frac{1}{2}x + 3, y = -\frac{1}{2}x + 5$

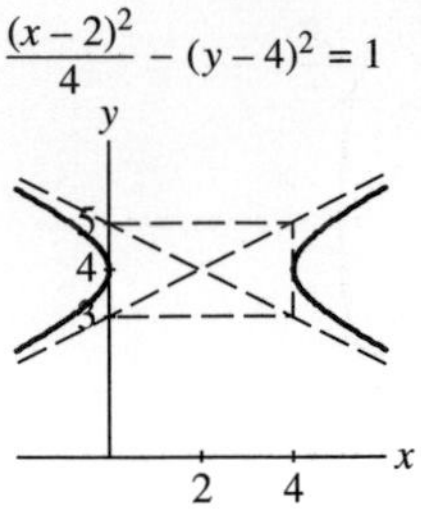

29. Hyperbola **31.** Ellipse **33.** Parabola **35.** Hyperbola

37. $x^2 + y^2 = 4$

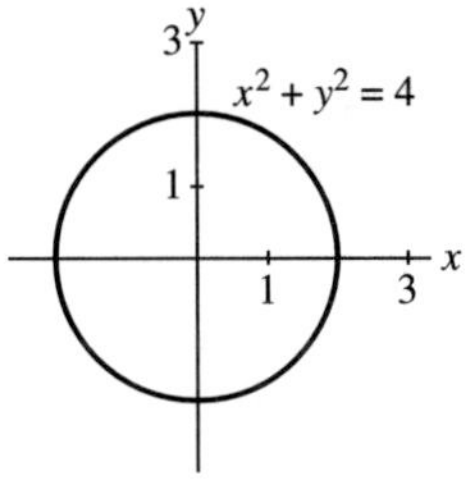

39. $y = \frac{1}{4}x^2 - 1$

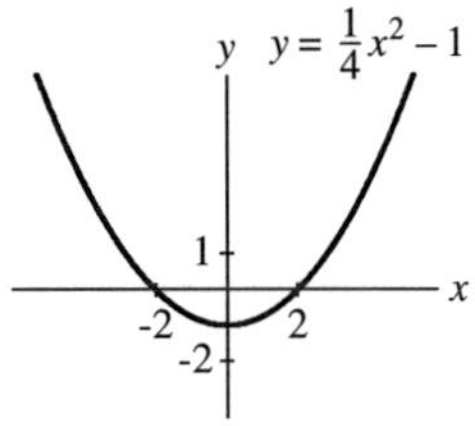

41. $\frac{x^2}{4} + y^2 = 1$

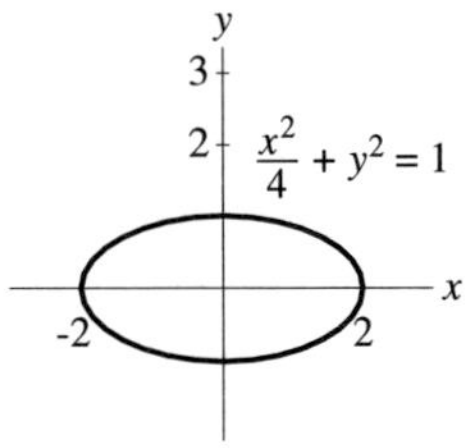

43. $y^2 + \frac{(x - 2)^2}{4} = 1$

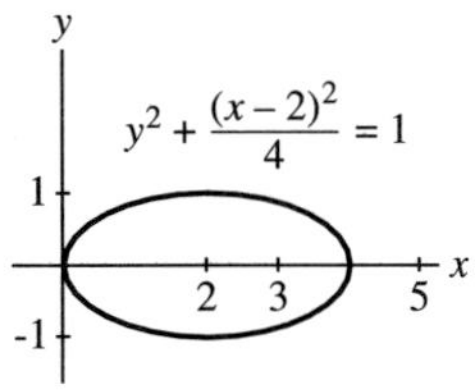

45. $x = (y - 3)^2 + \frac{3}{4}$ **47.** $\frac{x^2}{36} + \frac{y^2}{20} = 1$

49. $(x - 1)^2 + (y - 3)^2 = 20$ **51.** $\frac{x^2}{4} - \frac{y^2}{5} = 1$

53. $(x + 2)^2 + (y - 3)^2 = 9$ **55.** $\frac{(x + 2)^2}{9} + (y - 1)^2 = 1$

57. $\frac{(y - 1)^2}{9} - \frac{(x - 2)^2}{4} = 1$ **59.** $\frac{x^2}{100^2} - \frac{y^2}{120^2} = 1$

61. $\frac{x^2}{34^2} + \frac{y^2}{16^2} = 1$, 10.81 ft

Chapter 7 Test

1.

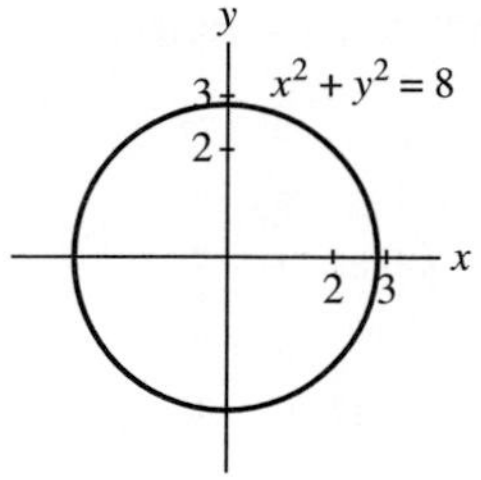

2.

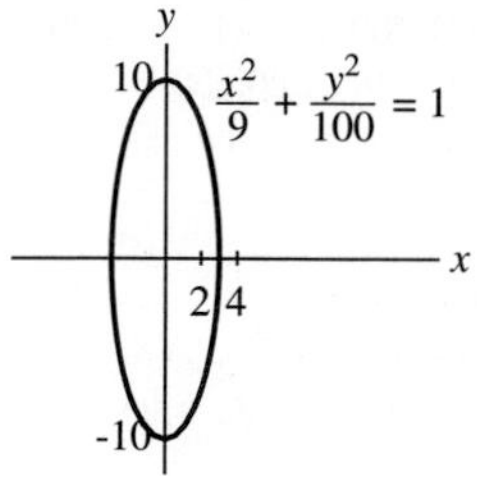

3.

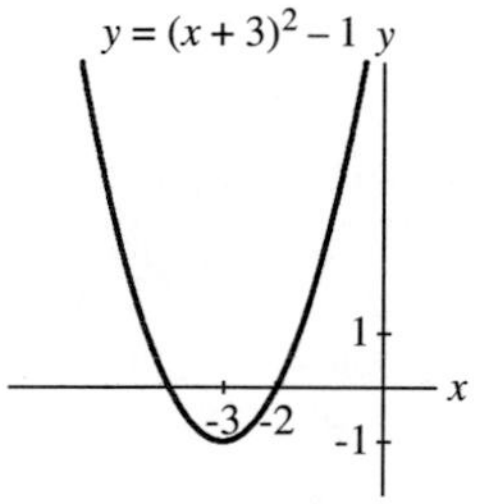

4.

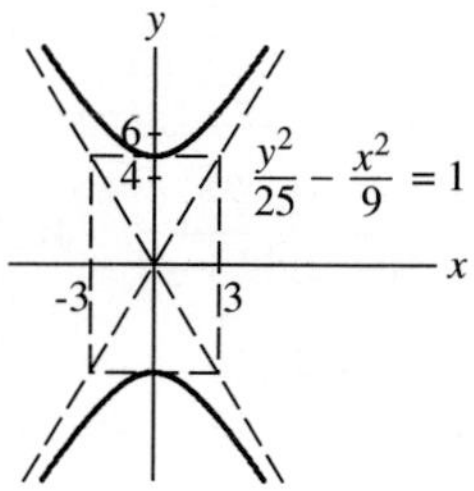

5.

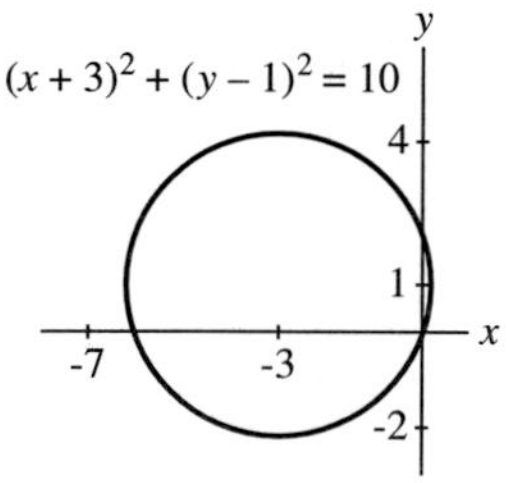

6.

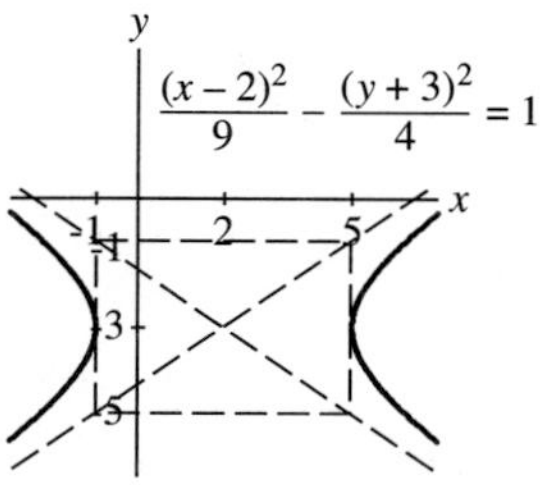

7. Hyperbola **8.** Parabola **9.** Circle **10.** Ellipse

11. $(x + 3)^2 + (y - 4)^2 = 12$ **12.** $x = \frac{1}{8}y^2$ **13.** $\frac{x^2}{4} + \frac{y^2}{10} = 1$

14. $\frac{x^2}{36} - \frac{y^2}{28} = 1$ **15.** $(2, -15/4)$, $y = -17/4$, $(2, -4)$, $x = 2$

16. $(\pm 2\sqrt{3}, 0)$, 8, 4 **17.** $(0, \pm\sqrt{17})$, $(0, \pm 1)$, $y = \pm\frac{1}{4}x$, 2, 8

18. $\left(-\frac{1}{2}, \frac{3}{2}\right)$, $\frac{3}{2}$ **19.** 24 cm

Tying It All Together Chapters P–7

1.

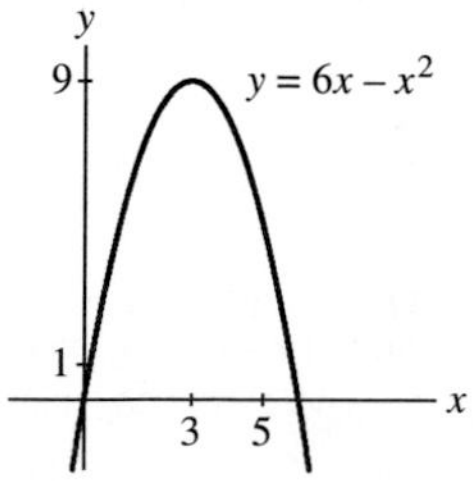

2.

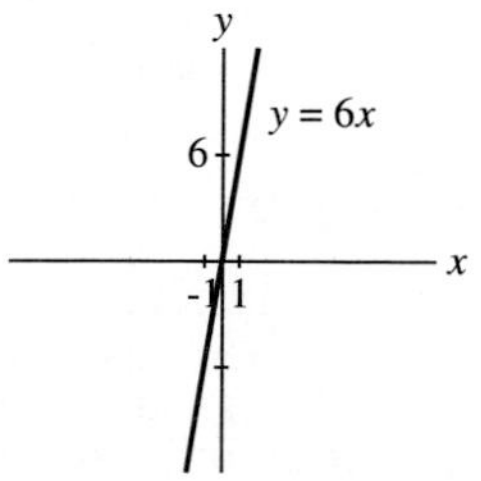

3.

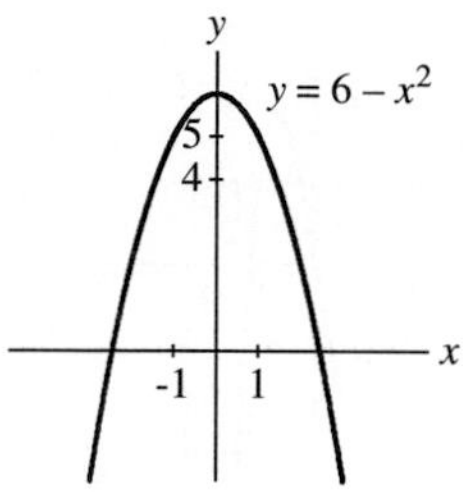

4.

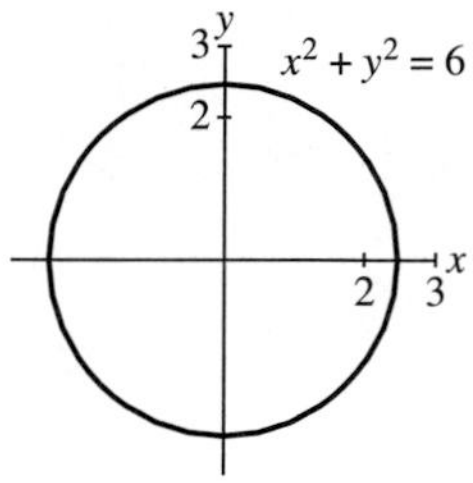

5.

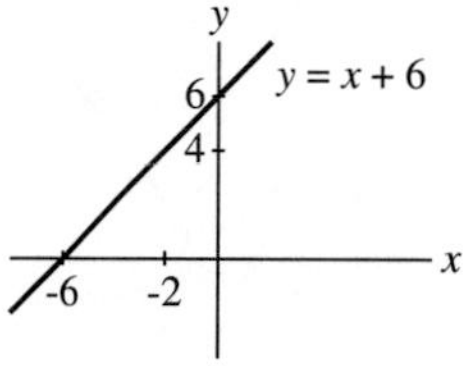

6.

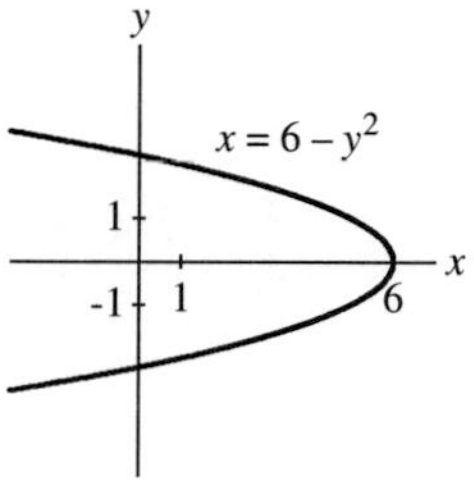

7.

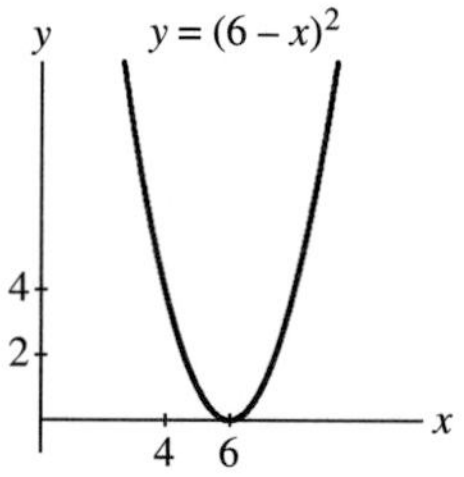

8.

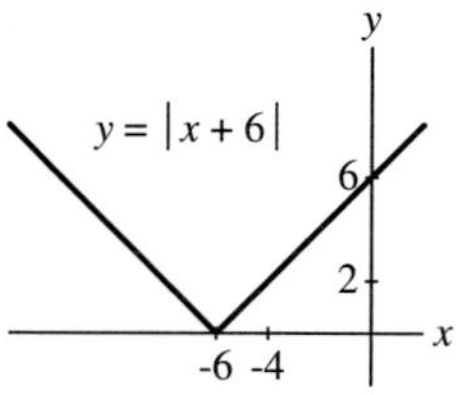

9.

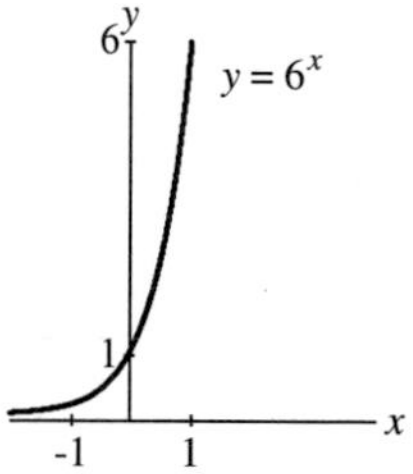

10.

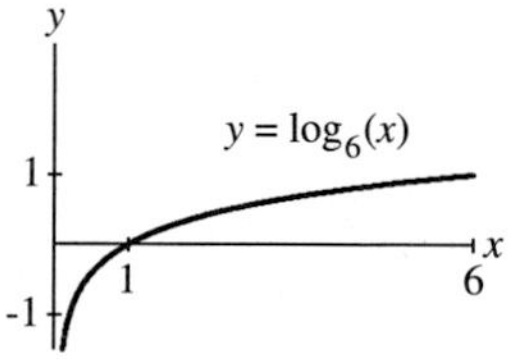

11.

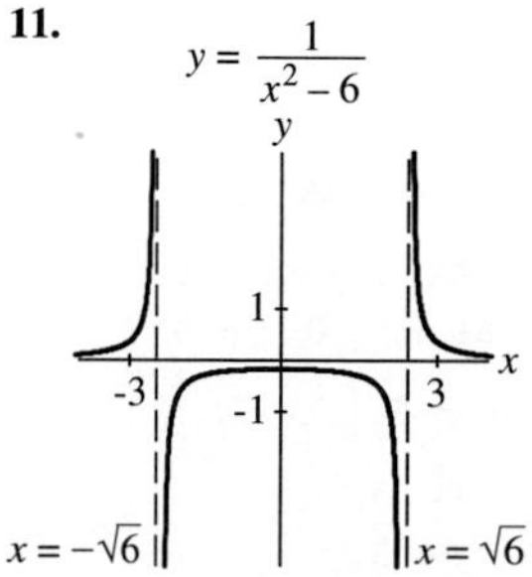

12.

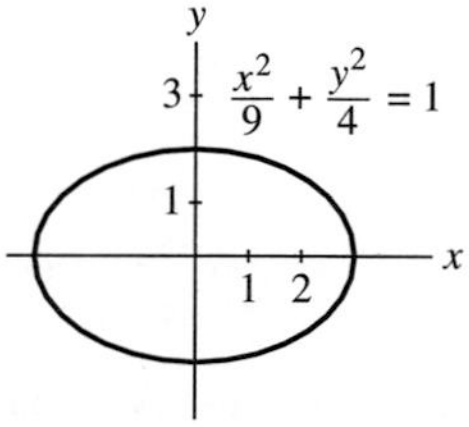

13.

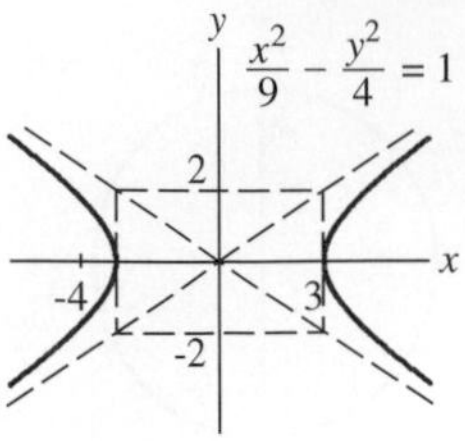

14.

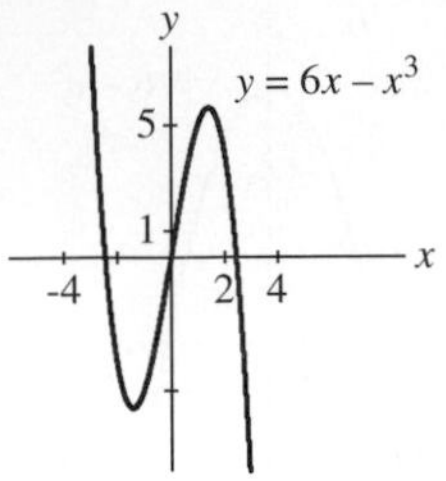

15.

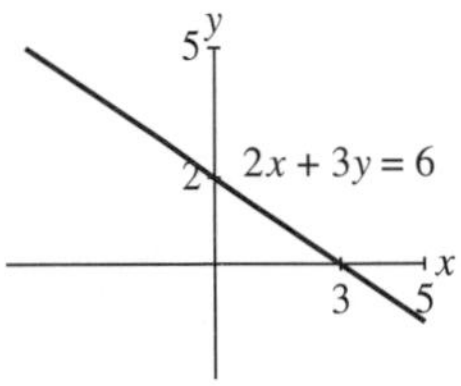

16.

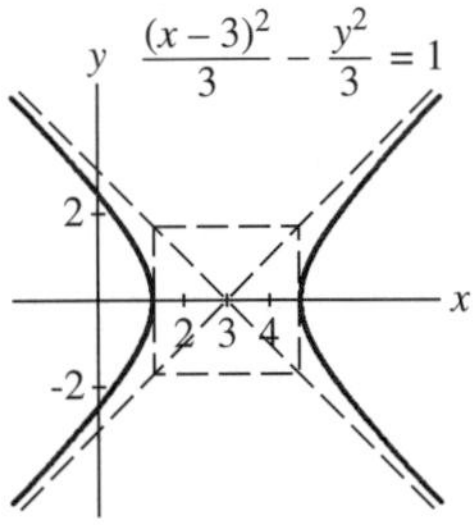

17. $\{-5/3\}$ **18.** $\{-5/3\}$ **19.** $\{11/15\}$ **20.** $\{0\}$ **21.** $\{22/3\}$
22. $\{50\}$ **23.** $\{-17, 3/2\}$ **24.** $\{-31/2, 0\}$
25. $\{17 - \sqrt{3}, 17 + \sqrt{3}\}$ **26.** $\{17 - i, 17 + i\}$

CHAPTER 8

Section 8.1

1. 1, 4, 9, 16, 25, 36, 49 **3.** $\frac{1}{2}, -\frac{1}{3}, \frac{1}{4}, -\frac{1}{5}, \frac{1}{6}, -\frac{1}{7}, \frac{1}{8}, -\frac{1}{9}$

5. 1, −2, 4, −8, 16, −32 **7.** $2, 1, \frac{1}{2}, \frac{1}{4}, \frac{1}{8}$

9. −6, −10, −14, −18, −22 **11.** 5, 5.5, 6, 6.5, 7, 7.5, 8

13. $1, 2, \frac{3}{2}, \frac{2}{3}, \frac{5}{24}$ **15.** 1, 1, 2, 6, 24, 120, 720

17. 8, 5, 2, −1; −19 **19.** $\frac{4}{3}, \frac{4}{5}, \frac{4}{7}, \frac{4}{9}; \frac{4}{21}$

21. $-\frac{1}{6}, \frac{1}{12}, -\frac{1}{20}, \frac{1}{30}; \frac{1}{132}$ **23.** $2, 2, \frac{4}{3}, \frac{2}{3}; \frac{4}{14{,}175}$

25. $-2, -8, -16, -\frac{64}{3}; -\frac{4096}{2835}$ **27.** 8.9, 8.8, 8.7, 8.6; 8

29. −4, −10, −28, −82; −6,562 **31.** 2, 1, −2, 1; 1

33. −15, −8, −1, 6; 34 **35.** $a_n = 2n$ **37.** $a_n = 2n + 7$

39. $a_n = (-1)^{n+1}$ **41.** $a_n = n^3$ **43.** $a_n = e^n$ **45.** $a_n = \frac{1}{2^{n-1}}$

47. Yes **49.** No **51.** No **53.** Yes **55.** $a_n = 5n - 4$

57. $a_n = 2n - 2$ **59.** $a_n = -4n + 9$ **61.** $a_n = 0.1n + 0.9$

63. $a_n = \frac{\pi}{6}n$ **65.** $a_n = 15n + 5$ **67.** 6, 3, 0, −3; −21

69. 1, 0.9, 0.8, 0.7; 0.1 **71.** $\frac{14}{3}, \frac{13}{3}, \frac{12}{3}, \frac{11}{3}; \frac{5}{3}$ **73.** 32 **75.** 27

77. 4.2 **79.** $a_n = 2.5n + 2.5$ **81.** $a_n = a_{n-1} + 9, a_1 = 3$

83. $a_n = 3a_{n-1}, a_1 = 1/3$ **85.** $a_n = \sqrt{a_{n-1}}, a_1 = 16$

87. \$21,709, \$23,011, \$24,392, \$25,856, \$27,407, $p = 20{,}480(1.06)^n$

89. 92 **91.** \$60,200

93. $C_n = 4, E_n = 8n - 8, I_n = 4n^2 - 8n + 4$ for $n = 1, 2, 3, \ldots$

Section 8.2

1. 55 **3.** 36 **5.** 15/32 **7.** 485 **9.** 1 **11.** 0 **13.** $\sum_{i=1}^{6} i$

15. $\sum_{i=1}^{5} (-1)^i(2i - 1)$ **17.** $\sum_{i=1}^{5} i^2$ **19.** $\sum_{i=1}^{5} \left(-\frac{1}{2}\right)^{i-1}$ **21.** $\sum_{i=1}^{3} \ln(x_i)$

23. $\sum_{i=1}^{11} ar^{i-1}$ **25.** $\sum_{j=0}^{31} (-1)^{j+1}$ **27.** $\sum_{j=1}^{10} (2j + 7)$

29. $\sum_{j=0}^{8} \frac{10!}{(j + 2)!(8 - j)!}$ **31.** $\sum_{j=5}^{9} \frac{5^{j-3}e^{-5}}{(j - 3)!}$

33. $0.5 + 0.5r + 0.5r^2 + 0.5r^3 + 0.5r^4 + 0.5r^5$

35. $a^4 + a^3b + a^2b^2 + ab^3 + b^4$ **37.** $a^2 + 2ab + b^2$ **39.** 74/3

41. 16 **43.** 14.151 **45.** 1128 **47.** −36 **49.** 406 **51.** 52.25

53. 360 **55.** 1.3 **57.** \$1,335,000, \$44,500 **59.** $\sum_{i=1}^{9} 12i = 540$

61. $\sum_{i=1}^{10} 1000(1.05)^i$ **63.** 455 mg **65.** 228.5 **67.** −1/2

Section 8.3

1. $a_n = \frac{1}{6}2^{n-1}$ **3.** $a_n = 0.9(0.1)^{n-1}$ **5.** $a_n = 4(-3)^{n-1}$

7. Arithmetic **9.** Geometric **11.** Neither

13. 2, 4, 6, 8; arithmetic **15.** 1, 4, 9, 16; neither

17. $\frac{1}{2}, \frac{1}{4}, \frac{1}{8}, \frac{1}{16}$; geometric **19.** 8, 32, 128, 512; geometric

21. 3, −9, 27, −81; geometric **23.** 3, −6, 12; −1536

25. 4, 0.4, 0.04; 4×10^{-9} **27.** 11 **29.** 3 **31.** ±3

33. $a_n = -3(-2)^{n-1}$ **35.** 242/27 **37.** −127.5 **39.** 31.8343

41. 1657.13 **43.** $\sum_{n=1}^{5} 3\left(-\frac{1}{3}\right)^{n-1}$ **45.** $\sum_{n=1}^{\infty} 0.6(0.1)^{n-1}$

47. $\sum_{n=1}^{\infty} -4.5\left(-\frac{1}{3}\right)^{n-1}$ **49.** 9/4 **51.** 1 **53.** −297/40

55. 34/99 **57.** No sum **59.** 2/3 **61.** 20 **63.** 2/45

65. 8172/990 **67.** 322.55 mg **69.** 1,073,741,823

71. $4000(1.02)^n$, \$8322.74 **73.** \$2561.87 **75.** \$183,074.35

77. 45 ft **79.** \$8,000,000

Section 8.4

1. 8 **3.** 6 **5.** 1024 **7.** 8 **9.** 210 **11.** 1 **13.** 126

15. 165 **17.** 1820 **19.** 109,736 **21.** 479,001,600 **23.** 2730

25. 6.3×10^{10} **27.** 4096 **29.** 72 **31.** 15,600 **33.** 30,000

35. 24 **37.** 131,072 **39.** 210

Section 8.5

1. 2,598,960 **3.** 13,983,816 **5.** 10 **7.** 27 **9.** 56 **11.** 1260 **13.** 497,001,600 **15.** 36 **17.** 144 **19.** 210 **21.** 2520 **23.** 3, 6, 10, $C(n, 2)$ **25.** 2520 **27.** $a^3 - 6a^2 + 12a - 8$
29. $8a^3 + 12a^2b^2 + 6ab^4 + b^6$
31. $x^4 - 8x^3y + 24x^2y^2 - 32xy^3 + 16y^4$
33. $x^8 + 4x^6 + 6x^4 + 4x^2 + 1$ **35.** $x^9 + 9x^8y + 36x^7y^2$
37. $4{,}096x^{12} - 24{,}576x^{11}y + 67{,}584x^{10}y^2$
39. $256s^8 - 512s^7t + 448s^6t^2$ **41.** $m^{18} - 18m^{16}w^3 + 144m^{14}w^6$
43. 56 **45.** −41,184 **47.** 13,860 **49.** 120

Section 8.6

1. 1/3, 2/3, 0 **3.** 2/3, 1, 5/6, 0, 1/6 **5.** 1/4, 3/4, 1/4, 3/4
7. 1/36, 11/36, 5/36, 35/36, 1/36 **9.** 3/13, 9/13, 9/13, 4/13
11. 1/72, 1/36, 1/18 **13.** 1/2,598,960, 1024/2,598,960 **15.** 0.999
17. 1/36, 35/36, 35/36 **19.** 0.05, 0.95 **21.** 4 to 1
23. 1 to 3, 3 to 1 **25.** 80% **27.** 1 to 9, 9/10 **29.** 3 to 5
31. 1 to 5 **33.** 1 to 1,999,999 **35.** 1/32 **37.** 78% **39.** 1/4
41. 4/13 **43.** 34%, 40%, 74%

Section 8.7

1. 1, 2, 3 **3.** 1, 2, 3 **5.** 1, 2 **7.** 2, 3
9. S_1: $2(1) = 1(1 + 1)$, S_k: $\sum_{i=1}^{k} 2i = k(k + 1)$,
S_{k+1}: $\sum_{i=1}^{k+1} 2i = (k + 1)(k + 2)$
11. S_1: $2 = 2(1)^2$, S_k: $2 + 6 + \cdots + (4k - 2) = 2k^2$,
S_{k+1}: $2 + 6 + \cdots + (4k + 2) = 2(k + 1)^2$
13. S_1: $2 = 2^2 - 2$, S_k: $\sum_{i=1}^{k} 2^i = 2^{k+1} - 2$,
S_{k+1}: $\sum_{i=1}^{k+1} 2^i = 2^{k+2} - 2$
15. S_1: $(ab)^1 = a^1b^1$, S_k: $(ab)^k = a^kb^k$,
S_{k+1}: $(ab)^{k+1} = a^{k+1}b^{k+1}$
17. S_1: if $0 < a < 1$ then $0 < a^1 < 1$, S_k: if $0 < a < 1$ then $0 < a^k < 1$, S_{k+1}: if $0 < a < 1$ then $0 < a^{k+1} < 1$

Chapter 8 Review Exercises

1. 1, 2, 4, 8, 16 **3.** −1, 1/2, −1/6, 1/24 **5.** 3, 1.5, 0.75
7. 3, 0, −3 **9.** 0.9375 **11.** 5450 **13.** 1/3 **15.** 33,065.9541
17. $a_n = \dfrac{(-1)^n}{n + 2}$ **19.** $a_n = 6\left(\dfrac{1}{6}\right)^{n-1}$ **21.** $\sum_{i=1}^{\infty} \dfrac{(-1)^{i+1}}{i + 1}$ **23.** $\sum_{i=1}^{14} 2i$
25. ±2 **27.** \$243.52 **29.** \$13,971.64 **31.** 1 million
33. $a^4 + 8a^3b + 24a^2b^2 + 32ab^3 + 16b^4$
35. $32a^5 - 80a^4b + 80a^3b^2 - 40a^2b^3 + 10ab^4 - b^5$
37. $a^{10} + 10a^9b + 45a^8b^2$ **39.** $256x^8 + 512x^7y + 448x^6y^2$
41. 715 **43.** 36,960 **45.** 24 **47.** 1,953,125 **49.** 210
51. 60 **53.** 56 **55.** 70, 5, 30 **57.** 180, 120 **59.** 128
61. 1/1024, 1/1024 **63.** 5/13, 8/13, 0, 1 **65.** 1 to 7 **67.** 9 to 1
69. 90% **71.** 40,320 **73.** 20 **75.** 1680 **77.** 28 **79.** 1680
81. 8

Chapter 8 Test

1. 2.3, 2.8, 3.3, 3.8 **2.** 20, 10, 5, 2.5 **3.** $a_n = (-1)^{n-1}(n - 1)^2$
4. $a_n = 3n + 4$ **5.** $a_n = \dfrac{1}{3}\left(-\dfrac{1}{2}\right)^{n-1}$ **6.** 4023 **7.** 13,050.5997
8. 49 **9.** $a_n = 1.5n - 4.5$ **10.** 3780 **11.** \$445
12. 10,000, 3/10,000 **13.** \$487,521.25, \$515,024.49, no
14. $a^5 - 10a^4x + 40a^3x^2 - 80a^2x^3 + 80ax^4 - 32x^5$
15. $x^{24} + 24x^{23}y^2 + 276x^{22}y^4$ **16.** $\sum_{i=0}^{30} \binom{30}{i} m^{30-i}y^i$
17. 1/6, 1 to 5 **18.** 220 **19.** 2970 **20.** 1/336

INDEX

A

Absolute value, 7
  equation, 89
  function, 195
  inequality, 138
  properties of, 8
ac-method, 57
Addition
  of complex numbers, 112
  of matrices, 501
  of polynomials, 43
  of radical expressions, 37
  of rational expressions, 69
Addition method, 426
Addition property
  of equality, 84
  of inequality, 134
Addition rule, 647
Additive identity
  of a matrix, 502
  of a real number, 4
Additive inverse
  of a matrix, 502
  of a real number, 4
Algebraic expression, 10
  domain, 11
Algebraic function, 361
Algorithm, division, 48
Analytical geometry, 170
Arithmetic expression, 9
Arithmetic sequence, 596
  common difference, 596
  *n*th term, 596
  sum of *n* terms, 605
Arithmetic series, 605
Array of signs for a matrix, 535
Associative properties, 4
Asymptotes
  horizontal, 340
  of a hyperbola, 577
  oblique, 342
  slant, 342
  vertical, 340
Augmented matrix, 486
Average rate of change, 183
Axis of symmetry, 282, 550

B

Base, 16, 362
Base-change formula, 396
Base-*e* exponential function, 371
Basic principle of rational numbers, 65
Binomial, 42
Binomial coefficient, 636
Binomial expansion, 635
Binomial theorem, 637
Bounded interval, 135
Break-even point, 321
Building up the denominator, 68

C

Carbon-14 dating, 371
Cartesian coordinate system, 164
Center
  of a circle, 191
  of an ellipse, 561
  of a hyperbola, 575
Change of base, 396
Circle, 191, 566
  center, 191, 566
  equation, 191, 567
  radius, 191, 566
Closed interval, 135
Closure property, 4
Coefficient, 11
  binomial, 636
  correlation, 96
  leading, 43
Coefficient matrix, 486
Column, 485
Column matrix, 485
Combination, 631
Combined variation, 250
Combining like terms, 11
Common denominator, 69
Common difference, 596
Common factor, 54
Common logarithm, 378
Common ratio, 611
Commutative properties, 4
Complementary event, 644
Completing the square, 120
Complex fraction, 71
Complex number, 110
  addition, 112
  conjugates, 113
  division, 113
  equal, 110
  imaginary part, 110
  multiplication, 112
  real part, 110
  standard form, 110
  subtraction, 112
Composition of functions, 225
Compound inequality, 135
Compound interest formula, 369
Compounding continuously, 370
Conditional equation, 87
Conic sections, 549
  circle, 566
  ellipse, 560
  hyperbola, 575
  parabola, 550
Conjugate pairs theorem, 305
Conjugate
  axis of a hyperbola, 578
  of a complex number, 113
  of a radical, 47
Constant, 42, 201
Constant function, 201, 264
Constant of variation, 247
Constant term, 42
Constraints, 472
Continuous compounding formula, 371
Coordinate geometry, 170
Coordinate plane, 164
Coordinates, 163
Coordinate system, 164
Correlation coefficient, 96
Counting numbers, 1
Cramer's rule, 530, 538
Cube root, 29
Cubed, 17
Cubes, sum and difference of, 59
Cubic polynomial, 43
Curve fitting, 440

D

Decreasing function, 201
Degenerate conic, 549
Degree, of a polynomial, 43
Denominator
  least common, 69
  rationalizing, 36
Dependent equations, 423
Dependent variable, 177
Descartes's rule of signs, 306
Determinants
  expansion by minors, 535
  of matrices, 528, 534
  minor, 533
Diagonal of a matrix, 489
Difference, 5
  of cubes, 59
  of squares, 58
Difference function, 223
Difference quotient, 184
Dimension, 486
Direct variation, 247
Directly proportional, 247
Directrix, 550
Discriminant, 124
Distance, 8
Distance formula, 170
Distance, rate, and time problems, 100
Distributive properties, 4
Divide out, 66
Dividend, 48
Division
  algorithm, 48
  of complex numbers, 113
  definition, 5
  of polynomials, 48
  of radical expressions, 37
  of rational expressions, 68
  of rational numbers, 68
  synthetic, 292
Division property of equality, 84
Divisor, 48
Domain
  of an algebraic expression, 11
  of a function, 179
  of a rational function, 339
  of a relation, 179
Dummy variable, 182

E

e, 370
Eccentricity, 568
Element
  of a matrix, 486
  minor of, 533
Elimination method, 447
Ellipse
  center, 561
  eccentricity, 568
  equation, 561
  focus, 560
  geometric definition, 560
  major axis, 561
  minor axis, 561
  vertices, 561
Empty set, 89
Entry of a matrix, 486
Equal matrices, 500
Equality
  addition property of, 84
  division property of, 84
  multiplication property of, 84
  subtraction property of, 84
Equally likely, 641
Equation, 83
  absolute value, 89, 320
  of a circle, 191
  conditional, 87
  dependent, 423
  of an ellipse, 561
  equivalent, 84
  exponential, 404
  identity, 86
  inconsistent, 87
  independent, 422
  linear, 84
  logarithmic, 402
  matrix, 511
  nonlinear, 446
  of a parabola, 280
  quadratic, 118
  of quadratic type, 319
  with radicals, 315
  with rational exponents, 316
  with rational expressions, 87
  root of, 83
  solution set, 83
  system of, 421
Equation of a line
  point-slope form, 267
  slope-intercept form, 265
  standard form, 166, 268
Equivalent
  augmented matrices, 487
  equations, 84
  expressions, 11
  inequalities, 133
  systems, 427
Evaluating expressions, 9
Even function, 215
Even root, 29
Event, 623
  complementary, 644
  equally likely, 641
  mutually exclusive, 647
Expansion by minors, 535
Experiment, 641
Exponent, 16
  base, 16
  definition of, 16
  negative, 18
  positive, 16
  power of a power rule, 20
  power of a product rule, 20
  power of a quotient rule, 20
  product rule, 19
  quotient rule, 20
  rational, 30
  rules for, 20
  zero, 20
Exponential equation, 367
Exponential expression, 16
Exponential function, 362
  base-*e*, 371
  one-to-one property, 367
  properties, 365
Expression
  algebraic, 10
  arithmetic, 9
  equivalent, 11
  exponential, 16
  radical, 33
  rational, 65
Extraneous root, 89

F

Factor, 11
  greatest common, 54
Factor theorem, 295
Factorial notation, 593
Factoring
  ac-method, 57
  completely, 60
  difference of cubes, 59
  difference of squares, 58
  greatest common factor, 54
  by grouping, 55
  perfect square trinomials, 58

polynomials, 54
by substitution, 59
sum of cubes, 59
trial and error, 58
trinomials, 56
Factoring out, 54
Feasible solution, 473
Finite sequence, 592
First coordinate, 163
f-notation, 181
Focal length, 550
Foci
of an ellipse, 560
of a hyperbola, 575
Focus of a parabola, 550
FOIL method, 45
Formula, 94
Fraction, complex, 71
Function, 177
absolute value, 195
composition of, 226
constant, 201, 264
decreasing, 201
domain of, 179
even, 215
exponential, 362
graphing, 191
greatest integer, 199
identity, 214
increasing, 201
inverse, 234
keys, 177
linear, 214, 264
logarithmic, 377
maximum value of, 282
minimum value of, 282
notation, 181
objective, 474
odd, 216
one-to-one, 235
operations, 222
piecewise, 198
polynomial, 263
quadratic, 279
range of, 179
rational, 339
transcendental, 361
vertical line test, 196
zero, 291
Fundamental counting principle, 624
Fundamental rectangle, 575
Fundamental theorem of algebra, 296
Future value, 369

G

Gauss, Carl F., 296
Gaussian elimination method, 488
General term, 592
Geometric problems, 100
Geometric sequence, 611
common ratio, 611
*n*th term, 611
Geometric series, 614
infinite, 616
sum of *n* terms, 615
Graph, 164
of a circle, 192
of an ellipse, 563
of a hyperbola, 577
of a line, 165
of a parabola, 279
Graphing, 2, 164
Greatest common factor, 54
Greatest integer function, 199
Grouping, factoring by, 55
Grouping symbols, 9

H

Half-closed interval, 133
Half-life, 406
Half-open interval, 133
Horizontal asymptote, 340
Horizontal component, 605
Horizontal line, 168
Horizontal line test, 236
Horner's method, 303
Hyperbola, 575
asymptotes, 577
branches, 574
center, 575
conjugate axis, 578
equation of, 576
foci, 575
fundamental rectangle, 577
transverse axis, 575
vertices, 575

I

i, 110
Identities, 86
Identity
additive, 4
multiplicative, 4
Identity function, 214
Identity matrix, 516
Identity properties, 4
Imaginary number, 110
Inconsistent equation, 87
Increasing function, 201
Independent variable, 177
Index of a radical, 33
Index of summation, 602
Indicated sum, 603
Induction, 653
Inequality, 132
absolute value, 138
addition property, 134
compound, 135
division property, 134
equivalent, 133
linear, 133, 463
multiplication property, 134
nonlinear, 466
properties of, 134
quadratic, 145
rational, 146
simple, 132
subtraction property, 134
systems of, 467
Infinite geometric series, 616
Infinite sequence, 592
arithmetic, 596
geometric, 611
Integers, 1
Integral exponents, 16
Interest, compound, 368
Intersection of intervals, 136
Intersection of sets, 136
Interval, 132
bounded, 135
closed, 135
half-closed, 133
half-open, 133
intersection, 136
notation, 132
open, 133
unbounded, 132
union, 136
Inverse
additive, 4
multiplicative, 4
Inverse function, 234
Inverse properties, 4
Inverse variation, 248
Inversely proportional, 248
Invertible function, 234
Invertible matrix, 517
Irrational numbers, 2
Irreducible polynomial, 60

J

Joint variation, 249
Jointly proportional, 249

L

Labeling problem, 633
Lambert, Johann Heinrich, 2
Leading coefficient, 43, 263
Leading coefficient test, 332
Least common denominator, 69
Like radicals, 37
Like terms, 11
Linear equation, 84, 166
Linear function, 214, 264
  in two variables, 474
Linear inequality, 133, 463
Linear polynomial, 43
Linear programming, 472
Linear regression, 96
Linear systems
  in three variables, 434
  in two variables, 421
  Cramer's rule for, 530, 538
Linear systems, solution of
  addition method, 426
  applications, 428
  determinant method, 530, 538
  elimination method, 447
  graphing method, 421
  substitution method, 423
Logarithm, 377
  changing the base, 396
  common, 378
  definition, 377
  inverse properties, 393
  Napierian, 378
  natural, 378
  power rule, 392
  product rule, 390
  properties, 394
  quotient rule, 391
Logarithmic equations, 381
Logarithmic function, 377
  one-to-one property, 382
  properties, 380
Lower bound, 306
Lowest terms, 66

M

Major axis of an ellipse, 561
Mathematical induction, 653
Matrix, 485
  addition, 501
  augmented, 486
  of coefficients, 486
  column, 485
  dimension, 486
  element, 486
  entry, 486
  equal, 500
  equation, 511
  identity, 516
  inverse, 517
  invertible, 517
  multiplication, 509
  multiplicative inverse, 517
  row, 485
  row operations, 488
  scalar, 503
  scalar multiplication, 503
  square, 486
  subtraction, 503
  zero, 502
Maximum value, 282
Mean, 605
Midpoint formula, 172
Minimum value, 282
Minor axis of an ellipse, 561
Minor of a matrix element, 533
Mixture problems, 101
Model, 94
Modulus of a complex number, 617
Monomial, 42
Multiplication
  of complex numbers, 112
  of matrices, 509
  of polynomials, 44
  of radical expressions, 37
  of rational expressions, 67
  of rational numbers, 67
  scalar, 603, 503
Multiplication property
  of equality, 84
  of inequality, 134
  of zero, 4
Multiplicative identity, 4
Multiplicative inverse, 4
Multiplicity, 303
Mutually exclusive events, 647

N

Napierian logarithm, 378
Natural constraints, 473
Natural logarithm, 378
Natural numbers, 1
Negative exponent, 18
Newton's law of cooling, 407
n factorial, 593
n-root theorem, 304
Nonlinear equation, 446
Nonlinear inequality, 466
Nonlinear systems of equations, 446
*n*th root, 29
  positive, 29
  principal, 29
*n*th term
  of an arithmetic sequence, 596
  of a geometric sequence, 611
  of a sequence, 592
Number line, 1
Numbers
  complex, 110
  counting, 1
  imaginary, 110
  integers, 1
  irrational, 2
  natural, 1
  rational, 2
  real, 3
  sets of, 3
  whole, 1

O

Objective function, 474
Oblique asymptote, 342
Odd function, 216
Odd root, 29
Odds, 646
One-to-one function, 235
  horizontal line test for, 236
Open interval, 133
Open sentence, 83
Operations
  on complex numbers, 112
  on functions, 223
  order of, 9
Opposites, 5
Order of a matrix, 486
Order of operations, 9
Ordered pair, 163
Ordered triple, 434
Origin, 164
  symmetric about, 216
Outcomes, 641

P

Parabola
 applications of, 555
 axis of symmetry, 282, 550
 directrix, 550
 equation of, 280
 focus, 550
 geometric definition, 550
 graphs of, 280
 opening, 280
 opening left and right, 554
 vertex, 281, 550
Parallel lines, 269
Parallelogram law, 604
Partial fractions, 457
 decomposition, 457
Pascal's triangle, 636
Perfect cube, 35
Perfect *n*th power, 35
Perfect square, 35
Perfect square trinomial, 58
Permutation, 626
Perpendicular lines, 269
pH, 388
Pi, 2
Piecewise functions, 198
Plotting points, 164
Point-slope form, 267
Polynomial, 42
 addition, 43
 binomial, 42
 cubic, 43
 degree, 43
 division, 48
 evaluating, 49
 factoring, 54
 function, 263
 inequalities, 335
 irreducible, 60
 linear, 43
 monomial, 42
 multiplication, 44
 prime, 60
 quadratic, 43
 quotient, 48
 remainder, 48
 subtraction, 43
 trinomial, 42
 in x, 42
 zero, 43
Polynomial function, 263
 zeros of, 291
Potassium-argon dating, 406
Power, 16
Power of a power rule, 20
Power of a product rule, 20
Power of a quotient rule, 20
Power rule for logarithms, 392
Present value, 369
Prime polynomial, 60
Principal *n*th root, 29
Principal square root, 114
Principle of mathematical induction, 653
Probability, 642
 addition rule, 647
Problem solving, 98
Product
 of complex numbers, 112
 of matrices, 509
 of polynomials, 44
 of a sum and a difference, 46
Product function, 223
Product rule for exponents, 19
Product rule for logarithms, 390
Product rule for radicals, 34
Properties of equality, 7, 84
Properties of inequality, 134
Properties of logarithms, 394
Properties of real numbers, 4
Proportionality constant, 247
Pythagorean theorem, 126

Q

Quadrant, 164
Quadratic equation, 118
 applications of, 125
 discriminant of, 124
 standard form, 118
 strategy for solving, 123
Quadratic formula, 123
Quadratic function, 279
Quadratic inequality, 145, 284
Quadratic polynomial, 43
Quadratic type, 317
Quotient, 5, 48
Quotient function, 223
Quotient rule for exponents, 20
Quotient rule for logarithms, 391
Quotient rule for radicals, 34

R

Radical equations, 314
Radical expression, 33
 simplified form, 35
Radical notation, 34
Radical sign, 33
Radicals, 33
 conjugate of, 47
 index of, 33
 rationalizing the denominator, 36
 rules for, 34
 simplifying, 35
Radicand, 33
Radius, 191
Range
 of a function, 179
 of a relation, 179
Rational exponents, 30
Rational expression, 65
 addition, 69
 building up, 68
 complex fraction, 71
 division, 68
 domain, 65
 least common denominator, 69
 multiplication, 67
 reducing, 65
 subtraction, 69
Rational function, 339
Rational inequalities, 347
Rational numbers, 2
 basic principle, 65
Rational zero theorem, 297
Rationalizing the denominator, 36
Real numbers, 3
 properties of, 4
Reciprocal, 4
Rectangular coordinate system, 164
Recursion formula, 595
Reflection, 208
Reflexive property, 7
Regression
 exponential, 376
 linear, 96
 quadratic, 290
Relation, 6, 176
Remainder, 48
Remainder theorem, 291
Richter scale, 400
Root
 bounds for, 308
 cube, 29
 of an equation, 83, 167
 even, 29
 extraneous, 89
 *n*th, 29
 odd, 29
 square, 29
Row, 485

Row matrix, 485
Row operations, 488
Rules for exponents, 20, 31
Rules for radicals, 34
  product, 34
  quotient, 34

S

Sample space, 641
Satisfying an equation, 83, 165
Scientific notation, 21
Second coordinate, 163
Semicircles, 197
Sequence, 592
  arithmetic, 596
  finite, 592
  general term, 592
  geometric, 611
  infinite, 592
  *n*th term, 592
Series, 603
  arithmetic, 605
  finite, 603
  geometric, 614
  infinite geometric, 616
Set
  empty, 89
  intersection, 136
  solution, 83, 165, 651
  union, 136
Sets of numbers, summary of, 3
Shrinking, 212
Sigma notation, 602
Sign array, 535
Sign graph, 145
Simple inequality, 132
Simplify, 11
Simplifying radicals, 35
Slant asymptote, 342
Slope
  formula, 264
  horizontal line, 264
  parallel lines, 269
  perpendicular lines, 269
  vertical line, 264
Slope-intercept form, 265
Solution of an equation, 83
Solution of an inequality, 134
Solution set, 83, 165, 421
Solving equations, 83
Solving for a specified variable, 95
Solving inequalities, 134
Solving word problems, 98
Special products, 46
Square of a binomial, 46
Square of a difference, 46
Square of a sum, 46
Square matrix, 486
Square root, 29
  of a negative number, 114
Square-root property, 118
Squared, 17
Standard form
  of a complex number, 110
  of a linear equation, 166
  of a quadratic equation, 118
Statement, 653
Stretching, 212
Substitution, factoring by, 59
Substitution method for systems, 423
Substitution property, 7
Subtraction
  of complex numbers, 112
  definition, 5
  of matrices, 503
  of polynomials, 43
  of radical expressions, 37
  of rational expressions, 69
Subtraction property of equality, 84
Sum
  of an arithmetic series, 606
  of cubes, 59
  of a geometric series, 615
  of an infinite geometric series, 617
  of two cubes, 59
Sum function, 223
Summation notation, 602
Symmetric property, 7
Symmetry
  about the origin, 216
  about the *y*-axis, 215
  axis of, 282
Synthetic division, 292
System of equations, 421
  dependent, 423
  equivalent, 427
  inconsistent, 423
  independent, 422
  linear, 421
  nonlinear, 446
  solution set of, 421
  in three variables, 434
  in two variables, 421
System of inequalities, 467

T

Term, 11
  coefficient of, 11
  constant, 42
  like, 11
Term of a sequence, 592
Test point method, 150
Transcendental function, 361
Transformation
  nonrigid, 211
  reflection, 208
  rigid, 212
  shrinking, 212
  stretching, 212
Transitive property, 7
Translation, 209
  of a circle, 567
  downward, 209
  of an ellipse, 565
  horizontal, 213
  of a hyperbola, 581
  to the left, 211
  of a parabola, 555
  to the right, 211
  upward, 209
  vertical, 213
Transverse axis of a hyperbola, 575
Tree diagram, 624
Trial and error, 58
Trichotomy property, 6
Trinomial, 42
  factoring, 56
  perfect square, 58

U

Unbounded interval, 132
Uniform motion problems, 100
Union of events, 647
Union of intervals, 136
Union of sets, 136
Unit, 2
Upper bound, 308

V

Value, 9
  of a polynomial, 49
Variable, 2
  dependent, 177

independent, 177
solving for a specified, 95
Variation
combined, 250
constant, 247
direct, 247
inverse, 248
joint, 249
of sign, 306
Vertex
of an ellipse, 561
of a hyperbola, 575
of a parabola, 281, 550
Vertical asymptote, 340
Vertical line, 168
Vertical line test, 196

W

Whole numbers, 1
Word problems, 99
Work problems, 102

X

x-axis, 164
x-intercept, 167
xy-plane, 164

Y

y-axis, 164
symmetric about, 215
y-intercept, 167

Z

Zero exponent, 20
Zero factor property, 119
Zero matrix, 502
Zero polynomial, 43
Zero of an equation, 167
Zero of a polynomial, 291

Algebra

Midpoint Formula

The midpoint of the line segment with endpoints (x_1, y_1) and (x_2, y_2) is

$$\left(\frac{x_1 + x_2}{2}, \frac{y_1 + y_2}{2}\right).$$

Slope Formula

The slope of the line through (x_1, y_1) and (x_2, y_2) is

$$\frac{y_2 - y_1}{x_2 - x_1} \qquad (\text{for } x_1 \neq x_2).$$

Linear Function

$f(x) = mx + b$ with $m \neq 0$

Graph is a line with slope m.

Quadratic Function

$f(x) = ax^2 + bx + c$ with $a \neq 0$

Graph is a parabola.

Polynomial Function

$f(x) = a_n x^n + a_{n-1} x^{n-1} + \cdots + a_1 x + a_0$ for $a_n \neq 0$ and n a nonnegative integer

Rational Function

$f(x) = \dfrac{p(x)}{q(x)}$, where p and q are polynomial functions with $q(x) \neq 0$

Exponential and Logarithmic Functions

$f(x) = a^x$ for $a > 0$ and $a \neq 1$

$f(x) = \log_a(x)$ for $a > 0$ and $a \neq 1$

Properties of Logarithms

Base-a logarithm: $y = \log_a(x) \Leftrightarrow a^y = x$

Natural logarithm: $y = \ln(x) \Leftrightarrow e^y = x$

Common logarithm: $y = \log(x) \Leftrightarrow 10^y = x$

One-to-one: $a^{x_1} = a^{x_2} \Leftrightarrow x_1 = x_2$

$\log_a(x_1) = \log_a(x_2) \Leftrightarrow x_1 = x_2$

$\log_a(a) = 1$ $\qquad \log_a(1) = 0$

$\log_a(a^x) = x$ $\qquad a^{\log_a(N)} = N$

$\log_a(MN) = \log_a(M) + \log_a(N)$

$\log_a(M/N) = \log_a(M) - \log_a(N)$

$\log_a(M^x) = x \cdot \log_a(M)$

$\log_a(1/N) = -\log_a(N)$

$$\log_a(M) = \frac{\log_b(M)}{\log_b(a)} = \frac{\ln(M)}{\ln(a)} = \frac{\log(M)}{\log(a)}$$

Compound Interest

P = principal, t = time in years, r = annual interest rate, and A = amount:

$$A = P\left(1 + \frac{r}{n}\right)^{nt} \text{ (compounded } n \text{ times/year)}$$

$$A = Pe^{rt} \text{ (compounded continuously)}$$

Variation

Direct: $y = kx$ $\quad (k \neq 0)$

Inverse: $y = k/x$ $\quad (k \neq 0)$

Joint: $y = kxz$ $\quad (k \neq 0)$

Straight Line

Slope-intercept form: $y = mx + b$

Slope: m $\qquad$ y-intercept: $(0, b)$

Point-slope form: $y - y_1 = m(x - x_1)$

Standard form: $Ax + By = C$

Horizontal: $y = k$ $\qquad$ Vertical: $x = k$